COLD SPRING HARBOR SYMPOSIA ON QUANTITATIVE BIOLOGY

VOLUME LIII

COLD SPRING HARBOR SYMPOSIA ON QUANTITATIVE BIOLOGY

VOLUME LIII

Molecular Biology of Signal Transduction

COLD SPRING HARBOR LABORATORY
1988

COLD SPRING HARBOR SYMPOSIA ON QUANTITATIVE BIOLOGY
VOLUME LIII

International Standard Book Number 0-87969-056-9
International Standard Serial Number 0091-7451
Library of Congress Catalog Card Number 34-8174

Printed in the United States of America

COLD SPRING HARBOR SYMPOSIA ON QUANTITATIVE BIOLOGY

Founded in 1933 by
REGINALD G. HARRIS
Director of the Biological Laboratory 1924 to 1936

Previous Symposia Volumes

I (1933) Surface Phenomena
II (1934) Aspects of Growth
III (1935) Photochemical Reactions
IV (1936) Excitation Phenomena
V (1937) Internal Secretions
VI (1938) Protein Chemistry
VII (1939) Biological Oxidations
VIII (1940) Permeability and the Nature of Cell Membranes
IX (1941) Genes and Chromosomes: Structure and Organization
X (1942) The Relation of Hormones to Development
XI (1946) Heredity and Variation in Microorganisms
XII (1947) Nucleic Acids and Nucleoproteins
XIII (1948) Biological Applications of Tracer Elements
XIV (1949) Amino Acids and Proteins
XV (1950) Origin and Evolution of Man
XVI (1951) Genes and Mutations
XVII (1952) The Neuron
XVIII (1953) Viruses
XIX (1954) The Mammalian Fetus: Physiological Aspects of Development
XX (1955) Population Genetics: The Nature and Causes of Genetic Variability in Population
XXI (1956) Genetic Mechanisms: Structure and Function
XXII (1957) Population Studies: Animal Ecology and Demography
XXIII (1958) Exchange of Genetic Material: Mechanism and Consequences
XXIV (1959) Genetics and Twentieth Century Darwinism
XXV (1960) Biological Clocks
XXVI (1961) Cellular Regulatory Mechanisms
XXVII (1962) Basic Mechanisms in Animal Virus Biology
XXVIII (1963) Synthesis and Structure of Macromolecules
XXIX (1964) Human Genetics
XXX (1965) Sensory Receptors
XXXI (1966) The Genetic Code
XXXII (1967) Antibodies
XXXIII (1968) Replication of DNA in Microorganisms
XXXIV (1969) The Mechanism of Protein Synthesis
XXXV (1970) Transcription of Genetic Material
XXXVI (1971) Structure and Function of Proteins at the Three-dimensional Level
XXXVII (1972) The Mechanism of Muscle Contraction
XXXVIII (1973) Chromosome Structure and Function
XXXIX (1974) Tumor Viruses
XL (1975) The Synapse
XLI (1976) Origins of Lymphocyte Diversity
XLII (1977) Chromatin
XLIII (1978) DNA: Replication and Recombination
XLIV (1979) Viral Oncogenes
XLV (1980) Movable Genetic Elements
XLVI (1981) Organization of the Cytoplasm
XLVII (1982) Structures of DNA
XLVIII (1983) Molecular Neurobiology
XLIX (1984) Recombination at the DNA Level
L (1985) Molecular Biology of Development
LI (1986) Molecular Biology of *Homo sapiens*
LII (1987) Evolution of Catalytic Function

All Cold Spring Harbor Laboratory publications may be ordered directly from Cold Spring Harbor Laboratory, Box 100, Cold Spring Harbor, New York 11724. Phone: 1-800-843-4388. In New York (516)367-8423.

Symposium Participants

ABRAHAM, IRENE, Dept. of Cell Biology, Upjohn Company, Kalamazoo, Michigan
ADAMS, DOLPH, Dept. of Pathology, Duke University Medical Center, Durham, North Carolina
ADARI, HEDY, Cetus Corporation, Emeryville, California
ADEREM, ALAN, Dept. of Cellular Physiology and Immunology, Rockefeller University, New York, New York
ADLER, JULIUS, Dept. of Biochemistry, University of Wisconsin, Madison
AHN, NATALIE, University of Washington, Seattle
ALCORTA, DAVID, Dept. of Cellular and Developmental Biology, Harvard University, Cambridge, Massachusetts
ALLENDE, CATHERINE, Dept. of Biology, Universidad de Chile, Santiago
ALLENDE, JORGE, Dept. of Biochemistry, Universidad de Chile, Santiago
APPLEBURY, MEREDITHE, University of Chicago, Eye Research Laboratories, Illinois
AXEL, RICHARD, Columbia University, Howard Hughes Medical Institute, New York, New York
BACKER, JOSPEH, American Cyanamid, Pearl River, New York
BAEUERLE, PATRICK, Whitehead Institute, Cambridge, Massachusetts
BALLOU, LISA, Friedrich Miescher Institut, Basel, Switzerland
BARRETT, RONALD, Dept. of Neuroscience, Abbott Laboratories, Abbott Park, Illinois
BAYLOR, DENIS, Dept. of Neurobiology, Stanford University School of Medicine, California
BELL, JOHN, Dept. of Biochemistry, McGill University, Montreal, Canada
BELL, ROBERT, Dept. of Biochemistry, Duke University Medical Center, Durham, North Carolina
BENOVIC, JEFFREY, Duke University, Durham, North Carolina
BERG, HOWARD, Dept. of Cellular and Developmental Biology, Harvard University Biological Laboratories, Cambridge, Massachusetts
BERKENSTAM, ANDERS, Dept. of Medical Nutrition, Karolinska Institute, Huddings University Hospital, Sweden
BERNINGER, MARK, Life Technologies Inc., Gaithersburg, Maryland
BERRIDGE, MICHAEL, Dept. of Zoology, Agricultural and Food Research Council, Cambridge, England
BILLINGSLEY, MELVIN, Dept. of Pharmacology, Milton S. Hershey Medical Center, Pennsylvania
BIRNBAUMER, LUTZ, Dept. of Cell Biology, Baylor College of Medicine, Houston, Texas
BIRNBAUMER, MARIEL, Dept. of Cell Biology, Baylor College of Medicine, Houston, Texas
BITENSKY, MARK, Los Alamos National Laboratory, New Mexico
BLENIS, JOHN, Dept. of Molecular Biology, Northwestern University Medical School, Chicago, Illinois
BLOCH, DONALD, Dept. of Genetics, Harvard Medical School, Boston, Massachusetts
BLUMER, KEN, Dept. of Biochemistry, University of California, Berkeley
BOBBITT, JESSE, Dept. of Biochemical Research, Lilly Research Laboratories, Indianapolis, Indiana
BOCCKINO, STEPHEN, Vanderbilt University, Howard Hughes Medical Institute, Nashville, Tennessee
BORASIO, GIAN DOMENICO, Dept. of Neurochemistry, Max-Planck Institute for Psychiatry, Martinsried, Federal Republic of Germany
BORST, JANNIE, Dept. of Immunology, Netherlands Cancer Institute, Amsterdam, The Netherlands
BOS, TIMOTHY, Dept. of Microbiology, University of Southern California, Los Angeles
BOTELHO, LYNNE, Sandoz Research Institute, East Hanover, New Jersey
BOTSTEIN, DAVID, Genentech Inc., South San Francisco, California
BOURNE, HENRY, Dept. of Pharmacology, University of California, San Franciso
BOYLE, WILLIAM, Dept. of Molecular Biology and Virology, Salk Institute, San Diego, California

BOWTELL, DAVID, University of California, Berkeley

BRADSHAW, SUZANNE, Dept. of Cell Biology, Baylor College of Medicine, Houston, Texas

BRASS, LAWRENCE, Hematology-Oncology Section, University of Pennsylvania Philadelphia

BRAVO, RODRIGO, European Molecular Biology Laboratory, Heidelberg, Federal Republic of Germany

BROWN, ARTHUR, Dept. of Physiology and Molecular Biophysics, Baylor College of Medicine, Houston, Texas

BROWN, DAVID, Dept. of Pharmacology, University College London, England

BROWN, KENNETH, Institute of Animal Physiology, Cambridge, England

BROWN, MICHAEL, Center for Genetic Disease, University of Texas Health Science Center, Dallas

BUCHOU, THIERRY, INSERM, Hôpital St. Antoine, Paris, France

BUELL, GARY, Pharmacia Genetic Engineering Inc., La Jolla, California

BURNS, DAVID, Dept. of Biochemistry, Duke University, Durham, North Carolina

BUTEL, JANET, Dept. of Virology, Baylor College of Medicine, Houston, Texas

BUTT, TAUSEEF, Smith Kline & French Laboratories, King of Prussia, Pennsylvania

CADD, GARY, Dept. of Pharmacology, University of Washington, Seattle

CAPON, DAN, Dept. of Molecular Biology, Genentech Inc., South San Francisco, California

CARON, MARC, Duke University Medical Center, Durham, North Carolina

CARROLL, DENNIS, Cold Spring Harbor Laboratory, New York

CARTER, ANTHONY, National Institutes of Health, Bethesda, Maryland

CASSATELLA, MARCO, Wistar Institute, Philadelphia, Pennsylvania

CATELLI, MARIA-GRAZIA, INSERM, Laboratoire Hormones, Montrouge, France

CATO-BIRO, LAURA, Dept. of Biochemistry, McGill University, Montreal, Canada

CHABRE, MARC, Centre d'Etudes Nucleaires, Laboratoire Biologie Moleculaire et Cellulaire, Grenoble, France

CHALEFF, DEBORAH, American Cyanamid Company, Princeton, New Jersey

CHAMBON, PIERRE, Institut de Chimie, Biologique, Strasbourg Faculty of Medicine, France

CHAO, MOSES, Dept. of Cell Biology and Anatomy, Cornell University Medical College, New York, New York

CHARDIN, PIERRE, INSERM, Faculté de Medecine Laboratory, St. Louis, Paris, France

CHEN, WILLIAMS S., University of California, Howard Hughes Medical Institute, San Diego

CLAPHAM, DAVID, Dept. of Pharmacology, Mayo Clinic, Rochester, Minnesota

CLARK, KAREN, University of Oregon Institute of Molecular Biology, Eugene

COFFEY, ROBERT, Dept. of Medicine, Division of Gastroenterology, Vanderbilt University, Nashville, Tennessee

CONNEELY, ORLA, Dept. of Cell Biology, Baylor College of Medicine, Houston, Texas

COOPER, J.A., Fred Hutchinson Cancer Center, Seattle, Washington

COTE, TOM, Uniformed Services University, Bethesda, Maryland

COURTNEIDGE, SARA, European Molecular Biology Laboratory, Heidelberg, Federal Republic of Germany

CURRAN, THOMAS, Dept. of Molecular Oncology, Roche Institute of Molecular Biology, Nutley, New Jersey

DAHLQUIST, FREDERICK, Institute of Molecular Biology, University of Oregon, Eugene

DALE, TREVOR, Imperial Cancer Research Fund, London, England

DANIEL, TOM, Vanderbilt University, Nashville, Tennessee

DARNELL, JAMES, Dept. of Molecular Biology, Rockefeller University, New York, New York

DAS MANJUSRI, Dept. of Biochemistry and Biophysics, University of Pennsylvania, Philadelphia

DASH, PRAMOD, Dept. of Neurobiology and Behavior, Columbia University, New York, New York

DETERRE, PHILIPPE, Centre d'Etudes Nucleaires, Laboratory Biophysique Moleculaire, Grenoble, France

DEVREOTES, PETER, Dept. of Biological Chemistry, Johns Hopkins University, Baltimore, Maryland

DIAMOND, IVAN, University of California, San Francisco General Hospital

DICKEY, BURTON, Pulmonary Center, Boston University School of Medicine, Massachusetts

DIDSBURY, JOHN, Dept. of Inflammation Research, Genentech Inc., South San Francisco, California

DILIBERTO, EMANUEL, Dept. of Medical Biochemistry, Burroughs Wellcome Company, Research Triangle Park, North Carolina

DIXON, RICHARD, Merck Sharp & Dohme Research Laboratories, West Point, Pennsylvania

DOSTMANN, WOLFGANG, Dept. of Chemistry, University of California, San Diego, La Jolla

DOWNWARD, JULIAN, Whitehead Institute, Cambridge, Massachusetts

DUDLEY, DAVID, Park-Davis Research Division, Ann Arbor, Michigan

EATOCK, RUTH ANN, Dept. of Physiology, University of Rochester, New York

EDWARDS, AL, Dept. of Biochemistry, McGill University, Montreal, Canada

ENDEMANN, GERDA, Dept. of Cell Biology, Stanford University, California

ERIKSON, RAYMOND, Dept. of Cellular and Developmental Biology, Harvard University Biological Laboratories, Cambridge, Massachusetts

ERNST, JOEL, University of California, San Francisco

ERTL, ADRIENNE, Dept. of Medicine, University of California, San Diego, La Jolla

ERUSALIMSKY, JORGE, Imperial Cancer Research Fund, London, England

ERCH, KEITH R., Dept. of Molecular Genetics, University of California, San Diego, La Jolla

ESPINO, PEARL, Dept. of Cellular Regulation, Integrated Genetics, Framingham, Massachusetts

EVANS, RON, Salk Institute, Howard Hughes Medical Institute Research Laboratories, La Jolla, California

EVANS, TONY, Dept. of Pharmacological Sciences, Genentech Inc., South San Francisco, California

FABBRO, DORIANO, University Clinic Medical School Laboratories, Basel, Switzerland

FANG, KATHY, Dept. of Biochemical Endocrinology, Rockefeller University, New York, New York

FARBER, JOSHUA, Dept. of Medicine, Johns Hopkins University, Baltimore, Maryland

FERAMISCO, JAMES, Cold Spring Harbor Laboratory, New York

FERRO, ARI, Dept. of Biology, University of Utah, Salt Lake City

FIRTEL, RICHARD, Center for Molecular Genetics, University of California, San Diego, La Jolla

FISCHER, J.A., University of Zurich, Klinik Balgrist, Switzerland

FISCHER, SIEGMUND, INSERM, Paris, France

FLANAGAN, JOHN, Dept. of Genetics, Harvard Medical School, Boston, Massachusetts

FLAVELL, RICHARD, Biogen Research Corporation, Cambridge, Massachusetts

FLINT, ANDREW, Dept. of Biochemistry, University of California, Berkeley

FLIRI, ANTON, Dept. of Medicinal Chemistry, Pfizer Central Research, Groton, Connecticut

FOLDES, ROBERT, Dept. of Cancer and Cell Biology, Mt. Sinai Hospital Research Institute, Toronto, Canada

FOSTER, DAVID, Dept. of Biological Sciences, Hunter College New York, New York

FOULKES, GORDON, Oncogene Science Inc., Manhasset, New York

FOWLKES, DANA, University of North Carolina, Chapel Hill

FRANZA, ROBERT, Cold Spring Harbor Laboratory, New York

KWOK-KEUNG FUNG, BERNARD, Jules Stein Eye Institute, University of California School of Medicine, Los Angeles

FURTH, MARK, Oncogene Science Inc., Manhasset, New York

GALIONE, ANTHONY, Dept. of Zoology, University of Cambridge, England

GARBERS, DAVID, Vanderbilt University, Howard Hughes Medical Institute, Nashville, Tennessee

GAROFALO, ROBERT, Dept. of Molecular Biology, Memorial Sloan Kettering Institute, New York, New York

GARRISON, JAMES, Dept. of Pharmacology, University of Virginia, Charlottesville

GATELL, SUSAN, Elsevier Publications, *Trends in Neurosciences,* Cambridge, England

GEORGOPAPADAKOU, NAFSIKA, Dept. of Chemotherapy, Hoffmann-La Roche, Nutley, New York

GIBBS, JAY, Dept. of Molecular Biology, Merck Sharp & Dohme Research Laboratories, West Point, Pennsylvania

GILL, GORDON, Dept. of Medicine, University of California, San Diego, La Jolla

GILMAN, ALFRED, Dept. of Pharmacology, University of Texas Health Science Center, Dallas

GILMAN, MICHAEL, Cold Spring Harbor Laboratory, New York

GINSBERG, MARK, Dept. of Immunology, Scripps Clinic and Research Foundation, La Jolla, California

GLASS, CHRISTOPHER, Dept. of Medicine, University of California, San Diego, La Jolla

GOLDSTEIN, JOSEPH, Dept. of Molecular Genetics, University of Texas Health Science Center, Dallas

GOMPERTS, BASTIEN, Dept. of Experimental Pathology, University College London, England

GONZATTI-HACES, MARY, NCI-Frederick Cancer Research Facility, Frederick, Maryland

GORDON, ADRIENNE, San Francisco General Hospital, University of California

GOSSE, MARILYN, Dept. of Pharmacology, Uniformed Services University, Bethesda, Maryland

GRAHAM, ROBERT, Cardiac Unit, Massachusetts General Hospital, Boston

GROPPI, VINCENT, Dept. of Cell Biology, Upjohn Company, Kalamazoo, Michigan

GUTKIND, SILVIO, LOPB, NIDR, National Institutes of Health, Bethesda, Maryland

HAIECH, JACQUES, CNRS, INSERM, Montpellier, France

HALL, ALAN, Institute of Cancer Research, London, Chester Beatty Laboratories, England

HALL, DAVID, Dana-Farber Cancer Institute, Boston, Massachusetts

HAMM, HEIDI, Dept. of Physiology and Biophysics, University of Illinois, Chicago

HAMPTON, RANDOLPH, Dept. of Biochemistry, University of Wisconsin, Madison

HANAFUSA, HIDESABURO, Rockefeller University, New York, New York

HANCOCK, JOHN, Institute of Cancer Research, London, Chester Beatty Laboratories, England

HANLEY, MICHAEL, University of Cambridge, MRC Molecular Neurobiology Unit, England

HANSBURG, DANIEL, Fox Chase Cancer Center, Philadelphia, Pennsylvania

HASLAM, R., Dept. of Pathology, McMaster University, Hamilton, Canada

HASSELL, JOHN, Dept. of Microbiology and Immunology, McGill University, Montreal, Canada

HAUBRUCK, HEINZ, Dept. of Molecular Genetics, Max-Planck Institute, Gottingen, Federal Republic of Germany

HE, XI, Dept. of Biology, University of California, San Diego, La Jolla

HEINRICH, JULIA, Dept. of Molecular Biology, Memorial Sloan Kettering Institute, New York, New York

HEMMINGS, BRIAN, Friedrich Miescher Institute, Basel, Switzerland

HEN, RENE, Columbia University, Howard Hughes Medical Institute, New York, New York

HENNER, DENNIS, Dept. of Cell Genetics, Genentech Inc., South San Francisco, California

HERSKOWITZ, IRA, Dept. of Biochemistry and Biophysics, University of California, San Francisco

HESS, J. FRED, Dept. of Biology, California Institute of Technology, Pasadena

HIGASHIDA, H., Dept. of Biophysics, Kanazawa University School of Medicine, Japan

HILL, DAVID, Genetics Institute, Cambridge, Massachusetts

HIRSCH, JEANNE, Dept. of Biology, Yale University, New Haven, Connecticut

HODGINS, M.B., Dept. of Dermatology, Glasgow University, Scotland

HOMCY, CHARLES, Harvard Medical School, Massachusetts General Hospital, Boston

HUGANIR, RICHARD, Dept. of Neuroscience, Johns Hopkins University, Baltimore, Maryland

HUNTER, TONY, Tumor Virus Laboratory, Salk Institute, La Jolla, California

HUTCHCROFT, JILL, Dept. of Biological Sciences, Purdue University, West Lafayette, Indiana

IMBRA, RICK, Dept. of Environmental Medicine, New York University School of Medicine, Tuxedo, New York

INGHAM, COLIN, Dept. of Biochemistry, University of Wisconsin Madison

IVES, HARLAN, Dept. of Medicine and Nephrology, University of California, San Francisco

IWASHITA, SHINTARO, Mitsubishi-kasei Institute, Tokyo, Japan

IYENGAR, RAVI, Dept. of Pharmacology, Mt. Sinai School of Medicine, New York, New York

IZUMO, SEIGO, Dept. of Cardiology, Harvard Medical School, Children's Hospital, Boston, Massachusetts

JACKOWSKI, SUZANNE, Dept. of Biochemistry, St. Jude Children's Research Hospital, Memphis, Tennessee

JACKSON, STEPHEN, Dept. of Biochemistry, University of California, Berkeley

JACQUET, MICHEL, Laboratoire d'Information Genetique et Development, Université de Paris-Sud, Orsay, France

JAMBOU, ROBERT, Dept. of Pathology, University of North Carolina, Chapel Hill, North Carolina

JASNY, BARBARA, *Science,* AAAS, Washington, D.C.

JAZIN, ELENA, New York State Department of Health, Wadsworth Center for Research, Albany, New York

JENNESS, DUANE, Dept. of Molecular Genetics and Microbiology, University of Massachusetts Medical School, Worcester, Massachusetts

JOHNSON, DOUGLAS, Dept. of Biology, University of Michigan, Ann Arbor, Michigan

JOHNSTON, ROGER, Dept. of Physiology and Biophysics, State University of New York, Stony Brook

JONES, DAVID, Dept. of Molecular Biology and Genetics, Johns Hopkins University, Baltimore, Maryland

JOSHI, LATA, Sandoz Research Institute, East Hanover, New Jersey

JULIUS, DAVID, Columbia University College of Physicians and Surgeons, Institute of Cancer Research, New York, New York

JURIVICH, DONALD, Duke University, Durham, North Carolina

KAHN, RICHARD, National Cancer Institute, NIH, Laboratory of Biological Chemistry, Bethesda, Maryland

KAJIJI, SHAMA, Pfizer Central Research, Neurosciences and Cancer, Groton, Connecticut

KANDEL, ERIC, Columbia University College of Physicians and Surgeons, Center for Neurobiology and Behavior, New York, New York

KATAN, MATILDA, Ludwig Institute for Cancer Research, London, England

KAUPP, BENJAMIN, Dept. of Biologie and Chemie, Universität Osnabruck, Federal Republic of Germany

KAZIRO, YOSHITO, Institute of Medical Sciences, University of Tokyo, Japan

KHACHATRIAN, LEILI, Dept. of Pathology, Boston University School of Medicine, Massachusetts

KHORANA, H. GOBIND, Dept. of Biology and Chemistry, Massachusetts Institute of Technology, Cambridge

KIBERSTIS, PAULA, *Cell* Editorial Office, Cambridge, Massachusetts

KIM, SUNG-HOU, Dept. of Chemistry, University of California, Berkeley

KIRSCH, DONALD, Dept. of Molecular Biology, E.R. Squibb Princeton, New Jersey

KLEIN, PETER, Dept. of Biological Chemistry, Johns Hopkins University, Baltimore, Maryland

KLEMENZ, ROMAN, Ludwig Institute for Cancer Research, Bern, Switzerland

KMIECIK, THOMAS, NCI-Frederick Cancer Research Facility, Basic Research Program, Frederick, Maryland

KNOPF, JOHN, Genetics Institute, Cambridge, Massachusetts

KNOX, BARRY, Dept. of Chemistry, Massachusetts Institute of Technology, Cambridge, Massachusetts

KOBILKA, BRIAN, Duke University Medical Center, Durham, North Carolina

KOSHLAND, DANIEL, Dept, of Biochemistry, University of California, Berkeley

KRAL, LEOS, Dept. of Biochemistry, Wayne State University, Detroit, Michigan

KREBS, EDWIN, University of Washington, Howard Hughes Medical Institute, Seattle

KUNG, CHING, Dept. of Molecular Biology, University of Wisconsin, Madison

KURJAN, JANET, Dept. of Biological Sciences, Columbia University, New York, New York

LACAL, JUAN CARLOS, National Cancer Institute, NIH, Bethesda, Maryland

LACOMBE, M.L., Institut Pasteur, Unité de Biochimie Cellulaire, Paris, France

LAI, CHUN-YEN, Dept. of Protein Biochemistry, Hoffmann-La Roche, Nutley, New Jersey

LANCET, DORON, Dept. of Membrane Research, Weizmann Institute of Science, Rehovot, Israel

LANCHO, TERESA, Laboratory of Molecular Microbiology, National Institutes of Health, Bethesda, Maryland

LANDIS, CLAUDIA, Dept. of Pharmacology, University of California, San Francisco

LARNER, JOSEPH, Dept. of Pharmacology, University of Virginia, Charlottesville

LEAR, JAMES, E.I. DuPont de Nemours & Company, Central Research & Development Experimental Station, Wilmington, Delaware

LEE, ANGEL, NHLBI, Clinical Hematology Branch, National Institutes of Health, Bethesda, Maryland

LEFKOWITZ, ROBERT, Duke University Medical Center, Durham, North Carolina

LEICHTLING, BEN, Dept. of Pediatrics, National Jewish Center for Immunology, Denver, Colorado

LEVY, DAVID, Rockefeller University, New York, New York

LI, CONGYI, Dept. of Biochemistry, University of Wisconsin, Madison

LIMBIRD, LEE, Dept. of Pharmacology, Vanderbilt University, Nashville, Tennessee

LIN, CHUN WEL, Abbott Laboratories, Abbott Park, Illinois

LITOSCH, IRENE, Dept. of Pharmacology, University of Miami, Florida

LLOYD, ALISON, Institute of Cancer Research, London, Chester Beatty Laboratories, England

LOW, MARTIN, Dept. of Physiology and Cellular Biophysics, Columbia University College of

Physicians and Surgeons, New York, New York

LOWE, DAVID, Dept. of Molecular Biology, Genentech Inc., South San Francisco, California

LUKAS, THOMAS, Dept. of Pharmacology, Vanderbilt University, Nashville, Tennessee

MACKENZIE, ROBERT, Abbott Laboratories, Abbott Park, Illinois

MACNAB, ROBERT, Dept. of Molecular Biophysics and Biochemistry, Yale University, New Haven, Connecticut

MAGEE, TONY, MRC National Institute for Medical Research, London, England

MAGGI, ADRIANA, University of Milan, Milano Molecular Pharmacology Laboratory, Italy

MAGUN, BRUCE, Dept. of Cell Biology and Anatomy, Oregon Health Sciences University, Portland

MARBACH, IRIT, Dept. of Biological Chemistry, Hebrew University of Jerusalem, Israel

MARRION, NEIL, Dept. of Neurobiology and Behavior, State University of New York, Stony Brook, Howard Hughes Medical Research Laboratory, New York

MARSH, LORRAINE, Dept. of Biochemistry and Biophysics, University of California, San Francisco

MARSHAK, DANIEL, Cold Spring Harbor Laboratory, New York

MARSHALL, MARK, Dept. of Molecular Biology, Merck Sharp & Dohme Research Laboratories, West Point, Pennsylvania

MARTIN-ZANCA, DIONISIO, NCI-Frederick Cancer Research Facility, Basic Research Program, Frederick, Maryland

MATSUI, HIROAKI, Dept. of Medicine, Division of Cardiology, Duke University Medical Center, Durham, North Carolina

MATSUI, YASUSHI, Biosciences Laboratory Research Center, Mitsubishi Chemical Ind., Yokohama, Japan

MATSUMOTO, KUNIHIRO, DNAX Research Institute, Palo Alto, California

MATTER, ALEX, Ciba-Geigy, Basel, Switzerland

MAYER, BRUCE, Rockefeller University, New York, New York

MCCORMICK, FRANK, Dept. of Molecular Biology, Cetus Corporation, Emeryville, California

MCKAY, IAN, Amersham International plc New Technologies, Buckinghamshire, United Kingdom

MCKNIGHT, STANLEY, G., Dept. of Pharmacology, University of Washington, Seattle

MCROBBIE, STUART, Saunders Scientific Publications, The Workshop, London, England

MEHMET, HUSEYIN, Imperial Cancer Research Fund, London, England

MERCKEN, LUC, Genetica, Paris, France

MEYER, TOBIAS, Dept. of Cell Biology, Stanford University, California

MEYERHOF, WOLFGANG, Universitäts Krankenhaus Eppendorf, Institut fur Zellbiochemie, Hamburg, Federal Republic of Germany

MICHELL, BOB, Dept. of Biochemistry, University of Birmingham, England

MICHITSCH, RICHARD, Oncogene Science Inc., Manhasset, New York

MIKI, TORU, Dept. of Cellular and Molecular Biology, National Cancer Institute, Bethesda, Maryland

MILLER, R. TYLER, Dept. of Pharmacology, University of California, San Francisco

MISKO, THOMAS, Dept. of Molecular Genetics and Mammalian Biology, Monsanto Company, St. Louis, Missouri

MISRA, RAVI, Dept. of Microbiology and Molecular Genetics, Harvard Medical School, Boston, Massachusetts

MOCHLY-ROSEN, MARIA, San Francisco General Hospital, University of California

MOE, GREG, Dept. of Biochemistry, University of California, Berkeley

MOELLING, KARIN, Max-Planck Institute of Molecular Genetics, Berlin, Federal Republic of Germany

MOLLER, WIM, Dept. of Medical Biochemistry, Sylvius Laboratoria, Leiden, The Netherlands

MOOLENAAR, WOUTER, Netherlands Cancer Institute, Amsterdam, The Netherlands

MOORE, DAVID, Dept. of Molecular Biology, Massachusetts General Hospital, Boston

MORRISON, DEBORAH, Dana-Farber Cancer Institute, Boston, Massachusetts

MOXHAM, CARY, Burroughs Wellcome, Research Triangle Park, North Carolina

MULHERKAR, RITA, NCI-Frederick Cancer Research Facility, Developmental Oncology Section, Frederick, Maryland

MURAD, FERID, Dept. of Medicine, Stanford University Medical Center, Palo Alto, California

MYERS, ALAN, Dept. of Biochemistry, Iowa State University, Ames

NAKAYAMA, NAOKI, Dept. of Molecular Biology, DNAX Research Institute, Palo Alto, California

NATHANS, DANIEL, Dept. of Molecular Biology and Genetics, Johns Hopkins University, Baltimore, Maryland

NATHANSON, NEIL, Dept. of Pharmacology, University of Washington, Seattle

NEER, EVA, Dept. of Medicine, Brigham and Women's Hospital, Boston, Massachusetts

NELKIN, BARRY, Oncology Center, Johns Hopkins University, Baltimore, Maryland

NIMAN, HENRY, PROGENX INC., San Diego, California

NISHIZUKA, YASUTOMI, Dept. of Biochemistry, Kobe University School of Medicine, Japan

NORTHUP, JOHN, Dept. of Pharmacology, Yale University, New Haven, Connecticut

NOVICK, PETER, Dept. of Cell Biology, Yale University School of Medicine, New Haven, Connecticut

NUMA, SHOSAKU, Dept. of Medical Chemistry, Kyoto University, Japan

O'MALLEY, BERT, Dept. of Cell Biology, Baylor College of Medicine, Houston, Texas

OESTREICHER, A. BEATE, Institute of Molecular Biology, University of Utrecht, The Netherlands

OOSAWA, KENJI, Dept. of Neurochemistry and Neuropharmacology, Osaka University Medical School, Japan

OTTO, HENNING, Dept. of Neurochemie, Free University Berlin, Federal Republic of Germany

PARK, MORAG, NCI-Frederick Cancer Research Facility, Basic Research Program, Frederick, Maryland

PARKER, CARL, Dept. of Chemistry, California Institute of Technology, Pasadena

PARKINSON, JOHN, Dept. of Biology, University of Utah, Salt Lake City

PARSONS, SARAH, Dept. of Microbiology, University of Virginia Medical School, Charlottesville

PAWSON, TONY, Dept. of Molecular and Developmental Biology, Mt. Sinai Hospital Research Institute, Toronto, Canada

PENN, LINDA, J.Z., Imperial Cancer Research Fund, London, England

PIGGOTT, JAMES, Dept. of Cellular Pharmacology, Smith Kline & French Research Ltd., Hertfordshire, England

POGGIOLI, JOSIANE, Université de Paris-Sud, Physiologie et Pharmacolgie Cellulaire, INSERM, Orsay, France

PONZETTO, CAROLA, Dept. of Medicine/Oncology, University of Turin, Italy

POUYSSEGUR, JACQUES, Dept. of Biochemistry, Division Research Center, CNRS, Nice, France

PRENTKI, MARC, Dept. of Medicine, Institut de Biochimie Clinique, Geneva, Switzerland

PRUSS, REBECCA, Dept. of Biochemical Sciences, Merrell Dow Research Institute, Cincinnati, Ohio

PRYWES, RON, Dept. of Biochemistry and Molecular Biology, Rockefeller, New York, New York

PTASHNE, MARK, Dept. of Biochemistry and Molecular Biology, Harvard University, Cambridge, Massachusetts

RABEN, DANIEL, Dept. of Physiology, Johns Hopkins University School of Medicine, Baltimore, Maryland

RACKER, EFRAIM, Dept. of Biochemistry, Cornell University, Ithaca, New York

RAETZ, CHRISTIAN, Dept. of Biochemistry, University of Wisconsin, Madison

RAJPUT, BHANA, Friedrich-Meischer Institute, Basel, Switzerland

RAKOWICZ-SZULCZYNSKA, EVA, The Wistar Institute, Philadelphia, Pennsylvania

RAPP, ULF, National Cancer Institute, Laboratory of Viral Carcinogenesis, Frederick, Maryland

RAYTER, SYDONIA, Oncogene Science Inc., Manhasset, New York

REDDY, PRASAD, Center for Chemical Physics, National Bureau of Standards, Gaithersburg, Maryland

REDISKE, JOHN, Dept. of Inflammation, CIBA-GEIGY Corporation, Summit, New Jersey

REED, RANDALL, Dept. of Molecular Biology and Genetics, Johns Hopkins University Medical School, Baltimore, Maryland

REED, STEVEN, Dept. of Molecular Biology, Scripps Clinic and Research Foundation, La Jolla, California

REGAN, JOHN, Duke University Medical School, Durham, North Carolina

RENEKE, JOHANNA, Dept. of Genetics, University of California, Berkeley

RIABOWOL, KARL, Cold Spring Harbor Laboratory

ROACH, ARTHUR, Dept. of Molecular Immunology and Neurobiology, Mt. Sinai Hospital Research Institute, Toronto, Canada

ROBERTS, BRUCE, Integrated Genetics, Framingham, Massachusetts

ROBERTS, THOMAS, Dept. of Pathology, Dana-

Farber Cancer Institute, Boston, Massachusetts

ROBERTSON, MIRANDA, *Nature,* London, England

RODER, JOHN, Dept. of Molecular Immunology and Neurobiology, Mt. Sinai Hospital Research Institute, Toronto, Canada

ROMERO, GUILLERMO, Dept. of Pharmacology, University of Virginia, Charlottesville

RONNETT, GABRIELE, Dept. of Neuroscience, Johns Hopkins University School of Medicine, Baltimore, Maryland

ROSEN, ANTONY, Dept. of Cellular Physiology and Immunology, Rockefeller University, New York, New York

ROSENBAUM, JAN, Procter & Gamble Company, Miami Valley Laboratories, Cincinnati, Ohio

ROSENFELD, MICHAEL, University of California School of Medicine, San Diego, La Jolla

ROSENTHAL, ARNON, Dept. of Molecular Biology, Genentech Inc., South San Francisco, California

ROSS, CHRISTOPHER, Dept. of Neuroscience, Johns Hopkins University, Baltimore, Maryland

ROSS, ELLIOTT, Dept. of Pharmacology, University of Texas Health Science Center, Dallas

ROTH, RICHARD, Dept. of Pharmacology, Stanford University Medical School, California

ROUSSEL, MARTINE, Dept. of Tumor Cell Biology, St. Jude Children's Research Hospital, Memphis, Tennessee

ROZENGURT, ENRIQUE, Imperial Cancer Research Fund, London, England

RUBIN, GERALD, Dept. of Biochemistry, University of California, Berkeley

RUDKIN, BRIAN, Dept. of Growth Factor, National Institutes of Health, NICHD, Bethesda, Maryland

RUGGIERI, ROSAMARIA, Dept. of Molecular Biology, DNAX Research Institute, Palo Alto, California

RYAN, TIMOTHY, Dept. of Physics, Cornell University, Ithaca, New York

SADOWSKI, HENRY, Dept. of Biophysics and Medicine, University of Rochester, New York

SAIMI, YOSHIRO, Dept. of Molecular Biology, University of Wisconsin, Madison

SAKMAR, THOMAS, Dept. of Chemistry, Massachusetts Institute of Technology, Cambridge, Massachusetts

SALOMON, YORAM, Dept. of Hormone Research, Weizmann Institute of Science, Rehovot, Israel

SALTIEL, ALAN, Dept. of Biochemistry and Endocrinology Laboratory, Rockefeller University, New York, New York

SASSONE-CORSI, PAOLO, Dept. of Molecular Biology and Virology Salk Institute, San Diego, California

SAXE, CHARLES, NIDDK/NIH, Dept. of Cellular and Developmental Biology, National Institutes of Health, Bethesda, Maryland

SCHLESSINGER, JOSEPH, Rorer Biotechnology Inc., Rockville, Maryland

SCHLONDORFF, DETLEF, Albert Einstein College of Medicine, Bronx, New York

SCHONTHAL, AXEL, Kenforschungszentrum Karlsruhe, Institute fur Genetik und Toxikologie, Federal Republic of Germany

SCHULTZ, GUNTER, Dept. of Pharmacology, Freie University of Berlin, Federal Republic of Germany

SCHULZ, JOHN, Dept. of Physiology, Tufts University, Boston, Massachusetts

SCHUTZ, GUNTHER, Institute of Cell and Tumor Biology, German Cancer Research Center, Heidelberg, Federal Republic of Germany

SCHWARTZ, JAMES, Columbia University, Howard Hughes Medical Institute, New York, New York

SEEDORF, KLAUS, Dept. of Developmental Biology, Genentech Inc., South San Francisco, California

SEGEV, NAVA, Dept. of Biochemistry, Stanford University, California

SEKIGUCHI, KAZUO, Dept. of Biochemistry, Kobe University School of Medicine, Japan

SELINGER, ZVI, Dept. of Biological Chemistry, Hebrew University of Jerusalem, Israel

SETH, PREM, Cancer Center, University of Rochester, New York

SEUWEN, KLAUS, CNRS, Centre de Biochimie, Nice, France

SHALABY, ISMAIL, Pfizer Central Research, Groton, Connecticut

SHAPIRO, ELI, Columbia University, Howard Hughes Medical Institute, New York, New York

SHEARMAN, MARK, Dept. of Biochemistry, Kobe University School of Medicine, Japan

SHERMAN-GOLD, RIVKA, BioGrowth Inc., Richmond, California

SHERR, CHARLES, Dept. of Tumor Cell Biology, St. Jude Children's Research Hospital, Memphis, Tennessee

SHI, WENYUAN, Dept. of Biochemistry, University of Wisconsin, Madison

SHIMIZU, YOSHIKO, Dept. of Molecular and Cellular Biology, University of Arizona, Tucson

SHOWELL, HENRY, Dept. of Immunology and Infectious Disease, Pfizer Central Research, Groton, Connecticut

SHUTTLEWORTH, TREVOR, Dept. of Physiology, University of Rochester, New York

SIGAL, IRVING, Merck Sharp & Dohme Research Laboratories, West Point, Pennsylvania

SIMON, MELVIN, Dept. of Biology, California Institute of Technology, Pasadena

SIMON, MICHAEL, Dept. of Biochemistry, University of California, Berkeley

SKINNER, RICHARD, Dept. of Molecular Sciences, Wellcome Foundation, Kent, England

SMART, JOHN, Dept. of Biological Chemistry, Hoffmann-La Roche, Nutley, New Jersey

SORBARA-CAZAN, Rockefeller University, New York, New York

SPRAGUE, GEORGE, University of Oregon, Institute of Molecular Biology, Eugene

STACEY, DENNIS, Dept. of Cell Biology, Roche Institute of Molecular Biology, Nutley, New Jersey

STADEL, JEFFREY, Dept. of Molecular Pharmacology, Smith Kline & French Laboratories, King of Prussia, Pennsylvania

STEARNS, TIM, Dept. of Biology, Massachusetts Institute of Technology, Cambridge, Massachusetts

STEPHENSON, JOHN, Oncogene Science Inc., Manhasset, New York

STERNWEIS, PAUL, Dept. of Pharmacology, University of Texas, Dallas

STEWART, RICHARD, University of Oregon, Institute of Molecular Biology, Eugene

STILES, CHARLES, Dana-Farber Cancer Institute, Harvard Medical School, Boston, Massachusetts

STOCK, ANN, Dept. of Molecular Biology, Princeton University, New Jersey

STOCK, JEFFRY, Dept. of Biochemical Sciences, Princeton University, New Jersey

STRASSER, RUTH, Dept. of Cardiology, University of Heidelberg, Federal Republic of Germany

STRUHL, KEVIN, Dept. of Biological Chemistry, Harvard Medical School, Boston, Massachusetts

STRULOVICI, BERTA, Syntex Research, Cancer and Developmental Biology, Palo Alto, California

STRYER, LUBERT, Dept. of Cell Biology, Stanford University Medical School, California

SULLIVAN, MARK, Eastman Kodak, Rochester, New York

SULTZMAN, LISA, Genetics Institute, Cambridge, Massachusetts

SWEATT, J. DAVID, Columbia University College of Physicians and Surgeons, Center for Neurobiology Behavior, New York, New York

TAYLOR, COLIN, Dept. of Zoology, University of Cambridge, England

TAYLOR, SUSAN, Dept. of Chemistry, University of California, San Diego, La Jolla California

TELERMAN, ADAM, Université Libre de Bruxelles, Faculte de Medecine, Belgium

THEIBERT, ANNE, Dept. of Neuroscience, Johns Hopkins University School of Medicine, Baltimore, Maryland

THOMAS, PAUL, Dept. of Protein Biochemistry, Hoffmann-La Roche, Nutley, New Jersey

THORNER, JEREMY, Dept. of Biochemistry, University of California, Berkeley

TIPPER, DONALD, Dept. of Molecular Genetics and Microbiology, University of Massachusetts Medical School, Worcester

TJIAN, ROBERT, Dept. of Biochemistry, University of California, Berkley

TREISMAN, RICHARD, MRC Laboratory of Molecular Biology, Cambridge, England

TSAI, MEN-HWEI, Dept. of Molecular Biology, Cleveland Clinic Foundation, Ohio

TSAI, SOPHIA, Dept. of Cell Biology, Baylor College of Medicine, Houston, Texas

TSIEN, ROGER, Dept. of Physiology and Anatomy, University of California, Berkeley

UI, MICHIO, Dept. of Physiological Chemistry, University of Tokyo, Japan

VAN DER VOORN, LOESJE, Netherlands Cancer Institute, Amsterdam, The Netherlands

VANKESSEL, GEURTS A., Dept. of Cell Biology, Erasmus University, Rotterdam, The Netherlands

VANNUCCI, SUSAN, Dept. of Physiology, Milton S. Hershey Medical Center, Pennsylvania

VANONI, MARCO, Sezione Biochimica Comparata, Fisiologia et Biochimica Generali, Milano, Italy

VERMA, INDER, Tumor Virology Laboratory, Salk Institute, San Diego, California

VICENTINI, LUCIA, Dept. of Pharmacology, University of Milan, Italy

VOGT, PETER, Dept. of Microbiology, University of Southern California, Los Angeles

WALTON, PAUL, University of California, San Diego, Theodore Gildred Cancer Facility, La Jolla

WANG, YUHUAN, University of Washington, Seattle

WARTMANN, MARKUS, University Clinic Medical School Laboratories, Basel, Switzerland

WATERFIELD, MICHAEL, Imperial Cancer Research Fund, Ludwig Institute for Cancer Research, London, England

WATSON, JAMES, Cold Spring Harbor Laboratory, New York

WATTERSON, MARTIN, Dept. of Pharmacology, Vanderbilt University, Nashville, Tennessee

WEIEL, JAMES, Dept. of Pharmacology, University of Washington, Seattle

WEINBERG, ROBERT, Whitehead Institute Center for Cancer Research, Cambridge, Massachusetts

WEINBERGER, OFRA, Dept. of Physiology, Columbia University, New York, New York

WEISS, ARTHUR, University of California, San Francisco, California

WELCH, MARY, Boehringer Mannheim Corporation, Indianapolis, Indiana

WESTHEAD, EDWARD, Dept. of Biochemistry, University of Massachusetts, Amherst

WHITEWAY, MALCOLM, National Research, Council, Biotechnology Research Institute, Montreal, Canada

WIGLER, MICHAEL, Cold Spring Harbor Laboratory, New York

WILLIAMS, LEWIS, University of California, Howard Hughes Medical Institute San Francisco

WILLIAMS, MEGAN, Dept. of Pathology, Brigham and Women's Hospital, Boston, Massachusetts

WISHART, WILLIAM, Dept. of Biotechnology, Sandoz Ltd., Basel, Switzerland

WITTERS, LEE, Dept. of Medicine and Biochemistry, Dartmouth Medical School, Hanover, New Hamshire

WODNAR-FILIPOWICZ, WODNAR, Friedrich Miescher Institute, Basel, Switzerland

WOLFF, HENRIK, Dept. of Pathology, University of Helsinki, Finland

WOLITZKY, BARRY, Dept. of Molecular Genetics, Hoffmann-La Roche, Nutley, New Jersey

WONG, STEPHEN, Dept. of Pharmacology, University of Texas Southwestern Medical Center, Dallas, Texas

WOOD, JOHN, Sandoz Institute for Medical Research, London, England

WRIGHT, TIMOTHY, Dept. of Medicine, Johns Hopkins University School of Medicine, Baltimore, Maryland

YAMAMOTO, KEITH, Dept. of Biochemistry and Biophysics, University of California, San Francisco

YANG, YU-CHUNG, Genetics Institute, Cambridge, Massachusetts

YU, VICTOR, Neurex Corporation, Menlo Park, California

ZABRECKY, JIM, Integrated Genetics, Framingham, Massachusetts

ZIFF, EDWARD, Dept. of Biochemistry, New York University School of Medicine, New York

First row: I. Verma, J. Schlessinger; J. Feramisco, J. Marin, M. Wigler
Second row: M. Berridge; E. Racker, E. Krebs; M. Brown
Third row: E. Kandel, J.D. Watson; F. Dahlquist, J. Goldstein, D. Koshland

First row: L. Stryer; M. Simon, I. Herskowitz; E. Neer
Second row: Coffee break at Grace Auditorium;
(*right*) W. Moller; D. Ryan; J. Inglis

First row: M. Gilman, R. Franza; Z. Selinger; E. Racker, M. Wigler
Second row: Break on Blackford lawn
Third row: D. Lancet; D. Botstein, J.D. Watson; K. Yamamoto
Fourth row: D. Koshland, I. Herskowitz, G. Rubin; Picnic

Foreword

For many decades, it has been clear that cells have a multitude of ways of sensing their environment and converting a plethora of external signals into measured intracellular responses. Already by 1965 Earl Sutherland had made the distinction between the "first-messenger" role of hormone-like signals and the "second-messenger" role of cyclic AMP, whose concentration was somehow influenced by signals from first messengers. In Sutherland's scheme, the first messenger binds to cellular receptors, with such binding events somehow leading to control of the concentration of the second messenger. Now we realize that many first messengers do not act directly through second messengers, but instead work at the genetic level by binding to cytoplasmically located receptors, which can then bind to DNA and turn on or off the functioning of specific genes. Today, we refer to the way that external signals are passed through various cellular components as signal transduction processes, with receptors and their associated molecules known as biological transducers. Because most transducer molecules are present in very limited amounts, their study at the biochemical level until recently was at best difficult, and hypotheses as to how they functioned were almost impossible to test rigorously.

Today, recombinant DNA techniques have dramatically changed the picture. Even very rare receptors are now open to analysis if their respective genes can be cloned, and virtually every month, the amino acid sequence of a new key biological transducer is established. The time was thus appropriate last June to hold a Cold Spring Harbor Symposium on the Molecular Biology of Signal Transduction. Our aim as always was to have the key players in the field present, and two members of our staff, Mike Wigler and Jim Feramisco, worked with me as the principal organizers. In so functioning, we contacted many outside advisors, and in particular we wish to thank Richard Axel, Howard Berg, Henry Bourne, Ron Evans, Alfred Gilman, Joe Goldstein, Ira Herskowitz, Tom Jessel, Eric Kandel, Dan Koshland, and Lubert Stryer for their help.

The final program consisted of 119 speakers, who spoke before an audience of 439, the largest ever yet to attend a Cold Spring Harbor Symposium. The crowd overflowed the Grace Auditorium, and a large closed-circuit screen in Bush was provided for those who tired of standing long hours in the side aisles. The meeting opened with a series of five virtually electric presentations by Howard Berg, Alfred Gilman, Mark Ptashne, Lubert Stryer, and Keith Yamamoto. A mode of high excitement prevailed throughout the subsequent 14 long sessions, which were concluded by a graciously thoughtful summary by Henry Bourne.

Super-large meetings like this require many major sources of financial backing, and we gratefully acknowledge our long-term Symposium backers, the National Cancer Institute of the National Institutes of Health, the National Science Foundation, and the Department of Energy. We are also indebted to the Lucille P. Markey Charitable Trust for major help provided for the third straight year. Essential funds from our Corporate Sponsors Program also provide key support for our meetings program: Abbott Laboratories; American Cyanamid Company; Amersham International plc; AMGen; Applied Biosystems; Becton Dickinson and Company; Boehringer Mannheim GmbH; Bristol Myers Company; Cetus Corporation; Ciba-Geigy Corporation; Diagnostic Products Corporation; E.I. du Pont de Nemours & Company; Eastman Kodak Company; Genentech, Inc.; Genetics Institute; Hoffmann-La Roche Inc.; Life Technologies, Inc.; Johnson & Johnson; Eli Lilly and Company; Millipore Corporation; Monsanto Company; Oncogene Science, Inc.; Pall Corporation; Pfizer Inc.; Pharmacia; Schering-Plough Corporation; Smith Kline & French Laboratories; Tambrands Inc.; The Upjohn Company; The Wellcome Research Laboratories; Burroughs Wellcome Co.; and Wyeth Laboratories.

Again our Meetings Office staff—Maureen Berejka, Barbara Ward, Diane Tighe, Karen Otto, and Michela McBride—performed at high level, overseeing the registration and housing of the participants as well as making them feel welcomed and special. Herb Parsons again ran a very competent audiovisual department, and my secretary, Andrea Stephenson, smoothly handled the massive telephone and written correspondence needed to put together a Symposium. The publication of these books was again ably overseen by Nancy Ford, Managing Director of our Publications Department, working with our Symposium editors Dorothy Brown, Patricia Barker, and Ralph Battey, assisted by Joan Ebert, Mary Cozza, and Inez Sialiano.

James D. Watson, Director
October 13, 1988

Contents

Part 1

Symposium Participants v
Foreword xix

Bacterial Chemotaxis

A Physicist Looks at Bacterial Chemotaxis *H.C. Berg* 1
Roles of Methylation and Phosphorylation in the Bacterial Sensing System *D.E. Koshland, Jr., D.A. Sanders, and R.M. Weis* 11
Osmotaxis in *Escherichia coli* *J. Adler, C. Li, A.J. Boileau, Y. Qi, and C. Kung* 19
Galvanotaxis in Bacteria *J. Adler and W. Shi* 23
Interaction of CheB with Chemotaxis Signal Transduction Components in *Escherichia coli*: Modulation of the Methylesterase Activity and Effects on Cell Swimming Behavior *R.C. Stewart, C.B. Russell, A.F. Roth, and F.W. Dahlquist* 27
Protein Phosphorylation and Bacterial Chemotaxis *J.F. Hess, R.B. Bourret, K. Oosawa, P. Matsumura, and M.I. Simon* 41
Phosphoproteins Involved in Bacterial Signal Transduction *A.M. Stock, D.C. Wylie, J.M. Mottonen, A.N. Lupas, E.G. Ninfa, A.F. Ninfa, C.E. Schutt, and J.B. Stock* 49
Structure-Function Studies of Bacterial Chemosensors *P. Ames, J. Chen, C. Wolff, and J.S. Parkinson* 59
The End of the Line in Bacterial Sensing: The Flagellar Motor *R.M. Macnab* 67

Protein Phosphorylation

Casein Kinase II as a Potentially Important Enzyme Concerned with Signal Transduction *E.G. Krebs, R.N. Eisenman, E.A. Kuenzel, D.W. Litchfield, F.J. Lozeman, B. Lüscher, and J. Sommercorn* 77
Microinjection of the Catalytic Subunit of cAMP-dependent Protein Kinase Induces Expression of the c-*fos* Gene *K.T. Riabowol, M.Z. Gilman, and J.R. Feramisco* 85
Regulating Cell Growth: Casein-kinase-II-dependent Phosphorylation of Nuclear Oncoproteins *D. Carroll, N. Santoro, and D.R. Marshak* 91
The Family of Protein Kinase C: Its Molecular Heterogeneity and Differential Expression *U. Kikkawa, K. Ogita, M.S. Shearman, K. Ase, K. Sekiguchi, Z. Naor, A. Kishimoto, Y. Nishizuka, N. Saito, C. Tanaka, Y. Ono, T. Fujii, and K. Igarashi* 97
Protein Kinase C Regulation by Sphingosine/Lysosphingolipids *R.M. Bell, C.R. Loomis, and Y.A. Hannun* 103
Expression of Wild-type and Mutant Subunits of the cAMP-dependent Protein Kinase *G.S. McKnight, G.G. Cadd, C.H. Clegg, A.D. Otten, and L.A. Correll* 111
cAMP-dependent Protein Kinase: A Framework for a Diverse Family of Enzymes *S.S. Taylor, J.A. Buechler, L.W. Slice, D.K. Knighton, S. Durgerian, G.E. Ringheim, J.J. Neitzel, W.M. Yonemoto, J.M. Sowadski, and W. Dospmann* 121
Targets for Signal-transducing Protein Kinase *T. Hunter, P. Angel, W.J. Boyle, R. Chiu, E. Freed, K.L. Gould, C.M. Isacke, M. Karin, R.A. Lindberg, and P. van der Geer* 131
Molecular Analyses of Gene Products Associated with the Response of Cells to Mitogenic Stimulation *R.L. Erikson, D. Alcorta, P.-A. Bedard, J. Blenis, H.-P. Biemann, E. Erikson, S.W. Jones, J.L. Maller, T.J. Martins, and D.L. Simmons* 143
Interactions between the Middle T Antigen of Polyomavirus and Host Cell Proteins *S.A. Courtneidge, R.M. Kypta, and E.T. Ulug* 153

Tyrosine Phosphorylation in Signal Transduction *T.M. Roberts, D. Kaplan, W. Morgan, T. Keller, H. Mamon, H. Piwnica Worms, B. Druker, B. Cohen, B. Schaffhausen, M. Whitman, L. Cantley, U. Rapp, and D. Morrison* 161
raf Family Serine/Threonine Protein Kinases in Mitogen Signal Transduction *U.R. Rapp, G. Heidecker, M. Huleihel, J.L. Cleveland, W.C. Choi, T. Pawson, J.N. Ihle, and W.B. Anderson* 173
Calmodulin and Calmodulin-regulated Protein Kinases as Transducers of Intracellular Calcium Signals *T.J. Lukas, J. Haiech, W. Lau, T.A. Craig, W.E. Zimmer, R.L.Shattuck, M.O. Shoemaker, and D.M. Watterson* 185
Regulation of the Vertebrate Cell Cycle by the *cdc2* Protein Kinase *G. Draetta, L. Brizuela, B. Moran, and D. Beach* 195

Mechanism of Action of G Proteins

Role of G Proteins in Transmembrane Signaling *P.J. Casey, M.P. Graziano, M. Freissmuth, and A.G. Gilman* 203
Structure of the Genes Coding for G-protein α Subunits from Mammalian and Yeast Cells *Y. Kaziro, H. Itoh, T. Kozasa, R. Toyama, T. Tsukamoto, M. Matsuoka, M. Nakafuku, T. Obara, T. Takagi, and R. Hernandez* 209
Mutations Probe Structure and Function of G-protein α Chains *H.R. Bourne, S.B. Masters, R.T. Miller, K.A. Sullivan, and W. Heideman* 221
Receptor-effector Coupling by G Proteins: Purification of Human Erythrocyte G_i-2 and G_i-3 and Analysis of Effector Regulation Using Recombinant α Subunits Synthesized in *Escherichia coli* *L. Birnbaumer, J. Codina, R. Mattera, A. Yatani, R. Graf, J. Olate, J. Sanford, and A.M. Brown* 229
Functions of G-protein Subunits *E.J. Neer, S.-Y. Kim, S.-L. Ang, D.B. Bloch, K.D. Bloch, Y. Kawahara, C. Tolman, R. Lee, D. Logothetis, D. Kim, J.G. Seidman, and D.E. Clapham* 241
Receptor- and G-Protein-mediated Modulations of Voltage-dependent Calcium Channels *W. Rosenthal, J. Hescheler, W. Trautwein, and G. Schultz* 247
Unique Properties of a New GTP-binding Protein with a Molecular Mass of 24,000 Daltons Purified from Porcine Brain Membranes *T. Katada and M. Ui* 255
Functional Role of Muscarinic Acetylcholine Receptor Subtype Diversity *A. Ashkenazi, E.G. Peralta, J.W. Winslow, J. Ramachandran, and D.J. Capon* 263
ras Oncogene Proteins: Three-dimensional Structures, Functional Implications, and a Model for Signal Transducer *S.-H. Kim, A.M. de Vos, L. Tong, M.V. Milburn, P.M. Matias, J. Jancarik, E. Ohtsuka, and S. Nishimura* 273

Mechanisms of Neurosensory Transduction

Molecular Basis of Visual Excitation *L. Stryer* 283
Molecular Basis of the Functional Heterogeneity of the Muscarinic Acetylcholine Receptor *S. Numa, K. Fukuda, T. Kubo, A. Maeda, I. Akiba, H. Bujo, J. Nakai, M. Mishina, and H. Higashida* 295
Life, Evolution, and the Pursuit of Single Photon Sensitivity *M.W. Bitensky, M.M. Whalen, and D.C. Torney* 303
Visual Signal Transduction: The Cycle of Transducin Shuttling between Rhodopsin and cGMP Phosphodiesterase *M. Chabre, J. Bigay, F. Bruckert, F. Bornancin, P. Deterre, C. Pfister, and T.M. Vuong* 313
Molecular Mechanics of the Cyclic-GMP-activated Channel of Retinal Rods *J.W. Karpen, A.L. Zimmerman, L. Stryer, and D.A. Baylor* 325
Inositol Lipid Cascade of Vision Studied in Mutant Flies *Z. Selinger and B. Minke* 333
Molecular Transduction in Smell and Taste *D. Lancet, D. Lazard, J. Heldman, M. Khen, and P. Nef* 343
Expression of G-protein α Subunits in Rat Olfactory Neuroepithelium: Candidates for Olfactory Signal Transduction *D.T. Jones, E. Barbosa, and R.R. Reed* 349

Studies on Light Transduction by Bacteriorhodopsin and Rhodopsin *M. Braiman, J. Bubis, T. Doi, H.-B. Chen, S.L. Flitsch, R.R. Franke, M.A. Gilles-Gonzalez, R.M. Graham, S.S. Karnik, H.G. Khorana, B.E. Knox, M.P. Krebs, T. Marti, T. Mogi, T. Nakayama, D.D. Oprian, K.L. Puckett, T.P. Sakmar, L.J. Stern, S. Subramaniam, and D.A. Thompson* 355
Direct Coupling of G Proteins to Ionic Channels *A.M. Brown, A. Yatani, Y. Imoto, G. Kirsch, H. Hamm, J. Codina, R. Mattera, and L. Birnbaumer* 365
Role of G-protein-coupled Phosphatidylinositol System in Signal Transduction in Vertebrate Neurons: Experiments on Neuroblastoma Hybrid Cells and Ganglion Cells *D.A. Brown, H. Higashida, P.R. Adams, N.V. Marrion, and T.G. Smart* 375
Functional Expression of the 5-HT_{1c} Receptor in Neuronal and Nonneuronal Cells *D. Julius, A.B. MacDermott, T.M. Jessell, K. Huang, S. Molineaux, I. Schieren, and R. Axel* 385
Molecular Convergence of Presynaptic Inhibition and Presynaptic Facilitation on Common Substrate Proteins of Individual Sensory Neurons of *Aplysia* *D. Sweatt, A. Volterra, S.A. Siegelbaum, and E.R. Kandel* 395
Electrical and Biochemical Properties of the cGMP-gated Cation Channel from Rod Photoreceptors *U.B. Kaupp, W. Hanke, R. Simmoteit, and H. Lühring* 407
Sevenless, a Gene Encoding a Putative Receptor for Positional Information *G.M. Rubin* 417
The Role of Arachidonic Acid Metabolites in Signal Transduction in an Identified Neural Network Mediating Presynaptic Inhibition in *Aplysia* *E. Shapiro, D. Piomelli, S. Feinmark, S. Vogel, G. Chin, and J.H. Schwartz* 425
Molecular Mechanisms of Phospholipid Signaling Pathways in Mammalian Nerve Cells *M.R. Hanley, T.R. Jackson, W.T. Cheung, M. Dreher, A. Gatti, P. Hawkins, S.I. Patterson, M. Vallejo, A.P. Dawson, and O. Thastrup* 435

Growth Factor and Hormone Receptors

Purification and Characterization of Placental Membrane Phosphotyrosine Phosphatases *C.J. Pallen, L. Sahlin, G. Panayotou, and M.D. Waterfield* 447
Signal Transduction by the Platelet-derived Growth Factor Receptor *L.T. Williams, J.A. Escobedo, M.T. Keating, and S.R. Coughlin* 455
Role of Intrinsic Protein Tyrosine Kinase in Function and Metabolism of the Epidermal Growth Factor Receptor *G.N. Gill, W.S. Chen, C.S. Lazar, J.R. Glenney, Jr., H.S. Wiley, H.A. Ingraham, and M.G. Rosenfeld* 467
Molecular Analysis of the Nerve Growth Factor Receptor *B. Hempstead, N. Patil, K. Olson, and M. Chao* 477
Structure-Function Analysis of the β-Adrenergic Receptor *R.A.F. Dixon, I.S. Sigal, and C.D. Strader* 487
Functional Domains in the β-Adrenergic Receptor *E.M. Ross, S.K.-F. Wong, R.C. Rubenstein, and T. Higashijima* 499
Molecular Biology of Adrenergic Receptors *R.J. Lefkowitz, B.K. Kobilka, J.L. Benovic, M. Bouvier, S. Cotecchia, W. Hausdorff, H.G. Dohlman, J.W. Regan, and M.G. Caron* 507
Signal Transduction by Epidermal Growth Factor Receptor *J. Schlessinger, A. Ullrich, A.M. Honegger, and W.H. Moolenaar* 515
Macrophage Colony-stimulating Factor, CSF-1, and Its Proto-oncogene-encoded Receptor *C.J. Sherr, C.W. Rettenmier, and M.F. Roussel* 521
Platelet-derived Growth Factor Generates at Least Two Distinct Intracellular Signals That Modulate Gene Expression *S.D. Jones, D.J. Hall, B.J. Rollins, and C.D. Stiles* 531
Insulin and Insulin-like Growth Factor Receptors and Responses *R.A. Roth, G. Steele-Perkins, J. Hari, C. Stover, S. Pierce, J. Turner, J.C. Edman, and W.J. Rutter* 537
Response and Binding Elements for Ligand-dependent Positive Transcription Factors Integrate Positive and Negative Regulation of Gene Expression *M.G. Rosenfeld, C.K. Glass, S. Adler, E.B. Crenshaw III, X. He, S.A. Lira, H.P. Elsholtz, H.J. Mangalam, J.M. Holloway, C. Nelson, V.R. Albert, and H.A. Ingraham* 545

Part 2

Signal Transduction in Unicellular Eukaryotes

From Membrane to Nucleus: The Pathway of Signal Transduction in Yeast and Its Genetic Control *L. Marsh and I. Herskowitz* 557
The Role of G Proteins in Yeast Signal Transduction *K. Matsumoto, M. Nakafuku, N. Nakayama, I. Miyajima, K. Kaibuchi, A. Miyajima, C. Brenner, K. Arai, and Y. Kaziro* 567
Analysis of the Role of *SCG1*, a G_α Homolog, and *SST2* in Pheromone Response and Desensitization in Yeast *J. Kurjan and C. Dietzel* 577
Function of the *STE4* and *STE18* Genes in Mating Pheromone Signal Transduction in *Saccharomyces cerevisiae* *M. Whiteway, L. Hougan, D. Dignard, L. Bell, G. Saari, F. Grant, P. O'Hara, V.L. MacKay, and D.Y. Thomas* 585
Functional Domains of a Peptide Hormone Receptor: The α-Factor Receptor (*STE2* Gene Product) of the Yeast *Saccharomyces cerevisiae* *K.J. Blumer, J.E. Reneke, W.E. Courchesne, and J. Thorner* 591
Recovery from Pheromone-induced Arrest of the Yeast Cell Cycle: α-Factor Binding and Mutants That Show Pheromone-independent Arrest of Cell Division *D. Blinder, P. Spatrick, S. Bouvier, C. Sullivan, and D. Jenness* 605
Response of Yeast α Cells to **a**-Factor Pheromone: Topology of the Receptor and Identification of a Component of the Response Pathway *K.L. Clark, N.G. Davis, D.K. Wiest, J.-J. Hwang-Shum, and G.F. Sprague, Jr.* 611
Isolation and Characterization of Two Genes Encoding Yeast Mating Pheromone Signaling Elements: *CDC72* and *CDC73* *S.I. Reed, J. Ferguson, and K.-Y. Jahng* 621
Diverse Biological Functions of Small GTP-binding Proteins in Yeast *D. Botstein, N. Segev, T. Stearns, M.A. Hoyt, J. Holden, and R.A. Kahn* 629
Regulation of Vesicular Traffic by a GTP-binding Protein on the Cytoplasmic Surface of Secretory Vesicles in Yeast *P.J. Novick, B. Goud, A. Salminen, N.C. Walworth, J. Nair, and M. Potenza* 637
Studies of *RAS* Function in the Yeast *Saccharomyces cerevisiae* *M. Wigler, J. Field, S. Powers, D. Broek, T. Toda, S. Cameron, J. Nikawa, T. Michaeli, J. Colicelli, and K. Ferguson* 649
cAMP Receptor and G-protein Interactions Control Development in *Dictyostelium* *M. Pupillo, P. Klein, R. Vaughan, G. Pitt, P. Lilly, T. Sun, P. Devreotes, A. Kumagai, and R. Firtel* 657
Ion Channels in *Paramecium*, Yeast, and *Escherichia coli* *Y. Saimi, B. Martinac, M.C. Gustin, M.R. Culbertson, J. Adler, and C. Kung* 667
A Molecular Analysis of G Proteins and Control of Early Gene Expression by the Cell-surface cAMP Receptor in *Dictyostelium* *A. Kumagai, S.K.O. Mann, M. Pupillo, G. Pitt, P.N. Devreotes, and R.A. Firtel* 675

Signal Transduction in the Nucleus

jun: Oncogene and Transcriptional Regulator *A.R. Ball, Jr., T.J. Bos, C. Löliger, L.P. Nagata, T. Nishimura, H. Su, H. Tsuchie, and P.K. Vogt* 687
Transcriptional Regulation by the AP-1 Family of Enhancer-binding Proteins: A Nuclear Target for Signal Transduction *D. Bohmann, A. Admon, D.R. Turner, and R. Tjian* 695
Transcriptional Activation by Yeast GCN4, a Functional Homolog to the *jun* Oncoprotein *K. Struhl, C.J. Brandl, W. Chen, P.A.B. Harbury, I.A. Hope, and S. Mehadevan* 701
YAP1 Encodes a Yeast Homolog of Mammalian Transcription Factor AP-1 *W.S. Moye-Rowley, K.D. Harshman, and C.S. Parker* 711
Analysis of Serum Response Element Function In Vitro *C. Norman and R. Treisman* 719

Gene Regulation by Growth Factors *R. Metz, J. Gorham, Z. Siegfried, D. Leonard, E. Gizang-Ginsberg, M.A. Thompson, D. Lawe, T. Kouzarides, R. Vosatka, D. MacGregor, S. Jamal, M.E. Greenberg, and E.B. Ziff* 727
Transcriptional Regulation of c-*fos* *R. Prywes, T.M. Fisch, and R.G. Roeder* 739
Regulation of Proto-oncogene *fos*: A Paradigm for Early Response Genes *P. Sassone-Corsi, W.W. Lamph, and I.M. Verma* 749
Intracellular Mediators of c-*fos* Induction *M.Z. Gilman, L.A. Berkowitz, J.R. Feramisco, B.R. Franza, Jr., R.M. Graham, K.T. Riabowol, and W.A. Ryan, Jr.* 761
Beyond the Second Messenger: Oncogenes and Transcription Factors *T. Curran, F.J. Rauscher III, D.R. Cohen, and B.R. Franza, Jr.* 769
Nuclear Oncoproteins Determine the Genetic Program in Response to External Stimuli *A. Schönthal, S. Gebel, B. Stein, H. Ponta, H.J. Rahmsdorf, and P. Herrlich* 779
Phorbol-ester-induced Activation of the NF-κB Transcription Factor Involves Dissociation of an Apparently Cytoplasmic NF-κB/Inhibitor Complex *P.A. Baeuerle, M. Lenardo, J.W. Pierce, and D. Baltimore* 789
Transcriptional Regulation of Interferon-stimulated Genes: A DNA Response Element and Induced Proteins That Recognize It *D. Levy, N. Reich, D. Kessler, R. Pine, and J.E. Darnell, Jr.* 799

Steroid Hormone Receptors

Ligand-regulated Nonspecific Inactivation of Receptor Function: A Versatile Mechanism for Signal Transduction *K.R. Yamamoto, P.J. Godowski, and D. Picard* 803
Cooperative and Positional Independent *trans*-Activation Domains of the Human Glucocorticoid Receptor *R.M. Evans and S.M. Hollenberg* 813
Expression and Function of the Human Estrogen Receptor in Yeast *J.H. White, D. Metzger, and P. Chambon* 819
Cooperative Interactions of Steroid Hormone Receptors with Their Cognate Response Elements *M.-J. Tsai, S.Y. Tsai, L. Klein-Hitpass, M. Bagchi, J.F. Elliston, J. Carlstedt-Duke, J.K. Gustafsson, and B.W. O'Malley* 829
Cooperative Action of the Glucocorticoid Receptor and Transcription Factors *U. Strähle, A. Münsterberg, R. Mestril, G. Klock, W. Ankenbauer, W. Schmid, and G. Schütz* 835

Growth Control in Mammalian Cells

Anti-oncogenes and the Negative Regulation of Cell Growth *J.M. Horowitz, S.H. Friend, R.A. Weinberg, P. Whyte, K. Buchkovich, and E. Harlow* 843
Interaction of *ras* p21 Proteins with GTPase Activating Protein *F. McCormick, H. Adari, M. Trahey, R. Halenbeck, K. Koths, G.A. Martin, W. Crosier, K. Watt, B. Rubinfeld and G. Wong* 849
Analysis of Mammalian *ras* Effector Function *A. Hall, C. Calés, J.F. Hancock, A. Lloyd, A. Self, S. Gardener, M.D. Houslay, M.J.O. Wakelam, and C.J. Marshall* 855
Structure/Function Studies of the *ras* Protein *I.S. Sigal, M.S. Marshall, M.D. Schaber, U.S. Vogel, E.M. Scolnick, and J.B. Gibbs* 863
Critical Role of Cellular *ras* Proteins in Proliferative Signal Transduction *D.W. Stacey, M.-H. Tsai, C.-L. Yu, and J.K. Smith* 871
Influence of $p21^{ras}$ on Phosphatidylinositol Turnover *J. Downward and R.A. Weinberg* 883
Genomic Response to Growth Factors *D. Nathans, L.F. Lau, B. Christy, S. Hartzell, Y. Nakabeppu, and K. Ryder* 893
Identification of Growth-factor-inducible Genes in Mouse Fibroblasts *R. Bravo, M. Zerial, L. Toschi, M. Schürmann, R. Müller, S.I. Hirai, M. Yaniv, J.M. Almendral, and R.-P. Ryseck* 901
Characterization of $p47^{gag\text{-}crk}$, a Novel Oncogene Product with Sequence Similarity to a Putative Modulatory Domain of Protein–Tyrosine Kinases and Phospholipase C *B.J. Mayer, M. Hamaguchi, and H. Hanafusa* 907

Phospholipase C-148: Chromosomal Location and Deletion Mapping of Functional Domains *A. Bristol, S.M. Hall, R.W. Kriz, M.L. Stahl, Y.S. Fan, M.G. Byers, R.L. Eddy, T.B. Shows, and J.L. Knopf* 915
Primary Structure of PLC-154 *M. Katan, R.W. Kriz, N. Totty, R. Philp, E. Meldrum, R.A. Aldape, J. Knopf, and P.J. Parker* 921

Second Messenger Systems

Inositol Trisphosphate and Calcium Signaling *M.J. Berridge and C.W. Taylor* 927
Agonist-induced Calcium Oscillations in Depolarized Fibroblasts and Their Manipulation by Photoreleased Ins(1,4,5)P_3, Ca^{++}, and Ca^{++} Buffer *A.T. Harootunian, J.P.Y. Kao, and R.Y. Tsien* 935
Signal Transduction in Mitogenesis: Further Evidence for Multiple Pathways *E. Rozengurt, J. Erusalimsky, H. Mehmet, C. Morris, E. Nånberg, and J. Sinnett-Smith* 945
Role of Glycosyl Phosphoinositides in Insulin Action *A.R. Saltiel, D.G. Osterman, and J.C. Darnell* 955
Insulin Mediators: Structure and Formation *J. Larner, L.C. Huang, G. Tang, S. Suzuki, C.F.W. Schwartz, G. Romero, Z. Roulidis, K. Zeller, T.Y. Shen, A.S. Oswald, and L. Luttrell* 965
Gram-negative Endotoxin: A Biologically Active Lipid *C.R.H. Raetz, K.A. Brozek, T. Clementz, J.D. Coleman, S.M. Galloway, D.T. Golenbock, and R.Y. Hampton* 973
GTP-binding Proteins in the Control of Exocytosis *B.D. Gomperts and P.E.R. Tatham* 983
The Membrane Form of Guanylate Cyclase *D.L. Garbers, D.G. Lowe, L.J. Dangott, M. Chinkers, D.S. Thorpe, J.K. Bentley, C.S. Ramarao, D.V. Goeddel, and S. Singh* 993
Effects of Nitrovasodilators, Endothelium-dependent Vasodilators, and Atrial Peptides on cGMP *F. Murad, D. Leitman, S. Waldman, C.-H. Chang, M. Hirata, and K. Kohse* 1005
Molecular Cloning of the Growth-factor-activatable Human Na^+/H^+ Antiporter *C. Sardet, A. Franchi, and J. Pouysségur* 1011

Summary

Summary: Signals Past, Present, and Future *H.R. Bourne* 1019

Author Index 1033

Subject Index 1037

COLD SPRING HARBOR SYMPOSIA ON QUANTITATIVE BIOLOGY

VOLUME LIII

A Physicist Looks at Bacterial Chemotaxis

H.C. Berg
Department of Cellular and Developmental Biology, Harvard University, Cambridge, Massachusetts 02138; Rowland Institute for Science, Cambridge, Massachusetts 02142

In this paper, a survey is given of the ways in which the enteric bacterium *Escherichia coli* (or *Salmonella typhimurium*) senses and responds to chemical stimuli. If one is to appreciate the molecular biology of signal transduction, it helps to understand the problems that the molecular biology is meant to solve. *E. coli* has managed to find its way in the world in the face of daunting physical constraints. These constraints arise because the organism is microscopic and immersed in an aqueous environment. In the realm of sensory transduction, not everything that is true for the colon bacillus is true for the elephant.

Matters of Size

To better appreciate the subject of scale, consider the differences in size of *Homo sapiens*, *E. coli*, and the flagellar basal apparatus of *E. coli* (Fig. 1). *E. coli* lives in the human gut. It outnumbers all other motile organisms attending this Symposium by a very wide margin. It is remarkably small (10^{-4} cm in diameter, some 200,000 times smaller in diameter than *Homo sapiens*). It swims by rotating about six thin helical filaments (generally longer than those shown), which arise at random points on its surface. The motor that drives each filament is embedded in the cell wall and the cytoplasmic membrane and is smaller in diameter than the cell body by a factor of about 50.

A Biased Random Walk

In the absence of a stimulus, *E. coli* simply wanders around, as illustrated in Figure 2. It swims smoothly by rotating its flagella counterclockwise (CCW). In this case, the filaments work together in a bundle that pushes the cell along a gently curved path, called a run. Runs are of random duration, distributed exponentially with a mean and standard deviation of about 1 second; the shortest runs are most probable. Runs are terminated by chaotic events, called tumbles, that arise when the flagella rotate clockwise (CW). In this event, the filaments work independently, moving the cell errati-

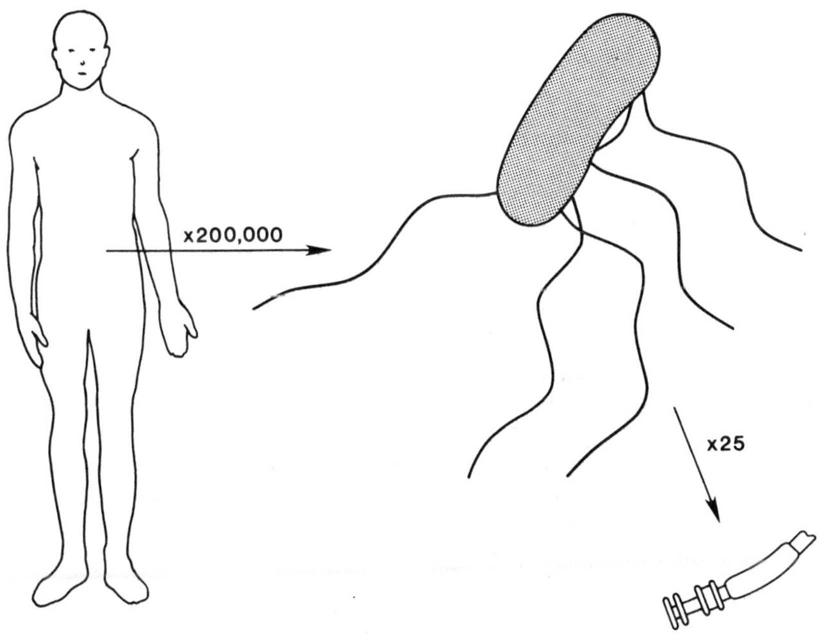

Figure 1. Comparison of the size of man, *E. coli*, and part of *E. coli*'s flagellar motor. The flagellar motor is some five million times smaller in diameter than one's fist. The sources of the diagrams are as follows: Man is adapted from Fig. 144 of Gray (1954). The bacterium is a tracing of Fig. 1 of Adler (1965), the publication that heralded modern work on bacterial chemotaxis. The motor fragment is a schematic outline of Fig. 1a of DePamphilis and Adler (1971), which showed the proximal end of a flagellum extracted from a cell by a neutral detergent and negatively stained with uranyl acetate. The curved segment is the proximal hook, which is on the outside of the cell. The adjacent pair of rings is a bushing that gets the drive shaft through the outer cell membrane. The inner pair of rings comprise part of the rotor and stator. The innermost ring is embedded in the cytoplasmic membrane.

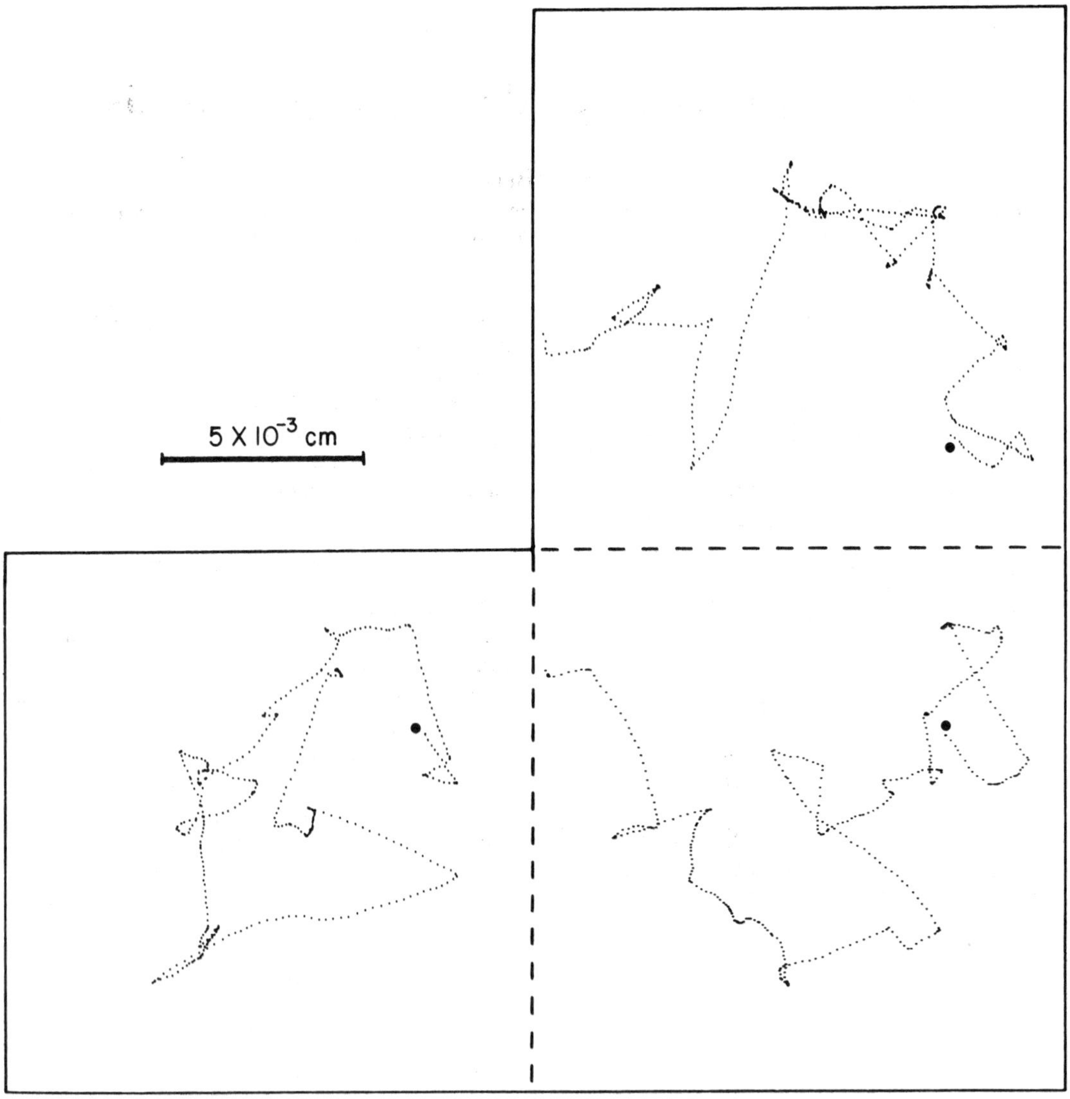

Figure 2. A digital plot (12.6 points/sec) of the displacement of a wild-type cell (*E. coli* strain AW405) executing a random walk in a homogeneous, isotropic medium. These are planar projections of a three-dimensional track: If the left and upper panels are folded out of the page along the dotted lines, the projections appear in proper orientation on three adjacent faces of the cube. Tracking began at the large dot and continued for about 30 sec. The cell swam at the speed of $\sim 2 \times 10^{-3}$ cm/sec. There were 26 runs and tumbles. (Reprinted, with permission, from Fig. 6.4 of Berg 1983, which was based on Fig. 1 of Berg and Brown 1972.)

cally with little net displacement. Tumbles also are of random duration, distributed exponentially, with a mean and standard deviation of about 0.1 second. Following a tumble, the cell runs again, picking a new direction, more or less at random. When the cell swims in a spatial gradient of a chemical attractant, runs that happen to carry it up the gradient are extended, whereas those that happen to carry it down the gradient are not. Thus, the cell drifts in a favorable direction by executing a biased random walk. The bias is positive, not negative; the runs get longer, not shorter. *E. coli* is an "optimist."

Limits Set by Physics

Physical constraints influence this sensory strategy in several ways: (1) The flagellar filaments are long, thin, and helical because motion is dominated by viscous rather than inertial forces. It sounds contradictory, but thrust must be generated by viscous drag. (2) A cell is unable to run in a straight line, because rotational perturbations due to Brownian movement knock it off its path. *E. coli* forgets where it is going in about 10 seconds. (3) A cell cannot improve its lot locally by swimming or stirring, because transport of small molecules is effected by diffusion rather than bulk flow. Local displacements of the fluid do not increase the cell's nutritional intake. (4) A cell must sense a chemical gradient temporally rather than spatially, because comparisons between concentrations in front or behind are overwhelmed by diffusive currents due to its rapid motion. Although the net increase in intake is very small, a cell always absorbs more molecules in front than behind. Therefore, with a spatial mechanism, any new direction is deemed favorable. Finally, (5) the precision with which a cell can make temporal comparisons is limited by statistical fluctuations. The counting statistics improve with the square root of the product of the concentration and the integration time. Therefore, a chemical cannot be sensed at an arbitrarily low concentration, because the integration time required would be prohibitively long. For further discussion, see Berg and Purcell (1977), or for treatments that are less mathematical, see Purcell (1977) and Berg (1985).

Mapping Responses in Time and Space

The time scale on which *E. coli* measures changes in its environment and the spatial scale on which signals are propagated internally from the chemoreceptors to the flagella have been determined by delivering chemical attractants (e.g., aspartate) or repellents (e.g., benzoate) with iontophoretic pipettes, as shown in Figure 3. In panel a, a cell is tethered to a glass surface by the stub of a single flagellar filament (following mechanical shear and the addition of antifilament antibody; cf. Silverman and Simon 1974). In panel b, an inert marker (a cell with abnormally long proximal hooks, called polyhooks, killed by fixation with glutaraldehyde) is attached to a polyhook of a living filamentous cell (by addition of antihook antibody; cf. Ishihara et al. 1983). The tethered cell or the inert marker spin alternately CW or CCW, reversing their directions at random times, on average about once per second. In a strain that is wild type for chemotaxis, the fraction of time that the cell or marker spins CCW (called the bias) is roughly one-half. Control of the addition of the attractant or repellent can be effected with a precision of a few milliseconds by passing a current through the pipette. In Figure 3a, the tip of the pipette is fixed a few micrometers away from the tethering axis. In Figure 3b, it is moved through a series of positions off the end of the filamentous cell or along the cell body (at similar distances from the marker, as shown by the dashed inverted "V").

Impulse Responses

When the experiment of Figure 3a is done with short impulsive stimuli (e.g., with pulses of aspartate of small amplitude and width), a biphasic response is obtained, as shown in Figure 4. The bias rises rapidly, returns to the baseline after about 1 second, falls below the baseline, and then returns to the baseline again after another 3 seconds; the areas of the positive and negative lobes are the same. If one uses a pulse of similar width but smaller amplitude, the response follows the same time course; only its amplitude is smaller. Evidently, a transient increase in occupancy of the aspartate receptor sets in motion a train of events that affect the direction of rotation of the flagellar motor for a span of 4 seconds. If one uses a mutant deleted for the genes *cheR* and *cheB*, whose products are enzymes that carboxymethylate and demethylate the receptor, respectively (see below), the second lobe of the response is missing. If one uses an amber mutant of *cheZ*, whose wild-type product is another cytoplasmic protein, the whole process slows down by a factor of about 10 (cf. Segall et al. 1986).

What *E. coli* Measures

Given a linear system, e.g., one without thresholds that is exposed to small stimuli, one can use the impulse response (Fig. 4) to predict the behavior expected with any stimulus. The impulse response is a weighting function, in the following sense. To discover how the pres-

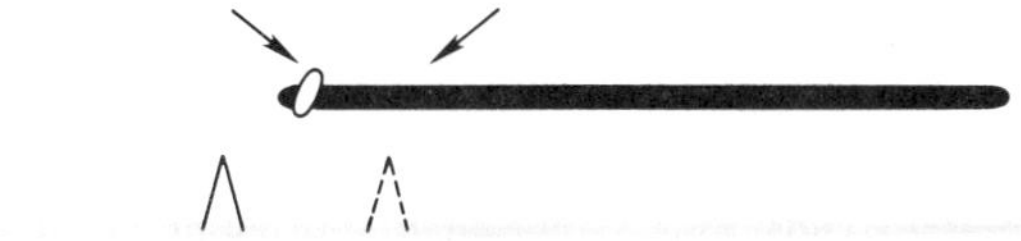

Figure 3. A schematic illustration of experiments in which a tethered cell (*a*) or a filamentous cell (*b*) is simulated by ejection of a charged chemical from the tip of an iontophoretic pipette. (*a*) The pipette is fixed; (*b*) it is moved to a series of alternative positions, one of which is shown (dashed inverted "V"). The tethered cell is driven by one of its own flagellar motors. The marker is inert, but it spins because it is linked to one of the motors of the filamentous cell. Filamentous cells can be prepared from nonseptate mutants or by growing ordinary cells in the presence of a β-lactam antibiotic (see Block et al. 1982; Ishihara et al. 1983; Segall et al. 1985).

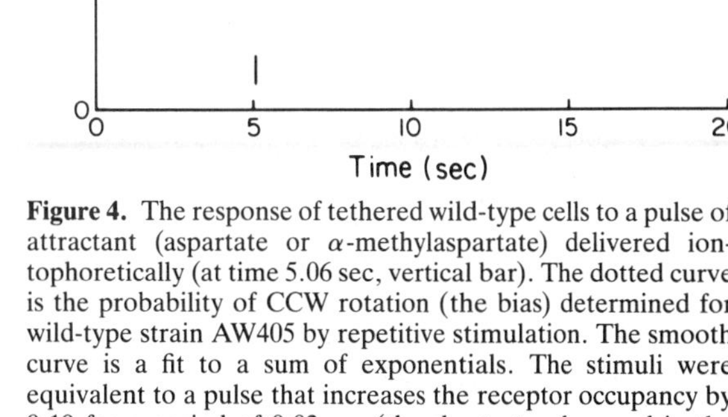

Figure 4. The response of tethered wild-type cells to a pulse of attractant (aspartate or α-methylaspartate) delivered iontophoretically (at time 5.06 sec, vertical bar). The dotted curve is the probability of CCW rotation (the bias) determined for wild-type strain AW405 by repetitive stimulation. The smooth curve is a fit to a sum of exponentials. The stimuli were equivalent to a pulse that increases the receptor occupancy by 0.19 for a period of 0.02 sec (the shortest pulse used in the experiments). (Reprinted, with permission, from Fig. 1 of Segall et al. 1986.)

ent bias of a cell deviates from its baseline, proceed as follows: (1) Plot the concentration of the attractant for the past several seconds on the same time scale as that in Figure 4, setting the present time at the vertical bar but going backwards in time from left to right. (2) Multiply the two curves together, calling the baseline of the impulse response zero. (3) Integrate to find the area of the product. The shift in bias is proportional to this number. Because the impulse response (for the wild type) remains at its baseline (the zero value) for times greater than 4 seconds, the present bias depends only on events that have occurred during the past 4 seconds. Because the areas of the two lobes of the impulse response are the same, a shift in bias will occur only if the concentration has changed over that time. In brief, wild-type cells exposed to stimuli in the physiological range make short-term temporal comparisons extending 4 seconds into the past: the past 1 second is given a positive weighting, the previous 3 seconds are given a negative weighting, and the cells respond to the difference.

The impulse response for repellent stimuli is similar to that for attractant stimuli (Fig. 4) but inverted (cf. Fig. 3 in Block et al. 1982). However, experiments in which the concentrations of attractants or repellents are varied over long periods of time by programmed mixing indicate that repellent stimuli are subject to a substantial threshold. Beyond a certain point, cells simply ignore gradual increases in the concentrations of repellents (or gradual decreases in the concentrations of attractants) (see Block et al. 1983). This mechanism underlies *E. coli*'s optimism.

Temporal Sensing

These temporal comparisons are made continuously, regardless of whether the cells run or tumble. This was evident early on from experiments in which an attractant (glutamate) was generated or destroyed enzymatically (by the action of alanine aminotransferase) in a solution containing swimming cells (of a mutant insensitive to alanine) that was otherwise homogeneous and isotropic (Brown and Berg 1974). When the attractant was generated, the cells swam in a random fashion, as before (i.e., as in Fig. 2), but all the runs were longer, regardless of the direction in which a cell happened to swim. When the attractant was destroyed (at a rate below the threshold noted above), no changes in behavior were observed.

The changes in bias found in experiments with programmed mixing are directly proportional to the time rate of change in receptor occupancy (Block et al. 1983). As noted by Delbrück and Reichardt (1956), comparisons between the recent and more distant past require at least two biochemical processes, one that occurs rapidly (e.g., within 1 sec of the specified change in receptor occupancy) and the other more slowly (e.g., during the following 3 sec). We know relatively little about the rapid process (or processes). The slow process, as we have seen from the disappearance of the second lobe of the impulse response in *cheR cheB* mutants, involves receptor carboxymethylation.

Saturation of the Response

It has proved convenient to separate rapid from slow changes that occur during signal processing by exposing cells suddenly to large changes in the concentrations of attractants or repellents. In this event, the response saturates rapidly (the bias goes to 1 or to 0 within about 0.2 sec; cf. Segall et al. 1982), and it remains saturated for a very long time (up to several min). The interval required for the bias to saturate has been called an "excitation time," and the interval required for the bias to come back on scale has been called a "recovery time" (Spudich and Koshland 1975) or a "transition time" (Berg and Tedesco 1975). But note that these excitation, or recovery, times are not related in anyze simple way to the times that characterize signal processing of stimuli in the normal physiological range. Cells swimming in the wild never see such abrupt and large temporal stimuli, because discontinuities in concentration are smoothed out by diffusion. Cells are sensitive to gradients only when the bias is on scale.

Gain of the System

The gain of the signal-processing system is prodigious. A step change that increases the occupancy of the aspartate receptor by 1 molecule (assuming 600 receptor molecules per cell) increases the rotational bias transiently by about 0.1. A ramp that increases the receptor occupancy by one molecule per second leads to a steady-state increase in bias of a similar magnitude.

An Optimum Strategy

The impulse response (Fig. 4) is matched to the random walk (Fig. 2). In trying to decide whether life is getting "better" or "worse," the cell averages concentrations for as long as it can (~1 sec). It does this to avoid being overwhelmed by fluctuations. However, if this integration time were much longer than 1 second, several tumbles could occur before a decision were reached, and the results would no longer be relevant. Note that the cell cannot solve this problem by making the mean run length longer, because rotational Brownian movement carries it off course by as much as 90° within 10 seconds. The cell sets the mean run length at 1 second to provide ample head room for a response, and it extends runs rather than shortening them to minimize changes in direction that might occur during the integration time.

A Simulation

One can put all of this together in a simulation of the sort shown in Figure 5. The slide used at the Sym-

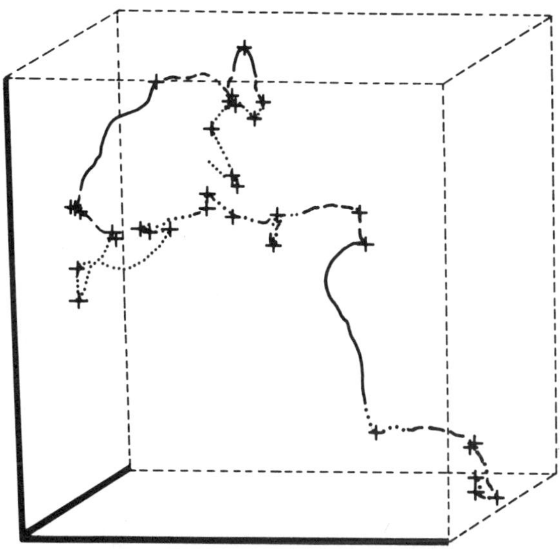

Figure 5. Simulation of the behavior of a wild-type cell in an exponential gradient of attractant (aligned in the vertical direction, positive upward, corresponding to an increment in receptor occupancy of 1 part in $10^4/\mu m$ at a mean occupancy of 0.5). The coordinate axes are ~200 μm long. The cell started off in a direction chosen at random near the bottom right edge of the figure. It ran at a fixed speed of 20 μm/sec. There were 35 runs and 34 tumbles. The value of the internal parameter setting the probability of a tumble is indicated by the nature of the line: large (dotted here; blue in the original), medium (dashed; yellow), small (solid; red). Other details are given in the text. (Figure courtesy of S.M. Block.)

posium was in color, which was more dramatic. The color (mimicked here by dots, dashes, and solid lines) indicates the value of an internal parameter that one might call "happiness," the signal that sets the bias, i.e., that determines the probability that the cell runs or tumbles. This parameter is computed by using the impulse response (Fig. 4) to weight the recent and more distant past in the manner described earlier. If the cell is running and the output of this computation (called a convolution) is negative, i.e., if the concentration has been decreasing for some time, the result is ignored, and the cell decides whether it should tumble by picking from an exponential distribution with a mean of 1 second. If the cell is running and the output of the convolution is positive, the cell picks from an exponential distribution with a larger mean (one with an exponent decreased in linear proportion to that output). On the other hand, if the cell is tumbling, it decides whether it should run by picking from an exponential distribution with a mean of 0.1 second. If a new run is called for, the cell heads in a direction that differs from that at the end of the previous run by a random choice from a distribution of angles peaked somewhat in the forward direction (cf. Fig. 3 in Berg and Brown 1972). Brownian rotation is included by giving the cell a small kick in angle every iteration (every 0.01 sec). For the simulation shown here, the gain of the system was chosen to be optimum, i.e., so that the probability of terminating a run would just go to zero if a cell were to swim unerringly straight up the gradient. Simulations of 100 cells for 10 minutes each gave a mean draft rate of 1.7 μm per second (~8% of the run speed).

When watching such tracks evolve on a computer screen, one gets the impression of a bloodhound following a scent. The cell sniffs about (dotted or blue), picks up the spoor (dashed or yellow), and then howls up the gradient (solid or red). Eventually, rotational Brownian motion carries it off the track, and it is forced to sniff about again. Note that for this cell (Fig. 5), most of the progress up the gradient occurred in two long runs (solid lines). The reason for playing this game is more serious than reference to bloodhounds might suggest. First, we would like to know whether we really understand the behavior of wild-type cells. Second, we would like to predict, from measurements of impulse responses of tethered chemotactic mutants, how pathological their swimming behavior might be.

Biochemistry and Genetics

What is the machinery that enables a creature of such small size to manage such remarkable sensory feats? The biochemical landscape is sketched in Figure 6. For more information, see the recent and extensive reviews of Stewart and Dahlquist (1987) and Macnab (1987a,b). Three major systems for chemoreception have been recognized. The first, noted at the top of Figure 6, detects serine and certain repellents such as leucine (mediated by Tsr = MCPI [*m*ethyl-accepting *c*hemotaxis *p*rotein I]), aspartate and certain other repellents such as Ni^+ (mediated by Tar = MCPII), dipeptides (mediated by Tap), and ribose and galactose (mediated by Trg = MCPIII). The second, shown at the left, detects sugars transported by the sugar phosphotransferase system. The third, shown at the bottom, detects oxygen and certain other electron acceptors. The first of these systems has been studied most extensively. For literature on the second and third systems, see Macnab (1987b), Vogler and Lengeler (1987), and Shioi et al. (1987). The chemotaxis genetic map is shown in Figure 7.

Transporting Information Across the Cytoplasmic Membrane

The proteins Tsr, Tar, Tap, and Trg bind ligands either directly or indirectly and mediate responses for several different kinds of inputs; therefore, they have been called transducers. Their genes appear to be members of one family, and they are composed of discrete structural and functional domains (cf. Krikos et al. 1983; Bollinger et al. 1984). The periplasmic domains are divergent, whereas the cytoplasmic domains are highly conserved. The former contain binding sites for attractants or attractant-binding proteins, whereas the latter contain two sets of clustered sites for methylation and a region thought to be involved in relaying the signal to the flagella. One transducer, the Tar protein, is shown schematically in Figure 8. An

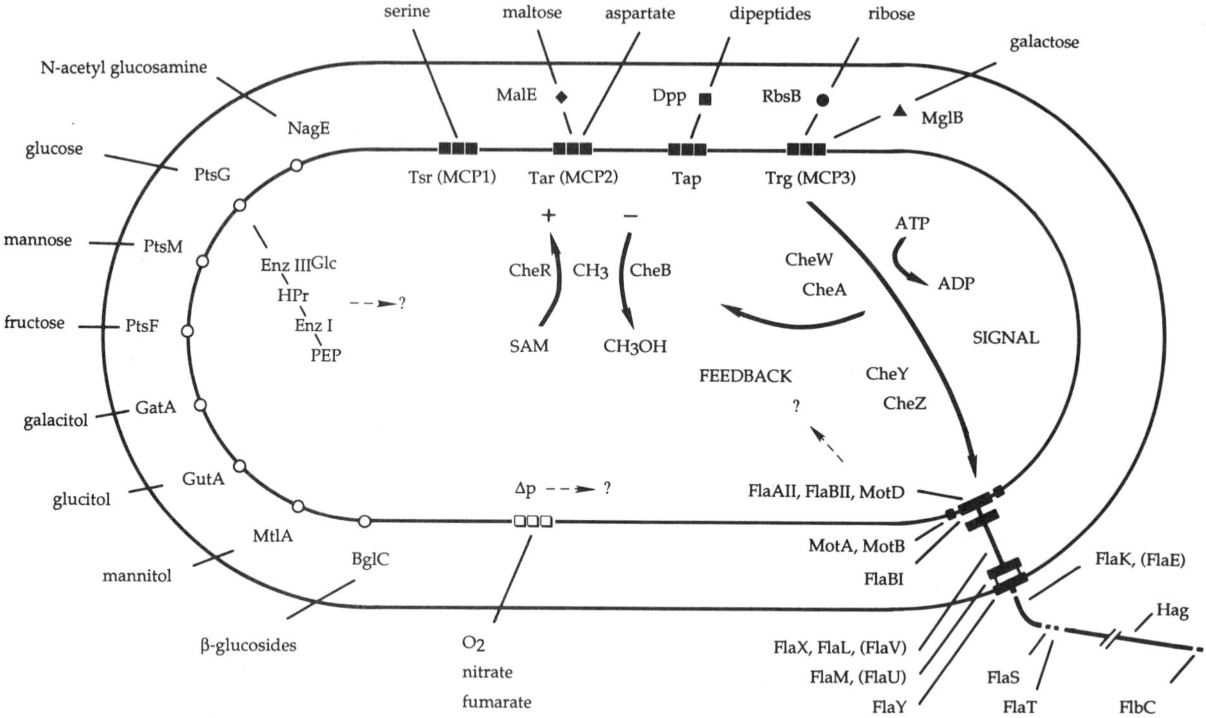

Figure 6. Machinery for chemotaxis in *E. coli* that has been identified thus far. The chemoreceptors are binding proteins found in the periplasmic space, such as MalE (which binds maltose), or proteins that span the cytoplasmic membrane, such as Tsr (which binds serine) or the enzymes II of the sugar phosphotransferase system, such as MtlA (which binds mannitol). Information about receptor occupancy is passed across the cytoplasmic membrane and eventually to the base of the flagellar motors, where it changes the probabilities that the motors turn CW or CCW. This signalling occurs through a cascade of events involving the cytoplasmic proteins CheW, CheA, CheY, and CheZ. A negative feedback loop counteracts the effects of this signal, enabling the cells to adapt. In the system studied most extensively, which involves the transducers Tsr, Tar, Tap, and Trg (also called MCPs), this feedback is effected by carboxymethylation of cytoplasmic domains. When more attractant is bound, methyl groups (CH_3) are transferred from *S*-adenosylmethionine (SAM) via a methyltransferase (CheR). When less attractant is bound, these methyl groups are removed by a methylesterase (CheB), liberating methanol (CH_3OH). In the Pts and O_2 systems, the signal and feedback pathways are not known, although signaling in the latter system might involve changes in protonmotive force (Δp). Components of the flagellar motor, hook, and filament are shown at the lower right. Fla components are required for flagellar assembly. FlaAII and FlaBII comprise the switch that enables a motor to change direction. The gene products shown in parentheses are not structural components per se but are required for proper assembly, e.g., FlaE controls the length of the hook (FlaK). Hag is the structural protein of the filament (flagellin). FlaS, FlaT, and FlbC are the so-called hook-associated proteins. MotA and MotB are required for force generation.

unanswered question, relevant to many other receptor systems, is how sensory information moves from the periplasmic domain to the cytoplasmic domain.

Linking the Transducers to the Flagella

The information-processing pathway is convergent. Information flows from many receptor molecules of several different kinds to a few flagellar motors (Fig. 6). Measurements of external concentration can be made most effectively (integration times can be shortened) if these receptors are scattered widely over the surface of the cell and if each contributes its bit to the signal that controls the direction of flagellar rotation (Berg and Purcell 1977). How is this coupling achieved? One approach to answering this question is illustrated in Figure 3b. When the pipette was close to the cell surface (as indicated by the dashed inverted "V"), the response of the motor driving the marker was larger than when the pipette was off the end of the cell (as indicated by the solid inverted "V"). This would be expected if there were an internal signal, because more receptors are influenced by the pipette when it is close to the cell surface. However, this effect faded rapidly as the pipette was moved farther away (beyond a few microns), indicating that the range of the internal signal is short. These data ruled out a number of possible mechanisms, including those involving changes in membrane potential (see also Margolin and Eisenbach 1984). Mutants of *cheZ* yielded a range longer by a factor of about 3. This inspired a hypothesis in which a small protein, CheY, is activated at the receptors and inactivated by CheZ as it diffuses through the cytoplasm. Active CheY binds to the components of the flagellar switch and enhances CW rotation.

CheY has been shown to have just such an effect (Clegg and Koshland 1984; Ravid et al. 1986; Wolfe et al. 1987). Indeed, the coupling pathway for a response to aspartate can be restored by adding back the compo-

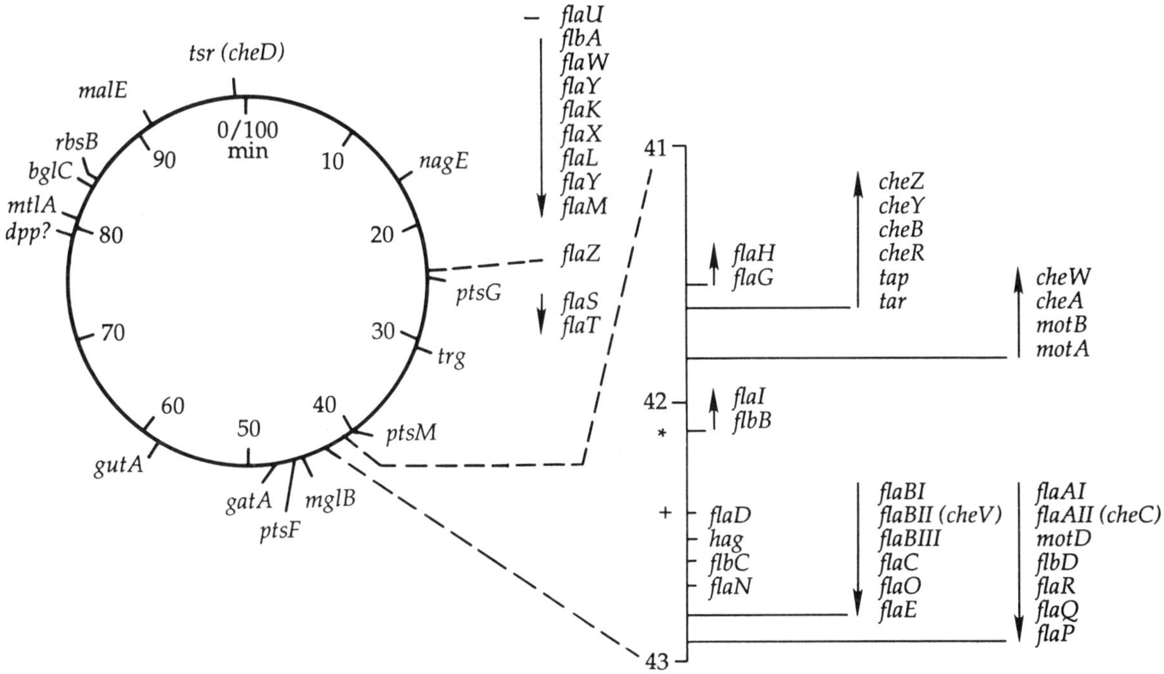

Figure 7. Genetic map of chemotaxis loci in *E. coli*. The numbers indicate map position in minutes. The arrows indicate the direction and extent of transcriptional units. (*) Master operon, turned on by cAMP and a catabolite activator protein (cap); (+) positive regulator of late operons; (−) negative regulator of late operons; late operons include hook-associated proteins (FlaS, FlaT, FlbV), Mot, Che, Hag (via FlaZ). The only *pts* genes shown are those for EnzII. Genes required for appearance of first incomplete flagellan structure: *flbA*, *flaW*, *flaX*, *flaL*; *flaZ*; *flaG*, *flaH*; *flbB*, *flaI*; *flaD*; *flaN*; *flaBI*, *flaBII*, *flaBIII*, *flaC*, *flaO*, *flaAI*, *flaAII*, *motB*, *flbD*, *flaR*, *flaW*, *flaP*. (Adapted from Fig. 1 of Parkinson [1981], using data tabulated by Macnab [1987b].)

nents Tar, CheW, CheA, and CheY to a cell that has been "gutted" for all of the transducers (Tsr, Tar, Tap, and Trg) and all of the soluble *che* gene products (CheW, CheA, CheR, CheB, CheY, and CheZ; Wolfe et al. 1987). Both CheW and CheA appear to be required (A.J. Wolfe et al., in prep.).

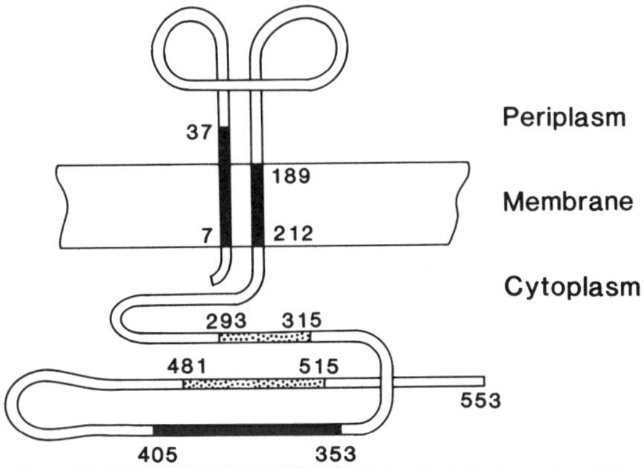

Figure 8. A schematic representation of the transducer Tar. The numbers refer to the positions of the amino acids in the peptide chain (shown as a ribbon), beginning at the amino terminus. Methylation of glutamyl residues occurs in two regions, between amino acids 293 and 315 (called K_1) and 481 and 515 (called R_1). The region between amino acids 405 and 353 is identical in Tsr, Tar, Tap, and Trg and is thought to be involved in signaling (cf. Mutoh et al. 1986). (Adapted from Fig. 6 of Krikos et al. [1983] and Fig. 5 of Russo and Koshland [1983].)

Phosphorylation

This discussion has been enriched enormously by the recent in vitro findings of Simon and co-workers (reviewed by Parkinson 1988) that CheA autophosphorylates (Hess et al. 1987), that phosphorylated CheA, in turn, phosphorylates CheY and CheB (Hess et al. 1988), and that CheZ accelerates the dephosphorylation of CheY (Hess et al. 1988). In addition, mutants of *cheA* defective in chemotaxis are defective in phosphorylation (Oosawa et al. 1988). Work of a similar nature on CheA and CheY has been reported by Stock et al. (1988) and Wylie et al. (1988). Thus, it appears that information transfer occurs via a cascade of reactions involving protein phosphorylation.

Other Modifications

In pursuing work on gutted cells containing CheY, A.J. Wolfe and M.P. Conley stumbled onto a surprising effect in which acetate but not benzoate enhances CW rotation. Normally, weak acids act by lowering the cytoplasmic pH, which perturbs both Tsr (leading to a CW response) and Tar (leading to a CCW response); benzoate is more effective than acetate. The acetate effect in gutted cells containing CheY has been traced to the intermediate acetyladenylyate and the enzyme acetate-CoA synthetase. Mutants defective in the latter enzyme, which grow if given oleic acid, fail to respond in capillary assays to serine, aspartate, or maltose.

Therefore, it is possible that phosphorylation will not be the only modification: Other kinds of chemical modification, such as acetylation or adenylylation, might occur at other stages in the signaling pathway (Wolfe et al. 1988).

Feedback

A branch in the pathway generating negative feedback via CheB occurs between CheA and CheY (Fig. 6). This branch has been studied extensively by Dahlquist and his co-workers, who have labeled cells with radioactive methionine and followed the evolution of radioactive methanol (cf. Kehry et al. 1985; Stewart and Dahlquist 1987). Because chemotaxis cannot occur if the rotational basis is 0 or 1, one also would expect feedback at the level of the flagellar motor (indicated by the dashed arrow and question mark in Fig. 6), but this branch remains to be identified.

Stochastic Output

What does the signal do when it reaches the flagellar motors? Experiments with filamentous cells carrying multiple markers (Ishihara et al. 1983) or with cells of normal size viewed with high-intensity light (Macnab and Han 1983) indicate that the motors do not change direction synchronously. The signal determines the rates at which transitions occur between CW and CCW states, not the transition times per se. Our current thinking is that activated CheY binds to FlaAII and/or FlaBII, changing these rates, and that CheY, once bound, is sensitive to attack by CheZ. But this is probably naive, because so little is known about how the motor controls its direction of rotation. Evidently, the coordination required for formation of a flagellar bundle arises from external interactions between different flagellar filaments.

The Rotary Motor

The flagellar motor is a remarkable machine. For one suggestion as to how it might work, see Berg and Khan (1983) or Khan and Berg (1983). Its specifications are given in Table 1. The speed is the value recorded in swimming cells at room temperature (Lowe et al. 1987). The power output seems miniscule, until one divides by the weight; internal combustion engines do no better (cf. McMahon and Bonner 1983). The motor is driven by an inward flux of protons (or hydronium ions); about 1000 are required per revolution (Meister et al. 1987). Proceeding more systematically than possible in earlier work (Block and Berg 1984), D.F. Blair has induced the synthesis of wild-type MotA and/or MotB in *mot*$^{-}$ strains of *E. coli* and found several examples in which tethered cells reach their top speed in eight steps; therefore, we now believe that there are only eight independent force-generating units (D.F. Blair and H.C. Berg, in prep.). For more information about motor components, see Macnab (this volume). It is the gears that are critical for chemotaxis.

Table 1. Motor Specifications (Approximate)

Feature	Specifications
Diameter	1 micro inch
Speed	6000 rpm
Power output	1/10 micro micro micro hp
Power output per unit weight	10 hp per pound
Power source	proton current
Cylinders	8
Number of different kinds of parts	30
Gears	forward and reverse

SUMMARY

What is distinctive about bacterial chemotaxis, as compared to, for example, taste in the elephant, is the time over which decisions must be made. The lower limit is set by diffusion of chemicals to and from the cell surface, which demands long times for statistically significant counts. The upper limit is set by diffusion of the cell itself, which demands short times for well-defined swimming paths. For an organism the size of *E. coli*, temporal comparisons of the concentrations of chemicals in the environment must be made within a few seconds. Although such short time spans might be difficult for the biochemist, they are not so difficult for *E. coli*, because diffusion can carry a small molecule across the cell in about 1 msec. *E. coli* has the opposite problem: How does it integrate inputs from many receptors over periods 1000 times as long?

The mechanisms for this signal processing are beginning to be understood. We know how most chemical attractants are identified, how temporal comparisons might be made, and how the behavioral output is effected. We know less about how sensory information crosses the cytoplasmic membrane, how the reactions that link the receptors to the flagella generate such high gain, and what actually controls the direction of flagellar rotation.

One thing is quite clear: *E. coli* demands our admiration and respect.

ACKNOWLEDGMENTS

I thank Karen Fahrner for her comments on the manuscript. This work was supported by grants from the National Science Foundation (DMB-8518257) and the National Institute of Allergy and Infectious Diseases (AI-16478).

REFERENCES

Adler, J. 1966. Chemotaxis in *Escherichia coli*. *Cold Spring Harbor Symp. Quant. Biol.* **30:** 289.

Berg, H.C. 1983. *Random walks in biology*. Princeton University Press, Princeton, New Jersey.

———. 1985. Physics of bacterial chemotaxis. In *Sensory perception and transduction in aneural organisms* (ed. G. Colombetti et al.), p. 19. Plenum Press, New York.

Berg, H.C. and D.A. Brown. 1972. Chemotaxis in *Escherichia coli* analysed by three-dimensional tracking. *Nature* **239:** 500.

Berg, H.C. and S. Khan. 1983. A model for the flagellar rotary motor. In *Mobility and recognition in cell biology* (ed. H. Sund and C. Veeger), p. 485. Walter de Gruyter, Berlin.

Berg, H.C. and E.M. Purcell. 1977. Physics of chemoreception. *Biophys. J.* **20:** 193.

Berg, H.C. and P.M. Tedesco. 1975. Transient response to chemotactic stimuli in *Escherichia coli. Proc. Natl. Acad. Sci.* **72:** 3235.

Block, S.M. and H.C. Berg. 1984. Successive incorporation of force-generating units in the bacterial rotary motor. *Nature* **309:** 470.

Block, S.M., J.E. Segall, and H.C. Berg. 1982. Impulse responses in bacterial chemotaxis. *Cell* **31:** 215.

———. 1983. Adaptation kinetics in bacterial chemotaxis. *J. Bacteriol.* **154:** 312.

Bollinger, J., C. Park, S. Harayama, and G.L. Hazelbauer. 1984. Structure of the Trg protein: Homologies with and differences from other sensory transducers of *Escherichia coli. Proc. Natl. Acad. Sci.* **81:** 3287.

Brown, D.A. and H.C. Berg. 1974. Temporal stimulation of chemotaxis in *Escherichia coli. Proc. Natl. Acad. Sci.* **71:** 1388.

Clegg, D.O. and D.E. Koshland, Jr. 1984. The role of a signaling protein in bacterial sensing: Behavioral effects of increased gene expression. *Proc. Natl. Acad. Sci.* **81:** 5056.

Delbrück, M. and W. Reichardt. 1956. System analysis for the light growth reactions of *Phycomyces*. In *Cellular mechanisms in differentiation and growth* (ed. D. Rudnick), p. 3. Princeton University Press, Princeton, New Jersey:

DePamphilis, M.L. and J. Adler. 1971. Fine structure and isolation of the hook-basal body complex in flagella from *Escherichia coli* and *Bacillus subtilis. J. Bacteriol.* **105:** 384.

Gray, H. 1954. *Anatomy of the human body*, 26th ed. (ed. C.M. Goss), p. 135. Lea and Febiger, Philadelphia, Pennsylvania.

Hess, J.F., K. Oosawa, N. Kaplan, and M.I. Simon. 1988. Phosphorylation of three proteins in the signaling pathway of bacterial chemotaxis. *Cell* **53:** 79.

Hess, J.F., K. Oosawa, P. Matsumura, and M.I. Simon. 1987. Protein phosphorylation is involved in bacterial chemotaxis. *Proc. Natl. Acad. Sci.* **84:** 7609.

Ishihara, A., J.E. Segall, S.M. Block, and H.C. Berg. 1983. Coordination of flagella on filamentous cells of *Escherichia coli. J. Bacteriol.* **155:** 228.

Kehry, M.R., T.G. Doak, and F.W. Dahlquist. 1985. Sensory adaptation in bacterial chemotaxis: Regulation of demethylation. *J. Bacteriol.* **163:** 983.

Khan, S. and H.C. Berg. 1983. Isotope and thermal effects in chemiosmotic coupling to the flagellar motor of *Streptococcus. Cell* **32:** 913.

Krikos, A., N. Mutoh, A. Boyd, and M.I. Simon. 1983. Sensory transducers of *E. coli* are composed of discrete structural and functional domains. *Cell* **33:** 615.

Lowe, G., M. Meister, and H.C. Berg. 1987. Rapid rotation of flagellar bundles in swimming bacteria. *Nature* **325:** 637.

Macnab, R.M. 1987a. Flagella. In Escherichia coli *and* Salmonella typhimurium*: Cellular and molecular biology* (ed. F.C. Neidhardt et al.), vol. 1, p. 70. American Society for Microbiology, Washington, D.C.

———. 1987b. Motility and chemotaxis. In Escherichia coli *and* Salmonella typhimurium*: Cellular and molecular biology* (ed. F.C. Neidhardt et al.), vol. 1, p., 732. American Society for Microbiology, Washington, D.C.

Macnab, R.M. and D.P. Han. 1983. Asynchronous switching of flagellar motors on a single bacterial cell. *Cell* **32:** 109.

Margolin, Y. and M. Eisenbach. 1984. Voltage clamp effects on bacterial chemotaxis. *J. Bacteriol.* **159:** 605.

McMahon, T.A. and J.T. Bonner. 1983. *On size and life*, p. 60. Scientific American Books, New York.

Meister, M., G. Lowe, and H.C. Berg. 1987. The proton flux through the bacterial flagellar motor. *Cell* **49:** 643.

Mutoh, N., K. Oosawa, and M.I. Simon. 1986. Characterization of *Escherichia coli* chemotaxis receptor mutants with null phenotypes. *J. Bacteriol.* **167:** 992.

Oosawa, K., J.F. Hess, and M.I. Simon. 1988. Mutants defective in bacterial chemotaxis show modified protein phosphorylation. *Cell* **53:** 89.

Parkinson, J.S. 1981. Genetics of bacterial chemotaxis. *Symp. Soc. Gen. Microbiol.* **31:** 265.

———. 1988. Minireview: Protein phosphorylation in bacterial chemotaxis. *Cell* **53:** 1.

Purcell, E.M. 1977. Life at low Reynolds number. *Am. J. Phys.* **45:** 3.

Ravid, S., P. Matsumura, and M. Eisenbach. 1986. Restoration of flagellar clockwise rotation in bacterial envelopes by insertion of the chemotaxis protein, CheY. *Proc. Natl. Acad. Sci.* **83:** 7157.

Russo, A.F. and D.E. Koshland, Jr. 1983. Separation of signal transduction and adaptation functions of the aspartate receptor in bacterial sensing. *Science* **220:** 1016.

Segall, J.E., S.M. Block, and H.C. Berg. 1986. Temporal comparisons in bacterial chemotaxis. *Proc. Natl. Acad. Sci.* **83:** 8987.

Segall, J.E., A. Ishihara, and H.C. Berg. 1985. Chemotactic signaling in filamentous cells of *Escherichia coli. J. Bacteriol.* **161:** 51.

Segall, J.E., M.D. Manson, and H.C. Berg. 1982. Signal processing times in bacterial chemotaxis. *Nature* **296:** 855.

Shioi, J., C.V. Dang, and B.L. Taylor. 1987. Oxygen as attractant and repellent in bacterial chemotaxis. *J. Bacteriol.* **169:** 3118.

Silverman, M. and M. Simon. 1974. Flagellar rotation and the mechanism of bacterial motility. *Nature* **249:** 73.

Spudich, J.L. and D.E. Koshland, Jr. 1975. Quantitation of the sensory response in bacterial chemotaxis. *Proc. Natl. Acad. Sci.* **72:** 710.

Stewart, R.C. and F.W. Dahlquist. 1987. Molecular components of bacterial chemotaxis. *Chem. Rev.* **87:** 997.

Stock, A., T. Chen, D. Welsh, and J. Stock. 1988. CheA protein, a central regulator of bacterial chemotaxis, belongs to a family of proteins that control gene expression in response to changing environmental conditions. *Proc. Natl. Acad. Sci.* **85:** 1403.

Vogler, A.P. and J.W. Lengeler. 1987. Indirect role of adenylate cyclase and cyclic AMP in chemotaxis to phosphotransferase system carbohydrates in *Escherichia coli* K-12. *J. Bacteriol.* **169:** 593.

Wolfe, A.J., M.P. Conley, and H.C. Berg. 1988. Acetyladenylate plays a role in controlling the direction of flagellar rotation. *Proc. Natl. Acad. Sci.* (in press).

Wolfe, A.J., M.P. Conley, T.J. Kramer, and H.C. Berg. 1987. Reconstitution of signaling in bacterial chemotaxis. *J. Bacteriol.* **169:** 1878.

Wylie, D., A. Stock, C.-Y. Wong, and J. Stock. 1988. Sensory transduction in bacterial chemotaxis involves phosphotransfer between Che proteins. *Biochem. Biophys. Res. Commun.* **151:** 891.

Roles of Methylation and Phosphorylation in the Bacterial Sensing System

D.E. KOSHLAND, JR., D.A. SANDERS, AND R.M. WEIS*
Department of Biochemistry, University of California, Berkeley, California 94720

Bacterial chemotaxis offers an interesting example of interactions between two second-messenger pathways activated by the same receptor. Because these two paths involve methylation and phosphorylation, they provide convenient analytical tools to explore the quantitative relationships between the pathways.

The methylation system was the first to be uncovered, and it was correlated in its timing with the adaptation system. By manipulating the concentrations of receptors, esterase and transferase, and using a truncated receptor that could not be methylated, the time required to reach a new steady state in methylation correlated with adaptation (Goy et al. 1977; Springer et al. 1977; Stock and Koshland 1981; Russo and Koshland 1983). Recently, it was found that the excitation pathway involved phosphorylation (Hess et al. 1987, 1988; Oosawa et al. 1988) in a system that has analogies with other bacterial and mammalian systems (J.B. Stock et al. 1985). As a first approximation, therefore, it could be stated that the phosphorylation pathway is involved with excitation, and the methylation pathway is involved with adaptation.

The importance of this division is revealed by the primary understanding of bacterial sensing. Bacteria sense gradients by a memory mechanism, which compared their present with the immediate past (Macnab and Koshland 1972). It was postulated that this memory could be achieved if a single effector molecule initiated two processes: a fast process leading to formation of a molecule X, referred to as the response regulator (Koshland 1977), and a second step leading to the degradation of the response regulator, which was slow. (The central point is one rapid step and one slow step. The degradation could be rapid and the formation could be slow and give a satisfactory mechanism.) Stimulus (or chemoeffector) would require the property of enhancing both processes by an equal amount but on different time scales. The reason for this complexity is that the bacterial system had two properties, a comparator, which compares the past with the present, and absolute adaptation (Russo and Koshland 1985), i.e., the return to the initial behavioral pattern, even though the external stimulus had increased. A system involving one slow step, which was essentially a recording of the past, and a fast step, which was a recording of the present concentration of the chemoeffector, provides the comparator function, and the fact that both enhancements were equal would lead to a final concentration of response regulator, which was the same at all external levels of a constant stimulus. This adaptation feature was explained as a reset-to-zero mechanism (Koshland 1982) to allow the bacterium to adjust and prepare to respond to new changes, which were a small increment of the chemicals in the external environment. This feature of adaptation is characteristic of our senses of smell, taste, vision, and so forth.

The response regulator had been shown to be a useful concept for explaining a wide variety of phenomena (Koshland 1977, 1980). The remaining question was the identification of the chemical species involved and their interactions (Macnab 1987; Stewart and Dahlquist 1987; Koshland 1988). The simplest explanation would be for the phosphorylation, i.e., excitation, pathway to be totally independent of the methylation, i.e., adaptation, pathway. The studies reported below indicate that this is not true and that the methylation pathway has feedback on the phosphorylation system and vice versa.

The Two Pathways

The complete resolution of all the steps in the two pathways is not yet delineated, but Figure 1 shows a good approximation of our present knowledge. The receptor can exist in a number of behavioral states: one that generates smooth-swimming signals, one that generates tumbling signals, and a third that generates a random mixture of each. Signals to cause tumbling ("change direction"), caused by a decrease in attractant gradient or an increase in repellent gradient, activate the CheA protein, which appears to be a kinase. The phosphorylated kinase acts on the CheY protein, causing it to be phosphorylated. The phosphorylated CheY protein (the response regulator or tumble generator) binds to the motor by means of two sensing proteins, CheC and CheV (Clegg and Koshland 1984). A decrease in the concentration of the phosphorylated Y protein caused by attractant increases or repellent decreases will generate smooth swimming. The CheZ protein is believed to be a phosphatase that hydrolyzes CheY phosphate (Hess et al. 1988). Smooth swimming or tumbling is thus determined by the amount of phosphorylated Y protein, which, in turn, is determined by the activation or inhibition of the CheA kinase. There-

*Present address: Department of Chemistry, University of Massachusetts, Amherst, Massachusetts 01003.

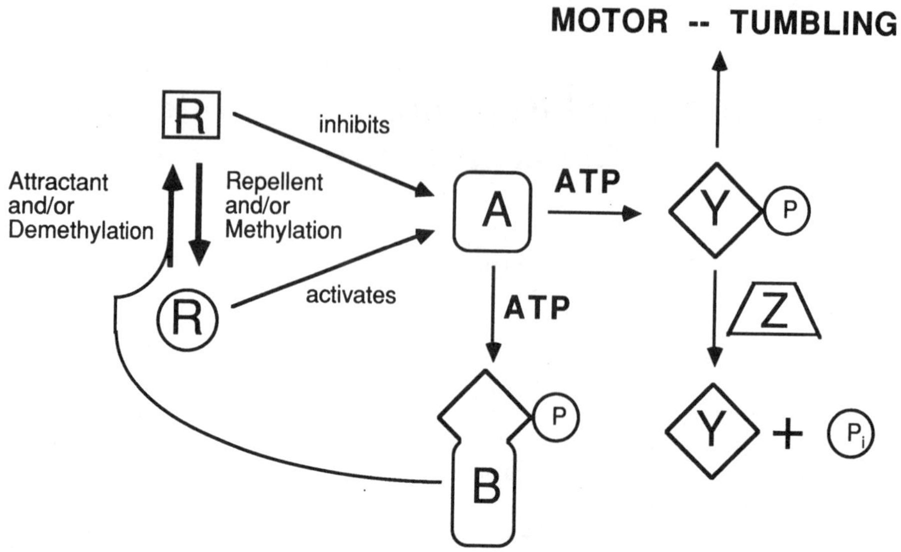

Figure 1. A model of the methylation and phosphorylation system network. Receptors (R) exist in two conformational states: One is favored by the binding of attractants and undermethylation (symbolized by the square), and the other is favored by the binding of repellents and overmethylation (symbolized by the circle). The attractant-bound form inhibits the kinase activity of the CheA protein (A), whereas the repellent-bound form activates it. The phosphorylated CheY protein (Y-P) binds to the flagellar motor and causes it to rotate clockwise, thus enhancing tumbling. The CheZ protein (Z) dephosphorylates CheY. CheA also phosphorylates the CheB protein, the methylesterase (B). The phosphorylated CheB protein is more active as a methylesterase. The phosphorylated domains of the CheB protein and the CheY protein share extensive amino acid similarity.

fore, the CheY phosphate protein in this formulation is the response regulator.

The process of adaptation can also be explained in molecular terms, as shown in Figure 2. The receptor binds the attractant aspartic acid in this model, causing a rapid conformation change, which causes smooth swimming. The new shape of the protein allows a change in the rate of the methyltransferase (CheR) and the methylesterase (CheB), thus leading to a new level of methylation. When the methylation has gone to a sufficient extent, it will compensate for the distortion of the protein caused by the binding of aspartate and return the signaling mode to what is termed the zero state in that figure. The zero state is defined in the bacterial behavior as random swimming. Removal of the stimulus would then generate the opposite state, indicated as the minus signaling state, which would then lead to demethylation in a slow step, thus regenerating the zero signaling state in the absence of a stimulus. At intermediate levels of stimulus, one would have intermediate levels of receptor occupancy and methylation.

The methylation system is thus essentially a feedback mechanism that dampens the excitation signal and resets the system to zero so that it can respond to other stimuli. The excitation system is one that generates levels of the response regulator. In essence, the pathways described above are sufficient for excitation and adaptation, and there is no apparent need for feedback of the excitation system onto the methylation system. Yet, this occurs; its existence and quantitation are indicated below.

Feedback of the Phosphorylation System on the Methylation System

That the excitation pathway affects the level of methylation has been suggested in several studies (Toews et al. 1979; Kehry et al. 1984, 1985; Springer and Zanolari 1984). A method of quantitating and verifying this feedback effect was difficult to obtain because of the difficulty of separating the two major receptors, the aspartate receptor and the serine receptor. Studies using localized mutagenesis, which introduced sulfhydryl groups into the receptors and the oxidation of these sulfhydryl groups to lead to cross-linked dimers (Falke and Koshland 1987), opened a possible route for such a study (Sanders and Koshland 1988). The theory of such an experiment is shown in Figure 3. Using the activation pathway described above, a cytoplasmic component might act equally on any receptor involved in methylation, because the transferase and esterase have both been identified as the common enzymes for a number of different receptors. The experiment therefore involved stimulating the aspartate receptor with aspartate and observing whether the methylation of the serine receptor was affected, or vice versa. If so, it would suggest a common signal that was feeding back on the methylation system in general. The experiment was carried on, allowing a sufficient period for adaptation and studying the time course during the interval of adaptation itself. At the completion of the experiment, receptor proteins were cross-linked through copper-phenanthroline-catalyzed oxidation. In this way, the aspartate receptor

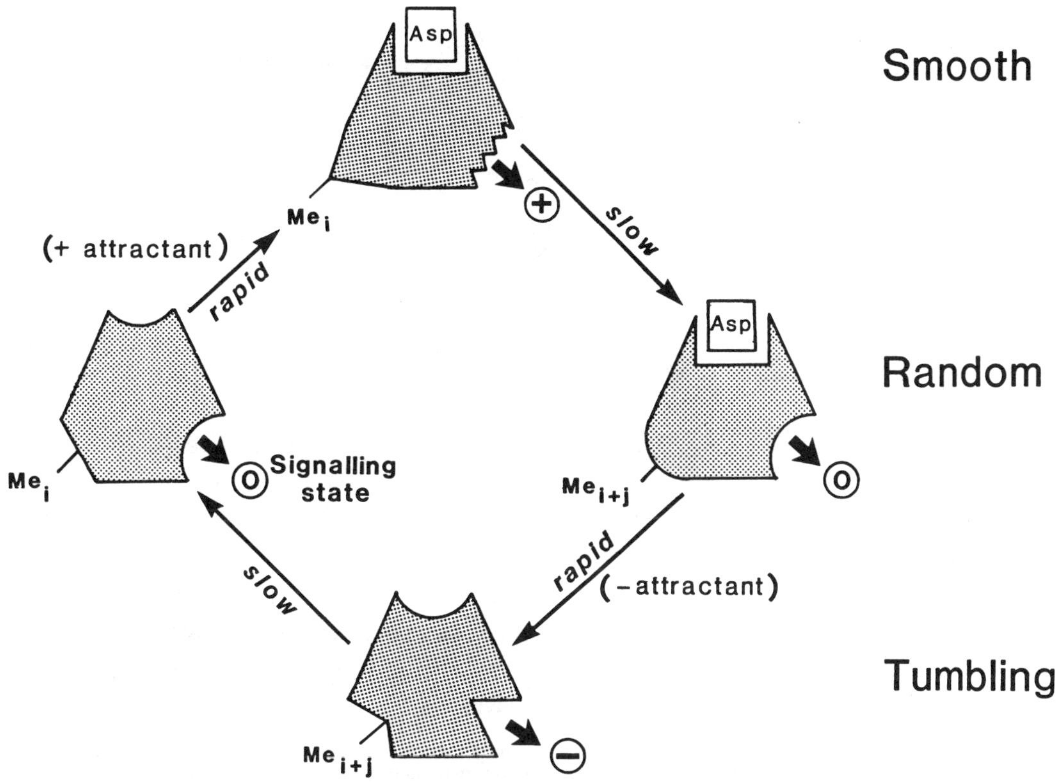

Figure 2. A model of the relation of methylation changes to attractant concentration changes. In the absence of attractant, receptors have a given methylation level (Me_i) and send a net signal to the flagella that promotes random motility (0-signaling state). Upon the binding of the attractant aspartate, a rapid receptor conformation change occurs, producing a signal that promotes smooth swimming (+ signaling state). A change in the methylation site also occurs, which leads to increased methylation. In a slow adaptation phase, the methylation of the receptor increases to a new steady-state level designated Me_{i+j}, at which point the receptors stop producing a smooth-swimming signal, and random motility ensues (0-signaling state). Upon attractant removal, the receptor is converted rapidly to a state in which it signals flagellar reversal to cause tumbling (− signaling state). After this adaptation, the receptor is demethylated until it attains the level of methylation that it had before attractant was added (Me_i). The receptor-generated signal to tumble is eliminated, and random motility ensues (return to the 0-signaling state).

containing sulfhydryl groups could be separated easily from the serine receptor, and the methyl groups in each receptor could be calculated.

The results were straightforward. The addition of aspartate, when both receptors were present, changed the methylation level of not only the aspartate receptor, but also the serine receptor. If no aspartate receptor was present, the addition of aspartate had no effect on the serine receptor, indicating that there was no general metabolic influence of aspartate. The addition of serine altered the methylation level of the aspartate receptor, as well as modifying the level of methylation of its own receptor. If the *cheA* and *cheW* genes were removed, stimulation of the aspartate receptor caused a change in the methylation level of the aspartate receptor but had no influence on the serine receptor. This shows that the *cheA* and *cheW* genes had a role in the feedback between the excitation and the methylation systems. This agreed with previous, more indirect experiments by Dahlquist and Springer and by our own laboratory (Kehry et al. 1984, 1985; Springer and Zanolari 1984).

Before concluding that the feedback effect between phosphorylation and methylation is caused by a cytoplasmic component, it is necessary to eliminate two other potential explanations. (1) If there were mixed dimers in which one peptide chain came from the aspartate receptor and one from the serine receptor, induced conformational changes in the aspartate receptor might affect the serine receptor, even without any indirect cytoplasmic component. Through the use of sulfhydryl cross-linking, it has been shown that there are no mixed dimers of this sort; therefore, this hypothesis is excluded (Milligan and Koshland 1988). (2) The ligand-bound receptor preferentially monopolizes transferase or esterase, or both and therefore tilts the pattern of methylation. In this way, the alternate receptor could have a changed level of methylation without an added cytoplasmic component. Two arguments can be made against this hypothesis. One is that the indirect effect is eliminated by the deletion of *cheA* and *cheW*, and the products of these genes have been shown to be unnecessary for the attractant-stimulated methylation of the receptor in vitro in the presence of transferase and esterase (Bogonez and Koshland 1985). The second argument is that the addition of aspartate should decrease the amount of methylation of the serine receptor, and the results shown here indicated just the

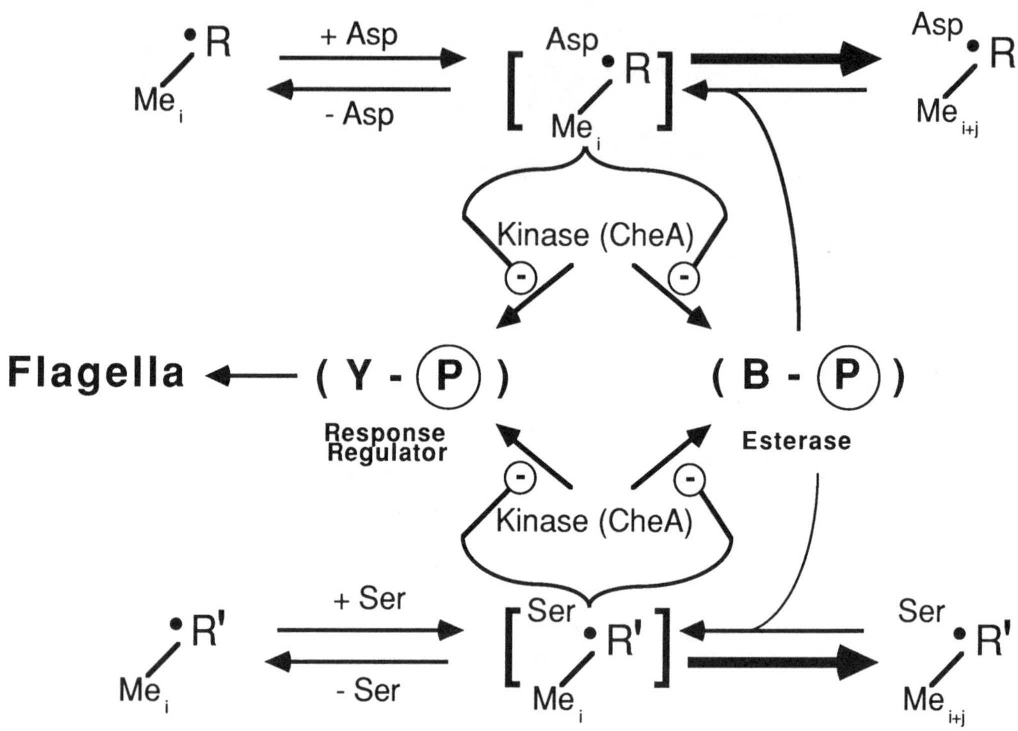

Figure 3. Two different receptors will interact if there is a common cytoplasmic component that reacts with either receptor. Applying the scheme of Fig. 1, attractants serine and aspartate are postulated to bind to their receptors and inhibit the kinase CheA. That inhibition leads to less Y-P (decreasing tumbling) and less B-P (decreasing demethylation). Because the esterase acts equally on either receptor, a serine- or aspartate-generated signal will feed back on the other receptor. The effect of the addition of attractants on the subsequent steps is indicated by minus (−) if it decreases the step.

opposite. Hence, it appears quite clear that an independent cytoplasmic component must be responsible for the feedback effect of the phosphorylation system on the methylation. The phosphorylation of the esterase by the CheA kinase fits in perfectly with these in vivo results and therefore strengthens the case that the in vitro phosphorylation experiment applies significantly in vivo.

A pulse-chase experiment confirmed that the feedback effect was on the methylesterase, which was inhibited by the addition of attractant (Sanders and Koshland 1988), in agreement with previous studies by Kehry et al. (1984, 1985). Thus, there are clearly two effects: One is a direct conformational effect caused by the ligand binding of its own receptor and causing a conformational change in the protein; this has been documented in in vitro experiments. In addition, there is a feedback effect through a cytoplasmic component that alters the level of methylation and works through the esterase. This feedback component, which has been called heterologous desensitization in some mammalian systems, appears to be of the same order of magnitude as the direct conformational effect.

Reasons for the Feedback on Methylation by the Phosphorylation System

At first glance, the feedback from the excitation pathway to the adaptation pathway appears curious. It is not necessary for adaptation. Moreover, the similarity of the sequence of the CheY protein to the sequence of the amino-terminal portion of the esterase (A. Stock et al. 1985) strongly indicates that the esterase is phosphorylated in its regulatory domain. Deletion of its amino-terminal region activates the esterase, and phosphorylation would be expected to—and does—activate the esterase (A. Lupas and J. Stock, pers. comm.). If this is so, it means that a signal that increases the amount of Y phosphate, and thereby increases tumbling, also increases the activation of the esterase, which will decrease tumbling. This system is quite different from other activation systems, such as in the phosphorylase cascade. In the phosphorylase system, phosphorylation of phosphorylase activates one pathway, whereas phosphorylation of glycogen synthase inhibits the opposing path; thus, the net effect of all the phosphorylations is to increase the velocity of glycogen to glucose (Stryer 1988). In the chemotaxis system, it would appear that the two phosphorylations are working at cross purposes.

A possible explanation arises from the purpose of tumbling and smooth swimming in bacteria. When bacteria are swimming in a favorable direction, their memory system tells them to shut off tumbling and keep going as long as possible. When swimming is in an unfavorable direction, they receive a signal to change direction. If one considers this carefully, there is no reason to prolong the negative signal. Once the signal to change direction has been generated, there is no real reason to accumulate further negative signal. Thus, it would be advantageous to have an asymmetry in the two response times. Swimming in a favorable direction should be enhanced as long as possible; swimming in an unfavorable direction should be cut off as soon as the signal to tumble has been received. Hence, it is probable that the feedback of phosphorylation of the esterase is designed to shorten the period of memory in an unfavorable response relative to that of a favorable response. Berg and his co-workers have given an indication that essentially all "wrong-direction" responses are fixed in a constant time frame (Berg and Brown 1972). Only if one gives large negative temporal gradients does one see an increase in tumbling frequency (Tsang et al. 1973). In the shallow gradients in which the bacteria proceed most of the time, there is little change in the tumble frequency. Hence, the most logical explanation for the excitation feedback on the adaptation pathway is to generate asymmetry in the chemotactic response. Therefore, there are three time constants: an initial rapid response, generated by the excitation system; a slow response, generated by the methylation-adaptation system; and an asymmetry between responses to an unfavorable direction from those in the favorable direction, caused by phosphorylation feedback on the methylation. As the feedback effects in more complex mammalian cells are unraveled; it will be interesting to observe whether the cross talk has a particular functional reason.

Role of Methylation

These findings bring into focus the role of methylation in chemotaxis. The early studies on methylation postulated that the methylation of the transmembrane proteins, later identified as receptors, was required for chemotaxis. Mutations in either the methyltransferase, which methylated receptors, or the methylesterase, which demethylated the receptors, eliminated chemotaxis. It therefore came as something of a shock to find that the double mutant, lacking both transferase and esterase, was capable of chemotaxis (J.B. Stock et al. 1981, 1985). The chemotactic ability of the transferase/esterase double mutant was not nearly as efficient as the wild type, but it was still able to migrate in a gradient. The resolution of this dilemma has revealed a great deal about chemotaxis and the processing of sensory systems. It involves two key logical features: (1) the reset-to-zero mechanism (Koshland et al. 1982) and (2) the timing of the sensory system (Koshland 1977). Both are involved in explaining how a double mutant can recover wild-type properties that are not present in single mutants, and this compensatory effect may well be observed in many other systems. In fact, some studies in our laboratory on phosphorylation in the Krebs cycle (LaPorte et al. 1985) have revealed exactly the same result, i.e., a double mutant in two different enzymes of a pathway restores vitality not present in either single mutant.

The reset-to-zero aspect of the problem developed from the finding that the initial receptor contained glutamines at two of the four positions that are ultimately methylated. These glutamine residues are hydrolyzed to glutamates by the esterase enzyme, which is therefore both an esterase and an amidase. Once they have been deamidated, they are reversibly methylated, as are the two other glutamic acid residues. However, it was found that the glutamine residue provides the same type of signaling properties as a glutamic acid methylester. The normal wild-type enzyme, which has been completely deamidated, exists with approximately two of the glutamates methylated in the absence of any stimulus. That cell exhibits random swimming in the absence of a gradient. Small changes in gradient cause migration because the behavioral pattern of the bacterium shifts from a random walk to a biased random walk, i.e., biased in the direction of a favorable gradient. When the bacterium has migrated to the more favorable environment and there is no further gradient, its methylation system restores the bacterial behavior to random swimming, i.e., it resets to zero. It is this resetting to zero that enables the bacterium to respond to any of a large number of gradients. It would be impractical for the bacterium if a very high concentration of aspartic acid prevented if from responding to a developing deficiency in oxygen in the environment.

It is a fortuitous property (perhaps selected by evolution) that made it possible for the double mutant to respond to gradients, whereas each single mutant could not. The methylesterase mutant was found to give no deamidation, and the remaining two glutamate positions were then methylated by the transferase. The overmethylated receptor generated a tumbling signal, and the strain tumbles incessantly. Because it tumbles incessantly, it is not able to sense gradients, and it is so far removed from the zero point that it cannot respond to small gradients of the type the bacterium usually encounters. On the other hand, the mutant lacking transferase does not have the capacity to methylate the receptor, and the esterase deamidates the two glutamines that are initially present and produces a receptor of all glutamates, which send out constant signals for smooth swimming. This bacterium therefore is as far removed from the zero point in the smooth-swimming direction as the esterase mutant is in the tumbling direction. It cannot respond to small stimuli and therefore is nonchemotactic. The double mutant, on the other hand, because of the two glutamines and lack of either transferase or esterase, is poised at the zero point and can respond to small gradients because it has partial adaptation. It does not respond as well as the wild type, but it can undergo appreciable chemotaxis. One could therefore imagine that this was the primordial system, but it is not optimal, because (1) it can only respond over a limited range, as it must be poised for random swimming before responding, and (2) strong stimuli overwhelm the system and are incapable of adaptation; hence, no chemotaxis.

This combination of properties raised the question as

to the actual role of methylation in chemotaxis. A number of hypotheses are possible. One is that methylation is essential and that the mutants uncovered previously actually have a small amount of methylation activity, which is sufficient to allow chemotaxis. The small amount of methylation could be due to the fact that either the mutants were leaky or there was a nonspecific methylating enzyme that could give a small amount of methylation to the receptor. Careful tests of a double deletion mutant that had the same phenotype and a sensitive assay for modification of the central methylated peptide eliminated all of these possibilities (Weis and Koshland 1988). A second hypothesis that the chemotaxis observed is caused by response to other systems that do not require methylation, such as repellents or oxygen, can also be eliminated.

A more interesting hypothesis is that the role of methylation is simply to reset to zero and that it is really not needed in the sensing of the gradient, i.e., it is not involved in memory. This is probably the system that is utilized in vision, where seeing an object is achieved through the cGMP pathway and occurs rapidly without any significant change in phosphorylation of the receptor. Receptor phosphorylation occurs subsequently and is a mechanism for correcting for the background light intensity or turning off the signal. The timing of phosphorylation in rhodopsin is not essential for the vision process. The argument could be made that reset to zero in chemotaxis must be done frequently along the gradient range to have chemotaxis respond to small changes in gradient. A test of this hypothesis was obtained by placing the bacteria in a gentle exponential gradient and observing whether there was any point at which the double mutant responded like wild type. If the methylation system only involved reset to zero, there should be some concentration of attractant in which the appropriate zero setting was appropriate for optimum chemotaxis. However, no such normal chemotaxis occurred over the entire range.

A possible deficiency in this experiment was that the esterase or transferase might have dual functions, i.e., that the absence of the esterase perturbed a second feature of the excitation pathway, which therefore made it nonoptimal for chemotaxis at any concentration. As described above, the esterase does have a dual function, but the finding that it is involved in a minor adjustment of the asymmetry of the response essentially eliminates this hypothesis.

This leaves the final alternative, that methylation is involved not only in resetting the system to zero, but also in the real time of the memory. It is largely, if not entirely, responsible for the slow step that is so important to the comparator. The partial adaptation that was observed in the double mutant can now be identified with the phosphorylation system, and the previous explanation that the response regulator system would have the mathematics of partial adaptation seems to be strongly supported. The absence of methylation produces a system that can allow chemotaxis, but not as efficiently, and over a much more limited range than the wild-type methylation system that exists today. Methylation therefore probably arose as a further improvement during evolution of the chemotaxis system. This conclusion would seem to be supported by the fact that the phosphorylation system is observed in so many different bacterial organisms with a combination of kinase and substrate similar to that in chemotaxis.

SUMMARY AND CONCLUSIONS

Chemotaxis is an intriguing model system for the study of second-messenger pathways. One of the puzzles of second-messenger pathways in eukaryotic cells has been that many of these pathways interact, with one pathway either desensitizing or sensitizing an alternate messenger pathway. The chemotaxis system offered a particularly interesting chance to analyze such systems, because one of them is a methylation pathway and the other a phosphorylation pathway. The above description indicates that these two pathways interact with each other in a highly sophisticated way to produce an extremely important survival system. The stimulus on a receptor activates an excitation system that operates through phosphorylation, or the inhibition of phosphorylation, depending on whether the stimulus is a repellent or an attractant. That system ultimately generates or inhibits phosphorylation of a small peptide, the CheY protein, which apparently is the response regulator. The instant that this fast excitation is generated by a change in gradient, the change in conformation of the protein sets in motion a second process, i.e., the adaptation. That is a device which, over a longer period of time, has two functions: It serves as the comparator, which allows the comparison of the past with the present, essential for deletion of a gradient; it also sets in motion the reset to zero, so that the bacterium will not be overwhelmed by any one stimulus but can use all of its receptors to optimize its environment. These two systems by themselves are adequate for chemotaxis, but there is a further elegent complexity: The excitation system feeds back into the adaptation system to produce an asymmetry in the responses. The reason for that asymmetry is that the bacterium wishes to travel in the wrong direction only long enough to produce a detectable signal that it is migrating incorrectly. It wishes to keep swimming in the positive direction as long as the signals indicate that the direction is favorable. Hence, the feedback between the phosphorylation and methylation system involves further fine tuning.

ACKNOWLEDGMENTS

We acknowledge financial support from the National Institutes of Health and the National Science Foundation.

REFERENCES

Berg, H.C. and D.A. Brown. 1972. Chemotaxis in *Escherichia coli* analysed by three-dimensional tracking. *Nature* **239:** 500.

Bogonez, E. and D.E. Koshland, Jr. 1985. Solubilization of a vectorial transmembrane receptor in functional form: Aspartate receptor of chemotaxis. *Proc. Natl. Acad. Sci.* **82:** 4891.

Clegg, D.O. and D.E. Koshland, Jr. 1984. The role of a signaling protein in bacterial sensing: Behavioral effects of increased gene expression. *Proc. Natl. Acad. Sci.* **81:** 5056.

Falke, J.F. and D.E. Koshland, Jr. 1987. Global flexibility in a sensory receptor: A site-directed crosslinking approach. *Science* **237:** 1596.

Goy, M.G., M.S. Springer, and J. Adler. 1977. Sensory transduction in *Escherichia coli*: Role of a protein methylation reaction in sensory adaptation. *Proc. Natl. Acad. Sci.* **74:** 4964.

Hess, J.F., K. Oosawa, N. Kaplan, and M.I. Simon. 1988. Phosphorylation of three proteins in the signaling pathway of bacterial chemotaxis. *Cell* **53:** 79.

Hess, J.F., K. Oosawa, P. Matsumura, and M.I. Simon. 1987. Protein phosphorylation is involved in bacterial chemotaxis. *Proc. Natl. Acad. Sci.* **84:** 7609.

Kehry, M.R., T.G. Doak, and F.W. Dahlquist. 1984. Stimulus-induced changes in methylesterase activity during chemotaxis in *Escherichia coli. J. Biol. Chem.* **259:** 11828.

———. 1985. Sensory adaptation in bacterial chemotaxis: Regulation of demethylation. *J. Bacteriol.* **163:** 983.

Koshland, D.E., Jr. 1977. A response regulator model in a simple sensory system. *Science* **196:** 1055.

———. 1980. Bacterial chemotaxis in relation to neurobiology. *Annu. Rev. Neurosci.* **3:** 43.

———. 1988. Chemotaxis as a model second messenger system. *Biochemistry* **27:** 5829.

Koshland, D.E., Jr., A. Goldbeter, and J.B. Stock. 1982. Amplification and adaptation in regulatory and sensory systems. *Science* **217:** 220.

LaPorte, D.C., P.E. Thorsness, and D.E. Koshland, Jr. 1985. Compensatory phosphorylation of isocitrate dehydrogenase: A mechanism for adaptation to the intracellular environment. *J. Biol. Chem.* **260:** 10563.

Macnab, R.M. 1987. Motility and chemotaxis. In Escherichia coli *and* Salmonella typhimurium: *Cellular and molecular biology* (ed. F.C. Neidhardt), p. 732. American Society for Microbiology, Washington, D.C.

Macnab, R.M. and D.E. Koshland, Jr. 1972. The gradient-sensing mechanism in bacterial chemotaxis. *Proc. Natl. Acad. Sci.* **69:** 2509.

Milligan, D.L. and D.E. Koshland, Jr. 1988. Site-directed cross-linking: Establishing the dimeric structure of the aspartate receptor of bacterial chemotaxis. *J. Biol. Chem.* **263:** 6268.

Oosawa, K., J.F. Hess, and M.I. Simon. 1988. *cheA* mutants defective in chemotaxis show modified protein phosphorylation. *Cell* **53:** 89.

Russo, A.F. and D.E. Koshland, Jr. 1983. Separation of signal transduction and adaptation functions of the aspartate receptor in bacterial sensing. *Science* **220:** 1016.

———. 1985. Receptor modification and absolute adaptation in bacterial sensing. In *Sensing and response in microorganisms* (ed. M. Eisenbach et al.), p. 27. Elseivier Science Publishers, Amsterdam.

Sanders, D.A. and D.E. Koshland, Jr. 1988. Receptor interactions via bacterial phosphorylation and methylation pathways in chemotaxis. *Proc. Natl. Acad. Sci.* (in press).

Springer, M.S. and B. Zanolari. 1984. Sensory transduction in *Escherichia coli:* Regulation of the demethylation rate by the *CheA* protein. *Proc. Natl. Acad. Sci.* **81:** 5061.

Springer, M.S., M.F. Goy, and J. Adler. 1977. Sensory transduction in *Escherichia coli:* A requirement for methionine in sensory adaptation. *Proc. Natl. Acad. Sci.* **74:** 183.

———. 1979. Protein methylation in behavioral control mechanisms and in signal transduction. *Nature* **280:** 279.

Stewart, R.C. and F.W. Dahlquist. 1987. Molecular components of bacterial chemotaxis. *Chem. Rev.* **87:** 997.

Stock, A., D.E. Koshland, Jr., and J. Stock. 1985. Homologies between the *Salmonella typhimurium* CheY protein and proteins involved in the regulation of chemotaxis, membrane protein synthesis, and sporulation. *Proc. Natl. Acad. Sci.* **82:** 7989.

Stock, A., D.E. Koshland, Jr., and J. Stock. 1985. Homologies between the *Salmonella typhimurium* CheY protein and proteins involved in the regulation of chemotaxis, membrane protein synthesis, and sporulation. *Proc. Natl. Acad. Sci.* **82:** 7989.
bacterial chemotaxis. *Cell* **41:** 683.

Stock, J.B., A.M. Maderis, and D.E. Koshland, Jr. 1981. Bacterial chemotaxis in the absence of receptor carboxymethylation. *Cell* **27:** 37.

Stryer, L. 1988. *Biochemistry*, 3rd edition, p. 449. W.H. Freeman, San Francisco.

Toews, M.L., M.F. Goy, M.S. Springer, and J. Adler. 1979. Attractants and repellents control demethylation of methylated chemotaxis proteins in *Escherichia coli. Proc. Natl. Acad. Sci.* **76:** 5544.

Tsang, N., R.M. Macnab, and D.E. Koshland, Jr. 1973. Common mechanisms for repellents and attractants in bacterial chemotaxis. *Science* **181:** 60.

Weis, R.M. and D.E. Koshland, Jr. 1988. Reversible receptor methylation is essential for normal chemotaxis of *Escherichia coli* in gradients of aspartic acid. *Proc. Natl. Acad. Sci.* **85:** 83.

Osmotaxis in *Escherichia coli*

J. ADLER,*† C. LI,‡ A.J. BOILEAU,§ Y. QI,* AND C. KUNG†**

**Department of Biochemistry, †Department of Genetics, ‡Program in Cellular and Molecular Biology §Neuroscience Training Program, and **Laboratory of Molecular Biology, University of Wisconsin, Madison, Wisconsin 53706*

Osmotaxis is the movement of organisms in search of an environment containing optimum water content. All organisms keep in check the concentration of cellular water. On the one hand, they seek a source of water, but too much water leads to undesired hydration. On the other hand, they avoid water wherein the concentration of solute is too high (such as high salinity). Thus, each organism must straddle these two extremes carefully: too low osmolarity and too high osmolarity. This drive is surely among the most basic in biology.

In the case of bacteria, osmotaxis was discovered in the nineteenth century by Massart (1889, 1891). Osmotaxis in *Escherichia coli* was indicated briefly by Tso and Adler (1974). We have now investigated osmotaxis in *E. coli* further. (Part of this work was presented in a preliminary form [Adler 1987, 1988].)

RESULTS

The osmotactic behavior of *E. coli* was studied in four ways.

Osmotaxis measured by capillary assays. Bacteria swimming in a pond were exposed to capillaries containing the chemical to be tested (Adler 1973). (The pond and the capillary both contained 0.01 M K^+ phosphate [pH 7.0] and 0.0001 M EDTA, needed for optimum motility.) In the capillary, the osmotic agent, ribitol or sucrose, was added up to 0.8 M. Both of these chemicals had been found previously to be neither attractants nor repellents for *E. coli* when tested below 0.1 M (Adler et al. 1973; Tso and Adler 1974).

Figure 1 shows that the bacteria seek an optimum osmolarity, about 0.2 M chemical in the capillary (lower concentrations diffuse out of the capillary to be active there.) At concentrations higher than 0.2 M, the attraction fell off; this suggests that the bacteria avoid such high concentrations. When the capillary contained only water, the bacteria also avoided it (see Fig. 1); this indicates osmotaxis away from pure water.

The attraction shown in Figure 1 appears to result from osmotaxis, rather than from chemotaxis. This was determined by finding a wild-type response to ribitol and sucrose in each of the known chemoreceptor mutants: *tsr* ($MCPI^-$) strain AW518, *tar* ($MCPII^-$) strain AW539, *trg* ($MCPIII^-$) strain AW701, *tap* ($MCPIV^-$) strain RP3525 and *ptsI* (enzyme I^-) strain AW509.

Osmotaxis measured by plug assays. Bacteria swimming in soft agar in a petri plate were exposed to test chemical diffusing out of a hard-agar plug (Tso and Adler 1974). Present in both the soft and hard agar were 0.01 M K^+ phosphate (pH 7.0) and 0.0001 M EDTA.

Figure 2A shows that *E. coli* cleared the zone near a plug containing 1 M ribitol or higher. Similar results were obtained with sucrose and with 15 other organic and 5 inorganic chemicals tested at 1 M or higher, whether known attractants or repellents or neither (Mesibov and Adler 1972; Adler et al. 1973; Tso and Adler 1974).

Even chemoattractants repel at high concentrations, as shown in Figure 2B. Here, bacteria were repelled from plugs containing 1 M or more concentrated D-ribose, but they formed an attractant ring to low concentrations of D-ribose (most prominent around plugs at 0.03 and 0.1 M). A *trg* mutant missing methyl-accepting chemotaxis protein III (MCP III), which is needed for attraction to D-ribose, could still form the osmotaxis ring, but it failed to form the ribose chemotaxis ring, as expected (Fig. 2C). In the case of L-aspartate, L-serine, and maltose, it was necessary to use the appropriate MCP mutant, *tar, tsr,* and *tar*, respectively, to show the osmotaxis rings: It seems that these chemicals are such powerful chemoattractants in wild-type bacteria, due to

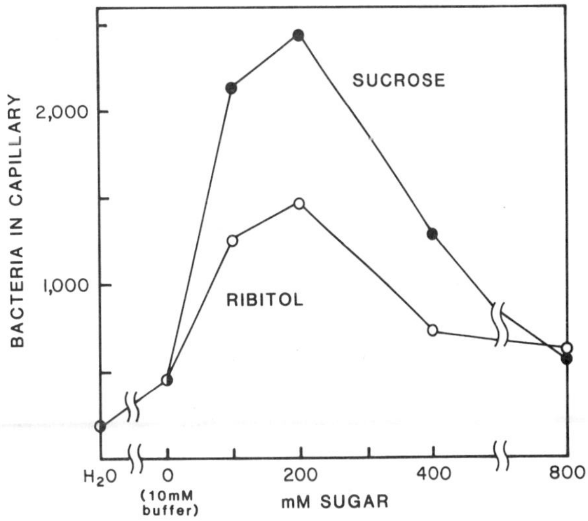

Figure 1. Bacteria are attracted by an optimum osmolarity. Capillary assays for attraction to ribitol and sucrose were carried out at 30°C for 30 min, according to the procedure described by Adler (1973). Bacteria in the capillaries were counted as colonies after incubating the capillary contents on tryptone plates overnight.

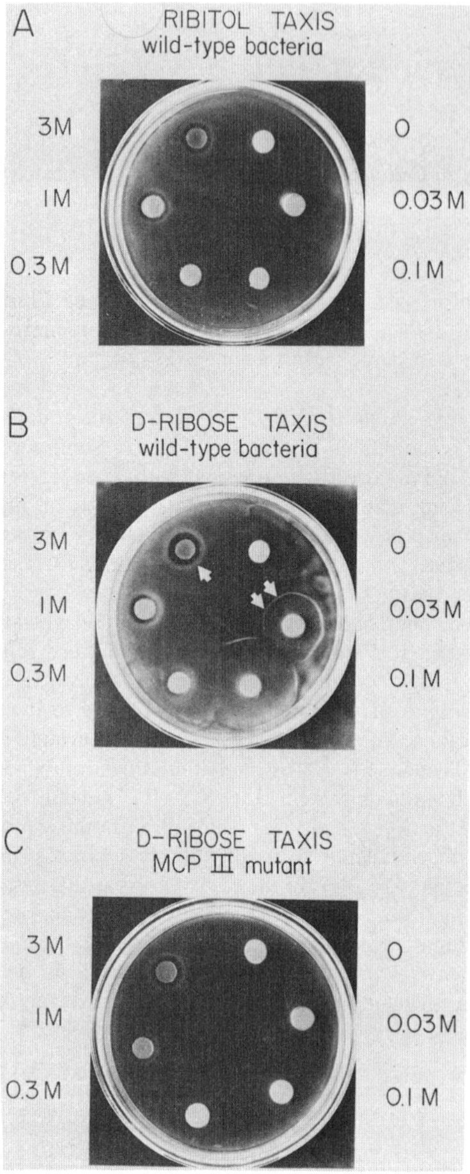

Figure 2. Osmotaxis in chemotactically wild-type and mutant *E. coli* was tested by chemical diffusion from a plug (Tso and Adler 1974). (*A*) A chemotactically wild-type strain (AW405) was repelled by 1 and 3 M ribitol plugs. (*B*) Wild-type AW405 was tested with D-ribose. The inner ring outside a zone of clearance (see 1 and 3 M, single arrow at 3 M) is believed to be due to osmorepulsion, whereas the outer ring (best apparent at 0.03 and 0.1 M, double arrow at 0.03 M) is believed to be due to chemoattraction. (*C*) *trg* (MCP III) mutant (AW701), known to be missing attraction to D-ribose, was tested with D-ribose. Only the inner ring (see 1 and 3 M), believed to be due to osmorepulsion, was observed.

the high concentration of MCP I and MCP II, that osmotaxis is obscured.

Which components of the chemotaxis machinery are necessary for osmotaxis? Table 1 (column 3) shows that normal osmotaxis was obtained in the plug assay with mutants missing each of the relevant MCPs or enzyme I needed for taxis to phosphotransferase sugars. Thus, any single receptor for chemotaxis does not appear to be the receptor for osmotaxis. However, a mutant that was missing all four MCPs did not perform osmotaxis in the plug assay.

Osmotaxis did not occur in plug assays when generally nonchemotactic (*che*) mutants were used (Table 1, column 3). Thus, by this criterion, it appears that osmotaxis proceeds by way of the general chemotaxis pathway (see Discussion).

Osmotactic behavior studied by observations of free-swimming cells. Wild-type *E. coli* run and tumble. The addition of an attractant causes them to run and then adapt; i.e., they return to running and tumbling. The addition of a repellent makes them tumble and then adapt (Berg and Brown 1972; Macnab and Koshland 1972; Tsang et al. 1973). Such direct addition of a chemical is the defining feature of a "temporal assay" as opposed to adding the chemical in a gradient coming out of a capillary or plug, which is a "spatial assay." Tumbling was observed through the microscope after addition of osmotic agent and recorded on videotape. Visual observations are reported.

Here, we find that the addition of an osmotic agent such as 0.05–0.1 M KCl or NaCl or 0.1–0.2 M ribitol or sucrose causes wild-type *E. coli* to tumble and then adapt. At higher concentrations between 0.2 M and 0.4 M KCl or NaCl or between 0.4 M and 0.8 M ribitol or sucrose, the bacteria became nonmotile, tumbled, and then adapted. (Studies at very low concentrations of osmotic agent have not yet been carried out by temporal assays, either with wild-type bacteria or with mutants.)

The same is true (Table 1, column 4) of all known receptor mutants and smooth mutants (*cheA, cheR, cheW, cheY, flaAII* smooth, *flABII* smooth). Even the tumbly mutants (*cheB, cheZ, flaII* tumbly, *flaBII* tumbly) lost their smooth stretches to become more tumbly when high osmolarity was added. Finally, two mutants (HCB326 and HCB437) lacking all four MCPs and all six Che proteins—normally perfectly smooth strains—tumbled when high osmolarity was added. They did not tumble when the chemorepellent 0.03 M benzoate was added. In every case, adaptation occurred within minutes, unless the concentration of osmotic agent was too high (e.g., 0.5 M).

To gain a direct comparison of chemotaxis and osmotaxis in temporal assays, we tested the MCP I repellents benzoate (0.03–0.08 M) and L-leucine (0.03 M). These tests showed little (compared with osmotaxis) or no response by the six *che* gene mutants, which leads to the conclusion that the *che* gene products are not required for osmotic response. (*CheR* usually does tumble to repellents.)

Osmotactic behavior studied by observations of tethered cells. Motile *E. coli* were tethered to glass; i.e., a flagellum was tied down with flagellar antibody so that the bacterial body rotated instead of the flagella (Silverman and Simon 1974). Tethered wild-type bacteria rotate both counterclockwise and clockwise, corre-

Table 1. Osmotactic Behavior of Wild-type and Mutant *E. coli*

Genotype	Strain number	Plug assays[a]	Assays on free-swimming cells[b]
Wild type	AW405	+	+
Receptor mutants			
tsr (MCP I⁻)	AW518	+	+
tar (MCP II⁻)	AW539	+	+
trg (MCP III⁻)	AW701	+	+
tap (MCP IV⁻)	RP3525	+	+
ptsI (enzyme I⁻)	AW509	+	+
tsr tar trg tap	CP362	−	+
Generally nonchemotactic mutants			
cheB (methylesterase⁻)	AW689	−	+
cheR (methyltransferase⁻)	AW688	−	+
cheA	AW690	−	+
cheW	RP4009	−	+
cheY	AW691	−	+
cheZ	AW692	−	+
flaAII (*cheC*) smooth	RP1581	−	+
flaAII (*cheC*) tumbly	RP2634	−	+
flaBII (*cheV*) smooth	RP1592	−	+
flaBII (*cheV*) tumbly	RP2638	−	+
Mutant missing four MCPs and six *che*s	HCB326	−	+

[a]Osmotaxis was studied by the plug assay described in the legend to Fig. 2. (+) Bacteria vacated the region around plugs containing 1 or 3 M osmotic agent, as shown in Fig. 2; (−) no apparent response. For each receptor mutant, the osmotic agent was a chemical detected by that receptor: L-serine, L-acetate, and benzoate for *tsr;* L-aspartate, maltose, Co^{2+}, and Ni^{2+} for *tar;* D-galactose and D-ribose for *trg;* D-mannose and D-sorbitol for *ptsI.* We could not test the *tap* mutant with 1 or 3 M dipeptides because of their low solubility. In every case, we also tested with D-glucose, ribitol, D-ribose, and $NaNO_3$.

[b]Osmotactic behavior was studied by adding osmotic agent to the swimming bacteria. In most cases, 0.1 to 0.15 M KCl was used (but NaCl, ribitol, and sucrose were also tested, see text). (+) A tumbling response; at such concentrations (0.05 M KCl), the bacteria recovered from tumbling to return to their unstimulated mode. (See text for effect of higher concentrations.) Tumbly mutants (*cheB, cheZ, flaAII* tumbly, and *flaBII* tumbly) became more tumbly (see text) with high osmolarity (0.3 M KCl).

sponding to the counterclockwise and clockwise rotation, respectively, of the flagellum (Berg 1974; Silverman and Simon 1974). The addition of an attractant is known to increase counterclockwise rotation corresponding to runs, whereas the addition of a repellent increases clockwise rotation corresponding to tumbles (Larsen et al. 1974).

Khan and Macnab (1980) reported that the addition of NaCl, KCl, or sucrose at high osmolarity to wild-type tethered cells causes counterclockwise rotation. This has now been confirmed by us (using 0.8 M sucrose or 0.1–0.6 M KCl) and by H. Berg and co-workers (pers. comm.). Our laboratory also found that a tethered tumbly *cheZ* mutant (RP1616) changed from its primarily clockwise rotation to counterclockwise rotation when an osmotic agent (0.2 M KCl) was added. We also tested the tethered, partly tumbly mutant HCB482, lacking all four MCPs and *cheA, cheB, cheR, cheW,* and *cheZ*, but containing CheY; it rotated both counterclockwise and clockwise, but when an osmotic agent (0.2 M KCl) was added, it rotated counterclockwise. Also, the tethered mutant HCB437 lacking all four MCPs and all six Che proteins remained counterclockwise when an osmotic agent (0.2 M KCl) was added. As mentioned, high osmolarity caused tumbling of the free-swimming cells in these various cases (Table 1, column 4). High osmolarity caused tumbling, but counterclockwise rotation of the flagella. In the past, counterclockwise rotation was considered to be the equivalent of running. (Counterclockwise-rotating flagella form a bundle that pushes the cell forward to produce a run.) We theorize the following reconciliation: The tumbling caused by high osmolarity could result from counterclockwise rotation of flagella, because the flagella may be pulled into the cell by the shrinking of the cytoplasm (brought about by the high osmolarity) to force apart the bundle of counterclockwise-rotating flagella. This would represent the first case reported of tumbling that results from counterclockwise rotation of the flagella.

DISCUSSION

On the basis of studies of free-swimming cells (Table 1, column 4), we conclude that osmotaxis does not use either the known chemoreceptors or the *che* machinery. Rather, osmotaxis could feed in at a point close to the base of the flagella (Fig. 3). An "osmosensor"

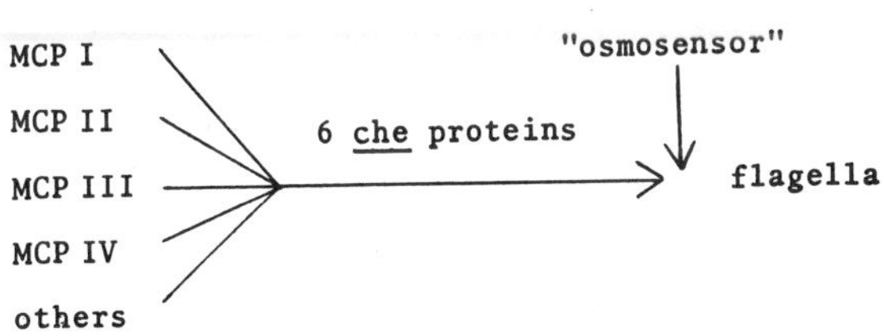

Figure 3. Scheme of chemotaxis mechanism to illustrate the point of entry of osmotaxis by way of a possible osmosensor.

might exist, and we are attempting to isolate mutants that lack it, i.e., mutants that fail in osmotaxis but carry out chemotaxis normally; or osmotic agents might act on the flagellar switch directly without mediation of an osmosensor.

The conclusion derived from temporal assays that osmotaxis takes place without participation of MCPs or *che* gene products (Table 1, column 4) is not consistent with the conclusion derived from spatial assays (Table 1, column 3). The inconsistency most likely relates to the steepness of the gradient—very steep in the temporal assay and relatively shallow in a spatial assay. Smooth mutants or tumbly mutants can be highly biased in the smooth or tumbly direction, so it may take a very steep gradient to overcome this bias.

CONCLUSIONS

Bacteria are repelled by both low and high osmolarity, i.e., they seek an optimum osmolarity. Here, repulsion by high osmolarity has been emphasized more than repulsion by low osmolarity. Free-swimming cells tumble to high osmolarity, but this tumbling seems to coincide unexpectedly with counterclockwise rotation of the flagella.

Osmotaxis appears to proceed in the absence of *che* gene products. If this proves to be true, osmotaxis may represent a primitive form of behavior that was present even before chemotaxis evolved. That idea would correspond with the universal, crucial need of organisms to control the water content of their cells.

ACKNOWLEDGMENTS

We thank J.S. Parkinson for strains numbered with RP and H.C. Berg for those numbered with HCB. We are very grateful to Matthew Buechner and Colin Ingham for help in experiments, and to the forementioned, Anne Delcour, Boris Martinac, Brian Sager, and Wenyuan Shi for discussions. This research was supported by grants from the National Institutes of Health (DK-39121) and from the Lucille P. Markey Charitable Trust to J.A. and C.K.

REFERENCES

Adler, J. 1973. A method for measuring chemotaxis and use of the method to determine optimum conditions for chemotaxis by *Escherichia coli*. *J. Gen. Microbiol.* **74:** 77.

———. 1987. How motile bacteria are attracted and repelled by chemicals: An approach to neurobiology. *Biol. Chem. Hoppe-Seyler* **368:** 163.

———. 1988. Chemotaxis: Old and new. *Botanica Acta* **101:** 93.

Adler, J., G.L. Hazelbauer, and M.M. Dahl. 1973. Chemotaxis toward sugars in *Escherichia coli*. *J. Bacteriol.* **115:** 824.

Berg, H.C. 1974. Dynamic properties of bacterial flagellar motors. *Nature* **249:** 77.

Berg, H.C. and D.A. Brown. 1972. Chemotaxis in *Escherichia coli* analysed by three-dimensional tracking. *Nature* **239:** 500.

Khan, S. and R.M. Macnab. 1980. The steady-state counterclockwise/clockwise ratio of bacterial flagellar motors is regulated by proton motive force. *J. Mol. Biol.* **138:** 563.

Larsen, S.H., R.W. Reader, E.N. Kort, W.-W. Tso, and J. Adler. 1974. Change in direction of flagellar rotation is the basis of the chemotactic response in *Escherichia coli*. *Nature* **249:** 74.

Macnab, R.M. and D.E. Koshland, Jr. 1972. The gradient-sensing mechanism in bacterial chemotaxis. *Proc. Natl. Acad. Sci.* **69:** 2509.

Massart, J. 1889. Sensibilité et adaptation des organismes à la concentration des solutions salines. *Arch. Biol.* **9:** 515.

———. 1891. Recherches sur les organismes inférieurs (1). *Bull. Acad. R. Med. Belg.* **22:** 148.

Mesibov, R. and J. Adler. 1972. Chemotaxis toward amino acids in *Escherichia coli*. *J. Bacteriol.* **112:** 315.

Silverman, M. and M. Simon. 1974. Flagellar rotation and the mechanism of bacterial motility. *Nature* **249:** 73.

Tsang, N., R. Macnab, and D.E. Koshland, Jr. 1973. Common mechanism for repellents and attractants in bacterial chemotaxis. *Science* **181:** 60.

Tso, W.-W. and J. Adler. 1974. Negative chemotaxis in *Escherichia coli*. *J. Bacteriol.* **118:** 560.

Galvanotaxis in Bacteria

J. ADLER*† AND W. SHI†

*Department of Biochemistry and †Department of Genetics, University of Wisconsin, Madison, Wisconsin 53706

Wonderful as are the laws and phenomena of electricity when made evident to us in inorganic or dead matter, their interest can bear scarcely any comparison with that which attaches to the same force when connected with the nervous system and with life.

Michael Faraday, 1839

Since Luigi Galvani applied the first electric shock to a frog leg to make it twitch in 1791, bioelectric phenomena ("galvanism") have been studied intensively.

Galvanotaxis is the movement of a motile organism or cell in direct response to an electric current. This migration is toward the cathode or the anode.

Many eukaryotes show galvanotaxis, e.g., amoebae, paramacia, hydrae, rotifers, shrimp, salamanders, and fish; motile cells of organisms also do this, such as vertebrate cells grown in tissue culture. For some research articles and reviews on galvanotaxis in eukaryotes, see Ludloff (1895), Verworn (1899), Jennings (1906), Loeb (1918), Umrath (1959), Carlile (1975), Machemer and De Peyer (1977), Hinrichsen et al. (1984), and Nuccitelli (1986a).

In bacteria, the first observation of galvanotaxis was made by Verworn in 1889: He noted briefly that some kinds of bacteria in a crude mixture of species swam toward the cathode, whereas other kinds swam toward the anode (Verworn 1889). Since then, galvanotaxis in prokaryotes has been studied in species of *Bacillus* and *Bacterium termo* (Abbott and Life 1908) and in blue-green algae (Häder 1977) and similar cyanobacteria (Murvanidze and Glagolev 1982). In this paper, we describe the first work on galvanotaxis, to our knowledge, in *Escherichia coli* and *Salmonella typhimurium*.

The mechanism of galvanotaxis is largely unknown. Because knowledge of the biochemistry and genetics of *E. coli* and *S. typhimurium* is very advanced relative to that of other organisms used for the study of galvanotaxis, we intend to use these bacteria for learning the mechanism. Furthermore, the function of galvanotaxis can be explored using these bacteria. It may be useful to try to apply what can be learned with bacteria to higher organisms. On the other hand, it is possible that prokaryotes and eukaryotes have different mechanisms and different functions for galvanotaxis. It is also possible that galvanotaxis is purely a physical phenomenon without any function in the bacteria, but even so, understanding how it works may lead to biological insights.

RESULTS

Bacteria swim in random directions when there is no current (Fig. 1, left); however, when the current is turned on, the bacteria migrate in a rather straight line toward one electrode or the other (Fig. 1, center). When the polarity is changed, the bacteria migrate in the opposite direction. This demonstrates galvanotaxis in bacteria. Upon change in polarity, each cell makes a U-turn, rather than backing up (Fig. 1, right). This indicates that the cells have a fixed dipole at that moment, i.e., one end is positively (+) charged relative to the negative (−) charge on the other end; the alternative idea of an induced dipole would allow the bacteria simply to back up when the polarity is changed.

Galvanotaxis by *E. coli* K-12 (strain AW405) is exclusively toward the anode. The speed is approximately 10 μm/sec at 3 V/cm. We have found the same result with three additional independent wild-type *E. coli* mutants: B (strain B62), C (strain B64), and W (strain AW490). In contrast, migration exclusively toward the cathode was found with *S. typhimurium* LT2 (strain AW621); this occurs at about the same speed as with *E. coli*. This same result has been obtained with six additional independent wild-type *S. typhimurium* mutants: UCD (strain SL1479), FIRN (strain SL3235), WRAY (strain SL3261), ovine (strain SL5489), C5 (strain SL5524), and St21 (strain SL5756). (Each contains an *aroA*, *p*-amino-acid-dependent mutation to make it nonvirulent. [Courtesy of B.A.D. Stocker].) This opposite result for *E. coli* and *S. typhimurium* demonstrates that the direction of migration is not an artifact of the assay. To determine the mechanism of galvanotaxis, it is important to discover the basis for the direction of migration.

Can mutants be obtained with altered direction of migration or with no flipping around (no fixed dipole) when the current is changed? If so, studies of these mutants to reveal the factors that determine the nature of the galvanotactic response would be helpful. We have now discovered that *S. typhimurium* SL7141 is a mutant that migrates toward the anode instead of the cathode (at about the same speed as wild-type *S. typhimurium*). With the help of B.A.D. Stocker, we are tracing the ancestors of this LT2 strain to determine the origin of this mutation and the cause of the altered galvanotaxis. In addition, we are now attempting to isolate our own galvanotaxis mutants of *S. typhimurium* and *E. coli*.

To determine which of the known gene products for chemotaxis and motility are required for galvanotaxis, we tested *E. coli* mutants lacking each product. It

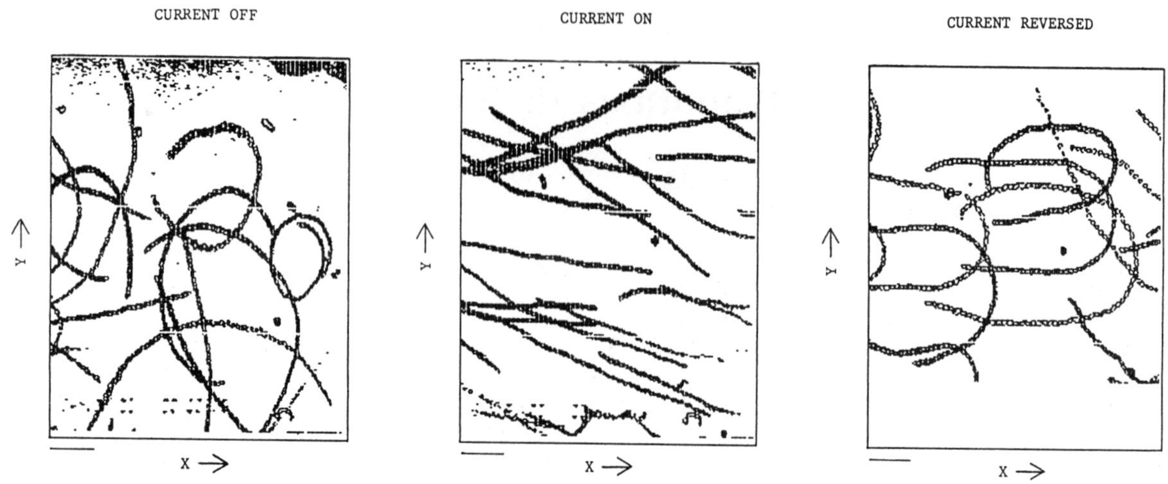

Figure 1. Galvanotaxis by motile *E. coli* strain AW405. (*Left*) 5-sec plot of paths of bacteria before current was turned on; (*center*) 5-sec plot of path of bacteria migrating toward the anode; (*right*) as in *center*, but polarity was switched at 2 sec; note resulting U-turns. Bars beneath art equal 7 μm. Galvanotaxis was observed through the microscope and recorded on videotape for study (as above) with a motion analysis system (Sager et al. 1988). Bacteria at $OD_{590} = 0.2$ were suspended in a 2 × 2.5-cm × 1-mm chamber containing 5×10^{-3} M potassium phosphate buffer at pH 7.0, with 10^{-4} M EDTA to enhance motility (Adler 1973) and were then covered with a coverslip. At each end of this chamber was a 5 × 2.5-cm × 5-mm compartment of 1.0% agarose (Bethesda Research Laboratories, electrophoresis grade) containing, distally to the bacteria, a platinum electrode hooked up to an adjustable voltage and current source with choice of polarity (Hoefer Scientific Instruments, model PS101). Galvanotactic responses were studied at 3 V/cm. The threshold depended on the conductance of the solution: 2 V/cm for 10^{-4} M, 4 V/cm for 10^{-2} M, and >20 V/cm for 1 M potassium phosphate.

appears that none of these known gene products are needed for galvanotaxis (Table 1). Thus, the information leading to galvanotaxis enters the flagellar apparatus at a point beyond the known machinery for chemotaxis (Fig. 2). In some manner, the electric current must affect the flagella: When *E. coli* and *S. typhimurium* are in the same electric field, the bundle of flagella must be located at opposite ends in order to bring about swimming in opposite directions. Galvanotaxis might be mediated by current flowing through ion channels, or the effect could be a purely physical one on the fixed dipole.

Although *E. coli* mutants with paralyzed flagella or lacking flagella and consequently lacking motility (e.g., a *motA* mutant or a *flaI* mutant) cannot carry out galvanotaxis (by definition, taxis requires motility), they do respond to electric current. They align in the current and flip around when the polarity is changed, as

Table 1. Galvanotaxis in Chemotaxis Mutants

Mutants[a]	Strain number	Galvanotaxis[b]
Wild type	AW405	+
Receptor mutants		
tsr (MCP I⁻)	AW518	+
tar (MCP II⁻)	AW539	+
trg (MCP III⁻)	AW701	+
tap (MCP IV⁻)	RP3525	+
tsr tar trg tap	CP362	+
ptsI (enzyme I⁻)	AW509	+
Chemotaxis-machinery mutants		
cheR (methyltransferase⁻)	AW688	+
cheB (methylesterase⁻)	AW689	+
cheA	AW690	+
cheW	RP4009	+
cheY	AW691	+
cheZ	AW692	—[c]
Mutant missing 4 MCPs and 6 *ches*	HCB326	+

[a]Strains that run and tumble (*tsr, tar, trg,* and *tap*) migrate in galvanotaxis by running and tumbling, although tumbling can be partially inhibited by use of voltages higher than normal. Mutants (*tsr tar trg tap, cheR, cheA, cheY,* and *cheW*) migrate in a smooth (running) form. Tumbly bacteria show poor galvanotaxis. The tumbly mutant *cheB* was made smooth by the addition of an attractant (10^{-2} M L-serine).

[b]Galvanotaxis was measured as described in Fig. 1; + indicates a positive response.

[c]It is difficult to judge galvanotaxis for *cheZ* because it makes very slow progress in the electric field due to its tumbliness.

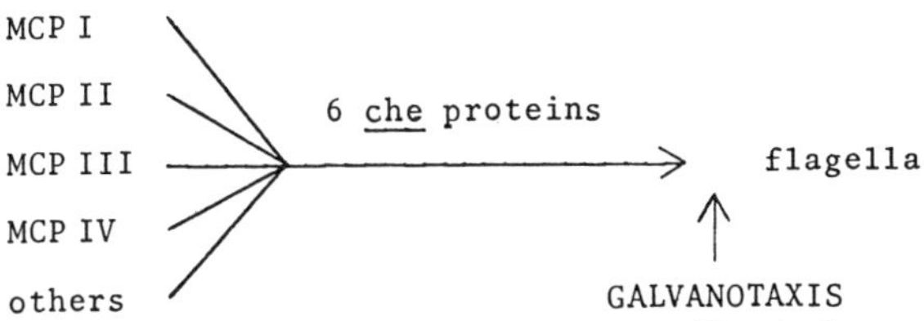

Figure 2. Galvanotaxis does not appear to utilize the known receptors and the known Che proteins of the chemotaxis mechanism. Although galvanotaxis requires flagella because it is an aspect of motility, nonmotile mutants lacking flagella align and flip around when the current is changed (see text).

cells do that have been killed by ether or by boiling. This shows that these bacteria have a dipole independent of flagella, that they survive death of cells, and that dipole is due at least, in part, to remaining macromolecules because materials of low molecular weight would have left such treated cells. A goal of this work will be to determine the molecular basis for this dipole. We have found that the nonmotile *Klebsiella planticola* also aligns in electric current but fails to flip around when the polarity is changed; therefore, this flipping requires some special, as yet unknown, structural quality.

It was found that motile bacteria migrate in an electric field at least 10 times faster than nonmotile bacteria; if "galvanotaxis" were simply electrophoresis, then motile and nonmotile forms would move at the same speed. It is still possible that electric force acts on the cells' dipole structure and net charge to align the cells and bring about movement toward one electrode, and then the flagellar bundle could just fall in behind to make the cells move faster.

CONCLUSIONS

Motile bacteria have an aligned swimming direction in an electric current. This direction is opposite for *E. coli* and *S. typhimurium*, and it is therefore not an artifact of the assay. We have discovered a mutant of *S. typhimurium* with opposite direction from its parent. None of the known gene products for chemotaxis is necessary for galvanotaxis, and it is not yet known how the electric current affects cell migration. Mutants with paralyzed flagella or lacking flagella align and flip around when the polarity is changed, as do dead cells.

PROSPECTUS

The fixed dipole ("natural electric field") plays an important role throughout eukaryotic biology, e.g., in determining cell polarity and in causing developmental events (Jaffe 1986; Nuccitelli 1986b). Now it can be studied by molecular genetics in *E. coli* and *S. typhimurium* to learn its mechanisms and discover its functions. If we find a mechanism similar to that in eukaryotes, the fixed dipole in bacteria may play a role in determining cell polarity and in controlling cell division. Even if bacterial galvanotaxis proves to be a functionless, physical by-product of the structure of the cell, a fixed dipole would still set up an electric field that could affect the cell's biological activities. Most promising is the fact that *E. coli* and *S. typhimurium* migrate in opposite directions in an electric field and that a reversed mutant exists: We wish to at least learn the basis for those facts.

ACKNOWLEDGMENTS

We thank Kevin Clark, Michael Lentz, and Brian Sager for contributions to experiments, and we are also indebted to Howard Berg, Ching Kung, Kim Lewis, and Boris Martinac for valuable discussions. B.A.D. Stocker kindly provided strains of wild-type *E. coli* and *S. typhimurium*. We thank Howard C. Berg for the HCB strains, G.L. Hazelbauer for the CP strain, and J.S. Parkinson for the RP strains. This research was supported by a grant from the National Science Foundation.

REFERENCES

Abbott, J.F. and A.L. Life. 1908. Galvanotropism in bacteria. *Amer. J. Physiol.* **22:** 202.

Adler, J. 1973. A method for measuring chemotaxis and use of the method to determine optimum conditions for chemotaxis by *Escherichia coli*. *J. Gen. Microbiol.* **74:** 77.

Carlile, M.J. 1975. Taxes and tropisms: Diversity, biological significance, and evolution. In *Primitive sensory and communication systems: The taxes and tropisms of microorganisms and cells* (ed. M.J. Carlile), p. 9. Academic Press, London.

Häder, D.P. 1977. Influence of electric fields on photophobic reactions in blue-green algae. *Arch. Microbiol.* **114:** 83.

Hinrichsen, R.D., Y. Saimi, and C. Kung. 1984. Mutants with altered Ca^{2+}-channel properties in *Paramecium tetraaurelia:* Isolation, characterization, and genetic analysis. *Genetics* **108:** 545.

Jaffe, L.F. 1986. Ion currents in development: An overview. In *Ionic currents in development* (ed. R. Nuccitelli), p. 351. Alan R. Liss, New York.

Jennings, H.S. 1906. *Behavior of the lower organisms*. Indiana University Press, Bloomington.

Loeb, J. 1918. *Forced movements, tropisms, and animal conduct*, p. 32. Lippincott, Philadelphia.

Ludloff, K. 1895. Untersuchungen über den Galvanotropismus. *Pflugers Arch. Gesamte. Physiol. Menschen Tiere* **59:** 525.

Machemer, H. and J. De Peyer. 1977. Swimming sensory cells: Electrical membrane parameters, receptor properties and motor control in ciliated Protozoa. *Verh. Dtsch. Zool. Ges.* p. 86.

Murvanidze, G.V. and A.N. Glagolev. 1982. Electrical nature of the taxis signal in Cyanobacteria. *J. Bacteriol.* **150:** 239.

Nuccitelli, R., ed. 1986a. *Ionic currents in development*, p. 263–347. Alan R. Liss, New York.

———. 1986b. An introduction by the editor. In *Ionic currents in development*. Alan R. Liss, New York.

Sager, B.M., J.J. Sekelsky, P. Matsumura, and J. Adler. 1988. Use of a computer to assay motility in bacteria. *Anal. Biochem.* **173:** 1.

Umrath, K. 1959. Galvanotaxis. In *Encyclopedia of plant physiology* (ed. W. Ruhland), vol. 17, p. 164. Springer-Verlag, Berlin.

Verworn, M. 1889. Die polare Erregung der Protisten durch den galvanischen Strom. *Pflugers Arch. Ges. Physiol. Menschen Tiere* **46:** 267.

———. 1899. The actions of electrical stimuli. In *General physiology, an outline of the science of life* (translated and edited by F.S. Lee), p. 455. Macmillan, London.

Interaction of CheB with Chemotaxis Signal Transduction Components in *Escherichia coli*: Modulation of the Methylesterase Activity and Effects on Cell Swimming Behavior

R.C. Stewart, C.B. Russell, A.F. Roth, and F.W. Dahlquist
Institute of Molecular Biology, University of Oregon, Eugene, Oregon 97403

Escherichia coli cells are capable of alternating between two swimming behaviors: smooth swimming (which moves the cells in a slightly curved path) and tumbling (which somersaults the cells and points them in new, random directions when smooth swimming resumes). Chemotaxis by *E. coli* results from the ability of a cell to alter the relative time spent performing these two swimming behaviors (Berg and Brown 1972; Macnab and Koshland 1972; Larsen et al. 1974; Silverman and Simon 1974) and to make this alteration in response to concentration changes of specific chemical stimuli, as detected by several receptor proteins (Adler 1969; Parkinson and Hazelbauer 1983). The components and processes that link receptor changes to changes of swimming behavior are beginning to be understood at a molecular level (Hess et al. 1988; Oosawa et al. 1988). All of the genes required specifically for chemotaxis have probably been identified (for review, see Parkinson 1981; Macnab 1987) and assigned to their corresponding proteins (Ridgway et al. 1977; Macnab 1987). Of these, four cytoplasmic proteins (CheA, CheW, CheY, and CheZ) are thought to play essential roles in the signal transduction process (Parkinson 1981; Parkinson and Hazelbauer 1983; Wolfe et al. 1987). In addition, the proteins encoded by the *cheR* and *cheB* genes are essential for allowing productive swimming behavior in the presence of unchanging levels of stimuli; e.g., they enable sensory adaptation of the chemotaxis system (Goy et al. 1978; Parkinson and Revello 1978; Springer et al. 1979; Yonekawa et al. 1983). The chemotaxis receptor proteins (Tsr, Tar, Trg, and Tap) that span the bacterial inner membrane are necessary for response to different groups of stimuli (Alder 1979; Parkinson and Hazelbauer 1983).

The receptor proteins convey extracellular information (ligand concentration) to the intracellular signal transduction components by an unknown mechanism that probably involves a ligand-induced conformation change by the receptor (Wang and Koshland 1980; Stock and Koshland 1981; Krikos et al. 1983). For this reason, these receptor proteins are frequently referred to as transducer proteins. These proteins appear to be capable of two distinct "signaling modes" or conformations: one that signals for smooth swimming and one that promotes tumbling (Parkinson 1980; Callahan and Parkinson 1985; Mutoh et al. 1986; Park and Hazelbauer 1986). The conformation adopted by a particular transducer protein is affected by two factors: (1) the extracellular concentration of specific attractant and repellent molecules and (2) the number of glutamate residues in the protein that have been methylated (converted to γ-glutamyl methyl esters). The methylation of the transducer proteins occurs only at several specific glutamate residues (four to six sites per protein molecule, depending on the particular transducer class) (Kehry and Dahlquist 1982a,b; Terwilliger and Koshland 1984; Nowlin et al. 1987). The fraction of these four to six sites that is methylated (the "methylation level") reflects the balance between the activities of the chemotaxis-specific methyltransferase (CheR) (Springer and Koshland 1977) and methylesterase (CheB) (Stock and Koshland 1978).

Transducer methylation levels change in characteristic ways in response to the addition or the removal of attractants and repellents that are bound by the transducer proteins (Kort et al. 1975; Silverman and Simon 1977; Springer et al. 1977; Kondoh et al. 1979). Increased attractant concentration results in a higher level of transducer methylation, whereas increased repellent concentration (or decreased attractant concentration) causes a lowering of the methylation level. These changes are specific to the class of transducer to which a particular attractant or repellent binds. Changes of methylation level occur relatively slowly (requiring some seconds to minutes, depending on stimulus strength) compared with the rapid interactions between ligands and the transducer proteins (which probably take place on a millisecond time scale). The methylation level of a transducer protein is important in determining its signaling status, promoting either smooth swimming or tumbling. In general, methylated transducer proteins promote tumbling, whereas unmethylated transducer proteins signal for smooth swimming (Parkinson and Hazelbauer 1983). The situation is somewhat more complicated in that a transducer protein's signaling mode is actually determined by some mechanism that "compares" the current ligand-occupancy status (reflecting the current environment) with the methylation level (which responds relatively slowly to stimuli and therefore reflects the cell's environment in the recent past). This comparison mechanism is thought to provide the bacterial cell with some form of "memory" to facilitate its determination of the appropriate swimming behavior.

Two distinct processes appear to account for the observed changes of transducer methylation in response to chemotactic stimuli. First, the conformation of the transducer protein appears to change upon binding attractants/repellents in such a way that attractant-bound transducers are better substrates for the methyltransferase than free or repellent-bound transducer proteins are (Kleene et al. 1979; Wang and Koshland 1980; Stock and Koshland 1981). A distinct mechanism modulates the activity of CheB (the methylesterase) in a manner that also promotes these changes of transducer methylation level. By an unknown sequence of events, attractants transiently decrease methylesterase activity; repellents transiently increase methylesterase activity (Toews et al. 1979; Kehry et al. 1984). This modulation of CheB activity appears to involve integration of information provided by all of the transducer proteins (Kehry et al. 1985). As discussed below in Results, this modulation requires CheA and CheW and does not result directly from altered accessibility of methylation sites in response to attractants and repellents.

We are interested in understanding the modulation of methylesterase activity of CheB at a molecular level. Specifically, we are attempting to define the cellular component(s) responsible for this modulation and to uncover details of the interactions that alter this activity. For example, we hope to discover what region of the CheB protein is involved in such interactions, to understand the events that modulate the methylesterase, and to determine whether modulation results from transient chemical modification of CheB or from association of CheB with an activator or inhibitor. In this paper, we summarize our recent progress toward these goals and discuss our results in the context of the model presented recently by Hess et al. (1988), who demonstrated that CheA autophosphorylates in vitro and functions as a CheB and CheY kinase.

RESULTS

CheA and CheW Are Necessary for Modulation of CheB Methylesterase Activity

For the work discussed in this paper, we used the flow-chase assay developed by Kehry et al. (1984) to measure methylesterase activity in intact *E. coli* cells. In this assay, [^{3}H]methanol is generated by prelabeled cells that are maintained on a filter over which Tris-buffered succinate solution is passed. During the labeling period, these cells have been provided with [^{3}H]methionine, which is converted to [^{3}H]*S*-adenosylmethionine, and this serves as the methyl group donor for transducer protein methylation by CheR. Removal of the methyl groups by CheB generates [^{3}H]methanol. By adding various attractants or repellents to the flow medium, we can observe the effect on methylesterase activity reflected as [^{3}H]methanol evolution as a function of time. These experiments are performed and interpreted most easily when carried out as cold chase experiments following the initial labeling period. Therefore, the apparent methanol evolution rate decreases with time as nonradioactive methyl groups replace the [^{3}H]methyl groups. Figure 1 shows such a decrease for unstimulated cells and the typical modulation of CheB activity in response to attractant addition (resulting in transient activity decrease) and attractant removal (resulting in transient activity increase) for wild-type chemotactic cells.

As discussed above, CheA, CheW, CheY, and CheZ are thought to play essential roles in the chemotaxis signal transduction pathway and are therefore reasonable candidates as modulators of CheB methylesterase activity. We have determined whether genetic deletion of these components (individually) affects the ability of cells to modulate CheB. These results (shown in Fig. 2 and summarized in Table 1) indicate that CheA is necessary for modulation of CheB activity in response to both positive and negative stimuli, whereas CheW is essential for modulation in response to negative stimuli. CheY is required for neither aspect of modula-

Table 1. Effects of Deleting and Overproducing Chemotaxis Components on CheB Modulation and Swimming Bias

	Methylesterase modulation[a]				Swimming behavior[b]	
	deletion		overproduction			
	pos.	neg.	pos.	neg.	deletion	overproduction
CheA	–	–	–	–	sm	sm
CheW	+	–	+	–	sm	sm
CheY	+	+	+	–	sm	sm
CheZ	+	+	+	+	tm	sm
CheB	n.a.	n.a.	–	–	tm	sm
CheR	n.a.	n.a.	+	+	sm	tm
Tar	+	+	+	+	wt	wt/sm
Tsr	+	+	+	+	wt	wt/sm

[a]The presence (+) or absence (–) of a normal response to positive (pos.) or negative (neg.) stimuli is indicated. (n.a.) Not applicable for *cheB* and *cheR* deletions because in vivo methylesterase activity cannot be measured in these mutants.

[b]Swimming behavior of unstimulated cells grown in minimal media is denoted as smooth (sm), tumbly (tm), or random (wt [wild type]).

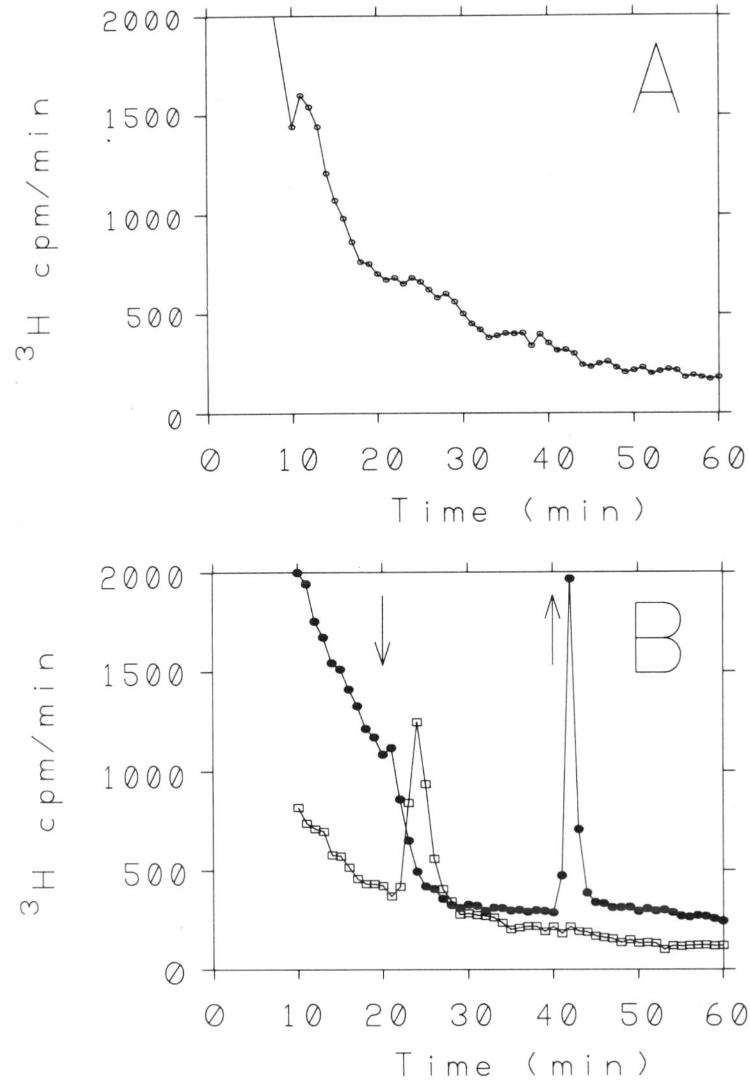

Figure 1. Results of flow-chase experiments demonstrating the response of CheB activity in wild-type chemotactic cells (RP487) to attractant and repellent stimuli. (*A*) A plot of apparent methylesterase activity (expressed as volatile ^{3}H counts generated per minute) as a function of time following initiation of a cold chase, as described in the text. (*B*) The data are for cells treated identically to those used in *A*, except that they were subjected to stimulus addition (downward arrow) and removal (upward arrow) at the indicated time points. The stimuli used were 50 μM L-serine (●) or 10 mM L-leucine (□). There is a characteristic "trough" in the activity curve following attractant addition and an activity "spike" following attractant removal or repellent addition.

tion. Deletion of *cheZ* has a subtle effect on the kinetics of the response to attractant addition, but the CheZ protein is not required for the full extent of modulation in response to either attractants or repellents.

Our results, using several *cheA* point mutants (data not shown), differ somewhat from those obtained with a *cheA* deletion mutant. With several point mutants, we observe loss of modulation in response to negative chemotactic stimuli, whereas modulation in response to positive stimuli appears to be essentially normal. These results are in agreement with those reported by Springer and Zanolari (1984), who monitored CheB activity in *cheA* point mutants by observing the disappearance of transducer methyl groups on polyacrylamide gels. These results may indicate that modulation of CheB activity in response to positive and negative stimuli involves distinct mechanisms, although both mechanisms require CheA.

Overproduction of CheB, CheA, CheW, and CheY Affect Modulation of the Methylesterase

We have also determined the effects on CheB modulation caused by overproducing the possible signal transduction components CheA, CheW, CheY, and CheZ, as well as the effects of overproducing CheB itself, the methyltransferase (CheR), and the transducer proteins Tsr and Tar. Some of the results are shown in Figure 3 and summarized in Table 1. These experiments involve placing the genes of interest into

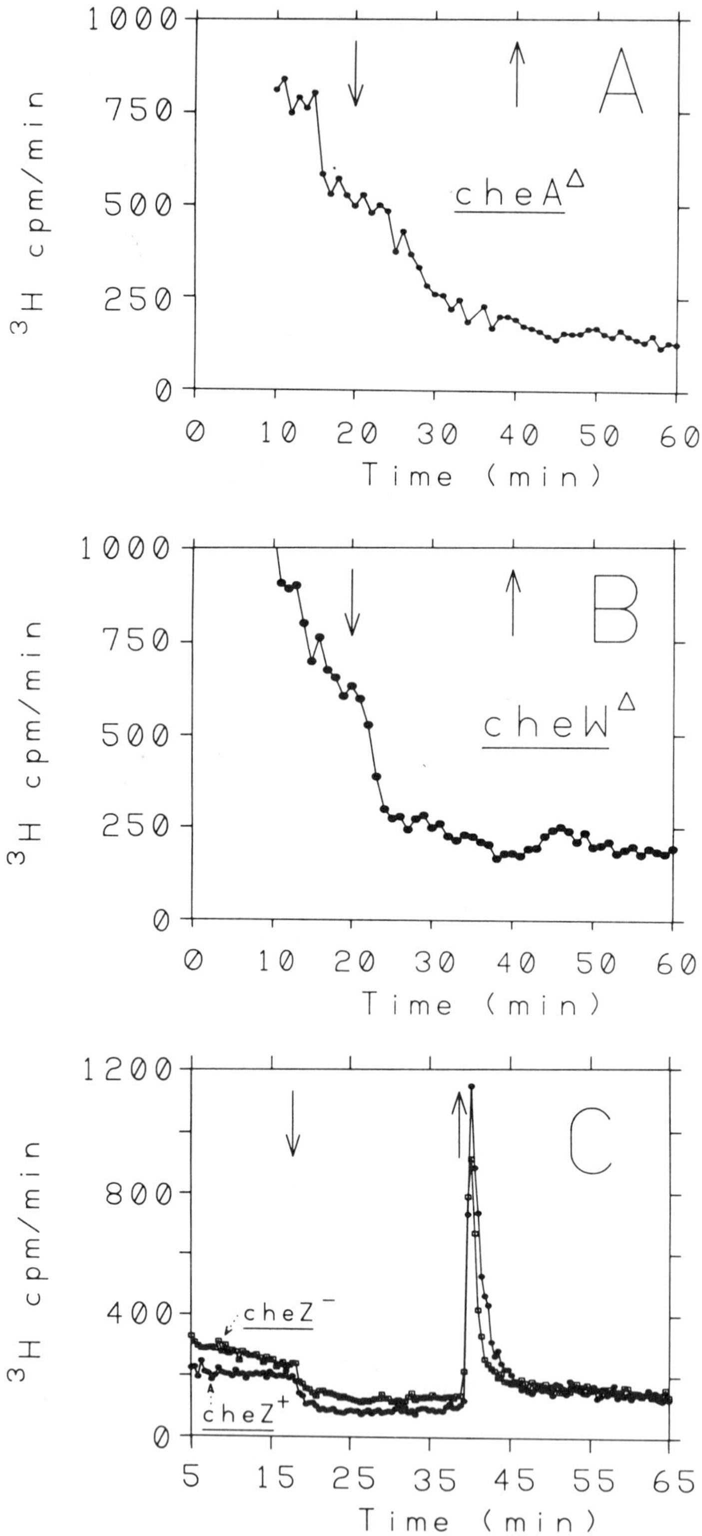

Figure 2. Results of flow-chase experiments demonstrating defective modulation of methylesterase activity in *cheA, cheW,* and *cheZ* mutants. A stimulus of 50 μM L-serine was added (downward arrow) at $t \sim 20$ min and removed (upward arrow) at $t \sim 40$ min, as indicated. (*A*) Results obtained with the *cheA*$^{\Delta}$ strain RP1788; (*B*) results obtained with the (*cheW-tar*)$^{\Delta}$ strain RP1078. The lack of response with these mutants is in striking contrast to the response in wild-type cells shown in Fig. 1B. (*C*) The somewhat slow response to attractant addition for the *cheZ*$^{\Delta}$ strain RP1616 (□) compared to that obtained with the same cells containing a plasmid that expresses CheZ at approximately wild-type levels (●).

plasmid vectors downstream from an inducible promoter (P_{lac}, P_{tac}, or P_{araB}) to enable variable levels of expression by adding appropriate concentrations of inducers (isopropyl-β-D-thiogalactoside [IPTG] or arabinose) to the growth medium. The multicopy versions of these plasmids also carry the genes for the appropriate regulatory proteins (*lacI*^{Q} or *araC*) to facilitate controlled expression of the chemotaxis gene. As shown in Figure 3A and Table 1, overproduction of CheB itself eliminates modulation of its methylesterase activity. We believe that this indicates the existence of some modulation component or process that can be saturated by excess CheB such that the modulation element can no longer interact with other components in response to chemotactic stimuli. CheA, CheW, and CheY may also interact with such a component, as indicated by the loss of CheB modulation, when they are overproduced (see Fig. 3B–D). The effect of CheY overproduction on CheB modulation appears to require the presence of CheZ, perhaps indicating the importance of rapid cycling between the phosphorylated and dephosphorylated forms of CheY enabled by CheZ, which can serve as a CheY phosphatase (Hess et al. 1988). Overproduction of CheR, CheZ, or the transducer proteins Tar and Tsr had no noticeable effects on CheB modulation.

In the experiments described above, the intracellular levels of the overproduced components were sufficient to impair the chemotactic ability of the cells significantly (as judged by swarm plate assays) and to alter the unstimulated swimming behavior of the cells dramatically. This resulted in either a markedly smooth swimming bias (for CheA, CheW, CheB, and CheZ overproduction) or a markedly tumbly bias (for CheY and CheR overproduction), as summarized in Table 1. It is interesting to note that the smooth-swimming phenotype associated with CheA and CheW overproduction is the same as that observed for *cheA* and *cheW* null mutants, whereas overexpression of *cheB, cheR, cheY,* and *cheZ* produce the opposite swimming bias of that observed with the respective null mutants. This is consistent with CheA and CheW each having multiple roles that affect swimming behavior differently, depending on the concentration of these key proteins.

Our investigations of CheB modulation in various deletion and overexpression strains indicate that CheA and CheW are necessary for the flow of information from the transducer proteins to CheB to enable appropriate modulation of methylesterase activity. Although not required for this information flow, CheY may interact with some required component in such a way that excess CheY inhibits the interaction of CheB with its activating component or process. These findings are summarized in Figure 4. In this scheme, CheA and CheW play a central role serving as the link between the transducer proteins and CheB. Because CheY overproduction diminishes CheB modulation, we believe that CheY and CheB may compete for the modulating component (CheA or a CheA-CheW complex). Further evidence for such competition is found in the following section. The interpretation of our results presented in Figure 4 is consistent with the recent results of Hess et al. (1988) who demonstrated (in vitro) the existence of CheB-CheA and CheY-CheA interactions involving phosphate transfer from CheA to CheB and CheY. This will be addressed in more detail below in the Discussion.

The Effect of CheB on Swimming Behavior

It has long been known that CheB affects the swimming behavior of unstimulated cells. *cheB*$^{\Delta}$ mutants tumble incessantly (Parkinson 1976, 1978; Yonekawa et al. 1983); complementation of the deletion with *cheB* carried on phage or plasmid vectors returns the swimming behavior to the normal pattern of alternating between smooth swimming and tumbling. We have found that progressive production of wild-type CheB above its normal intracellular level results in a progressively extreme smooth-swimming bias (see Fig. 5). Because this overexpression also results in dramatic reduction of transducer methylation levels, we initially assumed that this smooth bias resulted entirely from the consequent alteration of the transducers' signaling status. If the transducer proteins are always undermethylated significantly (as is the case in the presence of excess CheB), they would be expected to signal constantly for smooth swimming. However, we have found that neither the methylesterase activity of CheB nor the presence of transducer proteins is necessary to enable CheB overproduction to impart a smooth swimming bias to cells (see Fig. 5), indicating that CheB may influence swimming behavior by some mechanism in addition to that involving transducer demethylation. For example, overproduction (to ~10–20-fold of the wild-type concentration) of a CheB point mutant that lacks methylesterase activity can impart a smooth-swimming bias to otherwise wild-type cells, although the extent of transducer methylation is not affected by this overproduction. A similar (but less dramatic) effect is obtained by overproducing a CheB-LacZ fusion protein that includes only the amino-terminal half of CheB and therefore has no methylesterase active site (see below). This fusion protein and wild-type CheB can also influence the swimming behavior of cells that have been deleted for the genes encoding all four of the known transducer proteins. Because the unstimulated swimming behavior of these transducer$^-$ cells is extremely smooth, we have overproduced CheY in these cells to make them tumbly and have then determined whether increased expression of wild-type CheB and various CheB mutants can reverse this tumbly bias (as depicted in Fig. 5). The plasmid construction enabling such experiments has two inducible promoters (P_{tac} and P_{araB}) and the genes encoding their respective control elements separated by a kanamycin resistance cartridge. By adding IPTG to the growth media of transducer$^-$ cells carrying this plasmid, we can produce sufficient CheY to make the transducer$^-$ cells tumbly. By adding the appropriate concentrations of arabinose at the same IPTG level, we can elevate expression of *cheB* and observe the effect on the swimming behavior of the cells. The results are summarized in Figure 5. Such experiments indicate that CheB can reverse the tumbly effects of excess CheY in a transducer$^-$ cell. Control experiments in which a pBR322 fragment has replaced *cheB* in this plasmid indicate that arabinose itself or increasing expression off of the *araB* promoter does not reverse the effects of CheY. Therefore, CheB appears to be capable of influencing swimming behavior by some mechanism that does not involve transducer methylation levels. The effect of various mutant

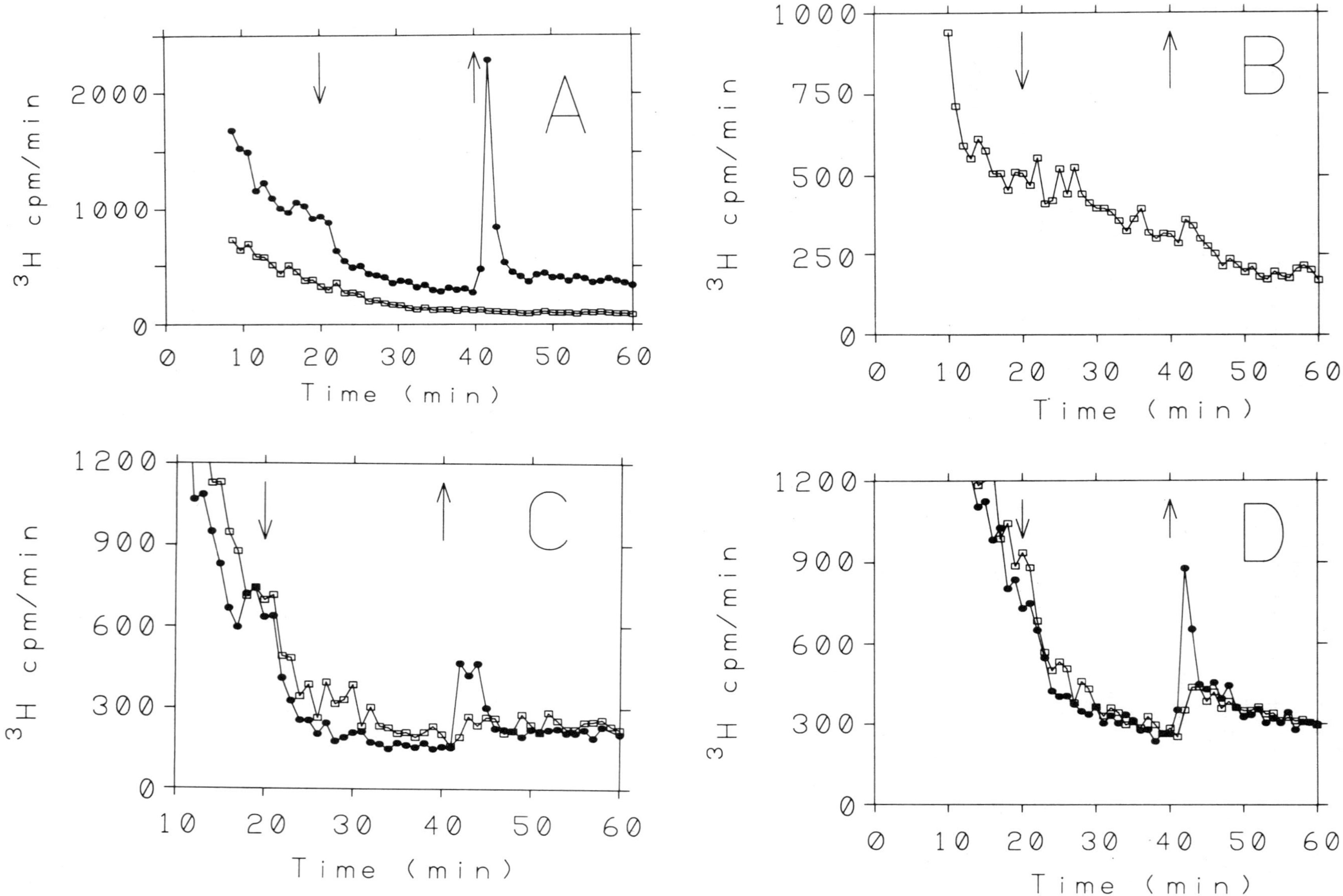

Figure 3. (*See facing page for legend.*)

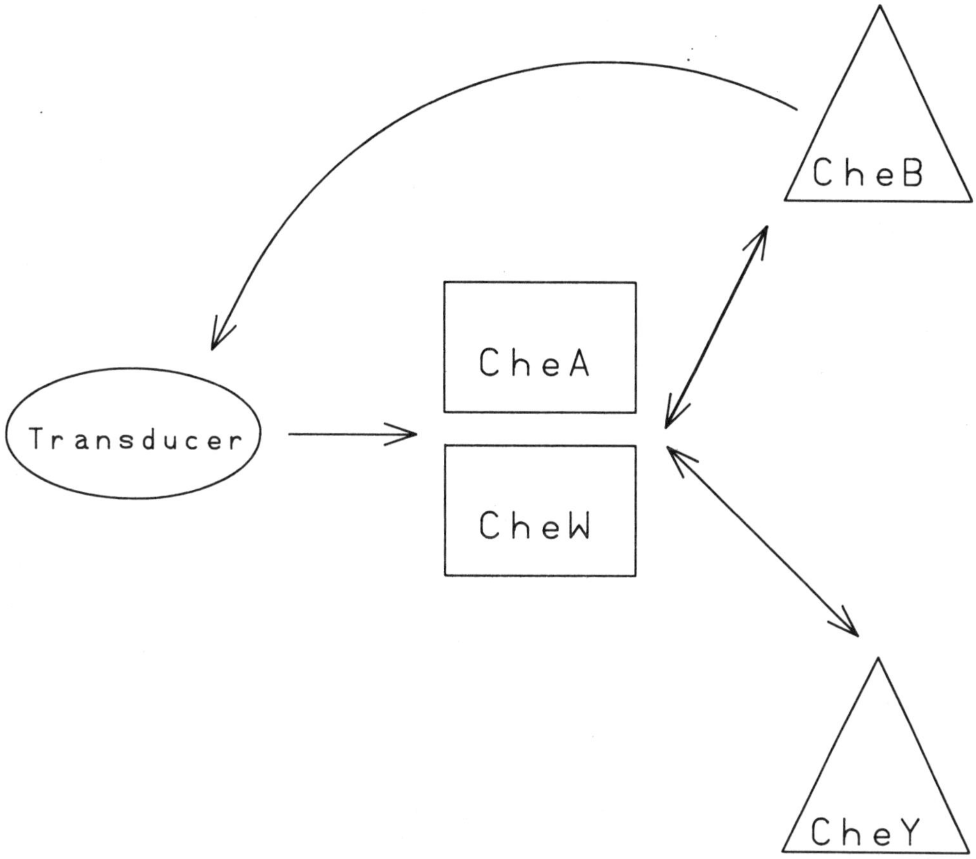

Figure 4. Schematic representation of the interactions that enable or affect modulation of methylesterase activity. The interactions indicated by double arrows summarize the results of the flow-chase experiments shown in Figs. 1–3. CheA is essential for both the activation of CheB in response to negative stimuli and the activity decrease in response to positive stimuli, as detected by the transducer proteins. CheW is necessary for the activation response. CheY is not necessary for either aspect of methylesterase modulation but may interact with a common modulating component (possibly CheA and/or CheW), as indicated by the loss of CheB activation in the presence of excess CheY. The changes in CheB methylesterase activity affect the methylation state and signaling intensity of the transducer proteins, thereby forming a feedback loop that eventually resets the methylesterase activity to its prestimulus level.

forms of CheB on the swimming behavior of such cells is summarized in Table 2 and will be discussed in detail in the following section.

The Amino-terminal Region of CheB Is Involved in Interactions That Modulate Methylesterase Activity

We have generated numerous *cheB* mutations in vitro and have used them to obtain information about what portion of the CheB protein molecule may be involved in the interactions that enable (1) modulation of its methylesterase activity and (2) its ability to influence swimming behavior. Among these mutations are the intragenic deletion constructions and the missense point mutations summarized in Figure 6. Our discussion will focus first on the deletion mutants before considering the effects of the various point mutations.

The two amino-terminal deletions shown in Figure 6

Figure 3. Overproduction of CheB, CheA, CheW, or CheY alters the ability of cells to modulate methylesterase activity. Results are from flow-chase experiments in which an attractant (50 μM L-serine) was added at $t = 20$ min (downward arrow) and removed at $t = 40$ min (upward arrow). (*A*) The normal response of the wild-type chemotactic strain RP487 at normal levels of CheB (●) and the lack of response at elevated levels of CheB (10- to 20-fold of the wild-type level □) achieved using a *cheB* expression plasmid. (*B*) The lack of response of wild-type chemotactic cells in which CheA has been overproduced. (*C*) The *cheW*$^{\Delta}$ strain RP1078 gives essentially normal responses to attractant addition and removal when provided with a plasmid that directs production of wild-type levels of CheW (●). The response to attractant removal is lost when CheW is overproduced in these cells (□). (*D*) The data are for the wild-type chemotactic strain RP487 containing a *cheY* expression plasmid. In the absence of IPTG, these cells have essentially normal levels of CheY and exhibit normal responsiveness to stimuli (●), but the presence of sufficient IPTG in the growth medium causes these cells to overproduce CheY significantly, resulting in a loss of the methylesterase activation in response to attractant removal (□). IPTG has no effect on the responsiveness of cells carrying a similar plasmid vector that lacks the *cheY* gene.

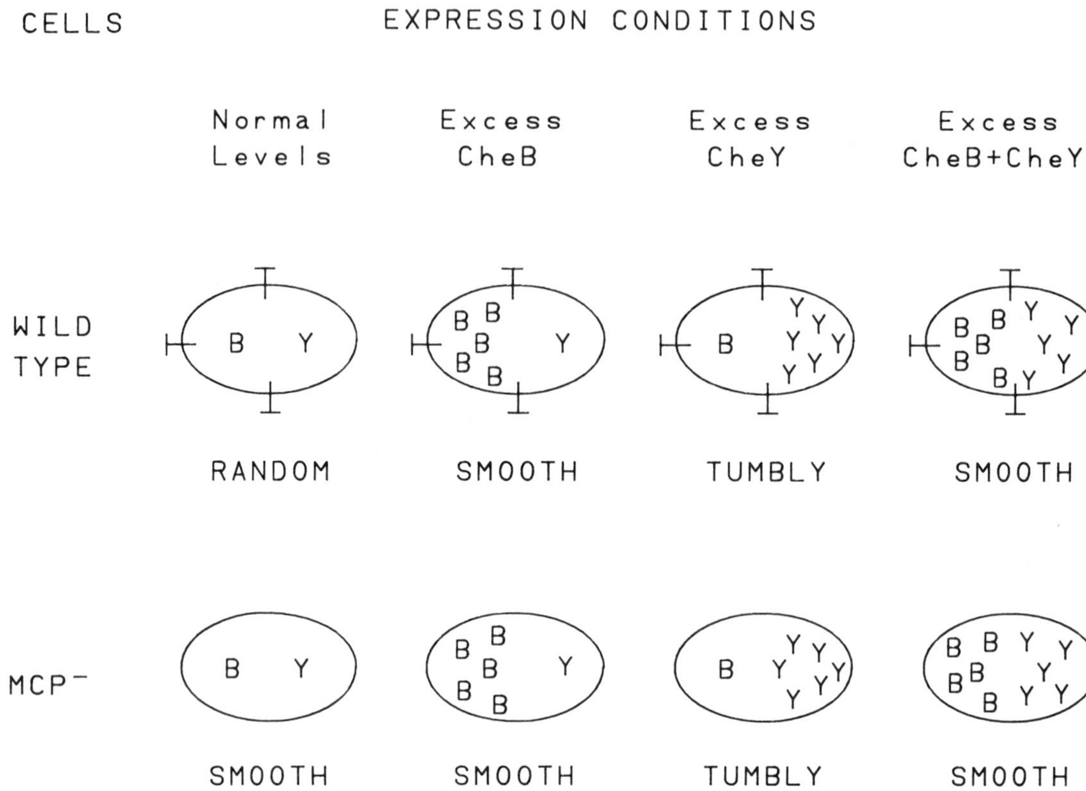

Figure 5. CheB can reverse the tumbly effects of excess CheY in both a wild-type chemotactic strain (RP487) and a strain that lacks the four known transducer proteins (CP362). The level of CheB (indicated schematically as the number of intracellular Bs) was controlled by adding appropriate concentrations of arabinose to the growth medium to induce expression from the *cheB* gene situated downstream from the *araB* promoter on a multicopy plasmid. Similarly, the CheY level (indicated by the number of Ys) was controlled by adding IPTG to the growth medium to induce expression of the *cheY* gene located downstream from the *tac* promoter located on the same plasmid as that expressing *cheB*. Wild-type cells have transducer proteins that span the bacterial inner membrane, as indicated by the Ts, whereas transducer⁻ cells lack these receptors.

result in removal of 25% and 50%, respectively, of the *cheB*-coding sequences while maintaining the reading frame for the remainder of the coding sequences. Immunoblots performed using anti-CheB antibodies indicate that expression of these deleted *cheB* genes (on multicopy plasmids) results in production of relative stable shortened forms of CheB that lack the deleted segments of the amino-terminal portion of this protein. When produced at appropriate levels, these deleted forms of CheB confer some swarming ability to $cheB^{\Delta}$ hosts and maintain significant methylesterase activity. The latter result confirms previous reports by Simms et

Table 2. Effects of Various *cheB* Mutations on Methylesterase Response and on Unstimulated Swimming Behavior

		Methylesterase modulation[b]		Swimming bias when overproduced[c]	
cheB allele	Mutation class	pos.	neg.	wt cells	MCP^-CheY^{+++} cells
Wild type	wild type	+	+	sm	sm
CheB-627	null	no activity		sm	sm
CheB(*Alu*I)	N-t deletion	+	−	sm	sm
CheB(*Hae*III)	N-t deletion	+	−	sm	sm
CheB-LacZ	C-t deletion	no activity		sm	sm
E58→K	dominant	+/−	−	sm	n.d.
E91→K	dominant	+/−	−	sm	n.d.
E125→K	dominant	+/−	−	sm	n.d.
R42→H	weak titrator	+/−	−	wt/sm	n.d.
R73→H	weak titrator	+/−	−	wt/sm	n.d.

[a](N-t) Amino-terminal; (C-t) Carboxy-terminal.
[b]The presence (+) or absence (−) of a normal response to positive (pos.) or negative (neg.) stimuli is indicated.
[c]Swimming behavior of cells in which various CheB mutant proteins are overproduced is indicated as smooth (sm) or random (wt [wild type]) for a chemotactic wild-type strain (RP487) and for a transducer (MCP) null strain (CP362) in which CheY is overproduced, as described in the text. (n.d.) Not determined.

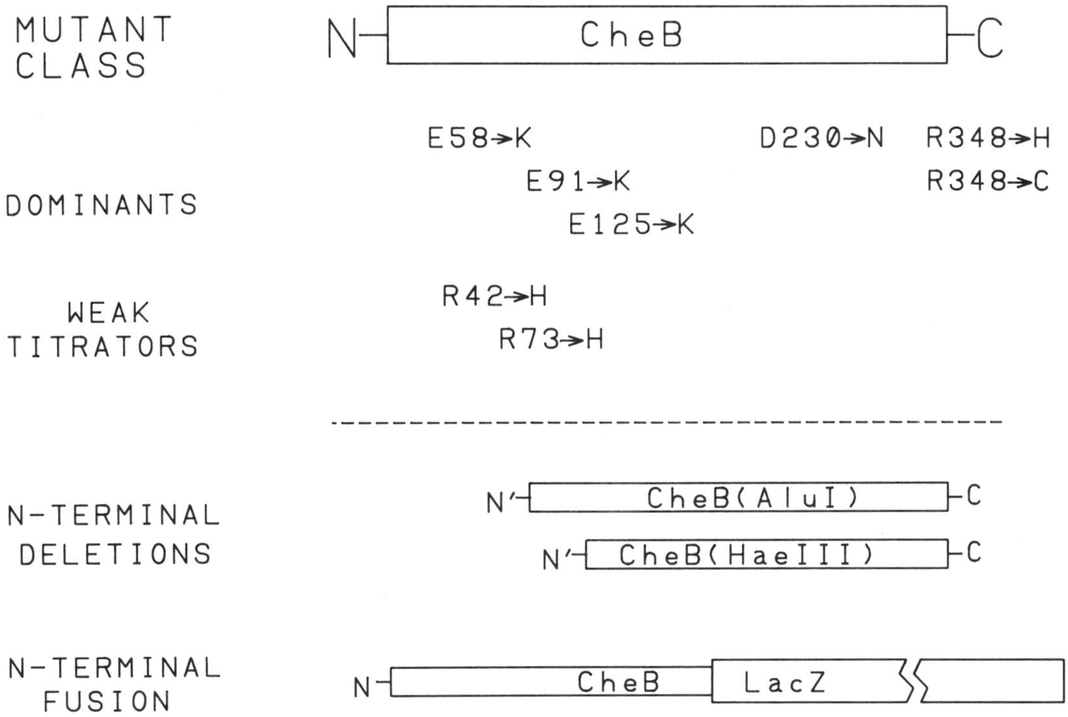

Figure 6. Mutations affecting the properties of CheB. The approximate extent of deletions and the general locations and amino acid changes of point mutants are shown for representative examples of each of four classes of mutants: amino-terminal deletions, carboxy-terminal deletions, dominant mutations, and weak titrator mutations. The properties of these CheB variants are described in the text.

al. (1985), who established that methylesterase activity requires only the carboxy-terminal three fifths of the protein and that a probable active-site cysteine lies in this portion of the protein.

As shown in Figure 7B, removal of the amino-terminal region of CheB dramatically impairs the ability of the methylesterase activity to respond to negative chemotactic stimuli (attractant removal or repellent addition), whereas the ability to respond to a positive stimulus is maintained. The flow-chase experiment shown in Figure 7B demonstrates the very limited response of one of the amino-terminal deletion mutants to serine removal and leucine addition. This is in striking contrast to the dramatic activity increase that is the main feature of the response of the wild-type protein. A similar lack of response to negative stimuli is observed with four other amino-terminal deletion mutants that removed as few as 7 to as many as 140 amino acids from the amino terminus of CheB. These results suggest that the amino-terminal portion of CheB is involved in activity modulation in response to negative chemotactic stimuli.

Results obtained with a *cheB-lacZ* fusion (depicted in Fig. 6) support this conclusion further. This translational fusion encodes a protein that includes the amino-terminal 206 residues of CheB fused to virtually the entire LacZ protein. This protein has no methylesterase activity. However, when it is overproduced in a wild-type cell (containing normal levels of the wild-type CheB protein), the ability of the cell to modulate (in response to negative stimuli) the methylesterase activity of the wild-type protein is lost (see Fig. 7A). We believe that this loss of modulation results from competition of the overproduced CheB-LacZ fusion protein for the component(s) responsible for modulating the activity of the wild-type CheB. Therefore, it appears that the amino-terminal portion of CheB is sufficient to enable such an interaction with the modulating component(s).

We have also determined the ability of the CheB amino-terminal deletion mutants and the amino-terminal CheB-LacZ fusion to affect the swimming behavior of cells (see Table 2). When overproduced, each of the proteins depicted in Figure 6 changes the swimming behavior of wild-type cells from random to smooth. Each of these proteins is also capable of reversing CheY-induced tumbliness in transducer$^-$ cells, as described above for the wild-type protein. These results suggest that both the amino-terminal portion and the carboxy-terminal portions of CheB are capable of altering the swimming bias of cells by some mechanism that does not involve transducer demethylation. We have no data to distinguish whether the amino-terminal and carboxy-terminal portions of CheB interact with a common component or with different components to counteract the effect of CheY. It should be noted that the amino-terminal 55 amino acid residues of CheB-(*Hae*III) are also present in the CheB-LacZ fusion and could account for the similar effects of overproduction of these two proteins. We are currently attempting to

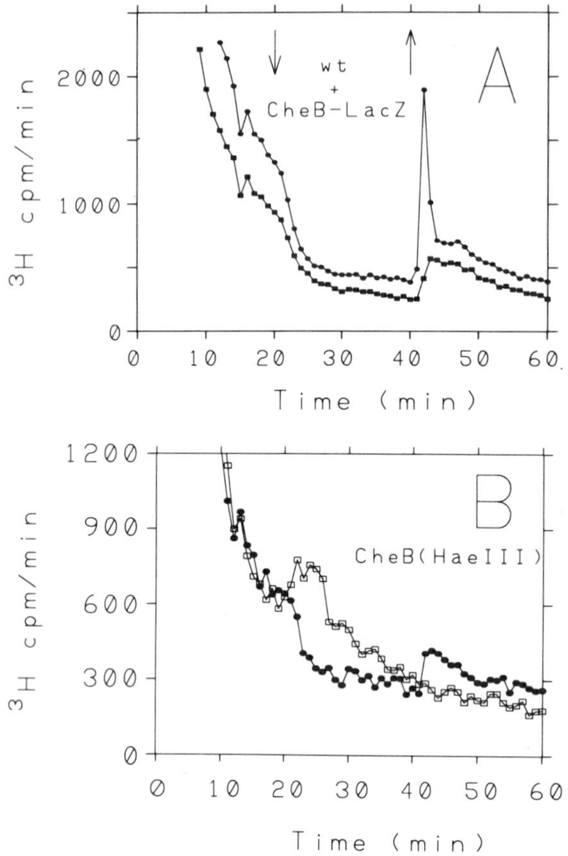

Figure 7. Deletion or overproduction of the amino-terminal portion of CheB greatly diminishes activation of the methylesterase activity in response to negative stimuli. (*A*) The results of flow-chase experiments for a wild-type chemotactic strain (RP487) transformed with a plasmid that directs expression of the *cheB-lacZ* translational fusion under control of the *lacuv*5 promoter. (●) Cells grown in the absence of IPTG; (■) cells grown in the presence of 1 mM IPTG to cause overproduction of the fusion protein. This overproduction results in the loss of the characteristic activity spike in response to attractant removal. (*B*) The results of flow-chase experiments for the $cheB^{\Delta}$ strain RP4972 transformed with a plasmid expressing the amino-terminal deletion construction *cheB*(*Hae*III). Although the response to serine addition appears essentially normal with this mutant, its response to attractant (serine) removal (●) and repellent (leucine) addition (□) is markedly diminished. Stimuli were added (downward arrow) and removed (upward arrow) at the indicated time points.

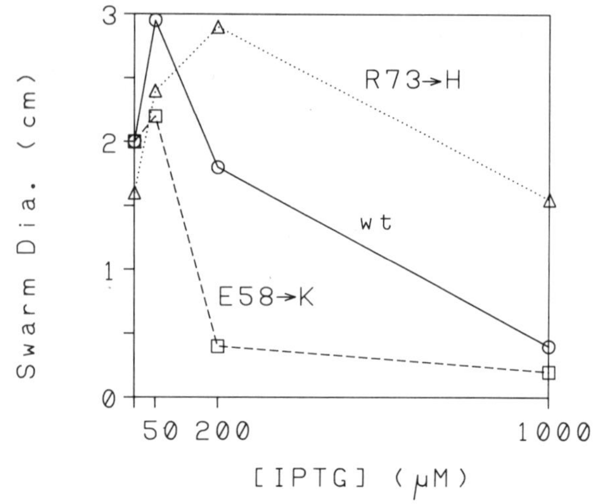

Figure 8. The effect of expression level of various representative *cheB* alleles on swarming ability. Multicopy plasmid were used to express the indicated *cheB* alleles under control of the *lacuv*5 promoter in the $cheB^{\Delta}$ strain RP4972. The indicated concentrations of IPTG were present in the λ tryptone swarm agar in which cells were grown to enable determination of chemotactic ability, as indicated by swarm diameter. CheB levels were 10–20% of the wild-type level at 0 μM IPTG, approximately equal to the wild-type level at 50 μM IPTG, and approximately ten times the wild-type level at 1 mM IPTG, as indicated by Western blots. Overproduction of the wild-type protein hinders swarming ability, as shown (○). This effect is enhanced with the dominant mutant E58→K (□) but is not very pronounced with the weak titrater mutant R73→H (△).

identify the component(s) necessary for CheB to reverse the tumbly effects of CheY.

Point Mutations in the Amino-terminal Portion of CheB Alter Modulation of CheB and Its Ability to Affect Swimming Behavior

As mentioned previously, overproduction of CheB is detrimental to the chemotactic ability of *E. coli*. With cells carrying a plasmid enabling inducible expression of *cheB*, this effect is easily observed as a diminished swarming ability with increasing concentration of inducer in the swarm plate (see Fig. 8). We have isolated two classes of *cheB* missense mutants (generated by random in vitro mutagenesis) that affect the ability of the overproduced mutant CheB portion to diminish swarming ability. These *cheB* alleles were placed downstream from the *lacuv*5 promoter in a multicopy plasmid vector. The amino acid sequence changes and properties of some of these mutants are summarized in Figure 6. The first class of mutants (called dominants) is composed of alleles that encode CheB variants that are significantly better than the wild-type protein at diminishing swarming ability when overproduced to varying levels. An example of the effects of such a mutant is shown in Figure 8. The second class of mutants (called weak titrators) are point mutants of CheB that are much less effective than the wild-type protein at hindering swarming when overproduced (for an example of the effects of one such mutant, see Fig. 8). The amino acid changes caused by these mutations are summarized in Figure 6. As judged by immunoblots, the mutations in both classes do not appear to affect the size or stability of CheB, nor do they affect the level of CheB produced at specific concentrations of the inducer IPTG. Each of the mutant forms of CheB is a functional methylesterase that can complement a $cheB^{\Delta}$ strain (make it swarm) to some extent at some level of expression. We believe that the altered abilities of these proteins to affect swarming behavior arise from altered abilities to interact with additional components of the chemotaxis machinery, such as those responsible for modulating methylesterase activity and/or those that enable CheB to influence swimming behavior. Although we are still studying the ability of these mutant forms of CheB to be modulated, it seems clear that they do not respond well, if at all, to negative chemotactic stimuli (see Table 2). The majority of these mutations fall in the amino-terminal portion of CheB, which we believe is involved in modulating methylesterase activity, as discussed above. It is intriguing that many of these mutations generate changes to charged amino acid side chains: Three of the dominant mutations result from Glu→Lys changes, whereas two of the weak titrator mutations result from Arg→His mutations. These charges may play some role in facilitating or hindering interactions with additional chemotaxis components.

DISCUSSION

In this section, we attempt to integrate our results into the rapidly emerging model of chemotaxis signal transduction. Our interpretation of this model is depicted schematically in Figure 9 and is based largely on those proposed recently by Hess et al. (1988), Parkinson (1988), and us (Stewart and Dahlquist 1987). This model accounts for CheA's kinase activity and resembles those proposed for several bacterial systems that regulate transcription of several genes in response to environmental conditions (Ninfa and Magasanik 1986; Nixon et al. 1986; Ronson et al. 1987; Kofoid and Parkinson 1988). For this discussion, it is instructive to emphasize several features of this model by following the signal transduction pathway for a negative stimulus through three "frames," as shown in Figure 9. We see the essential components in frame 1: transducer protein (methyl-accepting chemotaxis protein [MCP]), CheA, CheW, CheB, CheY, and the switch that regulates the direction of flagellar rotation. The sequence of events that results in response and adaptation is initiated by a conformational change in the transducer protein as it binds to a repellent molecule (as depicted in frame 2 of Fig. 9). By an unknown mechanism that may involve CheW (Hess et al. 1988; Oosawa et al. 1988), this conformational change enhances the ability of CheA to phosphorylate CheY and CheB (Hess et al. 1988; Wylie et al. 1988). These phosphorylation changes occur rapidly (in less than a second). Phosphorylation of CheY enables it to interact with the flagellar switch to cause tumbling (Parkinson et al. 1983; Clegg and Koshland 1984; Ravid et al. 1986; Yamaguchi et al. 1986; Wolfe et al. 1987; Smith et al. 1988). Phosphorylation of CheB increases its methylesterase activity and therefore results in a lower methylation level for the transducer protein. This decrease in methylation level takes place over a considerably longer time than the changes in CheB and CheY phosphorylation do, and it counteracts the effect of having repellent bound to the transducer protein such that the transducer no longer stimulates CheA's kinase activities, as shown in frame 3 of Figure 9. CheB and CheY then return to their nonphosphorylated states as a result of their autophosphase activities and/or the phosphatase activity of CheZ. This resets the methylesterase activity to its prestimulus level and causes the cells to resume smooth swimming.

Because the amino-terminal half of CheB shares

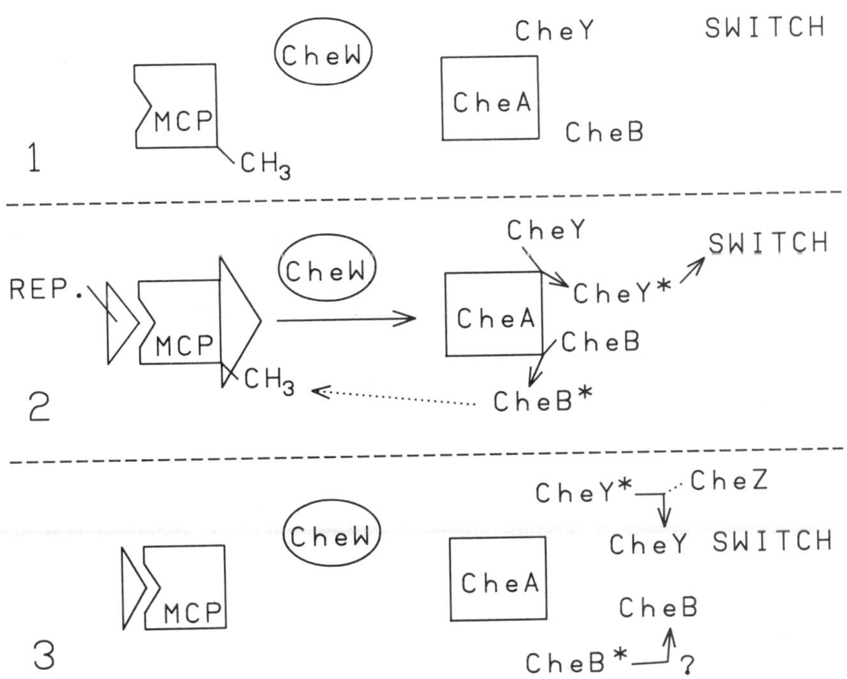

Figure 9. Model for the interactions of chemotaxis components that enable alteration of swimming behavior and modulation of methylesterase activity in response to chemotactic stimuli.

amino acid sequence homology with CheY and several bacterial transcriptional regulatory components that are phosphorylated (Stock et al. 1985; Nixon et al. 1986; Ronson et al. 1987), it is likely that CheB phosphorylation involves an amino acid residue in its amino-terminal half. This would imply a regulatory function for this part of CheB, which has been shown to be unnecessary for methylesterase catalytic activity (Simms et al. 1985). These homologous segments have been named "receiver modules" (Kofoid and Parkinson 1988; Parkinson 1988) or "regulator" components (Ronson et al. 1987) of signal transduction systems. At least one other receiver/regulator protein is phosphorylated: the transcriptional activator NtrC (Ninfa and Magasanik 1986; Ninfa et al. 1987). Interestingly, NtrB (the kinase responsible for phosphorylating NtrC) has significant amino acid sequence homology with the chemotaxis kinase CheA and other "transmitter components" of sensory transduction systems (Kofoid and Parkinson 1988; Parkinson 1988; Stock et al. 1988).

The results we have presented in this paper are consistent with several aspects of this model. First, we have shown that CheA and CheW are required for normal modulation of methylesterase activity. Second, we have shown that overproduction of CheA, CheW, or CheB eliminates the major features of this modulation, suggesting that the relative stoichiometries of these components may be crucial for enabling proper interactions. In addition, we have demonstrated that modulation in response to negative stimuli involves the amino-terminal portion of CheB, which probably encompasses the phosphorylation site or affects it in some way. And, finally, we have shown that CheY overproduction affects CheB modulation, whereas CheB overproduction can reverse the tumbly effects of CheY overproduction by some mechanism that does not require the transducer proteins. Taken together, these last two points suggest that CheB and CheY may compete for a common component (possibly CheA), as suggested by Figure 9.

There are several aspects of our results that have yet to be explained within the framework of this model. For example, our studies of the CheB amino-terminal deletion mutants suggest that inhibition of methylesterase activity in response to positive chemotactic stimuli is accomplished by a molecular mechanism distinct from the activation of esterase in response to negative stimuli. This transient activity decrease in response to attractants does not require the amino-terminal portion of CheB and is not affected by overexpression of CheY or the CheB-LacZ fusion protein. It is possible that this type of modulation does not involve changing the phosphorylation level of CheB, although it does require CheA. Some additional interaction between CheB and CheA may be responsible for this aspect of modulation and may account for the ability of native CheB and the various deletion mutant proteins to affect swimming behavior. It remains to be seen whether this "second" modulation mechanism is responsible for the residual response of the CheB amino-terminal deletion mutants to repellent stimuli (see Fig. 7). Alternatively, these deletions may not eliminate the phosphorylation site(s) but rather remove a region of the protein that facilitates the phosphorylation reaction.

The role of CheW in signal transduction is not well-defined by the model presented in Figure 9 because this protein has not yet been shown to affect CheA autophosphorylation or CheA kinase activity in vitro (Hess et al. 1988). But the general nonresponsiveness of *cheW*$^-$ cells strongly suggests that CheW must have some central role in signal transduction. Our results indicate that CheW is necessary for communicating negative chemotactic stimuli to the CheB modulation system but not for communicating positive stimuli. It therefore seems unlikely that CheW serves as an obligatory link between the transducer proteins and CheA for all stimuli. The effects of CheW overproduction may provide some insight into its normal mode of action. We have found that CheW overproduction causes a significant increase in the level of transducer methylation (C.B. Russell and F.W. Dahlquist, in prep.), yet the swimming behavior of cells containing excess CheW has a markedly smooth bias. The effect on transducer methylation could be accounted for by a diminished steady-state level of methylesterase activity (i.e., a lower steady-state level of activated CheB). If CheB activation arises from its phosphorylation by CheA, excess CheW could act by diminishing the rate of CheB phosphorylation or by enhancing the rate of CheB dephosphorylation. This would indicate a role for CheW either as a modifier of CheA kinase activity or as a CheB phosphatase. Similarly, the smooth-swimming bias would indicate a parallel effect on phosphorylation of CheY. Alternatively, CheW may normally participate in CheB modulation by binding to the transducer proteins and modulating the accessibility of methylated glutamate residues to CheB. In such a scheme, negative stimuli would cause release of CheW, thereby resulting in an apparent increase of methylesterase activity. The presence of excess CheW capable of complexing with the transducer proteins even in the presence of negative stimuli would eliminate this aspect of modulation.

In addition to providing new insight into interactions of chemotaxis signal transduction components, the in vitro work of Hess and co-workers (1988) has provided a testable model for the mechanism of CheB modulation. From our perspective, the most important question yet to be addressed is whether phosphorylation of CheB by CheA is the in vivo mechanism of modulating CheB methylesterase activity in response to chemotactic stimuli. We are currently determining the effects of various *che* gene deletions on the phosphorylation of CheB in intact cells and are investigating whether mutant forms of CheB (including those discussed in this paper) have altered phosphorylated levels in vivo.

ACKNOWLEDGMENTS

The transducer null strain CP362 was kindly provided by G.L. Hazelbauer. We are indebted to J.S.

Parkinson for providing a *cheW* expression vector, numerous chemotaxis mutant strains, and frequent helpful advice. R.S. was supported by grant DRG-863 from the Damon Runyon-Walter Winchell Cancer Fund and by the American Heart Association. This work was supported by U.S. Public Health Service grant AI-17807 from the National Institutes of Health.

REFERENCES

Adler, J. 1969. Chemoreceptors in bacteria. *Science* **166:** 1588.

Berg, H.C. and D.A. Brown. 1972. Chemotaxis in *Escherichia coli* analysed by three-dimensional tracking. *Nature* **239:** 500.

Callahan, A.M. and J.S. Parkinson. 1985. Genetics of methyl-accepting chemotaxis proteins in *Escherichia coli: cheD* mutations affect the structure and function of the Tsr transducer. *J. Bacteriol.* **161:** 96.

Clegg, D.O. and D.E. Koshland, Jr. 1984. The role of a signaling protein in bacterial sensing: Behavioral effects of increased gene expression. *Proc. Natl. Acad. Sci.* **81:** 5056.

Goy, M.F., M.S. Springer, and J. Adler. 1978. Failure of sensory adaptation in bacterial mutants that are defective in a protein methylation reaction. *Cell* **15:** 1231.

Hess, J.F., K. Oosawa, N. Kaplan, and M.I. Simon. 1988. Phosphorylation of three proteins in the signaling pathway of bacterial chemotaxis. *Cell* **53:** 79.

Kehry, M.R. and F.W. Dahlquist. 1982a. Adaptation in bacterial chemotaxis: *cheB*-dependent modification permits additional methylations of sensory transducer proteins. *Cell* **29:** 761.

———. 1982b. The methyl-accepting chemotaxis proteins of *E. coli*: Identification of multiple methylation sites on MCPI. *J. Biol. Chem.* **257:** 10378.

Kehry, M.R., T.G. Doak, and F.W. Dahlquist. 1984. Stimulus-induced changes in methylesterase activity during chemotaxis in *Escherichia coli. J. Biol. Chem.* **259:** 11828.

———. 1985. Sensory adaptation in bacterial chemotaxis: Global regulation of methylesterase activity. *J. Bacteriol.* **163:** 983.

Kleene, S.J., A.C. Hobson, and J. Adler. 1979. Attractants and repellents influence methylation and demethylation of methyl-accepting chemotaxis proteins in an extract of *Escherichia coli. Proc. Natl. Acad. Sci.* **76:** 6309.

Kofoid, E.C. and J.S. Parkinson. 1988. Transmitter and receiver modules in bacterial signaling proteins. *Proc. Natl. Acad. Sci.* **85:** 4981.

Kondoh, H., C.B. Ball, and J. Adler. 1979. Identification of a methyl-accepting chemotaxis protein for the ribose and galactose chemoreceptors of *Escherichia coli. Proc. Natl. Acad. Sci.* **76:** 260.

Kort, E.N., M.F. Goy, S.H. Larsen, and J. Adler. 1975. Methylation of a protein involved in bacterial chemotaxis. *Proc. Natl. Acad. Sci.* **72:** 3939.

Krikos, A., N. Mutoh, A. Boyd, and M.I. Simon. 1983. Sensory transducers of *E. coli* are composed of discrete structural and functional domains. *Cell* **33:** 615.

Larsen, S.H., R.W. Reader, E.N. Kort, W.-W. Tso, and J. Adler. 1974. Change in direction of flagellar rotation is the basis of the chemotactic response in *Escherichia coli. Nature* **249:** 74.

Macnab, R.M. 1987. Motility and chemotaxis. In Escherichia coli *and* Salmonella typhimurium: *Cellular and molecular biology* (ed. J.I. Ingraham et al.), p. 732. American Society for Microbiology, Washington, D.C.

Macnab, R.M. and D.E. Koshland, Jr. 1972. The gradient-sensing mechanism in bacterial chemotaxis. *Proc. Natl. Acad. Sci.* **69:** 2509.

Mutoh, N., K. Oosawa, and M. Simon. 1986. Characterization of *Escherichia coli* chemoreceptor mutant with null phenotypes. *J. Bacteriol.* **167:** 992.

Ninfa, A.J. and B. Magasanik. 1986. Covalent modification of the *glnG* product NRI by the *glnL* product NRII regulates the transcription of the *glnALG* operon in *Escherichia coli. Proc. Natl. Acad. Sci.* **83:** 5909.

Ninfa, A.J., L.J. Reitzer, and B. Magasanik. 1987. Initiation of transcription at the bacterial glnAp2 promoter by purified *E. coli* components as facilitated by enhancers. *Cell* **50:** 1039.

Nixon, B.T., C.W. Ronson, and F.M. Ausubel. 1986. Two-component regulatory systems responsive to environmental stimuli share strongly conserved domains with the nitrogen assimilation regulatory genes *ntrB* and *ntrC. Proc. Natl. Acad. Sci.* **83:** 7850.

Nowlin, D.M., J. Bollinger, and G.L. Hazelbauer. 1987. Sites of covalent modification in Trg, a sensory transducer of *Escherichia coli. J. Biol. Chem.* **262:** 6039.

Oosawa, K., J.F. Hess, and M.I. Simon. 1988. Mutants defective in bacterial chemotaxis show modified protein phosphorylation. *Cell* **53:** 89.

Park, C. and G.L. Hazelbauer. 1986. Mutation plus amplification of a transducer gene disrupts general chemotactic behavior in *Escherichia coli. J. Bacteriol.* **168:** 1378.

Parkinson, J.S. 1976. *cheA, cheB,* and *cheC* genes of *Escherichia coli* and their role in chemotaxis. *J. Bacteriol.* **126:** 758.

———. 1978. Complementation analysis and deletion mapping of *Escherichia coli* mutants defective in chemotaxis. *J. Bacteriol.* **135:** 45.

———. 1980. Novel mutations affecting a signaling component for chemotaxis of *Escherichia coli. J. Bacteriol.* **142:** 953.

———. 1981. Genetics of bacterial chemotaxis. *Symp. Soc. Gen. Microbiol.* **37:** 265.

———. 1988. Protein phosphorylation in bacterial chemotaxis. *Cell* **53:** 1.

Parkinson, J.S. and G.L. Hazelbauer. 1983. Bacterial chemotaxis: Molecular genetics of sensory transduction and chemotactic gene expression. In *Gene function in prokaryotes* (ed. J. Beckwith et al.), p. 293. Cold Spring Harbor Laboratory, Cold Spring Harbor, New York.

Parkinson, J.S. and P.T. Revello. 1978. Sensory adaptation mutants of *E. coli. Cell* **15:** 1221.

Parkinson, J.S., S.R. Reader, P.B. Talbert, and S.E. Houts. 1983. Interactions between chemotaxis genes and flagellar genes in *Escherichia coli. J. Bacteriol.* **155:** 265.

Ravid, S., P. Matsumura, and M. Eisenbach. 1986. Restoration of flagellar clockwise rotation in bacterial envelopes by insertion of the chemotaxis protein CheY. *Proc. Natl. Acad. Sci.* **83:** 7157.

Ridgway, H.F., M. Silverman, and M.I. Simon. 1977. Localization of proteins controlling motility and chemotaxis in *Escherichia coli. J. Bacteriol.* **132:** 657.

Ronson, C.W., B.T. Nixon, and F.M. Ausubel. 1987. Conserved domains in bacterial signaling proteins. *Cell* **49:** 579.

Silverman, M. and M.I. Simon. 1974. Flagellar rotation and the mechanism of bacterial motility. *Nature* **249:** 73.

———. 1977. Chemotaxis in *Escherichia coli:* Methylation of *che* gene products. *Proc. Natl. Acad. Sci.* **74:** 3317.

Simms, S.A., M.G. Keane, and J.B. Stock. 1985. Multiple forms of the CheB methylesterase in bacterial chemosensing. *J. Biol. Chem.* **260:** 10161.

Smith, J.M., E.H. Roswell, J. Shioi, and B.L. Taylor. 1988. Identification of site of ATP requirement for signal transduction in bacterial chemotaxis. *J. Bacteriol.* **170:** 2698.

Springer, W.R. and D.E. Koshland, Jr. 1977. Identification of a protein methyltransferase as the *cheR* gene product in the bacterial sensing system. *Proc. Natl. Acad. Sci.* **74:** 533.

Springer, M.S. and B. Zanolari. 1984. Sensory transduction in *Escherichia coli:* Regulation of the demethylation rate by CheA protein. *Proc. Natl. Acad. Sci.* **81:** 5061.

Springer, M.S., M.F. Goy, and J. Adler. 1977. Sensory transduction in *Escherichia coli:* Two complementary pathways of information processing that involve methylated proteins. *Proc. Natl. Acad. Sci.* **74:** 3312.

———. 1979. Protein methylation in behavioral control mechanisms and signal transduction. *Nature* **280:** 279.

Stewart, R.C. and F.W. Dahlquist. 1987. Molecular components of bacterial chemotaxis. *Chem. Rev.* **87:** 997.

Stock, A., D.E. Koshland, Jr., and J. Stock. 1985. Homologies between the *Salmonella typhimurium* CheY protein and proteins involved in the regulation of chemotaxis, membrane protein synthesis, and sporulation. *Proc. Natl. Acad. Sci.* **82:** 7989.

Stock, A., T. Chen, D. Welsh, and J. Stock. 1988. CheA protein, a central regulator of bacterial chemotaxis, belongs to a family of proteins that control gene expression in response to changing environmental conditions. *Proc. Natl. Acad. Sci.* **85:** 1403.

Stock, J.B. and D.E. Koshland, Jr. 1978. A protein methylesterase involved in bacterial sensing. *Proc. Natl. Acad. Sci.* **75:** 3659.

———. 1981. Changing reactivity of receptor carboxyl groups during bacterial sensing. *J. Biol. Chem.* **256:** 10826.

Terwilliger, T.C. and D.E. Koshland, Jr. 1984. Sites of methyl esterification and deamination on the aspartate receptor involved in chemotaxis. *J. Biol. Chem.* **259:** 7719.

Toews, M.L., M.F. Goy, M.S. Springer, and J. Adler. 1979. Attractants and repellents control demethylation of methylated chemotaxis proteins in *Escherichia coli. Proc. Natl. Acad. Sci.* **76:** 5544.

Wang, E.A. and D.E. Koshland, Jr. 1980. Receptor structure in the bacterial sensing system. *Proc. Natl. Acad. Sci.* **77:** 7157.

Wylie, D., A. Stock, C.-Y. Wong, and J. Stock. 1988. Sensory transduction in bacterial chemotaxis involves phosphotransfer between Che proteins. *Biochem. Biophys. Res. Commun.* **151:** 891.

Wolfe, A.J., P. Conley, T. Kramer, and H.C. Berg. 1987. Reconstitution of signaling in bacterial chemotaxis. *J. Bacteriol.* **169:** 1878.

Yamaguchi, S., S.-I. Aizawa, M. Kihara, M. Isomura, C.J. Jones, and R.M. Macnab. 1986. Genetic evidence for a switching and energy-transducing complex in the flagellar motor of *Salmonella typhimurium. J. Bacteriol.* **168:** 1172.

Yonekawa, H., H. Hayashi, and J.S. Parkinson. 1983. Requirement of the *cheB* function for sensory adaptation in *Escherichia coli. J. Bacteriol.* **156:** 1228.

Protein Phosphorylation and Bacterial Chemotaxis

J.F. HESS,* R.B. BOURRET,* K. OOSAWA,* P. MATSUMURA,† AND M.I. SIMON*
*Division of Biology, California Institute of Technology, Pasadena, California 91125;
†Department of Biological Sciences, University of Illinois at Chicago, Chicago, Illinois 60680

Bacteria are able to respond to changes in concentration of a large variety of chemicals and to changes in physical parameters, including viscosity, osmolarity, and temperature, by swimming toward a more favorable location (for review, see Stewart and Dahlquist 1987). Most chemotactic responses are mediated by a series of transmembrane receptor proteins that interact with or bind specific chemicals and thus monitor environmental conditions. Changes in ligand concentration detected by these receptors are transduced into two kinds of intracellular signals, adaptation and excitation signals. The adaptation signal acts as feedback, resulting in the covalent modification of specific glutamic acid residues in the cytoplasmic portion of the transmembrane receptor protein and thus in the modulation of receptor function (Springer et al. 1979). The excitation signal regulates the direction of rotation of bacterial flagella, causing changes in the "tumbling" or "smooth-swimming" behavior of the cell (Larsen et al. 1974).

Over the past 10 years, work in a number of laboratories has resulted in the identification and characterization of many of the genes and proteins required for the signal transduction process. The *cheB* and *cheR* gene products correspond to the methylesterase and the carboxymethyltransferase that are responsible for reversible demethylation and methylation of the transmembrane receptor proteins (Springer and Koshland 1977; Stock and Koshland 1978). The *cheA, cheW, cheZ,* and *cheY* genes are required to transmit and regulate the excitation signal that directly controls flagellar rotation (Parkinson 1978). CheA appears to be involved in generating the adaptation signal, as well as the excitation signal, because the *cheA* gene product apparently plays a role in the regulation of the activity of the CheB methylesterase (Springer and Zanolari 1984). Genetic evidence suggests that CheY and CheZ may interact with the "switch" that controls the direction of flagellar rotation (Parkinson et al. 1983; Yamaguchi et al. 1986). Furthermore, recent experiments suggest that the CheY protein plays a critical role in the excitatory signal and that the CheZ protein antagonizes CheY function (Clegg and Koshland 1984; Segall et al. 1985; Kuo and Koshland 1987; Wolfe et al. 1987).

How do the chemotaxis gene products generate an excitatory signal that can act within a fraction of a second to modify flagellar function? How does this system maintain a relatively stable behavioral pattern? We postulated that rapid and transient covalent modification of the chemotaxis gene products could function to transmit information from the receptor by regulating protein-protein interaction between the chemotaxis gene products. To test this idea, we purified the proteins corresponding to the *cheA, cheY, cheZ, cheW,* and *cheB* genes and tested the purified polypeptides to determine whether they could be covalently modified and whether they would interact with each other in vitro.

MATERIALS AND METHODS

The chemotaxis proteins CheA, CheB, CheW, CheY, and CheZ were overexpressed and purified (see Hess et al. 1987, 1988). CheA phosphorylation and phosphate transfer reactions were performed as described previously (Hess et al. 1988).

Purification of phosphorylated CheY. Phosphorylated CheY was prepared by incubating 0.5 nmole of purified CheA-phosphate with 1 nmole of CheY for 15 seconds in the presence of 5 mM $MgCl_2$. The phosphotransferase reaction was stopped by the addition of EDTA to a final concentration of 100 mM. CheY-phosphate was then purified on a Superose-12 FPLC column in buffer containing 70 mM EDTA. Reactions examining the dephosphorylation of CheY-phosphate in 70 mM EDTA were initiated by the addition of 95 mM $MgCl_2$ from a stock solution containing 1 M $MgCl_2$ in 1 M Tris (pH 9.0). CheZ and CheA were added in concentrations equimolar to CheY, where indicated.

Tryptic digestion of CheA. Partial proteolytic digestion of CheA-phosphate was performed by the addition of trypsin to purified phosphorylated CheA (w/w ratio of 1:100). The digestion was terminated at various times by the addition of SDS sample buffer, and the products were analyzed on SDS-PAGE. The phosphorylated 18-kD amino-terminal fragment of CheA was prepared by a 90-minute digestion of CheA-phosphate with trypsin. The digestion was stopped by the addition of soybean trypsin inhibitor, and the CheA fragment was purified on a Superose-12 FPLC column.

Mutagenesis of cheA. The *cheA* mutants listed in Table 1 were isolated and characterized as described previously (Oosawa et al. 1988). Site-specific mutations in *cheA* were constructed using the procedure of Kunkel et al. (1987) and were confirmed by DNA sequencing.

Table 1. Characteristics of Some Representative *cheA* Mutations

Region[a]	Amino acids	Allele number	Product[b]	Phosphorylation[c]	Dephosphorylation by CheB	Dephosphorylation by CheY	Swarm[d]
		wild type	+	++	++	++	++++
I	1–171	510	+	+	++	++	–
		516	+	++	–	+	–
		526	+	–			–
II	172–292	525	30 kD	–			–
III	293–429	535	+	+	+	++	–
		540	+	+	++	++	–
		541	+	+	++	++	+
IV	430–546	501	+	–			–
		546	+	++	++	++	++
V	547–654						

[a]Mutations were mapped to one of five restriction fragments in *cheA*. The amino acid numbers corresponding to each fragment are given. No mutations have been characterized in region V.
[b]*cheA* gene products were detected by immunoblotting with anti-CheA antibody following SDS-PAGE. (+) Wild-type pattern; (30 kD) a product of ~30 kD was detected.
[c]The level of phosphorylation of the CheA proteins was estimated from autoradiography following SDS-PAGE of the purified CheA proteins after reaction with [γ-^{32}P]ATP. (++) Similar to wild-type level; (+) <50% of wild-type level; (–) no phosphorylation detected.
[d]Swarm diameters after 12 hr at 30°C on tryptone swarm agar plates were 32 mm (++++), 24 mm (+++), 16 mm (++), 10–12 mm (+), <4 mm (–).

RESULTS

Phosphorylation of Chemotaxis Gene Products

Purified tetrameric CheA is phosphorylated rapidly when [γ-^{32}P]ATP is added, leading to the formation of a relatively stable phosphorylated protein (Fig. 1A). The phosphorylated CheA protein can be purified by gel filtration and contains approximately 1 mole of phosphate per polypeptide chain (Hess et al. 1987). The CheA-phosphate is in a high-energy form because the addition of ADP to this complex will reverse the reaction, leading to the formation of ATP (Hess et al. 1987).

The addition of specific chemotaxis gene products to purified phosphorylated CheA leads to the rapid transfer of phosphate to the added protein. Thus, when CheY is added to purified CheA-phosphate, the rapid dephosphorylation of CheA is accompanied by the concomitant phosphorylation of CheY (Fig. 1B). The reaction kinetics suggest that CheY-phosphate is an obligate intermediate in the CheY-mediated dephosphorylation of CheA. Maximal phosphorylation of CheY corresponds to between 1 and 2 moles of phosphate for each CheY molecule (Hess et al. 1988). Subsequently, CheY-phosphate hydrolyzes slowly and spontaneously, releasing inorganic phosphate. Similarly, the addition of the CheB protein results in the hydrolysis of CheA-phosphate via a CheB phosphoryl intermediate (Hess et al. 1988). However, the addition of CheW or of various proteins not related to chemotaxis has no effect on the stability of CheA-phosphate (Hess et al. 1987).

A very slow (100 times slower) hydrolysis of CheA-phosphate is found in the presence of CheZ, and no phosphorylated CheZ intermediate is detected (Hess et al. 1988). On the other hand, when CheA-phosphate is hydrolyzed by CheY, the addition of CheZ accelerates dephosphorylation, suggesting that CheZ can hydrolyze CheY-phosphate specifically (Hess et al. 1988). To test this possibility, CheY was added to CheA-phosphate, and the reaction was stopped by the addition of EDTA, which chelates the required magnesium. The low levels of the CheY-phosphate intermediate trapped by this procedure are relatively stable in the absence of Mg^{++} and were isolated by gel-filtration chromatography. The addition of $MgCl_2$ results in hydrolysis of CheY-phosphate (Fig. 2). The addition of CheA does not effect the rate of hydrolysis, whereas the addition of CheZ increases the rate markedly. Subsequent addition of CheA has no effect. Thus, CheZ alone appears to accelerate the rate of CheY dephosphorylation; rate increases ranging from five- to tenfold have been observed.

The phosphorylation reactions described can be assembled to suggest a pathway for information transfer in bacterial chemotaxis (Fig. 3). Although nothing is currently known about the interaction of the receptors with the *cheA* gene product, it is nonetheless plausible to postulate that the receptor may regulate the rate of CheA phosphorylation. In our current model, the addition of a repellent to the cells results in the rapid phosphorylation of CheA. In turn, this would generate a feedback signal by phosphorylating the *cheB* gene product. Indeed, preliminary experiments suggest that the phosphorylation of CheB results in a modest enhancement of the rate at which it demethylates receptors in vitro (J.F. Hess and M.I. Simon, unpubl.). Thus, phosphorylation could result in the rapid transient demethylation of receptors (i.e., the adaptation signal). Furthermore, the CheA-phosphate would also generate CheY-phosphate. The transfer of the phos-

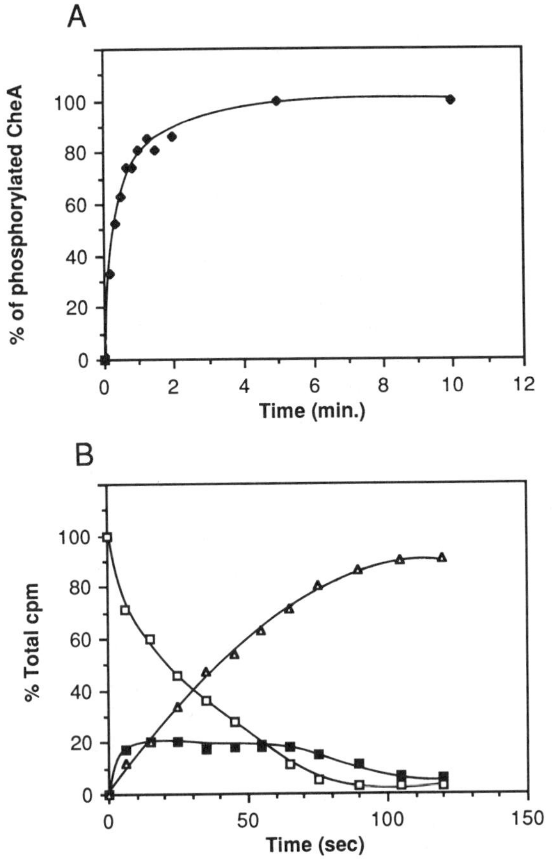

Figure 1. Rate of CheA autophosphorylation and phosphate transfer to CheY. (*A*) Time course of the formation of CheA-phosphate in the presence of [γ-^{32}P]ATP, $MgCl_2$, and KCl. (*B*) Time course of CheY-mediated dephosphorylation of CheA-phosphate. Purified phosphorylated CheA was present in a tenfold molar excess with respect to CheY. The amount of phosphate (cpm) present in CheA, CheY, or P_i is expressed relative to the amount of phosphate present in CheA prior to the initiation of the reaction by the addition of CheY. (□) Percentage of cpm in CheA; (■) percentage of cpm in CheY; (△) percentage of cpm in P_i.

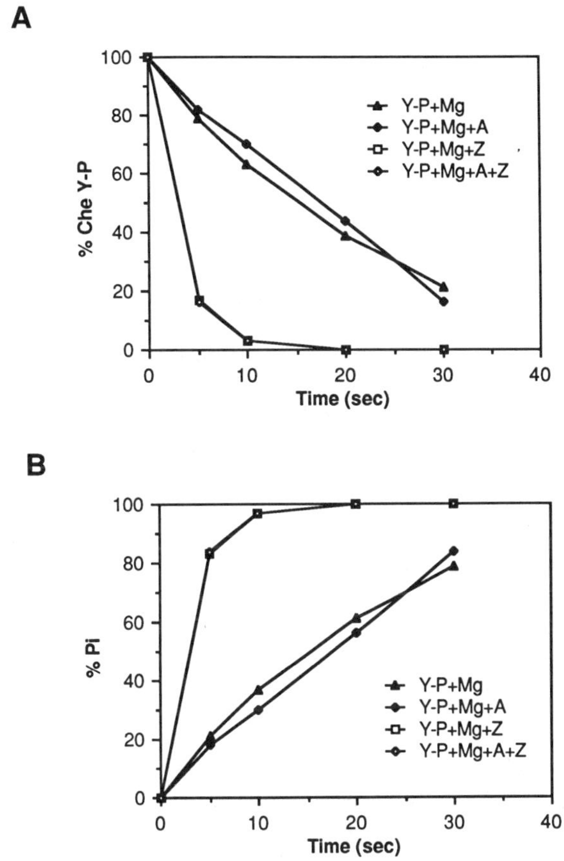

Figure 2. Dephosphorylation of purified [^{32}P]CheY-phosphate. (*A*) Time course of CheY dephosphorylation. The percentage of cpm in CheY is expressed relative to the amount of CheY-phosphate prior to the initiation of the reaction by the addition of $MgCl_2$. (*B*) Time course of P_i production during CheY dephosphorylation. The percentage of cpm present as P_i is expressed relative to the amount of CheY-phosphate prior to the initiation of the reaction. (▲) CheY-phosphate with $MgCl_2$; (◆) CheY-phosphate with $MgCl_2$ and CheA; (□) CheY-phosphate with $MgCl_2$ and CheZ; (◇) CheY-phosphate with $MgCl_2$, CheZ, and CheA.

phoryl group to CheY is very rapid. The transient nature of CheY-phosphate in vitro suggests that it may be the active intermediate that interacts with the region at the base of the flagellar motor to regulate the frequency of flagellar motor reversal. One role for the *cheZ* gene product in this scheme would be to hydrolyze CheY-phosphate rapidly. This would allow the system to remain responsive to new information from the transmembrane transducers, and it would regulate the effective lifetime of an excitatory CheY-phosphate signal.

CheA Mutants

We isolated mutants that are defective in CheA function in vivo to determine whether phosphorylation is required for chemotaxis (Oosawa et al. 1988). The properties of some of these mutants are summarized in Table 1. Mutations that map in the 5′ region of *cheA* and result in a full-length protein that assembles to form a tetramer exhibit three different phenotypes with respect to phosphorylation. Some of the mutations decrease or increase the rate of autophosphorylation. Others have little effect on autophosphorylation but decrease the ability of the phosphorylated protein to transfer the phosphoryl group to CheY or to CheB. A third class of mutations results in the absence of phosphorylation. Thus, the amino-terminal region of the molecule plays a primary role in the ability of CheA to autophosphorylate and to transfer phosphate.

Mutations in the central region of *cheA* result in two phenotypes. The proteins that carry these mutations are either unstable and degraded intracellularly or, if capable of forming stable tetramers, show little or no autophosphorylation. Finally, a mutation (*cheA*546) isolated as partially defective in chemotaxis was found to map in the 3′ region of the gene. This mutant shows aberrant but residual chemotaxis; however, the CheA

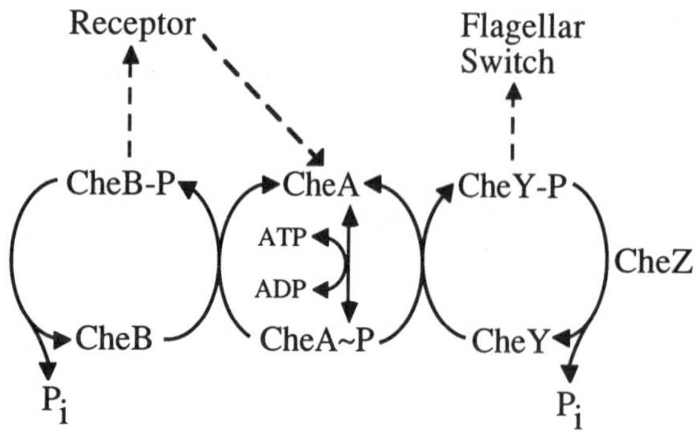

Figure 3. Model of the role of phosphate transfer reactions in the information pathway of bacterial chemotaxis. The reactions indicated by solid arrows have been shown to occur in vitro. The interactions indicated by dashed arrows are hypothesized to occur in vivo, as discussed in the text.

product is able to autophosphorylate and to transfer phosphate to CheB and CheY.

Our interpretation of these results and their effects is based on the notion that the CheA protein is made up of a number of different functional regions. The carboxy-terminal region may be responsible for the regulation of CheA interaction with receptors. This portion of the molecule could modulate autophosphorylation in response to interactions with transmembrane receptors or receptor complexes. The central region of CheA could function to stabilize the tetrameric CheA structure and may contain the catalytic activity responsible for autophosphorylation. The amino terminus could act as an intramolecular substrate for phosphorylation and as the site active in phosphate transfer to CheY and CheB. This hypothesis is further supported by previous work in the laboratory of J.S. Parkinson showing intracistronic complementation in *cheA* (Smith and Parkinson 1980). Thus, CheA proteins bearing amino-terminal mutations are able to interact with and compensate for carboxy-terminal mutants, suggesting that the molecule functions as a polymer of separate domains.

Functional Domains of CheA

The purified protein was treated with trypsin, and the distribution of peptide fragments after partial proteolysis was examined by SDS-PAGE to test the hypothesis that CheA is composed of discrete domains (Fig. 4A). A relatively stable 18-kD fragment was found to retain all of the phosphate when purified phosphorylated CheA was digested (Fig. 4B). The phosphorylated fragment was purified by separation from the rest of the fragments, using gel-filtration chromatography, to test its ability to transfer phosphate to CheB and CheY. The fragment behaved on gel filtration as a dimer, with an apparent molecular mass of 40–50 kD. The purified fragment retained the phosphoryl group and transferred phosphate rapidly and efficiently to CheY or to CheB (Fig. 5A,C). Furthermore, CheZ increased the rate of dephosphorylation of CheY markedly (Fig. 5B). The purified 18-kD fragment was identified as corresponding to the amino terminus of the CheA protein by amino acid sequence determination. The amino-terminal amino acid sequence of the fragment, starting with serine at position 2 and extending through threonine at position 11, was identical to the amino terminus of intact CheA, as determined previously. Thus, we can conclude that the phosphorylated amino-terminal domain of CheA extending approximately through amino acid 160 (i.e., the 18-kD fragment) is sufficient to transfer phosphate specifically to CheY and CheB. We have not been able

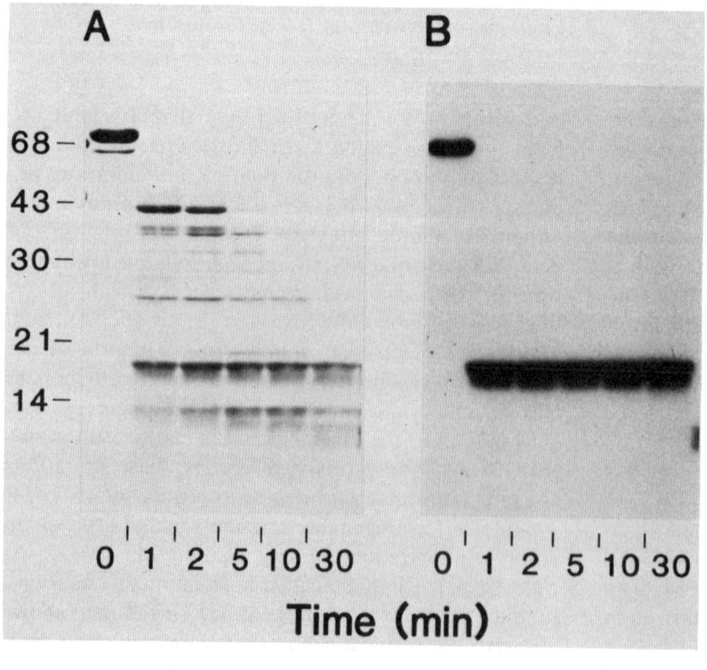

Figure 4. Partial proteolytic digestion of [^{32}P]CheA-phosphate. Purified phosphorylated CheA was digested with trypsin. The digestion was stopped at the indicated times by the addition of SDS sample buffer, and the products were analyzed by SDS-PAGE. (*A*) Coomassie blue stain of a polyacrylamide gel of the time course of partial tryptic digestion of CheA. (*B*) Autoradiograph of the gel in *A*. The position of molecular-mass standards is indicated on the left in kilodaltons.

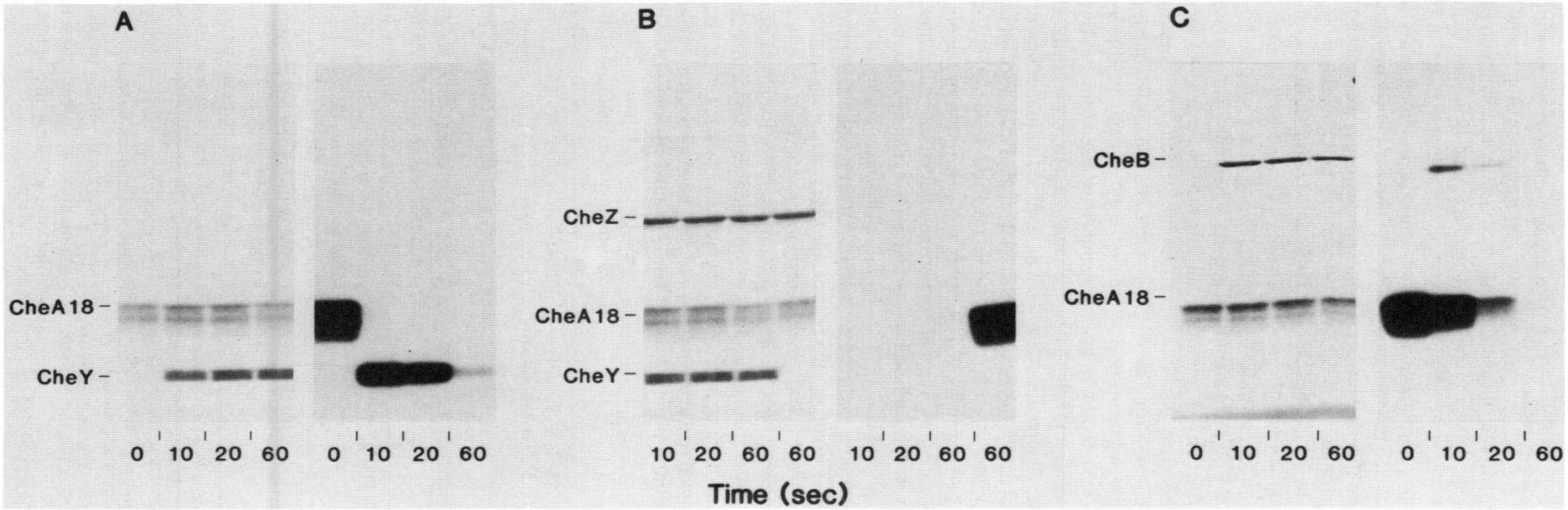

Figure 5. Phosphotransferase activity of the amino-terminal domain of CheA. Coomassie blue stain of a polyacrylamide gel (*A–C*, *left*) and the autoradiogram of that gel (*A–C*, *right*), following SDS-PAGE of reactions containing the purified ^{32}P-labeled phosphorylated CheA 18-kD fragment with CheY (*A*); CheZ and CheY (except for the last lane) (*B*); and CheB (*C*). The phosphorylated CheA fragment was present in a twofold molar excess with respect to CheY and a fourfold molar excess with respect to CheZ and CheB. The reactions were stopped at the indicated times by the addition of SDS sample buffer.

to demonstrate autophosphorylation by the purified 18-kD fragment in preliminary experiments.

The Nature of the Phosphorylated Residue

The amino-terminal 18-kD fragment was digested further by chymotrypsin, and the fragments were separated by high-performance liquid chromatography (HPLC) to determine the nature of the phosphorylated residue in CheA. The peptide that carried the radioactive phosphoryl group was isolated and digested further by trypsin. The products were separated again, the labeled peptides were isolated, and their sequence was determined. The phosphorylated peptide has the sequence arginine, alanine, alanine, histidine, serine, isoleucine, lysine (RAAHSIK) (Fig. 6). Which amino acid is the actual site of phosphorylation? A minor phosphorylated peptide with the sequence AAHSIK was also observed, eliminating arginine as a candidate. Trypsin is known to cleave polypeptides following basic residues. Because this peptide was generated by cleavage following the lysine residue, it seems unlikely that this peptide contains phospholysine. There is ample procedence for high-energy phosphohisitidine residues in bacterial proteins, notably in the phosphotransferase system of sugar transport (Postma and Lengeler 1985). On the other hand, it is possible that the serine residue could be phosphorylated and activated by adjacent amino acid residues to transfer the phosphate.

In an attempt to determine which residue is directly involved in phosphorylation, a number of site-specific mutations were introduced into *cheA*, and the chemotaxis and protein phosphorylation properties of the mutants were determined. His-48 was changed to a glutamine residue. Bacteria carrying this change in place of the wild-type *cheA* gene showed no chemotaxis. Furthermore, when the Gln-48 mutant protein was isolated, it showed no autophosphorylation. On the other hand, when the serine residue (Ser-49) was changed to threonine or alanine, both changes also resulted in the loss of chemotaxis. However, both of the corresponding Ala-49 and Thr-49 mutant proteins showed some residual autophosphorylation activity. Therefore, Ser-49 is not required for the phosphorylation. It is interesting that the modified genes were not effective in chemotaxis, although modification of Ser-49 allowed phosphorylation. These results suggest that the sequence adjacent to His-48 may also be important in CheA function (e.g., autophosphorylation or phosphate transfer). In a further series of experiments, His-48 was modified to each of ten different amino acids. None of the modified genes showed chemotaxis activity when they were present in a strain that was missing the wild-type *cheA* gene. Thus, His-48 appears to be the site of phosphorylation in CheA.

A series of pairs of bacterial proteins involved in monitoring changes in the concentrations of specific ligands have been found to show significant amino acid sequence homology with the CheA/CheY pair (Stock et al. 1985; Mutoh and Simon 1986; Nixon et al. 1986; Ronson et al. 1987). Protein phosphorylation is known to be involved in one other such system, in the regulation of transcription of the *glnA* gene. NtrB is rapidly phosphorylated and subsequently transfers phosphate to NtrC (Ninfa and Magasanik 1986; Keener and Kustu 1988). The phosphorylated NtrC protein is involved in DNA binding and activating RNA polymerase (Ninfa et al. 1987). The other pairs of sensor and regulator proteins may also share a similar mechanism of phosphoryl group exchange. Most of these systems appear to function by regulating gene transcription. The chemotaxis system acts to regulate cellular behavior and may have evolved to function rapidly and to maintain continual responsiveness to changes in transmembrane receptor activity. Thus, it will be informative to determine the extent to which these systems are analogous and the factors that regulate the relative rates of phosphate group transfer.

44 52
Phe Arg Ala Ala His Ser Ile Lys Gly
TTT CGG GCT GCC CAC TCG ATC AAA GGA

A G A
Gln Ala Thr

cheA Allele	Chemotaxis	Phosphorylation
Wild-type	+++	+++
His48 → Gln	-	-
Ser49 → Ala	-	+
Ser49 → Thr	+/-	++

Figure 6. Region of phosphorylation in CheA. (*Top*) Amino acid sequence of the region of CheA containing the site of phosphorylation and the corresponding DNA sequence. The changes made by site-specific mutagenesis are shown. (*Bottom*) The results of chemotaxis assays with intact bacteria on swarm plates and of in vitro autophosphorylation reactions with purified CheA proteins are tabulated for the various *cheA* alleles.

CONCLUSIONS

The picture that emerges from this work regarding signal transduction in *Escherichia coli* is summarized in Figure 7. We suggest that ligand binding at the receptor, in some fashion that we do not understand, modifies the rate of CheA phosphorylation. Receptor/CheA interaction may also involve the participation of CheW. An increase in phosphorylation of CheA leads to the rapid transfer of phosphate to CheY and CheB. Phosphorylated CheB demethylates the receptors rapidly and selective remethylation occurs, depending on the state of occupancy of the receptor. These reactions contribute to the ability of the cell to adapt to different levels of attractants or repellents. Phosphorylated CheY can interact with the switch at the base of the flagellum and generate a reversal of flagellar rotation (tumble). This interaction might require the simul-

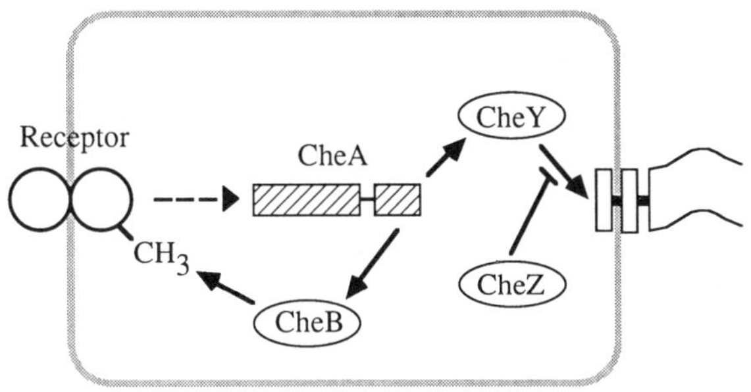

Figure 7. Model of signal transduction in *E. coli* chemotaxis. Information about the external environment is transmitted across the cell membrane by receptor/transducer molecules and then to CheA in a manner unknown at present (dashed arrow). The carboxy-terminal domain of the CheA protein presumably receives the signal generated by the receptor and modulates the phosphorylation of the amino-terminal domain. Phosphate is transferred from CheA to CheB and CheY to generate adaptation and excitation signals, respectively. Phosphorylation of the CheB-methylesterase enhances its activity. CheY-phosphate acts as a tumble signal at the flagellar switch but can be dephosphorylated by CheZ.

taneous participation of a number of CheY phosphate molecules; thus, the longer the lifetime of CheY-phosphate, the higher the probability that there will be a sufficient concentration to generate tumbles. The lifetime of CheY-phosphate is controlled by its slow endogenous dephosphorylation and by the enhanced dephosphorylation resulting from the presence of CheZ. The interaction of CheZ and phosphorylated CheY plays a critical role in determining the cells "bias" or relative frequency of flagellar reversal. We suggest further that the CheA protein acts as an input-output element to coordinate activity of a number of receptors and to transduce that activity to an output signal by autophosphorylation and phosphate transfer to the CheY and CheB proteins.

This model leaves a great many unanswered questions. For example, how do the receptors control CheA function? What is the role of CheW? Does the *cheZ* gene product act in any other capacity, or is its primary function the hydrolysis of CheY-phosphate? Because CheY does not contain histidine, what are the residues on CheY that act as phosphate acceptors? To what extent are the mechanisms proposed for CheA in general, and will they be found in other information-transducing systems? To what extent does the phosphorylation system alone regulate chemotaxis? Are there alternative pathways? These questions can be answered in the context of the bacterial chemotaxis system and may provide us with an understanding of a relatively simple behavior that can be used to test ideas about the mechanism involved in information processing in other biological systems.

ACKNOWLEDGMENTS

We thank F.W. Dahlquist and J.S. Parkinson for providing plasmids and bacterial strains, and N. Kaplan for purification of CheB. This work was supported by grant AI-19296-06 from the National Institutes of Health (to M.I.S.), by Damon Runyon-Walter Winchell Cancer Fund fellowship DRG-915 (to J.F.H.), and by National Institutes of Health fellowship AI-07798-01 (to R.B.B.).

REFERENCES

Clegg, D.O. and D.E. Koshland, Jr. 1984. The role of a signaling protein in bacterial sensing: Behavioral effects of increased gene expression. *Proc. Natl. Acad. Sci.* **81:** 5056.

Hess, J.F., K. Oosawa, N. Kaplan, and M.I. Simon. 1988. Phosphorylation of three proteins in the signaling pathway of bacterial chemotaxis. *Cell* **53:** 79.

Hess, J.F., K. Oosawa, P. Matsumura, and M.I. Simon. 1987. Protein phosphorylation is involved in bacterial chemotaxis. *Proc. Natl. Acad. Sci.* **84:** 7609.

Keener, J. and S. Kustu. 1988. Protein kinase and phosphoprotein phosphatase activities of nitrogen regulatory proteins NTRB and NTRC of enteric bacteria: Roles of the conserved amino-terminal domain of NTRC. *Proc. Natl. Acad. Sci.* **85:** 4976.

Kunkel, T.A., J.D. Roberts, and R.A. Zakour. 1987. Rapid and efficient site-specific mutagenesis without phenotypic selection. *Methods Enzymol.* **154:** 367.

Kuo, S.C. and D.E. Koshland, Jr. 1987. Roles of *cheY* and *cheZ* gene products in controlling flagellar rotation in bacterial chemotaxis of *Escherichia coli*. *J. Bacteriol.* **169:** 1307.

Larsen, S.H., R.W. Reader, E.N. Kort, W.-W. Tso, and J. Adler. 1974. Change in the direction of flagellar rotation is the basis of the chemotactic response in *Escherichia coli*. *Nature* **249:** 74.

Mutoh, N. and M.I. Simon. 1986. Nucleotide sequence corresponding to five chemotaxis genes in *Escherichia coli*. *J. Bacteriol.* **165:** 161.

Ninfa, A.J. and B. Magasanik. 1986. Covalent modification of the *glnG* product, NRI, by the *glnL* product, NRII, regulates the transcription of the *glnALG* operon in *Escherichia coli*. *Proc. Natl. Acad. Sci.* **83:** 5909.

Ninfa, A.J., L.J. Reitzer, and B. Magasanik. 1987. Initiation of transcription at the bacterial *glnAp2* promoter by purified *E. coli* components is facilitated by enhancers. *Cell* **50:** 1039.

Nixon, B.T., C.W. Ronson, and R.M. Ausubel. 1986. Two component regulatory systems responsive to environmental stimuli share strongly conserved domains with the nitrogen assimilation regulatory genes *ntrB* and *ntrC*. *Proc. Natl. Acad. Sci.* **83:** 7850.

Oosawa, K., J.F. Hess, and M.I. Simon. 1988. Mutants defective in bacterial chemotaxis show modified protein phosphorylation. *Cell* **53:** 89.

Parkinson, J.S. 1978. Complementation analysis and deletion mapping of *Escherichia coli* mutants defective in chemotaxis. *J. Bacteriol.* **135:** 45.

Parkinson, J.S., S.R. Parker, P.B. Talber, and S.E. Houts. 1983. Interactions between chemotaxis genes and flagellar genes in *Escherichia coli*. *J. Bacteriol.* **155:** 265.

Postma, P.W. and J.W. Lengeler. 1985. Phosphoenolpyruvate: Carbohydrate phosphotransferase system of bacteria. *Microbiol. Rev.* **49:** 232.

Ronson, C.W., B.T. Nixon, and F.M. Ausubel. 1987. Conserved domains in bacterial regulatory proteins that respond to environmental stimuli. *Cell* **49:** 579.

Segall, J.E., A. Ishihara, and H.C. Berg. 1985. Chemotactic signaling in filamentous cells of *Escherichia coli*. *J. Bacteriol.* **161:** 51.

Smith, R.A. and J.S. Parkinson. 1980. Overlapping genes at the *cheA* locus of *Escherichia coli*. *Proc. Natl. Acad. Sci.* **77:** 5370.

Springer, M.S. and B. Zanolari. 1984. Sensory transduction in *Escherichia coli:* Regulation of the demethylation rate by the CheA protein. *Proc. Natl. Acad. Sci.* **81:** 5061.

Springer, M.S., M.F. Goy, and J. Adler. 1979. Protein methylation in behavioral control mechanisms and in signal transduction. *Nature* **280:** 279.

Springer, W.R. and D.E. Koshland, Jr. 1977. Identification of a protein methylesterase as the *cheR* gene product in the bacterial sensing system. *Proc. Natl. Acad. Sci.* **74:** 533.

Stewart, R.C. and F.W. Dahlquist. 1987. Molecular components of bacterial chemotaxis. *Chem. Rev.* **87:** 997.

Stock, A., D.E. Koshland, Jr., and J.B. Stock. 1985. Homologies between the *Salmonella typhimurium* CheY protein and proteins involved in the regulation of chemotaxis, membrane protein synthesis, and sporulation. *Proc. Natl. Acad. Sci.* **82:** 7989.

Stock, J.B. and D.E. Koshland, Jr. 1978. A protein methylesterase involved in bacterial sensing. *Proc. Natl. Acad. Sci.* **75:** 3659.

Wolfe, A.J., P. Conley, T.J. Kramer, and H.C. Berg. 1987. Reconstitution of signaling in bacterial chemotaxis. *J. Bacteriol.* **169:** 1878.

Yamaguchi, S., S. Aizawa, M. Kihara, M. Isomura, C.J. Jones, and R.M. Macnab. 1986. Genetic evidence for a switching and energy-transducing complex in the flagellar motor of *Salmonella typhimurium. J. Bacteriol.* **168:** 1172.

Phosphoproteins Involved in Bacterial Signal Transduction

A.M. STOCK, D.C. WYLIE, J.M. MOTTONEN, A.N. LUPAS, E.G. NINFA, A.J. NINFA, C.E. SCHUTT, AND J.B. STOCK

Departments of Molecular Biology and Chemistry, Princeton University, Princeton, New Jersey 08544

Cells adjust their behavior continuously in response to changing environmental conditions. Stimuli include a wide range of chemicals, such as amino acids, peptides, sugars, oxygen, phosphate, ammonia, pheromones, hormones, and neurotransmitters, as well as parameters such as electrochemical potential, temperature, pressure, and light. Responses include regulation of gene expression, changes in metabolism, altered membrane permeability, and cell motility. One of the major unsolved problems in molecular biology is to understand how information from such a large number of sensory inputs is used to formulate an appropriate array of response strategies.

A number of specific stimulus-response systems have been investigated in bacteria. These include (Table 1) the chemotaxis system (Che) that mediates motor responses to attractant and repellent chemicals, the nitrogen regulatory system (Ntr) that controls induction of glutamine synthetase when cells are starved for nitrogen, the phosphorus system (Pho) that induces alkaline phosphatase when phosphorous is limiting, the system that controls expression of outer membrane proteins (Omp) in response to changes in osmotic pressure, the sporulation system (Spo0) that activates the developmental switch from vegetative growth to endospore formation, and the virulence system (Vir) that mediates bacterial infectivity of damaged plant tissues. Each of these systems has been investigated genetically, and many of the essential genes have been cloned and sequenced over the past few years. Surprisingly, all of these systems show a common set of components. In each case, the signal transduction proteins include members of two homologous families, which appear to comprise a cascade: Sensory information feeds into the first component, which activates the second component that, in turn, modulates a target activity within the cell. In this paper, we present evidence that the communication between the two components involves a phosphotransfer mechanism that is common to all of these regulatory systems.

METHODS

Materials. [γ-^{32}P]ATP (sp. act., 3000 Ci/mmole) was obtained from New England Nuclear/Du Pont. Ecoscint A scintillation solution was obtained from National Diagnostics, and Immobilon polyvinylidene difluoride membranes were obtained from Millipore. Phosphoramidate was a kind gift from N. Meadow and S. Roseman. Phosphohistidines were synthesized from phosphoramidate and L-histidine as described by Hulquist et al. (1966) and modified by Hays et al. (1973). All other reagents were obtained from standard commercial sources.

Protein isolation. *Salmonella typhimurium* CheA, CheB, and CheY were purified to greater than 95% homogeneity, as described previously (Simms et al. 1985; Stock et al. 1985, 1988). Preparations of *Escherichia coli* NR_I and NR_{II}, prepared as described

Table 1. A Family of Homologous Bacterial Signal Transduction Proteins

Stimulus	Signal transduction proteins I	Signal transduction proteins II	Response	References
Chemoeffector gradients	CheA	CheY CheB	chemotaxis	Stock et al. (1985); Stock et al. (1988)
α-Ketoglutarate : glutamine	NR_{II} (NtrB)	NR_I (NtrC)	nitrogen assimilation	Nixon et al. (1986)
Phosphate	PhoR	PhoB	phosphate assimilation	Wanner (1987)
Osmotic pressure	EnvZ	OmpR	porin expression	Mizuno et al. (1982); Wurtzel et al. (1982); Comeau et al. (1985)
Anaerobiosis	CpxA	ArcA (SfrA, Dye)	anaerobic repression	Iuchi and Lin (1988)
Nutrient deprivation	?	Spo0A Spo0F	sporulation	Ferrari et al. (1985); Trach et al. (1985)
Plant exudates	VirA	VirG	virulence	Winans et al. (1986); Leroux et al. (1987)
C4-dicarboxylates	DctB	DctD	dicarboxylic acid transport	Ronson et al. (1987)

previously (Ninfa et al. 1986, 1987), were greater than 90% pure, as determined by Coomassie blue staining of electrophoretograms. Phospho-CheA and phospho-NR_{II} were prepared by incubating the purified proteins with 0.1 mM [γ-^{32}P]ATP in 50 mM Tris-HCl, 50 mM KCl, 5 mM $MgCl_2$, 0.5 mM EDTA, and 0.1 mM dithiothreitol (pH 8.0) at 30°C. The phosphorylated proteins were separated from ATP by gel-filtration chromatography, using a 1.0 × 30-cm Sephadex G-50 column. Electrophoresis of proteins was performed in 90 × 50 × 0.75-mm SDS-polyacrylamide gels according to the method of Laemmli (1975). Samples were loaded in SDS-PAGE buffer containing 2% SDS, 10% glycerol, 0.002% bromophenol blue, 1% β-mercaptoethanol, and 63 mM Tris-HCl (pH 6.8).

RESULTS

Signal Transduction in Bacterial Chemotaxis

Chemotaxis in *E. coli* and *S. typhimurium* provides a well-characterized system in which to study signal transduction mechanisms (for a review, see Macnab 1987). Membrane receptor proteins at the cell surface monitor extracellular attractant and repellent chemicals. This information is processed by a set of six cytoplasmic components, the Che proteins, whose output controls flagellar activity. Two of these proteins, CheR and CheB, catalyze the reversible methylation of membrane receptors to control their sensitivity. The other four are more central to the signal transduction mechanism. CheW, CheA, and CheY are obligatory for communication between receptors and flagella, and CheZ acts to antagonize the effects of CheY. DNA sequence analysis indicates that CheA belongs to the first family of homologous signal transduction proteins, and both CheB and CheY belong to the second family (Table 1).

CheY is a soluble 128-amino-acid monomeric protein (Stock et al. 1985) that appears to interact directly with components at the flagellar motor to regulate swimming behavior (Parkinson et al. 1983; Clegg and Koshland 1984; Ravid et al. 1986; Yamaguchi et al. 1986; Wolfe et al. 1987). CheB is a soluble 349-amino-acid monomeric protein composed of two distinct domains: an amino-terminal region of approximately 140 residues that is homologous to the CheY protein, and a 200-residue carboxy-terminal domain that catalyzes the demethylation and deamidation of membrane chemoreceptor-transducer proteins (Simms et al. 1985). CheA is a soluble 300,000-molecular-weight homotetramer (Stock et al. 1988) that is required for stimulus-response coupling of both CheY-mediated motor responses (Clegg and Koshland 1984; Wolfe et al. 1987) and CheB-mediated decreases in receptor methylation (Springer and Zanolari 1984).

The signal transduction mechanism in chemotaxis appears to involve transfer of a phosphoryl moiety from CheA to CheY and CheB (Hess et al. 1988; Wylie et al. 1988). The CheA protein first undergoes an ATP-dependent autophosphorylation with a maximum of 1 phosphoryl group per 73,000-molecular-weight CheA subunit (Hess et al. 1987; Wylie et al. 1988). Phospho-CheA is stable at neutral pH and may be separated from ATP and other small molecules by gel-filtration chromatography. The phosphoryl moiety in CheA is readily transferred to the CheB and CheY proteins (e.g., see Fig. 1a). The phospho-CheB and phospho-CheY produced in these reactions dephosphorylate spontaneously to produce inorganic phosphate. CheB appears to be phosphorylated within its amino terminal regulatory domain, the portion homologous to CheY. We have shown previously that proteolytic cleavage of this domain causes a dramatic increase in demethylase

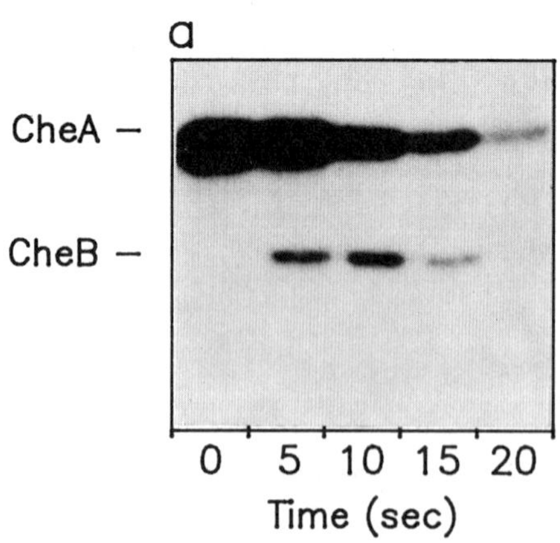

b

	CheB	$CheB_C$
CheB protein [0.2μM]	1.0	4.3
+ CheA [2.5μM]	0.4	-
+ ATP [5mM]	1.3	6.4
+ CheA + ATP	4.5	6.0
+ CheA + ATP + CheY [1.2μM]	0.8	-

Figure 1. (*a*) Phosphotransfer from CheA-P to CheB. Purified phospho-CheA (final concentration 0.5 μM) and CheB (2.0 μM) were mixed in 60 μl of buffer (0.10 M potassium phosphate, 5.0 mM $MgCl_2$ at pH 7.5) at 20°C. At the indicated times, 9.0-μl aliquots were removed, added to 2 × SDS-PAGE buffer, and electrophoresed in 12% polyacrylamide. An autoradiograph of the electrophoretogram is shown. (*b*) CheB methylesterase activity in vitro. *E. coli* membranes containing *methyl*-^{3}H-labeled *S. typhimurium* aspartate receptor, Tar, were incubated at 20°C with the indicated proteins. Aliquots were removed after 2, 4, and 8 min and mixed with glacial acetic acid to quench the reaction. Released [^{3}H]methanol was measured by vapor phase equilibrium as described previously (Simms et al. 1985). Data were normalized to the activity of the full-length CheB protein, which hydrolyzed 0.02% of the added Tar methylester groups per minute when present alone.

activity (Simms et al. 1985). Phosphorylation of the amino terminus appears to produce a similar activation (Fig. 1b). The CheA protein causes a five- to tenfold increase in esterase activity in the presence of ATP. This effect is antagonized by high concentrations of CheY, which compete with CheB for phospho-CheA. Moreover, CheA has no effect on the activity of the isolated carboxy-terminal catalytic domain. These results with purified components are consistent with the effect of CheA on demethylation that has been deduced from in vivo studies (Springer and Zanolari 1984). Thus, CheA acts as a central regulatory kinase in the chemotaxis system, controlling motility via phosphorylation of CheY and acting via CheB phosphorylation to control receptor activity by modulating the level of transducer methylation.

The other components of the chemotaxis system probably function by modulating the rates of phosphotransfer. CheW appears to provide the connection between membrane receptor-transducer proteins and CheA (Clegg and Koshland 1984; Stock et al. 1987; Wolfe et al. 1987), and the CheZ protein may function as a phosphatase to control the lifetime of phospho-CheY (Hess et al. 1988).

CheA is Phosphorylated at a Histidine Residue

CheA is autophosphorylated in an ATP-dependent reaction that produces ADP and phospho-CheA (Hess et al. 1987; Wylie et al. 1988). The equilibrium constant for the forward reaction is approximately 6×10^{-3}, corresponding to a $\Delta G°$ of +3 kcal/mole for phosphotransfer from ATP to CheA (Wylie et al. 1988). It is possible to push the reaction almost to completion by adding high concentrations of pure ATP, but in intact cells, where the ratio of ATP to ADP is maintained at approximately 10 (Bochner and Ames 1982), the level of CheA phosphorylation may never exceed a few percent of the total pool of CheA protein. High-energy phosphoproteins commonly involve phosphorylation at histidinyl side chains. Phosphohistidine intermediates have been identified in succinyl thiokinase (Suelter et al. 1961), nucleoside diphosphate kinase (Edlund et al. 1969), pyruvate phosphate dikinase (Spronk et al. 1976), fructose-2,6-bisphosphatase (Pilkis et al. 1987), and the phosphorylated proteins of the bacterial phosphoenolpyruvate:sugar phosphotransferase system, or PTS (Weigel et al. 1982). Phospho-CheA exhibits the characteristic alkaline stability and acid lability of a phosphoramidate (Wylie et al. 1988). Further studies have confirmed this finding and established that the group is a phosphohistidine (Fig. 2).

Recent studies of the CheA homolog involved in nitrogen regulation, NR_{II}, indicate that it is also autophosphorylated at a histidine residue (Ninfa et al. 1988). NR_{II} has been shown previously to function as a kinase to phosphorylate and activate NR_I (Ninfa and Magasanik 1986). NR_I is the component of the nitrogen regulation system that is homologous to CheY and CheB (Table 1). The NR_I protein has an overall design similar to that of CheB, with an amino-terminal domain

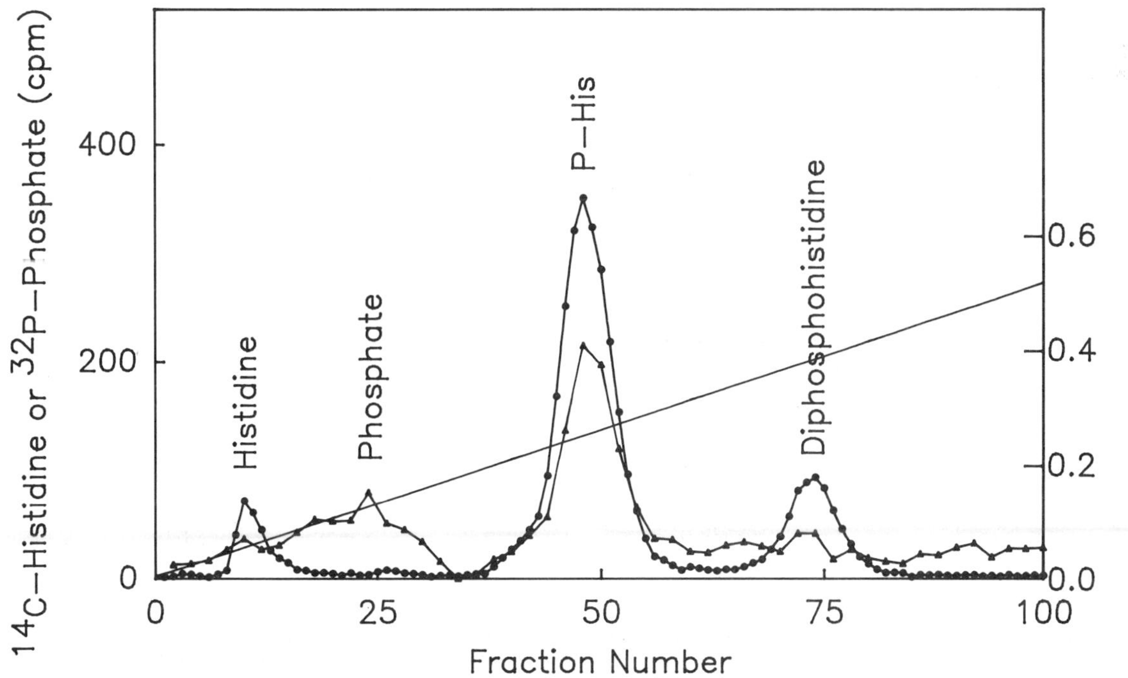

Figure 2. Analysis of hydrolyzed phospho-CheA by ion-exchange chromatography. CheA was phosphorylated with radiolabeled ATP (0.97 mole phosphate/mole CheA), resolved from free ATP, hydrolyzed in 3.0 M KOH for 6 hr at 100°C, and neutralized with perchloric acid (Steiner et al. 1980). An aliquot of the supernatant (1000 cpm) was resolved by anion-exchange chromatography on a Dowex 1-X8, 200 to 400-mesh column (1 × 7 cm), using a 10 mM to 1.0 M $KHCO_3$ gradient (800 ml), as described by Hays et al. (1973). The column had been calibrated previously with a [^{14}C]phosphohistidine standard. (●) ^{32}P-labeled phospho-CheA; (▲) [^{14}C]phosphohistidine standard. The volume of each fraction was 4.8 ml.

of approximately 120 amino acids that is homologous to CheY and a carboxy-terminal region with specific effector functions. Whereas the effector domain is a receptor demethylase in CheB, the effector function involves DNA binding and transcriptional activation in NR_I. In both cases, it seems likely that phosphotransfer to the Y domain is the key regulatory step in activation. Thus, the two component signal transduction systems may function by a common mechanism involving the initial formation of a high-energy phosphohistidinyl intermediate with subsequent phosphotransfer to the second component.

CheY Is Phosphorylated at an Acidic Residue

Phospho-CheY and its homologs are considerably less stable than the phosphorylated kinases CheA and NR_{II}. Halftimes for hydrolysis range from a few seconds in the case of phospho-CheB to several minutes for phospho-NR_I. If the proteins are denatured by heating in SDS or by precipitation in trichloroacetic acid, they become significantly more stable. In a denatured state, phospho-CheY exhibits a pattern of hydrolysis (Fig. 3) that is characteristic of an acyl phosphate (Koshland 1951; Di Sabato and Jenks 1961; Anthony and Spector 1972). Unlike phosphoramidates, acyl phosphates are relatively unstable in alkali and hydroxylamine. There are only six amino acid residues that are completely conserved between all proteins of the CheY family (Fig. 4). Two of these, corresponding to residues 13 and 57 in CheY, are aspartyl groups. Preliminary results indicate that the stoichiometry of CheY phosphorylation can exceed one phosphate per CheY monomer (Hess et al. 1988). Thus, it is possible that both of the conserved aspartate residues can function as phosphoacceptors.

Structure of CheY

We have recently solved the crystal structure of the *S. typhimurium* CheY protein to 2.7-Å resolution. CheY is a typical α/β protein composed of a central doubly wound five-stranded parallel β sheet flanked by

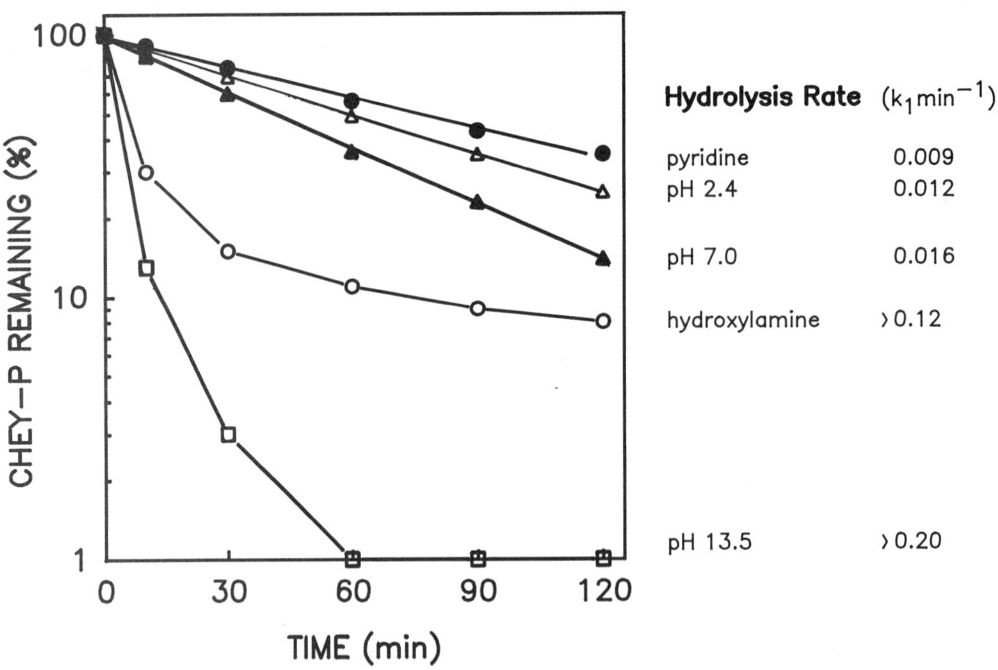

Figure 3. Hydrolysis of phospho-CheY. Phospho-CheA was prepared by incubating CheA with 10.0 μM [γ-^{32}P]ATP in 50 mM sodium phosphate, 5 mM $MgCl_2$, 0.5 mM EDTA, and 0.1 mM dithiothreitol at 20°C for 45 min. CheY was added (CheA:CheY molar ratio, 5:1), and the reaction was quenched 30 sec later by the addition of an equal volume of 2 × SDS-PAGE buffer containing 0.5 mg/ml cytochrome *c*. Phospho-CheY was isolated by electrophoresis in a 15% polyacrylamide gel at 200 V for 45 min. The protein was electroblotted onto polyvinylidene difluoride membranes in 20% methanol and 20 mM Tris-HCl (pH 7.5) for 1 hr at 20 mA, using a Sartorius semidry electroblotter SM 17556 at 4°C. The region of the membrane containing CheY, determined by the position of the cytochrome *c* marker, was excised. Membrane strips containing ~4000 cpm were rinsed in buffer and transferred to vials containing 0.6 ml of the following solutions: 0.4 M hydroxylamine (pH 7.6) at 20°C (○); 0.1 M pyridine at 20°C (●); 0.2 M sodium citrate (pH 2.4) at 45°C (△); 50 mM potassium phosphate (pH 7.0) at 45°C (▲); and 2.0 M NaOH (pH 13.5) at 45°C (□). At the indicated times, membranes were transferred to fresh solutions, 5 ml of Ecoscint scintillation fluid was added to the incubation solutions, and radioactivity was assayed in a Beckman LS-230 liquid scintillation spectrometer. Phospho-CheY remaining at each time was calculated as the difference between total and released radioactivity. Initial rates are indicated for each curve.

```
                                13
CheY     M A D K E L K F L V V D|D|F S T M R R I V R N L L K E L G F N N V E E A E - D G V D A L
CheB         M S K I R V L S V D|D|S A L M R Q I M T E I I N S H S D M E M V A T A P D P L V A R
SfrA         M Q T P H I L I V E|D|E L V T R N T L K S I F E A E G Y - D V F E A T - D G A E M H
OmpR       M Q E N Y K I L V V D|D|D M R L R A L L E R Y L T E Q G F - Q V R S V A - N A E Q M D
SpoA         M E K I K V S V A D|D|M R E L V S L L S E Y I E G Q E D M E V I G V A Y N G Q E C L
SpoF         M N E K I I L I V D|D|Q Y G I R E L L N E V F N K E G Y - Q T F Q A A - N G L Q A L
NtrCKp       M Q R G I A W I V D|D|D S S I R W V L E R A L T G A G L - S C T T F E - S G N E V L
NtrCBp       M P A G S I L V A D|D|D T A I R T V L N Q A L S R A G Y - E V R L T G - N A A T L W
VirG         E P L K H V L L V D|D|D V A M R H L I I E Y L T I H A F - K V T A V A - D S T Q F T
DctD       M D T L M P V A L I D|D|D K D L R R A T A Q T L E L A G F - S V S A Y D - G A K A A L
PhoB           M A R R I L V V E|D|E A P I R E M V C F V L E Q N G F - Q P V E A E - D Y D S A V
ORF2         M Q R E T V W L V E|D|E Q G I A D T L V Y M L Q Q E G F - A V E V F E - R G L P V L

                                57
CheY     N K L Q A G G F G F I I S|D|W N M P N M D G L E L L K T I R A D S A M S A L P V L M V T
CheB     D L I K K F N P D V L T L|D|V E M P R M D G L D F L E K L M R L R P M P V V M V S S L T
SfrA     Q I L S E Y D I N L V I M|D|I N L P G K N G L L L A R E L R E Q A N - - - V A L M F L T
OmpR     R L L T R E S F H L M V L|D|L M L P G E D G L S I C R R L R S Q S N - - P M P I I M V T
SpoA     S L F K E K D P D V L V L|D|I I M P H L D G L A V L E R L R E S D L K K Q P N V I M L T
SpoF     D I V T K E R P D L V L L|D|M K I P G M D G I E I L K R M K V I D E - - N I R V I I M T
NtrC     D A L T T K T P D V L L S|D|I R M P G M D G L A L L K Q I K Q R H P - - M L P V I I M T
NtrC     R W V S Q G E G D L V I T|D|V V M P D E N A F D L L P R I K K M R P - - N L P V I V M S
VirG     R V L S S A T V D V V V V|D|L N L V R E D G L E I V R N L A A K S D - - I P I I I I S G
DctD     A D L P A D F A G P V V T|D|I R M P E I D G L Q L F A T L Q G M D V - - D L P V I L M T
PhoB     N Q L N E P W P D L I L L|D|W M L P G G S G I Q F I K H L K R E S M T R D I P V V M L T
ORF2     D K A R K Q V P D V M I L|D|V G L P D I S G F E L C R Q L L A L H P - - A L P V L F L T

                              102 103     109 110
CheY     A E A K K E N I I A A A Q A|G A|S G Y V V|K P|F T A A T L E E K L N K I F E K L G M
CheB     G K G S E V T - L R A L E L|G A|I D F V T|K P|Q L G I R E G M L A Y S E A E K V R T ...
SfrA     G R D N E V D K I L G L E I|G A|D D Y I T|K P|F N P R E L T I R A R N L S R T M N L ...
OmpR     A K G E E V D R I V G L E I|G A|D D Y I P|K P|F N P R E L L A R I R A V R R Q A N E ...
SpoA     A F G Q E D V T K K A V D L|G A|S Y F I L|K P|F D M E N L V G H I R Q V S G N A S S ...
SpoF     A Y G E L D M I Q E S K E L|G A|L T H F A|K P|F D I D E I R D A V K K Y L P L K S N
NtrC     A H S D L D A A V S A Y Q Q|G A|F D Y L P|K P|F D I D E A V A L V D R A I S H Y Q E ...
NtrC     A Q N T F M T A I R P S E R|G A|Y E Y L P|K P|F D L K E L I T I V G R A L A E P K E ...
VirG     D R L E E T D K V V A L E L|G A|S D F I A|K P|F S I R E F L A R I R V A L R V R P N ...
DctD     G H G D I P M A V Q A I Q D|G A|Y D F I A|K P|F A A D R L V Q S V R R A S E K R R L ...
PhoB     A R G E E E D R V R G L E T|G A|D D Y I T|K P|F S P K E L V A R I K A V M R R I S P ...
ORF2     A R S E E V D R L L G L E I|G A|D D Y V A|K P|F S P R E V C A R V R T L L R R V K K ...
```

Figure 4. Amino acid sequence comparison of homologous response regulators. Residues that are totally conserved in all proteins are boxed. The complete sequences of CheY and Spo0F are aligned with the corresponding amino-terminal domains of the other proteins. Gaps (−) have been introduced to maintain alignment. Amino acid sequences were deduced from the nucleotide sequences of the corresponding genes: *S. typhimurium cheY* and *cheB* (Simms et al. 1985; Stock et al. 1985); *E. coli arcA, ompR, phoB,* and ORF2 (Wurtzel et al. 1982; Comeau et al. 1985; Drury and Buxton 1985; Amemura et al. 1986; Makino et al. 1986); *Bacillus subtilis spo0A* and *spo0F* (Ferrari et al. 1985; Trach et al. 1985); *Klebsiella pneumoniae ntrC* (Buikema et al. 1985); *Bradyrhizobium parasponiae ntrC* (Nixon et al. 1986); *Agrobacterium tumefaciens virG* (Winans et al. 1986); and *Rhizobium leguminosarum dctD* (Ronson et al. 1987).

five α helices (Fig. 5). This parallel α/β motif is common to a large number of proteins, including several dehydrogenases, reductases, and kinases (Richardson 1981). The active site is invariably located near the carboxyl termini of the β strands in these proteins (Hol et al. 1978; Brändén 1980), corresponding to the top of the molecule, as shown in Figure 5. This region of CheY contains four of the six totally conserved residues boxed in Figure 4. Asp-13 and Asp-57 lie close to each other at the carboxyl termini of β strands 1 and 3, respectively. Lys-109 and Pro-110 are positioned at the carboxyl terminus of strand 5, adjacent to the conserved aspartates in the three-dimensional structure. The other two totally conserved residues, Gly-102 and Ala-103, form a reverse turn on the opposite side of the molecule. The region at the carboxyl termini of the β strands contains an additional highly conserved acidic residue, Asp-12. This residue, together with Asp-13 and Asp-57, forms a pocket of acidic side chains that may function as the phosphoacceptor site. In addition to this region of highly conserved residues, other sequence similarities among the family of response regulators are explicable in terms of the CheY structure. All of these proteins have hydrophobic sequences that are consistent with the corresponding internal β strands in CheY. This type of analysis of the amino acid sequences of the CheY homologs in relation to the CheY structure strongly supports the hypothesis that all of the corresponding domains in these proteins have a common structural motif. The cluster of conserved residues

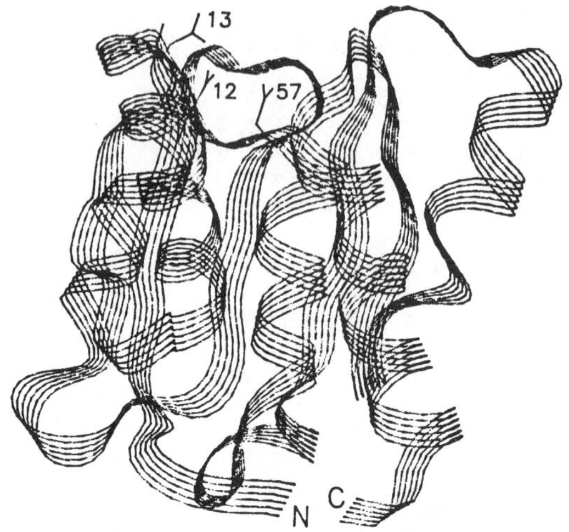

Figure 5. Structure of *S. typhimurium* CheY. The backbone is represented as a continuous flat ribbon, with coils representing α helices and straight runs representing β strands. The amino and carboxyl termini are indicated by N and C, respectively. Three aspartate residues are superimposed on the backbone at the positions of Asp-12, Asp-13, and Asp-57. This cluster of aspartate residues is thought to be the site of phosphorylation.

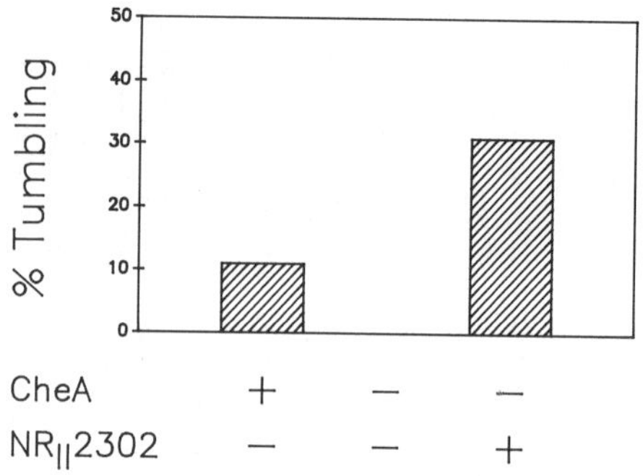

Figure 7. Activated NR$_{II}$ induces tumbly behavior in a CheA mutant. Wild-type *S. typhimurium*, PS1 (CheA +, NR$_{II}$2302 −) and a null *cheA* mutant, PS34, containing the plasmid pBR322 (CheA −, NR$_{II}$2302 −), or a pBR322-derived plasmid that overproduces a mutant highly active NR$_{II}$ protein, pTH814 (CheA −, NR$_{II}$2302 +) were grown in L broth at 37°C to mid-logarithmic phase. Cells were diluted 1:10 into motility buffer (50.0 mM KCl, 0.10 mM EDTA, 0.50 μM L-methionine, 10.0 mM potassium phosphate at pH 7.0) and incubated for 15 min at 20°C. Cell motility was recorded at 400 × magnification with a Zeiss phase-contrast microscope, Ikegami ITC-510 video camera, and a Panasonic NV8950 video recorder. Swimming behavior was analyzed, and the percentage of time the cells spent tumbling is indicated.

at the carboxyl termini of the β strands further suggests a conserved active site and thus a common mechanism of activation.

Crosstalk between Phosphotransfer Systems

Further evidence for a common mechanism in bacterial signal transduction has been obtained (Fig. 6) by demonstrating phosphotransfer from the first component of the chemotaxis system (phospho-CheA) to the second component of the nitrogen system (NR$_I$) and, conversely, from the first component of the nitrogen system (phospho-NR$_{II}$) to the second component of chemotaxis (CheY). Cross-reactivities between the chemotaxis and nitrogen regulators are inefficient compared with transfer between components of a single system. Nevertheless, CheA will function via NR$_I$ to activate transcription of the glutamine synthetase gene (Ninfa et al. 1988), and NR$_{II}$ can activate CheY in place of CheA to cause tumbly swimming behavior in vivo (Fig. 7).

DISCUSSION

The results presented here indicate that phosphotransfer reactions play a central role in signal transduc-

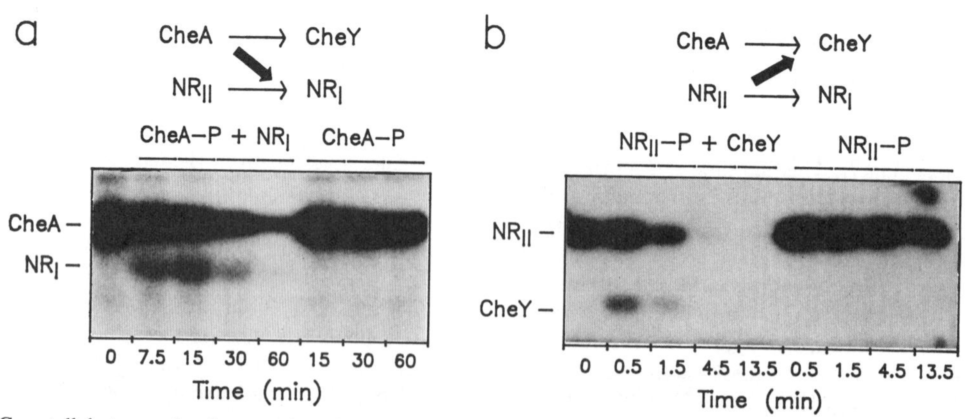

Figure 6. Crosstalk between the chemotaxis and nitrogen regulatory signal transduction proteins. (*a*) Purified phospho-CheA (85 nM) was incubated with NR$_I$ (12 μM) or alone at 37°C in Tris/phosphate buffer (32% glycerol, 82 mM Tris-HCl, 82 mM NaCl, 12 mM potassium phosphate, 1.6 mM dithiothreitol, 0.16 mM EDTA, 10 mM MgCl$_2$ at pH 7.5). At the indicated times, 25-μl aliquots were removed, added to 8.3 μl of 4 × SDS-PAGE buffer containing 20.0 mM EDTA, heated for 1 min at 60°C, and electrophoresed in 10% polyacrylamide. An autoradiograph of the electrophoretogram is shown. (*b*) Purified phospho-NR$_{II}$ (300 nM) was incubated with CheY (71 μM) or alone at 30°C in 50 μl of 0.10 M potassium phosphate and 5 mM MgCl$_2$ (pH 7.0). At the indicated times, 10-μl aliquots were removed, added to 10 μl of 2 × sample buffer containing 20.0 mM EDTA, heated 45 sec at 60°C, and electrophoresed in 15% polyacrylamide. An autoradiograph of the electrophoretogram is shown.

tion during bacterial chemotaxis. CheA is a kinase that phosphorylates both CheY and CheB via a high-energy phosphohistidinyl-CheA intermediate. Phospho-CheB is activated to demethylate the membrane chemoreceptor proteins, and phospho-CheY interacts with the flagellar motor to cause tumbly swimming behavior. Amino acid sequence conservations between homologous proteins indicate that residues corresponding to Asp-12, Asp-13, and/or Asp-57 in CheY may be the phosphorylated groups. An analysis of the tertiary structure of CheY supports this conclusion.

Signal transduction in the nitrogen regulatory system appears to function by the same mechanism as that involved in chemotaxis. The CheA homolog, NR_{II}, is a kinase that phosphorylates the CheY homolog, NR_{I}. Phosphorylated NR_{I} activates transcription at the nitrogen-regulated glutamine synthetase promoter. Moreover, NR_{II} can act as a CheY kinase to produce active phospho-CheY, and CheA can act as an NR_{I} kinase to produce active phospho-NR_{I}.

It seems likely that all members of the first family in Table 1 are protein kinases and that all members of the second family have phosphoaccepting Y domains. Studies of the tertiary structure of CheY have allowed us to begin to understand the sequence similarities between different Y domains in terms of a common set of structure-function relationships. The conserved amino acids at the carboxy-terminal end of the β sheet almost certainly represent critical active site residues. Presumably, these groups participate in the phosphotransfer reactions associated with Y activity.

It is not understood how sensory information regulates the activity of the kinases. Auxillary proteins are clearly involved in the Che and Ntr systems, most directly CheW in chemotaxis, and P_{II}, the product of *glnB*, in the nitrogen system. The mechanism of Y-domain regulation also remains to be determined. Although a subsequent phosphotransfer event has not been excluded, the current evidence suggests that a conformational change associated with phosphorylation is all that is required. The clearest example is the CheB methylesterase, where proteolytic cleavage of the Y domain appears to mimic the effects of phosphorylation.

The physiological significance of crosstalk between phosphotransfer systems remains to be determined. There are several examples where a mutational deficiency in a kinase appears to be suppressed partially by cross-reacting activities. For example, mutants deficient in NR_{II} still show NR_{I}-dependent glutamine synthetase expression (Backman et al. 1983), and strains lacking the phosphate-regulated kinase, PhoR, still show PhoB-dependent expression of alkaline phosphatase (Wanner et al. 1988). In the latter case, a cross-reactive kinase has been identified. This protein is homologous to CheA, NR_{II}, and PhoR. Although designated PhoM, it does not respond to changing levels of phosphate. Instead, its activity appears to be affected by extracellular glucose. The *phoM* gene is located in an operon next to an open reading frame, ORF2, that encodes a protein with a typical amino-terminal Y domain (Amemura et al. 1986). PhoM and ORF2 probably function together in a signal transduction pathway with unknown target activities. As more DNA sequence data become available, many more of these types of proteins will be discovered. Extrapolating from the current number, there may be over 50 sets of these proteins encoded in the *E. coli* genome. This phosphotransfer circuitry could constitute an intracellular neural network that provides coordinated response strategies to the complexities of a changing environment.

ACKNOWLEDGMENTS

We thank Greg Petsko for his support and guidance. We also thank Norman Meadow and Saul Rossman for providing the phoshoramidate used for synthesis of phosphohistidine, and Raymond Salemme and his group at Du Pont for providing access to the programs used for diagraming the CheY structure. This work was supported by grants from the U.S. Public Health Service (AI-20980) and the American Cancer Society (NP-515). D.C.W. and A.M.S. were supported by fellowships from the National Cancer Institute and the Damon Runyon-Walter Winchell Cancer Research Fund (DRG-933), respectively.

REFERENCES

Amemura, M., K. Makino, H. Shinagawa, and A. Nakata. 1986. Nucleotide sequence of the *phoM* region of *Escherichia coli*: Four open reading frames may constitute an operon. *J. Bacteriol.* **168:** 294.

Anthony, R.S. and L.B. Spector. 1972. Phosphorylated acetate kinase: Its isolation and activity. *J. Biol. Chem.* **247:** 2120.

Backman, K.C., Y.M. Chen, S. Ueno-Nishio, and B. Magasanik. 1983. The product of *glnL* is not essential for regulation of bacterial nitrogen assimilation. *J. Bacteriol.* **154:** 516.

Bochner, B.R. and B.N. Ames. 1982. Complete analysis of cellular nucleotides by two-dimensional thin layer chromatography. *J. Biol. Chem.* **257:** 9759.

Brändén, C.-I. 1980. Relation between structure and function of α/β-proteins. *Q. Rev. Biophys.* **13:** 337.

Buikema, W.J., W.W. Szeto, P.V. Lemley, W.H. Orme-Johnson, and F.M. Ausubel. 1985. Nitrogen fixation specific regulatory genes of *Klebsiella pneumoniae* and *Rhizobium meliloti* share homology with the general nitrogen regulatory gene *ntrC* of *K. pneumoniae*. *Nucleic Acids Res.* **13:** 4539.

Clegg, D.O. and D.E. Koshland, Jr. 1984. The role of a signaling protein in bacterial sensing: Behavioral effects of increased gene expression. *Proc. Natl. Acad. Sci.* **81:** 5056.

Comeau, D.E., K. Ikenaka, K. Tsung, and M. Inouye. 1985. Primary characterization of the protein products of the *Escherichia coli ompB* locus: Structure and regulation of synthesis of the OmpR and EnvZ proteins. *J. Bacteriol.* **164:** 578.

Di Sabato, G. and W.P. Jencks. 1961. Mechanism and catalysis of reactions of acyl phosphates. I. Nucleophilic reactions. *J. Am. Chem. Soc.* **83:** 4393.

Drury, L.S. and R.S. Buxton. 1985. DNA sequence analysis of the *dye* gene of *Escherichia coli* reveals amino acid homology between the Dye and OmpR proteins. *J. Biol. Chem.* **260:** 4236.

Edlund, B., L. Rask, P. Olsson, O. Walinder, O. Zetterqvist, and L. Engstrom. 1969. Preparation of crystalline nucleoside diphosphate kinase from Baker's yeast and identification of 1-[^{32}P]phosphohistidine as the main phosphorylated product of an alkaline hydrolysate of enzyme incubated with adenosine [^{32}P]triphosphate. *Eur. J. Biochem.* **9:** 9451.

Ferrari, F.A., K. Trach, D. LeCoq, J. Spence, E. Ferrari, and J.A. Hoch. 1985. Characterization of the Spo0A locus and its deduced product. *Proc. Natl. Acad. Sci.* **82:** 2647.

Hays, J.B., R.D. Simoni, and S. Roseman. 1973. Sugar transport: A trimeric lactose-specific phosphocarrier protein of the *Staphylococcus aureus* phosphotransferase system. *J. Biol. Chem.* **248:** 941.

Hess, J.F., K. Oosawa, N. Kaplan, and M.I. Simon. 1988. Phosphorylation of three proteins in the signaling pathway of bacterial chemotaxis. *Cell* **53:** 79.

Hess, J.F., K. Oosawa, P. Matsumura, and M.I. Simon. 1987. Protein phosphorylation is involved in bacterial chemotaxis. *Proc. Natl. Acad. Sci.* **84:** 7609.

Hol, W.G.J., P.T. Van Duijnen, and H.J.C. Berendsen. 1978. The α-helix dipole and the properties of proteins. *Nature* **273:** 443.

Hulquist, D.E., R.W. Moyer, and P.D. Boyer. 1966. The preparation and characterization of 1-phosphohistidine and 3-phosphohistidine. *Biochemistry* **5:** 322.

Iuchi, S. and E.C.C. Lin. 1988. *arcA* (*dye*) a global regulatory gene in *Escherichia coli* mediating repression of enzymes in aerobic pathways. *Proc. Natl. Acad. Sci.* **85:** 1888.

Koshland, D.E., Jr. 1951. Effect of catalysts on the hydrolysis of acetyl phosphate. Nucleophilic displacement mechanisms in enzymatic reactions. *J. Am. Chem. Soc.* **74:** 2286.

Laemmli, U.K. 1975. Cleavage of structural proteins during the assembly of the head of bacteriophage T4. *J. Biol. Chem.* **227:** 680.

Leroux, B., M.F. Yanofsky, S.C. Winans, J.E. Ward, S.F. Ziegler, and E.W. Nester. 1987. Characterization of the *virA* locus of *Agrobacterium tumefaciens:* A transcriptional regulator and host range determinant. *EMBO J.* **6:** 849.

Macnab, R.M. 1987. Motility and chemotaxis. In Escherichia coli *and* Salmonella typhimurium (ed. F.C. Neidhart et al.), vol. 2, p. 732. American Society for Microbiology, Washington, D.C.

Makino, K., H. Shinagawa, M. Amemura, and A. Nakata. 1986. Nucleotide sequence of the *phoB* gene, the positive regulatory gene for the phosphate regulon of *Escherichia coli* K-12. *J. Mol. Biol.* **190:** 37.

Mizuno, T., E.T. Wurtzel, and M. Inouye. 1982. Osmoregulation of gene expression: II. DNA sequence of the *envZ* gene of the *ompB* operon of *Escherichia coli* and characterization of its gene product. *J. Biol. Chem.* **257:** 13692.

Ninfa, A.J. and B. Magasanik. 1986. Covalent modification of the *glnG* product, NR_I, by the *glnL* product, NR_{II}, regulates the transcription of the *glnALG* operon in *Escherichia coli*. *Proc. Natl. Acad. Sci.* **83:** 5909.

Ninfa, A.J., L.J. Reitzer, and B. Magasanik. 1987. Initiation of transcription at the bacterial *glnAP2* promoter by purified *E. coli* components is facilitated by enhancers. *Cell* **50:** 1039.

Ninfa, A.J., S. Ueno-Nishio, T.P. Hunt, B. Robustell, and B. Magasanik. 1986. Purification of nitrogen regulator II the product of the *glnL* (*ntrB*) gene of *Escherichia coli*. *J. Bacteriol.* **168:** 1002.

Ninfa, A.J., E.G. Ninfa, A.N. Lupas, A. Stock, B. Magasanik, and J. Stock. 1988. Crosstalk between bacterial chemotaxis signal transduction proteins and regulators of transcription of the Ntr regulon: Evidence that nitrogen assimilation and chemotaxis are controlled by a common phosphotransfer mechanism. *Proc. Natl. Acad. Sci.* **85:** 5492.

Nixon, B.T., C.W. Ronson, and F.M. Ausubel. 1986. Two-component regulatory systems responsive to environmental stimuli share strongly conserved domains with the nitrogen assimilation regulatory genes *ntrB* and *ntrC*. *Proc. Natl. Acad. Sci.* **83:** 7850.

Parkinson, J.S., S.R. Parker, P.B. Talbert, and S.E. Houts. 1983. Interactions between chemotaxis genes and flagellar genes in *Escherichia coli*. *J. Bacteriol.* **155:** 265.

Pilkis, S.J., M.O. Lively, and M.R. el-Maghrabi. 1987. Active site sequence of hepatic fructose-2,6-bisphosphatase: Homology in primary structure with phosphoglycerate mutase. *J. Biol. Chem.* **262:** 12672.

Ravid, S., P. Matsumura, and M. Eisenbach. 1986. Restoration of flagellar clockwise rotation in bacterial envelopes by insertion of the chemotaxis protein CheY. *Proc. Natl. Acad. Sci.* **83:** 7157.

Richardson, J.S. 1981. The anatomy and taxonomy of protein structure. *Adv. Protein Chem.* **34:** 167.

Ronson, C.W., P.M. Astwood, B.T. Nixon, and F.M. Ausubel. 1987. Deduced proteins of C4-dicarboxylate transport regulatory genes of *Rhizobium leguminosarum* are homologous to nitrogen regulatory gene products. *Nucleic Acids Res.* **15:** 7921.

Simms, S., M. Keane, and J. Stock. 1985. Multiple forms of the CheB methylesterase in bacterial chemotaxis. *J. Biol. Chem.* **260:** 10161.

Springer, M.S. and B. Zanolari. 1984. Sensory transduction in *Escherichia coli*: Regulation of the demethylation rate by the CheA protein. *Proc. Natl. Acad. Sci.* **81:** 5061.

Spronk, A.M., H. Yoshida, and H.G. Wood. 1976. Isolation of 3-phosphohistidine from phosphorylated pyruvate, phosphate dikinase. *Proc. Natl. Acad. Sci.* **73:** 4415.

Steiner, A.W., E.R. Helander, J.M. Fujitaki, L.S. Smith, and R.A. Smith. 1980. High-performance liquid chromatography of acid-stable and acid-labile phosphoamino acids. *J. Chromatogr.* **202:** 263.

Stock, A.M., D.E. Koshland, Jr., and J.B. Stock. 1985. Homologies between the *Salmonella typhimurium* CheY protein and proteins involved in the regulation of chemotaxis, membrane protein synthesis, and sporulation. *Proc. Natl. Acad. Sci.* **82:** 7989.

Stock, A., T. Chen, D. Welsh, and J. Stock. 1988. CheA protein, a central regulator of bacterial chemotaxis, belongs to a family of proteins that control gene expression in response to changing environmental conditions. *Proc. Natl. Acad. Sci.* **85:** 1403.

Stock, A., J. Mottonen, T. Chen, and J. Stock. 1987. Identification of a possible nucleotide binding site in CheW, a protein required for sensory transduction in bacterial chemotaxis. *J. Biol. Chem.* **262:** 535.

Suelter, C.H., M. DeLuca, J.B. Peter, and P.D. Boyer. 1961. Detection of a possible intermediate in oxidative phosphorylation. *Nature* **192:** 43.

Trach, R.A., J.W. Chapman, P.J. Piggot, and J.A. Hoch. 1985. Deduced product of the stage 0 sporulation gene Spo0F shares homology with the Spo0A, OmpR, and SfrA proteins. *Proc. Natl. Acad. Sci.* **82:** 7260.

Wanner, B.L. 1987. Phosphate regulation of gene expression in *Escherichia coli.* In Escherichia coli *and* Salmonella typhimurium (ed. F.C. Neidhart et al.), vol. 2, p. 1326. American Society for Microbiology, Washington, D.C.

Wanner, B.L., M.R. Wilmes, and D.C. Young. 1988. Control of bacterial alkaline phosphatase synthesis and variation in an *Escherichia coli* K-12 mutant by adenyl cyclase, the cyclic AMP receptor protein, and the *phoM* operon. *J. Bacteriol.* **170:** 1092.

Weigel, N., M.A. Kukuruzinska, A. Nakazawa, E.B. Waygood, and S. Roseman. 1982. Sugar transport by the bacterial phosphotransferase system: Phosphoryl transfer reactions catalyzed by enzyme I of *Salmonella typhimurium*. *J. Biol. Chem.* **257:** 14477.

Winans, S.C., P.R. Ebert, S.E. Stachel, M.P. Gordon, and E.W. Nester. 1986. A gene essential for *Agrobacterium* virulence is homologous to a family of positive regulatory loci. *Proc. Natl. Acad. Sci.* **83:** 8278.

Wolfe, A.J., M.P. Conley, T.J. Kramer, and H.C. Berg. 1987.

Reconstitution of signaling in bacterial chemotaxis. *J. Bacteriol.* **169:** 1878.
Wurtzel, E.T., M.-Y. Chou, and M. Inouye. 1982. Osmoregulation of gene expression: I. DNA sequence of the *ompR* gene of the *ompB* operon of *Escherichia coli* and characterization of its gene product. *J. Biol. Chem.* **157:** 13685.
Wylie, D., A. Stock, C.-Y. Wong, and J. Stock. 1988. Sensory transduction in bacterial chemotaxis involves phosphotransfer between Che proteins. *Biochem. Biophys. Res. Commun.* **151:** 891.
Yamaguchi, S., H. Fujita, A. Ishihara, S.-I. Aizawa, and R.M. Macnab. 1986. Subdivision of flagellar genes of *Salmonella typhimurium* into regions responsible for assembly, rotation, and switching. *J. Bacteriol.* **166:** 187.

Structure-Function Studies of Bacterial Chemosensors

P. AMES, J. CHEN, C. WOLFF, AND J.S. PARKINSON
Biology Department, University of Utah, Salt Lake City, Utah 84112

Many of the chemotactic responses exhibited by *Escherichia coli* are mediated by a family of inner membrane proteins known as methyl-accepting chemotaxis proteins (MCPs). These relatively simple sensory transducers offer useful models for investigating stimulus detection and transmembrane signaling events at the molecular level. In this paper, we summarize genetic studies aimed at identifying the structural features of MCP molecules that are critical to their various signaling activities.

E. coli swims by rotating its flagellar filaments: counterclockwise (CCW) rotation produces forward swimming and clockwise (CW) rotation causes abrupt turns or tumbles. MCPs elicit chemotactic movements by modulating the rotational behavior of the flagellar motors in response to chemical stimuli. Four different MCP species have been identified in *E. coli*. Each monitors a different set of chemoeffector compounds through specific ligand-binding (chemoreceptor) sites arrayed on the periplasmic side of the membrane. Information about changes in receptor occupancy is transmitted across the membrane to the cytoplasmic portion of the transducer, which interacts with soluble chemotaxis proteins (CheA, CheB, CheR, CheW, CheY, CheZ) to generate internal signals that control flagellar rotation. Subsequent changes in MCP methylation state, catalyzed by the CheR and CheB enzymes, bring about sensory adaptation, enabling the cells to respond to further changes in chemoeffector levels.

MCP transducers are approximately 550 amino acid residues in length and are organized into two discrete structural domains (Fig. 1). The amino-terminal half of the molecule comprises the periplasmic receptor domain, which is flanked at each end by membrane-spanning segments (TM1 and TM2). The carboxy-terminal half of the molecule comprises the cytoplasmic signaling domain, which is flanked at each end by segments (K1 and R1) containing the methylation sites. This portion of MCPs is highly conserved in primary structure (Krikos et al. 1983; Russo and Koshland 1983; Bollinger et al. 1984), consistent with physiological evidence that all MCP transducers modulate the same internal signal (Berg and Tedesco 1975; Spudich and Koshland 1975). In contrast, the chemoreceptor domains, as well as the "linker" segment that joins the periplasmic and cytoplasmic portions of the molecule, vary considerably in primary structure.

To explore the roles of these various features in transducer function, we isolated and characterized

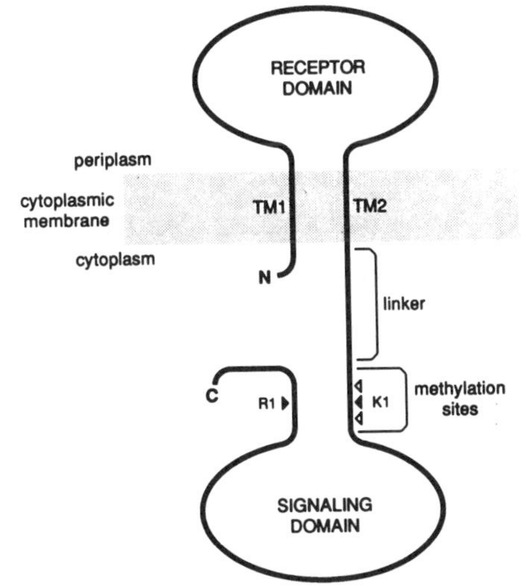

Figure 1. Membrane topology and domain structure of MCP transducers. This model is based primarily on DNA sequence analyses (Krikos et al. 1983; Russo and Koshland 1983; Bollinger et al. 1984), genetic studies (Krikos et al. 1985; Manoil and Beckwith 1986), and biochemical characterizations (Wang and Koshland 1980; Kehry and Dahlquist 1982). The Tsr molecule shown has four methylation sites: Two (△) are synthesized as glutamine residues and are posttranslationally converted to glutamic acids by a CheB-dependent deamidation reaction (Kehry et al. 1983).

MCP mutants with specific defects in ligand binding, transmembrane signaling, flagellar signaling, or sensory adaptation. These initial studies have enabled us to devise testable hypotheses about the nature of MCP signals and how they are modulated by changes in receptor occupancy and methylation state. Our preliminary findings suggest that transmembrane signaling by MCP molecules probably involves direct transmission of conformational changes between the periplasmic and cytoplasmic domains through the linker segment, rather than by ligand-dependent changes in subunit aggregation state.

EXPERIMENTAL PROCEDURES

Plasmid DNA carrying a structural gene for either the Tar or Tsr transducer was subjected to random mutagenesis with hydroxylamine in vitro. Treated DNA was introduced by transformation into recipient strains deleted for *tar* or *tsr*, and mutants with transducer defects were isolated by screening individual transformants for altered chemotactic behavior. The screening strategies used are described in Results and Discussion.

Transducer mutations with appropriate phenotypes were localized within the MCP-coding region by conventional deletion mapping. Mutations unsuited to this approach were mapped physically by recombining restriction fragments from mutant and wild-type genes and testing their resultant phenotype. Appropriate oligonucleotide primers were then chosen to determine the DNA sequence changes in the mutants, using the dideoxy method on either single- or double-stranded plasmid templates. Mutants are designated by their amino acid replacements (in one-letter code) and the position of the mutant residue. For example, AV123 indicates a change from alanine to valine at residue 123.

RESULTS AND DISCUSSION

Transducer Mutants with Chemoreception Defects

To better understand receptor site structure and the initial events of transmembrane signaling, we sought to isolate transducer mutants with specific defects in chemoreceptor function. A few such mutants have been described for the Tsr and Trg transducers (Hedblom and Adler 1980; Park and Hazelbauer 1986). The Tar transducer was chosen for this work because it handles two very different chemoeffector inputs—aspartate and maltose. Aspartate molecules are detected by direct binding to the transducer (Wang and Koshland 1980), whereas maltose is detected indirectly via maltose-binding protein (MBP) (Koiwai and Hayashi 1979; Richarme 1982), a periplasmic protein also involved in maltose transport. Competition experiments indicated that saturation of either the aspartate or MBP receptor site in Tar does not block subsequent responses to the other compound, demonstrating that the sites are not identical (Wolff 1983; Mowbray and Koshland 1987). Accordingly, we screened for *tar* mutants that were defective in aspartate, but not in maltose chemotaxis, as a means of identifying portions of the Tar molecule that are specifically involved in its aspartate-binding function.

A total of 17 such mutants were isolated and characterized. Their unstimulated swimming behavior and flagellar rotation patterns were identical to those of wild-type controls, as were their responses to serine stimuli, indicating that their transducer alterations do not have a generalized effect on chemotactic behavior. The mutants exhibited residual chemotactic responses to aspartate, but with elevated detection thresholds ranging from approximately 10^{-5} M to 10^{-3} M, in contrast to a 10^{-7} M threshold in wild type. Thus, the mutant transducers are capable of detecting aspartate stimuli but with reduced sensitivity. These findings indicate that the mutations affect the affinity of the aspartate receptor for ligand but do not eliminate its ability to generate flagellar signals. Maltose responses in the mutants ranged from 10% to 80% of wild-type, but unlike the aspartate results, the response thresholds were not demonstrably shifted to higher concentrations, indicating that the affinity of the mutant transducers for MBP is not substantially altered. These properties suggest that the mutants have specific defects in the aspartate-binding site of Tar.

The mutational changes in the mutants affect three closely spaced codons, which specify arginine residues at positions 64, 69, and 73 in Tar (Fig. 2). Each mutation results in replacement of one of the arginines with an uncharged amino acid. The proximity and periplasmic location of these arginine residues imply that they may play critical roles in aspartate detection. On the one hand, they could be directly involved in ligand binding, for example, through electrostatic interactions with the negatively charged carboxyl groups of aspartate. On the other hand, they could be responsible for maintaining proper conformation of the aspartate-binding site, which might be located elsewhere in the receptor domain.

In general, the receptor domains of different MCP transducers vary greatly in primary structure. It is therefore striking that the Tsr transducer, which serves as the chemoreceptor for serine, alanine, and glycine (Mesibov and Adler 1972), contains an arginine segment very similar to that of Tar. Arginine residues are found at the same positions in both transducers, and many of the flanking residues are identical or chemically similar as well (see Fig. 2). Tap and Trg, transducers that interact exclusively with periplasmic binding proteins (Hazelbauer and Adler 1971; Aksamit and Koshland 1974; Manson et al. 1986), show little sequence similarity to either Tar or Tsr in this portion of the receptor domain, suggesting that the arginine segment motif may be specifically involved in amino acid sensing.

Tar molecules undergo substantial conformational

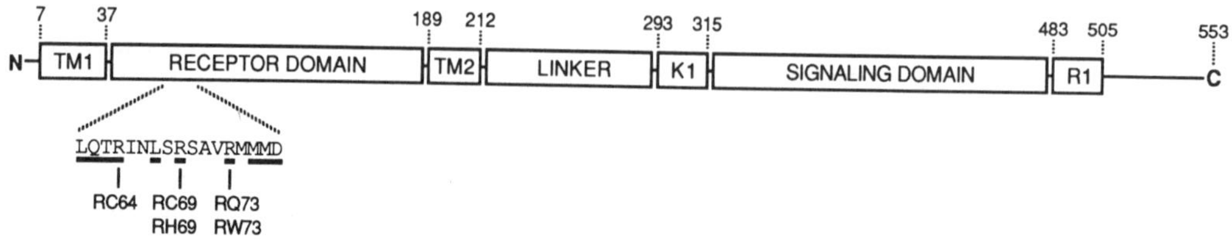

Figure 2. Mutational changes in aspartate affinity mutants of the Tar transducer. Numbers indicate the residue positions of important landmarks in the primary structure. The mutations fall into a small region of the receptor domain that may comprise the ligand-binding site. The underlined residues are also found in Tsr, the serine chemoreceptor.

changes upon ligand binding (Falke and Koshland 1987). It seems likely that changes in receptor site occupancy first trigger conformational shifts in the periplasmic domain, which, in turn, are propagated to the cytoplasmic domain to modulate its signaling activity. The arginine segment of the Tar transducer may represent a major conformational control point in the receptor domain that serves to amplify or propagate local structural changes triggered by ligand binding. For example, ligand-free receptor might be held in a "strained" conformation by electrostatic repulsion between the positively charged arginine residues. If aspartate detection occurred primarily through compensating charge interactions with these arginines, it could serve to alleviate some of those forces. Thus, aspartate binding might enable the receptor site to assume a less strained conformation, which, in turn, could precipitate conformational changes throughout the molecule.

Transducer Mutants with Locked Signal Output

To identify regions of the transducer molecule that are critically involved in controlling its signal output, we isolated MCP mutants that were "locked" in an active signaling state. Defects of this sort should cause a general loss of chemotactic ability if the offending signal is strong enough to drown out signals from other transducers (Parkinson 1980; Callahan and Parkinson 1985). In contrast, lesions that simply inactivate the transducer should eliminate a subset of responses but have no effect on other signaling pathways. Consequently, we looked for MCP mutations that conferred a dominant, generally nonchemotactic phenotype on wild-type cells. The *tsr* locus, whose product handles serine inputs, was chosen for this work because it is the most highly expressed of the MCP transducers.

Sequence changes were determined for a total of 58 independent mutations, yielding 39 different changes in 35 codons (Fig. 3). With the exception of one alteration in the first membrane spanner, all of the mutations affected the Tsr cytoplasmic domain, and all but three of them involved residues that are identical in all four MCP transducers of *E. coli*. The high degeee of sequence conservation throughout this portion of the transducer and the evidently critical nature of the conserved sites imply that relatively minor perturbations in transducer structure can have profound effects on its signaling activity.

Because transducer signals modulate flagellar rotation, a constantly signaling transducer should impart a permanent CW or CCW bias to the flagellar motors. All of the mutants exhibited aberrant flagellar rotation patterns: 43/58 had substantial CCW biases, with 70–100% of the cells rotating exclusively in the CCW direction; 11/58 had CW biases, with 30–70% of the cells rotating exclusively in the CW direction; and 4/58 (dubbed "slow switchers") exhibited individual cells of both types. Reversal episodes were much less frequent in the mutant strains than in wild-type controls, suggesting that MCP molecules may generate two sorts of signals, one that promotes CW rotation and one that augments CCW rotation. Because both types of mutants are dominant, it is simplest to assume that both represent active "signaling" conditions, although the possibility of dominant null ("spoiler") effects cannot be excluded rigorously as yet.

By definition, locked transducers must be refractory to the sensory adaptation system, implying that their signaling properties are unaffected by methylation changes. We found that nearly all of the mutant Tsr proteins were capable of accepting methyl groups, but their steady-state methylation levels varied greatly, ranging from much lower to much higher than wild type. MCP methylation levels reflect the relative activities of the CheR methyltransferase and the CheB methylesterase. Changes in methylation state during the adaptation phase of a chemotactic response are largely mediated through feedback regulation of CheB activity by transducer signals: CCW signals inhibit CheB activity and elicit methylation increases, whereas CW signals have the opposite effects (Kehry et al. 1985a). In general, the CCW-biased transducer mu-

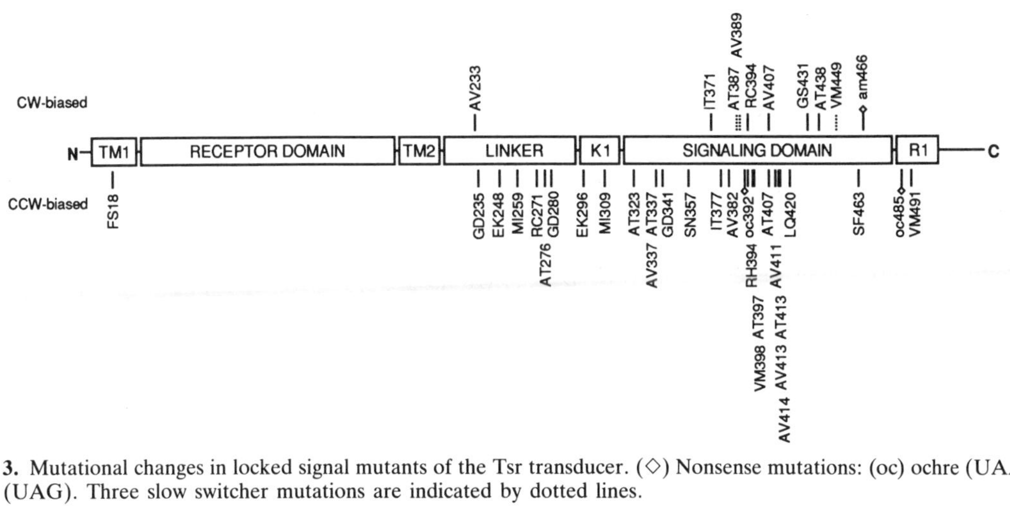

Figure 3. Mutational changes in locked signal mutants of the Tsr transducer. (◇) Nonsense mutations: (oc) ochre (UAA); (am) amber (UAG). Three slow switcher mutations are indicated by dotted lines.

tants exhibited high methylation levels, whereas the CW-biased mutants had low methylation levels. These anomalous methylation patterns probably result from uncompensated feedback control of the sensory adaptation machinery (Kehry et al. 1985b) and are consistent with the notion that the mutant transducers are locked in an active signaling state.

On the basis of these findings, we suggest that MCP transducers can assume alternative conformations that correspond to the CCW and CW signaling states and that the majority of the locked transducer mutants are simply stuck in one conformation. The slow switcher mutants may represent defects in a portion of the transducer molecule that is directly involved in these conformational transitions but is not essential to the function of either signaling state. Two of the slow switcher mutations, AT387 and AV389, are located approximately in the center of the highly conserved signaling domain (see Fig. 3). Because this region yielded many CW and CCW mutants as well, it may be a conformational switch or hinge. Consistent with this idea, different amino acid replacements at residues 394 or 407 resulted in mutants locked in either rotational mode, suggesting that the CW–CCW transition is delicately poised.

Three of the locked transducer mutants proved to have nonsense mutations, indicating that truncation of the signaling domain can alter its signaling properties. A stop codon at residue 392, in the middle of the presumptive switch region, confers a CCW bias, demonstrating that only the amino-terminal half of the signaling domain is needed for CCW output. A stop codon at residue 466 creates a CW-biased transducer, implying that the region between residues 392 and 466 is needed for CW rotation. Because a stop codon at residue 485 results once again in CCW-biased behavior, it appears that the carboxy-terminal portion of the transducer plays a role in controlling the equilibrium between CW and CCW conformations but is not essential for either type of signal output. Because the R1 methylation segment lies within this region of the molecule, it seems likely that the methylation segments are involved primarily in modulating transducer signaling properties.

Transducer Mutants with Biased Signal Output

Because locked transducers are insensitive to the sensory adaptation system, they may represent major structural alterations. To isolate mutants with less dramatic signaling defects, we looked for transducers with biased signal output that nevertheless retained the ability to mediate transmembrane signaling and respond to the adaptation machinery. Such mutants were obtained as phenotypic suppressors of *cheR* or *cheB* mutants. For example, *cheR* strains, which cannot methylate their MCPs, are unable to form chemotactic swarms on soft agar plates due to a CCW rotational bias. Modest swarming ability can be restored by mutations that impart some CW rotational behavior to the mutant cells. MCP alterations that generate a CW signal in the absence of methylation represent a common source of such suppressors. Conversely, selecting for suppression of *cheB* defects yields MCP mutants with CCW-biased signaling properties. In wild-type genetic backgrounds, the biased signal transducers obtained in this fashion appear to function normally, indicating that their signaling output can be modulated by the sensory adaptation machinery.

The mutational changes in 102 biased signal mutants of Tsr are summarized in Figure 4: 52 CW-biased isolates yielded 40 different changes at 37 sites and 50 CCW-biased isolates yielded 24 different changes at 23 sites. Unlike locked transducer mutations, the biased signal mutations generally altered nonconserved residues, consistent with the premise that those positions are less critical to transducer function. We discerned no pattern in the chemical nature of the amino acid replacements; however, their clustered distribution within the Tsr molecule may reflect the relative importance of various regions in modulating transducer signal output.

In striking contrast to the locked transducer mutants, a number of the biased signal mutants had changes in

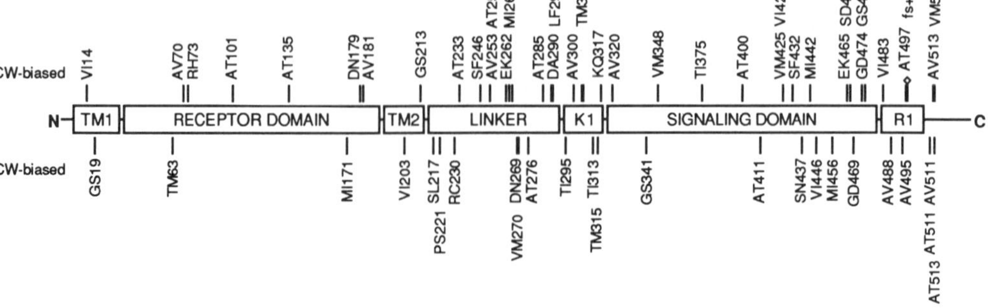

Figure 4. Mutational changes in biased signal mutants of the Tsr transducer. (◇ at position 498) A + 1 frameshift mutation that substitutes 27 new residues in place of 40 wild-type residues at the carboxyl terminus of the molecule.

the chemoreceptor domain, including three within the Tsr arginine segment, which may be the serine-binding site. These receptor alterations presumably mimic the action of chemoeffectors, generating a persistent transmembrane signal when the adaptation system is nonfunctional. It should be interesting to examine the ligand-binding properties of these mutants, which could prove useful in identifying major conformational control points in the receptor domain.

Many of the biased signal changes occur within or near the K1 or R1 methylation segments. Relatively few are located in the putative switch region identified with locked signal mutants. These findings support the idea that the methylation regions are involved in fine-tuning transducer signal output. Amino acid replacements within these segments probably elicit shifts in the distribution of CW and CCW transducer conformations in much the same manner as do changes in methylation state.

The Nature of Transducer Signals

Four cytoplasmic proteins—CheA, CheW, CheY, and CheZ—are involved in the transmission of sensory signals from the chemoreceptors to the flagellar motors. CheY and CheZ control the direction of flagellar rotation, perhaps through interactions with the switching mechanism of the motor (Parkinson et al. 1983). CheY is needed for CW rotation; CheZ seems to antagonize or inhibit CheY action, resulting in enhanced CCW bias. Thus, locomotor responses are probably carried out by regulating the activity of CheY and possibly CheZ. The roles of CheA and CheW are less clear. Mutants defective in either function are CCW-biased and appear to lack all signal-related activities, suggesting that CheA and CheW may interact with transducer molecules to generate the signal that is related to CheY and CheZ.

Recent biochemical studies indicate that sensory signaling in the chemotaxis system may involve protein phosphorylation cascades (Hess et al. 1987, 1988; Oosawa et al. 1988). CheA is known to autophosphorylate (Hess et al. 1987); its phosphate groups are subsequently transferred to CheB and CheY and probably serve to regulate their functional activities (Hess et al. 1988). Phosphorylated CheY may produce CW rotation of the flagellar motors; phosphorylated CheB may have elevated methylesterase activity. CheZ is not a phosphate acceptor but rather serves to accelerate dephosphorylation of CheY. Because loss of phosphate from CheY and CheB is relatively slow, the steady-state levels of the phosphorylated forms should reflect the rate at which phosphate groups can be cycled through CheA. Thus, the CheA autophosphorylation reaction is the likely site of chemoreceptor control. Several lines of circumstantial evidence indicate that the signaling domain of MCP transducers may modulate CheA function through interactions with the carboxyl terminus of the CheA molecule (Kofoid and Parkinson 1988; Oosawa et al. 1988). It is conceivable that the CheW protein is involved at this step because its function is also needed for CW rotation.

Mutant strains lacking all MCPs are extremely CCW-biased, indicating that the transducers play an active role in producing CW rotation, presumably through formation of CheY phosphate. Thus, MCP molecules may somehow activate the autophosphorylation capability of CheA molecules. We suggest that in the CW signaling mode, transducers bind both CheA and CheW, which may catalyze or otherwise facilitate CheA activation, perhaps culminating in its release from the transducer (Fig. 5). To ensure rapid and sensitive responses to stimuli, it is necessary to propose that the activated form of CheA has a short lifetime. Perhaps it is activated by a covalent or noncovalent modification that can be removed with a characteristic half-life, similar to the behavior of GTP-binding proteins in eukaryotic signaling systems.

According to this model, the CCW form of the transducer might simply fail to activate CheA. However, the existence of dominant CCW mutants implies that a transducer locked in the CCW mode interferes actively with CW rotation. Thus, we suggest that the CCW signaling mode represents a transducer conformation that is capable of binding or inactivating CheA, reducing the flow of phosphate through the signaling cascade. Transitions between the CCW and CW states, which can be thought of as closed and open forms, respectively, could be modulated by direct interactions between the methylation regions (Fig. 5). The possible mechanism by which transmembrane signals influence these transitions to elicit behavioral responses is discussed below.

The model shown in Figure 5 makes several predictions that are currently being tested. Most importantly, it predicts that CCW and CW signal output do not depend on the chemoreceptor domain and linker region. Accordingly, it should be possible to detect these signaling forms in truncated molecules consisting only of the signaling region and methylation sites. More-

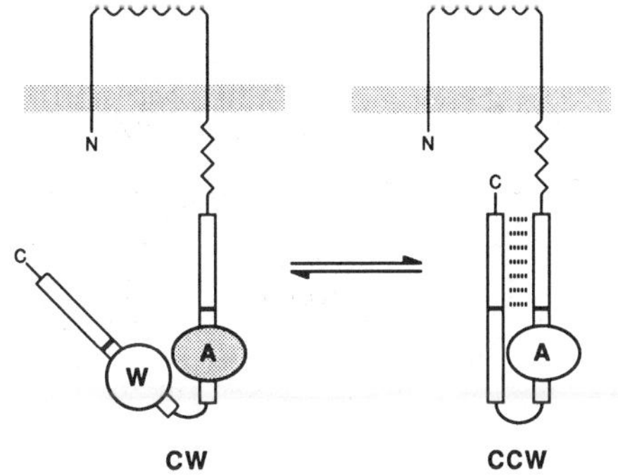

Figure 5. Model of the CW and CCW signaling conformations of the Tsr transducer. The periplasmic receptor domain and cytoplasmic linker segment are indicated by wavy lines. The methylation sites lie at the ends of each boxed region and may interact to control the equilibrium distributions of the two signaling forms. The segment joining the two arms is postulated to be a conformational switch or hinge.

over, if transitions between the CCW and CW signaling forms are regulated by interactions between the methylation segments, removal of either or both of them should result in CW signal output. By subcloning appropriate portions of the *tsr* coding region, we have isolated a soluble peptide consisting of residues 290–470 from the Tsr signaling domain that confers CW rotation on cells devoid of all other MCP transducers. The functional properties of this transducer fragment are at least consistent with these initial predictions. We hope to exploit the modular nature of the signaling domain to investigate some of its biochemical properties, particularly possible interactions with CheA and CheW molecules.

A Model of Transmembrane Signaling

The functions of various parts of the MCP transducers based on our mutant analyses (summarized in Fig. 6) suggest that transmembrane signaling may occur by a relatively simple mechanism in which the TM1 and TM2 membrane spanners and the linker segment play vital roles. A few transducers with locked or biased signaling properties had mutational changes in the membrane spanners (Figs. 3 and 4). Mutants of this type may be rare because only a few specific changes in these segments can alter transmembrane signaling activity without blocking membrane insertion. Oosawa and Simon (1986) constructed a similar mutation in TM1 of Tar and showed that mutations in TM2 were among its suppressors, leading to the suggestion that these membrane-spanning segments may interact with one another. We propose that such interactions are involved in transmitting conformational changes between the receptor and signaling domains. For example, ligand binding may cause TM1 and TM2 to shift relative orientations, leading to a push or pull on the linker segment, which, in turn, influences the signaling domain. Mutational changes in the linker can lead to either CW- or CCW-biased signal output, but this portion of the transducer is clearly not essential for production of either type of signal, suggesting that it could mediate transitions between the CW and CCW signaling states as part of the transmembrane signaling mechanism. We suggest that the linker modulates signal output by influencing the relative alignment of the methylation segments, which, in turn, control conformation of the signaling domain.

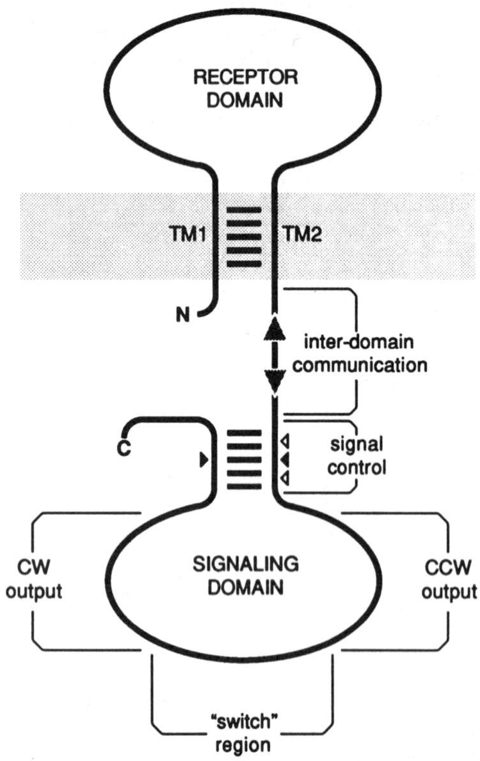

Figure 6. Structure-function relationships in MCP transducers. Interactions between TM1 and TM2 may initiate transmembrane signals, which are then propagated through the linker segment to the methylation regions, which, in turn, modulate signal output.

ACKNOWLEDGMENTS

This work was supported by research grant GM-19559 from the National Institutes of Health. C.W. was a recipient of fellowships from the Deutscher Akademischer Austauschdienst and the Hanns-Seidel Stiftung.

REFERENCES

Aksamit, R.R. and D.E. Koshland, Jr. 1974. Identification of the ribose binding protein as the receptor for ribose chemotaxis in *Salmonella typhimurium*. *Biochemistry* **13:** 4473.

Berg, H.C. and P.M. Tedesco. 1975. Transient response to chemotactic stimuli in *Escherichia coli*. *Proc. Natl. Acad. Sci.* **72:** 3235.

Bollinger, J., C. Park, S. Harayama, and G.L. Hazelbauer. 1984. Structure of the Trg protein: Homologies with and differences from other sensory transducers of *Escherichia coli*. *Proc. Natl. Acad. Sci.* **81:** 3287.

Callahan, A.M. and J.S. Parkinson. 1985. Genetics of methyl-accepting chemotaxis proteins in *Escherichia coli*: *cheD* mutations affect the structure and function of the Tsr transducer. *J. Bacteriol.* **161:** 96.

Falke, J.J. and D.E. Koshland, Jr. 1987. Global flexibility in a sensory transducer: A site-directed cross-linking approach. *Science* **237:** 1596.

Hazelbauer, G.L. and J. Adler. 1971. Role of the galactose binding protein in chemotaxis of *Escherichia coli* toward galactose. *Nature New Biol.* **230:** 101.

Hedblom, M.L. and J. Adler. 1980. Genetic and biochemical properties of *Escherichia coli* mutants with defects in serine chemotaxis. *J. Bacteriol.* **144:** 1048.

Hess, J.F., K. Oosawa, N. Kaplan, and M.I. Simon. 1988. Phosphorylation of three proteins in the signaling pathway of bacterial chemotaxis. *Cell* **53:** 79.

Hess, J.F., K. Oosawa, P. Matsumura, and M.I. Simon. 1987. Protein phosphorylation is involved in bacterial chemotaxis. *Proc. Natl. Acad. Sci.* **84:** 7609.

Kehry, M.R. and F.W. Dahlquist. 1982. The methyl-accepting chemotaxis proteins of *Escherichia coli*: Identification of the multiple methylation sites on methyl-accepting chemotaxis protein I. *J. Biol. Chem.* **257:** 10378.

Kehry, M.R., T.G. Doak, and F.W. Dahlquist. 1985a. Sensory adaptation in bacterial chemotaxis: Regulation of demethylation. *J. Bacteriol.* **163:** 983.

———. 1985b. Aberrant regulation of methylesterase activity in *cheD* mutants of *Escherichia coli. J. Bacteriol.* **161:** 105.

Kehry, M.R., M.W. Bond, M. Hunkapiller, and F.W. Dahlquist. 1983. Enzymatic deamidation of methyl-accepting chemotaxis proteins in *Escherichia coli* catalyzed by the *cheB* gene product. *Proc. Natl. Acad. Sci.* **80:** 3599.

Kofoid, E.C. and J.S. Parkinson. 1988. Transmitter and receiver modules in bacterial signaling proteins. *Proc. Natl. Acad. Sci.* **85:** 4981.

Koiwai, O. and H. Hayashi. 1979. Studies on bacterial chemotaxis. IV. Interaction of maltose receptor with a membrane-bound chemosensing component. *J. Biochem.* **86:** 27.

Krikos, A., N. Mutoh, A. Boyd, and M.I. Simon. 1983. Sensory transducers of *E. coli* are composed of discrete structural and functional domains. *Cell* **33:** 615.

Krikos, A., M.P. Conley, A. Boyd, H.C. Berg, and M.I. Simon. 1985. Chimeric chemosensory transducers of *Escherichia coli. Proc. Natl. Acad. Sci.* **82:** 1326.

Manoil, C. and J. Beckwith. 1986. A genetic approach to analyzing membrane protein topology. *Science* **233:** 1403.

Manson, M.D., V. Blank, G. Brade, and C.F. Higgins. 1986. Peptide chemotaxis in *E. coli* involves the Tap signal transducer and the dipeptide permease. *Nature* **321:** 253.

Mesibov, R. and J. Adler. 1972. Chemotaxis toward amino acids in *Escherichia coli. J. Bacteriol.* **112:** 315.

Mowbray, S.L. and D.E. Koshland, Jr. 1987. Additive and independent responses in a single receptor: Aspartate and maltose stimuli on the Tar protein. *Cell* **50:** 171.

Oosawa, K. and M.I. Simon. 1986. Analysis of mutations in the transmembrane region of the aspartate chemoreceptor in *Escherichia coli. Proc. Natl. Acad. Sci.* **83:** 6930.

Oosawa, K., J.F. Hess, and M.I. Simon. 1988. Mutants defective in bacterial chemotaxis show modified protein phosphorylation. *Cell* **53:** 89.

Park, C. and G.L. Hazelbauer. 1986. Mutations specifically affecting ligand interaction of the Trg chemosensory transducer. *J. Bacteriol.* **167:** 101.

Parkinson, J.S. 1980. Novel mutations affecting a signaling component for chemotaxis of *Escherichia coli. J. Bacteriol.* **142:** 953.

Parkinson, J.S., S.R. Parker, P.B. Talbert, and S.E. Houts. 1983. Interactions between chemotaxis genes and flagellar genes in *Escherichia coli. J. Bacteriol.* **155:** 265.

Richarme, G. 1982. Interaction of the maltose-binding protein with membrane vesicles of *Escherichia coli. J. Bacteriol.* **149:** 662.

Russo, A.F. and D.E. Koshland, Jr. 1983. Separation of signal transduction and adaptation functions of the aspartate receptor in bacterial sensing. *Science* **220:** 1016.

Spudich, J.L. and D.E. Koshland, Jr. 1975. Quantitation of the sensory response in bacterial chemotaxis. *Proc. Natl. Acad. Sci.* **72:** 710.

Wang, E.A. and D.E. Koshland, Jr. 1980. Receptor structure in the bacterial sensing system. *Proc. Natl. Acad. Sci.* **77:** 7157.

Wolff, C. 1983. "Genetic and biochemical studies of maltose chemotaxis in *Escherichia coli*." Diplom thesis, University of Konstanz, Federal Republic of Germany.

The End of the Line in Bacterial Sensing: The Flagellar Motor

R.M. MACNAB
Department of Molecular Biophysics and Biochemistry, Yale University, New Haven, Connecticut 06511

Sensory and Motor Subsystems of the Bacterial Behavioral System

Bacteria experience a multiplicity of environmental inputs, including pH, temperature, ionic strength, osmotic strength, and the concentration of various organic metabolites (Macnab 1987b). Some of these parameters affect the cell in more than one fashion. As examples, a sudden change in pH affects proton-motive force and also tactic behavior (Kihara and Macnab 1981; Repaske and Adler 1981; Shimada and Berg 1987). The external concentration of an amino acid can affect the kinetic properties of a transport system for it and therefore the cytoplasmic concentration; it can also affect the cell's tactic behavior to that amino acid. Thus, one can define in a broad way the sensory system of bacteria as the interaction of environmental information and the physiological state of the cell.

Bacteria, however, also possess a more narrowly and clearly defined sensory system. It is one in which sensory receptors (e.g., the aspartate chemoreceptor of *Escherichia coli* or *Salmonella typhimurium* [Moe and Koshland 1986; Oosawa et al. 1986], or the sensory rhodopsin photoreceptors of *Halobacterium halobium* [Spudich and Bogomolni 1988]), together with a limited set of central components for analysis of the sensory information, exist as a dedicated system whose sole function—as far as we know—is to generate and analyze sensory data. Where is this stream of information being directed, and for what reason? The answers are clear: The information is being directed to the cellular organelle for motility, the bacterial flagellum, and is affecting its modus operandi in a fashion that results in migration of the cell toward a more beneficial environment. A temporal change in receptor occupancy, as a result of the cell swimming through spatial gradients, affects the probability of directional changes and introduces a drift velocity to an otherwise random walk in three-dimensional space (Berg and Brown 1972; Macnab and Koshland 1972). The behavioral system of tactic migration therefore includes both the sensory and the motor subsystems, and an understanding of both is necessary before we can say that we have achieved the goal of understanding it as a whole. Of the two subsystems, the motor subsystem is by far the more complex, at least if we define complexity in terms of the number of genetic components. There are about 40 flagellar components (Fig. 1), whereas there are only 6 central sensory components and approximately 10 or so distinct receptors (this latter number may increase with further study, but unless they have a different class of mechanism, the actual number of receptor types does not really reflect a great increase in complexity). For more extensive reviews of flagella, motility, and chemotaxis, see Macnab (1987a,b), Stewart and Dahlquist (1987), and Macnab and DeRosier (1988).

Bacterial Flagellum

The bacterial flagellum has several characteristics that make it different from other mechanoenzymes and are therefore worth emphasizing. Its external portion consists largely of the flagellar filament, a macroscopically helical structure that is quite thin (20 nm) and simple in construction (Trachtenberg and DeRosier 1987), comparable to a microtubule, yet is rigid enough to act as an effective structure for propelling the cell. This filament is passive; i.e., it has no capacity to convert cellular metabolic energy into mechanical work. Instead, it receives mechanical energy transmitted from the basal structure at the cell surface. The mechanical energy is rotational (Berg 1974; Silverman and Simon 1974), which (because the filament is helical) transmits both torque and thrust to the medium. The rotary character of the motor is truly that—the filament can be made to rotate indefinitely in a given direction. However, such unmodulated operation of the motor would not be very useful in terms of generating a migrational response, and indeed a wild-type cell shows a spontaneous alternation between the counterclockwise (CCW) and clockwise (CW) senses of rotation. Thus, the motor contains a binary switch. This switch is the target of the sensory information stream, which can modulate the relative probabilities of the CCW and CW senses (Larsen et al. 1974). The final point to note about the flagellar motor is its energy source, which is the transmembrane proton-motive force (Manson et al. 1977) or, in some species, sodium-motive force (Hirota and Imae 1983).

Besides the filament, the external structure (Fig. 2) contains the hook and several minor junction proteins, called hook-associated proteins or HAPs (Homma et al. 1984). The basal body structure at the cell surface has a rod and two pairs of rings. The outermost ring (L, for lipopolysaccharide) is in the outer membrane, the next (P, for peptidoglycan) is in the periplasmic space, and the inner pair (S, for supramembrane, and M, for

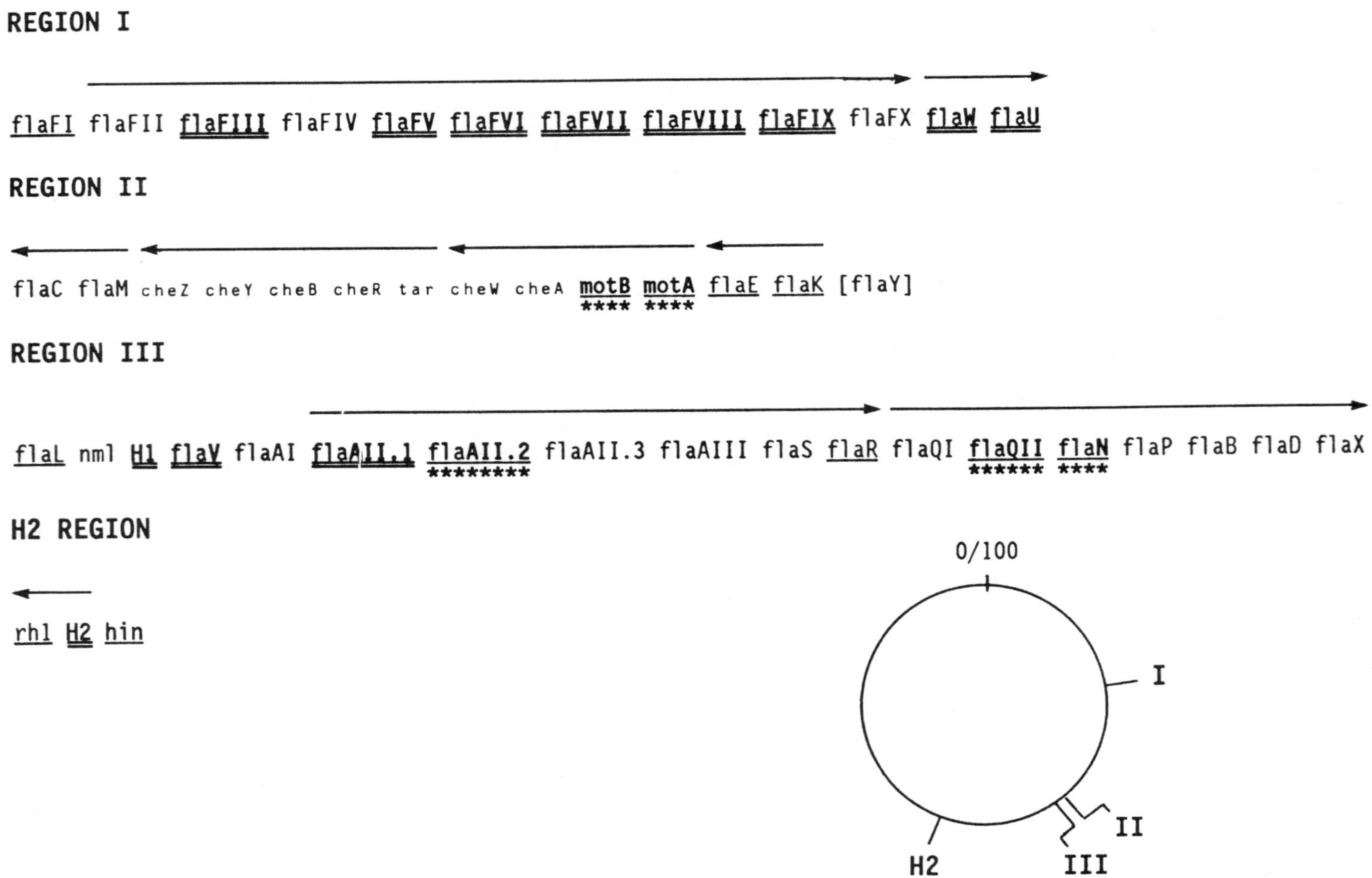

Figure 1. Flagellar and motility genes of *S. typhimurium*. The genes cluster in three main regions, located as shown on the circular map. Operons are indicated by arrows over the corresponding genes. Genes coding for known structural components of the filament, hook, and basal-body complex are indicated by double underlining; those that are involved in energy transduction or switching, on the basis of mutant phenotype, are indicated by asterisks; those with a role in regulation of gene expression or flagellar assembly are shown by single underlining; sensory (chemotaxis) genes are indicated in smaller letters. (Reprinted, with permission, from Macnab and DeRosier 1988.)

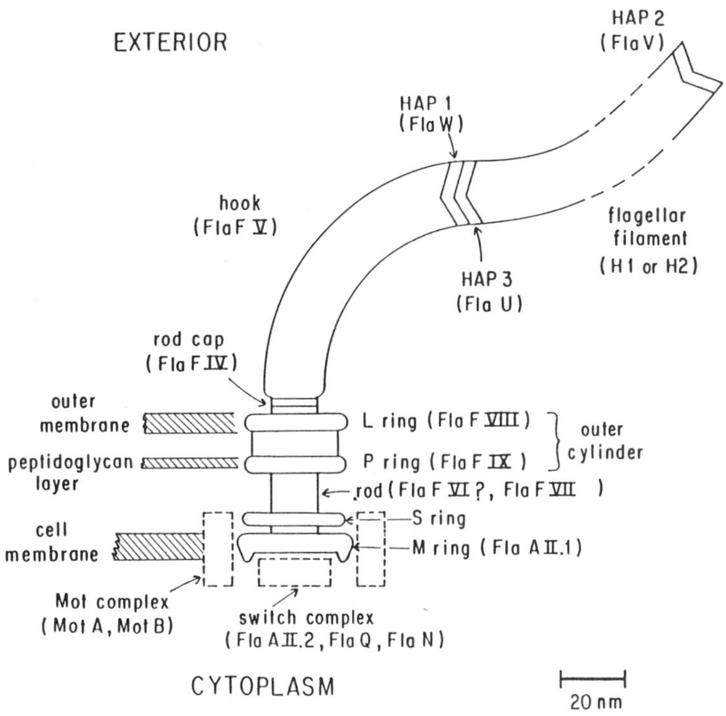

Figure 2. Current knowledge of the structure of the flagellum of *E. coli* or *S. typhimurium*, as deduced from electron microscopy and genetic and biochemical analysis. Principal features are the helical flagellar filament (truncated in this representation), the hook, and the basal body. The basal body is embedded in the layers of the cell surface, as shown. When the identity of a gene product is known or suspected, it is given in parentheses (using *S. typhimurium* nomenclature). The switch and Mot complexes have not been detected by electron microscopy and are therefore indicated in dashed outline; there are a number of lines of evidence to suggest that they are in the locations shown. (Modified, with permission, from Macnab 1987a.)

membrane) is at the cell membrane (DePamphilis and Adler 1971a,b). The proteins comprising these features and the genes coding for these proteins have, for the most part, been identified (Fig. 2) (Komeda et al. 1978; Bartlett and Matsumura 1984; Aizawa et al. 1985; Homma et al. 1987a,c,d, 1988; Ikeda et al. 1987; Jones et al. 1987).

The above description represents the bacterial flagellum as it is currently recognized morphologically; however, there are at least two reasons to suspect that matters are more complicated than this. The first is simply the total number of flagellar genes, about 40 (Fig. 1), compared to the much smaller number (~11) accounted for by the basal body, hook, filament, and HAPs (Homma et al. 1984; Aizawa et al. 1985). The second has to do with the mutant phenotype associated with a particular gene: Although mutants that are defective in rotation (paralyzed or Mot$^-$ phenotype) or in switching (nonchemotactic or Che$^-$ phenotype) have been isolated, none of the mutations map to basal-body genes (Yamaguchi et al. 1986a).

Two *mot* genes, *motA* and *motB*, are known (Matsumura et al. 1977). Their products are integral membrane proteins, which are not necessary for assembly of the basal body. Indirect evidence suggests that MotA and MotB may surround the motor (Ridgway et al. 1977; Block and Berg 1984; Dean et al. 1984; Stader et al. 1986; Wilson and Macnab 1988). Three flagellar genes (*flaAII.2*$_S$, *flaQII*$_S$, and *flaN*$_S$ in *S. typhimurium*; *flaBII*$_E$, *flaAII*$_E$, and *motD*$_E$ in *E. coli*)[1] give rise to multiple mutant phenotypes, not only nonflagellate (Fla$^-$, the null or deletion phenotype), but also Mot$^-$ and Che$^-$; in the latter case, both CW- and CCW-defective examples are found (Yamaguchi et al. 1986a). Genetic evidence (Parkinson et al. 1983; Yamaguchi et al. 1986b) suggests that these proteins form a peripheral membrane complex mounted to the M ring of the basal body, exposed to the cytoplasm, and in contact with the sensory system, specifically with CheY and CheZ, which are thought to modify the state of this complex and hence the rotational state of the motor (Ravid et al. 1986; Kuo and Koshland 1987; Wolfe et al. 1987). The putative structure has been called the "switch complex," but because some switch gene mutants are paralyzed, it should be recognized that the structure probably also plays a part in energy transduction.

Flagellar Regulon

The 40 or so flagellar genes, the genes for the 6 central sensory components, and a few of the chemoreceptor genes are all clustered into three regions of the chromosome (Fig. 1) and are organized into about 14 operons ranging in size from single-gene operons such as *flaL*$_S$ up to the *flaFII*$_S$ operon, which contains 9 genes (Kutsukake et al. 1980; references contained in Macnab 1987a). There are no nonmotility-linked genes within the clusters. A hierarchy exists with one master operon at the top and about three levels beneath that (Fig. 3) (Komeda 1986). The entire system constitutes a regulon, or coordinately expressed genetic unit under a single overall controlling signal, cAMP (Komeda et al. 1975). Although not quite a match for the ribosomal gene system (Jinks-Robertson and Nomura 1987), the

[1]Because *S. typhimurium* and *E. coli* will be discussed interchangeably, subscripts S and E, respectively, will be used to distinguish the genes and gene products from the two species where the symbols differ, as is the case with the flagellar genes.

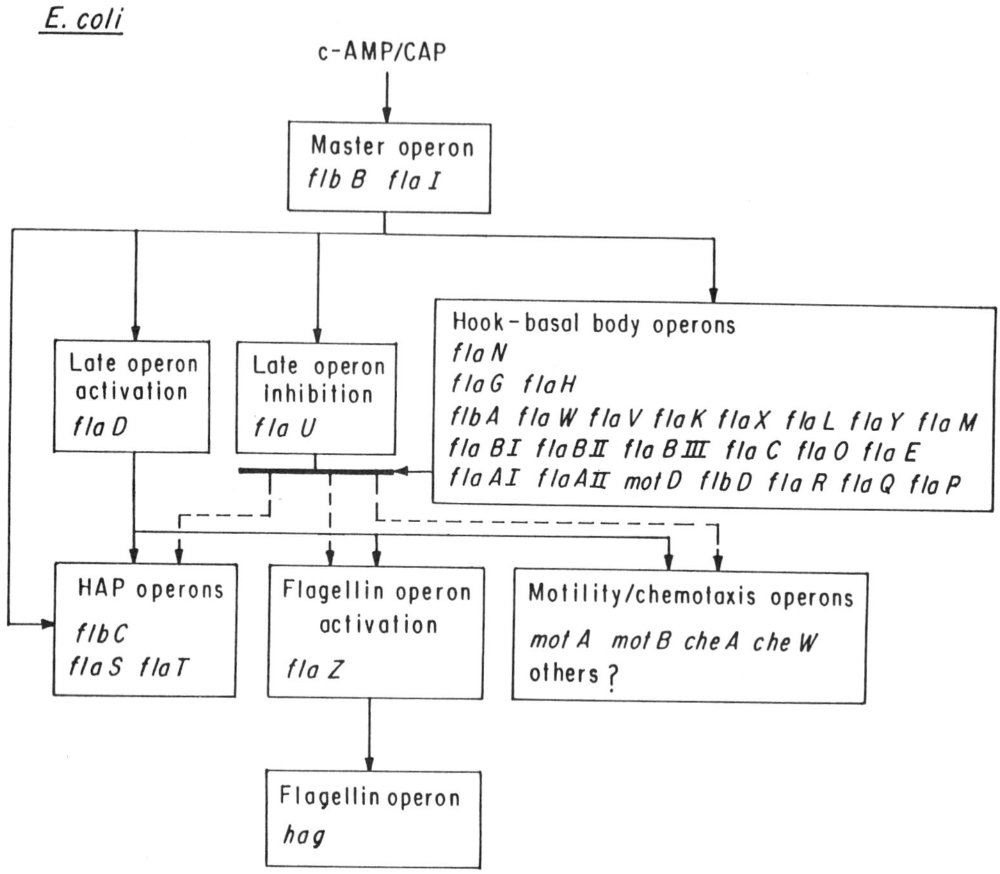

Figure 3. Control of gene expression within the flagellar/motility regulon of *E. coli*. (1) cAMP binds to cap protein, and the complex binds in the promoter region of the master operon, acting as a positive regulator. (2) Expression of the $flbB_E$ and $flaI_E$ genes activates expression of all other flagellar/motility genes, directly or indirectly. Direct activation occurs with most of the hook-basal body operons, and with two one-gene operons ($flaD_E$ and $flaU_E$). (3) $flaD_E$ expression activates operons for later components of the flagellar apparatus (hook-associated proteins, flagellin [with $flaZ_E$ expression as an intermediate], and Mot and Che proteins). (4) $flaU_E$ acts on the same operons as $flaD_E$ does, but as a negative regulator. (5) Where all of the basal body operons are being expressed, it is postulated that FlaU protein is incorporated in the basal body, preventing it from acting as an inhibitor of late gene expression. (Based on data from Komeda 1986; reprinted, with permission, from Macnab 1987a.)

flagellar gene system is certainly among the most complex in the genome.

Some Indications of the Complexities of Flagellar Gene Regulation

The flagellar regulon is likely to be subject to intricate patterns of regulation. A large number of the flagellar, motility, and chemotaxis genes, together with their controlling regions, have now been sequenced, and a number of intriguing regulatory features are beginning to emerge from these data. Some of these features have already been tested experimentally; others are still at the level of tantalizing clues indicated by the sequence information. Presented below is a collection of these features, some from the literature and others from recent observations from this laboratory.

Flagellar-specific promoters. Early efforts to spot putative promoters of flagellar operons were not particularly successful. The reason for this has recently become evident; namely, that flagellar promoters have consensus sequences that differ from the primary ones in the chromosome. These sequences are TTGACA for the −35 region and TATAAT for the −10 region in primary promoters, whereas they are TAAA and GCCGATAA in flagellar promoters (Helmann and Chamberlin 1987). There are indications (Bartlett et al. 1988) that a more extensive −10 consensus, TT(A/T)GCCGATAACG, may apply to operons classified at level 3 (Fig. 3) in the scheme of Komeda (1986). On the basis of its translated gene sequence, one of the two genes in the master operon ($flbB_E$) encodes a protein that is thought to be an alternate transcription initiation (σ)-factor responsible for recognizing flagellar promoters. Not all flagellar promoters have an obvious −35 consensus of either the primary or the flagellar-specific type; thus, the $flaBI_E$ operon (Bartlett et al. 1988) and the homologous $flaAII.1_S$ operon (C. Jones et al., in prep.) have AGTG at that position.

An intriguing example of the use of mixed promoter

consensus sequences exists in the *S. typhimurium* *flaQI*$_S$ operon (Fig. 4) (M. Kihara et al., in prep.). For both this operon and the homologous *E. coli* operon, *flaAI*$_E$ (P. Matsumura, pers. comm.), there are extensive stem-loop sequences that are likely to act as transcription terminators of the upstream operon (*flaAII.1*$_S$, or *flaBI*$_E$). However, the stem-loop sequences are not the same in the two species. In both cases, the 3′ half of the stem overlaps with the −35 region of the promoter and places constraints on the sequence that can be used there. In the case of *E. coli*, TCAA exists as a three-quarters match to the flagellar consensus. In *S. typhimurium*, T conflicts with the stem-loop requirement, and the sequence in register is a poor one-quarter match, ACAG. However, the stem-loop sequence permits the sequence TGGACA, which is a five-sixths match with the primary −35 consensus. Whether this means that the *S. typhimurium* promoter can be recognized by the primary σ-factor (in spite of the flagellar-specific −10 sequence) remains to be seen.

An invertible operon. A classic case of a major genetic rearrangement exists in the *H2* operon, one of the two operons for the flagellin genes of *S. typhimurium*. The promoter region of this operon contains an invertible element that, in one orientation, preserves the *H2* promoter and permits expression of the *H2* gene and the gene for the repressor of the other flagellin operon *H1* (Simon et al. 1980; Iino and Kutsukake 1983). This provides the cell with an opportunity to stochastically switch between expression of the two distinct flagellin genes and may provide an additional weapon against the host's immune system.

Overlapping operons? The *flbB*$_E$/*flaI*$_E$ operon (which is the master operon of the regulon; Komeda 1986) lies just upstream of the *mocha* operon (which contains *motA, motB, cheA,* and *cheW*). A classic ρ-independent transcription termination sequence lies just after the end of *flaI*$_E$ (Fig. 5) (Dean et al. 1984; Bartlett et al. 1988). Moderately good matches (four-sixths in each case) exist to the primary −10 and −35 consensus promoter sequences. But, surprisingly, there is an excellent match to the flagellar-specific promoter sequences, well within the coding region of the *flaI*$_E$ gene. The significance of this in terms of transcription is a mystery at present. It might mean that under some conditions, transcription of the *mocha* operon proceeds from within the *flaI*$_E$ gene and manages to bypass the transcription termination site of the *flbB*$_E$/*flaI*$_E$ operon.

Overlapping genes within an operon. The genes within the flagellar operons are quite tightly packed. Although examples with an intergenic spacing of 10–20 bases are not unusual, there are a number of examples where adjacent genes are so close that either the ribosome-binding site of one must lie within the coding region of the other or the coding regions themselves overlap. Examples ranked with an increasing degree of overlap are presented in Table 1. These arrangements make it likely that translation of the adjacent genes is coupled (cf. similar arrangements in other gene systems such as the *trp* operon; Yanofsky and Crawford 1987).

Potential secondary structure between genes in an operon. In contrast to the tightly packed situation described above, there is at least one example where there is a considerable distance (54 bp) between the termination of one gene and the initiation of another within the same operon and where this intergenic region has the potential for secondary structure (C.J. Jones et al., in prep.). This is the *flaFVII*$_S$–*flaFVIII*$_S$ region, which is shown schematically in Figure 6. FlaFVII$_S$ and FlaFVIII$_S$ fall into different functional categories. FlaFVII$_S$ is a rod protein (Aizawa et al. 1985; S.-I. Aizawa, pers. comm.), and FlaFVIII$_S$ is a ring protein (Jones et al. 1987); they have different export pathways (Homma et al. 1987c,d), and they have different dependencies on other components for their stable existence (C.J. Jones and R.M. Macnab, unpubl.). Punctuation of the operon at this point may

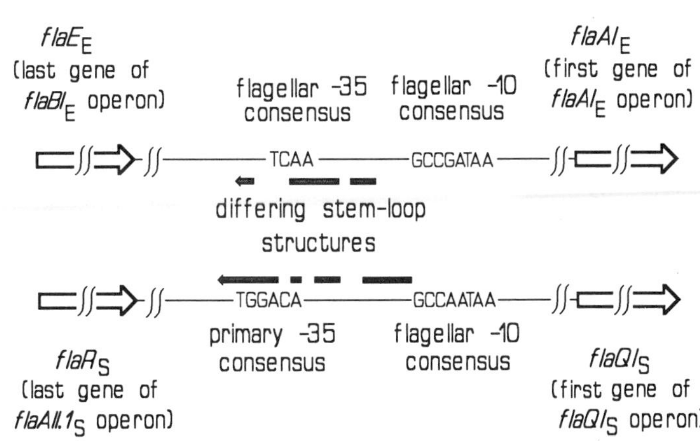

Figure 4. Comparison of the *flaAI* (*E. coli*) and *flaQI* (*S. typhimurium*) promoter regions. The −35 regions overlap with the right-hand portion of stem-loop structures that are probably termination sequences for the upstream operon (*flaBI* for *E. coli*, *flaAII.1* for *S. typhimurium*). In the case of *S. typhimurium*, the stem-loop sequence prevents the use of the flagellar consensus but permits the primary consensus instead (see text for details). (Based on data from Kuo and Koshland 1986; M. Kihara et al., in prep.)

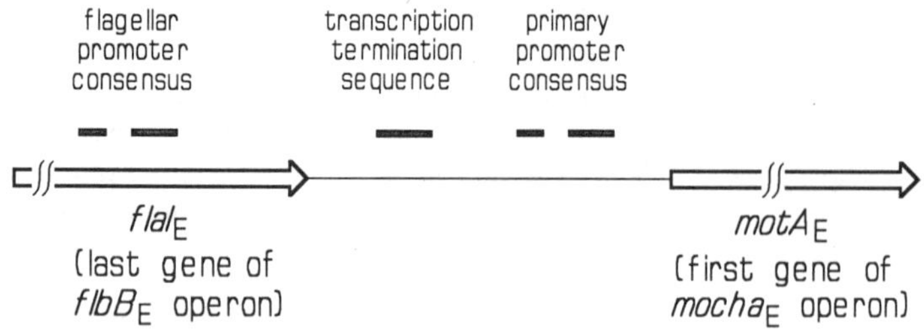

Figure 5. The promoter region for the *E. coli mocha* operon, which starts with the *motA* gene and lies downstream from the *flbB* operon. Although a moderately good primary promoter consensus exists in the noncoding region between the operons, shortly after the putative transcription terminator of the *flbB* operon, a much better flagellar-specific one exists toward the 3′ end of the *flaI* gene. (Based on data from Dean et al. 1984, Bartlett et al. 1988.)

therefore reflect different needs in terms of gene expression. Two possible functions of the stem-loop structure are to act as a partial terminator of transcription or to retard degradation of the mRNA for the genes up to and including $flaFVII_S$ in the operon.

Potential secondary structure within genes. There are also examples of potential secondary structure within genes. Within the 5′ end of the $flaAII.2_S$ gene lies a 20-bp palindrome (which probably functions as a stem-loop structure, because it seems rather long to be a binding site with dyad symmetry and because the palindrome is broken at two central positions in *E. coli*; P. Matsumura, pers. comm.). The $FlaAII.1_S$ protein is the structural component of the M ring, whereas the $FlaAII.2_S$ protein is part of the flagellar switch. They may require different control of gene expression.

***Alternative translational starts in* CheA.** Finally, we note that *cheA* is the first demonstrated example of a bacterial gene that has two translational start sites in the same frame (Smith and Parkinson 1980; Parkinson and Hazelbauer 1983). Both the large and the small forms of CheA are synthesized, with the latter containing an additional 95 amino-terminal amino acids (J.S. Parkinson, pers. comm.). Potential secondary structure exists in the vicinity of both ribosome-binding sites, which may affect the relative probabilities of translation initiation for the two forms.

Export and Assembly Process

This section considers how the flagellar components get to their natural location within the apparatus. Most proteins that cross the bacterial cell membrane are characterized by the presence of an amino-terminal signal peptide that is recognized by a specific export apparatus and is cleaved in the export process (Oliver 1985). Many of the flagellar proteins—the filament, hook, HAPs, rod, and the P and L rings—are external to the cell membrane. However, of these, only the P and L rings utilize the conventional pathway; they possess consensus amino-terminal sequences, which are experimentally found to be cleaved (Homma et al. 1987c,d). The translated gene sequences of the others fail to reveal such consensus sequences, and no evidence of cleavage has been obtained. This raises the issue of how these external flagellar components arrive without the help of the conventional export pathway.

At least for flagellin (the filament protein), assembly of new subunits occurs at the distal end of the structure (Iino 1969; Emerson et al. 1970). Because bulk diffusion is ruled out by a lack of cell density dependence of growth rate (and the tremendous losses that such a bulk

Table 1. Overlapping Sites of Translational Termination and Initiation in the Flagellar Gene System

Sequence	Examples	References
TAANNNATG	$flaAI_E$–$flaAII_E$ $flaQI_S$–$flaQII_S$	Kuo and Koshland (1986) M. Kihara et al. (in prep.)
TA(A/G)NNATG	$motD_E$–$flbD_E$ $flaN_S$–$flaP_S$	J. Malakooti et al. (in prep.) M. Kihara et al. (in prep.)
TGATG	$flaFIX_S$–$flaFX_S$ $flaAII.3_S$–$flaAIII_S$	C.J. Jones et al. (in prep.) M. Kihara et al. (unpubl.)
ATGA	$motA_E$–$motB_E$	Dean et al. (1984); Stader et al. (1986)
	$flaAII_E$–$motD_E$	Kuo and Koshland (1986); J. Malakooti et al. (in prep.)
	$flaQII_S$–$flaN_S$ $flaS_S$–$flaR_S$	M. Kihara et al. (in prep.) M. Kihara et al. (unpubl.)
ATGNNT(A/G)A	$flaBI_E$–$flaBII_E$–$flaBIII_E$ $flaAII.1_S$–$flaAII.2_S$–$flaAII.3_S$	P. Matsumura (pers. comm.) M. Kihara et al. (in prep.)

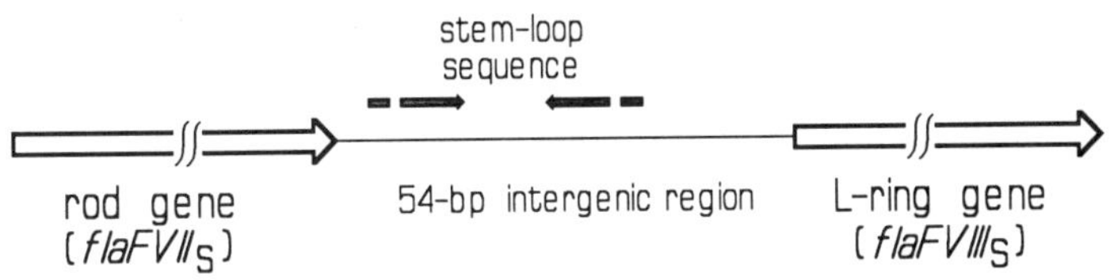

Figure 6. Stem-loop sequence in an intergenic region within the *flaFII* operon of *S. typhimurium*. The distance between the genes, one encoding a rod protein and the other a ring protein, is much larger than usual for the flagellar gene system, and the sequence may exist to permit different expression patterns for the two genes (see text). (Based on data from C.J. Jones et al., in prep.)

diffusion process would entail), the most probable mechanism involves extrusion through a central core. Mutant analysis has suggested that the carboxyl terminus of flagellin is important for its export (Homma et al. 1987b). This might be for reasons of three-dimensional structure or it might reflect a more local feature that was acting as a signal for the flagellar-specific export pathway. Next to flagellin, hook protein is the most abundant external flagellar protein, and it seemed to us at least possible that hook protein and flagellin might contain a common signal for export by the flagellar-specific pathway. The *S. typhimurium* hook gene, *flaFV*$_S$, has been sequenced recently (D.J. DeRosier et al., in prep.). Careful comparison of its translated sequence with that of flagellin fails to reveal any distinctive common motif at the carboxyl terminus, the amino terminus, or anywhere else. Thus, the basis for the recognition of flagellin and hook protein as molecules destined for export by the flagellar-specific pathway does not seem to be at the level of a local primary sequence signal; or if it is, the signal must be one with a low level of specificity.

Axial Protein Subfamily

We have found clear evidence that among flagellar proteins as a whole, flagellin and hook proteins are members of a subfamily. Although they are both major external structural proteins and occupy quite similar lattices, they are not physically adjacent in the structure, but instead are separated by two of the junction proteins, HAP1 and HAP3. The two rod proteins (products of the *flaFVI*$_S$ and *flaFVII*$_S$ genes), hook protein, HAP1, HAP3, flagellin, and HAP2 constitute a set of proteins, serially organized in that order along the axis of the flagellum (Fig. 2). The sequence of flagellin was already available (Wei and Joys 1985), and the sequences of the other genes in the family have recently been obtained (M. Homma et al., unpubl.). Comparison of the translated sequences of the axial proteins reveals a graded similarity (most evident at the amino and carboxyl termini), in which adjacent members show a higher degree of similarity than nonadjacent ones. For example, the FlaFVI$_S$ protein, which is thought to be a rod protein at the cytoplasmic-proximal end (S.-I. Aizawa, pers. comm.), shows in its first 40 amino-terminal residues, 13 identities with the FlaFVII$_S$ protein, which is a rod protein at the hook-proximal end (Aizawa et al. 1985; S.-I. Aizawa, pers. comm.). FlaFVII$_S$, in turn, shows 19 identities with hook protein, and hook protein shows 14 identities with HAP1. FlaFVI$_S$, however, shows only 8 identities with hook protein. Thus, the lack of any close similarity between flagellin and hook protein probably reflects the fact that they are somewhat distant members of the serial set. It seems likely that the graded similarities reflect a requirement for structural compatibility and may explain why the HAPs are necessary to create an indirect hook-filament junction, in spite of the superficial similarity of the lattices of the two proteins. The results also suggest that rod (acting as a rigid transmission shaft) and hook (acting as a flexible one) are related structures.

ACKNOWLEDGMENTS

I thank Shin-Ichi Aizawa, Phil Matsumura, and Sandy Parkinson for the provision of unpublished data, and the members of my own laboratory for their contributions to the unpublished work referred to in the text.

REFERENCES

Aizawa, S.-I., G.E. Dean, C.J. Jones, R.M. Macnab, and S. Yamaguchi. 1985. Purification and characterization of the flagellar hook-basal body complex of *Salmonella typhimurium*. *J. Bacteriol.* **161:** 836.

Bartlett, D.H. and P. Matsumura. 1984. Identification of *Escherichia coli* region III flagellar gene products and description of two new flagellar genes. *J. Bacteriol.* **160:** 577.

Bartlett, D.H., B.B. Frantz, and P. Matsumura. 1988. Flagellar transcriptional activators FlbB and FlaI: Gene sequences and 5′ consensus sequences of operons under FlbB and FlaI control. *J. Bacteriol.* **170:** 1575.

Berg, H.C. 1974. Dynamic properties of bacterial flagellar motors. *Nature* **249:** 77.

Berg, H.C. and D.A. Brown. 1972. Chemotaxis in *Escherichia coli* analysed by three-dimensional tracking. *Nature* **239:** 500.

Block, S.M. and H.C. Berg. 1984. Successive incorporation of force-generating units in the bacterial rotary motor. *Nature* **309:** 470.

Dean, G.E., R.M. Macnab, J. Stader, P. Matsumura, and C. Burks. 1984. Gene sequence and predicted amino acid sequence of the *motA* protein, a membrane-associated protein required for flagellar rotation in *Escherichia coli*. *J. Bacteriol.* **159:** 991.

DePamphilis, M.L. and J. Adler. 1971a. Fine structure and isolation of the hook-basal body complex of flagella from *Escherichia coli* and *Bacillus subtilis*. *J. Bacteriol.* **105:** 384.

———. 1971b. Attachment of flagellar basal bodies to the cell envelope: Specific attachment to the outer, lipopolysaccharide membrane and the cytoplasmic membrane. *J. Bacteriol.* **105:** 396.

Emerson, S.U., K. Tokuyasu, and M.I. Simon. 1970. Bacterial flagella: Polarity of elongation. *Science* **169:** 190.

Helmann, J.D. and M.J. Chamberlin. 1987. DNA sequence analysis suggests that expression of flagellar and chemotaxis genes in *Escherichia coli* and *Salmonella typhimurium* is controlled by an alternative σ factor. *Proc. Natl. Acad. Sci.* **84:** 6422.

Hirota, N. and Y. Imae. 1983. Na^+-driven flagellar motors of an alkalophilic *Bacillus* strain YN-1. *J. Biol. Chem.* **258:** 10577.

Homma, M., T. Iino, and R.M. Macnab. 1988. Identification and characterization of the products of six region III flagellar genes (flaAII.3-flaQII) of *Salmonella typhimurium. J. Bacteriol.* **170:** 2221.

Homma, M., S.-I. Aizawa, G.E. Dean, and R.M. Macnab. 1987a. Identification of the M-ring protein of the flagellar motor of *Salmonella typhimurium. Proc. Natl. Acad. Sci.* **84:** 7483.

Homma, M., H. Fujita, S. Yamaguchi, and T. Iino. 1987b. Regions of *Salmonella typhimurium* flagellin essential for its polymerization and secretion. *J. Bacteriol.* **169:** 291.

Homma, M., Y. Komeda, T. Iino, and R.M. Macnab. 1987c. The *flaFIX* gene product of *Salmonella typhimurium* is a flagellar basal body component with a signal peptide for export. *J. Bacteriol.* **169:** 1493.

Homma, M., K. Kutsukake, T. Iino, and S. Yamaguchi. 1984. Hook-associated proteins essential for flagellar filament formation in *Salmonella typhimurium. J. Bacteriol.* **157:** 100.

Homma, M., K. Ohnishi, T. Iino, and R.M. Macnab. 1987d. Identification of flagellar hook and basal body gene products (FlaFV, FlaFVI, FlaFVII and FlaFVIII) in *Salmonella typhimurium. J. Bacteriol.* **169:** 3617.

Iino, T. 1969. Polarity of flagellar growth in *Salmonella. J. Gen. Microbiol.* **56:** 227.

Iino, T. and K. Kutsukake. 1983. Flagellar phase variation in *Salmonella*: A model system regulated by flip-flop DNA inversions. In *Nucleic acid research. Future developments* (ed. K. Mizobuchi et al.), p. 395. Academic Press, New York.

Ikeda, T., M. Homma, T. Iino, S. Asakura, and R. Kamiya. 1987. Localization and stoichiometry of hook-associated proteins within *Salmonella typhimurium* flagella. *J. Bacteriol.* **169:** 1168.

Jinks-Robertson, S. and M. Nomura. 1987. Ribosomes and tRNA. In Escherichia coli *and* Salmonella typhimurium: *Cellular and molecular biology* (ed. F.C. Neidhart et al.), vol. 2, p. 1358. American Society for Microbiology Publications, Washington, D.C.

Jones, C.J., M. Homma, and R.M. Macnab. 1987. Identification of proteins of the outer (L and P) rings of the flagellar basal body complex of *Escherichia coli. J. Bacteriol.* **169:** 1489.

Kihara, M. and R.M. Macnab. 1981. Cytoplasmic pH mediates pH taxis and weak-acid repellent taxis of bacteria. *J. Bacteriol.* **145:** 1209.

Komeda, Y. 1986. Transcriptional control of flagellar genes in *Escherichia coli* K-12. *J. Bacteriol.* **168:** 1315.

Komeda, Y., M. Silverman, P. Matsumura, and M. Simon. 1978. Genes for the hook-basal body proteins of the flagellar apparatus in *Escherichia coli. J. Bacteriol.* **134:** 655.

Komeda, Y., H. Suzuki, J. Ishidsu, and T. Iino. 1975. The role of cAMP in flagellation of *Salmonella typhimurium. Mol. Gen. Genet.* **142:** 289.

Kuo, S.C. and D.E. Koshland, Jr. 1986. Sequence of the *flaA* (*cheC*) locus of *Escherichia coli* and discovery of a new gene. *J. Bacteriol.* **166:** 1007.

———. 1987. Roles of *cheY* and *cheZ* gene products in controlling flagellar rotation in bacterial chemotaxis of *Escherichia coli. J. Bacteriol.* **169:** 1307.

Kutsukake, K., T. Iino, Y. Komeda, and S. Yamaguchi. 1980. Functional homology of *fla* genes between *Salmonella typhimurium* and *Escherichia coli. Mol. Gen. Genet.* **178:** 59.

Larsen, S.H., R.W. Reader, E.N. Kort, W.-W. Tso, and J. Adler. 1974. Change in direction of flagellar rotation is the basis of the chemotactic response in *Escherichia coli. Nature* **249:** 74.

Macnab, R.M. 1987a. Flagella. In Escherichia coli *and* Salmonella typhimurium: *Cellular and molecular biology* (ed. F.C. Neidhart et al.), vol. 1, p. 70. American Society for Microbiology Publications, Washington, D.C.

———. 1987b. Motility and chemotaxis. In Escherichia coli *and* Salmonella typhimurium: *Cellular and molecular biology* (ed. F.C. Neidhart et al.), vol. 1, p. 732. American Society for Microbiology Publications, Washington, D.C.

Macnab, R.M. and D.J. DeRosier. 1988. Bacterial flagellar structure and function. *Can. J. Microbiol.* **34:** 442.

Macnab, R.M. and D.E. Koshland, Jr. 1972. The gradient-sensing mechanism in bacterial chemotaxis. *Proc. Natl. Acad. Sci.* **69:** 2509.

Manson, M.D., P. Tedesco, H.C. Berg, F.M. Harold, and C. van der Drift. 1977. A proton-motive force drives bacterial flagella. *Proc. Natl. Acad. Sci.* **74:** 3060.

Matsumura, P., M. Silverman, and M. Simon. 1977. Synthesis of *mot* and *che* gene products of *Escherichia coli* programmed by hybrid ColE1 plasmids in minicells. *J. Bacteriol.* **132:** 996.

Moe, G.R. and D.E. Koshland, Jr. 1986. Transmembrane signalling through the aspartate receptor. In *Microbial energy transduction: Genetics, structure, and function of membrane proteins* (ed. D.C. Youvan and F. Daldal), p. 163. Cold Spring Harbor Laboratory, Cold Spring Harbor, New York.

Oliver, D. 1985. Protein secretion in *Escherichia coli. Annu. Rev. Microbiol.* **39:** 615.

Oosawa, K., N. Mutoh, and M.I. Simon. 1986. Genetic studies of the structure and function of the Tar chemoreceptor. In *Microbial energy transduction: Genetics, structure, and function of membrane proteins* (ed. D.C. Youvan and F. Daldal), p. 169. Cold Spring Harbor Laboratory, Cold Spring Harbor, New York.

Parkinson, J.S. and G.L. Hazelbauer. 1983. Bacterial chemotaxis: Molecular genetics of sensory transduction and chemotactic gene expression. In *Gene function in prokaryotes* (ed. J. Beckwith et al.), p. 293. Cold Spring Harbor Laboratory, Cold Spring Harbor, New York.

Parkinson, J.S., S.R. Parker, P.B. Talbert, and S.E. Houts. 1983. Interactions between chemotaxis genes and flagellar genes in *Escherichia coli. J. Bacteriol.* **155:** 265.

Ravid, S., P. Matsumura, and M. Eisenbach. 1986. Restoration of flagellar clockwise rotation in bacterial envelopes by insertion of the chemotaxis protein CheY. *Proc. Natl. Acad. Sci.* **83:** 7157.

Repaske, D.R. and J. Adler. 1981. Change in intracellular pH of *Escherichia coli* mediates the chemotactic response to certain attractants and repellents. *J. Bacteriol.* **145:** 1196.

Ridgway, H.F., M. Silverman, and M.I. Simon. 1977. Localization of proteins controlling motility and chemotaxis in *Escherichia coli. J. Bacteriol.* **132:** 657.

Shimada, K. and H.C. Berg. 1987. Response of the flagellar rotary motor to abrupt changes in extracellular pH. *J. Mol. Biol.* **193:** 585.

Silverman, M. and M. Simon. 1974. Flagellar rotation and the mechanism of bacterial motility. *Nature* **249:** 73.

Simon, M., J. Zieg, M. Silverman, G. Mandel, and R. Doolittle. 1980. Phase variation: Evolution of a controlling element. *Science* **209:** 1370.

Smith, R.A. and J.S. Parkinson. 1980. Overlapping genes at the *cheA* locus of *Escherichia coli. Proc. Natl. Acad. Sci.* **77:** 5370.

Spudich, J.L. and R.A. Bogomolni. 1988. Sensory rhodopsins of halobacteria. *Annu. Rev. Biophys. Biophys. Chem.* **17:** 193.

Stader, J., P. Matsumura, D. Vacante, G.E. Dean, and R.M. Macnab. 1986. Nucleotide sequence of the *Escherichia coli motB* gene and site-limited incorporation of its product into the cytoplasmic membrane. *J. Bacteriol.* **166:** 244.

Stewart, R.C. and F.W. Dahlquist. 1987. Molecular components of bacterial chemotaxis. *Chem. Rev.* **87:** 997.

Trachtenberg, S. and D.J. DeRosier. 1987. Three-dimensional structure of the frozen-hydrated flagellar filament. The left-handed filament of *Salmonella typhimurium*. *J. Mol. Biol.* **195:** 581.

Wei, L.-N. and T.M Joys. 1985. Covalent structure of three phase-1 flagellar filament proteins of *Salmonella*. *J. Mol. Biol.* **186:** 791.

Wilson, M.L. and R.M. Macnab. 1988. Overproduction of the MotA protein of *Escherichia coli* and estimation of its wild-type level. *J. Bacteriol.* **170:** 588.

Wolfe, A.J., M.P. Conley, T.J. Kramer, and H.C. Berg. 1987. Reconstitution of signaling in bacterial chemotaxis. *J. Bacteriol.* **169:** 1878.

Yamaguchi, S., H. Fujita, A. Ishihara, S.-I. Aizawa, and R.M. Macnab. 1986a. Subdivision of flagellar genes of *Salmonella typhimurium* into regions responsible for assembly, rotation, and switching. *J. Bacteriol.* **166:** 187.

Yamaguchi, S., S.-I. Aizawa, M. Kihara, M. Isomura, C.J. Jones, and R.M. Macnab. 1986b. Genetic evidence for a switching and energy-transducing complex in the flagellar motor of *Salmonella typhimurium*. *J. Bacteriol.* **168:** 1172.

Yanofsky, C. and I.P. Crawford. 1987. The tryptophan operon. In Escherichia coli *and* Salmonella typhimurium*: Cellular and molecular biology* (ed. F.C. Neidhart et al.), vol. 2, p. 1453. American Society for Microbiology Publications, Washington, D.C.

Casein Kinase II as a Potentially Important Enzyme Concerned with Signal Transduction

E.G. KREBS,† R.N. EISENMAN,‡ E.A. KUENZEL,‡* D.W. LITCHFIELD,†
F.J. LOZEMAN,† B. LÜSCHER,‡ AND J. SOMMERCORN†

†Howard Hughes Medical Institute and Department of Pharmacology, University of Washington, Seattle, Washington 98195; ‡Division of Basic Sciences, Fred Hutchinson Cancer Center, Seattle, Washington 98104

Several of the effects of insulin and other growth factors whose receptors are protein tyrosine kinases are made manifest as a result of changes in the state of protein serine (threonine) phosphorylation involving certain cellular proteins (Denton 1986). It is even possible that *all* of the effects of such agents might be mediated in this manner (Krebs 1985). Such a scheme would provide an attractive mechanism that could explain the pleotypic action of these growth factors inasmuch as the protein serine (threonine) phosphorylation-dephosphorylation network is complex, involving a vast array of kinases and phosphatases that impinge on potential protein substrates in all parts of the cell (Edelman et al. 1987). Protein tyrosine phosphorylation is much less extensive than protein serine (threonine) phosphorylation (Hunter and Cooper 1981), but, as a result of coupling between the two systems, signals that give rise to changes in tyrosine phosphorylation could be amplified and widely disseminated within the cell.

Two general approaches are being taken to establish the nature of the linkages that exist between protein tyrosine phosphorylation and protein serine (threonine) phosphorylation. One of these involves "working forward" from the growth factor receptor-kinases themselves in an attempt to find substrates that are phosphorylated on tyrosine residues and that might be involved in linking the two systems. Such substrates could include enzymes involved in giving rise to second messengers, which might interact with protein serine (threonine) kinases or phosphatases, or the substrates might be the kinases (Yu et al. 1987) or phosphatases themselves. The second overall approach for determining the mechanism for the coupling of protein tyrosine and serine (threonine) phosphorylation is to "work back" from a particular growth-factor-stimulated protein serine (threonine) phosphorylation event in an effort to establish the linkage. The latter approach is the one being stressed in this laboratory.

A number of specific proteins have been identified that are either phosphorylated or dephosphorylated in reactions involving serine or threonine residues as a result of exposure of cells to insulin (Denton 1986). Among the proteins exhibiting increased phosphorylation, the following can be cited: ribosomal protein S6, acetyl-CoA carboxylase, ATP citrate lyase, and the insulin receptor itself. Proteins that appear to be dephosphorylated in response to insulin include glycogen synthase, pyruvate dehydrogenase, and HMG-CoA reductase. In general, all of these changes in protein serine (threonine) phosphorylation are rapid, occurring within minutes after treatment of cells with insulin. In some instances, information is available with respect to the particular protein serine (threonine) kinase or phosphatase that appears to be the first target of the cascade initiated as a result of activation of the insulin receptor, but relatively little is known concerning the mechanisms involved. It is known, however, that new protein synthesis is not required for these changes.

At least two of the proteins that undergo changes in serine (threonine) phosphorylation as a result of insulin action are substrates for the protein serine kinase known as casein kinase II (CK-II). This enzyme phosphorylates skeletal muscle glycogen synthase on what is referred to as site 5 (Cohen et al. 1982). Although phosphorylation of site 5 on the synthase by CK-II does not in itself affect the activity of the synthase, it is necessary for the subsequent phosphorylation of the enzyme at site(s) 3 by a different kinase, glycogen synthase kinase 3 (GSK-3), which causes inactivation of the enzyme (Picton et al. 1982; DePaoli-Roach et al. 1983; Woodgett and Cohen 1984). The molecular mechanism for the synergistic action of CK-II and GSK-3 has been established recently (Fiol et al. 1987). There is evidence that insulin regulates the activity of glycogen synthase through its effect on the state of phosphorylation of site(s) 3 (Parker et al. 1983). The other CK-II substrate whose serine (threonine) phosphorylation state is altered by insulin is acetyl-CoA carboxylase. Phosphorylation by CK-II does not alter the activity of the carboxylase (Witters et al. 1983), but this phosphorylation appears to facilitate dephosphorylation of a cAMP-dependent phosphorylation site by protein phosphatase-1, or phosphatase-2a (Sommercorn et al. 1987a). The latter reaction is one that leads to activation of the enzyme. It is significant that CK-II phosphorylates acetyl-CoA carboxylase at what is referred to as the insulin-sensitive site, as shown in experiments utilizing intact adipocytes (Witters et al. 1983; Brownsey and Denton 1985; Haystead and Hardie 1988).

The probable role of CK-II as a target for insulin in the regulation of glycogen synthase and acetyl-CoA

*Deceased, January 20, 1988.

carboxylase, as noted above, together with its potential role in the regulation of nuclear function, prompted initiation of an in-depth study of this particular protein kinase in our laboratories.

Review of the Properties of CK-II

CK-II is a ubiquitous messenger-independent protein serine (threonine) kinase whose properties have been described in several review articles (Hathaway and Traugh 1982; Pinna et al. 1985; Edelman et al. 1987). Briefly, it derives its name from the fact that it elutes after another casein kinase, casein kinase I, from DEAE-cellulose. Casein is not believed to be a physiological substrate for CK-II, inasmuch as a different casein kinase is responsible for the phosphorylation of casein in the Golgi apparatus of mammary gland cells. CK-II has an $\alpha_2\beta_2$ structure. The molecular weight of the α or catalytic subunit is reported as being in the range of 37,000–44,000 and that of the β subunit 24,000–28,000. No function has been assigned to the β subunit. The α subunit contains the catalytic domain, which is homologous to the catalytic domains of other protein kinases (Saxena et al. 1987), but it exhibits a special relationship to yeast *CDC28* in the carboxy-terminal regions (Saxena et al. 1987; Takio et al. 1987). The sequence of the β subunit, determined directly for the enzyme from beef lung (Takio et al. 1987) and derived from its cDNA for the *Drosophila* enzyme, is unique and shows no homology with other proteins in the data base. At least two forms of the α subunit, α and α', have been detected. CK-II is found primarily in the cytosol, but a significant amount of the enzyme is also present in the nucleus. Heparin is a potent inhibitor of CK-II, and the enzyme can be activated by polyamines; the physiological significance of the activation by polyamines has not been determined. CK-II utilizes either ATP or GTP as a nucleotide substrate, a property that is not shared by most of the protein kinases. The peptide substrate specificity of the enzyme will be discussed in a later section.

Regulation of CK-II Levels

Although the existence of CK-II has been known for a number of years, relatively little work has been done on its regulation. With respect to control of its cellular levels under different conditions, it has been reported that CK-II activity is significantly enhanced during early embryogenesis and in fetal liver (Schneider et al. 1986; Perez et al. 1987). Elevated levels were also seen in tumor cell cultures (Prowald et al. 1984; Schneider et al. 1986). The administration of dexamethasone to rats was found to increase the levels of liver nucleolar CK-II, an effect that was blocked by cycloheximide (Suzuki et al. 1987). In our own laboratory, we have studied the induction of CK-II during the differentiation of 3T3-L1 cells to fat cells (Sommercorn and Krebs 1987) and have also examined the tissue distribution of the enzyme.

To measure the levels of CK-II in crude tissue extracts, a synthetic peptide, Arg-Arg-Arg-Glu-Glu-Glu-Thr-Glu-Glu-Glu, was used as a substrate (Kuenzel and Krebs 1985; Sommercorn and Krebs 1987). When 3T3-L1 cells were stimulated to undergo differentiation into adipocytes, there was an increase in CK-II between days 2 and 4, which was followed by a decrease between days 4 and 7 (Fig. 1). Between day 0 and day 4, CK-II activity and insulin binding showed parallel changes. It will be noted, however, that the insulin receptor levels remained elevated after CK-II levels were decreasing. Extracts of differentiated and undifferentiated 3T3-L1 cells were subjected to Western analysis, and there was close agreement between the amount of immunoreactive CK-II and kinase activity estimated by peptide phosphorylation. This showed that the activity measurements served as a true index of the amount of CK-II protein that was present. Overall, these experiments supported the concept that CK-II induction contributes to the process of differentiation of the 3T3-L1 cells into adipocytes, rather than being a phenotypic change similar to what occurs with the insulin receptor.

A survey was made of the concentration of CK-II in extracts made from various tissues of adult male rats. The results are shown in Figure 2. The highest concentrations were found in testis and brain and the lowest concentration in skeletal muscle. Although great care was taken to determine that the assays would provide valid estimates of the amount of CK-II in the various extracts, there is always the possibility that interfering substances in the crude fraction could give misleading results. In addition, it should be noted that in the experiment reported here, no special precautions were used to prevent the possible lysis of nuclei. Thus, the levels shown in Figure 1 might reflect a mixture of cytosolic and nuclear enzyme. Singh and Huang (1985)

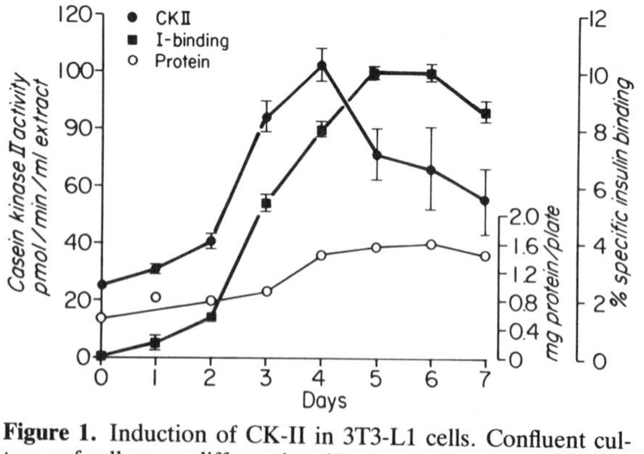

Figure 1. Induction of CK-II in 3T3-L1 cells. Confluent cultures of cells were differentiated by incubating them with 0.25 μM dexamethasone and 0.5 mM IBMX starting on day 0. On day 3, the medium was changed to fresh medium without dexamethasone and IBMX. CK-II activity was measured using 2 mM Arg-Arg-Arg-Glu-Glu-Glu-Thr-Glu-Glu-Glu and 100 mM [γ-^{32}P]ATP as substrates. Insulin binding and protein concentration were measured by conventional procedures. (Reprinted, with permission, from Sommercorn and Krebs 1987.)

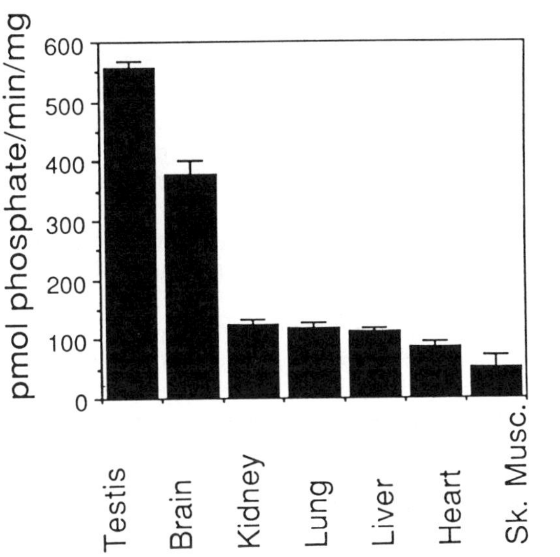

Figure 2. Distribution of CK-II in rat tissues. The tissues of adult male rats were immediately chilled in ice after removal and homogenized in 3 volumes of 100 mM Tris (pH 7.4), 0.3 M sucrose, 1 mM dithiothreitol, and 2 mM EDTA. The homogenates were centrifuged at 100,000*g*, and the supernates were assayed for kinase activity toward 2 mM Arg-Arg-Arg-Glu-Glu-Glu-Thr-Glu-Glu-Glu in the presence of 150 mM [γ-^{32}P] ATP, 10 mM $MgCl_2$, and 150 mM NaCl.

carried out a survey of the distribution of CK-II in the soluble fraction from rat tissue homogenate and found a somewhat different profile from that shown in Figure 2. The method that these workers used was different from that employed here, however, in that they carried out an initial fractionation step on the crude extracts before carrying out their assays. Thus, it is possible that variable recovery of enzyme from different tissues would have influenced their results. These workers also found very high levels of CK-II in the testis, second only to spleen, which was not analyzed in the present study. Another study on the tissue distribution of CK-II (Nakajo et al. 1986) also indicated that the level of the enzyme is high in the testis.

Short-term or Acute Regulation of CK-II

CK-II is commonly viewed as belonging to that large set of protein serine (threonine) kinases for which no second messenger has been defined, i.e., the so-called "independent" kinases. The activation of CK-II by polyamines (see above) has not generally been embraced as being of physiological significance, although this possibility should probably be left open. Our own interest in the idea that CK-II might be subject to short-term regulation was stimulated by the report of Witters et al. (1983), which showed that in vitro CK-II catalyzed the phosphorylation of acetyl-CoA carboxylase either at the same site or at one that is very close to the insulin-dependent phosphorylation site that had been identified in experiments with intact cells. Accordingly, we undertook a study to determine whether insulin and other growth factors could stimulate this kinase.

In experiments (Sommercorn et al. 1987b) utilizing fully differentiated 3T3-L1 cells, it was found that insulin treatment resulted in the rapid activation of CK-II, as measured in cell extracts utilizing the specific synthetic peptide substrate (see above) as a probe (Fig. 3). The extent of the response varied between 30% and 150% in different experiments and generally peaked within 10 minutes. Activation was very sensitive to insulin with a half-maximal response occurring at about 0.15 nM (Fig. 4). Preliminary kinetic measurements carried out on the crude enzyme, as present in the cell extracts, showed that activation of the enzyme was due to an increase in V_{max} and not a change in the K_m for the peptide substrate. Activation of CK-II in 3T3-L1 cells was not blocked by cycloheximide, indicating that protein synthesis is not required for the effect. When extracts from insulin-stimulated and nonstimulated cells were subjected to immunoblot analysis, it was found that the apparent concentration of CK-II was the same in both instances. This indicated that it was unlikely that translocation of the kinase from a bound form to a soluble form could explain the nature of the insulin effect. Finally, extracts from insulin-stimulated cells still exhibited higher CK-II activity when subjected to gel-filtration chromatography on Sephadex G-25, making it unlikely that the presence of a low-molecular-weight activator could account for the stimulation that is seen. Although the mechanism of activation of CK-II by insulin remains unknown, the data obtained thus far would be consistent with covalent modification of the enzyme.

Activation of CK-II by insulin was also found in

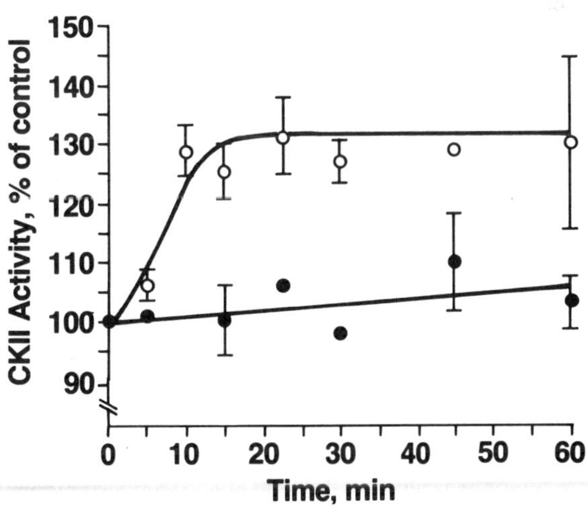

Figure 3. Activation of CK-II in 3T3-L1 adipocytes. 3T3-L1 mouse cells were fully differentiated (7–8 days) to adipocytes as described in Fig. 1. At various times, extracts were prepared from control (●) and insulin-treated (○) cells and assayed for CK-II activity with synthetic peptide substrate. The figure represents three separate experiments. Error bars represent S.E.M., $n = 3$–6; data points without error bars are the means of two values. (Reprinted, with permission, from Sommercorn et al. 1987b.)

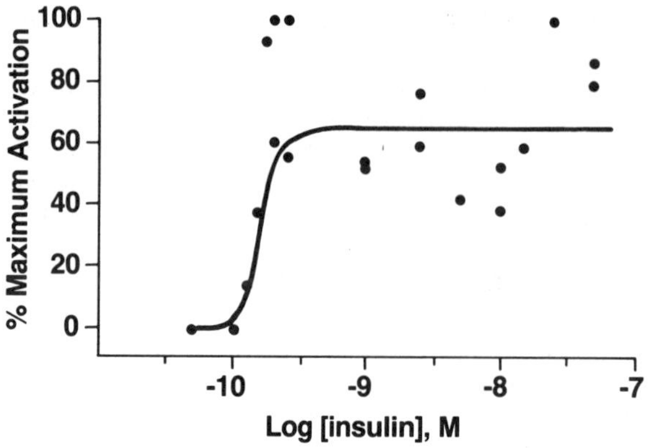

Figure 4. Dose response for the activation of CK-II by insulin in 3T3-L1 adipocytes. 3T3-L1 cells were incubated for 30 min in the presence of fresh control medium or medium supplemented with various concentrations of insulin. CK-II activities in cell extracts (see Figs. 1 and 3) were assayed, and the data within each experiment were standardized as described previously (Sommercorn et al. 1987b). (Reprinted, with permission, from Sommercorn et al. 1987b.)

H4-IIE cells (Sommercorn et al. 1987b). In addition, CK-II activation by epidermal growth factor was observed in A431 cells. Klarlund and Czech (1988) have reported independently that insulin and insulin-like growth factor 1 rapidly increase a CK-II-like activity in BALB/c-3T3 fibroblasts. Collectively, these results suggest that activation of CK-II may be a general response to growth factors. It should be emphasized, however, that in some cell types and/or animal tissues, we have failed to observe CK-II responses under conditions in which insulin or other growth factors do cause stimulation of ribosomal protein S6 kinase.

Specificity of CK-II

CK-II has been reported to catalyze the phosphorylation of a great many protein substrates, the number of which is ever-increasing. In a review written several years ago, as many as 16 different proteins that could be phosphorylated by this enzyme were listed (Pinna et al. 1985). A number of these proteins had been shown to serve as substrates in the intact cell as well as in vitro. Since that time, a number of new CK-II substrates have been described, including DNA topoisomerase II (Ackerman et al. 1985), nucleolar protein B23 (Chan et al. 1986; Caizergues-Ferrer et al. 1987), and calmodulin (Meggio et al. 1987). It is of interest that in keeping with the presence of CK-II in the nucleus, as well as in the cytoplasm, many of its substrates that have been identified are nuclear in origin.

It has been known for a number of years that CK-II phosphorylates serine (threonine) residues located in highly acidic regions of proteins and that glutamate or aspartate residues carboxy-terminal to the serine (threonine) are especially important (for review, see Hathaway and Traugh 1982). Examples of CK-II phosphorylation site sequences taken from proteins known to be phosphorylated by CK-II in the intact cell are shown in Table 1. Studies with synthetic peptide substrates have provided additional detail on the specificity of the enzyme (Marin et al. 1986; Kuenzel et al. 1987). It has been determined that, in general, serine residues are phosphorylated considerably more rapidly than threonine residues. Aspartic acid residues are somewhat better specificity determinants than are glutamic acid residues, although the latter are usually found in naturally occurring CK-II phosphorylation sites in proteins (Table 1). An acidic residue in the third position toward the carboxyl terminus is critical as a substrate determinant. The exact number of glutamates or aspartates carboxy-terminal to the serine (threonine) that are optimal for the lowest K_m and highest V_{max} values has not been determined, but synthetic peptide work and consideration of natural phosphorylation site sequences in protein substrates suggest that five or more of these residues may be efficacious. Finally, acidic residues amino-terminal to the phosphorylatable serine (threonine) are beneficial but not required.

Computer Search for Potential Substrates of CK-II

It was of interest to utilize available information on the molecular basis for CK-II peptide substrate specificity to identify new potential substrates for this enzyme. To this end, the National Biomedical Research Foundation (1987) Protein Sequences Database was searched for proteins that would contain Ser-Glu-Glu-Glu-Glu-Glu or Ser-Asp-Asp-Asp-Asp-Asp with

Table 1. Casein Kinase II Specificity

Substrate	Sequence	Reference
Type II regulatory subunit of cAMP-dependent protein kinase (two sites)	-Asp-Ala-Val-Ala-Asp-<u>Ser</u>-Glu-<u>Ser</u>-Glu-Asp-Glu-Glu-Asp-	Carmichael et al. (1982); Hemmings et al. (1982)
Inhibitor-2 of ATP-Mg dependent phosphoprotein phosphatase (three sites)	-Ile-Arg-Glu-Gln-Glu-<u>Ser</u>-<u>Ser</u>-Gly-Glu-Glu-Asp-Ser-Asp- -Asp-Asp-Asp-Ala-Tyr-<u>Ser</u>-Asp-Thr-Glu-Thr-Thr-Glu-Ala-	Holmes et al. (1986)
Glycogen synthase	-Ser-Ser-Pro-His-Gln-<u>Ser</u>-Glu-Asp-Glu-Glu-Glu-Pro-	Cohen et al. (1982)
Nucleolar protein B23	-Glu-Glu-Asp-Ala-Glu-<u>Ser</u>-Glu-Asp-Glu-Asp-Glu-Glu-Asp	Chan et al. (1986)
High mobility group protein 14	-Glu-Glu-Ser-Pro-Ala-<u>Ser</u>-Asp-Glu-Ala-Glu-Glu-Lys	Walton et al. (1985)

Sequence surrounding sites in proteins for which there is evidence for phosphorylation by casein kinase II in the intact cell. The residue(s) shown to be phosphorylated by casein kinase II are underlined.

up to two mismatches. Approximately 250 proteins were identified by this approach, but many of these could be dismissed because of mismatches at the serine site. The proteins that remained included many of the known substrates for CK-II, e.g., type II regulatory subunit of the cAMP-dependent protein kinase, troponin T, α-S2 casein, calmodulin, and ornithine decarboxylase, and, in addition, a number of others not previously implicated as possible substrates for CK-II. Noteworthy among the latter were most of the known nuclear proteins that are encoded by oncogenes; namely, Myc protein, Myb protein, Fos protein, large T antigen, ErbA protein, N-Myc protein, and EA1-transforming protein (E.A. Kuenzel et al., in prep.). This search also revealed a number of the same proteins that contain high "PEST-scores" and are known to be degraded rapidly in cells (Rogers et al. 1986).

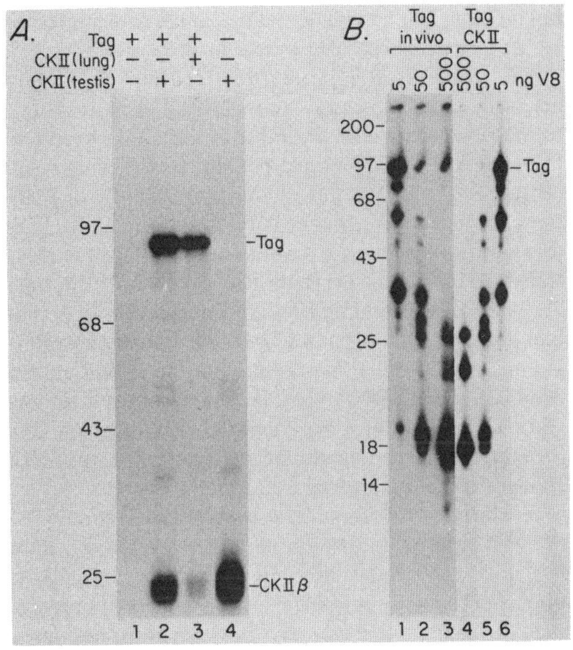

Figure 5. Phosphorylation of SV40 large T antigen by CK-II. (*A*) Purified large T antigen (500 ng/reaction) was incubated in 40 mM Tris-HCl (pH 7.6), 10 mM $MgCl_2$, 100 mM NaCl, and [γ-^{32}P]ATP as described previously (E.A. Kuenzel et al., in prep.) in the absence of any enzyme (lane *1*), in the presence of 50 ng CK-II from bovine testis (lane *2*), and in the presence of 110 ng CK-II from bovine lung (lane *3*). In lane *4*, 50 ng CK-II from bovine testis was incubated alone. (*B*) Large T antigen isolated from [^{32}P]orthophosphate-labeled CDS cells (lanes *1–3*) or phosphorylated in vitro by CK-II (lanes *4–6*) were compared by one-dimensional V8-protease mapping. Samples were analyzed on a 20% SDS-polyacrylamide gel.

Phosphorylation of Nuclear Oncoproteins by CK-II

The finding that a number of oncogene-encoded nuclear proteins appear to satisfy the substrate specificity requirements for CK-II (see above), coupled with the fact that most, if not all, of these proteins are known to be phosphoproteins, made it appear possible that they would serve as CK-II substrates. Accordingly, we examined several of them to see whether or not purified CK-II would phosphorylate them in vitro. Immune complexes containing c-*myc*, v-*myc*, or c-*myb* proteins as well as purified SV40 large T antigen (Tag) all served as substrates for the enzyme.

The SV40 large T antigen is heavily phosphorylated, and extensive phosphopeptide mapping has localized phosphorylated residues to two clusters of serine and threonine residues, one toward the amino terminus and one near the carboxyl terminus (Scheidtmann et al. 1982). Tag has been described to possess a number of biochemical functions (for review, see Butel 1986), some of which are regulated, at least in part, by phosphorylation. Removal of phosphate from Tag is accompanied by an increased ability to support DNA replication in vitro, whereas the ATPase activity remained unchanged (Graesser et al. 1987; Mohr et al. 1987). In addition, the interaction of Tag with DNA may be regulated, at least in part, by differential phosphorylations (Scheidtmann et al. 1984; Mohr et al 1987). We determined that purified Tag (obtained from J. Roberts, FHCRC) served as a substrate for two different preparations of pure bovine CK-II (Fig. 5A). Tag incubated on its own had no autophosphorylation capacity under the experimental conditions used (Fig. 5A). To determine if the regions phosphorylated in Tag by CK-II were phosphorylated in vivo, one-dimensional V8-protease phosphopeptide maps from Tag labeled in vivo with [^{32}P]orthophosphate and in vitro by CK-II with [γ-^{32}P]ATP were compared. As shown in Figure 5B, these maps were identical. There are two CK-II consensus sites within the Tag protein sequence, one within the amino-terminal cluster of phosphorylation (amino acids 111–115) and one within the carboxy-terminal cluster of phosphorylation (amino acids 632–638). From the lack of any differences in the phosphopeptide maps, we would predict that both these sites are phosphorylated in vitro and probably in vivo. Therefore, CK-II may play a role in the regulation of one or several of the activities attributed to Tag.

In comparison to Tag, very little is known about the function of both c-*myc* and c-*myb* oncoproteins. Both are implicated to play a role in growth control (Eisenman and Thompson 1986). Immune complexes containing c-*myc* and v-*myc* proteins from various avian and human cell lines (Hann and Eisenman 1984; Lüscher and Eisenman 1988) were incubated, as described for Tag, with CK-II. All the various forms of *myc* were found to be phosphorylated (E.A. Kuenzel et al., in prep.). Comparative phosphopeptide maps revealed that CK-II in vitro phosphorylated a subset of peptides found to be labeled in vivo by [^{32}P]orthophosphate (E.A. Kuenzel et al., in prep.). Thus, it is probable that one of the protein kinases that is active toward *myc* protein within the cell is, indeed, CK-II. Two peptides containing the consensus sequences for CK-II in *myc* were synthesized. Both peptides (amino acids 222–238 and 246–263) proved to be excellent substrates for pure bovine lung CK-II with K_m values in the low micromolar range (E.A. Kuenzel et al., in prep.).

The region of *myc* protein that contains the amino acid sequences present in the two peptides corresponds to the region that contains most of the total phosphate found in the MC29 v-*myc* protein in the intact cell (Bister et al. 1987). This region is apparently of functional significance, inasmuch as MC29 mutants lacking this major phosphorylation domain are altered in their transforming potential (Ramsay et al. 1980; Enrietta et al. 1983; Bister et al. 1987). No information is available, however, as to what might be the role of phosphorylating this region. It is possible that the ability of *myc* protein to bind DNA may be altered by phosphorylation. In addition, the rate of breakdown of *myc* protein in the cell could be affected as a result of changes in its state of phosphorylation.

Similarly, by using immune complexes containing c-*myb* (Lüscher and Eisenman 1988), we have established its phosphorylation by CK-II. Again, CK-II phosphorylated a subset of peptides when compared to peptides obtained from *myb* protein phosphorylated within the intact cell (B. Lüscher et al., in prep.). Preliminary data suggest that the *myb* CK-II site is conserved in evolution from man to *Drosophila* (B. Lüscher, unpubl.). Therefore, this site may be of importance in regulating *myb* protein function.

CONCLUSIONS

Although CK-II has been studied for a number of years, its physiological functions are not well understood (Hathaway and Traugh 1982; Pinna et al. 1985; Edelman et al 1987). Lately, however, a number of developments, some of which are reviewed in this paper, have stimulated a renewed interest in this particular protein kinase. Additional examples of enzymes whose activities are affected directly or indirectly as a result of phosphorylation have been found. The enzymes reported to be affected directly now include hepatoma RNA polymerases I and II (Rose et al. 1983), DNA topoisomerase II (Ackerman et al. 1985), and DNA topoisomerase I (Durban et al. 1985; Kaiserman et al. 1988). Thus, it may no longer be correct to think that CK-II is a protein kinase that catalyzes only "silent phosphorylations" that have no functional effects on its substrates. The indirect actions of CK-II on enzyme activity that have been reported are perhaps even more exciting than the direct effects, in that they open a new approach for studying protein kinases.

As indicated earlier, CK-II is now known to work synergistically with the glycogen synthase inactivating (phosphorylating) enzyme known as GSK-3. In this reaction, prior phosphorylation of the enzyme by CK-II, which by itself does nothing to change the activity, is necessary for subsequent phosphorylation by GSK-3, which does alter function. Three examples of the occurrence of this pattern of regulation are known. The substrates involved include (1) glycogen synthase (discussed earlier), (2) the regulatory subunit (inhibitor 2) of the form of protein phosphatase-1 that is activated by ATP (DePaoli-Roach 1984), and (3) type II regulatory subunit of the cAMP-dependent protein kinase (Hemmings et al. 1982). No regulatory significance with respect to the last example is known; however, the possible physiological significance of the synergistic action of CK-II and GSK-3 in protein phosphatase-1 activation has received considerable attention. The finding that the phosphatase regulatory subunit (inhibitor 2) is phosphorylated in fat cells in response to insulin (Lawrence et al. 1988) lends credence to the concept that this may be an important regulatory mechanism. Furthermore, it has been shown very recently (Chan et al. 1988) that protein phosphatase-1 undergoes activation in cells stimulated by growth factors. The mechanism of activation has not been determined but is consistent with covalent modification. Thus, it is possible that a cascade system exists in which CK-II is activated in response to growth factors (Sommercorn et al. 1987b), and activated CK-II then phosphorylates the regulatory subunit (inhibitor 2) of the protein phosphatase. This in turn leads to phosphorylation of the phosphatase by GSK-3 and activation of the enzyme. Such a scenario could explain the activation of glycogen synthase by growth factors.

Interest in CK-II as a potentially vital regulatory enzyme is also heightened by the facts that it is located in the nucleus and that many of its substrates are nuclear in origin and known to be involved in control of the cell cycle and cell replication. The short-term regulation of the enzyme by growth factors points to a possible role of CK-II as an intermediate step in the transduction of signals from growth factor receptors to effector proteins within the cell nucleus. However, the regulation of proteins such as *myc* and *myb* may be more complex, since both proteins seem to be phosphorylated by at least one additional kinase (E.A. Kuenzel et al., B. Lüscher et al.; both in prep.). As described above, CK-II may work synergistically by rendering its substrates (such as *myc* and *myb*) more susceptible to additional phosphorylation, allowing fine regulation.

CK-II phosphorylates sites within acidic regions. These may be rather extended, as in the case of chicken c-*myc*, which spans about 40 amino acids. Such regions, also called "acid blobs," are believed to play a role in transcriptional activation (for review, see Sigler 1988). Therefore, CK-II phosphorylation, introducing additional negative charges into an acidic region, may regulate the interactions of such proteins on the transcription machinery. Clearly, more work will be required to establish functional evidence for CK-II in the cell nucleus. However, its unique features described above make CK-II a candidate in controlling aspects of gene transcription.

ACKNOWLEDGMENTS

We thank Dr. Patrick Chou, Mr. Henry Zebroski, and Ms. Maria Harrylock for the preparation and purification of synthetic peptides, and Mrs. Christina Boyd for typing the manuscript.

REFERENCES

Ackerman, P., C.V.C. Glover, and N. Osheroff. 1985. Phosphorylation of DNA topoisomerase II by casein kinase II: Modulation of eukaryotic topoisomerase II activity *in vitro*. *Proc. Natl. Acad. Sci.* **82:** 3164.

Bister, K., C. Trachmann, H.W. Hansen, B. Schroler, and T. Patschinsky. 1987. Structure of mutant and wild-type MC 29 v-*myc* alleles and biochemical properties of their protein products. *Oncogene* **1:** 97.

Brownsey, R.W. and R.M. Denton. 1985. Evidence that insulin activates acetyl-CoA carboxylase by increased phosphorylation of a specific site. *Biochem J.* **202:** 77.

Butel, J.S. 1986. SV40 large T antigen: Dual oncogene. *Cancer Surv.* **5:** 343.

Caizergues-Ferrer, M., P. Belenguer, B. Lapeyre, F. Amalric, M.O. Wallace, and M.O.J. Olson. 1987. Phosphorylation of nucleolin by a nucleolar type NII protein kinase. *Biochemistry* **26:** 7876.

Carmichael, D.F., R.L. Geahlen, S.M. Allen, and E.G. Krebs. 1982. Type II regulatory subunit of cAMP-dependent protein kinase. *J. Biol. Chem.* **257:** 10440.

Chan, C.P., S.J. McNall, E.G. Krebs, and E.H. Fischer. 1988. Stimulation of protein phosphatase activity by insulin and growth factors in 3T3 cells. *Proc. Natl. Acad. Sci.* **85:** 6257.

Chan, P.-K., M. Aldrich, R.G. Cook, and H. Busch. 1986. Amino acid sequence of protein B23 phosphorylation site. *J. Biol. Chem.* **261:** 1868.

Cohen, P., D. Yellowlees, A. Aitken, A. Donella-Deana, B.A. Hemmings, and P.J. Parker. 1982. Separation and characterisation of glycogen synthase kinase 3, glycogen synthase kinase 4 and glycogen synthase kinase 5 from rabbit skeletal muscle. *Eur. J. Biochem.* **124:** 21.

Denton, R.M. 1986. Early events in insulin action. *Adv. cyclic nucleotide protein phosphorylation res.* **20:** 293.

DePaoli-Roach, A.A. 1984. Synergistic phosphorylation and activation of ATP-Mg-dependent phosphoprotein phosphatase by F_A/GSK-3 and casein kinase II. *J. Biol. Chem.* **259:** 12144.

DePaoli-Roach, A.A., Z. Ahmad, M.Carnici, J.C. Lawrence, Jr., and P.J. Roach. 1983. Multiple phosphorylation of rabbit skeletal muscle glycogen synthase. *J. Biol. Chem.* **258:** 10702.

Durban, E., M. Goodenough, J. Mills, and H. Busch. 1985. Topoisomerase I phosphorylation *in vitro* and in rapidly growing Novikoff hepatoma cells. *EMBO J.* **4:** 2921.

Edelman, A.M., D.A. Blumenthal, and E.G. Krebs. 1987. Protein serine and threonine kinases. *Annu. Rev. Biochem.* **56:** 567.

Eisenman, R.N. and C.B. Thompson. 1986. Oncogenes with potential nuclear function: *myc*, *myb* and *fos*. *Cancer Surv.* **5:** 309.

Enrietto, P.J., M.J. Hayman, G.M. Ramsay, J.A. Wyke, and L.N. Payne. 1983. Altered pathogenicity of avian myelocytomatosis (MC29) viruses with mutations in the v-*myc* gene. *Virology* **124:** 164.

Fiol, C.J., A.M. Mahrenholz, Y. Wang, R.W. Roeske, and P.J. Roach. 1987. Formation of protein kinase recognition sites by covalent modification of the substrate. *J. Biol. Chem.* **262:** 14042.

Graesser, F.A., K. Mann, and G. Walter. 1987. Removal of serine phosphates from simian virus 40 large T antigen increases its ability to stimulate DNA replication *in vitro* but has no effect on ATPase and DNA binding. *J. Virol.* **61:** 3373.

Hann, S.R. and R.N. Eisenman. 1984. Proteins encoded by the human c-*myc* oncogene: Differential expression in neoplastic cells. *Mol. Cell. Biol.* **4:** 2486.

Hathaway, G.M. and J.A. Traugh. 1982. Casein kinases—multipotential protein kinases. *Curr. Top. Cell. Regul.* **21:** 101.

Haystead, T.A.J. and D.G. Hardie. 1988. Evidence for casein kinase-2 mediated phosphorylation of acetyl-CoA carboxylase in insulin-treated adipocytes. *FASEB J.* **2:** A4054.

Hemmings, B.A., A. Aitken, P. Cohen, M. Rymond, and F. Hofmann. 1982. Phosphorylation of the type-II regulatory subunit of cyclic-AMP-dependent protein kinase by glycogen synthase kinase 3 and glycogen synthase kinase 5. *Eur. J. Biochem.* **127:** 473.

Holmes, C.F.B., J. Kuret, A.A.K. Chisholm, and P. Cohen. 1986. Identification of the sites on rabbit skeletal muscle protein phosphatase inhibitor-2 phosphorylated by casein kinase-II. *Biochim. Biophys. Acta* **870:** 408.

Hunter, T. and J.A. Cooper. 1981. Epidermal growth factor induces rapid tyrosine phosphorylation of proteins in A431 human tumor cells. *Cell.* **24:** 741.

Kaiserman, H.B., T.S. Ingebritsen, and R.M. Benbow. 1988. Regulation of *Xenopus laevis* DNA topoisomerase I activity by phosphorylation *in vitro*. *Biochemistry* **27:** 3216.

Klarlund, J.K. and M.P. Czech. 1988. Insulin and insulin-like growth factor I rapidly increase casein kinase II activity in BALB/c 3T3 fibroblasts. *J. Biol. Chem.* **263:** (in press).

Krebs, E.G. 1985. The phosphorylation of proteins: A major mechanism for biological regulations (Fourteenth Sir Frederick Gowland Hopkins Memorial Lecture). *Biochem. Soc. Trans.* **13(5):** 813.

Kuenzel, E.A. and E.G. Krebs. 1985. A synthetic peptide substrate specific for casein kinase II. *Proc. Natl. Acad. Sci.* **82:** 737.

Keunzel, E.A., J.A. Mulligan, J. Sommercorn, and E.G. Krebs. 1987. Substrate specificity determinants for casein kinase II as deduced from studies with synthetic peptides. *J. Biol. Chem.* **262:** 9136.

Lawrence, Jr., J.C., J. Hiken, B. Burnette, and A.A. De-Paoli-Roach. 1988. Phosphorylation of phosphoprotein phosphatase inhibitor-2 (I-2) in rat fat cells. *Biochem. Biophys. Res. Commun.* **150:** 197.

Lüscher, B. and R.N. Eisenman. 1988. C-*myc* and c-*myb* protein degradation: Effect of metabolic inhibitors and heat shock. *Mol. Cell. Biol.* **8:** 25040.

Marin, O., F. Meggio, F. Marchiori, G. Borin, and L.A. Pinna. 1986. Site specificity of casein kinase-2 (TS) from rat liver cytosol. *Eur. J. Biochem.* **160:** 239.

Meggio, F., A.M. Brunati, and L.A. Pinna. 1987. Polycation-dependent, Ca^{2+}-antagonized phosphorylation of calmodulin by casein kinase-2 and a spleen tyrosine protein kinase. *FEBS Lett.* **215:** 241.

Mohr, I.J., B. Stillman, and Y. Gluzman. 1987. Regulation of SV40 DNA replication by phosphorylation of T antigen. *EMBO J.* **6:** 153.

Nakajo, S., T. Hagiwara, K. Nakaya, and Y. Nakamura. 1986. Tissue distribution of casein kinases. *Biochem. Int.* **13:** 701.

Parker, P.J., F.B. Caudwell, and P. Cohen. 1983. Glycogen synthase from rabbit skeletal muscle; effect of insulin on the state of phosphorylation of the seven phosphoserine residues in vivo. *Eur. J. Biochem.* **130:** 227.

Perez, M., J. Grande, and E. Itarte. 1987. Developmental changes in rat hepatic casein kinases 1 and 2. *Eur. J. Biochem.* **170:** 493.

Picton, C., J.R. Woodgett, B.A. Hemmings, and P. Cohen. 1982. Multisite phosphorylation of glycogen synthase from rabbit skeletal muscle. *FEBS Lett.* **150:** 191.

Pinna, L.A., F. Meggio, A. Donella-Deana, and A.M. Brunati. 1985. Type-2 casein kinases: Structure, metabolic involvements and regulation. *FEBS Proc. Meet.* **16:** 155.

Prowald, K., H. Fischer, and O.-G. Issinger. 1984. Enhanced casein kinase II activity in human tumour cell cultures. *FEBS Lett.* **176:** 479.

Ramsay, G.M., T. Graf, and M.J. Hayman. 1980. Mutants of avian myelocytomatosis virus with smaller *gag* gene-related proteins have an altered transforming ability. *Nature* **288:** 170.

Rogers, S., R. Wells, and M. Rechsteiner. 1986. Amino acid sequences common to rapidly degraded proteins: The PEST hypothesis. *Science* **234:** 364.

Rose, K.M., D.A. Stetler, and S.T. Jacob. 1983. Phosphorylation of RNA polymerases; specific association of protein kinase NII with RNA polymerase I. *Philos. Trans. R Soc. Lond. B Biol. Sci.* **302:** 135.

Saxena, A., R. Padmanabha, and C.V.C. Glover. 1987. Isolation and sequencing of cDNA clones encoding alpha and beta subunits of *Drosophila melanogaster* casein kinase II. *Mol. Cell. Biol.* **7:** 3409.

Scheidtmann, K.H., B. Echle, and G. Walter. 1982. Simian virus 40 large T antigen is phosphorylated at multiple sites clustered in two separate regions. *J. Virol.* **44:** 116.

Scheidtmann, K.H., M. Hardung, B. Echle, and G. Walter. 1984. DNA-binding activity of simian virus 40 large T antigen correlates with a distinct phosphorylation state. *J. Virol.* **50:** 1.

Schneider, H.R., G.H. Reichert, and O.G. Issinger. 1986. Enhanced casein kinase II activity during mouse embryogenesis. *Eur. J. Biochem.* **161:** 733.

Singh, T.J. and K.-P. Huang. 1985. Glycogen synthase (casein) kinase-1: Tissue distribution and subcellular localization. *FEBS Lett.* **190:** 84.

Sigler, P.B. 1988. Acid blobs and negative noodles. *Nature* **333:** 210.

Sommercorn, J. and E.G. Krebs. 1987. Induction of casein kinase II during differentiation of 3T3-L1 cells. *J. Biol. Chem.* **262:** 3839.

Sommercorn, J., S.J. McNall, E.H. Fischer, and E.G. Krebs. 1987a. Phosphorylation of acetyl-CoA carboxylase by casein kinase II enhances the rate of dephosphorylation of the cAMP-dependent protein kinase site. *Fed. Proc.* **46:** 2003.

Sommercorn, J., J.A. Mulligan, F.J. Lozeman, and E.G. Krebs. 1987b. Activation of casein kinase II in response to insulin and to epidermal growth factor. *Proc. Natl. Acad. Sci.* **84:** 8834.

Suzuki, N., T. Saito, and T. Hosoya. 1987. *In vivo* effects of dexamethasone and cycloheximide on the phosphorylation of 110-kDa proteins and the protein kinase activities of rat liver nucleoli. *J. Biol. Chem.* **262:** 4696.

Takio, K., E.A. Kuenzel, K.A. Walsh, and E.G. Krebs. 1987. Amino acid sequence of the beta subunit of bovine lung casein kinase II. *Proc. Natl. Acad. Sci.* **84:** 4851.

Walton, G.M., J. Spiess, and G.N. Gill. 1985. Phosphorylation of high mobility group protein 14 by casein kinase II. *J. Biol. Chem.* **260:** 4745.

Witters, L.A., J.P. Tripper, and G.W. Bacon. 1983. Stimulation of site-specific phosphorylation of acetyl-CoA carboxylase by insulin and epinephrine. *J. Biol. Chem.* **258:** 5643.

Woodgett, J.R. and P. Cohen. 1984. Multisite phosphorylation of glycogen synthase: Molecular basis for the substrate specificity of glycogen synthase kinase-3 and casein kinase-II (glycogen synthase kinase-5). *Biochim. Biophys. Acta.* **788:** 339.

Yu, K-T., N. Khalaf, and M.P. Czech. 1987. Insulin stimulates a membrane-bound serine kinase that may be phosphorylated on tyrosine. *Proc. Natl. Acad. Sci.* **84:** 3972.

Microinjection of the Catalytic Subunit of cAMP-dependent Protein Kinase Induces Expression of the c-*fos* Gene

K.T. Riabowol,* M.Z. Gilman,* and J.R. Feramisco†
Cold Spring Harbor Laboratory, Cold Spring Harbor, New York 11724; †Cancer Center, University of California-San Diego, California 92103

Changes in the concentration of cAMP have been shown to regulate cell shape (Porter et al. 1974) and metabolism (Schacter et al. 1984) and have been implicated in the inhibition of cell growth (Otten et al. 1972), although it is now believed that a sustained increase in the cellular level of cAMP constitutes a growth-promoting signal in mammalian (Rozengurt 1986) and yeast (Matsumoto et al. 1985) cells. In *Escherichia coli*, the expression of genes encoding enzymes involved in sugar catabolism is activated following binding of cAMP to a catabolite gene-activator protein (cap) and subsequent binding of the complex to specific DNA sequences of cAMP-regulated genes (de Crombrugghe et al. 1984). In contrast, the major way by which cAMP exerts cellular effects in eukaryotes is thought to be through the cAMP-dependent protein kinase (A-kinase) (Beavo et al. 1974). The holoenzyme of A-kinase exists as an enzymatically inactive tetramer, composed of two catalytic (C) and two regulatory (R) subunits (Krebs and Beavo 1979). When levels of cAMP are elevated in eukaryotic cells, cAMP binds to the R subunits and the holoenzyme dissociates, yielding an R subunit dimer and two active C subunits. Two major types of A-kinases have been described previously (Corbin et al. 1975). These kinases share common C subunits but differ in their R subunits. The R subunits can be distinguished by their affinity for cAMP, their ability to be phosphorylated by the C subunit (Johnson et al. 1987), and their primary amino acid sequences. Recent studies have shown that there are at least four types of R subunits ($R_{I\alpha}$, $R_{I\beta}$, R_{II51}, and R_{II54}) and two forms of C subunits (C_α and C_β) (Oyen et al. 1988).

Both the C and R subunits of A-kinase have been implicated in the transcriptional regulation of cAMP-responsive genes. The C subunit of A-kinase could alter gene expression by phosphorylation of chromosomal or other nuclear proteins (Waterman et al. 1985), an idea that is supported by the observation that the C subunit migrates from the Golgi complex into the nucleus, following treatment of cells with agents that increase intracellular levels of cAMP (Nigg et al. 1985). An alternative hypothesis is that the R subunits could function in a manner similar to that of bacterial cap and, upon binding cAMP, act directly on a target DNA site in cAMP-responsive genes (Montminy et al. 1986). In support of this hypothesis, one of the R subunits (R_{II}) is homologous in amino acid sequence to cap (Weber et al. 1982). Although the homology between R_{II} and cap does not occur within the putative DNA-binding domain, R_{II} does contain sequences related to other DNA-binding proteins (Nagamine and Reich 1985) and has been shown to interact with DNA in a sequence-specific manner (Jungmann et al. 1986). The cAMP-responsive sequences (CREs) in some eukaryotic genes are also homologous to bacterial cap-binding sequences. Furthermore, it has been reported that the phosphorylated form of R_{II}, when bound with cAMP, can alter the superhelicity of several DNA substrates via an intrinsic topoisomerase activity (Constantinou et al. 1985). Because topoisomerase I may be involved in transcriptional activity in some eukaryotic systems, this activity could provide a mechanism through which R_{II} could induce the expression of cAMP-regulated genes. Recent studies have indicated, however, that topoisomerase activity can be separated from R_{II} (Shabb and Granner 1988), so a direct effect of R_{II} on gene transcription has yet to be proven.

Two recent studies have reported that the active C subunit of A-kinase was necessary for the cAMP-mediated stimulation of gene expression (Grove et al. 1987; Buchler et al. 1988). In one study, cotransfection of cells with a construct encoding an active protein kinase inhibitor peptide (Cheng et al. 1985) decreased the cAMP-induced expression of a reporter construct composed of the cAMP-responsive promoter element of the human proenkephalin gene, fused to the bacterial chloramphenicol acetyltransferase gene (Grove et al. 1987). The second study reported that expression of endogenous cAMP-regulated genes was inhibited when cultured rat hepatocytes were incubated with combinations of cAMP and three different inhibitors of A-kinase (Buchler et al. 1988). Although these studies indicate that the active C subunit of A-kinase is required for expression of cAMP-responsive genes, they do not rule out the possibilities that a cAMP-binding protein is also required for the response or that the R subunits may be responsible for activation of transcription by cAMP, following phosphorylation by a C subunit. To identify the subunits of A-kinase that were sufficient to activate the transcription of a cAMP-regulated gene, we microinjected purified preparations of the C, R_I, and R_{II} subunits of A-kinase into fibroblasts and assayed the expression of the endogenous c-*fos* gene.

METHODS

Cell culture. 208F rat fibroblasts (Quade 1979) were grown in Dulbecco's modified Eagle's medium (DMEM) containing 10% fetal calf serum (FCS), 100 units/ml of penicillin, and 100 μg/ml of streptomycin. Other culture conditions were carried out as described previously (Riabowol et al. 1988a).

Immunofluorescence. For the determination of c-*fos* expression, cells grown in DMEM containing 0.5% FCS for 24 hours were microinjected, incubated for various lengths of time, and fixed by sequential immersion for 10 minutes in phosphate-buffered saline (PBS) containing 3.7% formaldehyde, 0.5% Triton X-100, and 0.05% Tween 20. Immunofluorescent detection of c-*fos* was done by incubating fixed cells for 1 hour at 37°C in PBS containing 1% bovine serum albumin and 50 μg/ml of affinity-purified rabbit anti-*fos* antibody, followed by incubation with fluorescein-conjugated goat anti-rabbit immunoglobulin (IgG) (Riabowol et al. 1988a).

Immunohistology. Identification of injected cells using biotinylated goat anti-rabbit antibodies, Strepavidin-horseradish peroxidase, and development in 4-chloro-1-napthol-H_2O_2 was done as described previously (Riabowol 1988).

Proteins microinjected. The C and R_{II} subunits of A-kinase were purified from bovine heart (Fletcher et al. 1986). The R_I subunit and *ras* (T-24, cloned from a bladder carcinoma, containing a glycine to valine substitution at amino acid 12) proteins were overexpressed and purified from bacteria. Purified proteins were stored on ice as stock solutions containing 10 mM TRIS (pH 7.0), 100 mM NaCl, 0.1 mM EDTA, and 1 mM β-mercaptoethanol (BME). Immediately prior to use, they were diluted to the desired concentration in the same buffer, but lacking EDTA and BME.

RESULTS

Biological Activity of Microinjected Proteins

The activity of purified *ras* protein and the C subunit of A-kinase used in this study was determined by microinjection into cells in control experiments. As shown in Figure 1, the C subunit (0.5 mg/ml) resulted in the

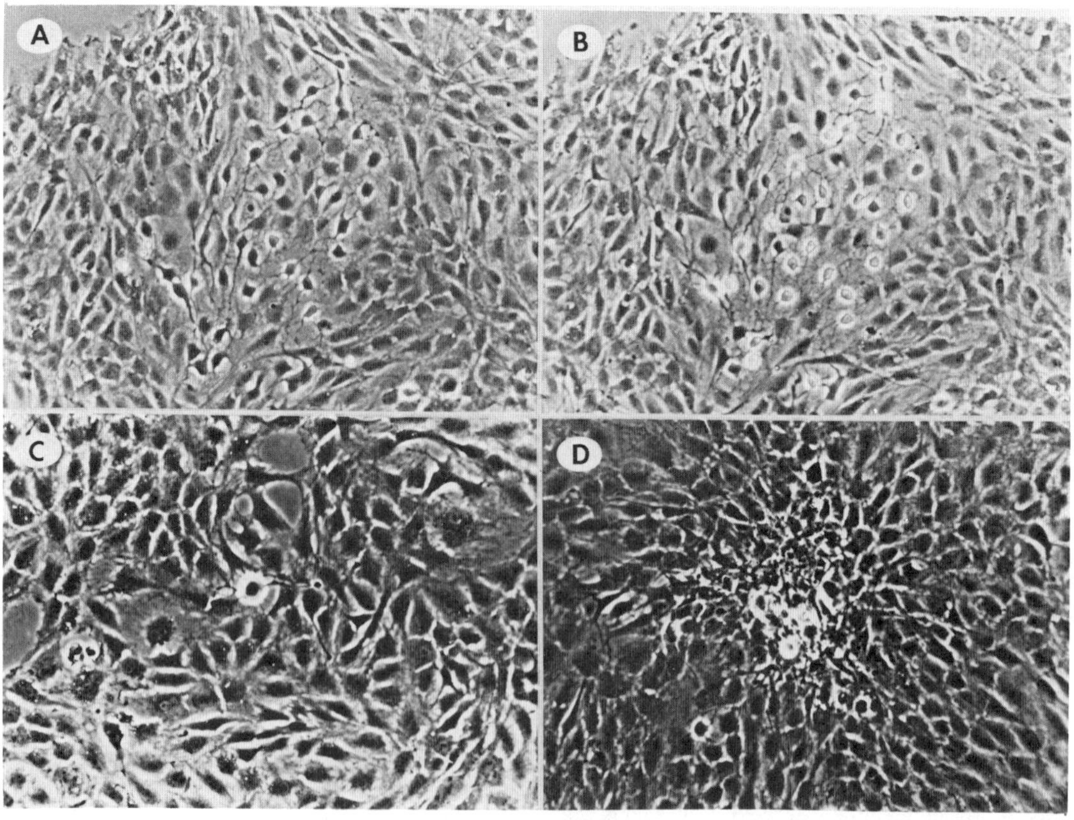

Figure 1. Determination of the biological activity of purified *ras* protein and the C subunit of A-kinase. Cells were microinjected using glass capillary needles (Feramisco et al. 1984). The *ras* protein (T-24) was injected at a concentration of 1 mg/ml, and the R_I, R_{II}, and C subunits of A-kinase were injected at a concentration of 0.5 mg/ml. Previous calculations indicate that microinjection of this concentration of A-kinase subunits elevates the endogenous levels of these proteins by two- to fourfold (Lamb et al. 1988). Cells injected with the C subunit of A-kinase were photographed 20 min (*A*) and 1 hr (*B*) after injection. Cells injected with the *ras* protein were photographed 1 hr (*C*) and 24 hr (*D*) following microinjection using phase-contrast optics.

collapse of actin microfilaments (Lamb et al. 1988) within 20 minutes following injection (Fig. 1A). This effect was more pronounced 1 hour after microinjection (Fig. 1B) and persisted for 18–22 hours (data not shown). The *ras* protein efficiently induced focus formation (Feramisco et al. 1984) when injected into fibroblast cells. Figure 1, C and D, shows a field of fibroblasts 1 hour and 24 hours, respectively, after injection of the *ras* protein at a concentration of 1 mg/ml. The R subunits of A-kinase were also biologically active, as discussed below. Aliquots of each of the protein preparations, which are estimated to be more than 95% pure, are shown in Figure 2.

Expression of the c-*fos* gene is known to be induced by increasing the intracellular levels of cAMP in macrophages (Bravo et al. 1987) and BALB/c-3T3 fibroblasts (Gilman 1988) but is apparently unresponsive to cAMP addition in NIH-3T3 (Bravo et al. 1987) and 208F (K. Riabowol, unpubl.) fibroblasts. Although the DNA sequences responsible for this regulation have not been fully defined for c-*fos*, regions 5′ of the transcribed region of this gene have been identified that contribute to basal levels of expression (Gilman et al. 1986). At least one of these sequences, which binds a protein factor and contributes to basal promoter activity, is cAMP-responsive (L.A. Berkowitz and M. Gilman, in prep.). Regions within the c-*fos* gene that show homology with a consensus response element (CRE) are shown in Figure 3A. Sequences located 5′ of several other genes whose expression is regulated by cAMP have been identified and shown to be essential for transcriptional induction in response to elevated levels of cAMP. Representatives of these cAMP response elements are shown in Figure 3B.

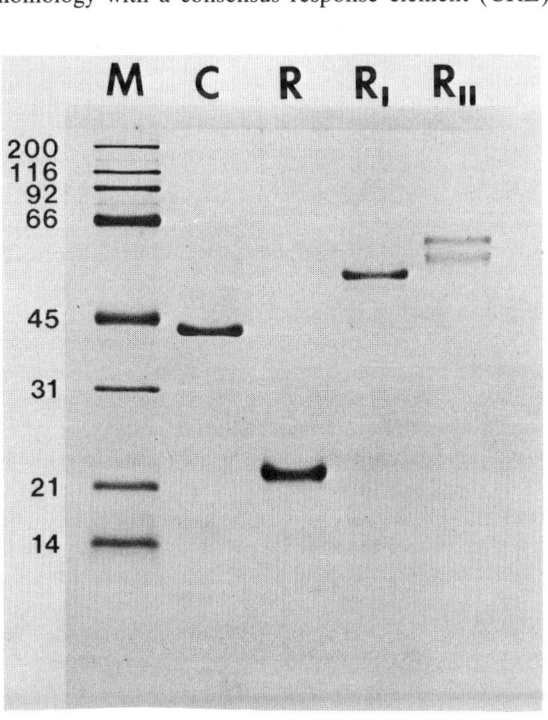

Figure 2. Proteins microinjected. Aliquots of *ras* (R), A-kinase catalytic subunit (C), type I regulatory subunit (R_I), and type II regulatory subunit (R_{II}) were boiled in Laemmli sample buffer and electrophoresed through a 10% polyacrylamide gel containing SDS. Following staining of the gel with Coomassie blue dye, two bands were observed for the R_{II} subunit, representing posttranslationally modified forms of the protein. (Reprinted, with permission, from Riabowol et al. 1988b.)

Expression of c-*fos* Is Induced by Microinjection of the C Subunit of A-kinase

Rodent fibroblasts rendered quiescent by incubation in medium containing 0.5% FCS for 24 hours were microinjected with the C subunit, *ras* protein, or mixtures of C and R_{II} subunits. Figure 4 shows that by 90 minutes, expression of c-*fos* was markedly induced in cells injected with the C subunit (Fig. 4A), and by 3 hours, very high levels of nuclear *fos* staining were observed (Fig. 4B). Figure 4D shows that a mixture of equal amounts of the C and R_{II} subunits failed to induce c-*fos* expression in these cells. Injection of the *ras* protein, which is also known to induce *fos* expression (Stacey et al. 1987), was included as a positive control and is shown in Figure 4C; however, in contrast to what is observed with the C subunit, no effect was seen earlier than 3 hours following injection (data not shown).

All Cells Injected with the C Subunit Express Elevated Levels of c-*fos*

To determine the proportion of fibroblast cells that responded to injection of the C subunit by increased expression of c-*fos*, we coinjected pure C subunit (0.5 mg/ml) with 0.25 mg/ml of nonspecific IgGs, so that the latter could serve as a marker for identification of injected cells. Coinjection of up to 5 mg/ml of nonspecific IgGs has been demonstrated to have no detectable effect upon cell metabolism by a variety of criteria (Riabowol 1988). Following injection, cells were processed to detect levels of c-*fos* by indirect immunofluorescence and stained for the presence of IgGs. All cells that were injected also showed increased levels of c-*fos* (data not shown).

DISCUSSION

Expression of the c-*fos* gene can be induced through several signal transduction pathways. One pathway requires a functional SRE (Treisman 1985, 1986) and is activated in response to serum, platelet-derived growth factor, and protein kinase C activators, such as phorbol myristate acetate (Greenberg and Ziff 1984; Curran 1988). Transcriptional activation through this pathway does not require the presence of other enhancer sequences (Treisman 1986; Gilman 1988) and is blocked by microinjection of oligodeoxynucleotides, corresponding to the SRE, 2 hours prior to serum stimulation (Gilman et al., this volume). Increased expression of the c-*fos* gene can also be elicited by treating cells with agents that increase intracellular levels of cAMP

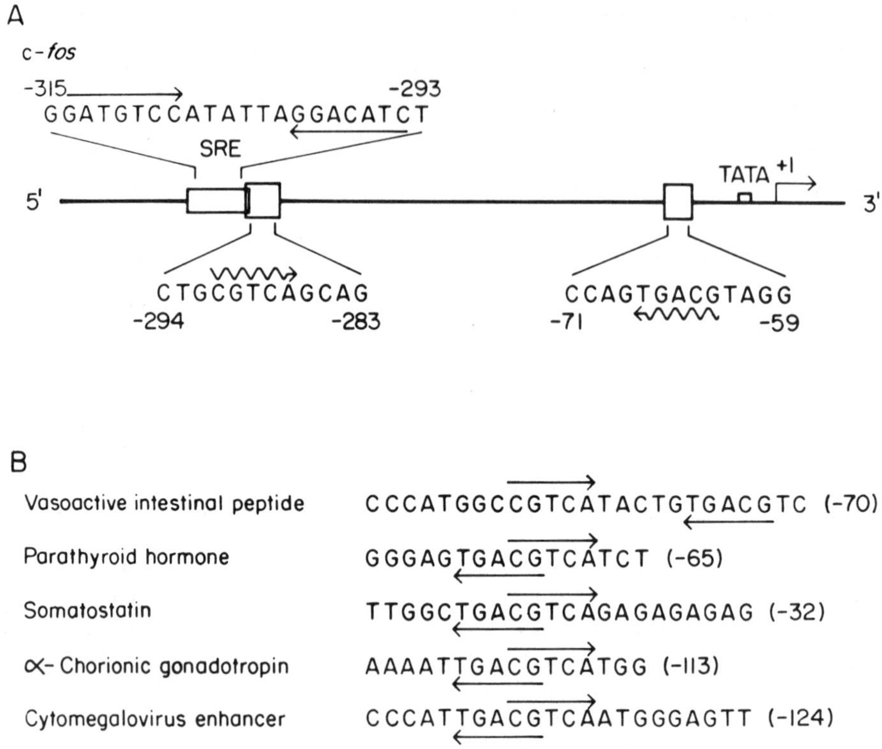

Figure 3. Sequences necessary for transcriptional induction by cAMP. (*A*) The mouse c-*fos* gene contains a region of dyad symmetry termed the serum response element (SRE). Two regions showing identity with a consensus cAMP-CRE of the sequence 5′-CGTCA-3′ are indicated by wavy arrows. (*B*) Upstream regions of genes containing sequences that show identity to a consensus CRE are shown. Several of these sequences have been shown to be necessary for cAMP-mediated expression of constructs containing these elements (Montminy et al. 1986; Delegeane et al. 1987).

(Bravo et al. 1987; Gilman 1988), although transcriptional response to cAMP varies markedly between cell types and among different lines of human and rodent fibroblasts (K. Riabowol, unpubl.). The regulation of gene expression by cAMP in prokaryotes involves the direct binding of a cAMP-cap to specific DNA sequences 5′ of cAMP-responsive genes. In mammalian cells, transcription of cAMP-responsive genes has been shown to require the active C subunit of A-kinase (Grove et al. 1987; Buchler et al. 1988). Data presented here indicate that injection of the C subunit of A-kinase is sufficient to induce expression of the c-*fos* gene. We have also tested the ability of the C subunit to activate the expression of a fusion gene consisting of the 5′ region of the cAMP-regulated human vasoactive intestinal peptide (VIP) gene linked to a bacterial β-galactosidase reporter gene. Sequences derived from the VIP gene contain two inverted copies of a highly conserved CGTCA motif (see Fig. 3), mutation of which abolishes response of the construct to cAMP in transient expression assays (Fink et al. 1988). Similar to the results seen for c-*fos*, we found that injection of the C subunit of A-kinase induces expression of the construct, in this case assayed by incubation of fixed cells in the presence of the chromogenic substrate X-gal (K. Riabowol et al., in prep.). Furthermore, the fact that injection of R_I or R_{II} individually or in the presence of a lower concentration of C subunit was unable to induce transcription implies that these R subunits do not play a crucial role in the expression of c-*fos*. Since these microinjection experiments were carried out in the absence of agents that increase intracellular levels of cAMP, they suggest also that endogenous cAMP-binding proteins were not involved in the transcriptional induction of this gene.

The c-*fos* gene contains at least four regions homologous to a consensus CRE. Although the region located between −71 and −59 of the mouse c-*fos* gene has been shown to enable fusion genes containing this sequence to respond to elevated cAMP levels, other sequences also contribute to cAMP-responsive gene expression (L. Berkowitz and M. Gilman, in prep.). To test whether transcription of the c-*fos* gene in response to cAMP requires the binding of factors specific for the CRE sequence, we coinjected different synthetic oligonucleotides with the C subunit of A-kinase. We found that the ability of the C subunit to induce c-*fos* expression was blocked by coinjection of a wild-type CRE (−71 to −59, Fig. 3), whereas coinjection of a CRE point mutant or of a wild-type SRE (Fig. 3) had minimal effects on C-subunit-induced expression of c-*fos* (Gilman et al., this volume).

Injection of the C subunit of A-kinase into several fibroblast lines (rat 208F [Quade 1979]; REF52 [McClure et al. 1982]; NIH-3T3 [obtained from M. Wigler]; BALB/c-3T3 [Gilman 1988]; human primary

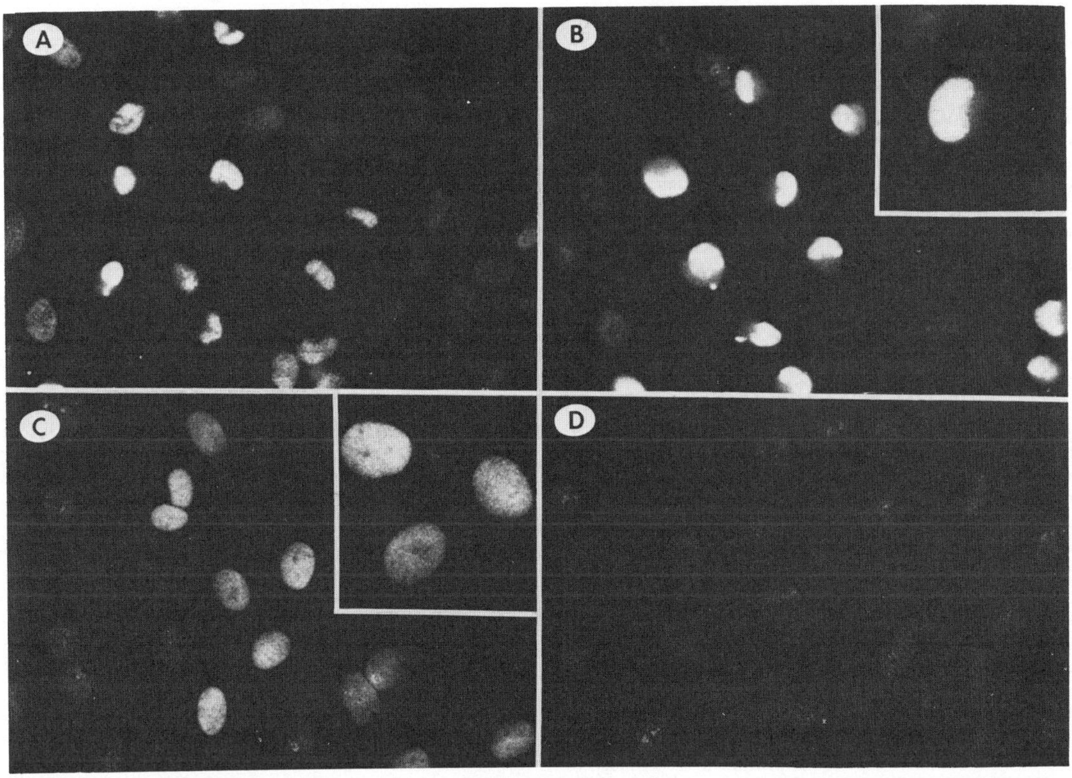

Figure 4. Microinjection of the C subunit of A-kinase induces expression of c-*fos*. Logarithmically growing rat 208F fibroblasts were made quiescent by incubation in medium containing 0.5% FCS for 24 hr prior to injection. Purified C subunit of A-kinase was injected, and the cells were incubated for 90 min (*A*) or 3 hr (*B*) prior to fixation and staining for the presence of c-*fos*. *ras* protein (*C*) and a mixture of equal amounts of C and R_{II} subunits (*D*) were also injected, and the cells were fixed and processed for immunocytochemistry with *fos* antibody following 3 hr (*ras*) or 6 hr ($C + R_{II}$) of incubation at 37°C. No nuclear fluorescence was observed in cells injected with the mixture of C and R_{II} when cells were fixed and processed for immunofluorescence 1, 2, or 4 hr following injection (data not shown). Insets in *B* and *C* show higher-magnification micrographs of the nuclei of C subunit and *ras*-injected cells, respectively. (Reprinted, with permission, from Riabowol et al. 1988b.)

foreskin fibroblasts [obtained from S. Cohen]) resulted in increased nuclear staining for c-*fos*. Conversely, some fibroblast lines failed to respond to treatment with 8-bromo cAMP or isobutyl methylxanthine and forskolin (Bravo et al. 1987; K. Riabowol, unpubl.). Since all fibroblast lines tested responded to injection of the catalytically active C subunit, the failure of some cell lines to respond to cAMP-elevating agents could be due to line-specific differences in permeability to these agents, differing endogenous levels of A-kinase, or differing activities of the enzymes largely responsible for regulating intracellular cAMP levels, adenylate cyclase, and cyclic nucleotide phosphodiesterase.

These results show that the C subunit of A-kinase is an important intermediate in the expression of genes in response to cAMP and suggest that phosphorylation of subsequent intermediates of this signal transduction pathway, possibly transcription factors, is required for transcriptional activation of genes in response to cAMP. Data presented here and in an accompanying paper (Gilman et al., this volume) also suggest that stimulation of transcription by cAMP requires binding of factors specific for the CRE. By injecting purified proteins or specific antibodies into tissue-culture cells, it may be possible to identify other cellular factors involved in regulating enhancer sequences.

ACKNOWLEDGMENTS

We thank Dr. D. Walsh for the kind gift of purified preparations of the C and R_{II} subunits of A-kinase and Dr. Susan Taylor for supplying the R_I subunit. The technical assistance of Christine Arrigo and the expert photographic services of Phillip Renna are also gratefully acknowledged.

REFERENCES

Beavo, J.A., P.J. Bechtel, and E.G. Krebs. 1974. Activation of protein kinase by physiological concentrations of cyclic AMP. *Proc. Natl. Acad. Sci.* **71:** 3580.

Bravo, R., M. Neuberg, J. Burckhardt, J. Almendral, R. Wallich, and R. Muller. 1987. Involvement of common and cell type-specific pathways in c-*fos* gene control: Stable induction by cAMP in macrophages. *Cell* **48:** 251.

Buchler, W., U. Walter, B. Jastorff, and S.M. Lohmann. 1988. Catalytic subunit of cAMP-dependent protein kinase is essential for cAMP-mediated mammalian gene expression. *FEBS Lett.* **228:** 27.

Cheng, H.C., S.M. Van Patten, A.J. Smith, and D.A. Walsh.

1985. An active twenty-amino-acid-residue peptide derived from the inhibitor protein of the cyclic AMP-dependent protein kinase. *Biochem. J.* **231:** 655.

Constantinou, A.I., S.P. Squinto, and R.A. Jungmann. 1985. The phosphoform of the regulatory subunit RII of cyclic AMP-dependent protein kinase possesses intrinsic topoisomerase activity. *Cell* **42:** 429.

Corbin, J.D., S.L. Keely, and C.R. Park. 1975. The distribution and dissociation of cyclic adenosine 3′:5′-monophosphate-dependent protein kinases in adipose, cardiac, and other tissues. *J. Biol. Chem.* **250:** 218.

Curran, T. 1988. *fos*. In *The oncogene handbook* (ed. E.P. Reddy et al.) Elsevier, Amsterdam. (In press.)

de Crombrugghe, B., S. Busby, and H. Buc. 1984. Cyclic AMP receptor protein: Role in transcription activation. *Science* **224:** 831.

Delegeane, A.M., L.H. Ferland, and P.L. Mellon. 1987. Tissue-specific enhancer of the human glycoprotein hormone α-subunit gene: Dependence on cyclic AMP-inducible elements. *Mol. Cell Biol.* **7:** 3994.

Feramisco, J.R., M. Gross, T. Kamata, M. Rosenberg, and R.W. Sweet. 1984. Microinjection of the oncogene form of the human H-*ras* (T24) protein results in rapid proliferation of quiescent cells. *Cell* **38:** 109.

Fink, J.S., M. Verhave, S. Kasper, T. Tsukada, G. Mandel, and R.H. Goodman. 1988. The CGTCA sequence motif is essential for biological activity of the vasoactive intestinal peptide gene cAMP-regulated enhancer. *Proc. Natl. Acad. Sci.* **85:** 6662.

Fletcher, W.H., S.M. Van Patten, H.C. Cheng, and D.A. Walsh. 1986. Cytochemical identification of the regulatory subunit of the cAMP-dependent protein kinase by use of fluorescently labeled catalytic subunit. *J. Biol. Chem.* **261:** 5504.

Gilman, M.Z. 1988. The c-*fos* serum response element responds to protein kinase C-dependent and -independent signals but not to cyclic AMP. *Genes Dev.* **2:** 394.

Gilman, M.Z., R.N. Wilson, and R.A. Weinberg. 1986. Multiple protein-binding sites in the 5′-flanking region regulate c-*fos* expression. *Mol. Cell Biol.* **6:** 4305.

Greenberg, M.E. and E.B. Ziff. 1984. Stimulation of 3T3 cells induces transcription of the c-*fos* proto-oncogene. *Nature* **311:** 433.

Grove, J.R., D.J. Price, H.M. Goodman, and J. Avruch. 1987. Recombinant fragment of protein kinase inhibitor blocks cyclic AMP dependent gene transcription. *Science* **238:** 530.

Johnson, K.E., S. Cameron, T. Toda, M. Wigler, and M.J. Zoller. 1987. Expression in *Escherichia coli* of BCYI, the regulatory subunit of cyclic AMP-dependent protein kinase from *Saccharomyces cerevisiae*. *J. Biol. Chem.* **262:** 8636.

Jungmann, R.A., A.I. Constantinou, S.P. Squinto, J. Kwast-Weifeld, and J.S. Schweppe. 1986. Regulation of lactate dehydrogenase gene expression by c-AMP-dependent protein kinase subunits. *Ann. N.Y. Acad. Sci.* **478:** 147.

Krebs, E.G. and J.A. Beavo. 1979. Phosphorylation and dephosphorylation of enzymes. *Annu. Rev. Biochem.* **48:** 923.

Lamb, N.J.C., A. Fernandez, M.-A. Conti, R.S. Adelstein, D.B. Glass, W.J. Welch, and J.R. Feramisco. 1988. Regulation of actin microfilament integrity in living nonmuscle cells by the cAMP-dependent protein kinase and the myosin light chain kinase. *J. Cell Biol.* **106:** 1955..

Matsumoto, K., I. Uno, and T. Ishikawa. 1985. Genetic analysis of the role of cAMP in yeast. *Yeast* **1:** 15.

McClure, D.B., M.J. Hightower, and W.C. Topp. 1982. Effect of SV40 transformation on the growth-factor requirements of the rat embryo cell line REF52 in serum-free medium. *Cold Spring Harbor Conf. Cell Proliferation* **9:** 345.

Montminy, M.R., K.A. Sevarino, J.A. Wagner, G. Mandel, and R.H. Goodman. 1986. Identification of a cyclic-AMP-responsive element within the rat somatostatin gene. *Proc. Natl. Acad. Sci.* **83:** 6682.

Nagamine, Y. and E. Reich. 1985. Gene expression and cAMP. *Proc. Natl. Acad. Sci.* **82:** 4606.

Nigg, E.A., H. Hilz, H.M. Eppenberger, and F. Dutly. 1985. Rapid and reversible translocation of the catalytic subunit of cAMP-dependent protein kinase type II from the golgi to the nucleus. *EMBO J.* **4:** 2801.

Otten, J., G.S. Johnson, and I. Pastan. 1972. Regulation of cell growth by cyclic adenosine 3′:5′-monophosphate. *J. Biol. Chem.* **247:** 7082.

Oyen, O., M. Sandberg, W. Eskild, F.O. Levy, G. Knutsen, S. Beebe, V. Hansson, and T. Jahnsen. 1988. Differential regulation of messenger ribonucleic acids for specific subunits of cyclic adenosine 3′:5′-monophosphate (cAMP)-dependent protein kinase by cAMP in rat sertoli cells. *Endocrinology* **122:** 2658.

Porter, K.R., T.T. Puck, A.W. Hale, and D. Kelly. 1974. An electron microscope study of the effects of dibutyryl cyclic AMP on Chinese hamster ovary cells. *Cell* **2:** 145.

Quade, K. 1979. Transformation of mammalian cells by avian myelocytomatosis virus and avian erythroblastosis virus. *Virology* **98:** 461.

Riabowol, K.T. 1988. Identification of microinjected cells using biotinylated antibodies and strep-avidin-conjugated horseradish peroxidase (HRP). *Anal. Biochem.* (in press).

Riabowol, K.T., R.J. Vosatka, E.B. Ziff, N.J. Lamb, and J.R. Feramisco. 1988a. Microinjection of *fos*-specific antibodies blocks DNA synthesis in fibroblasts. *Mol. Cell Biol.* **8:** 1670.

Riabowol, K.T., J.S. Fink, M.Z. Gilman, D.A. Walsh, R.H. Goodman, and J.R. Feramisco. 1988b. The catalytic subunit of cyclic AMP-dependent protein kinase induces expression of genes containing cyclic AMP-responsive enhancer elements. *Nature* (in press).

Rozengurt, E. 1986. Early signals in the mitogenic response. *Science* **234:** 161.

Schacter, E., P. Boon Chock, and E.R. Stadtman. 1984. Regulation through phosphorylation/dephosphorylation cascade systems. *J. Biol. Chem.* **259:** 12252.

Shabb, J.B. and D.K. Granner. 1988. Separation of topoisomerase I activity from the regulatory subunit of type II cyclic adenosine monophosphate-dependent protein kinase. *Mol. Endocrinol.* **2:** 324.

Stacey, D.W., T. Watson, H.F. Kung, and T. Curran. 1987. Microinjection of transforming *ras* protein induces c-*fos* expression. *Mol. Cell Biol.* **7:** 523.

Treisman, R. 1985. Transient accumulation of c-*fos* RNA following serum stimulation requires a conserved 5′ element and c-*fos* 3′ sequences. *Cell* **42:** 889.

———. 1986. Identification of a protein-binding site that mediates transcriptional response of the c-*fos* gene to serum factors. *Cell* **46:** 567.

Waterman, M., G.H. Murdoch, R.M. Evans, and M.G. Rosenfeld. 1985. Cyclic AMP regulation of eukaryotic gene transcription by two discrete molecular mechanisms. *Science* **229:** 267.

Weber, I.T., K. Takio, K. Titani, and T.A. Steitz. 1982. The cAMP-binding domains of the regulatory subunit of cAMP-dependent protein kinase and the catabolite gene activator protein are homologous. *Proc. Natl. Acad. Sci.* **79:** 7679.

Regulating Cell Growth: Casein-kinase-II-dependent Phosphorylation of Nuclear Oncoproteins

D. CARROLL, N. SANTORO, AND D.R. MARSHAK

Cold Spring Harbor Laboratory, Cold Spring Harbor, New York 11724

The ability of normal animal cells to respond to mitogenic signals depends on a signaling system that propagates external growth signals across the plasma membrane to the nucleus. This signaling network reflects a complex system of positive and negative pathways, which, collectively, are responsible for controlling cell growth. When there are lesions in this signaling process, neoplasia can occur, leading to the uncontrolled growth of cells. The recognition that signal transducer molecules, such as those encoded by cellular oncogenes, play a central role in normal and cancerous cell growth has led to considerable efforts on the part of many laboratories to eludicate the different components of the signaling pathways. In our own laboratory, we have begun to examine a set of nuclear proteins whose biological functions are linked to growth control. Those under study include viral oncogene and cellular proto-oncogene products, as well as transcriptional activators. In other laboratories, mutational analysis of the genes that encode these nuclear proteins has begun to identify distinct structural domains associated with their cellular functions. The relationship between protein structure and transforming activity is illustrated by the adenovirus E1A gene product, a 289-amino-acid nuclear oncoprotein (the 13S mRNA translation product) in which transformation and *trans*-activation functions have been mapped to distinct domains within the primary structure of the protein (for review, see Moran and Mathews 1987). Similar structure/function studies of the nuclear oncoprotein T antigen from SV40 (Kalderon and Smith 1984; Paucha et al. 1986; Sompayrac and Danna 1988), *fos* (Jenuwein and Muller 1987), and *myc* (Heaney et al. 1986; Stone et al. 1987) have mapped various oncogenic functions to relatively distinct areas of amino acid sequence.

Acidic Domains and Growth Control

The primary structures of the domains linked to transformation in the E1A proteins, T antigen, *fos*, and *myc* share a motif that is highly conserved among different viral and cellular forms of these proteins (Ralston and Bishop 1983; Stable et al. 1985; Figge et al. 1988). This motif contains a secondary structure predicted to be a β-turn followed by an acidic α-helix. Within this conserved motif are serine and threonine residues that, if phosphorylated, would further increase the negative charge of the region. In Figure 1, we show a comparison of the primary and predicted secondary structures of this region from several oncoproteins.

The presence of serine and threonine residues within this conserved acidic domain suggests a role for phosphorylation in the regulation of protein function by altering the net negative charge of the region. At physiological pH, each additional phosphate would add two negative charges. In E1A and T antigen, certain serine residues in these acidic domains appear to be phosphorylated in vivo (Scheidtmann et al. 1982; Grässer et al. 1988; C. Stephens et al., pers. comm.). This phosphorylation may be transient, however, because Tremblay et al. (1988) find no phosphorylation at this site in E1A at 8 hours after infection. In fact, Harlow and colleagues (Whyte et al. 1988) suggest that the region spanning these phosphorylatable residues may function to augment the efficiency of the core transformation domain. This hypothesis is supported by the observation that mutations that decrease the negative charge or delete the residues in this region can reduce transformation (Kalderon and Smith 1984; Lillie et al. 1986, 1987; Moran et al. 1986; Kuppuswamy and Chinnadurai 1987; Schneider et al. 1987) but do not completely obliterate the transforming function. Taken together, the data presently available indicate that transient phosphorylation of the serine and threonine residues within the acidic portions of the transformation-associated domain serves to enhance the function of this domain. These observations have led us (D. Carroll et al., in prep.) and others (Figge et al. 1988) to speculate that at least some of the activities of this

PROTEIN	SPECIES		P.(S/T) β-TURN		D/E ACIDIC DOMAIN
E1A	Ad2, Ad5	^{124}CHEAGF -	PPS	-	DDEDEEGEE
LTAg	SV40	^{105}CSEEM -	PSS	-	DDEATADS
FOS	FBJ MOUSE HUMAN	^{129}VEQL -	SP	-	EEEE
MYC	HUMAN	^{241}HEE	TPPTTSS		DSEEEQEDEEE

Figure 1. Conservation of primary and predicted secondary structure within the transformation-associated region of the nuclear oncoproteins. Transformation-associated regions of the adenovirus E1A proteins, SV40 LTAg, and *fos* and *myc* proteins were compared for conservation of primary and predicted secondary structure. The predicted β-turn and acidic α-helix are enclosed in an open box.

diverse group of nuclear proteins may be subject to a common mechanism of cellular regulation, involving protein phosphorylation.

Regulating Transformation by Casein Kinase II

On the basis of the primary and predicted secondary structures of these domains, a likely candidate kinase for these domains is the enzyme casein kinase II (CK-II). This kinase is found in both the cytoplasm and the nucleus of mammalian cells (Hathaway and Traugh 1982). The substrate specificity of CK-II is for serine or threonine residues adjacent to acidic amino acids. Studies using synthetic peptide substrates (Maggio et al. 1984; Kuenzel and Krebs 1985; Kuenzel et al. 1987) have shown that CK-II preferentially phosphorylates serine or threonine residues on the amino-terminal side of aspartic or glutamic acid residues. CK-II also appears to prefer to phosphorylate residues adjacent to predicted β turns (Hathaway and Traugh 1982). On the basis of the similarity of the CK-II substrate motif to the acidic domains found in the products of the nuclear oncogenes, we predicted that these domains would serve as substrates for the CK-II (D. Carroll et al., in prep.). Synthetic peptides representing these domains from E1A and T antigen were shown to be uniquely phosphorylated by CK-II with K_m values in the low micromolar range (D. Carroll et al., in prep.; see also Table 1). Similarly, we found that peptides derived from the transformation/immortalization domains of *myc* and *fos* proteins are high-affinity sites for CK-II phosphorylation (Table 1). Furthermore, we have found that this acidic motif in *myc* and *fos* is uniquely phosphorylated by CK-II (Table 2).

For the intact oncoproteins, Grässer et al. (1988) have demonstrated that CK-II phosphorylates the intact T-antigen protein in vitro at the serine residues within the acidic transformation domain. In our laboratory, we have found that both SV40 T antigen and *myc* proteins are readily phosphorylated by CK-II. In the case of CK-II-dependent phosphorylation of *myc* protein, 6 moles of phosphate are incorporated per mole of protein with a K_m of 150 nM.

Table 1. Kinase-specific Phosphorylation of Synthetic Peptides

Kinase	Substrate	Phosphate incorporated (pmole)
cAK	casein	2800
	Kemptide	2100
	E1A-001	<1
	LTAg-006	<1
	fos-029	<1
	myc-019	<1
	RRREEETEEE	<1
Phosphorylase	phosphorylase b	2000
	Kemptide	250
	E1A-001	<1
	LTAg-006	<1
	fos-029	<1
	myc-019	<1
	RRREEETEEE	<1
CK-I	casein	1750
	Kemptide	300
	E1A-001	<1
	LTAg-006	<1
	fos-029	<1
	myc-019	<1
	RRREEETEEE	<1
CK-II	casein	2100
	Kemptide	<1
	E1A-001	4150
	LTAg-006	4650
	fos-029	1875
	myc-019	4980
	RRREEETEEE	1250

Four different serine/threonine kinases (cAMP-dependent protein kinase [cAK], phosphorylase kinase, CK-I, and CK-II) were tested for their ability to phosphorylate the synthetic peptides spanning the transformation-associated regions for E1A proteins, T antigen from simian virus, and *fos* and *myc* proteins: peptides E1A-001, LTAg-006, FOS-029, and MYC-019, respectively. The primary sequence for each of these peptides is as follows:

E1A-001 HEAGFPPSDDEDEEG
LTAg-006 SEEMPSSDDEATAD
FOS-029 VEQLSPEEEE
MYC-019 HEETPPTTSSDSEEEQEDEEE

Preparation of the synthetic peptides is identical to the protocol described elsewhere (D. Carroll et al., in prep.) At the same time, control reactions were carried out with a known substrate for each kinase in place of a synthetic peptide. The amount of phosphate transferred by each kinase to the synthetic peptide substrate or the control substrate was measured as pmole phosphate incorporated. In addition, the CK-II-specific peptide substrate RRREEETEEE (Kuenzel and Krebs 1985) was similarly tested as a measure for CK-II specificity.

CK-II Motif and Transcriptional Regulation

In addition to identifying a CK-II motif within the transformation-associated domains of nuclear oncoproteins, we have found that the primary sequence of the transcriptional regulator proteins Jun, GCN4, and GAL4 all have acidic domains similar to the putative transformation regions of the nuclear oncoproteins. Other investigators (Ma and Ptashne 1987a; Hope et al. 1988; Struhl 1988) have demonstrated that it is the acidic property spanning this domain that defines the transcriptional activity of the regions. For example, within the region of amino acids 15–59 of Jun, there are four short sequences consisting of serine and/or threonine residues adjacent to both acidic residues and potential β turns. The net charge of this region is -9 and could be changed to -21 if fully phosphorylated at these sites. Current thinking would suggest that such an increase in the net negative charge of this region would enhance the transcriptional activity of the proteins (Ma and Ptashne 1987b). Thus, regulation of negative charge in this acidic motif by CK-II represents a potential mechanism for the regulation of transcriptional activators. We expect that similar domains within the transcriptional activators GCN4 and GAL4 will be substrates for CK-II.

Table 2. Kinetics of Peptide Phosphorylation

Peptide	K_m (μM)	V_{max} (nmole/mg)
E1A-001	25	230
LTAg-006	35	280
fos-029	2	60
myc-019	27	310
RRREEETEEE	720	38

The K_m and V_{max} for CK-II-dependent phosphorylation of each of the peptides are given. The protocol is the same as reported previously (Carroll et al., in prep.).

CK-II and Growth

On the basis of the observation that the transformation-associated regions of nuclear oncogenes are substrates for CK-II, we anticipate that a variety of biological signals that regulate nuclear oncogene expression might alter CK-II activity in cells. We have shown that serum stimulation of quiescent WI38 cells, a human lung fibroblast line, results in a sixfold increase in CK-II activity within 30 minutes of stimulation (D.Carroll et al., in prep.). CK-II activity returns to a level near that of quiescent cells by 2 hours after stimulation. Thus, there is a transient pulse in CK-II activity concomitant with transient rises in *fos* expression as seen in fibroblasts and pheochromocytoma cells (Greenberg and Ziff 1984; Greenberg et al. 1985). Additional cycles of CK-II activation and deactivation are seen at 12 and 24 hours after stimulation. The kinetics of CK-II cycles in WI38 cells have led us to suggest (D. Carroll et al., in prep.) that CK-II stimulation might be linked with the transitions from G_0 to G_1 and from G_1 to S phase. The possibility that CK-II plays a role at specific points in the cell cycle is supported by the observation made by Mulner-Lorillon et al. (1988) linking CK-II activity to the transition between prophase and metaphase of meiotic cell division in full-grown *Xenopus laevis* oocytes.

In separate experiments, we have shown that CK-II levels are stimulated upon adenovirus infection of baby rat kidney (BRK) epithelial cells in primary culture (Carroll et al. 1987). The kinetics of this activation are similar to those seen in serum stimulation of WI38 cells, with a greater than 12-fold elevation in CK-II activity by 15 minutes after viral infection. The parallel response of CK-II activity during normal cell growth and viral infection suggests that activation of CK-II may represent a central step in the regulation of cell growth.

Regulating CK-II Activity

The kinetics of serum and adenovirus activation of CK-II underscore the fact that CK-II activity can be coupled to extracellular signals. Both serum and adenovirus infection, however, represent highly complex signals with a large number of factors potentially involved in the activation of CK-II.

The rapid activation of CK-II upon adenovirus infection is consistent with a plasma membrane receptor-mediated event. In fact, we have found that purified fiber protein, the major coat protein from adenovirus (Ad5), stimulates CK-II activity in intact BRK cells more than twofold, with half-maximal stimulation occurring at 5×10^{-19} M fiber protein. The apparent high affinity of this activation suggests that there is a specific receptor-mediated mechanism that involves CK-II as part of a signaling pathway. The possibility that coat proteins might act as ligands to provoke rapid and complex changes in host biochemistry has been reported previously in the case of the polyomavirus coat protein, VP_1 (Zullo et al. 1987).

CK-II activation may result from the activation of other, well-established signal transduction pathways. The interaction between a protein-kinase-C-mediated pathway and CK-II activity is demonstrated in Figure 2. The tumor-promoting phorbol ester, TPA, a potent activator of protein kinase C, stimulates a threefold elevation in cellular CK-II activity. These results are

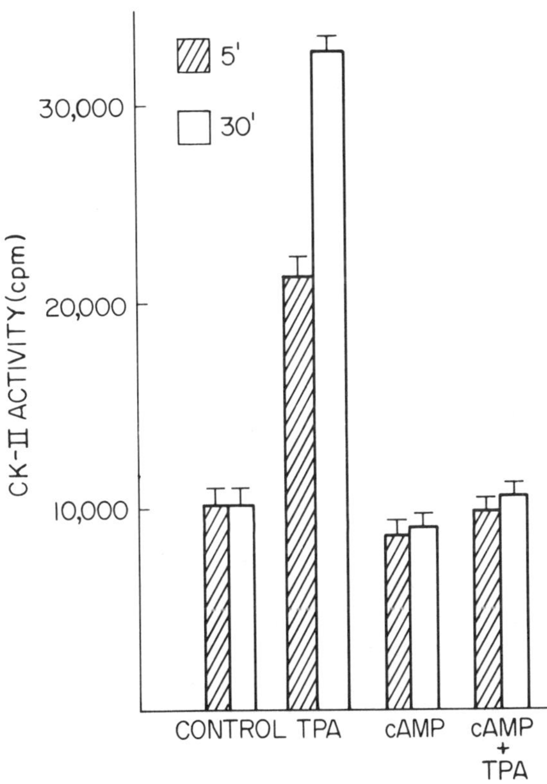

Figure 2. Activation and inhibition of CK-II. Primary cultures from 7-day-old BRK cells ($\sim 2 \times 10^7$ cells per dish) were incubated in the presence of 100 nM TPA or 100 μM dibutyryl cAMP. At 5 and 30 min after the addition of TPA and dibutyryl cAMP, cells were harvested, homogenized, and assayed for CK-II-dependent phosphorylation of the synthetic peptide E1A-001. The protocol describing the CK-II assay has been reported elsewhere (D. Carroll et al., in prep.). When measuring for the effect of elevated cAMP-dependent kinase activity on phorbol-ester stimulation of CK-II, cells were preincubated for 30 min with 100 μM dibutyryl cAMP prior to TPA treatment. Cells were harvested and analyzed exactly as before. Data are presented as the amount of radioactive phosphate incorporated into E1A-001.

consistent with the hypothesis that protein kinase C mediates the activation of CK-II.

The potential involvement of CK-II in a cAMP-dependent signaling pathway is indicated by recent work from our laboratory. We found that serum stimulation of WI38 cells and adenovirus infection of BRK cells resulted in a 10- and 20-fold inhibition of cAMP-dependent protein kinase (cAK) activity, respectively (Carroll et al. 1987 and in prep.). The inhibition of cAK activity was concomitant with the initial activation of CK-II. This reciprocal relationship is present only at the early phase of CK-II activation (0–4.0 hr) and not at the later activation phases. The inhibition of cAK activity was not surprising, since several reports have indicated that decreased cellular levels of cAMP are a requisite for the transiting of cells out of a quiescent cell state during growth stimulation (McConlogue et al. 1983; Trepel and Neckers 1987). To determine if lower cAK activity levels are necessary for the activation of CK-II, we tested the effect of elevated cAMP levels on TPA-dependent stimulation of CK-II. We preincubated intact BRK cells with dibutyryl cAMP for 30 minutes prior to activation with TPA. As shown in Figure 2, dibutyryl cAMP effectively inhibits the ability of TPA to activate CK-II. These results suggest that both cAK- and protein-kinase-C-dependent pathways act to regulate the cellular activity of CK-II.

FUTURE DIRECTIONS

It is clear from the recent work in our laboratory (D. Carroll et al., in prep.), as well as from work in other laboratories (Sommercorn et al. 1987), that CK-II activity is regulated in response to signals that provoke changes in oncogene expression and the transcription of other genes. The real challenge for future research is to show a necessary linkage between CK-II-dependent phosphorylation of these nuclear substrates and the regulation of their activities. To meet this challenge, new strategies will be needed. Current protocols utilizing genetic strategies have been useful in assigning the gross boundaries of the functional domains of proteins but are not useful in assessing more subtle regulatory properties, such as the effects of posttranslational modification on protein activities. In our laboratory, we are currently studying the consequences of CK-II-dependent phosphorylation on protein function. By exploiting recent developments in microchemical analysis of phosphoproteins and in vitro transcription assays, we hope to determine the effects that the phosphorylation of transcriptional activator proteins within the CK-II motif have on the kinetics of transcription.

ACKNOWLEDGMENTS

We thank Dr. J. Figueiredo for her helpful discussions; and Jim Duffy, Dave Greene, and Michelle Carroll for the preparation of the art work. This work has been supported by National Institutes of Health grant CA-13106 and training grant CA-09311.

REFERENCES

Carroll, D., E. Moran, and D.R. Marshak. 1987. Adenovirus regulation of protein kinases. In *Abstracts from DNA Tumor Virus Meeting*, Cambridge University, Cambridge, England.

Figge, J., T. Webster, T. Smith, and E. Paucha. 1988. Prediction of similar transforming regions in simian virus 40 large T antigen, adenovirus E1A and *myc* oncoproteins. *J. Virol.* **62:** 1814.

Grässer, F.A., K.H. Scheidtmann, P.T. Tuazon, J.A. Traugh, and G. Walter. 1988. *In vitro* phosphorylation of SV40 large T antigen. *Virology* **165:** 13.

Greenberg, M.E. and E.B. Ziff. 1984. Stimulation of 3T3 cells induces transcription of the c-*fos* proto-oncogene. *Nature* **311:** 433.

Greenberg, M.E., L.A. Greene, and E.B. Ziff. 1985. Nerve growth factor and epidermal growth factor induce rapid transient changes in proto-oncogene transcription in PC12 cells. *J. Biol. Chem.* **260:** 14101.

Hathaway, G.M. and J.A. Traugh. 1982. Casein kinases: Multipotential protein kinases. *Curr. Top. Cell. Regul.* **21:** 101.

Heaney, M., J. Pierce, and T. Parsons. 1986. Site-directed mutagenesis of the *gag-myc* gene of avian myelocytomatosis virus 29: Biological activity and intracellular localization of structurally altered proteins. *J. Virol.* **60:** 167.

Hope, I.A., S. Mahadevan, and K. Struhl. 1988. Structural and functional characterization of the short acidic transcriptional activation region of yeast GCN4 protein. *Nature* **333:** 635.

Jenuwein, T. and R. Muller. 1987. Structure-function analysis of *fos* protein: A single amino acid change activates the immortalizing potential of v-*fos*. *Cell* **48:** 647.

Kalderon, D. and A.E. Smith. 1984. *In vitro* mutagenesis of a putative DNA binding domain of SV40 large T. *Virology* **139:** 109.

Kuenzel, E.A. and E.G. Krebs. 1985. A synthetic peptide substrate specific for casein kinase II. *Proc. Natl. Acad. Sci.* **82:** 737.

Kuenzel, E.A., J.A. Mulligan, J. Sommercorn, and E.G. Krebs. 1987. Substrate specificity of casein kinase II as deduced from studies with synthetic peptides. *J. Biol. Chem.* **262:** 9136.

Kuppuswamy, M.N. and G. Chinnadurai. 1987. Relationship between the transforming and transcriptional regulatory functions of adenovirus 2 E1A oncogene. *J. Virol.* **159:** 31.

Lillie, J.W., M.Green, and M.R. Green. 1986. An adenovirus E1A protein region required for transformation and transcriptional repression. *Cell* **46:** 1043.

Lillie, J.W., P.M. Loewenstein, M.R. Green, and M. Green. 1987. Functional domains of adenovirus type 5 E1A proteins. *Cell* **50:** 1091.

Ma, J. and M. Ptashne. 1987a. Deletion analysis of GAL4 defines two transcriptional activating segments. *Cell* **48:** 847.

———. 1987b. A new class of yeast transcriptional activators. *Cell* **51:** 113.

Maggio, F., F. Marchiori, G. Borin, G. Chessa, and L. Pinna. 1984. Synthetic peptides including acidic clusters as substrates and inhibitors of rat liver casein kinase II. *J. Biol. Chem.* **259:** 14576.

McConlogue, L.S., L.J. Marton, and P. Coffino. 1983. Growth regulatory effects of cyclic AMP and polyamine depletion are dissociable in mouse lymphoma cells. *J. Cell Biol.* **96:** 762.

Moran, E., and M.B. Mathews. 1987. Multiple functional domains in the adenovirus E1A gene. *Cell* **48:** 177.

Moran, E., B. Zerler, T.M. Harrison, and M.B. Mathews. 1986. Separate domains in the adenovirus E1A gene for immortalization activity and the activation of early virus genes. *Mol. Cell. Biol.* **6:** 3480.

Mulner-Lorillon, O., J. Marot, X. Cayla, R. Pouhle, and R.

Belle. 1988. Purification and characterization of a casein kinase II-type enzyme from *Xenopus laevis* ovary: Biological effects on the meiotic cell division of full-grown oocyte. *Eur. J. Biochem.* **171:** 107.

Paucha, E., D. Kalderon, R.W. Harvsy, and A.E. Smith. 1986. Simian virus 40 origin DNA binding domains in large T antigen. *J. Virol.* **57:** 50.

Ralston, R. and J.M. Bishop. 1983. The protein products of the *myc* and *myb* oncogenes and adenovirus E1A are structurally related. *Nature* **306:** 803.

Schneider, J. and E. Fanning. 1988. Mutations in the phosphorylation sites of simian virus 40 T antigen alter its origin DNA binding specificity for sites I or II and effect SV40 DNA replication activity. *J. Virol.* **62:** 1598.

Schneider, J.F., F. Fisher, C.R. Goding, and N.C. Jones. 1987. Mutational analysis of the adenovirus E1A gene; the role of transcriptional regulation in transformation. *EMBO J.* **6:** 2053.

Scheidtmann, K.-H., B. Echle, and G. White. 1982. Simian virus 40 large T antigen is phosphorylated at multiple sites clustered into two separate regions. *J. Virol.* **44:** 116.

Sommercorn, J., J.A. Mulligan, F.J. Lozeman, and E.G. Krebs. 1987. Activation of casein kinase II in response to insulin and to epidermal growth factor. *Proc. Natl. Acad. Sci.* **84:** 8834.

Sompayrac, L. and K.J. Danna. 1988. A new SV40 mutant that encodes a small fragment of T antigen transforms established rat and mouse cells. *J. Virol.* **163:** 391.

Stable, S., P. Argos, and L. Philipson. 1985. The release of growth arrest by microinjection of adenovirus E1A DNA. *EMBO J.* **4:** 2329.

Stone, J., T. deLange, G. Ramsay, E. Jakobovits, J.M. Bishop, H. Varmus, and W. Lee. 1987. Definition of regions in human c-*myc* that are involved in transformation and nuclear localization. *Mol. Cell. Biol.* **7:** 1697.

Struhl, K. 1988. The JUN oncoprotein, a vertebrate transcription factor, activates transcription in yeast. *Nature* **332:** 649.

Trepel, J.B. and L.M. Neckers. 1987. Transcriptional inactivation of c-*myc* and the transferrin receptor in dibutyryl-cyclic AMP-treated HL-60 cells. *Mol. Cell. Biol.* **7:** 2644.

Tremblay, M.L., C.J. McGlade, G.E. Gerber, and P.E. Branton. 1988. Identification of the phosphorylation sites in early region 1A proteins of adenovirus type 5 by amino acid sequencing of peptide fragments. *J. Biol. Chem.* **263:** 6375.

Whyte, P., H.E. Ruley, and E. Harlow. 1988. Two regions of the adenovirus early region 1A proteins are required for transformation. *J. Virol.* **62:** 257.

Zullo, J., C.D. Stiles, and R.L. Garcea. 1987. Regulation of c-*myc* and c-*fos* mRNA levels by polyomavirus: Distinct roles for the capsid protein VP_1, and the viral early proteins. *Proc. Natl. Acad. Sci.* **84:** 1210.

The Family of Protein Kinase C: Its Molecular Heterogeneity and Differential Expression

U. KIKKAWA,* K. OGITA,* M.S. SHEARMAN,* K. ASE,* K. SEKIGUCHI,* Z. NAOR,* A. KISHIMOTO,* Y. NISHIZUKA,* N. SAITO,† C. TANAKA,† Y. ONO,‡ T. FUJII,‡ AND K. IGARASHI,‡

**Department of Biochemistry and †Department of Pharmacology, Kobe University School of Medicine, Kobe 650, Japan; ‡Biotechnology Laboratories, Central Research Division, Takeda Chemical Industries, Osaka 532, Japan*

Although protein kinase C (PKC) was once considered as a single entity (Nishizuka 1984), molecular cloning and enzymological analysis have revealed the existence of multiple subspecies of PKC in a number of mammalian tissues. These subspecies have a common structure closely related to, yet slightly different from, one another. The relative distribution of PKC subspecies varies markedly with different areas of tissues and cell types examined. It is tempting to surmise that each member of this family has a specific function in the various cellular processes that occur during cell-to-cell communication. The heterogeneity of the members of the PKC family is briefly summarized in this paper. The integrated nomenclature used herein for the multiple PKC subspecies reported from various laboratories is given in Table 1.

EXPERIMENTAL PROCEDURES

cDNA clones and sequence analysis. Sprague-Dawley rats were employed. βI- and βII-cDNA clones were obtained from a rat brain library as described previously (Ono et al. 1986). α- and γ-cDNA clones were obtained by sequential plaque screening using synthetic oligonucleotide probes (Kikkawa et al. 1987b). δ-, ϵ-, and ζ-cDNAs were isolated by using a mixture of α-, βII-, and γ-cDNAs as probes under low stringency conditions (Ono et al. 1987a, 1988). Nucleotide sequence analysis was performed using the enzymatic chain termination method in conjugation with M13-derived vectors as described previously (Ono et al. 1986).

Expression in COS cells. Fresh monolayers of COS-7 cells with Dulbecco's modified Eagle's medium containing 5% fetal calf serum were transfected with the cDNAs in plasmids using the calcium phosphate coprecipitation method as described previously (Ono et al. 1987b).

PKC assay and fractionation. PKC was routinely assayed with calf thymus H1 histone as phosphate acceptor (Kikkawa et al. 1987a; Sekiguchi et al. 1988). The enzyme was extracted from tissues and purified by DE-52 followed by hydroxyapatite column chromatography. The detailed procedures were described elsewhere (Kikkawa et al. 1987a; Shearman et al. 1987; Kosaka et al. 1988).

Table 1. List of cDNA Clones of PKC Family

cDNA library	cDNA clones							References
Rat brain	α	type I	type II	γ	δ	ε	ζ	Kikkawa et al. (1987a,b) Ono et al. (1986, 1987a,b, 1988)
Rat brain		PKC III	PKC II	PKC I				Knopf et al. (1986)
Rat brain		RP41				RP16		Housey et al. (1987)
Bovine brain	α		β	γ				Parker et al. (1986) Coussens et al. (1986)
Human brain	α	β2	β1	γ				Coussens et al. (1986, 1987)
Human spleen		βI	βII					Kubo et al. (1987)
Rabbit brain	γ	β	α	δ		nPKC		Ohno et al. (1987, 1988a,b)
Nomenclature	α	βI	βII	γ	δ	ε	ζ	

Antibodies and immunological procedures. The monoclonal antibody that specifically recognizes γ-PKC was prepared as described previously (Kitano et al. 1987; Hashimoto et al. 1988). Polyclonal antibodies against βI- and βII-PKC were prepared by using synthetic oligopeptides that correspond to the carboxy-terminal end sequences of each subspecies (Ase et al. 1988). Western blot analysis, Northern blot analysis, and immunocytochemistry were performed as described in detail elsewhere (Kitano et al. 1987; Ono et al. 1987b; Ase et al. 1988; Kose et al. 1988; Saito et al. 1988).

RESULTS AND DISCUSSION

Members of the Family

Initially, four cDNA clones that encode α-, βI-, βII-, and γ-subspecies were found in the bovine (Coussens et al. 1986; Parker et al. 1986), rat (Knopf et al. 1986; Ono et al. 1986; Kikkawa et al. 1987b), rabbit (Ohno et al. 1987, 1988b), and human (Coussens et al. 1986) brain, and subsequently in human spleen (Kubo et al. 1987) cDNA libraries. Although a comparison of the complete structures of these four subspecies can be made for the rat and rabbit brain PKCs, the animal species difference in each enzyme subspecies is extremely small. The genes for the α-, β- (βI plus βII), and γ-subspecies are located on different chromosomes (Coussens et al. 1986). Partial genomic analysis has indicated that βI- and βII-subspecies are derived from a single RNA transcript by alternative splicing. More recently, another group of cDNA clones, encoding at least three further subspecies designated δ-, ε-, and ζ-PKC, have been isolated from a rat brain library by using a mixture of α-, βII-, and γ-cDNAs as probes under low stringency conditions (Ono et al. 1987a, 1988). The cDNA clone encoding ε-sequence has also been isolated from a rabbit brain library (Ohno et al. 1988a). These subspecies have a common structure closely related to, but clearly distinct from, the four subspecies initially described.

Figure 1 shows schematically the structures of PKC subspecies cloned to date. They are all composed of a single polypeptide chain, with the group of α-, βI-, βII-, and γ-subspecies each having four conserved (C_1–C_4) and five variable (V_1–V_5) regions. The βI- and βII-subspecies differ from each other only in a short range of about 50 amino acid residues at their carboxy-terminal end regions, and even in this area, they possess a high degree of sequence homology. Although the second group of δ- and ε-subspecies lack the conserved region C_2, the molecular mass of the enzyme molecules is, nevertheless, similar. The cDNA clone encoding presently available for the ζ-subspecies does not contain a complete reading frame for the enzyme.

The conserved region C_1 contains a tandem repeat of a cysteine-rich sequence, $CX_2\,CX_{13(14)}CX_2CX_7CX_7C$, where C is cysteine and X represents any amino acid. This sequence is similar to the consensus sequence of a "cysteine-zinc DNA-binding finger" that is found in many metalloproteins and DNA-binding proteins related to transcriptional regulation (Berg 1986). There is no evidence presently, however, indicating that PKC will bind to DNA. In general, the enzyme is composed of two domains, one of which, the carboxy-terminal half containing the regions C_3 and C_4, appears to be the protein kinase domain, since it shows large clusters of sequence homology with many other protein kinases.

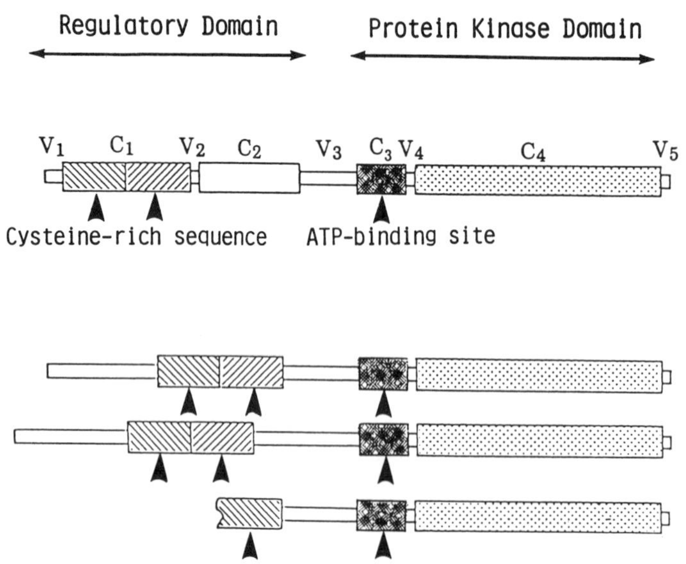

Clone	Amino Acids	Molecular weight
α	672	76,799
βI	671	76,790
βII	673	76,933
γ	697	78,366
δ	673	77,517
ε	737	83,474
ζ	> 492	——

Figure 1. Structures, amino acid residues, and deduced molecular weights of PKC subspecies. The conserved regions (C_1–C_4) and variable regions (V_1–V_5) are shown by thick and thin boxes, respectively. Each conserved region in the PKC family is emphasized. Cysteine-rich sequences and ATP-binding sites are indicated by triangles. The numbers of the amino acid residues and deduced molecular weights are summarized.

The conserved region C_3 has an ATP-binding sequence, GXGXXG----K, where G, K, and X represent glycine, lysine, and any amino acid, respectively. Although the conserved region C_4 contains an additional sequence similar to the ATP-binding site, GXGXXG----(X), the significance of this repeat remains unknown (Kikkawa et al. 1987b).

On the other hand, the conserved regions C_1 and C_2 apparently include the regulatory domain, but as yet the sites involved in calcium-, diacylglycerol-, and phospholipid-binding have not been identified with any certainty. Rather, the mechanism of PKC activation appears to be a complex interaction between these components (Nishizuka 1984, 1986).

PKC was originally detected as an undefined protein kinase, which was present in many tissues and could be activated by limited proteolysis with the Ca^{++}-dependent neutral protease, calpain (Kishimoto et al. 1983). This hydrolytic cleavage occurs at a site in the region V_3, resulting in the release of a catalytically fully active fragment, which subsequently is rapidly removed from the cell (A. Kishimoto et al., unpubl.). The physiological significance of this proteolysis has not been unequivocally established.

Differential Expression and Tissue Localization

Early studies using autoradiographic analysis after tritiated phorbol-12,13-dibutyrate binding have shown an uneven distribution of PKC in the brain (Nagle and Blumberg 1983; Worley et al. 1986). Northern blot analysis with specific oligonucleotide probes has also suggested a tissue-specific expression of some PKC subspecies (Coussens et al. 1986; Knopf et al. 1986; Ohno et al. 1987). In situ hybridization analysis has shown a distinct distribution of the transcripts of PKC subspecies in the rat brain and spleen (Brandt et al. 1987). Using a combination of biochemical, immunological, and cytochemical procedures with subspecies-specific antibodies, the relative activity and individual pattern of expression of multiple PKC subspecies in several tissues have recently been examined extensively and clarified in more detail in this laboratory and in others (Wood et al. 1986; Huang et al. 1987a,b; Kitano et al. 1987; Mochly-Rosen et al. 1987; Shearman et al. 1987; Ase et al. 1988; Hashimoto et al. 1988; Kose et al. 1988; Saito et al. 1988). Table 2 shows regional distribution of PKC subspecies in the rat brain.

PKC with γ-sequence is apparently expressed solely in the brain and spinal cord and is not found in other tissues and cell types. The highest enzyme activity is found in the hippocampus, cerebral cortex, amygdaloid complex, and cerebellar cortex. In cerebellar cortex, heavy immunoreaction of PKC with γ-sequence is observed in the cell bodies, dendrites, and axons in Purkinje cells (Fig. 2C). On the single cell level, γ-subspecies appear to be localized in the nerve endings of the Purkinje cell axon that terminate in the deep cerebellar nuclei, but not in the postsynaptic cell body itself. Immunoelectron microscopy analysis reveals that the γ-subspecies is associated with most membranous structures present throughout the cell, except for the nucleus. Mitochondria generally lack or poorly express PKC. The γ-subspecies develops postnatally and reaches maximum activity in the rat around 3 weeks after birth (Hashimoto et al. 1988). Although it seems premature to discuss the relationship of this PKC subspecies with any specific functions, it may play a role in specialized neuronal processes, such as long-term potentiation in the hippocampus.

Table 2. Distribution of PKC in Rat Nervous Tissue

	Specific activity (nmol/min/g tissue)			
	α	βI	βII	γ
Whole brain	4.9	1.2	8.5	5.2
Cerebrum	2.8	0.3	10.3	3.2
Cerebellum	6.0	3.2	10.3	21.5
Hippocampus	10.8	0.3	12.9	8.3
Spinal cord	3.2	1.1	2.4	0.2

PKC subspecies in various rat nervous tissue were separated and assayed as described in the text. The ratios of βI- and βII-subspecies were determined immunochemically using the antibodies specific to each subspecies as described previously (Shearman et al. 1987; Ase et al. 1988).

PKCs with βI- and βII-sequence also display differential expression in the brain and many other tissues, including endocrine tissues such as pancreatic islets and the pituitary gland. Normally, the activity of βII-subspecies far exceeds that of the βI-subspecies. These two subspecies are very similar, differing from each other only in the region V_5 as noted above. However, cytochemical analysis with polyclonal antibodies raised against synthetic peptides specific to each of the PKC subspecies indicates that, in certain tissues, a clear, distinct cellular expression is apparent (Ase et al. 1988). The rat cerebellar cortex illustrated in Figure 2 is an example of such. PKC with βI-sequence is localized mainly in the granular layer (Fig. 2B), whereas βII-subspecies is found primarily in the molecular layer, apparently in the presynaptic nerve endings terminating on the dendrites and cell body of the Purkinje cell (Fig. 2A). Immunoelectron microscopy analysis shows that, in some cells, βII-subspecies is associated primarily with the Golgi membrane, inferring a possible involvement in protein processing or transport (C. Tanaka et al. unpubl.).

On the other hand, PKC with α-sequence is widely distributed in many tissues and cell types. Most tissues, including liver, kidney, spleen, and testis, additionally contain β-subspecies in variable ratios. Some tissues such as heart, lung, and platelets appear to possess several undefined subspecies (Kosaka et al. 1988). In general, one cell type contains one or more than one subspecies of PKC. For instance, recent observation in our laboratory and others has shown that T and B lymphocytes and various cloned cell lines express α-, β-, and βII-subspecies in different ratios (Makowske et al. 1988; N. Berry et al. unpubl.). These subspecies, by light microscopy resolution, show a distinct intracellular location, depending on the state of differentiation or

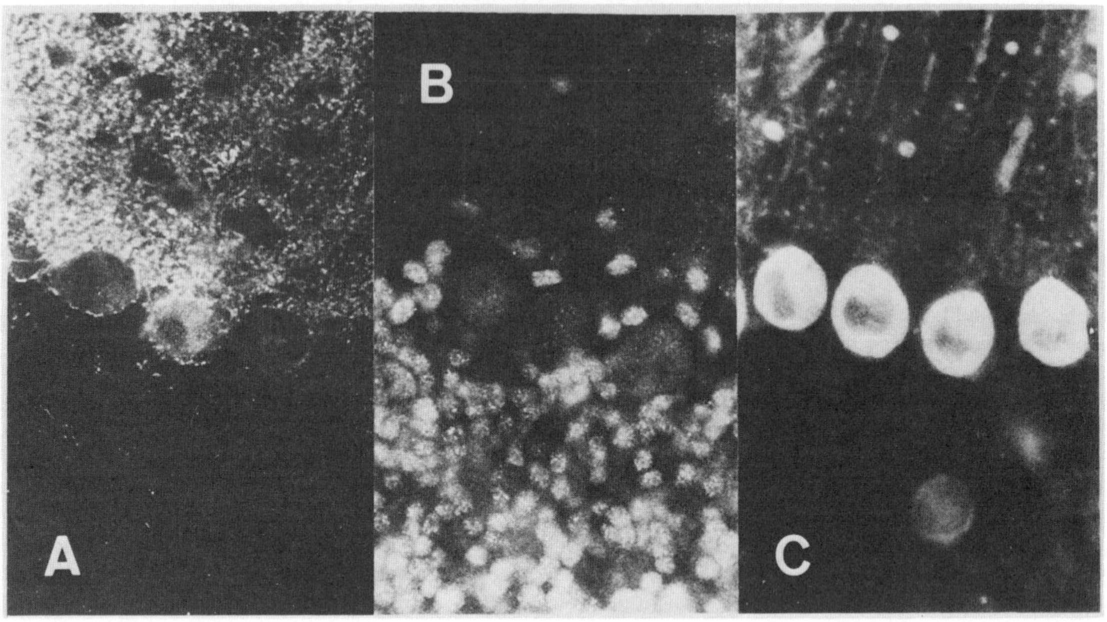

Figure 2. Immunostaining of the rat cerebellar cortex. The cerebellar sections were stained with the subspecies-specific antibodies against PKC by the immunofluorescent method. (A,B,C) Stained with the antibodies specific to βII-, βI-, and γ-subspecies, respectively. The detailed experimental conditions are described elsewhere (Ase et al. 1988).

proliferation. Cultured fibroblasts apparently express the α- but not β-subspecies (McCaffrey et al. 1987).

At present, the distribution and biochemical properties of the enzymes encoded by δ-, ϵ-, and ζ-subspecies have not been determined, but Northern blot analysis suggests that these members of the family are expressed in several tissues and organs in different ratios (Ono et al. 1988). Such a distinct regional localization and intracellular expression of various PKC subspecies per se may be an important factor in determining a variety of responses of different tissues and cells to external stimuli.

Individual Enzymological Properties

Direct enzymatic analysis of each PKC subspecies is a prerequisite for determining the potential roles of this enzyme family in signal transduction. The variable regions seen in each subspecies may play key roles in governing the individual enzymological characteristics and possibly localization and function. The diversity of the sequence in these regions allows separation of this enzyme into several subfractions after chromatography on a hydroxyapatite column (Huang et al. 1986; Jaken and Kiley 1987; Kikkawa et al. 1987a; Ono et al. 1987b; Pelosin et al. 1987). The structure and genetic identity of some of these subfractions have been determined by comparison with the enzymes expressed in COS-7 cells after transfection with individual PKC cDNA-containing plasmids. To date, three subfractions, types I, II, and III, have been shown to correspond to γ-, β- (βI- and βII-), and α-subspecies, respectively. PKCs with βI- and βII-sequence show nearly identical kinetic and catalytic properties and can be distinguished from each other only by immunochemical procedures (Ono et al. 1987b; Shearman et al. 1987; Ase et al. 1988). The subspecies having δ- and ϵ-sequence can be partially resolved using chromatography (Ono et al. 1988), but their correspondence to the subfractions chromatographically obtained from tissues remains to be clarified.

The PKC subfractions obtained thus far from various tissues are subtly different (Huang et al. 1986; Jaken and Kiley 1987; Sekiguchi et al. 1987, 1988; Asaoka et al. 1988). PKC with γ-sequence (type I) shows less activation by diacylglycerol but is significantly activated by the micromolar range of free arachidonic acid, as shown in Figure 3 (Sekiguchi et al. 1988; Z. Naor et al. unpubl.). Activation by arachidonic acid does not require Ca^{++}, nor does it depend on phospholipid and diacylglycerol. PKCs with βI- and βII-sequence (type II) show substantial activity without added Ca^{++} in the presence of diacylglycerol and phospholipid but respond much less to arachidonic acid. PKC with α-sequence (type III) shows properties apparently similar to γ-subspecies and, in addition, responds to high concentrations of free arachidonic acid only when Ca^{++} is increased. It is possible that some PKC subspecies may be activated at different phases of cellular responses by a series of phospholipid metabolites, such as diacylglycerol, and arachidonic acid and its metabolites, which successively appear subsequent to stimulation of the receptor.

In addition to these well-defined PKC subspecies, structurally undefined enzymes obtained from some tissues such as heart, lung, and platelets respond to phospholipid, diacylglycerol, and Ca^{++} in different ways (Kosaka et al. 1988). For instance, one enzyme obtained from human platelets is not sensitive to Ca^{++}

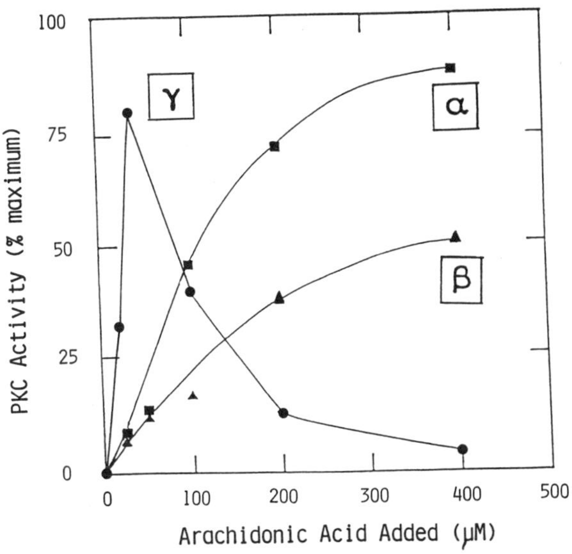

Figure 3. Activation of PKC subspecies by arachidonic acid. Each subspecies of PKC was assayed with various concentrations of arachidonic acid using calf thymus H_1 histone as phosphoacceptor. (●) PKC of γ-subspecies (type I); (▲) β-subspecies (type II); (■) α-subspecies (type III). The detailed experimental conditions are described elsewhere (Kikkawa et al. 1987a; Sekiguchi et al. 1988).

(M. Tsukuda et al., unpubl.). The exact relation of this enzyme to the members of the PKC family described above remains unclear. It is a general problem in protein phosphorylation research that, following disruption of the cell, most protein kinases show an activity to phosphorylate many proteins, which may or may not represent physiological substrates. In fact, a number of proteins, including cell-surface receptors, enzymes, and cytoskeletal and contractile proteins, have been proposed to serve as phosphate acceptors of PKC in cell-free systems. The facts that PKC subspecies may display different responsiveness to activators in vivo and that PKC subspecies are differentially distributed both in various cell types and within the same cell greatly complicate the search for physiological substrates for each member of the PKC family. Presumably, the PKC subspecies show different preferences for substrate proteins located in specific intracellular compartments. In an in vitro experiment, the EGF receptor of the A431 epidermoid carcinoma cells was found to be phosphorylated most rapidly by the ubiquitous α-subspecies of PKC, whereas it is more slowly phosphorylated by the brain-specific γ-subspecies (Ido et al. 1987). Thus, important questions remain to be answered concerning both the in vivo mechanism of activation of PKC and the true physiological target proteins for each subspecies.

ACKNOWLEDGMENTS

The investigations reported from the Department of Biochemistry, Kobe University School of Medicine, were supported by research grants from the Research Fund of the Ministry of Education, Science and Culture, Japan; Muscle Dystrophy Association, U.S.; Yamanouchi Foundation for Research on Metabolic Disorders; Merck Sharp & Dohme Research Laboratories; Ajinomoto Central Research Laboratories; Meiji Institute Health Sciences; and New Lead Research Laboratories of Sankyo Company.

REFERENCES

Asaoka, Y., U. Kikkawa, K. Sekiguchi, M.S. Shearman, Y. Kosaka, Y. Nakano, T. Satoh, and Y. Nishizuka. 1988. Activation of a brain-specific protein kinase C subspecies in the presence of phosphatidylethanol. *FEBS Lett.* **231:** 221.

Ase, K., N. Saito, M.S. Shearman, U. Kikkawa, Y. Ono, K. Igarashi, C. Tanaka, and Y. Nishizuka. 1988. Distinct cellular expression of βI- and βII-subspecies of protein kinase C in rat cerebellum. *J. Neurosci.* (in press).

Berg, J.M. 1986. Potential metal-binding domains in nucleic acid binding proteins. *Science* **232:** 485.

Brandt, S.J., J.E. Niedel, R.M. Bell, and W.S. Young III. 1987. Distinct patterns of expression of different protein kinase C mRNAs in rat tissues. *Cell* **49:** 57.

Coussens, L., L. Rhee, P.J. Parker, and A. Ullrich. 1987. Alternative splicing increases the diversity of the human protein kinase C family. *DNA* **6:** 389.

Coussens, L., P.J. Parker, L. Rhee, T.L. Yang-Feng, E. Chen, M.D. Waterfield, U. Francke, and A. Ullrich. 1986. Multiple, distinct forms of bovine and human protein kinase C suggest diversity in cellular signaling pathways. *Science* **233:** 859.

Hashimoto, T., K. Ase, S. Sawamura, U. Kikkawa, N. Saito, C. Tanaka, and Y. Nishizuka. 1988. Postnatal development of a brain specific protein kinase C in rat. *J. Neurosci.* **8:** 1678.

Housey, G.M., C.A. O'Brian, M.D. Johnson, P. Kirschmeier, and I.B. Weinstein. 1987. Isolation of cDNA clones encoding protein kinase C: Evidence for a protein kinase C-related gene family. *Proc. Natl. Acad. Sci.* **84:** 1065.

Huang, K.-P., H. Nakabayashi, and F.L. Huang. 1986. Isozymic forms of rat brain Ca^{2+}-activated and phospholipid-dependent protein kinase. *Proc. Natl. Acad. Sci.* **83:** 8535.

Huang, F.L., Y. Yoshida, H. Nakabayashi, and K.-P. Huang. 1987a. Differential distribution of protein kinase C isozymes in the various regions of brain. *J. Biol. Chem.* **262:** 15714.

Huang, F.L., Y. Yoshida, H. Nakabayashi, J.L. Knopf, W.S. Young III, and K.-P. Huang. 1987b. Immunochemical identification of protein kinase C isozymes as products of discrete genes. *Biochem. Biophys. Res. Commun.* **149:** 946.

Ido, M., K. Sekiguchi, U. Kikkawa, and Y. Nishizuka. 1987. Phosphorylation of the EGF receptor from A431 epidermoid carcinoma cells by three distinct types of protein kinase C. *FEBS Lett.* **219:** 215.

Jaken, S. and S.C. Kiley. 1987. Purification and characterization of three types of protein kinase C from rabbit brain cytosol. *Proc. Natl. Acad. Sci.* **84:** 4418.

Kikkawa, U., Y. Ono, K. Ogita, T. Fujii, Y. Asaoka, K. Sekiguchi, Y. Kosaka, K. Igarashi, and Y. Nishizuka. 1987a. Identification of the structures of multiple subspecies of protein kinase C expressed in rat brain. *FEBS Lett.* **217:** 227.

Kikkawa, U., K. Ogita, Y. Ono, Y. Asaoka, M.S. Shearman, T. Fujii, K. Ase, K. Sekiguchi, K. Igarashi, and Y. Nishizuka. 1987b. The common structure and activities of four subspecies of rat brain protein kinase C. *FEBS Lett.* **223:** 212.

Kishimoto, A., N. Kajikawa, M. Shiota, and Y. Nishizuka.

1983. Proteolytic activation of calcium-activated, phospholipid-dependent protein kinase by calcium-dependent neutral protease. *J. Biol. Chem.* **258:** 1156.

Kitano, T., T. Hashimoto, U. Kikkawa, K. Ase, N. Saito, C. Tanaka, Y. Ichimori, K. Tsukamoto, and Y. Nishizuka. 1987. Monoclonal antibodies against rat brain protein kinase C and their application to immunocytochemistry in nervous tissues. *J. Neurosci.* **7:** 1520.

Knopf, J.L., M.-H. Lee, L.A. Sultzman, R.W. Kriz, C.R. Loomis, R.M. Hewick, and R.M. Bell. 1986. Cloning and expression of multiple protein kinase C cDNAs. *Cell* **46:** 491.

Kosaka, Y., K. Ogita, K. Ase, H. Nomura, U. Kikkawa, and Y. Nishizuka. 1988. The heterogeneity of protein kinase C in various rat tissues. *Biochem. Biophys. Res. Commun.* **151:** 973.

Kose, A., N. Saito, H. Ito, U. Kikkawa, Y. Nishizuka, and C. Tanaka. 1988. Electron microscopic localization of type I protein kinase C in rat Purkinje cells. *J. Neurosci.* (in press).

Kubo, K., S. Ohno, and K. Suzuki. 1987. Primary structure of human protein kinase C βI and βII differ only in their C-terminal sequences. *FEBS Lett.* **223:** 138.

Makowske, M., R. Ballester, Y. Cayre, and O.M. Rosen. 1988. Immunochemical evidence that three protein kinase C isozymes increase in abundance during HL-60 differentiation induced by dimethyl sulfoxide and retinoic acid. *J. Biol. Chem.* **263**: 3402.

McCaffrey, P.G., M.R. Rosner, U. Kikkawa, K. Sekiguchi, K. Ogita, K. Ase, and Y. Nishizuka. 1987. Characterization of protein kinase C from normal and transformed cultured murine fibroblasts. *Biochem. Biophys. Res. Commun.* **146:** 140.

Mochly-Rosen, D., A.I. Basbaum, and D.E. Koshland, Jr. 1987. Distinct cellular and regional localization of immunoreactive protein kinase C in rat brain. *Proc. Natl. Acad. Sci.* **84:** 4660.

Nagle, D. and P.M. Blumberg. 1983. Regional localization by light microscopic autoradiography of receptors in mouse brain for phorbol ester tumor promoters. *Cancer Lett.* **18:** 35.

Nishizuka, Y. 1984. The role of protein kinase C in cell surface signal transduction and tumour promotion. *Nature* **308:** 693.

———. 1986. Studies and perspectives of protein kinase C. *Science* **233:** 305.

Ohno, S., Y. Akita, Y. Konno, S. Imajoh, and K. Suzuki. 1988a. A novel phorbol ester receptor/protein kinase, nPKC, distantly related to the protein kinase C family. *Cell* **53:** 731.

Ohno, S., H. Kawasaki, Y. Konno, M. Inagaki, H. Hidaka, and K. Suzuki. 1988b. A fourth type of rabbit protein kinase C. *Biochemistry* **27:** 2083.

Ohno, S., H. Kawasaki, S. Imajoh, K. Suzuki, M. Inagaki, H. Yokokura, T. Sakoh, and H. Hidaka. 1987. Tissue-specific expression of three distinct types of rabbit protein kinase C. *Nature* **325:** 161.

Ono, Y., T. Fujii, K. Ogita, U. Kikkawa, K. Igarashi, and Y. Nishizuka. 1987a. Identification of three additional members of rat protein kinase C family: δ-, ϵ-, and ζ- subspecies. *FEBS Lett.* **226:** 125.

———. 1988. The structure, expression, and properties of additional members of the protein kinase C family. *J. Biol. Chem.* **263:** 6927.

Ono, Y., T. Kurokawa, T. Fujii, K. Kawahara, K. Igarashi, U. Kikkawa, K. Ogita, and Y. Nishizuka. 1986. Two types of complementary DNAs of rat brain protein kinase C, heterogeneity determined by alternative splicing. *FEBS Lett.* **206:** 347.

Ono, Y., U. Kikkawa, K. Ogita, T. Fujii, T. Kurokawa, Y. Asaoka, K. Sekiguchi, K. Ase, K. Igarashi, and Y. Nishizuka. 1987b. Expression and properties of two types of protein kinase C: Alternative splicing from a single gene. *Science* **236:** 1116.

Parker, P.J., L. Coussens, N. Totty, L. Rhee, S. Young, E. Chen, S. Stabel, M.D. Waterfield, and A. Ullrich. 1986. The complete primary structure of protein kinase C—the major phorbol ester receptor. *Science* **233:** 853.

Pelosin, J.-M., I. Vilgrain, and E.M. Chambaz. 1987. A single form of protein kinase C is expressed in bovine adrenocortical tissue, as compared to four chromatographically resolved isozymes in rat brain. *Biochem. Biophys. Res. Commun.* **147:** 382.

Saito, N., U. Kikkawa, Y. Nishizuka, and C. Tanaka. 1988. Distribution of protein kinase C-like immunoreactive neurons in rat brain. *J. Neurosci.* **8:** 369.

Sekiguchi, K., M. Tsukuda, K. Ase, U. Kikkawa, and Y. Nishizuka. 1988. Mode of activation and kinetic properties of three distinct forms of protein kinase C from rat brain. *J. Biochem.* **103:** 759.

Sekiguchi, K., M. Tsukuda, K. Ogita, U. Kikkawa, and Y. Nishizuka. 1987. Three distinct forms of rat brain protein kinase C: Differential response to unsaturated fatty acids. *Biochem. Biophys. Res. Commun.* **145:** 797.

Shearman, M.S., Z. Naor, U. Kikkawa, and Y. Nishizuka. 1987. Differential expression of multiple protein kinase C subspecies in rat central nervous tissue. *Biochem. Biophys. Res. Commun.* **147:** 911.

Wood, J.G., P.R. Girard, G.J. Mazzei, and J.F. Kuo. 1986. Immunocytochemical localization of protein kinase C in identified neuronal compartments of rat brain. *J. Neurosci.* **6:** 2571.

Worley, P.F., J.M. Baraban, and S.H. Snyder. 1986. Heterogeneous localization of protein kinase C in rat brain: Autoradiographic analysis of phorbol ester receptor binding. *J Neurosci.* **6:** 199.

Protein Kinase C Regulation by Sphingosine/Lysosphingolipids

R.M. BELL,* C.R. LOOMIS,* AND Y.A. HANNUN†

*Department of *Biochemistry and †Medicine, Duke University Medical Center, Durham, North Carolina 27710*

Investigations on protein kinase C, actually a superfamily of seven or more closely related kinases (Ono et al. 1988), have led to connections between diverse areas of biology as a consequence of its central role in signal transduction processes (Nishizuka 1986) and its intimate regulation by cellular lipids and breakdown products of these lipids. This paper describes, in part, these biological connections, with emphasis on the role of protein kinase C in cell growth, regulation, differentiation, and oncogenesis. The main emphasis of the paper is on the connections and involvement of cellular lipids, both glycerolipids (Bell 1986) and sphingolipids (Hannun et al. 1986b), in the regulation of protein kinase C. The discovery that sphingolipid breakdown products inhibit protein kinase C has led to a unifying hypothesis with broad implications for cell regulation, pathogenesis, and medicine (Hannun et al. 1986b; Hannun and Bell 1987; Bell et al. 1988). The insights afforded by the connection of the sphingolipids to the major cellular signal transduction pathways may ascribe, in part, function to the diversity of the sphingolipids.

Discoveries of Protein Kinase C and Diacylglycerol Second Messengers

Within one decade from its discovery, protein kinase C permanently altered contemporary thinking about cellular regulation, unified once-divergent fields of research, and ignited interest in the cellular lipids and their breakdown products (Nishizuka 1986). Protein kinase C was discovered in rat brain cytosol as a noncyclic nucleotide-dependent protein kinase (Inoue et al. 1977; Takai et al. 1977 a,b). Protein kinase C could be activated by proteolysis or by factors present in the membrane and Ca^{++} (Takai et al. 1979a,b). The membrane factors proved to be lipids (Takai et al. 1979a,b). At first, phospholipids (notably phosphatidylserine [PtdS]) and Ca^{++} were considered sufficient to activate this novel kinase. When crude lipid extracts from brain proved more effective than PtdS in their ability to activate protein kinase C at reduced levels of Ca^{++}, a neutral lipid, *sn*-1,2-diacylglycerol (DAG), was discovered to increase the enzyme's affinity for both Ca^{++} and PtdS (Kishimoto et al. 1980). DAG was inferred to be an intracellular "second messenger" because DAG could activate the enzyme at resting levels of Ca^{++} and because DAG was produced in response to extracellular agents including neurotransmitters, growth factors, and hormones from the breakdown of phosphatidylinositol (PtdIns) (Kishimoto et al. 1980; Berridge and Irvine 1984; Nishizuka 1984a,b) (Fig. 1). There is much evidence in support of the second-messenger functions of DAG (Nishizuka 1984a,b; Ganong et al. 1986, 1987). Thus, protein kinase C provided three connections to cellular lipids: (1) phospholipid as cofactor, (2) DAG as cofactor and second messenger, and (3) the PtdIns as reservoirs of DAG. Other connections have now been made to other cellular glycerolipids as sources of DAG second messengers (for review, see Bishop and Bell 1988).

Protein Kinase C Is the Phorbol Ester Receptor

The involvement of protein kinase C in tumor promotion was proved by the demonstration that protein kinase C is the principal, if not exclusive, intracellular receptor for phorbol esters and other tumor promoters (Castagna et al. 1982; Niedel et al. 1983; Ashendel 1985; Nishizuka 1986). DAG and phorbol esters bind to protein kinase C at common or overlapping sites and activate it. Phorbol esters are more potent and, unlike DAGs, are not rapidly metabolized. Cell-permeable DAGs such as 1-oleoyl-2-acetylglycerol (OAG) and *sn*-1,2-dioctanoylglycerol (diC_8) mimic most, but not all, of the effects of phorbol esters (Ganong et al. 1987).

Transmembrane Signaling, Protein Kinase C, and Oncogene Products

The transmembrane signaling events depicted in Figure 1 summarize the key molecular events and make the connections between protein kinase C, DAG, and oncogene products (Bell 1986). In brief, extracellular agents interact with specific cell-surface receptors that are coupled through G-protein (GTP-dependent) activation of phospholipases to produce DAG. The cleavage of phosphatidylinositol 4,5-bisphosphate ($PtdInsP_2$) to DAG and inositol 1,4,5-trisphosphate ($InsP_3$) is depicted, but other glycerolipids can be converted to DAG (Bishop and Bell 1988). $InsP_3$ functions to mobilize intracellular Ca^{++} (Berridge 1987). Several oncogene products, *sis, erbB, ras,* and *crk* (Bell 1986; Mayer et al. 1988), resemble the elements involved in transmembrane signaling. Evidence demonstrating that oncogene products function to alter the levels of DAG and $InsP_3$, critical second messengers, continues to

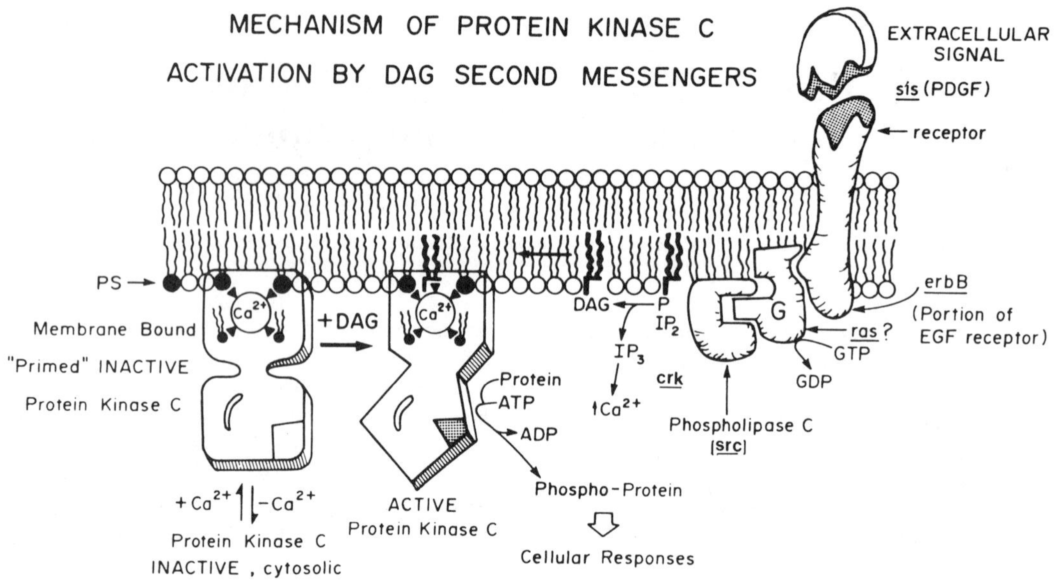

Figure 1. Transmembrane signaling. The model depicts transmembrane signaling events, the involvement of oncogene products, and the importance of protein kinase C in cellular regulation. This is an updated version. (Reprinted, with permission, from Bell 1986.)

mount (Preiss et al. 1986; for review, see Bishop and Bell 1988). Evidence for the critical roles of DAGs in cell proliferation and in cellular regulation (Nishizuka 1984a; Rozengurt et al. 1984; Davis et al. 1985; Ganong et al. 1987) also continues to mount using cell-permeable DAGs.

Protein Kinase C cDNA Cloning Reveals a Large Family of Closely Related Kinases

Sequence data from tryptic peptides of rat and bovine protein kinase C permitted oligonucleotides to be synthesized and cDNA libraries to be probed (Coussens et al. 1986; Knopf et al. 1986; Ono et al. 1986a,b; Parker et al. 1986; Housey et al. 1987). Such work led to the discovery of four forms of protein kinase C, which displayed tissue and cell-type expression. The βI and βII forms were derived by alternate splicing; three genes were involved and were mapped to different human chromosomes (Coussens et al. 1986). Three forms of protein kinase C could be isolated by hydroxyapatite chromatography (Huang et al. 1986). Subsequently, additional cDNA, δ, ϵ, ϵ', and ζ, were isolated using probes hybridized under reduced levels of stringency (Ono et al. 1988).

Expression of Protein Kinase C Alters Growth Regulation and Enhances Tumorigenicity

Protein kinase C-I (γ) cDNA, which is an isoform expressed normally in brain but not in fibroblasts, was transfected into NIH-3T3 cells. Expression of the γ-form of protein kinase C, as evidenced by Northern and Western analyses, enhanced [^{3}H]phorbol dibutyrate (PDBu) binding, and kinase activity caused altered growth regulation (Persons et al. 1988). The 3T3 transfectants had a lessened dependence on serum for growth, grew to a high saturation density, and were more refractile; they were not capable of forming colonies in soft agar. Similar results were obtained using protein kinase C-II (Housey et al. 1988). The transfected cell lines proved more tumorigenic than vector-transfected cell lines when injected into nude mice. Cell lines derived from the tumor retained protein kinase C-I in four of six cases. Interestingly, protein kinase C-γ also caused elevated DAG levels; the reason for this elevation is not understood (Persons et al. 1988).

Domain Structure and Structural Comparisons

The four protein kinase C cDNAs, α, βI, βII, and γ, from rat brain, encode single polypeptides that have five variable and four conserved regions (Fig. 2) (Ono et al. 1988). The carboxy-terminal two-thirds shows homology with the superfamily of protein kinases, contains an ATP-binding sequence, and is almost certainly the catalytic domain that could be released from the enzyme by mild proteolysis. The amino-terminal region contained two tandem repeated cysteine-rich regions C_1 and C_2; this region corresponds to the regulatory domain that interacts with phospholipid, calcium, and DAG (phorbol esters). A 32-kD fragment, released upon mild proteolysis, bound phorbol esters with high affinity in a phospholipid- and calcium-dependent manner (Lee and Bell 1986; Hoshijima et al. 1986). Interestingly, the cysteine residues fit the zinc finger motif. Two of the newest members of the protein kinase family, ϵ and ϵ', possess only one of the cysteine-rich domains. Near the amino terminus of the α, βI, and βII forms, a pseudosubstrate site exists that is believed

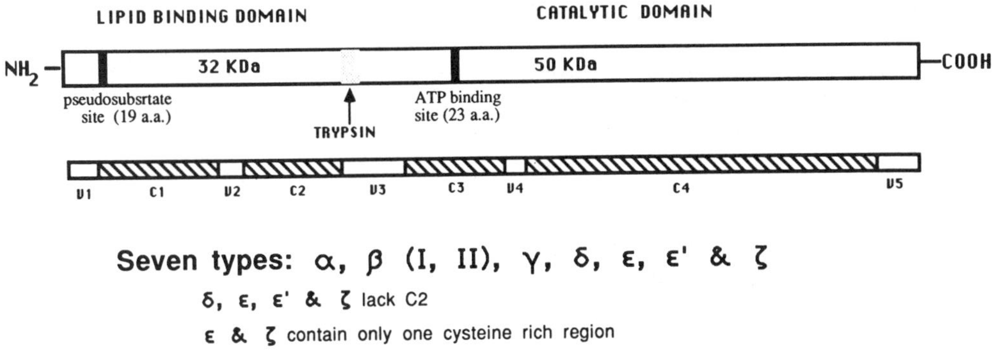

Figure 2. Protein kinases C. The figure depicts the domain structure of protein kinase C and compares the seven types of protein kinase C.

to occupy the active site until displaced upon activation (House and Kemp 1987).

Mechanism of Protein Kinase C Regulation by PtdS, Ca^{++}, and DAG (Phorbol Esters)

Studies on protein kinase C have generally employed sonically dispersed preparations of phospholipids and DAGs/phorbol esters (Nishizuka 1984a; Bell 1986). The physical properties of the lipids in such aggregates precluded efforts to determine the number of phospholipids and DAG required for activation and the oligomeric state of protein kinase C. Mixed micellar assay methods for protein kinase C were developed by our laboratory to overcome these problems (Hannun et al. 1985, 1986a). Mixed micellar methods permit the number of PtdS and DAG molecules to be systematically and independently varied within Triton X-100, B-octylglycoside, or other detergent micelles (Hannun et al. 1986a). Mixed micellar methods were also developed for [^{3}H]PDBu binding (Hannun and Bell 1986). Activation by DAG showed Hill numbers of 1, whereas activation by PtdS was highly cooperative (Hill numbers near 4.8). Molecular sieve chromatography demonstrated that monomeric (80 kD) protein kinase C was the active species and that it was the phorbol-dibutyrate-binding species. The dependencies strongly suggested that a single molecule of DAG was required for activation. Furthermore, one or more calciums were required, and four or more molecules of PtdS were required (Hannun et al. 1985, 1986a).

DAG structure-activity relationships revealed that activation required a micelle or membrane-bound DAG; activation was stereospecific and occurred with a high degree of specificity (Ganong et al. 1986). Three points of contact were inferred: one bond to the 3-hydroxyl and two bonds to the 1 and 2 oxygen esters (Ganong et al. 1986).

The stoichiometries, the three points of contact to DAG, and phospholipid structure-activity relationships suggested a model where four molecules of PtdS on the surface of the micelle or membrane-bound Ca^{++} through the four carboxyl groups (Ganong et al. 1986). Protein kinase C binds to the 4-PtdS-Ca^{++} complex to become membrane-bound but not active. Then DAG binds to the 4-PtdS-Ca^{++} · protein kinase C complex, occupying a position beneath the Ca^{++} coordination sphere such that a direct bond between DAG and Ca^{++} occurs as well as two bonds directly to protein kinase C (Ganong et al. 1986). This model is speculative but has proven useful in searching for inhibitors.

Protein Kinase C Inhibitors

Interpretation of studies with putative protein kinase C inhibitors is plagued with difficulties because of the necessity to maintain the lipid cofactors, essential for protein kinase C activity, in the face of inhibitors that are often amphipathic and/or hydrophobic in character. The mixed micellar methods can be extended to amphipathic inhibitors wherein the number of inhibitors can be systematically and independently varied. Since protein kinase C activity is independent of micelle number, an inhibitor whose potency is demonstrated in proportion to the number of micelles present is surface-active. Inhibition, like activation, must then be expressed as surface concentration (mole%), rather than bulk concentration (Bell et al. 1988).

The critical role of protein kinase C in oncogenesis and tumor promotion led us to investigate whether antitumor agents inhibited protein kinase C. The mixed micellar methods have been used to study the mechanism of inhibition. Table 1 shows several inhibitors discovered by other investigators and several that our group has uncovered. The aminoacridine antitumor agents were discovered because of the model in which they could bridge the 4-PtdS · Ca^{++} · protein kinase C complex and block DAG/phorbol ester binding

Table 1. Protein Kinase C Inhibitors

Inhibitors	Reference
H-7	Hidaka et al. (1984)
Staurosporine	Tamaoki et al. (1986)
Adriamycin	Wise et al. (1982)
Chlorpromazine	Mori et al. (1980)
Sangivamycin	Loomis and Bell (1988)
Aminoacridines	Hannun and Bell (1988)
Sphingosine/lysosphingolipids	Hannun and Bell (1987)

(Hannun and Bell 1988). Sangivamycin was discovered to inhibit competitively with respect to ATP and to be the sole protein kinase C inhibitor among numerous purine/pyrimidine analogs tested (Loomis and Bell 1988). Perhaps the most interesting class of inhibitors discovered to date are sphingosine and lysosphingolipids because of their potential as cellular regulators (Hannun et al. 1986b). The remainder of this paper discusses work and implications emanating from the discovery that sphingosine inhibits protein kinase C.

Sphingosine/Lysosphingolipids Inhibit Protein Kinase C

Sphingosine was discovered to be a potent and reversible inhibitor of protein kinase C as a result of looking at the molecular constituents of ceramide, which has structural features similar to DAG. Ceramide had no effect on protein kinase C activity, but sphingosine potently inhibited about as well as DAG activated. Inhibition was strongly dependent on mixed micelle composition and was subject to surface dilution. Therefore, the surface concentration of sphingosine (mole%) is the appropriate concentration unit (Hannun et al. 1986b).

Using the mixed micellar methods, detailed studies revealed that sphingosine interacted with the lipid-binding domain of protein kinase C, did not inhibit the catalytic fragment in the presence of phospholipids or Triton X-100 micelles, was a competitive inhibitor with respect to DAG (Fig. 3), and blocked [^{3}H]PDBu binding to protein kinase C but not the association of protein kinase C with the micellar complex (Fig. 4). A variety of lysosphingolipids were found to mimic the effects of sphingosine when analyzed in vitro using the mixed micellar methods (Hannun and Bell 1987).

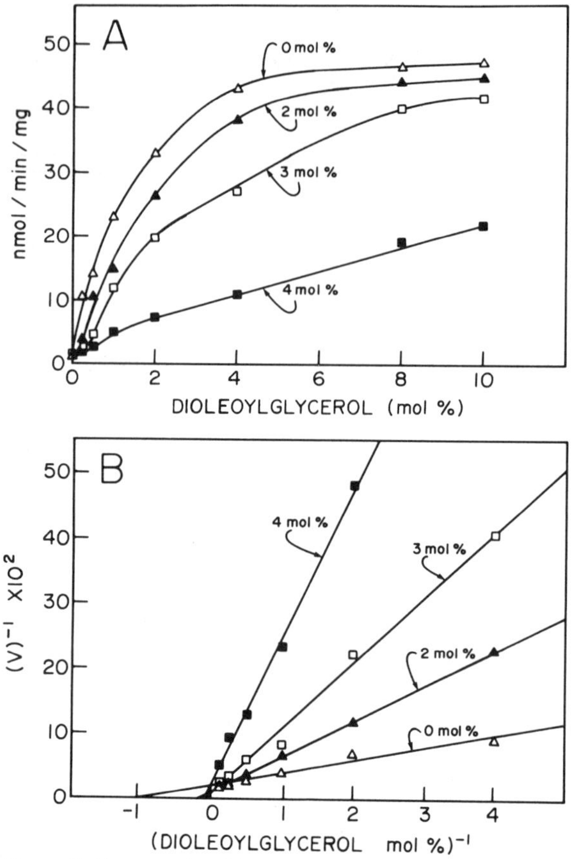

Figure 3. Pattern of sphingosine inhibition with respect to DAG. Protein kinase C was determined at 6 mole% PtdS, 50 μM $CaCl_2$, and variable $diC_{18:1}$ (*sn* 1,2-dioleoglycerol). Concentrations are shown in mole%. (Reprinted, with permission, from Hannun et al. 1986b.)

Sphingosine Inhibits Protein Kinase C in Cells

Sphingosine but not *N*-acetylsphingosine inhibited protein kinase C activity and [^{3}H]PDBu binding in numerous cell types, including human platelets, HL60 cells, and neutrophils (Hannun et al. 1986b, Merrill et al. 1986; Wilson et al. 1986). In platelets and neutrophils, sphingosine inhibition occurred without cell disruption and without blocking normal calcium transients or the generation of DAG (Hannun et al. 1987; Lambeth et al. 1988). In platelets, sphingosine blocked the second wave of aggregation in response to agonists and secretion. The addition of diC_8 or phorbol esters to sphingosine-inhibited platelets triggered secretion and aggregation, demonstrating that sphingosine inhibition was reversed by protein kinase C activators (Hannun et al. 1987). In neutrophils, sphingosine blocked the oxidative burst, secretion from azurophillic but not specific granules (Wilson et al. 1987), and the production of platelet-activating factor (PAF) and leukotrine B_4 (LTB_4) (McIntyre et al. 1987). Sphingosine also blocked protein kinase C translocation (Speizer et al. 1987) and down-regulation (Grove and Mastro 1988). Table 2 presents some of the pharmacologic effects of sphingosine when added to cells.

The effects of sphingosine have not rigorously been shown to be specific. Use of sphingosine in any cell system requires careful investigation of its selectivity and sites of action. To date, some enzymes in certain systems have been shown to be unaffected by sphingosine. These include phospholipase C, DAG kinase, cAMP- and cGMP-dependent protein kinases, and the calcium calmodulin-dependent myosin light-chain kinase. High concentrations of sphingosine did inhibit some of these enzymes (Y. Hannun and R.M. Bell, in prep.). To date, few studies have been done with lysosphingolipids.

Sphingosine/Lysosphingolipids: A Role in Pathogenesis

Sphingosine, psychosine, and other long-chain amino bases are cytotoxic to some cell lines (Merrill 1983). Sphingosine shows profound cytotoxic effects on CHO cells and, if added at concentrations greater than needed to inhibit protein kinase C, caused the death of platelets, neutrophils, macrophages, and HL60 cells (Y. Hannun, R.M. Bell, in prep.). Thus, the possibility

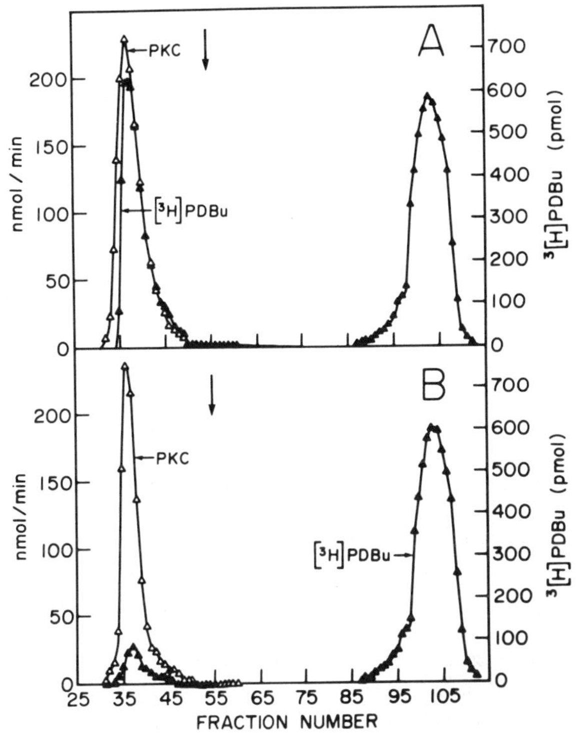

Figure 4. Molecular sieve chromatographic analysis of protein kinase C (PKC) interaction with PDBu. (*A*) Mixed micelles contained 16 mole% of PtdS. A fraction of [^{3}H]PDBu (▲) coelutes with protein kinase C (△) at an $M_r = 200{,}0000$. (*B*) Mixed micelles contained 16 mole% of PtdS and 12 mole% of sphingosine. Most of the bound [^{3}H]PDBu was displaced. Arrows indicate elution position of blue dextran protein kinase C when it is not bound to mixed micelles. (Reprinted, with permission, from Bell et al. 1988.)

that uncontrolled production of sphingosine/lysosphingolipids might function pathologically caused us to search for diseases where this might occur.

The sphingolipidoses, a family of diseases where sphingolipids accumulate as a consequence of defective lysosomal breakdown, were an initial point of inquiry. The same lysosomal enzyme that degrades the parental sphingolipid also functions, where it has been studied, to degrade the lysosphingolipid (the *N*-deacylated parental sphingolipid). Thus, cells genetically trapped to accumulate parental molecules would also be genetically trapped to accumulate the lysosphingolipid (Hannun and Bell 1987). In three cases, Tay-Sachs, Neimann-Pick, and Krabbe's diseases, accumulation of lysosphingolipids was reported (for review, see Hannun and Bell 1987). We suggested that lysosphingolipids represented the missing link in the pathogenesis of these disorders. Accumulation of lysosphingolipids occurs in these disorders to an extent where inhibition of protein kinase C may occur. At low concentrations of lysosphingolipids, signal transduction processes involving protein kinase C would be impaired; at higher concentrations, these molecules become cytotoxic. The best evidence for this hypothesis comes from the work of Miyatake and Suzuki (1972), who showed that psychosine rather than its parent cerebroside was the pathogenic agent.

Since protein kinase C is known to be involved in the differentiation of human promyelocytic leukemic cells (HL60 cells), we wondered whether the developmental trapping of these cells may not be a consequence of too much sphingosine (Hannun et al. 1987). In fact, sphingosine levels in HL60 cells are higher than in other cells

Table 2. Effects of Sphingosine on Cells

Cells and effects	References
Human neutrophils:	
inhibit oxidative burst and secretion of specific granules	Wilson et al. (1986); Wilson et al. (1987);
inhibit PAF and LTB_4 production	McIntyre et al. (1987)
HL60 promyelocytic leukemic cells:	Merrill et al. (1986)
inhibit differentiation	
Human platelets:	Hannun et al. (1986b);
inhibit secretion and aggregation	Hannun et al. (1987)
S49 lymphocytes:	Speizer et al. (1987)
block protein kinase C translocation	
Lymphocytes:	Grove and Mastro (1988)
prevent down-regulation	
3T3-L1 cells:	Nelson and Murray (1986)
inhibit insulin-dependent 2-deoxyglucose uptake	
PC12 cells:	Hall et al. (1988)
inhibit NGF-induced neurite outgrowth	
A431 cells:	Faucher et al. (1988)
inhibit EGF receptor phosphorylation	
increase affinity of EGF receptor for EGF	Hanai et al. (1987)
CHO cells:	Merrill (1983)
cytotoxic effects	
Mouse skin:	Fisher et al. (1988)
block induction of ornithine decarboxylase	
inhibit phorbol-ester-induced inflammation	

(Merrill et al. 1986); this is consistent with the hypothesis, but the absence of measurements from appropriate control cells makes a firm conclusion impossible at this time.

HYPOTHESIS AND SPECULATIONS

There is little knowledge of the concentrations of sphingosine and lysosphingolipids in cells or of their roles in cellular regulation. The diversity of the sphingolipids had suggested potentially important functional roles (the diversity of lysosphingolipids would match those of the parental molecules). Given that the effects of sphingosine on cellular regulation appear to be opposite those of DAG or phorbol esters, the question of physiological significance is worth considering. The question is whether sphingosine/lysosphingolipids exist normally in cells and function in cellular regulation. Do these molecules function as second messengers, and, as described above, are they involved in pathogenesis?

The hypothesis that sphingosine/lysosphingolipids function as second messengers, as negative effectors of protein kinase C, demands that the molecules be produced in response to extracellular agents. Furthermore, the hypothesis predicts the existence of a "sphingolipid cycle" that would be analogous to the well-established phosphatidylinositol cycle. The scheme for such a cycle is depicted in Figure 5, where lysosphingolipids can be produced directly by deacylation (Hannun and Bell 1987; Bell et al. 1988; Hirabayashi et al. 1988). Such enzymes have not been reported but would be analogous to phospholipase A_2 cleavage of glycerophosphatides. Sphingosine could be produced either from lysosphingolipids or from ceramide produced from the parental sphingolipids.

Although meager evidence exists today in support of such a cycle, some evidence is beginning to accumulate. A specific antibody was observed to cause sphingomyelin to undergo deacylation/reacylation in L-929 fibroblasts (Ulrich and Shearer 1984). In 3T3-L1 cells, sphingosine mimics the effects of dexamethasone on 2-deoxyglucose uptake (Nelson and Murray 1986). Dexamethasone induces a neutral sphingomyelinase activity in 3T3-L1 cells and causes an elevation in cellular sphingosine levels (Nelson and Murray 1988; Ramachandran et al. 1988). A. Merrill et al. (unpubl.) have detected enhanced production of sphingosine in neutrophils following stimulation. Methods to quantitate and assay for these sphingolipid breakdown products are in their early stages of development.

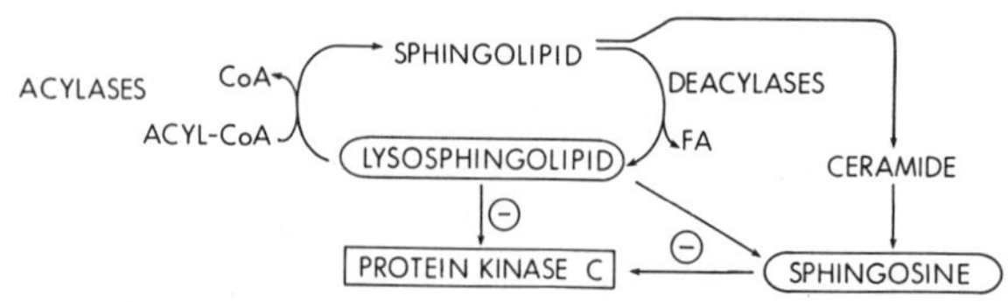

Figure 5. Hypothetical sphingolipid cycle. The figure depicts a scheme for the breakdown and resynthesis of sphingolipids to produce sphingosine/lysosphingolipids. (Reprinted, with permission, from Bell et al. 1988.)

Clearly, sphingosine and lysosphingolipids comprise a little-studied class of cellular molecules. It is premature to conclude that their cellular effects are limited to protein kinase C inhibition. These molecules may function in numerous ways that cannot be fully anticipated. Much work is needed to determine the concentrations of these molecules in cells and to define the enzymes involved in their formation and degradation.

The fact that protein kinase C is not active in resting cells that contain measurable DAG suggests that sphingosine/lysosphingolipids could prevent activation by resting levels of DAG. This physiological function would allow the ordinary roles of DAG in intermediary metabolism to be carried out without affecting protein kinase C activity. The concept of positive and negative regulators of protein kinase C is most attractive.

The possible relationship between positive (DAG) and negative (sphingosine/lysosphingolipids) effects on protein kinase C is depicted in relation to effector functions in oncogenesis in Figure 6. DAG produced by oncogene products would become the endogenous promoter. Following clonal expansion and hyperproliferation, secondary changes occur during the subsequent phase of oncogenesis and are referred to as progression. During this phase, a correlation with the increased presence of glycosphingolipids and tumor antigens has been observed (Hakomori 1986). The hypothesis that glycosphingolipids become major tumor antigens as a consequence of recessive oncogene prod-

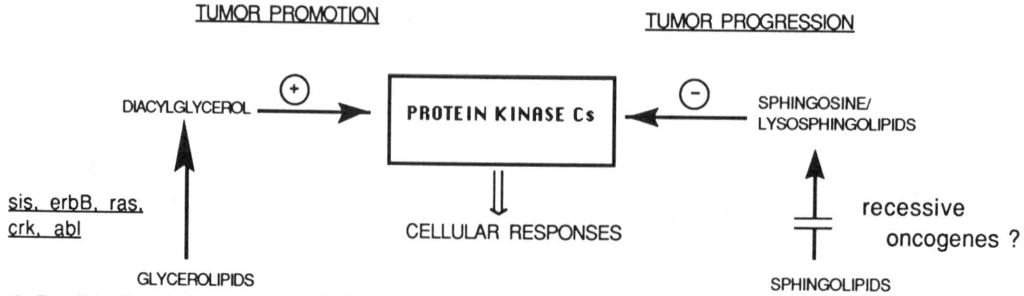

Figure 6. Possible link between sphingolipid tumor antigens and signal transduction. The figure illustrates the concept of reciprocal regulation of protein kinase C by positive (DAG) and negative (sphingosine/lysospholysis) effectors. The possible involvement of recessive oncogene products is shown.

ucts that fail to produce sphingosine/lysosphingolipids provides a plausible molecular connection. Further work is needed to understand the relationship between glycosphingolipids and altered growth regulation (Y. Hannun and R.M. Bell, in prep.).

Finally, protein kinase C has made important connections to the glycerolipids as cofactors (PtdS) and as second messengers (DAG) and to the sphingolipids as potential reservoirs for negative effectors, the "molecular brakes." The functions of sphingolipid breakdown products in cellular regulation are totally speculative at this time. New information is certain to emerge as a consequence of the existence of clear and testable hypotheses.

REFERENCES

Ashendel, C. 1985. The phorbol ester receptor: A phospholipid-regulated protein kinase. *Biochim. Biophys. Acta* **822:** 219.

Bell, R. 1986. Protein kinase C activation by diacylglycerol second messengers. *Cell* **45:** 631.

Bell, R., Y. Hannun, and C. Loomis. 1988. Regulation of protein kinase C by sphingosine/lysosphingolipids. In *Cellular and molecular aspects of inflammation.* (ed. G. Poste and S. Crooke), p. 265. Plenum Press, New York.

Berridge, B. 1987. Inositol trisphosphate and diacylglycerol: Two interacting second messengers. *Annu. Rev. Biochem.* **56:** 159.

Berridge, M. and R. Irvine. 1984. Inositol trisphosphate, a novel second messenger in cellular signal transduction. *Nature* **312:** 315.

Bishop, W. and R. Bell. 1988. Functions of diacylglycerol in glycerolipid metabolism, signal transduction and cellular transformation. *Oncogene Res.* **2:** 205.

Castagna, M., Y. Takai, K. Kaibuchi, K. Sano, U. Kikkawa, and Y. Nishizuka. 1982. Direct activation of calcium-activated, phospholipid-dependent protein kinase by tumor-promoting phorbol esters. *J. Biol. Chem.* **257:** 7847.

Coussens, L., P. Parker, L. Rhee, T. Yang-Feng, E. Chen, M. Waterfield, V. Francke, and A. Ullrich. 1986. Multiple, distinct, forms of bovine and human protein kinase C suggest diversity in cellular signaling pathways. *Science* **233:** 859.

Davis, R., B. Ganong, R. Bell, and M. Czech. 1985. *sn*-1,2-Dioctanoylglycerol. A cell-permeable diacylglycerol that mimics phorbol diester action on the epidermal growth factor receptor and mitogenesis. *J. Biol. Chem.* **260:** 1562.

Faucher, M., N. Girones, Y. Hannun, R. Bell, and R. Davis. 1988. Regulation of the epidermal growth factor receptor phosphorylation state by sphingosine in A431 human epidermoid carcinoma cells. *J. Biol. Chem.* **263:** 5319.

Fisher, G., A. Gupta, J. Elder, B. Nickoloff, and J. Voorhees. 1988. Sphingosine inhibits phorbol ester-induced inflammation ornithine decarboxylase activity and activation of protein kinase C in mouse skin. *FASEB Monogr.* **2:** A350.

Ganong, B., C. Loomis, Y. Hannun, and R. Bell. 1986. Specificity and mechanism of protein kinase C activation by *sn*-1,2-diacylglycerols. *Proc. Natl. Acad. Sci.* **83:** 1184.

———. 1987. Regulation of protein kinase C by lipid cofactors. In *Cell membranes: Methods and reviews* (ed. E. Elson et al.), vol. 3, p. 183. Plenum Press, New York.

Grove, D. and A. Mastro. 1988. Prevention of the TPA-mediated down-regulation of protein kinase C. *Biochem. Biophys. Res. Commun.* **151:** 94.

Hall, F., P. Fernyhough, D. Ishiii, and P. Vulliet. 1988. Suppression of nerve growth factor-directed neurite outgrowth in PC12 cells by sphingosine, an inhibitor of protein kinase C. *J. Biol. Chem.* **263:** 4460.

Hanai, N., G. Nores, C.-R. Torres-Mendez, and S. Hakomori. 1987. Modified ganglioside as a possible modulator of transmembrane signaling mechanism through growth factor receptors: A preliminary note. *Biochem. Biophys. Res. Commun.* **147:** 127.

Hannun, Y. and R. Bell. 1986. Phorbol ester binding and activation of protein kinase C on triton X-100 mixed micelles containing phosphatidylserine. *J. Biol. Chem.* **261:** 9341.

———. 1987. Lysosphingolipids inhibit protein kinase C: Implications for the sphingolipidoses. *Science* **235:** 670.

———. 1988. Aminoacridines, potent inhibitors of protein kinase C. *J. Biol. Chem.* **263:** 5124.

Hannun, Y., C. Greenberg, and R. Bell. 1987. Sphingosine inhibition of agonist-dependent secretion and activation of human platelets implies that protein kinase C is a necessary and common event of the signal transduction pathways. *J. Biol. Chem.* **262:** 13626.

Hannun, Y., C. Loomis, and R. Bell. 1985. Activation of protein kinase C by triton X-100 mixed micelles containing diacylglycerol and phosphatidylserine. *J. Biol. Chem.* **260:** 10039.

———. 1986a. Protein kinase C activation in mixed micelles: Mechanistic implications of phospholipid, diacylglycerol and calcium dependencies. *J. Biol. Chem.* **261:** 7184.

Hannun, Y., C. Loomis, A. Merrill, and R. Bell. 1986b. Sphingosine inhibition of protein kinase C activity and of phorbol dibutyrate binding *in vitro* and in human platelets. *J. Biol. Chem.* **261:** 12604.

Hakomori, S. 1986. Glycosphingolipids: The composition of these membrane molecules changes dramatically with cell differentiation and the onset of cancer. Exploiting such changes could lead to improved diagnosis and treatment of cancer. *Sci. Am.* **254:** 44.

Hidaka, H., M. Inagaki, S. Kawamoto, and Y. Sasaki. 1984. Isoquinolinesulfonamides, novel and potent inhibitors of cyclic nucleotide dependent protein kinase and protein kinase C. *Biochemistry* **23:** 5036.

Hirabayashi, Y., M. Kimura, M. Matsumoto, K. Yamamoto, S. Kadowaki, and T. Tochikura. 1988. A novel glycosphingolipid hydrolyzing enzyme, glycosphingolipid ceramide deacylase, which cleaves the linkage between the fatty acid and sphingosine base in glycosphingolipids. *J. Biochem.* **103:** 1.

Hoshijima, M., A. Kikuchi, T. Tanimoto, K. Kaibuchi, and Y. Takai. 1986. Formation of a phorbol ester binding fragment from protein kinase C by proteolytic digestion. *Cancer Res.* **46:** 3000.

House, C. and B. Kemp. 1987. Protein kinase C contains a pseudosubstrate prototope in its regulatory domain. *Science* **238:** 1726.

Housey, G., C. O'Brian, M. Johnson, P. Kirschmeier, and I. Weinstein. 1987. Isolation of cDNA clones encoding protein kinase C: Evidence for a protein kinase C-related gene family. *Proc. Natl. Acad. Sci.* **84:** 1065.

Housey, G., M.D. Johnson, W.L.W. Hsiao, C.A. O'Brien, J.P. Murphy, P. Kirschmeier, and I.B. Weinstein. 1988. Overproduction of protein kinase C causes disordered growth control in rat fibroblasts. *Cell* **52:** 343.

Huang, K., H. Nakabayshi, and F. Huang. 1986. Isozymic forms of rat brain Ca^{2+}-activated and phospholipid-dependent protein kinase. *Proc. Natl. Acad. Sci.* **83:** 8535.

Inoue, M., A. Kishimoto, Y. Takai, and Y. Nishizuka. 1977. Studies on cyclic nucleotide-independent protein kinase and its proenzyme in mammalian tissues. *J. Biol. Chem.* **252:** 7610.

Kishimoto, A., Y. Takai, T. Mori, U. Kikkawa, and Y. Nishizuka. 1980. Activation of calcium and phospholipid-dependent protein kinase by diacylglycerol, its possible relation to phosphatidylinositol turnover. *J. Biol. Chem.* **255:** 2273.

Knopf, J., M. Lee, L. Sultzman, R. Kritz, C. Loomis, R. Hewick, and R. Bell. 1986. Cloning and expression of multiple kinase C cDNAs. *Cell* **46:** 491.

Lambeth, J., D. Burnham, and S. Tyagi. 1988. Sphinganine effects on chemoattractant-induced diacylglycerol generation, calcium fluxes, superoxide production, and on cell viability in the human neutrophil: Delivery of sphinganine with bovine serum albumin minimizes cytotoxicity without affecting inhibition of the respiratory tract. *J. Biol. Chem.* **263:** 3818.

Lee, M. and R. Bell. 1986. The lipid binding, regulatory domain of protein kinase C. *J. Biol. Chem.* **261:** 14867.

Loomis, C. and R. Bell. 1988. Sangivamycin, a nucleoside analogue, is a potent inhibitor of protein kinase C. *J. Biol. Chem.* **263:** 1682.

Mayer, B., M. Hamaguchi, and H. Hanafusa. 1988. A novel viral oncogene with structural similarity to phospholipase C. *Nature* **332:** 272.

McIntyre, T., S. Reinhold, S. Prescott, and G. Zimmerman. 1987. Protein kinase C activity appears to be required for the synthesis of platelet-activating factor and leukotriene B_4 by human neutrophils. *J. Biol. Chem.* **262:** 15370.

Merrill, A. 1983. Characterization of serine palmitoyltransferase activity in Chinese hamster ovary cells. *Biochim. Biophys. Acta* **754:** 284.

Merrill, A., A. Sereni, V. Stevens, Y. Hannun, R. Bell, and J. Kinkade. 1986. Inhibition of phorbol ester dependent differentiation of human promyelocytic leukemic (HL60) cells by sphinganine and other long-chain bases. *J. Biol. Chem.* **261:** 12610.

Miyatake, T. and K. Suzuki. 1972. Globoid cell leukodystrophy: Additional deficiency of psychosine galactosidase. *Biochem. Biophys. Res. Commun.* **48:** 538.

Mori, T., Y. Takai, R. Minakuchi, B. Yu, and Y. Nishizuka. 1980. Inhibitory action of chlorpromazine, dibucaine and other phospholipid-interacting drugs on calcium-activated, phospholipid-dependent protein kinase. *J. Biol. Chem.* **255:** 8378.

Nelson, D.H. and D.K. Murray. 1986. Sphingolipids inhibit insulin and phorbol ester stimulated uptake of 2-deoxyglucose. *Biochem. Biophys. Res. Commun.* **138:** 4637.

———. 1988. Dexamethasone increases tissue sphingosine levels and blocks phorbol stimulated glucose transport. *FASEB Monogr.* **2:** A1417.

Niedel, J.E., L.J. Kuhn, and G.R. Vandenbark. 1983. Phorbol diester receptor copurifies with protein kinase C. *J. Biol. Chem.* **80:** 36.

Nishizuka, Y. 1984a. The role of protein kinase C in cell surface signal transduction and tumor promotion. *Nature* **308:** 693.

———. 1984b. Turnover of inositol phospholipids and signal transduction. *Science* **225:** 1365.

———. 1986. Studies and perspectives of protein kinase C. *Science* **233:** 305.

Ono, Y., T. Fujii, K. Ogita, U. Kikawa, K. Igarashi, and Y. Nishizuka. 1988. The structure, expression, and properties of additional members of the protein kinase C family. *J. Biol. Chem.* **263:** 69272.

Ono, Y., T. Kurokawa, K. Kawahara, R. Nishimura, K. Igarashi, and Y. Nishizuka. 1986a. Cloning of rat brain protein kinase C complementary DNA. *FEBS. Lett.* **203:** 111.

Ono, Y., T. Kurokawa, T. Fujii, K. Kawakasa, K. Igarashi, U. Kikkawa, K. Ogita, and Y. Nishizuka. 1986b. Two types of complementary DNAs of rat brain protein kinase C. *FEBS. Lett.* **206:** 347.

Parker, P., L. Coussens, N. Totty, L. Rhee, S. Young, E. Chen, S. Stabel, M. Waterfield, and A. Ullrich. 1986. The complete primary structure of protein kinase C—The major phorbol ester receptor. *Science* **233:** 853.

Persons, D.A., W.O. Wilkison, R.M. Bell, and O.J. Finn. 1988. Altered growth regulation and enhanced tumorigenicity of NIH 3T3 fibroblasts transfected with protein kinase C-I cDNA. *Cell* **52:** 447.

Preiss, J., C. Loomis, W. Bishop, R. Stein, J. Niedel, and R. Bell. 1986. Quantitative measurement of *sn*-1,2-diacylglycerols present in platelets, hepatocytes and *ras* and *sis*-transformed normal rat kidney cells. *J. Biol. Chem.* **261:** 8597.

Ramachandran, C.K., D.K. Murray, and D.H. Nelson. 1988. Increased activity of neutral sphingomyelinase in dexamethasone treated 3T3-L1 cells. *FASEB Monogr.* **2:** A581.

Rozengurt, E., A. Rodriquez-Pena, M. Coombs, and J. Sinett-Smith. 1984. Diacylglycerol stimulates DNA synthesis and cell division in mouse 3T3 cells: Role of Ca^{2+}-sensitive phospholipid-dependent protein kinase. *Proc. Natl. Acad. Sci.* **81:** 5748.

Speizer, L.A., M.J. Watson, and L.L. Brunton. 1987. Sphinganine inhibits translocation of protein kinase C induced by phorbol esters in S49 lymphoma cells. *FASEB Monogr.* **46:** 2065.

Takai, Y., A. Kishimoto, M. Inoue, and Y. Nishizuka. 1977a. Studies on a cyclic nucleotide-independent protein kinase and its proenzyme in mammalian tissues. *J. Biol. Chem.* **252:** 7603.

Takai, Y., M. Yamamoto, M. Inoue, A. Kishimoto, and Y. Nishizuka. 1977b. A proenzyme of cyclic nucleotide-independent protein kinase and its activation by calcium-dependent neutral protease from rat liver. *Biochem. Biophys. Res. Commun.* **77:** 542.

Takai, Y., A. Kishimoto, Y. Iwasa, Y. Kawahara, T. Mori, and Y. Nishizuka. 1979a. Calcium-dependent activation of a multifunctional protein kinase by membrane phospholipids. *J. Biol. Chem.* **254:** 3692.

Takai, Y., A. Kishimoto, Y. Iwasa, Y. Kawahara, T. Mori, Y. Nishizuka, A. Tamura, and T. Fujii. 1979b. A role of membranes in the activation of a new multifunctional protein kinase system. *J. Biochem.* **86:** 575.

Tamaoki, T., H. Nomoto, I. Takahashi, Y. Kato, M. Morimoto, and F. Tomita. 1986. Staurosporine, a potent inhibitor of phospholipid/Ca^{++}/dependent protein kinase. *Biochem. Biophys. Res. Commun.* **135:** 397.

Ulrich, R. and W. Shearer. 1984. Enhanced N-acylation of palmitic acid in sphingomyelin of antibody-stimulated L cells. *Biochem. Biophys. Res. Commun.* **121:** 605.

Wilson, E., M. Olcott, R. Bell, A. Merrill, and J. Lambeth. 1986. Inhibition of the oxidative burst in human neutrophils by sphingoid long chain bases: Role of protein kinase C in activation of the burst. *J. Biol. Chem.* **261:** 12616.

Wilson, E., W.G. Rice, J.M. Kinkade, Jr., A.H. Merrill, Jr., R.R. Arnold, and J.D. Lambeth. 1987. Protein kinase C inhibition by sphingoid long-chain bases: Effects on secretion in human neutrophils. *Arch. Biochem. Biophys.* **259:** 204.

Wise, B.C., D.B. Glass, C.-H.J. Chou, R.L. Raynor, N. Katoh, R.C. Schatzman, R.S. Turner, R.F. Kibler, and J.F. Kuo. 1982. Phospholipid-sensitive Ca^{2+}-dependent protein kinase from heart. II. Substrate specificity and inhibition by various agents. *J. Biol. Chem.* **257:** 8489.

Expression of Wild-type and Mutant Subunits of the cAMP-dependent Protein Kinase

G.S. McKnight, G.G. Cadd, C.H. Clegg, A.D. Otten, and L.A.Correll
Department of Pharmacology, University of Washington, Seattle, Washington 98195

Protein phosphorylation provides the cell with a rapidly reversible mechanism for altering the activity of specific proteins in response to hormones or other extracellular or intracellular signals. All of the well-characterized intracellular second messengers act either directly to modulate protein kinases or indirectly to affect the levels of other activators of cellular kinases or inhibitors of protein phosphatases. Both cAMP and cGMP directly activate specific protein kinases, and Ca^{++} and diacylglycerol activate the protein kinase C (PKC) family of kinases (Knopf et al. 1986). Ca^{++}, through interaction with calmodulin (CaM), also activates a small family of CaM-dependent kinases, some of which have a limited substrate spectrum, such as myosin light-chain kinase, and others, like Ca^{++}/CaM-dependent kinase II, that recognize many substrates (for review, see Edelman et al. 1987). Many growth factors and hormones interact with cell-surface receptors that have protein tyrosine kinase activity (e.g., EGF and insulin receptors), and their protein phosphorylating activity has been implicated in the mechanisms of growth factor and hormone action (Morgan et al. 1986). Although regulation of protein kinase activity by second messengers is clearly not the only important regulatory pathway in a cell's repertoire of responses to environmental stimuli, it is one of the most rapid and reversible ways of directly affecting the function of a protein and is a widely used mechanism in nature.

The discovery of cAMP by Sutherland and Rall (1957) led to a sustained interest in the mechanisms by which this newly recognized second messenger could regulate cellular events in animal cells. The important demonstration by Walsh et al. (1968) that a specific cAMP-dependent protein kinase (PKA) was the mediator of the effects of cAMP on glycogen breakdown led to the hypothesis that PKA might be responsible for all of the known effects of cAMP in higher animal cells (Krebs 1972). This hypothesis has proven correct in the vast majority of cAMP-regulated effects studied in detail, although recent evidence suggesting that cAMP can directly regulate ion channels in the olfactory system indicates that exceptions are possible (Nakamura and Gold 1987). There is also debate over the role of PKA in the very rapid cAMP-mediated induction of specific genes, such as those coding for somatostatin (Montminy et al. 1986), chorionic gonadotropin α subunit (Silver et al. 1987), and phosphoenolpyruvate carboxykinase (PEPCK) (Sasaki et al. 1984). By analogy to bacterial systems, gene regulation by cAMP could be accomplished by a DNA-binding cAMP receptor (similar to catabolite gene activator protein [cap] in *Escherichia coli*) that could act directly on promoter sequences to modify transcription (for review, see Adhya and Garges 1982). However, no evidence for such a cAMP-binding transcription factor has been found in higher animals, and recent reports favor an indirect mechanism involving phosphorylation of a specific transcription factor, which then binds to and modifies the activity of genes (Montminy and Bilezikjian 1987).

The structure of mammalian PKA has been studied in considerable detail, and the molecular mechanisms by which this enzyme responds to cAMP are the subject of numerous reports (for review, see Krebs and Beavo 1979). Chromatography on DEAE originally revealed the presence of two isoforms of PKA, termed PKA-I and PKA-II, on the basis of their order of elution from DEAE (Reimann et al. 1971). Each of these isoforms consists of a holoenzyme containing two molecules of a regulatory subunit (R) associated as a dimer and two molecules of a catalytic subunit (C). Both forms I and II appear to contain the same C subunit but differ in their R subunits (R_I and R_{II}). The holoenzyme is inactive until each regulatory subunit binds two molecules of cAMP, releasing the C subunits in an active form. Direct sequence analysis of the purified R proteins has helped to identify the cAMP-binding sites by their sequence similarity to the cAMP-binding site of cap protein, and the site on R_{II} (Arg-Arg-X-Ser) that is autophosphorylated by C and is therefore thought to interact directly with the catalytic domain of C in the holoenzyme complex (Takio et al. 1984). The R_I subunit also has a similar C recognition sequence (Arg-Arg-X-Ala), although the target serine for phosphorylation is replaced by an alanine, and the R_I protein cannot be autophosphorylated at this site (Titani et al. 1984). Another form of R_{II} has been characterized by specific antibody recognition, differential migration on SDS-polyacrylamide gels, and, most recently, by cloning and complete sequence analysis (Stein and Rubin 1985; Jahnsen et al. 1986). From our cloning studies detailed below, we have identified two active mouse genes coding for similar but distinct C subunits and two genes coding for R_I isoforms. These R and C isoforms show remarkable tissue-specific patterns of expression and may play distinctive roles in cellular responses to cAMP.

Although a great deal is known about this key regulatory enzyme complex, many interesting questions remain concerning the mechanisms of holoenzyme activation, regulation of specific subunit levels in the cell, differential function of the multiple R and C subunits, and the exact role of the PKA system in complex cellular responses in which other second-messenger systems, such as the PKC or Ca^{++} pathways, are also activated. Our efforts to use the techniques of molecular genetics first to isolate and characterize the genes for the various subunits of PKA and then to devise genetic approaches to help answer questions of biological interest are detailed below.

EXPERIMENTAL PROCEDURES

Most of the techniques used in this paper have been described previously and are cited within the text and figure legends.

In situ hybridization. Single-stranded ^{35}S-labeled RNA probes were synthesized from linearized plasmid vectors containing an SP6 promoter coupled to the 5′ region of either C_α or C_β. Newborn mice were anesthetized with ether and frozen on dry ice prior to sectioning with a cryostat. Sagittal sections (10 μm) were thaw mounted on poly-L-lysine-coated slides and hybridized as described by Rogers et al. (1987). Approximately 350,000 cpm of probe was applied to each slide in a volume of 60 μl in buffer containing 50% formamide, 1× Denhardt's solution, 10% dextran sulfate, 10 mM DTT, 250 μg/ml yeast RNA, 100 μg/ml denatured salmon sperm DNA, 0.6 M NaCl, 1 mM EDTA, and 10 mM Tris-Cl (pH 7.5). Slides were covered with Parafilm and incubated 16 hours at 50°C. The slides were then washed in 4× SSC and treated with 20 μg/ml RNase in 10 mM Tris-Cl (pH 8.0), 500 mM NaCl, and 1 mM EDTA for 30 minutes at 37°C. After further washing in 2× SSC followed by a wash with 0.1× SSC at 50°C, the slides were dehydrated with ethanol and exposed to X-ray film (Ultrofilm ^{3}H, LKB).

Ion exchange chromatography. Cell pellets were homogenized in 50 mM Tris-Cl (pH 7.6), 4 mM DTT, 1 mM EDTA, and 0.1 mM phenylmethylsulfonyl fluoride (PMSF), and the extract was sedimented at 100,000*g* for 1 hour at 4°C. For runs on the high-performance liquid chromatography (HPLC), the supernatant was filtered through a 0.2-μm acrodisc (Gelman Sci. Inc.), and about 4 mg of soluble protein was loaded onto a Bio-Gel DEAE–ion exchange column (75 × 7.5 mm). Conventional DEAE-cellulose columns were loaded with 15 mg of cytoplasmic protein and eluted with a linear salt gradient at 4°C. In both cases, fractions were stored on ice until assayed for kinase activity.

RESULTS AND DISCUSSION

Tissue-specific Expression of Endogenous R and C Subunit Genes

The cDNAs for six of the subunits of PKA have now been cloned and sequenced, and several of their corresponding genomic sequences have been isolated and characterized by mapping the exon/intron junctions and promoter elements ($R_{I\alpha}$: Lee et al. 1983; Nowak et al. 1987; $R_{I\beta}$: Clegg et al. 1988; $R_{I\alpha}$: Scott et al. 1987; $R_{II\beta}$: Jahnsen et al. 1986; C_α: Uhler et al. 1986b; C_β: Uhler et al. 1986a; Showers and Maurer 1986). To distinguish these various subunits from each other, we will use the nomenclature shown in Figure 1, which also summarizes our current understanding of the tissue specificity of expression of these mouse PKA subunits. This nomenclature has advantages over that which relies on apparent molecular weights on gels, since those values are not identical in different species, do not correspond to the predicted molecular weight from the primary sequence, and are affected by phosphorylation. In addition, we suspect that other forms of R and perhaps C subunits may be found, and the use of Greek letters will avert the confusing possibility of having an R_{IIC}.

There are two isoforms of each of the major subunits; we have referred to them as α and β isoforms. Figure 1 shows a comparison of the α and β subunits of R_I, R_{II}, and C, with differences in amino acid sequence noted by dark lines. All of the subunits are from mouse, with the exception of $R_{II\beta}$, which was cloned from rat granulosa cells (Jahnsen et al. 1986). In general, the α and β subunits are between 80% and 90% identical at the amino acid level, with many of the differences clustering in the amino-terminal domains.

The most striking difference between the α and β subunits is their tissue-specific expression. In general, the α subunits are constitutively expressed in most tissues, whereas the β subunits are highly expressed in brain with lesser levels in nonneuronal tissue. One exception to this is $R_{II\beta}$, which is highly inducible in granulosa cells by follicle-stimulating hormone (FSH) and estrogen (Hedin et al. 1987) and which is also actively expressed in adipose tissue and fetal liver (G. Cadd, unpubl.). The $R_{II\alpha}$ isoform can be detected in most tissues but is highly expressed in postmeiotic spermatids (Øyen et al. 1988) and in discrete regions of the brain (G. Cadd, unpubl.). Recently, we have used in situ hybridization to study the expression of R and C subunits in brain and in fetal tissues during development. Riboprobes specific for either C_α or C_β were hybridized to 10-μm sagittal sections of a newborn mouse, and the RNase-resistant hybrids were visualized by autoradiography. The results shown in Figure 2 demonstrate that the C_α subunit is expressed in many tissues during development, but that a specific C_β signal is only observed in brain and spinal cord. In the adult, the C_β isoform can be detected in virtually all tissues if sensitive Northern blots are used, but the levels of C_β

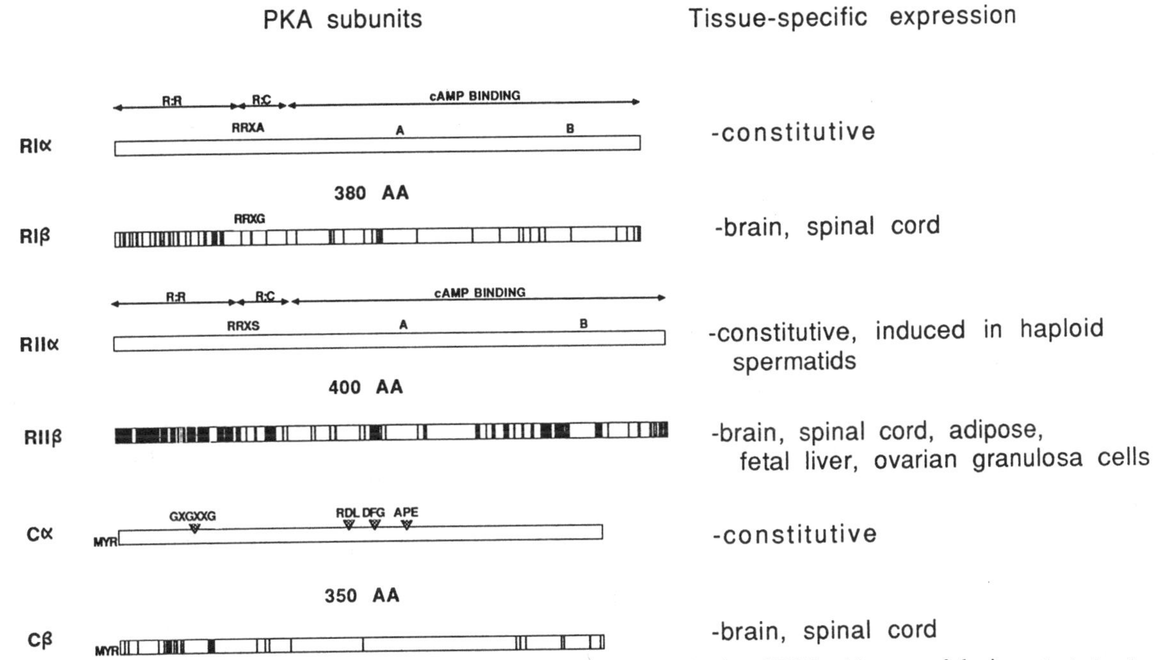

Figure 1. Subunits of PKA. Schematic representation of the six cloned subunits of PKA with some of the important structural features noted as well as a brief summary of expression in mouse tissues. For the R subunits, we have denoted the regions that correspond approximately to the dimerization domain (R:R), the hinge region that provides at least one site of interaction with C subunit (R:C) and contains the sequence RRXS/G/A, and the two carboxy-terminal cAMP-binding regions localized as A and B in the diagram. The β subunit amino acid sequences are compared to their corresponding α subunit, and amino acid differences are indicated by a black line. The C and R_I genes are isolated from mouse, and the R_{II} genes compared are from rat. There are only 21 amino acid changes between the mouse and rat $R_{II\alpha}$ sequences. For the C subunits, the ATP-binding region (GXGXXG) is noted, as are other amino acid sequences (RDL, DFG, APE) that are highly conserved in all protein kinases.

mRNA are very low in nonneuronal tissues (Uhler et al. 1986a). Although the discrete tissue localization of the various R and C subunits hints at discrete roles in intracellular regulation, biochemical analysis of the properties of the various R and C subunits has not yet indicated major differences in the ability of the R subunits to form active holoenzymes. Perhaps the presence of multiple R isoforms can be explained by their role in the subcellular localization of holoenzymes or by specific interactions between R subunits and cellular proteins other than C. The $R_{I\alpha}$ subunit appears to be generally soluble in most cells, but the $R_{II\beta}$ form specifically binds to a membrane-associated 75-kD protein in brain tissue (Sarkar et al. 1984). Both $R_{II\alpha}$ and $R_{II\beta}$ also appear to interact with microtubule-associated protein 2 (MAP 2), although the $R_{II\alpha}$ subunit has a higher affinity and capacity for MAP-2 binding (Leiser et al. 1986). The R_{II} subunits have been localized to Golgi and perinuclear membranes by immunofluorescence (Nigg et al. 1985b) and may serve to anchor the holoenzyme in this position until activation by cAMP occurs. C subunit released from the holoenzyme by cAMP has been observed to translocate into the nucleus in several cell lines (Nigg et al. 1985a); this may explain the rapid effects of cAMP on gene regulation via the phosphorylation of specific transcription factors.

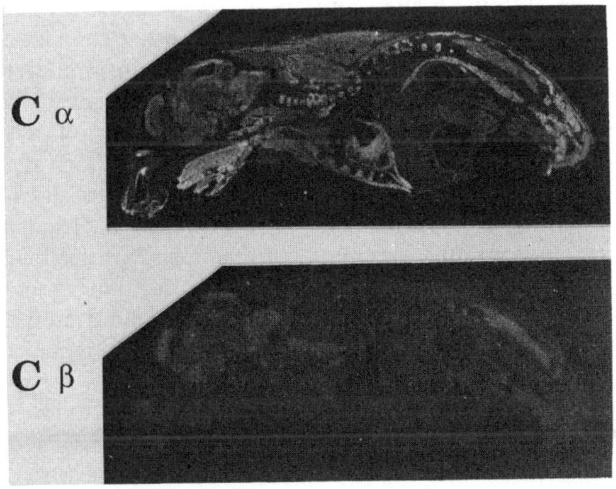

Figure 2. In situ hybridization of C_α- and C_β-specific riboprobes to sagittal sections from a newborn mouse. Adjacent 10-μm sagittal sections from the midline of a newborn mouse were hybridized to ^{35}S-labeled RNA probes as described in Experimental Procedures. After RNase digestion and extensive washing, the slides were exposed to Ultrofilm ^{3}H (LKB) for 8 days. The autoradiograms were printed as negatives.

Sequence of cAMP-binding Mutants in the R_I Subunit

Somatic cell genetic approaches have proven very useful in the analysis of the cAMP system in several cell lines that are growth-inhibited by cAMP. We have focused on the well-characterized series of mutants derived from S49 mouse lymphoma cells following mutagenesis and selection for growth in high concentrations of cAMP analogs (Coffino et al. 1976). Three classes of cAMP-resistant mutants have been described, including (1) kin^- cells that are lacking detectable C subunit activity or protein, (2) K_a cells that carry mutations in the R_I subunit that interfere with cAMP activation of the holoenzyme, and (3) V_{max} mutants that simply have decreased levels of an apparently normal holoenzyme.

The K_a mutants include a large family of cloned cell lines, and many have been mapped by charge shifts in peptide fragments (Murphy and Steinberg 1985). It is important to note that the K_a mutations selected are dominant and that the cells continue to synthesize the normal $R_{I\alpha}$ subunit in addition to the mutant protein. We have cloned $R_{I\alpha}$ cDNAs from cells that had defects that mapped close to either the site-A or site-B cAMP-binding domains and have sequenced the mutant forms (Clegg et al. 1987). The site-A mutant cell line contained a single G to A transition that changed the glycine at amino acid 200 to a glutamic acid (Fig. 3).

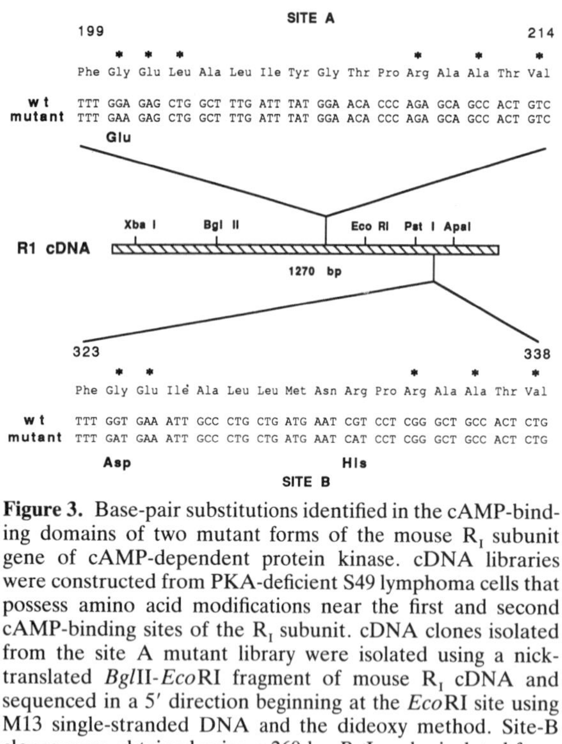

Figure 3. Base-pair substitutions identified in the cAMP-binding domains of two mutant forms of the mouse R_I subunit gene of cAMP-dependent protein kinase. cDNA libraries were constructed from PKA-deficient S49 lymphoma cells that possess amino acid modifications near the first and second cAMP-binding sites of the R_I subunit. cDNA clones isolated from the site A mutant library were isolated using a nick-translated *Bgl*II-*Eco*RI fragment of mouse R_I cDNA and sequenced in a 5′ direction beginning at the *Eco*RI site using M13 single-stranded DNA and the dideoxy method. Site-B clones were obtained using a 260-bp *Pst*I probe isolated from the 3′ end of mouse R_I cDNA. These clones were sequenced in a 3′ direction starting at the *Eco*RI site of R_I through to the stop codon of translation. The indicated sequences code for two highly conserved regions of the R_I protein implicated in cAMP binding. The predicted amino acid changes in the mutant R_I protein are presented in bold type. The asterisks above the amino acid sequence identify residues identical to *E. coli* cap protein. (Reprinted, with permission, from Clegg et al. 1987.)

The site-B mutant cell line contained three alleles: One was wild type, as expected, and the second contained a single mutation that changed Gly-324 to an Asp. These two independently selected mutations in site A and site B are highly homologous, changing the identical Gly to an acidic amino acid. Unexpectedly, we also found a third allele in the site-B mutant containing two changes: One was the Gly-324 to Asp described above, and the second was an Arg to His mutation several amino acids downstream from the Gly mutation. A summary of the mutations in site A and site B is shown in Figure 3.

Expression of the $R_{I\alpha}$ Mutations in Cultured Cells

The mutant genes that we have isolated from the S49 cells confer a dominant cAMP-resistant phenotype in those cells. The dominance of this mutation can best be understood by considering the interactions of R within the holoenzyme complex and possible effects on protein turnover. The mutant R protein still binds tightly to C to inhibit its activity, but this R subunit cannot be released by physiological levels of cAMP. Since it appears that R subunits are more likely to be degraded when released from the holoenzyme (see below), the levels of wild-type R subunit in mutant cells should decline and the proportion of mutant R increase. In addition, holoenzyme complexes containing both a wild-type and mutant R subunit may have a reduced ability to dissociate either C subunit, due to interactions within the complex. These phenomena may explain the rather striking inhibition of kinase activity seen in S49 cells expressing one wild-type and one mutant gene.

One of our goals was to use the isolated mutant R_I genes to confer a cAMP-resistant phenotype on other cell types. The type of selection used in S49 cells is not possible in most cells because they are not growth-arrested by cAMP. Therefore, we constructed regulatory subunit expression vectors (REV) that contained either the inducible mouse metallothionein promoter or the constitutive Harvey sarcoma virus LTR promoter. The structural domain of the mutant R_I cDNA was followed by a polyadenylation signal from the human growth hormone gene as shown schematically in Figure 4. We have transfected these vectors and similar constructs into a variety of cultured cells, using either $CaPO_4$ coprecipitation or electroporation. In all cases, stable transformants were selected using the bacterial neomycin-resistance gene and selection in G418.

Mouse adrenal cells (Y1) were chosen for some of these studies because this cell line has well-studied responses to cAMP that include differentiation and induction of steroid biosynthesis (Doherty et al. 1981). The REV-B vector (containing the site-B mutations) or the construct shown in Figure 4 containing both the site-A and the site-B mutation (REV-AB) was transfected into Y1 cells using $CaPO_4$ coprecipitation. Although there is some constitutive expression of mutant R_I in the stable transformants, this level can be aug-

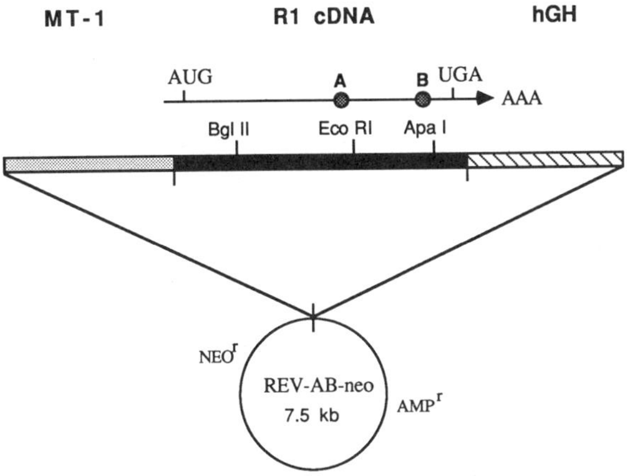

Figure 4. Expression vector containing the mutant $R_{I\alpha}$ structural sequence. REV-AB-*neo* contains the coding region of the R_I subunit gene of PKA flanked by the *Kpn*I-*Bam*HI fragment (~700 bp) of the mouse metallothionein-1 (MT) promoter, and 630 bp of the human growth hormone (hGH) 3′-untranslated and -flanking sequences.

mented dramatically by induction of the metallothionein promoter with Zn^{++}. When REV-B cells were treated with Zn^{++} concentrations sufficient to induce the metallothionein promoter (60–90 μM), a dramatic induction of mutant $R_{I\alpha}$ mRNA is seen as well as a five- to tenfold increase in $R_{I\alpha}$ protein (data not shown). Kinase activity in these cells was severely inhibited, even in the absence of Zn^{++}, and the ability of cAMP to activate PKA is further diminished after Zn^{++} treatment. Similar results were obtained with the REV-AB vector.

What are the biological consequences of inhibiting the cAMP activation of PKA in Y1 cells? We have examined four cAMP-mediated responses in these cells: growth inhibition, cell shape changes, steroid synthesis, and gene induction. In several independent cell lines expressing mutant forms of $R_{I\alpha}$, the growth inhibition normally seen in the presence of cAMP analogs was absent, and the cells no longer rounded up in response to cAMP as did wild-type Y1 cells. In Figure 5A, the time course of induction of steroid secretion by 150 μM 8-chlorophenylthio-cAMP (CPT-cAMP) is shown for wild type and a typical REV-B-expressing mutant. The wild-type Y1 cells synthesize large amounts of steroids, whereas the REV-B cells produce little steroid during the first 12 hours.

Several genes are induced in Y1 cells in response to cAMP, including plasminogen activator (PA), ornithine decarboxylase (ODC), and enzymes involved in steroid biosynthesis. In stable transformants expressing the REV-AB mutation, the induction of ODC mRNA is completely inhibited, as assayed by Northern blots (Fig. 5B), and the basal level of ODC mRNA has also declined severalfold in these mutant cells. The induction of PA is also decreased substantially, but this gene appears to remain partially inducible, even though the kinase levels are severely reduced. These results indi-

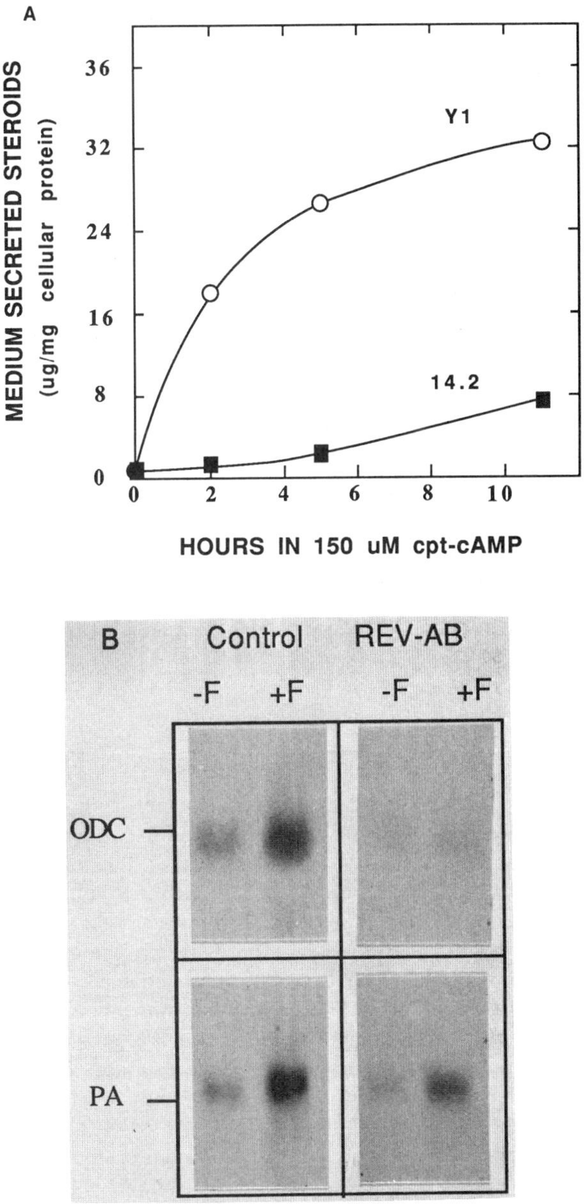

Figure 5. Expression of mutant R_I genes in Y1 cells inhibits the induction of steriod biosynthesis and gene activation by cAMP. (*A*) Cultures were fed medium containing 1.5% serum for 24 hr, and then refed with 2% serum plus 150 μM 8-CPT-cAMP. At the indicated times, culture medium and cells were collected. The accumulation of steroids in medium was measured using a fluorometric assay and normalized to cellular protein. Control Y1 cells (○) are compared to 14.2 cells (■) that express the B-site mutant form of $R_{I\alpha}$ shown in Fig. 3. Each data point represents an average taken from duplicate plates. (Adapted, with permission, from Clegg et al. 1987.) (*B*) The induction of ODC and PA is shown by hybridization of specific cDNA probes to Northern blots of RNA from control Y1 cells or a stable transformant expressing REV-AB-*neo*. The constitutive expression of REV-AB was sufficient to elicit the response seen, and induction of the vector with Zn^{++} was unnecessary. RNA was isolated after a 4-hr incubation in the absence (−F) or presence (+F) of 10 μM forskolin. The ODC mRNA migrates at 2.2 kb, and the PA mRNA migrates at 2.5 kb. The cDNA probes used were gifts from P. Coffino (ODC) and J. Degen (PA).

cate that the catalytic subunit is directly involved in gene regulation and suggest phosphorylation of regulatory transcription factors as a likely mechanism. The differential sensitivity of the PA and ODC genes might be due to differences in affinity for the limited level of phosphorylated transcription factors activated by cAMP in the mutant cells.

It is clear from these data that the mutant $R_{I\alpha}$ gene can confer a dominant cAMP-resistant phenotype on transfected cells that stably express relatively small amounts of the mutant protein. In addition to Y1 cells, we have transfected GH_3 pituitary cells, NIH-3T3 fibroblasts, and T84 human colon carcinoma cells, and in each case, we see a marked inhibition of PKA activity. Responses to cAMP in these cell lines are currently being studied. The mutant R_I genes should prove useful in producing a genetic lesion in the cAMP response pathway in cells that normally respond to hormones with the activation of multiple second-messenger systems. An examination of the biological responses affected by this mutation should help pinpoint which of the many effects of hormones and other extracellular signals are mediated by PKA.

Overproduction of C Leads to Compensation by R_I

One of the interesting aspects of the PKA holoenzyme system is the cell's ability to maintain two types of kinase, PKA-I and -II, within the cell and to prevent the C subunit from occurring in excess of R subunits. Clearly, a cell that overproduced free C subunit would lose the modulatory effect of cAMP on the kinase system and would have a constitutively active PKA. We have attempted to define the mechanisms involved in this type of cellular regulation by overexpressing C_α or C_β subunits and observing the cell's response (Uhler and McKnight 1987). The expression vectors used were very similar to that shown in Figure 4 for the R_I gene and contain the inducible metallothionein promoter to allow rapid changes in the levels of the C subunit. When C protein was overproduced in NIH-3T3 fibroblasts, we observed a large increase in kinase activity, but the majority of the induced kinase remained cAMP-dependent (Uhler and McKnight 1987). This suggested that the overproduced C has become complexed with an R subunit to maintain cAMP control over C activity.

To investigate this compensation phenomenon, we examined the effect of C overproduction on R subunit levels. Both R_I and R_{II} can be visualized by photoaffinity labeling with 8-azido-[^{32}P]cAMP and separation of the labeled R subunits on SDS-PAGE. The results obtained using cell extracts from several different cell types overexpressing C subunit demonstrated that R_I protein increased substantially but that R_{II} remained constant (data not shown). These results were further substantiated by Western blot analysis of C, R_I, and R_{II} in stable transformants overexpressing the C subunit. A fivefold increase in C and R_I protein was observed upon induction with Zn^{++}, but the levels of R_{II} protein remained unchanged. We conclude that only the R_I subunit is capable of increasing in response to C overexpression.

The mechanism by which the R_I compensation occurs was further examined by quantifying the levels of R_I and C mRNAs before and after C overexpression. We observed no change in $R_{I\alpha}$ mRNA in cells overexpressing the C subunit, indicating that the R_I compensation results from either an increased rate of translation of existing $R_{I\alpha}$ mRNA or a stabilization of the R_I protein. Experiments to rule out the former have not been completed, but we favor the latter explanation for several reasons. Earlier work from several laboratories has demonstrated that the R_I subunit is more susceptible to degradation when released from holoenzyme by cAMP than when complexed with C (Steinberg and Agard 1981). We have also observed that most of the cells examined contain free R_I subunit, which elutes between the type I and type II holoenzymes on a DEAE column. Since it is presumably important for cells to maintain the C subunits in a repressed state in the absence of cAMP, it appears likely that cells synthesize an excess of $R_{I\alpha}$ subunit which turns over rapidly if there is no C subunit with which to form holoenzyme.

Assembly of Type I Kinase in C Overexpressors

We have overexpressed C subunits in cells that do not have detectable levels of type I PKA when analyzed on DEAE columns. In control GH_3 rat pituitary cells, all of the kinase activity is found in the type II holoenzyme peak, although mRNA for R_I is present, and free R_I protein is readily detectable on Western blots. As described above, overproduction of C leads to a compensatory increase in R_I protein. The change in kinase subtypes as displayed on a DEAE column is quite dramatic, as shown in Figure 6. The GH_3 cells that overexpress C display little change in their levels of type II kinase but a large increase in type I kinase. We have observed similar results in NIH-3T3 cells, AtT20 pituitary cells, and Y1 adrenal cells, suggesting that this may be a common mechanism by which the cell ensures that the levels of R subunit will always at least equal the level of C in the cell. A simple interpretation of these results is that the type II kinase forms preferentially in cells, and the type I kinase is only seen when there exists an excess of C subunit over the available R_{II} subunit. Alternatively, it is possible that cells possess mechanisms for regulating the assembly of kinase subtypes that are independent of relative ratios of R_{II}, R_I, and C subunits.

Overexpression of R_{II} Decreases Type I Kinase Assembly

To test the hypothesis that the type II kinase forms preferentially under in vivo conditions, we have con-

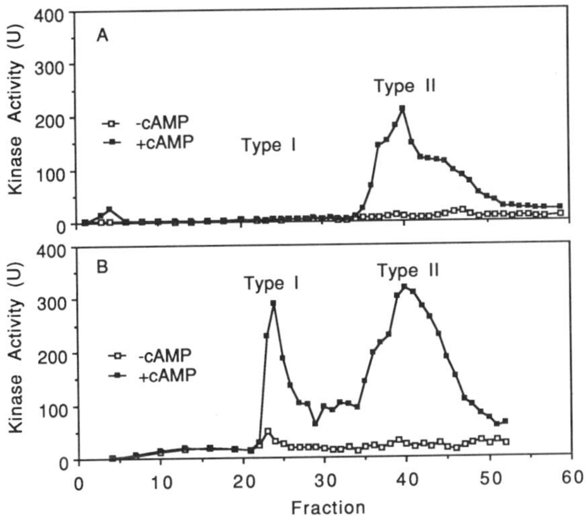

Figure 6. Increase in type I kinase in GH_3 pituitary cells overexpressing C_α. Cytoplasmic extracts from wild-type GH_3 rat pituitary cells (*A*) or C_α overexpressing GH_3 cells (*B*) were prepared and separated on a DEAE column developed with a linear gradient of 0–300 mM NaCl. The fractions were then assayed for kinase activity with Kemptide as substrate in the absence (□) or presence of 5 μM cAMP (■). The type I kinase eluted at 30–50 mM NaCl, the type II kinase eluted at about 130 mM NaCl, and free C subunit was not retained on this column.

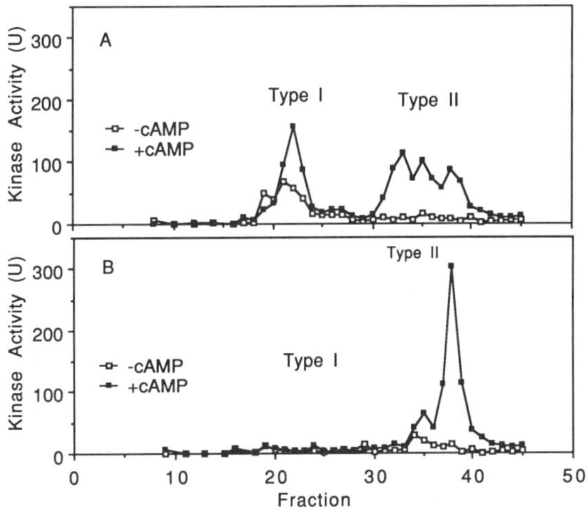

Figure 7. Overexpression of R_{II} causes a decrease in type I kinase in *ras*-transformed NIH-3T3 cells. Cytoplasmic extracts from a stable transformant of *ras*-transformed NIH-3T3 cells transfected with the neomycin resistance gene alone (*A*) or from $R_{II\alpha}$ overexpressing cells (*B*) were prepared and assayed as described in Fig. 6, except that the DEAE column was developed on an HPLC. The HPLC Bio-Gel TSK DEAE-5-PW column (Bio-Rad) produced a more complete separation of type I and II kinases and retained the free catalytic subunit that eluted slightly ahead of the type I peak.

structed an expression vector containing the structural sequence of mouse $R_{II\alpha}$ driven by the Harvey sarcoma virus LTR promoter to allow high constitutive expression. *ras*-transformed 3T3 cells were chosen as the recipient, since analysis on DEAE revealed that these cells contained nearly equal levels of type I and type II kinase. Stable transformants were obtained using the neomycin phosphotransferase gene as a selectable marker, and these transformants were analyzed for R_{II} mRNA and protein expression. Several clones were examined, and they expressed between 3 and 17 times more R_{II} protein than wild-type cells. Cytoplasmic extracts were prepared from control and R_{II}-transfected Ras-3T3 cells and applied to a DEAE-HPLC column. Fractions were assayed for kinase activity in the presence and absence of cAMP. Results are shown in Figure 7. Ras-3T3 cells overexpressing R_{II} at about three times the normal level displayed an 80% reduction in type I kinase levels, and the R_{II} overexpressors making 17-fold more R_{II} (shown in Fig. 7) had no detectable type I kinase and a greatly increased peak of type II enzyme compared to control cells. In experiments not shown, we have done the reciprocal experiment by overexpressing wild-type R_I in cells with predominantly type II kinase; in this case no change in kinase isoforms was observed on DEAE columns and the overexpressed R_I eluted in the position of free R_I (Clegg et al. 1988). These results indicate that R_{II} preferentially forms holoenzyme with the available catalytic subunit even in the presence of excess R_I protein and supports our hypothesis that type I kinase forms in cells only when an excess of C is synthesized compared to the available R_{II}.

CONCLUDING REMARKS

We have utilized the techniques of molecular genetics to devise new approaches to the study of the cAMP-dependent protein kinase system. To summarize briefly, two catalytic and four regulatory subunit genes have been characterized by us and other laboratories in mammalian systems. Three of these subunits ($R_{I\alpha}$, $R_{II\alpha}$, and C_α) are constitutively expressed in most tissues, although the levels of mRNA can vary significantly, most notably in male germ cells during spermatogenesis. The other three subunits (C_β, $R_{I\beta}$, and $R_{II\beta}$) are highly expressed in brain and spinal cord and are either absent or expressed at very low levels in many other tissues. Other sites of specific expression for the β subunits have been examined and, in particular, the $R_{II\beta}$ gene is highly inducible in several differentiating systems. We are just beginning to examine the differences in functional properties of the C_α and C_β subunits. Future work with specific antibodies should help resolve the potential for specific interactions between subtypes of R and C subunits. Studies on mutated forms of the R subunits have so far been limited to the cloning and expression of dominant cAMP-resistant R mutants from S49 cells. These studies have demonstrated that the R_I subunits harboring mutations in the cAMP-binding domains can be reintroduced into many cell types to impose a constitutive or inducible (when the metallothionein promoter is used) cAMP-resistant phenotype.

Overexpression of the C subunits in cultured cells has led to several insights into the regulation of holoenzyme subtypes and the coordinate regulation of R and

C subunit levels in cells. Many cultured cells are apparently overproducing the $R_{I\alpha}$ subunit and then preferentially degrading the R subunit that does not become associated with C subunit in a holoenzyme. In cells that can be induced to overexpress C, the $R_{I\alpha}$ subunit is stabilized, and its level can increase by at least fivefold. In untransfected NIH-3T3 cells, no type I holoenzyme is formed, despite the production of $R_{I\alpha}$ protein; however, when excess C is produced in these cells, it becomes associated with the free $R_{I\alpha}$ and forms a type I kinase. In cultured cells containing both type I and type II holoenzymes, we find that overexpression of modest levels of R_{II} protein leads to a loss of type I holoenzyme and a corresponding increase in type II. We suggest the possibility that cells preferentially assemble the type II holoenzyme and may only form the type I holoenzyme when the levels of C exceed the R_{II} subunits in the cell. Further studies addressing these and other basic questions regarding the role of PKA in intracellular regulation should continue to provide insights into this complex signaling system.

ACKNOWLEDGMENTS

We thank Thong To for expert technical assistance. This work was supported by grants from the National Institutes of Health (GM-32875) and the Cystic Fibrosis Foundation (Z-035-8).

REFERENCES

Adhya, S. and S. Garges. 1982. How cyclic AMP and its receptor protein act in *Escherichia coli*. *Cell* **29:** 287.

Clegg, C.H., G.G. Cadd, and G.S. McKnight. 1988. Genetic characterization of a brain specific form of the type-I regulatory subunit of cAMP-dependent protein kinase. *Proc. Natl. Acad. Sci.* **85:** 3703.

Clegg, C.H., L.A. Correll, G.G. Cadd, and G.S. McKnight. 1987. Inhibition of intracellular cAMP-dependent protein kinase using mutant genes of the regulatory type I subunit. *J. Biol. Chem.* **262:** 13111.

Coffino, P., H.R. Bourne, U. Friedrich, J. Hochman, P.A. Insel, I. Lemaire, K.L. Melmon, and G.M. Tomkins. 1976. Molecular mechanisms of cyclic AMP action: A genetic approach. *Recent Prog. Horm. Res.* **32:** 669.

Doherty, P.J., J. Tsao, B.P. Schimmer, M.C. Mumby, and J.A. Beavo. 1981. cAMP-dependent protein kinase and regulation of adrenocortical functions: a genetic evaluation. *Cold Spring Harbor Conf. Cell. Proliferation* **8:** 211.

Edelman, A.M., D.K. Blumenthal, and E.G. Krebs. 1987. Protein serine/threonine kinases. *Annu. Rev. Biochem.* **56:** 567.

Hedin L., G.S. McKnight, J. Lifka, J.M. Durica, and J.S. Richards. 1987. Tissue distribution and hormonal regulation of mRNA for regulatory and catalytic subunits of cAMP-dependent protein kinases during ovarian follicular development and luteinization in the rat. *Endocrinology* **120:** 1928.

Jahnsen, T., L. Hedin, V.J. Kidd, W.G. Beattie, S.M. Lohmann, U. Walter, J. Durica, T.Z. Schulz, E. Schiltz, M. Browner, C.B. Lawrence, S.L. Ratoosh, and J.S. Richards. 1986. Molecular cloning, cDNA structure and regulation of the regulatory subunit (R-II51) of type II cAMP-dependent protein kinase from rat granulosa cells. *J. Biol. Chem.* **261:** 12352.

Knopf, J.L., M.-H. Lee, L.A. Sultzman, R.W. Kriz, C.R. Loomis, R.M. Hewick, and R.M. Bell. 1986. Cloning and expression of multiple protein kinase C cDNAs. *Cell* **46:** 491.

Krebs, E.G. 1972. Protein kinases. *Curr. Top. Cell. Regul.* **5:** 99.

Krebs, E.G. and J.A. Beavo. 1979. Phosphorylation-dephosphorylation of enzymes. *Annu. Rev. Biochem.* **48:** 923.

Lee, D.C., D.R. Carmichael, E.G. Krebs, and G.S. McKnight. 1983. Isolation of a cDNA clone for the type I regulatory subunit of bovine cAMP-dependent protein kinase. *Proc. Natl. Acad. Sci.* **80:** 3608.

Leiser, M., C.S. Rubin, and J. Erlichman. 1986. Differential binding of the regulatory subunits (RII) of cAMP-dependent protein kinase II from bovine brain and muscle to RII-binding proteins. *J. Biol. Chem.* **261:** 1904.

Montminy, M.R. and L.M. Bilezikjian. 1987. Binding of a nuclear protein to the cyclic-AMP response element of the somatostatin gene. *Nature* **328:** 175.

Montminy, M.R., K.A. Sevarino, J.A. Wagner, G. Mandel, and R.H. Goodman. 1986. Identification of a cyclic-AMP-responsive element within the rat somatostatin gene. *Proc. Natl. Acad. Sci.* **83:** 6682.

Morgan, D.O., L. Ho, L.J. Korn, and R.A. Roth. 1986. Insulin action is blocked by a monoclonal antibody that inhibits the insulin receptor kinase. *Proc. Natl. Acad. Sci.* **83:** 328.

Murphy, C.S. and R.A. Steinberg. 1985. Hotspots for spontaneous and mutagen-induced lesions in regulatory subunit of cyclic AMP-dependent protein kinase in S49 mouse lymphoma cells. *Somatic Cell Mol. Genet.* **11:** 605.

Nakamura, T., and G.H. Gold. 1987. A cyclic nucleotide-gated conductance in olfactory receptor cilia. *Nature* **325:** 442.

Nigg, E.A., H. Hilz, H.M. Eppenberger, and F. Dutly. 1985a. Rapid and reversible translocation of the catalytic subunit of cAMP-dependent protein kinase type II from the Golgi complex to the nucleus. *EMBO J.* **4:** 2801.

Nigg, E.A., G. Schäfer, H. Hilz, and H.M. Eppenberger. 1985b. Cyclic-AMP-dependent protein kinase type II is associated with the Golgi complex and with centrosomes. *Cell* **41:** 1039.

Nowak, I., K. Seipel, M. Schwarz, D.A. Jans, and B.A. Hemmings. 1987. Isolation of a cDNA and characterization of the 5′ flanking region of the gene encoding the type I regulatory subunit of the cAMP-dependent protein kinase. *Eur. J. Biochem.* **167:** 27.

Øyen, O., J.D. Scott, G.G. Cadd, G.S. McKnight, E.G. Krebs, V. Hansson, and T. Jahnsen. 1988. A unique mRNA species for a regulatory subunit of cAMP-dependent protein kinase is specifically induced in haploid germ cells. *FEBS. Lett.* **229:** 391.

Reimann, E.M., D.A. Walsh, and E.G. Krebs. 1971. Purification and properties of rabbit skeletal muscle adenosine 3′,5′-monophosphate-dependent protein kinases. *J. Biol. Chem.* **246:** 1986.

Rogers, K.V., L. Vician, R.A. Steiner, and D.K. Clifton. 1987. Reduced preprosomatostatin messenger ribonucleic acid in the periventricular nucleus of hypophysectomized rats determined by quantitative in situ hybridization. *Endocrinology* **121:** 90.

Sarkar, D., J. Erlichman, and C.S. Rubin. 1984. Identification of a calmodulin-binding protein that co-purifies with the regulatory subunit of brain protein kinase II. *J. Biol. Chem.* **259:** 9840.

Sasaki, K., T.P. Cripe, S.R. Koch, T.L. Andreone, D.D. Petersen, E.G. Beale, and D.K. Granner. 1984. Multihormonal regulation of phosphoenolpyruvate carboxykinase gene transcription. The dominant role of insulin. *J. Biol. Chem.* **259:** 15242.

Scott, J.D., M.B. Glaccum, M. Zoller, M.D. Uhler, D.M. Helfman, G.S. McKnight, and E.G. Krebs. 1987. The molecular cloning of a type II regulatory subunit of the cAMP-dependent protein kinase from rat skeletal muscle and mouse brain. *Proc. Natl. Acad. Sci.* **84:** 5192.

Showers, M.O. and R.A. Maurer. 1986. A cloned bovine cDNA encodes an alternate form of the catalytic subunit of cAMP-dependent protein kinase. *J. Biol. Chem* **261:** 16288.

Silver, B.J., J.A. Bokar, J.B. Virgin, E.A. Vallen, A. Milsted, and J.H. Nilson. 1987. Cyclic AMP regulation of the human glycoprotein hormone alpha-subunit gene is mediated by an 18-base-pair element. *Proc. Natl. Acad. Sci.* **84:** 2198.

Stein, J.C. and C.S. Rubin. 1985. Isolation and sequence of a tryptic peptide containing the autophosphorylation site of the regulatory subunit of bovine brain protein kinase II. *J. Biol. Chem.* **260:** 10991.

Steinberg, R.A. and D.A. Agard. 1981. Turnover of regulatory subunit of cyclic AMP-dependent protein kinase in S49 mouse lymphoma cells. Regulation by catalytic subunit and analogs of cyclic AMP. *J. Biol. Chem.* **256:** 10731.

Sutherland, E.W. and T.W. Rall. 1957. The properties of an adenine ribonucleotide produced with cellular particles, ATP, Mg^{++}, and epinephrine or glucagon. *J. Am. Chem. Soc.* **79:** 3608.

Takio, K., S.B. Smith, E.G. Krebs, K.A. Walsh, and K. Titani. 1984. Amino acid sequence of the regulatory subunit of bovine type II adenosine cyclic 3′,5′-phosphate dependent protein kinase. *Biochemistry* **23:** 4200.

Titani, K., T. Sasagawa, L.H. Ericsson, S. Kumar, S.B. Smith, E.G. Krebs, and K.A. Walsh. 1984. Amino acid sequence of the regulatory subunit of bovine type I adenosine cyclic 3′,5′-phosphate dependent protein kinase. *Biochemistry* **23:** 4193.

Uhler, M.D. and G.S. McKnight. 1987. Expression of cDNAs for two isoforms of the catalytic subunit of cAMP-dependent protein kinase. *J. Biol. Chem.* **262:** 15202.

Uhler, M.D., J.C. Chrivia, and G.S. McKnight. 1986a. Evidence for a second isoform of the catalytic subunit of cAMP-dependent protein kinase. *J. Biol. Chem.* **261:** 15360.

Uhler, M.D., D.F. Carmichael, D.C. Lee, J.C. Chrivia, E.G. Krebs, and G.S. McKnight. 1986b. Isolation of cDNA clones coding for the catalytic subunit of mouse cAMP-dependent protein kinase. *Proc. Natl. Acad. Sci.* **83:** 1300.

Walsh, D.A., J.P. Perkins, and E.G. Krebs. 1968. An adenosine 3′,5′-monophosphate-dependent protein kinase from rabbit skeletal muscle. *J. Biol. Chem.* **243:** 3763.

cAMP-dependent Protein Kinase: A Framework for a Diverse Family of Enzymes

S.S. Taylor, J.A. Buechler, L.W. Slice, D.K. Knighton, S. Durgerian, G.E. Ringheim, J.J. Neitzel, W.M. Yonemoto, J.M. Sowadski, and W. Dospmann
Department of Chemistry, University of California, San Diego, La Jolla, California 92093

One of the first protein kinases to be discovered was the enzyme that is activated in response to cAMP, cAMP-dependent protein kinase (Walsh et al. 1968). In the intervening decades, many diverse kinases have been identified (Hanks et al. 1988). They include kinases that transfer phosphate to serine and threonine as well as those transferring phosphate to tyrosine. Some kinases are activated in response to second messengers, such as cAMP, Ca^{++}, and diacylglycerol, whereas others are activated by primary signals such as growth factors. The protein kinase family includes transforming proteins coded by oncogenes as well as growth factor receptors (for reviews, see Boyer and Krebs 1986).

In many ways, cAMP-dependent protein kinase serves as a prototype for the entire kinase family; however, its mechanism of activation is unusual and perhaps unique among the protein kinases. In the absence of cAMP, the enzyme is an inactive tetramer containing two regulatory (R) and two catalytic (C) subunits. With the exception of the transforming oncogenic kinases, the protein kinases are characteristically tightly regulated enzymes and are maintained in an inactive form in the absence of activating ligands. cAMP-dependent protein kinase is activated when intracellular cAMP levels are elevated. The major receptor for cAMP in eukaryotic cells is the regulatory subunit of cAMP-dependent protein kinase. Binding of cAMP induces conformational changes in the regulatory subunit that stabilize a structure having a lowered affinity for the catalytic subunit, thus promoting dissociation of the holoenzyme complex. The dissociated catalytic subunit ($M_r = 38{,}900$) is catalytically active, and its regulatory role in modulating the activity of many substrate proteins has been well documented (Krebs and Beavo 1979; Bramson et al. 1984).

Despite the diversity seen in the overall protein kinase family, many features of structure and function are shared. In particular, each protein kinase has a conserved catalytic core, indicating that these enzymes have evolved in part from a common origin. In this regard, cAMP-dependent protein kinase is not only one of the smallest protein kinases, but also one of the simplest, for the major regulatory elements are part of a distinct subunit that easily dissociates. The remaining catalytic subunit shares extensive sequence similarities with thc catalytic region of every kinase, and residues that we find to be essential for ATP binding or catalysis in the catalytic subunit are conserved throughout the kinase family. The catalytic subunit thus provides a valuable framework for viewing the entire family of protein kinases.

Homologous Protein Kinases

The protein kinases appear to have been assembled as modular structures by a mechanism that undoubtedly involves exon shuffling. General features of a few kinases are summarized in Figure 1. The closest relative to cAMP-dependent protein kinase is a cGMP-dependent protein kinase (Takio et al. 1984b). This enzyme also is activated by a cyclic nucleotide, but, in this case, the regulatory cyclic-nucleotide-binding region is part of a contiguous polypeptide chain that includes a catalytic domain at the carboxyl terminus. The Ca^{++}:calmodulin-dependent protein kinases, such as myosin light-chain kinase (Takio et al. 1986), contain a regulatory Ca^{++}:calmodulin-binding domain that is carboxy-terminal to the catalytic region. Protein kinase C, on the other hand, contains regulatory binding sites for Ca^{++}, diacylglycerol, and phosphatidyl serine in the region preceding the catalytic site (Parker et al. 1986). The epidermal growth factor (EGF) receptor and the insulin receptor both contain a hormone-binding extracellular domain, a single membrane-spanning segment, and a large intracellular domain having intrinsic protein kinase activity (Ullrich et al. 1984, 1985). The product of the oncogene in Rous sarcoma virus, pp60$^{v\text{-}src}$, also belongs to the protein kinase family (Hanks et al. 1988).

Although these protein kinases are striking for their diversity, each shares extensive sequence similarities throughout a conserved catalytic core that includes most of the catalytic subunit. Furthermore, if one considers the conserved sequences in conjunction with residues that have been shown to be associated with functional sites in the catalytic subunit, it is clear that the sequence similarities can be correlated with homologous functions.

Active Site

Mapping the nucleotide-binding site of the catalytic subunit with analogs of ATP initially provided a gener-

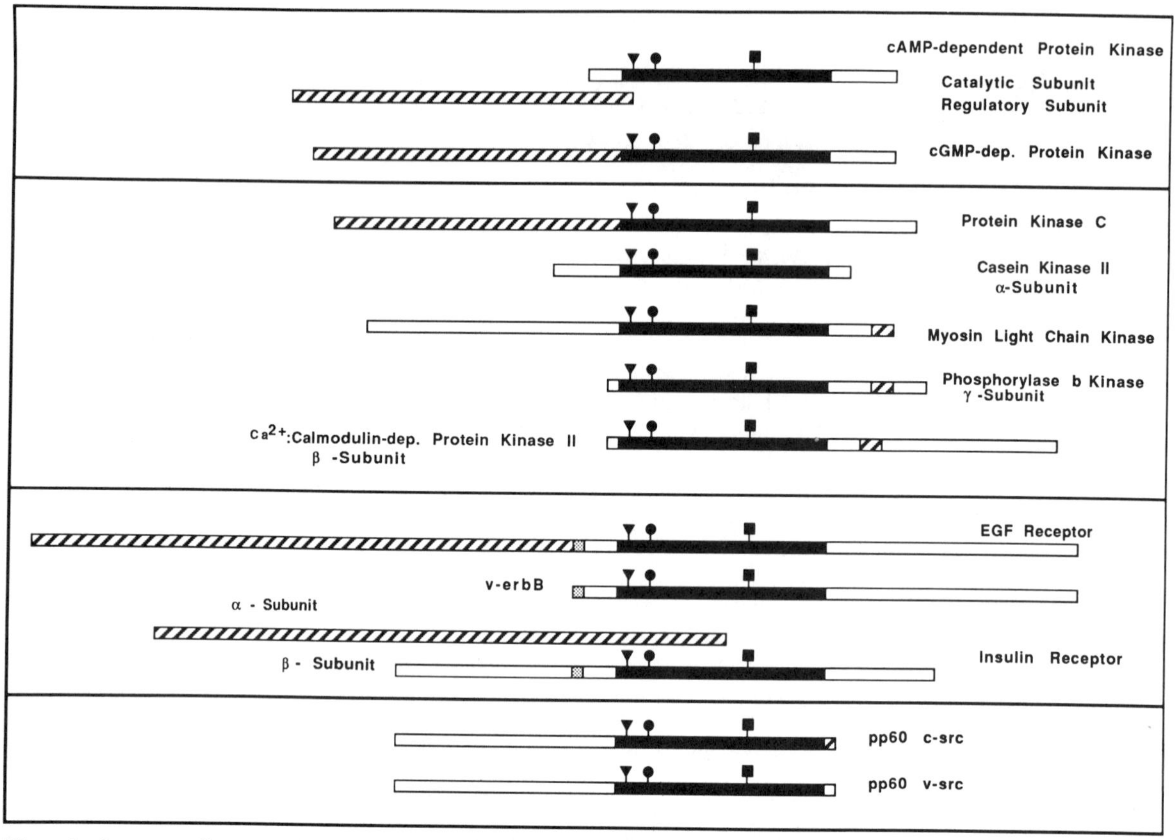

Figure 1. Sequence alignment of various protein kinases. Amino acid sequences were either determined directly or deduced from the DNA sequence. The conserved catalytic core is indicated in black, known regulatory regions are indicated by cross-hatching, and putative membrane-spanning segments are stippled. Conserved residues correspond to the glycine-rich segment, Gly-50-Gly-55 (▼), Lys-72 (●), and Asp-184 (■) in the catalytic subunit (Taylor et al. 1988).

al guideline for regions of the nucleotide that are important for interaction with the protein (Hoppe et al. 1978). On the other hand, the first definitive experimental evidence identifying a specific amino acid at the nucleotide-binding site was carried out by Zoller et al. (1981), who demonstrated that Lys-72 was covalently modified stoichiometrically by an affinity analog of ATP, *p*-fluorosulfonyl benzoyl 5′-adenosine (FSBA). Bhatnagar et al. (1984) later showed that the FSBA-modified catalytic subunit could no longer bind Mg-ADP or Mg-ATP but was capable still of binding AMP and adenosine, thus placing Lys-72 in close proximity to the β- and/or γ-phosphate of ATP.

Similar approaches have been used to localize and map the peptide recognition site. Synthetic peptides established the requirements for recognition by the catalytic subunit (for review, see Bramson et al. 1984). The catalytic subunit transfers phosphates from the γ-phosphate of ATP to the hydroxyl group of serine or threonine, and the site of phosphotransfer is typically preceded by two basic amino acids, usually arginines. Affinity labeling with a peptide analog led to inactivation that correlated with the covalent modification of Cys-199 (Bramson et al. 1983).

In addition to affinity labeling with substrate analogs, group-specific reagents have been used to target functional residues. The pH dependency of the reaction catalyzed by the catalytic subunit indicates that the mechanism involves a general base (Yoon and Cook 1987). In an effort to identify this group, a hydrophobic carbodiimide was used, dicyclohexylcarbodiimide (DCCD). DCCD partitions into hydrophobic regions and is an irreversible inhibitor of the catalytic subunit (Toner-Webb and Taylor 1987). A specific reactive carboxyl group in the catalytic subunit was not identified initially by modification with DCCD, presumably because of intramolecular cross-linking. To avoid this, the enzyme was treated first with acetic anhydride to block any reactive lysines. Treatment of this modified enzyme with DCCD, in the presence of [^{14}C]glycine ethyl ester, led to the covalent modification of one major residue, Asp-184, and Mg-ATP protected against this modification (Buechler and Taylor 1988b). Although it could also function in chelating Mg^{++}, Asp-184 is a good candidate for the general base that participates in catalysis.

Correlation of Conserved Sequences with Essential Functions

Chemical modifications have defined a general topographical arrangement for functional regions of the catalytic subunit. The amino-terminal region is associated with nucleotide binding, whereas peptide binding and catalysis are associated with the central portion of the molecule. Three amino acids, in particular, have been placed near the active site: Lys-72, Asp-184, and Cys-199. What criteria reinforce the likelihood that these residues contribute in an essential way to the functioning of the catalytic subunit? In general, if one considers the larger family of protein kinases, it is likely that the same general features of nucleotide binding and catalytic mechanism will be conserved. In addition, if one looks more selectively at the conserved sequences in the protein kinases, there are clearly specific regions that are highly conserved and some residues that are invariant. For example, Lys-72, which is thought to participate in ATP binding, is, in fact, invariant in every protein kinase. Furthermore, affinity labeling with FSBA has established that the homologous lysine is modified in v-*src* (Kamps et al. 1984). Finally, when Lys-296 in pp60$^{v\text{-}src}$ is selectively replaced, the enzyme activity is lost, as is its ability to transform (Kamps and Sefton 1986). Replacement of the homologous lysine in the EGF receptor and the insulin receptor by site-directed mutagenesis also generates an inactive enzyme, and the hormone-induced effects are lost (Chou et al. 1987; Honegger et al. 1987). Modification of this conserved lysine by site-directed mutagenesis has been used frequently to establish that intrinsic kinase activity is essential for function.

Asp-184 is another residue that is invariant throughout the family of protein kinases, and on this basis, it also is a good candidate for playing an essential role. As shown in Figure 2, the sequence flanking Asp-184 is highly conserved as well.

Unlike Asp-184 and Lys-72, Cys-199, which was targeted by the peptide analog, is not a highly conserved residue (Fig. 2). Thus, although Cys-199 may play a significant role in the catalytic subunit, this role is not conserved in the larger family of protein kinases. Given the variation in peptide recognition for each kinase, it is anticipated that regions important for peptide recognition will differ. A similar variability is seen in the family of serine proteases that conform to the chymotrypsin motif. Each has conserved the same mechanism, and residues essential for catalysis are invariant. In contrast, the residues lining the substrate binding pocket vary and presumably account for the unique specificity of each enzyme.

Catalytic Mechanism

A catalytic mechanism can be proposed for the catalytic subunit based on both kinetic and chemical evidence. One additional piece of information sheds light on the configuration of amino acid residues at the active site. When the catalytic subunit initially was treated with DCCD, inhibition was predicted to be due to an intramolecular cross-linking reaction. This cross-link has subsequently been identified. Asp-184 is the carboxyl group that initially reacts with DCCD to form a reactive acylisourea intermediate. This electrophilic intermediate then generates a cross-link between Asp-184 and the ϵ-amino group of Lys-72 (J.A. Buechler and S.S. Taylor, in prep.). Identification of this cross-link localizes both of these residues in close proximity to one another at the active site. As was emphasized earlier, these are two highly conserved regions of the catalytic subunit, and the lysine and aspartic acid, in particular, are invariant in every kinase.

A general mechanism for the catalytic subunit is proposed in Figure 3. This reaction proceeds by a direct transfer mechanism, resulting in an inversion of configuration of the γ-phosphate group (Ho et al. 1988). According to this mechanism, Mg-ATP preferentially binds first and interacts with the side chain of Lys-72.

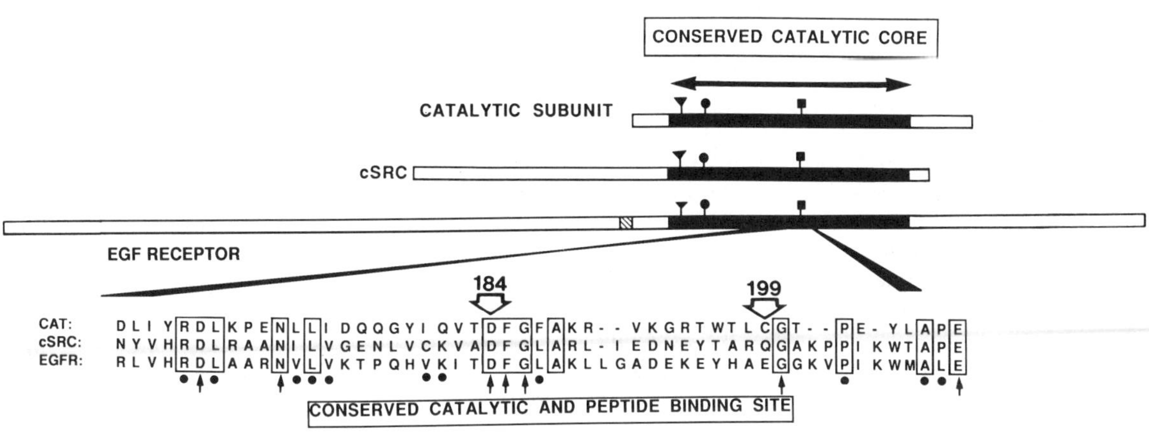

Figure 2. Comparison of kinase structures. The overall domain structure of the catalytic subunit is compared to v-*src* and the EGF receptor. The conserved catalytic core is designated in black, and the specific conservation of sequence in the region flanking the catalytic site is shown below. Residues that are invariant in all known protein kinases, so far, are indicated by the arrows, whereas residues that are highly conserved are designated by a filled circle. Asp-184, which is modified by DCCD and cross-linked to Lys-72, is indicated by the open arrow. Cys-199 is modified by affinity labeling with peptide analogs and is also indicated by an open arrow.

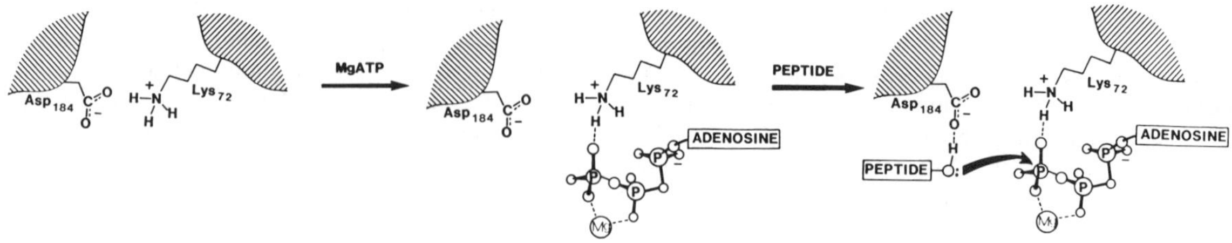

Figure 3. A proposed mechanism for the reaction catalyzed by the catalytic subunit.

Asp-184 is now free to act as a general base, thus making the oxygen on the serine hydroxyl group more nucleophilic for reacting with the γ-phosphate group of ATP. This ordered mechanism is consistent with the kinetic evidence of Whitehouse et al. (1983) and with the pH dependency described by Yoon and Cook (1987). An alternative role for Asp-184 may be to serve as a ligand for Mg^{++} in the Mg-ATP complex. This mechanism would also be consistent with a similar ordered binding pathway, but would require another assignment for the base.

Expression of Catalytic Subunit in *Escherichia coli*

One method for testing the proposed mechanism and for determining, in general, the specific role that any given amino acid plays is site-directed mutagenesis. Mutagenesis has been used extensively to identify functional regions of the oncogenic transforming proteins such as pp60^{src} (Hunter 1987); however, to genuinely understand the molecular consequences of any given mutation, it is necessary to isolate and characterize the mutant enzyme. The catalytic subunit provides a good model for correlating specific mutations with functional changes in the enzyme. A necessary prerequisite is to develop an expression vector that enables one to produce sufficient amounts of active or mutant enzyme. To achieve this, a cDNA clone for the mouse C_α-subunit, a kind gift from G. Stanley McKnight at the University of Washington, was used to construct a series of expression vectors.

A number of vectors have been used with varying degrees of success for the expression of eukaryotic proteins in *E. coli*. The most frequent difficulty encountered is the appearance of the recombinant protein in occluded particles that are neither soluble nor active. Expression of the catalytic subunit from a pUC vector containing either the *lac* promoter or the *tac* promoter yielded large amounts of insoluble protein. The strain of *E. coli* also can be a critical factor for achieving active, soluble enzyme. To overcome the solubility problems associated with expressing the catalytic subunit in *E. coli*, the cDNA coding for the catalytic subunit was genetically engineered to insert into the *Nde*I site of pT7-7. This vector, originally designed by Tabor and Richardson (1987), contains the T7 RNA polymerase promoter. This vector, containing the insert coding for the entire catalytic subunit, was transformed into a strain of *E. coli*, BL21 (DE-3) (Studier and Moffatt 1986), containing the T7 RNA polymerase gene incorporated into the bacterial genome under the regulation of the *lac* promoter. After the T7 polymerase is induced by isopropyl-D-thiogalactopyranoside, transcription proceeds from the T7 promoter. This expression system thus does not require elevated temperatures for induction.

Induction of the catalytic subunit using the pT7-7 vector in *E. coli* BL21 (DE-3) led to the expression of large quantities of protein, and, unlike other vectors and cells, a significant portion (20–30%) of the expressed protein was soluble and active (L. Slice and S.S. Taylor, in prep.). The enzyme was purified to homogeneity and is kinetically analogous to the enzyme that is purified from mammalian tissues. The catalytic subunit has one known posttranslational modification that does not occur in *E. coli*. The amino-terminal residue in the mammalian enzyme is myristylated (Carr et al. 1982), whereas the expressed enzyme from *E. coli* has a free α-amino group at the terminal glycine residue. The function of the myristyl group in the catalytic subunit is not known. The transforming protein, pp60^{src}, also is myristylated at its amino terminus, and, in this case, the myristyl group is thought to play a role in directing the enzyme to the plasma membrane (Kamps et al. 1986). Since, in general, the catalytic subunit is considered to be a cytosolic enzyme, the myristyl group on this enzyme presumably serves a different function. The only clear difference between the recombinant catalytic subunit and the mammalian enzyme is a slightly reduced stability. This was apparent when the enzyme was inactivated at 49°C. Under these conditions, the recombinant enzyme had a $t_{1/2}$ of 0.7 minutes in contrast with 3.9 minutes for the mammalian enzyme. The inclusion of 0.01% Triton X-100 in the buffer reduced the stability of the expressed enzyme further, whereas the mammalian enzyme was only marginally affected by Triton X-100. On this basis, the myristyl group may play a role in stabilization of the enzyme.

Several mutations subsequently have been introduced into the catalytic subunit. The first was intended to improve solubility. Specifically, the catalytic subunit contains two cysteine residues, Cys-199 and Cys-343 (Shoji et al. 1983). As indicated earlier, Cys-199 is thought to be at the active site of the enzyme. Nevertheless, both cysteines are protected against alkylation by iodoacetic acid. E.A. First et al. (in prep.) used fluorescence energy transfer techniques to measure the distance between Cys-199 and Cys-343. They established a minimum distance of 33 Å, and concluded that Cys-343 is not close to the active site but instead is protected by conformational changes that extend throughout a large portion of the protein. It was thought that replacement of Cys-343 with a serine could improve the solubility of the protein, since the insoluble aggregated protein found in *E. coli* migrates as high-molecular-weight aggregates on denaturing polyacrylamide gels run in the absence of reducing agents. The solubility of the expressed catalytic subunit was improved slightly, and the recombinant enzyme was fully active, confirming the nonessential nature of Cys-343. Changes are now being introduced into the active site regions of the enzyme to better understand the specific functional roles that key residues, such as Asp-184 and Lys-72, play. For example, although the lysine that is homologous to Lys-72 has been mutated in many protein kinases, there is no evidence indicating the structural consequences of that mutation. Does mutation of Lys-72 produce minor structural perturbations of the enzyme that simply render it inactive or, on the other hand, does it lead to major structural perturbations? This distinction is important to resolve, if results based on mutations are to be correctly interpreted.

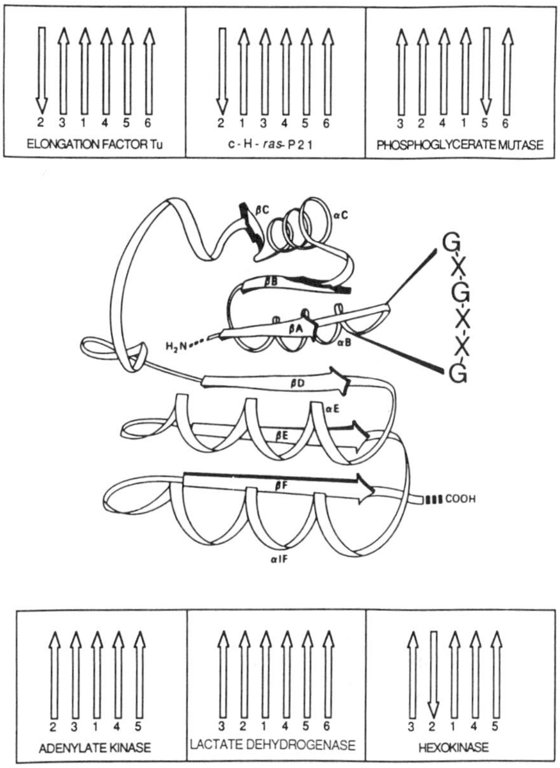

Figure 4. Schematic representation of the nucleotide folds for various adenine nucleotide-binding proteins. The helix and β-strand organization of lactate dehydrogenase, shown in the center, is based on the crystallographic coordinates. The alignments of the β-strands in the other proteins also were taken from the crystal structures of EF-Tu, *ras*-p21, phosphoglycerate mutase, adenylate kinase, and hexokinase (Taylor et al. 1988).

Nucleotide Fold

In addition to identifying residues that are essential for function, there are some general motifs that can be recognized, in particular, with regard to the nucleotide-binding portion of the enzyme. Affinity labeling localized the nucleotide-binding site to the amino-terminal region of the enzyme. A general nucleotide fold motif was first recognized by Rossmann et al. (1974) when they compared the crystal structures of lactate dehydrogenase and glyceraldehyde 3-phosphate dehydrogenase. Although there was little conservation of sequence, both enzymes had the same general features of secondary and tertiary structure in the nucleotide-binding domain. This motif, consisting of six β-strands forming a β-sheet and two helices above and below the plane of the sheet, has been found in most nucleotide-binding sites (Brändén 1980). In all cases, the nucleotide binds at the carboxy-terminal side of the β-strands. A comparison of several proteins having such a nucleotide fold motif is seen in Figure 4. It should be emphasized strongly that although the general motif is retained, details vary in each structure. For example, lactic dehydrogenase (LDH) has six parallel β-strands, whereas elongation factor-Tu (EF-Tu) (Jurnak 1985) and p21 *ras* (de Vos et al. 1988) have one antiparallel strand. The only consistent feature of the primary and secondary structure that is found in all cases is a triad of glycines, Gly X-Gly-X-X-Gly, which lies at a loop between the first β-strand and the first α-helix (Wierenga et al. 1986).

As seen in Figure 5, Lys-72 is preceded by a similar glycine-rich segment (50–55), consistent with this region of the catalytic subunit constituting a nucleotide fold motif. To probe the catalytic subunit structure in a more general way, lysine residues were differentially modified with acetic anhydride (J.A. Buechler et al., in prep.). This labeling identified two reactive regions of the enzyme, one at the carboxyl terminus and the other at the amino terminus. The intervening regions, Lys-92 to Lys-189, were unreactive under all conditions. In the presence of Mg-ATP, several lysines in the amino-terminal region were protected substantially. As seen more specifically in Figure 5, Lys-72 was protected to a very significant extent by Mg-ATP, which was predicted on the basis of the affinity labeling with FSBA. In addition, Mg-ATP afforded significant protection to Lys-47. If we build the sequence of the catalytic subunit into the glycine-rich loop region of LDH, the results shown in Figure 6 are consistent with the observed differential labeling by acetic anhydride. According to this model, the glycine-rich loop, as well as Lys-72 and Lys-47, would all be in close proximity to the ATP-binding site. This glycine-rich loop is predicted to play a significant role in the conformational changes that are induced by GTP binding to p21 *ras* (de Vos et al. 1988). As with p21 *ras*, this glycine-rich loop may be a factor in substrate-induced conformational changes that result from ATP binding in the protein kinases as well.

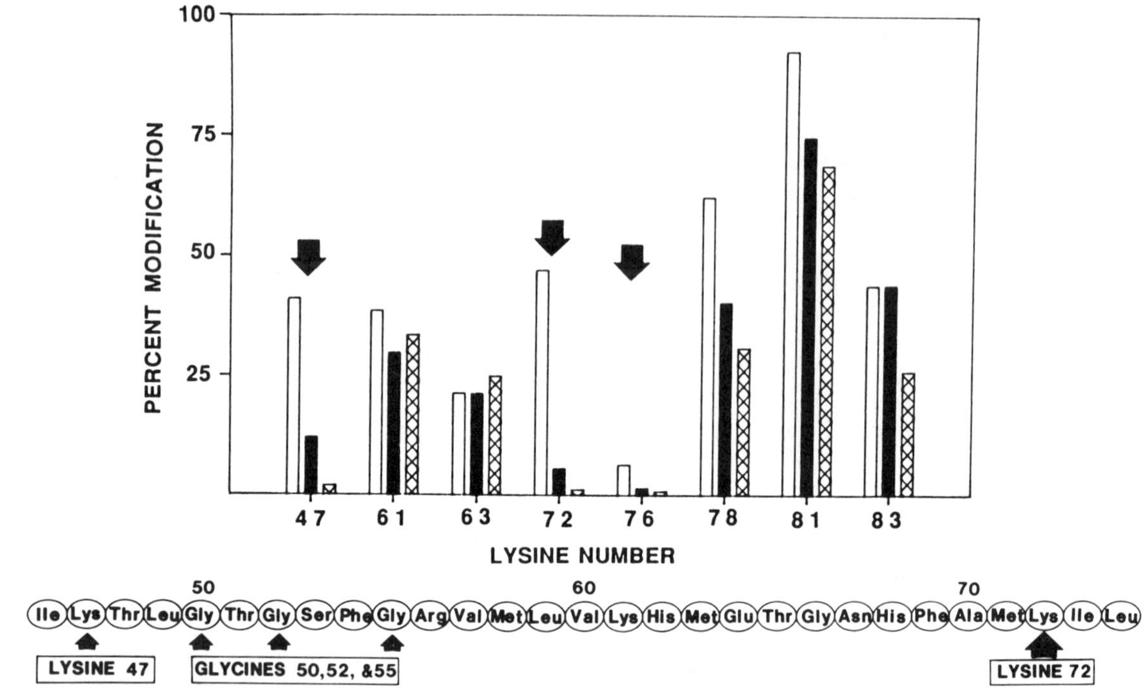

Figure 5. Reactivity of residues in the ATP-binding site. The graph indicates the reactivity of lysine residues toward acetylation with acetic anhydride. Reactivities in the apoenzyme (open rectangle) are compared with reactivity in the presence of 2 mM Mg-ATP (black rectangles), and Mg-ATP and inhibitor peptide (cross-hatched). The reactivity of all lysines was determined; however, only the region containing those residues protected significantly by Mg-ATP is shown. Lys-47, -72, and -76, indicated by the arrows, showed the greatest protection. The entire sequence spanning this region of the protein is indicated below. Lys-72 is affinity-labeled by FSBA, whereas the glycine-rich segment (50–55) is a conserved feature of most adenine nucleotide-binding sites.

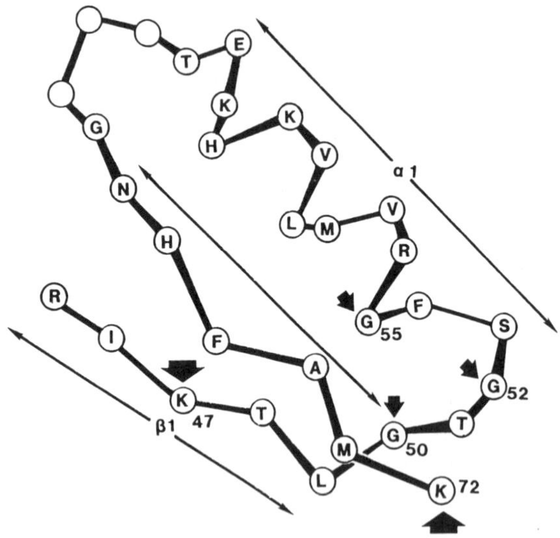

Figure 6. Proposed model for a portion of the ATP-binding site in the catalytic subunit. The amino acid sequence of the catalytic subunit has been fitted into the α-carbon backbone of lactate dehydrogenase (Rossmann et al. 1974; Wierenga et al. 1986). The glycine-rich segment most likely lies at a bend between the first β-strand (β1) and the first α-helix (α1). The region preceding Lys-72 in this model is shown as β-strand 2. According to this model, the glycines, Lys-72, and Lys-47 would all lie in close proximity to the bound nucleotide at the carboxyl terminus of the β-strands.

Inhibition of Catalytic Subunit

There are two known physiological mechanisms for inhibiting the catalytic subunit. One is interaction with the regulatory subunit in the absence of cAMP; the other is interaction with a heat-stable protein kinase inhibitor (PKI) (Walsh et al. 1971). Although the regulatory subunit and PKI represent two quite different molecules, the mechanism employed by each to inhibit the catalytic subunit is very similar. In fact, it is a mechanism that seems to be invoked frequently for inhibiting protein kinases. The only distinctive feature of cAMP-dependent protein kinase, in contrast to most other protein kinases, is that the regulatory region is part of a separate, dissociable subunit instead of being a contiguous part of a single polypeptide chain.

Several different forms of regulatory subunits are expressed. Some are constitutive, whereas other appear to be tissue-specific and inducible (Lee et al. 1983; Jahnsen et al. 1986; Scott et al. 1987). Despite these differences, each regulatory subunit shares a common domain structure. The carboxy-terminal two thirds of the polypeptide chain is highly conserved and consists of two tandem gene-duplicated segments representing two cAMP-binding sites. The variable amino-terminal region of the regulatory subunit contains the site of dimer interaction as well as a proteolytically sensitive "hinge" region that is essential for inhibition of the catalytic subunit. In some cases, this hinge region of the regulatory subunit contains an autophosphorylation site, whereas in others, it is a pseudophosphorylation site (Fig. 7). In all cases, this region contains two arginine residues that are essential for peptide recognition by the catalytic subunit. For the regulatory subunits that are classified as type II, these arginines are followed by a serine that is readily phosphorylated by the catalytic subunit (Potter and Taylor 1979; Takio et al. 1984a). In the case of type I regulatory subunits, the serine is generally replaced by an alanine so that autophosphorylation is not possible (Titani et al. 1984). The regulatory subunit is a competitive inhibitor with respect to peptide substrates (Granot et al. 1980), and this region of the regulatory subunit most likely occupies the peptide-binding site of the catalytic subunit in the holoenzyme complex. This mechanism for inhibition thus prevents access of other protein substrates to the active site of the catalytic subunit. PKI also appears to inhibit the catalytic subunit by a similar mechanism (Whitehouse and Walsh 1983). Like the type I regulatory subunits, PKI has a pseudophosphorylation site, and studies with synthetic peptide analogs have established that this peptide recognition site is essential for inhibition (Cheng et al. 1986). PKI also is a competitive inhibitor with respect to peptide substrates.

Several lines of evidence support this mechanism of inhibition. Limited proteolysis of porcine type II regulatory subunit from heart yielded two fragments differing by only three residues. The larger fragment was cleaved at Asp-91, just before the two arginine residues that lie at the hinge region and precede the autophoshorylation site, Ser-95 (Fig. 7). The resultant carboxy-terminal fragment retained the capacity to inhibit the catalytic subunit. On the other hand, a similar proteolytic fragment that began at Val-94 and lacked the two arginines no longer formed a high-affinity complex with the catalytic subunit (Weldon and Taylor 1985). Thus, the two arginine residues appear to be

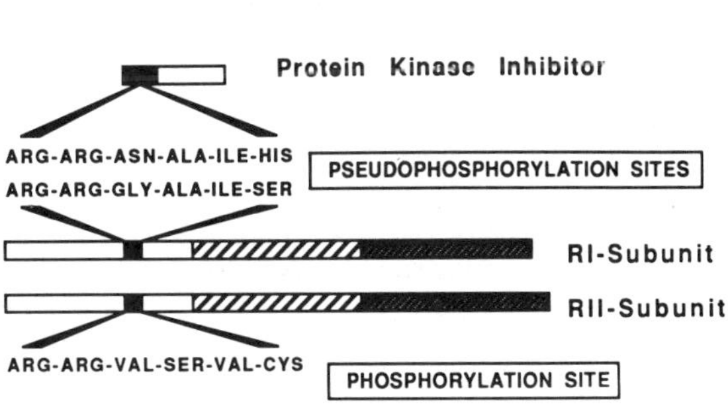

Figure 7. Domain structure and the catalytic subunit recognition site of inhibitors of the catalytic subunit. Two physiological inhibitors have been identified, the regulatory subunit and the heat-stable protein kinase inhibitor. PKI contains a pseudosubstrate site, whereas each regulatory subunit contains either a pseudosubstrate site or an autophosphorylation site. The sequences of these sites are indicated. The two tandem cAMP-binding sites at the carboxyl terminus of each regulatory subunit are indicated by cross-hatching.

essential for interaction between the regulatory subunit and the catalytic subunit.

In the presence of 5,5′-dithiobis 2-nitrobenzoic acid, or when the catalytic subunit is modified at its sulfhydryl groups with disulfide-forming compounds, reassociation of the type II regulatory subunit and the catalytic subunit leads to the formation of an intersubunit disulfide bond. This disulfide bond is formed between Cys-97 of the type II regulatory subunit and Cys-199 of the catalytic subunit, providing additional support for the hypothesis that the hinge region of the regulatory subunit occupies the peptide-binding site of the catalytic subunit (First et al. 1988).

To have greater flexibility in probing functional regions of the regulatory subunit, an expression vector was constructed for the bovine type I regulatory subunit (Saraswat et al. 1986). This pUC-derived vector leads to the expression of up to 30 mg of regulatory subunit per liter of culture, and unlike expression of the catalytic subunit, all of the expressed protein is soluble and fully active. Using this expression vector, pLST2, several mutant forms of the regulatory subunit have been constructed that lend further support to the inhibitory mechanism described above. These mutant subunits fall into two classes. First are those containing point mutations in the hinge region and showing altered properties of interaction with the catalytic subunit. The second class of mutants contain deletions at both the carboxyl terminus and at the amino terminus, allowing a minimum core region to be defined that is essential for recognition by the catalytic subunit.

A mutant type I regulatory subunit has been constructed that replaced the alanine at the hinge region with a serine, thus introducing a potential site of autophosphorylation (Fig. 7). This mutation yields a stable protein that is a good substrate for cAMP-dependent protein kinase. This mutant regulatory subunit now enables us to look at the specific consequences of introducing a single phosphorylation site. Two criteria were used to establish that this mutation does alter the interaction between the regulatory subunit and the catalytic subunit. For example, the native regulatory subunit is not dissociated in the presence of 0.5 M NaCl. In contrast, holoenzyme that was formed with the mutant regulatory subunit was mostly dissociated under these conditions. Formation of holoenzyme with the mutant regulatory subunit was not influenced significantly by the presence of Mg-ATP. Finally, autophosphorylation of the mutant regulatory subunit can occur readily in the holoenzyme complex and does not require dissociation of the holoenzyme, indicating that the portion of the regulatory subunit containing the autophosphorylation site must occupy the peptide-binding site in the holoenzyme (S. Durgerian and S.S. Taylor, in prep.).

Deletion mutants also have helped to define the regions of the regulatory subunit that are essential for interaction with the catalytic subunit. As indicated earlier, limited proteolysis established that the amino-terminal region preceding the arginines in the hinge region is not essential for interaction with the catalytic subunit. However, the hinge region itself, in particular Arg-94 and Arg-95, is essential. The deletion mutants shown in Figure 8 have been constructed in the type I regulatory subunit. The mutant indicated in the third panel introduced a stop codon at residue 260. This change deleted the entire second cAMP-binding domain. The resulting protein is stable and was purified to homogeneity, as indicated in Figure 8. It is also capable of forming a complex with the catalytic subunit in the presence of Mg-ATP, and this complex is dissociated by cAMP. Although the affinity of the regulatory subunit for the catalytic subunit is altered in this mutant, the entire second cAMP-binding domain is not essential for interaction with the catalytic subunit (Saraswat et al. 1988). The structure shown in the bottom panel is the minimal structure that can still form a complex with the catalytic subunit that can be dissociated by cAMP.

These types of mutations in the regulatory subunit have helped to define the structural criteria that are important for interaction with the catalytic subunit. Although other regions of the regulatory subunit must be important for forming a high-affinity complex with the catalytic subunit, an essential requirement for inhibition appears to be occupancy of the peptide-binding site in the catalytic subunit by a portion of the regulatory subunit. This region of the regulatory subunit corresponds to the hinge region that can be a site either of autophosphorylation or of pseudophosphorylation. This mechanism of inhibition appears to be common for many protein kinases, although in other protein kinases, the regulatory region is part of a contiguous polypeptide chain linked to the catalytic domain of the kinase (House and Kemp 1987; Kemp et al. 1987).

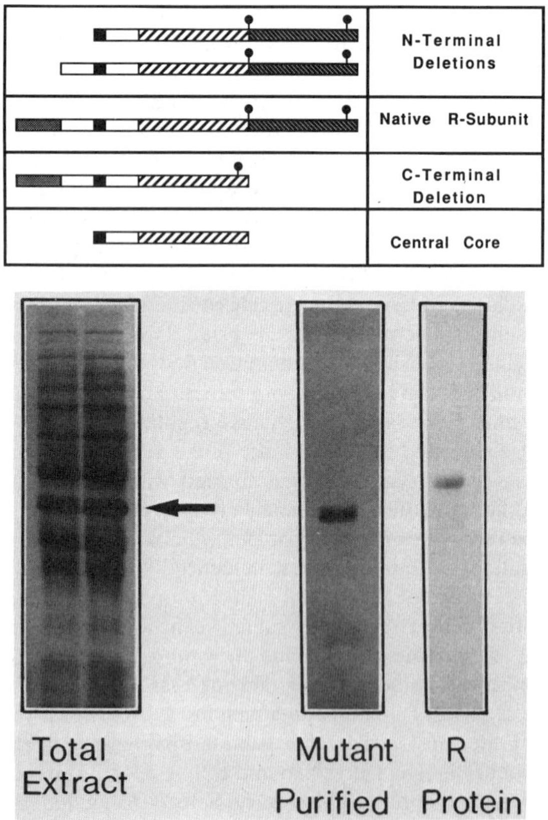

Figure 8. Deletion mutants of the type I regulatory subunit. A summary of the deletion mutants is shown (*top*), and expression of the mutant shown in panel 3 is indicated (*bottom*). This mutant regulatory subunit lacks the entire second cAMP-binding domain (Saraswat et al. 1988).

X-ray Crystallography

Affinity labeling, group-specific labeling, and site-directed mutagenesis can be used to identify specific residues that are essential for function. These approaches also enable one to identify functional regions topographically and to correlate specific functions with defined regions of the polypeptide chain. However, to fully understand the molecular structure of a molecule the size of the catalytic subunit, it is essential to have a high-resolution crystal structure. To achieve this, several different crystal forms of the catalytic subunit have been obtained. Crystallization of the apoenzyme yields cubic crystals that diffract to 3.2 Å. In contrast, crystallization of a ternary complex containing a catalytic subunit, Mg-ATP, and a 20-residue inhibitor peptide (Cheng et al. 1986) under identical conditions yields hexagonal crystals that diffracted to 3.0 Å (D.K. Knighton and J.M. Sowadski, unpubl.). It is likely that these represent two different conformational states of the enzyme. High-resolution data are being collected on the hexagonal crystals (D.K. Knighton and J.M. Sowadski, unpubl.).

ACKNOWLEDGMENTS

The authors wish to acknowledge the help of Dr. Thomas A. Vedvick for sequencing of peptides and Katherine Charles for purifying catalytic subunit. This work was supported in part by U.S. Public Health Service grants GM-19301 and GM-34921 and American Cancer Society grant CD-255 to S.S.T. J.A.B. and D.R.N. were supported by U.S. Public Health Service training grants AM-07233 and CA-09523, respectively.

REFERENCES

Bhatnagar, D., F.T. Hartle, R. Roskoski, Jr., R.A. Lessor, and N.J. Leonard. 1984. Adenosine cyclic 3′,5′-monophosphate-dependent protein kinase: Nucleotide binding to chemically modified catalytic subunit. *Biochemistry* **23:** 4350.

Boyer, P.D. and E.G. Krebs, eds. 1986. *The enzymes. Control by phosphorylation*, Part A. *General features, specific enzymes*. Vol. 17. Academic Press, San Diego.

Bramson, H.N., A.S. Mildvan, and E.T. Kaiser. 1984. Mechanistic studies of cAMP-dependent protein kinase action. *CRC Crit. Rev.* **15:** 93.

Bramson, H.N., N. Thomas, R. Matsueda, N.C. Nelson, S.S. Taylor, and E.T. Kaiser. 1983. Modification of the catalytic subunit of bovine heart cAMP-dependent protein kinase with affinity labels related to peptide substrates. *J. Biol. Chem.* **257:** 10575.

Brändén, C.-I. 1980. Relation between structure and function of α/β-proteins. *Quart. Rev. Biophys.* **13:** 317.

Buechler, J.A. and S.S. Taylor. 1988a. Dicyclohexylcarbodiimide crosslinks two essential residues at the active site of the catalytic subunit of cAMP-dependent protein kinase: Asp 184 and Lys 72. *Biochemistry* (in press).

———. 1988b. Identification of aspartate 184 as an essential residue in the catalytic subunit of cAMP-dependent protein kinase. *Biochemistry* **27:** 7356.

Carr, S.A., K. Biemann, S. Shoji, D.C. Parmelee, and K. Titani. 1982. n-Tetradecanoyl in the NH_2 terminal blocking group of the catalytic subunit of the cyclic AMP-dependent protein kinase from bovine cardiac muscle. *Proc. Natl. Acad. Sci.* **79:** 6128.

Cheng, H.-C., B.E. Kemp, R.B. Pearson, A.J. Smith, L. Misconi, S.M. VanPatten, and D.A. Walsh. 1986. A potent synthetic peptide inhibitor of the cAMP-dependent protein kinase. *J. Biol. Chem.* **261:** 989.

Chou, C.K., T.J. Dull, D.S. Russell, R. Gherzi, D. Lobwohl, A. Ullrich, and O. Rosen. 1987. Human insulin receptor mutated at the ATP binding site lacks protein tyrosine-kinase activity and fails to mediate post-receptor effects of insulin. *J. Biol. Chem.* **262:** 1842.

de Vos, A.M., T. Liang, M.V. Milburn, P.M. Matias, J. Jancarik, S. Noguchi, S. Nishimura, K. Miura, E. Ohtsuba, and S.-M. Kim. 1988. Three-dimensional structure of an oncogene protein: Catalytic domain of human c-H-*ras* p21. *Science* **239:** 888.

First, E.A., J. Bubis, and S.S. Taylor. 1988. Subunit interaction sites between the regulatory and catalytic subunits of cAMP-dependent protein kinase. Identification of a specific interchain disulfide bond. *J. Biol. Chem.* **263:** 5176.

Granot, J., A.S. Mildvan, K. Hiyama, H. Kondo, and E.T. Kaiser. 1980. Magnetic resonance studies of the effect of the regulatory subunit on metal and substrate binding to the catalytic subunit of bovine heart protein kinase. *J. Biol. Chem.* **255:** 4579.

Hanks, S.K., A.M. Quinn, and T. Hunter. 1988. Protein kinase family: Conserved features and deduced phylogeny of the catalytic domains. *Science* **241:** 42.

Ho, M.-F., H.N. Bramson, D.E. Hansen, J.R. Knowles, and E.T. Kaiser. 1988. Stereo-chemical course of the phospho group transfer catalyzed by cAMP-dependent protein kinase. *J. Am. Chem. Soc.* **110:** 2680.

Honegger, A.M., T.J. Dull, S. Felder, E. Van Obberghen, F. Bellot, D. Szapary, A. Schmidt, A. Ullrich, and J. Schlessinger. 1987. Point mutation at the ATP binding site of EGF receptor abolishes protein-tyrosine kinase activity and alters cellular routing. *Cell* **51:** 199.

Hoppe, J., W. Freist, R. Marutzky, and S. Shaltiel. 1978. Mapping the ATP binding site in the catalytic subunit of cAMP-dependent protein kinase: Spatial relationships with the ATP binding site of the undissociated enzyme. *Eur. J. Biochem.* **90:** 427.

House, C. and B.E. Kemp. 1987. Protein kinase C contains a pseudosubstrate prototype in its regulatory domain. *Science* **238:** 1726.

Hunter, T. 1987. A tail of two src's: mutatis mutandis. *Cell* **49:** 1.

Jahnsen, T., L. Hedin, V.J. Kidd, W.G. Beattie, S.M. Lohmann, V. Walter, J. Durica, T.Z. Schulz, E. Schlitz, M. Browner, C.B. Lawrence, D. Goldman, S.L. Ratoosh, and J.S. Rishards. 1986. Molecular cloning, cDNA structure, and regulation of the regulatory subunit of type II cAMP-dependent protein kinase from rat ovarian granulosa cells. *J. Biol. Chem.* **261:** 12352.

Jurnak, F. 1985. Structure of the GDP domain of EF-Tu and location of the amino acids homologous to *ras* oncogene proteins. *Science* **230:** 32.

Kamps, M.P. and S.M. Sefton. 1986. Neither arginine nor histidine can carry out the function of lys 295 in the ATP binding site of $pp60^{src}$. *Mol. Cell. Biol.* **6:** 751.

Kamps, M.P., J.E. Buss, and B.M. Sefton. 1986. Rous sarcoma virus transforming protein lacking myristic acid phosphorylates known polypeptide substrates without inducing transformation. *Cell* **45:** 105.

Kamps, M.P., S.S. Taylor, and B.M. Sefton. 1984. Oncogenic tyrosine kinases and cAMP-dependent protein kinases have homologous ATP binding sites. *Nature* **310:** 589.

Kemp, B.E., R.B. Pearson, V. Guerriero, Jr., I.C. Bagchi, and A.R. Means. 1987. The calmodulin binding domain of chicken smooth muscle myosin light chain kinase contains a pseudosubstrate sequence. *J. Biol. Chem.* **262:** 2542.

Krebs, E.G. and J.A. Beavo. 1979. Phosphorylation-dephosphorylation of enzymes. *Annu. Rev. Biochem.* **48:** 923.

Lee, D.C., D.F. Carmichael, E.G. Krebs, and G.S. McKnight. 1983. Isolation of a cDNA clone for the type I regulatory subunit of bovine cAMP-dependent protein kinase. *Proc. Natl. Acad. Sci.* **80:** 3608.

Nelson, N.C. and S.S. Taylor. 1981. Differential labeling and identification of the cysteine-containing tryptic peptides of catalytic subunit from porcine heart cAMP-dependent protein kinase. *J. Biol. Chem.* **256:** 3743.

Parker, P.J., L. Coussens, N. Tatty, L. Rhee, and A. Ullrich. 1986. The complete primary structure of protein kinase C—the major phorbol ester receptor. *Science* **233:** 853.

Peters, K.A., J.G. Demaille, and E.H. Fischer. 1977. Adenosine 3′:5′-monophosphate dependent protein kinase from bovine heart: Characterization of the catalytic subunit. *Biochemistry* **16:** 5691.

Potter, R.L. and S.S. Taylor. 1979. Correlation of the cAMP binding domain with a site of autophosphorylation on the regulatory subunit of cAMP-dependent protein kinase II from porcine skeletal muscle. *J. Biol. Chem.* **254:** 9000.

Rossmann, M.G., D. Moras, and K. Olsen. 1974. Chemical and biological evolution of nucleotide binding proteins. *Nature* **250:** 194.

Saraswat, L.D., M. Filutowics, and S.S. Taylor. 1986. Expression of the type I regulatory subunit of cAMP-dependent protein kinase in *Escherichia coli*. *J. Biol. Chem.* **261:** 11091.

Saraswat, L.D., G.E. Ringheim, J. Bubis, and S.S. Taylor. 1988. Deletion mutants as probes for localizing regions of subunit interaction in cAMP dependent protein kinase. *J. Biol. Chem.* (in press).

Scott, J.D., M.B. Glaccum, M.J. Zoller, M.D. Uhler, D.M. Helfman, G.S. McKnight, and E.G. Krebs. 1987. The molecular cloning of a type II regulatory subunit of the cAMP-dependent protein kinase from rat skeletal muscle and mouse brain. *Proc. Natl. Acad. Sci.* **84:** 5192.

Shoji, S., L.H. Ericsson, K.A. Walsh, E.H. Fischer, and K. Titani. 1983. Amino acid sequence of the catalytic subunit of bovine type II adenosine cyclic 3′,5′-phosphate dependent protein kinase. *Biochemistry* **22:** 3702.

Studier, F.W. and B.A. Moffatt. 1986. Use of bacteriophage T7 RNA polymerase to direct selective high-level expression of cloned genes. *J. Mol. Biol.* **189:** 113.

Tabor, S. and C.C. Richardson. 1987. DNA sequence analysis with a modified bacteriophage T7 DNA polymerase. *Proc. Natl. Acad. Sci.* **84:** 4767.

Takio, K., D.K. Blumenthal, K.A. Walsh, K. Titani, and E.G. Krebs. 1986. Amino acid sequence of rabbit skeletal muscle myosin chain kinase. *Biochemistry* **23:** 8049.

Takio, K., S.B. Smith, E.G. Krebs, K.A. Walsh, and K. Titani. 1984a. Amino acid sequence of the regulatory subunit of bovine type II adenosine cyclic 3′,5′-phosphate dependent protein kinase. *Biochemistry* **26:** 4200.

Takio, K., R.D. Wade, S.B. Smith, E.G. Krebs, K.A. Walsh, and K. Titani. 1984b. Guanosine cyclic 3′,5′-phosphate dependent protein kinase, a chimeric protein homologous with two separate protein families. *Biochemistry* **23:** 4207.

Taylor, S.S., J.A. Buechler, and D.R. Knighton. 1988. cAMP-dependent protein kinase: Mechanism for ATP-protein phosphoprotein. *CRC Crit. Rev.* (in press).

Titani, K., T. Sasagawa, L.H. Ericsson, S. Kumar, S.B.

Smith, E.G. Krebs, and K.A. Walsh. 1984. Amino acid sequence of the regulatory subunit of bovine type I adenosine cyclic 3′,5′-phosphate dependent protein kinase. *Biochemistry* **23:** 4193.

Toner-Webb, J. and S.S. Taylor. 1987. Inhibition of the catalytic subunit of cAMP-dependent protein kinase by dicyclohexylcarbodiimide. *Biochemistry* **26:** 7371.

Ullrich, A., J.R. Bell, E.Y. Chen, R. Herrera, L.M. Petruzzelli, T.J. Dull, A. Gray, L. Coussens, Y.-C. Liao, M. Tsubokawa, A. Mason, P.H. Seeburg, C. Grunfeld, O.M. Rosen, and J. Ramachandran. 1985. Human insulin receptor and its relationship to the tyrosine kinase family of oncogenes. *Nature* **313:** 756.

Ullrich, A., L. Coussens, J.A. Hayflick, T.J. Dull, A. Gray, A.W. Tam, Y. Lee, Y. Yarden, T.A. Libermann, J. Schlessinger, J. Downward, E.L.V. Mayes, N. Whittle, M. Waterfield, and P.H. Seeburg. 1984. Human epidermal growth factor receptor of cDNA sequence and aberrant expression of the amplified gene in S431 epidermal carcinoma cells. *Nature* **309:** 418.

Walsh, D.A., J.P. Perkins, and E.G. Krebs. 1968. An adenosine 3′,5′-monophosphate dependent protein kinase from rabbit skeletal muscle. *J. Biol. Chem.* **243:** 3763.

Walsh, D.A., C.D. Ashby, C. Gonzalez, D. Calkins, E.H. Fischer, and E.G. Krebs. 1971. Purification and characterization of a protein inhibitor of adenosine 3′,5′-monophosphate-dependent protein kinases. *J. Biol. Chem.* **246:** 1977.

Weldon, S.L. and S.S. Taylor. 1985. Monoclonal antibodies as probes for functional domains in cAMP-dependent protein kinases. *J. Biol. Chem.* **260**: 4203.

Whitehouse, S. and D.A. Walsh. 1983. $MgATP^{-2}$-dependent interaction of the inhibitor protein of the cAMP-dependent protein kinase with the catalytic subunit. *J. Biol. Chem.* **258:** 3682.

Whitehouse, S., J.R. Feramisco, J.E. Casnellic, E.G. Krebs, and D.A. Walsh. 1983. Studies on the kinetic mechanism of the catalytic subunit of the cAMP-dependent protein kinases. *J. Biol. Chem.* **258:** 3993.

Wierenga, R.K., P. Terpstra, and W.G.J. Hol. 1986. Prediction of the occurrence of the ADP-binding $\beta\alpha\beta$-fold in proteins, using an amino acid sequence fingerprint. *J. Mol. Biol.* **187:** 101.

Yoon, M.Y. and P.F. Cook. 1987. Chemical mechanism of the adenosine cyclic 3′,4′-monophosphate dependent protein kinase from pH studies. *Biochemistry* **26:** 4118.

Zoller, M.N., N.C. Nelson, and S.S. Taylor. 1981. Affinity labeling of cAMP-dependent protein kinase with IpR-fluorosulfonylbenzoyl adenosine. *J. Biol. Chem.* **256:** 10387.

Targets for Signal-transducing Protein Kinases

T. HUNTER,* P. ANGEL,† W.J. BOYLE,* R. CHIU,† E. FREED,* K.L. GOULD,* C.M. ISACKE,* M. KARIN,† R.A. LINDBERG,*‡ AND P. VAN DER GEER*

*Molecular Biology and Virology Laboratory, The Salk Institute, San Diego, California 92138; Departments of ‡Biology and †Pharmacology, University of California, San Diego, La Jolla, California 92093

The transduction of signals from the external medium across the plasma membrane into the cell is a process that is essential to the functioning of multicellular organisms. There are several basic mechanisms of signal transduction, many of which involve protein phosphorylation. Protein phosphorylation is ideally suited for signal transduction, since protein kinases have an inbuilt mechanism for signal amplification, and protein phosphatases provide a means of reversing the signal when the stimulus is removed. The most common type of signal-transducing protein kinase is the family of growth factor receptor protein-tyrosine kinases (PTK), such as the epidermal growth factor (EGF) receptor. These receptors have an intrinsic signaling mechanism in which binding of ligand to an external domain activates a cytoplasmic protein kinase domain to phosphorylate target proteins in the mitogenic signal pathway. Approximately half of all known protein-tyrosine kinase genes appear to encode ligand-regulated surface receptors (Table 1).

A second mitogenic signal transduction pathway that utilizes protein phosphorylation is protein kinase C (PKC). PKC is a family of Ca^{++}/phospholipid-dependent protein-serine kinases that are activated by binding diacylglycerol (DAG), which is produced as a consequence of the hydrolysis of phosphatidylinositol (PtdIns) by phospholipase C. Since the activity of phospholipase C is stimulated through specific surface receptors by a number of extracellular signals, including many mitogens, the activation of PKC is a common response to external signals. When activated, PKC is associated with the inner face of the plasma membrane, where it can phosphorylate target proteins in its signal transduction pathway. There is evidence to suggest that the growth factor receptor PTK pathway and the PKC pathway act in a parallel and independent fashion to trigger mitogenesis, although there is clearly cross-talk between these pathways.

We are interested in defining the components of these two signal pathways and the mechanisms by which they are regulated. To this end, we have studied protein kinases that act as the primary signal transducers and have attempted to identify the targets for these protein kinases. The response to mitogens is pleiotropic and includes a spectrum of cytoplasmic events, as well as a set of nuclear events. It seems likely that the pathways leading to these responses diverge at an early stage, perhaps starting with the primary signaling protein kinase, which may be able to phosphorylate and regulate a number of targets involved in different responses. In this paper, we cover several topics related to these issues, which include (1) the isolation of cDNAs for genes that encode novel PTKs that may be involved in signal transduction; (2) the regulation of $pp60^{c\text{-}src}$ activity by platelet-derived growth factor (PDGF) and its possible role in mitogenic signaling; (3) the identification of novel substrates for PKC; and (4) the characterization of possible nuclear targets for signal transduction including the *myb*, c-*jun* (Jun), and c-*fos* (Fos) gene products.

Isolation of Novel Protein-tyrosine Kinases

The preponderance of the growth factor receptors with PTK activity signifies that tyrosine phosphorylation is an important mitogenic signal pathway. This

Table 1. Protein-Tyrosine Kinase Subfamilies

src	
$pp60^{c\text{-}src}$-two forms generated by alternate splicing	
$pp62^{c\text{-}yes}$	
$p56^{lck}$	
c-*fgr* protein	
fyn protein	
lyn protein	
hck protein	
tkl protein	
fps/fes	
$p92^{c\text{-}fes}/p98^{c\text{-}fps}$	
NCP94 (FER gene?)	
TKR16	
abl	
$p150^{c\text{-}abl}$-two different amino termini generated by alternate splicing	
arg protein	
Growth factor receptors	
EGF.R	PDGF.R
c-*erb*B2 protein	CSF-1.R
	c-*kit* protein
INSULIN.R	*ret* protein
IGF-1.R	
met protein	*eph* protein
trk protein	*eck* protein (TKR11)
c-*ros* protein	
c-*sea* protein	
Others	
p75	
ltk protein	
flg protein	

is underscored by the existence of a second class of PTKs (Table 1; the *src* gene family) that may also be involved in signal transduction (see below). For this reason, we are interested in identifying novel PTKs that may be involved in signal transduction.

We have used two methods to isolate novel PTKs. The first method entails the use of antiphosphotyrosine (P.Tyr) antibodies to screen bacterial expression libraries. Our second approach is the homology-probing strategy devised by Hanks (1987), which consists of screening cDNA libraries with degenerate oligonucleotide probes designed to hybridize to two regions that are highly conserved in the catalytic domain of a specific subset of protein kinases.

The rationale for the first approach was as follows. Since *Escherichia coli* lacks detectable P.Tyr in proteins, a eukaryotic cDNA expressed in *E. coli* that encodes an active PTK might be detected if it were able to either autophosphorylate or phosphorylate bacterial proteins. We knew that it was possible to detect an active PTK in *E. coli* with anti-P.Tyr antibodies from the work of Kipreos et al. (1987), who had used immunoscreening of bacterial colonies to select temperature-sensitive mutants of the v-*abl* protein expressed from plasmids in *E. coli*. Even though we did not know if the signal obtained with a β-galactosidase fusion protein derived from a cDNA encoding a PTK catalytic domain would be strong enough to detect single plaques, we decided to screen a human fibroblast λgt11 library with affinity-purified anti-P.Tyr antibodies (obtained from Mark Kamps and Bart Sefton, Salk Institute). We expected a low frequency of positives, because the approach would rely not only on the complete catalytic domain's being expressed, but also on its being able to form a functional enzyme. In addition, enzymatic activity might require the absence of sequences that would normally have a negative regulatory function. Since this approach selects only for PTK function, in principle, it is less prejudiced than strategies relying on sequence homology. For this reason, we might obtain PTKs unrelated to the known PTKs, all of which show clear sequence relatedness.

In the first screening, nine clones were plaque-purified through three cycles and were found to fall into three restriction map groups. The first clone sequenced codes for a portion of a previously described PTK, *lyn* (Yamanashi et al. 1987). The *lyn* cDNA sequences comprise residues 130 to the carboxyl terminus, which would include the entire catalytic domain. The *lyn* cDNA insert has been transferred into two different inducible plasmid expression vectors. Following induction and examination of bacterial proteins by immunoblotting with anti-P.Tyr antibodies, we found that the *lyn* protein both autophosphorylates and phosphorylates various bacterial proteins. Our success validates the approach of using anti-P.Tyr antibodies for the cloning of active PTKs. The remaining two clones are now being sequenced, and their activities are being characterized.

The homology-probing approach was applied to a HeLa-cell-derived Okayama/Berg cDNA library (Hanks 1987). The oligonucleotide probes were designed to hybridize to regions that code for the conserved sequences: HRDLAAR and P(I/V)KW(T/M)APE, which are ~40 residues apart in the catalytic domain. The degeneracy of the oligonucleotide probes was limited to ~200-fold by amino acid choice and codon usage. Both probes were designed to detect cDNAs of PTK but not protein-serine/threonine kinases. Since we are primarily interested in identifying novel receptor-like PTKs, the first oligonucleotide was designed to detect receptor-like PTKs but not *src*-family PTKs, whereas the APE-region probe represents a composite of all PTKs. We screened 60,000 colonies, and 40 clones positive with either one or both probes representing several hybridization stringencies were picked for sequence analysis. The first two clones sequenced that appeared to be bona fide PTKs (by the criterion of containing sequences conserved in all PTKs) were both novel. The first of the two, *eck* (for epithelial cell kinase), was found with the HRDLAAR probe and has the receptor configuration in that region. It also has the DFG sequence common to all protein kinases, an autophosphorylation-site tyrosine, and an APE sequence, but differs from any known protein kinase in less-conserved regions (Fig. 1A). *eck* sequences hybridized to a 4.8-kb mRNA, which was expressed at the highest levels in cell lines of epithelial origin. Northern analysis of rat tissue RNAs also indicated that expression is greatest in proliferating epithelia; skin, intestine, lung, and ovary had the highest levels of *eck* mRNA, with nine other organs showing markedly less expression.

The deduced amino acid sequence for the kinase domain shows that *eck* is most closely related to the *eph* receptor-like PTK (see Hirai et al. 1987). There is 60% identity between *eck* and *eph*, whereas *eck* has ~30–45% identity with other reported PTKs. Like *eph* and other receptor PTKs, *eck* has a 24-residue hydrophobic transmembrane domain (followed by several basic residues) that is situated 61 amino acids amino-terminal to the kinase domain (Fig. 1B,C). The 97-amino-acid sequence carboxy-terminal to the kinase domain does not contain obvious similarities with any other PTK except *eph*, with which it has 48% identity. Although the entire structure of the ligand-binding domain has not been deduced, the sequence data indicate that *eph* and *eck* constitute a new family of receptor PTKs. *eck* appears to be a receptor PTK that may be involved in proliferation or differentiation of epithelial cells. Antisera to the *eck*-coding region are being raised to identify the *eck* gene product and to prove that it is a PTK.

The second cDNA clone (OB19) also has PTK consensus sequences in the probe target regions, but it is unique in not containing a tyrosine residue in the region in which all known kinases have an "autophosphorylation site" (Hanks et al. 1988). Northern analysis of cells and tissues shows that the gene encoding OB19 is ex-

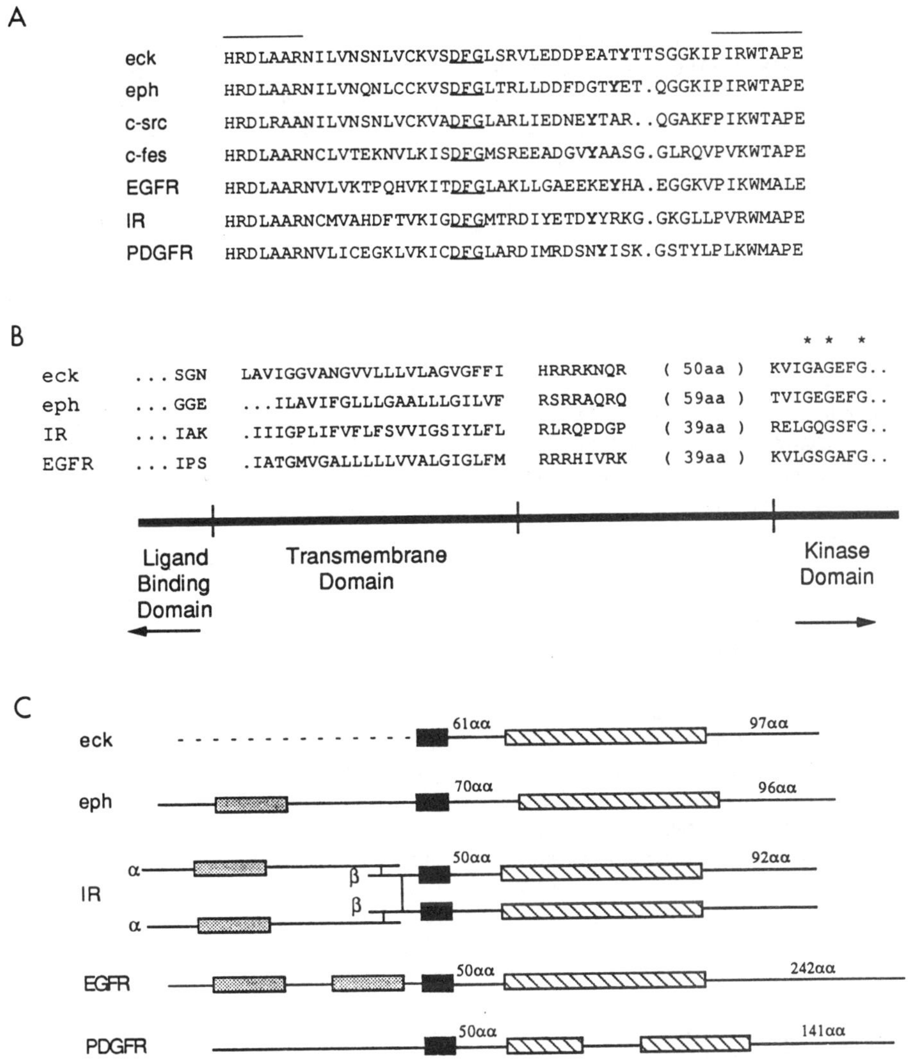

Figure 1. Structural comparison of *eck* with known PTKs. (*A*) Sequence comparison of *eck* with *eph*, c-*src*, c-*fes*, epidermal growth factor receptor (EGFR), insulin receptor (IR), and platelet-derived growth factor receptor (PDGFR) in the probe target region. Lines above the *eck* sequence correspond to the probes used for screening. Conserved DFG is underlined, and the autophosphorylation Tyr (Y) is in bold type. (*B*) Comparison of the transmembrane domain region of *eck* and other receptor PTKs. Asterisks denote the conserved glycines near the beginning of the kinase domain. (*C*) Domain structure of *eck* and other receptor PTKs. The structures are presented with the amino-terminal ligand-binding domain on the left. (Stippled boxes) Cysteine-rich region; (black boxes) transmembrane domain; (hatched boxes) protein kinase domain. The protein kinase domain is defined starting at the first glycine in the ATP-binding site and ending at the leucine corresponding to Leu-516 in pp60$^{c\text{-}src}$.

pressed as two RNAs, 3.0 and 3.3 kb. The 3.0-kb RNA is present in hematopoietic tissues and the 3.3-kb RNA is ubiquitous. Further studies to determine whether the product of this gene is a PTK are being carried out.

Phosphorylation of pp60$^{c\text{-}src}$

pp60$^{c\text{-}src}$ is a member of a family of closely related PTKs, which includes *src*, *yes*, *fgr*, *fyn*, *lyn*, *lck*, and *hck* (Cooper 1988). Each protein has a unique pattern of expression, with many being found only in specific hematopoietic lineages. These PTKs are all ~60 kD and have about 65% identity with each other over the carboxy-terminal 450 residues. All of them have an amino-terminal glycine residue, which can be myristylated. In the case of pp60$^{c\text{-}src}$, the myristyl group has been shown to be essential for membrane association. Although the true functions of the *src* family PTKs are not known, their localization to the inner face of cytoplasmic membranes, including the plasma membrane, has suggested that these proteins may have a signal-transducing function. Since, unlike the receptor PTKs, these PTKs do not span the membrane, they may act as subunits for ligand-binding proteins lacking their own catalytic domain.

We and other investigators have previously shown

that pp60$^{c\text{-}src}$ is negatively regulated by phosphorylation at a tyrosine six residues from the carboxyl terminus, Tyr-527 (Courtneidge 1985; Cooper et al. 1986; Cooper and King 1986; for review, see Hunter 1987). Phosphorylation at this site decreases the PTK-specific activity of pp60$^{c\text{-}src}$ five- to tenfold. In studying other phosphorylation events that might modulate the activity of pp60$^{c\text{-}src}$, we found that pp60$^{c\text{-}src}$ is a substrate for PKC, being rapidly and stoichiometrically phosphorylated at Ser-12 in cells treated with 12-tetradecanoylphorbol-13-acetate (TPA) (Gould et al. 1985). Rapid Ser-12 phosphorylation is also observed when quiescent mouse NIH-3T3 fibroblasts are treated with PDGF, prostaglandin $F_{2\alpha}$, serum, Na_3VO_4, bombesin, or basic fibroblast growth factor (FGF) (Gould and Hunter 1988). All these agents are mitogenic and are known to induce the turnover of PtdIns in NIH-3T3 cells, which leads to the production of DAG and the activation of PKC. So far, we have been unable to discern any effects of this modification on the activity or localization of pp60$^{c\text{-}src}$.

Ralston and Bishop (1985) first showed that treatment of resting fibroblasts with PDGF leads to the appearance of a more slowly migrating form of pp60$^{c\text{-}src}$ (p60$^+$), accounting for about 5% of the total population. p60$^+$ is modified in the amino-terminal 18 kD and, unlike pp60$^{c\text{-}src}$ in resting cells, contains P.Tyr in this amino-terminal region. The appearance of p60$^+$ is correlated with a two- to threefold increase in pp60$^{c\text{-}src}$ protein kinase activity toward an exogenous substrate. We have confirmed these observations and have shown that the appearance of p60$^+$ is a very rapid event, occurring within 2 minutes of PDGF treatment of quiescent NIH-3T3 cells (Gould and Hunter 1988). Both the PDGF AA and BB dimers (expressed in yeast and obtained from M. Murray, Zymogenetics) induce the formation of p60$^+$ as efficiently as the natural PDGF AB dimer. However, we have not observed the formation of p60$^+$ nor the enhancement of pp60$^{c\text{-}src}$ protein kinase activity with any other mitogen we have tested.

To determine the nature of p60$^+$, two-dimensional gel analysis of immunoprecipitates of [^{35}S]methionine-labeled pp60$^{c\text{-}src}$ from control and PDGF-treated cells was carried out. This analysis showed that p60$^+$ is highly phosphorylated and contains four to five phosphates per mole of pp60$^{c\text{-}src}$ (Gould and Hunter 1988). Dephosphorylation of p60$^+$ with potato acid phosphatase increases the mobility of p60$^+$ to that of the main form of pp60$^{c\text{-}src}$. Peptide mapping with V8 protease indicates that p60$^+$ differs from the normal form of pp60$^{c\text{-}src}$ in its amino-terminal 18 kD. Analysis of V8 protease fragments of ^{32}P-labeled p60$^+$ by peptide mapping shows that there is one novel phosphoserine-containing peptide derived from the amino-terminal 18 kD, as well as the expected Ser-17 and Ser-12 peptides. In addition to the novel serine phosphorylation, P.Tyr is detected in this 18-kD fragment from p60$^+$, but no tryptic peptide containing P.Tyr was recovered by our procedure. However, we were able to identify a P.Tyr-containing peptide following redigestion of the tryptic digest with thermolysin. There was no detectable difference in the level of phosphorylation of Tyr-527 in p60$^+$ compared with the main form of pp60$^{c\text{-}src}$ in PDGF-treated cells, nor was there any increase in the level of Tyr-416 phosphorylation, which is often an indication of increased pp60^{src} activity. Thus, the elevated pp60$^{c\text{-}src}$ PTK activity in PDGF-treated cells is not due to a relief of negative regulation at Tyr-527.

We are currently trying to map the new serine and tyrosine phosphorylation sites in the amino-terminal region and to determine which protein kinases are responsible for their phosphorylation. Preliminary evidence suggested that the tyrosine phosphorylation site was Tyr-149, which lies in the regulatory SH-2 domain of pp60$^{c\text{-}src}$. To determine whether this was the case, we synthesized a peptide corresponding to the predicted thermolysin peptide containing Tyr-149 and phosphorylated this peptide in vitro. Although this phosphopeptide migrated extremely close to the thermolysin P.Tyr-containing peptide derived from p60$^+$, it was clearly distinct. Alternative tyrosine phosphorylation sites in p60$^+$ would be Tyr-90 or Tyr-92, and we are currently exploring this possibility.

We have also attempted to find out whether this site is phosphorylated by the PDGF receptor, another PTK, or pp60$^{c\text{-}src}$ itself by autophosphorylation. Using purified PDGF receptor (obtained from C. Heldin, Uppsala), we have been unable to phosphorylate pp60$^{c\text{-}src}$ isolated in immunoprecipitates made with a variety of antibodies. This failure, however, could be a result of the use of immunoprecipitates, and we plan to repeat the experiment with purified pp60$^{c\text{-}src}$. We have also tested whether the PDGF-induced phosphorylation is a result of autophosphorylation, but so far we have not been able to find pp60$^{c\text{-}src}$ autophosphorylation conditions under which a similar P.Tyr-containing peptide is generated.

To summarize our results with PDGF and pp60$^{c\text{-}src}$, it seems likely that p60$^+$ is responsible for the increase in PTK activity and that the increased activity, as well as the mobility shift, of p60$^+$ is due to a combined effect of the additional serine and tyrosine phosphorylations. It is tempting to speculate that the activation of pp60$^{c\text{-}src}$ in PDGF-treated cells is important for the mitogenic response, but since a similar modification of pp60$^{c\text{-}src}$ is not seen with other mitogens, it may be involved in some response unique to PDGF.

There is an interesting parallel between the results obtained with p60$^+$ in PDGF-treated cells and pp60$^{c\text{-}src}$ in mitotic cells, where a novel threonine phosphorylation in the amino-terminal 18 kD correlates with the appearance of a more slowly migrating form of pp60$^{c\text{-}src}$ and an increase in PTK activity (Chackalaparampil and Shalloway 1988). In addition, elevated pp60$^{c\text{-}src}$ PTK activity is detected in neuronal cells, which contain a neuronal-specific form of pp60$^{c\text{-}src}$ that has an additional 6-amino-acid exon inserted at position 118, as well as yet another distinct serine phosphorylation site. All these observations suggest that the amino-terminal 18 kD of $pp60^{c\text{-}src}$ is likely to be a critical site for regulation by phosphorylation. However, in none of these three situations is there direct evidence that $pp60^{c\text{-}src}$ actually has increased PTK activity in the cell.

Novel PKC Substrates

PKC is activated by many mitogens through the stimulation of PtdIns turnover and the release of DAG. Furthermore, phorbol esters such as TPA, which mimic DAG in activating PKC, are mitogenic for some cells. For mitogens, such as bombesin, activation of PKC may constitute the major mitogenic signal pathway. In these cases, down-regulation of PKC by prolonged treatment with TPA blocks subsequent mitogenic stimulation by growth factor. In other cases, such as PDGF, although PKC is activated, down-regulation experiments indicate that the PKC pathway is not essential for mitogenesis. We have been interested in defining substrates for PKC with the aim of learning more about this mitogenic pathway. In the course of these studies, we have identified the EGF receptor (Hunter et al. 1984) and $pp60^{c\text{-}src}$ (Gould et al. 1985) as substrates for PKC.

In the past, we have generated a panel of monoclonal antibodies against surface proteins of human fibroblasts (Isacke et al. 1986). Four of these are directed against a 180-kD protein, which on intact cells displays a punctate immunostaining pattern that codistributes with clathrin heavy-chain staining, and which is localized to coated pits by immunoelectron microscopy. p180 is expressed on most cell types, but it is present at lower levels on epithelial cells. Labeling by surface iodination followed by proteolysis leads to the release of a 75-kD fragment, suggesting that a considerable portion of p180 might lie inside the cell. The biosynthesis of p180 was studied using pulse/chase labeling of FLOW 2000 human fibroblasts with [^{35}S]methionine. At short chase times, a 170-kD precursor is synthesized containing sugar chains that are sensitive to digestion with both endoglycosidases H and F, yielding a core protein of ~155 kD. This is consistent with the size of the product obtained in the presence of tunicamycin, an inhibitor of N-linked glycosylation. After longer chase times (1–2 hr) the mature protein is produced, which contains complex-type sugar chains sensitive to digestion with endoglycosidase F and neuraminidase, but resistant to digestion with endoglycosidase H. The mature protein has a half-life of 24–30 hours.

The phosphorylation of p180 has been studied both in vivo and in vitro. The protein is phosphorylated exclusively on serine residues in vivo. Phosphopeptide mapping indicates that p180 is phosphorylated constitutively on three to four sites in vivo. Treatment with TPA results in the rapid phosphorylation of two additional sites as well as an increase in the phosphorylation at one of the constitutively phosphorylated sites. Purified p180 is phosphorylated in vitro on serine residues by PKC, and the sites phosphorylated in vitro are identical to the major sites phosphorylated in response to TPA in vivo. Thus, p180 appears to be a bona fide substrate for PKC.

Because of its location to coated pits, we surmise that p180 might be a receptor of some sort. We have therefore undertaken to isolate molecular clones for the mRNA for p180. To do this, we have devised a purification scheme from human MG63 osteosarcoma cells, using one of the monoclonal antibodies to isolate p180 by affinity chromatography from detergent lysates of cell membranes. Essentially pure p180 can be obtained by elution from the antibody column at pH 11, followed by gel filtration over a Superose-6 column. Preliminary amino-terminal sequence data indicate that p180 is not a known protein, and in particular, it is not one of the growth factor receptor PTKs of this size, such as the EGF or PDGF receptors. In collaboration with R. Aebersold (Caltech), we have purified several tryptic peptides by microbore high-performance liquid chromatography and have obtained sequence data. On the basis of these sequences, degenerate oligonucleotides have been synthesized and are being used in conjunction with a polyclonal rabbit antiserum raised against purified p180 to screen human fibroblast and placental λgt11 libraries. The predicted sequence of p180 may show homology with a known receptor, which will give us a clue to its function. It seems unlikely that p180 will be a receptor PTK, since purified p180 lacks protein kinase activity. Indeed, its presence in coated pits might indicate that p180 is a receptor, like the low-density lipoprotein receptor, that constantly recycles. When the sequence of p180 becomes available from the cDNA cloning, we should be able to identify the PKC phosphorylation sites and assess their significance.

The fibronectin (FNR) and vitronectin (VNR) receptors and the platelet fibrinogen receptor are all members of the family of cell-surface receptors referred to as integrins, which play an important role in cellular adhesion processes. TPA causes rapid rounding of many cells, which could in part be due to a loss of cellular adhesion to the substratum. In collaboration with E. Ruoslahti (La Jolla Cancer Research Foundation), we have tested whether either the VNR or FNR is a substrate for PKC in order to assess whether phosphorylation of either of these receptors plays a role in this effect of TPA or in other aspects of receptor function. In human fibroblasts treated with TPA, there is a rapid and apparently stoichiometric phosphorylation of the smaller β subunit of the VNR on serine. A single major site is phosphorylated, and this same site is phosphorylated in vitro by PKC. In the course of this work, it has become apparent that the β-chain population is heterogeneous and that a subpopulation of β-chains with a greater electrophoretic mobility is the major phosphorylated form. On the basis of surface iodination experiments, both forms are present on the cell surface. Moreover, both will bind to and are specifically eluted from a column containing the cell attachment peptide Gly.Arg.Gly.Asp.Ser.Asp. We are attempting to determine whether this heterogeneity is due to posttranslational modification or whether it arises from the existence of multiple primary translation products. We are also assessing whether this phosphorylation affects the ability of the VNR to bind VN or some intracellular cytoskeletal protein such as talin.

In contrast to the VNR, we did not observe any change in the low level of phosphorylation of the FNR following TPA treatment. However, we have been able to extend the results of other investigators, who reported tyrosine phosphorylation of the FNR in Na_3VO_4-treated Rous sarcoma virus (RSV)-transformed chick cells (Hirst et al. 1986) by showing that there is increased tyrosine phosphorylation of the FNR in Na_3VO_4-treated RSV-transformed human A431 cells compared to that in normal A431 cells, as well as in PDGF-treated human fibroblasts in the presence of Na_3VO_4. We have also examined the phosphorylation of the related fibrinogen receptor in platelets in response to thrombin and TPA treatments. Here again the β (IIIa)-chain is phosphorylated, but this occurs mainly on threonine. Thus, despite the fact that platelet IIIa is believed to be very closely related to the VNR β-chain, it appears to be distinct in this respect. The basis for this difference is being pursued.

Nuclear Effectors of Mitogenic Signal Transduction

The ultimate fate of mitogenic signal transduction, triggered by protein kinases activated in response to signal reception, is to initiate a cascade of nuclear functions that culminate in cell division. The nuclear protein targets of signal transduction, once activated, may directly participate in crucial functions that regulate either early (G_1) and late (G_2) gene expression, replication and segregation of DNA, or disassembly and reassembly of the nuclear membrane/lamina. In addition to the receptor PTKs, there is a network of G proteins and phospholipases associated with the cell membrane that communicate via small messenger molecules with two major cellular protein kinase pathways, PKC and cAMP-dependent protein kinase. In addition, casein kinases I and II, which are activated in response to mitogens by an unknown mechanism (Sommercorn et al. 1987; Carroll et al., this volume), have isoforms that are active in both the cytoplasm and nucleus (Hathaway and Traugh 1982). Beyond this, however, it is not clear how signals are transmitted to the nucleus, to which proteins these signals are transmitted, and how many steps exist between the activation of cytoplasmic protein kinases and the subsequent activation of specific nuclear functions.

To learn more about nuclear signal transduction pathways, we are studying the phosphorylation of key nuclear proteins with a known role in cell growth control. Obvious choices for analysis are the viral and cellular oncogene products located in the nucleus, encoded by oncogenes such as *fos* (Curran et al. 1984), *myc* (Abrams et al. 1982), *myb* (Boyle et al. 1984), and a newly added member, *jun* (Bos et al. 1988). On the basis of their ability to act as dominant oncogenes during transformation, their products seem likely to play a major role in cell growth and development. This group of proteins could be the final nuclear recipients of transduced signals and could in turn act as direct effectors of mitogenesis. In agreement with this idea, each of these proteins has been found to be multiply phosphorylated by unidentified protein-serine/threonine kinase(s). Hence, direct analysis of the phosphorylation events of individual nuclear oncogene products should yield information important both for understanding their unique function(s) and for identifying adjacent proteins that regulate their function(s). We have investigated the role of nuclear protein phosphorylation in two systems: the *myb* protein during hematopoiesis and leukemogenesis, and Fos and AP-1/Jun during TPA-induced transcriptional activation.

Multisite Phosphorylation of *myb* Proteins during Normal and Abnormal Hematopoiesis

The first member of the *myb* gene family to be identified and characterized was the v-*myb* gene of avian myeloblastosis virus (AMV). AMV is an acutely oncogenic retrovirus that induces rapid and uniformly fatal myeloblastic leukemia in susceptible chickens (for review, see Lipsick and Baluda 1986). The transformation spectrum of AMV in vitro is restricted to immature chicken hematopoietic progenitor cells committed to the myeloid lineage. The second v-*myb* gene member, contained in the erythro- and myeloleukemogenic avian leukemia virus E26, is a structurally related oncogene encoding a fusion protein with an internal *myb* domain slightly smaller than v-*myb* of AMV (Nunn et al. 1983). The colony-forming activity of AMV, and presumably E26, requires an intact v-*myb* gene (Ibanez and Lipsick 1988) as well as the presence of exogenous myeloid-cell growth factor, cMGF (Leutz et al. 1984). Analysis of temperature-sensitive AMV and E26 mutants suggests that the v-*myb* gene functions during transformation by blocking the ability of cells to undergo normal differentiation (Ness et al. 1987). The v-*myb* gene of AMV encodes a 48-kD protein, $p48^{v\text{-}myb}$, whereas E26 encodes a larger hybrid protein, $p135^{myb\text{-}ets}$ (Nunn et al., 1983). Both proteins are located entirely within the nucleus and have short half-lives of ~30–60 minutes. Functional analysis of the v-*myb* gene of AMV has identified protein domains involved in DNA binding (Klempnauer and Sippel 1987), nuclear localization (Ibanez and Lipsick 1988), protein stability (Rogers et al. 1986), and associations with other proteins and subnuclear structures (Boyle et al. 1985) that determine its unique target cell specificity. Virtually all internal deletions disrupt the transforming activity of $p48^{v\text{-}myb}$, suggesting that the v-*myb* gene of AMV represents the smallest *myb* domain capable of transformation.

The v-*myb* genes and products are truncated versions of the highly evolutionarily conserved normal cellular gene c-*myb*, which encodes a 75-kD nuclear protein, p75$^{c\text{-}myb}$ (Klempnauer et al. 1983). p75$^{c\text{-}myb}$ shares many characteristics with p48$^{v\text{-}myb}$ but contains additional *myb* protein sequence at both its amino and carboxyl termini. During the genesis of AMV and E26, only an internal subset of the c-*myb* gene was transduced, resulting in the synthesis of structurally altered *myb* proteins with potentially aberrant activities. The c-*myb* gene itself is nontransforming, although its leukemogenic potential can be activated by deletion of sequences encoding amino- and carboxy-terminal c-*myb*-specific protein domains (F.A. Grässer and J.S. Lipsick, unpubl.). c-*myb* protein expression is regulated during the development of the immature hematopoietic stem cells of the lymphoid, myeloid, and erythroid lineages (Boyle et al. 1986), suggesting an important role in development and maintenance of the immune system. Expression of the c-*myb* gene is regulated during the cell cycle of exponentially replicating and serum-starved cells (Thompson et al. 1986). In resting peripheral blood T lymphocytes, there is no detectable expression of p75$^{c\text{-}myb}$, whereas treatment with mitogens such as phytohemagglutinin results in a late increase of p75$^{c\text{-}myb}$ that begins after the onset of DNA synthesis and extends into G_2 phase (Lipsick and Boyle 1987).

Recently, *myb*-related genes have been identified in two vastly divergent species, *Zea mays* (Paz-Ares et al. 1987) and *Drosophila* (Katzen et al. 1985), both of which lack an apparent counterpart to the vertebrate immune system. This suggests the c-*myb* protein may potentially have a more generalized function during growth of many cell types. All *myb* proteins from plant to man contain a highly conserved protein sequence of about 52 amino acids that is repeated two to three times in an equivalent position near the amino terminus (Klempnauer and Sippel 1987). The function of this domain is not known, but it is required in p48$^{v\text{-}myb}$ for both in vitro DNA-binding activity and in vivo nuclear localization. Although a DNA target-binding sequence has not been demonstrated, it is generally believed that this region of all *myb*-related proteins forms a unique type of DNA-binding structure, distinct from other binding motifs, including metal-binding fingers (Miller et al. 1985) and helix-turn-helix regions (Sauer et al. 1982). Interestingly, both p48$^{v\text{-}myb}$ and the v-*myb* sequence in p135$^{myb\text{-}ets}$ lack the first of the three tandem repeats present in p75$^{c\text{-}myb}$ (see Fig. 2).

Given a DNA-binding function, *myb* proteins could be transcription factors involved in cell growth by regulating developmental changes in gene expression. Many eukaryotic *trans*-acting factors, including GAL4, GCN4, AP-1/Jun, and Fos, are functionally divided into domains that are involved in DNA binding as well as in signal reception and regulation (see Vogt and Tjian 1988). By analogy, *myb* proteins may be organ-

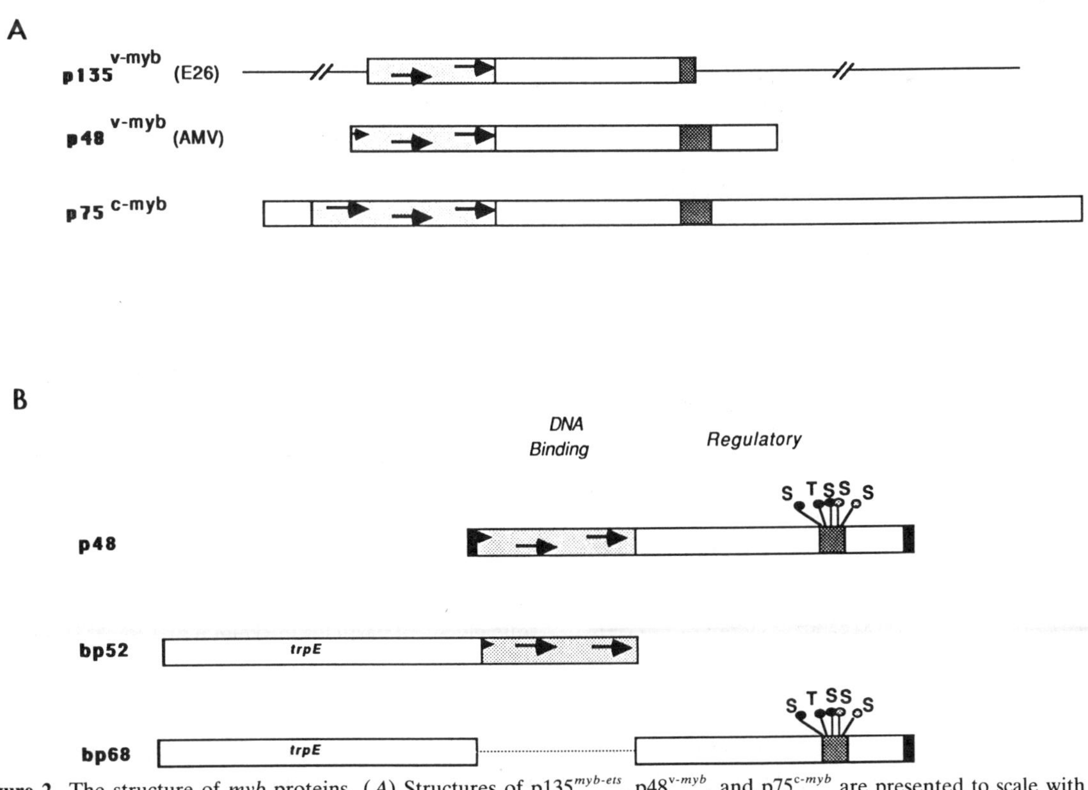

Figure 2. The structure of *myb* proteins. (*A*) Structures of p135$^{myb\text{-}ets}$, p48$^{v\text{-}myb}$, and p75$^{c\text{-}myb}$ are presented to scale with their amino termini on the left. (Tandem arrows) DNA-binding repeat; (dark-stippled boxes) PEST sequence. (*B*) Structures of p48$^{v\text{-}myb}$ and the bacterially expressed TrpE-v-*myb* fusion proteins are presented. (White boxes on left) trpE; (tandem arrows) DNA binding; (dark-stippled boxes) PEST sequence. The potential phosphorylation sites are shown above the PEST sequence.

ized in a similar manner, containing an amino-terminal DNA-binding domain and a carboxy-terminal regulatory domain. The amino-terminal repeat region of $p48^{v\text{-}myb}$ is necessary but not sufficient for transformation, indicating that its carboxyl terminus provides an important "regulatory" function (Ibanez and Lipsick 1988). Since transcription factors can respond to changes in signal transduction patterns, this class of proteins is likely to be regulated by protein modification, including phosphorylation. We therefore felt that the analysis of *myb* protein phosphorylation might allow us to identify a potential regulatory region and afford us the opportunity to study nuclear protein kinases that respond to growth stimuli.

To analyze *myb* protein phosphorylation, we chose to study phosphorylation of $p48^{v\text{-}myb}$ during AMV-induced transformation. This system has two features that are attractive. First, $p48^{v\text{-}myb}$ is expressed at higher levels than the other *myb* proteins and is therefore easier to analyze (Boyle et al. 1986). Second, there is available a G418-selectable myeloblastosis-associated virus (MAV) that expresses $p48^{v\text{-}myb}$ and transforms chicken myeloblasts in vitro (Ibanez and Lipsick 1988), which would ultimately allow us to determine the role of phosphorylation by analyzing $p48^{v\text{-}myb}$ phosphorylation site mutants. In both exponentially growing and serum-starved AMV-transformed myeloblasts, we find that $p48^{v\text{-}myb}$ is phosphorylated on both serine and threonine residues in a ratio of about 5:1. Tryptic digests of ^{32}P-labeled $p48^{v\text{-}myb}$ contain five discrete phosphopeptides that migrate in a pattern suggesting that they are phosphoisomers of the same peptide. Proteolysis of individual peptides of this group with V8 protease gives rise to identical products, reinforcing the idea that they are all forms of the same peptide differing by their extent of phosphorylation. The most acidic forms of this peptide contained both phosphoserine (P.Ser) and phosphothreonine, whereas the basic forms contain only P.Ser, indicating that serine phosphorylation occurs prior to threonine phosphorylation. The exact position of this peptide in $p48^{v\text{-}myb}$ was determined using parallel strategies including (1) the use of individual labeled amino acid tracers, (2) secondary cleavage of tryptic peptides using sequence-specific proteases, (3) determination of presence of candidate peptides in bacterially expressed v-*myb* deletion products labeled with appropriate amino acids, and (4) immunoprecipitation of the in-vivo-labeled tryptic phosphopeptides using a site-directed anti-*myb* monoclonal antibody that recognizes an internal sequence of this peptide (see Fig. 3A,B).

This peptide (residues 273–309), located in the sixth v-*myb*-specific exon, contains the bulk of phosphate detected in v-*myb*. The sequence of this peptide is highly conserved, being 80% identical between chicken and man. It contains an unusual amount of charged (Glu, Asp, Arg, and His; ~33%) and proline residues (15%) and has seven potential phosphorylation sites. This region of the protein has been identified as a "PEST" sequence because of its high proline, glutamate, serine, and threonine content. PEST sequences are frequently observed in proteins with short half-lives and are proposed to be primary determinants that specify rapid degradation (Rogers et al. 1986). The five residues (four serines and one threonine) that are conserved evolutionarily are boxed as potential targets (Fig. 3). However, we have not yet identified the individual residues that are phosphorylated in this sequence. Mutations that disrupt this structure change the biological activities of AMV (Ibanez and Lipsick 1988). A frameshift mutation that deletes the 3′ end of this sequence (LX3) has altered transformation properties and abolishes the colony-forming activities of AMV. Deletion of the entire 5′ end of this sequence abrogates transformation and disrupts nuclear localization of $p48^{v\text{-}myb}$. Thus, it is possible that phosphorylation events within this region are required for the leukemogenic functions of $p48^{v\text{-}myb}$. Interestingly, a portion of this region is missing in v-*myb* of E26 and is replaced by sequences derived from the c-*ets* proto-oncogene (Nunn et al. 1983). This may account in part for the differences in the biological activities of AMV and E26. $p75^{c\text{-}myb}$ is also phosphorylated in this region; however, it is also phosphorylated on at least two extra unknown serine residues.

The primary sequence of the region containing the *myb* protein phosphorylation sites is not characteristic of sites used by the major classes of protein-serine kinase. However, we noted that this sequence resembles in nature the region of glycogen synthase (GS) that is multiply phosphorylated by glycogen synthase kinase-3 (GSK-3) (see Woodgett and Cohen 1984). GSK-3 is thought to play a role in growth regulation and has several interesting properties. It phosphorylates sites 3a–c and 4 in GS that are sensitive to insulin treatment and inhibits GS activity. These serine residues are clustered together in GS and are phosphorylated in an ordered manner by GSK-3, but only if GS is phosphorylated at site 5. Thus, phosphorylation of the serine in site 5 by various protein kinases, notably casein kinase II, regulates phosphorylation by GSK-3. We find that purified GSK-3 can specifically phosphorylate v-*myb* when it is present in the bacterial *trp*E-*myb* fusion protein, bp68 (Fig. 2), at the same sites between residues 273–309 that are phosphorylated in vivo. We are currently testing whether other protein kinases, which may be involved in regulating *myb* function in vivo, cooperate during in vitro phosphorylation of *myb* proteins. It is not known whether GSK-3 exists within the nucleus, but it is normally found associated with cytosolic glycogen particles in skeletal muscle. It is possible that GSK-3, or a closely related enzyme, has nuclear forms that synergize with other protein kinases to regulate nuclear proteins, as is the case for GS. Using both recombinant *myb* proteins and synthetic peptides as abundant and specific substrates, we plan to identify the protein kinase(s) that phosphorylate *myb* proteins during normal and abnormal hematopoiesis.

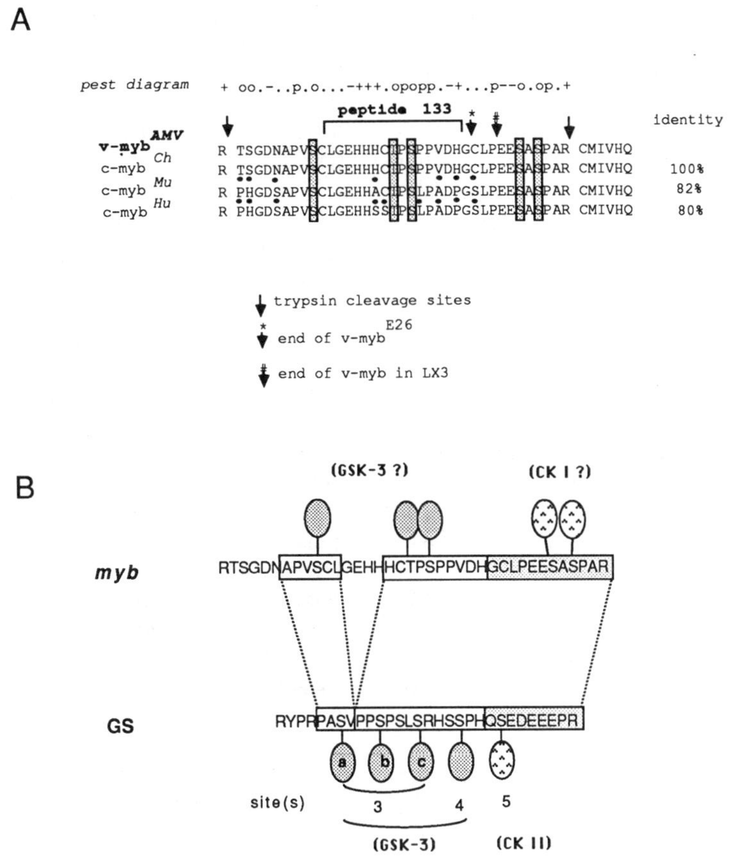

Figure 3. Phosphorylation sites in the v-*myb* protein. (*A*) Sequence of residues 272–315 of the v-*myb* protein are shown aligned with the corresponding regions in the chicken, mouse, and human c-*myb* proteins, with differences indicated by dots above the murine and human sequences. Potential trypsin cleavage sites are shown by arrows. Ends of the v-*myb* sequence in E26 and the LX3 mutant of AMV are indicated. Boxed residues designate serines and threonines conserved between all four v-*myb* sequences. A PEST diagram and the synthetic peptide (133) used to generate the anti-v-*myb* monoclonal antibody are shown above the predicted sequence. (*B*) Comparison of the sequences around the v-*myb* phosphorylation sites and the carboxy-terminal phosphorylation sites (sites 3a–c, 4 and 5) in glycogen synthase (GS).

Signal Transduction and the Transcriptional Activation of TPA Responsive Genes: Nuclear Phosphoproteins Fos and AP-1/Jun

TPA induces a variety of cellular metabolic and growth processes by directly affecting the activity of PKC. Included in the many growth-related processes stimulated are the rapid transcriptional activation of various genes, including SV40 early, collagenase, metallothionein IIA, and stromelysin genes (Angel et al. 1988). TPA responsiveness is mediated by a transcriptional activator protein, AP-1, that binds tightly to a specific oligonucleotide core 5′-TGACTCA-3′ that is contained within *cis*-acting TPA response elements (TRE). TRE-binding activity can be rapidly induced three- to fourfold upon the addition of TPA to serum-starved cells, even in the presence of protein synthesis inhibitors. This implies that the PKC pathway modulates AP-1 activity through a posttranslational mechanism, which is likely to involve phosphorylation. AP-1 purified from HeLa cells consists of two major polypeptides of 44 (p44) and 40 (p40) kD (Angel et al. 1988). The degree of structural similarity between p44 and p40 suggests that p40 is a cleavage product generated during purification (see below).

A potential role for AP-1 in transformation became apparent from analysis of members of the AP-1-binding protein family. We and other investigators noticed that the oligonucleotide-binding target of AP-1 was remarkably similar to the known binding site of the yeast transcriptional activator protein GCN4 (Angel et al. 1987). This suggested that GCN4 was a distant homolog of a family of AP-1-like proteins in higher vertebrates. At the same time, an acutely oncogenic, replication-defective avian retrovirus, ASV17, that induced fibrosarcomas in chickens was isolated. The oncogene responsible, v-*jun*, was novel and not structurally related to any of the known oncogenes (Vogt et al. 1987). Analysis of the predicted v-*jun* sequence indicated that it was very similar to a region of the yeast GCN4 protein that is required for site-specific DNA binding. Since GCN4 was predicted to be functionally similar to AP-1, this structural link between GCN4 and v-*jun* suggested that v-*jun* might be derived from the chicken AP-1 gene. When the human c-*jun* gene was cloned, using a chicken v-*jun* probe, a single long open reading frame was found, which could encode a 39-kD protein that is highly related to chicken v-*jun* (Angel et al. 1988). The amino acid sequences of four tryptic peptides isolated from purified AP-1 were found within the c-*jun* open reading frame, suggesting that AP-1 is indeed encoded by this gene (Bohmann et al. 1987). Furthermore, synthetic antipeptide antibodies to two separate v-*jun* domains recognized purified AP-1 on immunoblots (Bohmann et al. 1987; Angel et al. 1988). As final proof that the c-*jun* gene is functionally equivalent to AP-1, the c-*jun* product was shown to *trans*-activate reporter genes that utilize functional TREs (Angel et al. 1988). As predicted, a bacterial TrpE-*jun* fusion protein protects essentially identical sites on TPA-responsive genes as AP-1 (Angel et al. 1988).

To investigate how TPA induces a rapid increase in AP-1-binding activity and in transcription of TRE-containing genes, we have analyzed the phosphorylation of AP-1. For this purpose, we raised specific antibodies against a bacterially expressed TrpE-c-*jun* fusion protein, which specifically immunoprecipitated a 44-kD protein that comigrates with $p44^{AP-1}$ (Chiu et al. 1988). This protein was detected in a wide variety of cell lines from fibroblastic, epithelial, endothelial, and hematopoietic lineages originating from human, mouse, and chicken sources. This protein was not detected, however, in murine F9 teratocarcinoma cells. To determine if AP-1 phosphorylation changes after TPA stimulation, AP-1 immunoprecipitated from ^{32}P-labeled HeLa cells was subjected to tryptic peptide mapping. There was no change in the overall level of ^{32}P-labeled $p44^{AP-1}$ isolable by immunoprecipitation when serum-starved HeLa cells were treated with TPA for 30 minutes. $p44^{AP-1}$ from uninduced cells contained three P.Ser-containing peptides. TPA treatment did not lead to any change in this pattern, although one peptide was relatively decreased in intensity. This suggests that direct phosphorylation of AP-1 itself may not play a role in inducing transcription.

We then turned our attention to other nuclear proteins that may associate directly with AP-1 in response to TPA and in turn activate AP-1-binding activity. The Fos protein, the product of the c-*fos* gene, was a likely candidate to be involved in AP-1 interactions because its synthesis and multiple phosphorylations are induced in response to TPA treatment of quiescent fibroblasts (Barber and Verma 1987). Although the function of Fos was not known, a role for Fos protein in activating the adipocyte-specific gene aP2 became apparent when anti-Fos antibodies were found to immunoprecipitate a transcription complex assembled on the aP2 gene promoter (Distel et al. 1987). Since the promoter element contained a core TRE, it was possible that AP-1 protein was present in this complex together with Fos. Further analysis of the interaction between Fos and the aP2 promoter has shown that the formation of this complex requires the presence of an AP-1-binding site (Rauscher et al. 1988). Although purified AP-1 is capable of binding to specific DNA sequences, there is no evidence that Fos itself has sequence-specific DNA-binding activity. One explanation for the presence of Fos in complexes assembled on AP-1-binding sites is that Fos forms a complex with AP-1. This complex would be able to bind to TREs and stimulate transcription.

To test whether Fos acts as a specific transcriptional enhancer in cooperation with AP-1, we have utilized cell lines that harbor c-*fos* genes under the control of a heavy metal inducible promoter (Chiu et al. 1988). In 3T3MTfos cells treated with Cd^{++} to induce Fos expression, there is specific activation of AP-1-responsive reporter genes that is similar to the level of activation of endogenous TPA-responsive genes in TPA-treated HeLa cells. In contrast, no specific stimulation was seen when AP-2- and AP-3-responsive genes were tested. Induction of Fos expression in F9MTfos cells does not result in the *trans*-activation of AP-1-responsive genes, presumably because these cells do not contain AP-1. Transfection of F9MTfos cells with a functional c-*jun* or v-*jun* gene expression vector, however, restores the ability of Fos to *trans*-activate AP-1-responsive reporter genes. This implies that Fos functions as a transcriptional activator by interacting with a site-specific DNA-binding protein, AP-1.

To explore the possibility that Fos and AP-1 interact directly to mediate TPA-induced gene expression, we have tested whether they are physically associated. A Fos-protein complex that contains a Fos-associated protein (FAP), previously termed p39, is coimmunoprecipitated with anti-Fos antibodies together with Fos from a variety of cell types (Curran et al. 1984). Using both anti-AP-1 and anti-Fos antibodies, we have found that FAP p39 comigrates with AP-1 on one- and two-dimensional gels and is cross-reactive with AP-1 antibodies (Chiu et al. 1988). Two-dimensional tryptic peptide maps of [^{35}S]methionine-labeled FAP p39 and AP-1 are identical, proving that FAP p39 is AP-1 (Chiu et al. 1988). Transient expression of Fos and the c-*jun* protein in F9 cells, under conditions where we observe cooperative *trans*-activation, results in the formation of an exogenous Fos/AP-1 complex like that in TPA-treated HeLa cells.

A simplified model for the transcriptional activation of AP-1-responsive genes is presented in Figure 4. Although there are several alternatives, we propose that TPA acts at two levels during *trans*-activation. First, TPA causes modification of the activity of preexisting complexes, possibly through phosphorylation of Fos, and perhaps AP-1 as well, which results in increased DNA-binding activity. Second, TPA induces the expression of both AP-1 and Fos, leading to increased complex formation. The outcome will be a greater level of Fos/AP-1 complex bound to target TREs, which in turn will stimulate RNA polymerase-II-dependent transcription from these genes.

Fos appears to bind to AP-1 through its DNA-binding domain. In this regard, the nonspecific DNA-binding activity of Fos may be important in promoting the formation of this complex, since it will localize Fos to exposed regions of DNA corresponding to regulatory

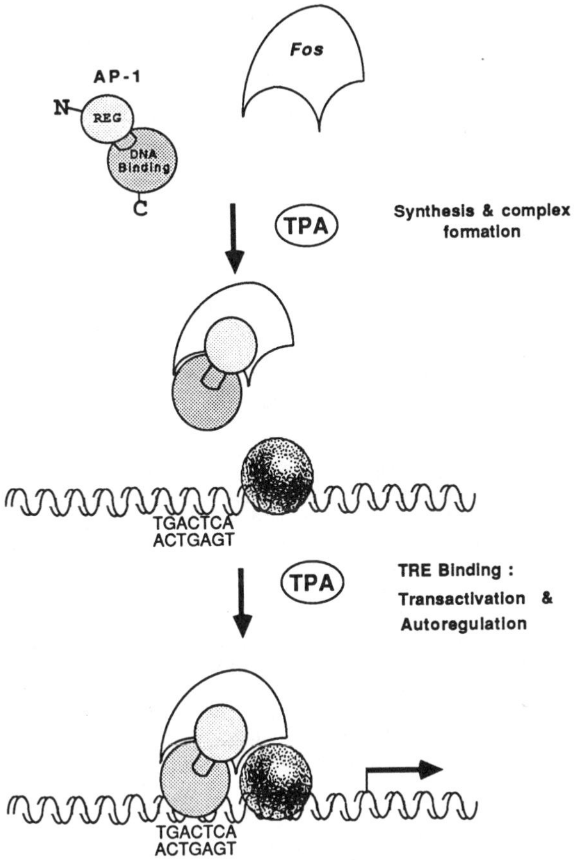

Figure 4. Model for transcriptional activation by Fos and AP-1.

regions where AP-1 may be bound. The mechanism of transcriptional stimulation is unknown, but the acidic nature of the clustered phosphorylation sites on Fos could constitute an "acid blob" of the type that is known to be essential for *trans*-activation by GAL4 (Gill and Ptashne 1987). It will obviously be important to determine whether phosphorylation of Fos regulates its ability to cooperate with AP-1, and to establish the nature of the protein kinase(s) that phosphorylates Fos. Although TPA activates PKC, it seems unlikely that PKC phosphorylates Fos directly in the nucleus. Instead, PKC may trigger a cascade of events that results in the activation of a nuclear protein kinase. The fact that TPA activates S6 ribosomal protein kinase provides a precedent for PKC-dependent activation of other protein kinases. In the long term, our goal will be to work backward from phosphorylations occurring in the nucleus to the signal-transducing protein kinases themselves and establish the number and nature of steps in this pathway.

This model for Fos action opens up the possibility that Fos may form complexes with other sequence-specific transcription factors, and thus act pleiotropically as a "third messenger" delivering membrane-integrated signals to the nucleus, where they affect the regulation of gene expression. Conversely, one or more of the Fos-related proteins (Franza et al. 1987) may associate with AP-1 or another factor for the same purpose. Clearly, we are just at the beginning of unraveling a complex regulatory network of protein-protein interactions in the nucleus. There is evidently considerable potential for protein phosphorylation in modulating this network via signals transduced from the cell surface through the cytoplasm.

ACKNOWLEDGMENTS

This work was supported by U.S. Public Health Service grants CA-39780 to T.H. and ES-0451 to M.K. P.A. is supported by a fellowship from the Deutscher Akademischer Austauschdienst; W.J.B. is a Howard Hughes Medical Institute fellow of the Life Sciences Research Foundation; R.C. was supported by a National Institutes of Health postdoctoral fellowship; E.F. is a student in the Graduate Program in Neurosciences at the University of California, San Diego; C.M.I. was the holder of a Damon Runyon-Walter Winchell postdoctoral fellowship (DRG-775); R.A.L. was supported by a National Institutes of Health predoctoral training grant to the University of Calfiornia, San Diego; P.v.d.G. is the holder of a Queen Wilhelmina fellowship.

REFERENCES

Abrams, H.D., L.R. Rohrschneider, and R.N. Eisenman. 1982. Nuclear location of the putative transforming protein of avian myelocytomatosis virus. *Cell* **24:** 427.

Angel, P., E.A. Allegretto, S. Okino, K. Hattori, W.J. Boyle, T. Hunter, and M. Karin. 1988. The *jun* oncogene encodes a sequence specific *trans*activator similar to AP-1. *Nature* **322:** 166.

Angel, P., M. Imagawa, R. Chiu, B. Stein, R.J. Imbra, H.J. Rahmsdorf, C. Jonat, P. Herrlich, and M. Karin. 1987. Phorbol ester-inducible genes contain a common *cis* element recognized by a TPA-modulated *trans*acting factor. *Cell* **49:** 729.

Barber, J.R. and I.M. Verma. 1987. Modification of c-*fos* proteins: Phosphorylation of c-*fos* but not v-*fos*, is stimulated by 12-tetradecanoyl-phorbol-13-acetate and serum. *Mol. Cell. Biol.* **7:** 2201.

Bohmann, D., T.J. Bos, A. Adom, T. Nishimura, P.K. Vogt, and R. Tjian. 1987. Human proto-oncogene c-*jun* encodes a DNA-binding protein with structural and functional properties of transcription factor AP-1. *Science* **238:** 1386.

Bos, T.J., D. Bohmann, H. Tsuchie, R. Tjian, and P.K. Vogt. 1988. v-*jun* encodes a nuclear protein with enhancer binding properties of AP-1. *Cell* **52:** 705.

Boyle, W.J., J.S. Lipsick, and M.A. Baluda. 1986. Antibodies to the evolutionarily conserved amino-terminal region of the v-*myb*-encoded protein detect the c-*myb* protein in widely divergent metazoan species. *Proc. Natl. Acad. Sci.* **83:** 4685.

Boyle, W.J., M.A. Lampert, A.C. Li, and M.A. Baluda. 1985. Nuclear compartmentalization of the v-*myb* oncogene product. *Mol. Cell. Biol.* **5:** 3017.

Boyle, W.J., M.A. Lampert, J.S. Lipsick, and M.A. Baluda. 1984. Avian myeloblastosis virus and E26 virus oncogene products are nuclear proteins. *Proc. Natl. Acad. Sci.* **81:** 4265.

Chackalaparampil, I. and D. Shalloway. 1988. Altered phosphorylation and activation of pp60$^{c\text{-}src}$ during fibroblast mitosis. *Cell* **52:** 801.

Chiu, R., W.J. Boyle, J. Meek, T. Smeal, T. Hunter, and M. Karin. 1988. The c-*fos* protein interacts with c-*jun*/AP-1 to stimulate transcription of AP-1 responsive genes. *Cell* **54:** 541.

Cooper, J.A. 1988. The *src*-family of protein-tyrosine kinases. In *Peptides and protein phosphorylation* (ed. B. Kemp and P.F. Alewood), CRC Press, Boca Raton, Florida. (In press.)

Cooper, J.A. and C.S. King. 1986. Dephosphorylation or antibody binding to the carboxy terminus stimulates pp60$^{c\text{-}src}$. *Mol. Cell. Biol.* **6:** 587

Cooper, J.A., K.L. Gould, C.A. Cartwright, and T. Hunter. 1986. Tyr^{527} is phosphorylated in pp60$^{c\text{-}src}$: Implications for regulation. *Science* **231:** 1431.

Courtneidge, S.A. 1985. Activation of the pp60$^{c\text{-}src}$ kinase by middle T antigen binding or by dephosphorylation. *EMBO J.* **4:** 1471.

Curran, T., A.D. Miller, L. Zokas, and I. Verma. 1984. Viral and cellular *fos* proteins: A comparative analysis. *Cell* **36:** 259.

Distel, R.J., H.-S. Ro, B.S. Rosen, D.L. Groves, and B.M. Speigelman. 1987. Nucleoprotein complexes that regulate gene expression in adipocyte differentiation: Direct participation of c-*fos*. *Cell* **49:** 835.

Gill, G. and M. Ptashne. 1987. Mutants of GAL4 protein altered in an activation function. *Cell* **51:** 121.

Gould, K.L. and T. Hunter. 1988. Platelet-derived growth factor induces multisite phosphorylation of pp60$^{c\text{-}src}$ and increases its protein-tyrosine kinase activity. *Mol. Cell. Biol.* **8:** 3345.

Gould, K.L., J.R. Woodgett, J.A. Cooper, J.E. Buss, D. Shalloway, and T. Hunter. 1985. Protein kinase C phosphorylates pp60^{src} at a novel site. *Cell* **42:** 849.

Franza, B.R., L.C. Sambucetti, D.R. Cohen, and T. Curran. 1987. Analysis of *Fos* protein complexes and *Fos*-related antigens by high-resolution two-dimensional gel electrophoresis. *Oncogene* **1:** 213.

Hanks, S. K. 1987. Homology probing: Identification of cDNA clones encoding members of the protein-serine kinase family. *Proc. Natl. Acad. Sci.* **84:** 388.

Hanks, S.K., A.M. Quinn, and T. Hunter. 1988. The protein kinase family: Conserved features and deduced phylogeny of the catalytic domains. *Science* **241:** 42.

Hathaway, G.M. and J.A. Traugh. 1982. Casein kinases-multipotential protein kinases. *Curr. Top. Cell Regul.* **211:** 101.

Hirai, H., Y. Maru, K. Hagiwara, J. Nishida, and F. Takaku. 1987. A novel putative tyrosine kinase receptor encoded by the *eph* gene. *Science* **238:** 1717.

Hirst, R., A. Horwitz, C. Buck, and L. Rohrschneider. 1986. Phosphorylation of the fibronectin receptor complex in cells transformed by oncogenes that encode tyrosine kinases. *Proc. Natl. Acad. Sci.* **83:** 6470.

Hunter, T. 1987. The tail of two *src*'s: Mutatis mutandis. *Cell* **49:** 3.

Hunter, T., N.C. Ling, and J.A. Cooper. 1984. Protein kinase C phosphorylation of the EGF receptor at a threonine residue close to the cytoplasmic face of the plasma membrane. *Nature* **311:** 480.

Ibanez, C.E. and J.S. Lipsick. 1988. Structural and functional domains of the *myb* oncogene: Requirements for nuclear transport, myeloid transformation, and colony formation. *J. Virol.* **62:** 1981.

Isacke, C. M., J. Meisenhelder, K.D. Brown, K.L. Gould, S.J. Gould, and T. Hunter. 1986. Early phosphorylation events following the treatment of Swiss 3T3 cells with bombesin and the mammalian bombesin-related peptide, gastrin-releasing peptide. *EMBO J.* **5:** 2889.

Katzen, A.L., T.B. Kornberg, and J.M. Bishop. 1985. Isolation of the proto-oncogene c-*myb* from *Drosophila melanogaster*. *Cell* **41:** 449.

Kipreos, E.T., G.J. Lee, and J.Y.J. Wang. 1987. Isolation of temperature-sensitive mutants of v-*abl* oncogene by screening with antibodies for phosphotyrosine. *Proc. Natl. Acad. Sci.* **84:** 1345.

Klempnauer, K.-H. and A. Sippel. 1987. The highly conserved amino-terminal region of the protein encoded by the v-*myb* oncogene functions as a DNA-binding domain. *EMBO J.* **6:** 2719.

Klempnauer, K.-H., G. Ramsay, J.M. Bishop, M.G. Moscovici, C. Moscovici, J.P. McGrath, and A.D. Levison. 1983. The product of the retroviral transforming gene v-*myb* is a truncated version of the protein encoded by the cellular oncogene c-*myb*. *Cell* **33:** 345.

Leutz, A., H. Beug, and T. Graf. 1984. Purification and characterization of cMGF, a novel chicken myelomonocytic growth factor. *EMBO J.* **3:** 3191.

Lipsick, J.S. and M.A. Baluda. 1986. The *myb* oncogene. In *Oncogenes: Gene amplification and analysis* (ed. T.S. Papas and G.F. Vande Woude), vol. 4, p. 73. Elsevier, New York.

Lipsick, J.S. and W.J. Boyle. 1987. c-*myb* protein expression is a late event during T-cell activation. *Mol. Cell. Biol.* **7:** 3358.

Lipsick, J.S., C.E. Ibanez, and M.A. Baluda. 1986. Expression of molecular clones of v-*myb* in avian and mammalian cells independent of transformation. *J. Virol.* **59:** 267.

Miller, J., A.D. McLachlan, and A. Klug. 1985. Repetitive zinc-binding domains in the protein transcription factor IIIA from *Xenopus* oocytes. *EMBO J.* **4:** 1609.

Ness, S.A., H. Berg, and T. Graf. 1987. V-*myb* dominance over v-*myc* in doubly transformed chick myelomonocytic cells. *Cell* **51:** 41.

Nunn, M.F., P.H. Seeburg, C. Moscovici, and P.H. Duesberg. 1983. Tripartate structure of the avian erythroblastosis virus E26 transforming gene. *Nature* **306:** 391.

Paz-Ares, J., D. Ghosal, U. Wienand, P.A. Peterson, and H. Seadler 1987. The regulatory *c1* locus of *Zea mays* encodes a protein with homology to *myb* proto-oncogene products and with structural similarities to transcriptional activators. *EMBO J.* **6:** 3553.

Ralston, R. and J.M. Bishop. 1985. The product of the protooncogene c-*src* is modified during the cellular response to platelet-derived growth factor. *Proc. Natl. Acad. Sci.* **82:** 7845.

Rauscher, F.J., L.C. Sambucetti, T. Curran, R.J. Distel, and B.M. Spiegelman. 1988. Common DNA-binding site for *Fos* protein complexes and transcription factor AP-1. *Cell* **52:** 471.

Rogers, S., R. Wells, and M. Rechsteiner. 1986. Amino acid sequences to rapidly degraded proteins: The PEST hypothesis. *Science* **234:** 364.

Sauer, R.T., R.R. Yocum, R.F. Doolittle, M. Lewis, and C.O.Pabo. 1982. Homology among DNA-binding proteins suggests use of a conserved supersecondary structure. *Nature* **298:** 447.

Sommercorn, J., J.A. Mulligan, F.J. Lozeman, and E.G. Krebs. 1987. Activation of casein kinase II in response to insulin and epidermal growth factor. *Proc. Natl. Acad. Sci.* **84:** 8834.

Thompson, C.B., P.B. Challoner, P.E. Neiman, and M. Groudine. 1986. Expression of the c-*myb* proto-oncogene during cellular proliferation. *Nature* **319:** 374.

Vogt, P.K. and R. Tjian. 1988. *Jun*: A transcriptional regulator turned oncogenic. *Oncogene* **3:** 3.

Vogt, P.K., T.J. Bos, and R.F. Doolittle. 1987. Homology between the DNA-binding domain of the GCN4 regulatory protein of yeast and the carboxy-terminal region of a protein coded for by the oncogene *jun*. *Proc. Natl. Acad. Sci.* **84:** 3316.

Woodgett, J.R. and P. Cohen. 1984. Multisite phosphorylation of glycogen synthase. *Biochim. Biophys. Acta* **788:** 339.

Yamanashi, Y., S. Fukushige, J. Sukegawa, N. Miyajima, K. Matsubara, T. Yamamoto, and K. Toyoshima. 1987. The *yes*-related cellular gene *lyn* encodes a possible tyrosine kinase similar to p56^{lck}. *Mol. Cell Biol.* **7:** 237.

Molecular Analyses of Gene Products Associated with the Response of Cells to Mitogenic Stimulation

R.L. ERIKSON,* D. ALCORTA,* P.-A. BEDARD,* J. BLENIS,‡ H.-P. BIEMANN,* E. ERIKSON,† S.W. JONES,* J.L. MALLER,† T.J. MARTINS,* AND D.L. SIMMONS*

*Department of Cellular and Developmental Biology, Harvard University, Cambridge, Massachusetts 02138; †Department of Pharmacology, University of Colorado Medical School, Denver, Colorado 80262; ‡Department of Molecular Biology, Northwestern University Medical School, Chicago, Illinois 60611

A description of the biochemical events in an animal cell that lead to division has recently begun to emerge. We have witnessed publication of numerous papers dealing with the early events that occur as cells enter G_1, such as the activation of protein kinase C (Nishizuka 1984). Perhaps an even greater effort has been expended on the study of gene expression, frequently on the proto-oncogenes c-*fos* and c-*myc* (Kelly et al. 1983). Genes such as c-*fos* and c-*myc* are presumed to be of importance because of their potential for malignant transformation. Other genes of potential relevance have been identified in cDNA libraries prepared from cells at various points in the cell cycle (Cochran et al. 1983; Lau and Nathans 1987). The lack of appropriate animal cell mutants limits the assignment of a specific function to most of these genes, however. There are, nevertheless, methods that partially compensate for the lack of mutants. These include the use of antisense RNA, overexpression of gene products with recombinant retroviruses or other vectors, and the use of purified gene products for in vitro characterization. Thus, we have initiated studies on cells in G_0 stimulated to enter G_1 with serum growth factors, tumor-promoting phorbol esters, or by the expression of pp60$^{v\text{-}src}$. We have focused on two areas during these studies; one concerns the characterization of a serine-specific protein kinase assayed in vitro by its capacity to phosphorylate the protein S6 in 40S ribosomal subunits, and the second concerns the identification of cDNAs corresponding to mRNAs unique to quiescent G_0 cells or dividing or transformed cells.

The activation of a vigorous serine-specific protein kinase and the concomitant phosphorylation of ribosomes is a highly conserved response of animal cells to mitogenic stimulation (Nielsen et al. 1982; Novak-Hofer and Thomas 1984). Nearly all of the phosphate is incorporated into a single protein, S6. The regulation of this enzyme activity is of interest because of the wide variety of agents that activate the kinase. Although these agents initially act through different receptors, the pathways apparently converge to activate a single enzyme or a limited number of distinct enzymes. To relate any findings on the activation of these enzymes to their structure, we have obtained molecular clones corresponding to mRNAs for these proteins. In our second area of interest, we have examined the expression of certain genes in quiescent, normal growing, or transformed cells. The nature of the products of these genes is under investigation.

RESULTS AND DISCUSSION

Stimulation of Protein Kinase Activity Specific for Ribosomal Protein S6

Previous studies have shown that S6 kinase activity is rapidly stimulated by serum, purified growth factors, phorbol ester, and pp60$^{v\text{-}src}$ (Trevillyan et al. 1984; Blenis and Erikson 1985). The mitogen-induced generation of the second messengers diacylglycerol (DAG) and inositol-1,4,5 trisphosphate has been proposed to mediate several growth-associated events. DAG presumably activates protein kinase C (PKC), as does phorbol ester, and thus it was possible that all the agents mentioned above utilize PKC to activate the S6 kinase. This possibility was tested in PKC-deprived cells after ligand-induced down-regulation of the enzyme. Such treatment strongly prevents activation of S6 kinase by phorbol ester (>90%), but only partially attenuates the ability of mitogens or pp60$^{v\text{-}src}$ to activate S6 kinase (Blenis and Erikson 1986). These results suggest that there is a PKC-dependent pathway of S6 kinase activation, as well as at least one PKC-independent pathway.

Phorbol ester is a mitogen for quiescent Swiss-3T3 cells. A variant cell line (3T3-TNR9) has been obtained that, when deprived of serum, does not respond mitogenically to phorbol ester but still displays many of the same responses observed in phorbol-ester-treated parental 3T3 cells (Blumberg et al. 1981; Butler-Gralla and Herschman 1981). Other than the absence of a mitogenic response to administration of a tumor promoter, the only other defective response of the TNR9 variant noted to date is the lack of increase in ornithine decarboxylase (ODC) activity (Butler-Gralla and Herschman 1983; Bishop et al. 1985). Serum, in contrast, does regulate ODC in these cells and indicates again the existence of a PKC-independent pathway leading to cell division. We examined the activity of the S6 kinase in TNR9 cells and found that although the response to serum was similar to that in the 3T3 parent, the response to phorbol ester was defective as com-

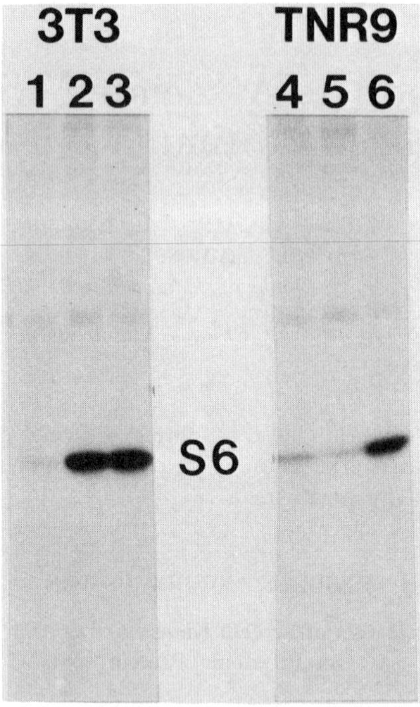

Figure 1. Differential regulation of S6 protein kinase activity in Swiss-3T3 cells and 3T3-TNR9 variants. Cell-free lysates were prepared from serum-deprived, quiescent cells or cells incubated for 1 hr with PMA (100 ng/ml) or fetal calf serum (FCS) (10% v/v) and equal amounts of lysate protein analyzed for S6 phosphotransferase activity as described previously (Blenis and Erikson 1986). (Lane *1*) Quiescent 3T3 cells (952 cpm); (lane *2*) 3T3 cells incubated with PMA for 1 hr (12,130 cpm); (lane *3*) 3T3 cells incubated with 10% (v/v) FCS for 1 hr (13,792 cpm); (lane *4*) quiescent 3T3-TNR9 cells (2063 cpm); (lane *5*) 3T3-TNR9 cells incubated with PMA for 1 hr (2100 cpm); (lane *6*) 3T3-TNR9 cells incubated with FCS (10% v/v) for 1 hr (8006 cpm). 3T3 and 3T3-TNR9 cells were cultured at 37°C in DMEM with 10% (v/v) FCS. For all experiments, cells were transferred at 0.75 to 1×10^6 cells per 100-mm Corning culture dish in 10 ml DMEM with 5% (v/v) FCS and cultured for 4 days. Medium was then changed to serum-free DMEM for an additional 24 hr to ensure quiescence. DMSO (final concentration 0.05% v/v), PMA 100 ng/ml (0.05% v/v DMSO), or FCS (10% v/v) was added for 1 hr, cell-free lysates were prepared, and S6 phosphotransferase activity was analyzed. Autoradiogram qualitatively indicates the extent of $^{32}P_i$ incorporation from [γ-^{32}P]ATP into the 40S ribosomal protein S6. For quantitative analysis, S6 was excised from the dried Coomassie-blue-stained gel and $^{32}P_i$ was determined by liquid scintillation spectrometry.

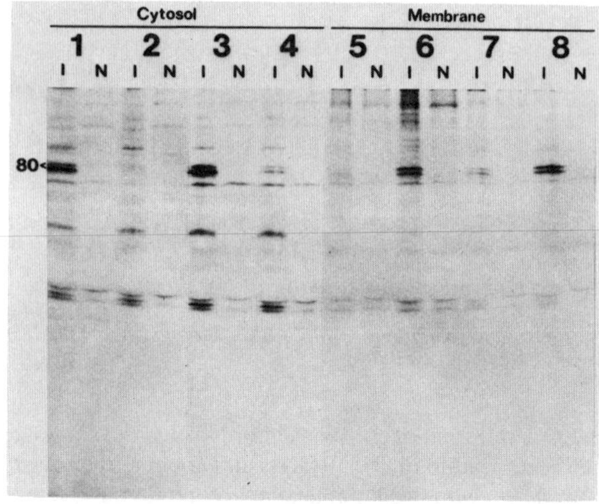

Figure 2. PMA-mediated membrane translocation of PKC in 3T3 and 3T3-TNR9 cells. Cytosolic (lanes *1–4*) and membrane (lanes *5–8*) fractions prepared from [^{35}S]methionine-labeled cells were immunoprecipitated with rabbit antibodies against PKC (lanes labeled I) or with normal (nonimmune) rabbit serum as control (lanes labeled N). PMA (100 ng/ml) incubations were carried out for 30 min. (*1*) Cytosol, untreated 3T3 cells; (*2*) cytosol, PMA-treated 3T3 cells; (*3*) cytosol, untreated 3T3-TNR9 cells; (*4*) cytosol, PMA-treated 3T3-TNR9 cells; (*5*) membrane, untreated 3T3 cells; (*6*) membrane, PMA-treated 3T3 cells; (*7*) membrane, untreated 3T3-TNR9 cells; (*8*) membrane, PMA-treated 3T3-TNR9 cells. Quiescent monolayers were labeled with 100 μCi/ml [^{35}S]methionine in DMEM containing 10% normal methionine concentration and 0.5% (v/v) heat-inactivated FCS for 26 hr prior to harvesting. Lysates were prepared at 4°C by washing monolayers with cold STE buffer (150 mM NaCl, 20 mM Tris [pH 7.2], 0.1 mM EDTA), gently scraping cells from the dish, centrifuging at 1500*g* for 3 min, resuspending the cell pellet in 1 ml hypotonic buffer (20 mM Tris [pH 7.5], 1 mM EGTA, 1 mM $MgCl_2$, 1 mM dithiothreitol, 40 μg/ml phenylmethylsulfonyl fluoride, 0.01% [v/v] leupeptin) and homogenizing cells in a 1-ml Dounce homogenizer with 30 strokes of a tight-fitting pestle. Cytosol and membrane fractions were prepared by centrifugation for 30 min at 100,000*g*. Proteins in the membrane pellet were solubilized by sonication in hypotonic buffer containing 0.5% Triton X-100. PKC antiserum (I) from a rabbit that had been immunized with bovine PKC and that produced antiserum specifically reacting with PKC (H.-P. Beimann and R.L. Erikson, unpubl.) or normal rabbit serum (N) was added to aliquots of lysate protein normalized for equivalent ^{35}S cpm. Immunoprecipitations were performed for 1.5 hr at 4°C in 10 mM Tris (pH 7.2), 150 mM NaCl, 1% sodium deoxycholate, 1% Triton X-100, and 0.1% SDS. Immune complexes were isolated, washed, and analyzed by SDS-PAGE and fluorography.

pared to the 3T3 parent cell line (Fig. 1). The lack of response was not due to a defective PKC as shown in Figure 2. Using rabbit antibodies, we can show a rapid phorbol-ester-stimulated translocation of this enzyme to the membrane fraction following treatment of both the 3T3 parent and the TNR9 variant. These data indicate that the variant has no obvious defect in normal early responses of PKC to phorbol ester. The S6 kinase is not defective, since it responds normally to other growth-promoting factors. Thus, the defect in the pathway must reside between PKC and S6 kinase. This defect occurs much earlier than the ODC defect and suggests a critical role for the stimulation of S6 protein kinase activity during mitogenesis. It is tempting to speculate that this role may be related to translation of ODC mRNA into functional protein.

Molecular Cloning of *Xenopus* Ribosomal Protein S6 Kinase, S6KII

To date, the most highly purified example of an S6 kinase, denoted S6KII, has been obtained from unfertilized *Xenopus* eggs (Erikson and Maller 1985, 1986). This enzyme has an apparent relative molecular weight of 92,000 as determined by SDS-PAGE. Antiserum raised against purified S6KII is capable of immunoprecipitating S6 kinase activity that can be measured in an immune complex. Studies with this antiserum have revealed that progesterone- or insulin-stimulated oocytes contain an antigenically related enzyme. Furthermore, anti-S6KII serum reacts with an S6 kinase activity found in extracts of chicken embryo fibroblasts (CEFs) that had been stimulated with serum or transformed by Rous sarcoma virus (RSV) (Erikson et al. 1987). These results indicate that S6KII or an antigenically related S6 kinase(s) is subject to mitogenic stimulation in various cell types.

A goal of our studies is to elucidate the mechanism of S6 kinase activation and to determine the relationship between various S6 protein kinases. These studies must take into account that the initial events during activation can involve tyrosine phosphorylation, activation of PKC, or, in the case of progesterone, a steroid hormone receptor linked to the adenylate cyclase system. To date, little information is available on the mechanism of activation of S6 kinase(s) in vivo, although in vitro *Xenopus* S6KII can be inactivated with serine phosphatases and can be phosphorylated and partially activated by microtubule-associated protein kinase (Sturgill et al. 1988). Whether this activation represents a pathway that is operative in vivo is unclear, but it raises the possibility that the pathways of activation converge not at the S6 kinase but at an S6 kinase kinase such as microtubule-associated protein kinase. Characterization of the relevant enzymes is necessary to describe the events in more detail. We have undertaken the molecular cloning of S6KII in order to relate any findings on its activation to the structure of the molecule and to attempt to determine the number of proteins that may contribute to S6 phosphorylation in cells.

Sequence analysis of cDNA for *Xenopus* S6KII indicates that it has an unusual structure when compared to other known protein kinases (Figs. 3 and 4)(Jones et al. 1988). It apparently consists of two domains: One that is similar to PKC, the catalytic subunit of cAMP-dependent protein kinase (cAK), and cGMP-dependent protein kinase (cGK) and one related to the catalytic subunit of phosphorylase *b* kinase (PhK) (Fig. 3). The two domains of S6KII also show internal homology that is particularly strong in the sequences 65–282 and 416–635. This is not unexpected, since the similarities to cAK and PhK include their putative ATP-binding sites and extend to a substantial fraction of the entire sequence of these kinases. This result raises the possibility that S6KII has two ATP-binding sites and two catalytic centers. The sequence in S6KII that begins at Gly-69 and ends at Lys-94 is very closely related to the consensus ATP-binding sites of PKC and cAK (GXGXFG[Xaa]16K) except for the insertion of three amino acids (Shoji et al. 1983; Parker et al. 1986). The S6KII sequence that begins at Gly-423 and ends at Lys-445 is related to the putative ATP-binding site of PhK (GXG[Xaa]19K) (Reimann et al. 1984). Which of the two kinase-like domains is functional or whether they are both functional awaits appropriate verification. In this regard, it should be noted that phosphorylase *b* does not serve as a substrate for S6KII, and, in addition, S6 is not phosphorylated by PhK. However, S6KII does phosphorylate glycogen synthase and troponin I in vitro (Erikson and Maller 1988), both of which can be phosphorylated by cAK and PhK. cAK, in contrast to PhK, phosphorylates S6 in vitro; however, this reaction is specifically inhibited by the heat-stable inhibitor of this enzyme, whereas the same inhibitor has no affect on S6KII.

At this time, there is no evidence that S6KII binds regulatory subunits as do cAK and PhK. It apparently has no second-messenger requirements for activity as do PKC, cAK, and cGK and, indeed, appears to be unrelated to the putative regulatory domains of these kinases. The S6 protein kinases do strongly respond to and become activated by mitogenic signals. It seems possible that enzymes such as S6KII, which are phosphoproteins and activated in the absence of protein synthesis, may be activated by phosphorylation. The closely related catalytic subunits of cAK and PhK are not, however, directly regulated by calcium or by phosphorylation. PhK consists of four subunits and is reg-

```
S6K    1 MPLAQLVNLWPEVAVVHEDPENGHGSPEEGGRHTSKDEVV..VKEFPITHHVKEGSEKADQSDFVLLKVLGQGSFGKVFLVRKITPPDANQLYAMKVLKKA.TLKVRDRVRTKMERDILADVHHPFVVRLHYAFQTEGKL 137
CAT    1 ....................GNAAAAKKGSEQESVKEFLAKAKEDFLKKWENPAQNTAHLDQFERIKTLGTGSFGRVMLVKHM...ETGNHYAMKILDKQKVVKLKQIEHTLNEKRILQAVNFPFLVKLEFSFKDNSNL 116

S6K  138 YLILDFLRGGDLFTRLSKEVMFTEEDVKFYLAELALGLDHLHSLGIIYRDLKPENILLDEEGHIKLTDFGLSKEAIDHEKKAYSFCGTVEYMAPEVVNRQGHSHSADWWSYGVLMFEMLTGSLPFQGKDRKETMTLILKA 277
CAT  117 YMVMEYVPGGEMFSHLRRIGRFSEPHARFYAAQIVLTFEYLHSLDLIYRDLKPENLLIDQQGYIQVTDFGFAKRVKG...RTWTLCGTPEYLAPEIILSKGYNKAVDWWALGVLIYEMAAGYPPFFADQPIQIYEKIVSG 253

S6K  278 KLGMPQFLSNEAQSLLRALFKRNPTNRLGSAMEGAEEIKRQPFFSTIDWNKLFRREMSPPFKPAVTQADDTYYFDTEFTSRTPKDSPGIPPSAGAHQLFRGFSFVAPALVEEDAKKTSS 396
CAT  254 KVRFPSHFSSDLKDLLRNLLQVDLTKRFGNLKDGVNDIKNHKWFATTDWIAIYQRKVEAPFIPKFKGPGDTSNFD.DYEEEEIRVSINEKCGKEFSEF.................... 350

S6K  397 PPVLSVPKTHSKNILFMDVYTVRETIGVGSYSVCKRCVHKGTNMEYAVKVIDKT.............KRDPSEEIEILRRYGQHPNIIALKDVYKEGNSIYVVTELMRGGELLDRILRQKFFSEREASSVLFTVCKTVEN 523
PhK    1 TRDAALPGSHSTHGFY.ENYEPKEILGRGVSSVVRRCIHKPTCKEYAVKIIDVTGGGSFSAEEVQELREATLKEVDILRKVSGHPNIIQLKDTYETNTFFFLVFDLMKKGELFDYLTEKVTLSEKETRKIMRALLEVICA 139

S6K  524 LHSQGVVHRDLKPSNILYVDESGDPESIRICDFGFAKQLRADNGLLMTPCYTANFVAPEVL......KRQGYDEGCDIWSLGILLYTMLAGYTPFANGLGDTPEEILARIGSGKFTLRGGNWNTVSAAAKDLVSRMLHVD 657
PhK  140 LHKLNIVHRDLKPENILLDDD....MNIKLTDFGFSCQLDPGEK.LREVCGTPSYLAPEIIECSMNDNHPGYGKEVDMWSTGVIMYTLLAGSPPF...WHRKQMLMLRMIMSGNYQFGSPEWDDYSDTVKDLVSRFLVVQ 271

S6K  658 PHKRLTAKQVLQHEWITKRDALPQSQLNRQD....VHLVKGAMAATYSALNSSKPTPLLQPIKSSILAQRRVKKLPSTTL................................ 733
PhK  272 PQKRYTAEEALAHPFFQQYVVEEVRHFSPRGKFKVICLTVLASVRIYYQYRRVKPVTREIVIRDPYALRPLRRLIDAYAFRIYGHWVKKGQQQNRAALFENTPKAVLFSLAEDDY 386
```

Figure 3. The sequence of S6KII has been aligned at the amino terminus with the sequence of the catalytic subunit of cAMP-dependent kinase and at the carboxyl terminus with the catalytic subunit of phosphorylase kinase.

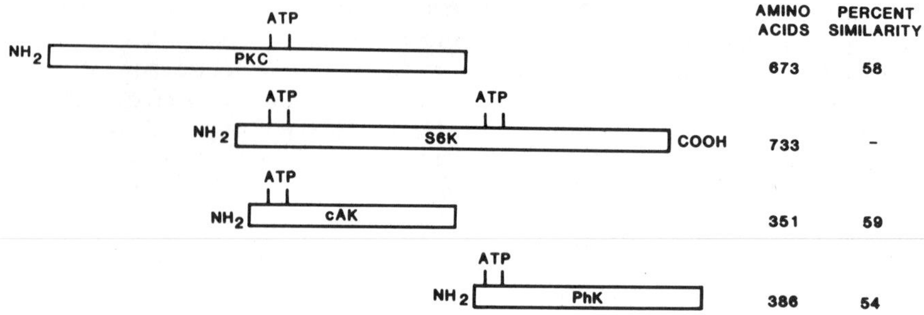

Figure 4. Diagram of S6KII sequence compared with other protein kinases.

ulated by phosphorylation, but in this case, the catalytic subunit does not become phosphorylated. Similarly, the regulatory subunit of cAK can be phosphorylated, and this may indirectly influence phosphotransferase activity.

The sequence of S6KII raises obvious questions concerning the evolution, regulation, and potential functions of this protein. A likely pathway for creation of the genes encoding S6KII would involve genomic rearrangements. The S6KII kinase genes may have arisen by an intragenic duplication event followed by divergent evolution of the duplicated segments. Alternatively, S6KII may be the product of a gene fusion event between a gene encoding a kinase resembling PKC, the catalytic subunit of cAK, or cGK and a gene encoding a kinase resembling the catalytic subunit of PhK by a process akin to exon shuffling. Such a mechanism has been proposed for the generation of cGK, in which sequences coding for a cyclic-nucleotide-binding region similar to the regulatory subunit of cAK have become fused to sequences coding for a catalytic region similar to the catalytic subunit of cAK. Further insight concerning this issue requires information about the genomic organization of the relevant protein kinases. Further analysis will be necessary to determine whether there are physiologically significant substrates for S6KII besides S6, whether such substrates are involved in growth control, and whether they are phosphorylated by a different catalytic domain in S6KII than is S6.

Cloning of pp60$^{v\text{-}src}$ Immediate Early Genes

A number of studies have focused on elucidating the signal transduction mechanisms accompanying transformation by v-*src*, yet little is known about the cellular genes that are activated by this process. A limited number of reports in the literature (Bedard et al. 1987a; Sugano et al. 1987) suggest that, like other mitogens, v-*src* induces some genes in an "immediate early" fashion, whereas others may be induced as secondary events. Genes in the latter group would be expected to require prior synthesis of regulatory or other proteins for induction to take place. In contrast, immediate early genes would be expected to be induced very early after expression of pp60$^{v\text{-}src}$ protein kinase activity, and moreover, induction would occur in the absence of protein synthesis. Using the temperature-sensitive mutant NY72-4 RSV, we have identified a set of immediate early genes whose mRNA levels are rapidly elevated by activation of the protein kinase activity of pp60$^{v\text{-}src}$.

To identify and clone pp60$^{v\text{-}src}$ immediate early genes, we constructed high-complexity cDNA libraries ($>1 \times 10^6$ recombinants) in λgt10 and λZAP (Bedard et al. 1987a). Libraries were constructed using mRNA isolated from NY72-4 RSV-infected CEFs held at the permissive temperature (35°C) for 4–6 hours in the presence of 75 μM cycloheximide. Differential screening of these libraries was done in two successive attempts. In the first screen, duplicate nitrocellulose filters were applied to 2.5×10^4 plaques plated at low density (2.5×10^3 plaques/150-mm plate) and were probed by differential hybridization. One set of filters was hybridized to random-primed, ^{32}P-labeled cDNA made from NY72-4 RSV-infected CEFs cultured at the nonpermissive temperature (41.5°C), and the other set was hybridized to cDNA from CEFs cultured at 35°C for 4–6 hours in the presence of cycloheximide. From this screen, 25 positive clones were plaque-purified and found by cross-hybridization and Northern blot analysis to fall into three groups, designated 5, 10, and 57. The combined incidence of these clones in the library was 1–2 per 100 plaques. To exclude these abundant cDNAs from being recloned, the second screen was done in triplicate, with the third filter being hybridized to a mixture of radiolabeled clones 5, 10, 57, and pCEF-4. The clone, pCEF-4, was included because we had shown previously that it hybridizes to an mRNA that is superinduced by pp60$^{v\text{-}src}$ in the presence of cycloheximide (Bedard et al. 1987a). Approximately 7.5×10^4 plaques were screened in the above manner, yielding nine new groups of clones—one of which was c-*fos*. Characteristics of the 11 novel clones isolated by both rounds of screening (i.e., neglecting c-*fos*) are presented in Table 1. Six of the clones (5, 10, 37, 101, 120, and 147) exhibit features similar to immediate early genes identified in other systems: (1) They are expressed as rare mRNAs in control cells; (2) they rise within 1 hour of temperature shift in NY72-4 RSV-

Table 1. Characteristics of Clones Isolated by Differential Screening of NY72-4RSV 35° and CHX Libraries

Clone	Clone size (kb)	mRNA size(s) (kb)	Cellular abundance[a]	Induction with temperature shift and CHX
5	3.0	3.0	low	>100-fold
10	1.9	1.9	low	>100-fold
37	1.0	3.0	low	>100-fold
57	3.6	1.7, 3.0, >6.0	moderate	>100-fold for >60-kb mRNA
68	0.8	0.5, 4.0	high	>2-fold
101	2.3	4, 3, 3.8, 3.6, and 2.1	low	>100-fold for 4.3-kb species
120	1.0	1.3, 3.7	1.3; moderate 3.7: low	~20-fold
134	0.7	0.7	low	3–4-fold
147	2.0	4.3, 3.8, 3.6	low	>100-fold for 4.3-kb species
157	1.0	2.0	high	2-fold
176	1.0	1.0	high	2-fold

[a]Cellular abundance in uninduced cells.

infected CEF cells; and (3) they are substantially superinduced by temperature shift in the presence of cycloheximide. The remaining five clones lack one or more of these features.

Figure 5 shows the induction patterns obtained for four of the $pp60^{v\text{-}sc}$ immediate early genes with chicken c-*fos*, c-*myc*, pCEF-4, and GAPDH. Clones 5 and 37 detect mRNAs that are induced with the same kinetic profile as c-*fos*, but are approximately five times more abundant than the latter mRNA (Fig. 5). All three of these mRNAs peak within 1 hour post-temperature-shift and decline to an intermediate level that is maintained out to 24 hours. The sequence of clone 5 (data not shown) shows it to be 70% identical to the murine EGR-1 (early growth response-1) gene, which is known to be concomitantly induced with c-*fos* (Sukhatme et al. 1988). Clone 147 detects a family of low-abundance mRNAs, of which only one species of approximately 5 kb (Fig. 5, arrow) is induced by temperature shift. Long exposures of autoradiographs demonstrate that this mRNA is induced biphasically, with an initial plateau being reached by 1 hour, and a second, tenfold higher plateau occurring at 4 hours and remaining elevated out to 24 hours. The exposure used in Figure 4 shows only the second plateau.

Activation of $pp60^{v\text{-}src}$ by temperature shift produced an extremely interesting induction pattern for clone 10, in which the mRNA reproducibly exhibited three peaks during the 24-hour time course. Clone 10 mRNA rose tenfold at 1 hour postshift, decreased to control or subcontrol levels by 2 hours, peaked again at 8 hours, fell 40% at 12 hours, and rose back to maximal levels at 24 hours. This phenomenon is also evident to a lesser extent in the expression of c-*myc*, which also shows a substantial decline at 2 hours (Fig. 5). The oscillating pattern observed for these mRNAs may reflect the interplay of multiple regulatory pathways stimulated by $pp60^{v\text{-}src}$.

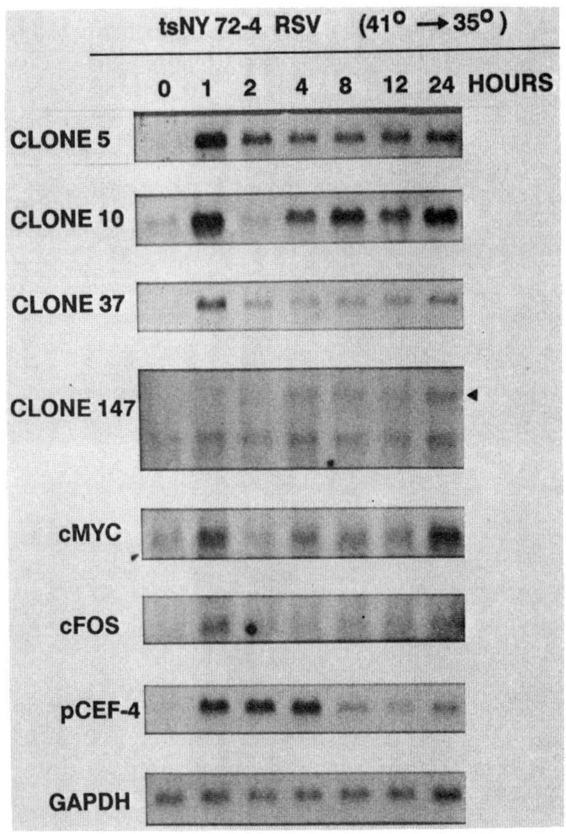

Figure 5. mRNA induction by $pp60^{v\text{-}src}$ in *ts*NY72-4 RSV-infected CEFs. Cells were seeded on 150-mm dishes at a density of 2×10^7 cells per plate. Cells were cultured for 48 hr in DMEM plus 5% serum, after which the cells were switched to low-serum-containing medium (DMEM plus 0.5% serum) for an additional 24 hr before temperature shift. At time zero, cells were shifted from the nonpermissive (41.5°C) to the permissive (35°C) temperature. Total RNA isolated at 0, 1, 2, 4, 8, 12, and 24 hr post-temperature-shift was electrophoresed on formaldehyde denaturing gels, blotted to nylon membranes, and hybridized to the probes indicated.

Effect of Temperature Shift Alone on pp60$^{v\text{-}src}$ Immediate Early Gene Expression

The shift of cells from one temperature to another in these experiments raised the possibility that the genes described above were induced by changing the ambient temperature from 41.5°C to 35°C and not by the activation of pp60$^{v\text{-}src}$. To address this question, temperature-shift experiments were performed on CEFs infected with the nontransforming, myristylation-defective virus NY315 RSV, and also with cells infected with transforming, wild-type Schmidt-Ruppin A virus (Fig. 6). In these experiments, clones 5, 37, c-*fos*, c-*myc*, and GAPDH revealed no change in mRNA levels following temperature shift. However, clone 10 showed nearly a linear increase following temperature shift, with the effect being most pronounced in NY315 RSV-infected cells. Clone 147 mRNA is also weakly induced in NY315-infected CEF (see Fig. 6) but not in wild-type transformed cells (data not shown). Therefore, for clones 10 and 147, the increase in mRNA observed is independent of the membrane-binding and transformation-stimulating activity of pp60$^{v\text{-}src}$, whereas pCEF-4 depends on these properties for a temperature shift to exert its inductive effect. We cannot determine from these experiments whether the elevation in mRNAs observed for clones 10, 147, and pCEF-4 is the result of changes in the activity of pp60$^{v\text{-}src}$ produced by temperature change or whether it is the result of changes in other cellular modulators. Nevertheless, it is clear that none of the changes in mRNA levels seen in Figure 6 are similar to those observed with NY72-4 RSV (Fig. 5), suggesting that the mRNA increases seen with the temperature-sensitive mutant are the result of activating the tyrosine kinase function of pp60$^{v\text{-}src}$.

Serum Stimulation of pp60$^{v\text{-}src}$ Immediate Early Genes

We have reported previously that serum induces pCEF-4 with kinetics quite different from those invoked by temperature shift of NY72-4 RSV-infected cells (Bedard et al. 1987a). Activation of the viral oncogene produces an immediate elevation of pCEF-4 mRNA that remains substantially above control levels out to 24 hours post-temperature-shift (Fig. 5). In contrast, treatment of serum-deprived CEFs with 10% calf serum produces an immediate, transient spike in

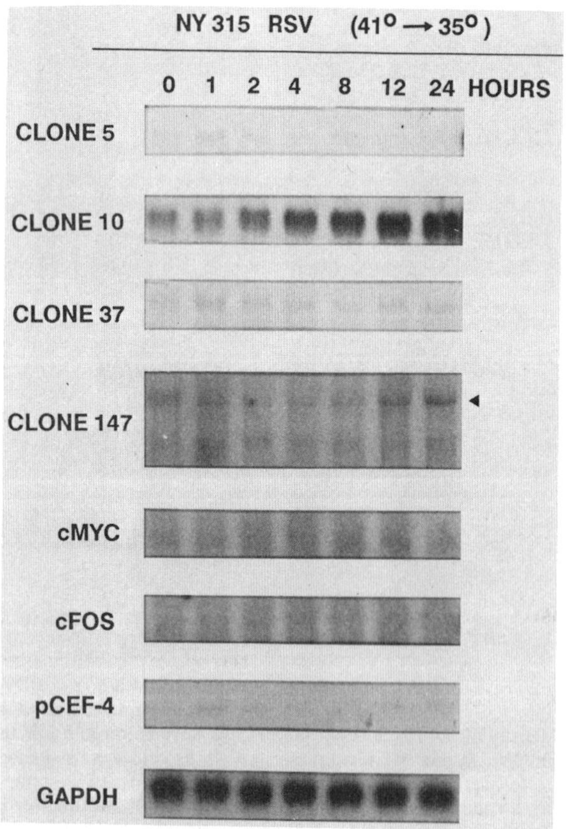

Figure 6. Effects of temperature shift on mRNA levels in NY315 RSV-infected CEFs. CEFs infected with the myristylation-defective mutant, NY315 RSV, were treated as described in Fig. 5.

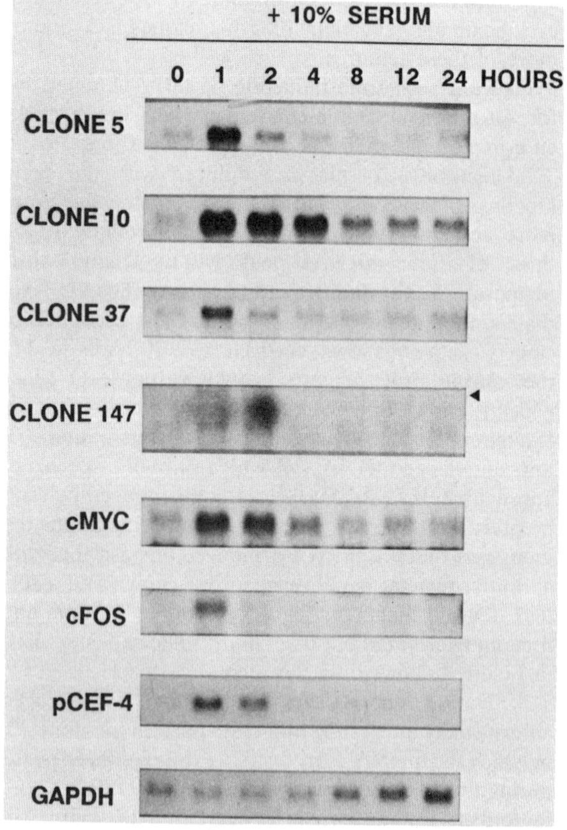

Figure 7. Serum induction of pp60$^{v\text{-}src}$ immediate early genes. CEFs were cultured as in Fig. 5, with the exception that cells were seeded at $\sim 1 \times 10^7$ cells per 150-mm plate. At time zero, cells were exposed to 10% calf serum and RNA was isolated and analyzed as described in Fig. 5. As can be discerned by hybridization to the GAPDH probe, increased amounts of mRNA appear to have been loaded for the last three time points; this must be taken into consideration in evaluating kinetics of induction.

pCEF-4 mRNA that rapidly declines within 2 hours of treatment to control levels (Fig. 7). Indeed, comparison of the results obtained with serum (Fig. 7) and NY72-4 RSV (Fig. 5) shows that (1) all pp60$^{v\text{-}src}$ immediate early genes thus far cloned are inducible by serum and (2) induction by serum in all cases is early and brief—rising and then declining to control levels within the first 8 hours after exposure to serum. Induction by serum is also monophasic (Fig. 7). Activation of pp60$^{v\text{-}src}$, on the other hand, generates a persistent induction of all the immediate early mRNAs out to 24 hours. For clones 10 and 147, this induction is multiphasic (Fig. 5).

Hence, our studies have not detected differences in the types of immediate early genes expressed upon mitogenic stimulation by pp60$^{v\text{-}src}$ versus nontransforming stimulators of cell proliferation, such as serum; but our analyses do show clear qualitative dissimilarities that distinguish these two types of processes.

Characterization of Immediate Early Gene Products

One of the major challenges we face after the preliminary characterization of the cDNAs identified above is the necessity of establishing their function. There are numerous approaches to this problem. One involves obtaining sufficient protein to establish function with assays with biological or biochemical tests. The sequence of pCEF-4 cDNA (Bedard et al. 1987a) predicts a protein that should be secreted from cells, but thus far we have been unable to purify it from medium conditioned by transformed cells. To obtain sufficient quantities of the protein to test biological activity, we turned to the use of the baculovirus *Autographa californía* (AcNPV) and the transplacement vectors from the laboratory of Summers (Summers and Smith 1987). The pCEF-4 cDNA was excised and cloned into the transplacement vector pVL941. Recombinant AcNPV that contain pCEF-4 were obtained by homologous recombination with a wild-type AcNPV polyhedrin gene. Putative recombinant viruses were initially identified by screening for occlusion-negative plaques. Confirmation that the viruses thus identified contained the pCEF-4 gene was established by dot-blot hybridization of DNA isolated from cells infected with plaque-purified virus. Recombinant virus was identified and determined to be expressing pCEF-4 mRNA by RNA blot analysis. When the supernatant from recombinant virus-infected cells was compared to that from wild-type-infected cells or uninfected cells, a specific low-molecular-weight polypeptide could be detected (Fig. 8). It is difficult to anticipate the mobility of proteins in this size range, but the 6-kD protein marked as the pCEF-4 gene product in Figure 8 is consistent with that predicted.

In anticipation that additional studies will confirm that the 6-kD protein is indeed the expected product, it has been extensively purified by conventional techniques and used for bioassays. The putative pCEF-4 protein product was tested for stimulation of DNA synthesis in chicken heart mesenchymal (CHM) cells (Bedard et al. 1987b). CHM cells were deprived of serum for 24 hours and then stimulated as shown in Table 2. DNA synthesis was measured by [^{3}H] thymidine uptake at 24 hours. As shown, the recombi-

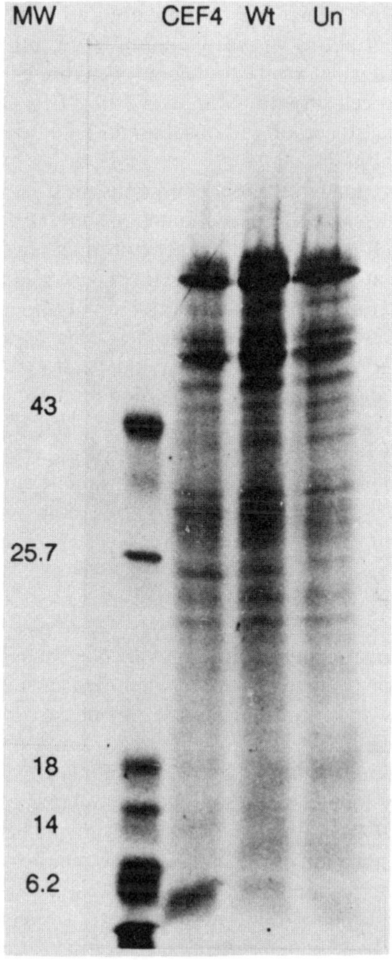

Figure 8. Expression of pCEF-4 gene product in baculovirus-infected insect cells. Recombinant baculovirus containing pCEF-4 under control of the polyhedrin promoter was used to infect insect cells. Small samples of supernatant taken from cells 48 hr postinfection were analyzed by SDS-PAGE. Recombinant virus-infected cells yielded a secreted protein at about 6 kD, whereas wild-type (Wt) and uninfected (Un) cell supernatants were negative for this protein.

Table 2. Chicken Heart Mesenchyme [^{3}H]Thymidine Uptake

Treatment	Quantity	Degree stimulation (%)
Serum	10%	100
CEF4	5–500 ng/ml	0
No addition	—	0
EGF/insulin	100/1000 ng/ml	48
EGF/insulin	10/100 ng/ml	26
CEF4 (EGF/insulin)	5–500 (10/100) ng/ml	23

nant DNA product had no influence on DNA synthesis at concentrations between 5 and 500 ng/ml, concentrations at which conventional mitogens are expected to influence cell growth. Moreover, the pCEF-4 protein did not act together with insulin and EGF to stimulate DNA synthesis.

Clearly, many different conditions must be tested to resolve the issue of pCEF-4 function, but these preliminary results suggest that the gene product may not be a mitogen at normal physiological concentrations. Thus, the induction of pCEF-4 expression may be irrelevant to the growth of RSV-transformed cells or to serum-stimulated cell division. It should be noted that the functional significance of expression of genes identified in this manner has seldom been established. Therefore, although they will undoubtedly be of interest in some area of development, they may not yield new insights in the context of their original identification.

SUMMARY

We have completed the molecular analysis of a ribosomal protein S6 kinase identified in unfertilized *Xenopus* eggs. Although some of our results support the notion that an antigenically related enzyme is present in animal cells, we are uncertain if such an enzyme accounts for all of the mitogen-stimulated phosphorylation of 40S subunits. Other experiments have led to the identification of mRNAs that are rapidly expressed in mitogen-stimulated cells. The relevance of such genes and their products is in doubt, however, until a function can be demonstrated for each of them. Preliminary experiments with the pCEF-4 gene product described here have failed to show that it acts as a mitogen for cells in culture, for example. Thus, although studies such as these may lead to the identification of novel genes, we may need to search elsewhere for a physiologically significant function.

ACKNOWLEDGMENTS

This research was supported by grant CA-42580 from the U.S. Public Health Service and a grant from the American Business Cancer Research Foundation. R.L.E. is an American Cancer Society Professor of Cellular and Developmental Biology.

REFERENCES

Bedard, P.A., D. Alcorta, D.L. Simmons, K.-C. Luk, and R.L. Erikson. 1987a. Constitutive expression of a gene encoding a polypeptide homologous to biologically active human platelet protein in Rous sarcoma virus-transformed fibroblasts. *Proc. Natl. Acad. Sci.* **84:** 6715.

Bedard, P.A., S.D. Balk, H.S. Gunther, A. Morisi, and R.L. Erikson. 1987b. Repression of quiescence-specific polypeptides in chicken heart mesenchymal cells transformed by the Rous sarcoma virus. *Mol. Cell. Biol.* **7:** 1450.

Bishop, R., R. Martinez, M.J. Weber, P.J. Blackshear, S. Beatty, R. Lim, and H.R. Herschman. 1985. Protein phosphorylation in a tetradecanoyl phorbol actetate-nonproliferative variant of 3T3 cells. *Mol. Cell. Biol.* **5:** 2231.

Blenis, J. and R.L. Erikson. 1985. Regulation of a ribosomal protein S6 kinase activity by the Rous sarcoma virus transforming protein, serum growth factors or phorbol ester. *Proc. Natl. Acad. Sci.* **82:** 7621.

———. 1986. Stimulation of ribosomal protein S6 kinase activity by pp60$^{v\text{-}src}$ or serum: Dissociation from phorbol ester-stimulated activity. *Proc. Natl. Acad. Sci.* **83:** 1733.

Blumberg, P.M., E. Butler-Gralla, and H.R. Herschman. 1981. Analysis of phorbol ester receptors in phorbol ester unresponsive 3T3 cell variants. *Biochem. Biophys. Res. Commun.* **102:** 818.

Butler-Gralla, E. and H.R. Herschman. 1981. Variants of 3T3 cells lacking mitogenic response to the tumor promoter tetradecanoyl-phorbol-acetate. *J. Cell. Physiol.* **107** :59.

———. 1983. Glucose uptake and ornithine decarboxylase activity in tetradecanoyl phorbol acetate nonproliferative variants. *J. Cell. Physiol.* **114:** 317.

Cochran, B.H., A.C. Reffel, and C.D. Stiles. 1983. Molecular cloning of gene sequences regulated by platelet-derived growth factor. *Cell* **33:** 939.

Erikson, E. and J.L. Maller. 1985. A protein kinase from *Xenopus* eggs specific for ribosomal protein S6. *Proc. Natl. Acad. Sci.* **82:** 742.

———. 1986. Purification and characterization of a protein kinase from *Xenopus* eggs highly specific for ribosomal protein S6. *J. Biol. Chem.* **261:** 350.

———. 1988. Substrate specificity of ribosomal protein S6 kinase II from *Xenopus* eggs. *Second Messengers and Phosphoproteins* (in press).

Erikson, E., D. Stefanovic, J. Blenis, R.L. Erikson, and J.L. Maller. 1987. Antibodies to *Xenopus* egg S6 kinase II recognize S6 kinase from progesterone- and insulin-stimulated *Xenopus* oocytes and from proliferating chicken embryo fibroblasts. *Mol. Cell. Biol.* **7:** 3147.

Jones, S.W., E. Erikson, J. Blenis, J.L. Maller, and R.L. Erikson. 1988. A *Xenopus* ribosomal protein S6 kinase has two apparent kinase domains that are each similar to distinct protein kinases. *Proc. Natl. Acad. Sci.* **85:** 3377.

Kelly, K., B.H. Cochran, C.D. Stiles, and P. Leder. 1983. Cell-specific regulation of the c-*myc* gene by lymphocyte mitogens and platelet-derived growth factor. *Cell* **35:** 603.

Lau, L.F. and D. Nathans. 1987. Expression of a set of growth-related immediate early genes in BALB/c 3T3 cells: Coordinate regulation with c-*fos* or c-*myc*. *Proc. Natl. Acad. Sci.* **84:** 1182.

Nielsen, P.J., G. Thomas, and J.L. Maller. 1982. Increased phosphorylation of ribosomal protein S6 during meiotic maturation of *Xenopus* oocytes. *Proc. Natl. Acad. Sci.* **79:** 2937.

Nishizuka, Y. 1984. The role of protein kinase C in cell surface signal transduction and tumor promotion. *Nature* **308:** 693.

Novak-Hofer, I. and G. Thomas. 1984. An activated S6 kinase in extracts from serum and epidermal growth factor-stimulated Swiss 3T3 cells. *J. Biol. Chem.* **259:** 5995.

Parker, P.J., L. Coussens, N. Totty, L. Rhee, S. Young, E. Chen, S. Stabel, M.D. Waterfield, and A. Ullrich. 1986. The complete primary structure of protein kinase C — The major phorbol ester receptor. *Science* **233:** 853.

Reimann, E.M., K. Titani, L.H. Ericsson, R.D. Wade, E.H. Fischer, and K.A. Walsh. 1984. Homology of the γ subunit of phosphorylase *b* kinase with cAMP-dependent protein kinase. *Biochemistry* **23:** 4185.

Shoji, S., L.H. Ericsson, K.A. Walsh, E.H. Fischer, and K. Titani. 1983. Amino acid sequence of the catalytic subunit of bovine type II adenosine cyclic 3′,5′-phosphate dependent protein kinase. *Biochemistry* **22:** 3702.

Sturgill, T.W., B.L. Ray, E. Erikson, and J.L. Maller. 1988. Insulin-stimulated MAP-II kinase phosphorylates and activates ribosomal protein S6 kinase II. *Nature* **334:** 715.

Sugano, S., M.Y. Stoeckle, and H. Hanafusa. 1987. Transformation by Rous sarcoma virus induces a novel gene with homology to a mitogenic platelet protein. *Cell* **49:** 321.

Sukhatme, V.P., X. Cao, L.C. Chang, C.H. Tsai-Morris, D. Stamenkovich, P.C.P. Ferreira, D.R. Cohen, S.A. Edwards, T.B. Shows, T. Curran, M.M. LeBeau, and E.D. Adamson. 1988. A zinc finger-encoding gene coregulated with c-*fos* during growth and differentiation, and after cellular depolarization. *Cell* **53:** 37.

Summers, M.D. and G.E. Smith. 1987. A manual of methods for baculovirus vectors and insect cell culture procedures. *Texas Agric. Exp. Stn. Bull.* **1555**.

Trevillyan, J.M., R.K. Kulkarni, and C.V. Byus. 1984. Tumor-promoting phorbol esters stimulate the phosphorylation of ribosomal protein S6 in quiescent reuber H35 hepatoma cells. *J. Biol. Chem.* **259:** 897.

Interactions between the Middle T Antigen of Polyomavirus and Host Cell Proteins

S.A. Courtneidge, R.M. Kypta, and E.T. Ulug
The European Molecular Biology Laboratory, 6900 Heidelberg, Federal Republic of Germany

Polyomavirus can transform cells to the neoplastic phenotype both in vivo and in vitro, an activity due to the early-region-encoded protein called middle T antigen (Tooze 1981). Middle T antigen (for review, see Smith and Ely 1983; Courtneidge 1986) is a 55-kD protein, phosphorylated in vivo predominantly on serine, but also to a lesser extent on threonine and tyrosine residues. It is associated with membranes, the plasma membrane being a particularly predominant location, via a stretch of hydrophobic amino acids near its carboxyl terminus and is exposed to the cytoplasm (although it is possible that a few carboxy-terminal amino acids are exposed on the surface of the cell). Analyses of mutants have shown that the membrane location is required for transformation. Middle T antigen was one of the first transforming proteins to be shown to have associated protein tyrosine kinase activity, which is also necessary for transformation.

Several studies had shown that middle T antigen did not behave as a monomeric protein; rather, it sedimented on density gradients in a heterodisperse manner, with the majority of the protein having a molecular mass greater than 100 kD (Schaffhausen and Benjamin 1981; Walter et al. 1982; Courtneidge et al. 1984). It was therefore proposed that middle T antigen existed either in multimers or in complexes with proteins of host origin. The latter proposal has proved to be a rather common occurrence. The first such complex to be demonstrated was between middle T antigen and the cellular tyrosine kinase pp60$^{c\text{-}src}$ (Courtneidge and Smith 1983), the homolog of the Rous sarcoma virus transforming protein, pp60$^{v\text{-}src}$. This complex accounted for, at least in part, the associated tyrosine kinase activity and was also shown to be required for transformation (Courtneidge and Smith 1984). Since that time, middle T antigen has also been shown to complex with other cellular proteins, both kinases and proteins without overt kinase activity. For example, complex formation has been demonstrated between middle T antigen and the product of the *yes* oncogene (Kornbluth et al. 1987), and there is also an abundant complex containing middle T antigen and a 61-kD protein of as yet unknown function (Grussenmeyer et al. 1985, 1987). We are particularly interested in understanding the relative importance of these interactions of middle T with host-cell proteins in the transformation process.

The complex containing middle T antigen and pp60$^{c\text{-}src}$ is perhaps the best studied to date. Middle T transformation causes a stimulation of pp60$^{c\text{-}src}$-kinase activity of approximately tenfold (Bolen et al. 1984; Cartwright et al. 1985; Courtneidge, 1985). However, since in a cell lysate of polyomavirus-transformed cells approximately 10–20% of the pp60$^{c\text{-}src}$ is complexed with middle T and only the complexed pp60$^{c\text{-}src}$ is activated (Bolen et al. 1984; Courtneidge 1985), the stimulation per molecule is approximately 50–100-fold. In normal cells, pp60$^{c\text{-}src}$ is phosphorylated on tyrosine at one major site, Tyr-527 (Cooper et al. 1986; Laudano and Buchanan 1986), and has very low kinase activity (Courtneidge 1985; Coussens et al. 1985; Iba et al. 1985). Dephosphorylation of this site, either in extracts (Courtneidge 1985) or in vitro (Cooper and King 1986), results in a stimulation of kinase activity. Middle T appears to activate pp60$^{c\text{-}src}$ by masking this regulatory phosphorylation site and thus constitutively activating the enzyme (Cartwright et al. 1986).

Within the complex of middle T antigen and activated pp60$^{c\text{-}src}$, we have recently identified a third protein of 81 kD (Courtneidge and Heber 1987). This p81 becomes phosphorylated, presumably by pp60$^{c\text{-}src}$, on tyrosine residues during in vitro kinase assays. The presence of this protein in complexes correlates completely with detection of phosphatidylinositol (PtdIns) kinase activity (Whitman et al. 1985), which suggests that p81 might be a component of a PtdIns kinase. Our recent findings concerning this protein are summarized below.

As mentioned above, middle T antigen can also bind to the product of the *yes* gene, although in this case kinase activity appears to be only slightly elevated (Kornbluth et al. 1987), and it is not yet clear if this complex is important for transformation. However, the *src* and *yes* proteins are but two members of a growing family of tyrosine kinases (for review, see Cooper 1988), all members of which have the potential to become associated with plasma membranes via myristylation of their amino-terminal glycine residues and to be regulated by phosphorylation:dephosphorylation of a tyrosine near the carboxyl terminus. It is possible that middle T antigen can, and perhaps to transform must, interact with all members of this family. We wished to investigate this hypothesis in more detail and have begun our analysis with the product of the *fyn* gene (previously called both *syn* and *slk*).

EXPERIMENTAL PROCEDURES

All of the procedures used in the studies presented in this paper that involve normal and polyomavirus-transformed cells, the use of antibodies specific for pp60$^{c\text{-}src}$ and middle T antigen, tyrosine kinase assays, PtdIns kinase assays, phosphoamino acid analysis, vanadate treatment of lysates, and SDS-gel electrophoresis have been described in previous publications from our laboratory and are cited at suitable points in the text.

Analysis of PtdIns turnover in normal and polyomavirus-transformed cells. NIH-3T3 cells and a derivative transformed with middle T antigen were plated at 2×10^5 in 35-mm dishes. After 24 hours, the cells were labeled with 2.5 μCi/dish [^{3}H]myoinositol in inositol- and serum-free Dulbecco's modified Eagle's medium supplemented with insulin and transferrin. Labeling was carried out for 48 hours, at which time lithium chloride was added to a final concentration of 10 mM. After a further 50 minutes, serum was added to a final concentration of 10% to half of the plates, and all plates were then incubated at 37°C for a further 10 minutes. Inositol-containing compounds were isolated essentially according to the method of Fry et al. (1985). Briefly, reactions were terminated by washing the cells with phosphate-buffered saline and adding 0.6 ml water and 0.12 ml 0.2 M HCl, into which the cells were scraped. Following the addition of 2.7 ml $CHCl_3$/MeOH (1:2, v/v) and incubation on ice for 15 minutes, 0.9 ml $CHCl_3$ and 0.9 ml water were added, the samples were mixed, and the phases were separated by centrifugation. Aliquots of the upper, aqueous phase were diluted 10× with water, and the inositol phosphates were separated using Dowex resin (Berridge 1983). Aliquots of the lower, organic phase were analyzed by thin-layer chromatography on oxalate-treated plates and contained predominantly PtdIns.

Generation of antipeptide antibodies specific for the fyn protein. The complete details of the generation and characterization of these antibodies will be published elsewhere (Kypta et al. 1988). Briefly, peptides corresponding to amino acid numbers 22–35 (fyn1) and 35–51 (fyn2) were synthesized by 9-fluoroenylmethoxycarbonyl (Fmoc) chemistry, coupled to keyhole limpet hemocyanin, emulsified with Freund's complete adjuvant, and injected into rabbits. Following two booster injections, rabbits were bled and serum was tested for the presence of antipeptide antibodies by enzyme-linked immunosorbent assay. Positive samples were then further tested for their ability to precipitate proteins from radiolabeled cell extracts. Both anti-fyn1 and anti-fyn2 antibodies recognized the same 59-kD protein in NIH-3T3 fibroblasts. We tested these antibodies for immune precipitation, kinase activity, etc., in exactly the same way as our previously published methods for pp60$^{c\text{-}src}$.

Phosphatase treatment of immune complexes. In essence, we followed the procedure of Cooper and King (1986), with two exceptions. First, phosphatase treatment was done in a buffer of pH 7.5; second, as a negative control, the immune complexes were treated with 2 μg bovine serum albumin (BSA) in phosphatase storage buffer.

RESULTS

p81 Protein Complexed with Middle T Antigen and pp60$^{c\text{-}src}$

In our earlier publications (Courtneidge and Smith 1983, 1984), we demonstrated the existence of a complex between middle T antigen and pp60$^{c\text{-}src}$. At this time, we only detected these two proteins, although the fact that the complex had a size of approximately 200–220 kD suggested that other proteins might be present. With the use of better reagents (e.g., monospecific antibodies), more sensitive kinase assays, and longer autoradiographic exposure times, it became clear that there was at least a third component in the complex: a protein of 81 kD, which became phosphorylated on tyrosine residues during immune complex kinase assays. An example of such an assay is shown in Figure 1A. When immune complexes were derived by treatment of lysates of normal fibroblasts with monoclonal antibody specific for pp60$^{c\text{-}src}$, autophosphorylated pp60$^{c\text{-}src}$ was the only product specifically detected (lane 2). In contrast, when lysates of middle-T-antigen-

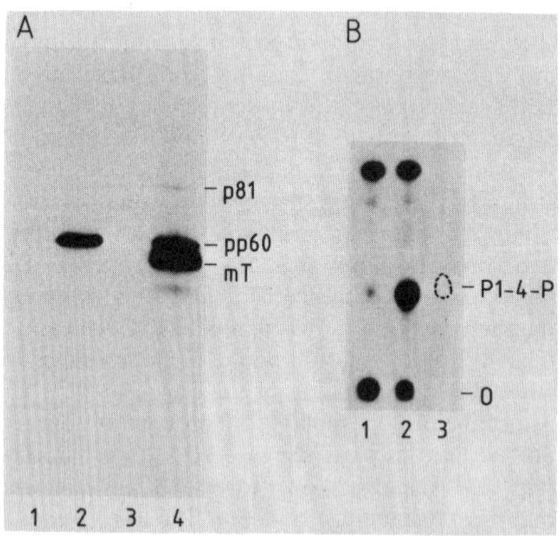

Figure 1. Kinase activities associated with middle T antigen. (*A*) Protein kinase assay. Lysates of normal (lanes *1* and *2*) or middle-T-antigen-transformed (lanes *3* and *4*) cells were immunoprecipitated either without (lanes *1* and *3*) or with (lanes *2* and *4*) monoclonal antibody specific for pp60$^{c\text{-}src}$, washed, and assayed for kinase activity as described previously (Courtneidge and Heber 1987). p81, pp60$^{c\text{-}src}$ (pp60), and middle T antigen (mT) are indicated. (*B*) PtdIns kinase assay. Lysates of middle-T-antigen-transformed cells were immunoprecipitated either without (lane *1*) or with (lane *2*) monoclonal antibody specific for pp60$^{c\text{-}src}$, washed, and assayed for PtdIns kinase activity as described previously (Courtneidge and Heber 1987). The positions of the origin (O) and authentic PtdIns4P are shown.

transformed cells were subjected to the same analysis, three specifically labeled proteins were seen (lane 4): pp60$^{c\text{-}src}$, middle T antigen, and an 81-kD protein. p81 was also precipitated from the polyomavirus-transformed cells using an antibody specific for middle T antigen (data not shown). Density gradient sedimentation analysis revealed that the three proteins also colocalized on sucrose gradients (Courtneidge and Heber 1987), strengthening the argument that they were complexed together.

We next wished to determine whether the presence of the p81 protein might be required for transformation by middle T antigen. Our approach was to screen several mutant middle T antigens for a coprecipitating p81 protein. There are many deletion mutants of middle T, which can be subdivided into three classes (Courtneidge and Heber 1987). Class I contains those middle T antigens, that, although sustaining deletions, are nevertheless wild-type in their transforming capacity, for example, *dl*8 and *dl*1014; these middle T antigens bind to and activate pp60$^{c\text{-}src}$. Class II comprises several mutants not able to transform cells fully, although they may form a few foci or very small colonies in soft agar. All members of this class are also able to bind to pp60$^{c\text{-}src}$ and activate its kinase activity, although perhaps not to wild-type levels. Examples in this class that we have analyzed include *dl*1015, *dl*23, MG13, and 82JF3 (Courtneidge et al. 1988). Class III mutants, e.g., NG59 and NS2, fail to associate with pp60$^{c\text{-}src}$ and fail completely to transform. Our analysis showed that all the class I mutants had an associated p81 protein, whereas none of the class III mutants did. The mutants in class II could be subdivided: *dl*1015 and MG13 had associated p81, but mutants *dl*23 and 82JF3 did not.

p81 Is Associated with PtdIns Kinase Activity

In searching for a function for p81, we recalled the description by Whitman et al. (1985) of a PtdIns kinase activity in immune complexes containing middle T antigen. We therefore tested our immunoprecipitates for such an enzyme activity, and an example is shown in Figure 1B. Immunoprecipitates were derived from polyomavirus-transformed cell lysates using either rabbit anti-mouse serum alone (lane 1) or a mouse monoclonal antibody specific for pp60$^{c\text{-}src}$ and rabbit anti-mouse serum (lane 2). Liposomes containing PtdIns were added as substrate, kinase assays were performed, and the lipids were subsequently extracted with chloroform:methanol and separated by thin-layer chromatography. The pp60$^{c\text{-}src}$ immunoprecipitates contained appreciable levels of a PtdIns kinase activity, whereas the control immunoprecipitates contained only very low levels. pp60$^{c\text{-}src}$ immunoprecipitates from normal cells showed no such activity (data not shown). Note also that the product of this reaction ran slightly slower than authentic PtdIns4P. Whitman et al. (1988) have recently reported that the product of the middle-T-associated PtdIns kinase is PtdIns3P. The data shown here are consistent with their finding.

To determine if there was any association between p81 and PtdIns kinase, we used the mutant middle T antigens described above. We found a complete correlation between the presence of p81 and detection of PtdIns kinase activity, suggesting that p81 may itself have PtdIns kinase activity, or it may regulate the activity of an as yet undetected protein in the complex.

PtdIns Kinase Activity May Be Regulated by Tyrosine Phosphorylation

The complex of middle T antigen and pp60$^{c\text{-}src}$ therefore contains a third protein of 81 kD, which appears to have the properties of a PtdIns kinase. p81 becomes phosphorylated on tyrosine(s) during kinase assays. We therefore reasoned that PtdIns kinase activity might be regulated by such phosphorylation. To test this hypothesis, we first examined the effect of dephosphorylation of immune complexes on kinase activity. We chose to use the acid phosphatase from potato (PAP), since it has been shown previously to dephosphorylate proteins in immune complexes (Cooper and King 1986). Under the conditions we used, proteins were dephosphorylated approximately 50–75% in 10 minutes at 30°C, and the enzyme dephosphorylated serine and tyrosine equally well (data not shown). Immune complexes were made from middle-T-antigen-transformed cells using an antipeptide antibody specific for middle T antigen, and incubated with either BSA (Fig. 2A, lanes 1 and 2) or PAP (lanes 3 and 4). Autophosphorylation assays subsequently performed showed that protein kinase activity was reduced approximately 50% by this treatment. When similar immune complexes were assayed for PtdIns kinase activity, we obtained the results shown in Figure 2B. Treatment with PAP reduced the PtdIns kinase activity to barely detectable levels compared to controls (compare lanes 2 and 4). These results are consistent with the hypothesis that the PtdIns kinase activity is positively regulated by phosphorylation. However, since PAP removes phosphate from all phosphorylated amino acids, we could not be sure that this effect was due to dephosphorylation at tyrosine.

Since we did not have a phosphotyrosine-specific phosphatase available, we turned to a more indirect approach to examine the effect of tyrosine phosphorylation on PtdIns kinase activity. We prepared lysates in the absence of sodium orthovanadate (a phosphotyrosine phosphatase inhibitor), under conditions that we have already shown allow phosphotyrosine phosphatases to be active (Courtneidge 1985), and compared the PtdIns kinase activity we obtained from these lysates to that present in immune complexes prepared in the usual way. (Control experiments carried out in parallel on normal cells showed that preparation of lysates without orthovanadate resulted in a 2.5-fold stimulation of pp60$^{c\text{-}src}$, in accord with previous reports.) Figure 2C shows that when dephosphorylation at tyrosine occurred during lysis, the PtdIns kinase activity subsequently measured was very low compared

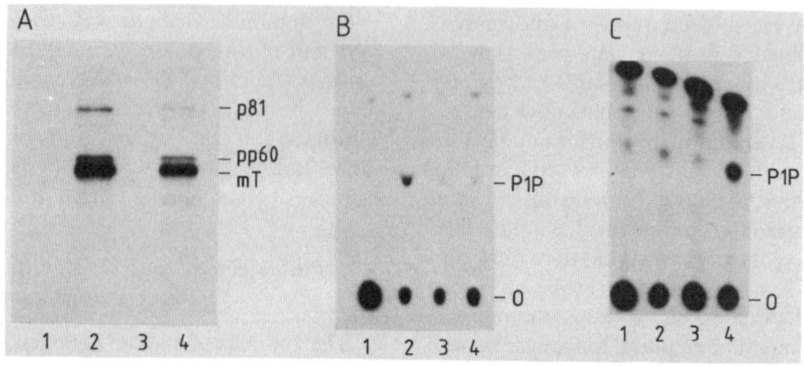

Figure 2. The effect of dephosphorylation on middle-T-antigen-associated tyrosine and PtdIns kinase activities. (*A*) Tyrosine kinase activity after dephosphorylation in vitro. Immune complexes derived from middle-T-antigen-transformed cells with either antipeptide antibody specific for middle T antigen (anti-mt.c, lanes *2* and *4*) or anti-mt.c preblocked with specific peptide (lanes *1* and *3*) were treated with either 2 μg BSA (lanes *1* and *2*) or 2 μg potato acid phosphatase (lanes *3* and *4*), washed, and assayed for tyrosine kinase activity. (*B*) PtdIns kinase activity after dephosphorylation in vitro. The immune complexes were derived and treated in exactly the same way as described in *A*, except that after washing they were assayed for PtdIns kinase activity. (*C*) PtdIns kinase activity after dephosphorylation during lysis. Polyomavirus-transformed cells were lysed in the absence (lanes *1* and *2*) or presence (lanes *3* and *4*) of 100 μM sodium orthovanadate as described by Courtneidge (1985). Immune complexes made with anti-mt.c. preblocked with specific peptide (lanes *1* and *3*) or anti-mt.c (lanes *2* and *4*) were subsequently prepared from these lysates, washed, and assayed for PtdIns kinase activity.

to control values. These results are therefore consistent with the model that PtdIns kinase activity is regulated by tyrosine phosphorylation.

Inositol Phosphate Production Appears Normal in Polyomavirus-transformed Cells

One of the early responses of a cell to many mitogens is the production of the two second messengers, diacylglycerol and inositol trisphosphate, by the agonist-stimulated breakdown of phosphatidylinositol 4,5-bisphosphate ($PtdInsP_2$) (for review, see Berridge 1984). Since PtdIns kinases act to produce $PtdInsP_2$, it was possible that in polyomavirus-transformed cells, increased activity of a PtdIns kinase could result in a stimulation of inositol phosphate production. To examine this, normal and polyomavirus-transformed cells were labeled with [^{3}H]myoinositol to equilibrium, and the levels of inositol phosphates were measured. A representative experiment is shown in Figure 3. We could detect no significant difference between the levels of inositol phosphates in the two cell lines (compare lanes 1 and 3). Our failure to detect a difference did not appear to be due to the methods used, since in both cell lines, serum stimulation produced an easily detectable increase in inositol phosphates. In this experiment, cells were treated with lithium for 1 hour prior to harvesting. We have also compared the amount and turnover of inositol phosphates in untreated cells and in actively growing cells and obtained the same results (data not shown). We conclude that either the activation of a PtdIns kinase in polyomavirus-transformed cells has no effect on the production of those inositol phosphates or the differences are too slight for us to detect with the methods used.

Identification of a New Tyrosine Kinase in Fibroblasts

Until now, our work has concentrated on the product of the *src* gene and its characterization in both normal and polyomavirus-transformed fibroblasts. However, other members of the *src* family of tyrosine protein kinases are also expressed in fibroblasts (for review, see Cooper 1988), and one of these, $pp62^{c\text{-}yes}$, has also

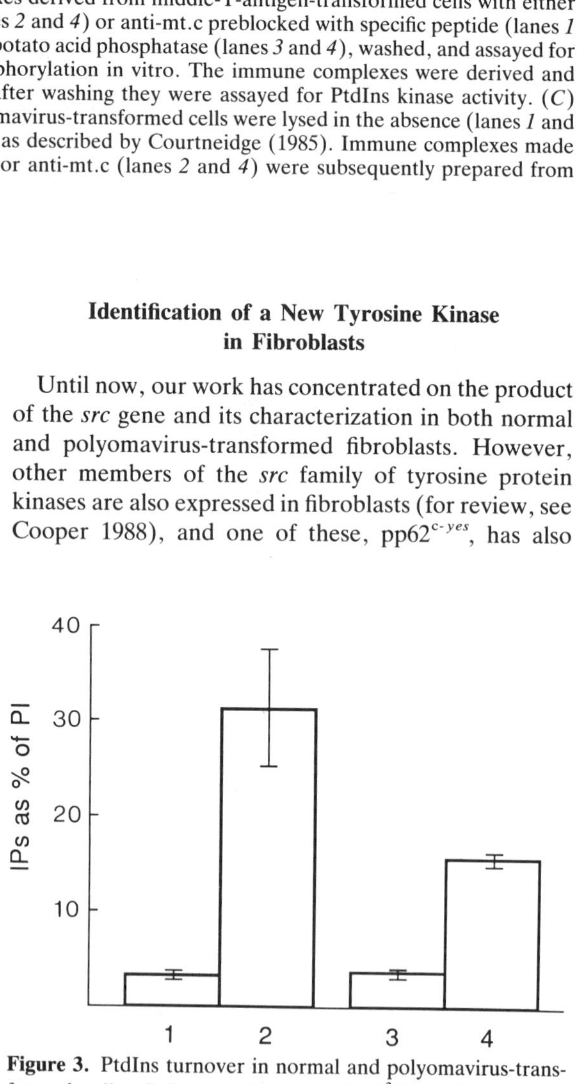

Figure 3. PtdIns turnover in normal and polyomavirus-transformed cells. Cells were labeled with [^{3}H]myoinositol and harvested, and inositol phosphates were isolated as described in Experimental Procedures . (Lane *1*) NIH-3T3 cells; (lane *2*) NIH-3T3 cells treated with 10% FCS; (lane *3*) middle-T-antigen-transformed NIH-3T3 cells; (lane *4*) middle-T-antigen-transformed cells treated with 10% FCS. The results are expressed as cpm in the inositol phosphates (IPs[InsP, $InsP_2$, and $InsP_3$]) as a percentage of the cpm in the organic phase (mostly PtdIns). Error bars show the range of values obtained.

been shown to bind to middle T antigen (Kornbluth et al. 1987). We wished to study the other members of this family in more detail, particularly with respect to their regulation by tyrosine phosphorylation and ability to bind to middle T antigen. We have begun our analysis with the *fyn*-encoded protein. This gene, also known as *syn* and *slk*, was identified as a *src*-related gene from human cDNA libraries (Kawakami et al. 1986; Semba et al. 1986). We first needed to identify the product of the *fyn* gene, to do which we have raised antibodies against peptides corresponding to unique sequences in *fyn*. These antibodies recognize a 59-kD protein in fibroblasts of human, rat, and mouse origin (Kypta et al., 1988). We used these antibodies in immune complex kinase assays, and the results of such an assay are shown in Figure 4A. A 59-kD band was common to both anti-fyn1 (lane 4) and anti-fyn2 (lane 6) immunoprecipitates, and in both cases its appearance was blocked by preincubation of the antibodies with their respective peptides (lanes 3 and 5). This protein was very similar in molecular weight to that of pp60$^{c\text{-}src}$ (lane 2). To investigate the relationship of the 59-kD protein recognized by the anti-fyn2 antibodies to pp60$^{c\text{-}src}$, we made one-dimensional V8-protease maps, which are shown in Figure 4B. As might be expected, the two proteins appeared very similar. In particular, the major peptide of p59 was very similar in size to that of the major peptide of pp60$^{c\text{-}src}$ (which represents the carboxy-terminal part of the molecule containing the autophosphorylation site). However, the maps were clearly distinct from each other, suggesting that the anti-fyn antibodies did not recognize pp60$^{c\text{-}src}$, but rather a closely related protein. In addition, phosphoamino acid analysis revealed that the 59-kD proteins were phosphorylated during kinase assays on tyrosine residues (Fig. 4C).

The *fyn* Product Is Also Regulated by Tyrosine Phosphorylation

We had previously shown that pp60$^{c\text{-}src}$ is regulated by tyrosine phosphorylation. This is conveniently done by testing for kinase activity using lysates prepared with and without sodium orthovanadate. To examine the effects of such dephosphorylation on the *fyn* protein, lysates were prepared in the same way and tested for kinase activity. As expected, dephosphorylation caused an approximately threefold increase in pp60$^{c\text{-}src}$ activity (Fig. 5, lanes 1 and 2). We also saw the same effect when measuring p59^{fyn} kinase activity, with dephosphorylation resulting in an approximately twofold increase in kinase activity (compare lanes 3 and 4), suggesting that the *fyn* protein, too, is regulated by tyrosine phosphorylation.

Middle T Antigen Also Complexes to the *fyn* Protein

To determine whether p59^{fyn} could also complex to middle T antigen, we compared immunoprecipitates made from normal and polyomavirus-transformed cells by kinase assay. When normal cells were analyzed, a single specific band corresponding to p59^{fyn} was detected (Fig. 6, lane 2). In contrast, analysis of transformed cells revealed two bands, p59^{fyn} and a protein that comigrated with middle T antigen (Fig. 6, lane 4).

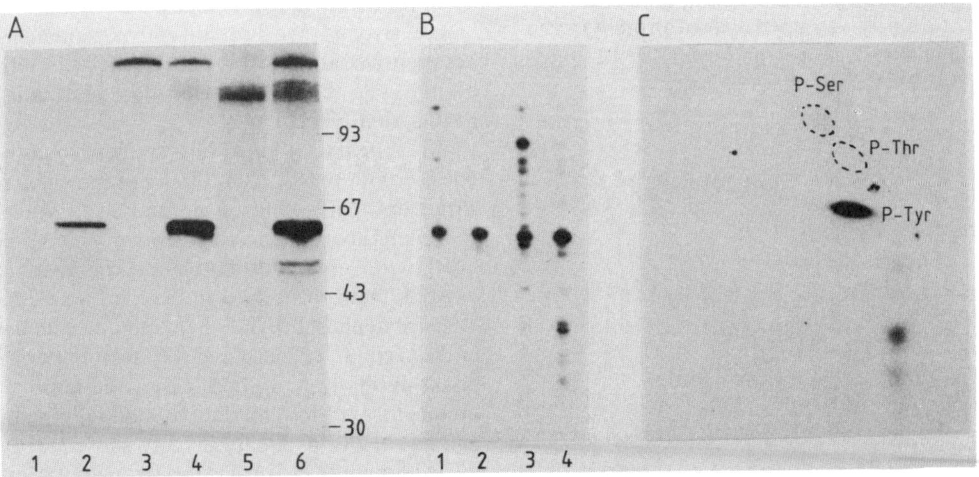

Figure 4. Kinase activity associated with p59^{fyn}. (*A*) Kinase assay. NIH-3T3 cells were lysed, immunoprecipitated as described below, and assayed for kinase activity. (Lane *1*) No specific antibody; (lane *2*) monoclonal anti-pp60$^{c\text{-}src}$; (lane *3*) anti-fyn1 preblocked with fyn1; (lane *4*) anti-fyn1; (lane *5*) anti-fyn2 preblocked with fyn2; (lane *6*) anti-fyn2. The positions of molecular-weight markers are indicated. (*B*) Peptide mapping. Bands from a kinase assay similar to the one shown in *A* were excised and digested with V8 protease as described previously (Courtneidge and Smith 1983). pp60$^{c\text{-}src}$ (lanes *1* and *2*) and pp59^{fyn} (lanes *3* and *4*) were digested with 20 (lanes *1* and *3*) and 100 (lanes *2* and *4*) μg V8 protease. (*C*) Phosphoamino acid analysis. p59^{fyn} immunoprecipitated using anti-fyn2 antibodies was labeled in a kinase assay, processed for autoradiography, and excised, and phosphopeptides were isolated and analyzed as described by Cooper et al. (1983).

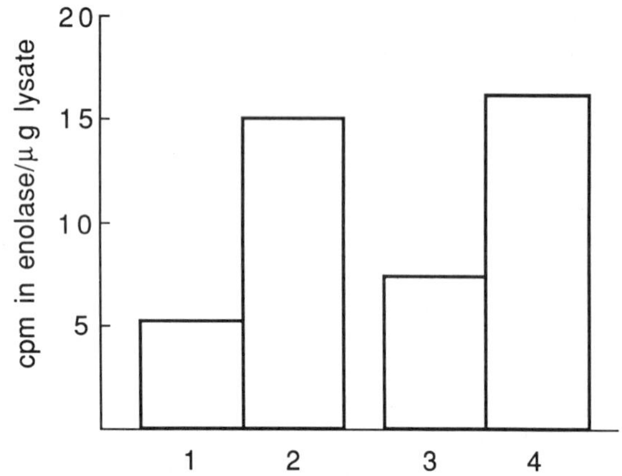

Figure 5. Kinase activities of pp60$^{c\text{-}src}$ and p59^{fyn} after dephosphorylation. Lysates were prepared in the presence (lanes *1* and *3*) or absence (lanes *2* and *4*) of sodium orthovanadate, and assayed for either pp60$^{c\text{-}src}$ (lanes *1* and *2*) or p59^{fyn} (lanes *3* and *4*) associated kinase activities.

These results strongly suggest that p59^{fyn} can also associate with middle T antigen. However, it is clear both from our unpublished observations and from a comparison of Figures 1 and 6 that little middle T antigen was detected complexed to p59^{fyn} compared to the amount complexed to pp60$^{c\text{-}src}$. It remains to be seen whether these differences reflect the in vivo situation or differences in the ability of the antibodies used to recognize complexes.

DISCUSSION

PtdIns Kinase Activity Associated with Middle T Antigen

We have previously described a third protein in the complex with middle T antigen and pp60$^{c\text{-}src}$. The presence of this protein of 81 kD is correlated with the detection of PtdIns kinase activity in middle T antigen immunoprecipitates. We have detected the p81 protein associated with transforming middle T antigens, suggesting that it may be important for transformation. However, although some of the nontransforming mutant middle T antigens do not have associated p81, others do. The fact that some middle T antigens are complexed with pp60$^{c\text{-}src}$ in the absence of p81 indicates that middle T binds directly to pp60$^{c\text{-}src}$. Since PtdIns kinase activity (Sugimoto et al. 1984) and also an 81-kD protein (S.A. Courtneidge, unpubl.) can be detected in pp60$^{v\text{-}src}$ immunoprecipitates, we believe that p81 also binds directly to the *src* protein in each case.

If p81 binding is critical to transformation by middle T antigen, then this could explain why those members of class II that lack p81 fail to transform. However, those mutants in which the complex appears wild-type remain an enigma. In our hands, none of the class II mutants are able to activate pp60$^{c\text{-}src}$ to wild-type levels, which may explain their defect. However, as already mentioned, middle T antigen is able to complex with many cell proteins, and the defect in these mutants may lie in these other complexes.

Our results show a complete correlation between detection of PtdIns kinase activity and presence of p81. Taking a different approach, using antiphosphotyrosine antibody immune complexes from PDGF-stimulated cells, Kaplan et al. (1987) have also recently demonstrated an 85-kD protein associated with PtdIns kinase activity. It seems likely that there is identity between p81 and this 85-kD protein. Since in neither of these studies were any other proteins detected, the most straightforward hypothesis is that p81 is itself the PtdIns kinase whose activity we detect. It remains possible, however, that it is instead a regulator of PtdIns kinase activity. We are currently examining the link between p81 and PtdIns kinase activity in more detail by purifying the p81 protein and assessing associated PtdIns kinase activity.

p81 appears to be a substrate for both activated pp60$^{c\text{-}src}$ and pp60$^{v\text{-}src}$, and perhaps may also be a substrate for the enzyme in normal cells. We therefore speculated that PtdIns kinase activity would be stimulated by tyrosine phosphorylation. To investigate this, we took two approaches: In the first, we examined the effect of dephosphorylation in immune complexes, and in the second, we measured the effect of dephosphorylation during lysis. In both cases, the same result was obtained: Dephosphorylation markedly reduced the amount of PtdIns kinase activity subsequently detected. Although these results are strongly suggestive that PtdIns kinase activity is positively regulated by tyrosine phosphorylation, they are not conclusive. We have to date been unable to rephosphorylate and reactivate PtdIns kinase activity in vitro; these experiments must await purified enzyme.

We next examined inositol phosphate production in normal and polyomavirus-transformed cells to determine whether the putative activation of a PtdIns kinase

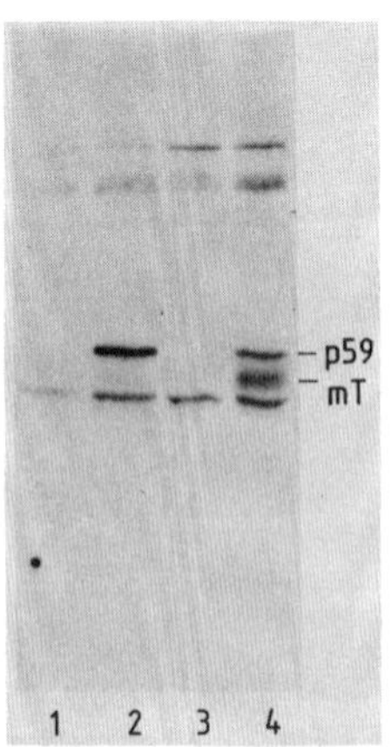

Figure 6. Analysis of p59^{fyn} in normal and middle-T-antigen-transformed cells. NIH-3T3 (lanes *1* and *2*) and middle-T-antigen-transformed NIH-3T3 (lanes *2* and *4*) cell lysates were immunoprecipitated with anti-fyn2 either preblocked with specific peptide (lanes *1* and *3*) or alone (lanes *2* and *4*), and kinase assays were performed.

would have an effect on PtdIns turnover in vivo. We could detect no differences in either the steady-state levels or the turnover rate of the inositol phosphates between the two cell types. We could show that the PtdIns pathway was functioning in the middle-T-antigen-transformed cells, however, since serum stimulation did increase the amount of inositol phosphates. These results contrast with published experiments with Abelson-virus-transformed cells, where the PtdIns pathway appears to be constitutively activated (Fry et al. 1985). Our results also contradict an earlier report of Kaplan et al. (1986), who showed increases in the basal levels of inositol phosphates in polyomavirus-infected cells. We do not understand the discrepancy between our results, although the fact that we looked at transformed cells and they looked at infected cells might be important. It is also possible that our detection methods were not sensitive enough. However, the recent demonstration by Whitman et al. (1988) that the product of the middle-T-antigen-associated PtdIns kinase is PtdIns3P, and not the more common PtdIns4P, should prompt a rethink, since PtdIns3P is not an intermediate of the PtdIns signaling pathway as described at present. Perhaps it acts as a precursor for the production of other molecules or as a regulator of PtdIns turnover. A more detailed comparison of inositol-containing compounds in normal and polyomavirus-transformed cells may be informative.

Another Tyrosine Kinase That Can Associate with Middle T Antigen

Fibroblasts express not just $pp60^{c\text{-}src}$, but also other members of the *src* family of tyrosine kinases. We are particularly interested in these proteins, especially in defining how they differ from and/or are similar to $pp60^{c\text{-}src}$. We have begun to approach this project by raising antisera that can recognize the product of the closely related *fyn* gene. We have succeeded in generating antibodies to two different synthetic peptides predicted from the sequence of the *fyn* gene. These antibodies both recognize the same 59-kD phosphoprotein, which we believe to be the authentic product of the *fyn* gene. A full description of these antibodies will be published elsewhere (Kypta et al. 1988). As shown in this communication, $p59^{fyn}$ immune complexes contained tyrosine kinase activity. The high degree of homology between *fyn* and *src*, together with the observation that the anti-fyn immunoprecipitable kinase activity sediments at about 60 kD on sucrose density gradients (R.M. Kypta unpubl.), suggests that this kinase activity is intrinsic to $p59^{fyn}$.

The tyrosine kinase activity of $pp60^{c\text{-}src}$ is down-regulated by phosphorylation at Tyr-527 (Courtneidge 1985). That this down-regulation is important in normal growth control is suggested by two observations: Mutation of Tyr-527 to phenylalanine generates a transforming protein (Cartwright et al. 1987; Kmiecik and Shalloway 1987; Piwnica-Worms et al. 1987), and masking of this site by middle T antigen binding also appears to be required for transformation (Cartwright et al. 1986). There is a tyrosine in an equivalent position (Tyr-531) in the *fyn* protein, suggesting that $p59^{fyn}$ kinase activity may be similarly regulated. Indeed, we show here that dephosphorylation of $p59^{fyn}$ at tyrosine also activates its kinase activity. Future studies will seek to determine whether the tyrosine being affected is indeed at position 531, and how perturbation of this phosphorylation site might affect the growth control of the cells.

Middle T antigen is able to bind to both $pp60^{c\text{-}src}$ and $pp62^{c\text{-}yes}$ proteins. Cheng et al. (1988) have recently shown that the binding of middle T antigen to $pp60^{c\text{-}src}$ involves residues near the tyrosine at position 527, although it is not yet clear whether other regions of $pp60^{c\text{-}src}$ are also in contact with middle T antigen. Since at least one recognition site occurs in the region common to all members of the *src* family, it may be that all members will bind middle T antigen. Indeed, we show here that $p59^{fyn}$ appears to also bind middle T antigen. However, it is clear from these and other studies that the amount of middle T antigen complexed to $p59^{fyn}$ is very low compared to that complexed with $p60^{c\text{-}src}$. In addition, we found little or no overall activation of $p59^{fyn}$ kinase activity in polyomavirus-transformed cells, under conditions where $p60^{c\text{-}src}$ was activated approximately 20-fold (Kypta et. al. 1988). It may be that the complexed form of $p59^{fyn}$ is poorly recognized by the antibodies we use, although we find this unlikely, since antibodies that bind to a very similar region of $pp60^{c\text{-}src}$ are able to recognize the complex, and we do detect some complexed $p59^{fyn}$ with these antibodies. An alternative, and in our view more likely, explanation is that middle T antigen does not bind equally well to all members of the family. Our current experiments address this issue.

ACKNOWLEDGMENTS

We thank Julie Childs and Tony Magee for the *fyn* peptides, Anna Hemming for help with the generation of anti-*fyn* antibodies, and our colleagues for advice and discussion.

REFERENCES

Berridge, M.J. 1983. Rapid accumulation of inositol trisphosphate reveals that agonists hydrolyse polyphosphoinositides instead of phosphatidylinositol. *Biochem. J.* **212:** 849.

———. 1984. Inositol trisphosphate and diacylglycerol as second messengers. *Biochem. J.* **220:** 345.

Bolen J.B., C.J. Thiele, M.A. Israel, W. Yonemoto, L.A. Lipsich, and J. S. Brugge. 1984. Enhancement of cellular *src* gene product associated tyrosyl kinase activity following polyoma virus infection and transformation. *Cell* **38:** 767.

Cartwright, C.A., M.A. Hutchinson, and W. Eckhart. 1985. Structural and functional modification of $pp60^{c\text{-}src}$ associated with polyoma middle tumor antigen from infected or transformed cells. *Mol. Cell. Biol.* **5:** 2647.

Cartwright, C.A., W. Eckhart, S. Simon, and P.L. Kaplan.

1987. Cell transformation by pp60$^{c\text{-}src}$ mutated in the carboxy-terminal regulatory domain. *Cell* **49:** 83.

Cartwright, C.A., P.L. Kaplan, J.A. Cooper, T. Hunter, and W. Eckhart. 1986. Altered sites of tyrosine phosphorylation in pp60$^{c\text{-}src}$ associated with polyoma virus middle tumor antigen. *Mol. Cell. Biol.* **6:** 1562.

Cheng, S.H., H. Piwnica-Worms, R.W. Harvey, T.M. Roberts, and A.E. Smith. 1988. The carboxy terminus of pp60$^{c\text{-}src}$ is a regulatory domain and is involved in complex formation with the middle T antigen of polyomavirus. *Mol. Cell. Biol.* **8:** 1736.

Cooper, J.A. 1988. The src-family of protein-tyrosine kinases. In *Peptides and protein phosphorylation* (eds. B. Kemp and P.F. Alewood). CRC Press, Boca Raton, Florida.

Cooper, J.A. and C.S. King. 1986. Dephosphorylation or antibody binding to the carboxy terminus stimulates pp60$^{c\text{-}src}$. *Mol. Cell. Biol.* **6:** 4467.

Cooper, J.A., B.M. Sefton, and T. Hunter. 1983. Detection and quantification of phosphotyrosine in proteins. *Methods Enzymol.* **99:** 387.

Cooper, J.A., K.L. Gould, C.A. Cartwright, and T. Hunter. 1986. Tyr^{527} is phosphorylated in pp60$^{c\text{-}src}$: Implications for regulation. *Science* **231:** 1431.

Courtneidge, S.A. 1985. Activation of pp60$^{c\text{-}src}$ kinase by middle T antigen binding or by dephosphorylation. *EMBO J.* **4:** 1471.

———. 1986. Transformation by polyoma virus middle T antigen. *Cancer Surv.* **5:** 173.

Courtneidge, S.A. and A. Heber. 1987. An 81 kd protein complexed with middle T antigen and pp60$^{c\text{-}src}$: A possible phosphatidylinositol kinase. *Cell* **50:** 1031.

Courtneidge, S.A. and A.E. Smith. 1983. Polyoma virus transforming protein associates with the product of the c-*src* gene. *Nature* **303:** 435.

———. 1984. The complex of polyoma virus middle T antigen and pp60$^{c\text{-}src}$. *EMBO J.* **3:** 585.

Courtneidge, S.A., B. Oostra, and A.E. Smith. 1984. Tyrosine phosphorylation and polyoma virus middle T protein. *Cancer Cells* **2:** 123.

Coutneidge, S.A., M. Read, J.B. Wilson, M. Fried. 1988. Cytoplasmic interaction between pp60$^{c\text{-}src}$ and a truncated polyoma virus middle T antigen. *Oncogene Res.* (in press).

Coussens, P.M., J.A. Cooper, T. Hunter, and D. Shalloway. 1985. Restriction of the in vitro and in vivo tyrosine protein kinase activities of pp60$^{c\text{-}src}$ relative to pp60$^{v\text{-}src}$. *Mol. Cell. Biol.* **5:** 2753.

Fry, M.J., A. Gebhardt, P.J. Parker, and J.G. Foulkes. 1985. Phosphatidylinositol turnover and transformation of cells by Abelson murine leukaemia virus. *EMBO J.* **4:** 3171.

Grussenmeyer, T., A. Carbone-Wiley, K.H. Scheidtmann, and G. Walter. 1987. Interaction between polyoma virus medium T antigen and three cellular proteins of 88, 61, and 37 kilodalton. *J.Virol.* **61:** 3902.

Grussenmeyer, T., K.H. Scheidtmann, M.A. Hutchinson, W. Eckhart, and G. Walter. 1985. Complexes of polyoma virus medium T antigen and cellular proteins. *Proc. Natl. Acad. Sci.* **82:** 7952.

Iba, H., F.R. Cross, E.A. Garber, and H. Hanafusa. 1985. Low level of cellular protein phosphorylation by nontransforming overproduced pp60$^{c\text{-}src}$. *Mol. Cell. Biol.* **5:** 1058.

Kaplan, D.R., M. Whitman, B. Schaffhausen, D.C. Pallas, M. White, L. Cantley, and T.M. Roberts. 1987. Common elements in growth factor stimulation and oncogenic transformation: 85 kd Phosphoprotein and phosphatidylinositol kinase activity. *Cell* **50:** 1021.

Kaplan, D.R., M. Whitman, B. Schaffhausen, L. Raptis, R.L. Garcea, D. Pallas, T.M. Roberts, and L. Cantley. 1986. Phosphatidylinositol metabolism and polyoma-mediated transformation. *Proc. Natl. Acad. Sci.* **83:** 3624.

Kawakami, T., C.Y. Pennington, and K.C. Robbins. 1986. Isolation and oncogenic potential of a novel human *src*-like gene. *Mol. Cell. Biol.* **6:** 4195.

Kmiecik, T.E., and D. Shalloway. 1987. Activation and suppression of pp60$^{c\text{-}src}$ transforming ability by mutation of its primary sites of tyrosine phosphorylation. *Cell* **49:** 65.

Kornbluth, S., M. Sudol, and H. Hanafusa. 1987. Association of the polyomavirus middle T antigen with c-*yes* protein. *Nature* **325:** 171.

Kypta, R.M., A. Hemming, and S.A. Courtneidge. 1988. Identification and characterization of p59^{fyn} (src-like protein tyrosine kinase) in normal and polyoma virus in transformed cell. *EMBO J.* **12:** (in press).

Laudano, A.P., and J.M. Buchanan. 1986. Phosphorylation of tyrosine in the carboxy-terminal tryptic peptide of pp60$^{c\text{-}src}$. *Proc. Natl. Acad. Sci.* **83:** 892.

Piwnica-Worms, H., K.B. Saunders, T.M. Roberts, A.E. Smith, and S.H. Cheng. 1987. Tyrosine phosphorylation regulates the biochemical and biological properties of pp60$^{c\text{-}src}$. *Cell* **49:** 75.

Schaffhausen, B.S. and T.L. Benjamin. 1981. Protein kinase activity associated with polyoma virus middle T antigen. *Cold Spring Harbor Conf. Cell Proliferation* **8:** 1281.

Semba, K., M. Nishizawa, N. Miyajima, M.C. Yoshida, J. Sukegawa, Y. Yamanashi, M. Sasaki, T. Yamamoto, and K. Toyoshima. 1986. *yes*-related protooncogene, *syn*, belongs to the protein-tyrosine kinase family. *Proc. Natl. Acad. Sci.* **83:** 5459.

Smith, A.E. and B.K. Ely. 1983. The biochemical basis of transformation by polyoma virus. *Adv. in Viral Oncol.* **3:** 3.

Sugimoto, Y., M. Whitman, L.C. Cantley, and R.L. Erikson. 1984. Evidence that the Rous sarcoma virus transforming gene product phosphorylates phosphatidylinositol and diacylglycerol. *Proc. Natl. Acad. Sci.* **81:** 2117.

Tooze, J., ed. 1981. *Molecular biology of tumor viruses*, 2nd edition, revised: *DNA tumor viruses*. Cold Spring Harbor Laboratory, Cold Spring Harbor, New York.

Walter, G., M.A. Hutchinson, T. Hunter, and W. Eckhart. 1982. Purification of polyoma virus middle T tumor antigen by immunoaffinity chromatography. *Proc. Natl. Acad. Sci.* **82:** 4568.

Whitman, M., C.P. Downes, M. Keeler, T. Keller, and L. Cantley. 1988. Type I phosphatidylinositol kinase makes a novel phospholipid, phosphatidylinositol-3-phosphate. *Nature* **332:** 644.

Whitman, M., D.R. Kaplan, B. Schaffhausen, L. Cantley, and T.M. Roberts. 1985. Association of phosphatidylinositol kinase activity with polyoma middle T competent for transformation. *Nature* **315:** 239.

Tyrosine Phosphorylation in Signal Transduction

T.M. ROBERTS,* D. KAPLAN,* W. MORGAN,* T. KELLER,* H. MAMON,* H. PIWNICA-WORMS,* B. DRUKER,* B. COHEN,† B. SCHAFFHAUSEN,† M. WHITMAN,* L. CANTLEY,† U. RAPP,‡ AND D. MORRISON*

**Division of Molecular and Cellular Biology, Dana Farber Cancer Institute, Harvard Medical School, Boston, Massachusetts 02115; †Department of Physiology, Tufts University Medical School, Boston, Massachusetts 02115; ‡National Cancer Institute, Frederick, Maryland 21701*

Our recent work has focused on the elucidation of the mechanisms by which membrane-bound tyrosine kinases transmit signals within the cell. To examine the role of tyrosine phosphorylation, we have employed the following strategy. First, we have utilized antibodies to phosphotyrosine (anti-P.Tyr) to identify candidate substrates of various tyrosine kinases, such as pp60$^{c\text{-}src}$, the CSF-1 receptor, or the platelet-derived growth factor (PDGF) receptor (Piwnica-Worms et al. 1987; Morrison et al. 1988). Second, we have attempted to characterize the biochemical properties of the putative substrates and to determine in what manner these properties are modified by phosphorylation on tyrosine residues (Kaplan et al. 1986, 1987; Piwnica-Worms et al. 1987; Morrison et al. 1988 and in prep.). In this endeavor, we are recapitulating the classic biochemical analysis used to study the effect of kinases on metabolism. The final portion of our work consists of using modern molecular biological strategies to clone the genes or cDNAs for the substrates and overproduce the relevant proteins for studies in vitro in defined systems. We can then confirm hypotheses based on biochemical experiments by making the appropriate mutations in the genes involved and studying the effects of such mutations in living cells (Piwnica-Worms et al. 1987, in prep., pers. comm.; Cheng et al. 1988). This paper describes the first and second aspects of our strategy, the identification and characterization of novel substrate molecules.

MATERIALS AND METHODS

Cells and mitogens. BALB/c-3T3 clone A31 (from C.D. Scher), v-*src*-transformed NIH-3T3 cells (Piwnica-Worms et al. 1987), polyomavirus middle-T-antigen-transformed NIH-3T3 cells (Cherington et al. 1986), v-*fms*-transformed NIH-3T3 cells (Morrison et al. 1988), v-*sis*-transformed BALB/c-3T3 cells (from C.D. Stiles), c-Ha-*ras*-transformed NIH-3T3 cells (activated for transformation by mutation at codon 61) (from M. Corbley), 45b cells (NIH-3T3 cells transfected with MMTV promoter-polyomavirus early region DNA) (Kaplan et al. 1986), BALB/c-3T3 cells transfected with MMTV promoter-c-*myc* DNA (Armelin et al. 1984), v-*fos*-transformed Rat-1 cells (from T. Curran), and polyomavirus large-T-antigen-expressing NIH-3T3 cells (D. Kaplan, unpubl.) were maintained in Dulbecco's modified Eagle's medium supplemented with 10% (v/v) newborn bovine serum. The polyoma middle T antigen, v-*src*-, v-*fms*-, c-*myc*-, v-*fos*-, and activated c-Ha-*ras*-expressing cell lines are all capable of anchorage-independent growth (Cherington et al. 1986; Kaplan et al. 1986; Piwnica-Worms et al. 1987; Zarbl et al. 1987; Morrison et al. 1988; B. Rollins and C.D. Stiles; M. Corbley; both pers. comm.). The v-*sis*-transformed cells are capable of forming foci on monolayers of 3T3 cells (D. Kaplan, unpubl.). PDGF and epidermal growth factor (EGF) (receptor grade) were obtained from Collaborative Research Inc. PDGF and fibroblast growth factor (FGF) were used at a concentration of 10 half-maximal units/ml, except where noted (1 unit induces half-maximal stimulation of [^{3}H]thymidine uptake by 3T3 cells). Acidic FGF (0.4 ng protein/unit) was obtained from P.A. D'Amore. Bovine pancreatic insulin, PMA, PBA, and dibutyryl cAMP were obtained from Sigma, and platelet-poor plasma was obtained from C.D. Stiles.

Radiolabeling of cells and immunoprecipitation. For ^{32}P labeling, cells were incubated at 37°C for 3 hours in phosphate-free medium supplemented with 2 mCi/ml [^{32}P]orthophosphate and 2% (v/v) platelet-poor plasma. For growth factor experiments, mitogens were added directly to the labeling medium. Assays were terminated by the addition of ice-cold phosphate-buffered saline (PBS) containing 100 μM sodium orthovanadate. Cell lysates were prepared as described previously (Kaplan et al. 1987), the lysates were normalized for protein content, and labeled proteins were immunoprecipitated for 3 hours at 4°C with anti-phosphotyrosine serum (White et al. 1985) or with anti-c-*raf* (anti-SP63) serum (Schultz et al. 1985). Immune complexes were precipitated with protein-A–Sepharose beads (Sigma), washed as described previously (Kaplan et al. 1987), and analyzed by SDS-PAGE. Phosphoproteins were visualized by autoradiography.

Phosphatase analysis. c-*raf* molecules were immunoprecipitated and washed as described above. The immunoprecipitated protein was treated with potato acid phosphatase (Cooper and King 1986) at 30°C for 20 minutes using 4 μg of phosphatase per 40 μl reaction. Reactions were terminated by washing in PBS containing 0.1% NP-40, 2 mM EDTA, and 100 μm sodium orthovanadate and by boiling in gel electrophoresis sample buffer.

Peptide mapping and phosphoamino acid analysis. ^{32}P-labeled proteins were resolved on SDS-polyacrylamide gels, extracted, and subjected to trypsin digestion as described previously (Pallas and Solomon 1982). Labeled peptides were separated in two dimensions on 10-cm thin-layer cellulose plates (EM Laboratories) by electrophoresis at pH 1.9 (formic acid:acetic acid:water; 1:3:16) for 20 minutes at 1000 V, followed by ascending chromatography as described previously (Hunter and Sefton 1980). Phosphoamino acid analysis of phosphorylated proteins was performed as described previously (Pallas and Solomon 1982).

Protein kinase assays in vitro. Immune complexes were prepared as described above. Immunoprecipitated c-*raf* protein was then assayed for the ability to phosphorylate histone type V-S in vitro as described previously (Moelling et al. 1984). Protein kinase assays were linear over the time of the assay.

Immunoblot analysis. For cytoplasmic protein extracts, 2×10^6 cells were rinsed twice with cold PBS and lysed in 400 μl of NP-40 lysis buffer (20 mM Tris [pH 8.0], 137 mM NaCl, 10% glycerol, 1% NP-40) containing 1 mM phenylmethylsulfonyl fluoride, 0.15 units/ml aprotinin, 10 mM EDTA, and 1 mM sodium vanadate at 4°C for 10 minutes. Insoluble material was removed by centrifugation at 4°C for 15 minutes at 10,000*g*. Total protein content of the lysate was determined by the Bio-Rad laboratories protein assay. Gel sample buffer (70 μl) (33% glycerol, 0.3 M dithiothreitol, 6.7% SDS and bromphenol blue) was added, and the extracts were heated to 100°C for 5 minutes prior to electrophoresis.

Lysates (150 μg/cm lane) were analyzed by SDS-PAGE using either 7.5% or 15% SDS-polyacrylamide gels. Proteins were electrophoretically transferred from the gel onto a 0.2-μm nitrocellulose filter (Schleicher and Schuell) in a buffer containing 25 mM Tris, 192 mM glycine, and 20% methanol at 0.4 mA for 4 hours at 4°C. Residual binding sites on the filter were blocked by incubating the nitrocellulose in TBS (10 mM Tris [pH 8.0]; 150 mM NaCl) containing 2% BSA for 1 hour at 25°C. All subsequent reactions were done at 25°C. The blots were then washed twice for 5 minutes in TBST (TBS with 0.05% Tween-20). Blots were incubated overnight with either anti-P.Tyr antibody or anti-c-*raf* antibody diluted 1:2000 in TBST. The primary antibody was removed, and the blots were washed three times for 5 minutes in TBST. To detect antibody reactions, the blots were incubated for 45–90 minutes with either anti-rabbit or anti-mouse alkaline phosphatase conjugated IgG diluted 1:2000 in TBST, washed three times in TBST, and then placed in a buffer containing 100 mM Tris (pH 9.5), 100 mM NaCl, 5 mM $MgCl_2$, 330 μg/ml 5-bromo-4-chloro-3-indolyl phosphate (BCIP), and 150 μg/ml nitro blue tetrazolium (NBT) for 5–60 minutes. Enzymatic color development was stopped by rinsing the filters in deionized water.

Phosphatidylinositol assays in vitro. Immune complex phosphatidylinositol (PtdIns) assays were conducted as described previously (Whitman et al. 1985, 1988; Kaplan et al. 1986). PtdIns assays were linear over the time of the assay.

RESULTS

Surveying Substrates of Activated Tyrosine Kinases with Anti-P.Tyr Sera

The use of anti-P.Tyr sera to study substrates of tyrosine kinases was pioneered by Frackelton and Williams (Frackelton et al. 1984), White (White et al. 1985), and Wang (Morla and Wang 1986). To identify targets of tyrosine kinases, we have utilized one-dimensional immunoblot analysis using anti-P.Tyr antibodies to examine cells expressing oncogene-encoded tyrosine kinases and cells in which growth factor receptors with intrinsic tyrosine kinase activity have been activated by their ligands. This method is considerably more sensitive than techniques in which tyrosine phosphorylation is detected by two-dimensional gel analysis of phosphate-labeled cell lysates. In our survey, we found that each activated tyrosine kinase gave a distinctive "fingerprint" of tyrosine phosphorylated proteins (Morrison et al. 1988). However, the pattern of apparent substrates was similar in certain cases. For example, the anti-P.Tyr sera recognized a similar set of proteins in lysates of v-*fms*- and v-*sis*-transformed cells and in lysates of monocytes and fibroblasts in which the CSF-1 and PDGF receptors had been activated, respectively, suggesting a potential common set of targets for the two receptors and oncogenes (Fig. 1).

We are currently attempting to identify these "substrate" molecules. One route to identifying these molecules involves biochemical purification of the individual substrates followed by peptide sequencing and molecular cloning of the genes that encode them. This type of analysis is in progress in our laboratory but has not yet yielded significant results. Another method of analysis has already proved fruitful and will be discussed here. In this case, we have examined immunoprecipitates made with anti-P.Tyr sera for enzymatic activities that we have reason to believe might be present. The logic is that to transmit signals effectively, the receptor must phosphorylate and modify the activity of cellular enzymes involved in mediating intracellular signaling (such as the proteins modulating PtdIns turnover and serine phosphorylation). We have studied in some detail two such activities, a PtdIns kinase, which we have tentatively identified as an 85-kD protein, and a serine kinase, which seems to be identical to the protein product of the proto-oncogene c-*raf* (Rapp et al. 1983; Moelling et al. 1984; Bonner et al. 1986).

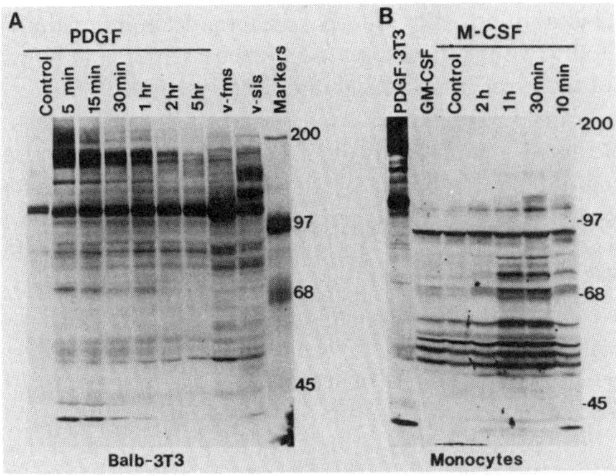

Figure 1. Proteins recognized by the anti-P.Tyr antibody in various cell lines. The extracted proteins were separated electrophoretically on 7.5% SDS-polyacrylamide gels and transferred to nitrocellulose filters. The blotted proteins were then probed with the anti-P.Tyr antibody. Molecular-weight markers are expressed in kilodaltons. (*A*) Cytoplasmic cell lysates were prepared from BALB/c-3T3 cells untreated (control), PDGF stimulated for 5 min, 15 min, 30 min, 1 hr, 2 hr, or 5 hr, and v-*sis* and v-*fms* cell lines. (*B*) Cytoplasmic cell lysates were prepared from BALB/c-3T3 cells stimulated for 5 min with PDGF, human monocytes untreated (control), GM-CSF stimulated for 10 min, M-CSF stimulated for 10 min, 30 min, 1 hr, or 2 hr. The substrate development time for the blots was 5 min.

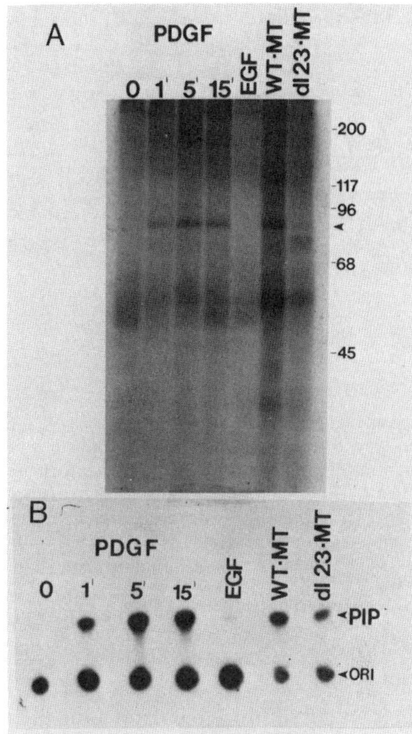

Figure 2. 85-kD phosphorylation of PtdIns kinase activities assayed in anti-P.Tyr immunoprecipitates from PDGF-stimulated and MTAg-expressing cells. Time course of the appearance of 85-kD phosphoprotein and PtdIns kinase activity in PDGF-stimulated quiescent BALB/c-3T3 cells as assayed in the anti-P.Tyr immunoprecipitates in vitro. Cells were treated for 0, 1, 5, or 15 min with 10 units/ml PDGF or 15 min with 10 units/ml FGF at 37°C. Cells expressing wild-type MTAg (WT-MT) or the *dl*23 mutant MTAg (*dl*23-MT) were also assayed. Proteins were analyzed by 10% SDS-PAGE and PtdIns kinase reactions by thin-layer chromatography (TLC). (PIP) Phosphatidylinositol 3-phosphate. (ORI) Origin.

The 85-kD Protein, a Candidate PtdIns Kinase

Some years ago, we identified a PtdIns kinase activity in immunoprecipitates of polyoma middle T antigen (MTAg), the presence of which appeared to correlate with the transformation competence of the MTAg gene (Kaplan et al. 1986). Subsequent work from our laboratory and others revealed that the activity was not intrinsic to MTAg nor to pp60$^{c\text{-}src}$, the cellular tyrosine kinase found in a complex with MTAg (MacDonald et al. 1985; Sugimoto and Erikson 1985; Kaplan et al. 1986; Piwnica-Worms et al. 1986). We hypothesized that the PtdIns kinase was a substrate for those pp60$^{c\text{-}src}$ molecules that had been activated as tyrosine kinases by means of their association with MTAg. The availability of anti-P.Tyr antibodies allowed us to test this hypothesis. We found that the PtdIns kinase activity was indeed immunoprecipitated by the P.Tyr-specific serum and that the presence of the PtdIns kinase correlated with the presence of an 85-kD protein (Fig. 2) (Courtneidge and Heber 1987; Kaplan et al. 1987). This same 85-kD protein associates with a number of activated tyrosine kinases such as the PDGF receptor (when stimulated by ligand), pp60$^{v\text{-}src}$, the MTAg/pp60$^{c\text{-}src}$ complex, and v-*fms* (Kaplan et al. 1987 and unpubl.).

Our current hypothesis is that the 85-kD protein is phosphorylated on tyrosine and activated as a PtdIns kinase by activated tyrosine kinases. Consistent with this hypothesis, PtdIns kinase activity is enhanced 100-fold in anti-P.Tyr immunoprecipitates from cells treated with PDGF, 20-fold in immunoprecipitates from MTAg-transformed cells, and 10-fold in immunoprecipitates from v-*fms*-transformed cells. The PtdIns kinase activity is unique in that it produces 1,3-PtdInsP, rather than the classic 1,4-PtdInsP, thereby generating a new PtdIns metabolite that may perform a novel second-messenger function (Fig. 3) (Whitman et al. 1987, 1988). The same kinase activity will apparently phosphorylate the 3 position on the inositol ring of 1,4-PtdInsP, yielding 1,3,4-PtdInsP$_2$ (L. Serunian and L. Cantley, unpubl.). The role(s) of these metabolites in cellular physiology is not known. However, genetic experiments using MTAg, v-*fms*, and c- and v-*src* indicate that this PtdIns kinase activity is necessary for transformation. Thus, we observe a close correlation between the ability of the oncogene products to elicit cellular transformation and their ability to associate with the PtdIns kinase in anti-P.Tyr immunoprecipitates. However, an MTAg mutant (*dl*1015) exists that is significantly defective in transformation but exhibits wild-type levels of PtdIns kinase in anti-P.Tyr immuno-

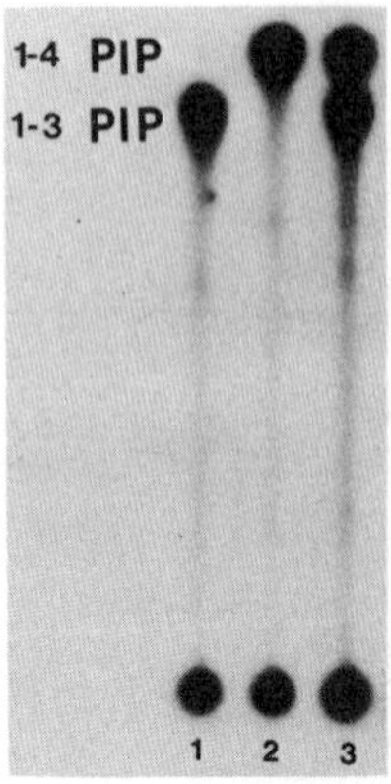

Figure 3. Type I PtdIns kinase produces a novel product. Type I and type II PtdIns kinases were prepared as described by Whitman et al. (1988) and used to phosphorylate PtdIns with [^{32}P]ATP. The reaction products were extracted and separated by TLC. Separations were done as described previously, except that the TLC plates were treated with 50 mM $CoCl_2$ (T. Keller and M. Whitman, unpubl.) to enhance separation. (Lane *1*) Product of type I PtdIns kinase. (Lane *2*) Product of type II PtdIns kinase. (Lane *3*) Mixture of the two products.

preciptitates prepared from cells expressing the mutant protein (Courtneidge and Heber 1987; Morgan et al. 1988). Therefore, this association may be necessary but is not sufficient to induce transformation. When the 85-kD protein was immunoprecipitated using anti-P.Tyr sera from PDGF-treated cells labeled in vivo with [^{32}P]orthophosphate, the protein was found to be phosphorylated on both tyrosine and serine residues. The 85-kD protein is also phosphorylated in roughly equivalent amounts on both tyrosine and serine residues when labeled in vitro by incubating anti-P.Tyr immunoprecipitates with Mn^{++} and [γ-^{32}P]ATP. This result indicates that one or more serine kinase activities are also present in the anti-P.Tyr immunoprecipitates.

The c-*raf* Protein Is Phosphorylated and Activated in Response to Membrane Signals

One source of serine-threonine kinase activity observed in anti-P.Tyr immunoprecipitates from cells expressing any of several activated, membrane-bound tyrosine kinases was identified as the c-*raf* protein, the cellular counterpart of the serine/threonine kinase v-Raf. The idea that c-Raf might be a tyrosine kinase substrate first became apparent from our studies of v-*fms*-transformed fibroblasts. Anti-*raf*-specific antibodies and anti-P.Tyr sera both immunoprecipitated proteins of similar molecular mass (roughly 74 kD) from lysates of v-*fms*-transformed cells. Proteins of similar mobility were detected in anti-P.Tyr immunoblots of lysates prepared from v-*src*-transformed cells and PDGF-stimulated cells. Quiescent cells stimulated with PDGF and v-*src*-transformed cells were incubated with [^{32}P]orthophosphate, and labeled proteins were immunoprecipitated with either the anti-c-*raf* antibody or anti-P.Tyr antibody. As seen in Figure 4A, the anti-P.Tyr antibody did not recognize a protein in unstimulated 3T3 cells that migrated with the same mobility as did the c-*raf* molecules found in these cells. However, in both PDGF-stimulated and v-*src*-transformed cells, the anti-P.Tyr antibody did precipitate a protein whose migration corresponded with at least some of the c-*raf* species found in these cells. To determine if the c-*raf* protein was indeed being phosphorylated on tyrosine residues, two methods of analysis were used. First, two-dimensional tryptic phosphopeptide mapping revealed that the proteins immunoprecipitated by the anti-P.Tyr and anti-c-*raf* sera from the v-*src*-transformed cells and by the anti-c-*raf* sera from unstimulated control cells shared a similar set of ^{32}P-labeled peptides (Fig. 4B). In the case of the comparison anti-*raf* band from control cells and the anti-P.Tyr band from v-*src* cells, the two sets of peptides are essentially the same. The identity of this set of peptides was confirmed by mixing ^{32}P-labeled peptides immunoprecipitated by the anti-P.Tyr and anti-c-*raf* sera from the v-*src*-transformed cells (data not shown). In addition, this set of peptides was observed in the tryptic maps of the c-*raf* protein immunoprecipitated from v-*src*-transformed cells and PDGF-treated cells (indicated with small arrowheads in Fig. 4B). The tryptic peptide maps also revealed that c-*raf* molecules from the PDGF-stimulated cells contained additional ^{32}P-containing peptides that were not observed in control or v-*src*-transformed cells (indicated with large arrowheads in Fig. 4B). Phosphoamino acid analysis demonstrated that the population of c-*raf* molecules recognized by the anti-P.Tyr serum was phosphorylated on both tyrosine and serine residues (Fig. 4C). However, the amount of ^{32}P-labeled c-*raf* protein in the anti-P.Tyr immunoprecipitates represented only a fraction of the total ^{32}P-labeled c-*raf* molecules observed in the anti-c-*raf* immunoprecipitates. In anti-c-*raf* immunoprecipitates from the PDGF-stimulated cells, we detected threonine phosphorylation and large increases in serine phosphorylation as compared to the quiescent 3T3 cells (Fig. 4C). Similar increases in serine and threonine phosphorylation were also observed in anti-c-*raf* immunoprecipitates from v-*src* cells (data not shown). We were unable to detect tyrosine phosphorylated residues on the c-*raf* protein precipitated with the anti-c-*raf* serum in these experiments. However, when cells overproducing the c-*raf* protein (tenfold overproduction) were stimulated with PDGF, a 74-kD protein was recognized by the anti-P.Tyr antibody when anti-c-*raf* immunoprecipitates were probed in immunoblot analysis (Fig. 4D). Furthermore, in immunoblot analysis, a 74-kD protein was recognized when anti-P.Tyr immunoprecipitates from v-*src*-transformed cells were probed on an immunoblot with anti-c-*raf* antibody (data not shown). Although our results indicate that a subpopulation of c-*raf* molecules are phosphorylated on tyrosine residues in response to PDGF stimulation and v-*src* transformation, we did not detect a band comigrating with c-*raf* in anti-P.Tyr immunoprecipitates from PMA-, FGF-, EGF-treated 3T3 cells or in

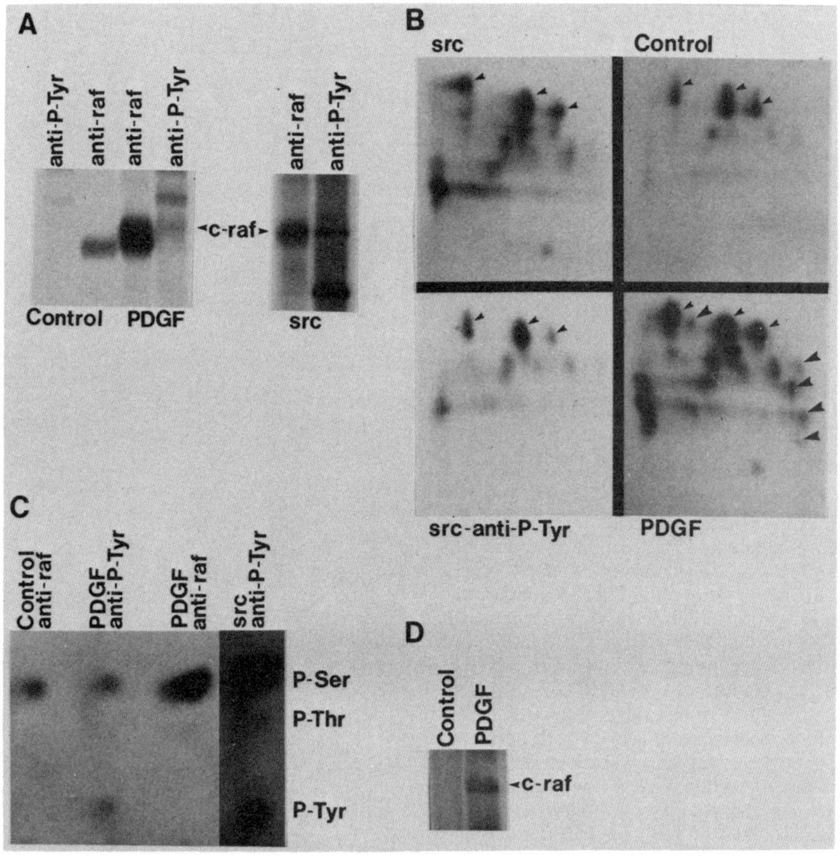

Figure 4. Enhancement of serine, threonine, and tyrosine phosphorylation in c-*raf* molecules from PDGF-treated and v-*src*-transformed cells. Cells were labeled with [^{32}P]orthophosphate for 3 hr. v-*src*-transformed cells, PDGF-treated cells (10 units/ml for 5 min), and quiescent cells were lysed and immunoprecipitated with either anti-c-*raf* antibody or an antibody that specifically recognizes phosphotyrosine (P.Tyr) residues. (*A*) Anti-P.Tyr and anti-c-*raf* immunoprecipitates from quiescent cells (control), PDGF-treated cells (PDGF), or v-*src*-transformed (src) cells. (*B*) Two-dimensional tryptic peptide mapping of phosphoproteins immunoprecipitated with anti-c-*raf* antibody from quiescent cells (control), PDGF-treated cells (PDGF), or v-*src*-transformed (src) cells. The panel labeled src-anti-P.Tyr is a map of the phosphoprotein comigrating with c-*raf* in anti-P.Tyr immunoprecipitates from v-*src*-transformed cells. The small arrowheads indicate common prominent phosphopeptides. The large arrowheads indicate phosphopeptides unique to c-*raf* molecules from PDGF-treated cells. Peptides were prepared as described in Materials and Methods and analyzed on thin-layer cellulose plates by electrophoresis (vertical axis) followed by chromatography (horizontal axis). (*C*) Phosphoamino acid analysis of the c-*raf* molecules immunoprecipitated by the anti-c-*raf* and anti-P.Tyr sera. (P-Ser) Phosphoserine. (P-Thr) Phosphothreonine. (P-Tyr) Phosphotyrosine. The phosphoamino acid analysis of the c-*raf* protein immunoprecipitated by anti-c-*raf* antibody was from 3×10^6 cells, whereas the analysis of c-*raf* protein immunoprecipitated by anti-P.Tyr antibody was from 10^7 cells. The exposure time was 3 days. (*D*) The c-*raf* protein was directly immunoprecipitated with anti-*raf* antibody from cells overproducing the c-*raf* protein (tenfold overproduction). Cells were either untreated (control) or treated with PDGF (10 units/ml) for 5 min. The *raf* immunoprecipitates were electrophoretically separated on a 7.5% polyacrylamide-SDS gel, transferred to a nitrocellulose filter, and probed with anti-P.Tyr antibody. The development time for the immunoblot was 2 min.

activated c-Ha-*ras*-transformed 3T3 cells (data not shown).

An additional observation from the above experiments was that the electrophoretic mobility of c-*raf* protein changed quickly in response to PDGF treatment (Fig. 4A). To pursue this observation, we examined immunoblots of whole-cell lysates with anti-*raf* sera to detect other perturbations in c-*raf* mobility. First, the c-*raf* protein was examined in 3T3 cells transformed by the v-*fms* and v-*src* oncogene products using immunoblot analysis. In this experiment, cytoplasmic cell lysates were prepared, and the proteins were separated on 7.5% SDS-PAGE before electrophoretic transfer to a nitrocellulose membrane. The c-*raf* protein was then detected using an antiserum (anti-SP63) that specifically recognizes the last 12 carboxy-terminal amino acids of the c-*raf* protein (Schultz et al. 1985). The binding of anti-SP63 to c-*raf* protein could be completely inhibited by the addition of a peptide containing these 12 amino acids to immunoprecipitates (data not shown). In both v-*fms*- and v-*src*-transformed cells, the c-*raf* molecules exhibited an altered electrophoretic mobility when compared to the mobility of the c-*raf* protein isolated from quiescent 3T3 cells. As seen in Figure 5, there is a population of c-*raf* molecules in the transformed cells that migrates more slowly

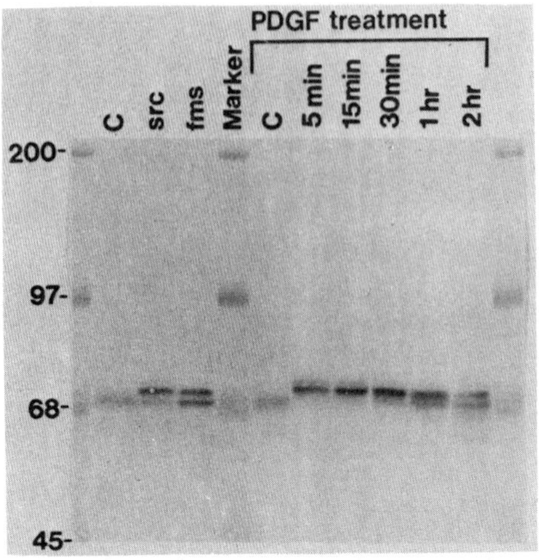

Figure 5. Migration patterns of c-*raf* protein in cells expressing membrane-bound oncogene products or treated with PDGF. Experiments were initiated on density-arrested monolayers as described previously (Kaplan et al. 1987). Growth factors were added directly to the medium, and incubation was continued at 37°C. Transformed cell lines were not fed with fresh medium for at least 48 hr prior to use. Cytoplasmic cell lysates were prepared and equalized for protein content before electrophoretic separation on 7.5% SDS-PAGE and transfer to nitrocellulose filters as described previously (Morrison et al. 1988). Protein blots were probed with anti-c-*raf* antibody, and immunological reactions were identified using an alkaline phosphatase detection system. c-*raf* proteins in 3T3 cells transformed by the v-*src* (src) or v-*fms* (fms) oncogenes or in quiescent BALB/c-3T3 cells (C) treated with PDGF for various amounts of time. Molecular size markers (in kilodaltons) are shown to the left.

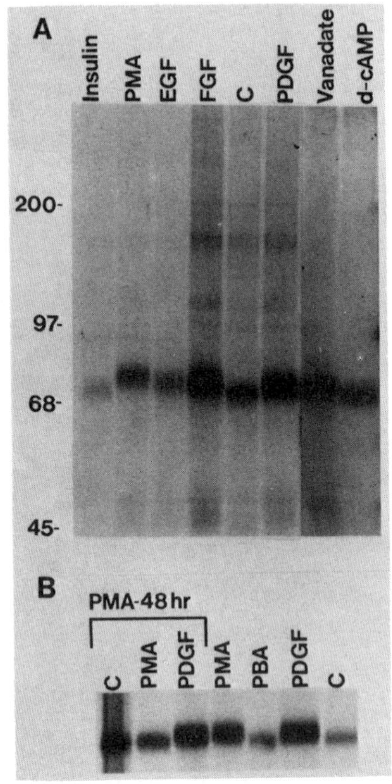

Figure 6. Changes in c-*raf* phosphorylation in response to mitogens, and PMA treatment. 3T3 cells were labeled with [^{32}P]orthophosphate for 3 hr prior to lysis. Anti-c-*raf* immunoprecipitates were prepared from cell lysates and analyzed on 7.5% SDS-PAGE. (*A*) Quiescent BALB/c-3T3 cells treated for 5 min with insulin (5 μg/ml), PMA (50 ng/ml), EGF (100 ng/ml), acidic FGF (10 units/ml), or PDGF (10 units/ml) prior to lysis. Cells were also treated with sodium orthovanadate (40 μM) or dibutyryl cAMP(d-cAMP) (1 mM) for 24 hr. (C) Control quiescent BALB/c-3T3 fibroblasts. Molecular size markers (in kilodaltons) are shown to the left. (*B*) Quiescent BALB/c-3T3 cells (C) chronically treated with 250 ng/ml PMA for 48 hr to remove PMA binding sites and protein kinase C activity. (PMA) Chronically treated quiescent cells treated for an additional 5 min with 50 ng/ml PMA. (PDGF) Chronically treated quiescent cells treated for 5 min with PDGF (10 units/ml). Also shown are quiescent BALB/c-3T3 cells (not pretreated with PMA) (C) treated for 5 min with PDGF (10 units/ml), PMA (50 ng/ml), or biologically inactive PMA (PBA) (50 ng/ml).

than the c-*raf* molecules in control cells. We next examined the mobility of the c-*raf* protein in lysates made from fibroblasts stimulated with the growth factor, PDGF (Fig. 5). When quiescent 3T3 cells were treated with PDGF (10 units/ml) for 5 minutes, all of the c-*raf* molecules exhibited a slower electrophoretic mobility when compared to the c-*raf* protein from unstimulated cells. This apparent molecular weight increase was transitory. The size increase was maximal 5 minutes following PDGF treatment; thereafter, multiple faster-migrating forms of the c-*raf* protein appeared. By 2 hours after PDGF treatment, approximately half of the c-*raf* molecules had returned to a mobility similar to that of the c-*raf* protein isolated from quiescent BALB/c-3T3 cells.

Phosphorylation of the c-*raf* Protein Increases in Response to Mitogens and PMA Treatment

To determine whether the changes in the mobility of the c-*raf* protein were accompanied by an increase in the phosphorylation of the c-*raf* protein, we labeled cells with [^{32}P]orthophosphate and examined the c-*raf* phosphoprotein in immunoprecipitates using the anti-c-*raf* antibody (Fig. 6). Alterations in c-*raf* protein phosphorylation were first examined in quiescent BALB/c-3T3 cells treated for 5 minutes with the following 3T3 cell mitogens: EGF, FGF, PDGF, and PMA (Stiles et al. 1979; Rozengurt 1986). In immunoprecipitates from mitogen-treated cells, the c-*raf* protein exhibited a decreased mobility and increased phosphate content as compared to untreated quiescent cells (Fig. 6A). Increased phosphate content was most dramatic in cells treated with PDGF and acidic FGF. Treatment for 24 hours with the phosphotyrosine phosphatase inhibitor sodium orthovanadate (40 μM) also increased the phosphorylation state and molecular weight of the c-*raf* protein (Fig. 6A). However, when quiescent BALB/c-3T3 cells were treated for 5 minutes with insulin, no size change or increased phosphorylation of the c-*raf* protein was observed. In addition, treatment of quiescent 3T3 cells with dibutyryl cAMP (1 mM), an activator of the cAMP-dependent protein kinase, produced no alterations in the c-*raf* molecule.

We next examined the effect of PMA on the migration and phosphorylation of the c-*raf* protein to determine if some of the changes in the c-*raf* molecules observed in response to oncogenic transformation or growth factor stimulation could be correlated with the activation of protein kinase C (Fig. 6B). Quiescent BALB/c-3T3 cells treated for 5 minutes with biologically inactive PMA (PBA) (50 ng/ml) showed no change in the c-*raf* protein, in contrast to the changes observed with biologically active PMA (50 ng/ml). If cells were chronically treated with 250 ng/ml PMA for 48 hours to remove PMA-binding sites and protein kinase C activity (Rozengurt 1986), the c-*raf* protein mobility and phosphorylation state returned to that observed in quiescent BALB/c-3T3 cells (Fig. 6B). Subsequent treatment of chronically treated cells with PMA (50 ng/ml for 5 min) still showed no change in the c-*raf* protein. However, treating these cells for 5 minutes with PDGF (10 units/ml) did increase the phosphate content and molecular weight of the c-*raf* molecules. These results suggest that although PMA activation of protein kinase C can affect the mobility and phosphorylation state of the c-*raf* protein, there are mechanisms independent of protein kinase C through which growth factors can bring about this change.

To confirm that the changes in c-*raf* protein mobility were due to phosphorylation, the c-*raf* protein immunoprecipitates from PDGF-stimulated and quiescent 3T3 cells were treated with potato acid phosphatase. The treated immunoprecipitates were electrophoresed on SDS-polyacrylamide gels and transferred to nitrocellulose filters. The blotted proteins were then probed with anti-c-*raf* antibody. Phosphatase treatment decreased the apparent molecular-mass of c-*raf* molecules from PDGF-stimulated cells from ~76 kD to ~72 kD. Phosphatase treatment also resulted in a molecular-mass size shift of the c-*raf* molecules from unstimulated cells from 74 kD to a molecular mass identical to that of the phosphatased c-*raf* molecules from PDGF-treated cells. Thus, the molecular-mass increase following PDGF stimulation can be accounted for solely by the enhanced phosphorylation of c-*raf* molecules (not shown).

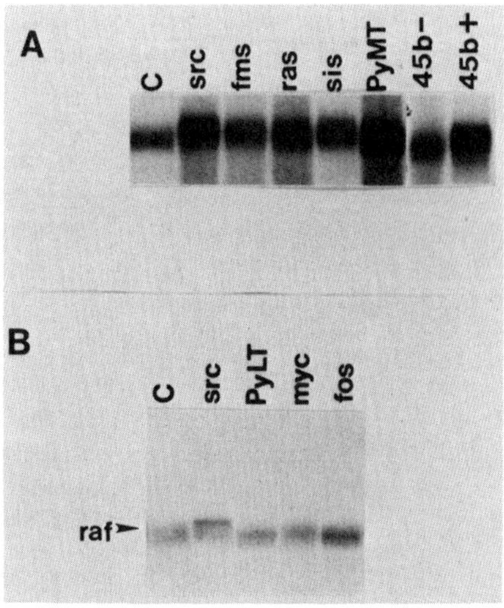

Figure 7. Changes in c-Raf in response to oncogene products. (*A*) 3T3 cells transformed with various membrane oncogenes were examined. 3T3 cells were labeled with [^{32}P]orthophosphate for 3 hr prior to lysis. Anti-c-*raf* immunoprecipitates were prepared from cell lysates and analyzed on 7.5% SDS-PAGE. The lanes show 3T3 cells transformed by v-*src*, v-*fms*, Ha-*ras*, v-*sis*, or polyomavirus middle T antigen (PyMT). 45b is an NIH-3T3 cell line expressing polyomavirus middle T antigen under the control of the dexamethasone-inducible MMTV promoter. (45b+) 45b cells treated for 16 hr with 5 μM dexamethasone to induce middle T antigen synthesis. (45b−) Untreated 45b cells. (*B*) The effects of several nuclear oncogene products are compared to those of v-*src*. In this case, the comparison is done via immunoblotting with the c-*raf* antibody, using the same technique as in Fig. 5. Shown are c-*raf* proteins in 3T3 cells transformed by v-*src* (src), Rat-1 cells transformed by v-*fos* (fos), and 3T3 cells transformed by c-*myc*. The c-*myc*-expressing cells are transformed due to c-*myc*-overexpression from an inducible promoter. PyLT is an NIH-3T3 cell line expressing polyomavirus large T antigen. (C) Control quiescent BALB/c-3T3 fibroblasts. Substrate development time for blots was 2 min.

Expression of Membrane but Not Nuclear Oncogene Products Alters c-*raf* Mobility

Expression of a battery of membrane-bound oncogene products affects changes in c-*raf* protein. As shown in Figure 7, the c-*raf* protein displayed an altered electrophoretic mobility and enhanced phosphate content in cells transformed by expression of the v-*src*, v-*sis*, v-*fms*, c-Ha-*ras* (activated for transformation by mutation at codon 61), or polyomavirus MTAg oncogenes. It was important to establish that the changes in the c-*raf* protein migration observed in transformed cells were not due to secondary effects unrelated to oncogene expression. Therefore, we examined the c-*raf* protein phosphorylation in cells where expression of the polyomavirus middle T oncogene is under the regulation of the dexamethasone regulatable MMTV promoter. Addition of dexamethasone (5 μM for 16 hr) to this cell line, 45b, results in the appearance of anchorage-independent growth concomitant with the induction of MTAg synthesis and MTAg-associated tyrosine kinase activity (Kaplan et al. 1986). Treatment of quiescent 45b cells with 5 μM dexamethasone for 16 hours resulted in an increase in phosphorylation and apparent molecular weight of the c-*raf* protein when compared to the c-*raf* protein from untreated 45b cells (Fig. 7A). The cells appeared more refractile, and the saturation density of the cultures increased following the addition of dexamethasone (data not shown). Dexamethasone treatment of quiescent 3T3 cells did not result in any alterations of the c-*raf* protein or cellular morphology (data not shown). These results suggest that the changes in c-*raf* protein phosphorylation and mobility observed in transformed cells are due to membrane oncogene expression. Furthermore, these changes correlate with the acquisition of the transformed state. When similar analyses were performed, via immunoblotting, using Rat-1 cells transformed by v-*fos* or 3T3 cells transformed by overexpression of the c-*myc* gene, no change in the mobility of c-Raf was observed (Fig. 7B). c-*raf* molecules were also examined in 3T3 cells expressing polyomavirus large T antigen, a protein with the capability of immortalizing primary rodent cells. The c-*raf* protein isolated from these cells exhibited an electrophoretic mobility indistinguishable from the c-*raf* protein isolated from quiescent 3T3 cells (Fig. 7B).

PDGF and PMA Treatment and v-*src* Expression Increase c-*raf*-associated Serine/Threonine Kinase Activity

To determine if the physical alterations of the c-*raf* molecules affect the serine/threonine kinase activity of the protein, the ability of immunoprecipitated c-*raf* molecules to phosphorylate histone type V-S was examined. Quantitative immune complex kinase assays were performed in vitro with c-*raf* proteins immunoprecipitated from quiescent BALB/c-3T3 cells, quiescent cells treated for 5 minutes with either PDGF (10 units/ml) or PMA (50 ng/ml), and v-*src*-transformed cells. To ensure that equivalent amounts of c-*raf* protein were present in each sample, cell lysates were prepared and normalized for protein content as in Figure 1. As shown in Figure 8, the ^{32}P labeling of histone-type V-S increased in the anti-c-*raf* immunoprecipitates from v-*src*-transformed cells and from cells treated with PMA and PDGF when compared to immunoprecipitates from quiescent control cells. Direct quantitation by Cerenkov counting of the ^{32}P radioactivity in the histone bands revealed that PMA treatment of cells resulted in a twofold increase in histone phosphorylation above basal levels; PDGF stimulation of cells, an approximate threefold increase; and v-*src*-transformation, a fivefold increase (as averaged from four experiments). Phosphoamino acid analysis of the in-vitro-phosphorylated histone molecules revealed that serine was the predominant phosphorylated residue. A slight amount of threonine phosphorylation was detected; however, no tyrosine phosphorylated residues were observed (data not shown). These data are consistent with the possibility that the increased phosphorylation of the c-*raf* molecules in response to mitogenic and oncogenic factors has modulated the associated serine/threonine kinase activity of the c-*raf* protein.

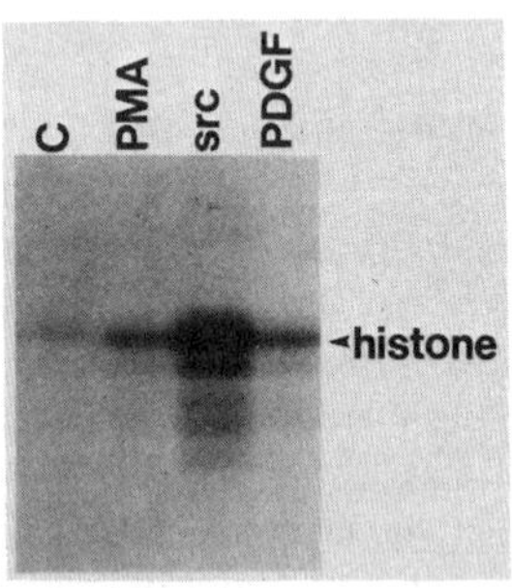

Figure 8. Enhancement of c-*raf*-associated serine/threonine kinase activity in anti-c-*raf* immunoprecipitates from v-*src*-transformed cells and in cells treated with PDGF or PMA. Anti-c-*raf* immunoprecipitates prepared from quiescent cells (C), PMA-treated cells (50 ng/ml for 5 min), v-*src*-transformed cells (src), or PDGF-treated cells (10 units/ml for 5 min) were assayed for the ability to phosphorylate histone type V-S in vitro. Protein kinase activites were tested in a total volume of 50 μl containing 50 mM Tris-HCl (pH 7.5), 50 mM NaCl, 1 mM dithiothreitol, 10 mM $MnCl_2$, 20 μCi of [γ-^{32}P]ATP, and 5 μg histone type V-S. Incubation time was 5 min at 25°C. Reactions were terminated by the addition of threefold concentrated electrophoresis buffer. Samples were analyzed by 12.5% SDS-PAGE, and phosphoproteins were detected by autoradiography. Protein kinase assays were linear over the time of the assay.

DISCUSSION

Immunoblotting whole-cell lysates with anti-P.Tyr sera has proved to be a powerful tool to identify potential substrates of tyrosine kinases. This technique has allowed the identification of P.Tyr-containing proteins in transformed cell lines, such as v-*fms* or polyoma MTAg, where they have been undetectable previously (Morrison et al. 1988). Several observations can be made from these types of data. First, each tyrosine kinase has its own distinct fingerprint of substrates (Morrison et al. 1988). Second, a receptor activated chronically by mutation displays a slightly different set of substrates from the same receptor activated transiently by its ligand (Fig. 1). Third, some receptors, such as c-*fms*, exhibit a slightly different profile of P.Tyr-containing proteins in immunoblot analysis when different cell types are examined (Fig. 1). This is to be expected in cases where the role of the kinase may vary from cell to cell.

After visualization of potential substrates, the identification of the biochemical properties of these substrates and determination of how these properties are modified by tyrosine phosphorylation must be addressed. More importantly, it must be determined what function the substrate performs in cellular growth control. This process is still in its infancy, a time subject to very rapid changes in any field. In particular, although the circuitry of intracellular signal transduction is obviously complex, our current hypotheses concerning the workings of the circuitry are quite simplistic. With that proviso, we will proceed to hypothesize, attempting as we do to warn the reader of the dividing line between correlations and proof. Our studies on the 85-kD protein and associated PtdIns kinase activity are a perfect case in point. For example, biochemical purification and genetic analysis suggest that this protein is a PtdIns kinase. However, only by isolating a molecular clone encoding the 85-kD protein can the required analysis be performed to determine definitively if this protein is a PtdIns kinase. Whether or not the 85-kD protein and PtdIns kinase are identical, the genetic and biochemical studies reported in this paper and elsewhere suggest that both protein and PtdIns kinase play a role in mitogenesis and oncogenesis. Similarly, although a model proposing regulation of PtdIns kinase activity through direct interaction with tyrosine kinases is consistent with our data, other models could also be considered. For instance, since the 85-kD protein/PtdIns kinase is phosphorylated on serine residues in vivo and in vitro, an indirect action of tyrosine kinases on serine kinases could occur. Finally and most importantly, we wish to point out our ignorance about the lipid kinase activity itself. We originally hypothesized that enhanced PtdIns kinase activity in vivo would stimulate intracellular PtdIns turnover, thereby increasing the intracellular concentrations of the PtdIns-derived second messengers, Ins(1,4,5)P_3 and diacylglycerol. Since this PtdIns kinase produces a PtdInsP isomer that is not in the pathway of Ins(1,4,5)P_3 production (Whitman et al. 1988), it is now quite possible that the metabolites generated by this kinase activity function independently of the classic PtdIns turnover pathway. Even if the type I PtdIns kinase enhanced PtdIns turnover, the roles of the many PtdIns-derived second messengers in mitogenesis have not as yet been clearly determined.

Perhaps the most novel aspects of our findings are those that suggest that c-Raf may play a broad role in signal transduction. Cytoplasmic serine/threonine kinases have been postulated to function downstream from membrane events in the mitogenic pathway (Smith et al. 1986). We have tested this hypothesis by examining the phosphorylation and protein kinase activity of c-*raf* protein isolated from growth-factor-stimulated and oncogene-transformed cells. Membrane-bound oncogene product expression increased the phosphate content and apparent molecular weight of c-*raf* protein, whereas nuclear oncogene product expression did not. Treatment of quiescent 3T3 cells with the mitogens PDGF, acidic FGF, EGF, and PMA also induced alterations in the c-*raf* protein. The phosphorylation and molecular-weight shift of the c-*raf* protein appear within minutes of growth factor addition, suggesting that this is a primary event in the mitogenesis response. In addition, c-*raf*-associated serine/threonine kinase activity was enhanced by growth factor treatment and membrane-bound oncogene product expression. These results suggest that c-*raf* molecules receive signals from membrane-bound receptors and oncogene products and transmit them to as yet unidentified cellular targets.

Our data suggest that the increase in the apparent molecular weight of c-*raf* protein may occur by at least two mechanisms. Phosphatase experiments, phosphoamino acid analysis, and phosphopeptide mapping all indicate that the shift in mobility of the c-*raf* protein is due to increased phosphorylation. The enhanced phosphorylation occurs primarily on serine residues. The increase in c-*raf* phosphorylation and apparent molecular weight in response to PMA treatment suggests that these alterations may occur as a result of serine phosphorylation directly mediated by protein kinase C. Consistent with this possibility, chronic treatment of cells with PMA, which reduces membrane protein kinase C activity (Rodriquez-Pena and Rozengurt 1984), prevents subsequent induction of c-*raf* protein phosphorylation by PMA. However, PDGF stimulation alters c-*raf* protein phosphorylation and mobility in cells chronically treated with PMA, and chronic treatment of v-*src*-transformed cells does not reverse the c-*raf* protein alterations observed in those cells. These experiments suggest that there are mechanisms, independent of protein kinase C, that modulate the c-*raf* protein phosphorylation. An alternative mechanism is suggested by experiments demonstrating that the c-*raf* protein is phosphorylated on tyrosine residues in response to PDGF stimulation and v-*src* expression. Treatment of cells with the phosphotyrosine phosphatase inhibitor, sodium orthovanadate, also results in increased serine and tyrosine phosphorylation of c-*raf* molecules. These results imply that increases in cellular tyrosine phosphorylation may affect c-*raf* protein migration and phosphorylation. Although tyrosine phosphorylation accounts for little of the total phosphate increase, it could stimulate *raf* phosphorylation by at least three mechanisms. First, c-*raf* protein phosphorylation may occur by the action of other serine/threonine kinases whose activity is induced by tyrosine kinases. Second, tyrosine phosphorylation of the c-*raf* protein by tyrosine kinases may induce c-*raf* protein kinase activity, resulting in enhanced autophosphorylation of the molecule. (Protein-kinase-C-mediated activation of c-Raf could be effected by either of these mechanisms as well.) Finally, c-*raf* protein phosphorylation may occur through a mechanism similar to that postulated for protein kinase C. In this case, effector molecules (such as diacylglycerol and Ca^{++}) generated by events occurring at the membrane could bind to c-Raf, thereby activating the autokinase activity.

Membrane-bound oncogene products and growth factor receptors might activate c-*raf* activity by the phosphorylation of amino acids residing in the amino-terminal portion of the protein. The amino-terminal domain of the c-*raf* protein is known to have a negative regulatory role based on experiments showing that c-Raf can be rendered a transforming protein by removal of this domain and that the v-*raf* product is truncated in this region (Rapp et al. 1988). Within the amino-terminal domain, there is a region rich in serine/threonine residues (Beck et al. 1987). The experiments presented here suggest that phosphorylation of some of these residues may activate c-*raf* protein kinase activity, thereby propagating proliferative signals to cellular targets mediating mitogenesis. The importance of the serine/threonine kinase activity to c-*raf* function and the nature of the targets of this activity are presently unknown. The precise role of tyrosine, threonine, and serine phosphorylation in modulating c-*raf* activity will only be determined by mutational analysis of the c-*raf* protein. However, it should be noted that the present experiments do not prove that phosphorylation of the c-Raf regulates its kinase activity; only that the correlation between the two phenomena has been demonstrated. We do know that tyrosine phosphorylation of c-Raf cannot be solely responsible for the increases in activity seen in Figure 8, since there is little or no tyrosine phosphate detectable in the anti-*raf* immunoprecipitates. Alternatively, tyrosine phosphorylation may activate c-*raf* autokinase activity. This possibility can be tested by using purified kinases produced by baculovirus expression systems (H. Piwnica-Worms et al., pers. comm.).

Our results indicate that there is a correlation between proliferative events generated at the membrane and alterations in the c-*raf* protein, including activation of the c-*raf*-associated serine/threonine kinase activity. However, expression of nuclear oncogene products did not have any apparent effect on the phosphorylation of the c-*raf* protein. These experiments imply that the c-*raf* protein is involved in the mediation of proliferative signals generated at the membrane but is not involved in the mediation of proliferative signals originating in the nucleus. These results are supported by the experiments of Smith et al. (1986) and Rapp et al. (1988) suggesting that cytoplasmic oncogene products with associated serine/threonine kinase activity, such as v-Raf and v-Mos, function downstream from membrane-associated oncogene products in the signal transduction pathway. In the studies by Smith et al. (1988), microinjected antibodies directed against c-Ras were able to inhibit the growth of cells transformed by v-*ras* and v-*src* but not those transformed by v-*raf* and v-*mos*. Rapp et al. (1988) observed that flat revertants of cells transformed by Kirsten murine sarcoma virus were able to suppress transformation by v-*ras* and v-*src* but not v-*raf* carrying viruses. These investigators concluded that the v-*raf* protein either acts independently of $p21^{c\text{-}ras}$ or functions downstream from $p21^{c\text{-}ras}$. The experiments presented here, showing c-*raf* protein phosphorylation in response to activated c-Ha-*ras* protein expression, suggest the latter possibility. We propose that the activation of c-*raf*-associated kinase activity by mitogenic stimuli at the membrane provides a new route of signal transduction to the nucleus. This route would function in addition to the well-established pathways mediated by protein kinases C and A, and the calcium-regulated kinases. The need for alternative mechanisms for signal transduction is suggested by the observation that a number of serum-inducible promoters cannot be activated by any combination of phorbol esters, cAMP analogs and calcium ionophores (Hall and Stiles 1987). Identification of the substrates of c-*raf* protein will test this hypothesis.

ACKNOWLEDGMENTS

We thank Michael Corbley for critical reading of the manuscript; Van Cherington, Chuck Stiles, Leda Raptis, and Michael Corbley for kindly providing cell lines; and Morris White for the anti-P.Tyr serum. This work was supported by grants from the U.S. Public Health Service to T.M.R. (CA-3002 and CA-43803), to B.S. (CA-34722), and to L.C. (GM-36622).

REFERENCES

Armelin, H.A., M.C.S. Armelin, K. Kell, T. Stewart, P. Leder, B.H. Cochran, and C.D. Stiles. 1984. Functional role for c-myc in mitogenic response to platelet-derived growth factor. *Nature* **310:** 655.

Beck, T.W., M. Huleihel, M. Gunnell, T.I. Bonner, and U.R. Rapp. 1987. The complete coding sequence of the human A-*raf*-1 oncogene and transforming activity of a human A-*raf* carrying retrovirus. *Nucleic Acids Res.* **15:** 595.

Bonner, T.I., H. Oppermann, P. Seeburg, S.B. Kerby, M.A. Gunnell, A.C. Young, and U.R. Rapp. 1986. The complete coding sequence of the human *raf* oncogene and the corresponding structure of the c-*raf*-1 gene. *Nucleic Acids Res.* **14:** 1009.

Cheng, S., H. Piwnica-Worms, R.W. Harvey, T.M. Roberts, and A.E. Smith. 1988. The carboxyl terminus of $pp60^{c\text{-}src}$ is a regulatory domain and is involved in complex formation with middle T antigen of polyoma virus. *Mol. Cell. Biol.* **8:** 1736.

Cherington, V., B. Morgan, B.M. Spiegelman, and T.M. Roberts. 1986. Recombinant retroviruses that transduce individual polyoma virus tumor antigens: Effects on growth and differentiation. *Proc. Natl. Acad. Sci.* **83:** 4307.

Cooper, J.A. and C.S. King. 1986. Dephosphorylation or antibody binding to the carboxyl terminus stimulates $pp60^{c\text{-}src}$. *Mol. Cel. Biol.* **6:** 4467.

Courtneidge, S.A. and A. Heber. 1987. An 81 kd protein complexed with middle T antigen and $pp60^{c\text{-}src}$: A possible phosphatidylinositol kinase. *Cell* **50:** 1031.

Frackelton, A.R., P.M. Tremble, and L.T. Williams. 1984. Evidence for the platelet-derived growth factor-stimulated tyrosine phosphorylation of the platelet-derived growth factor receptor in vivo. *J. Biol. Chem.* **259:** 7909.

Hall, D.J. and C.D. Stiles. 1987. Platelet-derived growth factor inducible genes respond differentially to at least two distinct intracellular second messengers. *J. Biol. Chem.* **262:** 15302.

Hunter, T. and B.W. Sefton. 1980. Transforming gene product of Rous sarcoma virus phosphorylates tyrosine. *Proc. Natl. Acad. Sci.* **77:** 1311.

Kaplan, D.R., M. Whitman, B. Schaffhausen, D.C. Pallas, M. White, L. Cantley, and T.M. Roberts. 1987. Common elements in growth factor stimulation and oncogenic transformation: 85 kd Phosphoprotein and phosphatidylinositol kinase activity. *Cell* **50:** 1021.

Kaplan, D.R., M. Whitman, B. Schaffhausen, L. Raptis, R.L. Garcea, D. Pallas, T.M. Roberts, and L. Cantley. 1986. Phosphatidylinositol metabolism and polyoma-mediated transformation. *Proc. Natl. Acad. Sci.* **83:** 3624.

MacDonald, M.L., E.A. Kuenzel, J.A. Glomset, and E.G. Krebs. 1985. Evidence from two transformed cell lines that the phosphorylations of peptide tyrosine and phosphatidylinositol catalyzed by different proteins. *Proc. Natl. Acad. Sci.* **82:** 3993.

Moelling, K., B. Heimann, P. Beimling, U.R. Rapp, and T. Sander. 1984. Serine- and threonine-specific protein kinase activities of purified gag-mil and gag-*raf* proteins. *Nature* **312:** 558.

Morgan, W.C., D. Kaplan, D. Pallas, and T.M. Roberts. 1988. Recombinant retroviruses that transduce middle T antigen cDNAs derived from polyoma virus mutants. *J. Virol.* (in press).

Morlan, A.O. and J.Y.J Wang. 1986. Protein tyrosine phosphorylation in the cell cycle of Balb/c 3T3 fibroblasts. *Proc. Natl. Acad. Sci.* **83:** 8191.

Morrison, D.K., P.J. Browning, M.F. White, and T.M. Roberts. 1988. Tyrosine phosphorylations in vivo associated with v-fms transformation. *Mol. Cell. Biol.* **8:** 176.

Pallas, D.P. and F. Solomon. 1982. Cytoplasmic microtubule-associated proteins: Phosphorylation at novel sites is correlated with their incorporation into assembled microtubules. *Cell* **30:** 407.

Piwnica-Worms, H., D. Kaplan, M. Whitman, and T.M. Roberts. 1986. Retrovirus shuttle vector for study of the kinase activities of pp60$^{c\text{-}src}$ synthesized in vitro and overproduced in vivo. *Mol. Cell. Biol.* **6:** 2033.

Piwnica-Worms, H., K.B. Saunders, T.M. Roberts, A.E. Smith, and S.H. Cheng. 1987. Tyrosine phosphorylation regulates the biochemical and biological properties of pp60$^{c\text{-}src}$. *Cell* **49:** 75.

Rapp, U.R., M.D. Goldsborough, G.E. Mark, T.I. Bonner, J. Groffen, F.H. Reynolds, Jr., and F.R. Stephenson. 1983. Structure and biological activity of v-*raf*, a unique oncogene transduced by a retrovirus. *Proc. Natl. Acad. Sci.* **80:** 4218.

Rapp, U.R., J.L. Cleveland, T.I. Bonner, and S.M. Storms. 1988. The *raf* oncogene. In *The oncogene handbook* (ed. P. Reddy et al.). Elsevier, Amsterdam. (In press.)

Rodriquez-Pena, A. and E. Rozengurt. 1984. Disappearance of Ca^{2+}-sensitive, phospholipid-dependent protein kinase activity in phorbol ester-treated 3T3 cells. *Biochem. Biophys. Res. Commun.* **3:** 1053.

Rozengurt, E. 1986. Early signals in the mitogenic response. *Science* **234:** 161.

Schultz, A.M., T.D. Copeland, G.E. Mark, U.R. Rapp, and S. Oroszlan. 1985. Detection of the myristylated gag-*raf* transformed protein with *raf*-specific antipeptide sera. *Virology* **146:** 78.

Smith, M.R., S.J. DeGudicbus, and D.W. Stacey. 1986. Requirement for c-ras proteins during viral oncogene transformation. *Nature* **320:** 540.

Stiles, C.D., G.T. Capone, C.D. Scher, H.N. Antoniades, J.J. van Wyk, and W.J. Pledger. 1979. Dual control of cell growth by somatomedins and platelet-derived growth factor. *Proc. Natl. Acad. Sci.* **76:** 1279.

Sugimoto, S. and R. Erikson. 1985. Phosphatidylinositol kinases active in normal and Rous sarcoma virus-transformed cells. *Mol. Cell. Biol.* **5:** 3194.

White, M.F., R. Maron, and C.R. Kahn. 1985. Insulin rapidly stimulates tyrosine phosphorylation of a M_r-185,000 protein in intact cells. *Nature* **318:** 183.

Whitman, M., D. Kaplan, T.M. Roberts, and L. Cantley. 1987. Fibroblasts contain two biochemically distinct PI kinases. *Biochem. J.* **247:** 165.

Whitman, M., C.P. Downes, M. Keeler, T. Keller, and L. Cantley. 1988. Type I PI kinase makes a novel inositol phospholipid, phosphatidylinositol-3 phosphate. *Nature* **322:** 644.

Whitman, M., D. Kaplan, B. Schaffhausen, L. Cantley, and T.M. Roberts. 1985. Association of PI kinase activity with polyoma middle T competent for transformation. *Nature* **315:** 239.

Zarbl, H., J. Latreille, and P. Jolicoeur. 1987. Revertants of v-fos-transformed fibroblasts have mutations in cellular genes essential for transformation by other oncogenes. *Cell* **51:** 357.

raf Family Serine/Threonine Protein Kinases in Mitogen Signal Transduction

U.R. Rapp,* G. Heidecker,* M. Huleihel,* J.L. Cleveland,* W.C. Choi,† T. Pawson,‡ J.N. Ihle,§ and W.B. Anderson†

**Laboratory of Viral Carcinogenesis, National Cancer Institute-Frederick Cancer Research Facility, Frederick, Maryland 21701; †Division of Cancer Biology and Diagnosis, National Cancer Institute, Bethesda, Maryland 20892; ‡Mount Sinai Hospital, Toronto, Canada; §Bionetics Research Institute, Frederick, Maryland 21701*

raf family genes encode signal transducers that first were identified in their altered form as oncogenes (Rapp et al. 1983a,b, 1988). Oncogenes are constitutively activated versions of genes encoding links in signal transduction pathways that feed into mitogenesis. In addition, they may regulate events unrelated to mitogenesis in terminally differentiated tissues. To date, *raf* family protein kinases remain the only widely expressed class of serine/threonine-specific kinases from which full-fledged oncogenes have emerged. The only other known serine/threonine protein-kinase-derived oncogene, *mos*, is expressed in only a few tissues (Seth and van de Woude 1988).

The general properties of *raf* family genes are summarized in Table 1. Briefly, there are three active genes known in man, A-*raf*, B-*raf*, and c-*raf*. Few data are available for B-*raf*, the most recently identified member, except that it is slightly closer to A-*raf* than to c-*raf* in sequence homology of the kinase domain (Ikawa et al. 1988). A-*raf* and c-*raf* encode cytoplasmic protein kinases of 68 and 74 kD, respectively (Bonner et al. 1986; Beck et al. 1987). All tissues we have analyzed so far express c-*raf* and/or A-*raf*. c-*raf* is ubiquitously expressed, as judged by examination of poly(A) RNA from entire tissues, whereas A-*raf* is more restricted in its expression, with highest levels in epididymis. The amounts of c-*raf* RNA vary approximately fivefold between various tissues, whereas A-*raf* levels show 100-fold differences (Heidecker et al. 1988). Neither c-*raf* nor A-*raf* appears to be transcriptionally regulated by mitogens or in a cell-cycle-dependent fashion (Rapp et al. 1988).

In this paper, we describe two approaches we have taken to elucidate how *raf* kinases function in signal transduction. The first focuses on structural features of the *raf* proteins that may be involved in receiving and transmitting signals; the other is a functional approach relating *raf* activity to other players in the signal network.

MATERIALS AND METHODS

Cells and viruses. NIH-3T3 clone 7 cells, a subline of NIH-3T3 cells selected for flat morphology, were grown in Dulbecco's modified Eagle's medium (DMEM) supplemented with 10% calf serum, 100 IU of penicillin, and 50 μg of streptomycin per milliliter. IL-3-dependent FDC-P1 cells were grown and infected

Table 1. Summary of *raf* Oncogene Characteristics

Three active genes in man: c-*raf*-1, B-*raf*, and A-*raf*-1
c-*raf*-1, on chromosome 3p25, site altered in many epithelial neoplasias, marker for von Hippel-Lindau disease
A-*raf*-1 on Xp21.3, expressed in select tissues
c-*raf* and A-*raf* have pseudogenes, c-*raf*-2 and A-*raf*-2; c-*raf*2 marks Huntington's chorea
D-*raf*-1 on X chromosome and D-*raf*-2 on chromosome 2 in *Drosophila melanogaster*
c-*raf*-1 and A-*raf*-1 encode 74- and 68-kD proteins, predominantly found in the cytoplasm, with associated Ser/Thr kinase activity
Function in signal transduction independent of *ras*
Synergize in transformation with *myc*
Involved in transcriptional control of polyoma and TGF-α enhancers
Distantly related to protein kinase C
Oncogenic activation can be achieved by truncation
Vaccination experiments with *raf* protein in a chemical mouse lung cancer induction system double latency

as described previously (Rapp et al. 1985a). Oncogene-carrying retroviruses included 3611MSV (Rapp et al. 1983b), J1–J5 (Rapp et al. 1985b,c), wild-type and temperature-sensitive Abelson murine leukemia virus (MLV), strains MO (wt), RK, DP, and TS (*ts*) (Engelman and Rosenberg 1987; Kipreos and Wang 1988), Moloney, Kirsten, and Harvey murine sarcoma virus (MSV) (Ihle et al. 1985), *trk-neo* MLV, *src*-MLV (Cleveland et al. 1988), and *fms*-MSV (Wheeler et al. 1987). Ki-*ras* revertant cell clone C-11 cells were fused with oncogene virus-transformed cells as reported by Noda et al. (1983). Virus infections and DNA transfections were carried out following standard procedures (Graham and van der Eb 1973; Rapp et al. 1983a). Where appropriate, transfected cells were selected in medium supplemented with 200 μg/ml of G418. Transformation was scored by focus induction and growth in semisolid medium.

Retroviral constructs. In general, procedures were carried out as described by Maniatis et al. (1982). Mutants were generated by BAL-31 resection, deletion of restriction fragments, and site-directed mutagenesis as described by Smith (1985). Mutations were verified by DNA sequencing. The retroviral vectors used were all MLV strain Leuk-2 (Rapp et al. 1983b) derived and carried on a pBR322 backbone. Plasmid DNAs were linearized by restriction digest prior to transfection onto NIH-3T3 cells.

RNA analyses. Total RNA was isolated from cultures of uninfected FDC-P1 cells and FDC-P1 cells infected with wild-type (MO-1) and temperature-sensitive *abl* retroviruses (RK-4, TS-1) at the times indicated in Figure 4 as described previously (Cleveland et al. 1987). RNA was electrophoresed, blotted, and hybridized, and blots were washed following standard procedures (Cleveland et al. 1987). In each lane, 20 μg of RNA was analyzed.

Immunohistochemical localization studies. NIH-3T3 clone 7 cells (1×10^2) were seeded into wells (1 cm^2) in Lab-Tek multiwell chamber slides and allowed to attach and grow for 24 hours under standard growth conditions. The attached cells then were washed thoroughly with serum-free DMEM and incubated for 24 hours at 37°C in the absence of serum. The serum-starved cells were incubated in the absence or presence of 5 ng/ml platelet-derived growth factor (PDGF) (PDGF, Inc.) or 0.25 μM 12-*O*-tetradecanoylphorbol-13-acetate (TPA) (Pharmacia P-L Biochemicals). At different time points of incubation, individual wells were rapidly washed with cold (4°C) phosphate-buffered saline (PBS), fixed by incubation at room temperature for 5 minutes with 2% formaldehyde in PBS, and then permeabilized by incubation for 5 minutes in cold (4°C) absolute methanol. The permeabilized cells were incubated at 4°C for 24 hours with rabbit anti-c-*raf* SP63 (1:300 dilution) sera (Schultz et al. 1988) in PBS containing 2% normal goat serum and 0.5% Triton X-100. The cells then were rinsed (three times) with PBS and c-*raf* protein was localized using a modification of the avidin-biotin-Texas Red method (Bayer and Wilchek 1979; Guesdon et al. 1979). The treated cells were incubated with biotin-labeled goat anti-rabbit IgG (2.5 mg/ml) in PBS for 2 hours at room temperature. Following this incubation, the cells were washed with PBS containing 50% glycerol, and slides were photographed with an Olympus Vanox microscope at 400 ×.

RESULTS AND DISCUSSION

Structural Analysis of *raf* Proteins

First clues to domains involved in the regulation of *raf* protein kinases came from a structural comparison of normal versus oncogenically activated *raf* genes. *raf* oncogenes isolated by retrovirus transduction, LTR promoter insertion, or DNA transfection all shared a minimal deletion of five 5′-coding exons (Rapp et al. 1983b; Molders et al. 1985; Ishikawa et al. 1986; Stanton and Cooper 1987). Inspection of the amino acid sequences of c-*raf* (and A-*raf*) in this region reveals two amino acid sequence motifs that may affect the structure and function of *raf* kinase. The first is a cluster of substrate sites for serine/threonine and tyrosine kinases (YQRRAS) present only in c-*raf* around position 40. This site may be phosphorylated either by *raf* kinase itself or by other serine/threonine or tyrosine kinases in the signal transduction network. The second obvious structure is a cysteine-rich sequence with the potential to fold into "zinc fingers" (Fig. 1) (Berg 1986). This potential metal-binding motif is part of the first of three highly conserved regions (CR1, CR2, and CR3) that become apparent by comparing the amino acid sequences of human A-*raf* and c-*raf* (Fig. 1) (Rapp et al. 1987; Heidecker et al. 1988). The CR2 domain is remarkable in that it is very rich in serine and threonines, and CR3 corresponds to the actual kinase domain.

The observation that removal of exons 1–5 activates kinase and transforming activity is consistent with a structure for c-*raf* (Fig. 2) similar to the one proposed by P. Parker for the only other family of cytoplasmic serine/threonine protein kinases with potential zinc fingers, the protein kinase C (PKC) family (see Parker et al. 1986; Kikkawa et al., this volume). In the case of c-*raf*, there are two arrays of cysteines and histidines in CR1. The first has the zinc finger consensus sequences of $C\text{-}X_2\text{-}C\text{-}X_9\text{-}C\text{-}X_2\text{-}C$ (position 150–166), and the second is degenerate ($H\text{-}X_2\text{-}C\text{-}X_7\text{-}C$, position 171-182). Molecular modeling (in collaboration with G. Michaels and R. Feldman, NIH) of the first finger indicates a rather hydrophobic surface that is likely to be buried within a folded structure such as illustrated in Figure 2.

We have investigated the role that specific structural features may play using *raf* mutants generated by gene fusion and deletion, linker insertion, and site-directed mutagenesis. These domains were mostly identified on the basis of homology of the *raf* family members with each other or with other protein kinases. They include

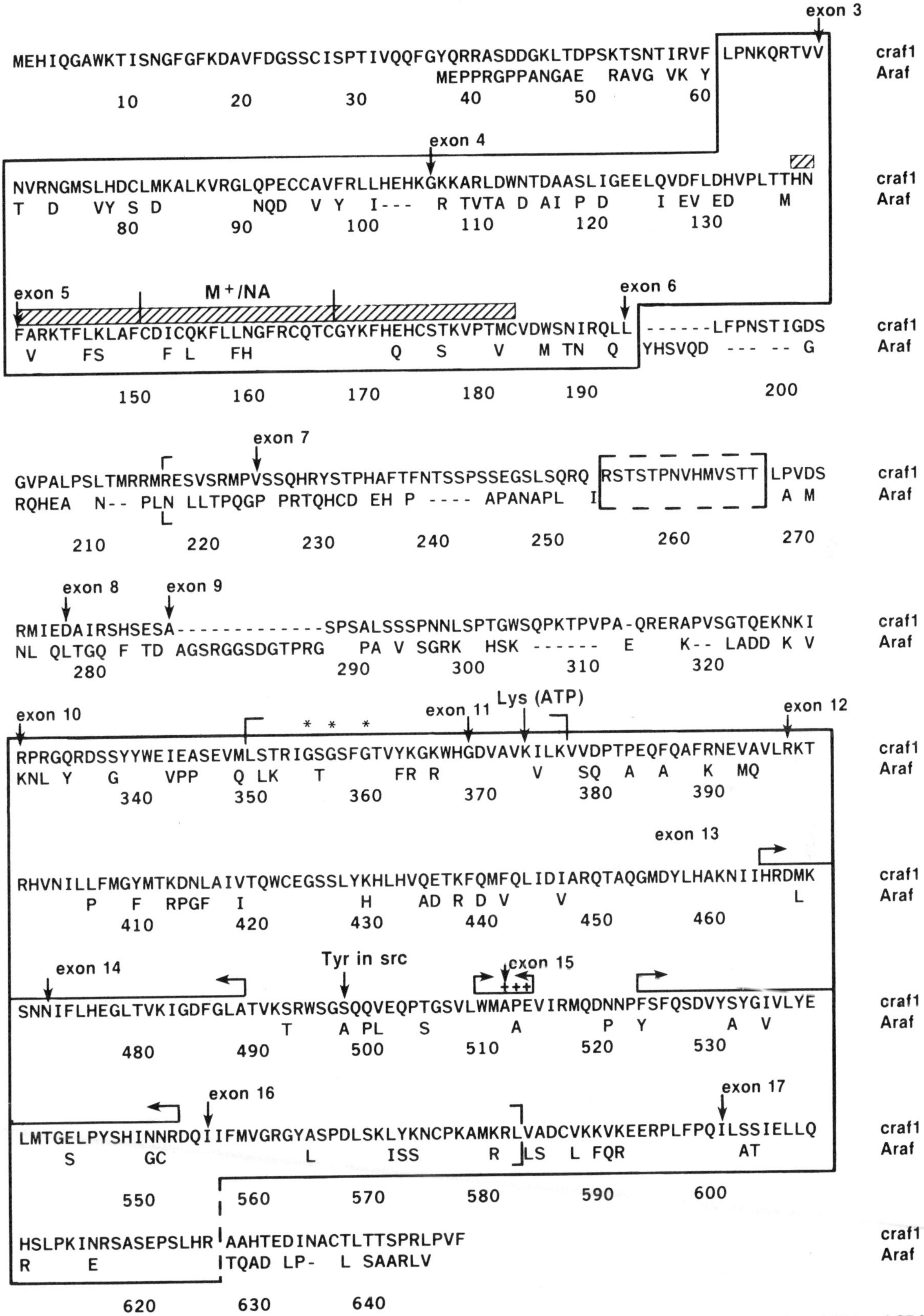

Figure 1. Comparison of full-length amino acid sequence of human c-*raf*-1 and A-*raf*-1. Highly conserved regions (CR1 and CR2 and CR3) between the two proteins are boxed; dashed lines indicate gaps introduced to align the sequences; open spaces represent amino acid identities with respect to c-*raf*-1. Sequences constituting the proposed zinc finger, the ATP-binding sites, and the site corresponding to the major phosphorylation site in *src* are indicated. (*) G-X-G-X-X-G sequence that is part of the ATP-binding consensus; (+) APE sequence of the kinase consensus. Hatched bars or brackets indicate sequences that have homology with PKC and cGMP protein kinase, respectively. Sequences are given in the standard single-letter abbreviation code.

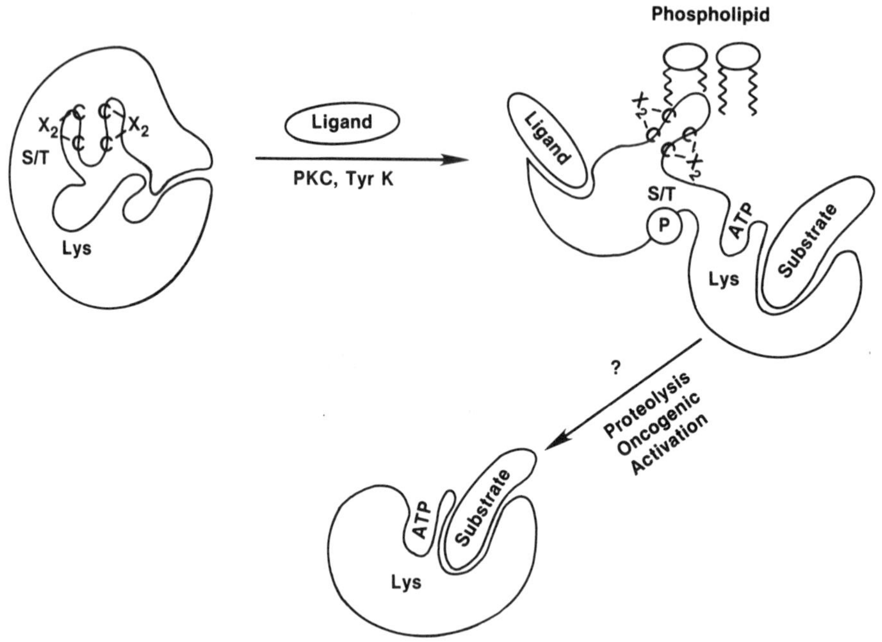

Figure 2. Model of c-*raf* protein activation. As yet, unidentified ligands bind to the regulatory domain, causing a conformational change of the *raf* protein. As a consequence, the protein molecule can bind ATP and phosphorylate unknown substrate(s). An alternative route of activation is proteolytic cleavage in the hinge region, which results in constitutive expression of a probably short-lived kinase moiety resembling oncogenically activated truncated *raf* protein.

the zinc finger domain, the amino-terminal "substrate" site, and the ATP-binding domain. Mutant *raf* genes were incorporated in eukaryotic expression vectors and assayed for their transforming activity in NIH-3T3 cells and for kinase activity in immunoprecipitation kinase assays (Huleihel et al. 1986; Beck et al. 1987; G. Heidecker et al., in prep.). The constructs analyzed are summarized in Figure 3. Several conclusions can be drawn from these analyses. First, amino-terminal truncation renders c-*raf* and A-*raf* transforming and constitutively activates their kinase activities. Second, high levels of expression of full-length *raf*, such as seen in cells carrying the constructs EC, EC1, EC3, and EC4, did not result in cell transformation, since these *raf* kinases still depend on mitogen treatment for activation (Roberts et al., this volume). Third, irritation of the protein structure by site-directed mutagenesis at several sites modulates *raf* activity. Mutation of Lys-56 (*raf* position 56 in v-*raf*) to glutamine enhances transforming activity, whereas a change to glutamic acid results in a less-transformed phenotype in cells carrying this allele. Lys-56 is located three residues downstream from the presumed ATP-binding lysine (position 53). Lys-53 was shown to be crucial for *raf* activity by conversion to tryptophan, which resulted in loss of both kinase and transforming activity. Full-length c-*raf* was moderately activated by a linker insertion into CR2 (Heidecker et al. 1988; Rapp et al. 1988).

The third mode of activating the *raf* kinase activity, besides truncation and point mutations, was achieved by fusing retroviral *gag* sequences to the intact amino terminus of c-*raf*. This activation is independent of *gag* myristilation. This fusion could, however, distort the structure of the amino terminus of *raf* and affect a potential kinase substrate site mentioned above. A pseudosubstrate sequence (KRGA) is present in the amino terminus of PKC, and peptides containing this region have been shown to inhibit the kinase activity of PKC (House and Kemp 1987). However, point mutations in the putative *raf* substrate region converting the serine into alanine or changing the two arginines into isoleucine and glutamic acid did not activate *raf* activity. We are currently testing whether any of these changes rendered the protein unresponsive to mitogen activation.

Analogous to a model for PKC activation, we speculate that normally *raf* kinase activity is regulated by binding of a ligand(s) (as yet unidentified) that triggers an opening of the protein structure rendering the kinase domain accessible to substrate(s). Activation and concomitant unfolding of the c-*raf* protein may also result from specific phosphorylations by other mitogen-regulated kinases (Rapp et al. 1987, 1988; Roberts et al., this volume). Activation-associated unfolding would likely expose the potential zinc finger, possibly allowing contacts to a variety of cellular structures (e.g., the plasma membrane), to other proteins with cysteine arrays via divalent cation (Zn) bridges (Frankel et al. 1988), or to nucleic acid (Berg 1986). Therefore, there may be something literal about the notion that mitogenic signals are "handed down" from the cell surface to the nucleus.

Signaling by *raf* Proteins

What other evidence indicates that c-*raf* protein kinase is involved in signal transduction, given that its oncogenic forms function as constitutive growth stimulators? We have used a variety of strategies to map *raf* function in signaling pathways; five of these will be briefly discussed here (Rapp et al. 1988). Two approaches utilize cells conditionally or permanently blocked in cellular *ras* activity. The conditional block was achieved by antibody microinjection, as described in detail by Stacey et al. (this volume). Stacey observed that NIH-3T3 cells stop growing in serum-containing medium following injection of a specific anti-*ras* monoclonal antibody (Mulcahy et al. 1985). Of all the oncogenes tested, only *raf* family and *mos* serine/threonine protein kinases could overcome this block (Huleihel et al. 1986; Smith et al. 1986; Beck et al. 1987). Earlier, a similar approach to classify oncogenes in functional categories had been taken by Bassin and co-workers (Noda et al. 1983), who isolated flat revertants of Ki-*ras*-transformed NIH-3T3 cells and then tested them for susceptibility or cross-resistance to transformation by other *ras* and non-*ras* oncogenes. In this system, the block to Ki-*ras* transformation is nonconditional, and therefore derivation of the suppressed line required that these cells retained their ability to grow in serum-containing medium. In other words, if *ras* function is normally obligatory for transmission of mitogenic signals from serum factors, the suppression in the revertants either had to be leaky or was associated with the opening of an alternate pathway(s). Oncogenes susceptible to cross-suppression in Ki-*ras* revertant cells included *src*, *fes*, v-Ki-*ras*, v-Ha-*ras*, and c-Ha-*ras*, whereas suppression was overcome by *sis*, *fms*, and *mos* as well as by c-*raf* and A-*raf* (Noda et al. 1983; Rapp et al. 1988; U.R. Rapp and R. Bassin, unpubl.). Thus, the data from the revertant cell transformation and the anti-*ras* antibody microinjection agree, with the exception of two oncogenes, *sis* and *fms*, that are thought to act analogously to ligand-activated PDGF and colony-stimulating factor 1 (CSF-1) receptor, respectively. Since PDGF is the major mitogen present in serum, the opening of an alternate pathway in the course of Ki-*ras* revertant cell isolation, if it occurs, would be expected to allow continued stimulation by PDGF. It is interesting to note in this context that both the PDGF and the CSF-1 receptors appear to activate a common signaling pathway, as judged by the coincident pattern of potential substrate proteins detectable by precipitation with anti-P.Tyr antibody in cells upon activation of either receptor (Roberts et al., this volume).

The third approach we took utilized cells that depend on the presence of a specific mitogen for growth, such as IL-3-dependent myeloid FDC-P1 cells or CSF-1-dependent primary macrophages. We tested the likelihood of *raf* and other protein kinase involvement in IL-3 (CSF-1) signal transduction by introducing exogenous *raf* and other oncogenes and challenging the cells to grow in the absence of growth factor (Table 2).

Table 2. Abrogation of Growth-factor Dependence

Virus	Oncogene	FDC-P1 cell −IL-3
None	—	−
src-MLV	v-*src*	+
abl-MLV	v-*abl*	+
fms-MLV	v-*fms*	+
*erb*B-MLV	v-*erb*B	+
3611MSV	v-*raf*	−
J-1	v-*raf*/*mil*	−
91V A-*raf* MSV	A-*raf*	−
Mo-MSV	v-*mos*	−
Ha-MSV	v-Ha-*ras*	−
J-2	v-*raf*/*mil* + *myc*	+
J-3	v-*myc*	+
J-5	v-*myc*	+
FBR	v-*fos*	−

FDC-P1 cells were infected with the indicated viruses in the presence of polybrene (25 μM) and maintained in media supplemented with IL-3 for 4 days. They were then cultured in media without IL-3 for 4 weeks. During this period the majority of cells died out, and factor-independent cells (indicated by +) emerged.

The addition of IL-3 to FDC-P1 cells results in a rapid tyrosine phosphorylation of a discrete set of proteins followed by induction of c-*myc* (Dean et al. 1987; Isfort et al. 1988b). One of the proteins consistently found phosphorylated is a 140-kD polypeptide that is also precipitable in immune complexes using antibodies directed against IL-3. This protein is therefore a likely candidate for the IL-3 receptor with intrinsic tyrosine kinase activity (Isfort et al. 1988a).

To test whether other activated tyrosine kinases will substitute for the triggered IL-3 receptor, a series of recombinant and natural isolates of defective murine retroviruses carrying different tyrosine kinase oncogenes were used to infect FDC-P1 cells. Virus-positive cells were identified by infectious cell center assays on NIH-3T3 indicator cells and challenged to grow in the absence of IL-3. The role of other oncogenes as signal transmitters downstream from the tyrosine kinase was similarly tested for *ras*, *myc*, and the serine/threonine kinase oncogenes *raf* and *mos*. Two classes of oncogenes were effective in relieving FDC-P1 cells (which normally express both c-*raf* and A-*raf*) from IL-3 requirements, first tyrosine kinases (efficiency of abrogation 10–30%), followed by *myc* (efficiency 1–2%). In contrast, *raf* genes had barely detectable effects (0.01%), although this low activity was clearly significant (about 100-fold above background). Since all tyrosine-kinase-abrogated cells, including those expressing temperature-sensitive *abl* under permissive conditions, had high constitutive levels of endogenous *myc*, it follows that *myc* activation is a downstream event (Fig. 4) (Cleveland et al. 1988). The fact that *raf* abrogates less efficiently than exogenous *myc* indicates that *raf* and *myc* are in different pathways.

In primary macrophages, others have shown that the CSF-1 tyrosine-kinase-receptor-derived oncogene *fms* alone abrogates CSF-1 dependence (Wheeler 1987). We observed that either *raf* or *myc* was ineffective in

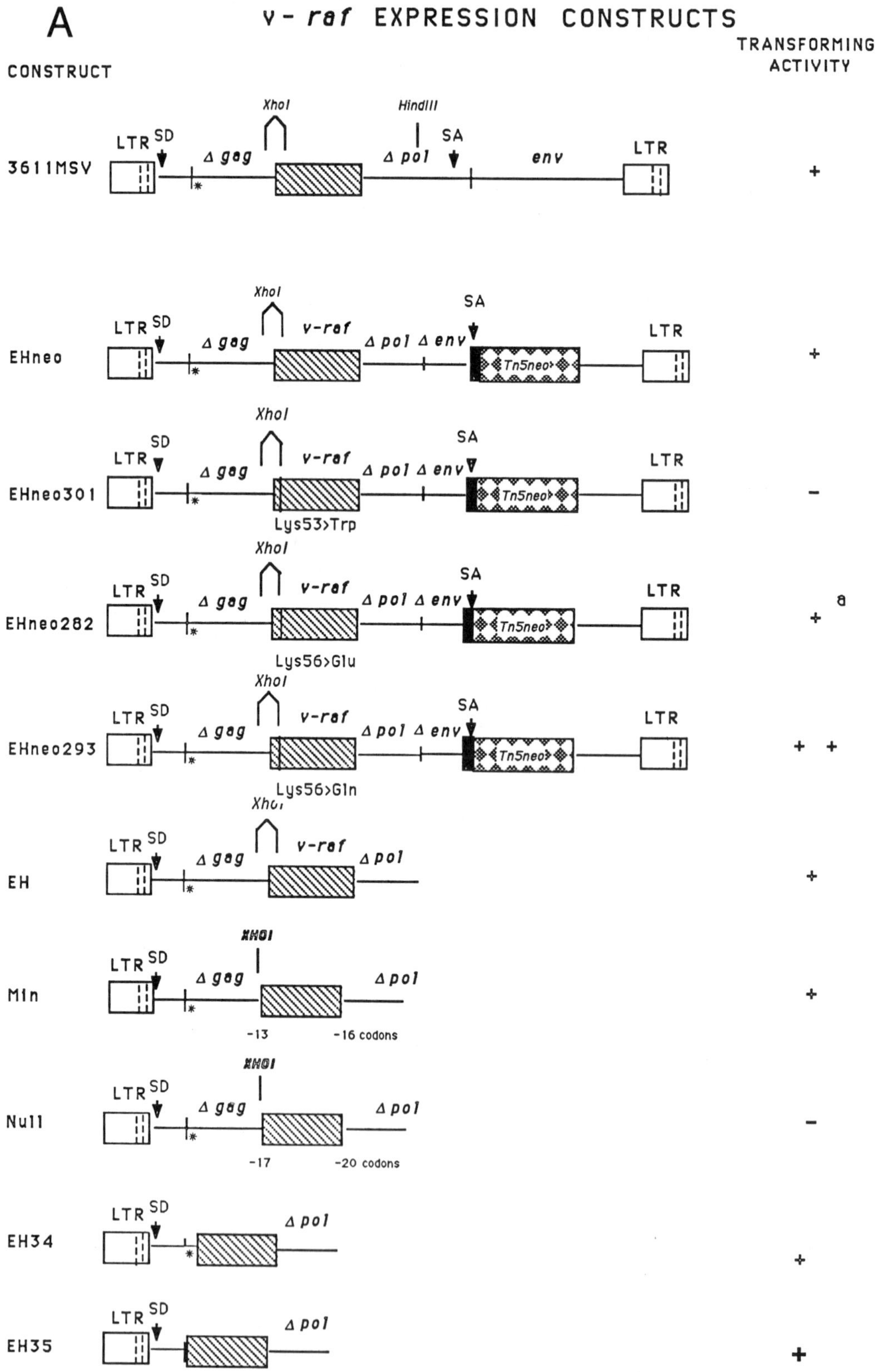

Figure 3. (*See facing page for part B.*)

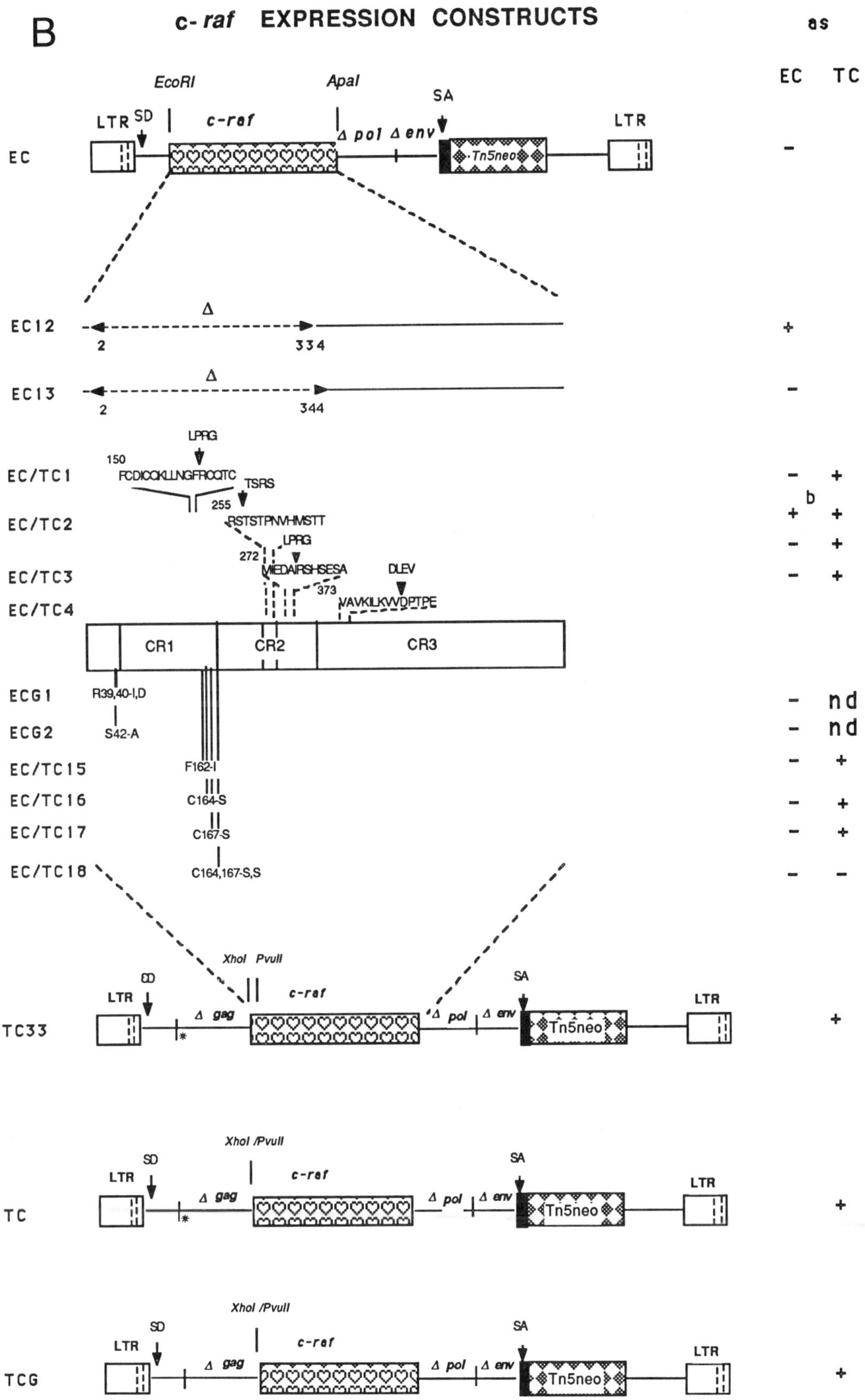

Figure 3. (*See following page for part C and legend.*)

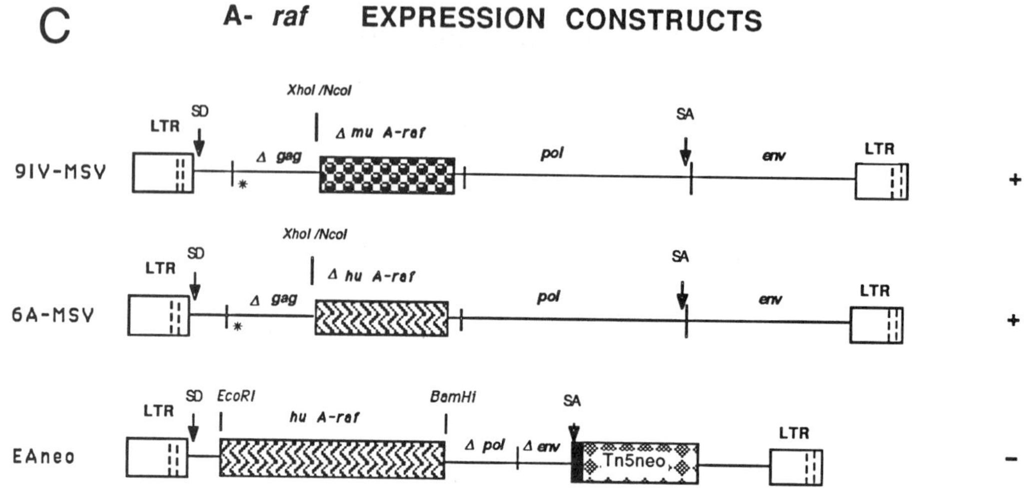

Figure 3. Transforming activity of *raf* expression vectors. Neo-vectors contain the Tn*5* neomycin-resistance gene, which is expressed as a subgenomic RNA after splicing at an RSV promoter inserted in front of it. Constructs that do not contain the *neo* gene were cotransfected with pSV2*neo* DNA (molar ratio 10:1) (Southern and Berg 1982). All constructs were assayed for induction of transformed foci without selection in addition to scoring the phenotype of *neo*-resistant foci in a parallel experiment. Expression of *raf* protein of the appropriate size in *neo*-resistant cultures was verified. Constructs that were scored as positive transformed more than 70% of all *neo* foci, both following transfection and, where possible, after infection with rescued virus. Special features of the constructs are as follows: SD and SA are splice donor and acceptor sites, respectively; the asterisk indicates the presence of the myristilation site. The drawings are to scale. Only 10%[a] and only 30%[b] of *neo* foci transformed.

abrogating CSF-1 dependence in these cells, but a combination of the two was very efficient (Blasi et al. 1985; Cleveland et al. 1986). These findings agree with *myc* and *raf* being in two separate pathways and demonstrate synergistic interaction between them in growth factor abrogation. We have yet to test whether a combination of *myc* and *raf* raises the abrogation efficiency in FDC-P1 cells to levels seen with tyrosine kinase oncogenes. We conclude from the abrogation experiments that the mitogenic IL-3 signal cannot be transmitted by *raf* alone, if *raf* is normally involved in this pathway at all. Similarly, if *raf* plays a role in CSF-1 signal transmission, it is insufficient by itself. The fact that *raf* synergizes with *myc* in abrogating CSF-1 dependence suggests that the CSF-1-activated tyrosine kinase activity of the CSF-1 receptor triggers minimally two pathways that are required jointly for induction of mitogenesis, a *myc* pathway and a *raf* pathway. Thus, tyrosine kinase receptors such as the CSF-1 receptor, although not belonging to the category of serpentine receptors (Marsh and Herskowitz, this volume), appear to speak with a forked tongue.

A schematic that depicts the channeling of signal transfer from a tyrosine kinase receptor and that is consistent with the *ras* suppressor cell experiments and the growth factor abrogation and synergism experiments is outlined in Figure 5. *raf* and *myc* have been placed in separate pathways because of the observed synergism in vivo (Rapp et al 1985c) and in vitro (Blasi et al. 1985). *raf* is located downstream from *ras* because *raf* kinases overcome a block to cellular *ras* function. It remains to be demonstrated whether the observed independence of *raf* from *ras* indicates a parallel path for *raf* or the indicated downstream position (Smith et al. 1986; Rapp et al. 1987). PKC was placed in the *myc*, or more generally competence pathway (see Jones, this volume) because PKC activation has been demonstrated to elevate levels of *myc* expression. Activation of PKC and *raf* kinases may occur in several ways, perhaps depending on the category of tyrosine kinase receptor, including recruitment of second messengers derived from tyrosine-stimulated lipid turnover (Berridge and Irvine 1984) and direct phosphorylation by tyrosine (see Roberts et al., this volume) and serine/threonine-specific protein kinases. Transcriptional activation of PKC and *raf* target genes is thought to occur by activation/inactivation of preexisting transcription factors via phosphorylation of such factors or their inhibitors (Rapp et al. 1986, 1987).

There are a number of predictions that this working hypothesis makes. One of them is that c-*raf* kinases are mitogen-regulated in activity. Mitogen stimulation of *raf* protein may alter its properties in several ways, including its kinase activity and subcellular distribution. Mitogen treatment has been demonstrated to activate c-*raf* kinase (Roberts et al., this volume). We have examined whether c-*raf* protein changes location in the cell upon mitogen or tumor promoter stimulation of serum-deprived cell populations using either PDGF or TPA. Rapid (maximal at 5 min with PDGF and 20 min with TPA) redistribution of *raf* was observed from the normally diffuse cytoplasmic fluorescence into a

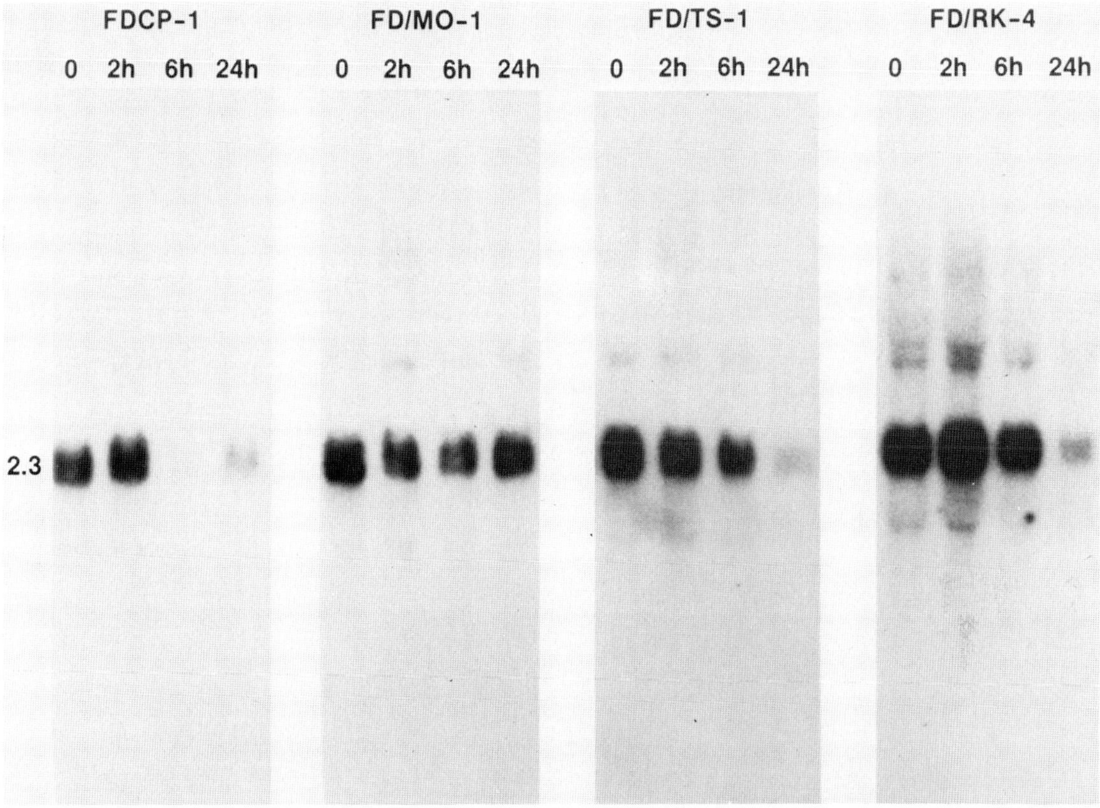

Figure 4. Effects of temperature shift on expression of c-*myc* in FD/*abl* virus clones. Cultures of FD/*abl* clones expanded at 32°C in the absence of IL-3 were shifted to 39°C, and, at indicated times, a portion of the culture was removed, and total RNA was isolated. As a control, parental FDC-P1 cells were also expanded at 32°C but in the presence of IL-3, washed twice and resuspended in medium without IL-3 and shifted to 39°C. Blots were exposed for 2 days.

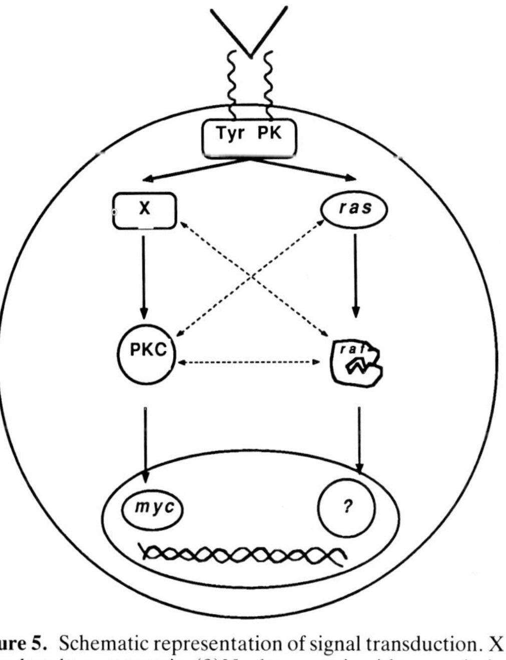

Figure 5. Schematic representation of signal transduction. X is thought to be a *ras* protein; (?) Nuclear protein with transcriptional regulatory activity. Lines denote primary pathways; dashed lines indicate less frequent cross-talk.

perinuclear area (Fig. 6). These results suggest that activation of PKC with TPA (Anderson et al. 1985; Nishizuka 1986), and perhaps with PDGF (Pelech et al. 1986), in turn can modulate *raf*, and that this may be part of the pathway for transduction of the mitogen signal to the nucleus.

In conclusion, *raf* family serine/threonine protein kinases have the hallmarks of signal transmitters involved in communicating between mitogen receptors at the cell surface and the nucleus. The types of approaches we have taken so far have provided us with a broad outline of the position in cellular growth control pathways where *raf* kinases function. We speculate that tyrosine kinase class receptors activate minimally two pathways, directly or indirectly, to trigger mitogenesis: a PKC (competence) and a *raf* (progression) pathway—at least one of which also includes *ras*, presumably upstream of *raf* (Fig. 4). We are now in the process of identifying activating ligands as well as substrates of *raf* kinases. Toward the latter goal, we are characterizing a set of characteristic phosphoproteins specifically associated with activated A-*raf* and c-*raf* kinases. These putative substrates might include those responsible for the transcriptional activation of experimental target genes (Table 1) (C. Wasylik et al., in prep.).

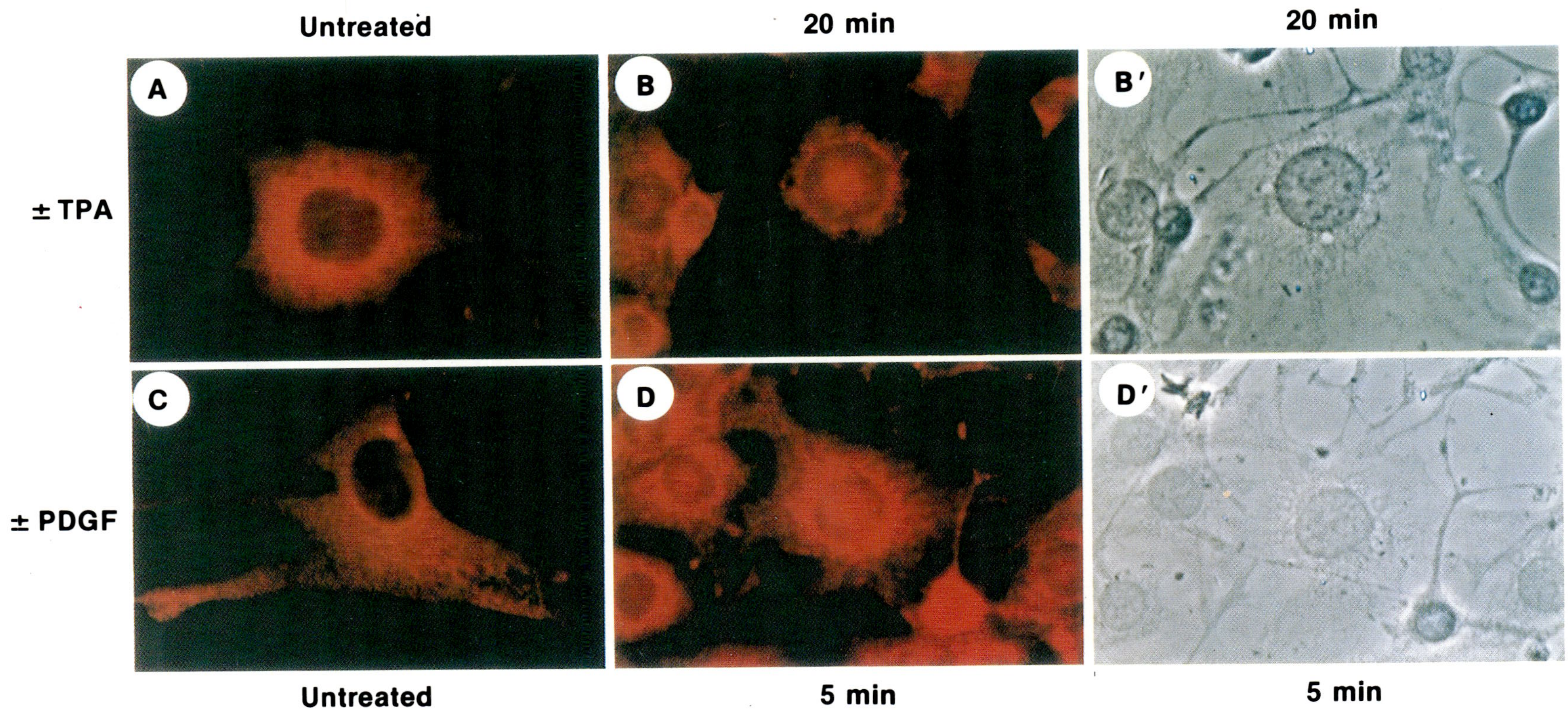

Figure 6. Immunohistochemical localization of c-*raf* kinase in NIH-3T3 clone 7 cells. (*A*,*C*) Untreated NIH-3T3 cells; (*B*) cells treated with PDGF for 5 min; (*D*) cells after TPA treatment for 20 min. Following the incubations as indicated, the cells were fixed and analyzed for c-*raf* kinase by immunohistochemical staining as described in Materials and Methods. (*B*′, *D*′) Phase-contrast photographs of the same cell field as depicted in panels *B* and *D*, respectively. No staining was observed when the antiserum was preabsorbed with SP63 peptide prior to use in the immunohistochemical localization of *raf* in cells treated with PDGF and TPA. Staining was also not detected with preimmune rabbit serum (data not shown).

REFERENCES

Anderson, W.B., A. Estival, H. Tapiovaara, and R. Gopalakristna. 1985. Altered subcellular distribution of protein kinase C (a phorbol ester receptor). *Adv. Cyclic Nucleotide Protein Phosphorylation Res.* **19:** 287.

Bayer, E.A. and M. Wilchek. 1979. The use of the avidin-biotin complex as a tool in molecular biology. *Methods Biochem. Anal.* **26:** 1.

Beck, T.W., M. Huleihel, M.A. Gunnell, T.I. Bonner, and U.R. Rapp. 1987. The complete coding sequence of the human A-*raf*-1 oncogene and transforming activity of a human A-*raf* carrying retrovirus. *Nucleic Acids Res.* **15:** 595.

Berg, J.M. 1986. Potential metal binding domains in nucleic acid binding proteins. *Science* **232:** 485.

Berridge, M.J. and R.F. Irvine. 1984. Inositol triphosphate, a novel second messenger in cellular signal transduction. *Nature* **312:** 315.

Blasi, E., B. Mathieson, L. Varesio, J.L. Cleveland, P.A. Borchert, and U.R. Rapp. 1985. Selective immortalization of murine macrophages from fresh bone marrow by a *raf/myc* recombinant murine retrovirus. *Nature* **318:** 667.

Bonner, T.I., H. Oppermann, P. Seeburg, S.B. Kerby, M.A. Gunnell, A.C. Young, and U.R. Rapp. 1986. The complete coding sequence of the human *raf* oncogene and the corresponding structure of the c-*raf*-1 gene. *Nucleic Acids Res.* **14:** 1009.

Cleveland, J.L., U.R. Rapp, and W.L. Farrar. 1987. Role of c-*myc* and other genes in interleukin-2 regulated CT6 T-lymphocytes and their malignant variants. *J. Immunol.* **138:** 3495.

Cleveland, J.L., Y. Weinstein, D.S. Askew, J.N. Ihle, and U.R. Rapp. 1986. Transformation and insertional mutagenesis in vitro of primary hematopoietic stem cell cultures. *Curr. Top. Microbiol. Immunol.* **132:** 44.

Cleveland, J.L., M. Dean, J.-Y Wang, A.-M Hedge, J.N. Ihle, and U.R. Rapp. 1988. Abrogation of IL-3 dependence of myeloid FDC-P1 by tyrosine kinase oncogenes is associated with induction of c-*myc*. *Curr. Top. Microbiol. Immunol.* (in press).

Dean, M., J.L. Cleveland, U.R. Rapp, and J.N. Ihle. 1987. Role of *myc* in the abrogation of IL-3 dependence of myeloid FDC-P1 cells. *Oncogene Res.* **1:** 279.

Engelman, A. and N. Rosenberg. 1987. Isolation of temperature sensitive Abelson virus mutants by site-directed mutagenesis. *Proc. Natl. Acad. Sci.* **84:** 8021.

Frankel, A.D., D.S. Bredt, and C.O. Pabo. 1988. Tat protein from human immunodeficiency virus forms metal-linked dimer. *Science* **240:** 70.

Graham, F.L. and A.J. van der Eb. 1973. A new technique for the assay of infectivity of human adenovirus 5 DNA. *Virology* **52:** 456.

Guesdon, J.L., T. Termynck, and S.Avremeas. 1979. The use of avidin-biotin interaction in immunoenzymatic techniques. *J. Histochem. Cytochem.* **27:** 1131.

Heidecker, G., J.L. Cleveland, T.W. Beck, M. Huleihel, W. Kolch, S.M. Storm, and U.R. Rapp. 1988. Role of *raf* and *myc* oncogenes in signal transduction. In *Genes and signal transduction in multistage carcinogenesis* (ed. N. Coburn) p. 339. Marcel Dekker, New York. (In press.)

House, C. and B.E. Kemp. 1987. Protein kinase C contains a pseudosubstrate prototype in its regulatory domain. *Science* **238:** 1726.

Huleihel, M., M. Goldsborough, J.L. Cleveland, M. Gunnell, T. Bonner, and U.R. Rapp. 1986. Characterization of murine A-*raf*, a new oncogene related to the v-*raf* oncogene. *Mol. Cell. Biol.* **6:** 2655.

Ihle, J.N., J. Keller, A. Rein, J.L. Cleveland, and U.R. Rapp. 1985. Interleukin-3 regulation of the growth of normal and transformed hematopoietic cells. *Cancer Cells* **3:** 211.

Ikawa, S., M. Fukui, Y. Ueyama, N. Tamaoki, T. Yamamoto, and K. Toyoshima. 1988. B-*raf*, a new member of the *raf* family, is activated by DNA rearrangement. *Mol. Cell. Biol.* **8:** 2651.

Isfort, R.J., D. Stevens, W.S. May, and J.N. Ihle. 1988a. IL-3 binding to a 140 kd phosphotyrosine containing cell surface protein. *Proc. Natl. Acad. Sci.* **87:**(in press).

Isfort, R.J., R. Abrahams, R.D. Huhn, A.R. Frankelton, Jr., and J.N. Ihle. 1988b. Interleukin-3 stimulation of factor dependent myeloid cell lines induces the tyrosine phosphorylation of several cellular substrates. *J. Biol. Chem.* (in press).

Ishikawa, F., F. Takaku, K. Hayashi, M. Nagao, and T. Sugimura. 1986. Activation of c-*raf* during transfection of hepatocellular carcinoma DNA. *Proc. Natl. Acad. Sci.* **83:** 3209.

Kipreos, E.D. and J.Y. Wang. 1988. Reversible dependence on growth factor interleukin-3 in myeloid cells expressing temperature sensitive v-*abl* oncogene. *Oncogene Res.* **2:** 277.

Maniatis, T., E.F. Fritsch, and J. Sambrook. 1982. *Molecular cloning. A laboratory manual.* Cold Spring Harbor Laboratory, Cold Spring Harbor, New York.

Molders, H., J. Defesche, D. Mueller, T.I. Bonner, U.R. Rapp, and R. Mueller. 1985. Integration of transfected LTR sequences into the c-*raf* proto-oncogene: Activation by promoter insertion. *EMBO J.* **4:** 693.

Mulcahy, L.S., M.R. Smith, and D.W. Stacey. 1985. Requirement for *ras* proto-oncogene function during serum-stimulated growth of NIH 3T3 cells. *Nature* **313:** 241.

Nishizuka, Y. 1986. Studies and perspectives of protein kinase C. *Science* **233:** 305.

Noda, M., Z. Selinger, E.M. Scolnick, and R.H. Bassin. 1983. Flat revertants isolated from Kirsten sarcoma virus-transformed cells are resistant to the action of specific oncogenes. *Proc. Natl. Acad. Sci.* **80:** 5602.

Parker, P.J., L. Coussens, L. Rhee, S. Young, E. Chen, S. Stabel, M.D. Waterfield, and A. Ullrich. 1986. The complete primary structure of protein kinase C, the major phorbol ester receptor. *Science* **233:** 853.

Pelech, S.L., K.E. Meier, and E.G. Krebs. 1986. Rapid microassay for protein kinase C translocation in Swiss 3T3 cells. *Biochemistry* **25:** 8348.

Rapp, U.R., T.I. Bonner, and J.L. Cleveland. 1986. The *raf* oncogene. In *Retroviruses and human pathology* (ed. R.C. Gallo et al.), p. 449. Humana Press, New Jersey.

Rapp, U.R., J.L. Cleveland, T.I. Bonner, and S.M. Storm. 1988. The *raf* oncogenes. In *Handbook of oncogenes.* (ed. P. Reddy et al.), p. 213. Elsevier Publishing, Amsterdam.

Rapp, U.R., F.H. Reynolds, and J.R. Stephenson. 1983a. New mammalian transforming retrovirus: Demonstration of a polypeptide gene product. *J. Virol.* **45:** 914.

Rapp, U.R., T.I. Bonner, K. Moelling, K. Bister, and J. Ihle. 1985a. Genes and gene products involved in growth regulation of tumor cells. *Recent Results Cancer Res.*, p. 221.

Rapp, U.R., J.L. Cleveland, K. Brightman, A. Scott, and J.N. Ihle. 1985b. Abrogation of IL-3 and IL-2 dependence by recombinant murine retroviruses expressing v-*myc* oncogenes. *Nature* **317:** 434.

Rapp, U.R., J.L. Cleveland, S.M. Storm, T.W. Beck, and M. Huleihel. 1987. Transformation by *raf* and *myc* oncogenes. In *Oncogenes and cancer* (ed. S.A. Aaronson et al.), p. 55. VNU Science Press, Utrecht, Netherlands.

Rapp, U.R., M.D. Goldsborough, G.E. Mark, T.I. Bonner, J. Groffen, F.H. Reynolds, and J. Stephenson. 1983b. Structure and biological activity of v-*raf*, a unique oncogene transduced by a retrovirus. *Proc. Natl. Acad. Sci.* **80:** 4218.

Rapp, U.R., J.L. Cleveland, T.N. Frederickson, K.L. Holmes, H.C. Morse, H.W. Jansen, T. Patschinsky, and K. Bister. 1985c. Rapid induction of hemapoietic neoplasms in newborn mice by a *raf(mil)/myc* recombinant murine retrovirus. *J. Virol.* **55:** 23.

Seth, A. and G.F. van der Woude. 1988. The mos oncogene. In *Genes and signal transduction in multistage carcinogenesis* (ed. N. Coburn). Marcel Dekker, New York.

Schultz, A.M., T.D. Copeland, S. Oroszlan, and U.R. Rapp. 1988. Identification and characterization of c-*raf* phosphoproteins in transformed murine cells. *Oncogene* **2:** 187.

Smith, M.R., S.J. DeGudicibus, and D.W. Stacey. 1986. Requirement for c-*ras* proteins during viral oncogene transformation. *Nature* **320:** 540.

Southern, P.J. and P. Berg. 1982. Transformation of mammalian cells to antibiotic resistance with a bacterial gene under control of SV40 early region promoter. *J. Mol. Appl. Genet.* **1:** 327.

Stanton, Jr., V.P. and G.M. Cooper. 1987. Activation of human *raf* transforming genes by deletion of normal amino-terminal coding sequences. *Mol. Cell. Biol.* **7:** 1171.

Wheeler, E.F., D. Askew, S. May, J.N. Ihle, and C.J. Sherr. 1987. The v-*fms* oncogene induces factor independent growth and transformation of the IL-3 dependent myeloid cell line, FDC-P1. *Mol. Cell. Biol.* **7:** 1673.

Calmodulin and Calmodulin-regulated Protein Kinases as Transducers of Intracellular Calcium Signals

T.J. LUKAS,* J. HAIECH,*† W. LAU,* T.A. CRAIG,* W.E. ZIMMER,*
R.L. SHATTUCK,* M.O. SHOEMAKER,* AND D.M. WATTERSON*
*Laboratory of Cellular and Molecular Physiology, Howard Hughes Medical Institute, and Department of Pharmacology, Vanderbilt University, Nashville, Tennessee 37232; †LP8402 CNRS-INSERM, Montpellier, France

Calcium serves as an intracellular signal transducer through its ability to reversibly interact with and modulate the structure and function of a class of proteins referred to as calcium-modulated proteins. One of these calcium-modulated proteins found in all eukaryotic cells is calmodulin (for review, see Van Eldik et al. 1982). Calmodulin is an integral subunit of a diverse array of enzymes and cytoskeletal structures and thus provides a common motif to one branch of the calcium signal transduction pathway. One class of enzymes that have calmodulin as the calcium-binding effector subunit are the protein kinases. It is through the proper expression and assembly of the calmodulin and protein kinase subunits that one set of signal transduction pathways are available to the eukaryotic cell. In this way, the chemical signal generated by a transient rise in the molar concentration of ionized intracellular calcium is transduced into a macromolecular conformational signal through its interaction with calmodulin; the signal is then amplified by the transduction of the calmodulin conformational signal into a protein kinase conformational signal, with a resultant increase in specific phosphotransferase activity within the cell.

In general, protein kinases are biological signal transducers and amplifiers that are required, like calmodulin, for cellular homeostasis. Although protein kinases vary widely in terms of their substrate specificity, regulation, and distribution, they can be classified in most cases as tyrosine kinases or serine/threonine kinases based on the amino acid in the protein substrate that is the phosphate acceptor (for review, see Edelman et al. 1987). Although all of the protein kinases efficiently use ATP as a nucleotide substrate, there is more selectivity in terms of the peptide substrate. The structural features of the peptide substrate required for effective phosphotransferase activity add an element of physiological specificity based on kinetic selectivity. In addition, mechanistic specificity is enhanced by the presence of effector regulation of certain protein kinases. In the case of calmodulin-dependent protein kinases, the calmodulin subunit is the calcium receptor that transduces a change in the intracellular concentration of ionized calcium into a biological response by modulating the activity of the kinase catalytic subunit (which has a substrate specificity for serine/threonine in all cases characterized to date).

As indicated in Table 1, most of the calmodulin(CaM)-dependent protein kinases are large multisubunit complexes in their active state. Notable exceptions are the myosin light-chain kinases (MLCKs). The simpler subunit structure of the active enzyme makes the MLCKs appealing, for technical reasons, as initial enzymes to examine the molecular basis of calmodulin regulation of a protein kinase. In terms of a defined role in cell function, smooth muscle MLCK is especially appealing because the regulation of enzyme activity is directly related to the role of the enzyme in smooth muscle cell function, i.e., regulation of contraction. Previous studies (Lukas et al. 1986) have identified a region of smooth muscle MLCK that contains a relatively high-affinity calmodulin-binding site. This information has been coupled more recently with site-specific mutagenesis studies (Craig et al. 1987) of calmodulin to gain further insight into the complementary structural features of calmodulin and smooth muscle MLCK that allow a specific, high-affinity interaction and conformational coupling between calmodulin and the kinase catalytic subunit. When this information is

Table 1. Comparison of Calmodulin-dependent Protein Kinases

Protein	Subunits	Molecular weight of holoenzyme
Phosphorylase kinase	$(\alpha_2\beta_2\gamma_2\delta_2)_2$	~1,300,000
Calmodulin-dependent protein kinase II	$(\alpha_x\beta_y)$ where $x + y = 12$	~750,000
Skeletal muscle myosin light-chain kinase	CaM + catalytic subunit	~100,000
Smooth muscle/nonmuscle myosin light-chain kinase	CaM + catalytic subunit	~150,000

coupled with similar studies of other calmodulin-regulated enzymes (e.g., adenylate cyclase), there is the possibility of identifying common structural motifs and similarities in mechanism, as well as obtaining insight into the unique features of each pathway that allow discrete responses to cellular signals while maintaining the appropriate level of integration or "biological cross-talk."

In this paper, we provide an update on some of our ongoing studies of the structural features of calmodulin and calmodulin-regulated enzymes that are focused on the long-term goal of elucidating the molecular mechanisms of calcium signal transduction through calmodulin. We also provide an update on models of how a eukaryotic cell is able to assemble a functional calcium signal transduction–amplification complex from calmodulin and a protein kinase.

RESULTS

Analysis of Protein Kinase Activation by Engineered Calmodulins

On the basis of the study by Craig et al. (1987), which showed that mutagenesis of a central helix charge cluster has a differential effect on three different types of calmodulin-dependent enzymes, a series of related mutant calmodulins were made and examined for their ability to activate the three calmodulin-dependent kinases (MLCK, γ subunit of phosphorylase kinase, and CaM-Kinase II). The activation of phosphodiesterase was examined for comparative purposes. A summary of the results of such a study is reported in Table 2. Although it was appealing to think that the mechanism of regulation might be similar for all the calmodulin-dependent kinases, these results show clearly that each enzyme has a set of unique features that characterize its activation by calmodulin. It appears that MLCK is affected by VU-8, VU-12, and VU-18 calmodulins where a cluster(s) of three acidic residues (glutamic or aspartic residues) has been replaced by a cluster(s) of lysine residues. The cluster mutations are located in different regions of the calmodulin molecule, as indicated in the model structure shown in Figure 1. The γ subunit of phosphorylase kinase is affected by the same set of mutants, but additionally, mutations in the carboxy-terminal end of the central helix (VU-7) of calmodulin have a deleterious effect on its activity. In contrast, CaM-Kinase II is sensitive only to the set of mutations in the carboxy-terminal part of the protein (VU-12 and VU-18). It should be noted that phosphodiesterase is neutral to most of these particular mutations of the calmodulin molecule. Finally, VU-11 calmodulin, which has a mutation in the amino-terminal end of the central helix, activates all the kinases similar to the standard VU-1 calmodulin. From this selected set of mutants, it appears that the effects of the VU-8 calmodulin on MLCK might be more a reflection of the sensitivity of MLCK to modifications of calmodulin that affect its electrostatic surface potential rather than a change in the central helix, per se. In contrast, CaM-Kinase II is sensitive to modifications of the carboxy-terminal part of the calmodulin central helix. Phosphorylase kinase is responsive to the same set of mutants as MLCK, but like CaM-Kinase II, to mutations of the carboxy-terminal end of the central helix. This suggests that the modulatory mechanisms are unique, although there may be some common elements. These initial findings imply that future studies should allow the dissection in more detail of the differences in the mechanism of activation of these calmodulin-dependent kinases.

Calmodulin-binding Peptides and Specificity

Recently, additional amino acid sequence data based on the analysis of cDNA structures (Fig. 2) (Guerriero et al. 1986; D.M. Watterson, unpubl.) have allowed a further analysis of calmodulin binding to and activation of MLCK. As an initial step, synthetic peptides based on the amino acid sequence in the vicinity of the calmodulin-binding site (Fig. 2) were examined for their effects on calmodulin stimulation of MLCK activity. The results obtained with synthetic peptides that antagonize the calmodulin activation of enzymes such as MLCK and phosphodiesterase are summarized in Table 3. These results suggested the relative importance of basic and hydrophobic residues at the amino terminus of the calmodulin-binding site and provided a foundation for current studies that utilize site-specific mutagenesis approaches. Additional parameters to be considered are the calcium dependence and specificity of the peptide binding to calmodulin and other members of the calcium-modulated protein family. As shown in Figure 3, complexes of RS20 and calmodulin are formed under these conditions, in the presence of

Table 2. Enzyme Activator Properties of Mutant Calmodulins

Protein	Mutation	Phosphodiesterase	MLCK	Calmodulin kinase II	Phosphorylase kinase
VU-7	(Ala-88 to Pro)	+	+	+	−
VU-8	(Glu-82–Glu-84 to Lys-82–Lys-84)	+	−	+	−
VU-11	(Lys-75 to Pro)	+	+	+	+
VU-12	(Asp-118–Glu-120 to Lys-118–Lys-120)	+	−	−	−
VU-18	(Glu-82–Glu-84 and Asp-118–Glu-120 to Lys-82–Lys-84 and Lys-118–Lys-120)	+/−	0	−	0

(+) Activation properties similar to VU-1 (both V_{max} and concentration required for half-maximal stimulation); (+/−) maximal activation properties similar to VU-1, but higher concentration required; (−) decreased maximal activation; (0) no activation.

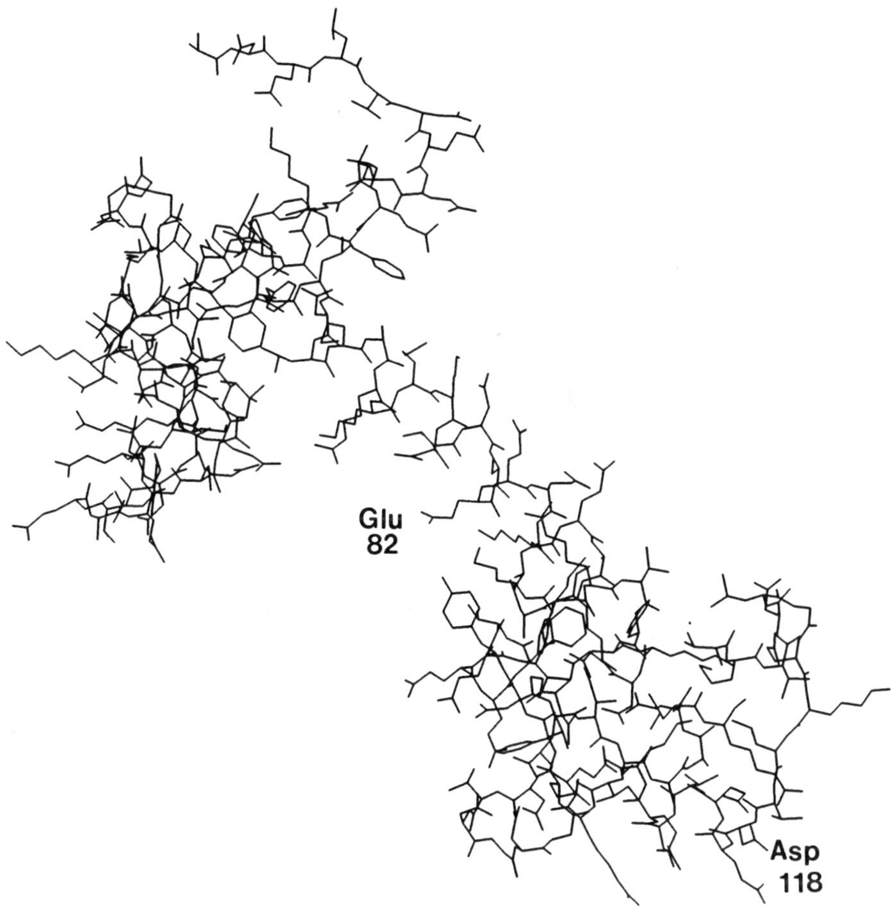

Figure 1. Three-dimensional model structure of VU-1 calmodulin. This structure was generated by Dr. Ray Salemme (E.I. du Pont de Nemours & Co.) using the α-carbon coordinates of rat testes calmodulin (Babu et al. 1985) as a backbone structure and generating side-chain coordinates from a crystallographic library. The final structure was regularized, and energy was minimized by techniques described elsewhere (Wendoloski et al. 1987).

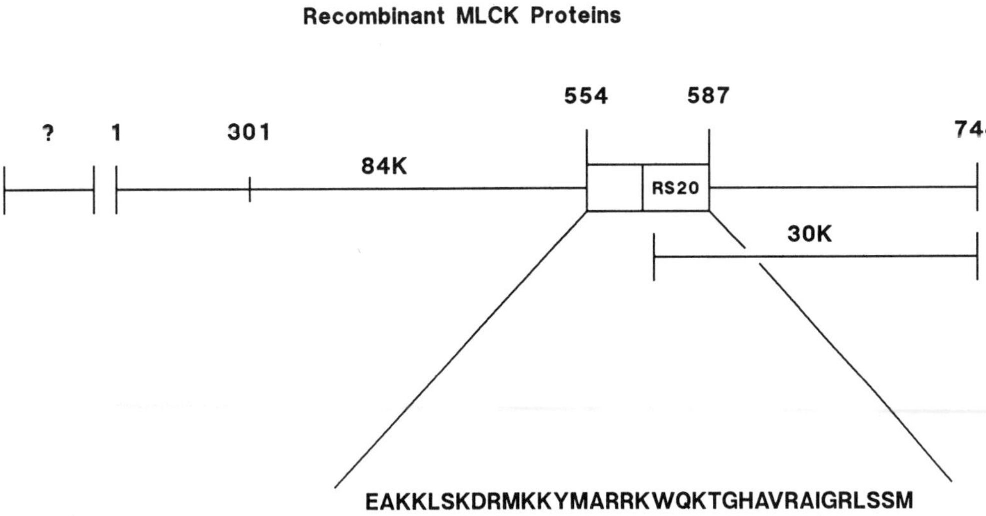

Figure 2. Schematic diagram of MLCK expression products. The 30K (relative SDS-gel mobility) protein was expressed using a cDNA clone inserted into the pVUCH-1 expression vector (Lukas et al. 1987). The 84K protein was expressed from a 2226-bp MLCK cDNA construct inserted into the pVUCH-1 expression system. Both proteins bind to calmodulin-Sepharose, but only the 84K protein exhibits calmodulin-dependent enzymatic activity. In the expanded area, the amino acid sequence of a calmodulin regulatory domain is shown. Amino acid sequence numbering is based on translation of the cDNA insert that contains the coding portion of the 84-kD protein.

Table 3. Synthetic Peptides Based on a Calmodulin-binding Site of smMLCK

Code name	Sequence	Calmodulin inhibitory activity[a]
RS20	R R K W Q K T G H A V R A I G R L S S S	+
KS25	K K Y M A R R K W Q K T G H A V R A I G R L S S S	+
YS23	Y M A R R K W Q K T G H A V R A I G R L S S S	+
KS18	K W Q K T G H A V R A I G R L S S S	+
KL12	K T G H A V R A I G R L	−
RL17	R R K W Q K T G H A V R A I G R L	+

[a]Peptides at various concentrations from 0 to 1 μM were tested in a pH 7.5 assay buffer containing 50 mM HEPES, 0.1 mM $CaCl_2$, 5 mM $MgCl_2$, 2 mM cAMP, and 8 nM vertebrate calmodulin. Reactions were initiated by addition of a fraction of calmodulin-activatable 3′, 5′ phosphodiesterase and incubated for 9 min at 25°C. Reactions were stopped by boiling, and the quantity of 5′-AMP produced was measured by HPLC as described previously (Watterson et al. 1980). (+) Inhibitor decreased the phosphodiesterase activity to basal levels at a concentration of 50 nM or less. (−) Inhibition was less than 50% at 1 μM peptide.

calcium but not in the presence of excess calcium chelator. Troponin-C, a naturally occurring calmodulin analog, does not form a complex with RS20 in the presence or absence of calcium. Thus, this 20-amino-acid sequence in smooth muscle MLCK is capable of qualitatively mimicking the calmodulin-binding properties of the whole enzyme (Burgess et al. 1984).

Analysis of MLCK Structure and Function by Using Recombinant DNA Engineered Enzyme Domains

On the basis of the amino acid sequence studies (Lukas et al. 1986) that resulted in the identification of a functional calmodulin-binding site, a series of synthetic oligonucleotide probes were made to screen a cDNA library produced from chicken gizzard polyadenylated RNA. Cloned DNAs were identified and characterized, and then used to screen a chicken embryo fibroblast cDNA library. This approach resulted in the isolation of a series of overlapping clones. The verification of the individual clones as coding for MLCK was done by correlating potential amino acid sequences translated from the DNA sequence with amino acid sequences actually determined for peptides generated by cleavage of MLCK isolated from chicken gizzard tissue and, as discussed below, by the ability to encode a calmodulin-regulated MLCK activity. The cDNA clones were then used to construct DNA inserts that would code for potential overlapping domains of MLCK. The approach used for construction of expression vectors containing these inserts and for the production of protein in *Escherichia coli* was similar to that used for the production of recombinant-DNA-encoded calmodulin (Roberts et al. 1987).

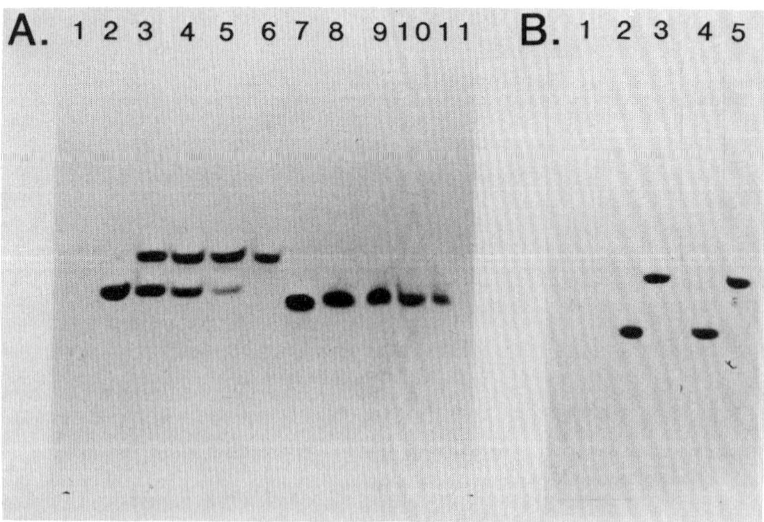

Figure 3. Polyacrylamide gel analysis of RS20 protein complexes. Samples were analyzed in 12.5% gels prepared as described previously (Watterson et al. 1976). (*A*) Complexes were prepared in 25 mM Tris (pH 7.5), containing 1 mM $CaCl_2$, 1 mM DTT at room temperature. The gel-running buffer contained 1 mM Ca^{++}. (Lane *1*) RS20 alone; (lane *2*) 154 pmoles of calmodulin; (lanes *3–6*) complexes of 77 pmoles of RS20 with 2×, 1.5×, 1×, and 0.5× molar ratios of calmodulin, respectively. (Lane *7*) 154 pmoles of troponin-C; and (lanes *8–11*) 2×, 1.5×, 1×, and 0.5× molar ratios of troponin-C in the presence of 77 pmoles of RS20. (*B*) Gels were prepared as in *A*, except that 1 mM EDTA was substituted for calcium and 1 mM EDTA was also in the electrophoresis buffer. (Lane *1*) RS20 alone. (Lanes *2* and *4*) Calmodulin alone (77 pmoles) and calmodulin + RS20 1:1 (mole/mole). (Lanes *3* and *5*) Troponin-C alone (77 pmoles) and troponin-C + RS20 1:1 (mole/mole), respectively.

One of the proteins produced by this approach, termed 30K on the basis of SDS-PAGE mobility, includes the carboxy-terminal region of MLCK and the RS20 calmodulin-binding region. Like the interaction of RS20 and calmodulin, the interaction of 30K protein and calmodulin is calcium-dependent. However, this protein does not exhibit protein kinase activity and does not have the various amino acid sequence elements often found in protein kinases. Another protein (84K) includes the amino acid sequence of 30K plus an additional amino acid sequence toward the amino terminus of the protein. The 84K protein has calmodulin-dependent MLCK activity. Thus, the amino acid sequence of 84K includes the elements necessary for proper folding and generation of a modulated enzyme activity. The linear relationship between these two genetically engineered proteins and native MLCK are diagrammatically depicted in Figure 2. In light of these results, it is not clear what biological function is served by the amino-terminal tail of MLCK. It is possible, however, that this region may be important in the in vivo targeting or subcellular localization of the kinase. Future studies that employ the various constructs generated from these studies and eukaryotic cell expression may allow a direct testing of this latter possibility.

Analysis of the Proposed Calmodulin-binding Domain of MLCK and Regulation of Enzymatic Activity

From studies with synthetic peptides, Kemp et al. (1987) proposed that part of the calmodulin-binding region is a "pseudo substrate" because some synthetic peptides based on these sequences are capable of inhibiting the enzyme. Competitive substrate inhibition, however, was not strictly observed, as evidenced by the fact that some peptides exhibited nonlinear plots. The reason for this behavior was not clear, but some of the reported inhibitors apparently interact directly with calmodulin and, under conditions where the calmodulin concentration was submaximal, exhibited dose-dependent calmodulin antagonism (Kemp et al. 1987). We attempted to further dissect this rather complicated relationship by establishing two types of peptide structures based on the amino acid sequence of this region (see Fig. 2). These included peptides that strongly antagonize the calmodulin activation of enzymes, such as MLCK and brain phosphodiesterase, and those that do not directly inhibit calmodulin activation. Some of the peptides in the first group of structures are given in Table 3. The second group of structures (see Table 4) includes synthetic peptides that do not significantly antagonize the calmodulin activation of phosphodiesterase under conditions of submaximal calmodulin concentration and peptide concentrations up to 100 μM. Although this screening does not preclude the potential direct antagonism of the calmodulin activation of MLCK under some conditions, it serves to operationally separate the high-affinity calmodulin interactions from those that are several orders of magnitude weaker and may involve interaction with the MLCK catalytic subunit.

As summarized in Table 4, selected peptides were tested for competitive inhibition of MLCK. The inhibition constants shown are those obtained from Lineweaver-Burke plots of the results. Although detailed analysis of substrate competition by the inhibitors was not done, double reciprocal plots (velocity vs. peptide substrate) at two concentrations of each inhibitor were characteristic of competitive inhibition. Peptides *1* and *2*, which contain the most amino-terminal sequence, are the poorest inhibitors. Addition of two nonbasic residues, as in peptide *3*, provided a significant increase in inhibitory potency. Two sets of peptides, *4;5* and *7;8*, contain common 4- or 10-residue sequences. The spacing of basic residues is similar to, but distinct from, the requirements for peptide substrates of smooth muscle MLCK (Kemp and Pearson 1985). The general substrate features, Bas-Bas-Bas-X-X-Arg-X-X-(Ser), found in the phosphorylatable myosin light chain, are mimicked only in peptides *5* and *8*,

Table 4. Peptide Inhibitors of smMLCK

Peptide[a]	Sequence	$K_i(\mu M)$[b]
1. MK(554–565)	E A K K L S K D R M K K	>250
2. MK(556–565)	K K L S K D R M K K	225
3. MK(A556, 557–567)	A K L S K D R M K K Y M	36
4. MK(564–571)	K K Y M A R R K	7
5. MK(568–574)	A R R K W Q K	6
6. MK(562–571)	R M K K Y M A R R K	1
7. MK(556–571)	K K L S K D R M K K Y M A R R K	0.4
8. MK(562–574)	R M K K Y M A R R K W Q K	0.2

[a]Numbers shown in parentheses are the residue numbers of 84K protein (see Fig. 3) produced by expression cloning and containing MLCK activity.

[b]All assays were done at 25°C, in 50 mM HEPES buffer (pH 7.5) containing 5 mM Mg^{++}, 0.1 mM Ca^{++}, 1 mM DTT, 0.2 mM ATP (200–500 cpm/pmole of γ-^{32}P), 0.1 mg/ml BSA, 1–2 nM MLCK, 1 μM calmodulin, and 5–100 μM peptide substrate (KKRPQRATSNVFAM). Assay time was 20 min. Aliquots of each reaction were spotted onto phosphocellulose filter paper, washed in 75 mM phosphoric acid solution, followed by 95% ethanol, and dried under a stream of air. Radioactivity was determined by counting the filters in 8 ml of Betafluor scintillation cocktail on a Beckman LS7500 instrument. Peptides were synthesized as described previously (Lukas et al. 1986). All methionine-containing peptides are in the sulfoxide form, which was used as a protecting group.

which are comparable to peptides *4* and *7* as inhibitors. Thus, potency of inhibitor peptides does not correlate simply with the number of basic residues, nor does only the length of the peptide necessarily relate directly to the potency of the inhibitor (e.g., *1–3* vs. *4–6*).

To test the kinase specificity of the inhibitors, two of the peptides were tested for competitive substrate inhibition of the catalytic subunit of cAMP-dependent protein kinase. Figure 4 illustrates that two peptides, which are quite good inhibitors of MLCK (K_i 0.2–0.4 μM), do not significantly inhibit the catalytic subunit of cAMP-dependent protein kinase when tested at concentrations as high as 50 μM inhibitor (i.e., sevenfold higher than the substrate). Thus, the MLCK inhibitory peptides based on sequences within MLCK appear to be more specific for MLCK than for the catalytic subunit.

DISCUSSION

Calmodulin-binding Versus Activation of Target Enzymes

As more is learned about calcium binding to calmodulin and calmodulin modulation of target proteins, early working models of molecular mechanism must be refined and reevaluated. The first working models used a two-step mechanism to depict the calmodulin regulation of a calmodulin-dependent enzyme. According to this model, calmodulin was able to bind one to four calcium ions following an increase in calcium concentration and to undergo a conformational change. The calmodulin/calcium complex then interacted with the target enzyme, and the tertiary complex calmodulin/calcium/enzyme was fully active. Subsequently, the study of the so-called anticalmodulin drugs and of the concentration of calcium needed to activate calmodulin-dependent enzymes led to a differentiation of the binding of calmodulin to a given enzyme from the activation of the calmodulin/enzyme complex. These and subsequent studies formed the basis of several popular working models that have been used in attempts to explain the physiological mechanism of calmodulin transduction of calcium signals in eukaryotic cells. Although additional data summarized here and elsewhere (Haiech and Watterson 1988) require further revision of our concepts, the more recent popular models have several general features in common:

1. The calcium regulation of a calmodulin-modulated enzyme is described by a minimal four-step mechanism (calcium binding to calmodulin, calcium-induced conformational change of calmodulin, calmodulin binding to the enzyme, activation of the enzyme). The calcium-induced conformational change is usually associated with the binding of two moles of calcium per mole of calmodulin, whereas the activation would need three to four calcium ions bound per molecule of calmodulin.
2. Calmodulin-dependent enzymes could bind to the same area of the calmodulin molecule, and a similar structural element would be present in calmodulin-regulated enzymes. This element would be the calmodulin-binding site and has been described as an amphiphilic helix.

An important implication of such models is that the response specificity of calmodulin inside a eukaryotic cell would be due to a rigid compartmentalization of calmodulin and calmodulin-dependent proteins and to a regulation of the calcium "wave" following external

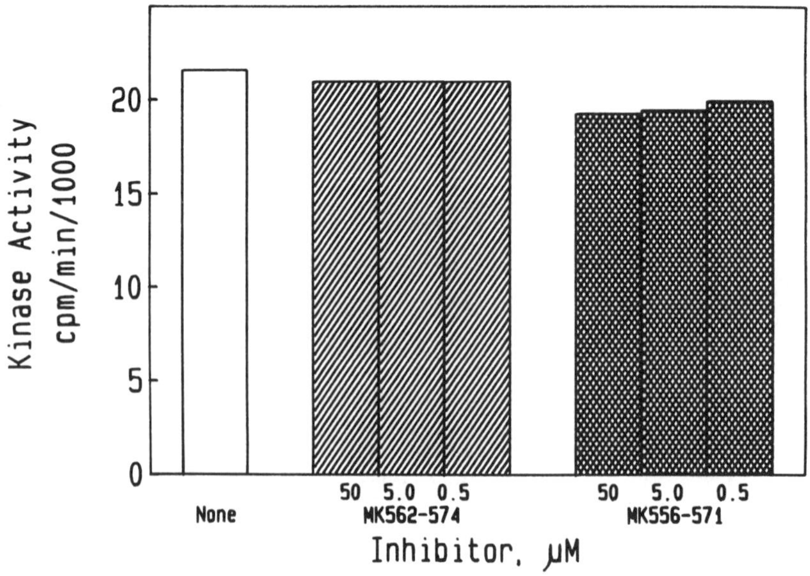

Figure 4. Test of peptide inhibitors based on MLCK with the catalytic subunit of cAMP-dependent protein kinase. The assay system is the same as described in Table 4 except that the substrate was 7.5 μM Kemptide (Peninsula Labs) and the reaction time was 1 min.

stimulation of the cell. As far as pathology is concerned, this model assumes that a mutation of the calmodulin molecule would be either neutral or deleterious for all calmodulin-dependent enzymes (i.e., lethal). In this type of model, probably no specific disease or altered phenotype would be associated with a mutated calmodulin. As discussed below, more recent data require a revision of these models.

Although as early as 1977 there was experimental evidence that some chemically modified calmodulins did not activate to the same extent all the calmodulin-dependent enzymes, the interpretation of these experiments was hampered by the fact that the population of chemically modified calmodulin was not homogeneous. More recently, the use of comparative activity and sequence analyses, as well as site-specific mutagenesis/protein engineering approach to analyze the relation of calmodulin structure to function, has allowed one to obtain chemically defined and homogeneous populations of precisely modified calmodulin molecules. Using the mutagenesis approach, Craig et al. (1987) demonstrated the feasibility of using engineered calmodulins to map areas of the calmodulin molecule involved in the regulation of a given calmodulin-dependent enzyme. The results summarized in Table 2 extend this first study and illustrate the differences in the mechanisms of activation of the different calmodulin-dependent kinases. It appears that each calmodulin-dependent enzyme interacts with calmodulin by a unique set of contact points, although specific contacts on calmodulin may be used by more than one enzyme (see Fig. 5A). The subsequent activation probably involves another set of contact points, although the interaction and activation contact points would clearly overlap (see Fig. 5B). Therefore, each enzyme would have two sets

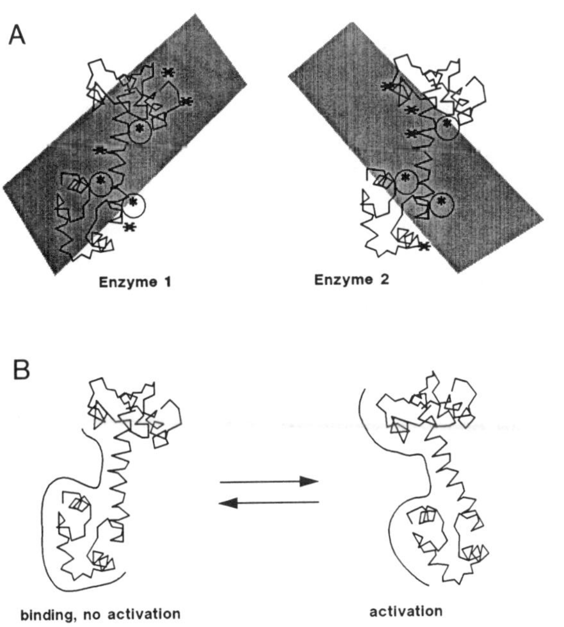

Figure 5. Summary of possible mechanisms of calmodulin binding and activation of a protein kinase.

of contact points associated with it: one set involved in the initial basal interaction (e.g., important in the biosynthesis of the signal transduction complex) and the other set involved in the activation by calmodulin (e.g., important in transduction of the rapid calcium signal). It is the object of future studies to determine these sets for specific calmodulin-dependent enzymes (e.g., MLCK and cyclases).

This evolving and increasingly detailed model of how the structural features of calmodulin are related to enzyme binding and activation has several implications for how a functional calcium signal transduction–amplification pathway is assembled. For example, the observation that formation of the calmodulin-MLCK complex is especially sensitive to changes in negative surface electrostatic potential suggests that the surface topology of negative charge in calmodulin is important in the biosynthetic formation of a functional complex. This is surprising because the past focus had been on how important the hydrophobic forces were in complex formation. In addition, studies using chemical modification of lysine residues had suggested that charge was not very important. The results of our mutagenesis of Lys-75 are consistent with the idea that the positive charge contribution of some lysines is not very critical, but the negative charge surface does appear to be critical. Hydrophobic interactions could still be important, especially in the stabilization of a complex once it is formed.

Another important biological implication of the in vitro mutagenesis/activity analysis studies described above is the demonstrated feasibility of selectivity in mutations of calmodulin that might occur in vivo. Recent studies (Schaefer et al. 1987) of the calmodulin from organisms with an altered calcium response phenotype are consistent with this concept. Briefly, the altered phenotype of pantophobiac A *Paramecia* could be restored to that of the wild-type organism by the microinjection of wild-type calmodulins, but not by the purified calmodulin from the mutant organism. Amino acid sequence analysis of the calmodulins from the wild-type and mutant organisms showed a single difference at residue 101 (Ser→Phe). These and other results are consistent with the model of calmodulin modulating potassium efflux activity and the altered phenotype of pantophobiac A being due to this single amino acid sequence change in calmodulin. Overall, these initial studies indicate the importance of examining the structure as well as the concentration of calmodulin when there is the possibility of an altered phenotype being due to a perturbation of calcium-calmodulin homeostasis.

Role of the Calmodulin Regulatory Domain in MLCK

The proposed calmodulin regulatory domain of smooth muscle MLCK (Fig. 3) appears to contain at least two overlapping properties: (1) binding of calmodulin and (2) inhibition of the enzyme. Possible amino acid sequences of the self-inhibiting region of

MLCK can be identified with the synthetic peptide, but it is not possible to demonstrate with this approach which specific part of the sequence within this domain of the kinase is involved in the self-inhibition of the enzyme. Although the peptides exhibit inhibition that is apparently competitive with the peptide substrate, kinetically indistinguishable mechanisms that modify the Michaelis constant of the substrate cannot be excluded. Similarly, the peptide inhibitors may also contribute to effects on ATP interaction when the nucleotide concentration is limiting (Ikebe et al. 1987; Kennelly et al. 1987). However, on the basis of results to date, several possible mechanisms may be elaborated.

1. The first mechanism assumes a true competition between substrate and inhibitory peptide (or internal inhibitory domain) at the same site. Either the inhibitory peptide uses the same set of contact points as the substrate (pseudo substrate theory) or the contacts are different but overlap (inhibitory domain theory). In the latter case, a phosphorylatable peptide based on the inhibitory domain will be a poor substrate for the enzyme. Both theories assume that the action of calmodulin binding to the enzyme removes the inhibitory domain from the region of the active site so that it can be occupied by the substrate.
2. The second mechanism assumes competition between substrate and inhibitory domain through two physically distinct sites, a substrate site and an inhibitory site. In this case, binding of the inhibitory domain to the inhibitory site will change the conformation of the substrate site and prevent the productive binding of substrate to the enzyme. Similarly, the binding of substrate will prevent the binding of the inhibitory domain. This mechanism is kinetically indistinguishable from the previous one. Following calmodulin binding, the conformation of inhibitory domains will be changed and, therefore, the inhibitory site will be released. The implication of such a model is that in the native enzyme, the virtual concentration of inhibitory domain is infinite, and productive substrate binding will not occur. A small conformational change will then release the inhibitory site, whereas in the first mechanism, a large movement of the inhibitory domain (a removal from the active site) is necessary to allow the substrate access to the active site. Both mechanisms explain the currently known data presented here and elsewhere. Variations of the interaction of calmodulin with the binding/inhibitory domain could then modulate the enzyme efficiency, as measured by maximal velocity, as well as the substrate binding potential. A key point to note is that if calmodulin only served to open the active site, then binding and activation events would be coincident. This is not consistent with what has been observed.

Obviously, other mechanisms may be postulated, but it is likely that they will be more complex than either of the two described here. Although synthetic peptides and MLCK fragments that contain the calmodulin-binding sequences have been localized, it is not clear within the context of the entire domain which residues are obligatory for high-affinity calmodulin binding versus those that may be necessary for conformational changes and those that participate directly in modulation of enzyme activity. Further work with the kinetic details of the activation of MLCK by calmodulin, the production of selected MLCK mutants, as well as physical studies of MLCK-calmodulin-substrate/inhibitor complexes should help to distinguish among the different mechanistic paths.

Biological Implications: Calmodulin Is Not a Simple Relay of the Calcium Signal

Results presented here illustrate that calmodulin may transfer calcium signals to different enzymes in the same area of the cell. Therefore, we might expect colocalization of calmodulin and several calmodulin-dependent enzymes. If true, the kinetic mechanism of activation of one specific enzyme by calmodulin and calcium is important to understand the calcium signal transduction in a given cell. In addition, since a mutation(s) in calmodulin may be associated with specific metabolic disease, a reappraisal of the role of a calmodulin in several pathological states of eukaryotic cells may be necessary.

Biological Implications: Ca^{++}-dependent Enzymes Are Probably Key Points in the Cross-talk between Second Messengers in Cell

The mechanism that we propose for regulation of MLCK by calmodulin may be generalized, in some form, to all calmodulin-dependent enzymes. As presented in Figure 6, the inhibition or activation of an enzyme is going to be driven by the coupling or uncoupling between an inhibitory site and the catalytic domain. Any phosphorylation, proteolytic modification, or effector binding to the enzyme may modify this relationship between the inhibitory site and the catalytic domain, thereby leading to a modulation of the enzymatic activity. In the normal state, the inhibitory site is occupied by an inhibitory domain coupled to the catalytic domain. Therefore, the enzyme is partially or totally inhibited. Following calmodulin binding, phosphorylation, or proteolytic degradation, the inhibitory site is uncoupled from the catalytic domain, either by removing the inhibitory domain from the inhibitory site (proximal to where calmodulin binds to the enzyme) or by physically uncoupling the inhibitory site from the catalytic domain, although the inhibitory domain may still bind to the inhibitory site.

This kind of general mechanism allows subtle modification of enzyme activity by several means without gross conformational change. Therefore, this mechanism is well suited for enzyme modulation by several second messengers. Clearly, a more detailed study of the molecular mechanism of several calmodulin-dependent enzymes is needed to understand the cross-talk

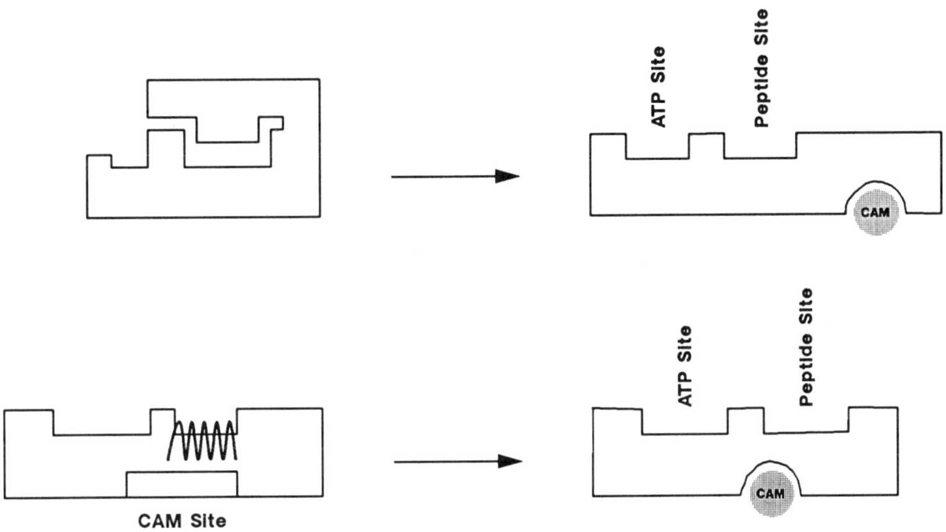

Figure 6. Possible mechanisms of activation of a calmodulin-dependent kinase.

between second messengers, and the outline presented here is meant only to serve as a useful framework for focusing such studies. Although the current emphasis is on the cross-talk between the cyclic nucleotide and calcium pathways via the calmodulin signal transduction pathways, this situation will become increasingly complex as the physiological significance of protein kinase C/lipid second messenger and G protein/guanine nucleotide interactions with calmodulin or calmodulin-regulated enzymes is investigated.

ACKNOWLEDGMENTS

The authors thank Cindy Reeder and Janis Elsner for their assistance with the preparation of this manuscript. These studies were supported in part by National Institutes of Health grant GM-30861 (D.M.W.).

REFERENCES

Babu, Y.S., J.S. Sack, T.G. Greenhough, C.E. Bugg, A.R. Means, and W.J. Cook. 1985. Three-dimensional structure of calmodulin. *Nature* **315:** 37.

Burgess, W.H., D.M. Watterson, and L.J. Van Eldik. 1984. Identification of calmodulin binding proteins in chicken embryo fibroblasts. *J. Cell. Biol.* **99:** 550.

Craig, T.A., D.M. Watterson, F.G. Prendergast, J. Haiech, and D.M. Roberts. 1987. Site-specific mutagenesis of the α-helices of calmodulin. *J. Biol. Chem.* **262:** 3278.

Edelman, A.M., D.K. Blumenthal, and E.G. Krebs. 1987. Protein serine/threonine kinases. *Annu. Rev. Biochem.* **56:** 567.

Guerriero, V., M.A. Russo, N.J. Olson, J.A. Putkey, and A.R. Means. 1986. Domain organization of chicken gizzard myosin light chain kinase deduced from a cloned cDNA. *Biochemistry* **25:** 8372.

Haiech, J. and D.M. Watterson. 1988. Site-specific mutagenesis and protein engineering approach to the molecular mechanism of calcium signal transduction by calmodulin. In *Calcium and calmodulin binding proteins* (ed. C.H. Gerday et al.), p. 191. Springer-Verlag, Berlin.

Ikebe, M., M. Stepinska, B.E. Kemp, A.R. Means, and D.J. Hartshorne. 1987. Proteolysis of smooth muscle myosin light chain kinase. *J. Biol. Chem.* **260:** 13828.

Kemp, B.E. and R.B. Pearson. 1985. Spatial requirements for location of basic residues in peptide substrates for smooth muscle myosin light chain kinase. *J. Biol. Chem.* **260:** 3355.

Kemp, B.E., R.B. Pearson, V. Guerriero, I.C. Bagchi, and A.R. Means. 1987. The calmodulin binding domain of chicken smooth muscle myosin light chain kinase contains a pseudosubstrate sequence. *J. Biol. Chem.* **262:** 2542.

Kennelly, P.J., A.M. Edelman, D.K. Blumenthal, and E.G. Krebs. 1987. Rabbit skeletal muscle myosin light chain kinase. *J. Biol. Chem.* **262:** 11958.

Lukas, T.J., T.A. Craig, D.M. Roberts, and D.M. Watterson. 1987. An interdisciplinary approach to the molecular mechanisms of calmodulin action: Comparative biochemistry, site-specific mutagenesis, and protein engineering. In *Calcium binding proteins in health and disease* (ed. A.W. Norman et al.), p. 533, Academic Press, Orlando.

Lukas, T.J., W.H. Burgess, F.G. Prendergast, W. Lau, and D.M. Watterson. 1986. Calmodulin binding domains: Characterization of a phosphorylation and calmodulin binding site from myosin light chain kinase. *Biochemistry* **25:** 1458.

Roberts, D.M., W.E. Zimmer, and D.M. Watterson. 1987. The use of synthetic oligodeoxyribonucleotides in the examination of calmodulin gene and protein structure and function. *Methods Enzymol.* **139:** 290.

Schaefer, W.H., R.D. Hinrichsen, A. Burgess-Cassler, C. Kung, I.A. Blair, and D.M. Watterson. 1987. A mutant *Paramecium* with a defective calcium-dependent potassium conductance has an altered calmodulin: A nonlethal selective alteration in calmodulin regulation. *Proc. Natl. Acad. Sci.* **84:** 3931.

Van Eldik, L.J., J.G. Zendegui, D.R. Marshak, and D.M. Watterson. 1982. Calcium-binding proteins and the molecular basis of calcium action. *Int. Rev. Cytol.* **77:** 1.

Watterson, D.M., D.B. Iverson, and L.J. Van Eldik. 1980. Rapid separation and quantitation of 3′:5′-cyclic nucleotides and 5′-nucleotides in phosphodiesterase reaction mixtures using high-performance liquid chromatography. *J. Biochem. Biophys. Methods* **2:** 139.

Watterson, D.M., W.G. Harrelson, P.M. Keller, F. Sharief, and T.C. Vanaman. 1976. Structural similarities between the Ca^{2+}-dependent regulatory proteins of 3′:5′-cyclic nucleotide phosphodiesterase and actomyosin ATPase. *J. Biol. Chem.* **251:** 4501.

Wendolski, J.J., J.B. Matthew, P.C. Weber, and F.R. Salemme. 1987. Molecular dynamics of a cytochrome C-cytochrome B5 electron transfer complex. *Science* **238:** 794.

Regulation of the Vertebrate Cell Cycle by the *cdc2* Protein Kinase

G. Draetta, L. Brizuela, B. Moran, and D. Beach
Cold Spring Harbor Laboratory, Cold Spring Harbor, New York 11724

A homolog of the *cdc2/CDC28* protein kinase of yeast is found in all vertebrate species that have been investigated. Human *cdc2* exists as a complex with a 13-kD protein that is homologous to the *suc1* gene product of fission yeast. In both human and fission yeast cells, the protein kinase also exists in a complex with a 62-kD polypeptide that has not been identified genetically but acts as a substrate in vitro. We have studied the properties of the protein kinase in rat and human cells, as well as in *Xenopus* eggs. We find that in baby rat kidney (BRK) cells, which are quiescent in cell culture, the *cdc2* protein is not synthesized. However, synthesis is rapidly induced in response to proliferative activation by infection with adenovirus. In human HeLa cells, the protein kinase is present continuously. It behaves as a cell-cycle oscillator that is inactive in G_1 but displays maximal enzymatic activity during mitotic metaphase. Mitotic activation of the kinase is associated with phosphorylation at threonine, tyrosine, and serine residues. Rearrangement of the subunit structure of the protein kinase also occurs. In *Xenopus* eggs, the *cdc2* protein kinase has been shown to be a component of the mitosis promoting factor (MPF), which has been difficult to characterize biochemically but has been known for many years to play a critical role in mitotic initiation. Taken together, these observations indicate that in a wide variety of vertebrate cells, the *cdc2* protein kinase is involved in regulating mitosis. Our approach taken toward study of the *cdc2* protein kinase highlights the possibilities that now exist for combining the advantages of ascomycete genetics with the cell-free systems of *Xenopus* and the biochemical advantages of tissue culture cells to investigate fundamental problems of the cell cycle.

The ascomycete fungi, *Saccharomyces cerevisiae* (Hartwell 1974), *Schizosaccharomyces pombe* (Nurse et al. 1976), and *Aspergillus nidulans* (Morris 1976), have been extensively used as model organisms for genetic investigation of the eukaryotic cell-cycle. One of the very first cell-cycle (*cdc*) genes to be characterized, initially by genetic analysis (Hartwell 1974) and subsequently at the molecular level (Reed and Nasmyth 1981), was the *CDC28* cell-cycle "start" gene of budding yeast. The product of this gene acts in G_1 prior to the initiation of DNA replication. After execution of the *CDC28* gene function, the yeast cell is committed to completion of the division cycle.

The budding yeast *CDC28* gene is homologous to the *cdc2*$^+$ gene of the fission yeast (Beach et al. 1982; Hindley and Phear 1984; Lorincz and Reed 1984). However, in fission yeast, *cdc2*$^+$ is required not only in G_1 prior to the initiation of DNA synthesis, but also in G_2 during the transition from interphase into mitosis (Nurse and Bissett 1981).

Both *cdc2*$^+$ and *CDC28* encode protein kinases of approximately 34 kD (Reed et al. 1985; Simanis and Nurse 1986). At present, the *cdc2*$^+$/*CDC28* protein kinase is poorly characterized. In fission yeast, its activity does not appear to alter during the division cycle, but in budding yeast, the enzyme is most active in G_1 and in vitro phosphorylates an endogenous complexed protein of 40 kD (Mendenhall et al. 1987). This substrate has not been identified genetically.

One protein, the 13-kD product of the *suc1*$^+$ gene, has been shown by both genetic and biochemical criteria to interact with the *cdc2* protein kinase. *suc1*$^+$ was originally identified as a plasmid-borne gene that could rescue some but not all temperature-sensitive alleles of *cdc2*$^+$ (Hayles et al. 1986). Subsequently, it was demonstrated by direct biochemical experiment that the p13^{suc1} gene product (Hindley et al. 1987) exists in vivo in a complex with at least a fraction of the *cdc2* protein kinase (Brizuela et al. 1987). By a variety of criteria, this interaction has been shown to be required for cell-cycle progression, but the precise role of the *suc1* protein is not fully clear at present. It does not act, for example, as a substrate of the *cdc2* protein kinase (Brizuela et al. 1987).

Genetic analysis has revealed large numbers of genes that appear to interact with *cdc2*$^+$ in the regulation of mitosis. In addition to *suc1*$^+$, these include *cdc13*$^+$ (Booher and Beach 1987), *cdc25*$^+$, *wee1*$^+$ (Fantes 1979; Russell and Nurse 1986, 1987), and a newly identified class of genes (*mcs1–6*, L. Molz et al., in prep.). It is not anticipated that each of these will interact directly with the *cdc2*$^+$ protein kinase, but some may turn out to be substrates. However, in none of these cases has the relationship between the products of these genes and the *cdc2*$^+$ protein kinase yet been elaborated in molecular detail.

Identification of Vertebrate *cdc2*

During the course of studies of the fission yeast *cdc2*$^+$ gene product, we had prepared both mono- and polyclonal antibodies against this protein. It was found that the polyclonal and some of the monoclonal antibodies would recognize the homologous *CDC28* gene product of *S. cerevisiae* (Draetta et al. 1987). Since the two yeast species are not closely related, we felt that these immunological reagents might also recognize a homologous protein in vertebrate cells. This turned out to be the case, and in human cells, a protein of 34 kD was found. Proteolytic cleavage of the fission yeast *cdc2*$^+$ protein, budding yeast *CDC28* protein, and human p34 with a chemical reagent that cleaves specifically at tryptophan residues revealed complete conservation of the four tryptophan residues in each of these polypeptides (Fig. 1) (Draetta and Beach 1988). The human *cdc2* gene has been isolated by rescue of a temperature-sensitive *cdc2* mutant of fission yeast (Lee and Nurse 1987). The predicted amino acid sequence of fission yeast *cdc2*$^+$, budding yeast *CDC28*, and the human protein each share approximately 60% identity.

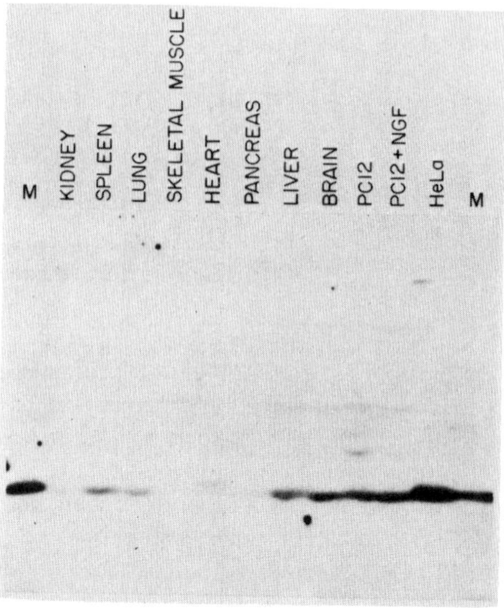

Figure 2. Immunological detection of p34 in rat organs and cell extracts. Extracts were prepared as described in the text. From each extract, 150 μg of protein was loaded per lane of a 7.5–15% SDS-polyacrylamide gel. Upon electrophoresis, proteins were transferred to nitrocellulose and probed with affinity-purified anti-p34^{cdc2} antibodies and ^{125}I-labeled Protein A. The marker (M) lane contains 200 ng of p34^{cdc2} expressed in *E. coli*. (Reprinted, with permission, from Draetta et al. 1988.)

Tissue Distribution of *cdc2* Protein

In yeast, the *cdc2*/*CDC28* protein kinase is involved in regulation of the cell cycle. If this is also the case in the cells of higher vertebrates, it is expected that the *cdc2* homolog will be present in actively dividing cells. We investigated the distribution of p34 in rat tissues by Western immunoblotting. The protein was relatively abundant in liver, spleen, and brain but was almost absent from skeletal muscle and kidney (Fig. 2) (Draetta et al. 1988). Furthermore, the protein was abundant in tissue-cultured rat PC-12 pheochromocytoma cells, both during active proliferation and during postmitotic differentiation induced by exposure to nerve growth factor (NGF) (Fig. 2). These observations indicate that the *cdc2* protein is not exclusively restricted to actively dividing cells but is found in cells that are dividing. In addition to the cell line tissues illustrated in Figure 2, we have detected the *cdc2* protein in human fibroblasts (Hs68), monkey cells (CV-1), mouse fibroblasts (NIH-3T3), and also *Xenopus* eggs (see below).

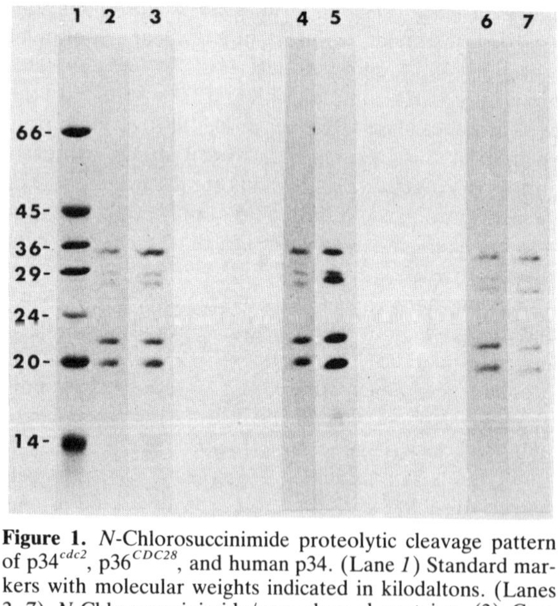

Figure 1. *N*-Chlorosuccinimide proteolytic cleavage pattern of p34^{cdc2}, p36^{CDC28}, and human p34. (Lane *1*) Standard markers with molecular weights indicated in kilodaltons. (Lanes *2–7*) *N*-Chlorosuccinimide/urea-cleaved proteins. (*2*) Coomassie-stained p34^{cdc2}; (*3*) ^{35}S-labeled p34^{cdc2}; (*4*) Coomassie-stained p34^{cdc2}; (*5*) ^{35}S-labeled p36^{CDC28}; (*6*) Coomassie-stained p34^{cdc2}; (*7*) ^{35}S-labeled human p34. Lanes *2* and *3*, *4* and *5*, *6* and *7* are different visualizations (Coomassie blue or autoradiography) of single gel lanes. In each lane, the uppermost band represents undegraded protein and the smallest major band (20 kD) carries the entire amino-terminal two thirds of the polypeptide. The intervening three bands are, in each case, partial cleavage products. (Reprinted, with permission, from Draetta et al. 1987.)

Induction of *cdc2* Synthesis in Adenovirus-infected Rat Cells

Since the *cdc2* protein is present at very low levels in certain nondividing cells, we tested whether proliferative stimulation of such cells is associated with synthesis of p34. We made use of BRK cells that rapidly cease cell division and have very low levels of p34 after plating in tissue culture, despite the presence of high levels of serum. These cells may be induced to proliferate by infection with adenovirus. Within a few hours of infection with wild type but not a mutant adenovirus (*dl*312, an E1A deletion strain), synthesis of the *cdc2* protein could be detected in immunoprecipitates of [^{35}S]methionine-labeled whole-cell protein (Fig. 3) (Draetta et al. 1988). Increase in p34 synthesis was accompanied by accumulation of the protein (Draetta et al. 1988). It should be noted that in BRK cells, the total rate of protein synthesis increases only approximately twofold following viral infection, and the increased synthesis of p34 therefore reflects specific activation of the *cdc2* gene (Draetta et al. 1988).

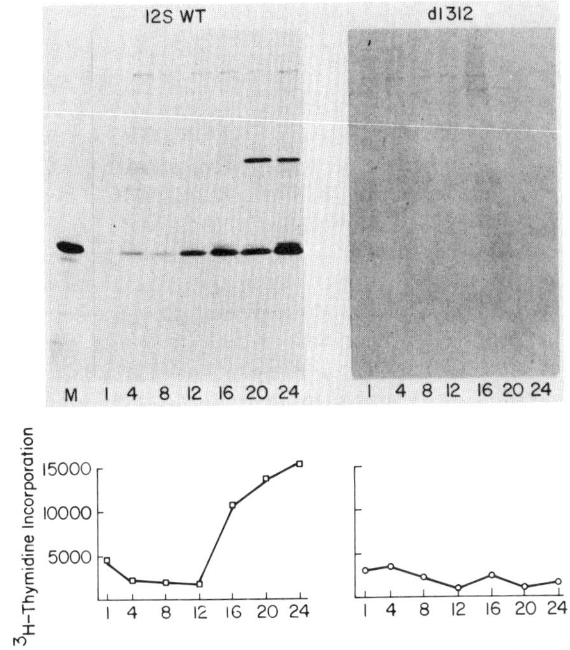

Figure 3. Induction of p34 synthesis by adenovirus infection of BRK cells. BRK cells infected with 12S-WT or the *dl*312 deletion mutant virus were labeled with Tran^{35}S-label or [^{3}H]thymidine for 1-hr pulses ending at the time points indicated below each panel. Samples of cell extracts normalized to contain equal amounts of precipitable ^{35}S cpm were immunoprecipitated with anti-p34 serum (*top* panels); the marker lane (M) contains ^{35}S-labeled p34^{cdc2} extracted from *E. coli*. The bottom panels show [^{3}H]thymidine incorporation plotted as trichloroacetic-acid-precipitable counts per 35-mm plate. (Reprinted, with permission, from Draetta et al. 1988.)

Human *cdc2* Is a Cell-cycle Oscillator

The preceding observations are consistent with the hypothesis that p34 is involved in the cell cycle of mammalian cells, but they give no indication about a specific role. We have used human HeLa cells to study the properties of the *cdc2* protein kinase at different stages of the continuous cell cycle, rather than during the transition from quiescence to activation.

Anti-*cdc2* immunoprecipitates of [^{35}S]methionine-labeled HeLa cells revealed that p34 exists in a complex with at least two other polypeptides in vivo (Fig. 4A). One is a protein of approximately 13 kD that has been shown to be the human homolog of the fission yeast *suc1*$^+$ gene product (Draetta et. al. 1987; Draetta and Beach 1988). The other is a protein of 62 kD that acts as a substrate of the *cdc2* protein kinase in vitro. The genetic identity of this protein has not been established, but in fission yeast, the *cdc2* protein is also found in a complex with a 62-kD protein that might be related to human p62 (R. Booher and D. Beach, unpubl.).

To investigate the properties of HeLa *cdc2* during the division cycle, suspension culture cells were synchronized by means of elutriation centrifugation. The smallest cells, those in early G_1, were reinoculated into fresh growth medium, and fractions were taken at 4-hour intervals. The *cdc2* protein kinase was immunoprecipitated and incubated with [γ-^{32}P]ATP. Casein was also added to the reaction as an artificial substrate. The activity of the *cdc2* kinase fluctuated dramatically during the division cycle. In G_1 cells, there was little or no activity with respect to either the endogenous 62-kD substrate or casein. However, kinase activity increased during the cycle and was maximal in cells in mitosis (Fig. 4B) (Draetta and Beach 1988). In addition, it was found that exposure of exponentially growing cells to the microtubule-disrupting drug, nocodazole (De Brabander et al. 1976), caused a massive activation of the protein kinase (Draetta et al. 1988). Since this drug arrests cells in mitotic metaphase, these results suggest that in HeLa cells, the protein kinase is maximally active in mitosis.

The molecular basis of the cell-cycle oscillation of the *cdc2* kinase activity has been investigated. It was found that the abundance of protein did not alter during the cell cycle but that its phosphorylation state and subunit structure were both subject to cell-cycle regulation. The protein kinase was not phosphorylated in G_1 cells but became progressively phosphorylated at two or more sites during G_2. Phosphoamino acid analysis revealed that the *cdc2* protein was predominantly phos-

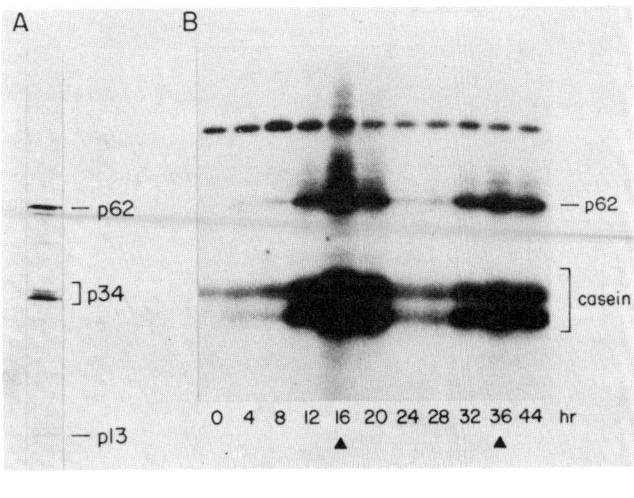

Figure 4. (*A*) Proteins associated with *cdc2* in HeLa cells. [^{35}S]Methionine-labeled HeLa cell lysate was immunoprecipitated with anti-*cdc2* polyclonal serum and analyzed by gel electrophoresis. (*B*) Periodicity of p34 protein kinase activity. G_1-enriched cells were obtained by elutriation, inoculated into fresh medium, and sampled at 4-hr intervals. Protein kinase activity in antipeptide precipitates was assayed in the presence of casein. (▲) Times at which the highest proportion of cells were in G_2/M, determined by flow cytometry.

phorylated on threonine residues but that there was also a significant fraction of tyrosine and serine phosphorylation (Fig. 5). Since *cdc2* is probably a serine/threonine kinase, this implies, among other things, that there exists a tyrosine kinase that phosphorylates *cdc2* in a cell-cycle-dependent manner. In addition to the cell-cycle-dependent phosphorylation of *cdc2*, it was also found that the association with the p62 endogenous substrate was subject to cell-cycle regulation. *cdc2* appears not to exist in a complex with p62 during G_1, but later in the cycle, the most highly phosphorylated fraction of the protein is found in association with p62 (Draetta and Beach 1988).

These observations indicate that in HeLa cells, the *cdc2* protein kinase exists in many different states. A hypothesized sequence of events is illustrated in Figure 6. In G_1 cells, *cdc2* is dephosphorylated and is inactive as a protein kinase. p34 phosphorylation occurs in conjunction with formation of the complex with the p62 substrate, and this event correlates with activation of the protein kinase. The p13^{suc1} homolog may exist in association with *cdc2* throughout the cell cycle, but the possibility that it is absent from the fully active mitotic form of the kinase has not been ruled out.

Cell-free Cell Cycle

The preceding series of observations indicate that in HeLa cells, the *cdc2* protein kinase displays exactly the biochemical properties expected if this enzyme plays a role in regulating the transition from interphase into mitosis. To investigate whether this is indeed the case, we have used lysates of *Xenopus* eggs in which many complex cell-cycle events have been mimicked in vitro. For example, morphologically normal nuclei may be reconstructed in vitro from naked DNA (Newport 1987). These interphase nuclei may fully replicate their DNA in such cell-free lysates (Blow and Laskey 1986; Newport 1987). Furthermore, following exposure to extract prepared from mitotic cells, interphase nuclei may undergo nuclear membrane breakdown accompanied by chromosome condensation (Lohka and Maller 1985; Miake-Lye and Kirschner 1985; Dunphy and Newport 1988). In extracts from *Xenopus* eggs that are very active in protein synthesis, nuclei may undergo full cell-cycle oscillation, consisting of rounds of DNA replication with intervening nuclear dissolution and chromosome condensation (Hutchinson et al. 1988).

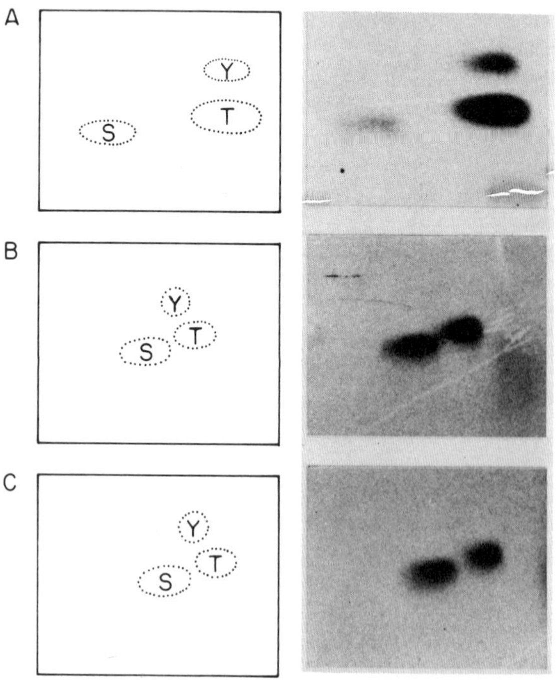

Figure 5. Phosphoamino acid analysis of p34 (*A*), p62 labeled in vivo (*B*), and p62 labeled in vitro (*C*).

cdc2 Is a Component of *Xenopus* MPF

The mitotic inducer that may now be directly assayed in vitro has been known for many years as MPF (Matsui and Markert 1971; Smith and Ecker 1971). This activity was first identified as a factor that could be transferred from *Xenopus* eggs to oocytes by microinjection to induce meiotic maturation in the oocytes (this process is normally triggered by steroid hormones). During maturation, oocytes that are arrested in late G_2/prophase enter meiosis and subsequently become arrested, as mature eggs prior to fertilization, in metaphase of the second meiotic division. It has been found that MPF activity may be detected in mitotic cells from a wide variety of eukaryotic species (Kishimoto et al. 1982; Adlakha et al. 1985). This universal mitotic inducer was also shown to be active only in mitotic cells and not during interphase (Gerhart and Kirschner 1984), although it appears to exist in a latent form during interphase (Cyert and Kirschner 1988; Dunphy and Newport 1988).

Since human $cdc2^+$ protein kinase has properties that suggest that it might act as a mitotic inducer, we have tested whether *Xenopus* eggs have a $cdc2^+$ homolog and whether this might be a component of the MPF activity. Western blotting revealed that *Xenopus* egg extracts do indeed contain a 34-kD protein that is recognized by a polyclonal serum prepared against the fission yeast $cdc2^+$ protein (Fig. 7) (Dunphy et al. 1988).

In preliminary investigations to test the possibility that *Xenopus cdc2* might play a role in mitotic conversion of nuclei in vitro, we found that highly purified fission yeast p13^{suc1} (expressed in *Escherichia coli*) could counteract *Xenopus* MPF activity in both microinjection and in vitro assays. This effect was highly specific, and full inhibition occurred at 2 μM p13 (Dunphy et al. 1988). Since p13 is known to interact with the *cdc2* protein kinase in both fission yeast (Brizuela et al. 1987) and human cells (Draetta and Beach 1988), it seemed most likely that this protein might inhibit MPF activity by associating with the *Xenopus cdc2* protein. To test this possibility, purified p13 was covalently coupled to an inert agarose matrix. It was found that if a *Xenopus* egg lysate was passed through this column,

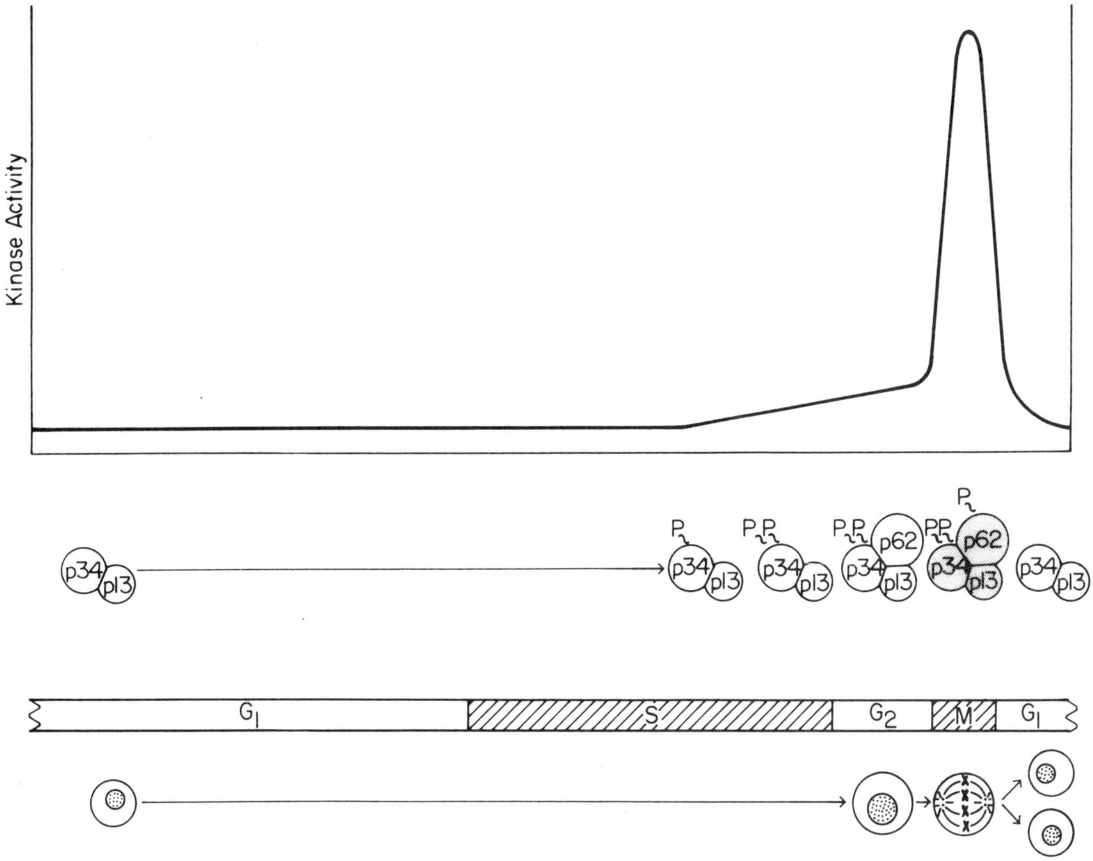

Figure 6. Relationship between the cell division cycle and the proposed cycle of p34 protein kinase. p34 is associated with p13. During the cell cycle, p34 becomes phosphorylated at two sites, associates with p62, and is finally activated by an unidentified event (represented by shading) during the transition from G_2 to mitosis. These steps are each reversed before cytokinesis. Direct contact between p34, p13, and p62 is shown only for simplicity in this illustration. The relative timing of G_1, S, G_2, and M in HeLa cells is shown approximately to scale. (Modified from Draetta and Beach 1988.)

the *cdc2* homolog was quantitatively retained (Fig. 7). In contrast, none of the protein was retained on a control column consisting only of agarose beads. This experiment demonstrates that fission yeast p13 may directly interact with *Xenopus cdc2*.

Finally, we tested whether the p13 column that depletes the *cdc2* protein from the *Xenopus* egg lysate might likewise deplete the MPF activity. This was exactly the result obtained. In contrast to a control column that did not deplete MPF activity, the flow-through fraction of the p13 column was devoid of activity (Fig. 8). This observation indicates that the *cdc2* protein kinase is one component of the MPF activity and that fission yeast p13 inhibits MPF by directly interacting with the frog *cdc2*. MPF appears to consist of two polypeptide species; one of approximately 34 kD, which is now shown to be the *Xenopus cdc2* homolog, and another of approximately 42 kD. This protein is also retained on the p13 affinity column (Dunphy et al. 1988). Using the techniques of classic biochemistry, very small quantities of MPF have been purified from frog eggs. The purest fractions consist of polypeptides of 34 and 45 kD. We assume that these are the same two proteins that are retained on the p13 affinity column.

PERSPECTIVE

The experiments outlined above utilized a variety of different vertebrate cell types to study the properties of the *cdc2* protein kinase. Despite this, a coherent picture has emerged. First, the protein is not restricted to cells that are actively dividing, but in at least one cell type that has low levels of *cdc2* (BRK), proliferative induction is accompanied by rapid induction of p34 synthesis. In HeLa cells, in which the protein is highly abundant, *cdc2* protein kinase behaves as a cell-cycle oscillator. Although the level of the protein does not vary during the division cycle, it is inactive as a protein kinase during G_1 and is maximally active during mitotic metaphase. Activation of the protein kinase involves a number of steps that have yet to be fully elaborated in molecular detail, but in which both phosphorylation of the protein kinase and also rearrangement of its subunit structure are implicated.

The properties of *cdc2* in HeLa cells are clearly compatible with the protein kinase playing a role in mitotic initiation. This has not been directly demonstrated in HeLa cells, but in *Xenopus* eggs, the protein kinase has been shown to be a component of the long-known but poorly characterized mitotic inducer known

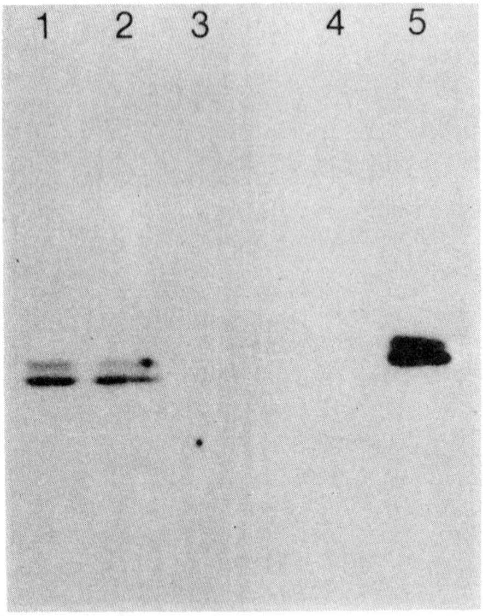

Figure 7. A p13 affinity column binds the *Xenopus cdc2* homolog. Crude MPF was applied to a column of either control- or p13-agarose. The flow through fraction was collected, and then each column was eluted with Laemmli buffer. (Lane *1*) Crude MPF preparation. Flow through fraction from either control-agarose (lane *2*) or p13-agarose (lane *3*). SDS eluate from control column (lane *4*) or p13-agarose (lane *5*). The immunoblot was reacted with anti-yeast *cdc2* antibodies. (Reprinted, with permission, from Dunphy et al. 1988.)

as MPF. Taken together, these observations indicate that the *cdc2* protein kinase does indeed play a role in the cell cycle of vertebrates. It acts as a critical component of the pathway that regulates the transition from interphase into mitosis.

These results raise many interesting new questions, some molecular and others evolutionary. First, what is the role of the p13^{suc1} protein in cell-cycle regulation? p13 does not appear to act as a conventional protein kinase inhibitor, such as the heat-stable inhibitor of cAMP-dependent protein kinase. In fission yeast, the *suc1*$^+$ gene is essential for cell-cycle progression, but gross overexpression is inhibitory to mitosis. p13 might be acting as an assembly factor, required for generating the appropriate subunit structure of the mitotically activated form of the *cdc2* protein kinase, but possibly not being a component of the final complex. This hypothesis would explain why p13 can inhibit *Xenopus* MPF activity and yet be essential for cell division in yeast.

Second, what is the relationship between the 62-kD protein that associates with *cdc2* in HeLa cells and the 42-kD protein in *Xenopus* and the 40-kD protein in budding yeast, and how do these proteins act in the regulation of the kinase? This remains entirely speculative until these proteins and their genes are characterized, but it is possible that these proteins are all homologs that happen to have different molecular weights. It has been assumed that the p62 and p42 polypeptides that associate with *cdc2* in HeLa and *Xenopus* cells, respectively, function primarily as activators of the kinase and only secondarily as substrates. However, critical evidence is lacking and will require the complex to be assembled in vitro. Also, since a complex between *cdc2* and a 42-kD protein constitutes MPF in *Xenopus*, it is a reasonable assumption that the *cdc2*/p62 complex represents HeLa MPF. This is also now testable.

Third, what are the substrates of the protein kinase? Apart from the p40, p42, and p62 substrates that exist in a complex with *cdc2*, no physiologically relevant substrates have been identified. The *cdc2* kinase does phosphorylate histone H1 kinase in vitro (G. Draetta et al., in prep.), and this might be a relevant substrate in vivo. Further substrates will probably be identified by studying the many proteins known to be specifically phosphorylated during mitosis (e.g., lamins; Gerace and Blobel 1982) and also from further genetic work in yeast.

Finally, is *cdc2* playing the same or different roles in cell-cycle regulation in evolutionarily diverse species? This is not fully clear at present. In fission yeast, *cdc2*$^+$ is required in both G_1 and G_2. In budding yeast, *CDC28* regulates the G_1 cell-cycle start event. However, due to the abnormally early initiation of mitotic events in this yeast (e.g., budding, which occurs close to the initiation of DNA synthesis, is the equivalent of cytokinesis in most cells), it is possible that start may encompass both the initiation of DNA synthesis and a parallel mitotic initiation pathway. In vertebrate cells, our results now allow the clear conclusion that the *cdc2* protein kinase is involved in the regulation of mitosis. However, it cannot be ruled out that there is also some role in the regulation of DNA replication. In HeLa cells, the enzyme is relatively inactive in G_1, but a low level of activity might be sufficient for a G_1 role. Alternatively, there might exist in mammalian cells some other element that performs the G_1 function of yeast *cdc2*/*CDC28*.

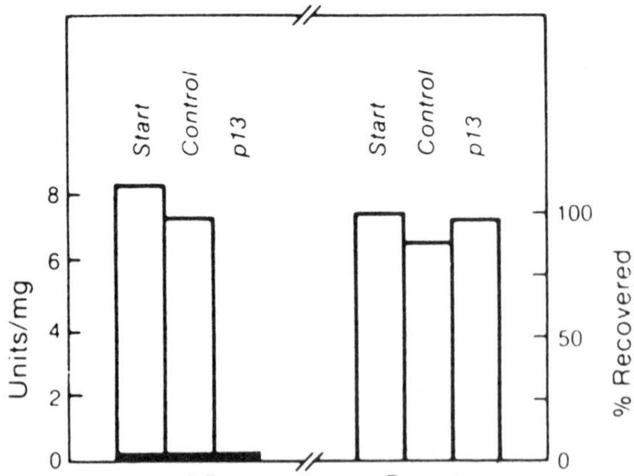

Figure 8. Affinity chromatography on p13-agarose depletes MPF. Data are from the same experiment as in Fig. 7. The flow through fractions from either a control column or a p13-agarose column were assayed for MPF activity (left portion of figure) or protein (right portion). The levels of MPF activity and protein in the initial MPF preparation (*Start*) are also indicated. There was no detectable MPF activity in the flow through fraction from the p13 column. (Reprinted, with permission, from Dunphy et al. 1988.)

The product of the *cdc2/CDC28* gene can no longer be regarded as a critical regulatory molecule per se. Rather, p34 is the catalytic subunit of a protein kinase that plays a critical cell-cycle role and which is subject to highly elaborate regulation at a variety of levels. The insight that *cdc2* is a component of the *Xenopus* MPF activity requires that the considerable body of phenomenological information concerning MPF be integrated with that concerning the *cdc2*$^+$ and *CDC28* protein kinases. For example, it has been shown that inactive MPF exists in a latent form throughout the cell cycle. This is entirely consistent with the finding in yeast and in HeLa cells that p34 is present at all stages of the division cycle, although it may be lost from nonproliferative cells. However, activation of MPF involves a process of autocatalytic activation, which can now be investigated in terms of the enzymatic properties of the protein kinase. Likewise, MPF is rapidly inactivated at the metaphase/anaphase transition. Again, this is consistent with the known properties of the *cdc2* complex in HeLa cells, and the mechanism of MPF inactivation is now open to full molecular investigation.

The ability to identify cell-cycle regulatory factors in yeast and to study their role in the cell cycle of higher organisms opens a powerful new approach to the study of the molecular basis of cell division in eukaryotes.

REFERENCES

Adlakha, R.C., D.A. Wright, C.G. Sahasrabuddhe, F.M. Davis, J. Prashad, H. Bigo, and P.N. Rao. 1985. Partial purification and characterization of mitotic factors from HeLa cells. *Exp. Cell Res.* **160:** 471.

Beach, D.H., B. Durkacz, and P.M. Nurse. 1982. Functionally homologous cell cycle control genes in budding and fission yeast. *Nature* **300:** 706.

Blow, J.J. and R.A. Laskey. 1986. Initiation of DNA replication in nuclei and purified DNA by a cell-free extract of *Xenopus* eggs. *Cell* **47:** 577

Booher, R. and D. Beach. 1987. Interaction between *cdc13*$^+$ and *cdc2*$^+$ in the control of mitosis in fission yeast; dissociation of the G_1 and G_2 roles of the *cdc2*$^+$ protein kinase. *EMBO J.* **6:** 3441.

Brizuela, L., G. Draetta, and D. Beach. 1987. p13^{suc1} acts in the fission yeast cell division cycle as a component of the p34^{cdc2} protein kinase. *EMBO J.* **6:** 3507.

Cyert, M. and M. Kirschner. 1988. Regulation of MPf activity in vitro. *Cell* **53:** 185.

De Brabander, M.J., R.M.L. Van de Veire, F.E.M. Aerts, M. Borgers, and P.A.J. Janssen. 1976. The effects of methyl [5-(2-thienylcarbonyl)-1H-benzimidazol-2-yl]carbamate, (R 17934; NSC 238159), a new synthetic antitumoral drug interfering with microtubules, on mammalian cells cultured *in vitro*. *Cancer Res.* **36:** 905.

Draetta, G. and D. Beach. 1988. Activation of *cdc2* protein kinase during mitosis in human cells: Cell-cycle dependent phosphorylation and subunit rearrangement. *Cell* **54:** 17.

Draetta, G., D. Beach, and E. Moran. 1988. Synthesis of p34, the mammalian homologue of the yeast *cdc2*$^+$/*CDC28* protein kinase, is stimulated during adenovirus-induced proliferation of baby rat kidney cells. *Oncogene* **2:** 553.

Draetta, G., L. Brizuela, J. Potashkin, and D. Beach. 1987. Identification of p34 and p13, human homologs of the cell cycle regulators of fission yeast encoded by *cdc2*$^+$ and *suc1*$^+$. *Cell* **50:** 319.

Dunphy, W.G. and J.W. Newport. 1988. Mitosis-inducing factors are present in a latent form during interphase in the *Xenopus* embryo. *J. Cell Biol.* **106:** 2047.

Dunphy, W.G., L. Brizuela, D. Beach, and J. Newport. 1988. The *Xenopus cdc2* protein is a component of *Xenopus* MPF, a cytoplasmic regulator of mitosis. *Cell* **54:** 423.

Fantes, P.A. 1979. Epistatic gene interactions in the control of division in fission yeast. *Nature* **279:** 428.

Gerace, L. and G. Blobel. 1982. Nuclear lamina and the structural organization of the nuclear envelope. *Cold Spring Harbor Symp. Quant. Biol.* **46:** 967.

Gerhart, J. and M. Kirschner. 1984. Cell cycle dynamics of an M-phase-specific cytoplasmic factor in *Xenopus laevis* oocytes and eggs. *J. Cell Biol.* **98:** 1247.

Hartwell, L.H. 1974. *Saccharomyces cerevisiae* cell cycle. *Bacteriol. Rev.* **38:** 164.

Hayles, J., S. Aves, and P. Nurse. 1986. *suc1* is an essential gene involved in both the cell cycle and growth in fission yeast. *EMBO J.* **5:** 3373.

Hindley, J. and G.A. Phear. 1984. Sequence of the cell division gene *cdc2*$^+$ from *Schizosaccharomyces pombe*, patterns of splicing and homology to protein kinases. *Gene* **31:** 129.

Hindley. J., G.A. Phear, M. Stein, and D. Beach. 1987. *suc1*$^+$ encodes a predicted 13-kilodalton protein that is essential for cell viability and directly involved in the division cycle of *Schizosaccharomyces pombe*. *Mol. Cell. Biol.* **7:** 504.

Hutchinson, C., R. Cox, and C. Ford. 1988. The control of DNA replication in a cell-free extract that recapitulates a basic cell cycle *in vitro*. *Development* **103:** 553.

Kishimoto, T., R. Kuriyama, H. Kondo, and H. Kanatani. 1982. Generality of the action of various maturation promoting factors. *Exp. Cell Res.* **137:** 121.

Lee, M. and P. Nurse. 1987. Complementation used to clone a human homologue of the fission yeast cell cycle control gene *cdc2*$^+$. *Nature* **327:** 31.

Lorincz, A.T. and S.I. Reed. 1984. Primary structure homology between the product of yeast cell division control gene *CDC28* and vertebrate oncogenes. *Nature* **307:** 183.

Masui, Y. and C.L. Markert. 1971. Cytoplasmic control of nuclear behavior during meiotic maturation of frog oocytes. *J. Exp. Zool.* **177:** 129.

Mendenhall, M.D., C.A. Jones, and S.I. Reed. 1987. Dual regulation of the yeast CDC28-p40 protein kinase complex: Cell cycle, pheromone and nutrient limitation effects. *Cell* **50:** 927.

Miake-Lye, R. and M.W. Kirschner. 1985. Induction of early mitotic events in a cell-free system. *Cell* **41:** 165.

Morris, N.R. 1976. Mitotic mutants of *Aspergillus nidulans*. *Genet. Res.* **26:** 237.

Newport, J. 1987. Nuclear reconstitution *in vitro*: Stages of assembly around protein-free DNA. *Cell* **48:** 205.

Nurse, P. and Y. Bissett. 1981. Gene required in G1 for commitment to cell cycle and in G2 for control of mitosis in fission yeast. *Nature* **292:** 558.

Nurse, P., P. Thuriaux, and K. Nasmyth. 1976. Genetic control of the cell division cycle in the fission yeast, *Schizosaccharomyces pombe*. *Mol. Gen. Genet.* **146:** 167.

Reed, S. and K.A. Nasmyth. 1981. Isolation of genes by complementation in yeast: Molecular cloning of a cell cycle gene. *Proc. Natl. Acad. Sci.* **77:** 2119.

Reed, S.I., J.A. Hadwiger, and A. Lorincz. 1985. Protein kinase activity associated with the product of the yeast cell cycle gene *CDC28*. *Proc. Natl. Acad. Sci.* **82:** 4055.

Russell, P. and P. Nurse. 1986. *cdc25*$^+$ functions as an inducer in the mitotic control of fission yeast. *Cell* **45:** 145.

———. 1987. Negative regulation of mitosis by *wee1*$^+$, a gene encoding a protein kinase homolog. *Cell* **49:** 261.

Simanis, V. and P. Nurse. 1986. The cell cycle control gene *cdc2*$^+$ of yeast encodes a protein kinase potentially regulated by phosphorylation. *Cell* **45:** 261.

Smith, L.D. and R.E. Ecker. 1971. Interaction of steroids with *Rana pipiens* oocytes in the induction of maturation. *Dev. Biol.* **25:** 232.

Role of G Proteins in Transmembrane Signaling

P.J. Casey, M.P. Graziano, M. Freissmuth, and A.G. Gilman
Department of Pharmacology, The University of Texas Southwestern Medical Center at Dallas, Texas 75235

Study of transmembrane signaling reactions that are initiated by hormones, neurotransmitters, and autacoids has been a major interest of molecular pharmacologists since the field was placed on firm footing in the late 1950s with the discovery of cyclic AMP by Earl Sutherland and Theodore Rall (Rall et al. 1957). More recent elucidation of the number and nature of the components that participate in these complex pathways has catalyzed the interest of investigators in a number of areas, and the field is now flourishing with the active participation of physiologists, biochemists, molecular biologists, and structural biologists.

An important prelude to this broadened interest was the demonstration by Rodbell, Birnbaumer, and their colleagues that GTP was an essential player in one transmembrane signaling system—hormone-sensitive adenylyl cyclase (Rodbell et al. 1971). One regulatory ligand (the receptor agonist) was found to be insufficient for activation of adenylyl cyclase. Two regulatory ligands were required—both a hormone and GTP. Thus, it seemed possible that there would be independent macromolecular sites of action for these ligands. Nevertheless, by the mid-1970s it still remained to be proven that the function of a hormone-sensitive adenylyl cyclase system was not fulfilled by a single polypeptide with an extracellularly oriented hormone-binding domain and intracellularly oriented regulatory and catalytic domains.

The development of radioligand-binding techniques for the identification and characterization of cell-surface receptors permitted the first dissection of a hormone-sensitive adenylyl cyclase system; the β-adrenergic receptor was resolved physically from the catalyst, adenylyl cyclase itself (Haga et al. 1977; Limbird and Lefkowitz 1977). An obvious next step was to learn how to reconstitute hormonal regulation of enzymatic activity by recombination of these crude components. Experiments designed to accomplish this goal took an unexpected twist when it was discovered that a third fraction was necessary (Pfeuffer 1977; Ross and Gilman 1977). Characterization and ultimate purification of the additional active component revealed that it was, in fact, a guanine-nucleotide-binding protein and that the essential mechanism for regulation of adenylyl cyclase activity was to promote the GTP-dependent association of this regulatory protein with adenylyl cyclase (Northup et al. 1980; Sternweis et al. 1981).

During the past decade, investigation of the role of guanine-nucleotide-binding regulatory proteins (G proteins) in the regulation of hormone-sensitive adenylyl cyclase activity, coupled with parallel studies of similar regulation of a rhodopsin-stimulated cyclic GMP-specific phosphodiesterase in retinal rod outer segments, has yielded a wealth of information on the mechanisms by which cell-surface receptors communicate with their respective intracellular effectors (Stryer and Bourne 1986; Gilman 1987). A model has evolved that appears to be applicable to a large number of such systems (Fig. 1): Activation of an extracellularly oriented receptor (by ligand or light) promotes interaction of the receptor with a specific G protein. As a result of this interaction, conformational alteration of the G protein permits exchange of tightly bound GDP (an inactive ligand) for GTP, the activating ligand. In the GTP-liganded state, the G protein (or components thereof) interacts with and influences the activity of appropriate effector molecules.

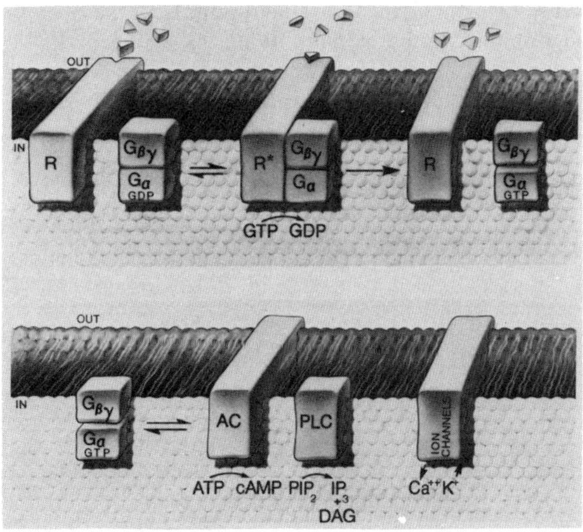

Figure 1. G-protein regulation of transmembrane signaling. (See text for detailed description.) The agonist-receptor complex catalyzes the dissociation of GDP from the α subunit of the oligomeric G protein. Subsequent binding of GTP is followed by the dissociation of G_α from $G_{\beta\gamma}$. Once dissociated, the G-protein subunits modulate the activity of a given effector(s). Hydrolysis of GTP to GDP, catalyzed by a GTPase intrinsic to the G_α subunit, is followed by reassociation of the G-protein subunits and cessation of the signal. (R) Receptor; (AC) adenylyl cyclase; (PLC) phosphoinositide-specific phospholipase C; (PIP_2) phosphatidylinositol-4,5-bisphosphate; (IP_3) inositol 1,4,5-trisphosphate; (DAG) 1,2-diacylglycerol. (Reprinted, with permission, from Graziano and Gilman 1987.)

G-protein Structure

Each of the G proteins that has been implicated as a participant in transmembrane signaling reactions is a heterotrimer, with subunits designated as α, β, and γ in order of decreasing mass (Table 1). The α subunit of each oligomer is unique and serves, for the moment, to distinguish one oligomer from another. The latest count reveals nine distinct genes that encode G-protein α subunits. In the case of the gene for the α subunit of G_s (G_s is the G protein that stimulates adenylyl cyclase), alternative splicing of mRNA appears to result in the production of four distinct polypeptides. Thus, 12 G-protein α subunits have been defined and sequenced as cDNAs. The corresponding proteins have been identified, and many of these have been purified to presumed homogeneity in quantities sufficient for biochemical characterization. Each G-protein α subunit contains a single high-affinity (nM) binding site for guanine nucleotides. In the basal state, this site is occupied by GDP. Activation of the G protein results from exchange of GDP for GTP. Deactivation of the G protein follows hydrolysis of GTP to GDP by a GTPase activity that is intrinsic to the α subunit. These subunits also contain one or two sites for NAD-dependent ADP-ribosylation—a covalent modification that is catalyzed by two distinct bacterial toxins. Cholera toxin catalyzes the ADP-ribosylation of G_s and G_t (one of the retinal G proteins) on an arginine residue; pertussis toxin catalyzes the ADP-ribosylation of the G_i proteins, G_o, and G_t on a cysteine residue near the carboxyl terminus of the subunit. When G_s or G_t is ADP-ribosylated by cholera toxin, the protein is activated; ADP-ribosylation of the G_i proteins or G_o by pertussis toxin interferes with the interaction between appropriate receptors and the G protein. The α subunits of G_i and G_o (but apparently not those of G_s and G_t) contain N-linked myristic acid, presumably in amide linkage at the amino terminus. Although the function of this N-myristoylation is not known, it is presumed to facilitate interactions between these G proteins and the membrane.

The primary amino acid sequences of the G-protein α subunits are extremely homologous; this fact is not surprising in view of their large number of shared functions (e.g., guanine nucleotide binding and hydrolysis, ADP-ribosylation, interaction with β and γ subunits). Variations of sequence, which are concentrated in three regions, are presumed to dictate the specificity of receptor and effector interactions for each individual G protein. Based in part on the similarity of the primary structures of the G-protein α subunits with that of elongation factor Tu, for which a crystal structure is available, Masters and co-workers have made predictions about the secondary structure of an "average" G-protein α subunit (Masters et al. 1986). Constraint of the four regions of α_{avg} that are believed to contribute to the guanine-nucleotide-binding site with the three-dimensional structure determined for this region of EF-Tu divides α_{avg} into three domains. Two of these, the amino and carboxyl termini, are obvious. The third domain results from a long insertion between the first two of the four regions involved in guanine nucleotide binding. Each of these three regions contains one of the stretches of greatest sequence variation mentioned above.

The terminology that is currently applied to G-protein α subunits is poor and reflects areas of ignorance. Names were chosen originally to evoke functional properties or, in some cases, to reflect the lack of

Table 1. Properties of G-protein Subunits

Subunit	M_r ($\times 10^{-3}$)[a]	Toxin	Receptor[b]	Role
$G_{s\alpha}$ (×4)	44.5–46	cholera	βAR ≫ α_2AR > Rho	activates adenylyl cyclase
$G_{i\alpha}$ (×3)	40.4–40.5	pertussis[c]	Musc, α_2AR, Rho > βAR[c,d]	inhibits adenylyl cyclase (weak)[d] P. tox. sensitive events[e]
$G_{o\alpha}$	39.9	pertussis	Musc, α_2AR, Rho	P. tox. sensitive events[e]
$G_{t\alpha,r}$	40	cholera + pertussis	Rho > α_2AR ≫ βAR	activates cGMP phosphodiesterase in retinal rods
$G_{t\alpha,c}$	40.5	cholera + pertussis[c]	cone opsins[c]	activates cGMP phosphodiesterase in cones[c]
$G_{z(\times)\alpha}$	40.9	unknown	unknown	P. tox insensitive events?
β (×2)	37.4	—	—	$\beta\gamma$ required for interaction of α with receptor; deactivates G_α; direct interactions with effectors
γ (×3?)	8–10	—	—	(?) $G_{t\beta\gamma}$ distinct from others

References. $G_{s\alpha}$: Bray et al. 1986; Robishaw et al. 1986; Gilman 1987. $G_{i\alpha}$: Gilman 1987; Jones and Reed 1987. $G_{o\alpha}$: Sternweis and Robishaw 1984; Gilman 1987. $G_{t\alpha,r}$: Fung 1983; Stryer 1986. $G_{t\alpha,c}$: Lerea et al. 1986. $G_{z(\times)\alpha}$: Fong et al. 1988; Matsuoka et al. 1988. $\beta\gamma$: Gilman 1987; Logothetis et al. 1987.

[a]The M_r values listed for α and β subunits are those calculated from their respective cDNAs.

[b]R·G interactions determined in reconstituted systems with respective G-protein oligomers (Gilman 1987; Logothetis et al. 1987). βAR, β-adrenergic; α_2AR, α_2-adrenergic; Musc, muscarinic cholinergic; Rho, rhodopsin.

[c]These properties are assumed for those proteins that are known only as cDNA-deduced sequences.

[d]The relative quantities of the three $G_{i\alpha}$ proteins in purified preparations of G_i used in functional studies are unclear.

[e]These proteins are candidates for involvement in the regulation of pertussis-toxin-sensitive cellular events (phosphoinositide hydrolysis, K^+ channels, Ca^{++} channels, etc.).

knowledge thereof. The discovery of related molecules and additional functions for known proteins has rendered the "system" obsolete, but it is probably premature to attempt to change it now.

G_s, named originally as the stimulatory regulator of adenylyl cyclase, is now also known to activate dihydropyridine-sensitive Ca^{++} channels (Yatani et al. 1987, 1988). This α subunit has the most distinctive structure, containing 380–395 amino acid residues, depending on the form. (All other α subunits contain 350–355 amino acid residues.) As mentioned above, four forms of $G_{s\alpha}$ are produced by alternative splicing of mRNA (Bray et al. 1986; Robishaw et al. 1986). The primary difference between these forms results from the inclusion or deletion of a stretch of 15 amino acid residues encoded by exon 3 of the gene for $G_{s\alpha}$ (Kozasa et al. 1988). It appears that any of the alternatively spliced forms of $G_{s\alpha}$ can activate both adenylyl cyclase and dihydropyridine-sensitive Ca^{++} channels.

G_i, as originally defined, is the G protein responsible for receptor-mediated inhibition of adenylyl cyclase. However, it is now known that the original "purified" preparations of G_i were heterogeneous. To date, three highly homologous cDNAs that encode $G_{i\alpha}$-like subunits have been isolated (Jones and Reed 1987). The use of antibodies generated against peptides synthesized according to amino acid sequences deduced from the cDNAs has permitted assignment of the $G_{i\alpha}$-1 cDNA sequence to a 41-kD polypeptide (found in particularly high concentrations in brain), the $G_{i\alpha}$-2 cDNA sequence to a relatively ubiquitous 40-kD polypeptide, and the $G_{i\alpha}$-3 cDNA sequence to another, distinct 41-kD polypeptide. All three $G_{i\alpha}$ polypeptides can be identified in a single cell type (the erythrocyte) and, thus, they do not represent cell-specific isoforms of the same protein (D.J. Carty et al., in prep.). The functional significance of this structural heterogeneity of the G_i proteins is unknown.

A related protein, G_o (o = "other" G protein), has also been purified, and cDNAs that encode its α subunit have been cloned (Van Meurs et al. 1987). G_o, which constitutes approximately 1% of brain membrane protein, possesses many of the properties associated with a G_i, although the sequence homology is not as great as that shared by the other three members of the group.

There are also multiple forms of $G_{t\alpha}$, the G protein involved in visual transduction. Two highly homologous cDNAs that encode different forms of this protein have been isolated. It appears that one form of this protein, $G_{t\alpha,r}$, is synthesized exclusively in retinal rods, whereas the other, $G_{t\alpha,c}$, is present only in cones (Lerea et al. 1986). The rod protein has been studied extensively and is the activator of a cGMP-specific phosphodiesterase; $G_{t\alpha,c}$ is presumed to have the same function.

A novel G-protein α subunit cDNA (termed $G_{z\alpha}$ or $G_{x\alpha}$, for lack of a better idea) has recently been described both by Simon and colleagues (Fong et al. 1988) and by Kaziro and associates (Matsuoka et al. 1988). Although the deduced amino acid sequence resembles that of the G_i proteins, there are significant differences, and certain of these have inspired considerable interest in the encoded product. First, the protein lacks a cysteine residue near the carboxyl terminus and, thus, is presumably not a substrate for pertussis toxin. This G protein thus becomes a candidate mediator of certain non-pertussis-toxin-sensitive pathways of signal transduction (e.g., inositol trisphosphate and diacylglycerol synthesis in several cell types). Of equal interest, this G-protein α subunit is the only one to differ in amino acid sequence in the highly conserved first region of the guanine-nucleotide-binding domain. Since the difference in sequence includes the residue that corresponds to the Gly-12 residue in *ras*, one assumes that the guanine-nucleotide-binding and/or hydrolytic properties of this protein will differ significantly from those of the other G-protein α subunits. The functional significance of these alterations remains to be studied on purified G_z protein.

The G-protein β and γ subunits form a high-affinity complex, and they have been resolved from each other only under denaturing conditions. The $\beta\gamma$ subunit complex of G_s, G_i, and G_o can be interchanged in functional assays, and it is assumed that this complex is shared among these different α subunits. Nevertheless, there are multiple forms of these polypeptides. Two cDNAs that encode β subunits have been cloned, sequenced, and expressed; the encoded proteins are 90% identical (Fong et al. 1987; Gao et al. 1987a). Preparations of β-subunit protein can be resolved into a doublet of bands by SDS-PAGE. The β_1 cDNA corresponds to the upper band of the doublet (β_{36}), whereas the β_2 cDNA encodes the β_{35} component (Gao et al. 1986b; Amatruda et al. 1988). All G-protein oligomers contain both β_{36} and β_{35} except for G_t, which contains only β_{36}. The evidence for heterogeneity of the γ subunit is based on analysis of the proteins, since only $G_{t\gamma}$ has been cloned to date. Peptide mapping indicates that $G_{t\gamma}$ is distinct from erythrocyte γ (Hildebrandt et al. 1985); furthermore, electrophoretic and immunological evidence indicates that there is more than one non-$G_{t\gamma}$ (Sternweis and Robishaw 1984; Evans et al. 1987).

G-protein-mediated Functions

Two G-protein-mediated events are understood in considerable detail: regulation of adenylyl cyclase activity and activation of retinal cGMP-specific phosphodiesterase. In both of these cases, all of the proteins involved in the primary pathway of signal transmission have been purified and reconstituted, and the major factors that regulate their interactions appear to be appreciated. Studies of these systems form the basis for the section on mechanisms, below.

In other cases, the involvement of a G protein has been clearly demonstrated, but important details remain to be discovered. There has been considerable recent interest in the involvement of G proteins in regulation of the activity of various ion channels. For example, muscarinic cholinergic receptors activate K^+ channels in the cardiac atrium. This process is dependent on GTP, is blocked by pertussis toxin, and can be mimicked by certain G-protein subunits. Thus, the involvement of a G protein seems certain. However, significant questions remain. Which G protein is involved? That G_i-3 can activate K^+ channels now seems clear (Codina et al. 1988), but it is not known if other G_i-like proteins can do so as well. Also unclear is the relative role of the α subunit versus the $\beta\gamma$-subunit complex in this response; conflicting reports have appeared in the literature (Codina et al. 1987; Logothetis et al. 1987). Finally, it is uncertain if the G protein interacts directly with the polypeptide(s) that constitutes the K^+ channel, or if other macromolecular intermediaries play a role. Recent evidence also indicates an important role for G proteins in the regulation of Ca^{++} channels (Hescheler et al. 1987; Ewald et al. 1988). G_o and perhaps some of the G_i proteins are capable of inhibiting neuronal Ca^{++} channels. As mentioned above, G_s can activate dihydropyridine-sensitive Ca^{++} channels in heart and skeletal muscle. The latter observation is particularly important, since it dispels the convenient notion that one G protein interacts with one type of effector molecule and thus establishes G proteins as branch points in pathways of transmembrane signaling.

Also of considerable interest is the potential involvement of G proteins in the regulation of various phospholipases (Cockcroft 1987). Hormonal regulation of the phospholipase C that is responsible for synthesis of inositol trisphosphate and diacylglycerol is dependent on GTP, and the system can be activated by nonhydrolyzable analogs of the nucleotide (e.g. GTPγS). In some cell types, the response is sensitive to pertussis toxin, and G_o or one of the G_i proteins may play a role. Confusing are the systems in which pertussis toxin is without effect, and the assumption is that a novel (non-pertussis-toxin substrate) G protein (such as G_z?) may be involved. Also confusing is the fact that no one has been able to stimulate phospholipase C activity with an activated G-protein preparation. This experimental paradigm is particularly powerful in all other situations. There is some evidence that a phospholipase A_2 activity may also be under the control of a G protein (Jelsema and Axelrod 1987). The role of G proteins in regulation of secretion and protein translocation is less clear. GTP-dependent steps in these pathways may reflect the activity of other macromolecules, such as members of the family of *ras*-like, ~21-kD GTP-binding proteins.

Mechanism of G-protein-mediated Signal Transduction

Purification and reconstitution of the components required for regulation of adenylyl cyclase and cGMP phosphodiesterase have permitted detailed studies of the molecular mechanism of G-protein-mediated transmembrane signaling (Gilman 1987; Casey and Gilman 1988). The crucial reactions are shown in Figure 2.

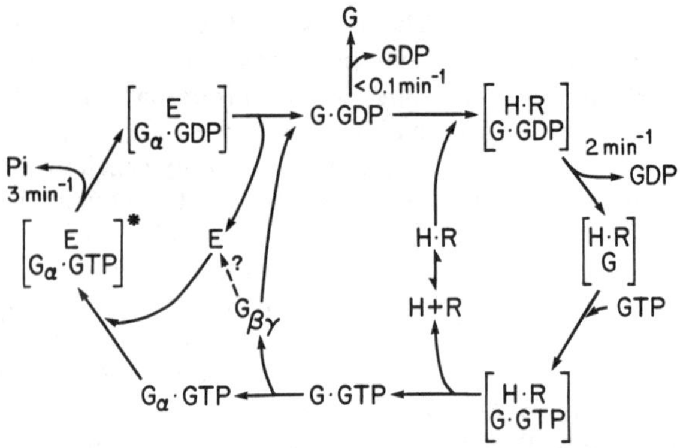

Figure 2. Interactions of receptor, G protein, guanine nucleotide, and effector. (G) G-protein oligomer ($\alpha\beta\gamma$); (H) hormone or agonist specific for (R) receptor; $[G_\alpha \cdot GTP \cdot E]^*$, activated form of effector. See text for explanation. (Reprinted, with permission, from Casey and Gilman 1988.)

In the basal state, G proteins exist as oligomers, and GDP is bound with high affinity to the α subunit. Interaction of the G protein with an appropriately liganded receptor (H·R) causes a conformational change that appears to "open" the guanine-nucleotide-binding site, such that the rate of dissociation of GDP is greatly enhanced. The H·R·G complex is apparently a relatively stable intermediate in the absence of guanine nucleotide. However, in the presence of cellular concentrations of GTP, the open guanine-nucleotide-binding site is filled rapidly with the triphosphate. The binding of GTP to H·R·G has two major consequences. The first is that the affinity of H for R (and R for G) is decreased. This negative heterotropic interaction between guanine nucleotide and hormone is a consequence of the utilization of the binding energy of R·H to drive G to the open form, thus necessarily weakening the observed binding of R for H. Dissociation of H from R and, therefore, of R from G·GTP permits R to recycle and thus function catalytically in the activation of G.

An additional significant consequence of the interaction of GTP with G is the activation of the G protein. There is considerable evidence that such activation greatly reduces the affinity of the GTP-liganded α subunit for the $\beta\gamma$ complex and that the resulting dissociation of subunits is an important component of their mechanism. $G_\alpha \cdot$GTP is the actual regulator of the target effector for both adenylyl cyclase and the cGMP phosphodiesterase. (There are, however, indications that the $\beta\gamma$ subunit complex may serve this role in some systems; see Logothetis et al. 1987.) Thus, the capacity of oligomeric G_s and G_t to activate adenylyl cyclase and the phosphodiesterase, respectively, can be fully accounted for by the activities of their α subunits.

In contrast to the results obtained with G_s and G_t, activation of G_i by nonhydrolyzable analogs of GTP and separation of the G_α-nucleotide and $\beta\gamma$ complexes revealed that the bulk of the capacity to inhibit adenylyl cyclase was in the $\beta\gamma$ subunit and that the inhibition was largely dependent on the presence of G_s. At least certain forms of $G_{i\alpha} \cdot GTP\gamma S$ appear to be able to inhibit adenylyl cyclase directly, but the activity has never been impressive. These observations led to the subunit exchange model for the activation of G_i. This hypothesis states that hormonal activation of G_i leads to dissociation of its subunits; the resultant increase in the concentration of free $\beta\gamma$ drives interaction with $G_{s\alpha}$; as a result, adenylyl cyclase is inhibited indirectly. This observation, along with the interchangeability of $\beta\gamma$ subunits mentioned previously, has led to the proposal that $\beta\gamma$ subunits may participate in a coordinate regulation of transmembrane signaling processes (Gilman 1987). The basic premise of the hypothesis is that receptor-mediated activation of a sufficient pool of a G protein will increase the concentration of $\beta\gamma$ to the point where it will inhibit some pathways, whereas the release of $\alpha \cdot GTP$ will initiate others.

Deactivation of the G protein appears to be due primarily to the GTPase activity of $\alpha \cdot GTP$. Reassociation of $\alpha \cdot GDP$ with $\beta\gamma$ is then presumed to follow. The rate constant for hydrolysis of GTP by various G-protein α subunits is approximately 3 min^{-1} at 20°C and 10 min^{-1} at 30°C. Thus, the lifetime of $G_\alpha \cdot GTP$ is many seconds, and this time is sufficient for considerable signal amplification due to prolonged activation of effector. This lifetime is very similar to that expected from the observed kinetics of deactivation of adenylyl cyclase. However, signals such as those initiated by light and rhodopsin are terminated much more rapidly; more information is needed to resolve this discrepancy.

Recombinant G-protein α Subunits

Given the existence of multiple forms of $G_{i\alpha}$, alternatively spliced forms of $G_{s\alpha}$, and the possibility that an apparently homogeneous G-protein preparation might be contaminated by a closely related but yet to be discovered member of the family, we have attempted to express cDNAs that encode G-protein α subunits in *Escherichia coli*. Our initial efforts have been directed at $G_{s\alpha}$, where the interaction with adenylyl cyclase has been well characterized (Graziano et al. 1987). We have utilized the expression system of Tabor and Richardson, where the cDNA of interest is placed under the control of the T7 RNA polymerase promoter (Tabor and Richardson 1985). Under typical culture conditions, $G_{s\alpha}$ accumulates to a concentration of approximately 1–2 mg/l of culture, which represents approximately 0.3% of total cell protein. Recent experiments suggest that this level can be increased by about fivefold if the culture conditions are more favorable. We have also devised a relatively simple purification procedure, and milligram quantities of homogeneous recombinant $G_{s\alpha}$ ($rG_{s\alpha}$) can be obtained in 2 days (M.P. Graziano et al., in prep.).

Using these procedures, we have first characterized the recombinant versions of two alternatively spliced forms of $G_{s\alpha}$ ($rG_{s\alpha\text{-}s}$ and $rG_{s\alpha\text{-}L}$), which differ by the deletion or inclusion of the 15 amino acid residues of exon 3 (Robishaw et al. 1986; Kozasa et al. 1988). As expressed in *E. coli*, both forms of $G_{s\alpha}$ bind $GTP\gamma S$ stoichiometrically. The proteins also hydrolyze GTP, although at different rates (0.13 min^{-1} and 0.34 min^{-1} for the short and long forms, respectively). These rates reflect differences in the rate of dissociation of GDP from the two proteins. Both forms of $rG_{s\alpha}$ have essentially the same K_{cat} for GTP hydrolysis, ~4 min^{-1} at 20°C.

$rG_{s\alpha}$ interacts functionally with G-protein $\beta\gamma$ subunits and with β-adrenergic receptors. The proteins can also be ADP-ribosylated by cholera toxin. Both forms of $rG_{s\alpha}$ can reconstitute GTP-, isoproterenol + GTP-, $GTP\gamma S$-, and fluoride-stimulated adenylyl cyclase activity in S49 cyc^- ($G_{s\alpha}$-deficient) membranes to maximal levels, although their specific activities for this reaction are lower than that observed for G_s purified from rabbit liver. Experiments with purified bovine brain adenylyl cyclase indicate that the affinity of $rG_{s\alpha}$ for adenylyl cyclase is 5–10 times lower than that of liver G_s; however, the intrinsic capacity of the recombinant protein to activate adenylyl cyclase is normal. These findings suggest that $G_{s\alpha}$, when synthesized in *E. coli*, may fail to undergo an unidentified posttranslational modification that is crucial for the high-affinity interaction of the G protein with adenylyl cyclase.

Despite the single apparent abnormality of $rG_{s\alpha}$—its relatively low affinity for adenylyl cyclase—the production of recombinant G-protein α subunits in *E. coli* does appear to constitute a valuable technique for analysis of the functional properties of individual α subunit polypeptides, including mutant and chimeric forms of the proteins. To date, we have successfully expressed and purified mutant $G_{s\alpha}$ proteins, as well as the α subunits of G_i-1, G_i-2, G_i-3, and G_o. Their characterization is in progress.

ACKNOWLEDGMENTS

This work was supported by U.S. Public Health Service grant GM-34497, American Cancer Society grant BC-555I, and the Raymond and Ellen Willie Chair of Molecular Neuropharmacology. We also acknowledge support from the Perot Family Foundation and the Lucille P. Markey Charitable Trust. P.J.C. was the recipient of National Institutes of Health fellowship GM-11982, and M.F. was the recipient of a Max Kade Foundation postdoctoral research exchange grant.

REFERENCES

Amatruda, T.T., III, H. Gautam, N.K.W. Fong, J.K. Northup, and M.I. Simon. 1988. The 35- and 36-kDa β subunits of GTP-binding regulatory proteins are products of separate genes. *J. Biol. Chem.* **263:** 5008.

Bray, P., A. Carter, C. Simons, V. Guo, C. Puckett, J. Kamholtz, A. Spiegel, and M. Nirenberg. 1986. Human cDNA clones for four species of $G_{s\alpha}$. *Proc. Natl. Acad. Sci.* **83:** 8893.

Casey, P.J. and A.G. Gilman. 1988. G protein involvement in receptor-effector coupling. *J. Biol. Chem.* **263:** 2577.

Cockcroft, S. 1987. Polyphosphoinositide phosphodiesterase:

Regulation by a novel guanine nucleotide binding protein, G_p. *Trends Biochem. Sci.* **12:** 75.

Codina, J., A. Yatani, D. Grenet, A.M. Brown, and L. Birnbaumer. 1987. The α subunit of the GTP binding protein G_k opens atrial potassium channels. *Science* **236:** 442.

Codina, J., J. Olate, J. Abramowitz, R. Mattera, R.G. Cook, and L. Birnbaumer. 1988. $\alpha_{1\text{-}3}$-cDNA encodes the α subunit of G_k, the stimulatory G protein of receptor-regulated K^+ channels. *J. Biol. Chem.* **263:** 6746.

Evans, T., A. Fawzi, E.D. Fraser, M.L. Brown, and J.K. Northup. 1987. Purification of a β_{35} form of the $\beta\gamma$ complex common to G-proteins from human placental membranes. *J. Biol. Chem.* **262:** 176.

Ewald, D.A., P.C. Sternweis, and R.J. Miller. 1988. G_o induced coupling of NPY receptors to calcium channels in sensory neurons. *Proc. Natl. Acad. Sci.* **85:** 3633.

Fong, H.K.W., T.T. Amatruda, B.W. Birren, and M.I. Simon. 1987. Distinct forms of the β subunit of GTP-binding regulatory proteins identified by molecular cloning. *Proc. Natl. Acad. Sci.* **84:** 3792.

Fong, H.K.W., K.K. Yoshimoto, P. Eversole-Cire, and M.I. Simon. 1988. Identification of a GTP-binding protein α subunit that lacks an apparent ADP-ribosylation site for pertussis toxin. *Proc. Natl. Acad. Sci.* **85:** 3066.

Fung, B.K.-K. 1983. Characterization of transducin from bovine retinal rod outer segments. I. Separation and reconstitution of subunits. *J. Biol. Chem.* **258:** 10495.

Gao, B., A.G. Gilman, and J.D. Robishaw. 1987a. A second form of the β subunit of signal transducing G proteins. *Proc. Natl. Acad. Sci.* **84:** 6122.

Gao, B., S.M. Mumby, and A.G. Gilman. 1987b. The G protein β_2 cDNA encodes the β_{35} subunit. *J. Biol. Chem.* **262:** 17254.

Gilman, A.G. 1987. G proteins: Transducers of receptor-generated signals. *Annu. Rev. Biochem.* **56:** 615.

Graziano, M.P. and A.G. Gilman. 1987. Guanine nucleotide-binding regulatory proteins: Mediators of transmembrane signaling. *Trends Pharmacol. Sci.* **8:** 478.

Graziano, M.P., P.J. Casey, and A.G. Gilman. 1987. Expression of cDNAs for G proteins in *Escherichia coli*: Two forms of $G_{s\alpha}$ stimulate adenylyl cyclase. *J. Biol. Chem.* **262:** 11375.

Haga, T., K. Haga, and A.G. Gilman. 1977. Hydrodynamic properties of the β-adreneregic receptor and adenylate cyclase from wild-type and variant S49 lymphoma cells. *J. Biol. Chem.* **252:** 5776.

Hescheler, J., W. Rosenthal, W. Trautwein, and G. Schultz. 1987. The GTP-binding protein, G_o, regulates neuronal calcium channels. *Nature* **325:** 445.

Hildebrandt, J.D., J. Codina, W. Rosenthal, L. Birnbaumer, E.J. Neer, A. Yamazaki, and M.W. Bitensky. 1985. Characterization by two-dimensional peptide mapping of the γ subunits of N_s and N_i, the regulatory proteins of adenylyl cyclase, and of transducin, the guanine nucleotide-binding protein of rod outer segments of the eye. *J. Biol. Chem.* **260:** 14867.

Jelsema, C.L. and J. Axelrod. 1987. Stimulation of phospholipase A_2 activity in bovine rod outer segments by the $\beta\gamma$ subunits of transducin and its inhibition by the α subunit. *Proc. Natl. Acad. Sci.* **84:** 3623.

Jones, D.T. and R.R. Reed. 1987. Molecular cloning of five GTP-binding protein cDNA species from rat olfactory neuroepithelium. *J. Biol. Chem.* **262:** 14241.

Kozasa, T., H. Itoh, T. Tsukamoto, and Y. Kaziro. 1988. Isolation and characterization of the human $G_{s\alpha}$ gene. *Proc. Natl. Acad. Sci.* **85:** 2081.

Lerea, C.L., D.E. Somers, and J.B. Hurley. 1986. Identification of specific transducin α subunits in retinal rod and cone photoreceptors. *Science* **234:** 77.

Limbird, L.E. and R.J. Lefkowitz. 1977. Resolution of β-adrenergic receptor binding and adenylate cyclase activity by gel exclusion chromatography. *J. Biol. Chem.* **252:** 799.

Logothetis, D.E., Y. Kurachi, J. Galper, E.J. Neer, and D.E. Clapham. 1987. The $\beta\gamma$ subunits of GTP-binding proteins activate the muscarinic K^+ channel in heart. *Nature* **325:** 321.

Masters, S.B., R.M. Stroud, and H.R. Bourne. 1986. Family of G protein α chains: Amphipathic analysis and predicted structure of functional domains. *Protein Eng.* **1:** 47.

Matsuoka, M., H. Itoh, T. Kozasa, and Y. Kaziro. 1988. Sequence analysis of cDNA and genomic DNA for a putative pertussis toxin-insensitive guanine nucleotide-binding regulatory protein α subunit. *Proc. Natl. Acad. Sci.* **85:** 5384.

Northup, J.K., P.C. Sternweis, M.D. Smigel, L.S. Schleifer, E.M. Ross, and A.G. Gilman. 1980. Purification of the regulatory component of adenylate cyclase. *Proc. Natl. Acad. Sci.* **77:** 6516.

Pfeuffer, T. 1977. GTP-binding proteins in membranes and the control of adenylate cyclase activity. *J. Biol. Chem.* **252:** 7224.

Rall, T.W., E.W. Sutherland, and J. Berthet. 1957. The relationship of epinephrine and glucagon to liver phosphorylase. IV. Effect of epinephrine and glucagon on the reactivation of phosphorylase in liver homogenates. *J. Biol. Chem.* **224:** 463.

Robishaw, J.D., M.D. Smigel, and A.G. Gilman. 1986. Molecular basis for two forms of the G protein that stimulates adenylate cyclase. *J. Biol. Chem.* **261:** 9587.

Rodbell, M., L. Birnbaumer, S.L. Pohl, and H.M.J. Krans. 1971. The glucagon-sensitive adenyl cyclase system in plasma membranes of rat liver. V. An obligatory role of guanyl nucleotides in glucagon action. *J. Biol. Chem.* **246:** 1877.

Ross, E.M. and A.G. Gilman. 1977. Resolution of some components of adenylate cyclase necessary for catalytic activity. *J. Biol. Chem.* **252:** 6966.

Sternweis, P.C. and J.D. Robishaw. 1984. Isolation of two proteins with high affinity for guanine nucleotides from membranes of bovine brain. *J. Biol. Chem.* **259:** 13806.

Sternweis, P.C., J.K. Northup, M.D. Smigel, and A.G. Gilman. 1981. The regulatory component of adenylate cyclase: Purification and properties. *J. Biol. Chem.* **256:** 11517.

Stryer, L. 1986. Cyclic GMP cascade of vision. *Annu. Rev. Neurosci.* **9:** 87.

Stryer, L. and H.R. Bourne. 1986. G proteins: A family of signal transducers. *Annu. Rev. Cell Biol.* **2:** 391.

Tabor, S. and C.C. Richardson. 1985. A bacteriophage T7 RNA polymerase/promoter system for controlled exclusive expression of specific genes. *Proc. Natl. Acad. Sci.* **82:** 1074.

Van Meurs, K.P., C.W. Angus, S. Lavu, H. Kung, S.K. Czarnecki, J. Moss, and M. Vaughan. 1987. Deduced amino acid sequence of bovine retinal $G_{o\alpha}$: Similarities to other guanine nucleotide-binding proteins. *Proc. Natl. Acad. Sci.* **84:** 3107.

Yatani, A., J. Codina, Y. Imoto, J.P. Reeves, L. Birnbaumer, and A.M. Brown. 1987. A G protein directly regulates mammalian cardiac calcium channels. *Science* **238:** 1288.

Yatani, A., Y. Imoto, J. Codina, S.L. Hamilton, A.M. Brown, and L. Birnbaumer. 1988. The stimulatory G protein of adenylyl cyclase, G_s, also stimulates dihydropyridine-sensitive Ca^{2+} channels. *J. Biol. Chem.* **263:** 9887.

Structures of the Genes Coding for G-protein α Subunits from Mammalian and Yeast Cells

Y. Kaziro, H. Itoh, T. Kozasa, R. Toyama, T. Tsukamoto, M. Matsuoka, M. Nakafuku, T. Obara, T. Takagi, and R. Hernandez*

Institute of Medical Science, University of Tokyo, 4-6-1, Shirokanedai, Minatoku, Tokyo 108, Japan

A superfamily of GTP-binding proteins consists of several families including (1) translational factors, (2) signal-transducing G proteins, (3) proto-oncogenic *ras* proteins, and (4) other low-molecular-weight GTP-binding proteins such as *rho*, *sec* (*sec4*, *YPT1*, etc.), and *arf* families. There are also other GTP-binding proteins such as tubulins and a group of metabolic GTP-binding proteins including succinate thiokinase and phosphoenolpyruvate carboxykinase, but they will not be discussed here. In this paper, we concentrate only on the family of signal-transducing G proteins that consists of three subunits, α (39–52 kD), β (35–36 kD), and γ (7–10 kD).

G proteins are involved in a variety of transmembrane signaling systems as transducers (for review, see Stryer and Bourne 1986; Gilman 1987). Two G proteins, G_s and G_i, are involved in hormonal stimulation and inhibition, respectively, of adenylate cyclase, whereas G_o (other G protein), which is present predominantly in brain tissues, may be involved in neuronal responses. Two transducins, G_t-1 and G_t-2, present in retinal rods and cones, respectively, regulate cGMP phosphodiesterase activity and mediate visual signal transduction. There is evidence suggesting the presence of additional G proteins, which may be involved in the activation of phospholipase C and phospholipase A_2, as well as in the gating of K^+ and Ca^{++} channels.

*Present address: Department of Molecular Biology, Institute of Investigaciones Biomédicas, Universidad Nacional Autónoma de México, 04510, México, D.F.

Isolation of cDNA Clones for G-protein α Subunits

Recently, cDNA sequences for α subunits of various G proteins have been reported (Table 1). These studies have revealed that there are several subtypes of $G_{i\alpha}$ that have closely related but distinct structures. The

Table 1. Molecular Cloning of G_s, G_i, G_o, and G_t α Subunit Genes and cDNAs

DNA library	Classification and nomenclature							Reference
	$G_{s\alpha}$	$G_{i1\alpha}$	$G_{i2\alpha}$	$G_{i3\alpha}$	$G_{o\alpha}$	$G_{t1\alpha}$	$G_{t2\alpha}$	
Genomic library								
human	$G_{s\alpha}$							Kozasa et al. (1988)
human		$G_{i1\alpha}$	$G_{i2\alpha}$	$G_{i3\alpha}$				Itoh et al. (1988b)
human			$G_{i2\alpha}$					Weinstein et al. (1988)
cDNA library								
human brain	$G_{s\alpha}$	αi-1						Bray et al. (1986, 1987)
human brain					$G_{o\alpha}$			Lavu et al. (1988)
human T-cells			αi2	αi3				Beals et al. (1987)
human monocytes (U-937)			$G_{i\alpha}$					Didsbury et al. (1987)
human granulocytes (HL-60)				$G_{x\alpha}$				Didsbury and Snyderman (1987)
human liver	$G_{s\alpha}$							Mattera et al. (1986)
human liver				αi − 3				Suki et al. (1987)
human liver				αi − 3				Codina et al. (1988)
bovine brain	$G_{s\alpha}$							Harris et al. (1985)
bovine adrenal gland	$G_{s\alpha}$							Robishaw et al. (1986b)
bovine cerebral cortex	$G_{s\alpha}$	$G_{i\alpha}$						Nukada et al. (1986a,b)
bovine pituitary gland		αi	αh					Michel et al. (1986)
bovine retina					$G_{o\alpha}$			Van Meurs et al. (1987)
bovine cerebellum					G_{39}			Ovchinnikov et al. (1987)
bovine retina						Tα		Tanabe et al. (1985)
bovine retina						Tα		Medynsky et al. (1985)
bovine retina						$G_{t\alpha}$		Yatsunami and Khorana (1985)
bovine retina							Tα	Lochrie et al. (1985)
rat glioma cells (C6)	$G_{s\alpha}$		$G_{i2\alpha}$	$G_{i3\alpha}$	$G_{o\alpha}$			Itoh et al. (1986, 1988b)
rat olfactory epithelium	$G_{\alpha s}$	$G_{\alpha i1}$	$G_{\alpha i2}$	$G_{\alpha i3}$	$G_{\alpha o}$			Jones and Reed (1987)
mouse macrophages (PU-5)	αs		αi					Sullivan et al (1986)
mouse lymphoma cells (S49)	$G_{s\alpha}$							Sullivan et al. (1987)
mouse lymphoma cells (S49)	$G_{s\alpha}$							Rall and Harris (1987)

presence of multiple $G_{i\alpha}$ subspecies had been suggested from molecular heterogeneity, immunological distinction, and functional differences of the pertussis toxin substrates in mammalian cells (for review, see Itoh et al. 1988a,b).

Our own studies on the cDNA cloning of G-protein α subunits from rat C6 glioma cells have revealed the structures of $G_{s\alpha}$, $G_{i2\alpha}$, $G_{i3\alpha}$, and $G_{o\alpha}$, consisting of 394, 355, 354, and 354 amino acid residues, respectively, with molecular weights of 45,663, 40,499, 40,522, and 40,068. $G_{i1\alpha}$, which is the most abundant among $G_{i\alpha}$ subtypes in brain tissues, was not expressed in C6 glioma cells, but we have recently obtained a $G_{i1\alpha}$ cDNA clone from the rat brain cDNA library (H. Itoh et al., unpubl.). The amino acid sequences of rat $G_{s\alpha}$, $G_{i2\alpha}$, $G_{i3\alpha}$, $G_{i1\alpha}$, $G_{o\alpha}$, and $G_{x\alpha}$ (a newly found G_α apparently insensitive to pertussis toxin, see below) deduced from the nucleotide sequences are shown in Figure 1.

Figure 2 shows a schematic representation of the structure of *Escherichia coli* EF-Tu, G-protein α subunits (G_α), yeast *RAS2* protein, and Ha-*ras* p21 protein. A remarkable homology was found in two regions, designated as P and G sites (Fig. 3), of all GTP-binding proteins. In EF-Tu, earlier biochemical studies indicated that the region around Cys-137 of EF-Tu (G site) is responsible for interaction with guanine nucleotides (Kaziro 1978), and later the four residues Asn-Lys-Cys-Asp were found by X-ray analysis to be situated close to the guanine ring (Jurnak 1985). On the other hand, it has been shown that the mutation of amino acid residue 12 of p21 from glycine to valine decreases GTPase activity and increases transforming activity. The sequence homologous to this region (P site) was found in all G_α proteins and also in translation factors, including EF-Tu. G′ site is a unique sequence highly conserved in all G proteins but absent in other GTP-binding proteins except *arf* (see Botstein et al.; Kahn, this volume).

Isolation of Human G_α Genes

We have screened human genomic libraries with the above-mentioned rat cDNA clones and have isolated human genes coding for $G_{s\alpha}$, $G_{i1\alpha}$, $G_{i2\alpha}$, $G_{i3\alpha}$, and $G_{o\alpha}$. So far, we have determined the gene organization and nucleotide sequences of total exons of $G_{s\alpha}$, $G_{i2\alpha}$, and $G_{i3\alpha}$ and have obtained the partial sequences (exons 1, 2, and 3) of $G_{i1\alpha}$ (see Fig. 4) (Itoh et al. 1988b; Kozasa et al. 1988). The human $G_{o\alpha}$ gene is a huge gene spanning at least 90 kb; however, the organization of exons is completely identical to that of the $G_{i\alpha}$ subfamily (T. Tsukamoto et al., unpubl.).

Structure of the Human $G_{s\alpha}$ Gene and Generation of Four $G_{s\alpha}$ cDNAs by Alternative Splicing

We have screened the human genomic libraries with rat $G_{s\alpha}$ cDNA and have isolated the human $G_{s\alpha}$ gene (Kozasa et al. 1988). We have also isolated three overlapping clones that covered the entire human $G_{s\alpha}$ gene. The gene contained 13 exons and 12 introns and spanned about 20 kb of genomic DNA (Fig. 4).

The presence of two species of $G_{s\alpha}$ protein with different molecular masses (45 and 52 kD) was known (Northup et al. 1980). Recently, Bray et al. (1986)

```
Gsα : MGCLGNSKTEDQRNEEKAQREANKKIEKQLQKDKQVYRATHRLLLLGAGESGKSTIVKQM  60
Gi2α: MGCTVSA-------EDKAAAERSKMIDKNLREDGEKAAREVKLLLLGAGESGKSTIVKQM  53
Gi3α: MGCTLSA-------EDKAAVERSKMIDRNLREDGEKAAKEVKLLLLGAGESGKSTIVKQM  53
Gi1α: MGCTLSA-------EDKAAVERSKMIDRNLREDGEKAAREVKLLLLGAGESGKSTIVKQM  53
Goα : MGCTLSA-------EERAALERSKAIEKNLKEDGISAAKDVKLLLLGAGESGKSTIVKQM  53
Gxα : MGCRQSS-------EEKEAARRSRRIDRHLRSESQRQRREIKLLLLGTSNSGKSTIVKQM  53

Gsα : RILHVNGFNGEGGEEDPQAARSNSDGEKATKVQDIKNNLKEAIETIVAAMSNLVPPVELA 120
Gi2α: KIIHEDGYSEE---------------ECRQYRAVVYSNTIQSIMAIVKAMGNLQIDFADP  98
Gi3α: KIIHEDGYSED---------------ECKQYKVVVYSNTIQSIIAIIRAMGRLKIDFGEA  98
Gi1α: KIIHEAGYSEE               ECKQYKAVVYSNTIQSIIAIIRAMGRLKIDFGDS  98
Goα : KIIHEDGFSGE---------------DVKQYKPVVYSNTIQSLAAIVRAMDTLGVEYGDK  98
Gxα : KIIHSGGFNLE---------------ACKEYKPLIIYNAIDSLTRIIRALAALKIDFHNP  98

Gsα : NPENQFRVDYILSVMNVPNFDFPPEFYEHAKALWEDEGVRACYERSNEYQLIDCAQYFLD 180
Gi2α: QRADDARQLFALSCAAEEQGMLPEDLSGVIRRLWADHGVQACFGRSREYQLNDSAAYYLN 158
Gi3α: ARADDARQLFVLAGSAEE-GVMTSELAGVIKRLWRDGGVQACFSRSREYQLNDSASYYLN 157
Gi1α: ARADDARQLFVLAGAAEE-GFMTAELAGVIKRLWKDSGVQACFNRSREYQLNDSAAYYLN 157
Goα : ERKADSKMVCDVVSRMEDTEPFSAELLSAMMRLWGDSGIQECFNRSREYQLNDSAKYYLD 158
Gxα : DRAYDAVQLFALTGPAESKGEITPELLGVMRRLWADPGAQACFGRSSEYHLEDNAAYYLN 158

Gsα : KIDVIKQADYVPSDQDLLRCRVLTSGIFETKFQVDKVNFHMFDVGGQRDERRKWIQCFND 240
Gi2α: DLERIAQSDYIPTQQDVLRTRVKTTGIVETHFTFKDLHFKMFDVGGQRSERKKWIHCFEG 218
Gi3α: DLDRISQTNYIPTQQDVLRTRVKTTGIVETHFTFKELYFKMFDVGGQRSERKKWIHCFEG 217
Gi1α: DLDRIAQPNYIPTQQDVLRTRVKTTGIVETHFTFKDLHFKMFDVGGQRSERKKWIHCFEG 217
Goα : SLDRIGAADYQPTEQDILRTRVKTTGIVETHFTFKNLHFRLFDVGGQRSERKKWIHCFED 218
Gxα : DLERIAAPDYIPTVEDILRSRDMTTGIVENKFTFKELTFKMVDVGGQRSERKKWIHCFEG 218

Gsα : VTAIIFVVASSSYNMVIREDNQTNRLQEALNLFKSIWNNRWLRTISVILFLNKQDLLAEK 300
Gi2α: VTAIIFCVALSAYDLVLAEDEEMNRMHESMKLFDSICNNKWFTDTSIILFLNKKDLFEEK 278
Gi3α: VTAIIFCVALSDYDLVLAEDEEMNRMHESMKLFDSICNNKWFTDTSIILFLNKKDLFEEK 277
Gi1α: VTAIIFCVALSDYDLVLAEDEEMNRMHESMKLFDSICNNKWFTDTSIILFLNKKDLFEEK 277
Goα : VTAIIFCVALSGYDQVLHEDETTNRMHESLMLFDSICNNKFFIDTSIILFLNKKDLFGEK 278
Gxα : VTAIIFCVELSGYDLKLYEDNQTSRMAESLRLFDSICNNNWFINTSLILFLNKKDLLSEK 278

Gsα : VLAGKSKIEDYFPEFARYTTPEDATPEPGEDPRVTRAKYFIRDEFLRISTASGDGRHYCY 360
Gi2α: IT--QSPLTICFPEYTGANKYDE-------------AASYIQSKFEDLNKRKDTKEIY-- 321
Gi3α: IK--RSPLTICYPEYTGSNTYEE-------------AAAYIQCQFEDLNRRKDTKEVY-- 320
Gi1α: IK--KSPLTICYPEYAGSNTYEE-------------AAAYIQCQFEDLNKRKDTKEIY-- 320
Goα : IK--KSPLTICFPEYPGSNTYED-------------AAAYIQTQF-ESKNRSPNKEIY-- 320
Gxα : IR--RIPLSVCFPEYKGQNTYEE-------------AAVYIQRQFEDLNRNKETKEIY-- 321

Gsα : PHFTCAVDTENIRRVFNDCRDIIQRMHLRQYELL 394
Gi2α: THFTCATDTKNVQFVFDAVTDVIIKNNLKDCGLF 355
Gi3α: THFTCATDTKNVQFVFDAVTDVIIKNNLKECGLY 354
Gi1α: THFTCATDTKNVQFVFDAVTDVIIKNNLKDCGLF 354
Goα : CHMTCATDTNNIQVVFDAVTDIIIANNLRGCGLY 354
Gxα : SHFTCATDTSNIQFVFDAVTDVIIQNNLKYIGLC 355
```

Figure 1. Deduced amino acid sequences of rat $G_{s\alpha}$, $G_{i2\alpha}$, $G_{i3\alpha}$, $G_{i1\alpha}$, $G_{o\alpha}$, and $G_{x\alpha}$. The sequences of $G_{s\alpha}$ and $G_{i2\alpha}$ are from Itoh et al. (1986); $G_{i3\alpha}$ is from Itoh et al. (1988b); $G_{i1\alpha}$ is from Jones and Reed (1987); $G_{o\alpha}$ is from Itoh et al. (1986) and Jones and Reed (1987); and $G_{x\alpha}$ is from Matsuoka et al. (1988).

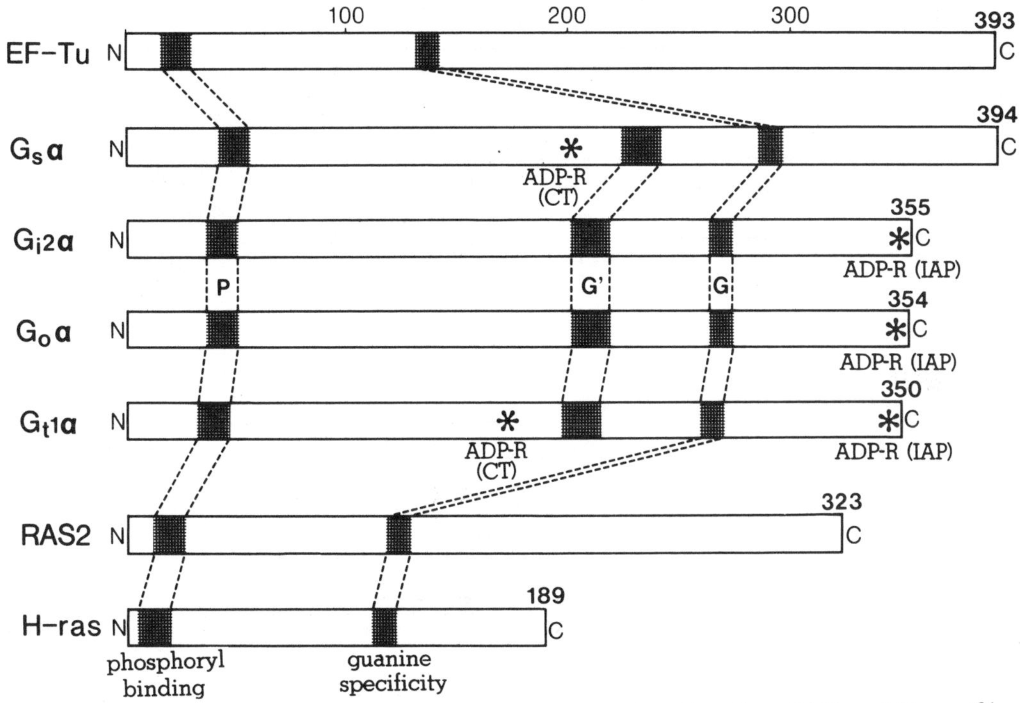

Figure 2. Schematic representation of structures of EF-Tu, $G_{s\alpha}$, $G_{i2\alpha}$, $G_{o\alpha}$, $G_{t1\alpha}$, *RAS2*, and Ha-*ras* p21.

isolated four different $G_{s\alpha}$ cDNAs ($G_{s\alpha}$-1 to -4) from human brain and characterized the partial structure. $G_{s\alpha}$-1 and $G_{s\alpha}$-3 are identical, except that $G_{s\alpha}$-3 lacks a single stretch of 45 nucleotides. $G_{s\alpha}$-2 and $G_{s\alpha}$-4 have three additional nucleotides (CAG) to $G_{s\alpha}$-1 and $G_{s\alpha}$-3 3′ to the above 45 nucleotides. Robishaw et al. (1986a) isolated two $G_{s\alpha}$ cDNAs from bovine adrenal that correspond to $G_{s\alpha}$-1 and $G_{s\alpha}$-4 and showed that these two cDNAs generated a 52-kD protein and a 45-kD protein when expressed in COS-m6 cells. Mattera et al. (1986) also isolated two $G_{s\alpha}$ cDNAs from human liver that correspond to $G_{s\alpha}$-1 and $G_{s\alpha}$-4.

Comparison of the four types of human $G_{s\alpha}$ cDNAs reported by Bray et al. (1986) with the sequence of the human $G_{s\alpha}$ gene of Kozasa et al. (1988) suggests that four types of $G_{s\alpha}$ mRNAs may be generated from a single $G_{s\alpha}$ gene by alternative splicing, as shown in Figure 5. $G_{s\alpha}$-1 has a sequence identical to exons 2, 3, and 4, whereas $G_{s\alpha}$-3 lacks exon 3. $G_{s\alpha}$-2 and $G_{s\alpha}$-4 have three additional nucleotides (CAG) to $G_{s\alpha}$-1 at the 3′ end of exon 4. This CAG sequence is found in the genomic sequence of the 3′ splice site of intron 3. The 5′ adjacent nucleotides to the CAG are TG and do not match with the 3′ splice consensus sequence AG. However, upstream of the splice site is the pyrimidine-rich sequence, which agrees with the rest of the 3′ splice

```
A   Gsα:  42-57  : R L L L L G A G E S G K S T I V
    Gi1α: 35-50  : K L L L L G A G E S G K S T I V
    Gi2α: 35-50  : K L L L L G A G E S G K S T I V
    Gi3α: 35-50  : K L L L L G A G E S G K S T I V
    Goα:  35-50  : K L L L L G A G E S G K S T I V
    Gxα:  35-50  : K L L L L G T S N S G K S T I V

B   Gsα:  223-240 : D V G G Q R D E R R K W I Q C F N D
    Gi1α: 200-217 : D V G G Q R S E R K K W I H C F E G
    Gi2α: 201-218 : D V G G Q R S E R K K W I H C F E G
    Gi3α: 200-217 : D V G G Q R S E R K K W I H C F E G
    Goα:  201-218 : D V G G Q R S E R K K W I H C F E D
    Gxα:  201-218 : D V G G Q R S E R K K W I H C F E G

C   Gsα:  285-296 : I S V I L F L N K Q D L
    Gi1α: 263-274 : T S I I L F L N K K D L
    Gi2α: 262-273 : T S I I L F L N K K D L
    Gi3α: 262-273 : T S I I L F L N K K D L
    Goα:  263-274 : T S I I L F L N K K D L
    Gxα:  263-274 : T S L I L F L N K K D L
```

Figure 3. Conserved sequences of GTP-binding proteins. Sequences of P site (*A*), G′ site (*B*), and G site (*C*) are shown.

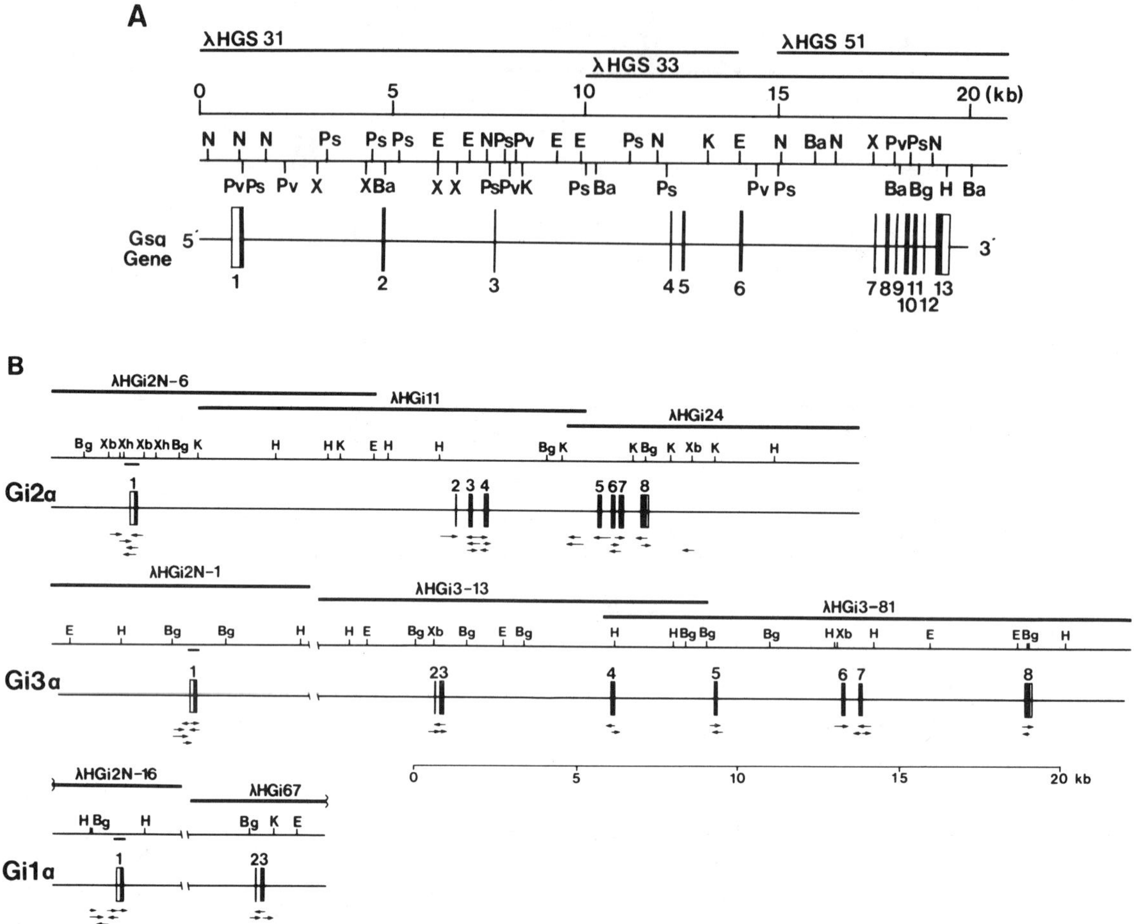

Figure 4. Organization of human genes for $G_{s\alpha}$ (*A*) and three $G_{i\alpha}$ subfamilies (*B*).

consensus sequence (Padgett et al. 1986). This 3′ splice site may be used for the production of $G_{s\alpha}$-2 and $G_{s\alpha}$-4.

An alternative splicing model proposed here to explain the generation of four different $G_{s\alpha}$ mRNAs implies the use of an unusual 3′ splice site sequence of TG. Since genomic Southern blot analysis indicates the presence of only a single $G_{s\alpha}$ gene per haploid genome, we consider that this unusual sequence TG at the 3′ splice site may probably be used. The amino acid sequence of exon 3, which is deleted in the differential splicing, is hydrophilic and may be present on the surface of the $G_{s\alpha}$ protein. One additional serine residue in $G_{s\alpha}$-2 and $G_{s\alpha}$-4 may be the potential site for phosphorylation by protein kinase C, and the alternative use of these splice sites may confer $G_{s\alpha}$ proteins with differential regulatory properties.

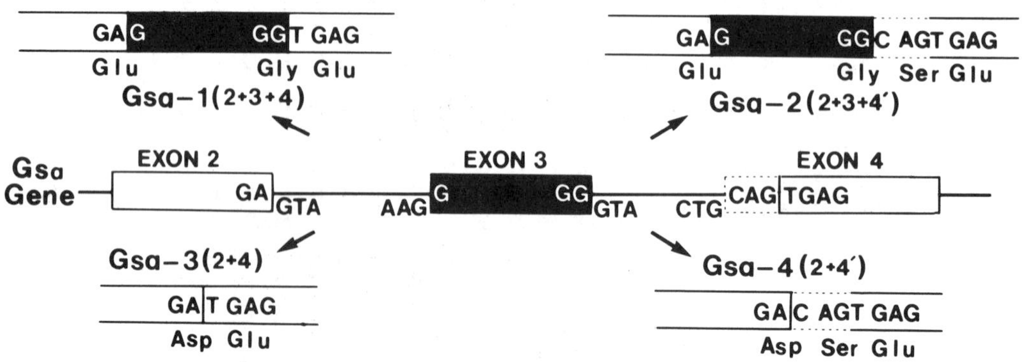

Figure 5. Generation of four different $G_{s\alpha}$ mRNAs by alternative splicing. The $G_{s\alpha}$ gene is shown in the center. $G_{s\alpha}$ mRNAs are indicated by $G_{s\alpha}$-1, -2, -3, and -4. For details see Kozasa et al. (1988).

Human Genes for $G_{i\alpha}$ Subtypes

To isolate genomic clones of the $G_{i\alpha}$ subfamily, we screened the human genomic DNA libraries under low-stringency conditions with rat $G_{i2\alpha}$ cDNA or rat $G_{i3\alpha}$ cDNA as a probe. Among a number of clones isolated, we identified three overlapping clones that cover the entire genomic sequence of $G_{i2\alpha}$, two overlapping clones covering exons 2–8 of $G_{i3\alpha}$, one nonoverlapping clone containing exon 1 of $G_{i3\alpha}$, and two nonoverlapping clones containing exon 1 and exons 2 and 3 of $G_{i1\alpha}$ (see Fig. 4).

The coding region of the human $G_{i2\alpha}$ and $G_{i3\alpha}$ genes splits into eight exons and seven introns (Itoh et al. 1988b). For human $G_{i1\alpha}$, we only have the sequences of exons 1–3 at present. Table 2 shows the sequences at the exon-intron boundaries of the human genes for $G_{i1\alpha}$, $G_{i2\alpha}$, and $G_{i3\alpha}$. All introns begin with the sequence GT at the 5′ end and terminate at the 3′ end with the sequence AG. Remarkably, the positions of the splice junctions on the sequence of cDNA for $G_{i2\alpha}$ and $G_{i3\alpha}$ were completely identical, although the lengths of introns are different (Itoh et al. 1988b). The same splice sites are also conserved in the partial sequence (exons 1, 2, and 3) of the human $G_{i1\alpha}$ gene as well as in the human $G_{o\alpha}$ gene (T. Tsukamoto, unpubl.). From the Southern blot analysis, it appears that each of the three $G_{i\alpha}$ genes occurs as a single copy per haploid human genome.

Organization of Human G_α Genes

The exon-intron organization of the $G_{s\alpha}$, $G_{i2\alpha}$, $G_{i3\alpha}$, and $G_{o\alpha}$ genes was compared with the predicted functional domain structure of proteins (Fig. 6). The amino-terminal domain encoded by exon 1 is hydrophilic and contains the site for limited tryptic digestions. Although this region may be involved in interaction with $\beta\gamma$ subunits, its precise function has not yet been shown. Exon 2 encodes a short-length region (24 and 14 amino acid residues, respectively, for $G_{s\alpha}$ and $G_{i\alpha}$s), which is the most conserved among all G_α proteins and responsible for GTP hydrolysis. Exon 3 of $G_{s\alpha}$ is unique to $G_{s\alpha}$. The most structurally divergent domain is encoded by exons 4–6 of $G_{s\alpha}$ and 3–4 of $G_{i\alpha}$. The amino acid sequences of residues 110–140 of $G_{s\alpha}$ and 90–130 of $G_{i\alpha}$ are remarkably diverse. Polypeptide secondary structure analysis (Chou and Fasman 1978) of this region in $G_{i2\alpha}$ is different from that of two other $G_{i\alpha}$ gene products (data not shown). Exon 8 of $G_{s\alpha}$ contains Arg-201, which is ADP-ribosylated in the presence of cholera toxin (Van Dop et al. 1984). ADP-ribosylation of $G_{s\alpha}$ by cholera toxin causes a decrease of affinity for $\beta\gamma$ subunits (Kahn and Gilman 1984). Arg-179 in exon 5 of $G_{i2\alpha}$ corresponds to this arginine residue. The domain encoded by exons 9–11 of $G_{s\alpha}$ and 6–7 of $G_{i\alpha}$ is strongly conserved among all G_α proteins. This domain is involved in formation of a core structure for GTP binding, together with that coded by exon 2

Table 2. Exon-Intron Organization of the Human $G_{i\alpha}$ Genes

Intron number	Species	5′ Boundary	Intron size (bp)	3′ Boundary	Exon size (bp)
1	$G_{i1\alpha}$	TGCTCG gtaagg ..	(>4000)	..tgtttatattttag GTGCTG	43
	$G_{i2\alpha}$	TGTTGG gtgagg ..	(~9800)	..gccctctgttccag GTGCTG	43
	$G_{i3\alpha}$	TACTCG gtgagg ..	(>7000)	..ttcttttcttacag GTGCTG	43
2	$G_{i1\alpha}$	GATGAA gtaagt ..	(96)	..ttttgtctcattag AATTAT	142
	$G_{i2\alpha}$	GATGAA gtaagt ..	(~350)	..gtcctggctatcag GATCAT	142
	$G_{i3\alpha}$	GATGAA gtaagt ..	(116)	..tctttgttttaaag AATCAT	142
3	$G_{i1\alpha}$	CGGGCG gtaagt ..			
	$G_{i2\alpha}$	AGAGCG gtatgt ..	(482)	..gccactgtgcccag GACGAC	161
	$G_{i3\alpha}$	AGGGCA gtaagt ..	(~5200)	..actgctttcttcag GATGAT	158
4	$G_{i2\alpha}$	TGCCTA gtgagt ..	(~3900)	..ccccccatccccag CTACCT	129
	$G_{i3\alpha}$	TTCATA gtaagt ..	(~3100)	..tgtcttttatttag TTATCT	129
5	$G_{i2\alpha}$	CTTCAA gtgagc ..	(~300)	..........ctgcag GATGTT	130
	$G_{i3\alpha}$	CTTCAA gtaagt ..	(~3600)	..ttccccttgcgcag GATGTT	130
6	$G_{i2\alpha}$	GAGATG gtgaga ..	(83)	..tattctacccccag AACCGC	154
	$G_{i3\alpha}$	GAGATG gtatgt ..	(~400)	..tatttctttttcag AACCGA	154
7	$G_{i2\alpha}$	ACACAG gtgtgg ..	(~500)	..tttctctcccccag GGGCCA	214
	$G_{i3\alpha}$	ACACAG gtaagg ..	(~5000)	..ggtgtccgttttag GTTCCA	212
8	$G_{i2\alpha}$	GGGATG gtgagc ..			
	$G_{i3\alpha}$	TGAAAG gtaaag ..			
Consensus sequence		AG:gtaagt		cccccccccc tttttttttt ncag:G	

The nucleotide sequences of exon-intron junctions were determined from genomic clones. For details see Itoh et al. (1988b).

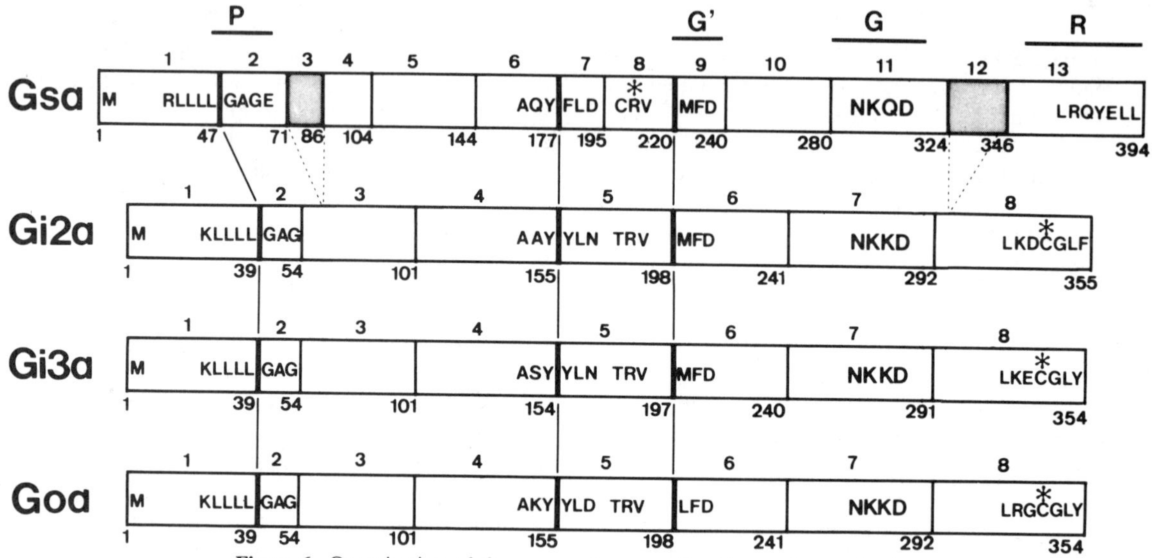

Figure 6. Organization of the exons of mammalian G-protein α subunits.

(Masters et al. 1986). The sequence, Asn-Lys-Xaa-Asp, consensus to all guanine-nucleotide-binding proteins, occurs in exon 11 of $G_{s\alpha}$ and exon 7 of $G_{i\alpha}$. The conserved Asp-223 in exon 9 of $G_{s\alpha}$ and Asp-201 in exon 6 of $G_{i2\alpha}$ may form a salt bridge to Mg^{++}, which is linked to the β phosphoryl group of GDP (Jurnak 1985). The exchange of GDP to GTP may result in displacement of the surrounding region residues 230–238 in exon 9 of $G_{s\alpha}$. A nonhydrolyzable GTP analog, but not GDP, prevents tryptic cleavage at Lys-210 in $G_{o\alpha}$ or Lys-205 in $G_{t1\alpha}$ (Hurley et al. 1984).

Exon 12 of $G_{s\alpha}$ is unique to $G_{s\alpha}$, and exon 13 of $G_{s\alpha}$ and exon 8 of $G_{i\alpha}$ encode the carboxy-terminal region. The domain may be involved in interaction with a receptor, since the cysteine residue that is ADP-ribosylated by pertussis toxin is present in this region of $G_{i\alpha}$, and, in addition the structure of this region is heterogeneous. Although in $G_{s\alpha}$, tyrosine instead of cysteine is present in this position, the replacement of arginine to proline at the −6 position gives rise to a mutant $G_{s\alpha}$, which is uncoupled with β-adrenergic receptor in S49 cells (Rall and Harris 1987; Sullivan et al. 1987; see also Bourne et al., this volume). Further studies, including site-directed mutagenesis and construction of chimeric genes, may shed more light on the structure-function relationship of G_α proteins.

Table 3. Conservation of G-protein α Subunit Sequences among Different Mammalian Species

Species	Amino acid sequences	Nucleotide sequences
$rG_{s\alpha}$ vs. $hG_{s\alpha}$	393/394 (99.7%)	1128/1182 (95.4%)
$bG_{i1\alpha}$ vs. $hG_{i1\alpha}$	354/354 (100%)	998/1062 (94.0%)
$rG_{i2\alpha}$ vs. $hG_{i2\alpha}$	350/355 (98.6%)	985/1065 (92.4%)
$rG_{i3\alpha}$ vs. $hG_{i3\alpha}$	349/354 (98.6%)	981/1062 (92.4%)
$rG_{o\alpha}$ vs. $bG_{o\alpha}$	348/354 (98.3%)	992/1062 (93.4%)
$rG_{x\alpha}$ vs. $hG_{x\alpha}$	349/355 (98.3%)	977/1065 (91.7%)

(h) Human; (r) rat; (b) bovine.

Comparison of the exon organization of the $G_{i\alpha}$ subfamily and $G_{o\alpha}$ with that of $G_{s\alpha}$ indicated that some of the exon junctions were conserved between the $G_{i\alpha}$ subfamily and $G_{s\alpha}$. Thus, 3 out of 12 splice sites of the human $G_{s\alpha}$ gene are shared with the human $G_{i\alpha}$ genes, and exon 1 and exons 7 and 8 of $G_{s\alpha}$ correspond to exon 1 and exon 5 of $G_{i\alpha}$, respectively.

Conservation of Primary Structure of Each G_α among Mammalian Species

Table 3 shows that, in addition to the remarkable homologies of the overall structures, there is a strong conservation of the amino acid sequence in each group of G-protein α subunit. The amino acid sequence of $G_{s\alpha}$ is strongly conserved between human and rat; only 1 out of 394 amino acids is different. The sequence of $G_{i1\alpha}$ is completely identical between bovine and human. For $G_{i2\alpha}$, $G_{i3\alpha}$, and $G_{o\alpha}$, more than 98% identity of amino acid sequences is maintained among different mammalian species.

The comparison of the deduced amino acid sequences of $G_{i1\alpha}$, $G_{i2\alpha}$, and $G_{i3\alpha}$ obtained from different mammalian sources is shown in Figure 7. The strong conservation of the amino acid sequence of each G-protein α subunit among distant mammalian species may reflect the presence of evolutionary pressure to maintain the specific physiological function of each G-protein gene product. Each G_α protein may be linked to a specific receptor and thereby be involved in a specific signal-transducing pathway.

An evolutionary tree of G-protein α subunits can be drawn on the basis of the homologies of the predicted amino acid sequences obtained from various mammalian sources (Fig. 8). It is remarkable that the homologies among three $G_{i\alpha}$ species are higher than that between rod ($G_{t1\alpha}$) and cone ($G_{t2\alpha}$) transducin α subunits (Lerea et al. 1986).

```
hGi2α:  1 MGCTVSAEDKAAAERSKMIDKNLREDGEKAAREVKLLLLGAGESGKSTIVKQMKIIHEDG
rGi2α:  1 ------------------------------------------------------------
mGi2α:  1 ------------------------------------------------------------
hGi3α:  1 ----L-------V-------R----------K----------------------------
rGi3α:  1 ----L-------V-------R----------K----------------------------
hGi1α:  1 ----L-------V-------R-------------------------------------A-
bGi1α:  1 ----L-------V-------R-------------------------------------A-

hGi2α: 61 YSEEECRQYRAVVYSNTIQSIMAIVKAMGNLQIDFADPSRADDARQLFALSCTAEEQGVL
rGi2α: 61 ---------------------------------------Q------------A-----M-
mGi2α: 61 ---------------------L----R------------Q------------A-----M-
hGi3α: 61 ---D--K--KV----------I--IR---R-K---GEAA---------V-AGS---*--M
rGi3α: 61 ---D--K--KV----------I--IR---R-K---GEAA---------V-AGS---*--M
hGi1α: 61 ------K--K-----------I--IR---R-K---G-SA---------V-AGA---*-FM
bGi1α: 61 ------K--K-----------I--IR---R-K---G-SA---------V-AGA---*-FM

hGi2α:121 PDDLSGVIRRLWADHGVQACFGRSREYQLNDSAAYYLNDLERIAQSDYIPTQQDVLRTRV
rGi2α:121 -E----------------------------------------------------------
mGi2α:121 -E----------------------------------------------------------
hGi3α:120 TPE-A---K---R-G------S-----------S------D--S--N-------------
rGi3α:120 TSE-A---K---R-G------S-----------S------D--S-TN-------------
hGi1α:120 TAE-A---K---K-S------N------------------D----PN-------------
bGi1α:120 TAE-A---K---K-S------N------------------D----PN-------------

hGi2α:181 KTTGIVETHFTFKDLHFKMFDVGGQRSERKKWIHCFEGVTAIIFCVALSAYDLVLAEDEE
rGi2α:181 ------------------------------------------------------------
mGi2α:181 ------------------------------------------------------------
hGi3α:180 ---------------Y---------------------------------D----------
rGi3α:180 -------------E-Y---------------------------------D----------
hGi1α:180 -------------------------------------------------D----------
bGi1α:180 -------------------------------------------------D----------

hGi2α:241 MNRMHESMKLFDSICNNKWFTDTSIILFLNKKDLFEEKITHSPLTICFPEYTGANKYDEA
rGi2α:241 ---------------------------------------Q--------------------
mGi2α:241 ---------------------------------------Q-S------------------
hGi3α:240 ---------------------E-----------------KR------Y-----S-T-E--
rGi3α:240 ---------------------------------------KR------Y-----S-T-E--
hGi1α:240 ---------------------------------------KK------Y---A-S-T-E--
bGi1α:240 ---------------------------------------KK------Y---A-S-T-E--

hGi2α:301 ASYIQSKFEDLNKRKDTKEIYTHFTCATDTKNVQFVFDAVTDVIIKNNLKDCGLF
rGi2α:301 -------------------------------------------------------
mGi2α:301 -------------------------------------------------------
hGi3α:300 -A---CQ-----R-------------------------------------E---Y
rGi3α:300 -A---CQ-----R------V------------------------------E---Y
hGi1α:300 -A---CQ------------------------------------------------
bGi1α:300 -A---CQ------------------------------------------------
```

Figure 7. Comparison of amino acid sequences of $G_{i\alpha}$ subfamily from different mammalian species. Hyphens represent the amino acid residue in each $G_{i\alpha}$ that is identical to the residue shown in human $G_{i2\alpha}$. For details see Itoh et al. (1988b).

Isolation and Characterization of a cDNA Clone Coding for a Pertussis-toxin-insensitive G-protein α Subunit

The α subunits of the G_i subfamily, G_o and G_t, are ADP-ribosylated by NAD in the presence of pertussis toxin. The cysteine residue at the fourth position from the carboxyl terminus is a feature common to all G proteins susceptible to modification by pertussis toxin, and the cysteine residue at this position of $G_{t\alpha}$ has been shown to be ADP-ribosylated by pertussis toxin (West et al. 1985). When ADP-ribosylated, these G proteins are uncoupled with their receptor and the signal transduction is inhibited. The sensitivity toward pertussis toxin has often been used as a criterion for the involvement of G proteins in certain cellular functions. Thus, it has been shown that the pertussis-toxin-sensitive G proteins are involved not only in inhibition of adenylate cyclase, but also in activation of phospholipase C and phospholipase A_2 as well as for gating of K^+ and Ca^{++} channels (for review, see Gilman 1987 and this volume). On the other hand, it has been suggested that a species of pertussis-toxin-sensitive G protein is involved in the activation of phospholipase C (Masters et al. 1985; for review, see Gilman 1987).

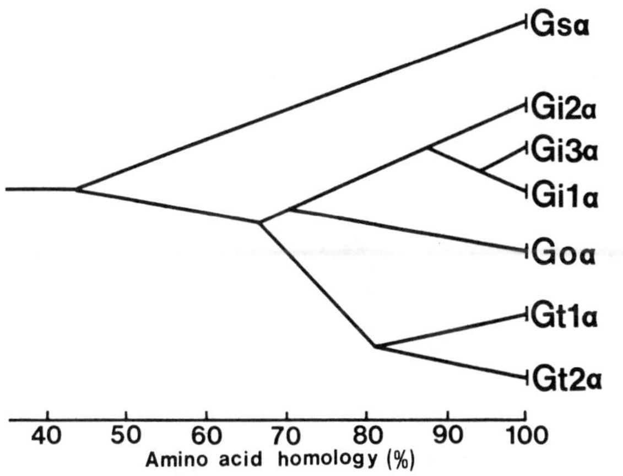

Figure 8. Relationships among the different mammalian G_α species. For details see Itoh et al. (1988b).

In the course of a study on the cloning of genomic DNAs for G-protein α subunits, we have isolated a clone whose partial sequence was similar to, but distinct from, any published G_α sequence. Furthermore, the sequence analysis indicated that the cysteine residue in the fourth position from the carboxyl terminus is replaced by isoleucine. The RNA blot analysis of various rat tissues with this human gene indicated that the counterpart of this new human G_α gene is predominantly expressed in brain. We therefore searched a rat brain cDNA library and isolated a cDNA clone corresponding to this new G_α species. We have tentatively designated this species as $G_{x\alpha}$, which turned out to be identical with $G_{z\alpha}$ isolated independently by Fong et al. (1988) from a human retina cDNA library. By RNA blot hybridization, the $G_{x\alpha}$ mRNA of 3.2 kb was detected mainly in brain and weakly in spleen.

Characteristics of the Predicted Amino Acid Sequence of $G_{x\alpha}$

The nucleotide sequence analysis of rat $G_{x\alpha}$ cDNA revealed the presence of a single open reading frame of 1065 nucleotides coding for a protein of 355 amino acids, with a calculated relative molecular weight of 40,879 (Matsuoka et al. 1988). The deduced amino acid sequence of rat $G_{x\alpha}$ (see Fig. 1) showed 40%, 67%, 66%, 66%, and 59% identity, respectively, with rat $G_{s\alpha}$, $G_{i1\alpha}$, $G_{i2\alpha}$, $G_{i3\alpha}$, and $G_{o\alpha}$. Therefore, $G_{x\alpha}$ is most closely related to the $G_{i\alpha}$ subfamily, but still distinct from it, since the similarity between $G_{x\alpha}$ and any of the $G_{i\alpha}$ subtypes is far less than that between the $G_{i\alpha}$ subtypes (>85%). Furthermore, the exon-intron organization of the human $G_{x\alpha}$ gene is quite different from that of human $G_{i\alpha}$ genes (M. Matsuoka, unpubl.).

Comparison of the predicted amino acid sequence of $G_{x\alpha}$ with those of known mammalian G_α proteins indicates several marked characteristics. As shown in Figure 3A, three amino acid residues in the consensus GTP-hydrolysis site of $G_{x\alpha}$ (Thr-Ser-Asn, at positions 41–43) are different from these residues in other G_α proteins (Ala-Gly-Glu). It remains to be seen whether the kinetics of the G_α-mediated signal transduction may be different from the other systems due to the replacement of glycine (which corresponds to Gly-12 of p21) to serine.

The most remarkable feature of the primary structure of the $G_{x\alpha}$ is that it lacks the cysteine residue in the fourth position from the carboxyl terminus, having instead an isoleucine (see Fig. 1). $G_{s\alpha}$, which is also resistant to modification by pertussis toxin, has a tyrosine in the fourth position from the carboxyl terminus instead of cysteine.

It has been suggested that a pertussis-toxin- and cholera-toxin-insensitive G protein is involved in activation of phospholipase C in a human astrocytoma cell line (Masters et al. 1985), a rat pituitary cell line (Lucas et al. 1986), and rat liver plasma membrane (Uhing et al. 1986). There are reports that a cholera-toxin-sensitive and pertussis-toxin-insensitive G protein is involved in the inositol phospholipid hydrolysis in a human pituitary cell line (Lo and Hughes 1987) and in a human T-lymphocyte line (Imboden et al. 1986). It is not known whether $G_{x\alpha}$ protein is ADP-ribosylated in the presence of cholera toxin, although it possesses an arginine residue at amino acid position 179. The arginine residue at this position is ADP-ribosylated by cholera toxin in $G_{s\alpha}$ and $G_{t\alpha}$ (Van Dop et al. 1984), but not in $G_{i\alpha}$ and $G_{o\alpha}$. Thus, $G_{x\alpha}$ could be a candidate for one of the proposed species of G_α.

G Proteins from *Saccharomyces cerevisiae*

A family of GTP-binding proteins, the *ras* family, is widely distributed among eukaryotes (for review, see Barbacid 1987) including yeast *S. cerevisiae* (DeFeo-Jones et. al. 1983; Powers et al. 1984) and *Schizosaccharomyces pombe* (Fukui and Kaziro 1985). It has been suggested that the *RAS2* gene in *S. cerevisiae* is involved in the activation of adenylate cyclase (Broek et al. 1985; Toda et al. 1985) and mimics the role of mammalian G_s. However, in view of the strong conservation of the amino acid sequences of each G protein species among different organisms (see Table 3 and Fig. 7), we speculated that the G protein may occur also in yeast. We have searched for G-protein homologous genes in yeast and have isolated two genes, *GPA1* (Nakafuku et al. 1987) and *GPA2* (Nakafuku et al. 1988), from *S. cerevisiae* that are homologous to cDNAs for mammalian G-protein α subunits (Fig. 9).

*GPA*1 and *GPA*2 code for the sequences of 472 and

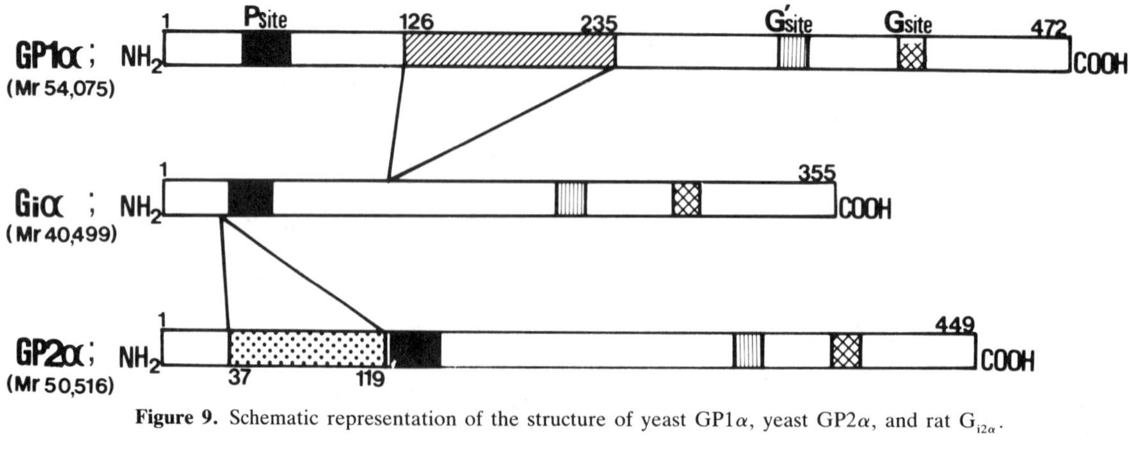

Figure 9. Schematic representation of the structure of yeast GP1α, yeast GP2α, and rat $G_{i2\alpha}$.

449 amino acid residues, respectively, with calculated relative molecular weights of 54,075 and 50,516. When aligned with the α subunit of mammalian G proteins to obtain maximal homology, GP1α (GPA1-encoded protein) and GP2α (GPA2-encoded protein) were found to contain the stretches of 110 and 83 additional amino acid residues, respectively, near the amino terminus.

Comparison of the Amino Acid Sequences of Yeast GP1α and GP2α with Those of Rat Brain $G_{i\alpha}$ and $G_{o\alpha}$

The deduced amino acid sequence of yeast GP1α and GP2α is highly homologous to those of rat brain $G_{i\alpha}$ and $G_{o\alpha}$. As shown in Figure 10, the homology is most remarkable in the region of GTP hydrolysis (amino acid residues 43–58 of GP1α, 125–140 of GP2α, and 35–50 of $G_{i2\alpha}$). The region responsible for GTP binding (amino acid residues 381–397 of GP1α, 358–374 of GP2α, and 263–279 of $G_{i2\alpha}$) was also highly homologous. Another region of homology was found in amino acid residues 321–334 of GP1α, 298–311 of GP2α, and 203–216 of $G_{i2\alpha}$, where a sequence of 14 contiguous amino acids was completely identical in yeast GP1α, GP2α, and rat $G_{i2\alpha}$.

The overall homology in nucleotide and amino acid sequences of yeast GP1α, yeast GP2α, rat $G_{i2\alpha}$, and rat $G_{s\alpha}$ is remarkable. Disregarding the unique sequences present in GP1α (residues 126–235) and GP2α (residues 37–119), the proteins are 60% homologous if conservative amino acid substitutions are considered to be homologous. The homology is smaller than that between rat $G_{i2\alpha}$ and $G_{o\alpha}$ (85%) but is comparable to that between rat $G_{i2\alpha}$ and $G_{s\alpha}$ (60%).

As described elsewhere in detail (Dietzel and Kurjan 1987; Miyajima et al. 1987; Nakafuku et al. 1988; see also Matsumoto et al.; Kurjan and Dietzel; both this volume), *GPA1* is a haploid-specific gene involved in the mating factor signal transduction. On the other hand, *GPA2* is expressed both in haploid and diploid cells and may be involved in the regulation of cAMP levels in *S. cerevisiae*. The demonstration of the occurrence of G proteins in yeast may open the way for a detailed genetic analysis of the function of G proteins in eukaryotic cells.

Reaction Mechanism of G Proteins

Table 4 lists G-protein α subunits present in mammalian cells and in yeast *S. cerevisiae*. Due to the limitation of time and space, we cannot discuss individual functions of each G protein in detail. How-

```
GP1α:   1- 37: MG-CTVSTQTIGDESDPFLQNK---RANDVIEQSLQLEKQR------------------
GP2α:   1- 55: MGLCA-SSEKNGSTPDT---QTASAGSDNV-GKAKVPPKQEPQKTVRTVNTANQQEKQQQ
Gi2α:   1- 29: MG-CTVSA------------EDKAAAERSKMIDKNLREDGEK-----------------
 Gsα:   1- 36: MG-CLGNSKTEDQRN----EEKAQREANKKIEKQLQKDKQV-----------------

GP1α:          ------------------------------------------------------------
GP2α:  56-115: RQQQPSPHNVKDRKEQNGSINNAISPTATANTSGSQQINIDSALRDRSSNVAAQPSLSDA
Gi2α:          ------------------------------------------------------------
 Gsα:          ------------------------------------------------------------

GP1α:  38- 78: ----DKNEIKLLLLGAGESGKSTVLKQLKLLHQGGFSHQ---------------ERLQYA
GP2α: 116-160: SSGSNDKELKVLLLGAGESGKSTVLQQLKILHQNGFSEQ---------------EIKEYI
Gi2α:  30- 70: ----AAREVKLLLLGAGESGKSTIVKQMKIIHEDGYSEE---------------ECRQYR
 Gsα:  37- 92: ----YRATHRLLLLGAGESGKSTIVKQMRILHVNGFNGEGGEEDPQAARSNSDGEKATKV

GP1α:  79-136: QVIWADAIQSMKILLIQAR-KLGIQLDCDDPINN-KDLFACKRILLKAKALDYINASVAG
GP2α: 161-210: PLIYQNLLEIGRNLI-QARTRFNVNLEPECELTQ-QDLSRTMSYEMPNNYTG--------
Gi2α:  71-118: AVVYSNTIQSIMAIVKAMG-NLQIDFADPQRADDARQLFALSCAAEEQG----------
 Gsα:  93-140: QDIKNNLKEAIETIVAAMS-NLVPPVELANPENQFRVDYILSVMNVPNF----------

GP1α: 137-197: GSDFLNDYVLKYSERYETRRRVQSTGRAKAAFDEDGNISNVKSDTDRDAETVTQNEDADR
GP2α:          ------------------------------------------------------------
Gi2α:          ------------------------------------------------------------
 Gsα:          ------------------------------------------------------------

GP1α: 197-255: NNSSRINLQDICKDLNQEGDDQMFVRKTSREIQGQNRRNLIHEDIAKAIKQLWNND-KGI
GP2α: 211-231: ---------------------------------------QFPEDIAGVISTLWALPSTQD
Gi2α: 119-137: ---------------------------------------MLPEDLSGVIRRLWA-D-HGV
 Gsα: 141-159: ---------------------------------------DFPPEFYEHAKALWE-D-EGV

GP1α: 256-314: KQCFARSNEFQLEGSAAYYFDNIEKFASPNYVCTDEDILKGRIKTTGITETEFNIGS-SK
GP2α: 232-291: LVNGPNASKFYLMDSTPYFMENFTRITSPNYRPTQQDILRSRQMTSGIFDTVIDMGSDIK
Gi2α: 138-196: QACFGRSREYQLNDSAAYYLNDLERIAQSDYIPTQQDVLRTRVKTTGIVETHFTFKD-LH
 Gsα: 160-218: RACYERSNEYQLIDCAQYFLDKIDVIKQADYVPSDQDLLRCRVLTSGIFETKFQVDK-VN

GP1α: 315-374: FKVLDAGGQRSERKKWIHCFEGITAVLFVLAMSEYDQMLFEDERVNRMHESIMLFDTLLN
GP2α: 292-351: MHIYDVGGQRSERKKWIHCFDNVTLVIFCVSLSEYDQTLMEDKNQNRFQESLVLFDNIVN
Gi2α: 197-256: FKMFDVGGQRSERKKWIHCFEGVTAIIFCVALSAYDLVLAEDEEMNRMHESMKLFDSICN
 Gsα: 219-278: FHMFDVGGQRDERRKWIQCFNDVTAIIFVVASSSYNMVIREDNQTNRLQEALNLFKSIWN

GP1α: 375-421: SKWFKDTPFILFLNKIDLFEEKVK--SMPIRKYFPDYQGRVGDAEAGL-----------K
GP2α: 352-417: SRWFARTSVVLFLNKIDLFAEKLR--KVPMENYFPDYTGG-SDINKAA-----------K
Gi2α: 257-302: NKWFTDTSIILFLNKKDLFEEKIT--QSPLTICFPEYTGANKYDEAAS------------
 Gsα: 279-338: NRWLRTISVILFLNKQDLLAEKVLAGKSKIEDYFPEFARYTTPEDATPEPGEDPRVTRAK

GP1α: 422-472: YFEKI-FLSLNK--TNKP--IYVKRTCATDTQTMKFVLSAVTDLIIQQNLKKIGII-
GP2α: 418-449: YILWR-FVQLNR--ANLS--IYFHVTQATDTSNIRLVFAAIKETILENTLKDSGVLQ
Gi2α: 303-355: -YIQSKFEDLNKRKDTKE--IYTHFTCATDTKNVQFVFDAVTDVIIKNNLKDCGLF-
 Gsα: 339-394: YFIRDEFLRISTASGDGRHYCYPHFTCAVDTENIRRVFNDCRDIIQRMHLRQYELL-
```

Figure 10. Alignment of the predicted amino acid sequence of yeast GP1α and GP2α with those of rat brain $G_{i2\alpha}$ and $G_{s\alpha}$. Identical or conservative amino acid residues are enclosed within solid lines. (Reprinted, with permission, from Nakafuku et al. 1988.)

Table 4. Family of G-protein α Subunit Gene

Species	Size of α subunit (kD)	Cholera toxin	Pertussis toxin	Expression	Involvement in
Mammalian					
G_s	45, 52	+	−	all tissues	A cyclase
G_{i1}	41	−	+	brain	A cyclase
G_{i2}	40	−	+	all tissues	PLC
G_{i3}	41	−	+	all tissues	K^+ channel
G_{to}	39	−	+	brain	Ca^{++} channel
G_{t1}	39	+	+	rod	cGMP PDE
G_{t2}	40	+	+	cone	cGMP PDE?
G_x	41	?	−	brain	PLC?
Yeast (*S. cerevisiae*)					
GP1	54	?	−	haploid cells	mating factor
GP2	51	?	−	also diploids	cAMP pathway?

ever, we do emphasize again that the basic reaction mechanism catalyzed by G proteins appears to be very much analogous to that proposed for translational elongation factors (Kaziro 1978). The GTP-bound form is an active conformation that transmits signals to the amplifiers (or effectors), and the hydrolysis of bound GTP to GDP is required to shift the conformation to an inactive form, i.e., to shut off the signal transduction.

This mechanism has now been extended to many other reactions dependent on either GTP or ATP. Other examples of the energy-dependent conformational change of proteins may be found in the transition of phosphorylated and unphosphorylated proteins, where the phosphate covalently bound to the protein, rather than the noncovalently bound nucleotide triphosphate, is effectively utilized in inducing the conformational alterations.

ACKNOWLEDGMENTS

We thank Kaoru Okuda for her help in preparing the manuscript. This work was supported by Grant in Aid for Scientific Research (62065008) from the Ministry of Education, Science, and Culture, Japan.

REFERENCES

Barbacid, M. 1987. *ras* genes. *Annu. Rev. Biochem.* **56:** 779.

Beals, C.R., C.B. Wilson, and R.M. Perlmutter. 1987. A small multigene family encodes G_i signal-transduction proteins. *Proc. Natl. Acad. Sci.* **84:** 7886.

Bray, P., A. Carter, V. Guo, C. Puckett, J. Kamholz, A. Spiegel, and M. Nirenberg. 1987. Human cDNA clones for an α subunit of G_i signal-transducing protein. *Proc. Natl. Acad. Sci.* **84:** 5115.

Bray, P., A. Carter, C. Simons, V. Guo, C. Puckett, J. Kamholz, A. Spiegel, and M. Nirenberg. 1986. Human cDNA clones for four species of $G_{\alpha s}$ signal transduction protein. *Proc. Natl. Acad. Sci.* **83:** 8893.

Broek, D., N. Samiy, O. Fasano, A. Fujiyama, F. Tamanoi, J. Northup, and M. Wigler. 1985. Differential activation of yeast adenylate cyclase by wild-type and mutant *RAS* proteins. *Cell* **41:** 763.

Chou, P.Y. and G.D. Fasman. 1978. Empirical predictions of protein conformation. *Annu. Rev. Biochem.* **47:** 251.

Codina, J., J. Olate, J. Abramowitz, R. Mattera, R.G. Cook, and L. Birnbaumer. 1988. αi-3 cDNA encodes the α subunit of Gk, the stimulatory G protein of receptor-regulated K^+ channels. *J. Biol. Chem.* **263:** 6746.

DeFeo-Jones, D., E.M. Scolnick, R. Koller, and R. Dhar. 1983. *ras*-Related gene sequences identified and isolated from *S. cerevisiae*. *Nature* **306:** 707.

Didsbury, J.R. and R. Snyderman. 1987. Molecular cloning of a new human G protein. Evidence for two $G_{i\alpha}$-like protein families. *FEBS Lett.* **219:** 259.

Didsbury, J.R., Y. Ho, and R. Snyderman. 1987. Human G_i protein α-subunit: Deduction of amino acid structure from a cloned cDNA. *FEBS Lett.* **211:** 160.

Dietzel, D. and J. Kurjan. 1987. The yeast *SCG1* gene: A Gα-like protein implicated in the *a*- and α-factor response pathway. *Cell* **50:** 1001.

Fong, H.K.W., K.K. Yoshimoto, P. Eversole-Cire, and M.I. Simon. 1988. Identification of a GTP-binding protein α subunit that lacks an apparent ADP-ribosylation site for pertussis toxin. *Proc. Natl. Acad. Sci.* **85:** 3066.

Fukui, Y. and Y. Kaziro. 1985. Molecular cloning and sequence analysis of a *ras* gene from *Schizosaccharomyces pombe*. *EMBO J.* **4:** 687.

Gilman, A.G. 1987. G proteins: Transducers of receptor-generated signals. *Annu. Rev. Biochem.* **56:** 615.

Harris, B.A., J.D. Robishaw, S.M. Mumby, and A.G. Gilman. 1985. Molecular cloning of complementary DNA for the alpha subunit of the G protein that stimulates adenylate cyclase. *Science* **229:** 1274.

Hurley, J.B., M.I. Simon, D.B. Teplow, J.D. Robishaw, and A.G. Gilman. 1984. Homologies between signal transducing G proteins and *ras* gene products. *Science* **226:** 860.

Imboden, J.B., D.M. Shoback, G. Pattison, and J.D. Stobo. 1986. Cholera toxin inhibits the T-cell antigen receptor-mediated increases in inositol triphosphate and cytoplasmic free calcium. *Proc. Natl. Acad. Sci.* **83:** 5673.

Itoh, H., T. Katada, M. Ui, H. Kawasaki, K. Suzuki, and Y. Kaziro. 1988a. Identification of three pertussis toxin substrates (41, 40 and 39 kDa proteins) in mammalian brain. *FEBS Lett.* **230:** 85.

Itoh, H., R. Toyama, T. Kozasa, T. Tsukamoto, M. Matsuoka, and Y. Kaziro. 1988b. Presence of three distinct molecular species of G_i protein α subunit. *J. Biol. Chem.* **263:** 6656.

Itoh, H., T. Kozasa, S. Nagata, S. Nakamura, T. Katada, M. Ui, S. Iwai, E. Ohtsuka, H. Kawasaki, K. Suzuki, and Y. Kaziro. 1986. Molecular cloning and sequence determination of cDNAs for α subunit of the guanine nucleotide-binding proteins G_s, G_i, and G_o from rat brain. *Proc. Natl. Acad. Sci.* **83:** 3776.

Jones, D.T. and R.R. Reed. 1987. Molecular cloning of five GTP-binding protein cDNA species from rat olfactory neuroepithelium. *J. Biol. Chem.* **262:** 14241.

Jurnak, F. 1985. Structure of the GDP domain of EF-Tu and

location of the amino acids homologous to *ras* oncogene proteins. *Science* **230:** 32.

Kahn, R.A. and A.G. Gilman. 1984. ADP-Ribosylation of G_s promotes the dissociation of its α and β subunits. *J. Biol. Chem.* **259:** 6235.

Kaziro, Y. 1978. The role of guanosine 5′-triphosphate in polypeptide chain elongation. *Biochim. Biophys. Acta* **505:** 95.

Kozasa, T., H. Itoh, T. Tsukamoto, and Y. Kaziro. 1988. Isolation and characterization of human $G_{s\alpha}$ gene. *Proc. Natl. Acad. Sci.* **85:** 2081.

Lavu, S., J. Clark, R. Swarup, K. Matsushima, K. Paturu, J. Moss, and H.-F. Kung. 1988. Molecular cloning and DNA sequence analysis of the human guanine nucleotide-binding protein $G_{o\alpha}$. *Biochem. Biophys. Res. Commun.* **150:** 811.

Lerea, C.L., D.E. Somers, J.B. Hurley, I.B. Klock, and A.H. Bunt-Milam. 1986. Identification of specific transducin α subunits in retinal rod and cone photoreceptors. *Science* **324:** 77.

Lo, W.W.Y. and J. Hughes. 1987. A novel cholera toxin-sensitive G-protein (G_c) regulating receptor mediated phosphoinositide signalling in human pituitary clonal cells. *FEBS Lett.* **220:** 327.

Lochrie, M.A., J.B. Hurley, and M.I. Simon. 1985. Sequence of the alpha subunit of photoreceptor G protein: Homologies between transducin, *ras*, and elongation factors. *Science* **228:** 96.

Lucas, D.O., S.M. Bajjalieh, J.A. Kawalchk, and T.F.J. Martin. 1986. Direct stimulation by thyrotropin-releasing hormone (TRH) of polyphosphoinositide hydrolysis in GH_3 cell membranes by a guanine nucleotide-modulated mechanism. *Biochem. Biophys. Res. Commun.* **132:** 721.

Masters, S.B., R.M. Stroud, and H.R. Bourne. 1986. Family of G protein α chains: Amphipathic analysis and predicted structure of functional domains. *Protein Eng.* **1:** 47.

Masters, S.B., M.W. Martin, T.K. Harden, and J.H. Brown. 1985. Pertussis toxin does not inhibit muscarinic-receptor-mediated phosphoinositide hydrolysis or calcium mobilization. *Biochem. J.* **227:** 933.

Matsuoka, M., H. Itoh, T. Kozasa, and Y. Kaziro. 1988. Sequence analysis of cDNA and genomic DNA for a putative pertussis toxin-insensitive guanine nucleotide-binding regulatory protein α subunit. *Proc. Natl. Acad. Sci.* **85:** 5384.

Mattera, R., J. Codina, A. Crozat, V. Kidd, S.L.C. Woo, and L. Birnbaumer. 1986. Identification by molecular cloning of two forms of the α-subunit of the human liver stimulatory (G_s) regulatory component of adenylate cyclase. *FEBS Lett.* **206:** 36.

Medynsky, D.C., K. Sullivan, D. Smith, C. Van Dop, F.-H. Chang, B.K.-K. Fung, P.H. Seeburg, and H.R. Bourne. 1985. Amino acid sequence of the α subunit of transducin deduced from the cDNA sequence. *Proc. Natl. Acad. Sci.* **82:** 4311.

Michel, T., J.W. Winslow, J.A. Smith, J.G. Seidman, and E.J. Neer. 1986. Molecular cloning and characterization of cDNA encoding the GTP-binding protein αi and identification of a related protein, αh. *Proc. Natl. Acad. Sci.* **83:** 7663.

Miyajima, I., M. Nakafuku, N. Nakayama, C. Brenner, A. Miyajima, K. Kaibuchi, K. Arai, Y. Kaziro, and K. Matsumoto. 1987. GPA1, a haploid-specific essential gene, encodes a yeast homolog of mammalian G-protein which may be involved in mating factor signal transduction. *Cell* **50:** 1011.

Nakafuku, M., H. Itoh, S. Nakamura, and Y. Kaziro. 1987. Occurrence in *Saccharomyces cerevisiae* of a gene homologous to the cDNA coding for the α subunit of mammalian G proteins. *Proc. Natl. Acad. Sci.* **84:** 2140.

Nakafuku, M., T. Obara, K. Kaibuchi, I. Miyajima, A. Miyajima, H. Itoh, S. Nakamura, K, Arai, K. Matsumoto, and Y. Kaziro. 1988. Isolation of a second yeast *Saccharomyces cerevisiae* gene (*GPA2*) coding for guanine nucleotide-binding regulatory protein: Studies on its structure and possible functions. *Proc. Natl. Acad. Sci.* **85:** 1374.

Northup, J.K., P.C. Sternweise, M.D. Smigel, L.S. Shleifer, E.M. Ross, and A.G. Gilman. 1980. Purification of the regulatory component of adenylate cyclase. *Proc. Natl. Acad. Sci.* **77:** 6516.

Nukada, T., T. Tanabe, H. Takahashi, M. Noda, T. Hirose, S. Inayama, and S. Numa. 1986a. Primary structure of the α-subunit of bovine adenylate cyclase-stimulating G-protein deduced from the cDNA sequence. *FEBS Lett.* **195:** 220.

Nukada, T., T. Tanabe, H. Takahashi, M. Noda, K. Haga, T. Haga, A. Ichiyama, K. Kangawa, M. Hiranaga, H. Matsuo, and S. Numa. 1986b. Primary structure of the α subunit of bovine adenylate cyclase-inhibiting G-protein deduced from the cDNA sequence. *FEBS Lett.* **197:** 305.

Ovchinnikov, Y.A., V.Z. Slepak, A.N. Pronin, A.B. Shlensky, N.B. Levina, V.L. Voeikov, and V.M. Lipkin. 1987. Primary structure of bovine cerebellum GTP-binding protein G_{39} and its effect on the adenylate cyclase system. *FEBS Lett.* **226:** 91.

Padgett, R.A., P.J. Grabowski, M.M. Konarska, S. Seiler, and P.A. Sharp. 1986. Splicing of messenger RNA precursors. *Annu. Rev. Biochem.* **55:** 1119.

Powers, S., T. Kataoka, O. Fasano, M. Goldfarb, J. Strathern, J. Broach, and M. Wigler. 1984. Genes in *S. cerevisiae* encoding proteins with domains homologous to the mammalian *ras* proteins. *Cell* **36:** 607.

Rall, T. and B.A. Harris. 1987. Identification of the lesion in the stimulatory GTP-binding protein of the uncoupled S49 lymphoma. *FEBS Lett.* **224:** 365.

Robishaw, J.D., M.D. Smigel, and A.G. Gilman. 1986a. Molecular basis for two forms of the G protein that stimulates adenylate cyclase. *J. Biol. Chem.* **261:** 9587.

Robishaw, J.D., D.W. Russell, B.A. Harris, M.D. Smigel, and A.G. Gilman. 1986b. Deduced primary structure of the α subunit of the GTP-binding stimulatory protein of adenylate cyclase. *Proc. Natl. Acad. Sci.* **83:** 1251.

Stryer, L. and H.R. Bourne. 1986. G proteins: A family of signal transducers. *Annu. Rev. Cell Biol.* **2:** 391.

Suki, W.N., J. Abramowitz, R. Mattera, J. Codina, and L. Birnbaumer. 1987. The human genome encodes at least three non-allelic G proteins with αi-type subunits. *FEBS Lett.* **220:** 187.

Sullivan, K.A., R.T. Miller, S.B. Masters, B. Beiderman, W. Heideman, and H.R. Bourne. 1987. Identification of receptor contact site involved in receptor–G protein coupling. *Nature* **330:**758.

Sullivan, K.A., Y.-C. Liao, A. Alborzi, B. Beiderman, F.-H. Chang, S.B. Masters, A.D. Levinson, and H.R. Bourne. 1986. Inhibitory and stimulatory G proteins of adenylate cyclase: cDNA and amino acid sequences of the α chains. *Proc. Natl. Acad. Sci.* **83:** 6687.

Tanabe, T., T. Nukada, Y. Nishikawa, K. Sugimoto, H. Suzuki, H. Takahashi, M. Noda, T. Haga, A. Ichiyama, K. Kanagawa, N. Minamino, H. Matsuo, and S. Numa. 1985. Primary structure of the α-subunit of transducin and its relationship to *ras* proteins. *Nature* **315:** 242.

Toda, T., I. Uno, T. Ishikawa, S. Powers, T. Kataoka, D. Broek, S. Cameron, J. Broach, K. Matsumoto, and M. Wigler. 1985. In yeast, *RAS* proteins are controlling elements of adenylate cyclase. *Cell* **40:** 27.

Uhing, R.J., V. Prpic, H. Jiang, and J.H. Exton. 1986. Hormone-stimulated polyphosphoinositide breakdown in rat liver plasma membranes. *J. Biol. Chem.* **261:** 2140.

Van Dop, C., M. Tsubokawa, H.R. Bourne, and J. Ramachandran. 1984. Amino acid sequences of retinal transducin at the site ADP-ribosylated by cholera toxin. *J. Biol. Chem.* **259:** 696.

Van Meurs, K.P., C.W. Angus, S. Lavu, H.-F. Kung, S.K. Czarnecki, J. Moss, and M. Vaughan. 1987. Deduced amino acid sequence of bovine retinal $G_{o\alpha}$: Similarities to other guanine nucleotide-binding proteins. *Proc. Natl. Acad. Sci.* **84:** 3107.

Weinstein, L.S., A.M. Spiegel, and A.D. Carter. 1988. Cloning and characterization of the human gene for the α-subunit of G_{i2}, a GTP-binding signal transduction protein. *FEBS Lett.* **232:** 333.

West, R.E., J. Moss, M. Vaughan, T. Liu, and T.-Y. Liu. 1985. Pertussis toxin-catalysed ADP-ribosylation for transducin. *J. Biol. Chem.* **260:** 14428.

Yatsunami, K. and H.G. Khorana. 1985. GTPase of bovine rod outer segments: The amino acid sequence of the α subunit as derived from the cDNA sequence. *Proc. Natl. Acad. Sci.* **82:** 4316.

Mutations Probe Structure and Function of G-protein α Chains

H.R. Bourne, S.B. Masters, R.T. Miller,
K.A. Sullivan, and W. Heideman
Departments of Pharmacology and Medicine and the Cardiovascular Research Institute, University of California, San Francisco, California 94143-0450

Recent investigations have identified new members of the G-protein family and have established similarities between G proteins and other GTP-binding proteins (for review, see Stryer and Bourne 1986; Gilman 1987). All signaling pathways mediated by G proteins are driven by a single kind of molecular machine, which is conserved within and beyond the G-protein family. In the G proteins, this GTP-dependent machine transduces chemical signals, detected by receptors for sensory or hormonal stimuli outside the cell, into regulation of effector proteins that control accumulation of intracellular second messengers.

This paper summarizes what we know of the structure and function of a key element of the signaling machinery, the G-protein α polypeptide chain. We begin with a brief description of key functions of G-protein α chains and features of a structural model of a composite α chain, α_{avg} (Masters et al. 1986). We then describe phenotypes produced by mutations in the α chain of G_s (termed α_s) and inferences that can be drawn from them.

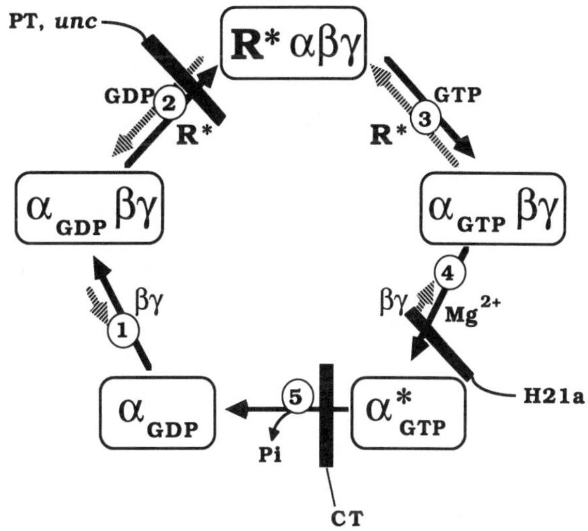

Figure 1. GTP-driven cycle of G-protein α subunits in transmembrane signaling. See text for details. α and $\beta\gamma$ represent the α subunit and the $\beta\gamma$ complex of the G-protein heterotrimer. R* represents the active (stimulated) form of the receptor, which is capable of stimulating the effector (e.g., adenylyl cyclase and retinal cGMP phosphodiesterase). Cross-hatched rectangles indicate sites at which the cycle is interrupted by mutations (*unc* and H21a) or by ADP-ribosylation catalyzed by bacterial toxins (CT, cholera toxin; PT, pertussis toxin).

Function of α Chains

Figure 1 illustrates the reactions that make up the cycle of activation and inactivation of a G-protein α chain (numbers refer to reactions shown in the diagram):

1. In its inactive GDP-bound state, the α chain binds with high affinity to the $\beta\gamma$ complex.
2. The $\alpha_{GDP}\beta\gamma$ complex can bind to the activated (i.e., hormone-bound) receptor (R*), although neither α nor $\beta\gamma$ alone can do so. Release of GDP from the $\alpha\beta\gamma$ complex causes the complex to bind much more tightly to R*; in effect, guanine nucleotide and R* heterotropically displace each other from binding to $\alpha\beta\gamma$.
3. Binding of GTP to the R*$\alpha\beta\gamma$ complex causes dissociation of R* and formation of a (transient) $\alpha_{GTP}\beta\gamma$ complex. (Note that binding of GDP would also cause dissociation of R*, as in reaction 2.)
4. In the presence of Mg^{++} ion, α_{GTP} has a very low affinity for $\beta\gamma$. Activated α^*_{GTP}, released from $\beta\gamma$, now can activate the effector molecule (e.g., adenylyl cyclase, cGMP phosphodiesterase, and K^+ channel).
5. The intrinsic GTPase activity of the α^*_{GTP} complex hydrolyzes the bound nucleotide to return to the inactive α_{GDP} state.

Two key elements of this cycle deserve special emphasis: One is the formation of α^*_{GTP} (4 in the diagram). The binding of GTP to α promotes a conformational change that results in (1) a marked decrease in affinity for binding $\beta\gamma$, (2) ability to bind and activate the effector, and (3) ability to hydrolyze GTP. These properties differ profoundly from those of α_{GDP}. How does the presence of the γ-phosphoryl on the bound guanine nucleotide produce these changes in α chains? Although poorly understood, this GTP-induced conformational change is a key element of the molecular mechanism common to all GTP-binding proteins and probably occurs in essentially the same way in each of them.

A second key element of the cycle is the heterotropic

displacement of R* by guanine nucleotide and of guanine nucleotide by R*. This displacement reaction (reactions 2 and 3) allows R* to catalyze replacement of GDP by GTP. The guanine-nucleotide-binding site of the α chain undergoes a conformational change, from "closed" (tight binding of either GDP or GTP) in the absence of R* to "open" (release of bound guanine nucleotide and, presumably, easy access for entry of a new GTP or GDP molecule into the site). Little is known of the mechanism of this conformational change, which appears common to all G-protein α chains, as well as other GTP-binding proteins—e.g., the guanine nucleotide exchange reaction of elongation factor (EF) Tu, catalyzed by EF-Ts. The $R^*\alpha\beta\gamma$ complex is relatively stable in the absence of guanine nucleotide. The receptor in this complex exhibits enhanced affinity for binding its activating ligand (hormonal agonist); this property is useful for assessing coupling of G proteins to their receptors.

To understand in molecular detail the GTP-dependent mechanism of signaling by G proteins, we will need to answer many questions: Which regions of the α subunits interact with $\beta\gamma$? With receptors? With effectors? Are these regions distinct, or do they overlap? Are they similarly located in α chains of different G proteins? How does the activated receptor open the guanine-nucleotide-binding site and accelerate dissociation of GDP? Conversely, how do guanine nucleotides reduce affinity of $\alpha\beta\gamma$ for receptor? How do the GTP- and GDP-bound conformations of a G protein differ? How does binding of GTP decrease the affinity of the α subunit for binding $\beta\gamma$ and increase its affinity for binding effector?

Satisfying answers to these questions will ultimately come from a combination of experimental approaches, utilizing the tools of classic biochemistry, immunologic probes, and X-ray crystallography, as well as manipulations of protein structure and expression via alterations in nucleic acid sequence. Such techniques are just beginning to provide strong hints regarding a few of the answers, as we describe below.

Model of α-chain Structure

Although still rudimentary, our present understanding of the structure and function of G-protein α chains greatly exceeds understanding of the β and γ chains. Molecular cloning has revealed the primary structures of one γ and two β polypeptides; further experimentation will be necessary to relate these structures to the two known functions of $\beta\gamma$, attachment of the G protein to membrane (Sternweis 1986) and presentation of the α chain to the receptor (Fung 1983; Stryer and Bourne 1986; Gilman 1987).

Comparison of the predicted primary structures of known G-protein α chains reveals a consistent pattern of conserved and divergent sequences. Halliday (1984) identified four stretches of primary sequence that are conserved among the GTP-binding proteins of ribosomal protein synthesis (e.g., EF-Tu) and the 21,000-dalton products ($p21^{ras}$) of the *ras* oncogenes. Halliday's prediction that these conserved sequences would play important roles in the structures of guanine-nucleotide-binding sites has been confirmed by the crystal structures of EF-Tu and $p21^{ras}$ (Jurnak 1985; La Cour et al. 1985; de Vos et al. 1988). Similar stretches of sequence are also conserved in the α chains of the G proteins.

We used the crystal structure of EF-Tu and similarities between regions of G-protein α chains and the Halliday sequences of EF-Tu to construct a model (Figs. 2 and 3) of the structure and function of a composite G-protein α chain, α_{avg} (Masters et al. 1986). Figure 2 illustrates locations of key structural features and mutations in the linear α_{avg} chain. Regions of sequence that comprise the putative guanine-nucleotide-binding site are designated by G in the diagram. Putative "domains" (I, II, and III), interspersed among portions of the guanine-nucleotide-binding site, may play key roles in interactions with $\beta\gamma$, receptors, and effectors. The lower portion of Figure 2 compares the amino acid sequences of the α chain of $G_s(\alpha_s)$ and the α chain (α_i-2) of one of the G_i proteins. The pattern in Figure 2 is representative of the pattern of conservation and divergence among primary structures of the nine G-protein α chains so far identified in mammals: All are highly conserved in their presumptive guanine-nucleotide-binding regions and diverge widely in domains I, II, and III.

Figure 3 illustrates the α_{avg} model in a different way:

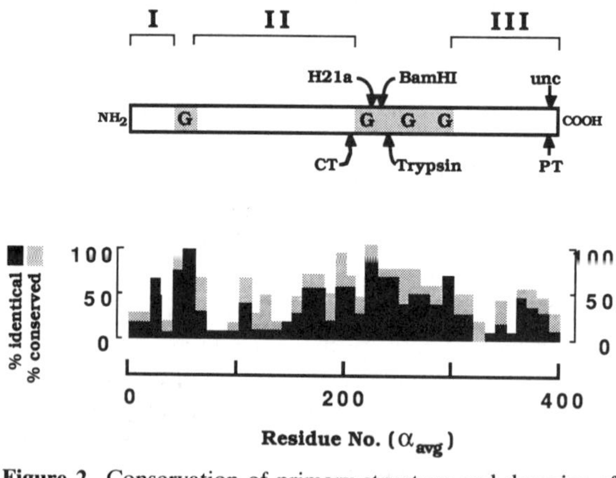

Figure 2. Conservation of primary structure and domains of G-protein α chains. The bar at the top represents the linear sequence of a composite G-protein α chain, α_{avg}. Letters and arrows indicate the sites ADP-ribosylated by cholera (CT) and pertussis (PT) toxins, the locations of four regions that contribute to the GDP-binding domain (G), and three additional presumed domains (I, II, II) of the α chains. Also shown are the sites of the *unc* and H21a mutations of α_s, the tryptic cleavage site that is protected in the GTP-bound active conformation of α chains, and the position of the *Bam*HI restriction site used for construction of α_i/α_s and α_t/α_s chimeras. The graph at the bottom compares the number of identical and conserved residues at each position between two α chains, murine α_s and α_{i2}. The scale (amino acid residue number in α_{avg}) is shown at the extreme bottom of the diagram.

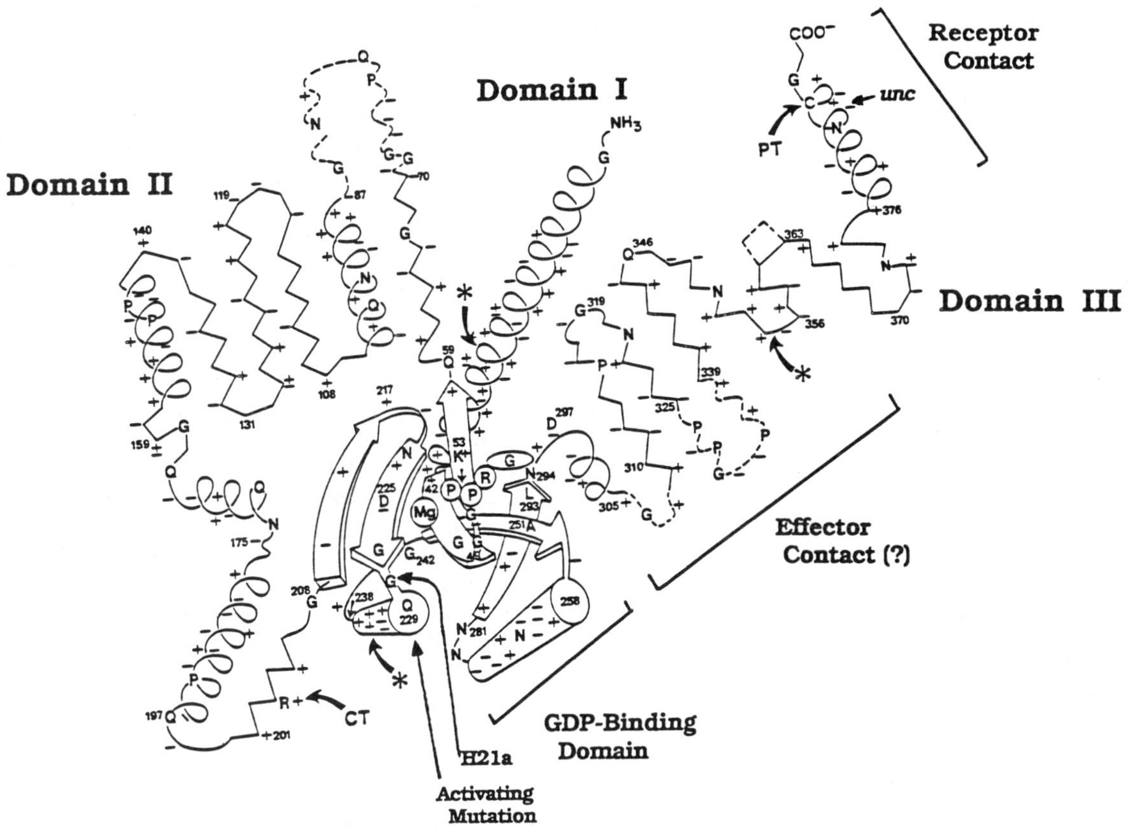

Figure 3. Structural model of a composite G-protein α chain, α_{avg} (Masters et al. 1988). The GDP-binding domain, (three-dimensional cartoon at the center of the model) is adapted from that of EF-Tu (Jurnak 1985). In this domain, the GDP molecule (G, guanine ring; R, ribose; P, phosphoryl) nestles in a pocket formed by turns between β strands and α helices. The rest of the model depicts predicted secondary structure (β strands, α helices, turns). Residues conserved among different α chains are depicted in the single-letter amino acid code, and numbers refer to positions of residues in α_{avg} (Masters et al. 1986). Asterisks with arrows indicate locations of tryptic cleavage sites in native α chains; the site protected in the GTP-activated conformation is located at residue 233 of α_{avg}. Sites of α_s mutations and toxin ADP-ribosylation sites are also indicated (see Fig. 2).

A three-dimensional cartoon of the guanine-nucleotide-binding region (center of the figure), based on that of EF-Tu (Jurnak 1985), is located in relation to predicted secondary structures of sequences in other domains of the protein (Masters et al. 1986). Numbers in the figure refer to residue numbers in α_{avg}, and letters designate some of the amino acid residues (single-letter code) that are conserved among three of the five different α chains used to make the model.

We have recently embarked on a series of experiments aimed at refining and testing elements of the model. To do so, we have used as a model system the hormone-sensitive adenylyl cyclase of S49 mouse lymphoma cells. The propensity of S49 cells to die upon prolonged elevation of intracellular cAMP has allowed selection of a variety of mutant strains with instructive defects in signaling through the cAMP pathway. Of these, one of the most useful is the S49 *cyc*⁻ cell, which is genetically deficient in transcription and translation products of the gene that encodes α_s. S49 *cyc*⁻ cells provide a unique "null" background in which to assess signaling phenotypes produced by expression of normal and genetically altered α_s chains. Details of the expression system, which utilizes a retroviral vector, have been reported (Sullivan et al. 1987). Here, we briefly describe the phenotypes produced by mutated α_s in S49 cells and sketch their relation to our developing model of α-chain structure and function.

Guanine Nucleotide Binding and GTPase Activity

If the model correctly identifies amino acid sequences that make up the guanine-nucleotide-binding region of α_{avg}, mutations that alter these sequences should alter binding or hydrolysis of GTP. A large number of mutations in the guanine-nucleotide-binding region of $p21^{ras}$ provide useful precedents. So far, we have assessed function of two α_s proteins with amino acid substitutions that correspond to mutation-induced substitutions that activate $p21^{ras}$. One of these proteins shows the expected "activated" phenotype; the other does not.

It should be noted that S49 *cyc*⁻ cells, like their wild-type parents, die upon continuous exposure to exogenous or endogenous cAMP. Although this property proved extremely useful in isolating mutants, including *cyc*⁻, it was important to prevent cAMP-induced death of cells transformed with mutant, potentially "activated" α_s chains. To make this possible, mutant α_s chains were expressed in an S49 cell bearing, in addition to the *cyc*$^-$ mutation, a mutation (*kin*$^-$) that causes inactivation of cAMP-dependent protein kinase, thereby preventing cAMP-induced cell death.

Near the amino terminus of α_{avg}, the first Halliday region forms a loop, located near the phosphoryl groups of GDP, that is thought to play a role in regulating and perhaps in catalyzing hydrolysis of bound GTP. In p21^{ras}, mutational replacement of Gly-12 (corresponding to Gly-49 in both α_s and α_{avg}; see Fig. 3) by valine (or almost any other amino acid) produces a protein with enhanced capacity to cause malignant transformation of fibroblasts (Seeburg et al. 1984). The same mutations also decrease the ability of the *ras* protein to hydrolyze bound GTP (Barbacid 1987; Trahey and McCormick 1987.)

Mutational replacement of Gly-49 by valine does not appear to produce a similar activation of α_s (S.B. Masters et al., unpubl.). Instead, adenylyl cyclase in *cyc*$^-$ cells expressing α_s-Val-49 responds quite poorly to stimulation by agents that act through α_s, including hormones, GTP analogs, and AlF_4^-. Indeed, this mutation produces an effect nearly opposite to that of the corresponding mutation in p21^{ras}. This discrepancy makes it unlikely that the mutated residue plays the same functional role in p21^{ras} and the G-protein α chains.

In EF-Tu, the next Halliday region is composed of a β strand connected by a loop to an amphipathic α helix. An aspartate residue in the β strand (Asp-225 in α_{avg}) interacts with an Mg^{++} ion that in turn is coordinated with the phosphoryl groups of GDP (Jurnak 1985). Transforming mutations in the third Halliday region of p21^{ras} have been identified as positions corresponding to residues 227, 229, and 231 of α_{avg} (for review, see Barbacid 1987). A mutation replacing one of these residues in α_s produces a constitutively activated phenotype quite similar to that of p21^{ras} bearing the corresponding mutation: Gln-227 of α_s, which corresponds to Gln-229 of α_{avg} (see Fig. 3), was replaced by leucine (S.B. Masters et al., unpubl.). This replacement corresponds to replacement by leucine of the glutamine residue at position 61 in the Ha-*ras* protein, a mutation that activates the protein's capacity for malignant transformation and diminishes its GTPase activity (Barbacid 1987).

Adenylyl cyclase is constitutively elevated, even in the absence of hormonal agonist, in *cys*$^-$ cells expressing α_s-Leu-227. cAMP is markedly elevated in intact cells expressing this mutant α chain, and the adenylyl cyclase activity of membrane preparations is high in the absence of exogenous stimulators. This activation is nearly equivalent to that produced in wild-type S49 membranes upon stimulation by hormonal agonist (e.g., the β-adrenoceptor agonist isproterenol) plus a hydrolysis-resistant GTP analog, such as GTPγS. Although we have not measured GPTase activity of the mutant α_s, we suspect that the constitutive activation of G_s in α_s-Leu-227 is due to slowing of the hydrolysis of GTP, because the corresponding mutation in p21^{ras} activates the protein by reducing GTPase activity. If this interpretation is correct, the Gln-227→Leu mutant phenotype constitutes strong evidence that this region of the guanine-nucleotide-binding domain of G protein α chains resembles those of p21^{ras} and EF-Tu.

The last two Halliday sequences together form a pocket for the guanine ring, including a conserved aspartate (Asp-297 in α_{avg}; see Fig. 3) that interacts with the 2-amino group of the guanine. The strong conservation of amino acid sequence in this region among the G proteins, EF-Tu, and p21^{ras} led to the prediction (Masters et al. 1986) that the guanine ring of the bound nucleotide is located in a pocket defined by corresponding amino acid residues in all three sets of proteins. The crystal structure of a p21^{ras} protein (de Vos et al. 1988) confirmed this prediction, at least with respect to EF-Tu and p21^{ras}. We are presently constructing mutant α_s DNAs designed to test functional similarities of these other portions of the putative GDP-binding region of G protein α chains to the corresponding sequences in other GTP-binding proteins.

GTP-induced Conformational Change

The crystal structures of p21^{ras} and EF-Tu provide only a static picture of the GDP-bound conformation of these proteins. Key questions, both for these proteins and for the G proteins, center on the change in conformation that occurs when bound GDP is replaced by GTP. The G proteins are converted to the "active" conformation upon binding GTP, synthetic GTP analogs, or GDP plus AlF_4^- ion. G protein activation is most easily measured by assessing the ability of the protein to regulate activity of effectors, such as adenylyl cyclase. Activation is accompanied by changes in many other properties, including altered binding affinity for receptors and $\beta\gamma$ as well as effectors (Gilman 1987); enhanced intrinsic tryptophan fluorescence of α chains (Higashijima et al. 1987a,b); altered reactivity of cysteine sulfhydryls (Winslow et al. 1987); susceptibility to ADP-ribosylation by cholera and pertussis toxins (Van Dop et al. 1984); and sensitivity to cleavage by trypsin (Fung and Nash 1983; Hurley et al. 1984).

The GTP-induced conformational change requires Mg^{++} ion. In the absence of this ion, G proteins do not activate effector enzymes or hydrolyze GTP. Measurements of the intrinsic tryptophan fluorescence of α_i and α_o highlight the importance of Mg^{++} for the GTP-induced conformational change: Binding of GTP in the absence of Mg^{++} produces a small change in fluorescence, which is greatly augmented upon addition of Mg^{++}; the fluorescence then decays in parallel with conversion of the bound GTP to GDP (Higashijima et al. 1987a,b). One attractive but untested interpretation of these observations is that the change in fluorescence reflects a change in conformation that results from entry of the Mg^{++} ion into a location corresponding to that of Mg^{++} in the crystal structure of EF-Tu—i.e., between the phosphoryl groups of GTP and an aspartate residue in the third Halliday region (Asp-225 of α_{avg}).

Useful clues regarding the GTP-dependent conformational change have come from studies of the phenotype of the S49 mutant cell line, H21a. The H21a phenotype (Bourne et al. 1981) results from replacement of a glycine by an alanine residue in α_s, at a position corresponding to Gly-228 of α_{avg} (R.T. Miller et al., unpubl.). This residue is located at the $\beta \rightarrow \alpha$ turn in the third Halliday region (see Fig. 3). Although G_s in H21a can bind GTP (as shown by the ability of GTP analogs to regulate the affinity of β-adrenoreceptors for agonists in H21a membranes), the mutant G_s cannot stimulate adenylyl cyclase in response to GTP analogs, AlF_4^-, hormones, or cholera toxin. In addition, the GTP- and AlF_4^--induced dissociation of α_s from $\beta\gamma$, detected by assessing altered migration of normal (S49 wild-type) α_s in sucrose density gradients, does not occur with G_s from H21a (R.T. Miller et al., unpubl.).

Studies of α_s cleavage by trypsin provide further evidence that G_s in H21a cannot undergo the conformational change that is normally induced by GTP. Activation of transducin and of (normal) G_s by GTP analogs protects the α chains of these proteins from cleavage by trypsin at a conserved site, corresponding to Arg-233 of α_{avg}. In contrast to the α_s of wild-type S49 cells, the mutant α_s of H21a is *not* protected from proteolysis by GTP analogs (R.T. Miller et al., unpubl.). Thus, the H21a mutation appears to have created an α chain that knows it has bound guanine nucleotide but that cannot distinguish GTP from GDP.

The functional defect of α_s in H21a can be precisely located in the GTP-driven activation cycle (see Fig. 1): The $\alpha\beta\gamma$ complex of H21a apparently is able to assume the "empty site" conformation (step 2), which binds tightly to the activated receptor and enhances the affinity of receptor for binding agonists. Furthermore, $\alpha\beta\gamma$ in H21a can be heterotypically displaced from association with the receptor upon binding guanine nucleotide (step 3), as shown by the ability of guanine nucleotide to lower receptor affinity for agonists in H21a membranes. The H21a mutation exerts its effect at the next step (step 4 in Fig. 1): Because the mutation prevents α_s from taking on its "active" conformation in the GTP-bound state, α_s does not dissociate from $\beta\gamma$ and cannot interact with the effector, adenylyl cyclase.

It is clear that the GTP-induced change from an inactive to an active conformation is common to all G proteins. We would like to know how the active GTP-bound conformation differs, in three dimensions, from the inactive GDP-bound conformation. If the guanine-nucleotide-binding pocket of G-protein α chains resembles those of EF-Tu and $p21^{ras}$, it is obvious that the γ phosphoryl of GTP must occupy space that is not available in the GDP-bound protein. One attractive possibility is that replacement of GDP by GTP somehow displaces the β strand and induces flexion of the $\beta \rightarrow \alpha$ turn in the third Halliday region. In EF-Tu, this turn begins at a conserved glycine residue that corresponds to the mutated residue in the H21a α chain; this glycine residue may serve as a hinge (see Fig. 3) that allows relative movement of separate domains of the protein. The Gly→Ala substitution in H21a could decrease flexibility of the presumed hinge in this region and thereby prevent a critical conformational change.

Interaction with Receptors

Several lines of evidence point to a carboxy-terminal region of the α chains (in domain III; see Fig. 2) as one of the sites involved in receptor interaction. The pertussis-toxin-catalyzed ADP-ribosylation of a cysteine four residues from the carboxyl terminus prevents G_i, G_o, and transducin from responding to stimulation by their respective receptors (Ui et al. 1984). Furthermore, pertussis-toxin-catalyzed ADP-ribosylation prevents transducin from binding to rhodopsin (Van Dop et al. 1984). The *unc* mutation in α_s creates a similar phenotype—uncoupling of G_s from interactions with the receptors that stimulate adenylyl cyclase (Haga et al. 1977). The *unc* mutation causes replacement by proline of an arginine located six residues from the carboxyl terminus of α_s (Rall and Harris 1987; Sullivan et al. 1987). Both the *unc* mutation of α_s and the cysteines modified by pertussis toxin in α_i, α_o, and α_t are located in a predicted α helix at the carboxyl terminus of the composite G-protein α chain, α_{avg} (Masters et al. 1986).

A retinal protein, arrestin, provides a third line of evidence implicating the predicted α helix at the carboxyl terminus as a contact site of α chains with receptors. Arrestin competes with transducin for binding to the phosphorylated form of photorhodopsin. Because arrestin contains a stretch of internal sequence very similar to that of the carboxyl terminus of α_t (Wistow et al. 1986), it is reasonable to propose that these conserved regions are both involved in rhodopsin binding.

Further evidence has come from phenotypes produced by recombinant chimeric α chains. Two such chimeras are composed of the amino-terminal 60% of either α_i (mouse α_i-2) or α_t (the α chain of bovine transducin) and the carboxy-terminal 40% of α_s; the chimeras were constructed by splicing together the corresponding portions of the cDNAs at the conserved *Bam*HI site near the middle of the coding region (see Fig. 2). As a result, the amino acid sequences of domains I and II in the resulting chimeras differ markedly from corresponding sequences of α_s. Expression of either the α_i-2/α_s chimera (Masters et al. 1988) or of the α_t/α_s chimera (S.B. Masters et al., unpubl.) in S49 *cyc*$^-$ cells produces cells in which β-adrenoceptors stimulate adenylyl cyclase. Other experiments showed that the α_i/α_s chimera does not interact with somatostatin receptors, which are coupled to G_i in S49 cells. These results imply that the carboxy-terminal portion of the chimera, which was contributed by α_s, confers on the α chain its specificity for interacting with receptors.

A third chimeric α chain, in which the carboxy-terminal 38 residues of α_s are replaced by the carboxy-terminal 36 residues of α_i-2, has been expressed in a fibroblast cell line (G.L. Johnson, pers. comm.). In comparison to normal cells or cells expressing recombinant normal α_s, adenylyl cyclase in the cells expressing this chimera behaves as if it is constitutively activated by receptor, even in the absence of receptor agonists: i.e., cAMP synthesis is elevated in the presence of GTP alone, and hydrolysis-resistant GTP analogs activate cAMP synthesis with a very short lag. In adenylyl cyclase systems coupled by normal α_s, the temporal lag in activation by GTP analogs is due to slow dissociation of GDP from α_s; hormone-receptor complexes normally abolish the lag by accelerating the rate of GDP dissociation from α_s. Accordingly, the simplest and most exciting interpretation of the phenotype produced by the chimera constructed by the Johnson group is that it is due to accelerated release of GDP and a consequent increase in the rate at which GTP binds to the α chain. In short, the effect of replacing the normal α_s carboxyl terminus with the cognate region of α_i-2 is to mimic the effect of hormone-receptor complex on normal G_s.

By combining the Johnson result with the other evidence cited above, we can construct a speculative but attractive and testable model of the mechanism by which hormone-receptor complexes are able to open the guanine-nucleotide-binding site, promote release of GDP, and set in motion the transduction of hormonal signals. The model suggests that the carboxy-terminal α helix of a G-protein α chain suppresses release of GDP when the G protein is in the inactive conformation. The hormone-receptor complex is imagined to bind to this α helix and to dislodge it from the position in which it suppresses release of GDP. This interpretation would account for the uncoupling effects of the Arg→Pro mutation in *unc* α_s and of the ADP-ribose attached to the helix by pertussis toxin. It would also account for the phenotype produced by the α_s/α_i-2 chimera, if we imagine that the substituted carboxyl terminus derived from α_i-2 fits poorly into the site at which the cognate portion of α_s normally suppresses GDP release.

Although further experiments will be required to support or refute this interesting interpretation, it is in keeping with the crystal structure of $p21^{ras}$ and the effect of a carboxy-terminal deletion in the *RAS2* gene product of the budding yeast, *Saccharomyces cerevisiae*. In the GDP-bound form of $p21^{ras}$ (de Vos et al. 1988), residues 152–165 form an α helix that lies directly over the GDP-binding domain, near the fourth Halliday region (corresponding to residues 290–300 of α_{avg}; see Fig. 3). In the yeast system, GTP-dependent stimulation of adenylyl cyclase is mediated by the *RAS2* protein, which has a primary structure (and therefore presumably a three-dimensional structure) very similar to that of $p21^{ras}$, especially in its guanine-nucleotide-binding domain. Like the mammalian G proteins, *RAS2* normally requires the presence of a GDP-GTP exchange protein, called *CDC25*, to stimulate cAMP synthesis. The requirement for *CDC25* is bypassed by deletion of a portion of *RAS2* near the carboxyl terminus (Marshall et al. 1987); the deleted portion of *RAS2* is located in the right place to serve as a homolog of the α helix (residues 152–165) in $p21^{ras}$. We propose, in turn, that this α-helical segment of $p21^{ras}$ is the functional homolog of the carboxy-terminal portion of G-protein α chains, in that both are critical regulators of GDP release.

Effector Contact Region

The location of the effector-binding region of G-protein α chains is not known. Masters et al. (1986) proposed that domain II (Fig. 2) would prove to mediate interactions with effectors. This proposal was based in part on the heterogeneity of amino acid sequence in this domain among the α chains that regulate different effectors and in part on suggestive evidence that the (much shorter) cognate regions of EF-Tu (Laursen et al. 1981) and $p21^{ras}$ (Sigal et al. 1986) may be involved in interactions with proteins analogous to the effectors of G-protein-mediated signaling systems.

The 60:40 α_i/α_s and α_t/α_s chimeras described above allowed a simple test of the notion that region II mediates effector interactions (Masters et al. 1988 and unpubl.). If this notion were correct, the chimeras would not be expected to mediate stimulation of adenylyl cyclase, because domain II in the chimeras is derived from α_i-2 or α_t rather than from α_s. Unexpectedly, the chimeric α chains behave like α_s, in that they mediate *stimulation* of adenylyl cyclase activity by receptors that stimulate cAMP synthesis in normal cells (e.g., β-adrenoceptors and PGE_1 receptors). Thus, the phenotypes produced by expression of these chimeras make it unlikely that domain II specifies stimulation of adenylyl cyclase. Instead, they suggest that domain III, and not domain II, contains the amino acid sequences that are critical for a specific stimulatory interaction with catalytic adenylyl cyclase (and, by implication, the sequences in other α chains that specify regulation of their effectors).

The precise location of the effector contact region(s) of G-protein α chains remains undefined, however. The α_i/α_s chimera suggests that such a region may be located in domain III; furthermore, the phenotype produced by another chimera, constructed by the Johnson laboratory (described above), suggests that the carboxy-terminal 38 residues of α_s are not part of the effector contact region (because their replacement with a quite different carboxyl terminus derived from α_i-2 does not disrupt stimulation of adenylyl cyclase). Consequently, one might speculate that the effector contact region resides in a relatively short stretch of domain III, corresponding to residues 300–355 of α_{avg}. Further experiments may show that this model is too simple—i.e., it is quite possible that sequences in domains I and II interact intimately with sequences in domain III, and that sequences from more than one domain may be required for interaction with effectors—or indeed with any other protein, including receptors and the $\beta\gamma$ complex as well.

Contact with the $\beta\gamma$ Complex

Proteolytic removal of the amino-terminal 21 amino acids of the α chain of rod cell transducin destroys its ability to bind to $\beta\gamma$ (Navon and Fung 1987). The inability to interact with $\beta\gamma$ prevents the truncated α_t chain from binding to rhodopsin and from undergoing photon-triggered exchange of guanine nucleotide. The simplest (but not the only) interpretation of these results is that $\beta\gamma$ binds directly to the extreme amino terminus of the α chain. Although observations that $\beta\gamma$ complexes can be functionally interchanged among several G proteins (Kanaho et al. 1984) suggest that the parts of α chains that interact with $\beta\gamma$ should be conserved, the primary structure of the α chain amino terminus (domain I) is *not* highly conserved (see Fig. 2, bottom). Masters et al. (1986) predicted a conserved α helical secondary structure for domain I in all G-protein α chains (see Fig. 3). Perhaps some feature of this conserved α helix, not dependent on strict conservation of a long stretch of amino acid sequence, specifies interaction with $\beta\gamma$. Alternatively, this region may be involved in regulating interactions with $\beta\gamma$ without contacting the complex directly.

SUMMARY

The molecular genetic approach has just begun to provide hints of answers to some of the questions posed at the outset of this paper. We have some idea of which portions of the α chain of G_s interact with receptors and effectors, and we guess that the same is true of corresponding regions of other α chains. We have a tantalizing hint that points to a key region of α_s that is necessary for the conformational change induced by binding GTP, and the vague outline of a hypothesis regarding the mechanism by which receptors release GDP from its binding site on the α chain. The α chain region (domain) that interacts with $\beta\gamma$ remains unknown.

The tenuous quality of all these hints and hypotheses is obvious, and at least in the short term, frustrating. Even the present level of our understanding, however, is impressive in comparison with what was known of G-protein function only 5 years ago. Now we can pose much more precise questions and hope that a combination of the molecular genetic approach with biophysical probes of structure will provide satisfying answers.

ACKNOWLEDGMENTS

Work from this laboratory is supported in part by research grants from the National Institutes of Health and from the March of Dimes.

REFERENCES

Barbacid, M. 1987. *ras* Genes. *Annu. Rev. Biochem.* **56:** 779.

Bourne, H.R., D. Kaslow, H.R. Kaslow, M. Salomon, and V. Licko. 1981. Hormone sensitive adenylate cyclase: Mutant phenotype with normally regulated beta-adrenergic receptors uncoupled from catalytic adenylate cyclase. *Mol. Pharmacol.* **20:** 435.

de Vos, A.M., L. Tong, M.V. Milburn, P.M. Matias, J. Jancarik, K. Miura, E. Ohtsuka, S. Noguchi, S. Nishimura, and S.-H. Kim. 1988. Three-dimensional structure of an oncogene protein: Catalytic domain of human c-H-*ras* p21. *Science* **239:** 888.

Fung, B.K.-K. 1983. Characterization of transducin from bovine retinal rod outer segments. *J. Biol. Chem.* **258:** 10495.

Fung, B.K.-K. and C.R. Nash. 1983. Characterization of transducin from bovine retinal rod outer segments. *J. Biol. Chem.* **258:** 10503.

Gilman, A.G. 1987. G proteins: Transducers of receptor-generated signals. *Annu. Rev. Biochem.* **56:** 615.

Haga, T., E.M. Ross, H.J. Anderson, and A.G. Gilman. 1977. Adenylate cyclase permanently uncoupled from hormone receptors in a novel variant of S49 mouse lymphoma cells. *Proc. Natl. Acad. Sci.* **74:** 2016.

Halliday, K. 1984. Regional homology in GTP-binding proto-oncogene products and elongation factors. *J. Cyclic Nucleotide Res.* **9:** 435.

Higashijima, T., K.M. Ferguson, M.D. Smigel, and A.G. Gilman. 1987a. The effect of GTP and Mg^{2+} on the GTPase activity and the fluorescent properties of G_o. *J. Biol. Chem.* **262:** 757.

Higashijima, T., K.M. Ferguson, P.C. Sternweis, E.M. Ross, M.D. Smigel, and A.G. Gilman. 1987b. The effect of activating ligands on the intrinsic fluorescence of guanine nucleotide-binding regulatory proteins. *J. Biol. Chem.* **262:** 752.

Hurley, J.B., M.I. Simon, D.B. Teplow, J.D. Robishaw, and A.G. Gilman. 1984. Homologies between signal transducing G proteins and *ras* gene products. *Science* **226:** 860.

Jurnak, F. 1985. Structure of the GDP domain of EF-Tu and location of the amino acids homologous to *ras* oncogene proteins. *Science* **230:** 32.

Kanaho, Y., S.C. Tsai, R. Adamik, E.L. Hewlett, J. Moss, and M. Vaughan. 1984. Rhodopsin-enhanced GTPase activity of the inhibitory GTP-binding protein of adenylate cyclase. *J. Biol. Chem.* **259:** 7378.

La Cour, T.F.M., J. Nyborg, S. Thirup, and B.F.C. Clark. 1985. Structural details of the binding of guanosine diphosphate to elongation factor Tu from *E. coli* as studied by X-ray crystallography. *EMBO J.* **4:** 2385.

Laursen, R.A., J.J. L'Italien, S. Nakarkatti, and D.L. Miller. 1981. The amino acid sequence of elongation factor Tu of *Escherichia coli*. The complete sequence. *J. Biol. Chem.* **256:** 8102.

Marshall, M.S., J.B. Gibbs, E.M. Scolnick, and I.S. Sigal. 1987. Regulatory function of the yeast *RAS* c terminus. *Mol. Cell. Biol.* **7:** 2309.

Masters, S.B., R.M. Stroud, and H.R. Bourne. 1986. Family of G protein α chains: Amphipathic analysis and predicted structure of functional domains. *Protein Eng.* **1:** 47.

Masters, S.B., K.A. Sullivan, R.T. Miller, B. Beiderman, N.G. Lopez, J. Ramachandran, and H.R. Bourne. 1988. Carboxy terminal domain of $G_{s\alpha}$ specifies coupling of receptors to stimulation of adenylyl cyclase. *Science* **241:** 448.

Navon, S.E. and B.K.-K. Fung. 1987. Characterization of transducin from bovine retinal rod outer segments. Participation of the amino-terminal region of T_α in subunit interaction. *J. Biol. Chem.* **262:** 15746.

Rall, T. and B.A Harris. 1987. Identification of the lesion in the stimulatory GTP-binding protein of the uncoupled S49 lymphoma. *FEBS Lett.* **224:** 365.

Seeburg, P.H., W.W. Colby, D.J. Capon, D.V. Goeddel, and A.D. Levinson. 1984. Biological properties of human c-Ha-*ras*1 genes mutated at codon 12. *Nature* **312:** 71.

Sigal, I.S., J.B. Gibbs, J.S. D'Alonzo, and E.M. Scolnick. 1986. Identification of effector residues and a neutralizing epitope of Ha *ras* p21. *Proc. Natl. Acad. Sci.* **83:** 4725.

Sternweis, P.C. 1986. The purified α subunits of G_o and G_i

from bovine brain require $\beta\gamma$ for association with phospholipid vesicles. *J. Biol. Chem.* **261:** 631.

Stryer, L. and H.R. Bourne. 1986. G proteins: A family of signal transducers. *Annu. Rev. Cell. Biol.* **2:** 391.

Sullivan, K.A., R.T. Miller, S.B. Masters, B. Beiderman, W. Heideman, and H.R. Bourne. 1987. Identification of receptor contact site involved in receptor-G protein coupling. *Nature* **330:** 758.

Trahey, M. and F. McCormick. 1987. A cytoplasmic protein stimulates normal N-*ras* p21 GTPase, but does not affect oncogenic mutants. *Science* **238:** 542.

Ui, M., T. Katada, T. Murayama, H. Kurose, M. Yajima, M. Tamura, T. Nakamura, and K. Nogimori. 1984. Islet-activating protein, pertussis toxin: A specific uncoupler of receptor-mediated inhibition of adenylate cyclase. *Adv. Cyclic Nucleotide Protein Phosphorylation Res.* **17:** 145.

Van Dop, C., G. Yamanaka, F. Steinberg, R.D. Sekura, C.R. Manclark, L. Stryer, and H.R. Bourne. 1984. ADP-ribosylation of transducin by pertussis toxin blocks the light-stimulated hydrolysis of GTP and cGMP in retinal photoreceptors. *J. Biol. Chem.* **259:** 23.

Winslow, J.W., J.D. Bradley, J.A. Smith, and E.J. Neer. 1987. Reactive sulfhydryl groups of α_{39}, a guanine nucleotide-binding protein from brain. Location and function. *J. Biol. Chem.* **262:** 4501.

Wistow, G.J., A. Katial, C. Craft, and T. Shinohara. 1986. Sequence analysis of bovine retinal S-antigen. Relationships with α transducin and G proteins. *FEBS Lett.* **196:** 23.

Receptor-Effector Coupling by G Proteins: Purification of Human Erythrocyte G_i-2 and G_i-3 and Analysis of Effector Regulation Using Recombinant α Subunits Synthesized in *Escherichia coli*

L. BIRNBAUMER,*† J. CODINA,* R. MATTERA,* A. YATANI,† R. GRAF,* J. OLATE,* J. SANFORD,* AND A.M. BROWN†

*Departments of *Cell Biology and †Physiology and Molecular Biophysics, Baylor College of Medicine, Houston, Texas 77030*

Molecular cloning has revealed the existence of many more G proteins than anticipated from purification studies, especially in the field of G_s, where what originally were thought to be two forms of α_s (Mattera et al. 1986; Robishaw et al. 1986) turned out to be four splice variants (Bray et al. 1986; Kozasa et al. 1988), and G_i, of which there are now known to exist three independent genes, as deduced from cDNA cloning (Jones and Reed 1987; Suki et al. 1987) and gene sequencing (Itoh et al. 1988). The abundance at which different α_s and α_i molecules are expressed varies from tissue to tissue, raising the question as to whether functional differences are associated with the structural differences, or whether the G_s proteins and, respectively, the G_i proteins should be thought of merely as isoforms. Although the final word on this question is not yet in, we have developed the method(s) required to answer this question. Thus, we concentrated on (1) purification of two of the G_i forms from each other so that we could test for differential biological functions, and (2) expression in bacteria of biologically active forms of the α subunits, designated here as recombinant α subunits.

We succeeded with both and used recombinant α subunits to answer two immediate questions. The first was to determine whether recombinant α_i-3 subunits stimulate K^+ channels. We had recently sequenced peptide fragments derived from the α subunit of human erythrocyte G_i, which we had discovered earlier to have G_K activity, i.e., to stimulate atrial and endocrine K^+ channels (Yatani et al. 1987b,c), and found them to be encoded by the α_i-3 gene (Codina et al. 1988) cloned by us in independent studies from a human liver cDNA library (Suki et al. 1987; Codina et al. 1988). However, no protein is ever pure, and we could not be absolutely sure that we might not have sequenced peptides from one of the contaminating α subunits, which in preparations used for sequencing may have comprised up to 10% of the total. The second question was to determine whether the stimulation of Ca^{++} channels obtained with purified human erythrocyte G_s (Yatani et al. 1987a, 1988) is indeed mediated by G_s, as opposed to being due to a contaminant. This point was of importance because up to now no G protein had been shown to regulate more than one effector function and the Ca^{++} channels stimulated by putative G_s are also stimulated by cAMP-mediated phosphorylation (Kameyama et al. 1986).

EXPERIMENTAL PROCEDURES

Urea gradient/SDS-PAGE. This procedure was carried out essentially as described by Scherer et al. (1987), with the exception that the acrylamide concentration was changed to 9%.

Purification of G_i-2 and G_i-3 from human erythrocytes. Preparation of erythrocyte membranes, extraction of G proteins with cholate, and chromatographies over DEAE-Sephacel (first DEAE), Ultrogel AcA 34, and heptylamine-Sepharose B were carried out as described previously (Codina et al. 1984a,b). For second and third DEAE chromatographies, DEAE-Toyopearl was substituted for DEAE-Sephacel. For separation of G_i proteins, their distribution in column eluates was monitored by ADP-ribosylation with ^{32}P-labeled NAD analyzing the migration of the labeled PTX substrates by urea gradient/SDS-PAGE and autoradiography to give the patterns reported in the text and figures. Heptylamine BI and BII columns were of the same size and developed in the same way as the heptylamine B column. Second DEAE-A and DEAE-B columns were 0.9 × 15 cm and eluted with a 0–4% Na-cholate gradient in buffer with 0.5% Lubrol PX. The second DEAE-AB column was 0.9 × 75 cm and eluted with a 600-ml 0–2% cholate gradient in buffer with 0.5% Lubrol PX. Third DEAE-A, DEAE-B, and DEAE-C columns had bed volumes of 0.9 × 15 cm and were eluted with a 600-ml 0–300 mM NaCl gradient in buffer with 0.5% Lubrol PX. All fractions were ~6 ml.

Bacterial expression of α subunits using plasmid pT7-7 cDNA carrier and M13-derived mGP1-2 phage as source of T7 RNA polymerase. The techniques of Tabor and Richardson (Tabor and Richardson 1985; Tabor et al. 1987) were followed. The various α-subunit cDNAs were first modified by site-directed mutagenesis to generate a unique *Nco*I site at the ATG initiation codon. The engineering was done in replicative forms of M13 phages. The open reading frames of the engineered cDNAs were then excised by digestion with *Nco*I and an appropriate second restriction enzyme. The excised cDNAs were filled with the Klenow fragment of DNA polymerase and subcloned by blunt-end ligation into the *Bam*HI site of the pT7-7 expression vector, giving recombinant plasmids designated pT7-α_x. Recombinant pT7-α_x plasmids with cDNAs inserted in either the antisense (3′ to 5′, open arrows on figures) or sense (5′ to 3′, filled arrows on figures) direction were transformed into *E. coli* K38 cells and expressed after infection of the cells with phage mGP1-2. For analysis of total cell protein, cells were harvested, dissolved in Laemmli's sample buffer, and analyzed by SDS-PAGE in 10% polyacrylamide gel slabs. The slabs were stained with Coomassie blue. For study of biological activity of recombinant α subunits, *E. coli* K38 cells were transformed with pT7-α_x plasmids, grown, infected with phage mGP1-2 in the presence of rifampicin and isopropyl-β-D-thiogalactoside (IPTG), harvested 60 minutes later, shocked with 2.4 M sucrose, and lysed with lysozyme in the cold. The lysed cells were diluted, treated with diisopropylfluorophosphate in the cold and GTPγS plus Mg, or AlF_4^- plus Mg and GDP at 32°C, and centrifuged at 100,000*g* to obtain a soluble extract containing GTPγS-activated (recomb α^*) or fluoride-activated (recomb $\alpha^{(F)}$) recombinant α subunits. These were partially purified by ion-exchange chromatography over DEAE-Sephacel, concentrated by ultrafiltration, and stored at −70°C until used. In some experiments, rather than providing the T7 RNA polymerase through infection with mGP1-2 phage, we supplied it by cotransformation of the K38 cells with the cDNA bearing pT7-α_x plasmid plus pGP1-2 plasmid with the T7 RNA polymerase gene under control of the λP_L promoter. RNA polymerase was then induced by heat inactivation of *c*I857 repressor.

Test for biological activity of recombinant α subunits. G_K activity was tested on inside-out atrial membrane patches from adult guinea pigs as described previously (Yatani et al. 1987c). G_s activity was tested by measuring cyc$^-$ reconstituting activity according to the method of Codina et al. (1984b). Ca^{++} channel stimulatory activity was measured on skeletal muscle T-tubule channels incorporated into planar lipid bilayers according to the method of Yatani et al. (1988).

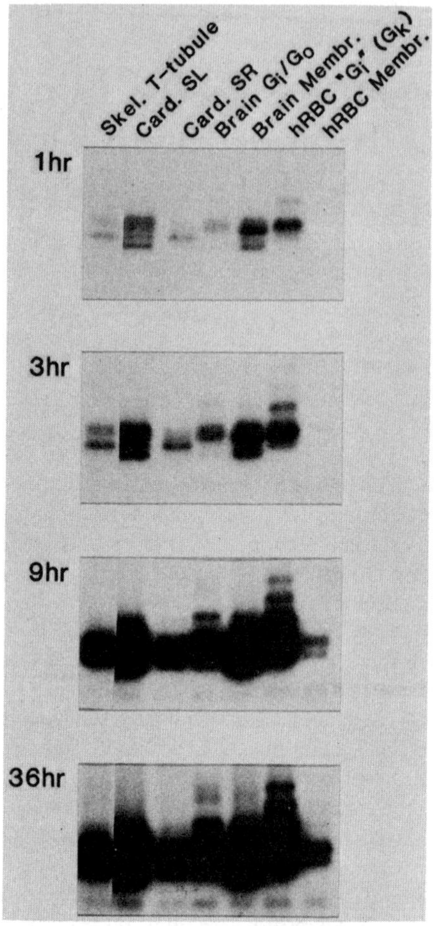

Figure 1. Membranes and "purified" proteins contain multiple PTX substrates. PTX-labeled membranes (1 μg/lane) or purified bovine brain mixture of G/G_o and human erythrocyte G_i (20 ng/lane) were separated by SDS-PAGE in 9% polyacrylamide gels in the presence of a 4–8-M urea gradient, the gel slabs were fixed, stained with Coomassie blue, destained (not shown), and then autoradiographed for the times indicated. Only the section of the autoradiograms with molecules between 35 and 45 kD are shown.

RESULTS

Purification of G_i-2 and G_i-3 from Human Erythrocytes: Use of Urea Gradient/SDS-PAGE to Follow the Progress of Purification

In 1985, we noticed that SDS-PAGE in the presence of a 4–8 M urea gradient in the separation gel results in resolution of what appear to be single, or at most double, bands of PTX-labeled substrates upon simple SDS-PAGE into many more (Fig. 1; Scherer et al. 1987). By this method of analysis, we found human erythrocyte membrane PTX substrate to be primarily of two types, an "upper" and a "lower," in addition to several minor bands of which one, constituting ~5% of the total and migrating in normal SDS-PAGE as a 43-kD protein (Iyengar et al. 1987), is the predominant one. On analyzing the preparations of PTX substrates purified to be free of G_s for studies of their G_K activity, we found that these consistently contained more than 95% of the upper band. On the other hand, the PTX substrate found as contaminant in G_s preparations was almost exclusively lower-band G_i. This indicated that the two predominating PTX substrates of human erythrocytes are separable by the fractionation techniques used in our laboratory.

We purify G proteins from cholate extracts of human erythrocyte membranes in their native unactivated form using sequential chromatographies over DEAE-Sephacel (first DEAE), Ultrogel AcA 34 (which splits G_s activity into a clearly defined peak [pool A] containing small amounts of PTX substrates [<15% of total] plus a less-defined second peak or tail [pool B] containing ~25–30% of the G_s activity and 85–90% of the PTX substrates), heptylamine-Sepharose (one for each, pool A: heptylamine A, and pool B: heptylamine B), and two successive DEAE anion exchange chromatographies: second DEAE (in 0.5% Lubrol PX and eluting with a 0.4–4.0% cholate gradient) and third DEAE (in 0.5% Lubrol PX and eluting with a NaCl gradient). Second and third DEAEs were originally done with DEAE-Sephacel (Codina et al. 1984a,b), but, following the lead of Katada et al. (1987a,b), were recently replaced by DEAE-Toyopearl (Codina et al. 1988; Yatani et al. 1988).

"Fine-tuning" of the purification strategy for separation of upper and lower PTX substrates by analyzing their elution from each of the columns using PTX labeling followed by urea gradient/SDS-PAGE allowed us to obtain not only the upper PTX substrate with G_K activity, free of significant contamination by the lower PTX substrate, but also the lower PTX substrate free of significant contamination by either upper-band or G_s activity. The progress of one such purification as seen by PTX labeling, urea gradient/SDS-PAGE, and autoradiography is shown in Figure 2. Figure 3 presents the analysis of the starting erythrocyte membranes and the resulting pooled fractions obtained from the purification shown in Figure 2. On the basis of the amino

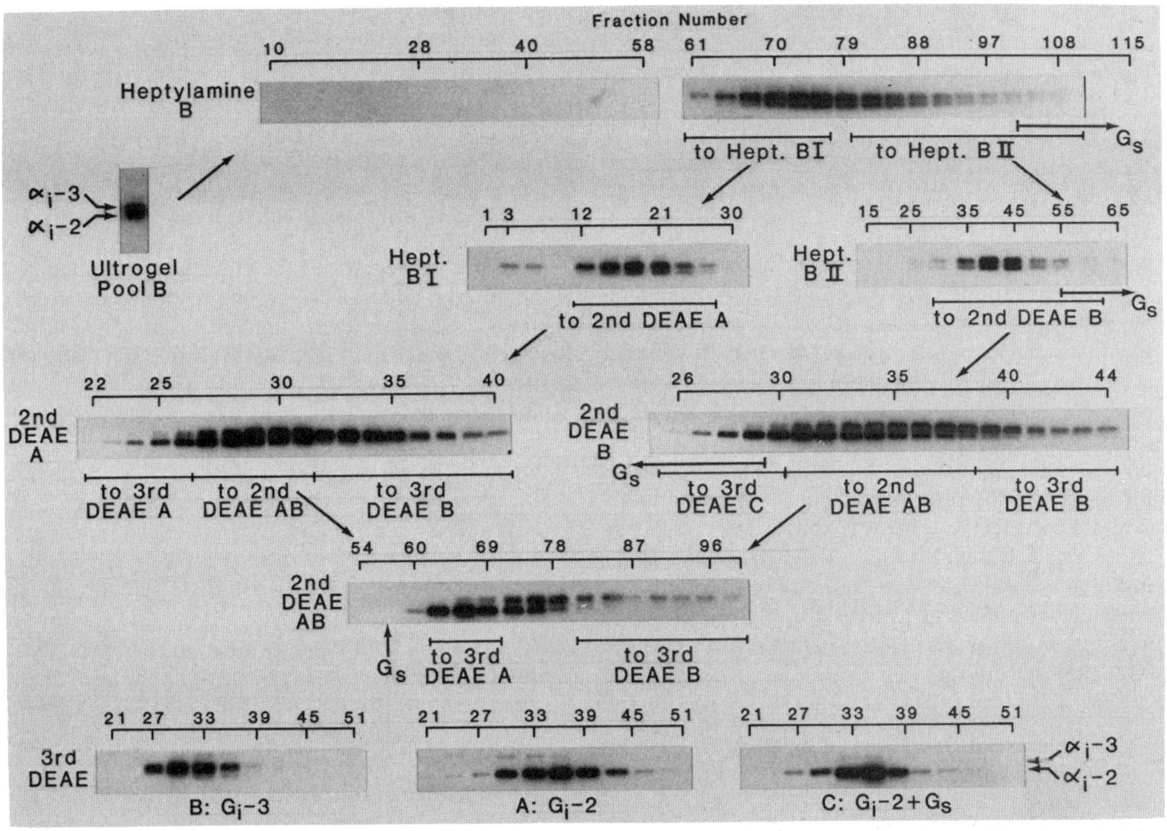

Figure 2. Strategy used to purify human erythrocyte G_i-2 ("lower" PTX substrate) and G_i-3 ("upper" PTX substrate). The progress of the purification is shown from the point of chromatographing pool B from the Ultrogel AcA 34 column over heptylamine-Sepharose (variant B) until upper and lower PTX substrates were separated and purified to better than 95% purity from other proteins. In the particular purification shown, the aliquots from fractions eluted off the heptylamine-Sepharose column were ADP-ribosylated with PTX and analyzed by urea gradient/SDS-PAGE to give the pattern shown for the heptylamine B column. Upper (leading) PTX substrate was partially separated from the lower (trailing) PTX substrate, the last part of which in turn overlapped with G_s. Fractions were pooled as shown and rechromatographed under identical conditions (heptylamine B-I and B-II chromatographies). The fractions from these two steps were then subjected to separate second DEAE chromatographies over DEAE-Toyopearl. This resulted in a further partial separation of lower (now leading) from upper (now trailing) PTX substrate, with the difference that whereas the leading lower PTX substrate eluting from the second DEAE-A was free of G_s, that eluting from the second DEAE-B contained an about equimolar amount of G_s. The eluates from the second DEAE columns were combined each into three pools of which the center pools (~50/50 upper and lower PTX substrates) were combined and rechromatographed over a 0.9 × 75-cm DEAE-Toyopearl column (second DEAE-AB). The fractions with the leading PTX substrate (lower) were combined with those from the second DEAE-A; the fractions with the trailing upper PTX substrate were combined with those from both the second DEAE-A and DEAE-B. These and the pool of leading lower PTX substrate plus contaminating G_s from the second DEAE-B pools were then subjected to a third DEAE chromatography (columns designated A, B, and C on the figure). Upper PTX substrate (G_i-3) elutes from the third DEAE column slightly earlier than the lower PTX substrate (G_i-2), which in turn essentially cochromatographs with G_s.

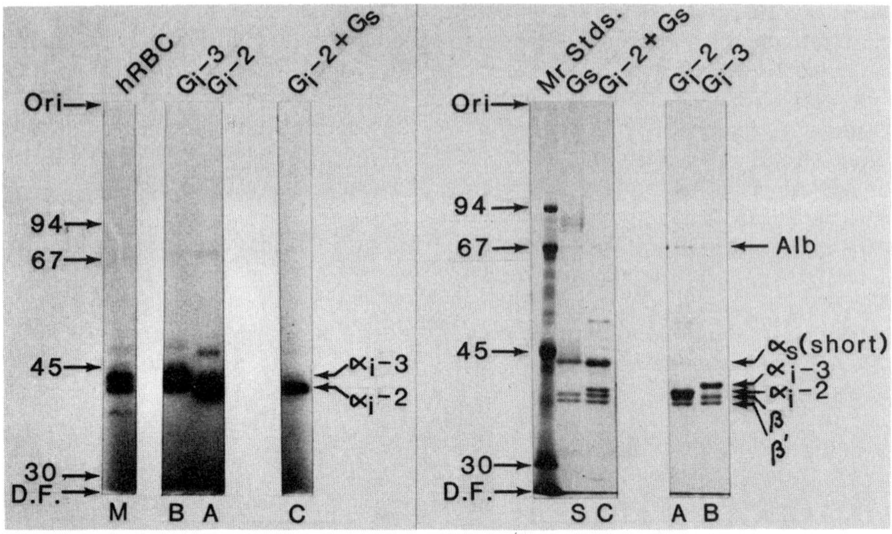

Figure 3. Urea gradient/SDS-PAGE analysis of fractions from the purification shown in Fig. 2. (*Left*) PTX-labeling pattern: (M) 7 μg human erythrocyte membranes; (B) 5 ng of protein obtained from the third DEAE-B; (A) 10 ng of protein obtained from the third DEAE-A; (C) 4 ng of protein obtained from the third DEAE-B. (*Right*) Coomassie blue stain: the same fractions analyzed for PTX labeling were analyzed; (B) 500 ng; (A) 250 ng; (C) 350 ng; (S) 200 ng G_s purified from Ultrogel pool A. Note the presence of two β subunits in each of the proteins.

acid sequence composition of two V8-protease fragments obtained upon digestion of fractions containing 80–90% upper PTX substrate, we assigned the G_i-3 identity to the human erythrocyte upper PTX substrate (Codina et al. 1988). Immunoblots carried out in A. Spiegel's laboratory with preparations now known to be enriched in lower PTX substrate showed strong positive reactivity with an antipeptide antibody specific for α_i-2 (A.M. Spiegel, pers. comm.). We have therefore assigned the G_i-2 identity to the lower PTX substrate obtained from the third DEAE-B column shown in Figures 2 and 3.

Availability of a highly purified G_i-2 preparation will allow us in future studies to characterize its intrinsic biological activity(ies).

Bacterial Expression of α Subunits: Uncertainties in Assignment of Function

Despite efforts such as those shown above, the question raised implicitly by the data of Figure 1 remains; i.e., how pure are purified G proteins made by the biochemist and hence how reliable are functional assignments?

For example, it is shown (A.M. Brown et al., this volume) that purified human erythrocyte G_s stimulates cardiac and skeletal muscle channels (see also Yatani et al. 1987c, 1988) and that a purified preparation of PTX substrate, containing predominantly G_i-3, stimulates K^+ channels (see also Codina et al. 1987a,b; Yatani et al. 1987b,c; Kirsch et al. 1988). Is it indeed G_s that stimulates the Ca^{++} channel or is it a contaminant? Did the proteolytic peptides obtained from G_K-containing preparations indeed come from the G protein that constituted >90% of the total or was one of the contaminants sequenced? Is the G_K activity obtained with the mostly upper PTX substrate-containing preparations indeed due to the upper presumed G_i-3? It is standard practice in the field of organic chemistry that structures of purified active principles deduced by analytical methods need confirmation by de novo synthesis. Although with proteins this was impossible in the past, the advent of recombinant DNA techniques and bacterial expression vectors opened the possibility to test and confirm the assignment of function to purified polypeptides. A search for an adequate bacterial expression vector to synthesize α subunits encoded in their respective cDNA molecules led Graziano et al. (1987), as well as us (see below), to the use of pT7 plasmids developed by Tabor and Richardson at Harvard (Tabor and Richardson 1985; Tabor et al. 1987). α subunits made in this way were then tested for function: α_s for reconstitution of the *cyc*$^-$ adenylyl cyclase system and stimulation of skeletal muscle Ca^{++} channels in lipid bilayers, and one of the α_i molecules (α_i-3) suspected of being an α_K, for stimulation of single channel K^+ currents in inside-out atrial membrane patches.

pT7-based Bacterial Expression Systems

These systems are double vector systems formed of a plasmid (pT7), into which the cDNA encoding the protein of interest is placed downstream from a T7 promoter sequence, and either a phage or a second plasmid that encodes the T7 RNA polymerase required for transcription of the cDNA is inserted into the first plasmid. In mGP1-2, a derivative of phage M13, the T7 RNA polymerase gene (gene 1), is under control of the *lac* promoter, allowing for induction of the T7 RNA polymerase by IPTG. In pGP1-2, a plasmid carrying the kanamycin resistance gene, the T7 RNA polymerase gene is under control of the leftward λ promoter (P_L) so that induction of the polymerase is contingent on inhibition of the λ repressor (*c*I gene product). Expression is carried out in *E. coli* K38 cells, which are constitutive producers of the temperature-sensitive *c*I857 form of the repressor. Currently, a family of the pT7 plasmids exists that are distinguished by suffix 1 through 7. All contain the β-lactamase selection marker, the ColE1 origin of replication for growth in host *E. coli* cells, and the promoter region of gene 10 of the T7 phage. They differ in nucleotide composition downstream from the gene 10 promoter, containing or not containing a ribosome-binding site (*rbs*) and a cloning cassette into which the cDNA of interest is cloned according to a desired reading frame.

Feasibility of α-subunit expression was first shown with pT7-5, which lacked a ribosomal-binding site as well as a true cloning cassette (Graziano et al. 1987). Experiments obtained with α-subunit cDNAs inserted into the pT7-7 plasmid, which has both the T7 gene 10 *rbs* and its natural ATG initiation codon followed by a multiple cloning cassette, are reported here. After insertion of an α-subunit cDNA, pT7-7 plasmids are referred to as pT7-α_x plasmids (Fig. 4). In initial experiments, we used the plasmid-phage system in which T7 RNA polymerase (gene 1 of the T7 phage) is supplied by infecting pT7-α_x-transformed host cells with phage mGP1-2; polymerase synthesis was induced with IPTG. In later experiments, we also used the two-plasmid system in which polymerase and α-subunit expression are induced by a temperature shift from 30°C to 42°C (Tabor et al. 1987). In both cases, the *Bam*HI restriction site was used to insert the α-subunit cDNAs into the pT7-7 vector, which led to addition of 27 nucleotides at the 5′ end of their open reading frame. These nucleotides encode nine extra amino acids at their amino termini, a modification that may have functional importance.

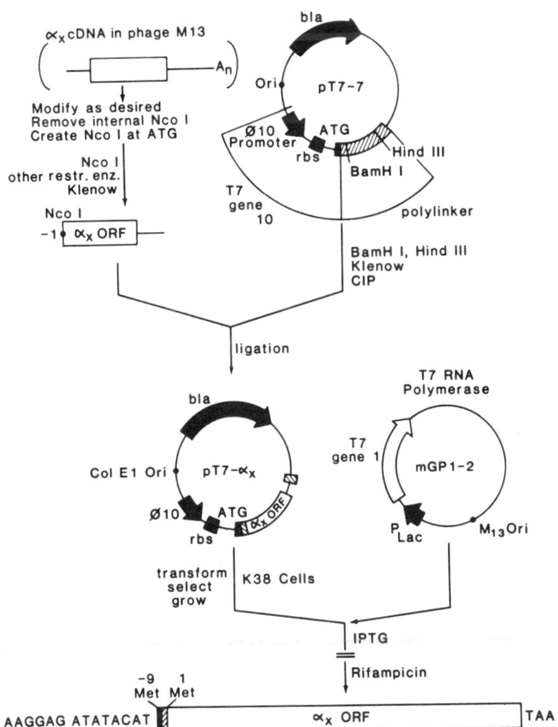

Figure 4. Tabor-Richardson strategy (plasmid/phage version) for bacterial expression of cloned proteins as applied to α subunits of G proteins (Tabor and Richardson 1985; Tabor et al. 1987).

Expression of α Subunits of G Proteins

As shown in Figure 5, analysis of total bacterial cell proteins after induction of expression revealed accumulation of polypeptides that depended on the composition and orientation of the cDNA inserted into the pT7-7 vector. The cDNAs inserted into the *Bam*HI site of the pT7-7 vector in the proper orientation yielded high levels of polypeptide expression (Fig. 5). In each case, the recombinant polypeptides accumulated within 60 minutes after induction and constituted ~5–8% of

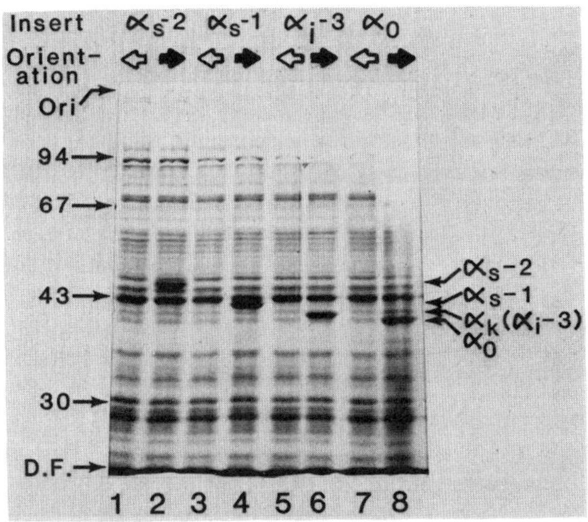

Figure 5. Bacterial expression of α-subunit fusion proteins according to the approach of Tabor and Richardson. cDNAs were inserted into the *Bam*HI site of pT7-7 plasmid, recombinant pT7-α_x plasmids with cDNAs inserted in either the antisense (3′ to 5′, open arrows) or sense (5′ to 3′, filled arrows) directions were isolated and transfected into *E. coli* K38 cells, and the open reading frame encoded downstream from the T7 promoter was expressed after infection of the cells with phage mGP1-2 and analyzed by SDS-PAGE in 10% polyacrylamide gel slabs. The slabs were stained with Coomassie blue and photographed. Note the appearance of insert-specific and orientation-dependent α-subunit bands that constitute between 5% and 8% of the total cell protein. (α_s-1) Short human α_s of 380 amino acids with Asp in position 71 and Ser in position 72 followed by Glu; (α_s-2) long human α_s of 394 amino acids with sequence Glu-Gly-Gly-Glu-Glu-Asp-Pro-Gln-Ala-Ala-Arg-Ser-Asn-Ser-Asp-Gly in the place of Asp-Ser; (α_i-3) PTX substrate of 354 amino acids cloned from a human liver cDNA library; (α_o) PTX substrate of 354 amino acids identical to rat brain and bovine retina α_o cloned from a rat heart cDNA library.

total bacterial cell protein, when the plasmid/phage system was used. Although the absolute amounts of recombinant α subunits made were similar using the two-plasmid system, they represented a smaller proportion of the total because of induction just prior to expression of pT7-α_x-encoded protein of several heat-shock proteins (not shown).

The difference in migration upon SDS-PAGE observed between the recombinant α_s-1 (apparent M_r = 42,000) and the recombinant α_s-2 (apparent M_r = 47,000) of about M_r 5000 correlates well with the apparent difference in migration of about M_r 5000–8000 observed when purified or cholera-toxin-labeled native α_s subunits are analyzed by similar means. This confirms that α_s subunits exhibit anomalous migration behavior on the basis of their primary amino acid composition (Robishaw et al. 1986). The other α subunits expressed in bacteria also migrate as would be expected from SDS-PAGE migrations of the natural polypeptides. For example, recombinant α_i-3 migrates with an apparent M_r of approximately 41,500, and recombinant α_o migrates with a slightly smaller apparent M_r of approximately 39,000 (Fig. 5).

Advantages and Drawbacks of α Subunits Synthesized in Bacteria

Drawbacks. Even though the levels of α-subunit protein synthesized by bacteria by the Tabor-Richardson two-vector system are quite high, making purification a relatively easy task, only a fraction (5–10%) can be recovered as a soluble protein after bacterial lysis, the remainder being soluble only in high urea and reducing agents such as DTT or β-mercaptoethanol. On assaying the soluble portion of α_s molecules for their *cyc*$^-$ reconstituting activity, we found them to have rather low intrinsic activity. Regardless of whether the recombinant α_s molecules were derived from a cDNA encoding a 394- or 380-amino-acid protein (Mattera et al. 1986; Robishaw et al. 1986), we obtained specific (intrinsic) activities that are only 3–5% of that of native human erythrocyte G_s.

Advantages. At this time, the principal advantage of synthesizing α subunits in bacteria is that the product obtained is absolutely free of any other α subunit. Despite the drawbacks listed above, of which the relatively low intrinsic activity is the more bothersome, the fact that a definable activity is associated with them makes them useful for identifying their functional role or roles as exemplified for one of the α_i and one of the α_s molecules.

Bacterially Expressed α_i-3 Stimulates the G-protein-gated Atrial K^+ Channel

Expression in bacteria and testing for K^+ channel stimulatory activity allowed us to test the hypothesis that α_i-3 has α_K activity. As shown in Figure 6, recombinant α_i-3, activated either by GTPγS or AlF_4^-, does indeed stimulate the atrial "muscarinic" K^+ channel, thus removing any doubt as to structural identity of the α subunit of G_K.

Bacterially Expressed Short Form of α_s Stimulates the Voltage-gated Ca^{++} Channel

As mentioned earlier, it has been well established that the cardiac voltage-gated dihydropyridine-sensitive Ca^{++} channel is positively regulated by cAMP-dependent phosphorylation (Kameyama et al. 1986). In agreement with this, it was found more recently that incubation of inside-out membrane patches from GH_3 cells can lead to recovery of Ca^{++} channel activity that had been lost on patch excision (Armstrong and Eckert 1987). The finding described above that G_s preparations stimulate the Ca^{++} channel directly was therefore surprising, for it means that these channels are apparently under dual and redundant regulation by G_s. The question of whether indeed it was G_s that caused the stimulation of Ca^{++} channel activity was thus an important one. Therefore, we prepared by bacterial synthesis one of the α_s subunits (380-amino-acid-long Asp-71, Ser-72 version) and tested whether its biological activity as seen with *cyc*$^-$ membranes had a correlate in stimulating the activity of Ca^{++} channels incorporated into lipid bilayers. Figure 7 illustrates that indeed recombinant α_s, but not recombinant α_i-3, stimulates the activity of the skeletal muscle T-tubule Ca^{++} channel.

These findings confirmed not only that Ca^{++} channels are controlled directly by G_s, but also that G_s, and potentially any other G protein, may interact with more than a single effector function.

DISCUSSION

About 80% of all known hormones and neurotransmitters that regulate cellular interactions, termed primary messengers, are now known to elicit cellular responses by combining to specific receptors that are coupled to effector functions by G proteins. Yet, even though the primary messengers are many, the number of distinct receptors that mediate their action is even larger. This is well exemplified by the existence of at least four types of adrenergic receptors (β1, β2, α1, and α2), at least five types of muscarinic acetylcholine (Ach) receptors (M1–M5), two types of dopamine receptors (D1 and D2), and several types each of adenosine, purinergic, and light (rhodopsin) receptors. Multiplicity of receptors is not restricted to neurotransmitters. There are at least three types of vasopressin receptors (V1a, V1b, and V2), and vasoactive intestinal peptide interacts with receptors that stimulate either adenylyl cyclase or phosphoinositide turnover. At this time, about 75 distinct receptors can be identified that recognize 30 hormones and neurotransmitters. It seems reasonable to assume that the total number of distinct receptors coupled by G proteins will be around 100.

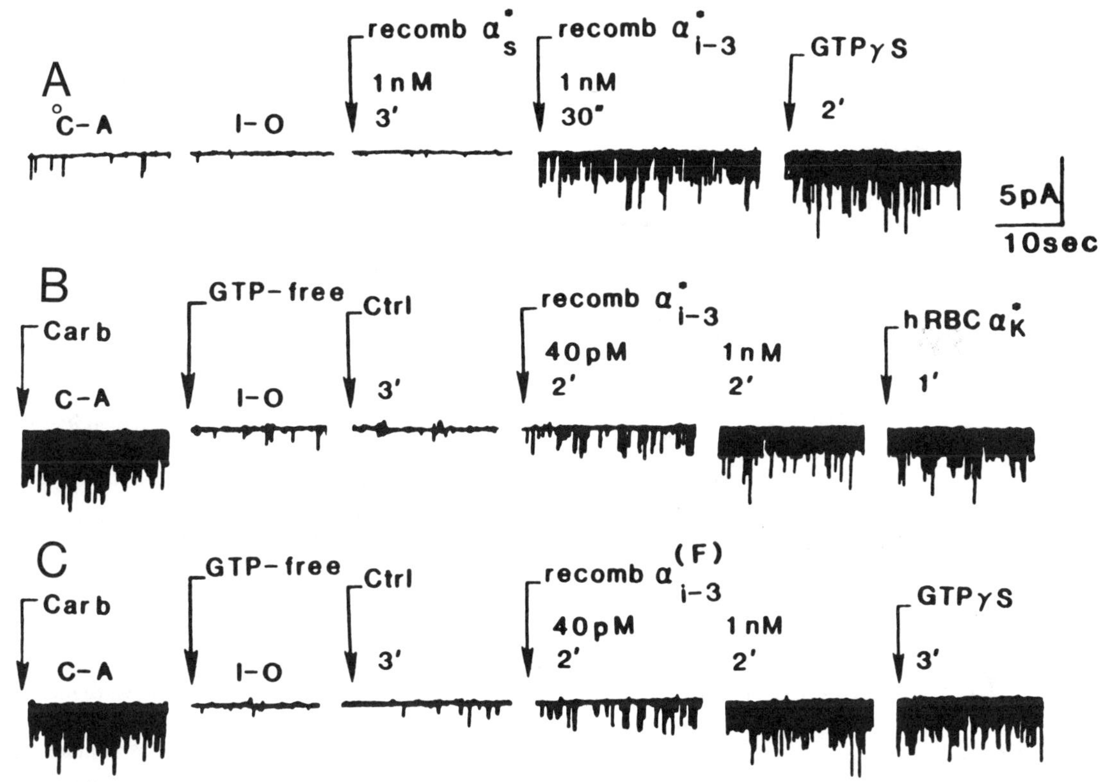

Figure 6. Stimulation of the atrial G-protein-gated K^+ channel by recombinant α_i-3. Recombinant α_i-3 and α_s subunits encoded in their respective cDNAs inserted into pT7-7 vector were expressed in *E. coli* by the one plasmid/one phage method of Tabor and Richardson (1985), the cells were lysed, and the recombinant α subunits were activated and purified partially by chromatography over DEAE-Sephacel (0–400 mM NaCl gradient). The partially purified recombinant GTPγS-activated α subunits were then added to inside-out membrane patches of guinea pig atria to test for possible K^+ channel stimulatory effects. (Experiment A) recombinant α_s does not, but recombinant α_i-3 does, stimulate the atrial K^+ channel. (Experiments B and C) Effects of two different doses of GTPγS-activated or fluoride-activated recombinant α_i-3. In these and repeat experiments, recombinant α_i-3 was about 3–10% as active as purified human erythrocyte GTPγS-activated G_K.

In contrast, both the final effector functions regulated by these receptors and the G proteins that provide for receptor-effector coupling are much fewer, probably not much more than ten each. Thus, at this time, we know of about eight to ten classes of G proteins, including G_s, several G_i proteins, a G_o, and two transducins; and of about as many effector functions, including adenylyl cyclase, the "original" effector system; the cAMP-specific phosphodiesterase of photoreceptor cells, polyphosphoinositide-specific phospholipase C; a phospholipase of the A_2-type that releases arachidonic acid, and two classes of ionic channels, one specific for K^+ and relatively independent of membrane potential for its activity, the other specific for Ca^{++} and strongly dependent on membrane potential for its activity. On the basis of this, the responses elicited through occupancy of G-protein-coupled receptors by primary messengers include changes not only of intracellular second-messenger levels—cyclic nucleotides, inositol phosphates, diacylglycerol, arachidonic acid, and Ca^{++}—but also of the cell membrane potential, which itself is a potent regulator of cellular function. Figure 8 summarizes this flow of information from primary messenger through receptor, G protein, and effector to elicit second-messenger or membrane potential changes.

Given the molecular diversity, several questions arise as to how intramembrane signal transduction pathways are set up and which of the interactions provides for specificity.

Ever since the discoveries in the late 1960s that up to five different hormone receptors can activate a single adenylyl cyclase system in an isolated membrane and, in the early and mid-1970s, that receptors can be transferred from one cell to another and that there are no species and/or tissue specificity restrictions as to the source of G_s for reconstitution of a hormonally stimulatable adenylyl cyclase system in mouse cyc^- membranes, it has been clear that single G proteins are designed to interact with classes of receptors as opposed to single receptor subtypes. The quantitative considerations mentioned above, indicating that not more than about ten G proteins receive the input of probably as many as 100 different receptors, actually require this to be so. Thus, it is clear that any given G protein can interact with several receptors of a class.

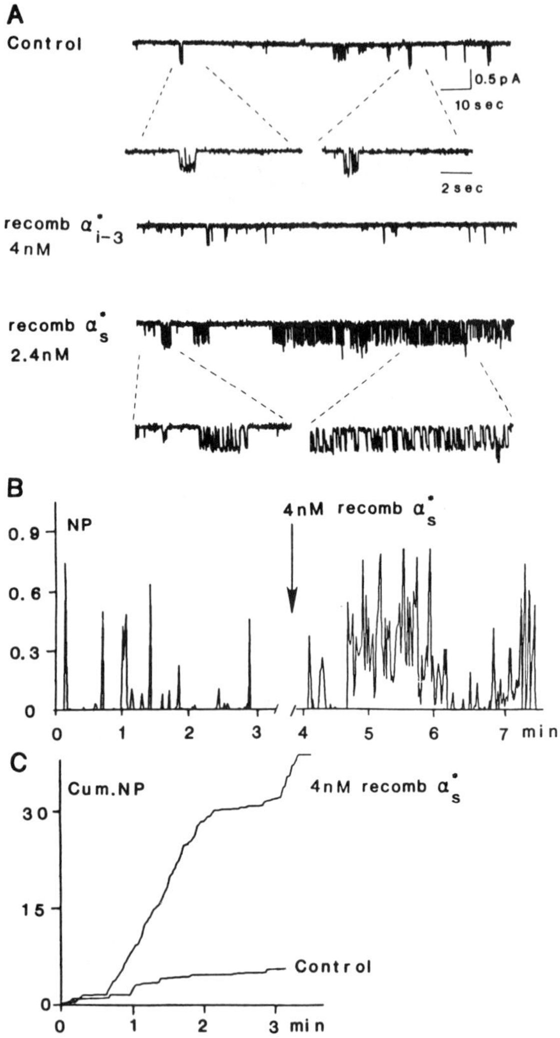

Figure 7. Stimulation of skeletal muscle T-tubule dihydropyridine-sensitive Ca^{++} channel incorporated into lipid bilayers by the short form of recombinant α_s. Recombinant α (Asp-71, Ser-72 version) and recombinant α_i-3 were prepared in their GTPγS-activated form and partially purified by DEAE-Sephacel chromatography, except that the two-plasmid approach was used to express the recombinant proteins (Tabor et al. 1987). (*A*) Single-channel Ba^{++} currents (*trans* to *cis*) before and after addition first of GTPγS-activated recombinant α_i-3 and then of GTPγS-activated recombinant α_s. (*B* and *C*) NPo diaries of Ca^{++} channel activities before and after addition of GTPγS-activated recombinant α_s to the *cis* chamber (*B*) and cumulative NPo values as a function of time of the same experiment (*C*).

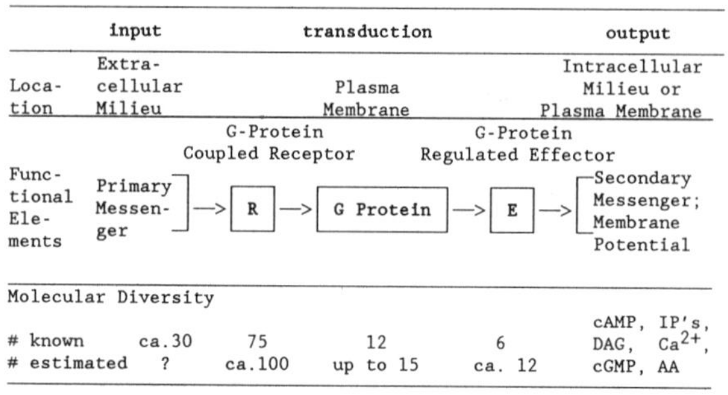

Figure 8. Flow of information through G-protein-dependent signal transduction systems as found in vertebrates.

The issue of how many different G proteins may interact with any given receptor is not so clear. Although the intuitive answer to this question might be *one*, indirect data from several different systems and a direct test in which a cloned receptor was transfected into a receptor-negative recipient cell indicate that receptors are likely to interact with more than one of the G proteins present in the membrane. Some examples stand out. Thrombin receptors in human platelets stimulate primarily phospholipase C activity, i.e., a G_p, but also inhibit adenylyl cyclase when tested in isolated membranes, i.e., couple to G_i (Crouch and Lapetina 1988). The same is true for thyrotropin-releasing hormone (TRH) receptors of GH_4C_1 cells (Straub and Gershengorn 1986; Toro and Birnbaumer 1986) and angiotensin II receptors in liver and adrenal glomerulosa cells that activate both G_p (PTX-insensitive) and G_i (PTX-sensitive). For all three, thrombin, TRH, and angiotensin II, activation of G_p appears to be the physiologically relevant response, because manipulations of the intact cells, e.g., incubation with phorbol esters and/or ionophores, which mimic phospholipase C activation, mimic the effects of the receptor ligands. With muscarinic ACh receptors of chick and mouse heart, the situation is reversed. ACh stimulates primarily G_i, causing a decrease in cAMP levels, but at higher concentrations also increases phosphoinositide turnover, indicative of G_p stimulation (Brown and Masters 1984). On the other hand, although thrombin stimulates both G_p and G_i in the human platelet, α_2-adrenergic stimulation of platelets seems to be restricted to G_i, there being no evidence for a G_p stimulatory action, as seen through mobilization of intracellular Ca^{++} (Crouch and Lapetina 1988). However, there is also evidence that receptors of the α_2-adrenergic type, such as somatostatin and "cardiac" or M2-type muscarinic ACh receptors, affect more than one function. Thus, in all cases so far measured, receptors that either stimulate K^+ channels acting through a G_K (e.g., somatostatin in pituitary cells, muscarinic in heart and pituitary cells) or inhibit Ca^{++} currents (e.g., α_2-adrenergic in dorsal root ganglia, opioid in neuroblastoma × glioma cells) also inhibit adenylyl cyclase via stimulation of a PTX-sensitive G_i. The finding in FRTL-5 cells that α_1-adrenergic stimulation activates independently both a phospholipase-C-specific and a phospholipase-A_2-specific G_p (Burch et al. 1986) provides yet another example of the possibility that single receptors may interact with more than one G protein. However, in none of these examples of one agonist eliciting activation of more than one G protein can the involvement of more than one receptor subtype be ruled out. Single cells can express multiple subtypes of the same receptor, as determined by Bockaert and collaborators for β_1- and β_2-adrenergic receptors in cultured cells (Homburger et al. 1981).

The potential ability of a single receptor to interact unequivocally with more than one G protein was first observed in in vitro reconstitution experiments, in which purified G proteins were coincorporated with purified receptors into phospholipid vesicles, and the ability of the receptors to drive the G proteins through their GTP-dependent regulatory cycle was assessed by measurement of GTP hydrolysis. Notably, rhodopsin was found to stimulate not only the GTP hydrolysis by transducin, the G protein with which it normally interacts, but also a mixture of human erythrocyte G_i-2 and G_i-3, of which the latter is now known to have G_K activity (Cerione et al. 1985). That a single receptor may indeed interact with more than one type of G protein in the normal setting of intact cell membranes was more recently shown with a cloned cardiac, M2-type muscarinic ACh receptor (Ashkenazi et al. 1987). On transfection into recipient cells, occupancy of the M2 receptor by ACh was found to cause in a dose-dependent manner both inhibition of adenylyl cyclase, i.e., activation of G_i, and stimulation of phosphatidylinositol bisphosphate hydrolysis, i.e., activation of a G_p. Interestingly, the agonist dose-response curves differ for the two effects, ACh or carbachol being more potent in inhibiting adenylyl cyclase than in promoting phosphoinositide hydrolysis. Since the same difference in potencies for these two effects had previously been found on stimulation of intact tissue with muscarinic agonists (Brown and Masters 1984), it is likely that, at least in this case, the two muscarinic effects were due to the action of a single receptor rather than to the independent effects of two separate receptor subtypes.

The second set of questions relates to the effector specificity of G proteins. The studies on the regulation of ionic channels by G proteins have led to the surprising discovery that one and the same G protein may affect more than one effector function. Presently, the example is only one, but it is likely that more such cases may be found once they are searched for.

Related to this is the question of whether each G protein indeed has separate functions or whether there are iso-G proteins, all of which fulfill the same function. Experimentally, this is best addressed by defining the functions of the three G_i-type PTX substrates. To date, G_i-3 has proven to have G_K activity, as seen with recombinant α_i-3. It will be of interest to determine if α_i-1 and/or α_i-2 have K^+ channel stimulatory activity, and if so, how potent they are. Furthermore, since all agonists that stimulate K^+ channels also inhibit adenylyl cyclase, one or the other of the α_i molecules should have G_i activity. Tests with G_i-3, which, as mentioned, has G_K activity, have thus far failed to show G_i effects on adenylyl cyclase, whereas preparations of brain G_i, probably mostly G_i-1, show some G_i activity. Furthermore, since G_i-2 is quite abundant in HL-60 cells, it may be this cell's PTK-sensitive stimulator of phospholipase C, i.e., a G_p. Recent immunoblotting studies with antibodies that distinguish between the three forms of α_i showed that all three α_i molecules can be expressed in a single cell type, the human erythrocyte (R. Iyengar et al., unpubl.). Yet the relative abundance of the three α_i molecules varies widely among cells. Resolution as to their true functional significance must await the testing of their individual properties, as can be done after bacterial expression. Thus, one receptor can couple to more than one G protein and one G protein can regulate more than one effector function.

A final question is with how many different types of G protein does any given effector interact? In the case of adenylyl cyclase, the answer is at present ambiguous because, clearly, both the short and the long α_s molecules can activate this enzyme. However, since to date no functional differences have been found for the various α_s molecules, and they are all derived from a single gene, it is possible to argue that to date only one G protein, G_s, has been found to stimulate the catalytic unit of the adenylyl cyclase system.

Given that receptor types far exceed the number of G proteins, yet effector systems are about as many as G proteins (Fig. 8); given, too, that many receptors interact with a single G protein, that a single receptor may interact with more than one G protein, that one G protein may interact with more than one effector system, that indeed several G proteins may interact with a single effector system, and that each cell type has a separate complement of receptors, G proteins, and effector systems, it is no longer possible to speak of a single or simple signal transduction pathway where a given receptor triggers a single effect. Figure 9 is therefore a modification of Figure 8 intended to include these complexities.

It is very likely that the norm will be that receptor stimulation leads to a complex set of G-protein stimulations and effector modulations, of which one or the other will dominate and vary from cell to cell depending on the presence or absence of given G proteins and/or effector systems. It will also depend on the specificity of receptor–G protein and G protein–effector interactions, since specificity does not appear to be absolute on selectivity parameters governing these interactions. As applied to any particular biological system, the unraveling of its G-protein-dependent signal

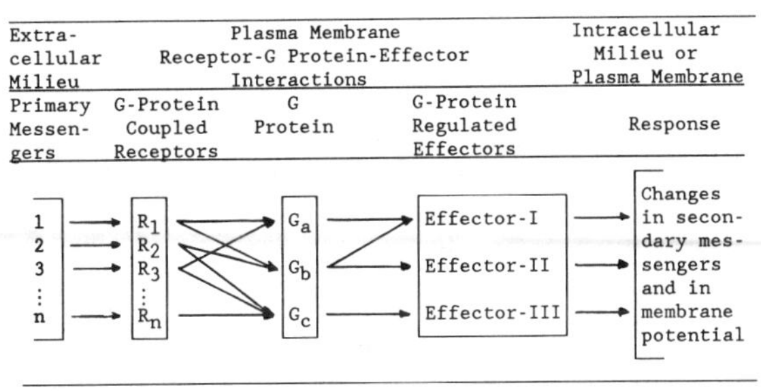

Figure 9. Flow of information through G-protein-dependent signal transduction systems of a cell with *n* receptors (R_1 through R_n), three G proteins (G_a through G_c), and three effector systems (Effector-I through Effector-III), each having separate and different selectivity for mutual interaction.

transductions, ultimately responsible for final cellular responses to agonists, be they stimulatory or inhibitory, will require first qualitative and then quantitative analyses of which receptors, which G proteins, and which effector systems are present in the cell systems under study, e.g., liver, heart, muscle, pancreatic β-cells, neurons, and testicular Leydig cells. Tools for nucleic acid hybridization and protein immunoblotting are becoming available to accomplish these goals. Finally, superimposed on these primary receptor–G-protein–effector connections is the fact that effector systems, G proteins, and receptors are themselves real or potential substrates for phosphorylation reactions triggered by the homologous activating pathway (feedback effect) or by the heterologous activating pathway (pathway cross-talk). These possibilities for feedback and cross-talk surely make for further difficulties in separating primary from secondary regulations of signal transductions, and also for new and, by definition, unexpected discoveries.

ACKNOWLEDGMENTS

We thank Dr. Ken Seamon (Food and Drug Administration, Bethesda, MD) for bringing to our attention the use of urea gradients in conjunction with SDS-PAGE. We are also deeply grateful to Dr. Ronald D. Sekura for his willingness to keep us supplied with pertussis toxin, without which most of the insights into many specific signal transduction pathways would not have been possible. Thanks are due also to Dr. Vincent Pirrotta and his students (Department of Cell Biology, Baylor College of Medicine) for acquainting us with the pT7 expression system and to Dr. Stanley Tabor for providing us with pT7 and pGP1 plasmids and mGP1-2 phage together with extensive documentation on their composition and use. The research reported here was supported in part by U.S. Public Health Service research grants DK-19318, HD-09581, HL-31164, and HL-37044 to L.B.; HL-39262 to A.M.B.; EY-06062 to H.E.H.; and DK-38761 and CA-44998 to R.I.; the Baylor College of Medicine Diabetes and Endocrinology Research Center grant DK-27685 (Dr. James B. Field, Director); and grant Q-1075 from the Welch Foundation.

REFERENCES

Armstrong, D. and R. Eckert. 1987. Voltage-activated calcium channels that must be phosphorylated to respond to membrane depolarization. *Proc. Natl. Acad. Sci.* **84:** 2518.

Ashkenazi, A., J.W. Winslow, E.G. Peralta, G.L. Peterson, M.I. Schimerlik, D.J. Capon, and J. Ramachandran. 1987. An M2 muscarinic receptor subtype coupled to both adenylyl cyclase and phosphoinositide turnover. *Science* **238:** 672.

Bray, P., A. Carter, C. Simons, V. Guo, C. Puckett, J. Kamholz, A. Spiegel, and M. Nirenberg. 1986. Human cDNA clones for four species of G α_s signal transducing protein. *Proc. Natl. Acad. Sci.* **83:** 8893.

Brown, J.H. and S.B. Masters. 1984. Muscarinic regulation of phosphatidylinositol turnover and cyclic nucleotide metabolism in the heart. *Fed. Proc.* **43:** 2613.

Burch, R.M., A. Luini, and J. Axelrod. 1986. Phospholipase A_2 and phospholipase C are activated by distinct GTP binding proteins in response to α_1 adrenergic stimulation in FRTL-5 cells. *Proc. Natl. Acad. Sci.* **83:** 7201.

Cerione, R.A., C. Staniszewski, J.L. Benovic, R.J. Lefkowitz, M.C. Caron, P. Gierschick, R. Somers, A. Spiegel, J. Codina, and L. Birnbaumer. 1985. Specificity of the functional interactions of the *beta*-adrenergic receptor and rhodopsin with guanine nucleotide regulatory proteins reconstituted in phospholipid vesicles. *J. Biol. Chem.* **260:** 1493.

Codina, J., D. Stengel, S.L.C. Woo, and L. Birnbaumer. 1987a. Beta-subunits of the human liver G_s/G_i signal-transducing proteins and those of bovine retinal rod cell transducin are identical. *FEBS Lett.* **207:** 187.

Codina, J., W. Rosenthal, J.D. Hildebrandt, R.D. Sekura, and L. Birnbaumer. 1984a. Updated protocols and comments on the purification without use of activating ligands of the coupling proteins N_s and N_i of the hormone sensitive adenylyl cyclase. *J. Recept. Res.* **4:** 411.

Codina, J., A. Yatani, D. Grenet, A.M. Brown, and L. Birnbaumer. 1987b. The *alpha* subunit of G_k opens atrial potassium channels. *Science* **236:** 442.

Codina, J., J. Olate, J. Abramowitz, R. Mattera, R.G. Cook, and L. Birnbaumer. 1988. α_i-3 cDNA encodes the α subunit of G_k, the stimulatory G protein of receptor-regulated K^+ channels. *J. Biol. Chem.* **263:** 6746.

Codina, J., J.D. Hildebrandt, R.D. Sekura, M. Birnbaumer, J. Bryan, C.R. Manclark, R. Iyengar, and L. Birnbaumer. 1984b. N_s and N_i, the stimulatory and inhibitory regulatory components of adenylyl cyclases. Purification of the human erythrocyte proteins without the use of activating regulatory ligands. *J. Biol. Chem.* **259:** 5871.

Crouch, M.F. and E.G. Lapetina. 1988. A role for G_i, in control of thrombin receptor-phospholipase C coupling in human platelets. *J. Biol. Chem.* **263:** 3363.

Graziano, M.P., P.J. Casey, and A.G. Gilman. 1987. Expression of cDNAs for G proteins in *Escherichia coli*. *J. Biol. Chem.* **262:** 11375.

Homburger, V., M. Lucas, E. Rosenbaum, G. Vassent, and J. Bockaert. 1981. Presence of both β_1 and β_2-adrenergic receptors in a single cell type. *Mol. Pharmacol.* **20:** 463.

Itoh, H., R. Toyama, T. Kozasa, T. Tsukamoto, M. Matsuoka, and Y. Kaziro. 1988. Presence of three distinct molecular species of G_i protein subunit. Structure of rat cDNAs and human genomic DNAs. *J. Biol. Chem.* **263:** 6656.

Iyengar, R., K.A. Rich, J.T. Herberg, D. Grenet, S.M. Mumby, and J. Codina. 1987. Identification of a new GTP-binding protein: An M_r = 43,000 substrate for pertussis toxin. *J. Biol. Chem.* **262:** 9239.

Jones, D.T. and R.R. Reed. 1987. Molecular cloning of five GTP-binding protein cDNA species from rat olfactory neuroepithelium. *J. Biol. Chem.* **262:** 14241.

Kameyama, M., J. Hescheler, F. Hofmann, and W. Trautwein. 1986. Modulation of Ca current during the phosphorylation cycle in the guinea pig heart. *Pfluegers Arch.* **407:** 123.

Katada, T., K. Kusakabe, M. Oinuma, and M. Ui. 1987a. A novel mechanism for the inhibition of adenylate cyclase via inhibitory GTP-binding proteins. Calmodulin-dependent inhibition of the cyclase catalyst by the beta-gamma subunits of GTP-binding protein. *J. Biol. Chem.* **262:** 11897.

Katada, T., M. Oinuma, K. Kusakabe, and M. Ui. 1987. A new GTP-binding protein in brain tissues serving as the specific substrate of islet-activating protein, pertussis toxin. *FEBS Lett.* **213:** 353.

Kirsch, G., A. Yatani, J. Codina, L. Birnbaumer, and A.M. Brown. 1988. The *alpha* subunit of G_k activates atrial K^+ channels of embryonic chick, neonatal rat and adult guinea pig. *Am. J. Physiol.* **254:** H1200.

Kozasa, T., H. Itoh, T. Tsukamoto, and Y. Kaziro. 1988. Isolation and characterization of the human G_s α gene. *Proc. Natl. Acad. Sci.* **85:** 2081.

Mattera, R., J. Codina, A. Crozat, V. Kidd, S.L.C. Woo, and L. Birnbaumer. 1986. Identification by molecular cloning of two forms of the *alpha* subunit of the human liver stimulatory (G_s) regulatory component of adenylyl cyclase. *FEBS Lett.* **206:** 36.

Robishaw, J.D., M.D. Smigel, and A.G. Gilman. 1986. Molecular basis for two forms of the G protein that stimulates adenylate cyclase. *J. Biol. Chem.* **261:** 9587.

Scherer, N.M., M.-J. Toro, M.L. Entman, and L. Birnbaumer. 1987. G protein distribution in canine cardiac sarcoplasmic reticulum and sarcolemma. Comparison to rabbit skeletal membranes and brain and erythrocyte G proteins. *Arch. Biochem. Biophys.* **259:** 431.

Straub, R.E. and M.C. Gershengorn. 1986. Thyrotropin-releasing hormone and GTP activate inositol trisphosphate formation in membranes isolated from rat pituitary cells. *J. Biol. Chem.* **261:** 2712.

Suki, W., J. Abramowitz, R. Mattera, J. Codina, and L. Birnbaumer. 1987. The human genome encodes at least three non-allelic G proteins with $alpha_i$-type subunits. *FEBS Lett.* **220:** 187.

Tabor, S. and C.C. Richardson. 1985. A bacteriophage T7 RNA polymerase/promoter system for controlled exclusive expression of specific genes. *Proc. Natl. Acad. Sci.* **82:** 1074.

Tabor, S., H.E. Huber, and C.C. Richardson. 1987. *Escherichia coli* thioredoxin confers processivity on the DNA polymerase activity of the gene 5 protein of bacteriophage T7. *J. Biol. Chem.* **262:** 16212.

Toro, M.-J. and L. Birnbaumer. 1986. Inhibitory regulation of GH_4 cell adenylyl cyclase (AC) by TRH and differentiation from that by somatostatin (SST) using cholera and pertussis toxins: Possible evidence of an action of another nucleotide regulatory component. *Endocrinology (Suppl. 1)* **118:** 415 (Abstr.).

Yatani, A., J. Codina, A.M. Brown, and L. Birnbaumer. 1987a. Direct activation of mammalian atrial muscarinic K channels by a human erythrocyte pertussis toxin-sensitive G protein, G_k. *Science* **235:** 207.

Yatani, A., J. Codina, R.D. Sekura, L. Birnbaumer, and A.M. Brown. 1987b. Reconstitution of somatostatin and muscarinic receptor mediated stimulation of K^+ channels by isolated G_k protein in clonal rat anterior pituitary cell membranes. *Mol. Endocrinol.* **1:** 283.

Yatani, A., J. Codina, Y. Imoto, J.P. Reeves, L. Birnbaumer, and A.M. Brown. 1987c. A G protein directly regulates mammalian cardiac calcium channels. *Science* **238:** 1288.

Yatani, A., Y. Imoto, J. Codina, S.L. Hamilton, A.M. Brown, and L. Birnbaumer. 1988. The stimulatory G protein of adenylyl cyclase, G_s, also stimulates dihydropyridine-sensitive Ca^{2+} channels. Evidence for direct regulation independent of phosphorylation by cAMP-dependent protein kinase or stimulation by a dihydropyridine agonist. *J. Biol. Chem.* **263:** 9887.

Functions of G-protein Subunits

E.J. Neer,* S.-Y. Kim,* ** S.-L. Ang,* D.B. Bloch,† K.D. Bloch,† Y. Kawahara,*††
C. Tolman,* R. Lee,* D. Logothetis,‡ D. Kim,§
J.G. Seidman,† and D.E. Clapham§

*Department of Medicine, Brigham and Women's Hospital and Harvard Medical School, Boston, Massachusetts 02115;
Departments of †Genetics and ‡Physiology, Harvard Medical School, Boston, Massachusetts 02115;
§Department of Pharmacology, Mayo Clinic Foundation, Rochester, Minnesota 55905

The signals from many receptors for hormones and neurotransmitters are sent through the plasma membranes via a set of heterotrimeric GTP-binding proteins composed of a GTP-binding α subunit and a $\beta\gamma$ subunit, which is a nondissociable dimer made up of two different polypeptides (for recent reviews, see Stryer and Bourne 1986; Gilman 1987; Lochrie and Simon 1988; Neer and Clapham 1988). The receptors on the cell surface are extremely specific for their agonists and can be differentiated from one another by their ligand-binding properties. However, at the cytoplasmic side of the membrane, the interactions appear to be less precise. Within a single cell, several separate receptors can often regulate each final pathway. For example, in the heart, acetylcholine and adenosine receptors appear to regulate the same K^+ channel (Kurachi et al. 1986).

Through the work of many laboratories, we have recently learned the deduced amino acid sequence of a number of cell-surface receptors, of close to a dozen G-protein α subunits, two G-protein β subunits, and even of some effectors. The results of these studies have revealed a surprising sequence conservation within each class of protein but a larger number of isoforms than might have been predicted. This abundance of similar proteins raises important questions about the mechanisms that a cell uses to assure that hormonal signals are correctly transmitted from receptor to effector without excessive cross-talk. Part of the specificity lies in the structure of the G proteins (e.g., only one class of G protein, G_s, stimulates adenylyl cyclase, although all the isoforms of G_s apparently can do so; Graziano et al. 1987). The structural similarity among the components makes it likely that additional factors, such as the relative concentrations of the subunits and their physical arrangement in the plasma membrane, will be important for the ability of the cells to respond correctly to external stimuli.

Structural Similarity of α Subunits

In this paper, we focus on the family of G proteins whose α subunits are substrates for covalent ADP-ribosylation by the bacterial toxin, pertussis toxin. These proteins are thought to mediate a variety of cellular functions, including inhibition of adenylyl cyclase, activation of phospholipase C, activation of phospholipase A_2, and regulation of ion channels. In the retina, similar proteins transduce signals from visual pigments to a cGMP phosphodiesterase (for reviews, see Stryer and Bourne 1986; Gilman 1987; Lochrie and Simon 1988; Neer and Clapham 1988).

At least four proteins other than those involved in visual transduction make up the group of pertussis toxin substrates. Jones and Reed (1987) identified all of these in a single species; many other laboratories have cloned cDNAs for various isoforms and are cited in the reviews noted above. These α subunits, α_i-1, α_i-2, α_i-3, and α_o, are more similar to each other than they are to α_s, the α subunit that mediates stimulation of adenylyl cyclase and other effectors (Jones and Reed 1987). Within the pertussis toxin substrates, the most closely related proteins are the three α_i-like α subunits. Figure 1 shows a comparison of the deduced amino acid sequence from the three α_i subtypes. The sequences were compared by taking segments of approximately ten amino acids and comparing the percentage of identity in each segment. The regions marked A, C, E, and G in

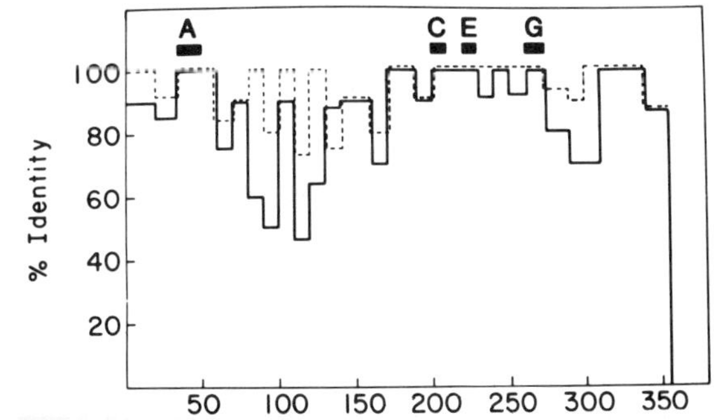

Figure 1. Comparison of amino acid sequences among α_i subtypes. (Broken line) α_i-1/α_i-3; (solid line) α_i-2/α_i-3. The number of identical amino acid residues in a given segment was expressed as a percentage and was plotted as a function of the amino acid residue number of three respective proteins. The length of segment (11 ± 3 residues) was determined to maximize the identity. The human α_i-2 has 355 amino acids, one more than all others. Therefore, one gap was introduced at position 123. The sequence of α subunits can be found in the following references: α_i-1, (Bray et al. 1987); α_i-2 (Didsbury et al. 1987); α_i-3 (Kim et al. 1988).

Present addresses: **Whitehead Institute for Biomedical Research, Cambridge, Massachusetts 02142; ††First Department of Internal Medicine, Kobe University School of Medicine, Kobe 650, Japan.

the figure denote segments of the sequence that are extremely similar in the signal-transducing G proteins, the oncogene *ras* protein, and bacterial elongation factor EF-Tu (Halliday 1984; Masters et al. 1986). In the latter two proteins, which have now been crystallized, it is known that these regions make up the GTP-binding site (Jurnak 1985; la Cour et al. 1985; de Vos et al. 1988). Therefore, it is assumed that these regions also make up the GTP-binding site of the G proteins. As shown in Figure 1, the putative GTP-binding region is identical among all of the three α_i subtypes. It is also identical in α_o (not shown). α_i-1 and α_i-3 are more similar to each other than either is to α_i-2. Sequence differences are scattered throughout the length of the proteins, although there is a region around amino acid number 100 where the difference among them is greatest. The function of this part of the molecule is not yet known, although it is tempting to speculate that it may be involved in interactions with effectors or with proteins that might target the α subunits to specific parts of the plasma membrane.

Genes for the G Protein Are Not Clustered

The strong similarity of sequences of the G-protein α subunits suggests that they may have evolved by gene duplication from a primordial guanine-nucleotide-binding protein. However, the genes for α subunits are no longer located at one place in the genome but are scattered over many chromosomes. Neer et al. (1987) found that genes for two α_i subtypes, α_i-1 and α_i-2, were located on different human chromosomes (7 and 12, respectively). Lochrie and Simon (1988) reported that genes for other G-protein α subunits are also scattered on multiple chromosomes. There are two apparent loci for α_i-2 (chromosome 12 and chromosome 3); it is possible that one of these is a pseudogene.

Distribution of G Proteins in α Subunits in Cells and Tissues

The identification of several extremely similar α_i-subtypes raises the question: Do each of these subtypes have distinct functions? If they do, how does the cell prevent cross-talk? One way to prevent mixed signals would be to segregate the proteins into two different cells. This solution has been adopted by the retinal rods and cones, which each contain one of two very closely related α subunits of transducin, the G protein that couples rhodopsin and other visual pigments to cGMP phosphodiesterase (Lerea et al. 1986). The rods and the cones also have different forms of opsins and phosphodiesterases, so that in the retina, the entire signaling system seems to be cell-specific (for review, see Stryer and Bourne 1986). However, matters are more complicated outside the visual system. Figure 2 shows that many human tissues and cell lines express mRNA for more than one subtype of α_i subunit. The mRNA from several permanent and primary human cell lines was tested with cDNA probes specific for each of the α_i subtypes. It is apparent from the figure that the cells analyzed contained multiple subtypes of α_i protein. Both α_i-2 and α_i-3 are expressed in all of the cell types we have analyzed. The α_i-1 protein has a more limited distribution and is, in general, absent from cells of myelocytic lineage such as T cells, B cells, HSB cells, and others (Kim et al. 1988).

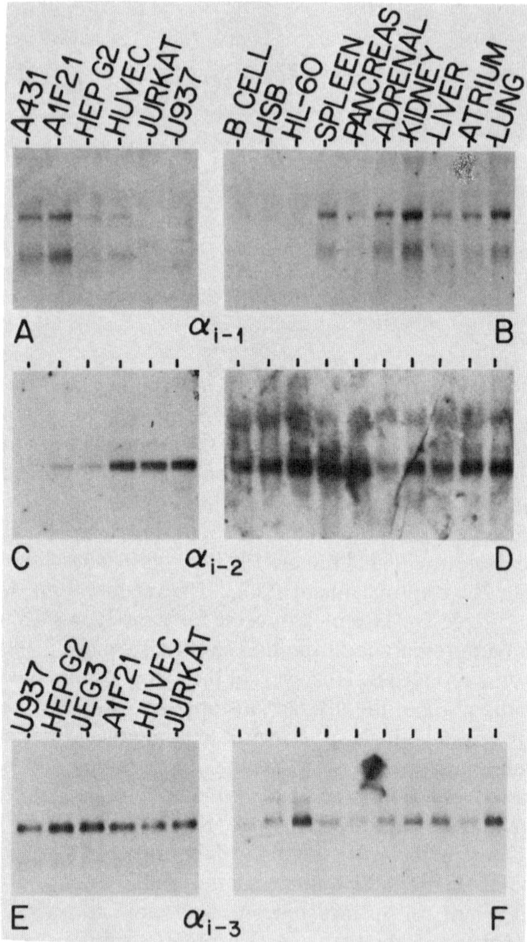

Figure 2. mRNA analysis from human fetal tissue and from human cell lines (Kim et al. 1988). Total cellular RNA was extracted from the human fetal tissues and cell lines. Blots were hybridized with probes for α_i-1, α_i-2, and α_i-3 as follows: (*A*,*B*) *Eco*RI-Bam fragment containing nucleotides 867–1344 of the published α_i-1 sequence and additional 3′-untranslated region corresponding to the larger mRNA. The bands are 3.9- and 2.2-kb. (*C*,*D*) Full-length *Eco*RI insert of human α_i-2. The predominant band is 2.7-kb. (*E*) 600-bp fragment derived from the 5′-flanking region of the α_i-3 gene as described in the text. (*F*) *Hin*dIII fragment of α_i-3 cDNA. The predominant band is 2.8-kb. The blots were washed at 65°C in 0.2 × SSC and 0.1% SDS. The cell lines are A431 (epithelial carcinoma); A1F21 (human dermal fibroblast); HepG-2 (hepatoma): HUVEC (human endothelial cells); Jurkat (T cell): U937 (monocyte); B cell (Epstein-Barr-virus-transformed peripheral blood B cells); HSB (T cell); HL-60 (monocyte).

Another protein that shows differential cell distribution is the 39-kD pertussis toxin substrate (called α_{39} or α_o), which is very abundant in brain (Neer et al. 1984;

Sternweis and Robishaw 1984), where it makes up about 0.5% of the total brain membrane protein. Immunocytochemical studies performed in collaboration with S. Snyder and his colleagues showed that α_o immunoreactivity was highly localized (Worley et al. 1986). Therefore, the average concentration underestimates the amount of α_o present in certain cells. It is not clear what function α_o serves in the nervous system; however, its distribution is similar to that of phorbol-ester-binding sites. This may provide a hint that α_o is involved in regulation of phosphoinositol metabolism. This subunit has also been proposed to regulate Ca^{++} channels (Hescheler et al. 1987).

The heterogeneity of G proteins in a single cell, revealed by analysis of mRNA, is corroborated by two-dimensional gel electrophoresis of pertussis toxin substrates in several clonal cell lines. Examples are shown in Figure 3. In each of the cell types, there are several ADP-ribosylated protein spots detected by this method. It is certainly possible that some of the heterogeneity in electrophoretic behavior may be attributed to posttranslational modification of the proteins. However, since we now know that there are multiple α subtypes, a simpler explanation is that at least some of the heterogeneity is due to the presence of different α subunits.

Although the subtype distribution of α_i is clearly different in different cells, is it really important to cell function which particular α_i subtypes are expressed, or can cells use them interchangeably? In the latter case, it might only matter what the total α_i subunit concentration is. The observation that α_i-1 is missing from certain cell types shows that this form of the α subunit is not obligatory either for inhibition of adenylyl cyclase or for pertussis-toxin-sensitive phosphoinositol turnover. The cells that lack this α_i subtype do have hormone-inhibited adenylyl cyclase activity and also have phospholipases that are apparently regulated by a pertussis toxin substrate (Bokoch and Gilman 1984; Brandt et al. 1985; Ohta et al. 1985).

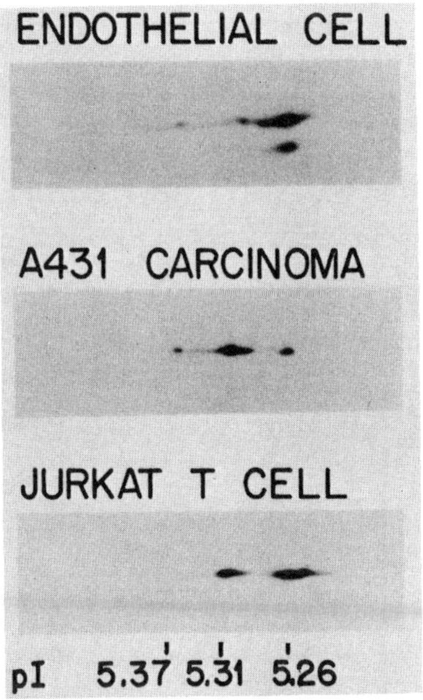

Figure 3. Membranes from human umbilical vein endothelial cells, A431 human carcinoma cells, or Jurkat, a transformed human T-cell line, were ADP-ribosylated by pertussis toxin as described by Huff et al. (1985). The membranes were solubilized and analyzed by isoelectric focusing followed by SDS-PAGE as described by O'Farrell et al. (1977). Isoelectric focusing was done using ampholytes of pH 5–7. Markers of known isoelectric points were used to calibrate the gels.

There is, as yet, not much evidence for specificity of action within α_i subtypes. In vitro reconstitution experiments reveal a considerable interchangeability (see Neer and Clapham 1988). Nevertheless, there are two lines of evidence that suggest that the particular kinds of α subunit expressed do matter to cell function. One is the extreme conservation of the α_i subtypes from species to species. α_i-1, for example, is identical in the human and in cows. There are, in general, only two to five amino acid differences in the α_i subtypes from species to species, so that 99% of the sequence is completely conserved (see Lochrie and Simon 1988). This conservation of sequence, in itself, argues for important differences in function.

Another kind of evidence for specificity of function is the demonstration that different α_i types are independently regulated by the cell. We have recently shown that mRNA for α_i-2 but not α_i-1, α_i-3, or α_s is increased threefold in human endothelial cells in response to interleukin-1 β (R.T. Lee et al., in prep.). However, we were not able to demonstrate any change in the ADP-ribosylatable α subunits in endothelial cells on treatment with interleukin-1 β. More sensitive assays, perhaps using type-specific antibodies, will allow us to determine whether the rise in mRNA is in fact reflected by a rise in protein. For the present, the point to be made is that mRNA levels for the different G proteins are independently regulated by different stimuli. This is a suggestive although not definitive argument for differential function.

Functions of α and $\beta\gamma$ Subunits

An essential feature of the membrane-bound heterotrimeric G proteins is that activation involves dissociation of the guanine-nucleotide-liganded α subunit from $\beta\gamma$.

$$\alpha\beta\gamma + G \rightleftarrows \alpha_G + \beta\gamma$$

(G is a guanine nucleoside triphosphate)

It should be pointed out that this dissociation has actually only been observed in solution with nonhydrolyzable analogs of GTP and not with GTP itself (Huff and Neer 1986). However, the failure to observe dissociation may be due to the rapid hydrolysis of GTP at the guanine-nucleotide-binding site so that in solution, the equilibrium is always shifted toward the GDP-liganded, associated form. Presumably, in the membrane where the G protein interacts with hormone receptors, the GTP-liganded state would be maintained longer.

The dissociation of the heterotrimer into α and $\beta\gamma$ subunits raises the possibility that both α and $\beta\gamma$ may be signaling molecules generated by a hormonal stimulus. There is ample evidence that the α subunit in its guanine-nucleotide-triphosphate-liganded form can be a direct activator of effectors (e.g., adenylyl cyclase [Northup et al. 1983], the muscarinic K^+ channel [Codina et al. 1987], and cGMP phosphodiesterase in the retina [Fung et al. 1981]).

The observation that the guanine-nucleotide-triphosphate-liganded α subunit can be an activator does not rule out the possibility that $\beta\gamma$ may also be a regulatory protein. In fact, Jelsema and Axelrod (1987) have shown that the $\beta\gamma$ subunit can activate phospholipase A_2 in rod outer segment membranes. Logothetis et al. (1987) have recently shown that $\beta\gamma$ subunits can activate the muscarinic K^+ channel in patch-clamp experiments using embryonic chick and adult rat heart cardiac atrial membranes. Since Codina et al. (1987) showed in guinea pig atrial patches that a GTPγS-liganded α subunit from human erythrocytes could activate the guinea pig atrial and rat pituitary cell membranes, both α and $\beta\gamma$ seem to be positive effectors in this system. Figure 4 shows that two kinds of α subunits, α_{40} from human erythrocytes (kindly provided by L. Birnbaumer) and α_{39} (α_o, the major G protein from bovine brain), both activate the K^+ channel at pmole concentrations in chick and rat atrial membranes. The $\beta\gamma$ subunit activates a channel with the identical electrical properties, although it does so at a higher concentration. Activation begins at approximately 200 pmoles $\beta\gamma$ subunit. We do not understand the reason for the difference in concentration, although the α subunit is a hydrophilic protein that can be applied in the absence of detergent, whereas $\beta\gamma$ is hydrophobic and, at the low concentrations of detergent required in order not to damage the patch, a large fraction (~80%) of $\beta\gamma$ subunits is aggregated (Logothetis et al. 1988).

Not only do the α and the $\beta\gamma$ subunits activate the same type of channel, but they seem to activate the same population of channels. Figure 5 shows that the effects of α and $\beta\gamma$ are not additive. Either one is sufficient maximally to activate all of the channels accessible in a given patch (Logothetis et al. 1988). It is not clear from these experiments that either α or $\beta\gamma$ is acting directly on the channel. That conclusion is not possible with reconstitution experiments done in intact membranes. One or both of the subunits may have intermediate steps.

Organizational Mechanisms

In vitro reconstitution experiments suggest that there is a considerable latitude in the ability of G proteins to interact with receptors and effectors. Receptors that probably do not interact with a particular G protein in vivo may do so in a reconstituted system (for review, see Neer and Clapham 1988). Some (but not all) of the reported cross-talk may only be apparent because it is likely that mixtures of subunits were sometimes used, blurring potential distinctions. However, even though the subunits may act interchangeably in vitro, in the cell, a variety of mechanisms may operate to enhance specificity.

One extremely important factor is the stoichiometry of the various α subunits. For example, in cardiac atria, the β-adrenergic receptor interacts with a heterotrimer of $\alpha_s.\beta\gamma$. Activation of the receptor by an agonist should cause dissociation of α from the $\beta\gamma$ subunit, and the $\beta\gamma$ subunit might then be expected to activate the K^+ channel. In fact, activation of the K^+ channel by β-adrenergic receptors is extremely weak. One explanation is that the amount of $\alpha_s.\beta\gamma$ in cardiac cells might

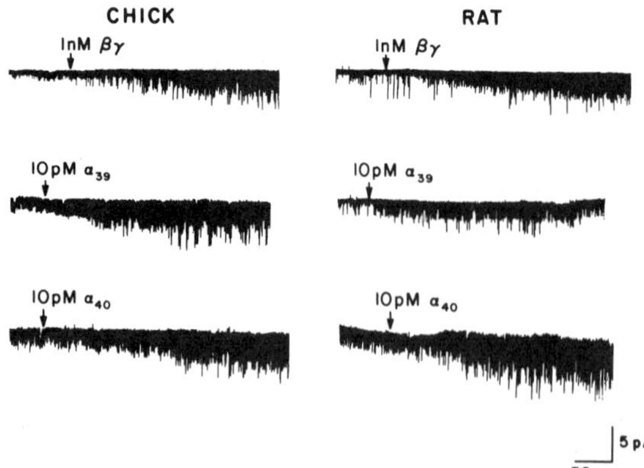

Figure 4. $\beta\gamma$, α_{39}-GTPγS, and α_{40}-GTPγS activate $i_{K.ACh}$ in both embryonic chick and neonatal rat atria. Each trace shows background single-channel activity (no GTP or ACh) before addition of the specified subunit (arrow). Holding potential is −80 mV. The conductance (35–40 pS) and kinetics (mean open time 1.0 ms at −80 mV) of rat and chick single-channel $i_{K.ACh}$ currents were indistinguishable.

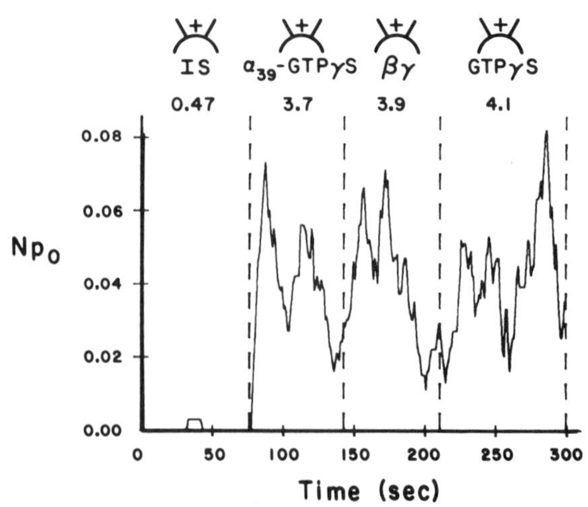

Figure 5. α_{39}-GTPγS (10 pM)-activated channel activity was not increased further by $\beta\gamma$ (10 nM) or GTPγS (100 μM). Likewise, α_{39}-GTPγS did not increase $\beta\gamma$-stimulated activity (not shown; $\overline{Np_o} = 1.7$ vs. 1.5).

be considerably lower than the amount of $\alpha_i.\beta\gamma$, so the amount of $\beta\gamma$ released from that source would be small in comparison to the amount of $\beta\gamma$ that could be released from other α components. Thus, although $\beta\gamma$s released from many receptors could, in principle, activate the K^+ channel, only receptors that could release larger amounts of $\beta\gamma$ might in fact do so.

Another factor is the relative affinity of different α subunits for $\beta\gamma$. All α subunits do not have the same affinity for $\beta\gamma$, and some may dissociate more readily than others (Huff et al. 1985). Therefore, one pool of $\alpha\beta\gamma$ heterotrimers may be a less effective source of $\beta\gamma$ than another. Finally, receptor number can affect the amount of G protein activated. If only a small number of receptors are present, only a small amount of G proteins may be activated. These G proteins would stimulate effectors with a high affinity for the subunits, but not effectors with a low affinity.

Considerations of stoichiometry and relative subunit affinity are particularly important if the components of the response system move freely in the plane of the lipid bilayer. There is evidence from several systems suggesting that free diffusion of components in the plasma membrane may occur (see, e.g., Tolkovsky et al. 1982; for review, see Neer and Clapham 1988). All of these experiments measure the ability of receptors and/or effector molecules to interact. None of them explicitly show that α and $\beta\gamma$ dissociate and float freely in the membrane separate from each other.

Another way to ensure the specificity of G-protein action is to constrain the location of the receptor G protein and/or effector. Within a limited area, subunits could dissociate, but they would not be free to diffuse throughout the membrane. There are several lines of evidence suggesting that this sort of compartmentation may in fact exist. Kinetic evidence suggests that there may be discrete pools of cAMP in cells that must originate from discrete populations of adenylyl cyclase (Dufau et al. 1978; Buxton and Brunton 1983). The components of the hormone-responsive system are differentially distributed in apical and basolateral membranes of polarized epithelia (Schlatz et al. 1975; Dominguez et al. 1987). There is evidence that the cytoskeleton influences the ability of hormones to stimulate adenylyl cyclase and of detergents to solubilize G protein (Insel and Kennedy 1978; Rasenick et al. 1981). Cytoskeletal components may associate with G-protein subunits (Carlson et al. 1986). Finally, there may be additional proteins that interact with the G-protein receptors and/or effectors and enhance the specificity or rates of the reaction. Recently, a cytosolic protein (Trahey and McCormick 1987) has been identified that acts as a nucleotide exchange factor for the *ras* protein. In this respect, the protein performs a function somewhat similar to that of a hormone receptor in the classic signaling system. Nevertheless, there may be cellular proteins that modulate G-protein function in addition to hormone receptors.

The various mechanisms we have outlined are not exclusive of each other, and each may apply to different parts of the same cell or have greater or lesser importance to different signaling systems. The challenge for the future will be to define how the G proteins actually function in situ and how the general principles developed through analysis of purified proteins are applied in living cells.

ACKNOWLEDGMENTS

The work from the authors' laboratories reviewed in this paper was supported by National Institutes of Health grants to E.J.N. (GM-36259), J.G.S. (AI-18436 and AI-19148), and D.E.C. (HL-34873), and by a Leukemia Society postdoctoral fellowship (D.B.B.), a Pfizer Pharmaceutical postdoctoral fellowship (K.D.B.), a Physician-Scientist award (R.L.), and a National Institutes of Health postdoctoral fellowship (C.T.).

REFERENCES

Bokoch, G.M. and A.G. Gilman. 1984. Inhibition of receptor-mediated release of arachidonic acid by pertussis toxin. *Cell* **39:** 301.

Brandt, S.J., R.W. Dougherty, E.G. Lapetina, and J.E. Niedel. 1985. Pertussis toxin inhibits chemotactic peptide-stimulated generation of inositol phosphates and lysosomal enzyme secretion in human leukemic (HL-60) cells. *Proc. Natl. Acad. Sci.* **82:** 3277.

Bray, P., A. Carter, V. Guo, C. Puckett, J. Kamholz, A. Spiegel, and M. Nirenberg. 1987. Human cDNA clones for a α subunit of G_i signal transduction protein. *Proc. Natl. Acad. Sci.* **84:** 5115.

Buxton, I.L., and L.L. Brunton. 1983. Compartments of cyclic AMP and protein kinase in mammalian cardiomyocytes. *J. Biol. Chem.* **258:** 10233.

Carlson, K.E., M.J. Woolkalis, M.G. Newhouse, and D.R. Manning. 1986. Fractionation of the β subunit common to guanine nucleotide-binding regulatory proteins with the cytoskeleton. *Mol. Pharmacol.* **30:** 463.

Codina, Y., A. Yatani, D. Grenet, A.M. Brown, and L. Birnbaumer. 1987. The α subunit of the GTP-binding protein G_k opens atrial potassium channels. *Science* **238:** 1288.

de Vos, A.M., L. Tong, M.V. Milburn, P.M. Matias, J. Jancarik, S. Noguchi, S. Nishimura, K. Miura, E. Ohtsuka, and S.-H. Kim. 1988. Three-dimensional structure of an oncogene protein: Catalytic domain of human c-H-*ras* p21. *Science* **239:** 888.

Didsbury, J.R., Y.-S. Ho, and R. Snyderman. 1987. Molecular cloning of a new human G protein. *FEBS Lett.* **211:** 160.

Dominguez, P., G. Velasco, F. Barros, and P.S. Lazo. 1987. Intestinal brush border membranes contain regulatory subunits of adenylyl cyclase. *Proc. Natl. Acad. Sci.* **84:** 6965.

Dufau, M.L., J.A. Horner, K. Hayashi, T. Tsuruhara, P.M. Conn, and K.J. Catt. 1978. Actions of choleragen and gonadotropin in isolated Leydig cells. *J. Biol. Chem.* **253:** 3721.

Fung, B.K.K., J.B. Hurley, and L. Stryer. 1981. Flow of information in the light-triggered cyclic nucleotide cascade of vision. *Proc. Natl. Acad. Sci.* **78:** 152.

Gilman, A.G. 1987. G-proteins: Transducers of receptor-generated signals. *Annu. Rev. Biochem.* **56:** 615.

Graziano, M.P., P.J. Casey, and A.G. Gilman. 1987. Expression of cDNAs for G-proteins in *E. coli*. *J. Biol. Chem.* **262:** 11375.

Halliday, K.R. 1984. Regional homology in GTP-binding

proto-oncogene products and elongation factors. *J. Cyclic Nucleotide Protein Phosphorylation Res.* **9:** 435.

Hescheler, J., W. Rosenthal, W. Trautwein, and G. Schultz. 1987. The GTP-binding protein G_o regulates neuronal calcium channels. *Nature* **325:** 445.

Huff, R.M., J.M. Axton, and E.J. Neer. 1985. Physical and immunological characterization of a guanine nucleotide-binding protein purified from bovine cerebral cortex. *J. Biol. Chem.* **260:** 10864.

Huff, R.M. and E.J. Neer. 1986. Subunit interactions of native and ADP-ribosylated α_{39} and α_{41}, two guanine nucleotide-binding proteins from bovine cerebral cortex. *J. Biol. Chem.* **261:** 1105.

Insel, P.A. and M.S. Kennedy. 1978. Colchicine potentiates β-adrenoreceptor-stimulated cyclic AMP in lymphoma cells by an action distal to the receptor. *Nature* **273:** 471.

Jelsema, C.L. and J. Axelrod. 1987. Stimulation of phospholipase A_2 activity in bovine rod outer segments by the $\beta\gamma$ subunits of transducin and its inhibition by the α subunit. *Proc. Natl. Acad. Sci.* **84:** 3623.

Jones, D.T. and R.R. Reed. 1987. Molecular cloning of five GTP-binding protein cDNA species from rat olfactory neuroepithelium. *J. Biol. Chem.* **262:** 14241.

Jurnak, F. 1985. Structure of the GDP domain of EF-Tu and location of the amino acids homologous to *ras* oncogene proteins. *Science* **230:** 32.

Kim, S.Y., S.-L. Ang, D.B. Bloch, K.D. Bloch, S.-L. Ang, Y. Kawahara, C. Tolman, R. Lee, J.G. Seidman, and E.J. Neer. 1988. Identification of cDNA encoding an additional α subunit of a human GTP-binding protein: Expression of three α_i subtypes in human tissues and cell lines. *Proc. Natl. Acad. Sci.* **85:** 4153.

Kurachi, Y., T. Nakajima, and T. Sugimoto. 1986. On the mechanism of activation of muscarinic K^+ channels by adenosine in isolated atrial cells: Involvement of GTP-binding proteins. *Pfluegers Arch. Eur. J. Physiol.* **407:** 264.

la Cour, T.F.M., J. Nyborg, S. Thirup, and B.F.C. Clark. 1985. Structural details of the binding of guanosine diphosphate to elongation factor Tu from *E. coli* as studied by X-ray crystallography. *EMBO J.* **4:** 2385.

Lerea, C.L. D.E. Somers, J.B. Hurley, I.B. Klock, and A.M. Bunt-Milam. 1986. Identification of specific transducin α subunits in retinal rod and cone photoreceptors. *Science* **234:** 77.

Lochrie, M.A. and M.I. Simon. 1988. G-protein multiplicity in eukaryotic signal transduction systems. *Biochemistry* **27:** 4957.

Logothetis, D.E., D. Kim, J. Northup, E.J. Neer, and D.E. Clapham. 1988. Specificity of action of guanine nucleotide-binding regulatory protein subunits on the cardiac muscarinic K^+ channels. *Proc. Natl. Acad. Sci.* **85:** 5814.

Logothetis, D.E., Y. Kurachi, J. Galper, E.J. Neer, and D.E. Clapham. 1987. The $\beta\gamma$ subunits of GTP-binding proteins activate the muscarinic K^+ channel in heart. *Nature* **325:** 321.

Masters, S.B., R.M. Stroud, and H.R. Bourne. 1986. Family of G-protein α chains: Amphipathic analysis and predicted structure of functional domains. *Protein Eng.* **1:** 47.

Neer, E.J. and D.E. Clapham. 1988. Roles of G-proteins in transmembrane signalling. *Nature* **333:** 129.

Neer, E.J., J.M. Lok, and L.G. Wolf. 1984. Purification and properties of the inhibitory guanine nucleotide regulatory unit of brain adenylate cyclase. *J. Biol. Chem.* **259:** 14222.

Neer, E.J., T. Michel, R. Eddy, T. Shows, and J.G. Seidman. 1987. Genes for two homologous G-protein α subunits map to different human chromosomes. *Hum. Genet.* **77:** 259.

Northup, J.K., M.D. Smigel, P.C. Sternweis, and A.G. Gilman. 1983. The subunits of the stimulatory regulatory component of adenylate cyclase. *J. Biol. Chem.* **258:** 11369.

O'Farrell, P.Z., H.M. Goodman, and P.H. O'Farrell. 1977. High-resolution two-dimensional electrophoresis of basic as well as acidic proteins. *Cell* **12:** 1133.

Ohta, H., F. Okajima, and M. Ui. 1985. Inhibition by islet-activating protein of a chemoactic peptide-induced early breakdown of inositol phospholipids and Ca^{2+} mobilization in guinea pig neutrophils. *J. Biol. Chem.* **260:** 15771.

Rasenick, M.M., P.J. Stein, and M.W. Bitensky. 1981. The regulatory subunit of adenylate cyclase interacts with cytoskeletal components. *Nature* **294:** 560.

Schlatz, L.J., I.L. Schwartz, E. Kinne-Saffran, and R. Kinne. 1975. Distribution of parathyroid hormone-stimulated adenylate cyclase in plasma membranes of cells of the kidney cortex. *J. Membr. Biol.* **24:** 131.

Sternweis, P.C. and J. Robishaw. 1984. Isolation of two proteins with high affinity for guanine nucleotides from membranes of bovine brain. *J. Biol. Chem.* **259:** 13806.

Stryer, L. and H.R. Bourne. 1986. G-proteins: A family of signal transducers. *Annu. Rev. Cell Biol.* **2:** 391.

Tolkovsky, A.M., S. Braun, and A. Levitski. 1982. Kinetics of interaction between the β-receptor, the GTP regulatory protein, and the catalytic unit of turkey erythrocyte adenylate cyclase. *Proc. Natl. Acad. Sci.* **79:** 213.

Trahey, M. and F. McCormick. 1987. A cytoplasmic protein stimulates normal N-*ras* p21 GTPase, but does not affect oncogene mutants. *Science* **238:** 542.

Worley, P.F., J.M. Baraban, C. Van Dop, E.J. Neer, and S. Snyder. 1986. G_o, a guanine nucleotide-binding protein: Immunohistochemical localization in rat brain resembles distribution of second messenger systems. *Proc. Natl. Acad. Sci.* **83:** 4561.

Receptor- and G-protein-mediated Modulations of Voltage-dependent Calcium Channels

W. Rosenthal,* J. Hescheler,† W. Trautwein,† and G. Schultz*

*Institut für Pharmakologie, Freie Universität Berlin, D-1000 Berlin 33;
†Physiologisches Institut, Universität des Saarlandes, D-6650 Homburg/Saar, Federal Republic of Germany

In eukaryotic cells, a 10,000-fold concentration gradient for Ca^{++} across the plasma membrane is established by the low Ca^{++} permeability of the plasma membrane at negative membrane potentials and by Ca^{++}-ATPases reducing the cytoplasmic Ca^{++} concentration to 10^{-7} M or less. Depolarization of the plasma membrane leads to an increase in Ca^{++} permeability by activation of voltage-dependent Ca^{++} channels and, consequently, to an elevated cytoplasmic Ca^{++} concentration. Besides regulation by the membrane potential, the activity of voltage-dependent Ca^{++} channels is modulated by extracellular signals, i.e., hormones and neurotransmitters, some of which may also stimulate Ca^{++} influx through voltage-independent cation channels.

The majority of extracellular signals, including hormones and neurotransmitters, bind to and activate membranous receptors that are coupled to transducer components located at the inner surface of the plasma membrane. These transducer molecules are represented by guanine-nucleotide-binding proteins (G proteins) composed of α subunits that are specific for the individual G protein and of functionally interchangeable and structurally very similar or identical β and γ subunits. Once activated by an agonist-occupied receptor, G proteins modify the activity of effectors generating intracellular signals. Enzymatic effectors include the hormone-sensitive adenylate cyclase, the retinal light-sensitive cGMP phosphodiesterase, and possibly the hormone-stimulated phospholipases C and A_2 found in virtually all cell types, and the NADPH oxidase, the superoxide-generating enzyme of neutrophils and macrophages. For the ubiquitously occurring adenylate cyclase and the retinal cGMP phosphodiesterase, stimulation of enzyme has been shown to be caused by direct interaction of G-protein α subunits and the respective enzyme (for recent reviews, see Iyengar and Birnbaumer 1987; Casey and Gilman 1988; Neer and Clapham 1988).

Hormonal Modulations of Voltage-dependent Ca^{++} Channel Activity

G proteins are also involved in hormonal modulations of transmembranous Ca^{++} fluxes through voltage-dependent Ca^{++} channels. One mechanism by which extracellular signals modulate Ca^{++} channel activity is phosphorylation of the channel protein or of unknown regulatory components. The protein kinases involved are stimulated by cytosolic signal molecules generated by G-protein-controlled effectors. Thereby, G proteins exert a distant control of Ca^{++} channel activity. β-Adrenergic agonists, which activate adenylate cyclase via G_s, a G protein sensitive to the exotoxin of *Vibrio cholerae* (cholera toxin), stimulate cardiac Ca^{++} currents by cAMP-dependent phosphorylation (Reuter 1983; Hofmann et al. 1987; Takahashi et al. 1987). Cyclic AMP-dependent phosphorylation of a Ca^{++} channel-forming protein and subsequent modulation of Ca^{++} channel activity have so far only been demonstrated for dihydropyridine-binding sites purified from skeletal muscle. After reconstitution into phospholipid bilayers, these sites form Ca^{++} channels that are stimulated by the addition of cAMP-dependent protein kinase and Mg-ATP (Flockerzi et al. 1986). Ca^{++} channels of other cells, e.g., hippocampal neurons (Gray and Johnston 1987), are also stimulated by cAMP-dependent phosphorylation, and in isolated membrane patches of GH_3 cells, Ca^{++} channel activity is only observed under conditions allowing cAMP-dependent phosphorylation (Armstrong and Eckert 1987). Other protein kinases, namely cGMP-dependent protein kinase in snail neurons (Paupardin-Tritsch et al. 1986) and protein kinase C in various cell types (Kaczmarek 1987), have also been reported to influence Ca^{++} channel activity. A direct phosphorylation of Ca^{++} channel-forming proteins by these protein kinases, resulting in an altered functional state of Ca^{++} channels, however, has not yet been demonstrated.

Within the last few years, experimental evidence has been provided for another mechanism by which extracellular signals acting via G-protein-coupled receptors influence the activity of voltage-dependent Ca^{++} channels. This mechanism may be membrane-confined, since there is no evidence for the participation of protein kinases regulated by cytosolic signal molecules such as cAMP. The only so far identified components of the signal transduction chain involved in the modulation of Ca^{++} channels are membranous receptors and G proteins.

G-protein-mediated Stimulation of Voltage-dependent Ca^{++} Channels

Receptor- and G-protein-mediated stimulations of voltage-dependent Ca^{++} channels have been described to occur in secretory cells but may not be restricted to

Table 1. Close Stimulatory Control of Voltage-dependent Ca^{++} Channels via G Proteins

Cell type	Activated receptors	Effects of guanine nucleotides	Effects of toxins	Reference
GH_3 cells (pituitary; rat)	LHRH, angiotensin II	n.d.	PT	Rosenthal et al. (1988)
Y1 cells (adrenal cortex; mouse)	angiotensin II	n.d.	PT	Hescheler et al. (1988a)
Glomerulosa cells (bovine)	angiotensin II	n.d.	n.d.	Cohen et al. (1988)
Cardiac myocytes (guinea pig)	β-adrenergic	GTPγS	CT[a]	Yatani et al. (1987)

(n.d.) Not determined; (GTPγS) guanosine 5′-*O*-(3-thio)triphosphate; (PT) pertussis toxin; (CT) cholera toxin.

[a]The hormonal stimulation was mimicked by the activated cholera-toxin-sensitive G protein, G_s. A direct (cAMP-independent) effect of cholera toxin has not been demonstrated.

these cellular systems (Table 1). On the basis of biochemical studies, evidence has been provided suggesting that angiotensin-II-induced secretion of aldosterone involves activation of voltage-dependent Ca^{++} channels (see references in Hescheler et al. 1988a). This hypothesis was recently verified by demonstrating that angiotensin II stimulates voltage-dependent Ca^{++} currents of the adrenocortical cell line, Y1 (Hescheler et al. 1988a), and in bovine adrenal glomerulosa cells (Cohen et al. 1988). The stimulation of Ca^{++} currents, as observed in the whole-cell-clamp configuration (Hamill et al. 1981), was seen within a few seconds and was rapidly reversed by removal of the hormone from the bath. Preliminary data confirm this effect of angiotensin II in isolated porcine (Hescheler et al. 1988a) and in pituitary GH_3 cells (Rosenthal et al. 1988).

Similar to angiotensin II, another secretion-stimulating hormone, luteinizing hormone-releasing hormone (gonadotropin-releasing hormone, LHRH), which has been under suspicion to act, at least in part, by activation of voltage-dependent Ca^{++} channels, has been shown to stimulate voltage-dependent Ca^{++} currents of GH_3 cells in a reversible fashion (Rosenthal et al. 1988).

Whereas Ca^{++} current stimulation in cardiac and skeletal muscle by cAMP-dependent phosphorylation is an established principle of Ca^{++} channel modulation, the stimulatory modulation of Ca^{++} currents in endocrine cells appears to be independent of cAMP for several reasons. Intracellularly applied cAMP did not stimulate Ca^{++} currents in Y1 or GH_3 cells as it did, for example, in cardiac myocytes (Hescheler et al. 1988a; Rosenthal et al. 1988); in addition, angiotensin II inhibited adenylate cyclase in all cell types tested so far (see references in Hescheler et al. 1988a) including GH_3 cells (Rosenthal et al. 1988). Finally, LHRH did not stimulate adenylate cyclase in membranes of GH_3 cells (Rosenthal et al. 1988).

The involvement of G proteins in the stimulatory hormonal modulation is evident from the effects of the exotoxin of *Bordetella pertussis* (pertussis toxin), which prevents functional coupling of activated receptors and a number of G proteins (e.g., G_i-type G proteins and G_o) by ADP-ribosylation of G-protein α subunits (Ui et al. 1984). Pretreatment of Y1 cells with pertussis toxin completely blocked hormonal stimulation of Ca^{++} currents (Hescheler et al. 1988a). This finding is consistent with a previously reported observation that stimulation of Ca^{++} influx by angiotensin II is abolished by the toxin (Kojima et al. 1986). Stimulation of Ca^{++} currents by LHRH in GH_3 cells (like somatostatin-induced inhibition of Ca^{++} currents in these cells; see below) was also abolished in cells pretreated with the toxin (Rosenthal et al. 1988).

The results indicate the involvement of a pertussis-toxin-sensitive G protein in the hormonal stimulations of Ca^{++} channel activity. Since angiotensin II and LHRH stimulate phosphatidylinositol 4,5-bisphosphate hydrolysis—and consequently protein kinase C—via a pertussis-toxin-*in*sensitive G protein in adrenocortical (Enyedi et al. 1986; Kojima et al. 1986) and pituitary (Naor et al. 1986) cells, it is unlikely that protein kinase C is involved in the Ca^{++} current stimulations by angiotensin II and LHRH. In preliminary experiments, a protein-kinase-C-activating phorbol ester did not affect Ca^{++} currents in Y1 cells (J. Hescheler et al., unpubl.).

The sensitivity of the stimulatory Ca^{++} current modulation toward pertussis toxin is another strong argument against the participation of a cAMP-dependent step, since stimulation of adenylate cyclase is mediated by G_s, a pertussis-toxin-*in*sensitive G protein.

Some conclusions regarding the identity of the G protein involved in the pertussis-toxin-sensitive stimulatory modulation of Ca^{++} currents can be drawn be relating hormonal effects on Ca^{++}currents to the pattern of G-protein α subunits in membranes of the cell types tested. Probing membranes of the adrenocortical cell line, Y1, with antisera raised against peptides corresponding to confined regions of G-protein α subunits revealed that Y1 cells contain G proteins of the G_i-type but are devoid of G_o (Hescheler et al. 1988a). Thus, the stimulation of Ca^{++} currents in this cell type is likely to be mediated by a G_i-type G protein. By analogy, the stimulatory effects of LHRH and angiotensin II in pituitary cells (GH_3) may occur via a G_i-type G protein.

Data obtained with isolated inside-out patches of cardiac myocytes and cardiac sarcolemmal vesicles incorporated into artificial phospholipid bilayers suggest that guanosine-5′-*O*-(3-thiotriphosphate) (GTPγS) stimulates Ca^{++} channel activity independently of cAMP-dependent protein kinase (Yatani et al. 1987). The effect of GTPγS was mimicked by activated G_s, the cholera-toxin-sensitive G protein mediating hormonal stimulation of adenylate cyclase, or by its activated α subunit, if either preparation was applied to the cytoplasmic face of the membrane patch or the appropriate side of the artificial phospholipid bilayer. The physiological significance of this apparently direct effect of G_s is not clear, since it has only been observed in cell-free systems at high concentrations (20–100 pM) of activated G_s or its α subunit. Data obtained with intact cells or reconstituted systems favor a cAMP-dependent mechanism as described above.

On the basis of electrophysiological and pharmacological properties, neuronal and possibly other voltage-dependent Ca^{++} channels can be differentiated into L-, N-, and T-type channels (Nowycky et al. 1985). At present, it is not clear which type of Ca^{++} channel is stimulated by LHRH or angiotensin II. In Y1 and GH_3 cells, hormone-sensitive currents inactivated slowly, exhibited a high apparent threshold potential, and were partially sensitive to Ca^{++} channel agonists and antagonists of the dihydropyridine type, suggesting that fluxes through L-type Ca^{++} channels contribute to these currents (Hescheler et al. 1988a; Rosenthal et al. 1988). Slowly inactivating Ca^{++} currents of GH_3 cells were also inhibited by the snake venom, ω-conotoxin (McCleskey et al. 1987), which blocks L- and N-type channels in neuronal cells (Suzuki and Yoshioka 1987). On the other hand, the angiotensin-II-sensitive Ca^{++} current in bovine glomerulosa cells apparently represents a current through T-type channels (Cohen et al. 1988). Thus, all three electrophysiological types of Ca^{++} channels are possibly stimulated via G-protein-coupled receptors.

G-protein-mediated Inhibition of Voltage-dependent Ca^{++} Channels

Noradrenaline, γ-aminobutyric acid (GABA) (via $GABA_B$ receptors), dopamine (via D_2 receptors), opioids (via κ and δ receptors), acetylcholine (via M_1 receptors), adenosine (via A_1 receptors), and somatostatin cause inhibition of Ca^{++} currents in various neuronal cells, including those of dorsal root ganglia of chick (Dunlap and Fischbach 1981; Deisz and Lux 1985; Forscher et al. 1986; Holz et al. 1986; Marchetti et al. 1986), rat (Dolphin et al. 1986; Scott and Dolphin 1986; Ewald et al. 1988), and mouse (Gross and MacDonald 1987), sensory and sympathetic ganglia of chick (Marchetti et al. 1986), rat sympathetic ganglia (Wanke et al. 1987), and neuroblastoma × glioma hybrid cells (Tsunoo et al. 1986; Hescheler et al. 1987) (Table 2). In these cells, the hormone-induced inhibition of Ca^{++} currents may be the molecular basis underlying inhibition of neurosecretion via presynaptic receptors. In nonneuronal cells, i.e., mouse (AtT-20; Lewis et al. 1986) and rat (GH_3; Rosenthal et al. 1988) pituitary cell lines, the G-protein-mediated inhibition of Ca^{++} currents by somatostatin and possibly dopamine may represent a molecular mechanism for the inhibition of pituitary secretion.

Similar to the stimulation of Ca^{++} currents (see above), the onset of the inhibitory modulation of Ca^{++} currents is rapid (within seconds) and easily reversed by removal of the receptor agonist (Holz et al. 1986; Lewis et al. 1986; Scott and Dolphin 1986; Tsunoo et al. 1986; Gross and MacDonald 1987; Hescheler et al. 1987; Wanke et al. 1987; Rosenthal et al. 1988).

The involvement of G proteins in the inhibitory Ca^{++} current modulation has been demonstrated by the effect of intracellularly applied analogs of GTP or GDP. Intracellular infusion of the GTP analog, GTPγS, at high concentrations (100–500 μM) drastically reduced Ca^{++} currents in neuronal (Holz et al. 1986; Scott and Dolphin 1986; Wanke et al. 1987; Hescheler et al. 1987, 1988b) and pituitary cells (Lewis et al. 1986). GTPγS applied intracellularly at a low concentration (1 μM) had no effect on Ca^{++} currents of neuroblastoma × glioma hybrid cells in the absence of a δ-receptor-activating synthetic peptide but caused a

Table 2. Close Inhibitory Control of Voltage-dependent Ca^{++} Channels via G Proteins

Cell type	Activated receptors	Effects of: guanine nucleotides	Effects of: toxins	Reference
DRG cells (chick)	α-adr., $GABA_B$, 5-HT	n.d.	n.d.	Dunlap and Fischbach (1981)
	$GABA_B$	n.d.	n.d.	Deisz and Lux (1985)
	α-adr., dopamin.	n.d.	n.d.	Marchetti et al. (1986)
	α-adr., $GABA_B$	GTPγS, GDPβS	PT	Holz et al. (1986)
	α-adr.	n.d.	n.d.	Forscher et al. (1986)
DRG cells (rat)	$GABA_B$	GTPγS, GDPβS	n.d.	Scott and Dolphin (1986)
	A_1-adenosine	n.d.	n.d.	Dolphin et al. (1986)
	neuropeptide Y	n.d.	PT[a]	Ewald et al. (1988)
DRG cells (mouse)	κ-opioid	n.d.	n.d.	Gross and MacDonald (1987)
SG cells (chick)	α-adr., dopamin.	n.d.	n.d.	Marchetti et al. (1986)
SG cells (rat)	M_1-muscarinic	GTPγS, GDPβS	PT	Wanke et al. (1987)
N × G hybrid cells (mouse/rat)	δ-opioid, SSt	n.d.	n.d.	Tsunoo et al. (1986)
	δ-opioid	GTPγS, GDPβS	PT[a]	Hescheler et al. (1987)
AtT20 cells (mouse pituitary)	SSt	GTPγS	PT	Lewis et al. (1986)
GH_3 cells (rat pituitary)	SSt	n.d.	PT	Rosenthal et al. (1988).

(n.d.) Not determined; (DRG cells) dorsal root ganglion cells; (SG cells) sympathetic ganglion cells; (N × G hybrid cells) neuroblastoma × glioma hybrid cells; (α-adr) α-adrenergic; (5-HT) 5-hydroxytryptamine; (dopamin.) dopaminergic; (SSt) somatostatin; (GTPγS) guanosine 5′-O-(3-thio)triphosphate; (GDPβS) guanosine 5′-O-(2-thio)diphosphate; (PT) pertussis toxin; (CT) cholera toxin.

[a]Following pretreatment of cells with pertussis toxin, the hormonal effect was efficiently restored by intracellular infusion of the α subunit of G_o.

marked *irreversible* inhibition of currents if the agonist was present in the extracellular medium (Hescheler et al. 1988b). On the other hand, intracellular application of the GDP analog, guanosine-5′-*O*-(2-thiodiphosphate) (GDPβS), prevented the inhibition of Ca^{++} currents by receptor agonists (Holz et al. 1986; Scott and Dolphin 1986; Hescheler et al. 1987, 1988b; Wanke et al. 1987). Thus, the effects of guanine nucleotide analogs on Ca^{++} currents resemble their respective effects on the hormone-sensitive adenylate cyclase.

The involvement of G proteins in the inhibitory hormonal modulation is also obvious from the effects of pertussis toxin. In neuroblastoma × glioma hybrid cells (Hescheler et al. 1987), dorsal root ganglia (Holz et al. 1986), and AtT-20 cells (Lewis et al. 1986), the toxin prevented hormonal inhibition of Ca^{++} currents. In analogy with data obtained on inhibition of adenylate cyclase, GTPγS-induced inhibition of Ca^{++} currents in neuroblastoma × glioma hybrid cells was not affected by the toxin (Hescheler et al. 1988b).

Most of the Ca^{++} current-reducing agonists inhibit adenylate cyclase via identical receptors coupled to a G_i-type G protein (e.g., adrenaline via α_2 receptors, GABA via $GABA_B$ receptors, dopamine via D_2 receptors, opioids via δ and κ receptors, and adenosine via A_1 receptors) (Jakobs et al. 1981). Inhibition of adenylate cyclase or a decrease in cytosolic cAMP is, however, not causal for the observed inhibition of Ca^{++} currents, since the tested inhibitory agonists remained active if cAMP was applied into the interior of the cell via the patch pipette (Holz et al. 1986; Lewis et al. 1986). Intracellular infusion of cAMP in the absence of a hormonal agonist (Hescheler et al. 1987; Wanke et al. 1987; Rosenthal et al. 1988) or extracellular application of the adenylate-cyclase-stimulating diterpene, forskolin (Hescheler et al. 1988a), did not modify neuronal, adrenocortical, or pituitary Ca^{++} currents.

Similarly, participation of protein kinase C in hormone-induced inhibition of Ca^{++} currents is unlikely. Although there are reports on the inhibition of neuronal Ca^{++} currents by protein-kinase-C-activating phorbol esters (see, e.g., Rane and Dunlap 1986), most of the Ca^{++} current-inhibiting receptor agonists do not stimulate phosphatidylinositol 4,5-bisphosphate hydrolysis (Berridge 1987) and therefore do not induce activation of protein kinase C via this pathway. In addition, protein kinase C activators and inhibitors did not influence Ca^{++} currents of rat sympathetic neurons (Wanke et al. 1987). In contrast, in smooth muscle cells, hormonal agonists may inhibit voltage-dependent Ca^{++} currents by activation of protein kinase C (see, e.g., Galizzi et al. 1987). Stimulation of phosphatidylinositol 4,5-bisphosphate hydrolysis and subsequent activation of protein kinase C may underlie the effect of intracellularly applied GTPγS in smooth muscle cells of the rabbit ear artery, in which the nucleotide renders the inhibition of Ca^{++} currents via α_1-adrenoceptors irreversible (Droogmans et al. 1987).

The identity of the pertussis-toxin-sensitive G protein involved in hormonal inhibition of voltage-dependent Ca^{++} channels in neuronal and endocrine cells is not clear. Brain contains at least two G proteins of the G_i-type (G_i-1 and G_i-2) (Mumby et al. 1988) and is highly enriched in G_o (Sternweis and Robishaw 1984). Similarly, neuroblastoma × glioma hybrid cells (Milligan et al. 1986) and the pituitary cell line, GH_3 (Rosenthal et al. 1988), contain G proteins of the G_i-type and G_o at a high concentration. Intracellular infusion of G_i-type G proteins and G_o into neuroblastoma × glioma hybrid cells reconstituted the inhibitory modulation of Ca^{++} currents by a synthetic opioid peptide after cells had become insensitive to the agonist following pretreatment with pertussis toxin (Hescheler et al. 1987). The isolated α subunit of G_o was still effective if applied at concentrations of 0.4 M via the patch pipette and was ten times more potent than was a G_i preparation from brain, probably representing mainly G_i-1. Similarly, infusion of the α subunit of G_o restored the inhibition of Ca^{++} current modulation by neuropeptide Y in rat dorsal root ganglion cells (Ewald et al. 1988). Although a physiological role of G_i-type G proteins in the inhibitory modulations cannot be excluded, the findings described above as well as the high abundance of G_o in cell types exhibiting inhibitory Ca^{++} current modulation favor a physiological role for G_o in functional coupling of inhibitory receptors to voltage-dependent Ca^{++} channels. This hypothesis is supported by the finding that in rat pituitary tumor cells that possess G_i-type G proteins but little or no α subunit of G_o, dopamine does not inhibit prolactin release as it does in G_o-containing pituitary tumor cells (Collu et al. 1988).

The occurrence of both pertussis-toxin-sensitive hormonal stimulation and inhibition of voltage-dependent Ca^{++} currents in GH_3 cells suggests that these opposite regulations are mediated by distinct G proteins and may be related to the occurrence of both G_i-type G proteins and G_o in this cell type (Rosenthal et al. 1988).

The properties of neuronal Ca^{++} currents modulated by inhibitory receptor agonists in mouse dorsal root ganglia (Gross and MacDonald 1987) and rat sympathetic ganglia (Wanke et al. 1987) indicate that these currents represent ion fluxes through N-type Ca^{++} channels. In addition, in contrast to L-type channel antagonists of the dihydropyridine type, ω-conotoxin abolished the inhibition of Ca^{++} currents by a synthetic opioid in neuroblastoma × glioma hybrid cells (J. Hescheler et al., unpubl.). In contrast, in dorsal root ganglia of chick, Ca^{++} currents and K^+-evoked transmitter (i.e., substance P) release are inhibited by L-type channel-blocking dihydropyridines, and noradrenaline and GABA appear to inhibit L-type currents (Rane et al. 1987 and references cited therein).

Possible Relevance of Receptor- and G-protein-mediated Effects on Voltage-dependent Ca^{++} Channels

In neuronal cells and skeletal and cardiac muscle, activation of voltage-dependent Ca^{++} channels, which is crucial for neurosection or contraction, is secondary to depolarization of the plasma membrane by activated fast Na^+ currents. Besides regulation of Ca^{++} channel activity by depolarization, agonists interacting with membranous receptors coupled to G proteins can stimulate protein kinases via cytosolic signal molecules; once stimulated, protein kinases activate Ca^{++} channels (at a given membrane potential) by phosphorylation of the channel-forming protein or a regulatory component. In neuronal cells, activation of neurosecretion-inhibiting, most likely presynaptic, receptors leads to G-protein-mediated inhibition of Ca^{++} channels, resulting in a decreased Ca^{++} influx during a depolarizing pulse (see Rane et al. 1987 and references cited therein).

In other cells, e.g., endocrine and possibly smooth muscle cells, the modulation of voltage-dependent Ca^{++} channels is mainly or exclusively achieved by activation of a group of cell-surface receptors coupled to G proteins. Typically, the membrane potential of these cells is lower than in neuronal cells and in striated myocytes and therefore close to the threshold potential for Ca^{++} channels (Schlegel et al. 1987; Hescheler et al. 1988a; Rosenthal et al. 1988). In these cells, fast Na^+ currents do not appear to contribute significantly to depolarization of the plasma membrane. In adrenocortical and pituitary cells, activation of G-protein-coupled stimulation or inhibition of secretion-conferring receptors leads to depolarization and hyperpolarization, respectively (Schlegel et al. 1987; Hescheler et al. 1988a; Rosenthal et al. 1988). The ionic basis of the changes in the membrane potential is not clear. In pituitary cells, somatostatin modulates the activity of voltage-dependent K^+ and Ca^{++} currents (Rosenthal et al. 1988 and references cited therein), and either effect may contribute to the effects of the hormone on the membrane potential.

Multiple Mechanisms of Receptor- and G-protein-mediated Ca^{++} Channel Regulations

Hormonal modulations of voltage-dependent Ca^{++} currents in vertebrate cells appear to follow two types of G-protein-involving mechanisms (Table 3). If protein kinases stimulated by intracellular signals are involved (e.g., cAMP-dependent protein kinase in the case of cardiac Ca^{++} channels), G proteins participate distantly and indirectly in channel modulation, since they mediate hormonal regulation of intracellular signal-generating enzymes. On the other hand, there is evidence for close stimulatory and inhibitory controls of Ca^{++} channels by G proteins. This mechanism may be membrane-confined, since there is so far no evidence for the involvement of cytosolic signal molecules and protein kinases activated by these signal molecules.

A stimulatory membrane-confined G-protein control of (cardiac) Ca^{++} channels has been proposed by Yatani et al. (1987), describing a cAMP-independent Ca^{++} channel stimulation by the activated α subunit of G_s in isolated membrane patches from cardiac ventricular cells. The pertussis-toxin-sensitive hormonal stimulation of Ca^{++} currents in an adrenocortical cell line (Y1), which possesses a G_i-type G protein but lacks G_o, argues for a role of a G_i-type G protein in hormonal stimulation of Ca^{++} channels in this and possibly other endocrine cells. The abundance of G_o in neuronal and pituitary cells as well as data obtained by reconstitution experiments suggests that this G protein is involved in hormonal inhibition of Ca^{++} channels in these two cell types.

With the exception of reconstitution experiments with isolated patches of cardiac sarcolemma and purified G_s (Yatani et al. 1987), experiments regarding the "close" control of voltage-dependent Ca^{++} channels by G proteins have been performed in the whole-cell-clamp configuration (Hamill et al. 1981). It remains to be shown whether activated α subunits of pertussis-toxin-sensitive G proteins modulate the activity of voltage-dependent Ca^{++} channels in plasma membrane patches or of channel proteins reconstituted into phospholipid bilayers. Although the involvement of cytosolic signal molecules is unlikely, as yet unknown membranous components may be required for functional coupling of G proteins and Ca^{++} channels. It is possible that G proteins directly regulate membrane-associated, not yet identified enzymes, e.g., protein kinases or phosphatases, which control channel activity by phosphorylation or dephosphorylation of the channel or of a regulatory protein. Alternatively, G proteins themselves may modify the channel protein and, as a consequence, its function by an inherent enzymatic activity.

The apparently close control of Ca^{++} channels by G proteins may be similar to that of the voltage-dependent K^+ channels in cardiac pacemaker and GH_3 cells.

Table 3. Mechanisms of G-protein-mediated, Hormonal Modulations of Voltage-dependent Ca^{++} Channels

Effect	G protein	Mechanism	Cell type	Hormonal agonist (examples)	References (examples)
Stimul.	(G_s)	cAMP-PK	card./skel. myoc.	β-adrenergic	Hofmann et al. (1987)
Stimul.	G_s	direct	card./skel. myoc.	β-adrenergic	Yatani et al. (1987)
Stimul.	G_i?	direct?	secretory cells	angiotensin II, LHRH	Hescheler et al. (1988a)
Inhibit.	G_o?	direct?	secret./neur. cells	SSt, opiates, α_2-adren.	Lewis et al. (1986); Hescheler et al. (1987)

(Stimul.) Stimulation; (Inhibit.) inhibition; (cAMP-PK) cAMP-dependent protein kinase, (card./skel. myoc.), cardiac and skeletal myocytes; (secret./neur. cells) secretory and neuronal cells; (SSt) somatostatin; (α_2-adren.), α_2-adrenergic.

In both systems, the activated α subunit of a G_i-type G protein purified from membranes of human erythrocytes (possibly G_i-3) is sufficient to enhance channel activity if applied to the cytosolic face of isolated membrane patches (Codina et al. 1987 and references cited therein). Therefore, the active G_i subtype has been termed G_K. As activation of K^+ channels (promoting hyperpolarization) antagonizes the activation of Ca^{++} channels, G_K is likely to represent a G_i subtype different from that mediating stimulatory modulation of Ca^{++} channels.

According to a recent report, G proteins not only act as transducers between activated receptors and voltage-dependent Ca^{++} channels, but also modify the effects of Ca^{++} channel ligands such as D-600, nifedipine, and diltiazem on Ca^{++} currents (Scott and Dolphin 1987). In rat dorsal root ganglia, the blocking effects of these compounds were turned into agonistic actions by intracellular infusion of GTPγS at high concentrations (500 μM). On the other hand, Ca^{++} channel ligands affect the binding of hormone receptor ligands to G-protein-coupled receptors (see, e.g., Glossmann and Hornung 1980; Nayler et al. 1982). These findings raise the question of whether Ca^{++} channel blockers interact exclusively with the channel protein or additionally with G proteins or other components participating in the modulation of voltage-dependent Ca^{++} channels.

According to recent electrophysiological and biochemical studies, the described modulations of the activity of voltage-dependent Ca^{++} channels represent just one mechanism by which extracellular signals regulate the influx of Ca^{++}. In isolated gonadotrophs from sheep, LHRH stimulated Ca^{++}-permeable cation channels exhibiting an ohmic conductance (Mason and Waring 1986). Similarly, in vascular smooth muscle cells of the rabbit ear artery, ion channels exhibiting an ohmic conductance and a moderate selectivity for Ca^{++} were activated via purinergic receptors of the P_2-type (Benham and Tsien 1987). Vasopressin has been proposed to stimulate Ca^{++} influx in a pertussis-toxin-sensitive manner through voltage-independent ion channels in hepatocytes (Hughes et al. 1986, 1987) and through electrophysiologically undefined, dihydropyridine-insensitive channels in vascular smooth muscle (Wallnöfer et al. 1987). Hormonal control of Ca^{++} influx through an unidentified cation channel may also occur in the rabbit pulmonary artery (Haeusler et al. 1986). In this system, an α_2-adrenergic agonist caused a contraction that required extracellular Ca^{++}, was independent of the membrane potential, and was inhibited by pretreatment of the isolated organ with pertussis toxin. Similarly, contraction of rat vas deferens by α_1-adrenergic agonists was sensitive to dihydropyridines (Han et al. 1987). The sensitivity of the hormonal stimulation of the described transmembranous Ca^{++} fluxes toward pertussis toxin found in at least two systems and the fact that activated receptors that stimulate Ca^{++} influx interact with G proteins suggest an additional role for G proteins in functional coupling of activated cell-surface receptors to Ca^{++}-permeable cation channels that differ from voltage-dependent Ca^{++} channels. Finally, Ca^{++} entry is apparently activated by interaction of cytosolic molecules with membranous components. Ca^{++}-permeable cation channels stimulated by elevated cytosolic Ca^{++} occur in neutrophils (von Tscharner et al. 1986) and other systems (Partridge and Swandulla 1988). Ca^{++} entry into sea urchin eggs (Irvine and Moor 1987) and acinar cells of mouse lacrimal glands (Morris et al. 1987) is accomplished by the combined action of inositol trisphosphates and inositol-1,3,4,5-tetrakisphosphate. Since both compounds are formed in many mammalian cell types following hormonal stimulation of phosphatidylinositol 4,5-bisphosphate hydrolysis, these inositol phosphates may be of general importance as cytosolic signal molecules mediating receptor-regulated Ca^{++} entry. The question of whether the observed stimulation of Ca^{++}-permeable cation channels in human T-lymphocytes by inositol-1,4,5-trisphosphate (Kuno and Gardner 1987) is a direct effect or due to the release of intracellular Ca^{++} remains to be solved.

ACKNOWLEDGMENTS

We are grateful to Rosemarie Krüger for help in typing the manuscript. Work reported herein was supported by the Deutsche Forschungsgemeinschaft, SFB 246, and the Fonds der Chemischen Industrie.

REFERENCES

Armstrong, D. and R. Eckert. 1987. Voltage-activated calcium channels that must be phosphorylated to respond to membrane depolarization. *Proc. Natl. Acad. Sci.* **84:** 2518.

Benham, C.D. and R.W. Tsien. 1987. A novel receptor-operated Ca^{2+}-permeable channel activated by ATP in smooth muscle. *Nature* **328:** 275.

Berridge, M.J. 1987. Inositol trisphosphate and diacylglycerol: Two interacting second messengers. *Annu. Rev. Biochem.* **56:** 159.

Casey, P.J. and A.G. Gilman. 1988. G protein involvement in receptor-effector coupling. *J. Biol. Chem.* **263:** 2577.

Codina, J., A. Yatani, D. Grenet, A.M. Brown, and L. Birnbaumer. 1987. The α-subunit of the GTP binding protein G_k opens atrial potassium channels. *Science* **236:** 442.

Cohen, C.J., R.T. McCarthy, P.Q. Barrett, and H. Rasmussen. 1988. Ca channels in adrenal glomerulosa cells: K^+ and angiotensin II increase T-type Ca channel current. *Proc. Natl. Acad. Sci.* **85:** 2412.

Collu, R., C. Bouvier, G. Lagace, C.G. Unson, G. Milligan, P. Goldsmith, and A. Spiegel. 1988. Selective deficiency of guanine nucleotide-binding protein G_o in two dopamine-resistant pituitary tumors. *Endocrinology* **122:** 1176.

Deisz, R.A. and H.D. Lux. 1985. v-Aminobutyric acid-induced depression of calcium currents of chick sensory neurons. *Neurosci. Lett.* **56:** 205.

Dolphin, A.C., S.R. Forda, and R.H. Scott. 1986. Calcium-dependent currents in cultured rat dorsal root ganglion neurons are inhibited by an adenosine analogue. *J. Physiol.* **373:** 47.

Droogmans, G., I. Declerck, and R. Casteels. 1987. Effect of adrenergic agonists on Ca^{2+}-channel currents in single vascular smooth muscle cells. *Pflugers Arch. Eur. J. Physiol.* **409:** 7.

Dunlap, K. and G.D. Fischbach. 1981. Neurotransmitters

decrease the calcium conductance activated by depolarization of embryonic chick sensory neurones. *J. Physiol.* **317:** 519.

Enyedi, P., I. Mucsi, L. Hunyady, K.J. Catt, and A. Spät. 1986. The role of guanyl nucleotide binding proteins in the formation of inositol phosphates in adrenal glomerulosa cells. *Biochem. Biophys. Res. Commun.* **140:** 941.

Ewald, D.A., P.C. Sternweis, and R.J. Miller. 1988. Guanine-nucleotide-binding protein Ca-induced coupling of neuropeptide Y receptors to Ca^{2+} channels in sensory neurons. *Proc. Natl. Acad. Sci.* **85:** 3633.

Flockerzi, V., H.-J. Oeken, F. Hofmann, D. Pelzer, A. Cavalié, and W. Trautwein. 1986. Purified dihydropyridine-binding site from skeletal muscle t-tubules is a functional calcium channel. *Nature* **323:** 66.

Forscher, P., G.S. Oxford, and D. Schulz. 1986. Noradrenaline modulates calcium channels in avian dorsal root ganglion cells through tight receptor-channel coupling. *J. Physiol.* **379:** 131.

Gallizi, J.-P., J. Qar, M. Fosset, C. Van Renterghem, and M. Lazdunski. 1987. Regulation of calcium channels in aortic muscle cells by protein kinase C activators (diaclyglycerol and phorbol esters) and by peptides (vasopressin and bombesin) that stimulate phosphoinositide breakdown. *J. Biol. Chem.* **262:** 6947.

Glossmann, H. and R. Hornung. 1980. Calcium- and potassium channel blockers interact with α-adrenoceptors. *Mol. Cell. Endocrinol.* **19:** 243.

Gray, R. and D. Johnston. 1987. Noradrenaline and β-adrenoceptor agonists increase activity of voltage-dependent calcium channels in hippocampal neurons. *Nature* **327:** 620.

Gross, R.A. and R.L. MacDonald. 1987. Dynorphin A selectivity reduces a large transient (N-type) calcium current of mouse dorsal root ganglion neurons in cell culture. *Proc. Natl. Acad. Sci.* **84:** 5469.

Haeusler, G., J.-E. De Peyer, M. Yajima, and G. Schultz. 1986. Vascular smooth muscle: Availability of calcium through α-adrenoceptor stimulation. *J. Cardiovas. Pharmacol.* **8:** S107.

Hamill, O.P., A. Marty, E. Neher, B. Sakman, and F.J. Sigworth. 1981. Improved patch-clamp techniques for high-resolution current recording from cells and cell-free membrane patches. *Pflugers Arch. Eur. J. Physiol.* **391:** 85.

Han, C., P.W. Abel, and K.P. Minneman. 1987. α_1-Adrenoceptor subtypes linked to different mechanisms for increasing intracellular Ca^{2+} in smooth muscle. *Nature* **329:** 333.

Hescheler, J., W. Rosenthal, W. Trautwein, and G. Schultz. 1987. The GTP-binding protein, G_o, regulates neuronal calcium channels. *Nature* **325:** 445.

Hescheler, J., W. Rosenthal, K.-D. Hinsch, M. Wulfern, W. Trautwein, and G. Schultz. 1988a. Angiotensin II-induced stimulation of voltage-dependent Ca^{2+} currents in an adrenal cortical cell line. *EMBO J.* **7:** 619.

Hescheler, J., W. Rosenthal, M. Wulfern, M. Tang, M. Yajima, W. Trautwein, and G. Schultz. 1988b. Involvement of the guanine nucleotide-binding protein, N_o, in the inhibitory regulation of neuronal calcium channels. *Adv. Second Messenger Phosphoprotein Res.* **21:**165.

Hofmann, F., W. Nastainczyk, A. Röhrkasten, T. Schneider, and M. Sieber. 1987. Regulation of the L-type calcium channel. *Trends. Pharmacol. Sci.* **8:** 393.

Holz IV, G.G., S.G. Rane, and K. Dunlap. 1986. GTP-binding proteins mediate transmitter inhibition of voltage-dependent calcium channels. *Nature* **319:** 670.

Hughes, B.P., S.E. Milton, G.J. Barritt, and A.M. Auld. 1986. Studies with verapamil and nifedipine for the presence in liver cell plasma membrane of two types of Ca^{2+} inflow transporters which are dissimilar to potential-operated Ca^{2+} channels. *Biochem. Pharmacol.* **35:** 3045.

Hughes, B.P., J.N. Crofts, A.M. Auld, L.C. Read, and G.J. Barritt. 1987. Evidence that a pertussis-toxin-sensitive substrate is involved in stimulation by epidermal growth factor and vasopressin of plasma-membrane Ca^{2+} inflow in hepatocytes. *Biochem. J.* **248:** 911.

Irvine, R.F. and R.M. Moor. 1987. Inositol-(1,3,4,5)tetrakisphosphate-induced activation of sea urchin eggs requires the presence of inositol trisphosphate. *Biochem. Biophys. Res. Commun.* **146:** 284.

Iyengar, R. and L. Birnbaumer. 1987. Signal transduction by G-proteins. *ISI Atlas Sci. Pharmacol.* **1:** 213.

Jakobs, K.H., K. Aktories, and G. Schultz. 1981. Inhibition of adenylate cyclase by hormones and neurotransmitters. *Adv. Cyclic Nucleotide Protein Phosphorylation Res.* **14:** 173.

Kaczmarek, L.K. 1987. The role of protein kinase C in the regulation of ion channels and neurotransmitter release. *Trends Neurosci.* **10:** 30.

Kojima, I., H. Shibata, and E. Ogata. 1986. Pertussis toxin blocks angiotensin II-induced calcium influx but not inositol trisphosphate production in adrenal glomerulosa cell. *FEBS Lett.* **204:** 347.

Kuno, M. and P. Gardner. 1987. Ion channels activated by inositol 1,4,5-trisphosphate in plasma membrane of human T-lymphocytes. *Nature* **326:** 301.

Lewis, D.L., F.F. Weight, and A. Luini. 1986. A guanine nucleotide-binding protein mediates the inhibition of voltage-dependent calcium current by somatostatin in a pituitary cell line. *Proc. Natl. Acad. Sci.* **83:** 9035.

Marchetti, C., E. Carbone, and H.D. Lux. 1986. Effects of dopamine and noradrenaline on Ca channels of cultured sensory and sympathetic neurons of chick. *Pflugers Arch. Eur. J. Physiol.* **406:** 104.

Mason, W.T. and D.W. Waring. 1986. Patch clamp recordings of single ion channel activation by gonadotrophin-releasing hormone in ovine pituitary gonadotrophs. *Neuroendocrinology* **42:** 205.

McCleskey, E.W., A.P. Fox, D.H. Feldman, L.J. Cruz, B.M. Olivera, R.W. Tsien, and D. Yoshikami. 1987. ω-Conotoxin: Direct and persistent blockade of specific types of calcium channels in neurons but not muscle. *Proc. Natl. Acad. Sci.* **84:** 4327.

Milligan, G., P. Gierschik, A.M. Spiegel, and W.A. Klee. 1986. The GTP-binding regulatory proteins of neuroblastoma × glioma, NG108-15, and glioma, C6, cells. *FEBS. Lett.* **195:** 225.

Morris, A.P., D.V. Gallacher, R.F. Irvine, and O.H. Petersen. 1987. Synergism of inositol trisphosphate and tetrakisphosphate in activating Ca^{2+}-dependent K^+ channels. *Nature* **330:** 653.

Mumby, S., I.-H. Pang, A.G. Gilman, and P.C. Sternweis. 1988. Chromatographic resolution and immunologic identification of the α_{40} and α_{41} subunits of guanine nucleotide-binding regulatory proteins from bovine brain. *J. Biol. Chem.* **263:** 2020.

Naor, Z., A. Azrad, R. Limor, H. Zakut, and M. Lotan. 1986. Gonadotropin-releasing hormone activates a rapid Ca^{2+}-independent phosphodiester hydrolysis of polyphosphoinositides in pituitary gonadotrophs. *J. Biol. Chem.* **261:** 12506.

Nayler, W.G., J.E. Thompson, and B. Jarrott. 1982. The interaction of calcium antagonists (slow channel blockers) with myocardial alpha adrenoceptors. *J. Mol. Cell. Cardiol.* **14:** 185.

Neer, E.J. and D.E. Clapham. 1988. Roles of G protein subunits in transmembrane signalling. *Nature* **333:** 129.

Nowycky, M.C., A.P. Fox, and R.W. Tsien. 1985. Three types of neuronal calcium channel with different calcium agonist sensitivity. *Nature* **316:** 440.

Partridge, L.D. and D. Swandulla. 1988. Calcium-activated non-specific cation channels. *Trends Neurosci.* **11:** 69.

Paupardin-Tritsch, D., C. Hammond, H.M. Gerschenfeld, A.C. Nairn, and P. Greengard. 1986. cGMP-dependent protein kinase enhances Ca^{2+} current and potentiates the serotonin-induced Ca^{2+} current increase in snail neurones. *Nature* **323:** 812.

Rane, S.G. and K. Dunlap. 1986. Kinase C activator 1,2-oleoylacetylglycerol attenuates voltage-dependent calcium current in sensory neurons. *Proc. Natl. Acad. Sci.* **83:** 184.

Rane, S.G., G.G. Holz IV, and K. Dunlap. 1987. Dihydropyridine inhibition of neuronal calcium current and substance P release. *Pflugers Arch. Eur. J. Physiol.* **409:** 361.

Reuter, H. 1983. Calcium channel modulation by neurotransmitters, enzymes and drugs. *Nature* **301:** 569.

Rosenthal, W., J. Hescheler, K.-D. Hinsch, K. Spicher, W. Trautwein, and G. Schultz. 1988. Cyclic AMP-independent, dual regulation of voltage-dependent Ca^{2+} currents by LHRH and somatostatin in a pituitary cell line. *EMBO J.* **7:** 1627.

Schlegel, W., B.P. Winiger, P. Mollard, P. Vacher, F. Wuarin, G. Zahnd, C.B. Wollheim, and B. Dufy. 1987. Oscillations of cytosolic Ca^{2+} in pituitary cells due to action potentials. *Nature* **329:** 719.

Scott, R.H. and A.C. Dolphin. 1986. Regulation of calcium currents by a GTP analogue: Potentiation of (−)-baclofen-mediated inhibition. *Neurosci. Lett.* **69:** 59.

———. 1987. Activation of G proteins promotes agonist responses to calcium channel ligands. *Nature* **330:** 760.

Sternweis, P.C. and J.D. Robishaw. 1984. Isolation of two proteins with high affinity for guanine nucleotides from membranes of bovine brain. *J. Biol. Chem.* **259:** 13806.

Suzuki, N. and T. Yoshioka. 1987. Differential blocking action of synthetic ω-conotoxin on components of Ca^{2+} channel current in clonal GH3 cells. *Neurosci. Lett.* **75:** 235.

Takahashi, M., M.J. Seager, J.F. Jones, B.F.X. Reber, and W.A. Catterall. 1987. Subunit structure of dihydropyridine-sensitive calcium channels from skeletal muscle. *Proc. Natl. Acad. Sci.* **84:** 5478.

Tsunoo, A., M. Yoshii, and T. Narahashi. 1986. Block of calcium channels by enkephalin and somatostatin in neuroblastoma-glioma hybrid NG108-15 cells. *Proc. Natl. Acad. Sci.* **83:** 9832.

Ui, M., T. Katada, T. Murayama, H. Kurose, M. Yajima, M. Tamura, T. Nakamura, and K. Nogimori. 1984. Islet-activating protein, pertussis toxin: A specific uncoupler of receptor-mediated inhibition of adenylate cyclase. *Adv. Cyclic Nucleotide Protein Phosphorylation Res.* **17:** 145.

von Tscharner, V., B. Prod'Hom, M. Baggiolini, and H. Reuter. 1986. Ion channels in human neurotrophils activated by a rise in free cytosolic calcium concentration. *Nature* **324:** 369.

Wallnöfer, A., C. Cauvin, and U.T. Rüegg. 1987. Vasopressin increases $^{45}Ca^{2+}$ influx in rat aortic smooth muscle cells. *Biochem. Biophys. Res. Commun.* **148:** 273.

Wanke, E., A. Ferroni, A. Malgaroli, A. Ambrosini, T. Pozzan, and J. Meldolesi. 1987. Activation of a muscarinic receptor selectively inhibits a rapidly inactivated Ca^{2+} current in rat sympathetic neurons. *Proc. Natl. Acad. Sci.* **84:** 4313.

Yatani, A., J. Codina, Y. Imoto, J.P. Reeves, L. Birnbaumer, and A.M. Brown. 1987. A G protein directly regulates mammalian cardiac calcium channels. *Science* **238:** 1288.

Unique Properties of a New GTP-binding Protein with a Molecular Mass of 24,000 Daltons Purified from Porcine Brain Membranes

T. KATADA* AND M. UI†

*Department of Life Science, Faculty of Science, Tokyo Institute of Technology, Yokohama 227, Japan; †Department of Physiological Chemistry, Faculty of Pharmaceutical Sciences, University of Tokyo, Tokyo 113, Japan

GTP-binding proteins (G proteins), so far identified as membrane signal transducers, are characterized by their common $\alpha\beta\gamma$-trimeric structure and unique susceptibility to bacterial toxins that catalyze the transfer of the ADP-ribosyl moiety of NAD to these proteins. These G proteins communicate between a variety of receptors and effectors, including adenylate cyclase, phospholipases C and A_2, K^+ and Ca^{++} channels, and cGMP-specific phosphodiesterase (for review, see Gilman 1984, 1987; Ui 1984, 1986, 1988b; Stryer and Bourne 1986; Casey and Gilman 1988).

G_s ($\alpha_{45}\beta\gamma$) derives from one genomic DNA and is the only G protein that can activate the adenylate cyclase catalyst directly and is ADP-ribosylated by cholera toxin but not by pertussis toxin, islet-activating protein (IAP). In contrast, G_i (G_i-1 [or $\alpha_{41}\beta\gamma$], G_i-2 [or $\alpha_{40}\beta\gamma$], and G_i-3) and G_o (or $\alpha_{39}\beta\gamma$) belong to a subfamily of G proteins that are, encoded by their respective genes (Itoh et al. 1988), probably involved in the phospholipase C and A_2 activation, K^+ channel opening, and Ca^{++} channel regulation, as well as in the adenylate cyclase inhibition. These G proteins serve as the substrates of IAP-catalyzed ADP-ribosylation; once being ADP-ribosylated, they are uncoupled from receptors and no signal is any longer transmitted to these effectors (Ui 1984, 1986, 1988b). Likewise, there are two isoforms of G_t (or transducin) encoded by different DNAs—one that activates the photosensitive cGMP phosphodiesterase of rod outer segments (G_t-1) in vertebrate retinal cells and one that plays the analogous role in cones (G_t-2).

These G proteins differ from one another in their α subunits with M_r values of 39,000–52,000, each having a binding site for guanine nucleotides and a site to be ADP-ribosylated by bacterial toxins, since the accompanying $\beta\gamma$ subunits are most common. GTP is hydrolyzed to GDP at the binding site in the α subunit. Thus, the α subunit appears to be an active component of the G protein, whereas the $\beta\gamma$ subunits may act primarily as the regulator of the α subunit or its membrane anchor, although $\beta\gamma$ subunits that have been resolved from the α subunit could play their own roles as signal transducers by interacting directly with the adenylate cyclase catalyst (Katada et al. 1986a), calmodulin (Asano et al. 1986; Katada et al. 1987a), K^+ channels (Logothetis et al. 1987), and phospholipase A_2 (Jelsema and Axelrod 1987).

The physiological role of GTP-binding proteins with lower molecular weights remains to be defined. They include p21, encoded by c-*ras* and v-*ras* oncogenes, G_p purified from human placenta (Evans et al. 1986) and bovine brain (Waldo et al. 1987) membranes, the ADP-ribosylation factor (ARF) to support the cholera-toxin-catalyzed reaction (Kahn and Gilman 1984, 1986), and the protein with $M_r = 24{,}000$ purified from bovine brain (Kikuchi et al. 1988). We have also purified a GTP-binding protein with a molecular mass of 24 kD (henceforth referred to as the 24-kD G protein) from porcine brain, which displayed unique properties distinctly different from those of the above-mentioned trimeric G proteins, as will be described below.

EXPERIMENTAL PROCEDURES

Purification of 24-kD G protein. The initial steps for purification of the 24-kD G protein from porcine brain membranes are analogous to those described previously for purification of IAP substrate G proteins (G_i-1 or $\alpha_{41}\beta\gamma$, G_i-2 or $\alpha_{40}\beta\gamma$, G_o or $\alpha_{39}\beta\gamma$) from rat (Katada et al. 1986b) and porcine (Katada et al. 1987b) brains. This allows good recovery of all of these proteins from the same preparation of the brain membrane extract. The buffer solutions used were 20 mM Tris-HCl (pH 8.0) (TED) and 50 mM Na 3-(*N*-molpholino)propanesulfonate (pH 7.4) (MED), both containing 1 mM EDTA, 1 mM 1 dithiothreitol, and 50 units/ml of aprotinin.

The first step was the chromatography of the extract on a column of DEAE-Sephacel that had been equilibrated with TED/25 mM NaCl/1% Na cholate and then eluted with a linear gradient of NaCl from 0 to 200 mM added to the same buffer. The second column was that of Ultrogel AcA-34 eluted with TED/100 mM NaCl/0.8% Na cholate. This allowed a rough separation of the fraction of IAP substrates from the fraction containing the 24-kD G protein. Both fractions were individually applied to the column of phenyl-Sepharose that was eluted with a linear gradient of (0.2–1.2)% Na cholate/(250–25) mM NaCl in TED.

Further purification of the 24-kD G protein was

achieved by the sequential application to columns of DEAE-Toyopearl 650(S) and Superose-12 HR10/30, which were eluted with a linear gradient of (250–25) mM NaCl in TED/0.7% CHAPS and TED/100 mM NaCl/0.7% CHAPS, respectively. The mostly homogeneous 24-kD G protein exhibiting the GTPγS-binding activity was thus obtained and stored at −80°C or on ice until use without loss of activity for several months (−80°C) or 2–3 weeks (on ice).

Activity assays. The activity of proteins to bind radioactive GTP, GDP, or GTPγS was determined by the use of a rapid filtration technique described previously (Murayama and Ui 1984; Katada et al. 1986b). ADP-ribosylation of proteins by preactivated IAP or the A-protomer of IAP (Ui et al. 1985; Ui 1988a) was performed in the presence of α-^{32}P-labeled NAD, and the labeled proteins were separated on an SDS-polyacrylamide gel electrophoretic plate (Katada and Ui 1982a,b; Oinuma et al. 1987).

Identification of guanine nucleotides bound to 24-kD G protein. The 24-kD G protein was incubated with radioactive GTP, GDP, or GTPγS and applied to a small column of Sephadex G-25. Over 90% of the protein was eluted from the column within the void volume that was contaminated with less than 0.01% of the unbound nucleotides. Rapid boiling of the thus separated protein with carrier guanine nucleotides for 3 minutes allowed more than 99% of the bound nucleotides to be released into the supernatant, which was then separated on a Mono Q HR5/5 anion exchanger column using a Pharmacia FPLC system. Neither GTP nor GDP was hydrolyzed during boiling.

RESULTS

Purification of 24-kD G Protein from Porcine Brain Membranes

The cholate extract of porcine brain membranes was first applied to a DEAE-Sephacel column, and a fraction of the eluate enriched with G proteins was next applied to a column of the Ultrogel AcA-34, which

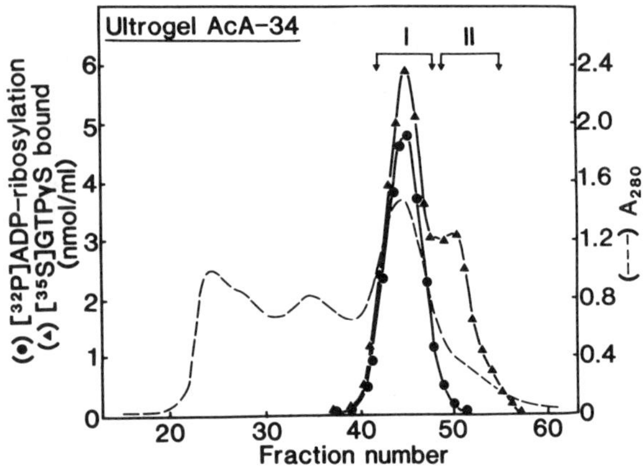

Figure 1. GTP-binding protein not serving as an IAP substrate eluted from an Ultrogel AcA-34 column.

produced the main sharp peak and an accompanying shoulder of the GTPγS-binding activity (Fig. 1). Essentially all of the proteins ADP-ribosylated by preactivated IAP were eluted as the main symmetrical peak (fraction 1 in Fig. 1), whereas the shoulder (fraction II in Fig. 1) was mostly free of the IAP-substrate activity. Most of G_s was found in fraction I, rather than in fraction II, revealing that G proteins in fraction II were new proteins different from G_s and any member of the IAP-substrate subfamily.

The application of fractions I and II to phenyl-Sepharose columns yielded three mutually separated peaks of G_s, IAP substrates, and the new G proteins, all of which served as good sources for the further purification of these different G-protein groups. The further FPLC procedure with the sequential uses of DEAE-Toyopearl and Superose 12 columns afforded the mostly homogeneous preparation of a new G protein (Fig. 2). The G protein thus purified and treated with *N*-ethylmaleimide moved as a single sharp band at the same rate as did a protein marker with $M_r = 24{,}000$ on SDS-polyacrylamide gels (inset in Fig. 2). The protein is hence referred to as the 24-kD G protein in this paper. The purification of the 24-kD G protein is shown in Table 1.

Table 1. Purification of G Proteins from Porcine Brain

Step	Protein (mg)	GTPγS binding		ADP-ribosylation	
		nmole(%)	nmole/mg	nmole(%)	nmole/mg
Extract (550 ml)	2420	1100 (100)	0.46	460 (100)	0.19
DEAE-Sephacel	360	610 (55)	1.7	340 (74)	0.94
Ultrogel AcA-34					
Peak I	92	310 (28)	3.4	270 (59)	2.9
Peak II	31	130 (12)	4.2	20 (4)	0.6
Peak I: IAP substrates					
Phenyl-Sepharose (I)	41	250 (20)	6.1	210 (46)	5.1
Peak II: 24-kD G protein					
Phenyl-Sepharose (II)	13	99 (9)	7.6	4 (0.9)	0.3
DEAE-Toyopearl	3.6	62 (6)	17	not detectable	
Superose-12	1.2	43 (4)	36		

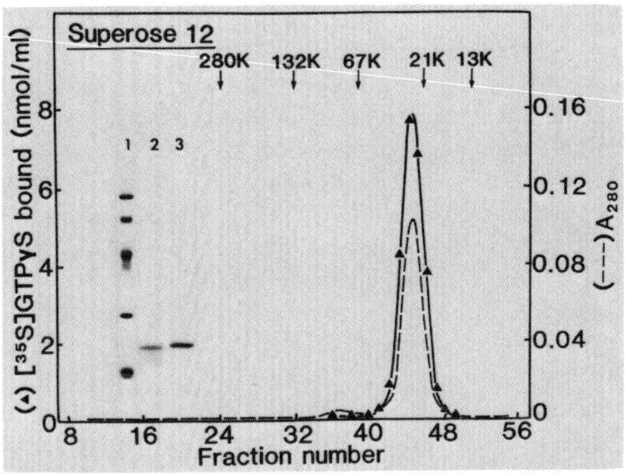

Figure 2. Homogeneous preparation of the 24-kD G protein eluted from a column of Superose 12. (*Inset*) Pattern of SDS-PAGE. (Lane *1*) Molecular-weight markers; (lane *2*) protein eluted from the column; (lane *3*) protein eluted from the column and treated with *N*-ethylmaleimide.

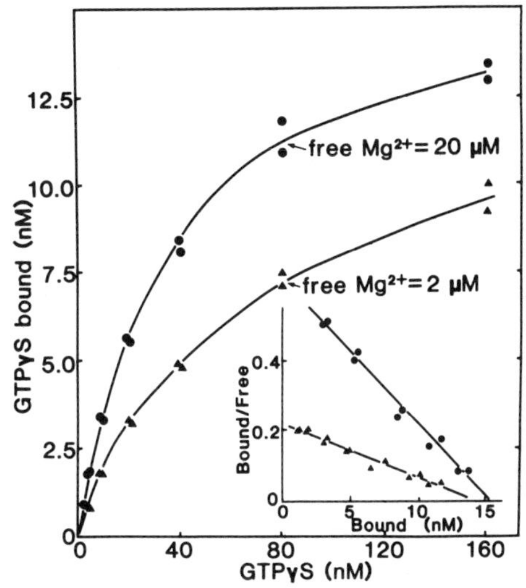

Figure 4. The concentration-dependent binding of [^{35}S]GTPγS to the 24-kD G protein at two levels of Mg^{++}. The binding of [^{35}S]GTPγS to the protein was determined as in Fig. 3A. (*Inset*) Scatchard plots of the binding.

Mg^{++}-dependent Binding of GTP and GDP to the 24-kD G Protein

GTPγS bound to the purified 24-kD G protein very rapidly in the presence of Mg^{++} at concentrations higher than 20 μM (Fig. 3A). The amount of GTPγS bound as well as the rate of the binding was strictly dependent on the concentration of Mg^{++} from 0.25 to 20 μM. Likewise, GTPγS, once bound in the presence of 1 mM Mg^{++}, was released when the Mg^{++} concentration was lowered to less than 20 μM (Fig. 3B). GTP or GDP displayed just the same Mg^{++}-dependent binding and releasing behavior as did GTPγS (data not shown).

Mg^{++} increased not only the capacity and rate, but also the affinity for the GTPγS binding to the 24-kD G protein (Fig. 4). The Scatchard analysis of the GTPγS binding revealed that the apparent dissociation constants (K_d) were 60 and 25 nM for GTPγS in the presence of 2 and 20 μM Mg^{++}, respectively (inset in Fig. 4). The binding sites on the 24-kD G protein were specific for GTP and GDP, since other nucleotides, including GMP, ATP, ADP, AMP, and UTP, did not inhibit the binding of GTPγS even at their higher concentrations (data not shown).

Lack of Guanine Nucleotide Exchange on the 24-kD G Protein

Radioactive GTPγS once bound to the 24-kD G protein was not released in the presence of Mg^{++} at concentrations higher than 100 μM (see Fig. 3). Such was the case with radioactive GTP or GDP. No radioac-

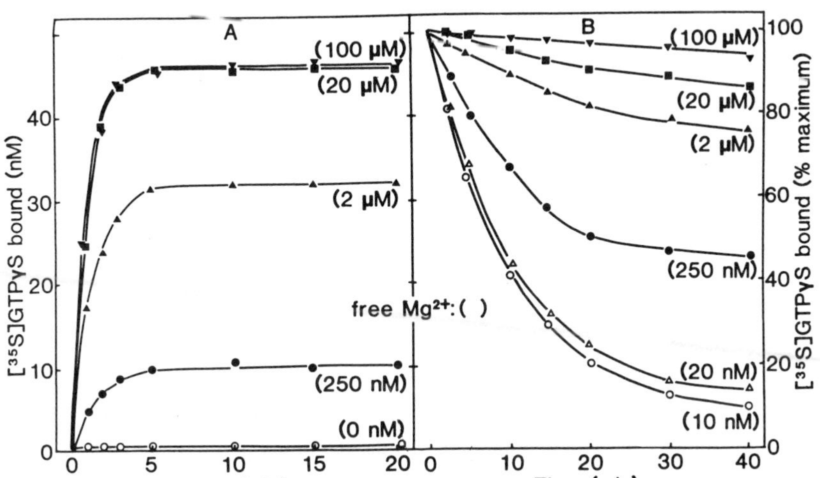

Figure 3. Mg^{++}-dependent binding of [^{35}S]GTPγS to, and its release from, the purified 24-kD G protein. (*A*) The protein was incubated with 0.2 μM [^{35}S]GTPγS before the analysis of the protein-bound radioactivity. (*B*) The [^{35}S]GTPγS-bound protein was incubated in the nucleotide-free medium to be analyzed for the radioactivity release into the medium. The concentration of Mg^{++} calculated from the concentration of EDTA and $MgCl_2$ added to the medium is shown in parentheses.

tive GTP or GDP was released significantly even when the reaction mixture was further fortified with nonradioactive guanine nucleotides, indicating that essentially no exchange occurred between the protein-bound guanine nucleotides and the external nucleotides.

The rate of [^{3}H]GDP-GDP exchange was compared between the 24-kD G protein and G_o ($\alpha_{39}\beta\gamma$) in Figure 5. The prebound [^{3}H]GDP was released in exchange for added nonradioactive GDP from G_o with the first order kinetic constant (k) of 0.248 (Fig. 5B). The rate of exchange was very low with the 24-kD G protein ($k = 0.017$; Fig. 5A). Thus, the GDP exchange reaction occurred on the 24-kD G protein only at a rate one-fifteenth lower than the rate of the same reaction occurring on G_o.

The addition of 0.25 M $(NH_4)_2SO_4$ to the reaction mixture caused the release of [^{3}H]GDP from the 24-kD G protein as well as from G_o, probably due to the conformational change in the vicinity of guanine-nucleotide-binding sites of these proteins (Ferguson et al. 1986).

Apparently Low GTPase Activity of the 24-kD G Protein Despite Rapid Hydrolysis of GTP to GDP at the Binding Sites

The GTPase activity (V_{max}) of the 24-kD G protein measured under the standardized conditions was 0.024 ± 0.002 mole of P_i released per minute per mole of protein (the mean ±S.E.M. from six observations). The activity was much lower than the activity of G_o measured under the identical conditions (0.35 ± 0.02). Since these GTPase activities were well correlated with the rate constants (k) for GDP exchange measured in Figure 5, it is very likely that the rate of the overall GTPase reaction is limited by the process of the release of GDP, the product of the reaction, from the enzymatic protein in exchange for the substrate, GTP.

The intrinsic conversion of GTP to GDP at the binding sites of the 24-kD G protein was then measured taking advantage of the quantitative discharge of the bound guanine nucleotides to the supernatant upon boiling of the G protein for only 3 minutes. The 24-kD protein labeled with [^{3}H]GTP or [^{3}H]GDP was quickly separated from the unbound nucleotides by gel filtration before boiling. The supernatant of the thus heat-denatured protein was then analyzed for the radioactivity content of the GTP, GDP, and GMP fractions subsequently separated on an FPLC system (Fig. 6). The radioactivity was detected in the GDP fraction exclusively, regardless of whether the protein had been labeled with [^{3}H]GTP or [^{3}H]GDP. Thus, GTP was rapidly converted to GDP on the 24-kD G protein due to intrinsic GTPase activity. The product, GDP, however, was not readily released to the supernatant in exchange for the substrate GTP, thereby minimizing the apparent GTPase activity.

Failure of the 24-kD G Protein to be Complexed with $\beta\gamma$ Subunits of IAP Substrates

One of the unique characteristics of the usual transducer G proteins is their $\alpha\beta\gamma$-trimeric structure (see above). The GDP-binding α subunits of these G proteins form trimeric complexes when combined with the $\beta\gamma$ subunits, a component common to these G proteins. The activities of the G-protein α subunits to bind GTP or GDP, to hydrolyze GTP, and to be ADP-ribosylated by IAP were profoundly affected by the $\alpha\beta\gamma$-complex formation (Katada et al. 1986b; Ui 1988b). In the case of the 24-kD G protein, however, the Mg^{++}-dependent binding of GDP (or its release) was not affected at all by the addition of the $\beta\gamma$ subunits of G_o to the reaction mixture (data not shown). In fact, no association prod-

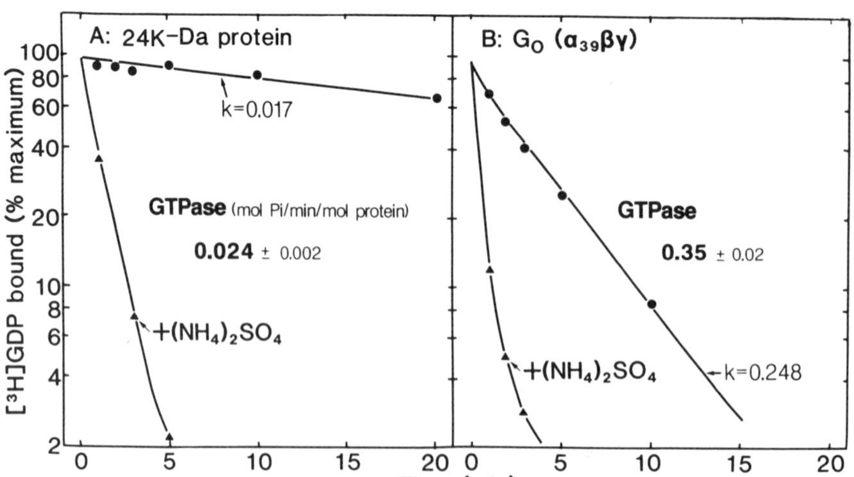

Figure 5. The release of prebound [^{3}H]GDP from the 24-kD G protein (A) or G_o (B) in exchange for nonradioactive GDP added to the reaction mixture in 2 mM Mg^{++}. The concentration of GDP added was 200 μM. Where indicated, the reaction mixture was supplemented with 250 mM $(NH_4)_2SO_4$. The radioactive GDP still bound at each incubation time is plotted as the percentage of the initial value at the start of incubation.

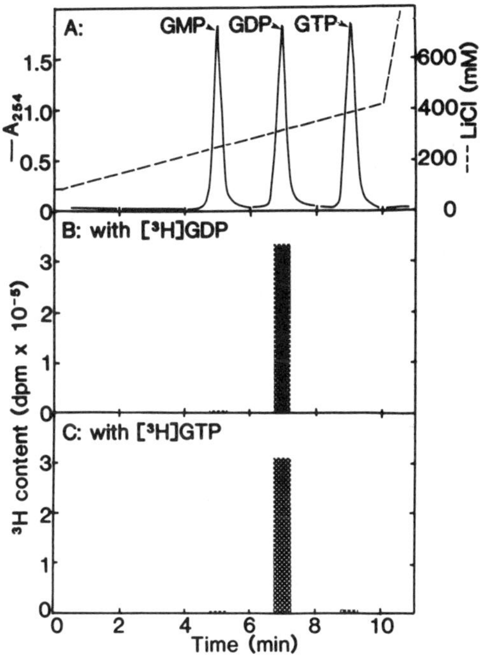

Figure 6. Identification of guanine nucleotides bound to the 24-kD G protein. Authentic samples of GTP, GDP, and GMP were separated by chromatography on an anion exchanger column as shown in *A*. The elution profile was not altered by boiling the samples before application. The 24-kD G protein prelabeled with [^{3}H]GDP (*B*) or [^{3}H]GTP (*C*) was boiled, and the supernatant (after being supplemented with carrier GDP and GTP) was analyzed for the radioactivity content of the fractions separated as in *A*.

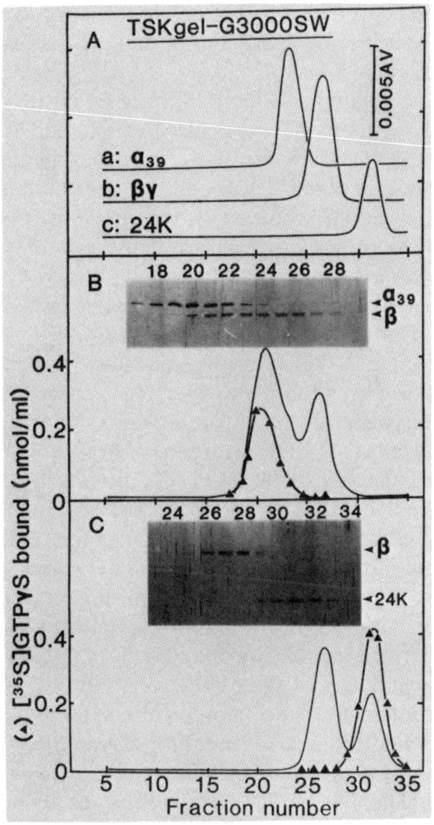

Figure 7. Failure of the 24-kD G protein to associate with the $\beta\gamma$ subunits resolved from G_o. The α subunit (α_{39}) of G_o, its $\beta\gamma$ subunits ($\beta\gamma$), and the 24-kD G protein (24K) were eluted individually from a column of TSK-3000SW as in *A*. The elution profiles of the mixture of α_{39} with $\beta\gamma$ (*B*) and the mixture of the 24-kD G protein with $\beta\gamma$ (*B*) are shown. (*Insets*) SDS-PAGE patterns of eluates. The GTPγS-binding activity of the eluate was measured as shown by solid triangles.

uct was formed when the $\beta\gamma$ subunits and the 24-kD protein were mixed, in contrast to the product formation between the $\beta\gamma$ subunits and the G_o α subunits (or α_{39}) (Fig. 7). The elution profile of the 24-kD G protein from a TSK gel permeation column was not altered by combination with the $\beta\gamma$ subunits (Fig. 7C), whereas the α_{39} subunit was eluted at a higher rate together with the $\beta\gamma$ subunits than was the same protein eluted alone (Fig. 7B).

DISCUSSION

The 24-kD G protein purified proved to be a new protein distinctly different from the GTP-binding proteins previously reported to have analogous molecular masses. The antibody raised against purified p21 did not interact with the 24-kD G protein, and there was a marked difference in Mg^{++}-dependency of the guanine nucleotide binding between these two proteins (data not shown). The 24-kD G protein differs from ARF (Kahn and Gilman 1984, 1986) or from G_p (Evans et al. 1986; Waldo et al. 1987); it neither enhances the cholera-toxin-induced ADP-ribosylation of G_s (not shown), nor is it complexed with the $\beta\gamma$ subunits of IAP substrate G protein (Fig. 7). The GTP-binding protein purified from bovine brains by Kikuchi et al. (1988) exhibited such a high GTPase activity as to be comparable with that of usual trimeric G proteins, making it unlikely that their protein is the same entity as the 24-kD G protein described in this paper. The content of the 24-kD G protein in porcine brain membranes was in the same order of magnitude as its content in bovine or rat brain membranes and attained roughly to 5–10% of the total GTP-binding proteins in the membrane extract (Table 1). This would be more than a reasonable quantity required for a protein that should be functionally coupled to receptors and effectors in membranes.

The unique properties of the 24-kD G protein are summarized as follows. First, the guanine-nucleotide-binding site in the protein is emptied of nucleotides during incubation in the medium containing no Mg^{++}. Since the purification was achieved in the presence of 1 mM EDTA, the binding sites on the purified 24-kD G protein were unoccupied by guanine nucleotides, in contrast with the previous trimeric G proteins to which GDP continues to bind tightly, even in the absence of Mg^{++}, during the purification procedure. The exposure of proteins to 250 mM $(NH_4)_2SO_4$ was the only means to deprive the trimeric G proteins of endogenous GDP and was effective on the 24-kD proteins in this regard as well. Guanine nucleotides bound to these GDP-depleted G proteins very rapidly; binding occurred only in the presence of Mg^{++} in the case of the 24-kD G proteins (Fig. 3A).

Second, guanine nucleotides, once bound to the 24-kD G protein, are never released (or released at only a slow rate), even in the presence of exogenous nucleotides, unless Mg^{++} was omitted from the medium. This forms a sharp contrast with the previously isolated trimeric G proteins, which are converted to a GTP-bound form from a GDP-bound form as a result of exchange of prebound GDP with the exogenous GTP in the presence of Mg^{++}.

Third, the 24-kD G protein does not exhibit a significant activity of GTPase when the activity measurement was achieved based on liberation of P_i or GDP from added GTP. In the presence of Mg^{++}, GTP binds to the vacant sites on the 24-kD G protein and is readily converted to GDP as such (Fig. 6), but GDP thus formed cannot be released in exchange for GTP in the reaction mixture. Hence, the apparent lack of the GTPase activity in the 24-kD G protein results from inaccessibility of the substrate to catalytic sites that are persistently occupied by the product, rather than a functional defect of the catalytic activity.

The interconversion between the GDP-bound and GTP-bound forms is essential for classic $\alpha\beta\gamma$-trimeric G proteins to play their significant role as signal transducers. The GDP-to-GTP conversion results from exchange of GTP with GDP prebound to the α subunits, whereas the GTP-to-GDP conversion reflects the hydrolysis of GTP on the α subunits. The former conversion is accelerated by coupling with agonist-occupied receptors, and the resultant GTP-bound α subunits and/or $\beta\gamma$ subunits resolved therefrom are responsible for the transmission of signal to effectors. The latter conversion regains the original GDP-bound $\alpha\beta\gamma$ trimers that are ready for receptor coupling. Thus, the 24-kD G protein appears to be lacking, as such, in the function essential as transducers in cells in which the Mg^{++} concentration is anticipated to be high. An extensive search is currently being made for membrane proteins or lipids that confer a transducer function to the 24-kD G protein by permitting exogenous GTP to exchange with prebound GDP and hence permitting continuous GTP hydrolysis to occur on the G protein. The relationship of the 24-kD G protein to proteins encoded by oncogenes other than *ras* genes, such as *rho* or *ral* genes, is also the subject of our current investigations.

ACKNOWLEDGMENTS

This work was supported in part by research grants from the Scientific Research Fund of the Ministry of Education, Science, and Culture, Japan, and by research grants from Yamada Scientific Foundation and from Toray Science and Technology grants in Japan.

REFERENCES

Asano, T., N. Ogasawara, S. Kitajima, and M. Sano. 1986. Interaction of GTP-binding proteins with calmodulin. *FEBS Lett.* **203:** 135.

Casey, P.J. and A.G. Gilman. 1988. G protein involvement in receptor-effector coupling. *J. Biol. Chem.* **263:** 2577.

Evans, T., M.L. Brown, E.D. Fraser, and J.K. Northrup. 1986. Purification of the major GTP-binding proteins from human placental membranes. *J. Biol. Chem.* **261:** 7052.

Ferguson, K.M., T. Higashijima, M.D. Smigel, and A.G. Gilman. 1986. The influence of bound GDP on the kinetics of guanine nucleotide binding to G proteins. *J. Biol. Chem.* **261:** 7393.

Gilman, A.G. 1984. G proteins and dual control of adenylate cyclase. *Cell* **36:** 577.

———. 1987. G proteins: Transducers of receptor-generated signals. *Annu. Rev. Biochem.* **56:** 615.

Itoh, H., T. Katada, M. Ui, H. Kawasaki, K. Suzuki, and Y. Kaziro. 1988. Identification of three pertussis toxin substrates (41, 40, and 39 kDa proteins) in mammalian brain. *FEBS Lett.* **230:** 85.

Jelsema, C.L. and J. Axelrod. 1987. Stimulation of phospholipase A_2 activity in bovine rod outer segments by the $\beta\gamma$ subunits of transducin and its inhibition by the α subunit. *Proc. Natl. Acad. Sci.* **84:** 3623.

Kahn, R.A. and A.G. Gilman. 1984. Purification of a protein cofactor required for ADP-ribosylation of the stimulatory regulatory component of adenylate cyclase by cholera toxin. *J. Biol. Chem.* **259:** 6228.

———. 1986. The protein cofactor necessary for ADP-ribosylation of G_s by cholera toxin is itself a GTP binding protein. *J. Biol. Chem.* **261:** 7906.

Katada, T. and M. Ui. 1982a. Direct modification of the membrane adenylate cyclase system by islet-activating protein due to ADP-ribosylation of a membrane protein. *Proc. Natl. Acad. Sci.* **79:** 3129.

———. 1982b. ADP-ribosylation of the specific membrane protein of C6 cells by islet-activating protein association with modification of adenylate cyclase activiity. *J. Biol. Chem.* **257:** 7210.

Katada, T., M. Oinuma, and M. Ui. 1986a. Mechanisms for inhibition of the catalytic activity of adenylate cyclase by the guanine nucleotide-binding proteins serving as the substrate of islet-activating protein, pertussis toxin. *J. Biol. Chem.* **261:** 5215.

———. 1986b. Two guanine nucleotide-binding proteins in rat brain serving as the specific substrate of islet-activating protein, pertussis toxin. Interaction of the α-subunits with $\beta\gamma$-subunits in the development of their biological activities. *J. Biol. Chem.* **261:** 8182.

Katada, T., K. Kusakabe, M. Oinuma, and M. Ui. 1987a. A novel mechanism for the inhibition of adenylate cyclase via inhibitory GTP-binding proteins. Calmodulin-dependent inhibition of the cyclase catalyst by the $\beta\gamma$-subunits of GTP-binding proteins. *J. Biol. Chem.* **262:** 11897.

Katada, T., M. Oinuma, K. Kusakabe, and M. Ui. 1987b. A new GTP-binding protein in brain tissues serving as the specific substrate of islet-activating protein, pertussis toxin. *FEBS Lett.* **213:** 353.

Kikuchi, A., T. Yamashita, M. Kawata, K. Yamamoto, K. Ikeda, T. Tanimoto, and Y. Takai. 1988. Purification of characterization of a novel GTP-binding protein with a molecular weight of 24,000 from bovine brain membranes. *J. Biol. Chem.* **263:** 2897.

Logothetis, D.E., Y. Kurachi, J. Galper, E.J. Neer, and D.E. Clapham. 1987. The $\beta\gamma$ subunits of GTP-binding proteins activate the muscarinic K^+ channel in heart. *Nature* **325:** 321.

Murayama, T. and M. Ui. 1984. [^{3}H]GDP release from rat and hamster adipocyte membranes independently linked to receptors involved in activation or inhibition of adenylate cyclase. Differential susceptibility to two bacterial toxins. *J. Biol. Chem.* **259:** 761.

Oinuma, M., T. Katada, and M. Ui. 1987. A new GTP-binding protein in differentiated human leukemic (HL-60) cells serving as the specific substrate of islet-activating protein, pertussis toxin. *J. Biol. Chem.* **262:** 8347.

Streyer, L. and H.R. Bourne. 1986. G proteins: A famıly of signal transducers. *Annu. Rev. Cell Biol.* **2:** 391.

Ui, M. 1984. Islet-activating protein, pertussis toxin: A probe for functions of the inhibitory guanine nucleotide regulatory component of adenylate cyclase. *Trends Pharmacol. Sci.* **5:** 277.

———. 1986. Pertussis toxin as a probe of receptor coupling to inositol lipid metabolism. In *Receptor biochemistry and methodology. Phosphoinositides and receptor mechanisms* (ed J.W. Putney), vol. 7, p. 163. Alan R. Liss, New York.

———. 1988a. The multiple biological activities of pertussis toxin. In *Pathogenesis and immunity in pertussis* (ed. A.C. Wardlaw and R. Parton), p. 121, John Wiley, London.

———. 1988b. G proteins identified as pertussis toxin substrates. In *G proteins and calcium mobilization* (ed. P.H. Nachache). CRC Press, Boca Raton. (In press.)

Ui, M., K. Nogimori, and M. Tamura. 1985. Islet-activating protein, pertussis toxin: Subunit structure and mechanism for its multiple biological actions. In *Pertussis toxin* (ed. R.D. Sekura et al.), p. 19. Academic Press, New York.

Waldo, G.L., T. Evans, E.D. Fraser, J.K. Northrup, M.W. Martin, and T.K. Harden. 1987. Identification and purification from bovine brain of a guanine-nucleotide-binding protein distinct from G_s, G_i and G_o. *Biochem. J.* **246:** 431.

Functional Role of Muscarinic Acetylcholine Receptor Subtype Diversity

A. Ashkenazi,* E.G. Peralta,* J.W. Winslow,† J.Ramachandran,‡ and D.J. Capon*
*Departments of *Molecular Biology and †Developmental Biology, Genentech, Inc., South San Francisco, California 94080;*
‡Neurex Corporation, Menlo Park, California 94023

Multicellular organisms use many of the same basic mechanisms as do unicellular organisms to convey information from their environment to the interior of the cell. It is reasonable to believe that although greater numbers of cellular signal transduction mechanisms are used by multicellular organisms, these mechanisms will have often evolved as a result of the duplication and divergence of genes encoding relatively few types of signal transduction molecules. Indeed, molecular cloning efforts have consistently revealed that the components of the signal transduction machinery (receptors, pumps, channels, G proteins, effector enzymes, etc.) are members of families of structurally related proteins. By using many variations on a relatively small number of signal transduction themes, multicellular organisms can fine-tune the specificity of a response for a given cell or tissue or at a particular time during development.

To gain insight into how diversity of a given signal transduction component is successfully exploited, we have studied muscarinic acetylcholine receptors (mAChRs). The diversity in the functions of muscarinic receptors was amply suggested by their wide distribution throughout the central and autonomic nervous systems, their considerable pharmacological complexity, and their specific regulation during development (Nathanson 1987). Moreover, the physiological tasks performed by muscarinic receptors are also diverse, including the regulation of muscle contraction, exocrine gland secretion, and learning and memory. At the cellular level, these actions are accomplished by the activation of G proteins capable of regulating ion channels as well as second-messenger-generating enzymes such as adenylyl cyclase and phospholipase C.

Here we summarize the isolation of human mAChR genes, a family of at least four members, and our efforts to characterize the differences in their expression and in the functions of the proteins they encode. These studies have revealed that mAChR subtypes are expressed in a highly specific manner in cells and tissues and interact differently with both G proteins and the adenylyl cyclase and phospholipase C second-messenger systems they regulate, portraying a highly evolved language of acetylcholine-mediated communication.

EXPERIMENTAL PROCEDURES

The experimental procedures used in this work have been described in detail elsewhere (Ashkenazi et al. 1987; Peralta et al. 1987a,b, 1988a; and references cited therein).

RESULTS AND DISCUSSION

Molecular Cloning of Four Human Muscarinic Receptor Subtypes

Previous studies by S. Numa and colleagues and by our laboratory have led to the isolation of cDNAs encoding the porcine brain M1 (PM1) and porcine heart M2 (PM2) mAChR subtypes (Kubo et al. 1986a,b; Peralta et al. 1987b). These two mAChR subtypes are the polypeptide products of different genes, sharing 38% overall amino acid identity. Despite these differences, hydropathicity analysis suggests that PM1 and PM2 have a common structure of seven hydrophobic membrane-spanning domains similar to that proposed for β-adrenergic receptors and visual rhodopsins (Dixon et al. 1986; Nathans et al. 1986; Yarden et al. 1986). PM1 and PM2 share greatest amino acid identity within the first five transmembrane segments. These subtypes have a segment virtually identical to the β-adrenergic receptors and rhodopsins extending from the first cytoplasmic loop to the second transmembrane domain—a region proposed to be a site of G-protein interaction (Zucker et al. 1985). To isolate the genes for human mAChRs, we therefore used a hybridization probe containing the first five transmembrane domains of PM2 as well as two pools of oligonucleotides encoding the highly conserved region spanning the first cytoplasmic loop and second transmembrane region. A human genomic library was screened, since we anticipated that human mAChR genes like PM2 would lack introns in their coding region. We obtained four classes of clones on the basis of differential hybridization with the various probes (Peralta et al. 1987a). Two of these clones encoded polypeptides virtually identical to PM1 and PM2 and were designated HM1 and HM2, respectively. The two other classes of clones encoded novel human mAChR subtypes, designated HM3 and HM4, which share a high degree of identity with the trans-

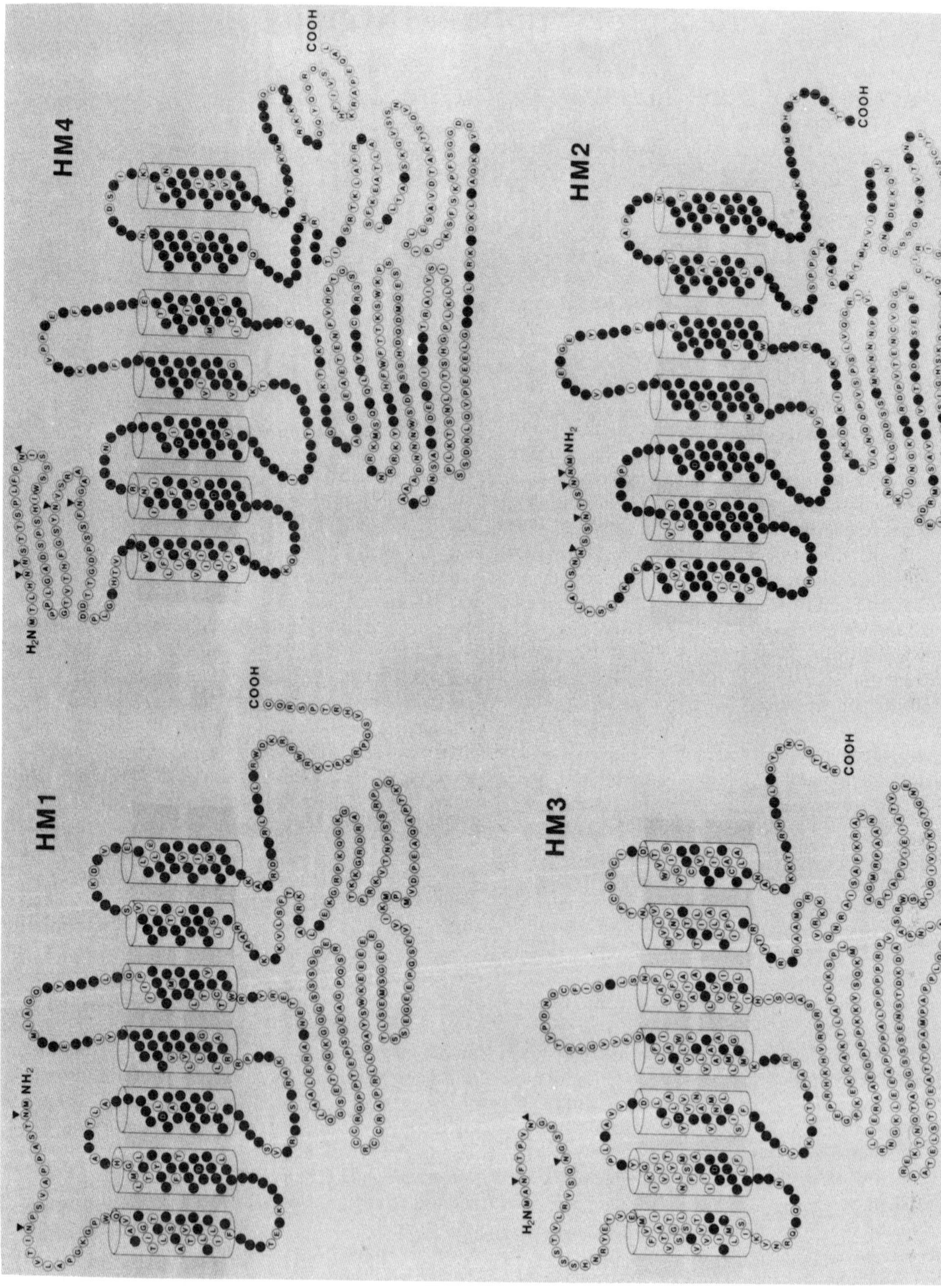

Figure 1. *(See facing page for legend.)*

membrane domains of HM1 and HM2, but are substantially different in other regions (Fig. 1).

Hydropathicity analysis supports the idea that all four human mAChR subtypes have a structure consisting of seven membrane-spanning domains, an extracellular amino terminus containing 2–5 potential N-linked glycosylation sites, and an intracellular carboxyl terminus (Fig. 1) (Peralta et al. 1988b). Two aspartic acid residues are conserved among all four subtypes within the otherwise hydrophobic second and third transmembrane domains. This suggests a model in which acetylcholine binds in a cleft of the receptor formed upon insertion of its hydrophobic domains into the lipid bilayer, and the complex is stabilized by an ionic interaction between the quaternary amine of acetylcholine and the negatively charged aspartic acid residues of the receptor.

HM1 and HM2 are proteins of 460 and 466 amino acids, respectively, with virtual amino acid identity (>97%) to their porcine homologs. HM3 is a polypeptide of 479 residues with 40%, 55%, and 37% identity to HM1, HM2, and HM4, respectively. The 184-amino-acid 5-6 loop of HM3 has significant identity (21%) with that of HM2 (181 residues) only (Fig. 1). HM4 is the largest subtype (590 amino acids) with 45%, 35%, and 37% overall identity to HM1, HM2, and HM3, respectively. The 241 residues of the HM4 5-6 loop region share 22% identity with HM1 (156 amino acids), but have little identity to HM2 and HM3 (Fig. 1). The greater similarity of the 5-6 loops between HM2 and HM3 and between HM1 and HM4 correlates with similarities in their functions (see below).

Selective Binding of Muscarinic Ligands to Muscarinic Receptor Subtypes

The classic pharmacological distinction between M_1 (high affinity) and M_2 (low affinity) mAChRs is based on binding to the antagonist pirenzepine (Hammer et al. 1980). Other compounds such as the cardioselective antagonist AF-DX116 have helped to identify still other subtypes of mAChRs (Hammer et al. 1986). Our identification of four mAChR gene products suggested that each could possess distinct ligand-binding properties; if so, this would provide a novel approach for elucidating the relationship of subtypes identified by pharmacological criteria. To characterize their ability to bind muscarinic ligands, we expressed each mAChR subtype individually in cells that lack endogenous mAChRs (Peralta et al. 1987a). Nonselective antagonists like atropine bound to all four subtypes with comparable affinities, whereas selective antagonists including pirenzepine and AF-DX116 bound to the four recombinant mAChR subtypes with significantly different affinities (Fig. 2). For example, pirenzepine is bound by HM1 with an affinity 25-fold greater than HM2, whereas HM2 binds AF-DX116 significantly bet-

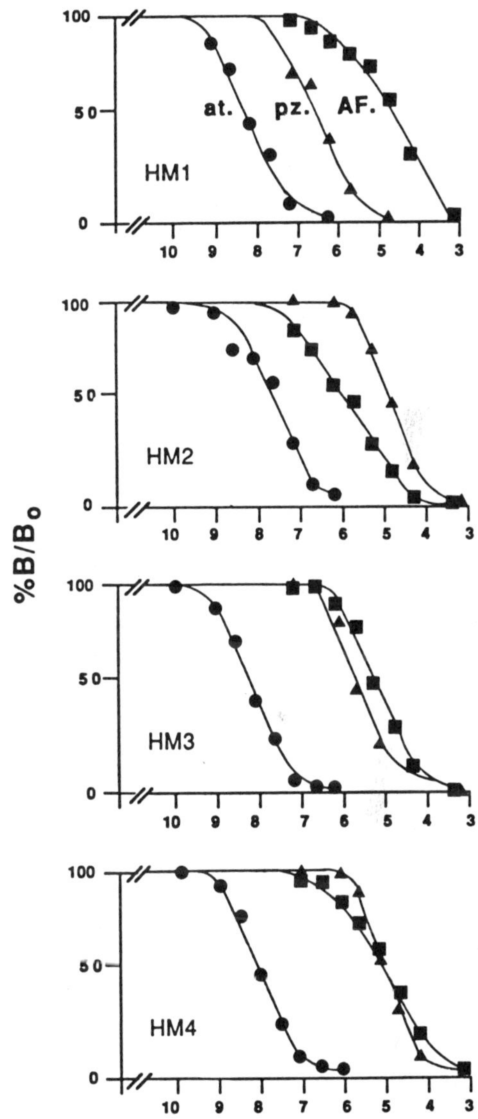

Figure 2. Antagonist-binding properties of recombinant human mAChR subtypes expressed in human kidney cell line. Competition displacement of [^{3}H]QNB binding to intact cells expressing individual recombinant mAChR subtypes by the muscarinic antagonists atropine (●); pirenzepine (▲), and AF-DX116 (■).

Figure 1. Predicted transmembrane domain structure and amino acid sequence comparison of the four human mAChR subtypes. The shaded residues superimposed on the structural models indicate the following comparisons: (HM1) amino acid identities shared between all four human mAChRs; (HM2) identities between HM2 and HM3; (HM3) identities shared between all four human mAChRs and the human, hamster, and avian β-adrenergic receptors; (HM4) identities shared between HM1 and HM4. (●) The two aspartic acid residues conserved between these four mAChR subtypes and β-adrenergic receptors. (▼) Potential N-linked glycosylation sites found in the predicted extracellular regions of the receptors.

ter than HM1. The binding of HM1 and HM2 to these antagonists is thus similar to that of the mAChRs found in the tissues from which these subtypes were cloned: M_1 receptors of cerebral cortex and M_2 receptors of atrium. HM3 and HM4 bind pirenzepine and AF-DX116 with intermediate affinities, consistent with the properties of mAChR subtypes suggested more recently by pharmacological studies.

Differences were similarly observed in the agonist-binding properties of the four mAChRs (Peralta et al. 1987a). Significantly, each subtype displayed both high- and low-affinity binding to a given muscarinic agonist. Because high-affinity binding is believed to result from G-protein interaction, this finding suggests that each subtype is distinct in its ability to interact with the G proteins resident in these cells. Indeed, treatment of cell homogenates with pertussis toxin (PTX) or with guanine nucleotides, either of which uncouples receptors from G proteins, was found to abolish the high-affinity agonist binding of PM2, confirming that high-affinity agonist binding is a property of individual mAChR subtypes resulting from G-protein interaction (Ashkenazi et al. 1987).

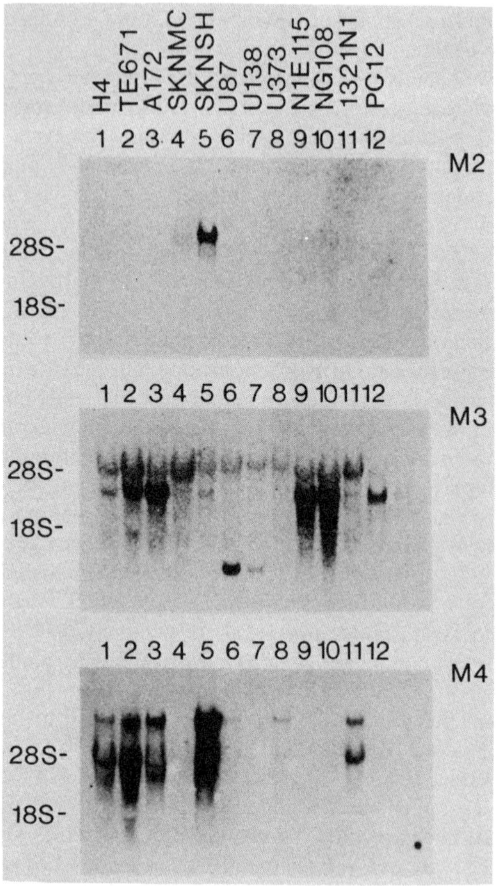

Figure 3. Cell-specific transcription of mAChR subtype genes. Approximately 5 μg of poly(A)$^+$ mRNA from each cell line was analyzed by Northern blot analysis with ^{32}P-labeled mAChR subtype-specific probes. The types of cell lines are as follows: (H4) human neuroglioma; (TE671) human medulloblastoma; (A172) human glioblastoma; (SK-N-MC and SK-N-SH) human neuroblastoma; (U87, U138, and U373) human glioblastoma; (1321N1) human astrocytoma; (N1E-115) mouse neuroblastoma; (NG108-15) mouse neuroblastomaX-rat glioma; (PC-12) rat adrenal pheochromocytoma. Autoradiographs are shown for M2-, M3-, and M4-probed blots. M1, which gave no detectable signal in these cell lines, is not shown. The mobility of ribosomal 28S and 18S RNA is indicated.

Tissue- and Cell-specific Transcription of Muscarinic Receptor Genes

We have examined the patterns of mAChR gene transcription in tissues in which mAChRs are known to regulate biological functions. Northern blot analysis of poly(A)$^+$ mRNA was carried out with subtype-specific cDNA probes encoding the unique 5-6 loop regions (Peralta et al. 1987a). Significant expression of all four mAChRs was observed in whole brain, whereas in the cerebral cortex only M1, M3, and M4 expression was detected. Specific expression of a single subtype was observed in heart (M2) and pancreas (M4). To determine whether mAChR subtype transcription can be highly specific at the cellular level, as it was found to be in bulk for these tissues, we carried out a similar analysis with a number of brain-derived cell lines (J.W. Winslow et al., in prep.). Most of these cell lines expressed either M3, M4, or both subtypes, whereas M1 expression was not detected, and M2 was expressed only in a single line (Fig. 3). Certain cell lines expressed a single subtype only, for example, M3 in NG108-15 neuroblastoma × glioma and N1E-115 neuroblastoma cells, and M4 in 1321N1 astrocytoma cells. In other cell lines, coexpression of subtypes was observed, including M3 and M4 in H4 neuroglioma, TE671 medulloblastoma, and A172 glioblastoma cells, and M2 and M4 in SK-N-SH neuroblastoma cells.

These results show that mAChR subtype expression can be highly specific in different cells of a given tissue such as brain. Given the different functions of mAChR subtypes (see below), this finding suggests that subtype-specific expression is important in determining the specific response of a given cell to acetylcholine. Moreover, certain cells appear to express more than one functional mAChR subtype, suggesting that acetylcholine can exert more than one subtype-defined effect, or set of effects, on a given cell.

Differential Regulation of Adenylyl Cyclase and Phosphatidylinositol (PtdIns) Hydrolysis by Muscarinic Receptor Subtypes

Signal transduction by mAChRs is carried out through activation of specific G proteins, which in turn regulate the activity of effector proteins such as enzymes and ion channels (Nathanson 1987). The structural diversity of mAChR subtypes, their distinct patterns of expression, and the number of functions they regulate suggest that the different mAChRs might have different functions. Indeed, experiments with tissue homogenates have provided evidence that different mAChR subtypes couple to different effectors; for example, the M_1-specific antagonist pirenzepine blocks muscarinic activation of PtdIns hydrolysis in rat brain more efficiently than muscarinic inhibition of adenylyl cyclase in rat heart (Gil and Wolfe 1985). However, the complex pattern of mAChR subtype expression in different cells and tissues limits the interpretation of such studies. We have approached this question by studying the functional properties of cloned mAChRs expressed individually in cells that ordinarily lack mAChRs.

We established cell lines stably transfected with vectors directing the expression of each of the four human mAChRs by cotransfection with a plasmid conferring neomycin resistance and selection using the neomycin analog, G418 (Peralta et al. 1988a). Cell lines express ing specific receptor numbers were isolated by single-cell cloning. Cell lines were obtained with comparable levels of each mAChR subtype (121,000–199,000 receptors/cell), as determined by whole-cell binding to the muscarinic antagonist [^{3}H]quinuclidinyl benzylate (QNB). Untransfected cells expressed fewer than 400 mAChRs per cell (Table 1).

To investigate the ability of each mAChR to inhibit adenylyl cyclase activity, we compared the effects of the muscarinic agonist carbachol on intracellular levels of cAMP following direct stimulation of adenylyl cyclase by forskolin (Fig. 4) (Peralta et al. 1988a). The phosphodiesterase inhibitor isobutylmethylxanthine (IBMX) was included to ensure that changes in cAMP levels were not due to changes in phosphodiesterase activity. In cells expressing HM2 or HM3, carbachol inhibited forskolin-stimulated accumulation of cAMP by up to 66% or 74%, respectively, with EC_{50} values of 3.7×10^{-8} M and 1.3×10^{-8} M (Table 1). No significant inhibition of cAMP accumulation was observed in HM1- or HM4-expressing cells; instead, at higher carbachol concentration, cAMP accumulation was stimulated above the forskolin-induced levels by 9.8- and 7.6-fold, respectively, with EC_{50} values of 2×10^{-6} M and 8×10^{-6} M (Fig. 4).

The same transfected cell lines expressing comparable levels of a single mAChR subtype were examined for the ability of each subtype to activate PtdIns hydrolysis (Peralta et al. 1988a). We analyzed the kinetics of carbachol-stimulated formation of inositol mono-, bis-, and trisphosphates ($InsP_1$, $InsP_2$, and $InsP_3$) resulting from phospholipase-C-catalyzed hydrolysis of PtdIns bisphosphate (Berridge 1987). All four mAChR subtypes significantly activated PtdIns hydrolysis (Figs. 5 and 6); however, the activation by HM1 and HM4 both resulted in substantially greater accumulation of all three inositol phosphates and occurred at much lower carbachol concentrations (Table 1). In addition, $InsP_2$ and $InsP_3$ formation stimulated by HM1 and HM4 had a prolonged duration; in contrast, levels of these inositol phosphates peaked within 2 minutes of carbachol treatment of HM2- and HM3-expressing cells (Figs. 5 and 6).

Notably, the EC_{50} values for PtdIns hydrolysis and cAMP stimulation mediated by HM1 and HM4 were

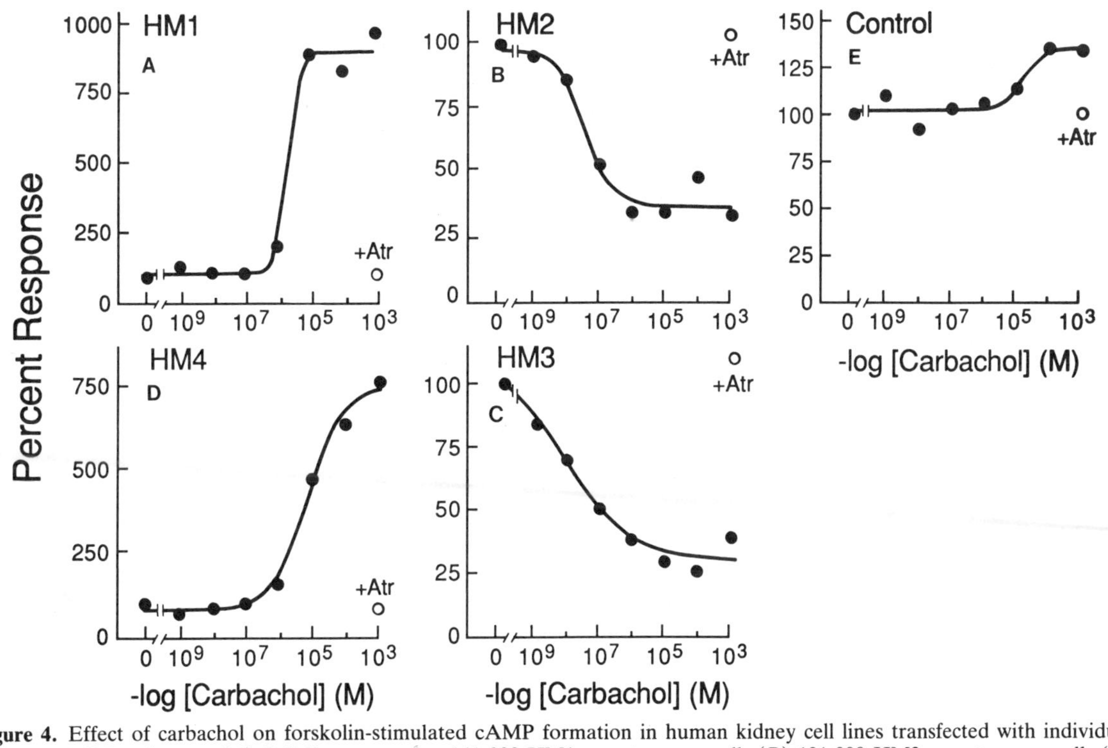

Figure 4. Effect of carbachol on forskolin-stimulated cAMP formation in human kidney cell lines transfected with individual human mAChR subtypes. (*A*) Cell line expressing 141,000 HM1 receptors per cell; (*B*) 121,000 HM2 receptors per cell; (*C*) 168,000 HM3 receptors per cell; (*D*) 199,000 HM4 receptors per cell; (*E*) control (untransfected) cells. (○) Response in the presence of 10 μM atropine.

Table 1. Effect of Carbachol on Inositol Phosphate and cAMP Formation

mAChR cell lines	[³H]QNB sites/cell ($\times 10^{-3}$)	[³H]QNB K_D(pM)	Maximal accumulation of inositol phosphates (cpm/10^6 cells)						EC_{50} carbachol (M)		
			$InsP_1$	time (min)	$InsP_2$	time (min)	$InsP_3$	time (min)	$InsP_1$ (30 min)	cAMP inhibition	cAMP stimulation
Control	0.3 ± 0.1	19 ± 8	240 ± 11	(1)	79 ± 4	(5)	80 ± 10	(0.25, 1)	7.0×10^{-6}		14.0×10^{-6}
HM1	141 ± 7.8	58 ± 12	8,906 ± 105	(30)	2,984 ± 25	(20)	1,378 ± 26	(20)	1.0×10^{-6}		2.0×10^{-6}
HM2	121 ± 3.3	44 ± 1	1,528 ± 343	(30)	154 ± 19	(2)	247 ± 13	(0.5–2)	16.0×10^{-6}	3.7×10^{-8}	
HM3	24 ± 0.8	17 ± 5	359 ± 5	(20)	80 ± 12	(1–5)	207 ± 50	(1)			
HM3	168 ± 28.6	75 ± 24	477 ± 85	(30)	158 ± 17	(0.5)	195 ± 26	(0.5)	4.0×10^{-6}	1.3×10^{-8}	
HM3	306 ± 30.7	94 ± 15	479 ± 2	(30)	177 ± 50	(1)	219 ± 46	(0.5, 1)			
HM4	13 ± 0.3	15 ± 1	1,075 ± 145	(30)	293 ± 30	(20)	239 ± 58	(1)			
HM4	199 ± 3.4	19 ± 3	14,112 ± 1621	(30)	4,075 ± 1093	(5)	1,390 ± 259	(5)	1.6×10^{-6}		8.0×10^{-6}

The number of recombinant mAChRs expressed is shown by the number of [³H]QNB-binding sites per cell. The maximal accumulation of inositol phosphates was determined at the incubation time points (shown in parentheses) resulting in the highest level of each hydrolysis product. The level of each inositol phosphate generated in the absence of carbachol was subtracted, and the data represent the mean ±S.D. of two experiments or the mean ±S.E.M. of three experiments performed in duplicate.

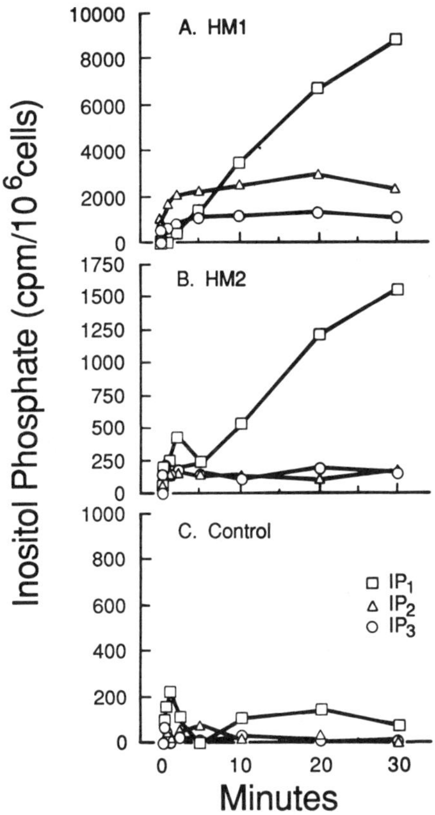

Figure 5. Time course of carbachol-stimulated accumulation of inositol phosphates in transfected human kidney cell lines expressing the HM1 and HM2 mAChR subtypes. (*A*) Cell line expressing 141,000 HM1 receptors per cell; (*B*) cell line expressing 121,000 HM2 receptors per cell; (*C*) control cells expressing ~400 endogenous mAChRs per cell. (□) $InsP_1$; (△) $InsP_2$; (○) $InsP_3$. The basal level of each inositol phosphate was subtracted from each time point.

comparable (Table 1). This suggests that the increases in cAMP by HM1 and HM4 may be a secondary effect of their efficient activation of PtdIns hydrolysis. Indeed, previous studies have demonstrated "cross-talk" between these two second-messenger systems: adenylyl cyclase can be stimulated by cellular factors such as protein kinase C and calcium-dependent calmodulin (Salter et al. 1981; Nishizuka 1986; Yoshimata et al. 1987), which are activated by PtdIns hydrolysis as a result of diacylglycerol formation and $InsP_3$-mediated calcium mobilization (Berridge 1987).

Although the EC_{50} values for HM1- and HM4-mediated PtdIns hydrolysis and cAMP stimulation were comparable, much higher carbachol levels were needed for HM2- and HM3-mediated stimulation of PtdIns hydrolysis than inhibition of adenylyl cyclase (Table 1). Similarly, a carbachol dose-response analysis revealed that the PM2 mAChR subtype is significantly more efficient in the inhibition of adenylyl cyclase activity than in the stimulation of PtdIns hydrolysis in Chinese hamster ovary (CHO) cells (Fig. 7) (Ashkenazi et al. 1987). Another indication that PM2 is coupled more efficiently to adenylyl cyclase inhibition than to PtdIns hydrolysis stimulation was revealed in studies with CHO cells expressing different PM2 levels (Fig. 8). The stimulation of PtdIns hydrolysis by PM2 was highly dependent on the receptor number reaching a plateau above 1.45×10^6 receptors per cell, whereas the inhibition of adenylyl cyclase was similar over a range of receptor numbers between 0.24×10^6 and 2.5×10^6 receptors per cell (Fig. 8).

These results demonstrate that the functions of HM2 and HM3 can be clearly distinguished from those of HM1 and HM4 by two independent criteria: the ability to regulate intracellular levels of the second messenger

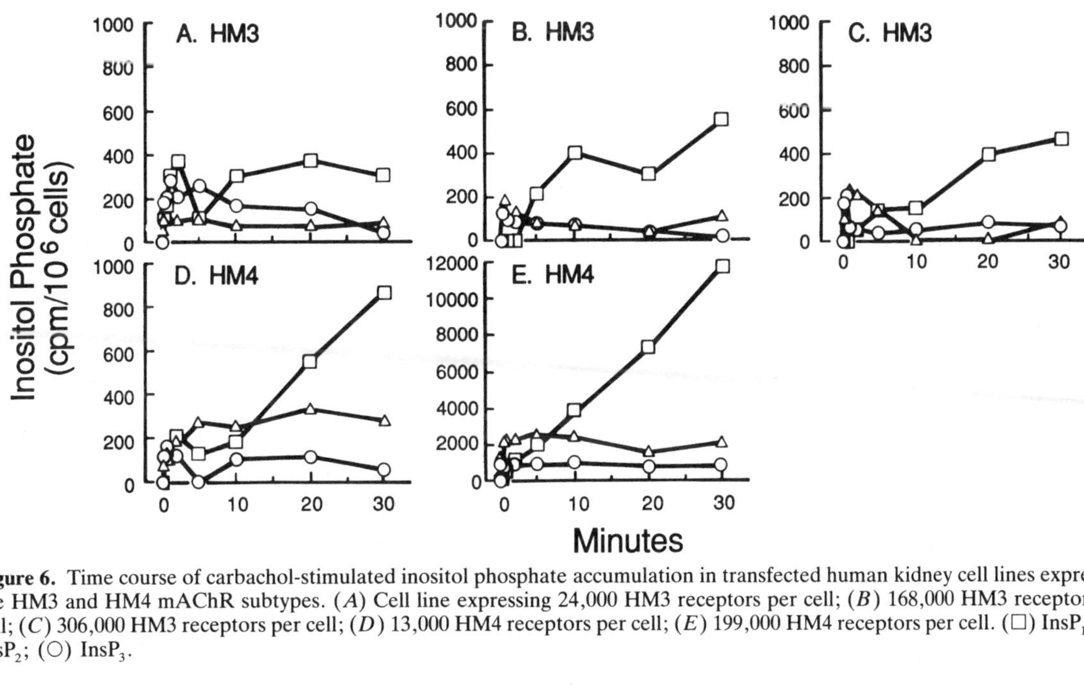

Figure 6. Time course of carbachol-stimulated inositol phosphate accumulation in transfected human kidney cell lines expressing the HM3 and HM4 mAChR subtypes. (*A*) Cell line expressing 24,000 HM3 receptors per cell; (*B*) 168,000 HM3 receptors per cell; (*C*) 306,000 HM3 receptors per cell; (*D*) 13,000 HM4 receptors per cell; (*E*) 199,000 HM4 receptors per cell. (□) $InsP_1$; (△) $InsP_2$; (○) $InsP_3$.

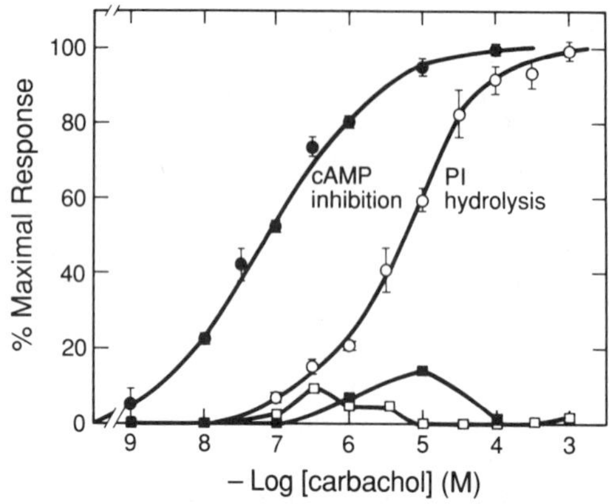

Figure 7. Comparison of carbachol dose response for inhibition of adenylyl cyclase and activation of PtdIns hydrolysis by recombinant PM2 mAChR expressed in CHO cells. Inhibition of forskolin-induced accumulation of cAMP (●, ■) and stimulation of inositol phosphate accumulation (○, □) by carbachol in CHO cells expressing 1,450,000 PM2 receptors per cell (●, ○), and in untransfected cells (■, □), which express less than 100 endogenous mAChRs per cell. The maximal inhibition of cAMP accumulation 76.4 ± 15.0%, and the maximal activation of inositol phosphate accumulation was 4.5 ± 0.5-fold.

cAMP and the ability to regulate phosphoinositide hydrolysis. These results also show that a given subtype can be coupled to more than one effector and can be more efficient in activating one response than another. This suggests that the exact response of a given cell to a given neurotransmitter, such as acetylcholine, will depend significantly on the concentration of the neurotransmitter, even if that cell expresses only a single subtype of the receptor that recognizes that neurotransmitter.

Muscarinic Receptor Subtypes Are Selectively Coupled by Different G Proteins to PtdIns Hydrolysis

The substantial difference in the magnitude of PtdIns hydrolysis activated by HM1 and HM4 compared to HM2 and HM3 in a given cell suggested the existence of different pathways for the coupling of these receptors to PtdIns hydrolysis. It has been shown previously that multiple G proteins (termed Gp) can couple receptors to PtdIns hydrolysis (Cockroft 1987); these Gp proteins can be distinguished by sensitivity to PTX, which ADP-ribosylates certain G proteins, thus uncoupling them from receptors. To determine whether distinct Gp proteins couple different mAChR subtypes to PtdIns hydrolysis in CHO cells, we investigated the effects of this toxin on PtdIns hydrolysis stimulated by carbachol, as well as by agonists to two G-protein-coupled receptors endogenous to CHO cells, thrombin, and cholecystokinin (CCK) (A. Ashkenazi et al., in prep.). Treatment of CHO cells with up to 100 ng/ml PTX, which is sufficient to ADP-ribosylate all detectable PTX substrate in these cells (Ashkenazi et al. 1987), had no effect on PtdIns hydrolysis stimulated by CCK receptors (Fig. 9). In contrast, the response stimulated by thrombin receptors was completely inhibited by 10–100 ng/ml PTX. These results reveal the existence of two types of Gp in CHO cells, distinguished by their sensitivity to PTX. PTX treatment of cells expressing PM2 mAChRs completely abolished carbachol-stimulated PtdIns hydrolysis, indicating that PM2 couples to a Gp similar or identical to that employed by thrombin receptors. In contrast, PTX treatment of HM1- or HM4-expressing cells resulted in partial inhibition (maximally 20–50%) of carbachol-stimulated PtdIns hydrolysis (Fig. 9). This partial in-

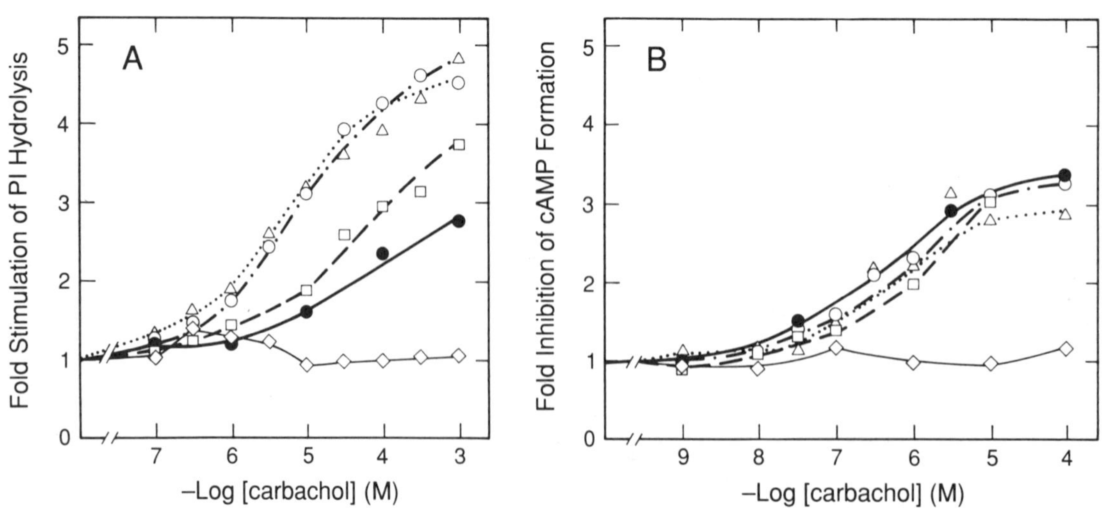

Figure 8. Effect of the expression level of PM2 mAChRs on the stimulation of PtdIns hydrolysis and inhibition of adenylyl cyclase. CHO cells expressing 2.4×10^5 (●); 6×10^5 (□); 1.45×10^6 (○); or 2.5×10^6 (△) PM2 receptors per cell and untransfected cells (◇) were analyzed for carbachol-stimulated PtdIns hydrolysis (*A*) or inhibition of adenylyl cyclase (*B*) as in Fig. 7.

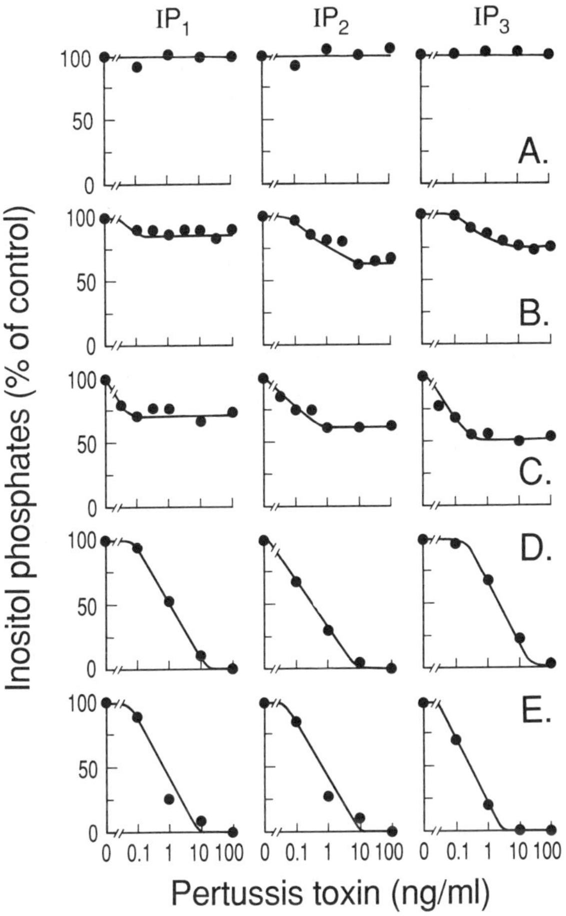

Figure 9. Effect of PTX on agonist-stimulated PtdIns hydrolysis in CHO cells transfected with mAChRs. Untransfected CHO cells (*A,E*) or cell lines expressing HM1 (3.8×10^5 receptors per cell) (*B*); HM4 (1.9×10^5 receptors per cell) (*C*); and PM2 (1×10^6 receptors per cell) (*D*) were treated with PTX and analyzed for accumulation of inositol phosphates stimulated by CCK (*A*), carbachol (*B–D*), or thrombin (*E*).

hibition was complete at 10 ng/ml PTX; increasing the concentration of PTX to 100 ng/ml had no further inhibitory effect. The biphasic inhibition curves seen for HM1 and HM4 suggest that more than one Gp protein is responsible for the coupling of these receptors to PtdIns hydrolysis: Each subtype may be coupled to both a PTX-sensitive and a PTX-insensitive Gp, or alternatively, to a third Gp whose activity is merely attenuated by PTX. These results indicate that HM1 and HM4 activate PtdIns hydrolysis by a Gp-protein-defined pathway(s) that is distinct from that used by PM2, HM2, and HM3. Thus, the ability of different receptors, or of different receptor subtypes, to activate different magnitudes of a given response in the same cellular context is facilitated by selective coupling by different G proteins.

CONCLUSION

We have studied muscarinic receptors to learn how the diversity of a specific component of the signal transduction machinery can be used to regulate diverse physiological functions in multicellular organisms. By utilizing heterologous expression of cloned genes in mammalian cells, we have demonstrated that individual mAChR subtypes are functionally distinct, that is, they regulate the formation of distinct cellular second messengers upon binding of acetylcholine. Thus, HM1 and HM4 are very efficient in stimulating the hydrolysis of polyphosphoinositides by phospholipase C, whereas HM2 and HM3 are very efficient in inhibiting the synthesis of cAMP by adenylyl cyclase. Further differences are to be expected between subtypes, for example, in their ability to regulate ion channels. This functional specialization strongly supports the idea that multiple subtypes of a given receptor have evolved to economically accomplish diversity in signal transduction.

The molecular cloning of mAChR subtypes has revealed significant structural diversity, which parallels the functional specification of these receptors. mAChR subtypes share a great degree of structural identity in specific regions, but diverge dramatically in others. Conserved structural features are likely to be involved in generic functions shared by all mAChR subtypes, such as the binding of acetylcholine, whereas diverged features are likely to be involved in determining functional specificity, such as coupling to different G proteins and effector enzymes. Indeed, subtypes that exhibit similarity in their coupling to phospholipase C or adenylyl cyclase share a relatively higher degree of identity within otherwise diverged regions, especially the 5-6 loop, a region predicted to be involved in interaction with G proteins. By the same token, significant structural differences between subtypes that exhibit similarity in function, such as HM2 or HM3, suggest that specific functional differences, perhaps more subtle, indeed exist between such subtypes.

In addition to the structural and functional diversity of mAChR subtypes, our investigation has demonstrated that the transcription of specific mAChR genes in tissues is highly specific. In certain tissues, such as heart or pancreas, only a single subtype is expressed, whereas in other tissues, such as brain, multiple subtypes are expressed. Even in different cells of a given tissue, such as brain, specific patterns of expression are observed. Some cell types express a single mAChR subtype, whereas others express multiple subtypes. These findings, together with the functional diversity of mAChR subtypes, indicate that the identity of mAChR subtype(s) expressed in a given cell is a most crucial determinant of the nature of the response that acetylcholine elicits in this cell. In cells that express a single mAChR subtype, the response will be relatively simple. In contrast, in cells that express more than one subtype, more than one type of response will be elicited by acetylcholine, and cross-talk between responses may serve to accomplish even greater diversity. The observation that a single mAChR subtype can regulate more than one second-messenger system indicates that cells that express a single subtype may also respond in a complex manner to acetylcholine. In such cases, because the efficiency of coupling of a single receptor subtype to different second messengers can differ, the level of this receptor in the cell and the local concentration of acetylcholine can also become critical determinants to the nature of the response. Yet another type of diversity is demonstrated by the ability of different receptor subtypes to activate a common second-messenger system, such as the phosphoinositide system, to different degrees by coupling to different regulatory G proteins.

The muscarinic receptor system is a revealing example of how duplication and divergence of genes encoding a given type of signal-transducing molecules are successfully used by multicellular organisms to accomplish complex regulation. Structural diversity within the muscarinic receptor family gives rise to functional specificity, which, together with differential cell-specific expression, creates a complex array of acetylcholine-mediated communication that facilitates the specific regulation and fine-tuning of diverse physiological functions at the level of tissues and of the whole organism.

ACKNOWLEDGMENTS

We thank Scot Marsters for assistance in tissue culture, Carol Morita and Kerrie Andow for preparation of figures, and Rebecca Ward for useful comments on this manuscript. This work was supported by Genentech, Inc. and by a National Institutes of Health grant to J.R.

REFERENCES

Ashkenazi, A., J.W. Winslow, E.G. Peralta, G.L. Peterson, M.I. Schimerlik, D.J. Capon, and J. Ramachandran. 1987. An M2 muscarinic receptor subtype coupled to both adenylyl cyclase and phosphoinositide turnover. *Science* **238:** 672.

Berridge, M.J. 1987. Inositol trisphosphate and diacylglycerol: Two interacting second messengers. *Annu. Rev. Biochem.* **56:** 159.

Cockroft, S. 1987. Polyphosphoinositide phosphodiesterase: Regulation by a novel guanine nucleotide binding protein, Gp. *Trends Pharmacol. Sci.* **12:** 75.

Dixon, R., B. Kobilka, D. Strader, J. Benovic, H. Dohlman, T. Ferielle, M. Bolanowski, C. Bennett, E. Rands, R. Diehl, R. Mumford, E. Slater, I. Sigal, M. Caron, R. Lefkowitz, and C. Strader. 1986. Cloning of the gene and cDNA for mammalian beta adrenergic receptor and homology with rhodopsin. *Nature* **321:** 75.

Gil, D.N. and B.J. Wolfe. 1985. Pirenzepine distinguishes between muscarinic receptor mediated phosphoinositide breakdown and inhibition of adenylate cyclase. *J. Pharmacol. Exp. Ther.* **232:** 608.

Hammer, R., C.P. Berrie, N.J.M. Birdsall, A.S.V. Burgen, and E.C. Hulme. 1980. Pirenzepine distinguishes between different subclasses of muscarinic receptors. *Nature* **283:** 90.

Hammer, R., E. Giraldo, G.B. Schiavi, E. Monferini, and H. Ladinski. 1986. Binding profile of a novel cardioselective muscarinic receptor antagonist, AF-DX 116, to membranes of peripheral tissues and brain in rat. *Life Sci.* **38:** 1653.

Kubo, T., A. Maeda, K. Sugimoto, I. Akiba, A. Mikami, H. Takahashi, T. Haga, K. Haga, A. Ichiyama, K. Kangawa, H. Matsuo, T. Hirose, and S. Numa. 1986a. Primary structure of porcine cardiac muscarinic acetylcholine receptor deduced from the cDNA sequence. *FEBS Lett.* **209:** 367.

Kubo, T., K. Fukuda, A. Mikami, A. Maeda, H. Takahashi, M. Mishina, T. Haga, K. Haga, A. Ichiyama, K. Kangawa, M. Kojima, H. Matsuo, T. Hirose, and S. Numa. 1986b. Cloning, sequencing and expression of cDNA encoding the muscarinic acetylcholine receptor. *Nature* **323:** 411.

Nathans, J., D. Thomas, and D.S. Hogness. 1986. Molecular genetics of human color vision. *Science* **232:** 193.

Nathanson, N.M. 1987. Molecular properties of the muscarinic acetylcholine receptor. *Annu. Rev. Neurosci.* **10:** 195.

Nishizuka, Y. 1986. Studies and perspectives of protein kinase C. *Science* **233:** 305.

Peralta, E.G., A. Ashkenazi, J.W. Winslow, J. Ramachandran, and D.J. Capon. 1988a. Differential regulation of PI hydrolysis and adenylyl cyclase by muscarinic receptor subtypes. *Nature* **334:** 434.

Peralta, E.G., A. Ashkenazi, J.W. Winslow, D.H. Smith, J. Ramachandran, and D.J. Capon. 1987a. Distinct primary structures, ligand binding properties and tissue specific expression of four human muscarinic acetylcholine receptors. *EMBO J.* **6:** 3923.

Peralta, E.G., J.W. Winslow, A. Ashkenazi, D.H. Smith, J. Ramachandran, and D.J. Capon. 1988b. Structural basis of muscarinic acetylcholine receptor subtype multiplicity. *Trends Pharmacol. Sci.* (Feb. suppl.), p.6.

Peralta, E.G., J.W. Winslow, G.L. Peterson, D.H. Smith, A. Ashkenazi, J. Ramachandran, M.I. Schimerlik, and D.J. Capon. 1987b. Primary structure and biochemical properties of an M2 muscarinic receptor. *Science* **236:** 600.

Salter, R.S., M.H. Krinks, C.B. Klee, and E.J. Neer. 1981. Calmodulin activates the isolated catalytic unit of brain adenylate cyclase. *J. Biol. Chem.* **256:** 9830.

Yarden, Y., H. Rodriguez, S.K.F. Wong, D.R. Brandt, D.C. May, J. Burnier, R.N. Harkins, E.Y. Chen, J. Ramachandran, and A. Ulrich. 1986. The avian beta adrenergic receptor: Primary structure and membrane topology. *Proc. Natl. Acad. Sci.* **83:** 6795.

Yoshimata, T.D., R. Sibley, M. Bouvier, R.J. Lefkowitz, and M.G. Caron. 1987. Cross talk between cellular signalling pathways suggested by phorbol ester induced adenylate cyclase phosphorylation. *Nature* **327:** 67.

Zucker, C., A. Cowman, and G. Rubin. 1985. Isolation and structure of a rhodopsin gene from *D. melanogaster*. *Cell* **40:** 851.

ras Oncogene Proteins: Three-dimensional Structures, Functional Implications, and a Model for Signal Transducer

S.-H. KIM,* A.M. DE VOS,* L. TONG,* M.V. MILBURN,* P.M. MATIAS,* J. JANCARIK,* E. OHTSUKA,† AND S. NISHIMURA‡

**Department of Chemistry and Lawrence Berkeley Laboratory, University of California, Berkeley, California 94720; †Faculty of Pharmaceutical Sciences, Hokkaido University, Sapporo, Japan; ‡Biology Division, National Cancer Center Research Institute, Chuo-ku, Tokyo, Japan*

ras oncogene products (for review, see Barbacid 1987) are of interest for at least two reasons: First, the most commonly found oncogenes isolated from human tumors or transformed human cell lines belong to the *ras* gene family (for a survey, see Nishimura and Sekiya 1987); second, *ras* gene products are thought to belong to the family of signal transducer proteins, G proteins (for review, see Stryer and Bourne 1986; Gilman 1987; Neer and Clapham 1988). In the human genome, there are three distinct cellular *ras* genes: c-Ha-*ras*, c-Ki-*ras*, and N-*ras*. Proteins encoded by these genes, p21 proteins, have a molecular weight of 21,000 with 188 or 189 amino acid residues, bind GDP/GTP (Scolnick et al. 1979), and have practically identical amino acid sequences, with the exception of the carboxy-terminal 25 residues. The difference between the normal *ras* proto-oncogenes and the activated (transforming) *ras* oncogenes is usually a single base change at one of a few critical positions, resulting in a single amino acid substitution in p21 (Reddy et al. 1982; Tabin et al. 1982; Taparowsky et al. 1982). Transgenic mice harboring an activated *ras* gene develop tumors (Quaife et al. 1987).

In addition to their roles in human and animal cancer, the *ras* genes have been implicated as playing crucial roles in cell proliferation, terminal differentiation, and other essential cellular processes, suggesting that *ras* proteins may participate as signal transducers, like G proteins, at the beginning of the cascade of reactions leading to these processes. Overall sequence homology between the p21 proteins and the α subunit of G proteins is rather limited, but the homology is high in a few regions that form a putative guanine-nucleotide binding site in both proteins (Hurley et al. 1984; Lochrie et al. 1985).

We have recently determined the crystal structures of the "catalytic domain" of a normal (de Vos et al. 1988) and two transforming c-Ha-*ras* p21 proteins (L. Tong et al., in prep.) lacking the flexible 18 residues at the carboxyl termini (Willumsen et al. 1985). We present here features of these crystal structures relevant to the known biochemical properties and functions, describe the differences between normal and transforming p21 proteins in the GDP-bound form, propose a model of the α subunit of G proteins derived on the basis of crystal structure of p21, and suggest a possible mechanism for conformational change on GDP/GTP exchange.

EXPERIMENTAL PROCEDURES

Expression and purification of soluble proteins. p21 proteins used for the structural studies were prepared according to the method of Miura et al. (1986). Water-soluble p21 proteins were expressed in *Escherichia coli* harboring plasmids that contained the synthetic genes coding for amino acid residues 1–171, and others coding for residues 1–189 of normal as well as transforming c-Ha-*ras* oncoproteins. The genetic codes used for the genes were those preferred by *E. coli*. The proteins were purified by essentially the same procedure as that reported by Gibbs et al. (1984). The purified proteins had the same amino-terminal sequence of Met-Thr-Glu-Tyr-Lys as the sequence predicted from the cDNA clones (Goldfarb et al. 1982; Pulciani et al. 1982; Reddy et al. 1982; Shih and Weinberg, 1982; Capon et al. 1983). As detected by complex formation with tritium-labeled GDP, catalytic domains of normal p21 and activated p21, as well as full-length p21 proteins, bind GDP. GTPase activity of the catalytic domain of normal p21 was found to be the same as that of full-length normal p21, whereas the catalytic domain and full-length p21 with mutations at residues 12 (Val-12) or 61 (Leu-61) showed, as predicted, almost no detectable GTPase activity (data not shown).

Crystallization and diffraction data collection. Single crystals of catalytic domains of both Val-12 and Leu-61 mutants were obtained by the same method (Jancarik et al. 1988) as that used for normal p21 protein, i.e., the sitting drop technique from solutions containing 10 mg/ml protein, 0.05 M $CaCl_2$, 0.075 M HEPES buffer (pH 7.5), 1 mM EDTA, 0.5 mM DTT, and 0.005% *n*-octyl glucoside equilibrated to 30% PEG 400. The space group of all three catalytic domain crystals is $P6_522$ with cell parameters a = b = 83.2 Å and c = 105.1 Å. There is one p21 catalytic domain per asymmetric unit, and 46% of the unit cell volume is protein. Diffraction data were collected on X-ray films using an Enraf-Nonius rotation camera installed on the 8-pole wiggler branch line at the Stanford Synchrotron Radiation Laboratory, Palo Alto, California.

Structure determination. For the determination of the crystal structure of the normal p21 protein (de Vos et al. 1988), five heavy-atom derivatives were used: $KAuI_4$, K_2HgI_4, and K_2PtBr_6 soaks, $KAuI_4$ and K_2HgI_4 double soak (Au/Hg), and $KAuI_4$ and K_2PtBr_6 double soak (Au/Pt). Difference Patterson maps for each derivative were calculated using reflection data between 10 and 4 Å resolution and interpreted for heavy-atom positions using the program HASSP (Terwilliger et al. 1987). Heavy-atom positions were refined using reflections between 40.0 and 3.0 Å resolution. rms lack-of-closure errors were calculated from centric reflections and updated after each refinement cycle. Reflections with lack-of-closure errors greater than 2.1 times the rms value were rejected from the phase refinement. Several minor heavy-atom sites were located from difference Fourier maps and included in the refinement process. A noise-filtering procedure (Wang 1985) was subsequently applied to improve the MIR phases, and the phasing was extended to 2.7 Å resolution. The resulting electron density map revealed most of the structure. A partial backbone model represented by polyalanine was fit into the electron density map by using the program package FRODO (Jones 1985) for an Evans and Sutherland PS300 computer graphics system. Phases calculated from the partial model were combined with the original MIR phases, and improved by the noise-filtering technique. The new map helped in the assignment of side chains, and a complete tracing was achieved. At this point, the majority of the side chains could be identified, thus confirming the correctness of the chain tracing. The backbone tracing was further supported by the location of the major heavy-atom positions: One site was found to lie between Cys-80 and Met-111, a second site near Cys-118, and a third between Met-67 and Met-72.

The crystal structures of the two transforming forms, one with Val substituting residue 12 (Val-12 mutant) and the other with Leu at position 61 (Leu-61 mutant), were solved by (2F[observed]-F[calculated]) maps. All three structures were refined independently. Current R factors range from 22% to 25%. The details will be published elsewhere (L. Tong et al., in prep.).

RESULTS

Overall Backbone Structure

The backbone structures of the Val-12 mutant and the Leu-61 mutant lacking the carboxy-terminal 18 residues are almost identical to that of the normal p21 protein (de Vos et al. 1988), shown in Figure 1. The topological arrangement of the secondary structure elements is schematically represented in Figure 2. The structure contains a six-stranded β-sheet, four α-helices, and nine connecting loops. Two α-helices ($\alpha 2$, $\alpha 3$) pack against the convex face of the β-sheet, and the remaining two ($\alpha 1$, $\alpha 4$) pack against the concave face. The carboxy-terminal helix $\alpha 4$ is the longest in the molecule, containing approximately seven turns, the last two turns of which form a protrusion from the otherwise globular shape of the protein. The β-α-β-α-β structure at the carboxy-terminal half of the protein is the same topology as the nucleotide-binding domain motif (Rao and Rossmann 1973), although the way GDP binds to this fold is very different from that found in other nucleotide-binding proteins. Five (L1, L2, L5, L7, and L9 in Fig.1) out of nine loops are on the "cytoplasmic side," the side of the molecule where GDP is bound; three (L3, L6, and L8) are on the "membrane side," the side of the carboxyl terminus known to attach to the membrane (Willumsen et al. 1984); and one long loop, L4, spans the two sides (Figs. 1 and 2).

Two-domain Structure

The structure appears to consist of two recognizable domains: The amino-terminal domain contains the first three β-strands and the carboxy-terminal domain is the "nucleotide-binding domain" containing the remaining three β-strands (see Fig. 2). This separation of domains is also clear from the distribution of temperature factors of residues in each domain. The residues in the amino-terminal domain have higher temperature factors, thus are more mobile than those of the carboxy-terminal domain. The high mobility of the amino-terminal domain may have a functional significance, in that it is this domain where the catalytic site of GTP hydrolysis and the putative effector region are located. The carboxy domain recognizes the guanine base of GDP.

Guanine Nucleotide Pocket

The GDP molecule is bound in a pocket formed by four loops (L1, L2, L7, and L9). Specifically, residues 12–17 of loop L1 surround the β-phosphate of GDP, and Asp-30 of loop L2 is near the hydroxyl groups of the ribose sugar. Residues 116, 117, 119, and 120 of loop L7 and residues 145–147 of loop L9 form a side of the pocket for the guanine base. At the current stage of refinement, Asn-116 appears to form hydrogen bonds to N7 and O6 of guanine, Asp-119 to N1 and N2, Ser-145 and the backbone NH of Ala-146 to O6, and Lys-117 to the ring oxygen of the ribose. Phe-28 is pointing almost perpendicularly to the face of the guanine base. A schematic drawing representing these interactions is shown in Figure 3.

P-Loop: The Catalytic Site of GTP Hydrolysis

Loop L1 forms a tight belt around the β-phosphate of GDP and would have straddled the phosphodiester bond between the β- and γ-phosphates of GTP if GTP were bound to the protein. Thus, loop L1 is most likely the catalytic site for the hydrolysis of GTP. The way this

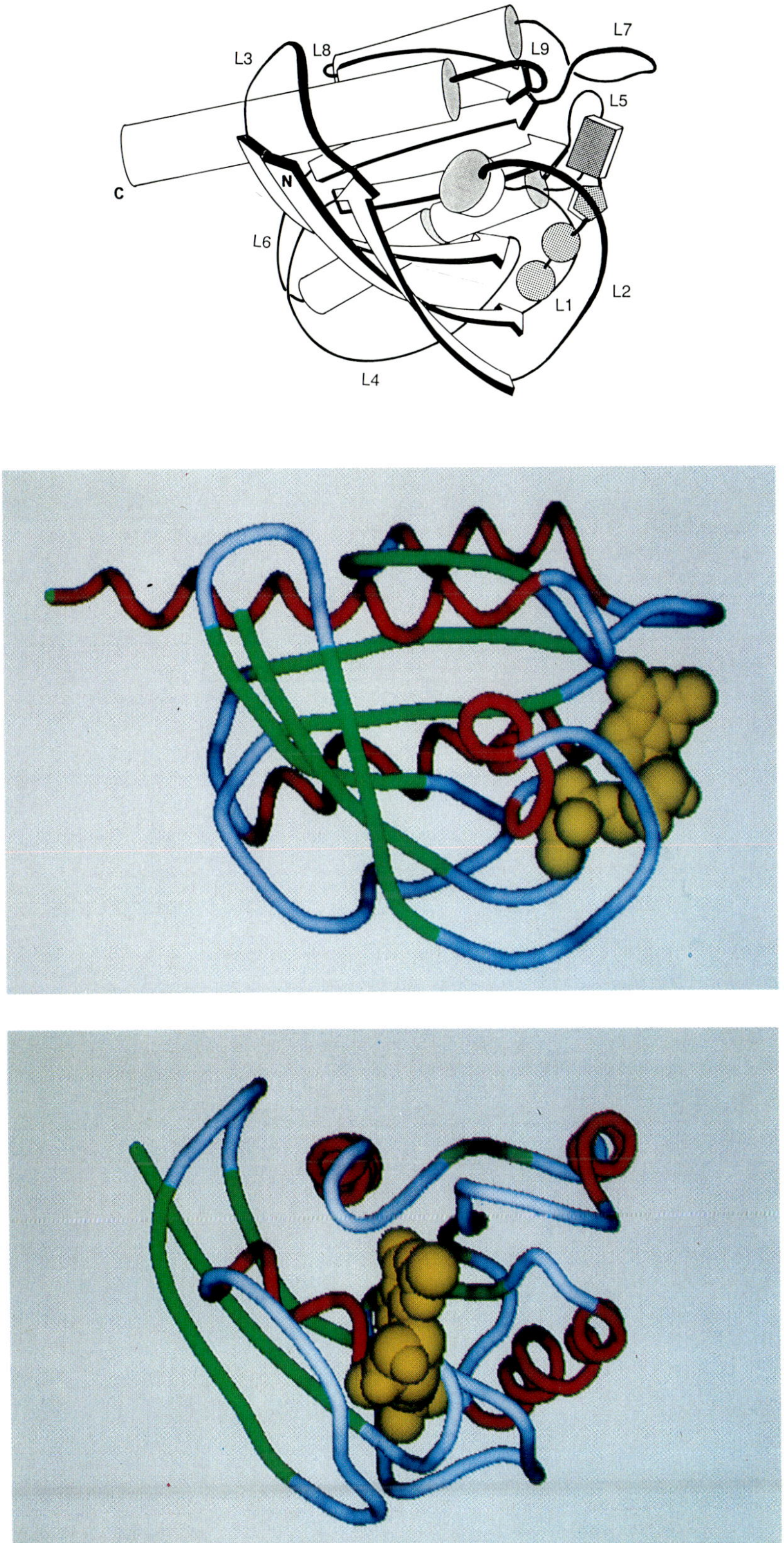

Figure 1. (*Top*) Artist's drawing of the backbone structure of the catalytic domain of human c-Ha-*ras* p21 protein. There are six β-strands (flat arrows) and four α-helices (cylinders) connected by nine loops labeled L1 to L9. The guanine base, the ribose, and the phosphates are represented as a stippled rectangular block, a pentagonal block, and spheres, respectively. The amino and carboxyl termini are indicated. (*Middle*) Three-dimensional structure showing the β-strands in green, the α helices in red, and the loops in light blue. The bound GDP is represented by a space-filling structure in gold. (*Bottom*) The same as (*Middle*) except the structure is rotated 90° around a vertical axis to show the GDP pocket.

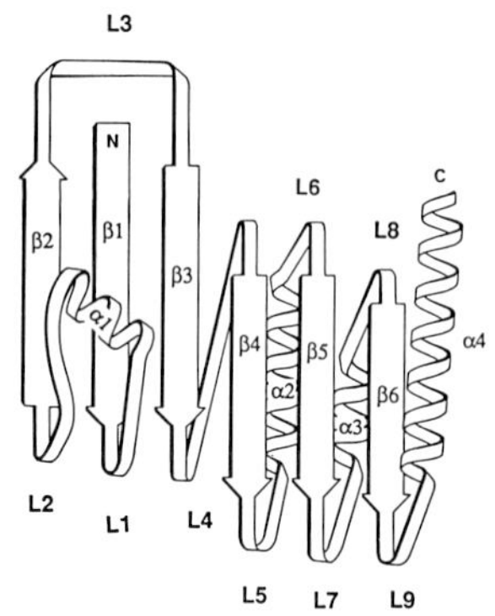

Figure 2. Topological structure of the catalytic domain of human c-Ha-*ras* p21 protein. The β-strands, α-helices, and loops are represented as wide arrowed ribbons, helical ribbons, and narrow ribbons, respectively. Loops, α-helices, and β-strands are numbered sequentially from amino to carboxyl termini.

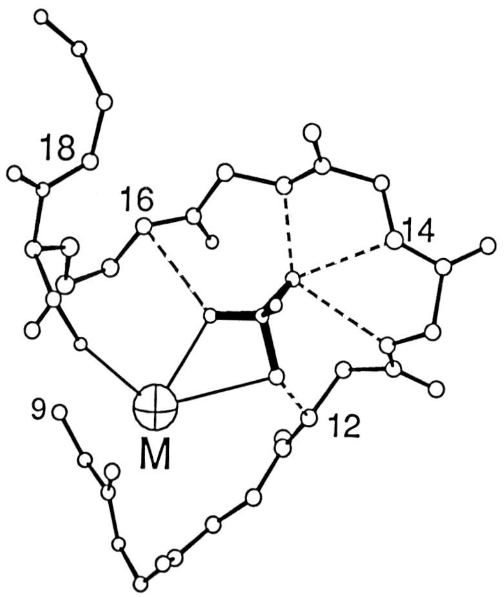

Figure 4. Hydrogen-bonded P-loop. NH groups of the backbone of catalytic loop L1 make several hydrogen bonds to the β-phosphate of GDP. A metal ion, M, also coordinates to the phosphate. Hydrogen bonds from Lys-16 to the β-phosphate are not shown for clarity.

tight loop holds and orients the β-phosphate is highly unusual. In contrast to most catalytic sites of enzymes, where the side chains of the catalytic site residues form hydrogen bonds to substrates, all residues in loop L1 have no functional groups to bond to hydrogen. Instead, the backbone NH groups of residues 12–16 and the side chain of Lys-16 are forming hydrogen bonds to the β-phosphate (Fig. 4). A possible mechanism of GTP hydrolysis suggested by the crystal structure is that the β-phosphate of GTP is held in a particular orientation by this unusual phosphate-binding loop, P-loop, and the metal ion and that a water molecule makes a nucleophilic attack on the phosphorus atom of the β-phosphate releasing the γ-phosphate. Another mechanism involving an attack on γ-phosphate has

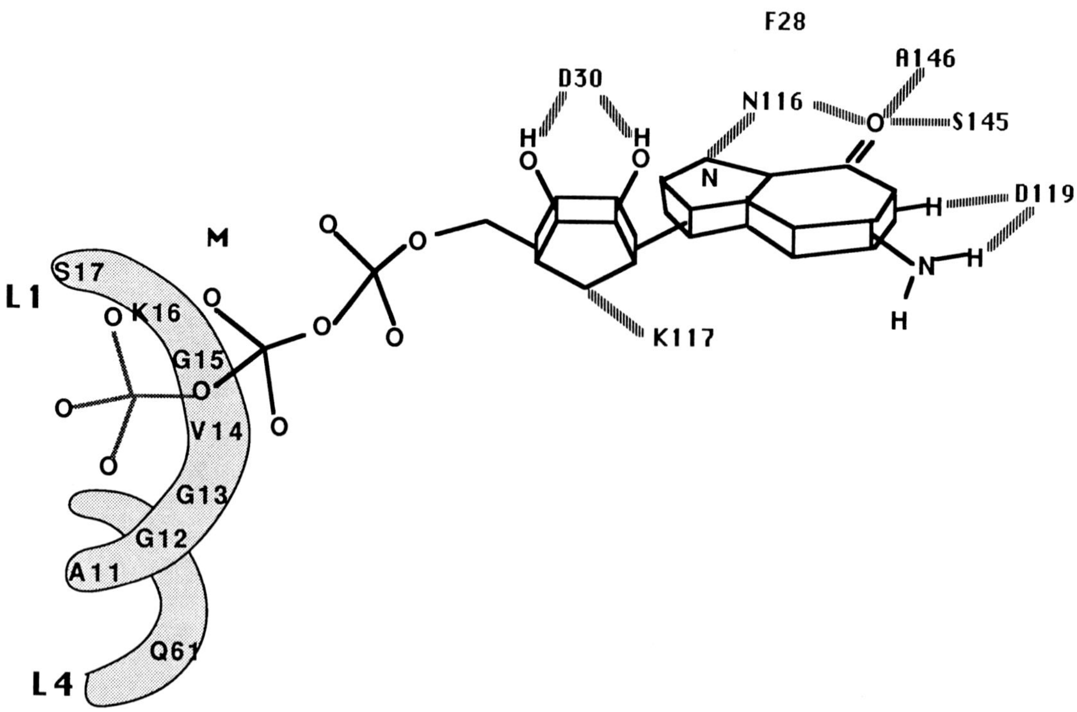

Figure 3. Guanine nucleotide pocket. The immediate neighborhood of the bound GDP is shown schematically. One-letter codes for amino acids are used. Loops L1 and L4 are shaded, and γ-phosphate is added to reveal the relative position of loop L1, which would straddle the phosphodiester bond between the β- and γ-phosphates. M stands for a metal ion, Mg^{++} or Ca^{++} ion, which is coordinated to Ser-17 and the β-phosphate. Many hydrogen bonds from loop L1 and Lys-16 side chain to the β-phosphate are not shown for clarity. For details, see the text.

been proposed for elongation factor Tu (Eccleston and Webb 1982). In addition, Ser-17 is coordinated to a bound metal ion, probably a Mg^{++} or Ca^{++} ion, which in turn is coordinated to the β-phosphate. Its highly constrained conformation also explains the critical nature of the residues in this loop; e.g., substitution at residue 12 by any amino acid other than proline endows the protein with transforming activity (Seeburg et al. 1984). The highly unusual composition (Gly/Ala-rich) of the catalytic loop, L1, may be a remnant example of a primitive enzyme for formation or cleavage of a nucleic acid backbone.

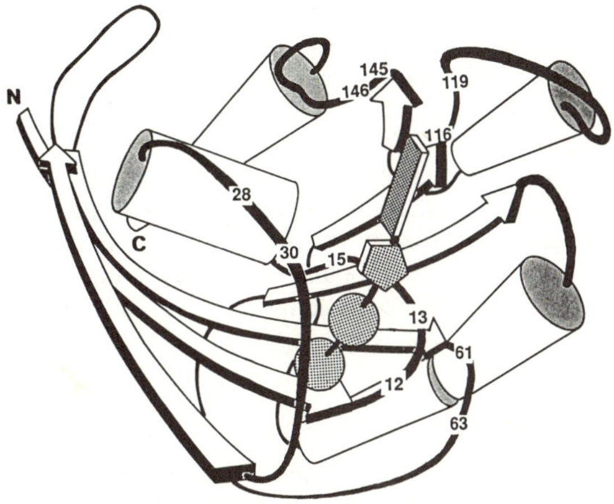

Figure 6. Schematic drawing of the backbone structure of human c-Ha-*ras* p21 protein looking to the GDP-binding pocket. Activating mutation sites as well as residues in contact with GDP are labeled.

Differences between Normal and Transforming p21 Structures

The crystal structures of the catalytic domains of two commonly found transforming p21 proteins, Val-12 and Leu-61 mutants, reveal that the significant differences between the normal protein and these two transforming forms are found in the conformation of the P-loop (L1) and of the long loop L4. In the Val-12 mutant, the P-loop became enlarged and loose (Fig. 5) and has lost two hydrogen bonds to the β-phosphate of GDP. This conformational change in the catalytic site immediately explains the drastic reduction of GTPase activity of the transforming p21 protein, thus keeping the proteins in a GTP-complexed conformation, which is thought to signal cell proliferation. Similar changes were observed also in Leu-61 mutant.

Correlation between Activating Mutation Sites and GDP Binding

All the known activating in vivo and in vitro point mutations (at positions 12, 13, 15, 59, 61, 63, 116, 119; for review, see Barbacid 1987) are localized within three loops of the GDP-binding site (see Fig. 6). These can be divided into three groups: The first four are mutations in the P-loop (L1); the next three in loop L4, which is in direct contact with the P-loop; and the last two in loop L7, which binds the guanine base. Activation by mutation of residues 12, 13, or 15 may be due to the inhibition of GTPase activity caused by the conformational change at the catalytic site and/or enhancement of GTP/GDP exchange by the reduction in the phosphate-binding strength of the P-loop. Mutation at residues 116 or 119 could reduce the guanine-binding strength, resulting in the increase of the GTP/GDP exchange rate. In all cases, mutation at these locations would increase the concentration of GTP-bound p21 over the GDP-bound form, thus prolonging the "signal-on" state of p21 molecules, and therefore in-

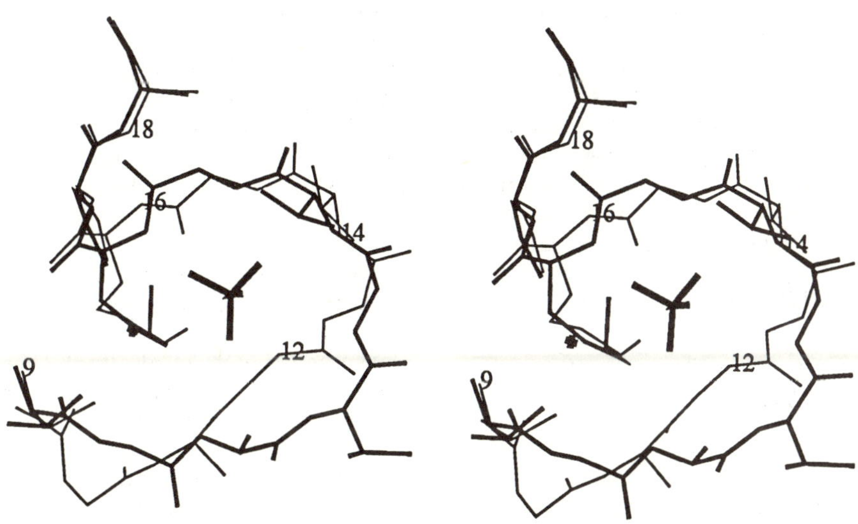

Figure 5. Differences of the catalytic loop L1 in normal p21(Gly-12) and transforming p21(Val-12). Thin lines are for p21(Gly-12). The β-phosphate is in the middle of the loop, and the metal ion is indicated as a cross. A few residues are labeled for reference.

ducing transformation. On the other hand, residues 59, 61, and 63 belong to loop L4, which is not in contact with the phosphate groups of GDP but is in direct contact with the catalytic loop L1. It seems possible that mutations at these positions in loop L4 may activate p21 through indirect conformational changes of the P-loop. Furthermore, their positions in the structure are in the pocket that the γ-phosphate would have occupied in the GTP-bound form; thus, a mutation in these positions may, in addition, affect GTP binding and/or GTPase activity.

Acceptor Site of Autophosphorylation

Transforming viral p21 proteins have an additional mutation, a threonine residue at position 59 (Shih et al. 1982). This residue is located in loop L4, adjacent to the catalytic loop L1. If we imagine the γ-phosphate of GTP extending out from the β-phosphate into the guanine-nucleotide-binding pocket, the only available space is toward the beginning of loop L4, and the hydroxyl of Thr-59 of viral protein would be the closest functional group to receive the γ-phosphate group with accompanying conformational changes of the loop L4. The phosphorylation of Thr-59 may freeze the conformation of p21 in GTP-bound form, thus making the protein transformation competent.

Longest Loop, L4, Binds a Neutralizing Antibody

The longest loop, L4, spans the two extreme ends of the structure, the cytoplasmic end and the "membrane end." Furthermore, this loop starts right after one of the most conserved sequences between p21 proteins and α subunits of G proteins, Asp-(Thr/Val)-(Ala/Gly)-Gly-Gln. This loop includes residues 63–73, which are known to be part of the epitope (Sigal et al. 1986) for a neutralizing monoclonal antibody, Y13-259 (Furth et al. 1982), that has been shown to inhibit morphological transformation induced by activated *ras* proteins (Mulcahy et al. 1985; Kung et al. 1986). A tentative explanation for the neutralizing effect may be either that the antibody binding induces a conformational change in loop L4, which in turn affects the catalytic loop L1 due to their direct contacts, reverting the catalytic site (loop L1) conformation from a transforming to a normal state, or that the large size of IgG covers the effector recognition site, preventing the continuous generation of a second message or altering the exchange rate of GTP/GDP.

Loop L2, a Candidate for Effector and/or GAP Recognition

Loop L2 is the largest loop on the cytoplasmic (GDP-binding) side of the molecule. Except for Phe-28 and Asp-30, the residues in this loop are "free" and well-exposed. Amino acid substitutions in the region between residues 35 and 40 reduce the transformation capacity of activated *ras* proteins without affecting their GTPase activity or guanine nucleotide binding (Sigal et al. 1986), suggesting that loop L2 may be a recognition site for the effector molecule. None of the residues between 35 and 40 are in direct contact with GDP, thus explaining their noninvolvement with GDP/GTP binding or GTP hydrolysis. A recent study (Adari et al. 1988; Calés et al. 1988) suggests that loop L2 may also be the binding site for a GTPase-activating protein, GAP (Trahey and McCormick 1987).

DISCUSSION

Comparison with Guanine-nucleotide-binding Domain of EF-Tu

Although the overall homology is limited, there are small regions of sequence homology separated by nonhomologous regions between p21 proteins and the guanine-nucleotide-binding domain of bacterial elongation factor EF-Tu (G-EFTu) if conservative substitutions are permitted (Halliday 1984; Leberman and Egner 1984). Comparison of the crystal structures of p21 (de Vos et al. 1988) and G-EFTu (Jurnak 1985; La Cour et al. 1985) shows that there is an overall similarity between the two. The similarity is very strong for the carboxy-terminal domain, but there are significant differences in the amino-terminal domain. Besides the presence of an extra α-helix in G-EFTu, the relative position of two β-strands, β1 and β3, are switched in the G-EFTu structure (Fig. 7A,B). Structural models (Jurnak 1985; McCormick et al. 1985) for p21 built based on the G-EFTu structure therefore inherited the same differences. Details of the differences will be revealed once the entire structure of EF-Tu is solved and refined at a reasonably high resolution.

Model for the α Subunit of the G Protein

The sequence homology, although limited, between p21 proteins and the α subunits of G proteins (Lochrie et al. 1985), as well as GDP/GTP binding (Scolnick et al. 1979) and GTP hydrolysis (Shih et al. 1980) properties of p21, invites speculation that both groups of proteins have similarities in their three-dimensional structures (Masters et al. 1986; Deretic and Hamm 1987). Furthermore, p21, like G proteins, might have surface regions specific to one or more target proteins and other region(s) specific to signal receptor molecule(s), and signal transduction may be mediated by a conformational change of p21 proteins. There are several different criteria one could use in aligning the two sequences (Masters et al. 1986; Gilman 1987). An alignment of the sequences based on the crystal structure of p21 allows one to make a reasonable prediction (Fig. 7C) for the tertiary structure of the α subunit of G proteins (S. Holbrook and S.-H. Kim, in prep.). Most of the differences between the two classes of proteins can be localized in four regions and two termini: In our model, region R2 (corresponding to the residues 57–152 in α subunit of G_i-2, $G_{i\alpha}$-2) of G_α has 84 more

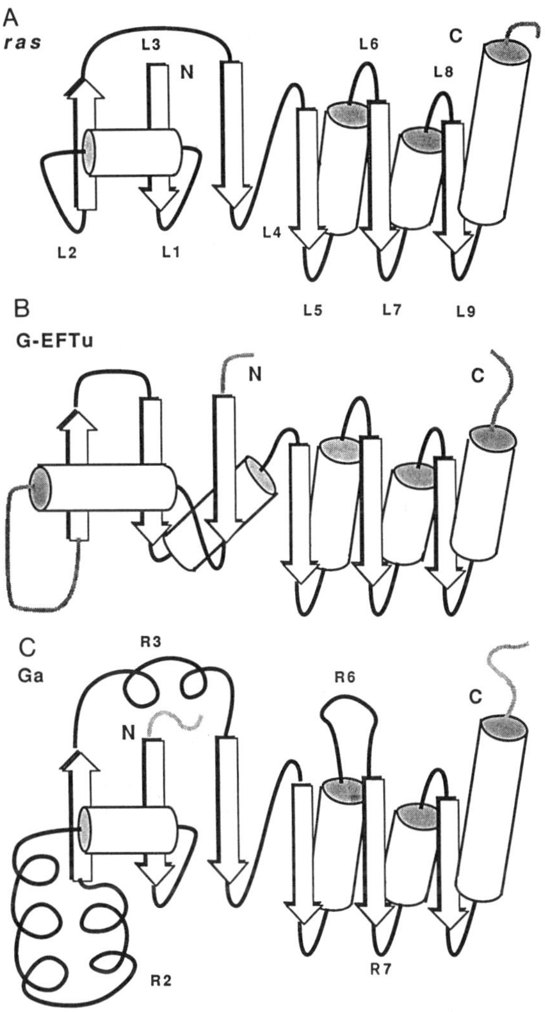

Figure 7. Topological drawings representing (*A*) the crystal structure of c-Ha-*ras* p21 protein, (*B*) the crystal structure of the GDP-binding domain of EF-Tu, and (*C*) a model for the α subunit of G protein. The loops in the p21 protein are labeled from L1 to L9, and three regions in G_α corresponding to the loops L2, L3, and L6 are labeled as R2, R3, and R6, respectively. Since it is known that the carboxy-terminal region of the p21 protein is palmitylated (Chen et al. 1985) and anchored to the cell membrane, one can assign the top of each figure as facing the cell membrane.

residues than loop L2 in the p21 structure, region R3 (the residues 158–197 in $G_{i\alpha}$-2) has 31 residues more than loop L3, region R6 (the residues 247–263 in $G_{i\alpha}$-2) has 13 residues more than loop L6, and R7 (residues 272–305 in $G_{i\alpha}$-2) has 25 residues more than L7 (Fig. 7C). The amino and carboxyl termini are 32 and 17 residues longer in G_α proteins than in p21 *ras* proteins. Tertiary structures of these "inserts" are difficult to determine. Our model predicts that the site for ADP ribosylation of G_α by cholera toxin will be in loop L3, which faces the membrane, suggesting how such modification may interfere with the interaction of the α subunit with the membrane-bound $\beta\gamma$ subunit of G proteins. The model also orients loops L3, L6, L8, and both amino- and carboxy-terminal regions toward the membrane side, suggesting that some of these may contain the receptor recognition region.

Conformational Switch System

Analogous to G proteins, p21 proteins are thought to become active when bound to GTP and are inactive in GDP-bound conformation. It is assumed that an effector protein recognizes the GTP-bound form and sends out a second message. On the basis of the crystal structure of the GDP-bound form, one can make a qualitative prediction of the structure of the GTP-bound form (see Fig. 8). In the crystal structure, there is a hole below the β-phosphate, just above the highly conserved sequence of Asp-(Thr-Ala)-Gly-Gln, which could accommodate the γ-phosphate. The β-phosphate is tightly hydrogen-bonded by the backbone NH groups

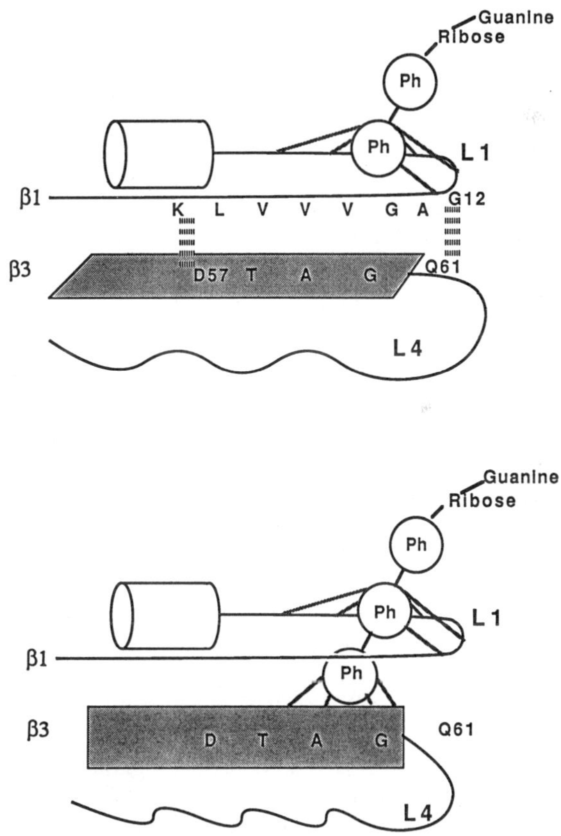

Figure 8. Conformational switch system model. (*Top*) Three-tier system, as appeared in the crystal structure of human c-Ha-*ras* p21, consists of β1-L1-α1, residues 57–61, and L4. One-letter codes for amino acids are used, and Ph stands for phosphate group. The first and second regions are highly conserved among p21 *ras* proteins and α subunits of G proteins, and the third within p21 proteins or within the α subunits. The β-phosphate is extensively hydrogen-bonded to the backbone NH groups of L1. Gln-61 is in direct contact with Gly-12, and Asp-57 and Lys-5 potentially form a charge pair. (*Bottom*) Presence of γ-phosphate would force conformational changes on the stretch of conserved sequence of the middle tier, which in turn changes the conformation of the leading strand, L4, and/or the lagging strand, β3, distorting the β-sheet that holds the two subdomains of the molecule.

of loop L1, but the γ-phosphate would be in a hydrophobic pocket if there are no conformational changes. One obvious way to resolve this conflict would be to rotate a part of the backbone of the second-tier conserved region to present the backbone NH groups of the region toward the γ-phosphate to form hydrogen bonds to it. Such rotation would induce conformational changes in the leading strand, presenting loop L4 with a different conformation at the surface of the molecule and/or in the lagging strand, β3, disrupting or distorting the β-sheet, possibly changing the relative orientation of the two subdomains of the molecule. Lys-16 is also likely to participate in hydrogen bonding to the γ-phosphate. Since this model includes two of the most conserved sequences of G proteins, a similar mechanism could be operative in G proteins as well.

OTHER COMMENTS

The crystal structure of p21 reveals that nine looped regions are exposed at the surface of the protein, of which four are pointing toward the cytoplasmic side and used for GDP/GTP binding and GTP hydrolysis. It is tempting to speculate that the remaining three loops on the cytoplasmic side, L2, L4, and L5, as well as the exposed α-helices, are important for recognition by particular target proteins, and that the two loops on the membrane side, L3 and L8, and the carboxy-terminal end may be involved in the recognition of signal receptor molecule(s).

The structure of the catalytic domain of p21 lacks 18 residues at the carboxyl terminus. The long carboxy-terminal α-helix in this structure suggests that the catalytic domain is well separated from the membrane attachment domain. This long α-helix could play an important role in transmitting a signal from the membrane attachment domain (receptor recognition domain?) to the catalytic domain, and in keeping the catalytic domain some distance away from the membrane surface to increase the probability for the catalytic domain to interact with target proteins.

ACKNOWLEDGMENTS

The research reported here has been partially supported by grants from the National Institutes of Health (GM-29287, CA-45593) and the Department of Energy to S.H.K.; a gift from Merck, Sharp & Dohme to the University of California; and grants-in-aid from the Japanese Ministry of Health and Welfare for a comprehensive 10-year strategy for cancer control to S.N. P.M.M. was a fellow of the Damon Runyon-Walter Winchell Cancer Fund.

REFERENCES

Adari, H., D.R. Lowy, B.M. Willumsen, C.J. Der, and F. McCormick. 1988. Guanosine triphosphatase activating protein (GAP) interacts with the p21 *ras* effector binding domain. *Science* **240:** 518.

Barbacid, M. 1987. *ras* Genes. *Annu. Rev. Biochem.* **56:** 779.

Calés, C., J.F. Hancock, C.J. Marshall, and A. Hall. 1988. The cytoplasmic protein GAP is implicated as the target for regulation by the *ras* gene product. *Nature* **332:** 548.

Capon, D.J., E.Y. Chen, A.D. Levinson, P.H. Seeburg, and D.V. Goeddel. 1983. Complete nucleotide sequences of the T24 human bladder carcinoma oncogene and its normal homologue. *Nature* **302:** 33.

Chen, Z.Q., L.S. Ulsh, G. DuBois, and T.Y. Shih. 1985. Posttranslational processing of p21 *ras* proteins involves palmitylation of the C-terminal tetrapeptide containing cysteine-186. *J. Virol.* **56:** 607.

Deretic, D. and H.E. Hamm. 1987. Topographic analysis of antigenic determinants recognized by monoclonal antibodies to the photoreceptor guanyl nucleotide-binding protein, transducin. *J. Biol. Chem.* **262:** 10839.

de Vos, A.M., L. Tong, M. Milburn, P.M. Matias, J. Jancarik, S. Noguchi, S. Nishimura, K. Miura, E. Ohtsuka, and S.-H. Kim. 1988. Three-dimensional structure of an oncogene protein: Catalytic domain of human c-H-*ras* p21. *Science* **239:** 888.

Eccleston, J.F. and M.R. Webb. 1982. Characterization of the GTPase reaction of elongation factor Tu. *J. Biol. Chem.* **257:** 5046.

Furth, M.E., L.J. Davis, B. Fleurdelys, and E.M. Scolnick. 1982. Monoclonal antibodies to the p21 products of the transforming gene of Harvey murine sarcoma virus and of the cellular *ras* gene family. *J. Virol.* **43:** 294.

Gibbs, J.B., I.S. Sigal, M. Poe, and E.M. Scolnick. 1984. Intrinsic GTPase activity distinguishes normal and oncogenic *ras* p21 molecules. *Proc. Natl. Acad. Sci.* **81:** 5704.

Gilman, A.G. 1987. G proteins: Transducers of receptor-generated signals. *Annu. Rev. Biochem.* **56:** 615.

Goldfarb, M., K. Shimizu, M. Perucho, and M. Wigler. 1982. Isolation and preliminary characterization of a human transforming gene from T24 bladder carcinoma cells. *Nature* **296:** 404.

Halliday, K.R. 1984. Regional homology in GTP binding protooncogene products and elongation factors. *J. Cyclic Nucleotide Protein Phosphorylation Res.* **9:** 435.

Hurley, J.B., M.I. Simon, D.B. Teplow, J.D. Robishaw, and A.G. Gilman. 1984. Homologies between signal transducing G proteins and *ras* gene products. *Science* **226:** 860.

Jancarik, J., A.M. de Vos, S.-H. Kim, K. Miura, E. Ohtsuka, S. Noguchi, and S. Nishimura. 1988. Crystallization of human c-H-*ras* oncogene products. *J. Mol. Biol.* **200:** 205.

Jones, T.A. 1985. FRODO: A molecular graphics program. University of Uppsala, Uppsala, Sweden. (Subsequently modified by B.L. Bush at Merck, Sharpe and Dohme, Rahway, New Jersey; and by J.W. Pflugrath, M.A. Saper, and J.S. Sack, Rice University, Austin, Texas.)

Jurnak, F. 1985. Structure of the GDP domain of EF-Tu and location of the amino acids homologous to *ras* oncogene proteins. *Science* **230:** 32.

Kung, H.F., M.R. Smith, E. Bekesi, V. Manne, and D.W. Stacey. 1986. Reversal of transformed phenotype by monoclonal antibodies against Ha-*ras* p21 proteins. *Exp. Cell. Res.* **162:** 363.

La Cour, T.F.M., J. Nyborg, S. Thirup, and B.F.C. Clark. 1985. Structural details of the binding of guanosine diphosphate to elongation factor Tu from *E. coli* as studied by X-ray crystallography. *EMBO J.* **4:** 2385.

Leberman, R. and U. Egner. 1984. Homologies in the primary structure of GTP-binding proteins: The nucleotide-binding site of EF-Tu and p21. *EMBO J.* **3:** 339.

Lochrie, M.A., J.B. Hurley, and M.I. Simon. 1985. Sequence of the alpha subunit of photoreceptor G protein: Homologies between transducin, *ras*, and elongation factors. *Science* **228:** 96.

Masters, S.B., R.M. Stroud, and H.R. Bourne. 1986. Family of G protein α chains: Amphipathic analysis and predicted structure of functional domains. *Protein Eng.* **1:** 47.

McCormick, F., B.F.C. Clark, T.F.M. la Cour, M. Kjeldgaard, L. Norskov-Lauritsen, and J. Nyborg. 1985. A model for the tertiary structure of p21, the product of the *ras* oncogene. *Science* **230:** 78.

Miura, K., Y. Inoue, H. Nakamori, S. Iwai, E. Ohtsuka, M. Ikehora, S. Noguchi, and S. Nishimura. 1986. Synthesis and expression of a synthetic gene for the activated human c-Ha-*ras* protein. *Jpn. J. Cancer Res. (Gann)* **77:** 45.

Mulcahy, L.S., M.R. Smith, and D.W. Stacey. 1985. Requirements for *ras* proto-oncogene function during serum-stimulated growth of NIH 3T3 cells. *Nature* **313:** 241.

Neer, E.J. and D.E. Clapham. 1988. Roles of G protein subunits in transmembrane signalling. *Nature* **333:** 129.

Nishimura, S. and T. Sekiya. 1987. Human cancer and cellular oncogenes. *Biochem. J.* **243:** 313.

Perucho, M., M. Goldfarb, K. Shimizu, C. Lama, J. Fogh, and M. Wigler. 1981. Human-tumor derived cell lines contain common and different transforming genes. *Cell* **27:** 467.

Pulciani, S., E. Santos, A.V. Lauver, L.K. Long, K.C. Robbins, and M. Barbacid. 1982. Oncogenes in human tumor cell lines: Molecular cloning of a transforming gene from human bladder carcinoma cells. *Proc. Natl. Acad. Sci.* **79:** 2845.

Quaife, C.J., C.A. Pinkert, D.M. Ornitz, R.D. Palmiter, and R.L. Brinster. 1987. Pancreatic neoplasia induced by *ras* expression in acinar cells of transgenic mice. *Cell* **48:** 1023.

Rao, S.T. and M.G. Rossmann. 1973. Comparison of super-secondary structures in proteins. *J. Mol. Biol.* **76:** 241.

Reddy, E.P., R.K. Reynolds, E. Santos, and M. Barbacid. 1982. A point mutation is responsible for the acquisition of transforming properties by the T24 human bladder carcinoma oncogene. *Nature* **300:** 149.

Scolnick, E.M., A.G. Papageorge, and T.Y. Shih. 1979. Guanine nucleotide-binding activity and an assay for *src* protein of rat-derived murine sarcoma viruses. *Proc. Natl. Acad. Sci.* **76:** 5355.

Seeburg, P.H., W.W. Colby, D.J. Capon, D.V. Goeddel, and A.D. Levinson. 1984. Biological properties of human c-Ha-*ras*1 genes mutated at codon 12. *Nature* **312:** 71.

Shih, C. and R.A. Weinberg. 1982. Isolation of transforming sequence from a human bladder carcinoma cell line. *Cell* **29:** 161.

Shih, T.Y., A.G. Papageorge, P.E. Stokes, M.O. Weeks, and E.M. Scolnick. 1980. Guanine nucleotide-binding and autophosphorylating activities associated with the p21 SRC protein of Harvey murine sarcoma virus. *Nature* **287:** 686.

Shih, T.Y., P.E. Stokes, G.W. Smythers, R. Dhar, and S. Oroszlan. 1982. Characterization of the phosphorylation sites and the surrounding amino acid sequences of the p21 transforming proteins coded for by the Harvey and Kirsten strains of murine sarcoma viruses. *J. Biol. Chem.* **257:** 11767.

Sigal, I.S., J.B. Gibbs, J.S. D'Alonzo, and E.M. Scolnick. 1986. Identification of effector residues and a neutralizing epitope of Ha-*ras*-encoded p21. *Proc. Natl. Acad. Sci.* **83:** 4725.

Stryer, L. and H.R. Bourne. 1986. G proteins: A family of signal transducer. *Annu. Rev. Cell Biol.* **2:** 391.

Tabin, C.J., S.M. Bradley, C.I. Bargmann, R.A. Weinberg, A.G. Papageorge, E.M. Scolnick, R. Dhar, D.R. Lowy, and E.H. Chang. 1982. Mechanism of activation of a human oncogene. *Nature* **300:** 143.

Taparowsky, E., Y. Suard, O. Fasano, K. Shimizu, M. Goldfarb, and M. Wigler. 1982. Activation of the T24 bladder carcinoma transforming gene is linked to a single amino acid change. *Nature* **300:** 762.

Terwilliger, T., S.-H. Kim, and D. Eisenberg. 1987. Generalized method of determining heavy-atom positions using the difference patterson function. *Acta Crystallogr.* **A43:** 1.

Trahey, M., and F. McCormick. 1987. A cytoplasmic protein stimulates normal N-*ras* p21 GTPase, but does not affect oncogenic mutants. *Science* **238:** 542.

Wang, B.C. 1985. Resolution of phase ambiguity in macromolecular crystallography. *Methods Enzymol.* **115:** 90.

Willumsen, B.M., A. Christensen, N.L. Hubbert, A.G. Papageorge, and D.R. Lowy. 1984. The p21 *ras* C-terminus is required for transformation and membrane association. *Nature* **310:** 583.

Willumsen, B.M., A. Papageorge, N. Hubbert, E. Bekesi, H.-F. Kung, and D.R. Lowy. 1985. Transforming p21 *ras* protein: Flexibility in the major variable region linking the catalytic and membrane-anchoring domains. *EMBO J.* **4:** 2893.

Molecular Basis of Visual Excitation

L. Stryer

Department of Cell Biology, Sherman Fairchild Center, Stanford University School of Medicine, Stanford, California 94305

Retinal rod cells are remarkable detectors. Psychophysical experiments carried out by Selig Hecht nearly half a century ago revealed that a rod cell can be excited by a single photon. Moreover, the sensitivity of rods is adjusted continuously according to the background light level, enabling them to perceive contrast over a 10^5-fold range of ambient light intensity. High-fidelity signaling is another noteworthy feature of rods. The amplitude and timing of their responses to light are highly reproducible. Much of the transduction process takes place in a discrete part of the cell, the outer segment. This miniaturized detector and transducer is a triumph of membrane design and protein targeting. Electrical engineers have regarded them with admiration and envy, and biochemists have viewed them as ripe for exploration. Indeed, the rod outer segment (ROS) of vertebrates is a gift of nature. It contains the transduction machinery in highly abundant and concentrated form, separate from the energy-generating and protein-synthesizing apparatus in the inner segment. The molecules mediating vision can be isolated, and their interplay can be unraveled. Equally important, rods and their outer segments are very attractive for electrophysiological study. A coherent picture of the molecular mechanism of visual transduction has recently come into view as a result of biochemical, biophysical, molecular genetic, and electrophysiological studies carried out in many laboratories. An overview of the process is presented here, from photon to nerve impulse; more detailed accounts of different facets are presented in several recent reviews (Pugh and Cobbs 1986; Stryer 1986; Liebman et al. 1987; Nathans 1987; Yau and Baylor 1989).

In essence, light triggers a nerve signal by activating an enzymatic cascade that leads to closure of channels in the plasma membrane of the outer segment (Fig. 1). This cascade is initiated by the photoisomerization of the retinal chromophore of rhodopsin (R). Photoexcited rhodopsin (R*) then activates transducin (T), a member of the G-protein family, by catalyzing the exchange of GTP for bound GDP. The GTP form of transducin (specifically, T_α-GTP, the activated α subunit), in turn, switches on a potent phosphodiesterase (PDE*) that rapidly hydrolyzes cGMP. In the dark, Na^+ and Ca^{++} enter outer segments rapidly through cation-specific channels that are opened by the binding of cGMP. The light-induced decrease in the cGMP level closes channels, thereby blocking the influx of cations. The resulting hyperpolarization of the plasma membrane is transmitted to the synaptic terminal at the other end of the cell. Recovery of the dark state is mediated by deactivation of the PDE and activation of guanylate cyclase. The cGMP level is restored, and membrane channels open again. Calcium ion plays a key feedback role in the recovery process by regulating guanylate cyclase.

Photoisomerization of Retinal Generates R*

Vision begins with the absorption of a photon by the 11-*cis* retinal chromophore of rhodopsin, a 40-kD transmembrane protein. Many years ago, G. Wald (1968) discovered that the primary event in visual excitation is the isomerization of 11-*cis* retinal to the all-*trans* form. Photoisomerization is an efficient process, with a quantum yield of about 0.6. Absorption of a photon dramatically alters the electronic distribution of retinal. The change in dipole moment of 11-*cis* retinal in going to the excited state is 12 debyes, which is equivalent to the movement of 0.2 electronic charge across the 1.2-nm length of the chromophore (Mathies and Stryer 1976). The interaction of this large excited-state dipole moment of retinal with a suitably located charged group or dipole on opsin probably controls the position of the absorption maximum of retinal. Moreover, the light-induced movement of charge toward the Schiff-based end of retinal facilitates isomerization by conferring single-bond character on the C11–C12 double bond. The highly dipolar excited state very likely drives conformational changes in retinal and surrounding groups in opsin. Indeed, an activated conformational state, bathorhodopsin, is formed within 6 psec of the absorption of a photon (Busch et al. 1972). Picosecond-resonance Raman studies have shown that much of the isomerization process occurs in less than 20 psec (Hayward et al. 1981; Carlsen 1984). In contrast, elec-

$$\mathrm{R} \xrightarrow{h\nu} \mathrm{R}^* \longrightarrow \mathrm{T}_\alpha\text{-GTP} \longrightarrow \mathrm{PDE}^* \longrightarrow \mathrm{cGMP}\downarrow \longrightarrow \text{Channel closure} \longrightarrow \text{Membrane hyperpolarization}$$

Figure 1. Flow of information in visual excitation. Light hyperpolarizes the plasma membrane of vertebrate photoreceptor cells by lowering the concentration of cGMP which, in turn, closes cation-specific channels. (R) Rhodopsin; (R*) photoexcited rhodopsin; (T_α-GTP) α subunit of transducin containing bound GTP.

trophysiological studies of rod cells in the dark indicate that spontaneous isomerization of the retinal chromophore of rhodopsin occurs about once in a millenium (Baylor et al. 1979). Thus, retinal changes its shape quickly and efficiently following absorption of light but keeps silent in the dark, patiently awaiting the arrival of a photon. The capacity of rods to detect single photons is critically dependent on the exceptional signal-to-noise characteristics of rhodopsin, the first element in the transduction chain.

The elucidation of the amino acid sequence of rhodopsin (Ovchinnikov 1982; Hargrave et al. 1983) led to a structural model that brought together a wealth of structural information about this photoreceptor protein (Dratz and Hargrave 1983). The presence of seven predominantly hydrophobic stretches suggested that rhodopsin contains seven transmembrane helices. The proposed threading of the 348-residue polypeptide chain (Fig. 2) is based on the results of numerous enzymatic labeling, proteolytic digestion, antibody-binding, and lectin-binding studies. The amino terminus is on the intradiskal side and the carboxyl terminus is on the cytosolic (interdiskal) side of the disk membrane. The postulated distribution of mass and orientation of α helices fits the findings of spectroscopic, neutron-scattering, and magnetic anisotropy studies (Chabre 1985). The two N-linked oligosaccharides, attached to Asp-2 and Asp-15, are located on the intradiskal side. Cys-322 and Cys-323 in the carboxy-terminal region are palmitylated; acylation of these adjacent cysteines may anchor the carboxy-terminal region to the disk membrane (Ovchinnikov et al. 1988a). The carboxy-terminal region on the cytosolic face is rich in serine and threonine residues, which become phosphorylated following illumination. Phosphorylation deactivates rhodopsin, as discussed below.

The 11-*cis* retinal chromophore of rhodopsin is bound to Lys-296 on the seventh helix by a protonated Schiff-base linkage. The long axis of retinal is nearly parallel to the plane of the membrane. Fluorescence energy-transfer studies showed that 11-*cis* retinal is located near the center of the membrane (Thomas and Stryer 1982). Some of the five charged residues within the membrane probably participate in tuning the absorption maximum of rhodopsin, which is at 500 nm, compared to 440 nm for an unperturbed protonated Schiff base. Solid-state ^{13}C-labeled nuclear magnetic resonance (NMR) spectra point to the presence of a negative charge near C-13 of retinal (Lugtenburg et al. 1988).

A series of distinctive spectral intermediates appear following absorption of a photon: bathorhodopsin in picoseconds, lumirhodopsin in nanoseconds, metarhodopsin I in microseconds, and metarhodopsin II in about a millisecond. In metarhodopsin II, the Schiff-base linkage is unprotonated, and the chromophore is in the all-*trans* form. In a few minutes, metarhodopsin II becomes converted into metarhodopsin III, which is then hydrolyzed to opsin and all-*trans* retinal. The cGMP cascade is triggered by metarhodopsin II (Emeis et al. 1982), which is called R*. This conformationally activated state stores about 27 kcal/mole of the energy

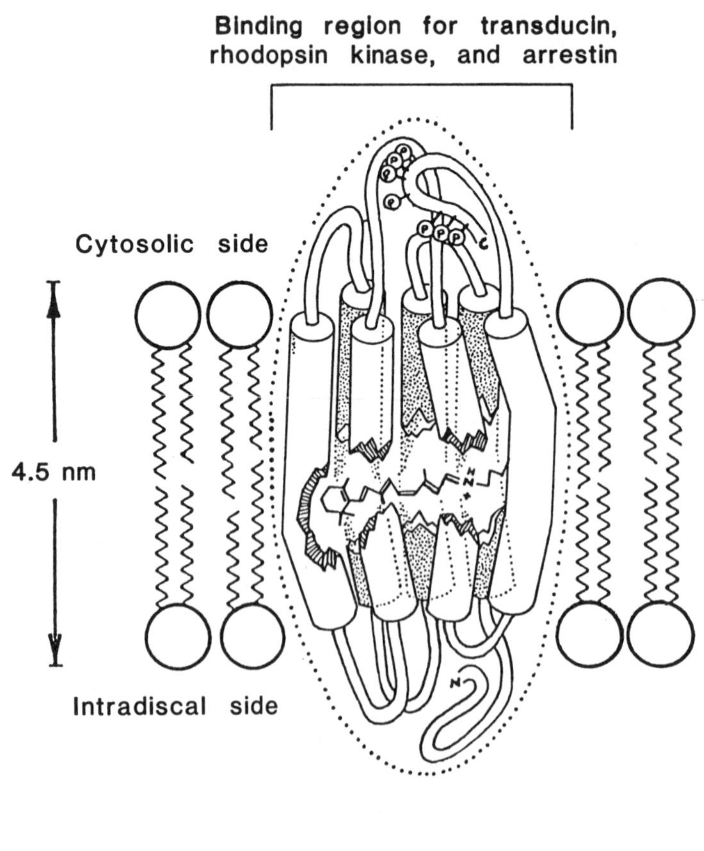

Figure 2. A model of the structure of vertebrate rhodopsin (Dratz and Hargrave 1983).

of the incident photon (Cooper 1981). The large activation barrier between rhodopsin and metarhodopsin II, more than 40 kcal/mole, assures that transducin is triggered only rarely in the absence of light. The photoisomerization of retinal leads to conformational changes that are transmitted from the center of the protein to its cytosolic face through a distance of at least 2 nm. The nature of these conformational changes has not yet been established, but their functional consequences are strikingly evident. R*, but not rhodopsin, interacts with transducin and induces the release of its bound GDP (Fung and Stryer 1980). Furthermore, R*, but not rhodopsin, binds rhodopsin kinase and becomes phosphorylated (Kühn 1984). The third cytosolic loop (the one joining the fifth and sixth helices) of R* is probably critical in binding transducin, as evidenced by the effects of limited proteolysis (Kühn 1984) and the outcome of site-specific mutagenesis (Franke et al. 1988). Photoisomerization of retinal may be communicated to this loop by a rotation and translation of helices, as has been proposed for the gating of ion channels (Unwin 1987).

R* Activates Transducin by Catalyzing GTP-GDP Exchange

The next step in visual excitation is the activation of transducin by R*. Transducin is a multisubunit peripheral membrane protein consisting of three polypeptide chains: α (39 kD), β (36 kD), and γ (8 kD). cDNAs encoding the α, β, and γ subunits of transducin have been cloned and sequenced (for references, see Table 1). The roles of these subunits have been determined by reconstitution experiments (Fung et al. 1981; Fung 1983). The α subunit of transducin (T_α) contains a guanyl-nucleotide-binding site and a catalytic site for the hydrolysis of GTP to GDP. T_α in the GTP form is the activator of the PDE. The β and γ chains form a $T_{\beta\gamma}$ subunit. T_α is associated with $T_{\beta\gamma}$ when GDP is bound (the quiescent dark state), whereas they are separate when GTP is bound (the light-activated state). The only known role of $T_{\beta\gamma}$ is to present T_α to R* so that GTP can be exchanged for GDP. A light-activated amplification cycle involving rhodopsin, transducin, and the PDE is shown in Figure 3. The major features of this cycle are as follows (Fung and Stryer 1980; Stryer 1983):

1. In the dark, nearly all the transducin is in the T-GDP form, which does not activate the PDE.
2. R* encounters T-GDP by lateral diffusion of both components in the plane of the membrane. An R*–T-GDP complex is formed. GDP is released, leaving an empty guanyl-nucleotide-binding site. GTP then binds to form an R*–T-GTP complex. The only role of R* is to induce the release of GDP from transducin.
3. The binding of GTP markedly diminishes the affinity of transducin for R*. Moreover, the affinity of T_α for $T_{\beta\gamma}$ is very much lowered. These changes in affinities are critical for excitation. The released R* is free to interact with another T-GDP, enabling R* to act catalytically. The dissociation of transducin allows T_α-GTP to carry the excitation signal to the PDE.
4. R* is a potent catalyst for GTP-GDP exchange. Light-scattering studies of magnetically oriented ROS have shown that a molecule of R* activates a molecule of transducin in 1 msec (Vuong et al. 1984). About 500 molecules of T_α-GTP are formed by the action of a single R*. The activation of transducin is the first stage of amplification in vision.
5. Under physiological conditions, T-GDP is membrane-bound, whereas most of the T_α-GTP formed during the activation process is released into the cytosol. About 5% of the transducin stays membrane-bound in both GDP and GTP states (Wensel and Stryer 1988). The membrane-bound form of T-GTP is located in the plasma membrane, whereas the soluble form of T-GTP comes from disk membranes (Molday and Molday 1987; R. Molday, pers. comm.).
6. T_α-GTP activates the PDE by overcoming inhibi-

Table 1. Visual Transduction cDNAs and Genes

Encoded protein	Source	References
Rhodopsin	bovine	Nathans and Hogness (1983)
Rhodopsin	human	Nathans and Hogness (1984)
Cone opsins	human	Nathans et al. (1986)
Photoreceptor opsins	*Drosophila*	O'Tousa et al. (1985); Zucker et al. (1985); Cowman et al. (1986); Montell et al. (1987)
Photoreceptor opsin	octopus	Ovchinnikov et al. (1988b)
T_α	bovine	Khorana (1985); Lochrie et al. (1985); Medynski et al. (1985); Tanabe et al. (1985); Yatsunami (1985) and Lerea et al. (1986)
T_β	bovine	Fong et al. (1986); Sugimoto et al. (1985)
T_γ	bovine	Hurley et al. (1984a)
PDE_α	bovine	Ovchinnikov et al. (1987)
PDE_γ	bovine	Ovchinnikov et al. (1986)
PDE_γ	mouse	Tuteja and Farber (1988)
Arrestin	bovine	Shinohara et al. (1987)

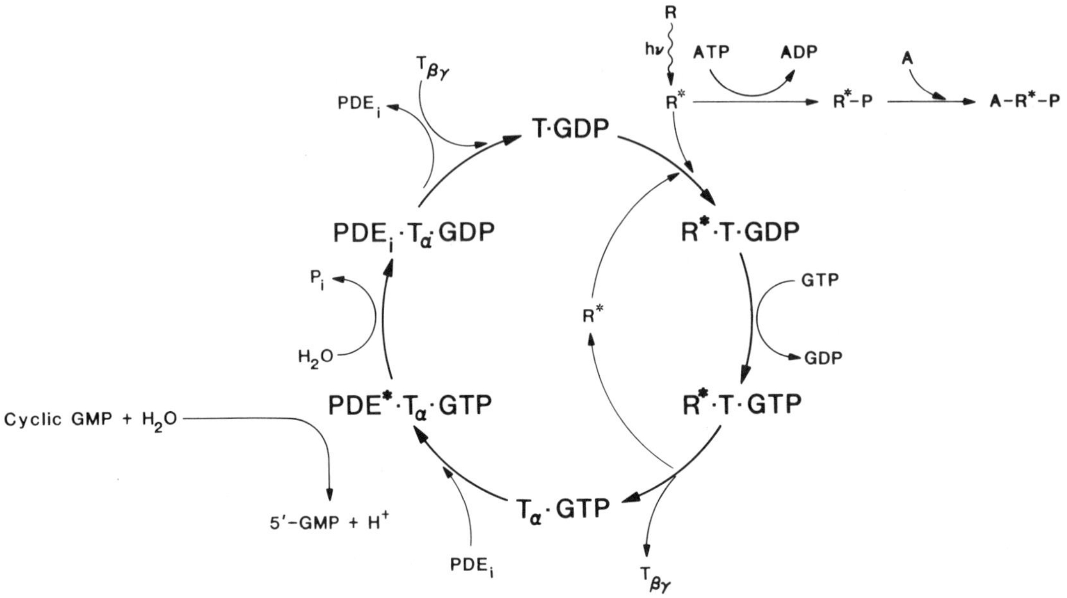

Figure 3. Light-activated transducin cycle. (R) Rhodopsin; (R*) photoexcited rhodopsin; (T) transducin; (PDE_i) inhibited phosphodiesterase; (PDE*) activated phosphodiesterase; (A) arrestin.

tory constraints imposed by two subunits of the PDE, as discussed below. Activated PDE hydrolyzes cGMP rapidly.

7. The GTPase activity inherent in the α subunit of transducin converts T_α-GTP to T_α-GDP, thereby deactivating the PDE. Recent studies suggest that this turnoff occurs in a second or two under physiological conditions (Dratz et al. 1987; Vuong and Chabre 1988), rather than in 30 seconds to 1 minute, as was observed in dilute ROS suspensions. T_α-GDP then rejoins $T_{\beta\gamma}$.

Rhodopsin Kinase and Arrestin Deactivate R*

The hydrolysis of GTP bound to T_α is necessary but not sufficient to restore the dark state. The release of all-*trans* retinal and binding of 11-*cis* retinal to resynthesize rhodopsin occur in minutes. R* must be switched off within a second or so to obtain a correspondingly transient neural signal. The capacity of R* to activate transducin is turned off by two proteins: rhodopsin kinase, a 68-kD cytosolic protein, and arrestin, a highly abundant 48-kD cytosolic protein (also known as the 48K protein or S antigen). Following illumination, rhodopsin kinase binds to the cytosolic face of R* (but not to rhodopsin) and catalyzes the phosphorylation of multiple serines and threonines in the carboxy-terminal region of R* (Wilden and Kühn 1982). Arrestin binds to phosphorylated R* but not to unphosphorylated rhodopsin. The capacity of phosphorylated R* to activate transducin is diminished markedly by arrestin (Miller et al. 1986; Wilden et al. 1986). In the absence of arrestin, R* must be phosphorylated at more than ten sites to become totally inactive (Bennett and Sitaramayya 1988). In the presence of arrestin, much less phosphorylation, perhaps only one or two phosphoryls per R*, is required for complete inactivation (Miller et al. 1986). Arrestin quenches phosphorylated R* by blocking the simultaneous binding of transducin. It is noteworthy that arrestin and the α subunit of transducin display a considerable degree of sequence similarity, particularly in their carboxy-terminal region (Shinohara et al. 1987). This sequence similarity implies that they recognize common groups on the cytosolic face of R*. Transducin and arrestin cannot be bound simultaneously to R* because their binding sites are overlapping.

Transducin Switches on PDE by Removing Its Inhibitory Subunits

T_α-GTP then carries the excitation signal to the PDE, a multisubunit peripheral membrane protein. The PDE consists of three kinds of subunits: α (88 kD), β (85 kD), and γ (9 kD) (Baehr et al. 1979). A key property of this enzyme is its very low catalytic activity in the dark state. The finding that limited tryptic digestion increases its catalytic activity markedly led to the proposal that native PDE is subject to an inhibitory constraint (Miki et al. 1975). We then found that the γ subunit, but not α or β, is rapidly degraded by trypsin (Hurley and Stryer 1982). The kinetics of activation of PDE paralleled the kinetics of degradation of γ. The trypsin-activated enzyme is an αβ complex. Purified γ subunit inhibits the catalytic activity of trypsin-activated PDE more than 99% by lowering its maximal velocity without appreciably changing its K_m. The binding of γ to αβ is very tight ($K_d = 0.13$ nM). These experiments demonstrated that the PDE consists of distinct catalytic and regulatory subunits. The maximal catalytic activity obtained by addition of T_α-GppNHp was nearly the same as that achieved by tryptic activation. This identity strongly suggested that T_α-GTP activates PDE by overcoming the inhibitory constraint imposed by γ. Recent studies (Deterre et al. 1988) have shown that the stoichiometry of fully inhibited PDE is $\alpha\beta\gamma_2$, not $\alpha\beta\gamma$, as was formerly presumed.

How does T_α-GTP reverse the inhibition of PDE imposed by its γ subunits? Three models seem plausible a priori. In model 1, T (denoting T_α-GTP) carries the γ subunits away from the catalytic subunits of PDE; in this scheme, $\alpha\beta$ is catalytically active. In model 2, T binds to the catalytic subunits and displaces the γ subunits, forming a catalytically active $T\alpha\beta$ complex. In model 3, T binds to the holoenzyme, forming a catalytically active $T\alpha\beta\gamma_2$ complex. Several lines of experimental evidence discriminate between these possibilities:

1. Light-triggered activation of PDE by transducin can be reversed by the addition of the γ subunit to ROS membranes (Wensel and Stryer 1986). As shown in Figure 4, nanomolar levels of γ subunit reverse the activation of PDE by micromolar levels of transducin. This result rules out model 3, which lacks a pathway enabling added γ to inhibit PDE activity.
2. Fluorescent-labeled γ subunit that retains inhibitory activity has proven to be a useful probe of the regulatory mechanism of the PDE (T.G. Wensel and L. Stryer, in prep.). The fluorescence polarization of this labeled γ-chain increases substantially when it binds to another protein. In the absence of T, labeled γ exchanges slowly (on the time scale of minutes) with unlabeled γ in PDE holoenzyme. In contrast, exchange is rapid in the presence of T. Thus, activated transducin displaces the γ subunits of PDE from their inhibitory sites. This experiment also rules out model 3.
3. Under physiological conditions, γ released from bovine PDE during the activation process stays on the disk membrane (Wensel and Stryer 1986).
4. A complex of T_α-GTPγS with γ can be eluted at low ionic strength from membranes that were activated by light in the presence of GTPγS (Deterre et al. 1986). Furthermore, $\alpha\beta$ and $\alpha\beta\gamma$ were isolated following low-ionic-strength elution of activated membranes, whereas unactivated membranes yielded $\alpha\beta\gamma_2$ (Deterre et al. 1988).
5. T_α-GTPγS, by itself, does not bind tenaciously to ROS membranes. Likewise, the γ subunits of PDE by themselves have little affinity for the membrane. However, the complex of T_α-GTPγS and γ formed in the activation process binds strongly to disk membranes. This leads to the prediction that T_α-GTPγS and γ added to disk membranes will become membrane-bound. Indeed, the amount of added T_α-GTPγS that becomes bound to reconstituted membranes devoid of PDE is proportional to the amount of added γ; the stoichiometry is nearly equimolar (B.K.-K. Fung, pers. comm.).

These experiments reveal that transducin activates the PDE by removing its inhibitory γ subunits (Fig. 5). The first step in activation is the binding of T_α-GTP to the PDE holoenzyme. Activation does not occur by dissociation of γ from the holoenzyme and the subsequent binding of γ to T_α-GTP. The subnanomolar dissociation constant of γ (Hurley and Stryer 1982) and the exchange reaction kinetics assayed by fluorescence polarization indicate that dissociation of γ in the absence of T_α-GTP is very slow, on the time scale of minutes, compared with PDE activation, which can occur in less than about 100 msec (Bennett 1982). T_α-GTP must actively remove γ from the catalytic subunits of PDE to activate the enzyme.

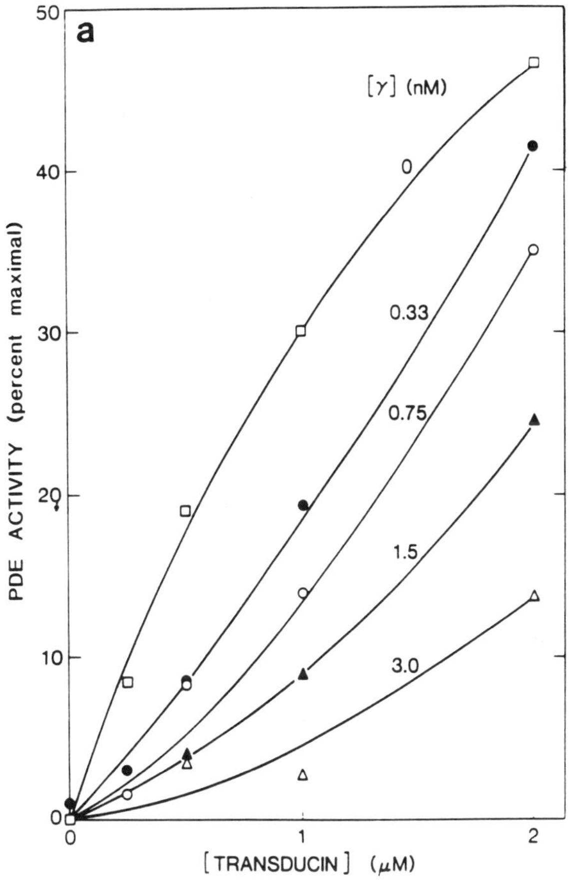

Figure 4. Reciprocal control of PDE by its γ subunit and transducin. Transducin (in the GTPγS form) activates the PDE, whereas added γ subunit inhibits it (Wensel and Stryer 1986).

The amino acid sequence of the γ subunit of PDE (Ovchinnikov et al. 1986) exhibits a striking distribution of charged residues. Of 15 basic residues, 12 are clustered in one region, and 5 of 9 are clustered in another. Each cluster is devoid of residues of the opposite charge. These highly charged sequences may form patches that interact with complementary patches on T_α and the catalytic subunits of PDE. Polyanions such as heparin and polycations such as protamine can activate PDE in the absence of activated transducin (Bitensky et al. 1981). These highly charged polymers probably turn on PDE activity by blocking the binding of γ to $\alpha\beta$.

Lane Brown, a graduate student in my laboratory, is

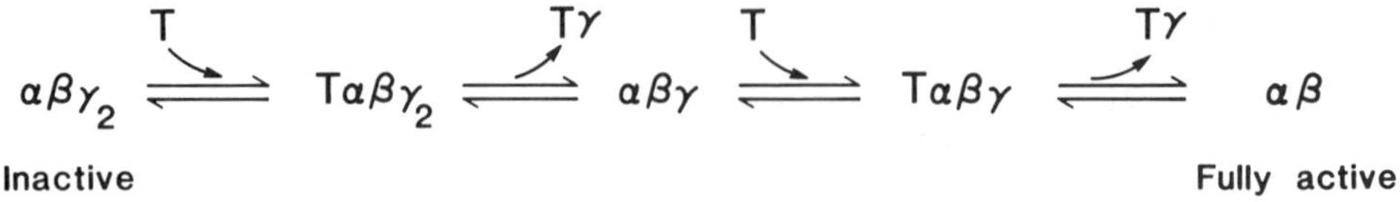

Figure 5. Transducin activates the PDE by removing its inhibitory subunits. T denotes T_α-GTP. The catalytic activity of the intermediate containing a single γ subunit is not known.

carrying out site-specific mutagenesis studies of the γ subunit. He has synthesized a gene encoding the γ chain of bovine PDE with codons optimized for expression in *Escherichia coli* and conveniently located unique restriction sites. This gene was inserted into an expression vector containing the p_L promoter and the gene for the first 31 amino acids of the *c*II protein of bacteriophage λ (Nagai and Thgersen 1984). The plasmid also encodes a four-residue linker between *c*II and the amino-terminal residue of γ that can be cleaved specifically by blood-clotting factor X_a. A substantial quantity of the *c*II-γ fusion protein was expressed in transfected *E. coli* cells. The fusion protein was then cleaved to yield a synthetic γ subunit that is identical to native γ, except for the absence of an acetyl group on the amino-terminal methionine. Synthetic γ is fully functional: It inhibits trypsin-activated PDE and native PDE, and its inhibition of native PDE is reversed by T_α-GTP. Surprisingly, the fusion protein, which carries an extra 35 amino acids at the amino terminus, is also fully functional. Hence, the amino-terminal region of γ probably does not interact intimately with either the catalytic subunits of PDE or the α subunit of transducin. This expression system should enable us to pinpoint sites on γ that are critical for its inhibitory action and its release by transducin.

Activated PDE Has Great Catalytic Power

The catalytic prowess of activated PDE is noteworthy. The ratio of the maximal velocity of the enzyme (4200/sec) to its Michaelis constant (70 μM), k_{cat}/K_m, is equal to 6×10^7 M/sec (Hurley and Stryer 1982). This value is very close to the limit set by the diffusion-controlled encounter of enzyme and substrate. The total concentration of cGMP in dark-adapted ROS is 60 μM, of which about 2 μM is free. Hence, [cGMP] $\ll K_m$ under physiological conditions, and so the rate of hydrolysis of fully activated PDE is given by the product of [PDE*] (30 μM) and k_{cat}/K_m, which is equal to 1.8×10^3/sec. Thus, cGMP is hydrolyzed to $1/e$ of its initial concentration by fully activated PDE in only 0.6 msec.

The Plasma Membrane Is a cGMP Electrode

How does the hydrolysis of cGMP lead to the closure of channels in the plasma membrane? The answer proved to be unexpectedly simple. Patch-clamp studies of excised pieces of ROS plasma membrane revealed that cGMP applied to the cytosolic side of the membranes directly opens channels, whereas calcium ion and nucleoside triphosphates have no effect (Fesenko et al. 1985). Channel opening by cGMP is reversible and depends on its concentration in a cooperative manner. The Hill coefficient for channel opening is 3.0; half the channels are open at 10 μM cGMP (Haynes et al. 1986; Zimmerman and Baylor 1986). Desensitization does not occur: The conductance depends only on the instantaneous concentration of cGMP. Under physiological conditions, the channel opens and closes in times of milliseconds in response to changes in cGMP level (Karpen et al. 1988a and this volume). The introduction of 8-bromo-cGMP, a hydrolysis-resistant analog of cGMP, into retinal rod cells blocks the normal light-induced decrease in membrane current, strongly suggesting that the channel is gated solely by cGMP (Zimmerman et al. 1985). Thus, the plasma membrane of the ROS behaves as a cGMP electrode. The channel protein, a 63-kD subunit, has been purified from bovine ROS and functionally reconstituted into liposomes and planar bilayer membranes (Cook et al. 1987; Hanke et al. 1988).

The channel is permeable to divalent cations, as well as to monovalent cations. Under physiological conditions, the fractions of inward current carried by Na^+, Ca^{++}, and Mg^{++} are estimated to be 0.70, 0.15, and 0.05, respectively (Nakatani and Yau 1988). Most of the light-induced hyperpolarization, the excitation signal, comes from blockage of Na^+ influx. In contrast, the blockage of Ca^{++} influx gives rise to a feedback signal that is central to recovery and adaptation.

Guanylate Cyclase Is Activated by the Light-induced Lowering of the Cytosolic Calcium Level

The cytosolic calcium level (Ca_i) of dark-adapted rods is estimated to be about 300 nM. Following illumination, Ca_i drops markedly in about 0.5 seconds, probably to less than 50 nM (Yau and Nakatani 1985; McNaughton et al. 1986). The cytosolic calcium level falls because the entry of Ca^{++} through the cGMP-gated channels stops, whereas its efflux by the sodium-calcium antiporter continues. The finding that the cGMP content of ROS increases severalfold when Ca_i is lowered to less than 10 nM pointed to a negative feedback loop between cGMP and Ca_i (Lipton and Dowling 1981).

How does Ca_i control the cGMP level? One possibility is that Ca_i modulates the activity of the PDE and thereby controls the rate of cGMP breakdown (Torre et al. 1986). High calcium levels do inhibit PDE, but the effect is unlikely to be important physiologically because millimolar levels are required (A.E. Barkdoll III et al., in prep.). Alternatively, the target could be guanylate cyclase, as suggested by electrophysiological experiments (Hodgkin et al. 1985; Hodgkin and Nunn 1988; Kondo and Miller 1988). Support came from biochemical experiments showing that guanylate cyclase in suspensions of toad ROS is strongly stimulated by lowering the calcium level from 10 μM to 10 nM if the sample is excited by light (Pepe et al. 1986).

We have recently found that the activity of guanylate cyclase in bovine ROS suspensions is exquisitely sensitive to the calcium level in the nanomolar range, whether illuminated or not (Koch and Stryer 1988). Cyclase activity increased more than fivefold when Ca_i was lowered from 120 to 20 nM (Fig. 6). The change in guanylate cyclase activity was half-maximal at 90 nM Ca_i, and the Hill coefficient was 3.9 ± 1.2. Calcium sensitivity was lost when ROS membranes were washed with a large volume of low-ionic-strength buffer. Guanylate cyclase stays with the membranes and exhibits a low level of enzymatic activity independent of Ca_i. Calcium sensitivity was partially restored when a soluble fraction derived from washing ROS with an equal volume of low-salt buffer was added back to the membranes. Four conclusions can be drawn: (1) Guanylate cyclase is regulated by a highly cooperative calcium switch that senses changes in Ca_i in the submicromolar range. (2) A membrane-bound protein that can be released at low ionic strength is an essential component of this switch. (3) Guanylate cyclase in the absence of the calcium-sensitive regulator has low activity at any calcium level. Hence, guanylate cyclase is stimulated by this regulatory protein at low Ca_i, rather than being inhibited by it at high Ca_i. (4) The stimulated activity of 15 nmole/min/mg rhodopsin in ROS suspensions corresponds to 60 μM cGMP/sec. ^{18}O studies of intact rabbit retinas showed that the cGMP flux increases to 67 μM/sec on illumination (Ames et al. 1986). Thus, the observed extent of stimulation of guanylate cyclase activity would suffice to replenish the cGMP store of moderately illuminated rods.

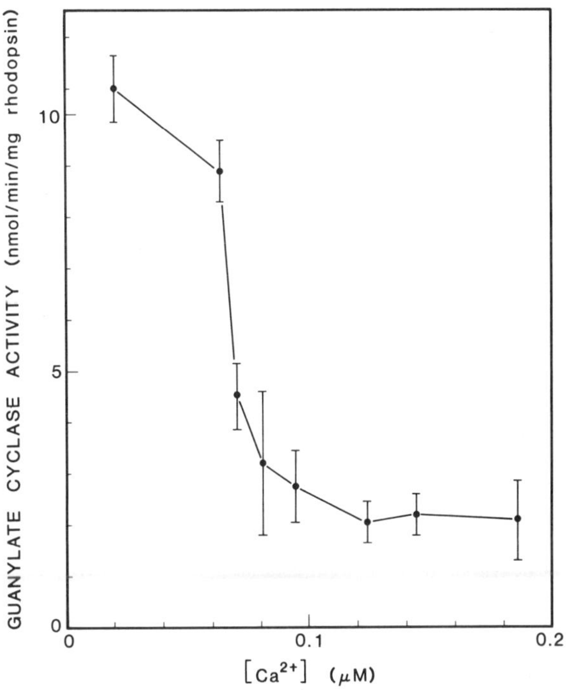

Figure 6. The guanylate cyclase of an ROS suspension is activated markedly when the concentration of calcium ion is lowered in the submicromolar range. The switching of catalytic activity by calcium ion is highly cooperative (Koch and Stryer 1988).

These biochemical and electrophysiological experiments establish that the light-induced decrease in the cytosolic level of calcium ion plays a key feedback role in initiating the recovery process (Fig. 7). Ca_i may be central to adaptation as well as recovery (Kondo and Miller 1988; Matthews et al. 1988; Nakatani and Yau 1988). In essence, calcium ion is the remembrance of past photons. Ca_i also serves as a signal to maintain a constant cGMP level in the dark state. A transient excess of cGMP would lead to increased entry of Ca^{++} that, in turn, would inhibit guanylate cyclase until the cGMP level is brought back to the standard level.

Kinetic and Thermodynamic Optimization of the Cascade

The single-photon sensitivity and reliability of visual transduction by retinal rods are consequences of the kinetic and thermodynamic optimization of each step in the cascade. The integration of components and the economy of design are equally impressive. Nature has chosen retinal as the chromophore of rhodopsin because it is unique among biomolecules in having a highly tunable absorption band, a high absorbance

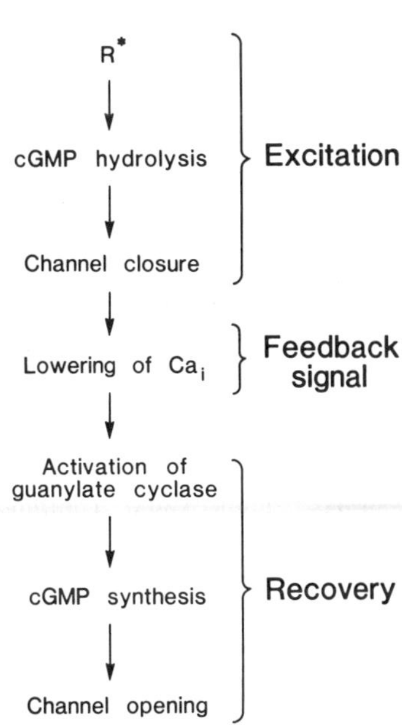

Figure 7. The light-induced lowering of Ca_i is the feedback signal that leads to the restoration of the dark state.

coefficient, a high quantum yield of photoisomerization, an extremely low thermal isomerization rate, a large light-induced conformational change, and a reactive group for attachment to opsin. The membrane setting of rhodopsin is optimized for a rapid response to light. The rate-limiting step in single-photon vision is the diffusion-controlled encounter of R* and transducin. The highly fluid disk membrane arising from the abundance of polyunsaturated fatty acyl groups enables R* to meet T-GDP in less than a millisecond (Liebman and Pugh 1981; Vuong et al. 1984).

The energetics of the transducin cycle are also noteworthy (Stryer 1987). The cycle is powered by the hydrolysis of GTP, rather than by the energy of the absorbed photon, which is used only to generate a trigger (R*). The phosphoryl potential of GTP is partitioned aptly between the formation of T_α-GTP and its subsequent hydrolysis to T-GDP. The formation of T_α-GTP is driven by a large decrease in free energy. However, about half of the phosphoryl potential of GTP is saved to make the hydrolysis step virtually irreversible to assure a very low level of T_α-GTP in the dark. Hence, the transducin cycle proceeds unidirectionally once triggered by R*. In the absence of R*, T-GDP undergoes GTP-GDP exchange very slowly, in times of hours, because there is a large activation for opening the guanyl-nucleotide-binding site. R* lowers the free energy of this transition state by about -10 kcal/mole, which accelerates the exchange reaction by a factor of 10^7. Like other efficient catalysts, R* is complementary to the transition state but not to the substrate (T-GDP) or product (T-GTP).

The PDE is also designed for speed. Its k_{cat}/K_m value of 6×10^7 M/sec is close to the limit set by its diffusion-controlled encounter with cGMP. The PDE is a member of an elite class of enzymes, exemplified by carbonic anhydrase, catalase, and acetylcholinesterase, which have achieved kinetic perfection. Equally significant, the PDE has low catalytic activity in the dark because it binds its inhibitory subunits tenaciously. Fewer than 1 in 1700 PDE molecules are catalytically active in situ in the dark (Wensel and Stryer 1986). The cGMP hydrolysis time in the dark is about 1 second, compared to less than 1 msec for the fully activated enzyme following illumination.

The amounts of the constituents of the cascade have been optimized to give activation before deactivation. R* interacts first with transducin and then with rhodopsin kinase because the concentration of T-GDP is about 100-fold higher than that of the kinase (Kühn 1984). Phosphorylation of rhodopsin begins only after much of the transducin has been activated. The binding of arrestin to phosphorylated R*, but not to unphosphorylated R*, gives transducin unhindered access to the trigger.

The cGMP-gated channel has several properties that are critical for effective phototransduction. The highly cooperative opening of the channel by cGMP enables it to be switched off by a small decrease in cGMP concentration (Yau and Baylor 1989). The millisecond closing of the channel following light-triggered cGMP hydrolysis gives a swift response at high light levels. Millisecond channel kinetics also help to make the singe-photon response stand out amid noise (Karpen et al. 1988). Reliability is also enhanced markedly by having many open channels, each with a low conductance, instead of a few channels, each with high conductance (Yau and Baylor 1989). The conductance of the cGMP-gated channel is lowered 250-fold by the high concentration of divalent cations in the extracellular milieu. At first glance, it seems curious that fewer than 5% of the channels are open in the dark state. Why did nature not choose to have a smaller total number of channels with a higher proportion open in the dark? A priori, this could have been accomplished by having a higher concentration of unbound cGMP or a higher binding affinity for cGMP. However, either alternative would have been deleterious. A higher binding affinity would slow the response of the channel; a higher cGMP level would be energetically prohibitive for the rod cell (Stryer 1987).

The ionic selectivity of the channel is well suited to the overall transduction process. By making the channel permeable to Ca^{++} as well as Na^+, nature has found an ingenious means of timing the onset of recovery precisely. The duration of the excitation phase is determined by the kinetics of the fall in Ca_i and the ensuing activation of guanylate cyclase. Excitation itself sets the recovery process in motion. The large store of bound cGMP, comprising more that 90% of the total cGMP, also aids the recovery process. It will be interesting to learn how this buffer is regulated.

Retinal Rods Exemplify Recurring Motifs of Signal Transduction

The transduction mechanism of vertebrate cones is very much like that of rods. The chromophore is again 11-*cis* retinal. The genes for the three human visual pigments mediating color vision have been cloned and sequenced (Nathans et al. 1986). The deduced amino acid sequences can be folded into the seven transmembrane helix motif of vertebrate rhodopsins (Nathans 1987). The Schiff-base lysine on the seventh helix, glycosylation sites in the amino-terminal region, and multiple phosphorylation sites in the carboxy-terminal region are found in the cone pigments, as in rhodopsin. About 45% of the sequence of each of the cone pigments is identical to that of rhodopsin. Likewise, the transducin in cones is similar to that in rods (Lochrie et al. 1985). Cones also contain a cGMP PDE that is like that of rods (Hurwitz et al. 1985). Finally, the light-sensitive channel in the membrane of cone outer segments is gated by cGMP and has many features in common with the channels of rods (Cobbs et al. 1985; Haynes and Yau 1985). Thus, the transduction machineries of rods and cones are variations of the same fundamental theme. It will be interesting to learn how differences between rod and cone molecules are related to the distinctive physiological roles of these two classes of sensory cells.

The photoreceptor cells of vertebrates, mollusks, and arthropods are anatomically different and are believed to have evolved independently. However, the chromophore in all three phyla is 11-*cis* retinal. The transduction machinery of the three phyla exhibit additional fundamental similarities. The genes for several *Drosophila* visual pigments have been cloned and sequenced, as has the cDNA of octopus R (for references, see Table 1). About 30% of their sequence closely resembles that of vertebrate rhodopsin. Arthropod, molluscan, and vertebrate visual pigments share the same fundamental architectural plan: seven transmembrane helices, amino-terminal oligosaccharide sites, carboxy-terminal phosphorylation sites, loops for interaction with cytosolic proteins, a retinal-binding lysine near the center of the membrane, and a conserved aspartate that may be the Schiff-base counterion. Indeed, R* from a molluscan retina can trigger the activation of vertebrate transducin. Furthermore, invertebrate photoreceptors contain a G protein akin to vertebrate transducin. This G protein plays a key role in invertebrate vision (Tsuda 1987). However, the transduction pathways appear to differ beyond transducin. The G protein of fly photoreceptors activates phospholipase C, which then generates inositol 1,4,5-trisphosphate and diacylglycerol (Devary et al. 1987). Injection of inositol 1,4,5-trisphosphate into *Limulus* photoreceptors mimics the action of light (Brown et al. 1984; Fein et al. 1984). It seems likely that the phosphoinositide cascade is central to invertebrate vision. Thus, rhodopsin and transducin are ancient molecules, more than 700 million years old, that have become coupled to different effector systems in vertebrate and invertebrate visual systems.

The molecules mediating the cGMP cascade of rods and cones are found in many other signal-transducing systems. The seven-helix rhodopsin motif is found in other membrane receptors, such as the β-adrenergic receptor (Dixon et al. 1986), the muscarinic acetylcholine receptor (Kubo et al. 1986), and yeast mating-factor receptor (Blumer et al., this volume). These receptors are deactivated by kinases resembling rhodopsin kinase and probably also by proteins like arrestin (Benovic et al. 1987). Transducin is a member of a large family of signal-coupling proteins, the G proteins, that play key roles in sensory transduction, hormone action, and growth control (Gilman 1987). G proteins participate in the adenylate cyclase cascade, the phosphoinositide cascade, and the regulation of membrane channels. The kinship of transducin to these G proteins is clearly established by multiple criteria: mechanistic similarity, sequence homology, susceptibility to toxins, and interchangeability of components (Stryer and Bourne 1986). The cGMP-gated channel also has its counterpart in another sensory system. Olfactory neurons contain an odorant-triggered cAMP cascade and cation-specific channels that are directly opened by cyclic nucleotides (Jones and Reed 1987; Lancet et al. 1987; Nakamura and Gold 1987). The study of the cGMP cascade of vision is providing rewarding views of recurring motifs of signal transduction in addition to revealing how the exquisite sensitivity of retinal rod cells is achieved.

ACKNOWLEDGMENTS

I am indebted to Denis Baylor, Lane Brown, Jeffrey Karpen, Karl-Wilhelm Koch, and Theodore Wensel for stimulating discussions. The research carried out in my laboratory was supported by research grants from the National Eye Institute and the National Institute of General Medical Sciences.

REFERENCES

Ames, A., T.F. Walseth, M. Barad, R.M. Graeff, and N.D. Goldberg. 1986. Light-induced increases in cGMP metabolic flux correspond with electrical responses of photoreceptors. *J. Biol. Chem.* **261:** 13034.

Baehr, W., M.J. Devlin, and M.L. Applebury. 1979. Isolation and characterization of cGMP phosphodiesterase from bovine rod outer segments. *J. Biol. Chem.* **254:** 11669.

Barkdoll, A.E. III, E.N. Pugh, Jr., and A. Sitaramayya. 1988. The calcium dependence of the activation and inactivation kinetics of the light-activated phosphodiesterase of retinal rods. *J. Gen. Physiol.* (in press).

Baylor, D.A., T.D. Lamb, and K.-W. Yau. 1979. Responses of retinal rods to single photons. *J. Physiol.* **288:** 613.

Bennett, N. 1982. Light-induced interactions between rhodopsin and the GTP-binding protein. Relation with phosphodiesterase activation. *Eur. J. Biochem.* **123:** 133.

Bennett, N. and A. Sitaramayya. 1988. Inactivation of photoexcited rhodopsin in retinal rods: The roles of rhodopsin kinase and 48-kDa protein (arrestin). *Biochemistry* **27:** 1710.

Benovic, J.L., H. Kühn, I. Weyand, J. Codina, M.G. Caron, and R.J. Lefkowitz. 1987. Functional desensitization of the isolated β-adrenergic receptor by the β-adrenergic receptor kinase: Potential role of an analog of the retinal protein arrestin (48-kDa protein). *Proc. Natl. Acad. Sci.* **84:** 8879.

Bitensky, M.W., G.L. Wheeler, A. Yamazaki, M.M. Rasenick, and P.J. Stein. 1981. Cyclic nucleotide metabolism in vertebrate photoreceptors: A remarkable analogy and an unraveling enigma. *Curr. Top. Membr. Transp.* **15:** 237.

Brown, J.E., L.J. Rubin, A.J. Ghalayini, A.P. Tarver, R.F. Irvine, M.J. Berridge, and R.E. Anderson. 1984. Myo-inositol polyphosphate may be a messenger for visual excitation in Limulus photoreceptors. *Nature* **311:** 160.

Busch, G.E., M.L. Applebury, A.A. Lamola, and P.M. Rentzepis. 1972. Formation and decay of prelumirhodopsin at room temperature. *Proc. Natl. Acad. Sci.* **69:** 2802.

Carlsen, W.F. 1984. "Picosecond resonance Raman spectroscopy of visual pigments." Ph.D. thesis, Stanford University, Stanford, California.

Chabre, M. 1985. Trigger and amplification mechanisms in visual phototransduction. *Annu. Rev. Biophys. Biophys. Chem.* **14:** 331.

Cobbs, W.A., A.E. Barkdoll III, and E.N. Pugh, Jr. 1985. Cyclic GMP increases photocurrent and light sensitivity of retinal cones. *Nature* **317:** 64.

Cook, N.J., W. Hanke, and U.B. Kaupp. 1987. Identification, purification, and functional reconstitution of the cyclic GMP-dependent channel from rod photoreceptors. *Proc. Natl. Acad. Sci.* **84:** 585.

Cooper, A. 1981. Rhodopsin photoenergetics: Lumirhodopsin and the complete energy profile. *FEBS Lett.* **123:** 324.

Cowman, A.F., C.S. Zuker, and G.M. Rubin. 1986. An opsin gene expressed in only one photoreceptor cell type of the *Drosophila* eye. *Cell* **44:** 705.

Deterre, P., J. Bigay, F. Forquet, M. Robert, and M. Chabre. 1988. cGMP phosphodiesterase of retinal rods is regulated by two inhibitory subunits. *Proc. Natl. Acad. Sci.* **85:** 2424.

Deterre, P., J. Bigay, M. Robert, C. Pfister, H. Kühn, and M. Chabre. 1986. Activation of retinal rod cyclic GMP-phosphodiesterase by transducin: Characterization of the complex formed by phosphodiesterase inhibitor and transducin α-subunit. *Proteins Struct. Funct. Genet.* **1:** 188.

Devary, O., O. Heichal, A. Blumenfeld, D. Cassel, E. Suss, S. Barash, C.T. Rubinstein, B. Minke, and Z. Selinger. 1987. Coupling of photoexcited rhodopsin to inositol phospholipid hydrolysis in fly photoreceptors. *Proc. Natl. Acad. Sci.* **84:** 6939.

Dixon, R.A., B.K. Kobilka, D.J. Strader, J.L. Benovic, H.G. Dohlman, T. Frielle, M.A. Bolanowski, C.D. Bennett, E. Rands, R.E. Diehl, R.A. Mumford, E.E. Slater, I.S. Sigal, M.G. Caron, R.J. Lefkowitz, and C.D. Strader. 1986. Cloning of the gene and cDNA for mammalian beta-adrenergic receptor and homology with rhodopsin. *Nature* **321:** 75.

Dratz, E.A. and P.A. Hargrave. 1983. The structure of rhodopsin and the rod outer segment disk membrane. *Trends Biochem. Sci.* **8:** 128.

Dratz, E.A., J.W. Lewis, L.E. Schaechter, K.R. Parker, and D.S. Kliger. 1987. Retinal rod GTPase turnover rate increases with concentration: A key to the control of visual excitation? *Biochem. Biophys. Res. Commun.* **146:** 379.

Emeis, D., H. Kühn, J. Reichert, and K.P. Hofmann. 1982. Complex formation between metarhodopsin II and GTP-binding protein in bovine photoreceptor membranes leads to a shift of the photoproduct equilibrium. *FEBS Lett.* **143:** 29.

Fein, A., R. Payne, D.W. Corson, M.J. Berridge, and R.F. Irvine. 1984. Photoreceptor excitation and adaptation by inositol 1,4,5-trisphosphate. *Nature* **311:** 157.

Fesenko, E.E., S.S. Kolesnikov, and A.L. Lyubarsky. 1985. Induction by cyclic GMP of cationic conductance in plasma membrane of retinal rod outer segment. *Nature* **313:** 310.

Fong, H.K., J.B. Hurley, R.S. Hopkins, R. Miake-Lye, M.S. Johnson, R.F. Doolittle, and M.I. Simon. 1986. Repetitive segmental structure of the transducin beta subunit: Homology with the CDC4 gene and identification of related mRNAs. *Proc. Natl. Acad. Sci.* **83:** 2162.

Franke, R.R., T.P. Sakmar, D.D. Oprian, and H.G. Khorana. 1988. A single amino acid substitution in rhodopsin (lysine 248→leucine) prevents activation of transducin. *J. Biol. Chem.* **263:** 2119.

Fung, B.K.-K. 1983. Characterization of transducin from bovine retinal rod outer segments. Separation and reconstitution of the subunits. *J. Biol. Chem.* **258:** 10495.

Fung, B.K.-K. and L. Stryer. 1980. Photolyzed rhodopsin catalyzes the exchange of GTP for GDP in retinal rod outer segment membranes. *Proc. Natl. Acad. Sci.* **77:** 2500.

Fung, B.K.-K., J.B. Hurley, and L. Stryer. 1981. Flow of information in the light-triggered cyclic nucleotide cascade of vision. *Proc. Natl. Acad. Sci.* **78:** 152.

Gilman, A.G. 1987. G proteins: Transducer of receptor-generated signals. *Annu. Rev. Biochem.* **56:** 615.

Hanke, W., N.J. Cook, and U.B. Kaupp. 1988. cGMP-dependent channel protein from photoreceptor membranes: Single-channel activity of the purified and reconstituted protein. *Proc. Natl. Acad. Sci.* **85:** 94

Hargrave, P.A., J.H. McDowell, D.R. Curtis, J.K. Wang, E. Juszczak, S.-L. Fong, J.K. Mohana Rao, and P. Argos. 1983. The structure of bovine rhodopsin. *Biophys. Struct. Mech.* **9:** 235.

Haynes, L. and K.-W. Yau. 1985. Cyclic GMP-sensitive conductance in outer segment membrane of catfish cones. *Nature* **317:** 61.

Haynes, L.W., A.R. Kay, and K.-W. Yau. 1986. Single cyclic GMP-activated channel activity in excised patches of rod outer segment membrane. *Nature* **317:** 61.

Hayward, G., W. Carlsen, A. Siegman, and L. Stryer. 1981. The retinal chromophore of rhodopsin photoisomerizes within picoseconds. *Science* **211:** 942.

Hodgkin, A.L. and B.J. Nunn. 1988. Control of light-sensitive current in salamander rods. *J. Physiol.* **403:** 439.

Hodgkin, A.L., P.A. McNaughton, and B.J. Nunn. 1985. The ionic selectivity and calcium dependence of the light-sensitive pathway in toad rods. *J. Physiol.* **358:** 447.

Hurley, J.B. and L. Stryer. 1982. Purification and characterization of the gamma regulatory subunit of the cyclic GMP phosphodiesterase from retinal rod outer segments. *J. Biol. Chem.* **257:** 11094.

Hurley, J.B., H.K. Fong, D.B. Teplow, W.J. Dreyer, and M.I. Simon. 1984a. Isolation and characterization of a cDNA clone for the gamma subunit of bovine retinal transducin. *Proc. Natl. Acad. Sci.* **81:** 6948.

Hurley, J.B., M.I. Simon, D.B. Teplow, J.D. Robishaw, and A.G. Gilman. 1984b. Homologies between signal transducing G proteins and *ras* gene products. *Science* **226:** 860.

Hurwitz, R.L., A.H. Bunt-Milam, M.L. Chang, and J.A. Beavo. 1985. cGMP phosphodiesterase in rod and cone outer segments of the retina. *J. Biol. Chem.* **260:** 568.

Jones, D.T. and R.R. Reed. 1987. Molecular cloning of five GTP-binding protein cDNA species from rat olfactory neuroepithelium. *J. Biol. Chem.* **262:** 14241.

Karpen, J.W., A.L. Zimmerman, L. Stryer, and D.A. Baylor. 1988. Gating kinetics of the cyclic-GMP-activated channel of retinal rods: Flash photolysis and voltage-jump studies. *Proc. Natl. Acad. Sci.* **85:** 1287.

Koch, K.-W. and L. Stryer. 1988. Highly cooperative feedback control of retinal rod guanylate cyclase by calcium ion. *Nature* **334:** 64.

Kondo, H. and W.H. Miller. 1988. Rod light adaptation may be mediated by acceleration of the phosphodiesterase-guanylate cyclase cycle. *Proc. Natl. Acad. Sci.* **85:** 1322.

Kubo, T., A. Maeda, K. Sugimoto, I. Akiba, A. Mikami, H. Takahashi, T. Haga, K. Haga, A. Ichiyama, K. Kangawa, H. Matsuo, T. Hirose, and S. Numa. 1986. Primary structure of porcine cardiac muscarinic acetylcholine receptor deduced from the cDNA sequence. *FEBS Lett,* **209:** 367.

Kühn, H. 1984. Interactions between photoexcited rhodopsin and light-activated enzymes in rods. In *Progress in retinal research* (ed. N. Osborne and J. Chader), p. 123. Pergamon, New York.

Lancet, D., Z. Chen, A. Ciobotariu, F. Eckstein, M. Khen, J. Heldman, D. Ophir, I. Shafir, and U. Pace. 1987. Toward a comprehensive molecular analysis of olfactory transduction. *Ann. N.Y. Acad. Sci.* **510:** 27.

Lerea, C.L., D.E. Somers, J.B. Hurley, I.B. Klock, and A.H. Bunt-Milam. 1986. Identification of specific transducin alpha subunits in retinal rod and cone photoreceptors. *Science* **234:** 77.

Liebman, P.A., and E.N. Pugh, Jr. 1981. Control of rod disk membrane phosphodiesterase and a model for visual transduction. *Curr. Top. Membr. Transp.* **15:** 157.

Liebman, P.A., K.R. Parker, and E.A. Dratz. 1987. The molecular mechanism of visual excitation and its relation to the structure and composition of the rod outer segment. *Annu. Rev. Physiol.* **49:** 965.

Lipton, S.A. and J.E. Dowling. 1981. The relation between Ca^{2+} and cyclic GMP in rod photoreceptors.. *Curr. Top. Membr. Transp.* **15:** 381.

Lochrie, M.A., J.B. Hurley, and M.I. Simon. 1985. Sequence of the alpha subunit of photoreceptor G protein: Homologies between transducin, *ras*, and elongation factors. *Science* **228:** 96.

Lugtenburg, J., R.A. Mathies, R.G. Griffin, and J. Herzfeld. 1988. Structure and function of rhodopsins from solid state

NMR and resonance Raman spectroscopy of isotopic retinal derivatives. *Trends Biochem. Sci.* **13:** 388.

Mathies, R.A. and L. Stryer. 1987. Retinal has a highly dipolar singlet excited state: Implications for vision. *Proc. Natl. Acad. Sci.* **73:** 2169.

Matthews, H.R., R.L.W. Murphy, G.L. Fain, and T.D. Lamb. 1988. Photoreceptor light adaptation is mediated by cytoplasmic calcium concentration. *Nature* **334:** 67.

McNaughton, P.A., L. Cervetto, and B.J. Nunn. 1986. Measurement of the intracellular free calcium concentration in salamander rods. *Nature* **322:** 261.

Medynski, D.C., K. Sullivan, D. Smith, C. Van Dop, F.H. Chang, B.K. Fung, P.H. Seeburg, and H.R. Bourne. 1985. Amino acid sequence of the alpha subunit of transducin deduced from the cDNA sequence. *Proc. Natl. Acad. Sci.* **82:** 4311.

Miki, N., J.M. Baraban, J.J. Keirns, J.J. Boyce, and M.W. Bitensky. 1975. Purification and properties of the light-activated cyclic nucleotide phosphodiesterase of rod outer segments. *J. Biol. Chem.* **250:** 6320.

Miller, J.L., D.A. Fox, and B.J. Litman. 1986. Amplification of phosphodiesterase activation is greatly reduced by rhodopsin phosphorylation. *Biochemistry* **25:** 4983.

Molday, R.S. and L.L. Molday. 1987. Differences in the protein composition of bovine retinal rod outer segment disk and plasma membranes isolated by a ricin-gold-dextran density gradient perturbation method. *J. Cell Biol.* **105:** 2589.

Montell, C., K. Jones, C. Zuker, and G. Rubin. 1987. A second opsin gene expressed in the ultraviolet-sensitive R7 photoreceptor cells of *Drosophila melanogaster*. *J. Neurosci.* **7:** 1558.

Nagai, K. and H.C. Thgersen. 1984. Generation of beta-globin by sequence-specific proteolysis of a hybrid protein produced in *Escherichia coli*. *Nature* **309:** 810.

Nakatani, K. and K.-W. Yau. 1988. Calcium and light adaptation in retinal rods and cones. *Nature* **334:** 69.

Nathans, J. 1987. Molecular biology of visual pigments. *Annu. Rev. Neurosci.* **10:** 163.

Nathans, J. and D.S. Hogness. 1983. Isolation, sequence analysis, and intron-exon arrangement of the gene encoding bovine rhodopsin. *Cell* **34:** 807.

———. 1984. Isolation and nucleotide sequence of the gene encoding human rhodopsin. *Proc. Natl. Acad. Sci.* **81:** 4851.

Nathans, J., D. Thomas, and D.S. Hogness. 1986. Molecular genetics of human color vision: The genes encoding blue, green and red pigments. *Science* **232:** 193.

Nukamura, T. and G.H. Gold. 1987. A cyclic nucleotide-gated conductance in olfactory receptor cilia. *Nature* **325:** 442.

O'Tousa, J.E., W. Baehr, R.L. Martin, J. Hirsh, W.L. Pak, and M.L. Applebury. 1985. The *Drosophila* ninaE gene encodes an opsin. *Cell* **40:** 839.

Ovchinnikov, Y.A. 1982. Rhodopsin and bacteriorhodopsin: Structure-function relationships. *FEBS Lett.* **148:** 179.

Ovchinnikov, Y.A., N.G. Abdulaev, and A.S. Bogachux. 1988a. Two adjacent cysteine residues in the C-terminal cytoplasmic fragment of bovine rhodopsin are palmitylated. *FEBS Lett.* **230:** 1.

Ovchinnikov, Y.A., N.G. Abdulaev, A.S. Zolotarev, I.D. Artamonov, I.A. Bespalov, A.E. Dergachev, and M. Tsuda. 1988b. Octopus rhodopsin. Amino acid sequence deduced from cDNA. *FEBS Lett.* **232:** 69.

Ovchinnikov, Y.A., V.V. Gubanov, N.V. Khramtsov, K.A. Ishenko, V.E. Zagranichny, K.G. Muradov, T.M. Shuaeva, and V.M. Lipkin. 1987. Cyclic GMP phosphodiesterase from bovine retina. *FEBS Lett.* **223:** 169.

Ovchinnikov, Y.A., V.M. Lipkin, V.P. Kumarev, V.V. Gubanov, N.V. Khramtsov, N.B. Akhmedov, V.E. Zagranichny, and K.G. Muradov. 1986. Cyclic GMP phosphodiesterase from cattle retina. Amino acid sequence of the gamma subunit and nucleotide sequence of the corresponding cDNA. *FEBS Lett.* **204:** 288.

Pepe, I.M., I. Panfoli, and C. Cugnoli. 1986. Guanylate cyclase in rod outer segments of the toad retinal. Effect of light and Ca^{2+}. *FEBS Lett.* **203:** 73.

Pugh, E.N., Jr. and W.H. Cobbs. 1986. Visual transduction in vertebrate rods and cones: A tale of two transmitters, calcium and cyclic GMP. *Vision Res.* **26:** 1613.

Shinohara, T., B. Dietzschold, C.M. Craft, G. Wistow, J.J. Early, L.A. Donoso, J. Horwitz, and R. Tao. 1987. Primary and secondary structure of bovine retinal S antigen (48-kDa protein). *Proc. Natl. Acad. Sci.* **84:** 6975.

Stryer, L. 1983. Transducin and the cyclic GMP phosphodiesterase: Amplifier proteins in vision. *Cold Spring Harbor Symp. Quant. Biol.* **48:** 841.

———. 1986. Cyclic GMP cascade of vision. *Annu. Rev. Neurosci.* **9:** 87.

———. 1987. Visual transducin: Design and recurring motifs. *Chemica Scripta* **27B:** 161.

Stryer, L. and H.R. Bourne. 1986. G. Proteins: A family of signal transducers. *Annu. Rev. Cell Biol.* **2:** 389.

Sugimoto, K., T. Nukada, T. Tanabe, H. Takahashi, M. Noda, N. Minamino, K. Kangawa, H. Matsuo, T. Hirose, S. Inayama, and S. Numa. 1985. Primary structure of the beta-subunit of bovine transducin deduced from the cDNA sequences. *FEBS Lett.* **191:** 235.

Tanabe, T., T. Nukada, Y. Nishikawa, K. Sugimoto, H. Suzuki, H. Takahashi, M. Noda, T. Haga, A. Ichiyama, K. Kangawa, N. Minamino, H. Matsuo, and S. Numa. 1985. Primary structure of the alpha-subunit of transducin and its relationship to *ras* proteins. *Nature* **315:** 242.

Thomas, D.D. and L. Stryer. 1982. The transverse location of the retinal chromophore of rhodopsin in rod outer segment disc membranes. *J. Mol. Biol.* **154:** 145.

Torre, V., H.R. Matthews, and T.D. Lamb. 1986. Role of calcium in regulating the cyclic GMP cascade of phototransduction in retinal rods. *Proc. Natl. Acad. Sci.* **83:** 7109.

Tsuda, M. 1987. Photoreception and phototransduction in invertebrate photoreceptors. *Photochem. Photobiol.* **45:** 915.

Tuteja, N. and D.B. Farber. 1988. γ-Subunit of mouse retinal cyclic-GMP phosphodiesterase: cDNA and corresponding amino acid sequences. *FEBS Lett.* **232:** 182.

Unwin, P.N.T. 1987. Design and action of cell communication channels. *Chemica Scripta* **27B:** 47.

Vuong, T.M., M. Chabre, and L. Stryer. 1984. Millisecond activation of transducin in the cyclic nucleotide cascade of vision. *Nature* **311:** 659.

Wald, G. 1968. The molecular basis of visual excitation. *Nature* **219:** 800.

Wensel, T.G. and L. Stryer. 1986. Reciprocal control of retinal rod cyclic GMP phosphodiesterase by its gamma subunit and transducin. *Proteins Struct. Funct. Genet.* **1:** 90.

———. 1988. Membrane-bound GTP-transducin efficiently activates retinal cGMP phosphodiesterase. In *Enzyme dynamics and regulation* (ed. P.B. Chock et al.), p. 102. Springer-Verlag, Berlin.

Wilden, U. and H. Kühn. 1982. Light-dependent phosphorylation of rhodopsin: Number of phosphorylation sites. *Biochemistry* **21:** 3014.

Wilden, U., S.W. Hall, and H. Kühn. 1986. Phosphodiesterase activation by photoexcited rhodopsin is quenched when rhodopsin is phosphorylated and binds the intrinsic 48-kDa protein of rod outer segments. *Proc. Natl. Acad. Sci.* **83:** 1174.

Yatsunami, K. and G.K. Khorana. 1985. GTPase of bovine rod outer segments: The amino acid sequence of the alpha-subunit as derived from the cDNA sequence. *Proc. Natl. Acad. Sci.* **82:** 4316.

Yau, K.-W. and D.A. Baylor. 1989. Cyclic GMP-sensitive conductance of retinal photoreceptor cells. *Annu. Rev. Neurosci.* **12:** (in press).

Yau, K.-W. and K. Nakatani. 1985. Light induced reduction of cytoplasmic free calcium in retinal rod outer segment. *Nature* **313:** 579.

Zimmerman, A.L. and D.A. Baylor. 1986. Cyclic GMP-sensitive conductance of retinal rods consists of aqueous pores. *Nature* **321:** 70.

Zimmerman, A.L., G. Yamanaka, F. Eckstein, D.A. Baylor, and L. Stryer. 1985. Interaction of hydrolysis-resistant analogs of cyclic GMP with the phosphodiesterase and light-sensitive channel of retinal rod outer segments. *Proc. Natl. Acad. Sci.* **82:** 8813.

Zucker, C.S., A.F. Cowman, and G.M. Rubin. 1985. Isolation and structure of a rhodopsin gene from *D. melanogaster. Cell* **40:** 851.

Molecular Basis of the Functional Heterogeneity of the Muscarinic Acetylcholine Receptor

S. NUMA,* K. FUKUDA,* T. KUBO,* A. MAEDA,* I. AKIBA,* H. BUJO,* J. NAKAI,* M. MISHINA,* AND H. HIGASHIDA†

*Departments of Medical Chemistry and Molecular Genetics, Kyoto University Faculty of Medicine, Kyoto 606, Japan; †Department of Biophysics, Neuroinformation Research Institute, Kanazawa University School of Medicine, Kanazawa 920, Japan

The muscarinic acetylcholine receptor (mAChR) mediates a variety of cellular responses, including inhibition of adenylate cyclase, breakdown of phosphoinositides, and modulation of potassium channels, through the action of guanine-nucleotide-binding regulatory proteins (G proteins) (Burgen 1984; Brown 1986; Gilman 1986). The question then arises as to whether multiple mAChR species exist that are responsible for the various biochemical and physiological effects. In fact, pharmacologically distinguishable forms of the mAChR occur in different tissues and have been provisionally classified into M_1 (I), M_2 cardiac (II), and M_2 glandular (III) subtypes on the basis of their difference in apparent affinity for antagonists (Hammer et al. 1980, 1986; Birdsall et al. 1986). Here, we have made attempts to understand the molecular basis of the functional heterogeneity of the mAChR, using recombinant DNA technology.

Primary Structure of mAChRs

DNAs complementary to porcine cerebral and cardiac mRNAs encoding mAChRs were cloned, and the complete amino acid sequences of the mAChRs (designated as mAChR I and mAChR II, respectively) were deduced from the nucleotide sequences of the cDNAs (Kubo et al. 1986a,b; Peralta et al. 1987b). By using cloned cDNA sequences as hybridization probes, two additional mammalian cerebral cDNAs (or genomic DNAs) encoding mAChRs (designated as mAChR III and mAChR IV) were subsequently isolated and sequenced (Bonner et al. 1987; Peralta et al. 1987a; Akiba et al. 1988); mAChR III and mAChR IV are defined as described by Bonner et al. (1987). The four mAChRs are homologous to one another in amino acid sequence, except that the region between the putative transmembrane segments V and VI (see below) is divergent; some sequence similarity is noted in this region, however, between mAChR I and mAChR III as well as between mAChR II and mAChR IV. The hydropathy profiles of the four mAChRs are similar, suggesting the presence of seven hydrophobic segments (I–VII) that represent putative transmembrane α helices; segment VII is less hydrophobic than the others. The only charged amino acid residues present in these segments are two aspartic acid residues (one each in segments II and III) that are conserved in all the mAChRs. The mAChRs are devoid of an amino-terminal signal sequence and contain potential *N*-glycosylation sites in the amino-terminal region preceding segment I. The four mAChRs are also similar in that their carboxy-terminal region contains several threonine or serine residues as potential sites of phosphorylation.

These structural features are generally shared by rhodopsin (Ovchinnikov 1982) and the β-adrenergic receptor (Dixon et al. 1986). This suggests that the mAChRs have a transmembrane topology similar to that of bacteriorhodopsin (Henderson and Unwin 1975). It is noteworthy that the nonhomologous segment present in the mAChRs (see above) corresponds to the region where an insertion is found in the mAChRs, as compared with the β-adrenergic receptor and rhodopsin.

Molecular Distinction between mAChR Subtypes

mRNAs specific for the individual mAChRs were synthesized by transcription in vitro of the respective cDNAs or genomic DNAs (containing no introns in the protein-coding region) linked with the bacteriophage SP6 promoter and were injected into *Xenopus* oocytes to yield functional mAChRs (Kubo et al. 1986b; Fukuda et al. 1987; Akiba et al. 1988). The ligand-binding properties of the four mAChRs expressed in oocytes were examined using cell extracts (Table 1). The apparent dissociation constant (K_d) for (−) [^{3}H]quinuclidinyl benzilate (QNB), estimated by Scatchard analysis, was similar for all the mAChRs. The apparent K_d values for selective antagonists were obtained by measuring displacement of (−)[^{3}H]QNB binding by increasing concentrations of the antagonists. Pirenzepine, selective for the M_1 (I) subtype (Hammer et al. 1980), showed the highest binding affinity for mAChR I (K_d = 18 nM). mAChR III and mAChR IV exhibited intermediate affinities for pirenzepine (K_d = 120–180 nM), and mAChR II exhibited the lowest affinity (K_d = 660 nM). These three classes of affinity for pirenzepine are similar to those reported originally for mAChR in brain (cerebral cortex and hippocampus), glands, and heart (Hammer et al. 1980). AF-DX 116, selective for the M_2 cardiac (II) subtype (Hammer et al. 1986), showed a higher binding affinity for mAChR

Table 1. Apparent Dissociation Constants (K_d) of mAChR Subtypes for Antagonists

Antagonist	K_d (M) mAChR I (porcine)	mAChR II (porcine)	mAChR III (porcine)	mAChR III (rat)	mAChR IV (rat)
(−) [^{3}H]QNB	8.4×10^{-11}	1.3×10^{-10}	1.2×10^{-10}	1.0×10^{-10}	9.5×10^{-11}
Pirenzepine	1.8×10^{-8}	6.6×10^{-7}	1.8×10^{-7}	1.3×10^{-7}	1.2×10^{-7}
AF-DX 116	2.5×10^{-6}	7.3×10^{-7}	2.3×10^{-6}	3.1×10^{-6}	2.3×10^{-6}
Hexahydrosiladifenidol	5.1×10^{-8}	2.8×10^{-7}	4.4×10^{-9}	4.0×10^{-9}	2.0×10^{-8}

Reprinted, with permission, from Akiba et al. (1988.)

II ($K_d = 0.73$ μM) than for the other mAChR species ($K_d = 2.3$–3.1 μM). Hexahydrosiladifenidol, selective for the M_2 glandular (III) subtype relative to the M_2 cardiac (II) subtype (Birdsall et al. 1986; Tacke and Zilch 1986; Ladinsky et al. 1988), exhibited the highest binding affinity for mAChR III ($K_d = 4.0$–4.4 nM). The affinity for this antagonist decreased in the order of mAChR IV ($K_d = 20$ nM), mAChR I ($K_d = 51$ nM), and mAChR II ($K_d = 280$ nM).

These results indicate that mAChR I, mAChR II, and mAChR III correspond most closely to the pharmacologically defined M_1 (I), M_2 cardiac (II), and M_2 glandular (III) subtypes, respectively. In support of this view is the differential tissue location of the mRNAs encoding these mAChR species, revealed by blot hybridization analysis (Kubo et al. 1986a,b; Peralta et al. 1987a). Thus, the mAChR heterogeneity in tissues with respect to antagonist binding can be accounted for by the presence of individual molecularly distinct mAChR species or various combinations of them. mAChR subtypes are defined hereafter as distinct gene products.

Selective Coupling of mAChR Subtypes with Effector Systems

To examine whether the different molecularly defined mAChR subtypes are coupled with distinct effector systems, we investigated agonist-induced responses mediated by the individual mAChR subtypes, using *Xenopus* oocytes and NG108-15 neuroblastoma-glioma hybrid cells (Nirenberg et al. 1983) as transient and stable expression systems, respectively (Kubo et al. 1986b; Fukuda et al. 1987, 1988). *Xenopus* oocytes injected with the mAChR-I-specific mRNA or the mAChR-II-specific mRNA exhibited acetylcholine (ACh)-activated inward currents when tested under voltage clamp at a holding potential of −70 mV; no detectable response was observed in control oocytes (noninjected or injected with water). The ACh-activated currents were abolished by atropine or (−)scopolamine in a reversible manner, whereas (+) tubocurarine and α-bungarotoxin exerted no effect.

The ACh responses mediated by mAChR I and mAChR II differed in nature from each other (Fukuda et al. 1987). Figure 1 shows typical ACh responses observed in oocytes injected with the mAChR-I-specific mRNA (Fig. 1A,a) or with the mAChR-II-specific mRNA (Fig. 1B,a). Oocytes implanted with

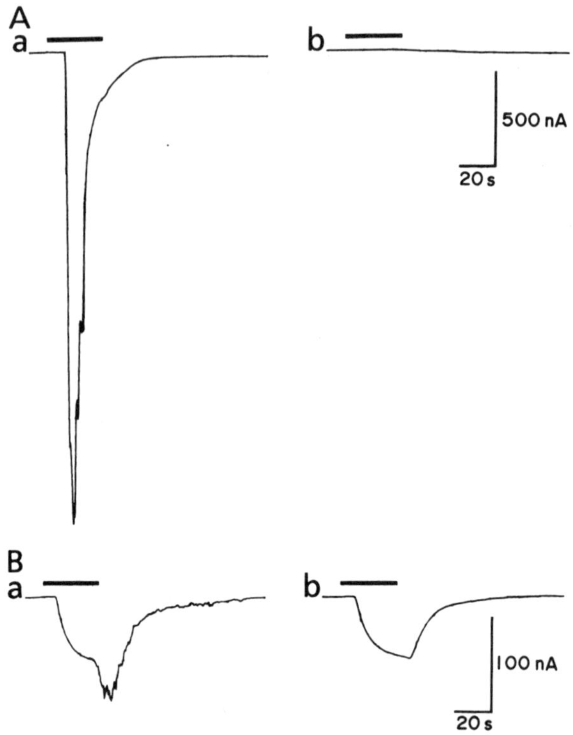

Figure 1. ACh-activated currents in *Xenopus* oocytes injected with the mAChR-I-specific mRNA (*A*) or with the mAChR-II-specific mRNA (*B*). Whole-cell currents activated by bath application of 1 μM ACh were recorded under voltage clamp at a holding potential of −70 mV before (*a*) and after (*b*) intracellular injection of EGTA. The duration of ACh application is indicated by bars without taking into account the dead-space time in the perfusion system (~7 sec). (Reprinted, with permission, from Fukuda et al. 1987.)

mAChR I exhibited an inward current, which was usually oscillatory in nature. In contrast, a typical ACh response induced in oocytes implanted with mAChR II comprised an initial smooth inward current followed by an oscillatory component. The oscillatory current evoked by activation of mAChR II varied in amplitude among the oocytes tested and was undetected in some of them. The average peak inward current measured in mAChR-II-implanted oocytes was much smaller than that measured in mAChR-I-implanted oocytes. The difference was 560-fold when the density of mAChRs expressed on the cell surface, estimated by (−) [^{3}H]*N*-methylscopolamine binding, was taken into account. The latency of the ACh response in mAChR-II-implanted oocytes was shorter than that in mAChR-I-implanted oocytes. Most of the latency of the ACh response in mAChR-II-implanted oocytes was due to the dead-space time in the perfusion system (~7 sec), whereas the current response in mAChR-I-implanted oocytes occurred after an additional delay of 3 ± 2 seconds (mean ±S.D.).

Figure 2 shows the membrane potential dependence of the ACh responses. The mean reversal potential of the ACh-activated current in mAChR-I-implanted oocytes, obtained in Ringer's solution (Fig. 2A, closed circles), was -25.6 ± 2.2 mV. This is close to the equilibrium potential of chloride ions observed in *Xenopus* oocytes (Barish 1983). In fact, the reversal potential shifted by 17.9 ± 1.6 mV to a more positive value when the external Cl^- concentration was halved by substitution with SO_4^{--} (Fig. 2A, open circles), but it did not significantly change when the external Na^+ concentration was decreased threefold by substitution with Tris or when the external K^+ concentration was increased threefold by addition of K_2SO_4. Thus, the ACh response mediated by mAChR I is due principally to an increase in the Cl^- conductance of the oocyte membrane.

The ACh-activated currents recorded from an mAChR-II-implanted oocyte in Ringer's solution at various holding potentials are shown in Figure 2B. The oscillatory current was reversed in polarity at a potential around −25 mV, which is close to the equilibrium potential of chloride ions in *Xenopus* oocytes. In contrast, the reversal potential of the smooth current was around 10 mV. This value does not correspond to the equilibrium potential of any single species of ions, indicating the involvement of more than one ion species in the generation of the smooth current. The ionic basis of the smooth current was further examined in EGTA-loaded oocytes (see below). The reversal potential in Ringer's solution was 11 ± 3 mV and was virtually unchanged even when external Cl^- was completely replaced by SO_4^{--}. In a K^+-free solution (120 mM Na^+), the reversal potential was 11 ± 1 mV and shifted to -2 ± 1 mV when the Na^+ concentration was reduced to one half by substitution with Tris (Fig. 2C). The reversal potential in a Na^+-free solution (120 mM K^+) was -2 ± 1 mV and changed to -15 ± 1 mV when the K^+ concentration was halved by substitution with Tris (Fig. 2D). These results suggest that the smooth current evoked by activation of mAChR II is carried principally by sodium and potassium ions, whereas the oscillatory current is carried mainly by chloride ions.

It is known that an elevation in intracellular Ca^{++} concentration causes activation of Cl^- channels in *Xenopus* oocytes (Barish 1983; Miledi and Parker 1984; Dascal et al. 1985). Replacement of extracellular Ca^{++} by Mg^{++} did not significantly affect the ACh-activated oscillatory current in oocytes implanted either with mAChR I or with mAChR II. However, intracellular injection of the calcium-chelating agent EGTA

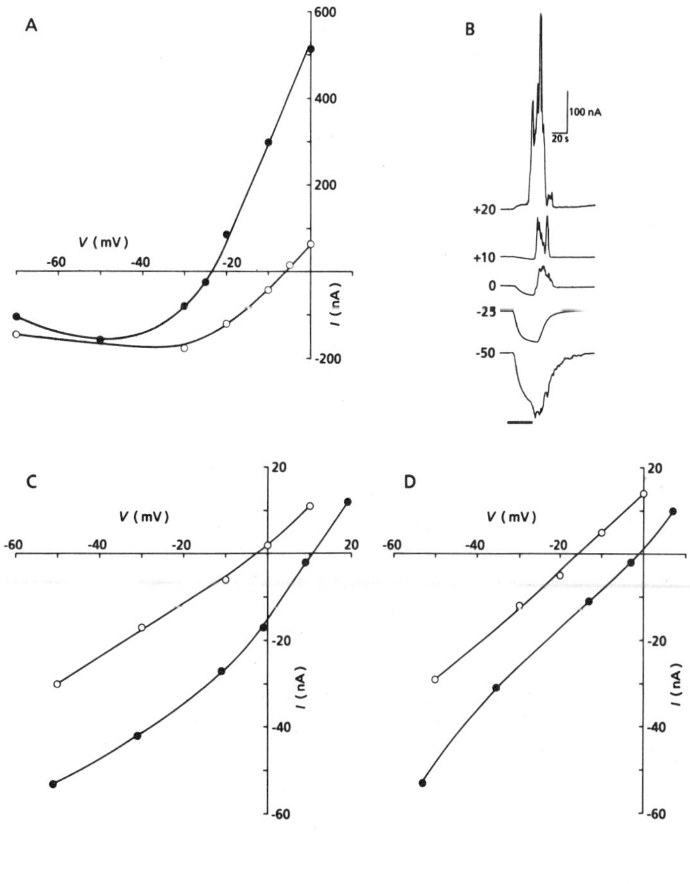

Figure 2. Membrane potential dependence of ACh-activated whole-cell currents in *Xenopus* oocytes injected with the mAChR-I-specific mRNA (*A*) or with the mAChR-II-specific mRNA (*B–D*). (*A*) *I-V* relationships (inward current shown as negative current) obtained from an mAChR-I-implanted oocyte in Ringer's solution (●) and in the solution in which half of the Cl^- was replaced by SO_4^{--} (○). Currents activated by ionophoretic application of ACh were recorded under voltage clamp at different holding potentials. (*B*) ACh-activated currents recorded from an mAChR-II-implanted oocyte in Ringer's solution under voltage clamp at different holding potentials (mV), as indicated on the left-hand side of each trace. Inward current is downward. The duration of bath application of 1 μM ACh is indicated as in Fig. 1. (*C*) *I-V* relationships obtained from an EGTA-loaded, mAChR-II-implanted oocyte in a K^+-free solution containing 120 mM Na^+ (●) or 60 mM Na^+ (○). (*D*) *I-V* relationships obtained from an EGTA-loaded, mAChR-II-implanted oocyte in a Na^+-free solution containing 120 mM K^+ (●) or 60 mM K^+ (○). (Reprinted, with permission, from Kubo et al. 1986b [*A*] and Fukuda et al. 1987 [*B*].)

abolished almost completely the ACh-activated current in mAChR-I-implanted oocytes (Fig. 1A,b), leaving only a small long-lasting inward current. The oscillatory current in mAChR-II-implanted oocytes similarly disappeared after this treatment, whereas the smooth component was virtually unaffected (Fig. 1B,b). Thus, our results provide evidence that mAChR I induces activation of a Ca^{++}-dependent Cl^- current, whereas mAChR II principally evokes activation of Na^+ and K^+ currents in a Ca^{++}-independent manner. Most likely, the activation of the Cl^- current mediated by mAChR I results from phosphoinositide hydrolysis leading to Ca^{++} release from intracellular stores (see below).

We next investigated ACh responses in NG108-15 neuroblastoma-glioma hybrid cells that stably expressed the individual mAChR subtypes (Fukuda et al. 1988). NG108-15 cells were transfected with cDNA (or genomic DNA) encoding each of the four mAChR subtypes, using an expression vector carrying the SV40 early gene promoter and the neomycin resistance marker gene. Neomycin-resistant clones that exhibited (−) [^{3}H]QNB-binding activities more than threefold higher than the activity of nontransfected cells were selected. Blot hybridization analysis of total cellular RNA using mAChR-subtype-specific probes indicated that the transformed clones expressed RNA species (with sizes expected from the expression constructs) encoding the corresponding mAChR subtypes. The mAChR-IV-specific probe revealed an additional hybridizable RNA species of ~3200 nucleotides, which was also found in nontransfected or vector-transformed control cells, thus representing the endogenous mAChR IV mRNA in NG108-15 cells (see also Peralta et al. 1987a). Control cells showed no detectable levels of RNA species hybridizable with any of the three other mAChR-subtype-specific probes.

The transformed clones were tested for electrophysiological response to focal application of ACh under voltage clamp at a holding potential of −30 mV. Cells transformed to express mAChR I or mAChR III exhibited a similar response to ACh. A typical response comprised an initial outward current, followed by a sustained inward current (Figs. 3A and 4A). Both these currents were abolished by atropine in a reversible manner. The initial outward current was accompanied by an increase in input conductance, as measured by the currents evoked by short voltage steps (Figs. 3A,B and 4A,B). The outward current was reversed to an inward current on hyperpolarizing cells to −73 to −85 mV, and the current-voltage (*I-V*) relation was apparently linear (Figs. 3C and 4C). The outward current response was reduced or abolished by apamin (0.4 μM) or (+) tubocurarine (0.2 mM), blockers of a voltage-independent Ca^{++}-activated K^+ current, but was resistant to tetraethylammonium (1 mM), a blocker of a voltage-dependent Ca^{++}-activated K^+ current. These results indicate that the initial outward current is principally attributable to activation of a subclass of Ca^{++}-

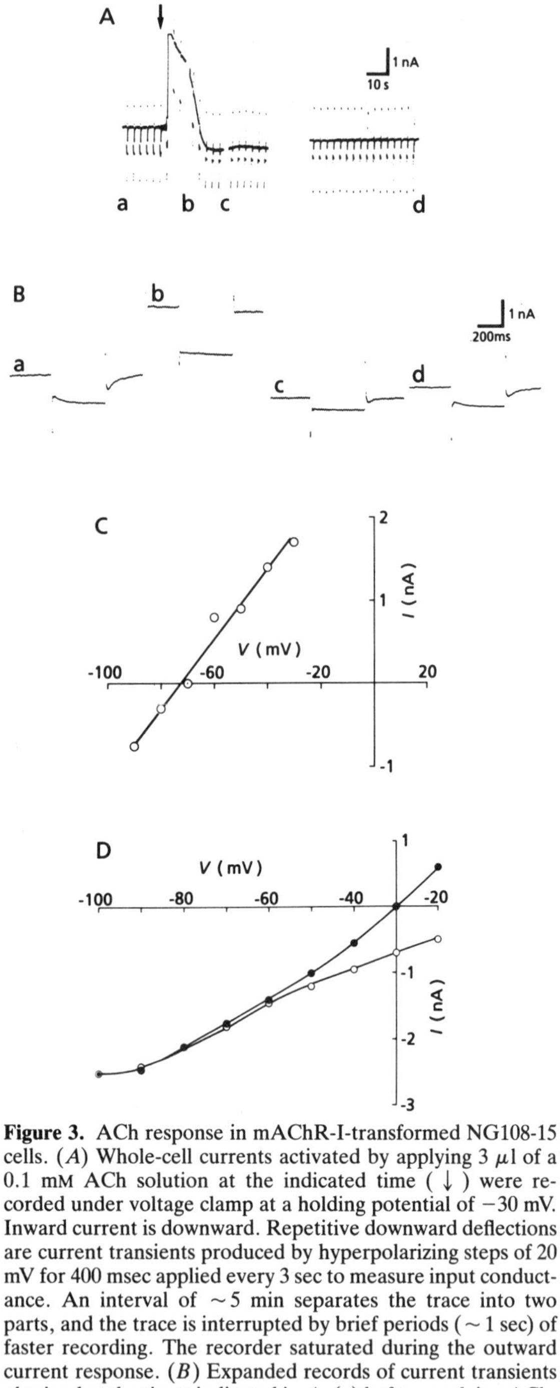

Figure 3. ACh response in mAChR-I-transformed NG108-15 cells. (*A*) Whole-cell currents activated by applying 3 μl of a 0.1 mM ACh solution at the indicated time (↓) were recorded under voltage clamp at a holding potential of −30 mV. Inward current is downward. Repetitive downward deflections are current transients produced by hyperpolarizing steps of 20 mV for 400 msec applied every 3 sec to measure input conductance. An interval of ~5 min separates the trace into two parts, and the trace is interrupted by brief periods (~1 sec) of faster recording. The recorder saturated during the outward current response. (*B*) Expanded records of current transients obtained at the times indicated in *A*: (*a*) before applying ACh; (*b*) during the ACh-induced outward current; (*c*) during the subsequent inward current; (*d*) after partial recovery. (*C*) Effect of changing clamp holding potential on the ACh-induced outward current. ACh-activated currents were recorded under voltage clamp at different holding potentials. The reversal potential is −73 mV. (*D*) *I-V* relationships obtained before applying ACh (●) and during the ACh-induced sustained inward current (○). The current was measured as the displacement from the original holding current at the end of a series of incrementing 400-msec voltage steps from a holding potential of −30 mV. (Reprinted, with permission, from Fukuda et al. 1988.)

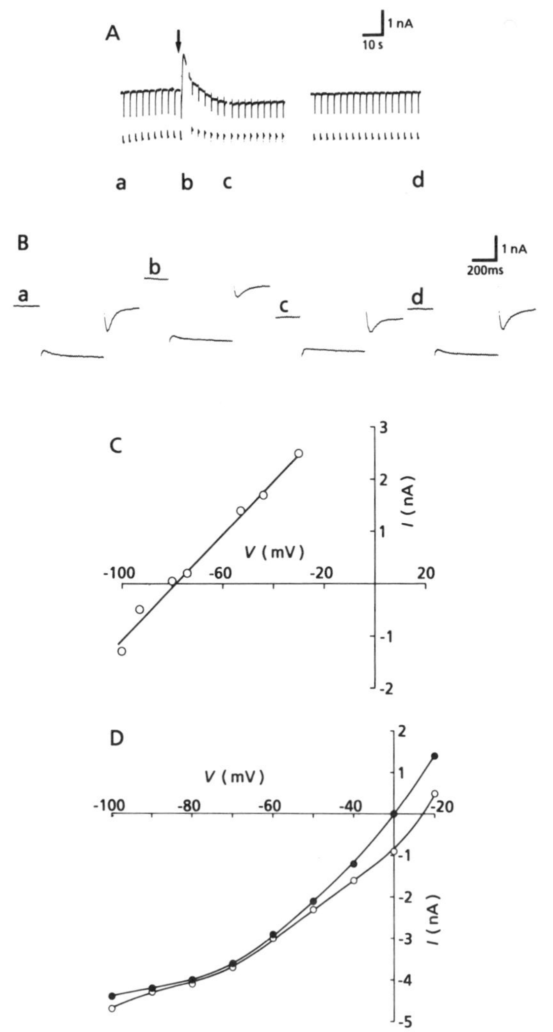

Figure 4. ACh response in porcine mAChR-III-transformed NG108-15 cells. The experimental conditions used were the same as those in Fig. 3, unless otherwise specified. (*A*) ACh-activated whole-cell currents. The duration of hyperpolarizing steps was 500 msec. An interval of ~1 min separates the trace into two parts. (*B*) Expanded records of current transients obtained at the times indicated in *A*. (*C*) Effect of changing clamp holding potential on the ACh-induced outward current. The reversal potential is −79 mV. (*D*) *I-V* relationships obtained before applying ACh (●) and during the ACh-induced sustained inward current (○). The duration of voltage steps was 500 msec. A similar ACh response was observed in rat mAChR-III-transformed cells. (Reprinted, with permission, from Fukuda et al. 1988.)

dependent K^+ currents analogous to I_{AHP} (Pennefather et al. 1985).

Unlike the initial outward current, the secondary inward current induced by ACh was usually accompanied by a decrease in input conductance and by a reduction in the time-dependent inward current relaxation observed during hyperpolarizing voltage steps (Figs. 3A,B and 4A,B). The steady-state *I-V* curve that was obtained while the sustained inward current was present showed a reduction in the outward rectification in a potential range less negative than approximately −70 mV (tested up to −20 mV), where the M current is activated (Figs. 3D and 4D) (Brown and Adams 1980; Higashida and Brown 1986). These data indicate that the secondary inward current results primarily from inhibition of the M current, which is known to be present in NG108-15 cells, some neuronal cells, and smooth muscle, and to be modulated by different neurotransmitters (Brown and Adams 1980; Halliwell and Adams 1982; Constanti and Galvan 1983; Sims et al. 1985; Higashida and Brown 1986).

The initial outward current was observed in 88% and 81% of the mAChR-I- and mAChR-III-transformed cells tested, respectively. The secondary inward current occurred in 59% and 64% of the mAChR-I- and mAChR-III-transformed cells examined, respectively. In contrast, the percentages of responsive cells among the mAChR-II- or mAChR-IV-transformed cells tested (12% for the outward current and 6% for the inward current), despite their (−) [^{3}H]QNB-binding activities being comparable to those of the mAChR-I- or mAChR-III-transformed cells, were not higher than those of nontransfected or vector-transformed control cells (14% and 6%, respectively). The ACh responses observed in mAChR-II- or mAChR-IV-transformed cells and in control cells were not characterized further.

It is known that the activation of the endogenous mAChR in NG108-15 cells leads to adenylate cyclase inhibition without apparent effect on phosphoinositide hydrolysis (Harden et al. 1986). Using [^{3}H]inositol-labeled cells, we examined the effect of carbachol on the formation of [^{3}H]inositol phosphates. In mAChR-I- or mAChR-III-transformed cells, a four- to sevenfold increase in the release of total inositol phosphates occurred in response to carbachol stimulation, as compared with control values obtained without stimulation. In mAChR-II- or mAChR-IV-transformed cells, as well as in nontransfected or vector-transformed control cells, no appreciable increase in inositol phosphate release was observed. Figure 5 shows the time courses of the carbachol effects on the formation of individual inositol phosphates in the different cell types. In mAChR-I- or mAChR-III-transformed cells, the release of inositol trisphosphate increased rapidly, reaching levels four- to sixfold higher than control values within 15 seconds (Fig. 5A). The formation of inositol bisphosphate (Fig. 5B) and inositol monophosphate (Fig. 5C) was elevated more slowly than that of inositol trisphosphate. These data suggest that mAChR I and mAChR III mediate stimulation of phospholipase C, thus enhancing release of inositol 1,4,5-trisphosphate, which is known to mobilize Ca^{++} from intracellular stores (Berridge 1987). The formation of inositol phosphates and the biphasic changes in membrane conductance induced by agonist in mAChR-I- or mAChR-III-transformed cells were not appreciably affected by prior incubation of the cells with pertussis toxin, whereas the inhibition by agonist of prostaglandin-E_1-induced cAMP accumulation observed in these cells

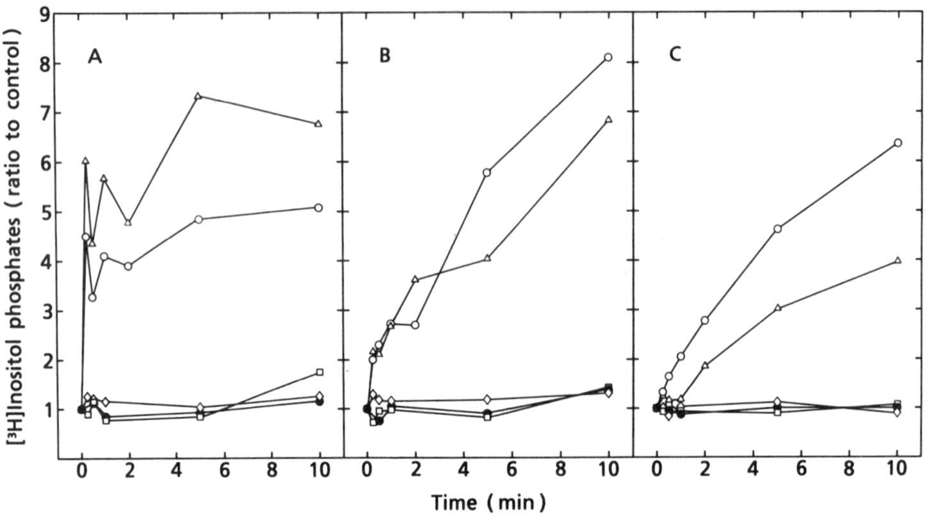

Figure 5. Effect of carbachol on the formation of [^{3}H]inositol trisphosphate (*A*), [^{3}H]inositol bisphosphate (*B*), and [^{3}H]inositol monophosphate (*C*) in nontransfected NG108-15 cells (●) and transformed cells with mAChR I (○), mAChR II (□), mAChR III (porcine) (△), or mAChR IV (◇). [^{3}H]Inositol-labeled cells were incubated at 37°C for the indicated times with or without 1 mM carbachol. The individual inositol phosphates were separated by anion-exchange chromatography and counted for radioactivity. Data are expressed as ratios to the respective controls without carbachol stimulation. (Reprinted, with permission, from Fukuda et al. 1988.)

and in nontransfected cells (Kurose et al. 1983) was strongly reduced by the same treatment. This suggests that the response mediated by mAChR I or mAChR III and the inhibition of adenylate cyclase mediated by the endogenous mAChR (probably the mAChR IV subtype; see above) occur through distinct G proteins.

Our results indicate that the molecularly defined mAChR I and mAChR III subtypes, but not the mAChR II and mAChR IV subtypes, efficiently mediate phosphoinositide hydrolysis and sequential changes in membrane K^+ conductance in NG108-15 cells. These conductance changes are attributable to activation of a Ca^{++}-dependent K^+ current, followed by inhibition of the M current. A similar current response has been shown to be induced by bradykinin through phosphoinositide hydrolysis in NG108-15 cells (Yano et al. 1985; Higashida and Brown 1986). It is suggested that the mAChR IV subtype mediates adenylate cyclase inhibition without affecting phosphoinositide hydrolysis in NG108-15 cells (see above; Harden et al. 1986). Recent work reports that the mAChR II subtype expressed in Chinese hamster ovary cells is coupled with adenylate cyclase inhibition, although this subtype also mediates phosphoinositide hydrolysis less efficiently in these cells (Ashkenazi et al. 1988). These findings, together with our cDNA expression studies using *Xenopus* oocytes, suggest that mAChR subtypes are selectively coupled with different effector systems, albeit not exclusively.

Some brain neurons have been reported to exhibit ACh-induced hyperpolarization, followed by depolarization with a time course similar to that of the sequential membrane conductance changes we have observed in mAChR-I- and mAChR-III-transformed NG108-15 cells (Segal 1982; McCormick and Prince 1987). Furthermore, slow excitatory postsynaptic potentials, resulting from blockade of K^+ currents, occur widely at many levels of vertebrate nervous systems (Jones and Adams 1987). Our finding that the mAChR I and mAChR III subtypes mediate inhibition of the M current suggests that these subtypes are involved in the cholinergic excitation of vertebrate neurons evoked by muscarinic activity.

CONCLUSIONS

The existence and the primary structures of multiple mAChR species have been revealed by cloning and sequence analysis of the cDNAs. The antagonist-binding properties of the individual mAChR species expressed from cDNAs (or genomic DNAs), together with the differential tissue location of the mRNAs encoding them, indicate that the mAChR heterogeneity in tissues with respect to antagonist binding can be accounted for by the presence of distinct mAChR gene products by themselves or in various combinations. The agonist-induced responses in NG108-15 neuroblastoma-glioma hybrid cells and *Xenopus* oocytes expressing the individual mAChR species provide evidence that the molecularly defined mAChR subtypes are selectively coupled with different effector systems, albeit not exclusively. The finding that the mAChR I and mAChR III subtypes mediate inhibition of the M current suggests the involvement of these subtypes in the muscarinic excitation of vertebrate neurons.

ACKNOWLEDGMENTS

This research was supported in part by research grants from the Ministry of Education, Science and

Culture of Japan, the Institute of Physical and Chemical Research, the Mitsubishi Foundation, and the Japanese Foundation of Metabolism and Diseases.

REFERENCES

Akiba, I., T. Kubo, A. Maeda, H. Bujo, J. Nakai, M. Mishina, and S. Numa. 1988. Primary structure of porcine muscarinic acetylcholine receptor III and antagonist-binding studies. *FEBS Lett.* **235:** 257.

Ashkenazi, A., J.W. Winslow, E.G. Peralta, G.L. Peterson, M.I. Schimerlik, D.J. Capon, and J. Ramachandran. 1988. An M2 muscarinic receptor subtype coupled to both adenylyl cyclase and phosphoinositide turnover. *Science* **238:** 672.

Barish, M.E. 1983. A transient calcium-dependent chloride current in the immature *Xenopus* oocyte. *J. Physiol.* **342:** 309.

Berridge, M.J. 1987. Inositol trisphosphate and diacylglycerol: Two interacting second messengers. *Annu. Rev. Biochem.* **56:** 159.

Birdsall, N.J.M., E.C. Hulme, M. Keen, E.K. Pedder, D. Poyner, J.M. Stockton, and M. Wheatley. 1986. Soluble and membrane-bound muscarinic acetylcholine receptors. *Biochem. Soc. Symp.* **52:** 23.

Bonner, T.I., N.J. Buckley, A.C. Young, and M.R. Brann. 1987. Identification of a family of muscarinic acetylcholine receptor genes. *Science* **237:** 527.

Brown, D. 1986. Acetylcholine and brain cells. *Nature* **319:** 358.

Brown, D.A. and P.R. Adams. 1980. Muscarinic suppression of a novel voltage-sensitive K^+ current in a vertebrate neurone. *Nature* **283:** 673.

Burgen, A.S.V. 1984. Muscarinic receptors—An overview. *Trends Pharmacol. Sci.* (suppl.) **5:** 1.

Constanti, A. and M. Galvan. 1983. M-current in voltage-clamped olfactory cortex neurones. *Neurosci. Lett.* **39:** 65.

Dascal, N., B. Gillo, and Y. Lass. 1985. Role of calcium mobilization in mediation of acetylcholine-evoked chloride currents in *Xenopus laevis* oocytes. *J. Physiol.* **366:** 299.

Dixon, R.A.F., B.K. Kobilka, D.J. Strader, J.L. Benovic, H.G. Dohlman, T. Frielle, M.A. Bolanowski, C.D. Bennett, E. Rands, R.E. Diehl, R.A. Mumford, E.E. Slater, I.S. Sigal, M.G. Caron, R.J. Lefkowitz, and C.D. Strader. 1986. Cloning of the gene and cDNA for mammalian β-adrenergic receptor and homology with rhodopsin. *Nature* **321:** 75.

Fukuda, K., T. Kubo, I. Akiba, A. Maeda, M. Mishina, and S. Numa. 1987. Molecular distinction between muscarinic acetylcholine receptor subtypes. *Nature* **327:** 623.

Fukuda, K., H. Higashida, T. Kubo, A. Maeda, I. Akiba, H. Bujo, M. Mishina, and S. Numa. 1988. Selective coupling with K^+ currents of muscarinic acetylcholine receptor subtypes in NG108-15 cells. *Nature* **335:** 355.

Gilman, A.G. 1986. Receptor-regulated G proteins. *Trends Neurosci.* **9:** 460.

Halliwell, J.V. and P.R. Adams. 1982. Voltage-clamp analysis of muscarinic excitation in hippocampal neurons. *Brain Res.* **250:** 71.

Hammer, R., C.P. Berrie, N.J.M. Birdsall, A.S.V. Burgen, and E.C. Hulme. 1980. Pirenzepine distinguishes between different subclasses of muscarinic receptors. *Nature* **283:** 90.

Hammer, R., E. Giraldo, G.B. Schiavi, E. Monferini, and H. Ladinsky. 1986. Binding profile of a novel cardioselective muscarine receptor antagonist, AF-DX 116, to membranes of peripheral tissues and brain in the rat. *Life Sci.* **38:** 1653.

Harden, T.K., L.I. Tanner, M.W. Martin, N. Nakahata, A.R. Hughes, J.R. Hepler, T. Evans, S.B. Masters, and J.H. Brown. 1986. Characteristics of two biochemical responses to stimulation of muscarinic cholinergic receptors. *Trends Pharmacol. Sci.* (suppl.) **7:** 14.

Henderson, R. and P.N.T. Unwin. 1975. Three-dimensional model of purple membrane obtained by electron microscopy. *Nature* **257:** 28.

Higashida, H. and D.A. Brown. 1986. Two polyphosphatidylinositide metabolites control two K^+ currents in a neuronal cell. *Nature* **323:** 333.

Jones, S.W. and P.R. Adams. 1987. The M-current and other potassium currents of vertebrate neurons. In *Neuromodulation* (ed. L.K. Kaczmarek and I.B. Levitan), p. 159. Oxford University Press, New York.

Kubo, T., A. Maeda, K. Sugimoto, I. Akiba, A. Mikami, H. Takahashi, T. Haga, K. Haga, A. Ichiyama, K. Kangawa, H. Matsuo, T. Hirose, and S. Numa. 1986a. Primary structure of porcine cardiac muscarinic acetylcholine receptor deduced from the cDNA sequence. *FEBS Lett.* **209:** 367.

Kubo, T., K. Fukuda, A. Mikami, A. Maeda, H. Takahashi, M. Mishina, T. Haga, K. Haga, A. Ichiyama, K. Kangawa, M. Kojima, H. Matsuo, T. Hirose, and S. Numa. 1986b. Cloning, sequencing and expression of complementary DNA encoding the muscarinic acetylcholine receptor. *Nature* **323:** 411.

Kurose, H., T. Katada, T. Amano, and M. Ui. 1983. Specific uncoupling by islet-activating protein, pertussis toxin, of negative signal transduction via α-adrenergic, cholinergic, and opiate receptors in neuroblastoma × glioma hybrid cells. *J. Biol. Chem.* **258:** 4870.

Ladinsky, H., E. Giraldo, E. Monferini, G.B. Schiavi, M.A. Vigano, L.D. Conti, R. Micheletti, and R. Hammer. 1988. Muscarinic receptor heterogeneity in smooth muscle: Binding and functional studies with AF-DX 116. *Trends Pharmacol. Sci.* (suppl.) **9:** 44.

McCormick, D.A. and D.A. Prince. 1987. Actions of acetylcholine in the guinea-pig and cat medial and lateral geniculate nuclei, *in vitro*. *J. Physiol.* **392:** 147.

Miledi, R. and I. Parker. 1984. Chloride current induced by injection of calcium into *Xenopus* oocytes. *J. Physiol.* **357:** 173.

Nirenberg, M., S. Wilson, H. Higashida, A. Rotter, K. Krueger, N. Busis, R. Ray, J.G. Kenimer, and M. Adler. 1983. Modulation of synapse formation by cyclic adenosine monophosphate. *Science* **222:** 794.

Ovchinnikov, Y.A. 1982. Rhodopsin and bacteriorhodopsin: Structure-function relationships. *FEBS Lett.* **148:** 179.

Pennefather, P., B. Lancaster, P.R. Adams, and R.A. Nicoll. 1985. Two distinct Ca-dependent K currents in bullfrog sympathetic ganglion cells. *Proc. Natl. Acad. Sci.* **82:** 3040.

Peralta, E.G., A. Ashkenazi, J.W. Winslow, D.H. Smith, J. Ramachandran, and D.J. Capon. 1987a. Distinct primary structures, ligand-binding properties and tissue-specific expression of four human muscarinic acetylcholine receptors. *EMBO J.* **6:** 3923.

Peralta, E.G., J.W. Winslow, G.L. Peterson, D.H. Smith, A. Ashkenazi, J. Ramachandran, M.I. Schimerlik, and D.J. Capon. 1987b. Primary structure and biochemical properties of an M_2 muscarinic receptor. *Science* **236:** 600.

Segal, M. 1982. Multiple actions of acetylcholine at a muscarinic receptor studied in the rat hippocampal slice. *Brain Res.* **246:** 77.

Sims, S.M., J.J. Singer, and J.V. Walsh, Jr. 1985. Cholinergic agonists suppress a potassium current in freshly dissociated smooth muscle cells of the toad. *J. Physiol.* **367:** 503.

Tacke, R. and H. Zilch. 1988. Sila-substitution—A useful strategy for drug design? *Endeavour* **10:** 191.

Yano, K., H. Higashida, H. Hattori, and Y. Nozawa. 1985. Bradykinin-induced transient accumulation of inositol trisphosphate in neuron-like cell line NG108-15 cells. *FEBS Lett.* **181:** 403.

Life, Evolution, and the Pursuit of Single Photon Sensitivity

M.W. Bitensky, M.M. Whalen, and D.C. Torney
Los Alamos National Laboratory, Los Alamos, New Mexico 87545

Evolution and Molecular Biology

Recent advances in understanding the informational macromolecules of heredity have captured our attention. The utility of restriction enzymes (Nathans and Smith 1975) and the recognition function of DNA sequences have permitted the design and synthesis of hybridization probes (Moyzis et al. 1987) that simplify gene mapping, ordering, and sequencing. Pulsed-field gel electrophoresis (Schwartz and Cantor 1984) and synthetic yeast chromosomes (Burke et al. 1987) allow separation and cloning of 5-megabase fragments of DNA. We are learning to unravel, understand, and manipulate the fragile threads that form the tapestry of life.

DNA's capacity for information storage, self-replication, and change has contributed new dimensions to our understanding of the mechanisms of molecular evolution. Gene function optimization and gene modification via exon shuffling (Holland and Blake 1987) or nucleotide substitutions have permitted a more molecular and quantitative understanding of the methodological devices of evolution. Data on the rates of econiche filling following extinctions (Sepkoski 1982) suggest that evolutionary dynamics can produce rapid change unexpectedly. Although the uncompromising Darwinian algorithms of competition and survival often unfold in a macroscopic world, the molecular program for these macroscopic characteristics is clearly shaped in the submicroscopic domain of introns and exons.

The unchanging intent of evolution is an optimized fit between organism and econiche. When a given capability provides advantage for a particular species, no amount of DNA modification is too extravagant in achieving and optimizing the needed function. Add to this the habits of parsimony, conservatism, and opportunism, and one has a dynamic picture of evolution's operational plan.

Conservatism appears as the tenacious preservation of successful gene products, their architectures and ensembles, and their application in ever-changing contexts. The conservative design features must also be perceived in the context of continuity. Gene innovations and modifications are superimposed upon collections of existing gene products. A species can experiment with the development of a new or replacement gene only while it maintains the viability and continuity of the existing gene pool within which the new feature will be assimilated. Parsimony manifests in the selection of the most frugal gene product arrays in achieving a particular function. Whatever the task at hand (be it single photon sensitivity, muscle contraction, or bone formation), the approach that accomplishes the desired result with the fewest demands on material, energy, and informational resources will be the approach that prevails. Opportunism is encountered because the evolutionary process is inherently adventuresome and unconstrained by rules of propriety. In its 3.6 billion years of tinkering, evolution has engaged in some unconventional experiments; e.g., crystalline glycogen has been utilized successfully as the conical lens material in each facet of the arthropod eye.

Another example of resourceful and determined optimization is found in the raptor eye. Raptors employ modified lens and foveal arrangements (Snyder and Miller 1978) and widened spacing of their photoreceptors (Miller 1979; Hirsch 1982) in order to achieve the image resolution needed to locate their prey.

Molecular Infrastructure of Signal Transduction

The three-segment, GTP-dependent architecture of signal transduction is an ancient evolutionary device that has met the test of time (Rodbell 1985; Bitensky et al. 1987a,b; Gilman 1987; Deterre et al. 1988). This architecture (Fig. 1) has found wide application in the transfer of information across the cell membrane. In its earliest forms, this ensemble of proteins probably alerted single-celled organisms to the presence of food molecules. Its initial utility as a chemodetection device is now exploited in olfactory (Lancet et al., this volume) and gustatory (Jones et al., this volume) signal transduction. Although our awareness of this mechanism was born in an era that featured cAMP as second messenger (Rall and Sutherland 1961), we have since learned that other molecules can serve in this capacity, e.g., inositol trisphosphate (Selinger and Minke, this volume), Ca^{++}, and diacylglycerol (Litosch and Fain 1986).

The success of the three-segment architecture of signal transduction is inherent in the capabilities offered by each segment. The receptor elements (segment 1) contribute stereospecific binding and recognition for an endless variety of ligands, including photons of specific wavelengths (Nathans et al. 1987). The receptor itself is conformationally modified by signal. In the photoreceptor system, receptor changes include phosphor-

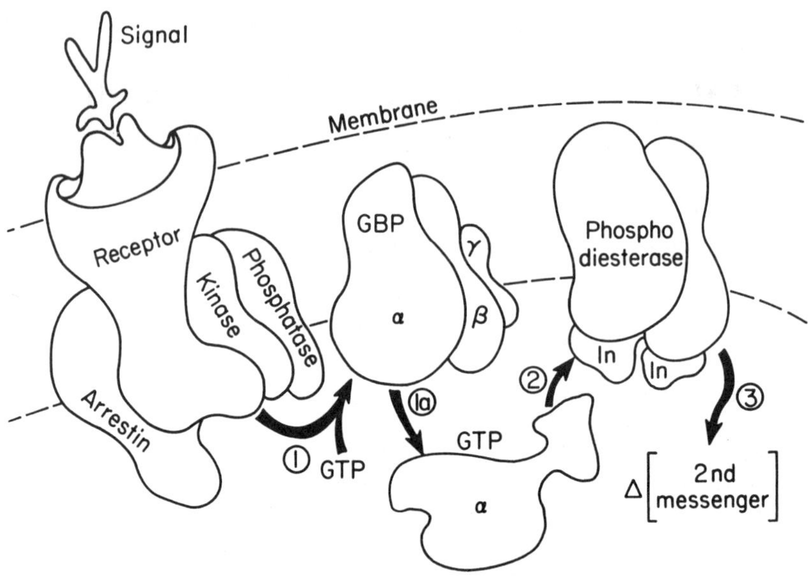

Figure 1. The three-segment architecture of signal transduction. This example is illustrated with rod photoreceptor components: Step *1* facilitates binding of GTP to $N_{h\nu}\alpha$. In step *1a*, the activator complex dissociates from $N_{h\nu}\beta\gamma$. Step *2* depicts the activator attack on the PDE inhibitory subunit, also called PDEγ. As a consequence of PDE displacement and activation, there is a decrease in rod cGMP concentrations. The kinase and arrestin participate in the inactivation of photoisomerized receptor. Both large subunits of PDE appear to have catalytic activity.

ylation, interaction with the inhibitory protein arrestin (Mangini and Pepperberg 1987), and formation of a series of time-dependent photoconformers. The latter result in the rapid appearance and, subsequently, the gradual disappearance of metarhodopsin II, the active photoconformer of opsin. From *Chlamydomonas* to cow, the retinal-opsin collaboration is exploited in photon transduction (Foster and Smyth 1980). A receptor must provide impeccable consistency and specificity in signal discrimination and must also participate in the processes of down regulation (Lefkowitz et al. 1983) and signal termination (Wheeler et al. 1977; Selinger and Minke, this volume).

The second segment has the capacity for signal amplification and can generate from a few to the order of a thousand GTP complexes or activators in response to a single liganded receptor (Liebman and Pugh 1982). Signal gain varies with the function of the receptor and its recent history of interaction against agonist. Although there are many potential loci to explain light adaptation, a particularly efficient one is the receptor/activator transition. By modifying the rates of receptor inactivation as a function of ambient light intensity, the yield of activators per liganded receptor could determine photoreceptor sensitivity. Evidence for a light-dependent change in immunoreactivity of arrestin is interesting in this context (Mangini and Pepperberg 1987). In the visual enzyme cascade, the activator takes the form of a complex between GTP and the α subunit of the heterotrimeric rod GTP-binding proteins (GTP $\cdot$ $N_{h\nu}\alpha$ [$N_{h\nu}$ = light-activated guanine-nucleotide-binding protein]) (Shinozawa et al. 1979; Fung and Stryer 1980; Uchida et al. 1981). The GTP-binding protein contributes a significant part of the amplification that arises in photon transduction.

The activator complex (GTP $\cdot$ $N_{h\nu}\alpha$) also provides a component of signal termination because it has the capacity for GTP hydrolysis, which converts the activating signal to a nonsignal (Wheeler et al. 1977; Yamazaki et al. 1987; E.A. Dratz et al., in prep.). The activator must communicate the signal to the effector segment of the cascade prior to the hydrolytic event that terminates activation.

The third or effector segment has the capability of changing the concentration of a second messenger. In the photoreceptor system, the effector is a phosphodiesterase (PDE), which preferentially hydrolyzes cGMP. In the case of the adenylyl cyclase and phospholipase C systems, the effector moiety can *increase* the concentration of other second messengers, including cAMP, inositol trisphosphate, and diacylglycerol. By utilizing an enzymatic function in the third segment of the cascade, the transduction process acquires additional opportunities for amplification that are inherent in the turnover number of the effector enzyme. In the case of photoreceptor PDE, this number approaches (in vitro) 10^3 cGMPs hydrolyzed per PDE per second (Miki et al. 1975; Sitaramayya 1986; M.M. Whalen and M.W. Bitensky, in prep). This value is misleading because the rate at which second-messenger molecules are synthesized or dismantled in vivo reflects the substrate concentrations selected by evolution, rather than the saturating concentrations selected by the enzymologist. Because the K_m of light-activated PDE is in the millimolar range (Robinson et al. 1980) at 5 μM cGMP, which is the estimated in vivo, dark, free cGMP pool concentration, rod PDE will hydrolyze approximately 10 cGMPs per second.

The mechanism by which the effector is activated in the visual system has been studied extensively. Here, the activator $GTP \cdot N_{h\nu}\alpha$ removes or displaces an inhibitory subunit from the effector enzyme PDE (Hurley and Stryer 1982; Yamazaki et al. 1983; Deterre et al. 1986). The effect upon inhibitor appears somewhat different with regard to the two catalytic subunits of rod PDE. It is now known that there are two inhibitor moieties (γ subunits) per high-molecular-weight PDE dimer (Deterre et al. 1986; M.M. Whalen and M.W. Bitensky, in prep.), designated as PDE $\alpha,\beta(\gamma)_2$. The inhibitor appears to have a lower affinity for one of its binding sites and is thus released more easily from PDE into the soluble (cytoplasmic) fraction. The inhibitor moiety also has a higher affinity for a second class of inhibitory sites (M.M. Whalen and M.E. Bitensky, in prep.). Even when PDE is transiently displaced from its high-affinity site (by high concentrations of activator), any lowering of activator concentrations permits a rapid restoration of inhibitory function. There is a defined sequence for PDE disinhibition, which exhibits two phases. The first phase, requiring less activator (micromolar range), is readily accompanied by solubilization of a GTP/activator/inhibitor complex from the enzyme ($GTP \cdot N_{h\nu}\alpha \cdot PDE\gamma$).

The second phase of disinhibition, which requires larger amounts of activator, is associated with a transient displacement of inhibitor. For this second class of high-affinity PDE inhibitor-binding sites, activator concentrations greater than 10 μM are needed for PDEγ solubilization (M.M. Whalen and M.W. Bitensky, in prep.). The cow and frog rod enzyme systems are conceptually similar. A principal difference is found in the amount of PDE activator unmasked by release of the first inhibitory moiety. This step unmasks about half of the PDE activity in the frog and only 10% in the cow (M.M. Whalen and M.W. Bitensky, in prep.). We suggest that these differences reflect the different physiological adaptations of amphibian and bovine species. The frog is nocturnal and feeds at twilight, whereas the cow is essentially diurnal. The observed arrangement may provide a threshold for rod vision in the cow, which is not encountered in the frog; i.e., the frog rod would appear capable of responding to very small bleaches even in a dimly lighted surround. The cow rod, on the other hand, must be more fully dark adapted before it can generate a significant response to bleaches of very low intensity (M.M. Whalen and M.W. Bitensky, in prep.).

Activation of the effector enzyme segment by the release of an inhibitory constraint appears as an attractive model (Bitensky et al. 1982) that might be generally employed in the three-segment molecular cascade of signal transduction. However, there is not yet conclusive evidence in support of the idea that other classes of GTP-liganded α subunits (of the GTP-binding protein family) also function by displacing an inhibitory moiety.

For that class of cyclase-linked forms of single transduction, where the signal has an inhibitory character and reduces cellular levels of cAMP (e.g., α-adrenergic agonists), an analogous mechanism is operating (i.e., the liganding of an inhibitory cyclase receptor is followed by the binding of GTP to an inhibitory heterotrimeric GTP-binding complex). In this case, however, the inhibitory GTP complex ($GTP \cdot N_i\alpha$ [N_i = inhibitory guanine-nucleotide-binding protein]) interferes with cAMP synthesis. The mechanism of cyclase inhibition appears to be somewhat more complex than the mechanism of activation. For one thing, there is evidence that the cyclase inhibitory complex can directly inhibit the effector segment (Hildebrandt et al. 1983). It would also appear that the N_i β and γ subunits, which are released in consequence of binding to the inhibitory nucleotide-binding subunit, will also exert an inhibitory influence by complexing with stimulatory ($N_s\alpha$ [N_s = stimulatory guanine-nucleotide-binding protein]) subunits (Gilman 1987). Thus far, GTP-binding proteins are known to interact with phospholipase C (Berridge and Irvine 1984), adenylyl cyclase, rod PDE, ion channels, and phospholipase A_2. It is expected that other effector enzymes will be identified as members of a larger family of GTP-dependent signal transduction systems.

In a memorable essay on the metabolic code, Tomkins (1975) observed that the earliest functional significance of cAMP was as a surrogate for hunger. The absence of glucose reliably elevates cAMP levels in the cytoplasm of vertebrate liver and coliform bacteria and in the extracellular environment of *Dictyostelium discoideum*. In *D. discoideum*, cAMP (as a herald of hunger) also bore witness to those early experiments in which evolution made a tentative transition from single to multicellular forms (Gabius et al. 1985). cAMP functions as a signal of glucose deprivation in single-celled animals and as a signal for substrate mobilization following the development of multicellular forms. Substrate storage and mobilization became central themes in such specialized cells as hepatocytes and adipocytes. In multicellular organisms, the three-segment molecular architecture of signal transduction made a further evolutionary transition to a more general role. cAMP could elicit those specialized products or services that a cell contributed altruistically to the whole organism, including thyroid hormone, insulin, enhanced force and rate of myocardial contraction, increased renal concentrating power, and the synthesis of melanin (Pober and Bitensky 1979).

Photoreceptor Cell

In the vertebrate photoreceptor system, the three-segment architecture of signal transduction has become specialized in a most unusual way. The achievement of single quantum sensitivity was a major shaping force in the evolution of photon transduction (Hecht et al. 1952). In addition, specializations that permitted resolution of light at different wavelengths (Nathans 1987) and specializations that allowed photoreceptors to respond over a wide range of signal intensities (Pugh and Cobbs 1986) have enhanced their ability to communicate detailed information about the external environment.

The rod must detect a single quantum and also manage to be relatively free of noise. The full gamut of noise-reducing techniques is not known, nor is it known precisely what fraction of noise reduction is accomplished by higher-order neuronal data processing. The visual system plays a fundamental role in predation and escape from predators, reproduction, and the location of food and shelter. It is not surprising that evolution has fashioned some unusual gene product ensembles in the photoreceptor.

The amphibian rod is a specialized neuron that can reduce its 40-pS membrane current by approximately 4% in response to 1 photon and virtually abolish membrane current in response to 10^5 photons. It is now known that light can rapidly reduce the rod cytoplasmic concentration of cGMP (Blazynski and Cohen 1986; Cote et al. 1986). Moreover, through the combined functions of the cation channel and Na^+/Ca^{++} exchanger, light can also lower cytoplasmic Ca^{++} concentrations (Yau and Nakatani 1985).

Earlier Problems

A series of early problems that arose in the course of studies on cyclic nucleotides in visual excitation are reviewed in this paper. In our analysis of these problems, we focus on the evolutionary perspective developed here and show how a number of these problems arose in consequence of the fact that our experimental designs have often been unmindful of the precisely shaped conditions that evolution has painstakingly fashioned for the functioning of photoreceptors.

Light activation of PDE in broken-rod suspensions. A perplexing problem that faced vision researchers in the early 1970s was the difficulty in documenting a light-dependent increase in rod PDE activity. In illuminated, broken-rod outer segment suspensions to which GTP was added, there was often little increase in the observed activity of the rod (cGMP preferring) PDE. In many such experiments, the suspensions of disk membranes were prepared at a substantial dilution when compared with actual protein concentrations in the living rod.

Because rhodopsin comprises roughly 50% of rod outer segment protein, the amphibian rod has a protein concentration approaching 400 mg/ml. We often use disk suspensions with protein concentrations less than 20 mg/ml. In addition to the parameter of dilution, the rod disk membrane suspensions are often "washed" by sedimentation and resuspension. Under these experimental conditions, there is a possibility that a portion of the $N_{h\nu}\alpha$ subunits are lost. To mount an attack of sufficient intensity to dislodge the PDEγ subunits bound to the low-affinity sites, concentrations of activator in the 1 μM range are needed. Such an activator contingent might unmask less than 10% of the potential PDE activity in the bovine system. Moreover, to dislodge the PDEγ subunits bound to the higher-affinity class of binding sites, a population of activators in the range of 10 μM is required. Thus, in disk membranes partially bleached and extensively diluted during preparation, the light and GTP-dependent activation of bovine rods might indeed appear to be insufficient.

Even in the beginning phase of rod cyclic nucleotide studies, there were indications that many properties of the rod system required extremely concentrated membrane suspensions for their observation (Bitensky et al. 1975). This was found empirically, although the actual molecular reasons were not understood initially. Robust activation of PDE by light and GTP were best observed in fully dark-adapted, unwashed, and concentrated disk membrane suspensions, using unilluminated membranes as a control. In view of the single photon sensitivity of the rod, the best preparations were those that had been processed under infrared light.

Effects of light on intact photoreceptors. A second and related problem arose when some investigators were unable to find a decrease in the concentration of rod outer segment cGMP using intact rod outer segments or whole retinas and short light flashes. More recent information suggests that the duration of the cGMP decrease induced by a flash of light is very brief and, in fact, must correspond closely to the duration of the light-induced decrease in photoreceptor current. Because early measurements did not achieve the time resolution sufficient to observe this decrease, what was actually measured was a cGMP level that had already been restored by guanylyl cyclase.

Moreover, because the problem of distinguishing between bound and free cGMP (Yamazaki et al. 1980) was not addressed, the total photoreceptor cGMP pool was being measured. Because bound cGMP may constitute as much as 90% of the total rod outer segment cGMP pool in the unstimulated photoreceptor, a significant change in the free pool would appear as only a modest change in the total, with very small bleaches (e.g., $< 10^{-8}$ of the rhodopsin pool) when only a modest fraction of the total PDEs were activated. For example, a 30% decrease in the free pool would translate into a 3% decrease in total cGMP, which was within the margin of error for these measurements. Moreover, such cGMP measurements were complicated not only by the rapid restoring force of guanylyl cyclase, but also by the fact that adjacent (and often contaminating) retinal neurons also contain cGMP and could thus be a source of obfuscation.

(Observed) rates of GTPase are too slow. A recurring concern has focused on the issue of why the measured rate constant for the GTPase activity of rod $N_{h\nu}\alpha$ was so slow. The observed hydrolysis of about 1 GTP/sec/ $N_{h\nu}\alpha$ would make it impossible to shut down rod PDE in the short time span ($\sim$1 sec) needed to accommodate the electrical response (Yamazaki et al. 1987). Following a saturating bleach, the rod PDE continues to be active some 30 seconds after the photocurrent has recovered (Miller 1982). The measured rate for the GTPase activity would still be too slow to accommodate this prolonged time span.

We may have markedly underestimated the rate of the GTP hydrolytic step, however, primarily because of excessively dilute conditions in vitro. This is because it is necessary for $GDP \cdot N_{h\nu}\alpha$ to interact with $N_{h\nu}\beta,\gamma$ and metarhodopsin II each time a GTP replaces its hydrolyzed antecedent. It is not sufficient to be mindful only of GTP concentrations in designing a GTPase experiment, especially because GTP binding to $N_{h\nu}\alpha$ readily saturates in the micromolar range.

It is also necessary to achieve or approach those concentrations of rhodopsin (3×10^{-3} M) and $N_{h\nu}\alpha$ (3×10^{-4} M) that would be encountered in vivo. It may not be necessary to consider undiscovered GTPase-activating factors (E.A. Dratz et al., in prep.) that must be present in high concentration, but rather to recognize the roles and concentrations of each of the known reactants as they exist in vivo. Thus, the rod GTPase is a rather atypical enzyme system insofar as it demonstrates high-concentration, macromolecular cofactor requirements (including a liganded receptor), which must act in concert for the GTP-binding step. One can account for the observed twofold increase in GTPase turnover number with each doubling of preparation concentration (E.A. Dratz et al., in prep.), using mass action:

$$\frac{dGTP}{dt} \sim GTP \cdot N_{h\nu}\alpha \sim K(N_{h\nu}\alpha)(N_{h\nu}\beta\gamma)$$

This equation predicts the observed (concentration-dependent) gain in rate, simply on the basis of concentrating those gene products known to be essential for GTP binding. Alternatively, Dratz' explanation that an additional factor increases GTPase rate in a concentration-dependent manner may be correct.

Trypsin activation of PDE. Trypsin activation of PDE was usually found to exceed that obtained with light and GTP. The reason was not that trypsin was altering the enzyme, but rather that conventional approaches to physiological activation did not generate those concentrations of activator (> 100 μM) that are likely to appear in vivo. When such concentrations are utilized in vitro, one can readily achieve activations of PDE that are equivalent to those found with trypsin (Fung et al. 1981; Sitaramayya 1986).

Comparisons between guanylyl cyclase and PDE. In earlier studies of the rod, it was not unusual to experience disappointment in the amount of guanylyl cyclase activity measured in vitro, especially when this activity was compared with the activity exhibited by rod PDE. This discrepancy arose because PDE was measured at substrate concentrations in the millimolar range, whereas the actual enzyme velocity in vivo is profoundly constrained by substrate levels in the 5 μM range (see above).

cGMP and Regulation of the Multiply Liganded Rod Cation Channels

A central question about the regulation of cyclic nucleotide-dependent channels in the rod concerns the concentration of free cGMP in the unstimulated rod and the change in that concentration occurring as a function of photon capture. Measurements of cGMP in rod outer segments are subject to some ambiguity because a significant fraction of the cGMP pool (~90%) is bound primarily to unstimulated PDE. Moreover, the actual dimension of the cytoplasmic space in which cGMP is found is not defined rigorously. However, there is a total pool of approximately 50 μM cGMP in the intact rod outer segments.

In reasoning from data obtained with "pipetted" membrane patches and the binding constant (~17 μM) calculated for the multiply liganded cation channel, one can arrive at a value for the dark, free cGMP pool in the vicinity of 5 μM.

At present, there is a rather compelling collection of observations that indicate rod cation channel patency is controlled by cytoplasmic cGMP levels. These data emerge from experiments demonstrating that inhibitors of rod PDE would increase rod dark current (Miller and Nicol 1979; Miller 1982). Moreover, infusions of cGMP also increase dark current. These rod recordings were followed by isolated outer segment membrane patch experiments, which have also demonstrated that cGMP regulates the flux of cations through the rod membrane (Fesenko et al. 1985). Moreover, studies with intact photoreceptors and membrane patches have provided evidence for a multiply liganded cGMP-regulated cation channel. Positive cooperativity among the ligands has also been proposed. This arrangement is especially sensitive, even to modest reductions in cytoplasmic cGMP levels such as those associated with single photon capture.

The dependence of the fraction of channels open, f, upon the cGMP concentration, as measured by Zimmerman and Baylor (1986), is approximately $f = (cGMP/K)^3/(1 + [cGMP/K]^3)$.

The form of this equation implies that all sites must be bound for a channel to be open and also implies positive cooperativity. This cooperativity increases the sensitivity further for concentrations of cGMP near 5 μM; i.e., the derivative df/dcGMP increases by a factor of approximately 2, as compared with a channel that behaves similarly but with three independent (noncooperative) sites. This function of cGMP as a rod cation channel regulator is consistent with the concept that cyclic nucleotides regulate the highly specialized functions of fully differentiated cells.

Hypothetical Subcellular Localizations of Rod Enzymes

Would an advantage be gained by sequestering the PDE and/or the guanylyl cyclase in proximity to the cation channels, i.e., near the disk margins? One anatomical feature that prompts such a localization is the enlarged, rounded disk rim margins that bring the disks in close proximity to the plasma membrane. Furthermore, the 240-kD integral disk protein "incisin" appears to be localized within the disk margins (Papermaster et al. 1978) and could serve as an organizing structure about which to array disk rim proteins. In this model, the extended plane of the disk would be dedicated to photon capture and the generation of activator complexes.

If one were to localize either the cyclase or PDE in the disk rim area, would the complimentary enzyme have to be placed nearby in order to achieve regulatory balance? Having the cyclase near the channels and the PDE dispersed throughout the disk plane would impose a greater demand on hydrolytic activity in the activation step, especially if one assigns a robust basal rate to the guanylyl cyclase. On the other hand, localizing the rod PDE adjacent to the channels and dispersing the cyclase uniformly in the plane of the disk membrane would create an increased demand on the cGMP synthetic activity needed for the recovery step because the PDE hydrolytic function would dominate in the channel vicinity. However, such an arrangement would provide a useful time lag between activation of the PDE and that of the guanylyl cyclase.

Experiments based on differential sedimentation link the guanylyl cyclase with the rod axoneme (Fleischman and Denisevich 1979). This specialized domain between inner and outer segments concentrates the microtubular elements of the rod. The microscopical anatomy of the rod exhibits microtubular projections that radiate outward from the axoneme toward the apex of the outer segment, confining their excursions within the channels formed by the vertical alignment of disk incisures.

A modified interpretation of the above data would affiliate the guanylyl cylase not only with axonemal structures, but also with outer segment projections of the microtubules. Such a distribution would locate the guanylyl cyclase in a region adjacent to the plasma membrane cation channels. The rounded disk rim margins could provide an adjacent locus for the hydrolytic activity of PDE. Having both the synthetic and hydrolytic enzymes adjacent to the cation channels provides more immediate impact upon the cGMP concentrations in the cytoplasmic space adjacent to the channels.

A possible advantage for the speculative confinement of cyclase and PDE adjacent to the membrane channels could arise from more effective PDE utilization in the single photon case. If PDE could diffuse rapidly within the rim margin, a single photon might influence all of the PDE molecules in one disk, rather than the half distributed in that disk membrane bilayer in which the photon was captured. In principle, in a fully dark-adapted rod, a single photolyzed rhodopsin can activate all of the $N_{h\nu}\alpha$ molecules residing in one plane of the disk membrane. This collection of activators would interact with the disk rims by diffusion in the narrow aqueous channel between two adjacent disks.

Generic Evidence for Gene Product Localization

There is ample precedent for subcellular localization of enzyme systems and organelles. Evolution has repeatedly demonstrated a capability that extends beyond regulating enzyme rates and substrate concentrations to a full capability for localizing gene products and their ensembles in particular subcellular domains. One need go no further than the rod to discover examples of subcellular localizations, which include clustering of mitochondria adjacent to the axoneme, neurotransmitter secretory granules adjacent to the synaptic terminus, segregation of rhodopsin and the Na^+/Ca^{++} exchanger in the outer segment, and the Na^+ pump in the inner segment. Yet another example of subcellular localization is the segregation of the apparatus for photon capture in a discrete organelle. These cases exemplify evolution's capacity for topological specialization, as well as its fine quantitative control over gene product expression.

A Light-dependent Activator of Guanylyl Cyclase

A recent study has suggested the utility of having a light-generated guanylyl cyclase activator to facilitate restoration of cGMP levels following the initial fall caused by photon capture (Kondo and Miller 1988). The possible advantages of such an activator would include a wider dynamic range of response, as well as enhanced timing for restoration of the dark current. Although the fall in Ca^{++} associated with channel closure and continuing activity of the Na^+/Ca^{++} exchanger may contribute to the light-dependent activation of guanylyl cyclase, this fall may not provide the regulatory capability needed to orchestrate the recovery phase of the light response by itself.

Documentation of a Rapid Light-sensitive Change in Rod cGMP Levels

Two recent studies have demonstrated rapid and sensitive light-mediated declines in total rod cGMP levels (Blazynski and Cohen 1986; Cote et al. 1986). In these experiments, the decline in cGMP concentrations does not parallel the size of the bleach over a wide range of light intensities. The magnitude of the guanylyl cyclase activation appears to reflect the magnitude of the PDE activation so that the amount of hydrolysis is compensated for by an appropriate quantity of synthesis. This exquisite coordination of the rod outer segment major-force enzymes would benefit from an additional light-sensitive regulator of the guanylyl cyclase, as discussed above. This coordination must also persist during the PDE activity observed after completion of the electrical response (Miller 1982).

Measurements of Flux through the cGMP Pathway

Much consideration has been given to the increased flux through the cGMP pathway measured in illuminated retinas, and rightly so. The microanalytic technique utilizes ^{18}O and mass spectroscopy to measure this flux. However, there is a minor ambiguity with regard to the specificity of these measurements because there is a background cGMP flux in adjacent, retinal higher-order neurons: Changes in the background are minimized by the use of aspartate (Goldberg et al. 1983).

The possibility has been raised that the flux itself could regulate rod cation channel dynamics. The light-associated flux through the cGMP pathway appears to reflect the dynamic capacity of a rod photoreceptor to restore itself rapidly to a functional status following bleaches of varying magnitude. To complete a light response, a burst of cGMP synthesis is needed to compensate effectively for the amount of light-induced cGMP hydrolysis. The magnitude of the synthetic response must correlate with the magnitude of the activation signal. Thus, the amount of flux must correlate with the size of the bleach rather than the kinetic features of the light response.

The rapid measurements of changes in rod cGMP levels clearly document that the cytoplasm of the rod does exhibit a reduced concentration of cGMP in response to light. They also demonstrate that the magnitude of the decline is not profoundly different for smaller and larger bleaches. This is in contrast to the observation that the size of the flux is more sensitive to the size of the bleach. Because the actual light response (membrane current drop) appears to follow the decrease in cGMP closely, these data may be interpreted to support the view that the rod converts cGMP flux changes of different magnitude into cGMP concentration changes of comparable magnitude as a feature of achieving its dynamic range. The flux data have focused attention on the need for light regulation of guanylyl cyclase. They have also indicated the different sizes of guanine nucleotide pools, as well as different turnover rates in these pools.

SUMMARY

Evolutionary forces have designed a large family of rod and cone photoreceptors, each member of which suits the lifestyle requirements and circadian patterns of a particular species. The three-segment architecture of signal transduction is conspicuous in the biochemistry of photoreceptors and supports their demonstrated properties of extreme sensitivity, low noise levels, extended dynamic range, and light adaptation. The designs elaborated by evolution reflect a gradual process of modification, with the sequential elaboration of layers of control and refinements in control. The end results of this long evolutionary labor are the functional efficiency and dynamic range that give the rod its utility. Our conceptual problems in deriving observed rod properties from the collective features of known rod gene products may well give way when we have learned more about the true composition and topology of the outer segment gene set and both bound and free nucleotide concentrations.

The invertebrates have developed alternative solutions to the problems of photoreceptor sensitivity and wide dynamic range. The vertebrate rod represents a truly optimized way to capture and interpret low-intensity photon signals. One may anticipate, with some enthusiasm, those molecular and kinetic data that will permit an understanding of how cones differ from rods and how release from the requirement for single photon detection has shaped the design of this wavelength-specific companion photoreceptor.

The utilization by evolution of the three-segment architecture of GTP-dependent signal transduction for other modalities of sensory perception, such as olfaction (Lancet et al., this volume) and gustation (Jones et al., this volume), is certainly a reasonable and successful choice. Here, the need for versatility in the repertoire of signal recognition moieties, as well as a capacity for sensitivity and signal amplification, strongly recommends the generic use of the GTP-dependent signal-activated macromolecular switch in neurosensory signal transduction mechanisms.

ACKNOWLEDGMENTS

We thank Alexandra Vigil for facile manuscript preparation. This work was supported by grant EY-06816 from the National Institutes of Health.

REFERENCES

Barkdoll, A.E. III, E.N. Pugh, Jr., and A. Sitaramayya. 1988. Kinetics of the hydrolysis of 8-bromo-cGMP by the light-activated PDE of toad rods. *J. Neurochem.* **50:** 839.

Berridge, M.J. and R.F. Irvine. 1984. Inositol trisphospate, a novel second messenger in cellular signal transduction. *Nature* **312:** 315.

Bitensky, M.W., D. Torney, A. Yamazaki, M.M. Whalen, and J.S. George. 1987a. A model of the light dependent regulation of retinal rod phosphodiesterase, guanylate cyclase and cation flux. In *Molecular mechanisms of neuronal responsiveness* (ed. Y.H. Ehrlish et al.), p. 107. Plenum Publishing, New York.

Bitensky, M.W., M.M. Whalen, D.C. Torney, M. Tatsumi, and A. Yamazaki. 1987b. A common algorithm for the transduction, amplification and cellular response to photons, hormones, and neurotransmitters. In *Progress in clinical and biological research: Mechanisms of signal transduction by hormones and growth factors* (ed. M.C. Cabat and W.L. McKeehan), p. 3. A.R. Liss, New York.

Bitensky, M.W., N. Miki, J.J. Keirns, J.A. Baraban, M.A. Wheeler, J. Lacy, and F.R. Marcus. 1975. Activation of photoreceptor disc membrane phosphodiesterase by light and ATP: Cyclic GMP as modulator of sense receptor functions. *Adv. Cyclic Nucleotide Res.* **5:** 213.

Bitensky, M.W., M.A. Wheeler, M.M. Rasenick, A. Yamazaki, P.J. Stein, K.R. Halliday, and G.L. Wheeler. 1982. Functional exchange of components between light activated (photoreceptor phosphodiesterase and hormone activated) adenylate cyclase systems. *Proc. Natl. Acad. Sci.* **79:** 3408.

Blazynski, C. and A.I. Cohen. 1986. Rapid declines in cGMP of rod outer segments of intact frog photoreceptors after illumination. *J. Biol. Chem.* **261:** 14142.

Burke, D.T., G.F. Carle, and M.V. Olson. 1987. Cloning of large segments of exogenous DNA into yeast by means of artificial chromosome vectors. *Science* **236:** 806.

Cote, R.H., G.D. Nicol, S.A. Burke, and M.D. Bownds. 1986. Changes in cGMP concentrations correlate with some but not all aspects of the light regulated conductance of frog rod photoreceptor. *J. Biol. Chem.* **261:** 12965.

Deterre, P., J. Bigay, F. Forquet, M. Robert, and M. Chabre. 1988. cGMP phosphodiestrase of retinal rods is regulated by two inhibitory subunits. *Proc. Natl. Acad. Sci.* **85:** 2424.

Deterre, P., J. Bigay, M. Robert, C. Pfister, H. Kuhn, and M. Chabre. 1986. Activation of retinal rod cGMP phosphodiesterase by transducin: Characterization of the complex formed by phosphodiesterase and the transducin alpha subunit. *Proteins Struct. Funct. Genet.* **1:** 188.

Fesenko, E.E., S.S. Kolesnikow, and A.L. Lubarsky. 1985. Induction of cyclic GMP of cationic conductance in plasma membrane of retinal rod outer segment. *Nature* **313:** 310.

Fleischman, D. and M. Denisevich. 1979. Guanylate cyclase of isolated bovine retinal rod axonemes. *Biochemistry* **18:** 5060.

Foster, K.W. and R.D. Smyth. 1980. Light antennas in phototactic algae. *Microbiol. Rev.* **44:** 572.

Foster, K.W., J. Saranak, N. Patel, G. Zarilli, M. Okabe, T. Kline, and K.A. Nakanishi. 1984. Rhodopsin is the functional photoreceptor for phototaxis in the unicellular eukaryote chlamydomonas. *Nature* **311:** 756.

Fung, B.K. and L. Stryer. 1980. Photolyzed rhodopsin catalyzes the exchange of GTP for bound GDP in retinal rod outer segments. *Proc. Natl. Acad. Sci.* **77:** 2500.

Fung, B.K., J.B. Hurley, and L. Stryer. 1981. Flow of information in the light triggered cyclic nucleotide cascade of vision. *Proc. Natl. Acad. Sci.* **78:** 152.

Gabius, H.J., W.R. Springer, and S.H. Barondes. 1985. Receptor for the cell binding site of discoidin I. *Cell* **42:** 449.

Gilman, A.G. 1987. G. proteins: Transducers of receptor-generated signals. *Annu. Rev. Biochem.* **56:** 615.

Goldberg, N.A., A. Ames III, J.E. Gander, and T. Walseth. 1983. Magnitude of increase in retinal cGMP metabolic flux determined by ^{18}O incorporation into nucleotide alpha phosphoryls corresponds with intensity of photic stimulation. *J. Biol. Chem.* **258:** 9213.

Hecht, S., S. Schlaer, and M.H. Pirenne. 1952. Energy, quanta and vision. *J. Gen. Physiol.* **25:** 819.

Hildebrandt, J.D., R.D. Sekura, J. Codina, R. Iyengar, C.R. Manclark, and L. Birnbaumer. 1983. Stimulation and inhibition of adenylyl cyclases mediated by distinct regulatory proteins. *Nature* **302:** 706.

Hirsch, J. 1982. Falcon visual sensitivity to grating contrast. *Nature* **300:** 57.

Holland, S.K. and C.C.F. Blake. 1987. Proteins, exons and molecular evolution. *Bio/Systems* **20:** 181.

Hurley, J.B. and L. Stryer. 1982. Purification and characterization of the g regulatory subunit of the cyclic GMP PDE from retinal rod outer segments. *J. Biol. Chem.* **257:** 11094.

Kondo, H. and W.H. Miller. 1988. Rod light adaptation may be mediated by acceleration of the phosphodiesterase-guanylate cyclase cycle. *Proc. Natl. Acad. Sci.* **85:** 1322.

Lefkowitz, R.J., J.M. Stadel, and J.G. Caron. 1983. Adenylate cyclase coupled β-adrenergic receptors: Structure and mechanisms of activation and desensitization. *Annu. Rev. Biochem.* **52:** 159.

Liebman, P.A. and E.N. Pugh, Jr. 1982. Gain, speed and sensitivity of GTP binding vs PDE activation in visual excitation. *Vision Res.* **22:** 1475.

Litosch, I. and J.N. Fain. 1986. Minireview: Regulation of phosphoinositide breakdown by guanine nucleotides. *Life Sci.* **39:** 187.

Mangini, N.J. and D.R. Pepperberg. 1987. Localization of Retinal "48K" (S-antigen) by electron microscopy. *Jpn. J. Ophthalmol.* **31:** 207.

Miki, N., J.M. Barban, J.J. Keirns, J.J. Boyce, and M.W. Bitensky. 1975. Purification and properties of the light activated cyclic nucleotide phosphodiesterase of rod outer segments. *J. Biol. Chem.* **250:** 6320.

Miller, W.N. 1979. Intraocular filters. In *Handbook of sensory physiology* (ed. H. Autrum), p. 70. Springer Verlag, Berlin.

———. 1982. Physiological evidence that light mediated decrease in cGMP is an intermediary process in retinal rod transduction. *J. Gen. Physiol.* **80:** 103.

Miller, W.N. and G.D. Nicol. 1979. Evidence that cyclic GMP regulates membrane potential in rod photoreceptor. *Nature* **280:** 64.

Moyzis, R.K., K.L. Albright, M.F. Barholdi, L.S. Cram, L.L. Deaven, C.E. Hildebrand, N.E. Juste, J.L. Longmire, J. Meyne, and T.S. Robinson. 1987. Human chromsome-specific repetitive DNA sequences: Novel markers for genetic analysis. *Chromosoma* **95:** 375.

Nathans, J. 1987. Molecular biology of visual pigments. *Annu. Rev. Neurosci.* **10:** 163.

Nathans, D. and H. Smith. 1975. Restriction endonucleases in the analysis and restructuring of DNA molecules. *Annu. Rev. Biochem.* **44:** 273.

Papermaster, D.S., B.G. Schneider, M.A. Zorn, and J.P. Kraehenbuhl. 1978. Immunocytochemical localization of a large intrinsic membrane protein to the incisures and margins of frog outer segment discs. *J. Cell Biol.* **77:** 415.

Pober, J.S. and M.W. Bitensky. 1979. Light regulated enzymes of vertebrate retinal rods: A critical review. *Cyclic Nucleotide Res.* **11:** 265.

Pugh, E.N., Jr. and W.H. Cobbs. 1986. Visual transduction in vertebrate rods and cones: A tale of two transmitters, calcium and cGMP. *Vision Res.* **26:** 1613.

Rall, T.W. and E.W. Sutherland. 1961. The regulatory role of adenosine-3′5′ phosphate. *Cold Spring Harbor Symp. Quant. Biol.* **25:** 347.

Robinson, P.R., S. Kawamura, B. Abramson, and M.D. Bownds. 1980. Control of the cyclic GMP phosphodiesterase of frog photoreceptor membranes. *J. Gen. Physiol.* **76:** 631.

Rodbell, M. 1985. Programmable messengers: A new theory of hormone action. *Trends Biochem. Sci.* **10:** 461.

Schwartz, D.C. and C.R. Cantor. 1984. Separation of chromosome-size DNAs by pulsed field gradient gel electrophoresis. *Cell* **37:** 67.

Sepkoski, J. 1982. Geologic implications of impacts of large asteroids and comets on the earth. In *Mass extinctions in the Phanerozoic oceans: A review* (ed. L.T. Silver and P.H. Schultz, p. 283. Special Paper 190. Geological Society of America, Boulder, Colorado.

Shinozawa, T., I. Sen, G.L. Wheeler, and M.W. Bitensky. 1979. Predictive value of the analogy between hormone-sensitive adenylate cyclase and light sensitive photoreceptor cyclic GMP phosphodiesterase: A specific role for light sensitive GTPase as a component in the activation sequence. *J. Supramol. Struct.* **10:** 185.

Sitaramayya, J.H., J.H. Parkes, C. Gonzalez-Oliva, and P.H. Liebman. 1986. Kinetic studies indicate that light activated cyclic GMP PDE is a complex with G-protein subunits. *Biochemistry* **25:** 651.

Snyder, A.W. and W.H. Miller. 1978. Telephoto lens system of falconiform eyes. *Nature* **275:** 127.

Tomkins, G.M. 1975. The metabolic code. *Science* **189:** 760.

Uchida, S., G.L. Wheeler, A. Yamazaki, and M.W. Bitensky. 1981. A GTP-protein activator of phosphodiesterase which forms in response to bleached rhodopsin. *J. Cyclic Nucleotide Res.* **7:** 95.

Wheeler, G.L., Y. Matuo, and M.W. Bitensky. 1977. A light activated GTPase in vertebrate photoreceptors. *Nature* **269:** 822.

Yamazaki, A., P.J. Stein, N. Chernoff, and M.W. Bitensky. 1983. Activation mechanism of rod outer segment cyclic GMP phosphodiesterase. *J. Biol. Chem.* **258:** 8188.

Yamazaki, A., M. Tatsumi, D.C. Torney, and M.W. Bitensky. 1987. The GTP binding protein of ROS (1) Role of each subunit in the GTP hydrolytic cycle. *J. Biol. Chem.* **262:** 9316.

Yamazaki, A., I. Sen, M.W. Bitensky, J.E. Casnellie, and P. Greengard. 1980. Cyclic GMP-specific, high affinity noncatalytic binding sites on light activated phosphodiesterase. *J. Biol. Chem.* **255:** 11619.

Yau, K.W. and K. Nakatani. 1985. Light induced reduction of cytoplasmic free calcium in retinal rod outer segment. *Nature* **313:** 579.

Zimmerman, A.L. and D.A. Baylor. 1986. cGMP sensitive conductance of retinal rods consists of aqueous pores. *Nature* **321:** 70.

Visual Signal Transduction: The Cycle of Transducin Shuttling between Rhodopsin and cGMP Phosphodiesterase

M. CHABRE, J. BIGAY, F. BRUCKERT, F. BORNANCIN, P. DETERRE, C. PFISTER, AND T.M. VUONG
Biophysique Moléculaire et Cellulaire, Unité associée 520 du CNRS Fédération de Biologie, DRF/CENG, BP 85, 38041 Grenoble, France

Among the numerous G proteins that couple membrane receptors to effectors of intracellular messengers, the system of transducin (T) shuttling between photoexcited rhodopsin (R*) and cGMP phosphodiesterase (PDE) constitutes a most interesting model. Three features of this system have made a large variety of in situ biophysical measurements possible and have greatly facilitated biochemical purification and reconstitution experiments: (1) the exceptionally high concentration of the components (rhodopsin, transducin, and PDE) in retinal rod outer segments (ROS); (2) the convenient excitability of the system by light flashes; and (3) the peculiarity of the solubility of transducin and PDE in the total absence of detergent.

The objection is sometimes made that the visual transduction cascade might be a very special case, particularly that transducin is an atypical G protein, due to its solubility. G proteins were long considered to be membrane proteins that would interact within the lipid phase with the receptors. This concept is not borne out by the sequence comparison between transducin and other G proteins: The extensive sequence homologies, particularly the absence of additional hydrophobic stretches in the other G proteins, preclude that some of them would be structurally very different from transducin and would thus have their functional interaction domains in the lipids, whereas those of transducin are necessarily cytoplasmic (Chabre 1987a). Among the three subunits making up all known signal-transducing G proteins, the 36-kD β polypeptide of G_s and G_i, for example, is strictly identical to T_β. The small T_γ subunit, which is not dissociable from T_β without denaturation of the complex, is indeed immunologically and electrophoretically different from the G_γs. The sequence of T_γ is quite hydrophilic, which can account for the fact that the $T_{\beta\gamma}$ complex is only weakly membrane bound and is solubilized from the membrane in media of low ionic strength. The sequences for other G_γs are not yet known and must be responsible for the much tighter attachment of the $G_{\beta\gamma}$ complex to the membrane, with G_γ acting as an anchor for G_β. As for the guanine-nucleotide-binding subunit T_α, its sequence does not differ from that of other G_αs in a way that could account for a deep membrane penetration of the G_αs. However, all G_α polypeptides are terminated on their amino ends by a glycin that is myristylated in $G_{o\alpha}$ and $G_{i\alpha}$ but remains free in $G_{s\alpha}$ and T_α (Muss et al. 1987). This posttranslational modification, which does not modify the structure of the protein part, is probably the main reason for the variation in membrane attachment of these basically cytoplasmic G_α subunits. It has been shown (Lynch et al. 1986) that when activated by GTPγS, $G_{s\alpha}$ dissociates from $G_{\beta\gamma}$ and becomes soluble in the absence of detergent, just as T_αGTPγS does.

On the receptor side, rhodopsin clearly stands as an archetype for a large and fast-growing class of "7 α-helix" receptors, which already includes the M_1 and M_2 muscarinic, the α_2, β_1, and β_2 adrenergic, and the substance K receptors (Dohlman et al. 1987). In contrast to these other receptors, rhodopsin is exceptionally abundant in the rod cell: Compared to the 10^4 to 10^5 receptors found in the plasma membrane of a hormone-responsive cell, there are nearly 10^8 rhodopsin molecules in a mammalian retinal rod. This large number is made necessary by the special physical character of the visual signal: Photons do not diffuse around in the extracellular medium until they are captured by a receptor as chemical transmitter molecules are. Whereas a limited number of receptors are sufficient to confer to a cell an efficiency close to the theoretical maximum for the capture of a diffusing molecule (Berg 1983), a maximum number of pigmented receptors is needed to increase the probability of absorbing photons that cross the cell only once in a straight line. The visual transduction system is thus atypical in that the total number of receptors far exceeds that of the G proteins to which these receptors can couple. However, the physiological range of response entails the photoexcitation of only a small fraction of the receptor population. As in other G-protein systems, transducin is then in large excess with respect to these activated receptors. That rhodopsin is present in excess over transducin, however, has two interesting implications: (1) Conceptually, any significant "precoupling" between the inactivated receptor and the G protein is excluded a priori; the large majority of the receptors cannot be precoupled due to the insufficient number of transducin molecules. (2) Experimentally, the total pool of transducin can be made to interact with an excess of R*. This was realized and exploited by Kühn (1980) to arrive at very elegant and efficient purification procedures for transducin and is of utmost interest in studying the key step of coupling between transducin and R*.

On the effector side, the cGMP-specific PDE that is

capable of activating by transducin is also peripherally membrane-bound. Two other major classes of enzymes activated by other G proteins are adenylyl cyclases, which seem to be intrinsic transmembrane proteins, and phosphatidylinositol-specific phospholipases C, which are soluble or peripherally attached to membranes. These three types of enzymes share the common feature of acting on phosphodiester links, and in all cases, the G protein's α subunit, in its GTP-binding state, seems solely responsible for the activation. In the case of transducin, T_αGTP activates the PDE by releasing the effect of an inhibitor molecule to which it binds and detaches it from the cGMP hydrolytic units. The other types of G_α probably activate their respective effector proteins by similar processes, although the inhibitory domains might not be physically dissociated from the activated enzyme in all cases.

This study follows the path of transducin as it shuttles between rhodopsin and PDE (Fig. 1). Recent experiments are discussed at each stage, focusing mostly on the interactions between the subunits of the three main proteins of the cascade and their relationships with the membrane. Rhodopsin, the three subunits of transducin, and those of the PDE have all been cloned and sequenced, but no one, to our knowledge, has succeeded in crystallizing any of the components of this system, although most are soluble. Detailed macromolecular models have been proposed only by analogies with other systems (bacteriorhodopsin for rhodopsin; EF-Tu for T_α). Our discussions of the coupling mechanisms of transducin alternatively with rhodopsin and with PDE can therefore only be in terms of subunit or, at best, domain interactions, and for the ligands, retinal in rhodopsin and guanine nucleotide in transducin, in terms of sites whose detailed structures are not actually known. It is still mainly the logistics, the kinetics, and the energetics of the shuttling process that are being explored here.

The Resting State: Lateral Diffusion of Holotransducin Uncoupled to Rhodopsin

When no rhodopsin is photoactivated, transducin is bound to the membrane as a heterotrimeric complex, T_α GDP-$T_{\beta\gamma}$, with an unexchangeable GDP locked in the nucleotide site of T_α. The membrane attachment is very sensitive to the ionic strength and composition of the solution. It is most probably not dependent on specific binding to membrane proteins, i.e., to rhodopsin, which would be the only intrinsic species present in sufficient amount. The surface charge of rhodopsin may, however, contribute unspecifically to the binding of transducin, as do the charged lipids. Even in the physiological medium where the binding is not very tight, an equilibrium exists with solubilized transducin. This can deplete the membrane-bound pool significantly if the membrane suspension is diluted to less than 0.1 mg total protein per milliliter, but the membrane concentration is so high in situ that 99.7% of the transducin is on the membrane, and solubilized holotransducin plays no role (Bruckert et al. 1988). Magnesium ions, which are required in conjunction with the nucleotide in its binding site, are also critical for the association of T_αGDP to $T_{\beta\gamma}$ (Deterre et al. 1984), as well as for the attachment of the heterotrimeric complex to the membrane (Kühn 1981). This multiplicity of effects, which also perturb the kinetics of the light-activated process involving transducin, sometimes leads to misunderstandings as to a putative regulatory role for magnesium on transducin and other G-protein-mediated processes. One must keep in mind that the magnesium concentration is buffered in the millimolar range in vivo. Effects that are only observable far above or below this range cannot have a physiologically significant regulatory function.

The lateral mobility of transducin on the membrane may appear critical for the transducin process through

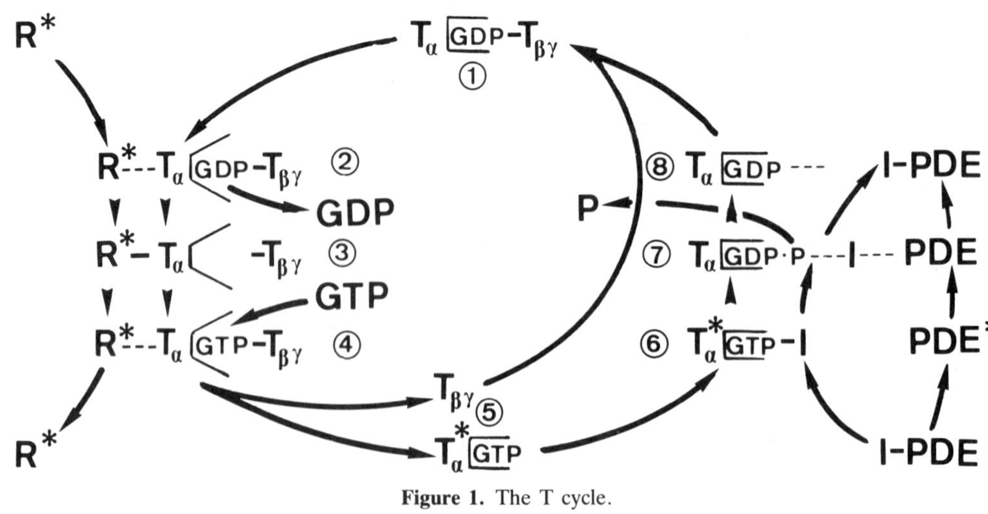

Figure 1. The T cycle.

its possible influence on the collision rate of transducins with the fewer R*s. However, it has never been measured directly, and it is often assumed implicitly that rhodopsin mobility, which is well known, is the only important one. To check this point, we have devised an experiment (F. Bruckert et al., in prep.), on the basis of the analysis of near infrared light-scattering transient observed on oriented rod suspension (Vuong et al. 1984), combined with photoexcitation by fringe interference patterns. The insensitivity of the kinetics to the fringe pattern suggests that the relative diffusion of rhodopsin with respect to transducin is not a rate-limiting factor for the rising phase of the cascade. This seems only compatible with a higher lateral mobility for transducin than which has been measured for rhodopsin. This could be expected since transducin moves in the cytoplasm on the membrane, and rhodopsin moves in the more viscous lipid medium of the membrane. It is noteworthy that rhodopsin appears fixed in quasicrystalline arrays in insect visual cells and, therefore, the G proteins must be the only mobile component. When discussing collision-coupling models for other G-protein systems, the emphasis is often only on the receptor mobility. As with the visual tranduction cascade, however, it is likely that the most mobile component is always the G protein: A G protein that is only attached to the membrane by a myristyl chain and whose entire polypeptide chain is in the cytoplasm probably has a faster lateral diffusion than that of a necessarily transmembranous receptor.

The Interaction between R* and Transducin

This capital interaction controls the amplification of the signal relayed by the receptor to the cell membrane's cytoplasmic surface. The photoexcited receptor binds to transducin and induces the opening of the nucleotide site on T_α. The formerly locked-in GDP becomes fast exchangeable. Its affinity remains on the order of 20 μM (Bennett and Dupont 1985), but it can be released and eventually replaced in the site by a new GTP. It is worth stressing that the replacement of GDP by GTP is not particularly favored over a simple GDP/GDP exchange: The kinetic binding constants of the open site for both types of nucleotides are comparable, as demonstrated by the inhibitory competition of GDP or GDPβS on the rate of GTP binding, which can be monitored on the millisecond time scale by near infrared light-scattering techniques (Vuong et al. 1984; Chabre 1985). However, the eventual binding of a GTP induces a transconformation of T_α: T_αGTP then loses its high affinity for R*, as well as for the other transducin subunit $T_{\beta\gamma}$, and it is instantly released in solution. The nucleotide site is again locked on the bound GTP, and further nucleotide exchange is prevented. What drives the reaction toward the GTP bound state is not a significantly higher affinity of the open site for GTP but the dissociation of the R*-T_αGTP-$T_{\beta\gamma}$ complex into three independent subunits, which hinders the reverse reaction. R* becomes free again to act on a new transducin, until it is inhibited by the combined action of rhodopsin kinase and arrestin (Wilden et al. 1986). The three successive stages (Fig. 1, 2–4) of this overall R* catalyzed GDP/GTP exchange reaction will be discussed in more detail, with special attention devoted to the intermediate state in which transducin is bound to R* with an open site that has lost its nucleotide.

R*-T binding and opening of the nucleotide site. The binding step may be studied under the condition that the exchange reaction is prevented by suppressing GTP from the medium: Upon illumination, each R* binds one transducin, the nucleotide site opens, GDP is released and dilutes into the medium, and the reaction can go no farther. If the number of R*s is sufficient on morphologically intact ROS (R*/R $\geq$ 1%), a modification of the disk membrane stacking in the ROS is observed, which can be monitored by light scattering as well as by neutron diffraction (Vuong et al. 1987). This is due to a change in electrostatic interaction between closely stacked disk membranes, which occurs upon the formation of the R*-T complexes. This fast signal does not originate from the change in rhodopsin conformation upon photoexcitation or from the binding to the membrane of transducins that were previously in solution, but clearly from the formation of R*-T complexes from transducins already membrane-bound. The charge modification does not simply result from the release of GDP from the open site, as it is observed even in the presence of levels of GDP or GDPβS that are high enough to prevent this release. It must therefore monitor the modification in the conformation of T_α, which is induced by its interaction with R* and results in the nucleotide site opening: Even though GDP is still present in the site in stage 2 (Fig. 1), the conformation of T_α is different from that in unbound transducin.

The tight intermediate R*-T_α-empty-$T_{\beta\gamma}$ complex. As long as GDP remains present in the open site, which is the case if a sufficient concentration of GDP is maintained in the medium, the binding of transducin to R* is not very tight. The R*-T complex is short-lived because transducins exchange fast on a given R*. However, if both GDP and GTP have been suppressed from the medium so that the "open" nucleotide site empties, the binding of T_α-empty to R* becomes very tight and the normally transient R*-T_α-empty-$T_{\beta\gamma}$ complex remains stable: Under this unphysiological condition, the system is "frozen" in an intermediate state of catalytic activation of transducin by the receptor. The binding of the two proteins is indeed extremely tight, and no significant extraction of the bound transducin is obtained upon extensive dilution and washing of the membranes in low-ionic-strength media in which GDP containing transducin would be readily soluble. Interestingly, in this strong R*-T_α-empty interaction, not only is the conformation of transducin modified by R*, but, conversely, the tight binding of T_α-empty acts on

R* at the level of the chromophore site. The active form of R* is the so-called metarhodopsin II state, in which the chromophore has isomerized to the all-*trans* conformation but is still bound in the protein site and has its spectral absorbance peak at 380 nm. This spectral state is normally transient at physiological temperature and evolves in tens of seconds toward another spectral state, metarhodopsin III, followed by the release of the chromophore from the protein part, the opsin. As observed previously (Pfister et al. 1983), the spectral decay was notably slower for R* bound to transducin as compared to free R*, but these measurements were done without totally eliminating the endogenous GDP. We have reinvestigated this problem and studied both the spectral and the functional evolutions of rhodopsin that is blocked in R*-T_{α}-empty-$T_{\beta\gamma}$ complexes (Fig. 2): After various incubation periods at 30°C, R* is liberated from the complex by the addition

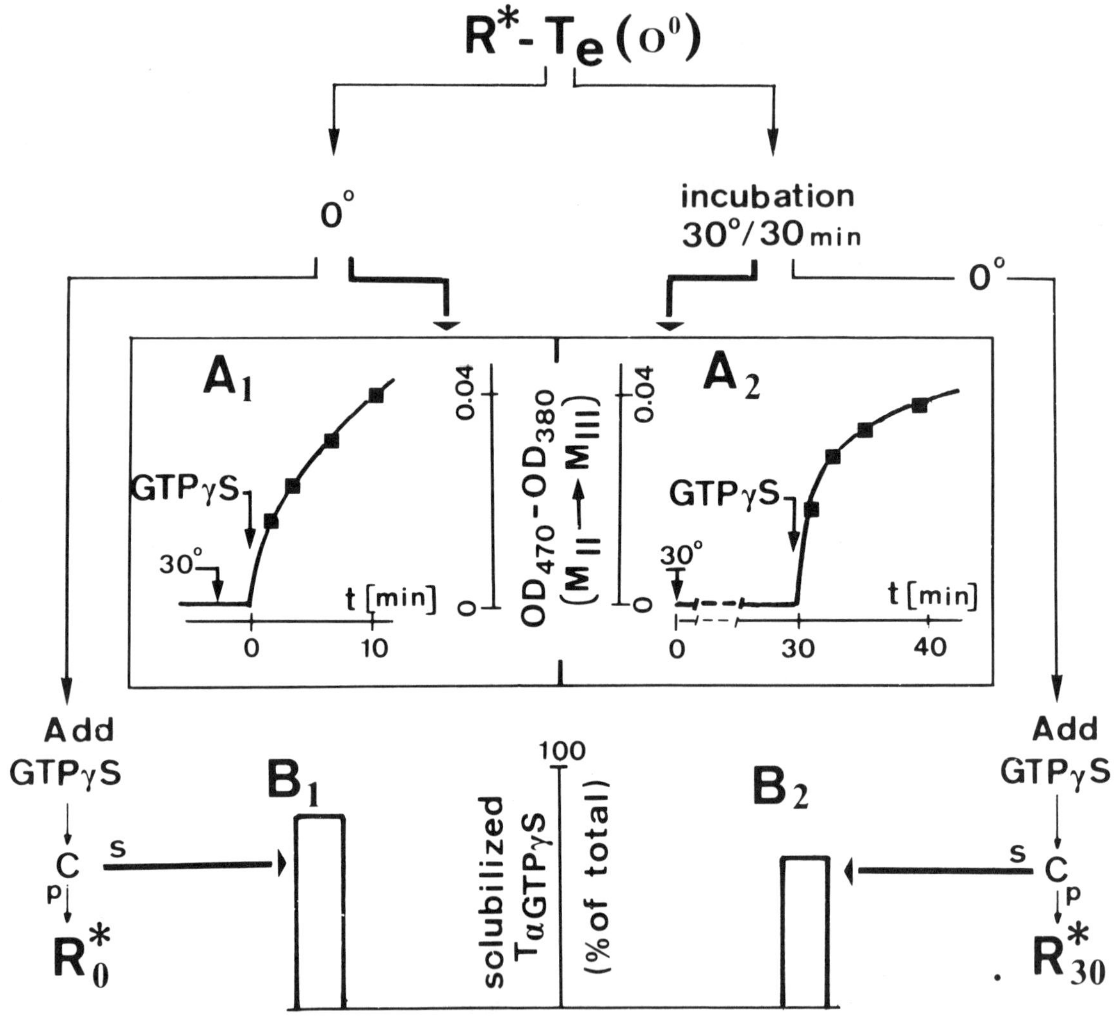

Figure 2. Perturbation of spectral and functional evolution of R* by its interaction with nucleotide-free transducin. The R*-T-empty complexes are obtained by illuminating ROS membranes to create R* in excess over the pool of transducin and by washing the membrane suspension extensively in low-ionic-strength buffer at 0°C. The pellets are resuspended in a medium-ionic-strength buffer (20 mM Tris-HCl, pH 7.4, 120 mM NaCl, 1 mM $MgCl_2$). (*A*) Spectral study. The suspension (0.5 mg/ml rhodopsin) is brought to 30°C and poured into two identical temperature-contolled cuvettes set in two beams of a Perkin-Elmer λ 5 dual beam spectrophotometer. A first fast scan of the difference spectrum determines that the absorption of the two samples is identical from 250 to 650 nM. After 1 min of incubation at 30°C (A_1), GTPγS (10 μM) is added to one of the cuvettes to liberate R* from its complex with transducin. Fast scanned difference spectra then show a decrease in absorption at 380 nm compensated by an increase at 480 nm. This is characteristic of the spectral transition of metarhodopsin II → metarhodopsin III, which occurs in the cuvette where GTPγS has been added. The time course is a few minutes for the decay of free R* from metarhodopsin II to metarhodopsin III at 30°C. The amplitude of the observed spectral difference is the same if GTPγS is added after a 30-min incubation at 30°C (A_2). This demonstrates that transducin-bound R* was maintained in the metarhodopsin II state over this incubation period. (*B*) Biochemical control of the release of transducin. Aliquots of the same ROS suspension are sedimented after GTPγS has been added after 1 or 30 min of incubation at 30°C. The released transducin is titrated in the supernatant according to the technique of Bradford and by Coomassie blue staining on gels and is normalized to the total transducin content measured by further extraction of the membrane at low ionic strength. The GTPγS-dependent release is the same in both samples.

of GTPγS, which allows the dissociation of transducin. The spectral decay of the newly liberated R* is measured by difference with that of an aliquot sample in which no GTPγS is added. One sees (Fig. 2A) that even after an incubation of 30 minutes at 30°C, which allows the total decay of all free R*, the decay pattern of the R* that had been maintained in R*-T complexes during the incubation is identical to that observed before the incubation. This proves that the T-bound R* has been maintained quantitatively in the spectral state metarhodopsin II. The retinal-binding site is located in the middle of the transmembrane core of rhodopsin, and the transducin-binding site is located on the cytoplasmic surface, because transducin is attached peripherally and does not penetrate the membrane (Fig. 3). These two distant sites therefore remain allosterically coupled after photoexcitation. Initially, it is the isomerization of retinal that induces the formation of the transducin-binding site, but later, the binding of transducin can block the release of the chromophore from its internal site.

These spectral and functional observations may seem anecdotal, as they are entirely conditioned by the artificial removal of nucleotide from the medium. Yet they are of interest when one draws on analogies made with other receptor-G-protein systems. First, one notices that in the physiological cycle, transducin is never found devoid of nucleotide unless it is tightly bound to rhodopsin. The guanine nucleotide seems necessary for the structural stability of isolated transducin: Our effort to separate T_α-empty-$T_{\beta\gamma}$ from R* resulted in very fragile and mostly denatured transducin. The spectral observation on the complex can be reworded in terms of receptor-G-protein coupling: The binding of transducin, a G protein, to R*, the equivalent of an agonist-liganded receptor, hinders the release of the chromophore that plays the role of a bound agonist. One cannot speak here of agonist affinity, as the chromophore was already bound, albeit in an inactive conformation before its isomerization by light. However, it was observed phenomenologically that the release of this agonist is slower in the absence of GTP, when the R*-T complex is blocked, than in its presence when R* is rapidly released. For a hormone receptor, a slower agonist release means a higher affinity. The GTP dependence of retinal decay in rhodopsin is therefore equivalent to the well-known GTP dependence of agonist affinity in other G-protein-coupled receptors. Here, the GTP dependence clearly results from the "postcoupling" (artificially enhanced in the absence of GTP) of transducin to already activated rhodopsin, rather than from a putative "precoupling" of transducin to inactive rhodopsin. The same is probably true in other G-protein-coupled receptor systems. The concept of the "affinity state" of a receptor for its agonist is misleading: An affinity only describes the equilibrium between the inactive receptor state, R, and the agonist liganded state R*-L. By definition, an agonist changes the conformation of its receptor, and the conformational states R and R* must be different; otherwise no signal would be transmitted. Therefore, for any hormonal or neurotransmitter receptor, the causal sequence must be as follows: (1) The ligand's binding to the receptor changes the receptor conformation, and

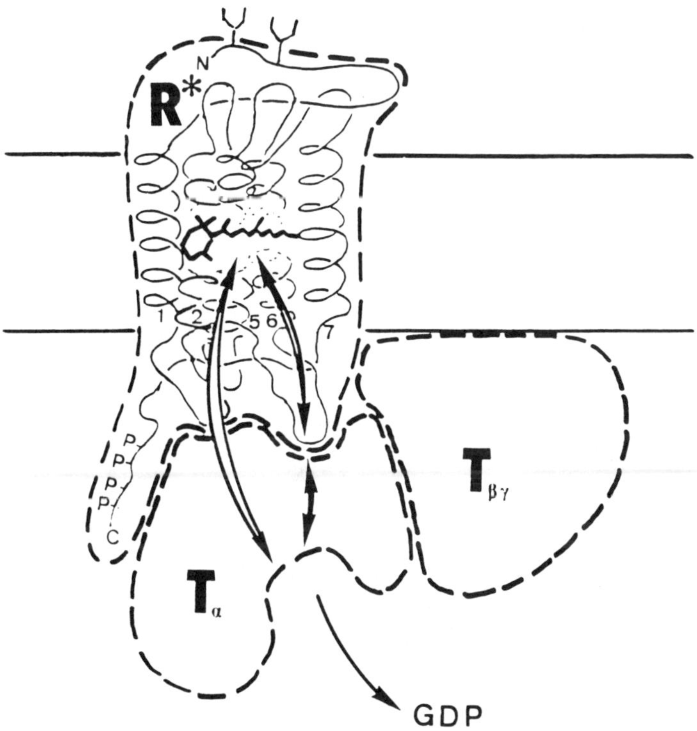

Figure 3. The R*-T-empty complex. The representation for rhodopsin takes into account what is known about its structure. The thick arrows symbolize interactions between distant sites.

(2) this induces the subsequent functional coupling of the G proteins to R*. Step 2 is a consequence of step 1, not the opposite. The so-called high affinity state observed when GTP is suppressed does not predate the binding of the agonist; it is an artificial consequence of the permanent binding of the G protein that is induced only when the agonist is added to assess its affinity (Chabre 1987b).

The "Activated" Form of T_α: T_αGTP and Its Analogs: T_αGDP-AlF_3 or T_αGDP-BeF_3

It was observed previously (Fung et al. 1981; Kühn 1981) that upon the R*-catalyzed binding of GTP, the two subunits T_αGTP and $T_{\beta\gamma}$ dissociate and become two independent entities. In a medium of ionic composition close to that of the cytoplasm, $T_{\beta\gamma}$ remains membrane-bound, but T_αGTP is released into solution. Light-scattering studies (Vuong et al. 1984; Bruckert et al., 1988) have demonstrated that this solubilization occurs in situ within less than 20 msec of the completion of the exchange reaction and therefore may be physiologically relevant as it allows the activated T_α to diffuse through the cytoplasm toward another disk membrane surface. The solubility of T_αGTP (or its hydrolysis-resistant analog T_αGTPγS) is also a good monitor of its functional state. One usually thinks of T_αGTP as the "activated" form of the G protein, because it can couple to the effector enzyme and activate it in this state. However, this does not imply that the bound GTP has transferred any free energy to the protein yet. This is clear from the fact that analogs such as GPP-NH-P or GTPγS, which cannot transfer any energy from phosphate bond cleavage, are perfectly able to activate T_α and lock the protein in this active conformation permanently.

Another artificial way of activating T_α, by fluorometallic complexes, has been elucidated recently and provides further insight into the role of GTP. Sternweis and Gilman (1982) demonstrated that the long known activating effect on G proteins of millimolar concentrations of F^- anions depends on the presence of micromolar concentrations of aluminum or beryllium etched from the glassware by the NaF solution. These workers suggested that the activator was the ion complex AlF_4^- or BeF_3^-. We measured the stoichiometric binding of one aluminum per transducin and noticed that the prerequisite for activation was the presence of GDP in the site of T_α. Analogs of GDP are also efficient, provided their β phosphate remains unsubstituted. Observing structural analogies between AlF_4^- and BeF_3^- and a phosphate group, we proposed that the fluorometallic complex binds next to the β phosphate of GDP and stimulates the presence of the γ phosphate of GTP, hence conferring its active conformation to T_α (Bigay et al. 1985). The validity of this model has now been amply confirmed (Bigay et al. 1987) and extends to other types of proteins interacting with nucleotides or simply with phosphates. For transducin and other G proteins, aluminofluoride activation bypasses the requirement for a catalyzing receptor, because the GDP does not have to be exchanged: It is "complemented" by the fluorometallic complex in the nonexchangeable site. The $T_{\beta\gamma}$ subunit is not needed either. Indeed, the addition of an excess $T_{\beta\gamma}$ hinders the activation by pushing the reversible equilibrium to the right:

$$T_\alpha\text{GDP-AlF}_3 + T_{\beta\gamma} \rightleftharpoons T_\alpha\text{GDP-}T_{\beta\gamma} + \text{AlF}_4$$

As written above, the aluminofluoride complex loses one fluoride upon entering the site so that the aluminum remains tetracoordinated by binding directly to the β-phosphate oxygen. This view is reinforced by the similar efficiency of beryllium: In contrast to aluminum, which might be penta- or hexacoordinated, beryllium is strictly tetracoordinated and must bind as such in the site (Fig. 4). That the binding of a tetrahedral beryllium confers to T_αGDP the conformation observed with bound GTP indicates that in this activated state, the γ phosphate of GTP is not constrained in a penta- or hexacovalent conformation, as often found for intermediate states of interaction between energy-rich nucleotide triphosphates and proteins. In the activated state, the γ phosphate of GTP must still

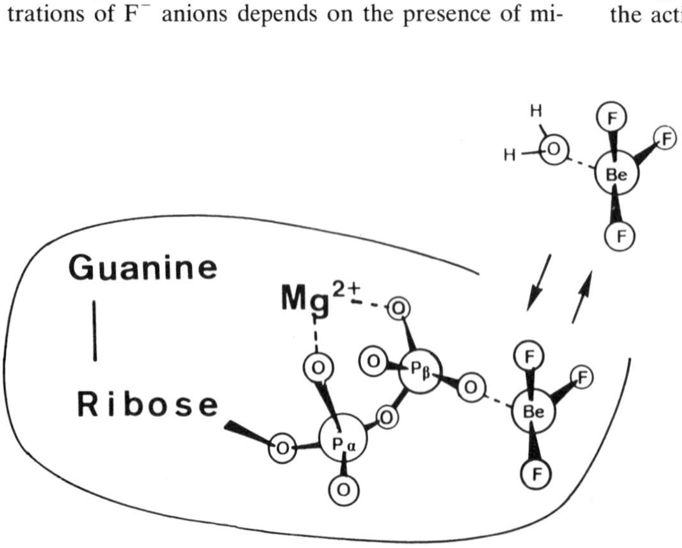

Figure 4. The reversible binding of tetrahedral BeF_3^- to GDP in the nucleotide-binding site of transducin.

be in the tetrahedral conformation of free GTP; hence, it has not yet conferred any chemical potential energy to the protein. That the binding of these phosphate analogs to T_αGDP is spontaneous and fast also confirms that the action of the receptor is purely to facilitate the release of the locked-in GDP.

Interaction of T_αGTP with the cGMP PDE

Binding of T_αGTP to the inhibitor of PDE and release of the T_α-I complex. The PDE is capable of being activated by trypsin (Miki et al. 1975), which suggests that it is maintained in its basal state by a proteolyzable inhibitor. Hurley and Stryer (1982) isolated a heat-stable 11-kD subunit (hereafter called I) that inhibits the trypsin-activated catalytic complex $PDE_{\alpha\beta}$ with high affinity. In the absence of solid data on the inhibitor's abundance, a 1:1:1 stoichiometry was usually assumed for the α, β, and I subunit in the inactive holoenzyme. The activation of PDE by T_αGTP corresponds to a release of inhibition, but it was not clear whether this involves the physical removal of the inhibitor from the catalytic complex, as for proteolytic activation. An analysis by ion-exchange chromatography of the products of interaction of purified T_αGTPγS with native I-$PDE_{\alpha\beta}$ (Deterre et al. 1987) provided strong evidence (Fig. 5) for a scheme where T_α binds to the inhibitor and forms a stoichiometric complex with it that dissociates from the catalytic units:

$$T_\alpha GTP + I\text{-}PDE_{\alpha\beta} \rightleftharpoons T_\alpha GTP\text{-}I\text{-}PDE_{\alpha\beta}$$
$$T_\alpha GTP + I\text{-}PDE_{\alpha\beta} \rightleftharpoons T_\alpha GTP\text{-}I + PDE_{\alpha\beta}$$

Gel-filtration studies of the same products suggested a molecular weight of about 50,000 and hence a 1:1 stoichiometry for the T_α-I complex. Gel filtration also demonstrated that the intermediate complex T_α-I-PDE is unstable, as no transducin is found associated with the high-molecular-weight peak containing the PDE catalytic subunits associated with the remaining inhibitor. However, this is only representative of the situation after the proteins have been extracted from the membrane by washing at low ionic strength. At physiological ionic strength, the soluble T_αGTPγS that interacts with the PDE becomes membrane-bound. As yet, we have no way of detecting whether the membrane-bound T_αGTPγS-I complex is separated physically from the activated PDE that is also on the membrane. However, the fact that a low-ionic-strength extraction is sufficient to separate T_αGTPγS-I from $PDE_{\alpha\beta}$ suggests that the inhibitor is already dissociated from the catalytic subunits on the membrane.

Two inhibitory subunits per PDE. Further analysis of the transducin-PDE interaction by ion-exchange chromatography demonstrates the existence of not one but two inhibitory subunits per native PDE (Deterre et al. 1988). The formation of T_α-I complexes should leave an equivalent amount of $PDE_{\alpha\beta}$ stripped of I. Indeed, in addition to the excess native, inactive PDE peak, not one but two peaks of active PDE were resolved. Electrophoresis gels of the peak fractions revealed that only the last peak was totally devoid of inhibitor. Quantitative analysis indicated that the I/$PDE_{\alpha\beta}$ ratio in the first active PDE peak was close to 50% of that found for native PDE. The simplest interpretation for the existence of three PDE peaks with a 2:1:0 relative amount of inhibitor for a normalized amount of $PDE_{\alpha\beta}$ is that they correspond, respectively, to inactive I_2-$PDE_{\alpha\beta}$ and to two different states of activated PDE: I-$PDE_{\alpha\beta}$ and $PDE_{\alpha\beta}$. The same two species of activated PDE are observed upon activation of native PDE by very short trypsin proteolysis, which degrades and eliminates the two inhibitory subunits successively. When tested in solution, the specific activities for the trypsin-activated species and for the corresponding transducin-activated PDE are the same. In the presence of membrane and at physiological ionic strength, however, the trypsin-activated species no longer binds to the membrane as the transducin-activated ones do, and this causes kinetic differences in their efficiencies (Wensel and Stryer 1986). This detachment from the membrane is due to the rapid cleavage by trypsin of a very short terminal peptide from the catalytic units, which precedes the removal of the inhibitory subunits (P. Deterre and P. Catty, in prep.).

The specific activity for the I-$PDE_{\alpha\beta}$ peak fractions is about half of that measured on totally stripped $PDE_{\alpha\beta}$. However, by mixing purified $PDE_{\alpha\beta}$ with native I_2-$PDE_{\alpha\beta}$, exchange of inhibitor is observed:

$$PDE_{\alpha\beta} + I_2\text{-}PDE_{\alpha\beta} \rightleftharpoons 2\,I\text{-}PDE_{\alpha\beta}$$

The possibility of this exchange prevents an unambiguous estimate of the actual activity of the intermediate complex I-$PDE_{\alpha\beta}$ as yet, because elution fractions containing initially pure I-$PDE_{\alpha\beta}$ may have dissociated and reequilibrated before the PDE activity measurement can be made.

A diffusion-controlled fast inactivation of PDE? The existence of two inhibitory subunits per PDE complex might allow for subtle modulations of PDE activity. The exchangeability of inhibitors between inactive I_2-$PDE_{\alpha\beta}$ and fully activated $PDE_{\alpha\beta}$ raises the possibility of a fast, diffusion-controlled switch-off mechanism of the PDE activity after a flash. This might bypass the "standard" inactivation due to GTP hydrolysis in T_α, which will be discussed later. All of the transducins surrounding an R* are activated within a few hundred milliseconds, in situ, and the local concentration of solubilized T_αGTP in the cytoplasmic cleft between two disks may reach 500 μM (Chabre 1985). This high concentration may be necessary to overcome the high affinity of the last inhibitor for the PDE and fully strip $PDE_{\alpha\beta}$. However, as soon as R* is blocked by phosphorylation and arrestin binding, the pool of T_αGTP dilutes into the cytoplasm and its concentration may drop too rapidly to produce any more $PDE_{\alpha\beta}$. The initially formed $PDE_{\alpha\beta}$ also diffuses away into the

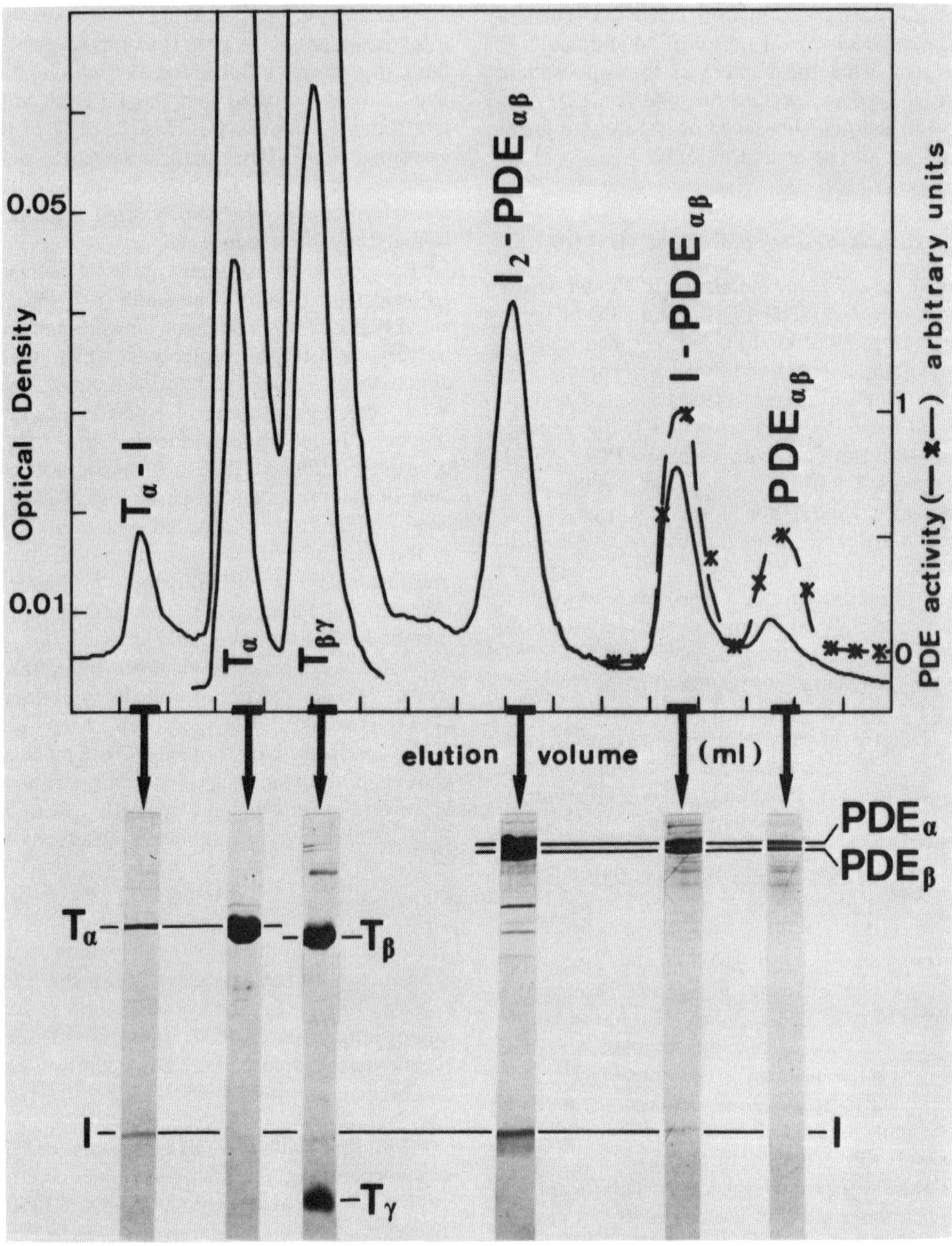

Figure 5. Ion-exchange chromatography of products resulting from the interaction between $T_{\alpha}GTP\gamma S$ and native I_2-PDE. Transducin and cGMP PDE are extracted from 10 mg of ROS membrane suspension after being made to interact by illumination in the presence of $GTP\gamma S$. The low-ionic-strength extract is loaded on an FPLC Polyanion SI column (Pharmacia) and eluted by a Na_2SO_4 gradient. The polypeptide compositions of the various peaks are shown on SDS gels under the corresponding peaks. The PDE activity of the two last peak fractions are given on the right scale.

membrane area where it encounters excess native I_2-PDE $_{\alpha\beta}$, from which it can regain one inhibitor, forming 2 I-$PDE_{\alpha\beta}$. If I-$PDE_{\alpha\beta}$ has a lower activity or a higher K_m than $PDE_{\alpha\beta}$, this process could provide a rapid quenching of the cGMP hydrolysis that would precede the permanent activation of transducin resulting from the GTP hydrolysis.

Lifetime of the Active $T_{\alpha}GTP$ State and Regeneration of Inactive Transducin

What is the significance of the GTPase rate? The "GTPase activity" stimulated by agonist-liganded receptors was the first characteristic identified of G proteins. In retinal rods in the dark, transducin has a very low basal GTPase because, as has been discussed, spontaneous GDP release from T_αGDP-$T_{\beta\gamma}$ is very slow. R* as an agonist-liganded receptor does not accelerate the GTP hydrolysis rate by increasing the intrinsic hydrolytic capacity of transducin on its bound GTP, but by accelerating the release of GDP, hence, allowing the faster binding and subsequent hydrolysis of GTP. Yet even upon strong illumination when the number of R* exceeds that of transducin, the turnover rate of GTP hydrolysis, measured in vitro remains surprisingly low: on the order of three GTPs per minute per transducin. This low turnover rate poses a major problem for the inactivation of the cascade, because at first sight, it seems to imply much too long a lifetime, of about 20 seconds, for the active T_αGTP state. This is to be compared to the shutoff time of less than 1 second for the physiological response to a flash. The hydrolytic step $T_\alpha GTP \rightarrow T_\alpha GDP + P$ is usually seen as the inactivation of T_α, because it correlates with a loss of affinity of T_α for the inhibitor of PDE, which will rebind to the catalytic units, and a recovery of its affinity for $T_{\beta\gamma}$. The association to $T_{\beta\gamma}$ is required for T_αGDP to be able to bind to R* and undertake a new cycle of GDP/GTP exchange. However, the number of $T_{\beta\gamma}$ is not rate limiting: Fung (1983) reported that the GTPase rate saturated in reconstituted systems when more than one $T_{\beta\gamma}$ was added for ten T_αs. However, the GTPase rate measured in vitro might be artifactually limited due to the dilution of $T_\alpha$in the solutions, where it is soluble in the GTP bound state: Large differences between GTPase rates measured by different groups—from one to nearly five GTPs hydrolyzed per transducin per minute—are probably related to variability in the dilutions and the ionic composition of the suspensions. Indirect measurements by light-scattering technique (Dratz et al. 1987) indeed suggest that higher membrane concentrations lead to faster GTPase rate. Another suggestion that could make the slow GTPase rate compatible with fast turnoff kinetics would be that T_αGTP becomes inactive, i.e., loses its affinity for the PDE inhibitor before the actual release of the γ phosphate of GTP. This occurs for other nucleotide-dependent systems such as actin, for example, where the fast hydrolysis, $ATP \rightarrow ADP + \text{bound P}$, is followed by a much slower release of the bound P_i (Korn et al. 1987). For transducin, the intermediate state T_αGDP-P would have lost its affinity for the PDE inhibitor but not yet regained its affinity for $T_{\beta\gamma}$. The total GTPase cycle time would then be divided in a short active time followed by a longer "dead time" during which T_αGDP-P would not be capable of reactivating. Recent data of Arshavski et al. (1987) do not favor this hypothesis, which may, however, deserve further investigation.

Measurement of T_αGTP lifetime by time-resolved microcalorimetry. One way of studying the lifetime of T_αGTP and looking for possible intermediate states during the GTPase cycle is to try to analyze the time course of enthalpy release in a population of T_αGTP formed by a short pulse of R* activation. One GTP is hydrolyzed to GDP + P and therefore 8 kcal is released for each complete cycle of GDP/GTP exchange followed by GTP hydrolysis and regeneration of T_αGDP-$T_{\beta\gamma}$. We have designed and built a fast time-resolved microcalorimeter based on the use of pyroelectric PVDF film (Fig. 6; T.M. Vuong and M. Chabre, in prep.). Each sample contains the rods extractable from two to four cattle retinas. The sensitivity of the device is on the order of 1 μcal/sec, and its impulse response, as measured by flashing an absorbing dye, is 0.5 seconds (full width at half maximum). When GTP is added to a rod suspension, illumination elicits a heat pulse that far exceeds the heat from the flash of light itself (which can be measured accurately and subtracted). The time course of the heat pulse depends not only on the lifetime of hydrolysis produced by T_αGTP, but also on the lifetime of the R*s: As long as the R*s are active, they will reload GTP on the transducins that have already hydrolyzed their first loaded GTP. To reach the real lifetime of T_αGTP, that of R* must be reduced to a negligible value. This is probably ensured in vivo by the combined action of rhodopsin kinase and arrestin that rapidly block the R*s. However, these soluble enzymes are very dilute in these broken rod preparations, making their action too slow. To hasten the decay of R*, we therefore add millimolar amounts of hydroxylamine in the suspension, which does not act on rhodopsin but is known to hydrolyze the Schiff base bond of the retinal specifically in the metarhodopsin II state (i.e., R*) and to release the chromophore rapidly, hence inactivating R*. Several representative experiments are shown in Figure 6, which demonstrate that the lifetime of the heat pulse resulting from the hydrolysis of GTP can be reduced to 3 seconds when sufficient hydroxylamine is present. Increasing the hydroxylamine to higher concentration (up to 50 mM) then reduces only the amplitude of the heat pulse, probably by destroying R* too fast to allow the loading of GTP on all of the transducins, but does not modify its time course. This 3-second limit must therefore represent the lifetime of hydrolysis of GTP in T_αGTP at 23°C. This is one order of magnitude faster than what would be expected from the known turnover rate of the GTPase, which, one should note, has always been measured in diluted membrane suspension. Our preliminary measurements of the GTPase turnover rate, as measured through the rate of GDP production upon saturating illumination in samples of the same concentration as that used for microcalorimetry, suggest a much faster rate than that published previously. Our preferred hypothesis is that dilution of the membrane sample slows down the turnover rate in vitro: The soluble T_αGTP dilutes into the suspension before GTP hydrolysis and is converted in solution into T_αGDP, which must recombine with $T_{\beta\gamma}$ on the membrane before it is able to undertake a new cycle of R*-catalyzed GDP/GTP exchange.

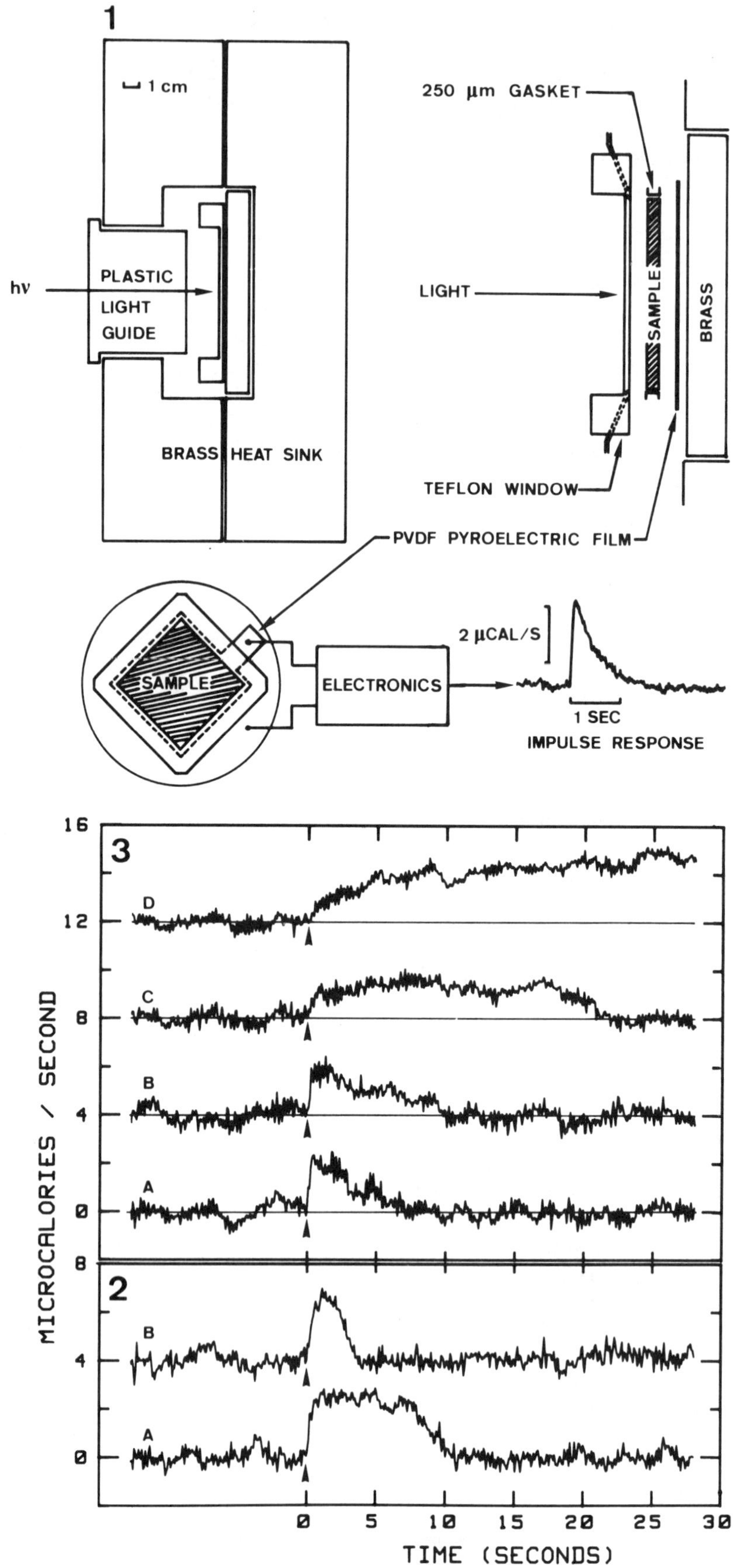

Figure 6. (*See facing page for legend.*)

CONCLUSION

The cycle of transducin shuttling between rhodopsin and PDE is more well known than that of any other G-protein system. However, the interactions between the three proteins at the molecular level are still not understood. This understanding would first require the determination of their three-dimensional structures. The analysis of the two key protein-protein interactions would also demand crystallographic studies of the two corresponding complexes: R^*-$T_\alpha$empty-$T_{\beta\gamma}$ and T_αGTP-I. The mechanism of action of the guanine nucleotide in the system must be analyzed further, perhaps by drawing not only on possible analogies with other GTP-binding proteins such as elongation factors and tubulin, but also on analogies with ATP-binding proteins like actin or other contractile proteins. Notwithstanding some peculiar characteristics related to the specific nature of the visual signal, the transducin cascade has proven and will continue to prove to be a very instructive model for G-protein transduction processes.

REFERENCES

Arshavski, Y.U., M.P. Antoch, and P.P. Philippov. 1987. On the role of transducin GTPase in the quenching of a phosphodiesterase cascade of vision. *FEBS Lett.* **224:** 19.

Bennett, N. and Y. Dupont. 1985. The G-protein of retinal rod outer segment (transducin) mechanism of interaction with rhodopsin and nucleotides. *J. Biol. Chem.* **260:** 4156.

Berg, H.C. 1983. *Random walks in biology*, p. 30. Princeton University Press, New Jersey.

Bigay, J., P. Deterre, C. Pfister, and M. Chabre. 1985. Fluoraluminates activate transducin-GDP by mimicking the γ-phosphate of GTP in its binding site. *FEBS Lett.* **191:** 181.

———. 1987. Fluoride complexes of aluminum and beryllium act on G-proteins as reversibly bound analogs of the γ-phosphate of GTP. *EMBO J.* **6:** 2907.

Bruckert, F., T.M. Vuong, and M. Chabre. 1988. Light and GTP dependence of transducin solubility in retinal rods. Further analysis by near infared light scattering. *Eur. Biophys. J.* (in press).

Chabre, M. 1985. Molecular mechanism of visual phototransduction in retinal rod cells. *Annu. Rev. Biophys. Biophys. Chem.* **14:** 331.

———. 1987a. The G protein connection: Is it in the membrane or the cytoplasm? *Trends Biochem. Sci.* **12:** 213.

———. 1987b. Receptor-G protein precoupling: Neither proven or neeeded. *Trend Neurosci.* **10:** 355.

Deterre, P., J. Bigay, C. Pfister, and M. Chabre. 1984. Guanine nucleotides and magnesium dependence of the association states of the subunits of transducin. *FEBS Lett.* **178:** 228.

Deterre, P., J. Bigay, F. Forquet, M. Robert, and M. Chabre. 1988. The cGMP phosphodiesterase of retinal rods is regulated by two inhibitory subunits. *Proc. Natl. Acad. Sci.* **85:** 2424.

Deterre, P., J. Bigay, M. Robert, C. Pfister, H. Kühn, and M. Chabre. 1987. Activation of retinal rod cyclic GMP-phosphodiesterase by transducin: Characterization of the complex formed by phosphodiesterase inhibitor and transducin α-subunit. *Proteins Struct. Funct. Genet.* **1:** 188.

Dohlman, H.G., M.G. Caron, and R.J. Lefkowitz. 1987. A family of receptors coupled to guanine nucleotide regulatory proteins. *Biochemistry* **26:** 2657.

Dratz, E.A., J.W. Lewis, L.E. Schaechter, K.R. Parker, and D.S. Kliger. 1987. Retinal rod GTP-ase turn over rate increases with concentration: A key to the control of visual excitation. *Biochem. Biophys. Res. Commun.* **148:** 379.

Fung, B.K.K. 1983. Characterization of transducin from bovine retinal rod outer segments. I. Separation and reconstitution of the subunits. *J. Biol. Chem.* **258:** 10495.

Fung, B.K.K., J.B. Hurley, and L. Stryer. 1981. Flow of information in the light-triggered cyclic nucleotide cascade of vision. *Proc. Natl. Acad. Sci.* **78:** 152.

Hurley, J.B. and L. Stryer. 1982. Purification and characterization of the γ regulatory subunit of the cyclic GMP phosphodiesterase from retinal rod outer segments. *J. Biol. Chem.* **257:** 11094.

Kühn, H. 1980. Light- and GTP-regulated interaction of GTP-ase and other proteins with bovine photoreceptor membranes. *Nature* **283:** 587.

———. 1981. Interactions of rod cell proteins with the disk membrane: Influence of light, ionic strength and nucleotides. *Curr. Top. Membr. Transp.* **15:** 171.

Korn, E.D., M.F. Carlier, and D. Pantaloni. 1987. Actin polymerization and ATP hydrolysis. *Science* **238:** 638.

Lynch, C.J., L. Morbach, P.F. Blackmore, and J.H. Exton. 1986. α-Subunits of N_s are released from the plasma membrane following cholera toxin activation. *FEBS Lett.* **200:** 333.

Miki, N., J.M. Baraban, J.J. Keirns, J.J. Boyce, and M.W. Bitensky. 1975. Purification and properties of the light-activated cyclic nucleotide phosphodiesterase of rod outer segments. *J. Biol. Chem.* **250:** 6320.

Muss, J.E., S.M. Mumby, P.J. Casey, A.G. Gilman, and B.M. Sefton. 1987. Myristylated α subunit of guanine nucleotide-binding regulatory proteins. *Proc. Natl. Acad. Sci.* **84:** 7493.

Pfister, C., H. Kühn, and M. Chabre. 1983. Interaction between photoexcited rhodopsin and peripheral enzymes in

Figure 6. Time-resolved microcalorimetry measurement of the GTPase rate of transducin. (*1*) Experimental apparatus: A 4 × 4-cm × 40-μm pyroelectric PVDF film (Solvay et Cie, Bruxelles) is glued with epoxy onto a 40-kg brass heat sink and, together with a Teflon window, makes up the sample compartment. The sample thickness of 0.25 mm is defined by a parafilm gasket. The apparatus is stabilized passively at 23°C by being enclosed in a Styrofoam box. A Plexiglas light guide permits illumination from the outside through the translucent Teflon window while also providing mechanical and acoustic isolation. The ROS suspension (4–8 mg/ml rhodopsin) is introduced via the lower opening (dotted line) of the window into the compartment, and excess air is evacuated through the upper opening. The sample volume is ~330 μl. (*2*) Data obtained for different hydroxylamine concentrations on aliquots of the same suspension. ROS concentration: 115 μM rhodopsin; photoexcitation: $R^*/R = 9.0\ 10^{-5}$; 200 μM GTP, 5 mM hydroxylamine for trace A and 20 mM for trace B. (*3*) Effects of successive flashes on one sample at low level of hydroxylamine. ROS concentration: 63 μM rhodopsin, photoexcitation: $R^*/R = 1.7\ 10^{-4}$ per flash; 300 μM GTP; 3 mM hydroxylamine. (Trace A) Average of flashes 1 and 2; (trace B) average of flashes 3 and 4; (trace C) average of flashes 6–8; (trace D) average of flashes 11–14. The first flashes produce signals that rise and decay quickly, but the signals rise more slowly and last longer with later flashes. Rhodopsin kinase (which also accepts GTP as substrate) and arrestin probably contribute to the fast quenching of R* in the first flash, but arrestin is present only at a very low concentration in these preparations and must soon be exhausted, as it remains permanently bound to R*. The hydroxylamine concentration is then too low to ensure a sufficiently fast quenching of R* by itself. The slowing down of the rising phase is not yet fully understood.

frog retinal rods. Influence of the post metarhodopsin II decay and phosphorylation rate of rhodopsin. *Eur. J. Biochem.* **136:** 489.

Sternweis, P.C. and A.G. Gilman. 1982. Aluminum: A requirement for activation of the regulatory component of adenylate cyclase by fluoride. *Proc. Natl. Acad. Sci.* **79:** 4888.

Vuong, T.M., M. Chabre, and L. Stryer. 1984. Millisecond activation of transducin in the cyclic nucleotide cascade of vision. *Nature* **311:** 659.

Vuong, T.M., C. Pfister, D.L. Worcester, and M. Chabre. 1987. The transducin cascade is involved in the light-induced structural changes observed by neutron diffraction on retinal rod outer segments. *Biophys. J.* **52:** 587.

Wensel, T.G. and L. Stryer. 1987. Reciprocal control of retinal rod cyclic GMP phosphodiesterase by its γ subunit and transducin. *Proteins Struct. Funct. Genet.* **1:** 90.

Wilden, U., S.W. Hall, and H. Kühn. 1986. Phosphodiesterase activation by photoexcited rhodopsin is quenched when rhodopsin is phosphorylated and binds the intrinsic 48-kDa protein of rod outer segments. *Proc. Natl. Acad. Sci.* **83:** 1174.

Molecular Mechanics of the Cyclic-GMP-activated Channel of Retinal Rods

J.W. KARPEN,* A.L. ZIMMERMAN,†‡ L. STRYER,* AND D.A. BAYLOR†

*Cell Biology and †Neurobiology Departments, Sherman Fairchild Science Center, Stanford University School of Medicine, Stanford, California 94305

Recent evidence indicates that the electrical response to light in retinal rod cells is generated by a member of a new class of ion channels that are directly gated by cyclic nucleotides (Fesenko et al. 1985, 1986; Nakatani and Yau 1985; Zimmerman et al. 1985; Haynes et al. 1986; Pugh and Cobbs 1986; Stern et al. 1986; Stryer 1986; Zimmerman and Baylor 1986; Matthews 1987; Matthews and Watanabe 1987; Yau and Baylor 1989). Channels somewhat like those of rods appear to participate in transduction by retinal cone cells (Cobbs et al. 1985; Haynes and Yau 1985) and olfactory receptor neurons (Nakamura and Gold 1987). In rods, channels are held open in darkness by cGMP. A reduction in cGMP concentration in light closes channels and lowers the inward Na^{+} current, producing the membrane hyperpolarization. A small Ca^{++} current through the open channel (Yau and Nakatani 1984; Hodgkin et al. 1985; Nakatani and Yau 1988) provides negative feedback control of cGMP synthesis (Lolley and Racz 1982; Pepe et al. 1986; Hodgkin and Nunn 1988; Koch and Stryer 1988).

The functional design of the channel seems well suited for its role in single photon detection (see Baylor 1987; Yau and Baylor 1989). The conductance behaves as a sensitive molecular switch that gives amplified responses to small changes in ligand concentration: At physiological concentrations, the current varies with the cube of the cGMP concentration (Haynes et al. 1986; Zimmerman and Baylor 1986; Cook et al. 1987). Although channels are continuously gated by cGMP in darkness, background electrical noise is minimized by a very low effective elementary conductance (Detwiler et al. 1982; Bodoia and Detwiler 1985; Gray and Attwell 1985; Zimmerman and Baylor 1985). Absorption of a single quantum, which produces a photocurrent about 1 pA in size (Baylor et al. 1979), is encoded by the abolition of a large number of small and brief single-channel currents, making the response highly reproducible. The functional role of the cGMP-activated conductance in phototransduction is diagramed in Figure 1.

We have studied the molecular mechanisms that underlie the functional behavior of the conductance. In particular, we have examined the kinetics of activation by cGMP as well as the mechanism of ion conduction through the channel.

‡Present address: Section of Physiology and Biophysics, Box G, Brown University, Providence, Rhode Island 02912.

METHODS

Inside-out patches were excised from the surface membrane of salamander (*Ambystoma tigrinum*) rod outer segments held by a suction electrode (Zimmerman et al. 1985) or stuck to the bottom of a chamber pretreated with 1 mg/ml polylysine. Unless otherwise indicated, the patch pipette and experimental chamber were filled with a low-divalent-cation solution (130 mM NaCl, 10 mM Tris [pH 6.8], 0.02 mM EDTA). cGMP was applied to the cytoplasmic side of the patch by chamber perfusion; in some experiments, the solution was changed by rapidly switching the patch between two streams emerging from a divided tube. The cGMP-induced current was determined from the difference in currents with and without cGMP. Patch currents were low-pass-filtered (eight-pole Bessel) and acquired directly by a PDP 11/73 computer (voltage-jump experiments) or stored by a PCM video recorder unit with a sampling rate of 44 kHz (rapid perfusion and flash-photolysis experiments). Recorded signals were replayed and analyzed with the computer.

The 4,5-dimethoxy-2-nitrobenzyl ester of cGMP ("caged" cGMP; Nerbonne et al. 1984) was obtained from Molecular Probes (Eugene, Oregon). Photolysis of caged cGMP to cGMP was driven by a Lambda Physics 102E xenon fluoride excimer laser, loaned by the San Francisco Laser Center (University of California, Berkeley). This instrument delivers 14-nsec flashes of wavelength 351 nm, with energies up to 75 mJ. Purification of the commercial caged compound, determination of the photolysis rate, and other details of the experiments with caged cGMP have been described previously (Karpen et al. 1988).

RESULTS

Lack of Desensitization in Response to cGMP

Studies of cGMP-gated ion fluxes in vesicles prepared from rod membranes have been interpreted to indicate desensitization of the cGMP-gated channel (Puckett and Goldin 1986; Koch et al. 1987), but we have not observed such an effect in excised patches.

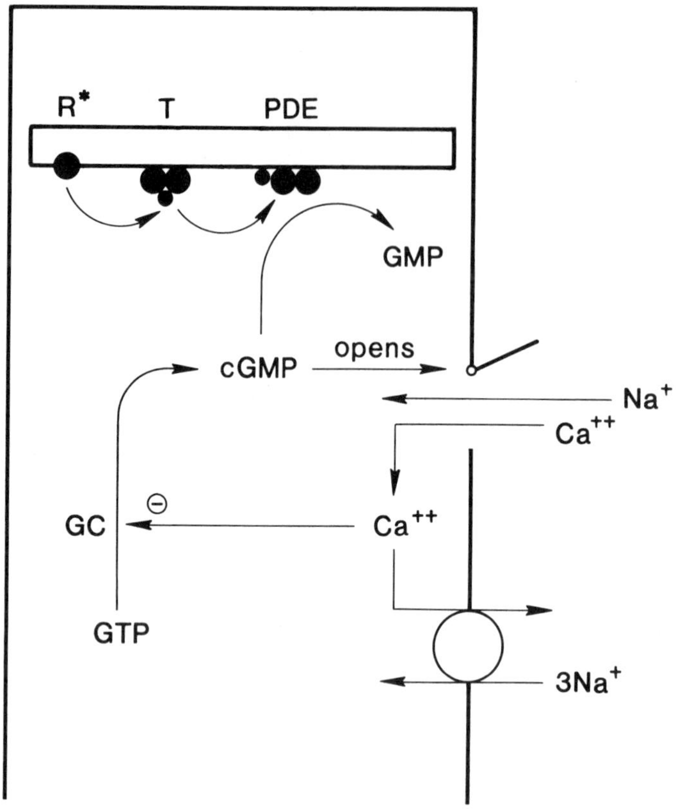

Figure 1. Role of the cGMP-gated channel in phototransduction by retinal rods. In darkness, the internal free concentration of cGMP is relatively high. cGMP binds to channels in the surface membrane of the outer segment and opens them, allowing Na^+ ions to enter. Light closes channels by activating a nucleotide cascade that lowers the concentration of cGMP: (1) Illumination of the outer segment photoisomerizes rhodopsin (R) in the disk membrane; (2) activated rhodopsin (R*) catalyzes exchange of GTP for bound GDP on many molecules of transducin (T); (3) active T, with GTP bound, removes an inhibitory constraint from the cGMP phosphodiesterase (PDE); and (4) activated PDE hydrolyzes cGMP to GMP, which cannot open the channel. Closure of channels generates a hyperpolarization that modulates the release of transmitter at the synaptic terminal of the rod. Ca^{++} also enters the rod through the open cGMP-activated channel. It is extruded by a Na/Ca exchanger, which transports three Na^+ ions inward for each Ca^{++} that moves outward. Light closes the cGMP-activated channel, blocking Ca^{++} entry. Continued extrusion by the exchanger then lowers the internal free Ca^{++} concentration. Internal Ca^{++} has little direct effect on the channel but controls it indirectly by inhibiting guanylate cyclase, which synthesizes cGMP from GTP. Ca^{++}, therefore, is a key participant in a negative feedback loop that assists recovery of the flash response and may mediate light adaptation.

When cGMP was applied by chamber perfusion, the cGMP-activated currents showed no significant decline for times of up to half an hour. In the experiment shown in Figure 2, no droop was evident even when the channels were rapidly opened with a saturating (500 μM) or half-saturating (10 μM) concentration of cGMP. At 500 μM, the response rose in 2.5 msec, and at 10 μM it rose in 35 msec. The absence of desensitization is remarkable, inasmuch as desensitization has been observed in most other ligand-controlled channels studied thus far (for review, see Aldrich et al. 1986). Lack of desensitization implies that the steady-state relation

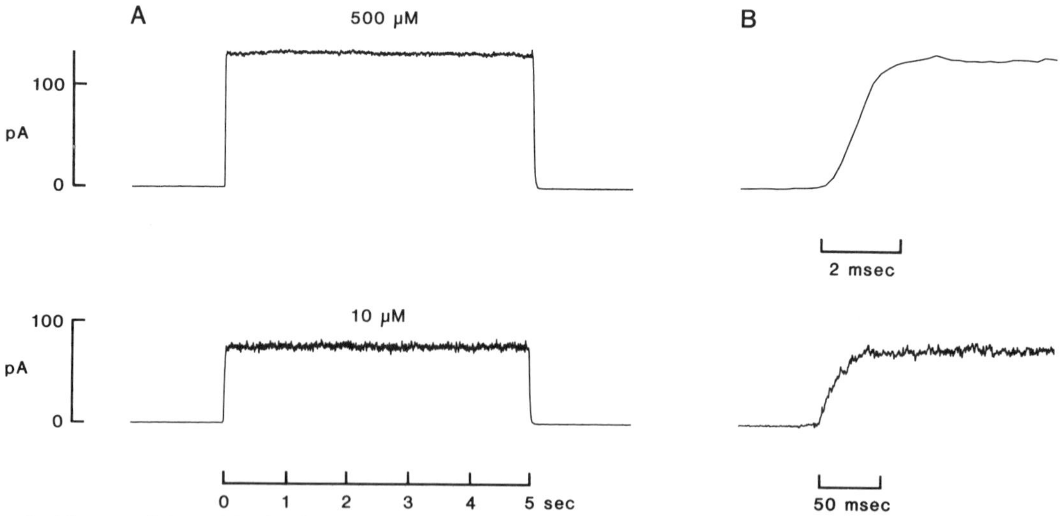

Figure 2. Absence of desensitization in the response of an excised patch to sudden application of cGMP. Patch current as a function of time. (*A*) In the record above, 500 μM cGMP was applied at time $t = 0$ by rapid perfusion. After 5 sec, the patch was switched to a solution lacking cGMP. In the lower record, 10 μM cGMP was applied. The holding potential was +20 mV, the bandwidth 0–100 Hz, and the sampling rate was 500 Hz. The bath and pipette solutions contained 130 mM NaCl, 2 mM Tris (pH 7.6), 0.02 mM EDTA. The seal resistance was 9 GΩ and the temperature was 24°C. (*B*) Rising phases of the responses in (*A*) on expanded time scales. The bandwidth was 0–1 kHz, and the sampling rate was 5 kHz.

between the free cGMP concentration and the number of channels open is independent of the history of exposure to cGMP.

Kinetics of Activation after a cGMP Concentration Jump

Rapid perfusion does not give fast enough cGMP changes to allow activation kinetics to be resolved, but flash photolysis of caged cGMP (Nerbonne et al. 1984) produces a cGMP concentration step that rises in less than 100 μsec (Karpen et al. 1988). This is much faster than the channel response over the accessible range of cGMP concentrations. Tests showed that the caged compound, as well as the aldehyde photolysis product, had no effect on the channel and that the rise in patch current after a flash was generated by the jump in cGMP concentration. Using this method, we have found that the channels open as rapidly as diffusion delivers cGMP to them. Figure 3 illustrates a response to a flash that caused the cGMP concentration to jump to 9.7 μM, as estimated from the patch's steady-state dose-response relation. The semilog plot in the lower part of Figure 3 shows that the response rose exponentially with a time constant of 2.1 msec. This is comparable to the time required for a molecule the size of cGMP to diffuse to a binding site.

After reaching the peak, the response slowly declined. We attribute the decline to diffusion of cGMP away from its site of creation in the excised patch (Karpen et al. 1988). The patch consists of plasmalemma and adherent membranous material (Zimmerman et al. 1988) into which caged cGMP partitions (Nerbonne et al. 1984); a flash will liberate cGMP, a charged species, at high concentration in the aqueous phase near the plasmalemma. Several other possible explanations for the decline were tested and ruled out (Karpen et al. 1988); desensitization is unlikely for the reasons described above.

If diffusion of cGMP to binding sites on the channels limits the rise of the photolysis response, the rise should be faster for higher peak concentrations. This was observed in several patches, when various concentrations of cGMP were added to the caged compound bathing the patch. At concentrations between 5 and 18 μM, the slope of the relation between reciprocal rise time and peak cGMP concentration was about $5 \times 10^7\ \text{M}^{-1}\text{sec}^{-1}$. We interpret this as a lower limit for any bimolecular rate constant in the activation of the channel by cGMP. Some patches had somewhat slower kinetics, possibly because of hindered diffusion of cGMP near the membrane (Zimmerman et al. 1988).

Kinetics of Responses to Membrane Voltage Jumps

Although channel gating has rather little dependence on membrane voltage, changes in voltage weakly perturb activation by cGMP and thus provide a useful tool for investigating the kinetics of activation. An advantage of this method, which gave results very similar to those obtained by flash photolysis, is that activation kinetics can be examined over the full range of cGMP concentrations. With caged cGMP, it was difficult to resolve responses at ligand concentrations higher than roughly 20 μM. The voltage dependence of activation is illustrated by the steady-state dose-response relations plotted in Figure 4. The relation at +50 mV lies to the left of that at −50 mV, indicating that activation was more effective at the positive potential. The respective half-saturating concentrations of cGMP were 12 and 16 μM. The continuous curves were calculated from the kinetic model shown below.

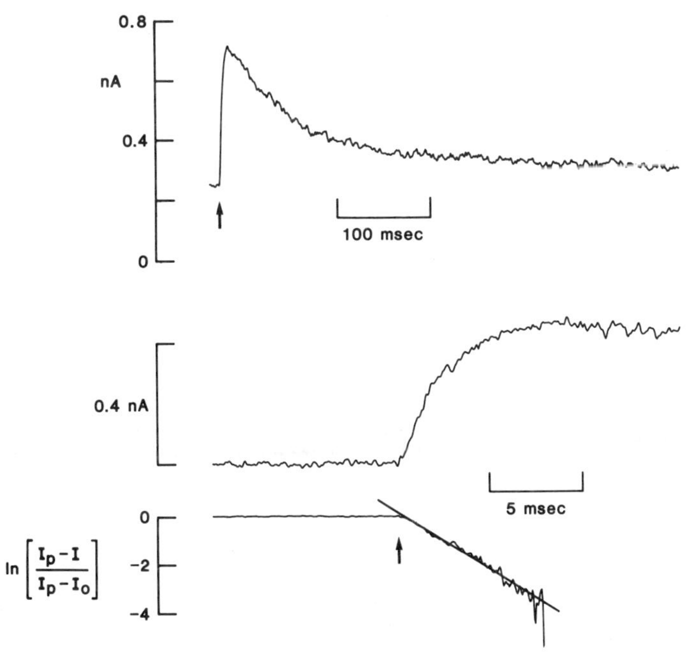

Figure 3. Change in patch current after flash photolysis of caged cGMP. The top and middle traces are the same response plotted on different time scales. (*Top*) Bandwidth 0–300 Hz, 2-kHz sampling; (*middle*) bandwidth 0–3 kHz, 20-kHz sampling; (→) time of flash. Solution contained 250 μM caged cGMP with no added cGMP. Holding potential was +50 mV; response to saturating cGMP was 1068 pA; and seal resistance was 10 GΩ. (*Bottom*) Semilog plot of rising phase of response. I is the current, I_o is the initial value, and I_p is the peak value. Straight line gives the rise time constant as 2.1 msec. (Reprinted, with permission, from Karpen et al. 1988.)

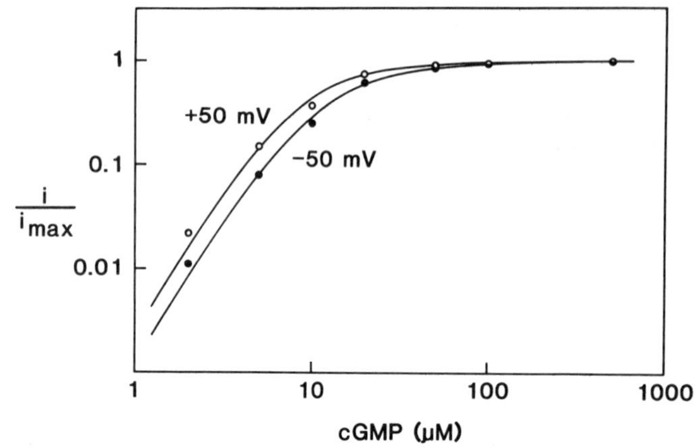

Figure 4. Double logarithmic plot of dose-response relations for cGMP at two voltages (patch of Figs. 5–7). Steady-state currents at +50 and −50 mV (pulse protocol of Fig. 5) were normalized by the saturating value (i_{max}). Continuous curves were calculated from the model in the text. (Reprinted, with permission, from Karpen et al. 1988.)

Activation kinetics were investigated by applying different steady concentrations of cGMP to the patch and switching the voltage between −50 and +50 mV while observing the current (Fig. 5). To determine the kinetics of channel gating after the voltage was switched, current transients were corrected for changes in ionic concentration that occurred near the membrane due to hindered diffusion at the cytoplasmic patch surface (Zimmerman et al. 1988). Gating transients obtained in this way are illustrated in Figure 6, which plots the normalized change in the number of open channels as a function of time after switching the voltage from −50 to +50 mV. The speed of channel activation depended strongly on cGMP concentration, increasing by roughly two orders of magnitude as the cGMP concentration increased from 2 to 50 μM. At high concentration, a limiting rate was reached; the transients at 50 and 500 μM rose with a time constant of about 60 μsec. Because the effect of cGMP on the channel is saturated at these concentrations, we interpret the fast relaxations to arise

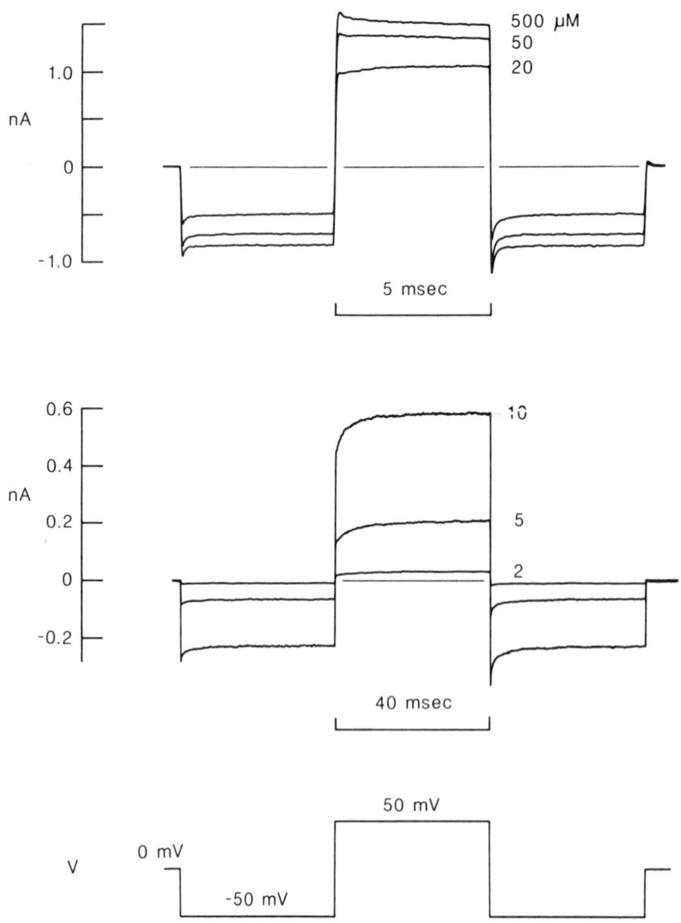

Figure 5. cGMP-activated currents during the voltage pulse protocol shown at the bottom of the figure, with cGMP concentrations as indicated. Each trace was averaged from 14–30 trials. Upper set of traces were low-pass-filtered at 6 kHz and sampled at 62 kHz. Similar kinetics were obtained from three other patches with 10 kHz filtering and 100 kHz sampling; hence, responses were not instrumentally limited. Middle set of traces were filtered at 3 kHz and sampled at 16 kHz, seal resistance 5 GΩ. (Reprinted, with permission, from Karpen et al. 1988.)

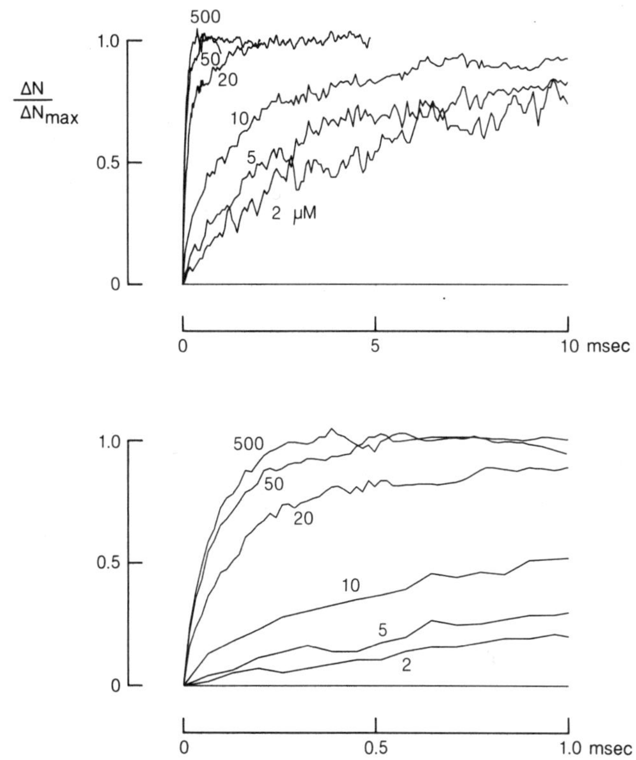

Figure 6. Channel relaxation kinetics as a function of cGMP concentration when voltage was switched from −50 to +50 mV. Fractional change in current vs. time at different cGMP concentrations on two time bases. Transients were corrected for changes in ionic concentration at the surface membrane (Zimmerman et al. 1988). (Reprinted, with permission, from Karpen et al. 1988.)

from voltage-dependent open-closed transitions in the fully liganded channel.

A working scheme of the activation process is shown below and compared with experimental results in Figure 7.

$$C + G \underset{4}{\overset{1}{\rightleftarrows}} GC + G \underset{5}{\overset{2}{\rightleftarrows}} G_2C + G \underset{6}{\overset{3}{\rightleftarrows}} G_3C \;\; {}_{8}\!\uparrow\!\downarrow_{7} \;\; G_3O$$

Here, C stands for the closed channel, G is cGMP, and O denotes the open channel. The microscopic rate constants (which for binding steps apply to a single site) are: k_1, k_2, and $k_3 = 1 \times 10^8$ M^{-1} sec^{-1}; k_4 and $k_5 = 1.8 \times 10^3$ sec^{-1}; $k_6 = 8 \times 10^2$ sec^{-1}; $k_7 = 1.44 \times 10^4$ sec^{-1}; $k_8 = 2.3 \times 10^3$ sec^{-1} (value at +50 mV, *e*-fold increase in −123 mV). At low concentrations of cGMP, the voltage-jump kinetics varied by as much as 50% between different patches, perhaps because of differences in the effective diffusion coefficient for cGMP (Zimmerman et al. 1988). Initially (Karpen et al. 1988), we attributed the entire voltage dependence of activation to the rate constant for channel closing. Subsequent analysis suggests that there may in addition be a weaker voltage dependence to the opening rate constant.

Relation of Gating and Permeation

Previous work has shown that the channel consists of aqueous pores that are strongly blocked under physiological conditions by external divalent cations, which carry small inward currents (Haynes et al. 1986; Zimmerman and Baylor 1986). Work described elsewhere (Zimmerman and Baylor 1988 and in prep.) suggests that the channel contains a single site to which cations must bind in order to permeate. Only one cation occupies the site at a time, and divalents bind more strongly than monovalents. Movements of ions to and from the binding site are voltage-dependent because the site is located within the transmembrane electric field.

A fundamental question is whether the channel can close with an ion bound inside it. Under physiological conditions, open channels are blocked 99% of the time by divalents. If a blocked channel cannot close, blockage should slow the response to a drop in free cGMP by a factor of 100. This does not appear to be the case. In experiments on intact rods, Cobbs and Pugh (1987) found that after a very bright flash, channels close with a 2-msec time constant. This is comparable to estimates from our kinetic experiments, which were performed without divalents and in which the channel was empty roughly half the time.

Evidence that a divalent cation can be bound in the closed channel is presented in Figure 8. The degree of blockage of Na^+ currents through the channel is plotted as a function of internal Mg^{++} concentration at low and high concentrations of cGMP. On these axes, blockage at a single site gives a straight line whose slope is the apparent association constant for the blocker. If only open channels can contain an Mg^{++} (model 1), blockage should be more effective at high cGMP concentration, where more channels are open. If Mg^{++} is

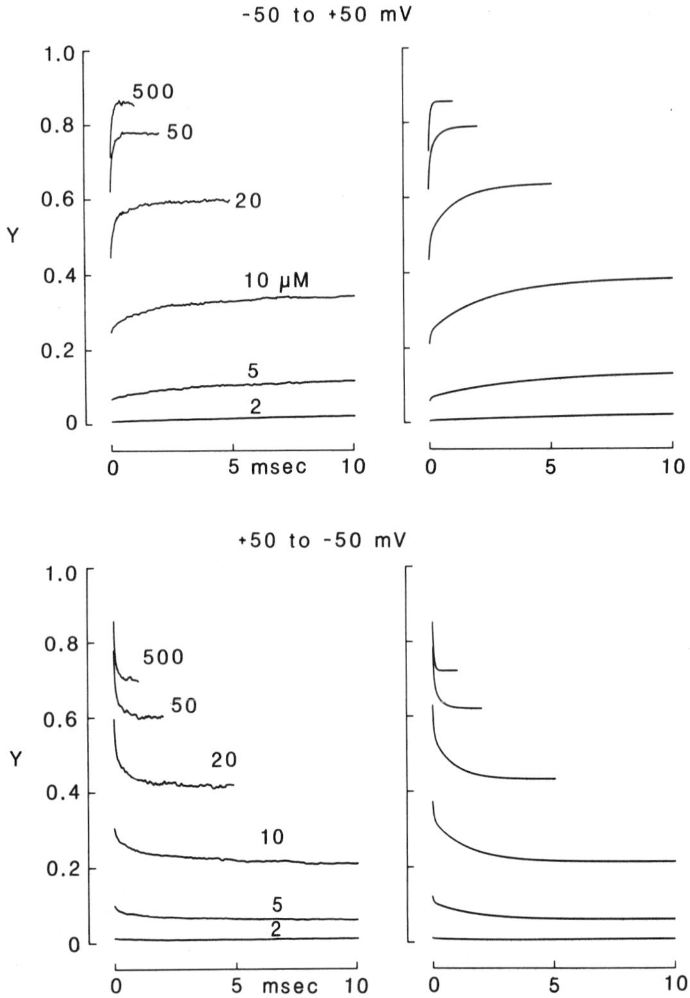

Figure 7. Comparison of voltage-jump relaxations with the predictions of a simple kinetic scheme. Traces at the left are current relaxations corrected for ionic changes and plotted as fraction of channels open vs. time; cGMP concentrations are indicated. Fraction of channels open was obtained by multiplying $i/[i_{max}$ at 500 μM cGMP] for each final voltage by the calculated fraction of channels open at 500 μM cGMP and +50 mV $[k_7/(k_7 + k_8]$; see scheme in text). Traces on the right are calculated from the scheme. (*Upper*) Change from −50 to +50 mV (same data as Fig. 6). (*Lower*) Change from +50 to −50 mV. (Reprinted, with permission, from Karpen et al. 1988.)

bound in open and closed channels equally strongly (model 2), the efficiency of blockage will be independent of cGMP concentration. Assuming an activation mechanism like that shown above, model 1 predicts that the ratio of currents in the absence and presence of Mg^{++} is given by

$$I/I_{Mg} = 1 + KF[Mg] \tag{1}$$

whereas for model 2 it is given by

$$I/I_{Mg} = 1 + K[Mg] \tag{2}$$

The constant K in the expressions above is the association constant for binding of Mg^{++} to all states of the channel (model 2) or to the open state (model 1); F is the fraction of open channels. In Equation 1, the apparent association constant KF is strongly dependent on the cGMP concentration, whereas in Equation 2 it is independent of the cGMP concentration. Given the observed blockage at 5 μM cGMP, model 1 predicts that the blockage at 100 μM cGMP should follow the dashed line in Figure 8. The blockage actually observed at 100 μM cGMP is slightly weaker than that at 5 μM. This behavior suggests that Mg^{++} is bound in the closed channel.

DISCUSSION

The scheme for cGMP activation presented above seems to be the simplest that can provide a reasonable description of the experimental results, but we regard it as incomplete. For example, the two conductance states observed in single-channel recordings (Haynes et al. 1986; Zimmerman and Baylor 1986) will have to be accounted for when more information becomes available. An attractive scheme that may be appropriate for explaining such behavior is a concerted allosteric model in which the mean conductance depends on the number of ligand molecules bound (Stryer 1987). It will also be important to determine the values for individual rate constants by other methods.

Several model-independent conclusions can be drawn. The speed of activation by cGMP points to a mechanism that involves the diffusion-controlled binding of ligand to a preformed channel unit in the membrane. Mechanisms involving the aggregation of protein subunits to form a conducting channel require unrealistically large subunit densities to account for the kinetics. Activation appears to involve at least three cGMP-binding reactions with bimolecular rate con-

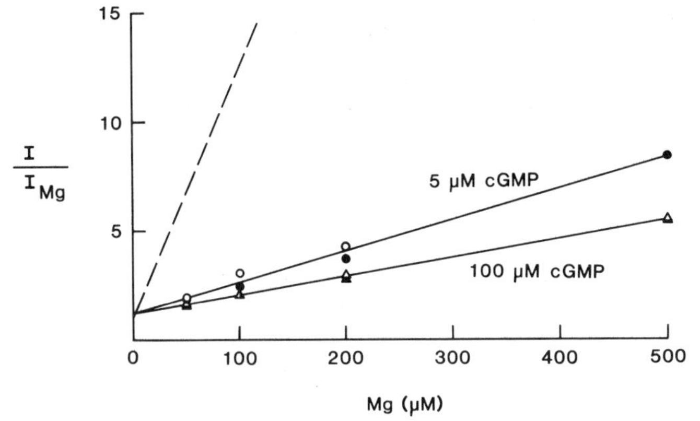

Figure 8. Evidence that closed channels can contain a Mg ion. Blockage of Na currents by internal Mg at different cGMP concentrations. The ratio of currents in the absence and presence of Mg is plotted vs. free Mg concentration. Steady-state currents were measured at +30 mV. Data are from one patch. (○,●) Data at 5 μM cGMP; (△,▲) data at 100 μM cGMP. Open and closed symbols indicate data collected at two different times during the experiment. (—) data obtained by linear regression; (- - - -) the expected behavior at 100 μM cGMP, based on the data at 5 μM cGMP, for a model in which Mg is only bound to the open channel (see text). The solution in the patch pipette and experimental chamber contained 130 mM NaCl, 5 mM sodium phosphate (pH 7.6), and 0.5 mM EDTA. $MgCl_2$ was added to the chamber perfusion solutions at the following concentrations (calculated free Mg concentrations in parentheses): 540.5 μM (50 μM); 595.2 μM (100 μM); 697.6 μM (200 μM); 999 μM (500 μM). Seal resistance was 4 GΩ.

stants near the diffusion-controlled limit. Fully liganded channels undergo very rapid, voltage-dependent open-closed transitions. Under physiological conditions, the channel's gating kinetics will be limited by the cGMP-binding steps, rather than the open-closed transitions.

The channel seems designed for speed of response. Rod responses to single photons will be limited by the kinetics of the enzymatic events that change the intracellular concentration of cGMP, rather than by channel gating. At first sight, the channel seems "over-engineered" in this respect, because the difference in time scales of channel gating and photon response is so large (roughly two orders of magnitude in amphibian rods). Rapid gating kinetics are useful in another respect, however, because they help the single photon effect to stand out from channel noise in darkness. The outer segment appears to contain a large number of channels, each with a small probability (about 2%) of being open in darkness. The number N_o of channels open in darkness therefore will be a Poisson-distributed random variable. If a photon blocks a fixed fraction, b, of the open channels for a mean time t_i, the signal/noise ratio for the photon effect (total number of channel openings prevented divided by standard deviation of dark fluctuation) will be

$$S/N = b\left(\frac{t_i}{\tau}\right)^{1/2} (\bar{N}_o)^{1/2}, \qquad (3)$$

where τ is the mean channel open time and $\bar{N}_o$ is the mean number of channels open in darkness. The difference in the time scales of single-channel current and the photon effect thus can be viewed as improving S/N by a factor of $(t_i/\tau)^{1/2}$. For $t_i = 1$ sec (Baylor et al. 1979) and τ = 2.5 msec, S/N is raised by a factor of 20. To realize this improvement, the rod's output should be integrated for a characteristic time t_i.

The results in Figure 8 suggest than an ion can be bound inside a closed channel. Further work is required to determine whether such a bound ion is accessible to one of the external solutions. Miller et al. (1987) have obtained evidence that a Ba^{++} ion can be trapped inside a Ca^{++}-activated K^+ channel when it is closed. The ability of the cGMP-gated channel to close on a divalent cation preserves rapid gating kinetics in the face of requirements for a low effective unit conductance to reduce quantizing noise and for a small inward current of Ca^{++} to regulate the gain of the cGMP cascade.

ACKNOWLEDGMENTS

This work was supported by grants EY01543 and EY02005 from the National Eye Institute and GM-30387 from the National Institute of General Medical Sciences. J.W.K. is the recipient of National Research Service award EY059020. We thank Dr. Markus Meister for comments on the manuscript and R. Schneeveis for excellent technical assistance.

REFERENCES

Aldrich, R.W., V.E. Dionne, E. Hawrot, and C.F. Stevens. 1986. Ion transport through ligand-gated channels. In *Physiology of membrane disorders* (ed. T.E. Andreoli et al.), p. 491. Plenum Press, New York.

Baylor, D.A. 1987. Photoreceptor signals and vision. *Invest. Opthal. Visual Sci.* **28:** 34.

Baylor, D.A., T.D. Lamb, and K.-W. Yau. 1979. Responses of retinal rods to single photons. *J. Physiol.* **288:** 613.

Bodoia, R.D. and P.B. Detwiler. 1985. Patch-clamp recordings of the light-sensitive dark noise in retinal rods from the lizard and frog. *J. Physiol.* **367:** 183.

Cobbs, W.H. and E.N. Pugh, Jr. 1987. Kinetics and components of the flash photocurrent of isolated retinal rods of the larval salmander, *Ambystoma tigrinum. J. Physiol.* **394:** 529.

Cobbs, W.H., A.E. Barkdoll III, and E.N. Pugh, Jr. 1985. Cyclic GMP increases photocurrent and light sensitivity of retinal cones. *Nature* **317:** 64.

Cook, N.J., W. Hanke, and U.B. Kaupp. 1987. Identification, purification, and functional reconstitution of the cyclic GMP-dependent channel from rod photoreceptors. *Proc. Natl. Acad. Sci.* **84:** 585.

Detwiler, P.B., J.D. Conner, and R.D. Bodoia. 1982. Gigaseal patch clamp recordings from outer segments of intact retinal rods. *Nature* **300:** 59.

Fesenko, E.E., S.S. Kolesnikov, and A.L. Lyubarsky. 1985. Induction by cyclic GMP of cationic conductance in plasma membrane of retinal rod outer segment. *Nature* **313:** 310.

———. 1986. Direct action of cGMP on the conductance of retinal rod plasma membrane. *Biochim. Biophys. Acta* **856:** 661.

Gray, P. and D. Attwell. 1985. Kinetics of light-sensitive channels in vertebrate photoreceptors. *Proc. R. Soc. Lond. B. Biol. Sci.* **223:** 379.

Haynes, L. and K.-W. Yau. 1985. Cyclic GMP-sensitive conductance in outer segment membrane of catfish cones. *Nature* **317:** 61.

Haynes, L.W., A.R. Kay, and K.-W. Yau. 1986. Single cyclic GMP-activated channel activity in excised patches of rod outer segment membrane. *Nature* **321:** 66.

Hodgkin, A.L. and B.J. Nunn. 1988. Control of light-sensitive current in salamander rods. *J. Physiol.* **403:** 439.

Hodgkin, A.L., P.A. McNaughton, and B.J. Nunn. 1985. The ionic selectivity and calcium dependence of the light-sensitive pathway in toad rods. *J. Physiol.* **358:** 447.

Karpen, J.W., A.L. Zimmerman, L. Stryer, and D.A. Baylor. 1988. Gating kinetics of the cyclic-GMP-activated channel of retinal rods: flash photolysis and voltage-jump studies. *Proc. Natl. Acad. Sci.* **85:** 1287.

Koch, K.-W. and L. Stryer. 1988. Highly cooperative feedback control of retinal rod guanylate cyclase by calcium ion. *Nature* **334:** 64.

Koch, K.-W., N.J. Cook, and U.B. Kaupp. 1987. The cGMP-dependent channel of vertebrate rod photoreceptors exists in two forms of different cGMP sensitivity and pharmacological behavior. *J. Biol. Chem.* **262:** 14415.

Lolley, R.N. and E. Racz. 1982. Calcium modulation of cyclic GMP synthesis in rat visual cells. *Vision Res.* **22:** 1481.

Matthews, G. 1987. Single-channel recordings demonstrate that cGMP opens the light-sensitive ion channel of the rod photoreceptor. *Proc. Natl. Acad. Sci.* **84:** 299.

Matthews, G. and S.-I. Watanabe. 1987. Properties of ion channels closed by light and opened by guanosine 3′,5′-cyclic monophosphate in toad retinal rods. *J. Physiol.* **389:** 691.

Miller, C., R. Lattore, and I. Reisin. 1987. Coupling of voltage-dependent gating and Ba^{++} block in the high-conductance, Ca^{++}-activated K^{+} channel. *J. Gen. Physiol.* **90:** 427.

Nakamura, T. and G.H. Gold. 1987. A cyclic nucleotide-gated conductance in olfactory receptor cilia. *Nature* **325:** 442.

Nakatani, K. and K.-W. Yau. 1985. cGMP opens the light-sensitive conductance in retinal rods. *Biophys. J.* **47:** 356a.

———. 1988. Calcium and magnesium fluxes across the plasma membrane of the toad rod outer segment. *J. Physiol.* **395:** 695.

Nerbonne, J.M., S. Richard, J. Nargeot, and H.A. Lester. 1984. New photoactivatable cyclic nucleotides produce intracellular jumps in cyclic AMP and cyclic GMP concentrations. *Nature* **310:** 74.

Pepe, I.M., I. Panfoli, and C. Cugnoli. 1986. Guanylate cyclase in rod outer segments of the toad retina. *FEBS Lett.* **203:** 73.

Puckett, K.L. and S.M. Goldin. 1986. Guanosine 3′,5′-cyclic monophosphate stimulates release of actively accumulated calcium in purified disks from rod outer segments of bovine retina. *Biochemistry* **25:** 1739.

Pugh, E.N., Jr. and W.H. Cobbs. 1986. Visual transduction in vertebrate rods and cones: A tale of two transmitters, calcium and cyclic GMP. *Vision Res.* **26:** 1613.

Stern, J.H., U.B. Kaupp, and P.R. Macleish. 1986. Control of the light-regulated current in rod photoreceptors by cyclic GMP, calcium, and l-*cis*-diltiazem. *Proc. Natl. Acad. Sci.* **83:** 1163.

Stryer, L. 1986. Cyclic GMP cascade of vision. *Annu. Rev. Neurosci.* **9:** 87.

———. 1987. Visual transduction: Design and recurring motifs. *Chem. Scri.* **27B:** 161.

Yau, K.-W. and D.A. Baylor. 1989. Cyclic GMP-activated conductance of retinal photoreceptor cells. *Annu. Rev. Neurosci.* **11:** (in press).

Yau, K.-W. and K. Nakatani. 1984. Cation selectivity of light-sensitive conductance in retinal rods. *Nature* **309:** 352.

Zimmerman, A.L. and D.A. Baylor. 1985. Electrical properties of the light-sensitive conductance of salamander retinal rods. *Biophys. J.* **47:** 357a.

———. 1986. Cyclic GMP-sensitive conductance of retinal rods consists of aqueous pores. *Nature* **321:** 70.

———. 1988. Ionic permeation in the cGMP-activated channel of retinal rods. *Biophys. J.* **53:** 472a.

Zimmerman, A.L., J.W. Karpen, and D.A. Baylor. 1988. Hindered diffusion in excised membrane patches from retinal rod outer segments. *Biophys. J.* **54:** 351.

Zimmerman, A.L., G. Yamanaka, F. Eckstein, D.A. Baylor, and L. Stryer. 1985. Interaction of hydrolysis-resistant analogs of cyclic GMP with the phosphodiesterase and light-sensitive channel of retinal rod outer segments. *Proc. Natl. Acad. Sci.* **82:** 8813.

Inositol Lipid Cascade of Vision Studied in Mutant Flies

Z. Selinger* and B. Minke†

*Departments of *Biological Chemistry and †Physiology, The Hebrew University of Jerusalem, Jerusalem 91904, Israel*

In the process of phototransduction, photoreceptor cells translate light stimuli into neuronal signals. A large body of evidence indicates that phototransduction is mediated by a cascade of enzyme reactions with feedback regulatory loops (Payne 1986; Stryer 1986). A common strategy to unravel such complex pathways has been to dissect the system by single point mutations, each affecting a discrete step of the enzyme cascade. With remarkable foresight, genetic analysis has been applied to the visual system of *Drosophila* flies, yielding over 100 photoreceptor potential mutants. These mutants fall into 15 mutual complementation groups, thus setting a minimal number of gene products that are involved in the phototransduction process (for review, see Pak 1979).

It is now well established that cGMP is the internal messenger of visual transduction in vertebrate rods and that a guanine-nucleotide-binding protein (the G protein transducin) couples the photoexcited rhodopsin to cGMP hydrolysis (for review, see Stryer 1986). Much less is known about phototransduction in the invertebrate photoreceptors, and even the identity of the second messenger of phototransduction is still in dispute. This is because both inositol trisphosphate ($InsP_3$) (Brown et al. 1984; Fein et al. 1984) and cGMP (Johnson et al. 1986) have been found to excite the *Limulus* ventral photoreceptors. We have initiated combined electrophysiological and biochemical studies of phototransduction in *Drosophila, Lucilia,* and *Musca* flies, since the extensive genetic and physiological background of *Drosophila* mutants can be used to assign physiological roles to biochemical reactions that unavoidably must be measured under in vitro conditions that disrupt the spatial arrangement of the photoreceptor cell. Furthermore, a particular phenotype in a mutant can direct us to explore parts of the phototransduction system that would otherwise remain unnoticed.

The approach that we use is to introduce into the photoreceptors products, activators, and inhibitors of putative intermediate steps in phototransduction and to analyze their electrophysiological response in the intact photoreceptor cell. Complementary biochemical studies are conducted in cell-free membrane preparations to identify light-dependent reactions, which are attenuated or missing in conditional and null mutants, respectively. Application of this interdisciplinary approach to *Drosophila* photoreceptor potential mutants should yield a comprehensive and detailed picture of phototransduction in the fly photoreceptor cell.

Photoexcited Rhodopsin Activates a G Protein

In vertebrate rods, photoisomerization of rhodopsin leads to dissociation of the retinal chromophore from the protein opsin. Regeneration of rhodopsin is achieved only after an extended period by a series of chemical reactions, some of which are carried out outside the photoreceptor cell (Dowling 1960). In contrast, in the invertebrates, absorption of light by rhodopsin (R) converts R to the stable metarhodopsin (M), whereas light absorption by metarhodopsin regenerates M back to R. In flies, R and M have maximal absorptions at ~490 nm and ~570 nm, respectively (Hamdorf 1979) (Fig. 1B). Thus, by the illumination with either blue (converting) or red (regenerating) lights, the content of M, the active form of the photopigment, can be easily manipulated. We used intense illumination, which brings the photopigment to photoequilibrium between the R and M states, to measure the fractional amount of M at each wavelength of light (the photoequilibrium spectrum of M) (Hillman et al. 1983). The fractional amount of M is given by the following equation.

$$f_M(\lambda) = \frac{\alpha_R(\lambda)}{\alpha_R(\lambda) + \alpha_M(\lambda) \cdot \phi}$$

$\alpha_R(\lambda)$ and $\alpha_M(\lambda)$ are the molecular extinction coefficients of R and M, respectively, and ϕ is the ratio of the quantum efficiencies of M and R, respectively (Fig. 1A).

In the intact photoreceptor of the fly, conversion of substantial amounts of R to M by blue light induces persistent excitation that far outlasts the light stimulus (Fig. 2B). The duration of this excitation, known as the prolonged depolarizing afterpotential (PDA), lasts about half an hour in *Musca* and as much as several hours in *Drosophila* (Hillman et al. 1983). The PDA can be suppressed to baseline at any time by photoconversion of the pigment back from M to R by a red light (Fig. 2B).

This same pattern of blue and red illumination was used to monitor GTPase activity in membrane preparations in order to test the relevance of the latter to the physiological response. Indeed, the same pattern of responses to colored lights was obtained (Fig. 2A). The finding that the PDA phenomenon can be reproduced by monitoring GTP hydrolysis in a cell-free membrane preparation is the first direct evidence that the process underlying the PDA takes place at an early stage of the

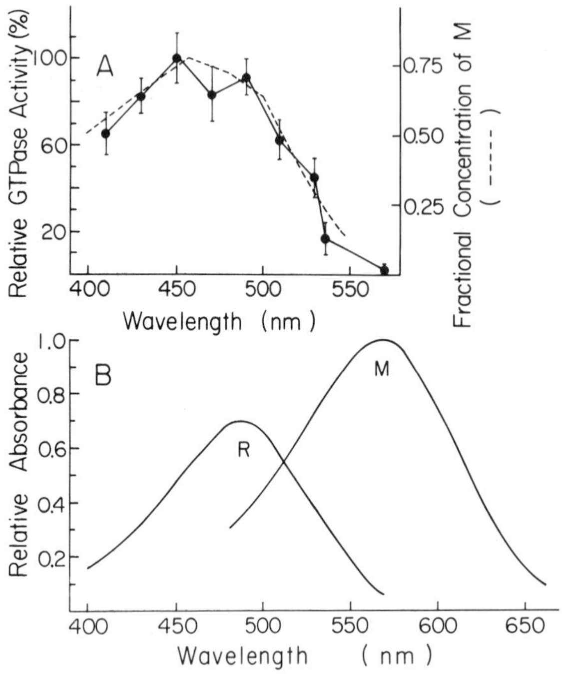

Figure 1. Spectral characteristics of fly photopigment and light-induced GTPase activity. (*A*) *Musca* eye membranes were illuminated for 1 min of GTPase reaction by 12-V, 100-W halogen lamp through interference filters (Ditric 3-cavity), KG3 heat filters, and neutral density filters that were adjusted to give 1.56×10^{16} photons $\cdot$ cm^{-2} at each wavelength. This amount of light is sufficient to bring the photopigment to photoequilibrium. The spectrum was obtained by plotting the normalized GTPase activity as a function of wavelength (solid line). Each point is the average of nine determinations. Fractional concentration of metarhodopsin (M) (dashed line) was measured by the early receptor potential in intact eyes (Minke and Kirschfeld 1979). (*B*) Relative absorption spectra of fly rhodopsin (R) and metarhodopsin (M) calculated from photometric measurements of the difference spectrum and the photoequilibrium spectrum (Hillman et al. 1983). Only white-eyed flies were used in this study.

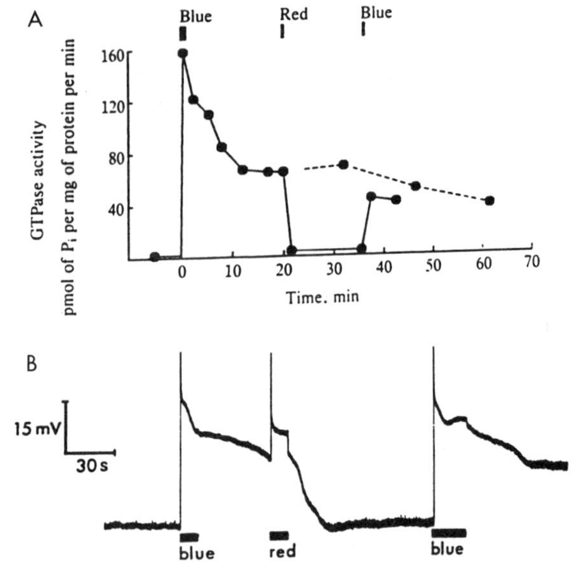

Figure 2. Effect of colored lights on the prolonged depolarizing afterpotential (PDA) and GTPase activity in *Musca* photoreceptors and in eye membrane preparations, respectively. (*A*) GTPase activity of *Musca* eye membranes (upper curve). *Musca* eye membranes suspended in a solution of 40 mM MOPS buffer (pH 6.7), 6 mM $MgC1_2$, and 1 mM dithiothreitol were incubated at 20°C and illuminated with saturating blue and red light as indicated. At each time point, an aliquot was removed for 1-min assay of GTPase activity at 4°C as described previously (Blumenfeld et al., 1985). (*B*) Intracellular recordings from *Musca* photoreceptor cells (lower curve). Intracellular recordings were performed on living *Musca* flies immobilized by wax; 2 M KC1-filled micropipettes of 60–100 MΩ resistance were lowered into the retina via a small hole in the cornea that had been covered by petroleum jelly. Blue and red illuminations were obtained by xenon light beam (XBO, 150 W) passed through either a blue filter (Ditric 480 nm cut-off filter) or a red-orange 610-nm edge filter (Schott OG-610) as indicated.

transduction process. Since the PDA is revealed by measurements of light-dependent GTPase, persistent activation must occur at either the photopigment or the G-protein level.

To test which state of the photopigment gives rise to light-dependent GTPase activity, we monitored GTP hydrolysis during illumination with saturating lights of different wavelengths. As shown in Figure 1A, the action spectrum of the light-dependent GTPase coincides with the photoequilibrium spectrum of metarhodopsin, indicating that metarhodopsin gives rise to the light-dependent GTPase activity.

We also determined the number of G-protein molecules in the photoreceptors by monitoring the time course of [^{35}S]GTPγS binding to eye membrane preparations, preilluminated with blue light. Binding assays over a range of GTPγS concentrations from 10^{-8} M to 10^{-7} M all saturated at about 10 pmoles GTPγS bound per milligram membrane protein, indicating that this value is the total number of binding sites in the membrane. Calculations based on the presence of 1.8×10^4 peripheral retinula cells in a compound *Musca* eye, an average yield of 5 μg membrane protein per eye and assuming that one GTPγS binds to one G-protein molecule, give a content of 1.6×10^6 molecules of G protein in a peripheral *Musca* photoreceptor cell.

To identify the G protein, we employed an [α-^{32}P] azidoanilido analog of GTP to label the light-dependent GTP-binding protein component in the *Musca* eye membranes. Polyacrylamide gel electrophoresis and autoradiography revealed in the blue illuminated membranes a labeled protein band of 41 kD that was not seen in the systems illuminated with red light (Devary et al. 1987).

$InsP_3$ Is a Second Messenger of Phototransduction

An indication that phosphoinositide hydrolysis is involved in phototransduction of invertebrates first came from experiments in which $InsP_3$, one of the two putative second messengers of the phosphoinositide transduction pathway, was injected into photoreceptors of the *Limulus* ventral eye (Brown et al. 1984; Fein et al. 1984). To study how light activates the enzyme that generates the putative second messenger, we have developed a membrane preparation in which light-dependent phosphoinositide hydrolysis can be studied under defined conditions and effects of activators and inhibitors can be analyzed. *Musca* eyes or *Drosophila* heads containing intact cells were preincubated with [^{3}H]inositol to label the phosphoinositides. The membrane prepared from these cells was used to analyze the hydrolysis of phosphoinositides. The *Musca* eye membrane preparation responded to illumination by an increase in the accumulation of $InsP_3$ and inositol bisphosphate ($InsP_2$), the respective products of polyphosphoinositide hydrolysis by a phospholipase-C-type enzyme (Fig. 3, left panel). In the illuminated system, there is a small increase in the concentration of $InsP_3$, which levels off after 1 minute. The greatest effect of light is seen, however, on the accumulation of $InsP_2$, which increases linearly for the first 5 minutes, then levels off and decreases slightly at longer incubation periods. The accumulation of inositol phosphate ($InsP_1$) (not shown) is characterized by a lag period, indicating that this product may arise by subsequent step-wise enzymatic dephosphorylation of polyphosphoinositols.

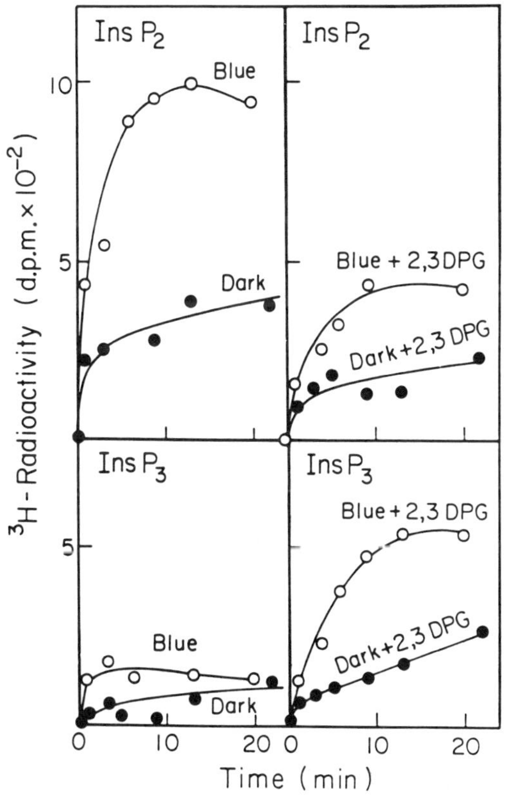

Figure 3. Light-induced hydrolysis of phosphoinositides in *Musca* eye membranes and effect of DPG. Equivalents of 100 *Musca* eyes each cut into two halves were incubated in the dark for 4 hr at 30°C with [^{3}H]inositol. Preparation of eye membranes and measurement of light-dependent phosphoinositide hydrolysis were carried out as described previously (Devary et al. 1987). The medium was adjusted with Ca-EGTA to give 50 nM free Ca^{++}. The reaction was initiated by the addition of membranes (500 μg protein/ml). The right column shows the effect of DPG (10 mM) added together with $MgCl_2$ (10 mM) (Reprinted, with permission, from Devary et al. 1987.)

The large accumulation of $InsP_2$ compared with that of $InsP_3$ in the *Musca* and *Drosophila* preparations and the ability of either $InsP_2$ or $InsP_3$ to excite *Limulus* ventral eye photoreceptors raise the question of which of the two inositol phosphates is the major product of light-induced phosphoinositide hydrolysis. To answer this question, we added 2,3-diphosphoglycerate (DPG), a known inhibitor of $InsP_3$ phosphatase (Downes et al. 1982). In the presence of 10 mM DPG, there is a substantial decrease in the accumulation of $InsP_2$ and a concurrently large increase in the accumulation of $InsP_3$, which now becomes the major product of the light-induced phosphoinositide hydrolysis (Fig. 3, right panel). Further evidence for the preference of the light-dependent phospholipase C enzyme for phosphatidylinositol 4,5-bisphosphate ($PtdInsP_2$) over phosphatidylinositol 4-phosphate (PtdInsP) comes also from our observation that, under assay conditions, the concentration of PtdInsP in the fly membranes is five times higher than that of $PtdInsP_2$, the immediate substrate for $InsP_3$ production. Yet, when hydrolysis of $InsP_3$ is blocked by DPG, more $InsP_3$ is accumulated than $InsP_2$ (Fig. 3). It is also apparent that fly membranes are endowed with the enzymatic system necessary to eliminate $InsP_3$ after it has been produced. Both the light-dependent preferential production of $InsP_3$ and a turn-off mechanism to stop its action are consistent with an internal messenger role of $InsP_3$ in fly phototransduction.

Two further criteria of an internal second messenger are (1) introduction of the putative messenger into the cell should reproduce the physiological response; and (2) agents that inhibit the inactivation of the messenger should augment the effect of the physiological stimulus. These criteria were tested for by light-induced introduction of $InsP_3$ + DPG into the *Musca* photoreceptor cells.

In the fly retina, illumination has been shown to facilitate the penetration of hydrophilic molecules into the photoreceptor cells. This has been shown for the dye Lucifer yellow (Wilcox and Franceschini 1984) and hydrolysis-resistant GTP analogs (Minke and Stephenson 1985). In this procedure, production of noisy depolarizing response made up of many unitary depolarizations is indicative of those agents that affect the phototransduction process.

Illumination with dim red light pulses of increasing intensity shows correspondingly increased depolarization with large noise during light (Fig. 4, bottom left). Likewise, introduction of increasing amounts of $InsP_3$ + DPG into the cells by extracellular injection combined with intense light results in a progressively increased noise in the dark (Fig. 4, traces 1,2). A combination of a weak excitation by $InsP_3$ + DPG together with dim light (Fig. 4, bottom right) markedly enhanced the response to dim lights, relative to the control (log I_{max} /I = 5,4). Power spectra calculated from the noise reveal that light and $InsP_3$ + DPG produce similar effects: Both cause an increased variance density throughout the relevant range of frequencies. Although $InsP_3$ + DPG usually produces an enhanced

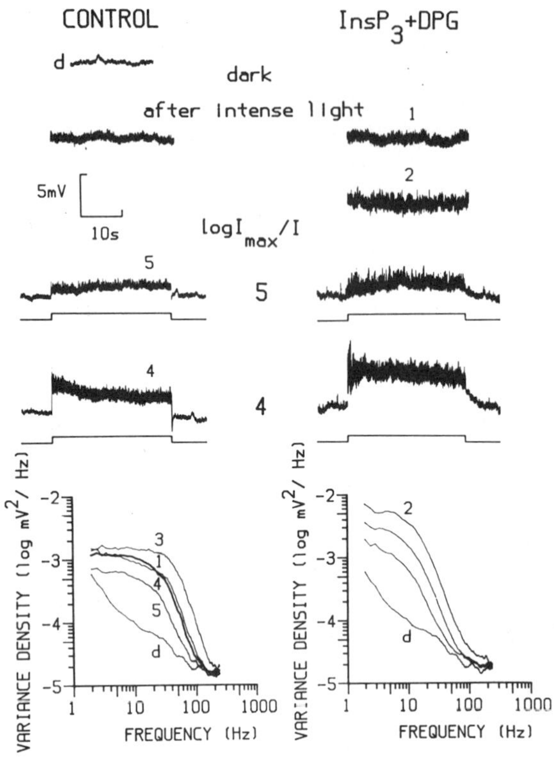

Figure 4. 1,4,5 $InsP_3$ + DPG excites *Musca* photoreceptors in the dark. (*Left*) The noise level in the dark-adapted cell (d) and 40 sec after intense (5 sec) white light (second trace). (*Right*) The noise in the dark recorded 40 sec (1) and 60 sec (2) after injections of $InsP_3$ (1 mM) + DPG (50 mM) in Ringer solution combined with the intense white light of 5 sec (1) and 20 sec (2) duration. The graphs below plot the power spectra calculated from the control responses to dim orange lights (*left*, 3,4,5); the dark noise after intense white lights combined with $InsP_3$ + DPG 60 sec (2), 3 min (*middle*), and 19 min after the bright white light was turned off (*right*). Curve 1 ($InsP_3$ + DPG) is similar to curve 4 (light).

noise relative to light, a weak excitation by $InsP_3$ + DPG in the dark has characteristics similar to dim light, as revealed by the similarity in their calculated power spectra (Fig. 4, traces and curves 1,4). After application of DPG alone and its introduction into the cell by intense light, there is not much change in the dark noise, but DPG does potentiate the effect of dim light, suggesting an endogenous production of $InsP_3$ during illumination (not shown).

Both the excitation of the photoreceptor cells by $InsP_3$ + DPG in the dark, which closely resembles the effect of light, and the potentiation of the effect of dim light by $InsP_3$ + DPG are consistent with a second-messenger role for $InsP_3$ in fly phototransduction.

Phospholipase C Is Under Stringent Control of a G Protein

In several transduction mechanisms, including phosphoinositide hydrolysis, coupling between the activated receptor and the effector enzyme is mediated by a G protein that is active in the GTP-bound form and reverses to the inactive state upon hydrolysis of the bound GTP to GDP (Cassel et al. 1977; Fung and Stryer 1980; Cockcroft and Gomperts 1985). Since omitting GTP from the incubation system had little, if any, effect on phosphoinositide hydrolysis, the involvement of a G protein was tested by the addition of GTPγS and GDPβS, the hydrolysis-resistant analogs of GTP and GDP, respectively (Eckstein et al. 1979). In the control systems incubated in the absence of guanine nucleotide analogs (Fig. 5, left panel), illumination with blue light for 30 seconds considerably increased the rate of phosphoinositide hydrolysis, resulting in a burst of $InsP_3$ production that leveled off after 1 minute, whereas the accumulation of $InsP_2$ proceeded for a longer period and achieved a greater extent than that of $InsP_3$. In the presence of GDPβS, there was a marked decrease in the accumulation of $InsP_2$ in the dark, whereas in the presence of GTPγS, light further increased both the length and extent of $InsP_2$ and $InsP_3$ accumulation (Fig. 5, right panel). These results indicate that the light-dependent phosphoinositide hydrolysis in the fly eye membrane is under the stringent control of a G protein that is inactive when it contains bound GDPβS and is persistently active when the bound nucleotide is the hydrolysis-resistant analog GTPγS.

Conditional Phospholipase C Activity in *Drosophila norpA* Mutants

A question that has not been settled yet is whether $InsP_3$ is involved in the light-induced excitation of invertebrate phosphoreceptors. This question was raised

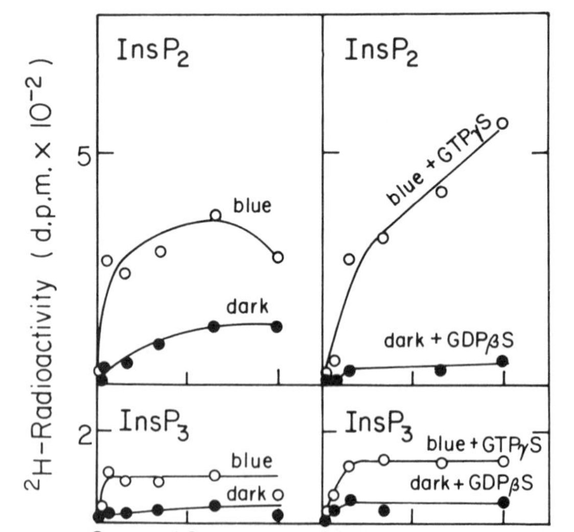

Figure 5. Effect of GTPγS and GDPβS on inositol phosphate production in *Musca* eye membranes. Incubation conditions were carried out as described in Fig. 3. Where indicated, the concentrations of GTPγS were 10 μM and GDPβS 100 μM. (●) Systems incubated in the dark; (○) systems illuminated with blue light.

by experiments in *Limulus* ventral photoreceptors in which introduction of the calcium chelator EGTA into the cell effectively blocked the response elicited by intracellular injection of $InsP_3$, but was ineffective in blocking the excitation by light (Payne et al. 1986). It was therefore argued that in addition to releasing $InsP_3$, light releases some other as yet unidentified transmitter that excites the cell by some means other than by raising calcium concentration.

The visual *Drosophila* mutant *norpA*, which in some alleles completely fails to produce receptor potential in response to light, is particularly useful to test the involvement of $InsP_3$ in photoexcitation. The *norpA* mutant has been found previously to have decreased diglyceride kinase and phospholipase C activities (Yoshioka et al. 1984; Inoue et al. 1985). It is not clear from these studies, however, if these enzymes are light-dependent and which of the two enzymes is the primary cause of the mutant phenotype. A decreased diglyceride kinase activity can give rise to persistent activation of protein kinase C, which in turn may cause secondary changes in phosphoinositide metabolism.

To overcome these difficulties, we have employed the temperature-sensitive allele $norpA^{H52}$. This mutant has normal receptor potential at the permissive temperatures of 19–30°C, but promptly loses its receptor potential when shifted to the restrictive temperature of >34°C (Deland and Pak 1973). It should be pointed out that $norpA^{H52}$ differs from other *norpA* alleles that are developmentally temperature sensitive (Inoue et al. 1985). These mutants reveal their characteristic phenotype (absence of light coincidence receptor potential) only if grown at the restrictive temperature throughout their life cycle, whereas $norpA^{H52}$ retains its temperature sensitivity even when grown at permissive temperatures. Light-dependent phosphoinositide hydrolysis was measured by monitoring the accumulation of $InsP_2$ in reaction mixtures containing either $norpA^{H52}$ or wild-type derived head membranes that were prelabeled with [^{3}H]inositol at the permissive temperature of 21°C. In the systems containing wild-type membranes, there was practically no difference in phosphoinositide hydrolysis at 21°C or 37°C. On the other hand, incubation of $norpA^{H52}$ membranes at the restrictive temperature of 37°C considerably reduced the rate and extent of $InsP_2$ accumulation, as compared with the system that was incubated at 21°C. This effect is greatly enhanced in $norpA^{H52}$ membranes that had been preincubated for 4 minutes at the restrictive temperature of 37°C, but not in preincubated wild-type membranes nor in mutant membranes that had been preincubated at 21°C (Fig. 6). Since the light-dependent phospholipase C activity is coupled to photoexcited rhodopsin by a G protein, and diglyceride kinase activity was reported to be greatly reduced in *norpA* mutants (Yoshioka et al. 1984), we have separately assayed the G protein by measuring its light-dependent GTPase activity and the diglyceride kinase by measuring incorporation of labeled phosphate from [γ-^{32}P]ATP to exogenously added diglyceride. None of these activities were diminished by temperature shift from 21°C to 37°C (not shown).

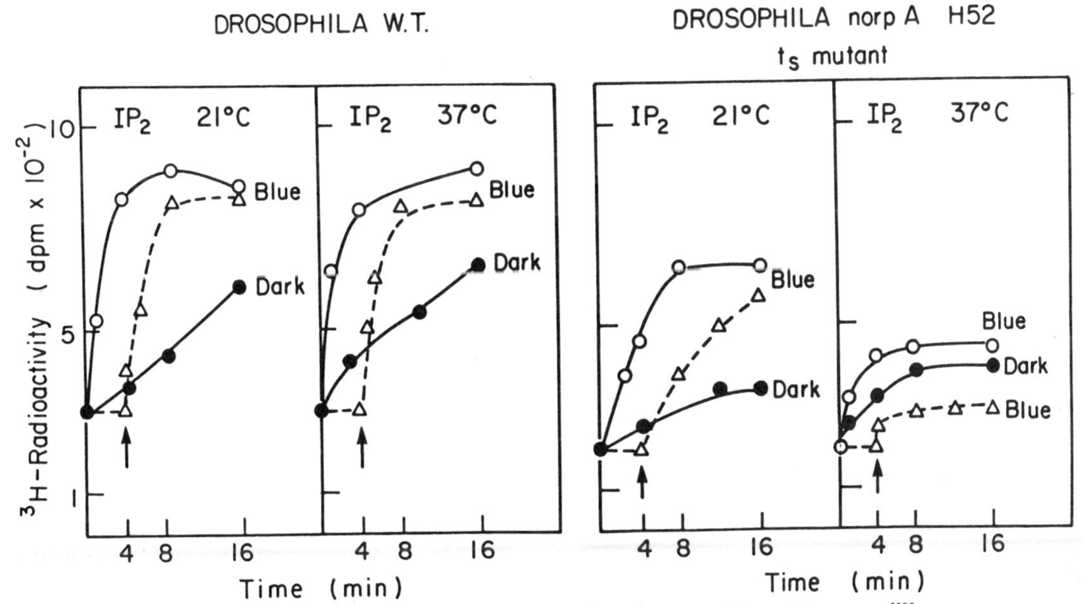

Figure 6. Light-dependent phosphoinositide hydrolysis by normal (W.T.) *Drosophila* and by $norpA^{H52}$ mutant eye membranes incubated at permissive and restrictive temperatures. *Drosophila* (normal and $norp^{H52}$ temperature-sensitive mutant) flies up to 5 days old were used. Incorporation of [^{3}H]inositol and preparation of crude *Drosophila* eye membranes were carried out as described previously (Devary et al. 1987). The reaction was initiated by addition of membranes to reaction media preequilibrated at the indicated temperatures and processed as described in Fig. 3. Systems depicted by dotted lines are preincubated for 4 min at the indicated temperatures in medium lacking ATP and regeneration system. These latter components were subsequently added (arrow) to initiate the reaction.

The *nss* Mutant of *Lucilia* Is Affected at a Stage Subsequent to $InsP_3$ Production

The transient receptor potential (*trp*) mutant of *Drosophila* and the no steady state (*nss*) mutant of the sheep blowfly *Lucilia* respond to a short light pulse like wild-type flies but are unable to maintain persistent excitation upon continuous illumination (Minke 1982; Howard 1984). In the present study, we used the *nss* mutant, which is technically more accessible to electrophysiological studies due to the much larger diameter of its photoreceptors. Our electrophysiological and biochemical studies have indicated that the two mutants have a very similar phenotype. Experiments on the *nss* and *trp* mutants using various combinations of dim background lights and prolonged, more intense, test light revealed that the rate at which the receptor potential decays, in response to continuous illumination with test light, is very sensitive to the presence of dim background light.

In both mutants, background light considerably accelerates the decay of the receptor potential. It was therefore possible to use chemical excitation equivalent to a dim background light to localize the phototransduction step that is modified by the *nss* mutation. Presumably, agents that mimic the effect of dim background light on the *nss* mutant must act on the transduction pathway at a site prior to that which is modified by the mutation. The experiments depicted in Figure 7 show that introduction of $InsP_3$ + DPG into the *nss* *Lucilia* photoreceptor cells accelerates the rate of decline of the receptor potential in response to subsequent continuous illumination in a manner similar to that of background light. It therefore appears that the *nss* gene product operates at a late stage of the phototransduction pathway, subsequent to production of $InsP_3$.

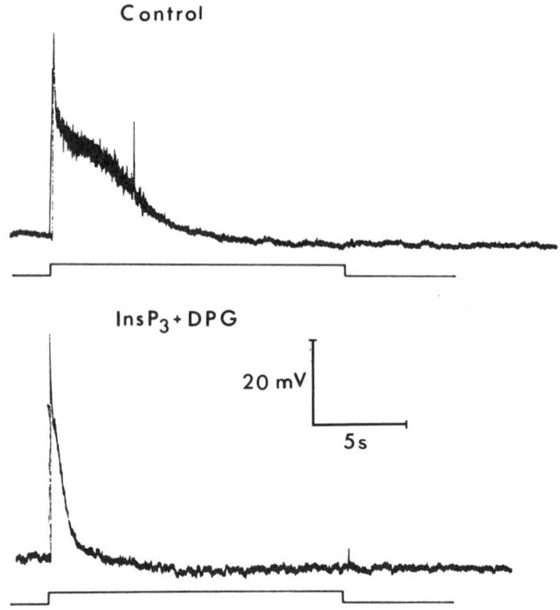

Figure 7. $InsP_3$ + DPG facilitates the light response of the *nss* mutant, accelerates its decline to baseline, and induces noise in the dark. Intracellular recordings from a photoreceptor of the *nss* mutant showing a response of a dark-adapted cell to continuous illumination with orange light of maximal intensity attenuated by 2 log units (upper trace). The bottom trace shows a response to the same orange light 5 min after 1 mM 1,4,5 $InsP_3$ together with 50 mM 2,3-DPG was injected into the eye during maximal intensity white illumination of 50-sec duration, given in pulses of 5-sec duration. The acceleration in the decay rate of the response (*bottom*) is typical for the effect of prolonged dim background illumination on the response to a test pulse of medium intensity.

Chemical Excitation Can Be Used to Localize the *rdg B* Mutation

The *rdg B Drosophila* is a retinal degeneration mutant that reveals its characteristic phenotype when the fly is exposed to light (Hotta and Benzer 1970). To this effect, the *rdg B* is a light-dependent conditional mutant that requires the operation of a phototransduction pathway to express its phenotype. From this and other experiments, it was inferred that some intermediate step along the phototransduction pathway is inactivated by the normal *rdg B* gene product, and when the latter is defective (as in the *rdg B* mutants) an unbalanced action of this particular but as yet unidentified intermediate brings about retinal degeneration (Harris and Stark 1977). Mosaic analysis showed that the *rdg B* mutant is autonomous to the eye (Hotta and Benzer 1970) and that in some *rdg B* alleles like the *rdg* B^{KS222}, the degeneration is confined to the peripheral R1–6 retinula cells, whereas the central R-7 and R-8 cells are spared and remain well preserved (Harris and Stark 1977). We have chosen the allele *rdg* B^{KS222} to study how chemical excitation affects retinal degeneration, since it has a built-in internal control. We considered any chemical agent that causes retinal degeneration in the dark to mimic the effect of light only if the degeneration is confined to the R1–6 cells while the R7 and R8 central cells are spared. It is argued that agents that activate intermediate stages in the phototransduction prior to the step with which the normal *rdg B* gene product interacts should bring about retinal degeneration, whereas activation of intermediate steps subsequent to this interaction should not cause retinal degeneration. Initial experiments aimed at localization of the step in the phototransduction pathway with which the *rdg B* gene product interacts are presented in Figure 8. In these experiments, GTPγS and control Ringer solution were applied once daily to the eyes of *rdg B* flies reared in the dark for 8 days from eclosion. Two other groups of flies, *rdg B* and wild type, were kept under standard conditions of a 12-hour light/dark cycle at 24°C. Upon eclosion, the *rdg B* flies have normal-looking photoreceptors (Fig. 8B) not much different from wild-type flies raised for 8 days under standard conditions (Fig. 8A). In contrast, *rdg B* flies treated for 8 days with GTPγS in the dark or exposed to 12-hour dark/light cycle for 8 days both demonstrate severe degeneration of the peripheral R1–6 retinula

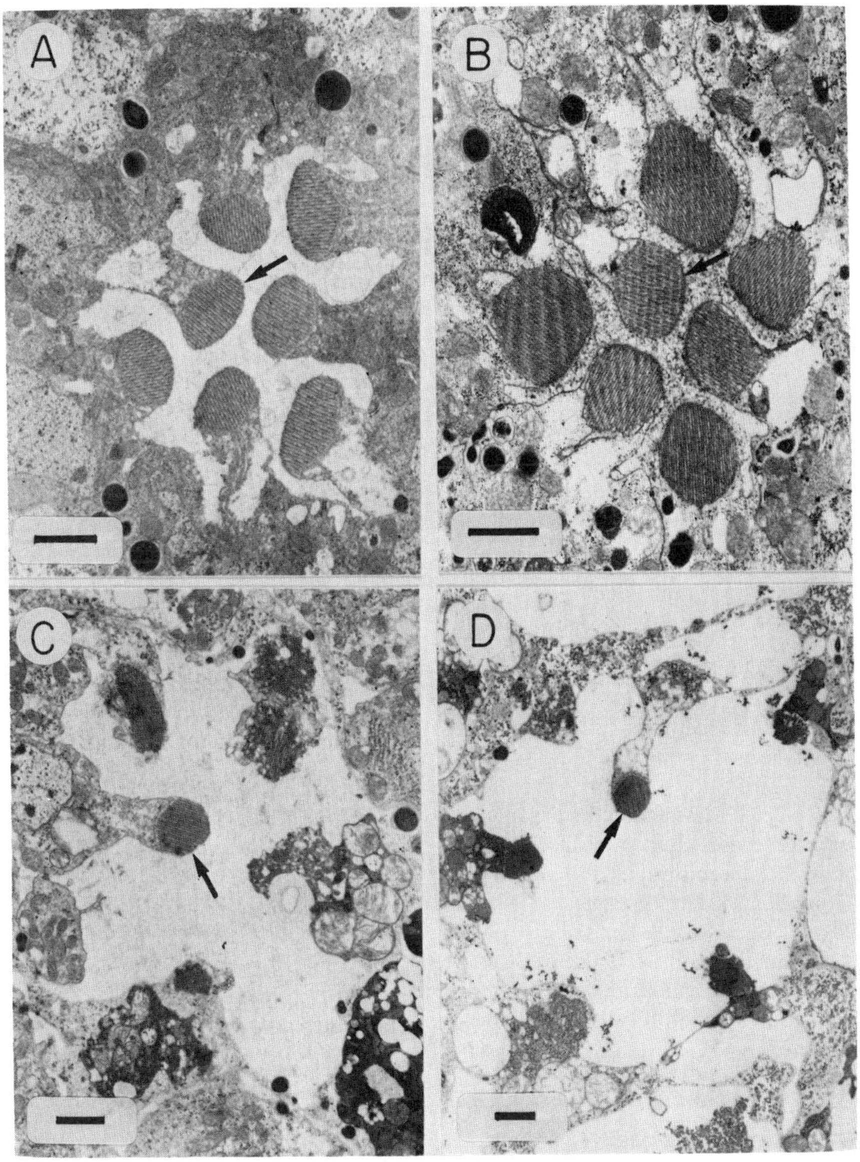

Figure 8. Ultrastructure of light- and GTPγS-induced retinal degeneration in *Drosophila rdg* B^{KS222} mutant. High-magnification electron micrographs of a cross section of *Drosophila* eyes. Each micrograph shows typical ommatidia representing the dominant picture in the field. Bars, 1 μm. (*A*) Preparation from an 8-day-old wild-type *Drosophila* reared under 12-hr light/dark cycle at 24°C. Normal-looking ommatidium is observed. (*B*) Preparation from 1-day-old *rdg B* fly reared in the dark. The typical structure of fly ommatidium with seven photoreceptor cells and their striated rhabdomers is observed. (*C*) Preparation from an 8-day-old *rdg B* fly reared under 12-hr light/dark cycle at 24°C. The characteristic features of degeneration of the peripheral R1–6 retinula cells and preservation of the central R7/8 cells (arrow) are observed. (*D*) Preparation from 8-day-old *rdg B* fly reared in the dark at 24°C and treated daily by application of a drop of 8 mM GTPγS solution to each eye. Clear manifestation of degeneration exists. The rhabdomers of the peripheral retinula cells R1–6 are absent and the cells contain vesicular and multilamellar bodies. Strikingly, the central retinula cells R7/8 (arrow) remain intact with well-preserved rhabdomeral structure. Treatment of the flies and processing were done according to the method of Rubinstein. (Modified from Rubinstein 1988.)

cells, although the central retinula cells R7 and R8 (arrow) remain well preserved (Fig. 8C,D). In control experiments, *rdg B* flies were reared in the dark, and wild-type flies were treated with GTPγS in the dark. Both of these groups of flies revealed little if any degeneration (not shown). Taken together, the present results suggest that the G protein precedes the step in the phototransduction with which the *rdg B* gene product interacts. Further studies along these lines should provide a powerful tool to investigate the molecular mechanism of light-induced retinal degeneration in the *rdg B* mutant.

DISCUSSION

The photoreceptor of the *Drosophila* fly is the only part of the nervous system for which extensive genetic background is available (Pak 1979). In part, this is due to the ease with which phototactic mutants can be isolated, but it is also due to the obvious physiological function of the photoreceptor cells (i.e., translation of light stimuli into neuronal signals), which lends itself to detailed electrophysiological characterization as well as to simplified extracellular recording that can be used for screening of mutants. What has been lacking for many years is biochemical characterization of the phototransduction process in terms of an assayable cascade of enzyme reactions. Superficially, the transduction process of invertebrate photoreceptors appears to have an inverse direction to that of vertebrate rod photoreceptors. The two photoreceptor systems have opposite signs of electrical responses to light, depolarization in the invertebrate, and hyperpolarization in the vertebrate as well as correspondingly light-induced increase and decrease in intracellular calcium concentrations. For these reasons, an inverse change in cGMP concentration was considered as a plausible second-messenger mechanism for both vertebrate and invertebrate photoreceptors.

The electrophysiological biochemical and genetic evidence that has accumulated in the last few years suggests, however, that the vertebrates and invertebrates have evolved profoundly different phototransduction mechanisms in their photoreceptors. It is clear now that in contrast to vertebrate photoreceptors that evolved a highly specific cGMP-mediated transduction mechanism unique to these cells, the invertebrate photoreceptors use the ubiquitous inositol lipid signaling system, a mechanism that operates in almost every eukaryotic cell (Berridge 1987). It is therefore likely that the information gained from studies of phototransduction in the fly photoreceptor will be widely relevant to a great number of cells and a variety of biological phenomena.

ACKNOWLEDGMENTS

We thank Dr. William L. Pak for generously supplying us with the *Drosophila* mutants used in this study. This research was supported by grants from the National Institutes of Health (EY-03529) and the U.S.–Israel Binational Science Foundation.

REFERENCES

Berridge, M.J. 1987. Inositol trisphosphate and diacylglycerol: Two interacting second messengers. *Annu. Rev. Biochem.* **56:** 159.

Blumenfeld, A., J. Erusalimsky, O. Heichal, Z. Selinger, and B. Minke. 1985. Light-activated guanosinetriphosphatase in *Musca* eye membranes resembles the prolonged depolarizing afterpotential in photoreceptor cells. *Proc. Natl. Acad. Sci.* **82:** 7116.

Brown, J.E., L.J. Rubin, A.J. Ghalayini, A.P. Tarver, R.F. Irvine, M.J. Berridge, and R.E. Anderson. 1984. Myo-inositol polyphosphate may be a messenger for visual excitation in *Limulus* photoreceptors. *Nature* **311:**160.

Cassel, D., H. Lefkovitz, and Z. Selinger. 1977. The regulatory GTPase cycle of turkey erythrocyte adenylate cyclase. *J. Cyclic Nucleotide Res.* **3:** 393.

Cockcroft, S. and B.D. Gomperts. 1985. Role of guanine nucleotide binding protein in the activation of polyphosphoinositide phosphodiesterase. *Nature* **314:** 534.

Deland, M.C. and W. Pak. 1973. Reversible temperature sensitive phototransduction mutant of *Drosophila melanogaster. Nature New Biol.* **244:** 184.

Devary, O., O. Heichal, A. Blumenfeld, D. Cassel, E. Suss, S. Barash, C.T. Rubinstein, B. Minke, and Z. Selinger. 1987. Coupling of photoexcited rhodopsin to inositol phospholipid hydrolysis in fly photoreceptors. *Proc. Natl. Acad. Sci.* **84:** 6939.

Dowling, J.E. 1960. Chemistry of visual adaptation in the rat. *Nature* **188:** 114.

Downes, C.P., M.C. Mussat, and R.H. Michell. 1982. The inositol trisphosphate phosphomonoesterase of the human erythrocyte membrane. *Biochem. J.* **203:** 169.

Eckstein, P., D. Cassel, H. Levkovitz, M. Lowe, and Z. Selinger. 1979. Guanosine 5′*O*-(2-thiodiphosphate): An inhibitor of adenylate cyclase stimulation by guanine nucleotides and fluoride ions. *J. Biol. Chem.* **254:** 9829.

Fein, A., R. Payne, D.W. Corson, M.J. Berridge, and R. Irvine. 1984. Photoreceptor excitation and adaptation by inositol 1,4,5-trisphosphate. *Nature* **311:** 157.

Fung, B.K.K. and L. Stryer. 1980. Photolysed rhodopsin catalyzes the exchange of GTP for GDP in rod outer segment membranes. *Proc. Natl. Acad. Sci.* **77:** 2500.

Hamdorf, K. 1979. The physiology of invertebrate visual pigment. In *Handbook of sensory physiology* (ed. H. Autumn), vol. 7, p. 145. Springer-Verlag, Berlin.

Harris, W.A. and W.S. Stark. 1977. Hereditary retinal degeneration in *Drosophila melanogaster*. A mutant defect associated with the phototransduction process. *J. Gen. Physiol.* **69:** 261.

Hillman, P., S. Hochstein, and B. Minke. 1983. Transduction in invertebrate photoreceptors: The role of pigment bistability. *Physiol. Rev.* **63:** 668.

Hotta, Y. and S. Benzer. 1970. Genetic dissection of the *Drosophila* nervous system by means of mosaics. *Proc. Natl. Acad. Sci.* **67:** 1156.

Howard, J. 1984. Calcium enables photoreceptors pigment migration in a mutant fly. *J. Exp. Biol.* **113:** 471.

Inoue, H., T. Yoshioka, and Y. Hotta. 1985. A genetic study of inositol trisphosphate involvement in phototransduction using *Drosophila* mutants. *Biochem. Biophys. Res. Commun.* **132:** 512.

Johnson, E.C., P.R. Robinson, and J.E. Lisman. 1986. Cyclic GMP is involved in the excitation of invertebrate photoreceptors. *Nature* **324:** 468.

Minke, B. 1982. Light-induced reduction in excitation efficiency in *trp* mutant of *Drosophila. J. Gen. Physiol.* **79:** 361.

Minke, B. and K. Kirschfeld. 1979. The contribution of a sensitizing pigment to the photosensitivity spectra of fly rhodopsin and metarhodopsin. *J. Gen. Physiol.* **73:** 517.

Minke, B. and R.S. Stephenson. 1985. The characteristics of chemically induced noise in *Musca* photoreceptors. *J. Comp. Physiol.* **156:** 339.

Pak, W.L. 1979. Study of photoreceptor function using *Drosophila* mutants. In *Neurogenetics: Genetic approaches to the nervous system* (ed. X. Breakefield), p. 67. Elsevier North Holland, New York.

Payne, R. 1986. Phototransduction by microvillar photoreceptors of invertebrates mediation of the visual cascade by inositoltrisphosphate. *Photobiochem. Photobiophys.* **13:** 373.

Payne, R., W.D. Corson, A. Fein, and M.J. Berridge. 1986. Excitation and adaptation of *Limulus* ventral photoreceptors by inositol 1,4,5 trisphosphate result from a rise in intracellular calcium. *J. Gen. Physiol.* **88:** 127.

Rubinstein, C.T. 1988. "Transient voltage responses induced

by light in invertebrate photoreceptors." Ph.D. thesis. Hebrew University of Jerusalem.

Stryer, L. 1986. Cyclic GMP cascade of vision. *Annu. Rev. Neurosci.* **9;** 87.

Wilcox, M. and N. Franceschini. 1984. Illumination induced dye incorporation in photoreceptor cells. *Science* **225:** 851.

Yoshioka, T., H. Inoue, and Y. Hotta. 1984. Absence of diglyceride kinase activity in the photoreceptor cells of *Drosophila* mutants. *Biochem. Biophys. Res. Commun.* **119:** 389.

Molecular Transduction in Smell and Taste

D. LANCET, D. LAZARD, J. HELDMAN, M. KHEN, AND P. NEF
Department of Membrane Research, The Weizmann Institute of Science, Rehovot 76100, Israel

In this paper we describe recent progress in our understanding of the molecular mechanism of smell and taste. Such new insight highlights the similarity of olfaction and gustation to another molecular sensory mechanism–visual photoreception. It also underlines the homology of recognition and transduction events in smell and taste with those that take place in hormone and neurotransmitter reception (Figs. 1 and 2) (Lancet and Pace 1987). An important theme in the present paper is the possibility that although the chemosensory cells utilize a molecular transduction apparatus akin to that of other sensory and nonsensory cells, the actual molecular components involved may be unique. Other reviews on molecular chemosensory mechanisms have appeared recently (Getchell et al. 1984; Lancet 1986, 1988a,b; Anholt 1987; Lancet and Pace 1987; Pace and Lancet 1987; Snyder et al. 1988). Other theories of olfaction have been described (Davies 1971).

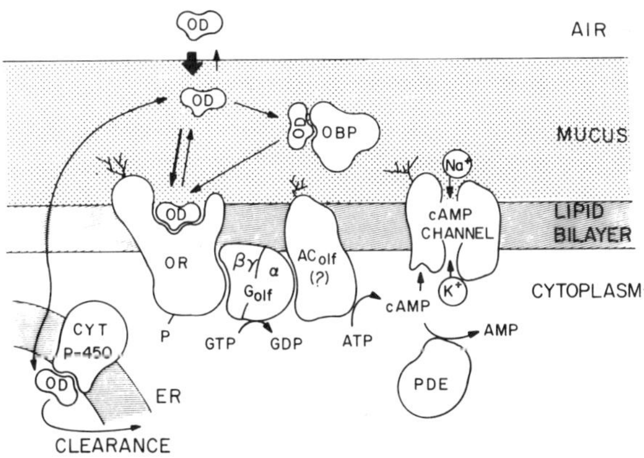

Figure 1. Scheme of olfactory reception and transduction. Odorants (OD) partition favorably from air to mucus (Lancet 1986) and bind to putative olfactory receptor (OR), a transmembrane glycoprotein. The odorant-binding site is shown to be within the transmembrane region, in analogy with β-adrenergic receptors. An odorant-binding protein (OBP) in the mucus serves as a putative carrier. OR (which may undergo phosphorylation [– P]) activates G_{olf}, the olfactory-specific G_s, composed of an $\alpha\beta\gamma$ heterotrimer, whose α chain catalyzes the hydrolysis of GTP to GDP. The same subunit also activates adenylate cyclase (AC), perhaps of an olfactory-specific type. AC generates the second-messenger cAMP, which activates a cation channel and is degraded by phosphodiesterase (PDE). AC and the cAMP channel are presumably transmembrane glycoproteins as well. Cytochrome P450 in the endoplasmic reticulum (ER) may serve in odorant clearance (see text for more details).

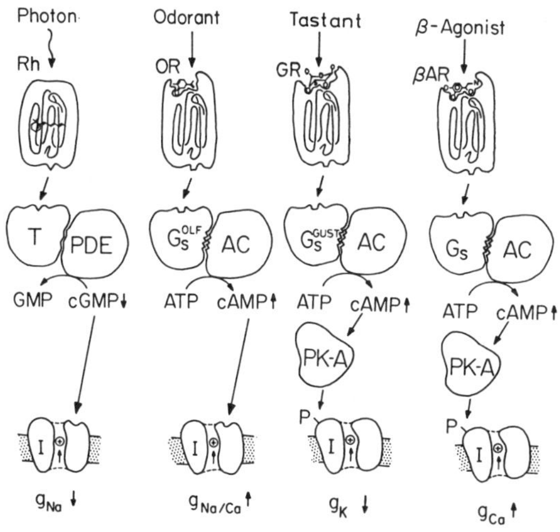

Figure 2. Functional homology among transduction mechanisms. Seven transmembrane receptors, rhodopsin (Rh), β-adrenergic receptor (βAR), and hypothetically also olfactory receptor (OR) and gustatory receptor (GR) are activated by their respective agonists. All are known to activate GTP-binding proteins: Rh activates transducin (T), and the other three receptors activate different forms of G_s (the identity and function of gustatory G_s are not established). Cyclic nucleotides (cGMP in visual photoreception, cAMP in the rest) act as second messengers. Ion channel (I) conductance is modulated directly by the cyclic nucleotide in photoreception and in olfaction, or via the action of cAMP-dependent protein kinase (PK-A) in βAR and gustatory reception. The conductance (g) to various ions is affected.

Olfactory Receptors: Expected Properties and Strategies of Identification

The demonstration that olfactory reception involves a stimulatory GTP-binding protein (G_s) and adenylate cyclase (Pace et al. 1985; Pace and Lancet 1986; Anholt 1987; Lancet and Pace 1987) suggests that olfactory receptors (OR, Figs. 1 and 2) share some properties with other G-protein-coupled receptors (cf. Dohlman et al. 1987). Most of the current strategies for attempting OR identification and isolation stem from such expected homology.

Two polypeptides that constitute receptor candidates have been identified in our laboratory. The first is gp95, a specific transmembranal glycoprotein enriched in the frog olfactory cilia preparation (Chen et al. 1986b,c). Glycoprotein gp95 has been immunohistochemically localized to the ciliary surface of olfactory epithelium and found to be absent in respiratory epithelium, retina, and brain (Chen et al. 1986a). In all, seven of nine criteria set for OR identification (Lancet 1986) are fulfilled by gp95.

The second polypeptide, gp56, has been identified in olfactory membrane preparations from rat and cow. It is also a transmembranal glycoprotein that is absent in control respiratory epithelial membranes (Kropf et al. 1987). We have recently obtained protein sequence information from several cyanogen bromide fragments of gp56 and are currently using this information to attempt the molecular cloning of gp56. Complete sequence information will allow one to ask whether gp56 shows any homology with other receptors with seven hydrophobic stretches that are known to activate G proteins.

Additional approaches utilized in our laboratory to identify OR proteins and the nucleic acid sequences coding for them are (1) studies of phosphorylation of olfactory cilia proteins by various protein kinases and its possible relation to chemosensory adaptation (Heldman and Lancet 1986; Pace et al. 1987); (2) screening of olfactory epithelial cDNA libraries to identify cDNA clones that bear homology with known sequences of G-protein-coupled receptors. This approach has been successfully implemented recently in the cloning of G-21, a member of the seven transmembrane domain receptor superfamily (Kobilka et al. 1987).

Olfactory G_s: A Unique Species

Olfactory cilia preparations from several species, which span almost all vertebrate classes from fish to mammals, contain odorant-sensitive adenylate cyclase. Significantly, odorant activation is seen only in the presence of GTP (Pace et al. 1985; Pace and Lancet 1986; Shirley et al. 1986; Sklar et al. 1986). Such synergism of physiological ligands with GTP serves as an indication for the involvement of a GTP-binding protein (G protein). Since odorants activate rather than inhibit adenylate cyclase, it appears that a G protein of the stimulatory type (G_s) is involved. The coupling of G_s to olfactory adenylate cyclase is also seen in that nonhydrolyzable guanine nucleotides, such as GTPγS and GppNHp, strongly activate adenylate cyclase in olfactory cilia preparations (Pace and Lancet 1986). Olfactory G_s has been identified by cholera-toxin-catalyzed ADP ribosylation (Pace et al. 1985; Pace and Lancet 1986) and by specific antibodies (Anholt et al. 1987; Lancet et al. 1988).

Early observations suggested that olfactory G_s is a unique species, different from G_s in other tissues and cells. These observations are given below.

1. G_s of olfactory cilia in rat and frog shows a considerably higher apparent affinity toward guanine nucleotides. For example, half-maximal activation of adenylate cyclase by GTPγS is attained at roughly 1 nM in olfactory cilia, whereas at least 100 times higher concentrations are required in control membranes from brain (Pace and Lancet 1986).
2. The mobility of $G_{s\alpha}$, visualized by cholera-toxin-catalyzed [^{32}P]ADP ribosylation, is about 1 kD higher than that of the respective species in other membranes (e.g., from liver). This difference is also seen both in the rat and in the frog (Pace et al. 1985; Pace and Lancet 1986; Lancet et al. 1988).
3. Olfactory G_s manifests a different guanine nucleotide pharmacology (Lancet et al. 1988; I. Shafir et al., in prep.). This is seen when employing two GTP thio analogs with sulfur atoms at the α phosphate position (GTPαS). Introducing a thio replacement at this position renders the α phosphorous atom asymmetric; hence, two optical isomers exist: R-GTPαS and S-GTPαS. We find that whereas in brain and liver membranes both optical isomers activate adenylate cyclase equally well (both being weak activators, like GTP), the R-GTPαS isomer is a much stronger activator than S-GTPαS in olfactory cilia. This, again, is true both for rat and frog. Such a difference in guanine nucleotide behavior suggests a different GTP-binding site configuration in olfactory cilia G_s, compared with G_s in other membranes.
4. Olfactory G_s shows a unique cellular distribution, being highly enriched in olfactory cilia, compared with the rest of the sensory cell (Pace et al. 1985).

Taken together, these results were interpreted to mean that at least the α chain of olfactory G_s is a different molecular species from other $G_{s\alpha}$ types. Until recently, it was known that $G_{s\alpha}$ has four variants, coded for by different mRNA species derived by alternative splicing from a single $G_{s\alpha}$ gene (Gilman 1987). The possibility exists that olfactory G_s corresponds to one of these variants, probably of the "small" type, which, like olfactory G_s, has an apparent molecular mass of about 42 kD. Alternatively, it was suggested that olfactory G_s could represent a different gene. The results reported by R. Reed and colleagues (Jones et al., this volume) strongly suggest that the latter possibility is correct. They were able to clone a new $G_{s\alpha}$-like molecule that appears to be specific to olfactory sensory neurons. These findings lend strong support to the biochemical data described above and to the notion that G_s is important in olfactory transduction.

Olfactory Adenylate Cyclase

Olfactory cilia preparations show extremely high specific activity of adenylate cyclase. The basal levels are 3–20 nmoles of cAMP per milligram of protein per minute. The rat and frog preparations tend toward the higher number; the level in the bovine preparation is lower and more variable. This level is ~10 times higher than that seen in control membranes such as from brain, deciliated olfactory epithelium (Pace et al. 1985), and olfactory sensory axons (Anholt 1988). The enzyme is classically activated by forskolin, as well as by GTP, nonhydrolyzable GTP analogs, and fluoride anions, as expected, in view of its coupling to G_s. It is not Ca^{++}/calmodulin-dependent, and it is inhibited by Ca^{++} at millimolar concentrations (Sklar et al. 1986).

There are several indications that olfactory adenylate cyclase represents a unique species:

1. Olfactory adenylate cyclase is activated by odorants. This property is not clearly seen in control membrane preparations (Pace et al. 1985), including membranes from deciliated olfactory epithelium (Chen et al. 1986c).
2. Olfactory adenylate cyclase is more heat stable than counterparts from other sources. Thus, whereas adenylate cyclase in other membranes is often completely inactivated after incubation at 37°C for 30 minutes, the enzyme in olfactory cilia remains almost fully active after a similar incubation and is not fully inactivated even at 47°C (Pace 1988).
3. Olfactory adenylate cyclase couples to olfactory G_s, which is coded by a $G_{s\alpha}$ gene different from the gene that couples to adenylate cyclase elsewhere (see above).
4. Olfactory adenylate cyclase segregated into the olfactory cilia membrane, being enriched in them compared with the rest of the epithelium. This suggests a possible association with membrane or cytoskeleton components of the sensory neuronal organelles.

Since it is Ca^{++}/calmodulin-independent, olfactory adenylate cyclase may be a variant of the high-molecular-mass (150 kD) species found in many tissues and distinct from the brain-specific Ca^{++}/calmodulin-sensitive 120-kD form (Mollner and Pfeuffer 1988). Studies using monoclonal antibodies and molecular genetics should soon help clarify the identity of olfactory adenylate cyclase.

Olfactory Ion Channels

One of the most exciting recent developments in our understanding of olfactory transduction arises from an electrophysiological study by Gold and co-workers (Nakamura and Gold 1987). This study shows that excised patches from the ciliary membranes of isolated toad olfactory sensory neurons contain a cAMP-gated cation conductance. The properties of the ion channel involved are highly reminiscent of those reported for cGMP-gated ion channels in retinal rod outer segments (for review, see Stryer 1986). Such properties include ion specificity for sodium and calcium, a significant decrease of single ion channel conductance by divalent cations, as well as positive cooperativity ($n_H > 1$) in the response to cyclic nucleotides.

Recently, Kaupp and colleagues successfully purified a 63-kD polypeptide, which, when incorporated into phospholipid vesicles, results in the functional reconstitution of the cGMP channels (Kaupp et al., this volume). The molecular cloning of this protein should reveal the polypeptide structural details of a new family of ion channels that are directly gated by intracellular ligands. The olfactory ion channel may belong to this family. An intriguing possibility is that cAMP-gated channels exist also in other cells, and they could represent a general pathway for cAMP action, parallel to the classic pathway of cAMP-dependent protein kinase.

Peri-receptor Molecules in the Olfactory System

The major pathway of signal transduction in the olfactory system appears to be from transmembrane receptors (yet to be identified) to G_s, adenylate cyclase, and cAMP-gated ion channels. Other pathways leading from reception to ion channel modulation may be present, such as the possible activation of the phosphatidylinositol mechanism (Huque and Bruch 1986) and direct ion channel gating by odorants (Price 1984; Labarca et al. 1987; Nomura and Kurihara 1987). Yet, while there is ample evidence obtained in intact cells for the cAMP mechanism (Kurihara and Koyama 1972; Minor and Sakina 1973; Menevse et al. 1977; Trotier and McLeod 1986; De Simone et al. 1988; Suzuki 1988; Vodyanoy 1988), practically no such evidence is available for any of the alternative mechanisms.

Given that the cAMP mechanism is the most probable one, there may still be peri-receptor components that are indirectly involved in signal transduction (cf. Getchell et al. 1984). This could include molecules that are involved in the onset and termination of odorant signals.

Odorant-binding proteins. One such component is odorant-binding protein (OBP), which is a low-molecular-weight, water-soluble, extracellular protein that constitutes a homodimer of 19-kD polypeptides. OBP is located in mucous glands and secretions of nasal epithelia. OBP binds odorants with micromolar affinities and has been suggested to serve as a carrier, possibly involved in facilitating the onset and/or turn-off kinetics of odorant interaction with the ciliary membrane receptors (cf. Bignetti et al. 1985; Pevsner et al. 1986; Lee et al. 1987; Snyder et al. 1988).

Cytochrome P450. Another set of relevant molecular components is the cytochrome P450 enzyme system (Fig. 1). This multigene family of mono-oxygenases (Adesnik and Atchison 1986; Black and Coon 1987; Nebert and Gonzalez 1987) includes several dozen members, which catalyze the stereospecific conversion of C–H bonds to C–OH (hydroxyl groups) in various lipophilic substrates, including many odorants. They play a major role in metabolic pathways, e.g., steroid biosynthesis, and in detoxification. Olfactory epithelium contains high concentrations of cytochrome P450 (Dahl et al. 1982; Dahl 1988). P450 molecules could be important in converting odorants to nonodorous compounds, thus serving as a signal turn-off mechanism. Alternatively, P450 could be important in detoxifying odorants that accumulate in the olfactory epithelium. Such accumulation may occur due to the presence of specific receptors and binding proteins in the mucus and in the ciliary membranes. A third possibility is that P450 converts chemicals for which only low-affinity receptors exist into derivatives that can be recognized with a higher affinity.

As suggested by Margolis (1987), the cytochrome P450 enzyme system also has several properties that make it an excellent model for ORs (for a review of the expected properties of these receptors, see Lancet 1986):

1. P450 enzymes are integral membrane proteins, a property believed to be shared by ORs.
2. P450 molecules are diverse, with multiple forms of binding sites to accommodate multiple ligands and with broad and overlapping specificities, as suggested for ORs.
3. P450 enzymes, like ORs, recognize and act on hydrophobic compounds. Many of these compounds are xenobiotic; i.e., they come from the outside world and are absent in the organism, similar to odorants.
4. Upon substrate binding, P450 molecules undergo a conformational change, which plays a role in triggering some of the steps in the enzymatic cycle. A similar allosteric mechanism has been suggested to underlie olfactory reception (Dodd and Persaud 1981; Lancet 1986).
5. All P450 subtypes interact with another macromolecule, the NADPH cytochrome P450 reductase. This molecule is unique and shows no diversity, i.e., it serves as a "common" component capable of interacting with diverse P450 polypeptides. Similarly, many OR types may interact with common transduction components, such as G_s and adenylate cyclase.
6. At least one subtype of rat P450 recently cloned in our laboratory (P. Nef et al., in prep.) is specific for olfactory epithelium and is absent in several control tissues, including respiratory epithelium.
7. Many P450 subtypes are inducible by their substrate. Thus, it is possible that the expression of some olfactory P450 species is enhanced by exposure to specific odorants. Such odorant modulation, which is yet to be demonstrated for OR expression, may underlie some forms of long-term plasticity in the olfactory system.

Future work should help establish whether P450 enzymes actually play a role in olfactory reception. Future results will also indicate how many different olfactory P450 enzymes exist (i.e., how diverse is the olfactory P450 subfamily), what are the substrates for different olfactory P450 species, what is their tissue localization, and whether their expression is controlled by odorant exposure.

Taste Transduction

The cAMP cascade appears to be a prime candidate for mediating taste responses to some amino acids and to sweeteners, and perhaps also to bitter compounds. At present, there is less evidence for cAMP involvement in gustatory responses as compared to the case of olfactory responses.

Recent results (Striem et al. 1986; Lancet et al. 1987; B.J. Striem et al., in prep.) support a role for adenylate cyclase and G_s in gustatory reception. We have shown that the rat lingual epithelium membrane contains adenylate cyclase, whose activity is enhanced by sucrose and other mono- and disaccharides in a guanine-nucleotide-dependent manner. Activation occurred in the presence of both GTP and its nonhydrolyzable analog GTPγS, suggesting that gustatory G_s is of the "closed" type, similar to that of turkey erythrocytes (Londos et al. 1974). In this, the G_s of the taste system may be different from olfactory G_s, which is of the "open" type, and is equally and fully activated by GTPγS in the presence or in the absence of odorants. It should be noted that F^- ions activate maximally and abolish agonist activation in tongue membranes (B.J. Striem et al., in prep.), just as they do in olfactory cilia (Pace 1988). This observation provides further support to the specificity of tastant activation.

When testing different sweet-tasting compounds, it was found that the half-maximal activation of the tongue adenylate cyclase occurred at concentrations (0.1–1.0 M) that adequately matched the physiological half-maximal concentrations. The order of potency in adenylate cyclase activation (sucrose > glucose > maltose) is in good agreement with published electrophysiological results (Hagstrom and Pfaffmann 1959; Hiji and Imoto 1980).

Recently, two independent electrophysiological studies demonstrated the involvement of cAMP as a gustatory second messenger (Avenet et al. 1988; Tonosaki and Funakoshi 1988). These studies suggest that activation of cAMP-dependent protein kinase leads to phosphorylation of potassium channels (Fig. 2). Such modifications result in channel closure, thereby leading to cell depolarization, the well-documented consequence of taste-cell stimulation by tastants. These results complement and strongly support the biochemical evidence for cAMP involvement in taste transduction.

REFERENCES

Adesnik, M. and M. Atchison. 1986. Genes for cytochrome P450 and their regulation. *CRC Crit. Rev. Biochem.* **19:**247.

Anholt, R.R.H. 1987. Primary events in olfactory reception. *Trends Biochem. Sci.* **12:** 58.

———. 1988. Functional reconstitution of the olfactory membrane: Incorporation of the olfactory adenylate cyclase in liposomes. *Biochemistry* **27:** 6464.

Anholt, R.R.H., S.M. Mumby, D.A. Stoffers, P.R. Girard, J.F. Kuo, and S.H. Snyder. 1987. Transductory proteins of olfactory receptor cells: Identification of guanine nucleotide binding proteins and protein kinase C. *Biochemistry* **26:** 788.

Avenet, P., F. Hofmann, and B. Lindemann. 1988. Transduction in taste receptor cells requires cAMP-dependent protein kinase. *Nature* **331:** 351.

Bignetti, E., A. Cavaggioni, P. Pelosi, K.C. Persaud, R.T. Sorbi, and R. Tirindelli. 1985. Purification and characterization of an odorant-binding protein from cow nasal tissue. *Eur. J. Biochem.* **149:** 227.

Black, S.D. and M.J. Coon. 1987. P450 cytochromes: Structure and function. *Adv. Enzymol.* **60:** 35.

Chen, Z., D. Ophir, and D. Lancet. 1986a. Monoclonal antibodies to glycoproteins of frog olfactory cilia. *Brain Res.* **368:** 329.

Chen, Z., U. Pace, D. Ronen, and D. Lancet. 1986b. Polypeptide gp95: A unique glycoprotein of olfactory cilia with transmembrane receptor properties. *J. Biol. Chem.* **261:** 1299.

Chen, Z., U. Pace, J. Heldman, A. Shapira, and D. Lancet. 1986c. Isolated frog olfactory cilia: A preparation of dendritic membranes from chemosensory neurones. *J. Neurosci.* **6:** 2146.

Dahl, A.R. 1988. The effect of cytochrome P450-dependent metabolism and other enzyme activities in olfaction. In *Molecular neurobiology of the olfactory system* (ed. F.L. Margolis and T.V. Getchell), p. 51. Plenum Press, New York.

Dahl, A.R., W.M. Hadley, F.F. Hahn, J.M. Benson, and R.O. McClellan. 1982. Cytochrome P 450-dependent monooxygenases in olfactory epithelium of dogs: Possible role in tumorigenicity. *Science* **216:** 57.

Davies, J.T. 1971. Olfactory theories. In *Handbook of sensory physiology* (ed. L.M. Beidler), vol. 4, p. 322. Springer Verlag, Berlin.

DeSimone, J.A., K.C. Persaud, and G.L. Heck. 1988. Voltage clamp studies of the isolated olfactory mucosa. In *Molecular neurobiology of the olfactory system* (ed. F.L. Margolis and T.V. Getchell), p. 159. Plenum Press, New York.

Dodd, G., J. Pevsner, P.B. Sklar, and S.H. Snyder. 1986. Odorant binding protein, localization to nasal glands and secretions. *Proc. Natl. Acad. Sci.* **83:** 4942.

Dohlman, H.G., M.G. Caron, and R.J. Lefkowitz. 1987. A family of receptors coupled to guanine nucleotide regulatory proteins. *Biochemistry* **26:** 2657.

Getchell, T.V., F.L. Margolis, and M.L. Getchell. 1984. Perireceptor and receptor events in vertebrate olfaction. *Prog. Neurobiol.* **23:**317.

Gilman, A.G. 1987. G-proteins: Transducers of receptor-generated signals. *Annu. Rev. Biochem.* **56:** 615.

Hagstrom, E.G. and C. Pfaffmann. 1959. The relative taste effectiveness of various sugars and alcohols for the rat. *J. Comp. Physiol. Psychol.* **52:** 259.

Heldman, J. and D. Lancet. 1986. Cyclic AMP dependent protein phosphorylation in chemosensory neurons: Identification of cyclic nucleotide regulated phosphoproteins in olfactory cilia. *J. Neurochem.* **47:** 1527.

Hiji, Y. and T. Imoto. 1980. Some electrophysiological properties of taste receptors measured by summated nerve recordings. *Biomedical Res.* (supp.), **1:** 124.

Huque, T. and R.C. Bruch. 1986. Odorant- and guanine-nucleotide stimulated phosphoinositide turnover in olfactory cilia. *Biochem. Biophys. Res. Commun.* **137:** 36.

Kobilka, B.K., T. Frielle, S. Collins, T. Yang-Feng, T.S. Kobilka, U. Francke, R.L. Lefkowitz, and M. Caron. 1987. An intronless gene encoding a potential member of the family of receptors coupled to guanine nucleotide regulatory proteins. *Nature* **329:** 75.

Kropf, R., D. Lancet, and D. Lazard. 1987. A bovine olfactory cilia preparation: Specific transmembrane glycoproteins and phosphoproteins. *Soc. Neurosci. Abstr.* **13:** 1410.

Kurihara, K. and N. Koyama. 1972. High activity of adenyl cyclase in olfactory and gustatory organs. *Biochem. Biophys. Res. Commun.* **48:** 30.

Labarca, P., S.A. Simon, and R.R.H. Anholt. 1987. Activation by odorants of a multistate cation channel from olfactory cilia. *Proc. Natl. Acad. Sci.* **85:** 944.

Lancet, D. 1986. Vertebrate olfactory reception. *Annu. Rev. Neurosci.* **9:** 329.

———. 1988a. Molecular components of olfactory reception and transduction. In *Molecular neurobiology of the olfactory system* (ed. F.L. Margolis and T.V. Getchell), p. 25. Plenum Press, New York.

———. 1988b. Receptor-activated chemosensory enzyme cascades. In: *Receptor and transduction mechanisms in taste and olfaction* (ed. J.G. Brand et al.). Marcel Dekker, New York. (In press.)

Lancet, D. and U. Pace. 1987. The molecular basis of odor recognition. *Trends Biochem. Sci.* **12:** 63.

Lancet, D., B.J. Striem, U. Pace, U. Zehavi, and M. Naim. 1987. Adenylate cyclase and GTP binding protein in rat sweet taste transduction. *Soc. Neurosci. Abst.* **13:** 361.

Lancet, D., Z. Chen, A. Ciobotariu, F. Eckstein, M. Khen, J. Heldman, D. Ophir, I. Shafir, and U. Pace. 1988. Towards a comprehensive molecular analysis of olfactory transduction. *Ann. NY Acad. Sci.* **510:** 27.

Lee, K.-H., R.G. Wells, and R.R. Reed. 1987. Isolation of an olfactory cDNA: Similarity to retinol-binding protein suggests a role in olfaction. *Science* **235:** 1053.

Londos, C., Y. Salomon, M.C. Lyn, J.P. Harwood, M. Schramm, and M. Rodbell. 1974. 5′-Guanylylimidodiphosphate, a potent activator of adenylate cyclase systems in eukaryotic cells. *Proc. Natl. Acad. Sci.* **71:** 3087.

Margolis, F.L. 1987. Olfactory receptor neurons: Specific gene expression and a hypothetical model for stimulus receptors. *Discuss. Neurosci.* **4:** 47.

Menevse, A., G. Dodd, and T.M. Poynder. 1977. Evidence for the specific involvement of cAMP in the olfactory transduction mechanism. *Biochem. Biophys. Res. Comm.* **77:** 671.

Minor, A.V. and N.L. Sakina. 1973. Role of cyclic adenosine-3′,5′-monophosphate in olfactory reception. *Neurophysiology* **5:** 415.

Mollner, S. and T. Pfeuffer. 1988. Two different adenylyl cyclases in brain distinguished by monoclonal antibodies. *Eur. J. Biochem.* **171:** 265.

Nakamura, T. and G.H. Gold. 1987. A cyclic nucleotide-gated conductance in olfactory receptor cilia. *Nature* **325:** 442.

Nebert, D.W. and F.J. Gonzalez. 1987. P450 genes: Structure, evolution and regulation. *Annu. Rev. Biochem.* **56:** 945.

Nomura, T. and K. Kurihara. 1987. Effect of changed lipid composition on responses of liposomes to various odorants. *Biochemistry* **26:** 6141.

Pace, U. 1988. " Transduction mechanisms in olfactory reception." Ph.D. thesis, The Weizmann Institute, Rehovot, Israel.

Pace, U. and D. Lancet. 1986. Olfactory GTP-binding protein: Signal transducing polypeptide of vertebrate chemosensory neurons. *Proc. Natl. Acad. Sci.* **83:** 4947.

———. 1987. Molecular mechanisms in vertebrate olfaction: Implications to pheromone biochemistry. In *Pheromone biochemistry* (ed. G.D. Prestwich and G.J. Blomquist), p. 529. Academic Press, New York.

Pace, U., E. Hanski, Y. Salomon, and D. Lancet. 1985. Odorant-sensitive adenylate cyclase may mediate olfactory reception. *Nature* **316:** 255.

Pace, U., J. Heldman, I. Shafir, G. Rimon, and D. Lancet. 1987. Molecular correlates of olfactory adaptation: Adenylate cyclase and protein phosphorylation. *Soc. Neurosci. Abstr.* **13:** 362.

Persaud, K. 1981. Biochemical mechanisms in vertebrate primary olfactory neurons. In *Biochemistry of taste and olfac-*

tion (ed. R.H. Cagan and M.R. Kare), p. 333. Academic Press, New York.

Price, S. 1984. Mechanisms of stimulation of olfactory neurons: An essay. *Chem. Sens.* **8:** 341.

Shirley, S.G., C.J. Robinson, K. Dickinson, R. Aujla, and G.H. Dodd. 1986. Olfactory adenylate cyclase of the rat. Stimulation by odorants and inhibition by Ca^{2+}. *Biochem. J.* **240:** 605.

Sklar, P.B., R.R.H. Anholt, and S.H. Snyder. 1986. The odorant-sensitive adenylate cyclase of olfactory receptor cells: Differential stimulation by distinct classes of odorants. *J. Biol. Chem.* **261:** 15538.

Snyder, S.H., P.B. Sklar, and J. Pevsner. 1988. Olfactory receptor mechanisms: Odorant binding protein and adenylate cyclase. In *Molecular neurobiology of the olfactory system* (ed. F.L. Margolis and T.V. Getchell), p. 3. Plenum Press, New York.

Striem, B.J., U. Pace, U. Zehavi, M. Naim, and D. Lancet. 1986. Is adenylate cyclase involved in sweet taste transduction? *Chem. Sens.* **11:** 669.

Stryer, L. 1986. Cyclic GMP cascade of vision. *Annu. Rev. Neurosci.* **9:** 87.

Suzuki, N. 1988. Voltage- and cyclic nucleotide-gated currents in isolated olfactory receptor cells. In *Receptor and transduction mechanisms in taste and olfaction.* (ed. J.G. Brand et al.). Marcel Dekker, New York. (In press.)

Tonosaki, K. and M. Funakoshi. 1988. Cyclic nucleotides may mediate taste transduction. *Nature* **331:** 354.

Trotier, D. and P. McLeod. 1986. cAMP and cGMP open channels and depolarize olfactory receptor cells. *Chem. Sens.* **11:** 674.

Vodyanoy, V. 1988. Cyclic nucleotide-gated electrical activity in olfactory receptors. In *Receptor and transduction mechanisms in taste and olfaction* (ed. J.G. Brand et al.). Marcel Dekker, New York. (In press.)

Expression of G-protein α Subunits in Rat Olfactory Neuroepithelium: Candidates for Olfactory Signal Transduction

D.T. Jones,* E. Barbosa,† and R.R. Reed*

*Howard Hughes Medical Institute Laboratory of Genetics, Department of Molecular Biology and Genetics, †Department of Neurology, Johns Hopkins University School of Medicine, Baltimore, Maryland 21205

The olfactory system is responsible for transducing a wide variety of odorant stimuli into neuronal action potentials. Biochemical evidence suggests that the initial events of signal transduction are mediated by a stimulatory receptor-G-protein-coupled adenylyl cyclase cascade within the cilia of the olfactory sensory neurons (Pace et al. 1985; Sklar et al. 1986). Recently, cyclic-nucleotide-gated channels that open in response to increasing concentrations of cAMP have been identified in patch-clamp recording studies of olfactory cilia (Nakamura and Gold 1987). Together, these observations imply a model for olfactory signal transduction in which odorants cause a local increase in ciliary cAMP concentration, leading to the opening of cation channels within the ciliary membrane and depolarization of the sensory cell. This depolarization then elicits an action potential in the axonal process leading to the olfactory bulb.

G proteins are heterotrimers composed of α, β, and γ subunits (Stryer and Bourne 1986; Gilman 1987). The α subunit ostensibly confers identity to the multimer and, in most systems, governs the specificity of the interaction with receptor and effector molecules. Accordingly, there are several different types of α subunits, including $G_{s\alpha}$ and $G_{i\alpha}$, which stimulate and inhibit adenylyl cyclase, respectively; $G_{o\alpha}$, a GTP-binding protein rich in brain not yet associated with known receptors or effectors; and transducin (T_{α}), a G protein specific for the visual system. The β and γ subunits copurify as a tightly associated complex that appears to function interchangeably between different α subunits.

The observation of increased adenylyl cyclase activity in response to odorants implicates G_s as the G-protein component of the olfactory system. Pace and Lancet (1986) reported that olfactory cilia are rich in a cholera toxin substrate of approximately the same relative molecular weight as the small form of $G_{s\alpha}$ (45,000) from other tissues. To identify the G protein(s) involved in odorant signal transduction, we screened a rat olfactory cDNA library with an oligonucleotide directed to a highly conserved G_{α}-subunit domain and isolated five distinct G_{α}-subunit cDNA species, $G_{s\alpha}$ (45- and 52-kD forms), $G_{o\alpha}$, and three forms of $G_{i\alpha}$ (Jones and Reed 1987). One might expect that the G_{α} subunit responsible for odorant signal transduction would be expressed abundantly within the sensory neurons, and perhaps its expression may even be confined to olfactory tissue. None of these five clones are expressed exclusively in olfactory tissue. However, the known properties of $G_{s\alpha}$ fulfill the biochemical criterion for adenylyl cyclase activation, and its message is expressed abundantly in olfactory tissue, suggesting a role for this α subunit in olfactory signal transduction.

The olfactory neuroepithelium is the only tissue in the adult mammal where mature neurons undergo continuous replacement (Graziadei and Monti-Graziadei 1979; Monti-Graziadei and Graziadei 1979). Mature olfactory neurons have a lifetime of approximately 40 days, after which they die and are regenerated from a population of pluripotent stem cells that reside near the basal lamina. As each olfactory neuron matures, it projects a single axon through the cribiform plate to the olfactory bulb. In contrast to the visual system, all of the neurons on a given side of the nasal septum project to the lobe of the olfactory bulb on that same side. Unilateral bulbectomy or axotomy leads to selective, unilateral degeneration and loss of the olfactory neurons on that side of the nasal septum. In this paper, we utilize this neuronal depletion technique to examine the expression of the five G_{α} subunits within the olfactory sensory neurons.

EXPERIMENTAL PROCEDURES

Immunochemistry. Olfactory marker protein (OMP) staining was performed on paraffin-embedded PLPG-fixed (PLPG = 9 mM sodium periodate, 0.96% glutaraldehyde, 0.48% paraformaldehyde, 72 mM lysine, and 36 mM sodium phosphate at pH 7.4) rat olfactory sections using primary goat anti-OMP antibody (1:500), followed by visualization with ABC-HRP, as described by the manufacturer (Vector Labs).

Northern analysis. Total RNA was isolated from the dissected tissue of 3- to 6-month-old Sprague-Dawley rats by guanidinium-CsCl centrifugation (Chirgwin et al. 1979). Treated RNA was isolated from dissected olfactory turbinates 8 days after bulbectomy, when OMP levels are at a minimum (Monti-Graziadei and Graziadei 1979). For Northern analysis, glyoxal-treated total RNA (10 μg/lane) was resolved on 1%

agarose gels and transferred to Nytran filters (Thomas 1980). Filters were prehybridized at 37°C for common annealing region (CAR) oligonucleotide probes (Largent et al. 1988) or 42°C for nick-translated probes for 6-12 hours in hybridization solution (50% deionized formamide, 5 × SSC [1 × = 150 mM NaCl, 15 mM sodium citrate], 5 × Denhardt's solution [5 × = Ficoll, bovine serum albumin, and polyvinylpyrrolidone, each at 1 mg/ml], 0.1% SDS, and 100 μg/ml boiled, sonicated salmon sperm DNA). Hybridizations were performed in hybridization solution containing ^{32}P-labeled CAR probes at 37°C ($G_{o\alpha}$, all $G_{i\alpha}$ and OMP blots) or ^{32}P-labeled nick-translated probes at 42°C ($G_{s\alpha}$, G_β, and tubulin blots) for 18–24 hours. Filters hybridized with CAR probes were washed at 45°C in 2 × SSC and 0.1% SDS for 40 minutes, and filters hybridized with nick-translated probes were washed at 65°C in 0.5 × SSC and 0.1% SDS for 40 minutes before apposing to X-ray film.

Immunochemical analysis. Total olfactory tissue was dissected from normal or 8-day postbulbectomy 3–6-month old Sprague-Dawley rats, homogenized with a polytron in 50 mM Tris-HCl, 10 mM $MgCl_2$, 1 mM EDTA, and 1 mM phenylmethylsulfonyl fluoride (PMSF), and centrifuged at 800*g* for 5 minutes. Proteins from the supernatant were denatured with SDS, resolved in 10% polyacrylamide gels using the Laemmli (1970) buffer system, and transferred to Immobilon. G_β protein was detected using U-49 peptide antiserum (1:5000; Mumby et al. 1986), followed by ^{125}I-labeled protein A and autoradiography.

Adenylyl cyclase assays. Adenylyl cyclase assays were performed as described by Salomon (1979) using the 800*g* olfactory tissue supernatant described above.

RESULTS AND DISCUSSION

OMP—A Neuronal-specific Marker

OMP is a cytoplasmic 19-kD protein of unknown function, that is abundantly and exclusively expressed by mature neurons of the olfactory neuroepithelium (Margolis 1972; Keller and Margolis 1975; Monti-Graziadei et al. 1977). Immunohistochemical localization of OMP near the septum (Fig. 1), which separates the left and right olfactory turbinates, demonstrates the dramatic loss of OMP in the bulbectomized side (right), compared with the levels in the untreated side (left). OMP mRNA levels parallel protein levels and drop at least 20-fold in the RNA isolated from affected epithelium (Fig. 2a). In contrast, tubulin mRNA levels remain relatively constant, despite bulbectomy, indicating that depletion of the olfactory neurons does not lead to gross metabolic changes of neighboring epithelial cells (Fig. 2b). Since OMP is expressed only by the mature sensory neurons, it serves as a useful marker for their presence in the olfactory epithelium after bulbectomy.

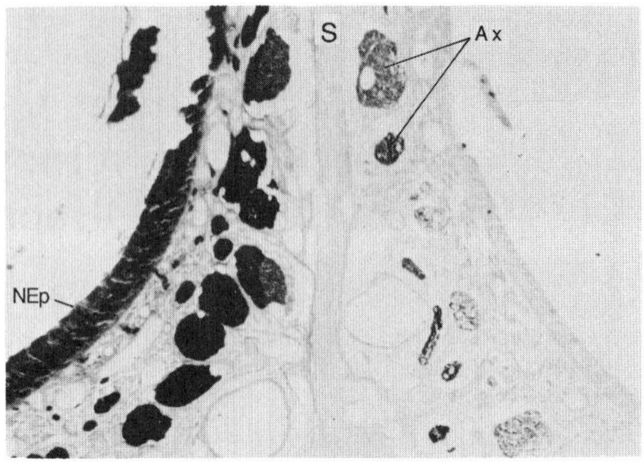

Figure 1. Immunohistochemical localization of OMP in rat olfactory neuroepithelium. Depicted is a coronal section through rat olfactory tissue. Removal of the right olfactory bulb 8 days prior to fixation has resulted in the morphologic disappearance of the sensory neurons on the right side of the septum. The left olfactory bulb was left intact. Note the intense OMP staining in the neuroepithelial layer (NEp) and the axonal bundles (Ax) of the untreated (left) side and the virtual absence of staining in these structures on the treated (right) side. S indicates nasal septum.

G_α-subunit Expression in Bulbectomized Tissue

To examine the expression of the five G_α subunits we previously identified within the olfactory neuroepithelium, we compared the message levels in normal

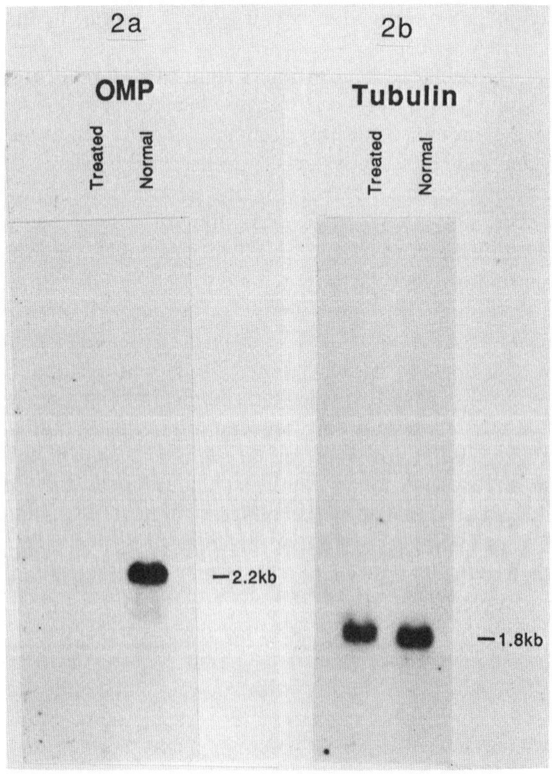

Figure 2. Northern analysis. Determination of message levels for OMP (*a*) or α-tubulin (*b*) in olfactory tissue derived from normal or bulbectomized animals. Autoradiographic exposure time varied from 1 to 2 days.

and bulbectomized RNA using monospecific probes (Fig. 3). Surprisingly, $G_{s\alpha}$, the most likely candidate for signal transduction on biochemical grounds, has increased levels of expression after bulbectomy. This increase could be due to a rapid surge of expression in the neuronal precursor cells as they differentiate into mature neurons or to increased expression by neighboring nonneuronal cells in response to neuronal loss. Alternatively, the increase may be attributed to overrepresentation of mRNA from nonneuronal cells because equal amounts of RNA were loaded in each lane.

The messages encoding the $G_{i\alpha}$ subunits, at least one of which inhibits adenylyl cyclase, continue to be expressed after bulbectomy (Fig. 3). Similar to $G_{s\alpha}$, the expression of each of the $G_{i\alpha}$'s increases after bulbectomy, suggesting that the corresponding messages may be expressed by neurons but are derived primarily from nonneuronal cells.

$G_{o\alpha}$ mRNA levels in olfactory tissue decrease after bulbectomy, suggesting that some of the message is derived from the olfactory neurons (Fig. 3). However, $G_{o\alpha}$ is unlikely to transduce odorant signals since message levels are low compared with $G_{s\alpha}$ (R.R. Reed et al., unpubl.), and the relative decrease in expression after bulbectomy is small. Furthermore, $G_{o\alpha}$ does not appear to interact with adenylyl cyclase (Roof et al. 1985).

In summary, although $G_{s\alpha}$ fulfills the biochemical criteria, analysis of the Northern blot data does not suggest unusually abundant expression of $G_{s\alpha}$ within the mature sensory neurons. The other G_α subunits neither fit the biochemical criteria nor, with the possible exception of $G_{o\alpha}$, seem to be expressed neuronally.

G_β Expression in Bulbectomized Tissue

In an attempt to interpret the conflicting biochemical and Northern hybridization results for $G_{s\alpha}$, we investigated the expression of the G_β subunit, which is common to all of the G-protein subtypes. A cDNA for rat G_β-1 has recently been isolated in our laboratory (R.S. Sosnowski et al., unpubl.). Although the message for G_β is not olfactory-specific, it is expressed abundantly in olfactory tissue. Northern analysis indicates that G_β mRNA levels decrease slightly (Fig. 4a), whereas immunochemical detection of G_β demonstrates a dramatic drop in protein levels after bulbectomy (Fig. 4b). This suggests that a considerable fraction of the G_β protein is contained in sensory neurons. The residual G_β message that remains after bulbectomy is probably used to synthesize G_β subunits that interact with the G_α subunits expressed in the remainder of the epithelium.

Adenylyl Cyclase Activity

Olfactory tissue is a rich source of adenylyl cyclase, with a specific activity tenfold higher than that found in

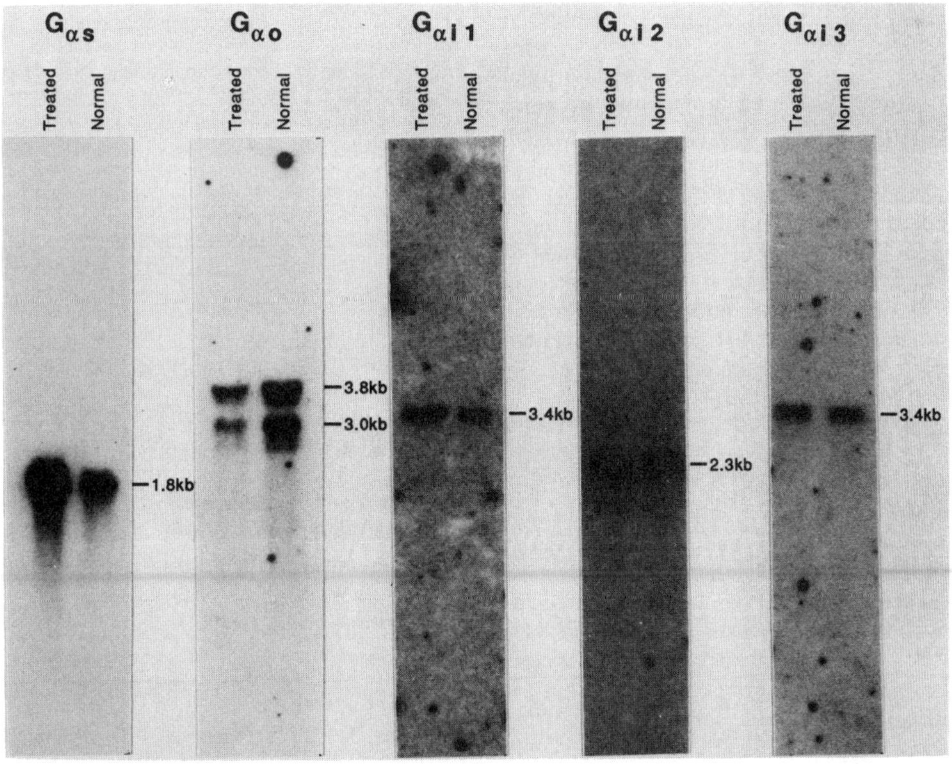

Figure 3. G_α subunit Northern analysis. Comparison of message levels for the five G_α subunits in olfactory tissue derived from normal or bulbectomized animals. Autoradiographic exposure time varied from 1 day to 3 weeks.

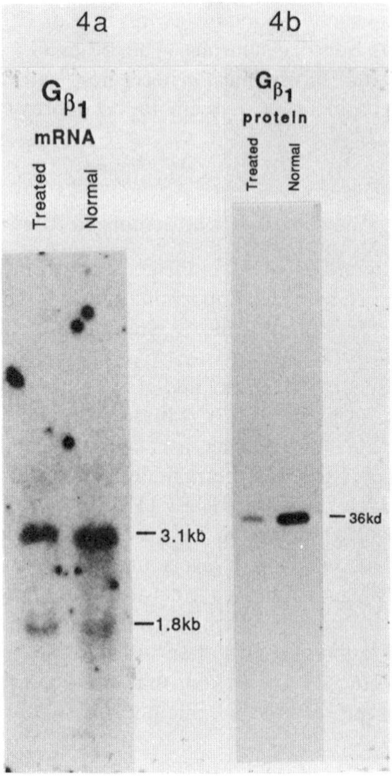

Figure 4. G_β subunit message and protein levels. Determination of G_β message (*a*) or protein (*b*) levels in olfactory tissue derived from normal or bulbectomized animals. Autoradiographic exposures were approximately 12 hr.

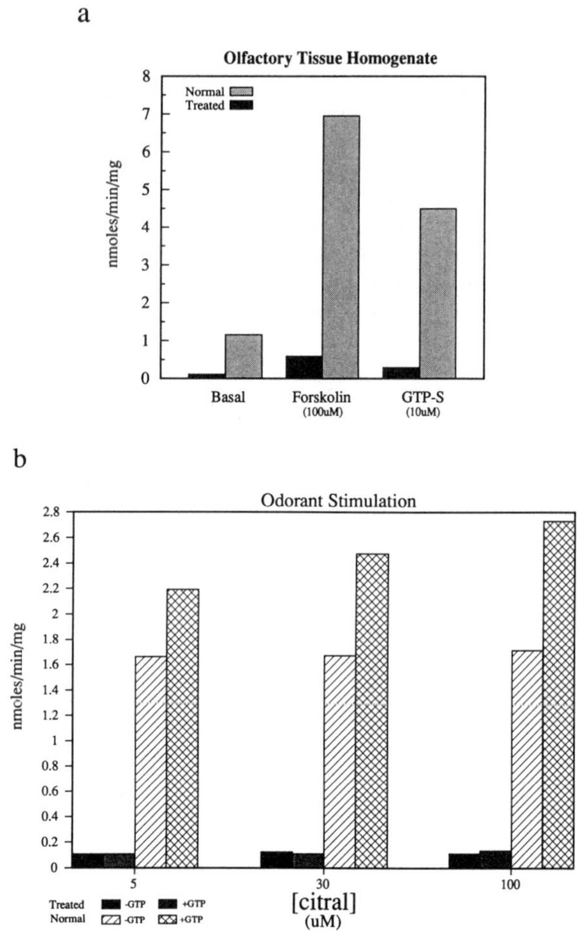

Figure. 5. Effect of bulbectomy on adenylyl cyclase activity in olfactory tissue. (*a*) Adenylyl cyclase activity activated by GTPγS (GTP-S) or forskolin in tissue derived from normal or bulbectomized animals. (*b*) Odorant-stimulated adenylyl cyclase activity in tissue derived from normal or bulbectomized animals.

other tissues (Pace et al. 1985). We measured adenylyl cyclase activity in total olfactory tissue homogenates after bulbectomy to determine whether the majority of the activity derives from olfactory sensory neurons. The specific activity of GTPγS or forskolin-stimulated catalytic subunit decreases approximately tenfold after bulbectomy (Fig. 5a). This argues that most of the adenylyl cyclase enzyme activity is contained within the sensory neurons. Furthermore, odorant stimulation of adenylyl cyclase activity disappears after bulbectomy (Fig. 5b), indicating that the neurons are responsible for odorant-dependent adenylyl cyclase activation. These results strongly support a model in which activation of neuronal adenylyl cyclase plays a pivotal role in olfactory signal transduction.

CONCLUSION

The decrease in G_β message and protein levels after bulbectomy suggests that a GTP-binding protein is expressed abundantly by the sensory neurons. The apparent paradox of increased or relatively unchanged G_α message levels and significantly decreased G_β levels is most easily explained by the presence of another G_α subunit in olfactory neurons. Recently, using low-stringency hybridization, we identified a novel G_α cDNA whose message levels drop dramatically after bulbectomy. The protein corresponding to this cDNA is therefore a likely candidate for the G protein involved in olfactory signal transduction. Apparently, this abundant G_α cDNA was missed in the original screen due to insufficient homology with the oligonucleotide probe.

ACKNOWLEDGMENTS

The authors are grateful to Dr. S. Mumby and Dr. A. Gilman for kindly providing the G_β antiserum and to Dr. F. Margolis for providing the OMP antiserum. This work was supported by grant PO1-CA16519 (R.R.R.) and grant NSPB 5K08NS01103-03 (E.B.) from the National Institutes of Health. D.T.J. is supported by a National Science Foundation predoctoral fellowship.

REFERENCES

Chirgwin, J.M., A.E. Przybyla, R.J. MacDonald, and W.J. Rutter. 1979. Isolation of biologically active ribonucleic acid from sources enriched in ribonuclease. *Biochemistry* **18:** 5294.

Gilman, A.G. 1987. G-proteins: Transducers of receptor-generated signals. *Annu. Rev. Biochem.* **56:** 615.

Graziadei, P.P.C. and G.A. Monti-Graziadei. 1979. Neurogenesis and neuron regeneration in the olfactory system of mammals. I. Morphological aspects of differentiation and structural organization of the olfactory sensory neurons. *J. Neurocytol.* **8:** 1.

Jones, D.T. and R.R. Reed. 1987. Molecular cloning of five GTP-binding protein cDNA species from rat olfactory neuroepithelium. *J. Biol. Chem.* **262:** 14241.

Keller, A. and F.L. Margolis. 1975. Immunocytochemistry of the olfactory marker protein. *J. Neurochem.* **24:** 1101.

Laemmli, U.K. 1970. Cleavage of structural proteins during the assembly of the head of bacteriophage T4. *Nature* **227:** 680.

Largent, B.L., D.T. Jones, R.R. Reed, R.C.A. Pearson, and S.H. Snyder. 1988. G-protein mRNA mapped in rat brain by *in situ* hybridization. *Proc. Natl. Acad. Sci.* **85:** 2864.

Margolis, F.L. 1972. A protein unique to the olfactory bulb. *Proc. Natl. Acad. Sci.* **69:** 1221.

Monti-Graziadei, G.A. and P.P.C. Graziadei. 1979. Neurogenesis and neuron regeneration in the olfactory system of mammals. II. Degeneration and reconstitution of the olfactory sensory neurons after axotomy. *J. Neurocytol.* **8:** 197.

Monti-Graziadei, G.A., F.L. Margolis, J.W. Harding, and P.P.C. Graziadei. 1977. Immunohistochemistry of olfactory marker protein. *J. Histochem. Cytochem.* **25(12):** 1311.

Mumby, S.M., R.A. Kahn, D.R. Manning, and A.G. Gilman. 1986. Antisera of designed specificity for subunits of guanine nucleotide-binding regulatory proteins. *Proc. Natl. Acad. Sci.* **83:** 265.

Nakamura, T. and G.H. Gold. 1987. A cyclic nucleotide-gated conductance in olfactory receptor cilia. *Nature* **325:** 442.

Pace, U. and D. Lancet. 1986. Olfactory GTP-binding protein: Signal-transducing polypeptide of vertebrate chemosensory neurons. *Proc. Natl. Acad. Sci.* **83:** 4947.

Pace, U., E. Hanski, Y. Salomon, and D. Lancet. 1985. Odorant-sensitive adenylate cyclase may mediate olfactory reception. *Nature* **316:** 255.

Roof, D.J., M.L. Applebury, and P.C. Sternweis. 1985. Relationships within the family of GTP-binding proteins isolated from bovine central nervous system. *J. Biol. Chem.* **260:** 16242.

Salomon, Y. 1979. Adenylate cyclase assay. *Adv. Cyclic Nucleotide Res.* **10:** 35.

Sklar, P.B., R.R.H. Anholt, and S.H. Snyder. 1986. The odorant-sensitive adenylate cyclase of olfactory receptor neurons. *J. Biol. Chem.* **261:** 15538.

Stryer, L. and. H.R. Bourne. 1986. G proteins: A family of signal transducers. *Annu. Rev. Cell Biol.* **2:** 391.

Thomas, P.S. 1980. Hybridization of denatured RNA and small DNA fragments transferred to nitrocellulose. *Proc. Natl. Acad. Sci.* **77:** 5201.

Studies on Light Transduction by Bacteriorhodopsin and Rhodopsin

M. Braiman, J. Bubis, T. Doi, H.-B. Chen,* S.L. Flitsch, R.R. Franke, M.A. Gilles-Gonzalez, R.M. Graham,† S.S. Karnik, H.G. Khorana, B.E. Knox, M.P. Krebs, T. Marti, T. Mogi, T. Nakayama, D.D. Oprian,‡ K.L. Puckett, T.P. Sakmar, L.J. Stern, S. Subramaniam, and D.A. Thompson

Departments of Biology and Chemistry, Massachusetts Institute of Technology, Cambridge, Massachusetts 02139

The visual photoreceptor pigments in vertebrates and invertebrates all use retinal (vitamin A aldehyde) as the light-absorbing molecule. Recently, Stoeckenius et al. (1979) discovered bacteriorhodopsin (bR) in the purple membrane of the extreme halophile, *Halobacterium halobium*, which also contains all-*trans* retinal as the chromophore. bR carries out light-dependent proton translocation from the inside to the outside of the *H. halobium* cell. Since the discovery of bR, *H. halobium* has been found to elaborate three more retinal-based light-transducing proteins. These are halorhodopsin, a chloride ion pump, and sensory rhodopsins I and II (Spudich and Bogomolni 1988).

The visual pigments as well as bR are all transmembrane proteins that are presumed to traverse the membrane seven times. The sole function of light in these retinal-based pigments appears to cause the isomerization of the protein-bound retinal molecules, 11-*cis* → all-*trans* for visual pigments and all-*trans* → 13-*cis* for bR. A series of structural changes ensues in the proteins, as evidenced by the photo intermediates (Wald 1968). We are carrying out structure-function studies of bacteriorhodopsin, bovine rhodopsin, and related proteins primarily by the technique of recombinant DNA; we summarize below the results we have obtained recently.

BACTERIORHODOPSIN

Bacteriorhodopsin as a Transmembrane Protein

bR contains a single polypeptide chain of 248 amino acids, whose sequence is known (Dunn et al. 1983). The protein is synthesized in the *H. halobium* cell as a precursor, with extra sequences of 13 amino acids at the amino terminus and an aspartic acid residue at the carboxyl terminus. The amino-terminal presequence appears to be removed cotranslationally, but the processing of the carboxy-terminal aspartic acid is not understood. The orientation of the mature protein in the membrane is such that the carboxyl terminus is at the cytoplasmic side with about 20 amino acid residues protruding into the cytoplasm, whereas the amino terminus is at the outside of the cell membrane (Fig. 1). All-*trans* retinal is attached as a Schiff base to Lys-216 in the putative helix G. From the mode of cross-linking of a photoactivatable retinal analog to the protein, the retinal appears to be oriented in the bilayer pointing toward the extracellular side of the protein (Huang et al. 1982). The latter result is not different from that obtained by other methods (Heyn et al. 1988).

The purple membrane can be solubilized to give bR as monomers in detergents such as Triton X-100. The bR monomers can be completely freed from endogenous phospholipids (Huang et al. 1980). The delipidated bR can be reconstituted into vesicles without denaturation or loss of the bound retinal. bR usually acquires inverted orientation in reconstituted vesicles irrespective of the phospholipids or the procedures used (Huang et al. 1980). However, largely right-side-out vesicles can be prepared from the *H. halobium* cell envelopes (MacDonald and Lanyi 1975). The most remarkable property of bR is its rapid and virtually quantitative renaturation following complete denaturation (Huang et al. 1981). Thus, the denatured apoprotein, bacterio-opsin (bO), refolds correctly and quantitatively to its native structure, binds retinal, and regenerates the bR chromophore. Purified proteolytic fragments of bR also reassociate to form the native bR-like structure (Liao et al. 1984).

Structure of bR and Intermediates in the Photocycle

A secondary structure model consistent with the electron and neutron diffraction data, the primary structure, and proteolytic cross-linking data is shown in Figure 1. Many uncertainties and questions remain in the model; for example, (1) correlation of the helices in the diffraction pattern with those in the primary structures; (2) the number of amino acids in the individual α helices; (3) the sizes and structures of the loops that connect the successive helices; and (4) the specific interactions between the membrane-embedded helices.

The photointermediates observed in the bR photocycle have been intensively studied (Stoeckenius et al.

Present addresses: *Shanghai Institute of Organic Chemistry, Academia Sinica, Shanghai, China; †Cellular and Molecular Research Laboratory, Massachusetts General Hospital and Harvard Medical School, Boston, Massachusetts 02114; ‡Department of Biochemistry, Brandeis University, Waltham, Massachusetts 02254.

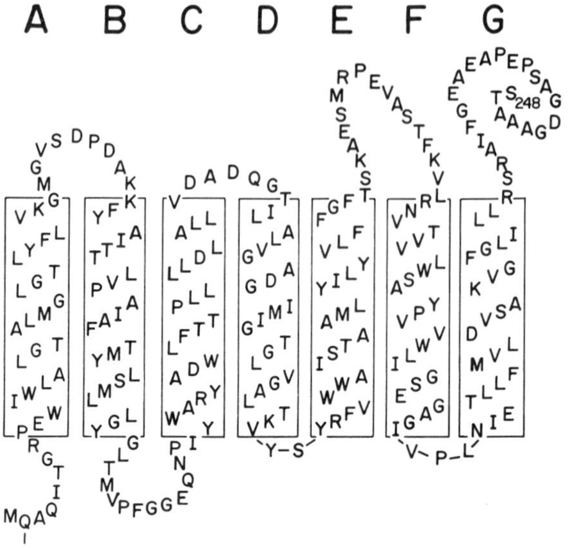

Figure 1. Secondary structure model of bacteriorhodopsin.

1979). An incisive synthesis of a great deal of conflicting work on the bR photocycle has been accomplished recently by Kouyama et al. (1988). Their model is consistent with the translocation of one proton per photocycle.

The photointermediates may reflect changes in the protein, internal motions, or conformational changes that correspond to steps in proton translocation. A number of questions arise; e.g., Is the transient deprotonation of the Schiff base at the M state directly coupled to H^+ translocation? The protonation-deprotonation of other groups such as tyrosine and aspartic acid side chains has been demonstrated recently (Braiman et al. 1988 and in prep.). The availability of defined mutants described below makes it possible to identify the specific amino acid residues involved.

Structure-function Studies by Amino Acid Replacements

The central question is the mechanism of proton translocation by bR. In particular, does it involve proton conductance along a hydrogen-bonded chain in which the side chains of certain specific amino acids participate? Second, the absorption characteristics of the chromophore in bR and the visual pigments display a large bathochromic shift relative to retinal Schiff base. How is this shift determined by protein-retinal interactions? Third, what is the nature of the specific interactions among the membrane-embedded helices and between the protein and the phospholipids that stabilize the folded structure of bR?

One approach to the above questions is specific amino acid replacements. Site-specific mutagenesis at the gene level provides a versatile approach for this purpose. The bR gene in *H. halobium* has been characterized (Dunn et al. 1981). To express the gene, two broad approaches can be considered. The first is transformation of *H. halobium* with a suitable vector carrying the gene or its mutants. This approach is currently under investigation. The second approach is the use of heterologous hosts such as *Escherichia coli* for expression of the bR gene and its mutants. This approach, which has been used in the present work, involves the following steps: (1) the use of a totally synthetic bR gene containing suitably placed unique restriction sites (Nassal et al. 1987) and the substitution of amino acids by replacement of restriction fragments encoding the native amino acid sequence by synthetic counterparts containing the altered codon(s); (2) development of a relatively efficient expression system for bO (Karnik et al. 1987); (3) rapid and efficient purification of the expressed bO mutants in the denatured form (Braiman et al. 1987); and (4) refolding of the bO mutants and regeneration of the bR-like chromophores in the presence of retinal.

To test if the H^+ translocation mechanism involves the participation of amino acids with side chains capable of forming H bonds, substitutions of single amino acids, such as Tyr→Phe, were carried out (Mogi et al. 1987).

Single Amino Acid Substitutions

The total substitutions made are shown in Table 1 (Fig. 1). All the mutants regenerated bR-like chromophores. However, the rates of chromophore regeneration for the different mutants varied widely (Table 2). Some mutants were very slow (e.g., Asp-212→Glu) in regenerating the chromophore, whereas an occasional mutant regenerated faster than native bO (e.g., Asp-85→Glu). Furthermore, on the basis of their effects on the absorption spectrum, the mutants could be divided into two classes (Table 3). Class I showed chromophores with λ_{max} values very similar to that of the wild-type bR. Examples are 8 of the 11 Tyr→Phe replacements (Mogi et al. 1987) and several Asp→Asn, Arg→Gln, and Pro→Ala substitutions. The chromophores in class II mutants showed significant to marked shifts in the λ_{max} from that of the wild type. Trp-86, -137, -182, and -189→Phe substitutions gave relatively large (20–70 nm) blue shifts (T. Mogi, unpubl.). The substitutions Pro-186→Leu (Val) and Thr-89→Val also gave large (70 nm) blue shifts, whereas some other substitutions (Asp-115→Glu, Asp-212→Ala) gave smaller (~20 nm) blue shifts. In contrast, the substitutions Asp-85→Asn, Asp-212→Glu, and Arg-82→Cys or Gln gave large (30–35 nm) red shifts.

We conclude that the spectral shifts in the mutants are caused by the substitution of those amino acids that interact with retinal in some way, whereas the amino acids corresponding to class I substitutions do not interact with the chromophore, at least in the ground state. It is noteworthy that the amino acids whose replacements cause spectral shifts are inside the membrane in the secondary structure model (Fig. 1).

Table 1. Single Amino Acid Substitution Mutants of bR

Amino acid[a] substitution	Positions in bR sequence
Arg→Gln	7, 82,[b] 134, 164, 175, 225, 227
Asp→Asn	36, 38, 85,[b] 96,[b] 102, 104, 115,[b] 212[b]
XXX→Cys	Gly-72, Arg-82, Thr-90,[b] Leu-92, Ser-169, Lys-216
Glu→Gln	9,[b] 74,[b] 161, 166, 194, 204
Pro→Ala	50,[b] 91,[b] 186[b]
Ser→Ala	183, 193
Thr→Val	17, 24, 46,47, 55, 89,[b] 90,[b]
	107, 121, 128, 142, 157, 178, 205
Trp→Phe	10, 12, 80, 86, 137, 138, 182,[b] 189[b]
Tyr→Phe	26, 43, 57, 64, 79, 83, 131, 133, 147, 150, 185[b]

Data from Hackett et al. (1987), Mogi et al. (1987, 1988), and T. Mogi et al. (unpubl.).
[a]Every amino acid of this type has been substituted as indicated.
[b]Other substitutions have also been made at this position.

Table 2. Chromophore Regeneration Rates of Selected Single Amino Acid Substitution Mutants of bR

Substitution	Regeneration rate $t_{1/2}$ (min)
(Wild type)	0.6
Arg-187→Gln	15
Asp-85→Asn	20
Asp-85→Glu	0.1>
Asp-212→Asn	31
Asp-212→Glu	38
Pro-50→Ala	31
Pro-91→Ala	29
Thr-89→Val	54
Tyr-57→Phe	22
Tyr-185→Phe	11

Table 4. Proton Pumping Activity of Selected Single Amino Acid Substitution

Substitution	Pumping rate[a]
(Wild type)	1.00
Asp-85→Asn	<0.001
Asp-85→Glu	0.32
Asp-96→Asn	0.03
Asp-96→Glu	0.82
Asp-212→Asn	0.15
Asp-212→Glu	0.06
Pro-186→Ala	1.09
Pro-186→Val	0.32
Trp-86→Phe	0.26
Trp-137→Phe	0.24
Tyr-185→Phe	0.36

[a]Fraction of wild-type proton pumping rate.

Amino Acid Substitutions That Affect Proton Pumping

Examples where proton pumping seemed to be partially affected were found (Table 4). Thus, in Tyr-185→Phe, it was 40% of normal. The meaning of partial blocks in proton pumping is not clear. Multiple interpretations are possible. Additional mutations were found that had strong effects on proton pumping. Replacement of Asp-85 or Asp-96 by asparagine abolished proton pumping completely or almost completely (Mogi et al. 1988). Asp-212 mutants (Asp-212→Glu and Asp-212→Asn) showed 10–20% proton pumping, whereas the mutant Asp-212→Ala was unstable to light and showed no detectable pumping. Therefore, it is certain that Asp-85 and Asp-96 in helix C (Fig. 1) are involved in proton translocation. Similarly, Asp-212 may also participate in proton translocation, or it may regulate protonation-deprotonation of the Schiff base. These results show for the first time that proton translocation occurs by a conductance pathway involving the side chains of several amino acids. An immediate task is the characterization of the photocycles of the above mutants. Fourier transform infrared spectroscopy and UV difference spectroscopy of the Tyr-185→Phe mutant have shown that the Tyr-185 residue normally changes its protonation state during the bR→K step and again during the L→M step (Ahl et at. 1988; Braiman et al. 1988). These and other time-resolved experiments (T. Marinetti et al., unpubl.) have begun to reveal steps at which Asp-85, -96, and -212 gain or lose protons.

Table 3. Absorbance Maxima of the Chromophores of Selected Single Amino Acid Substitution Mutants of bR

Substitution	λ_{max} (light adapted) (nm)
(Wild type)	561
Arg-82→Gln	580
Asp-85→Asn	594
Asp-85→Glu	560
Asp-212→Asn	548
Asp-212→Glu	581
Pro-186→Ala	567
Pro-186→Val	479
Thr-89→Val	493
Trp-182→Phe	491
Trp-189→Phe	521
Tyr-185→Phe	573

Projection Model for Bacteriorhodopsin

The projection model (Fig. 2) (Mogi et al. 1988) shows rough spatial relationships between the amino

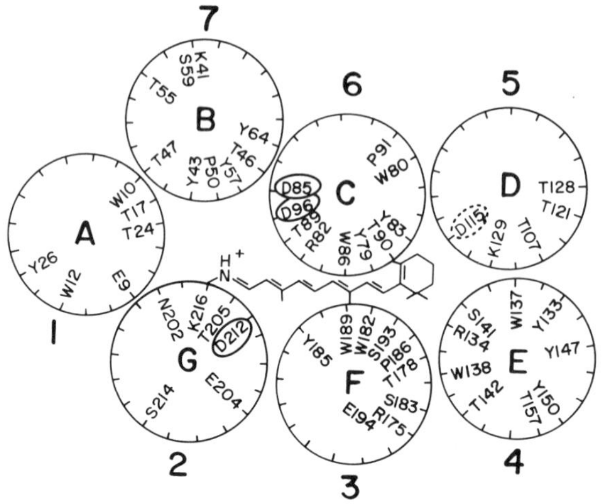

Figure 2. Helical wheel model of bacteriorhodopsin showing the relationships between the aspartic acid mutations that affect proton pumping and the amino acids that are believed to interact with the chromophore.

acids relevant to proton translocation. (Those amino acids that are believed to interact with retinal are shown in ovals.) One possible starting point for proton movement could be Asp-96. Another possible sequence of proton transfers could begin with the protonated Schiff base, which undergoes rapid deuterium/hydrogen exchange with water in the dark and perhaps maintains tight association with a water molecule (Hildebrandt and Stockburger 1984). If proton translocation begins at the Schiff base, how is the proton subsequently restored to this site? Is there a gated pore that makes the Schiff base transiently accessible to water from the cytoplasmic side? Such a light-dependent channel seems reasonable because of the known light-dependent bleaching of the protonated Schiff base by NH_2OH as well as its light-dependent reduction by $NaBH_4$.

Further Work

Although the results on mutants affecting proton translocation are encouraging, clearly we need to know a great deal more to understand the mechanism of the process. Site-specific mutagenesis can be taken further and, in addition, chemical and biophysical studies of the different groups of mutants must be comprehensive. Additional chemical approaches need to be tapped for understanding bR as a membrane protein. Furthermore, an essential requirement is the solution of the three-dimensional structure. This is most urgent, since we need to bring about a synthesis of the different types of results. Finally, we need to understand the energetics of coupling in the H^+ translocation process in bR.

VISUAL RHODOPSIN

Rhodopsin is the photoreceptor in the rod cell of the visual system. It is an integral membrane protein for which a secondary structure model (Fig. 3) has been proposed (Dratz and Hargrave 1983). The chromophore, 11-*cis*-retinal, is linked to Lys-296 as a Schiff base. The isomerization of retinal, 11-*cis* to all-*trans*, causes structural changes in the protein that result in the formation of a series of transient intermediates (Wald 1968). One intermediate, *meta*-rhodopsin II, seems to activate a guanine nucleotide-binding protein, transducin (T), and undergoes phosphorylation by rhodopsin kinase (Stryer 1986). Following activation, T_α exchanges bound GDP for GTP and dissociates from the $\beta\gamma$ subunits to form T_α-GTP. The latter then activates the rod cell cGMP phosphodiesterase. This results in a decrease in cGMP levels, which causes closure of plasma membrane cation channels, with consequent hyperpolarization of the rod cell. The above cascade is modulated by the intrinsic GTPase activity of G_α-GTP, by rhodopsin phosphorylation, and possibly by 48K protein binding to the phosphorylated rhodopsin.

Rhodopsin is an example par excellence of a large number of receptors (e.g., adrenergic, muscarinic, and neural peptide hormone receptors) containing seven transmembrane segments that undergo structural changes by light or ligand binding and activate G-binding proteins and are subsequently desensitized by phosphorylation and, possibly, other mechanisms (Sibley et al. 1987). Rhodopsin and related pigments of the visual system are attractive models for studies of the above general and important classes of signal transduction systems.

Orientation of Retinal in Rhodopsin

At one end, retinal is tethered to Lys-296. However, its orientation and the nature of its binding interactions within the protein are not known. One approach is the use of a retinal analog that carries a photoactivatable group. Following photolysis, the specific amino acid(s) to which the analog becomes cross-linked could be determined. Rhodopsin was reconstituted with the ^{3}H-labeled analog, 1,5-dimethyl-3-trifluoromethyl diazirinophenyl 11-*cis*-retinal (I).

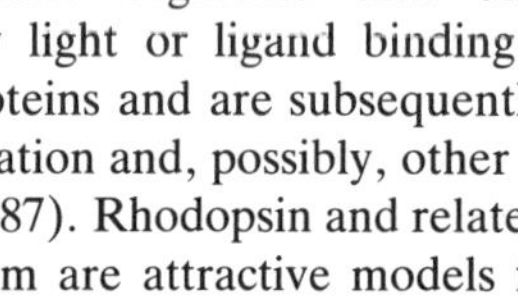

(I)

A rhodopsin-like pigment with λ_{max} at 458 nm was obtained. The linkage of the analog to Lys-296 via Schiff base was confirmed. Furthermore, on illumination, the reconstituted rhodopsin activated transducin in a light-dependent manner and also underwent phosphorylation by rhodopsin kinase. Photolysis of the reconstituted pigment at 365 nm (−15°C) led to the cross-linking of the analog to the opsin. After proteolysis and separation of the resulting peptides, two radioactive peptides were obtained. Sequence analysis showed that

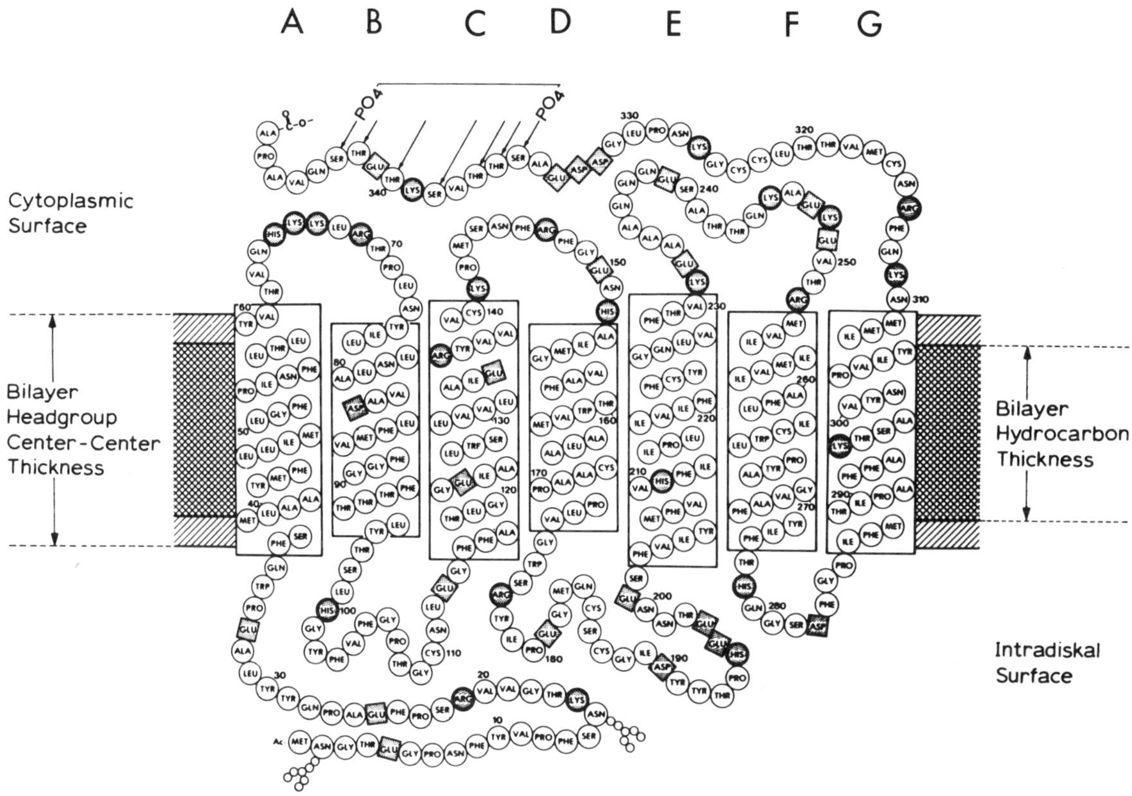

Figure 3. A secondary structure model of bovine rhodopsin (Reprinted, with permission, from Dratz and Hargrave 1983.)

the major cross-linking occurred within the sequences Gly-114-Glu-150 (putative helix C) and the minor within Ser-202-Gly-232 (putative helix E). Thus, the β-ionone ring of retinal appears to orient toward helices C and E from its point of attachment in helix G. Identification of sites of cross-linking should provide insight into specific interactions of the retinal with amino acid side chains that determine the spectral properties of the pigment.

Expression of a Rhodopsin Gene in COS-1 Cells

The first requirement for structure-function studies of rhodopsin is the satisfactory expression of rhodopsin in a functional form. We have previously reported on the total synthesis of a gene for rhodopsin (Ferretti et al. 1986). The chief motivation, as with bR, was site-specific mutagenesis by cassette replacement.

The vector for the expression of the rhodopsin gene in the monkey kidney cell line, COS-1, has been described elsewhere (Oprian et al. 1987). Isolation of more than 80% pure rhodopsin from the cells was accomplished by the use of an immunoaffinity procedure. Rhodopsin thus prepared showed the characteristic UV/Vis absorption spectrum and activated bovine transducin in a light-dependent manner. The specific activity of COS cell rhodopsin was identical to that of purified bovine rhodopsin.

Expression of Rhodopsin in *Xenopus* Oocytes and Coupling to Endogenous Channels

Xenopus oocytes provide a complementary expression system in which receptors and channels can be assayed in situ (Dascal 1987). The expression of the synthetic gene encoding bovine rhodopsin, its SP6 transcripts, and total retinal mRNA was demonstrated in *Xenopus* oocytes. Immunoprecipitation showed the presence of three forms of opsin, with an apparent relative molecular weight of 41K, 35K, and 30K. The 30K protein was unglycosylated opsin, whereas the higher-molecular-weight proteins, the major products, were differentially glycosylated forms. Addition of 11-*cis*-retinal to the cells expressing the above proteins regenerated rhodopsin, which, after purification by immunoaffinity chromatography, showed the expected UV/Vis absorption spectrum and characteristic light-dependent activation of T. Oocytes expressing rhodopsin exhibited light-dependent ionic currents that were detected by voltage clamp techniques (Fig. 4).

In Vivo Phosphorylation of Rhodopsin in COS Cells

Light-dependent desensitization of visual transduction resulting in impaired G-protein activation is thought to involve receptor phosphorylation (Kuhn et al. 1973; Liebman and Pugh 1980). Phosphorylation of

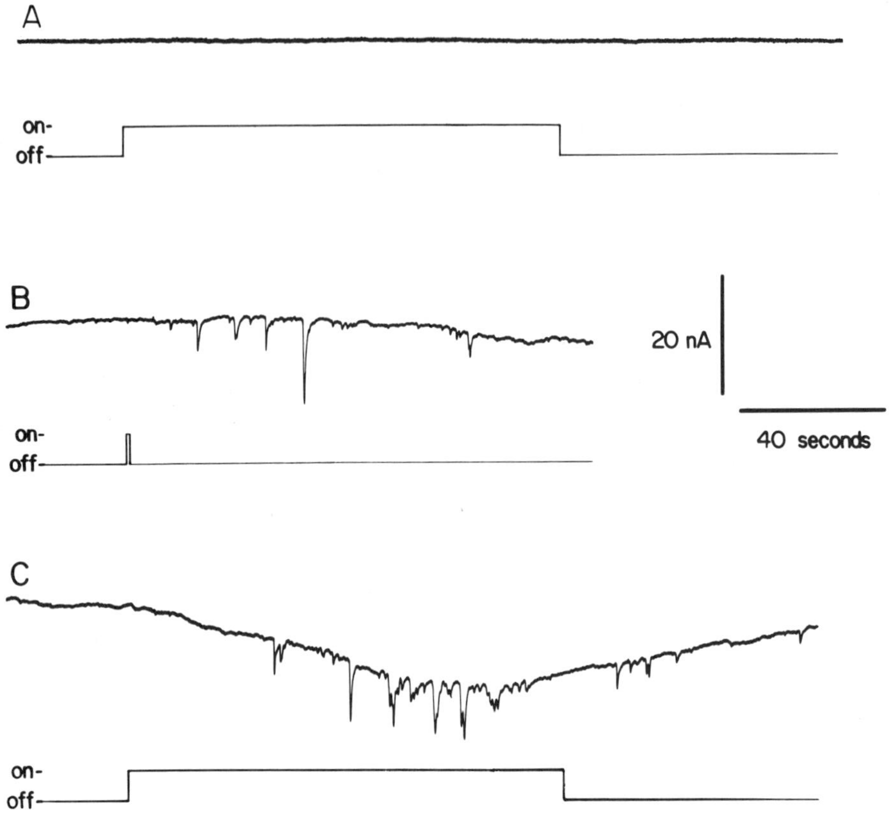

Figure 4. Cells were perfused with 11-*cis*-retinal, and the currents flowing through the plasma membrane were recorded. The membrane potential of the cells was clamped at −80 mV, and the current signal was low-pass-filtered at 10 Hz upon playback. In each panel, the upper trace is the current record and the lower trace is the light stimulus monitor. Inward current is represented as a downward deflection of the record. Upward deflections of the stimulus monitor (on) denote the time the cell was exposed to light. (*A*) Current response of an uninjected cell to 2 min of illumination. (*B*) Currents induced in an oocyte expressing rhodopsin in response to a 1-sec flash of white light. (*C*) Current response of the same cell to continuous (2 min) illumination.

rhodopsin as expressed in COS cells was tested as follows. The transfected COS cells were incubated with inorganic ^{32}P after the addition of 11-*cis*-retinal. Samples were either left in the dark or exposed to room light. Rhodopsin was then purified. SDS-PAGE, followed by silver-staining and autoradiography, showed that rhodopsin regenerated with 11-*cis*-retinal and exposed to light in COS cells was phosphorylated, whereas controls (dark, COS cell opsin lacking 11-*cis*-retinal) were not phosphorylated (Fig. 5). The results indicate light-dependent phosphorylation of rhodopsin in COS cells by endogenous receptor kinases such as β-adrenergic receptor kinases (Benovic et al. 1986). Furthermore, purified COS cell rhodopsin, following incorporation into phospholipid vesicles, was also phosphorylated in a light-dependent manner by rod outer segment extracts enriched for rhodopsin kinase activity.

Role of Cysteine Residues in Rhodopsin Structure-Function

The role of cysteine residues in integral membrane proteins has been studied by a number of workers, but clear-cut conclusions are largely lacking. Bovine rhodopsin contains ten cysteine residues in the primary structure (Fig. 3). These residues are conserved in human and bovine rhodopsin (Applebury and Hargrave 1986). The cysteines appear to be distributed into three topographical groups as follows: three cysteines (Cys-110, Cys-185, and Cys-187) are in the intradiskal loops, four cysteines (Cys-140, Cys-167, Cys-222, and Cys-264) are membrane-embedded, and three cysteines (Cys-316, Cys-322, and Cys-323) are located in the carboxy-terminal tail (Dratz and Hargrave 1983). Conflicting conclusions have been drawn by different workers regarding the involvement of different cysteine residues in disulfide bond formation.

We have constructed ten mutants of the opsin gene in which cysteines were replaced by serines as shown in Table 5. All of the mutants (CysI-CysVII) were expressed in COS cells. An immunoblot of the product is shown in Figure 6. The band patterns obtained with CysII (replacements of Cys-140, -167, -222, and -264), CysIII (replacements of Cys-316, -322, and -323), and CysV (replacements as in CysII and CysIII) are typical of the pattern obtained with the wild-type rhodopsin gene. However, the mutant groups (CysI, CysIV, CysVI, and CysVII) gave patterns that were similar to each other, but radically different from that obtained

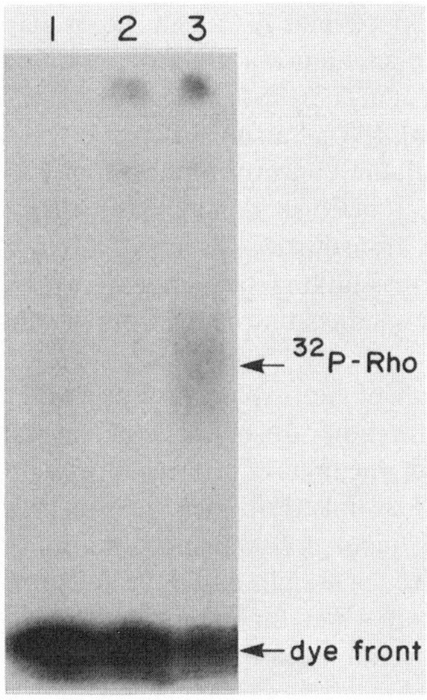

Figure 5. Light-dependent in vivo phosphorylation of rhodopsin in COS cells. The rhodopsin gene was expressed in COS cells; ^{32}P inorganic phosphate and 11-*cis*-retinal were added. A portion of the cells were exposed to light and the remainder was kept in the dark. After purification by immunoaffinity absorption, the cells were subjected to SDS-PAGE and autoradiography. (Lane *1*) Rhodopsin purifed from COS cells incubated with ^{32}P but without 11-*cis* retinal; (lane *2*) rhodopsin from cells preincubated with 11-*cis*-retinal and ^{32}P and kept in the dark; (lane *3*) rhodopsin from 11-*cis*-retinal-treated cells, labeled with ^{32}P under light exposure.

with the wild type. In each, there were faster-migrating bands, suggesting altered levels of glycosylation. The area above the band corresponding to the "normally" glycosylated rhodopsin was, in each case, strikingly devoid of slower-moving proteins (Fig. 7). In addition, whereas the class B (Table 5) mutants all regenerate the chromophore with 11-*cis*-retinal and activate transducin in a light-dependent manner, none of the class A (Table 5) mutants regenerate the chromophore or activate transducin.

Further study with single Cys→Ser substitution mutants showed (Table 5) that replacement of Cys-110 or

Table 5. Cys/Ser Substitutions in Rhodopsin

Mutants	Sites mutated (Cys to Ser)	Domain
Class A		
CysI	110, 185, 187	intradiskal
CysIV	110, 140, 167, 185, 187, 222, 264	intradiskal + membrane
CysVI	110, 185, 187, 316, 322, 323	intradiskal + membrane
CysVII	all ten	all
Cys-110	110	intradiskal
Cys-187	187	intradiskal
Class B		
CysII	140, 167, 222, 264	membrane
CysIII	316, 322, 323	cytoplasmic
CysV	140, 167, 222, 264, 316, 322, 323	membrane + cytoplasmic
Cys-185	185	intradiskal

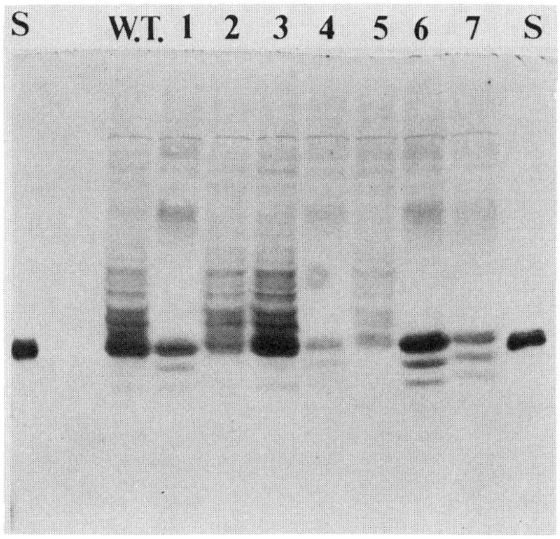

Figure 6. Immunoblot analysis of rhodopsin cysteine mutants. Transiently transfected COS cells were solubilized in CHAPS detergent, and soluble fractions were run on SDS-polyacrylamide gels. Description of the mutants is in Table 5. (Lane *S*) Bovine rhodopsin standard; (lane *W.T.*) wild-type COS cell rhodopsin; (lane *1*) mutant CysI; (lane *2*) mutant CysII; (lane *3*) mutant CysIII; (lane *4*) mutant CysIV; (lane *5*) mutant CysV; (lane *6*) mutant CysVI; (lane *7*) mutant CysVII.

Cys-187 conferred on these mutants the phenotype of class A mutants (CysI) (Fig. 7), whereas the substitution of Cys-185 gave the wild-type protein pattern.

These results show that, although all the other cysteine residues in rhodopsin are dispensable, Cys-110 and Cys-187 are necessary. Presumably, the formation of a specific disulfide bond between Cys-110 and Cys-187 is required for the formation of the correct tertiary structure. This enables the binding of 11-*cis*-retinal and the subsequent activation of transducin.

Transducin Interaction with Photolyzed Rhodopsin

To probe the rhodopsin domain(s) that interacts with transducin, rhodopsin mutants with alterations on the cytoplasmic surface have been prepared (Franke et al. 1988). The following three mutants in loop EF (Fig. 3) were prepared: mutant 1, Glu-239→Gln; mutant 2, Lys-248→Leu; and mutant 3, Glu-247→Gln, Lys-248→Leu, and Glu-249→Gln. Following expression in COS cells, all of the mutant proteins displayed wild-type visible absorption spectra. However, as seen in Figure 7, only mutants 1 and 3 stimulated the light-dependent GTPase activity of transducin. Mutant 2 failed to activate transducin. These results show that the loop EF is involved in some way in the interaction with transducin.

Deletions in the Intradiskal Loops

A series of mutants with varying deletions in the loops BC, DE, and FG were prepared. The amino acid sequences deleted were as follows: (1) 102–109, (2) 177–182, (3) 181–197, (4) 189–197, and (5) 280–283.

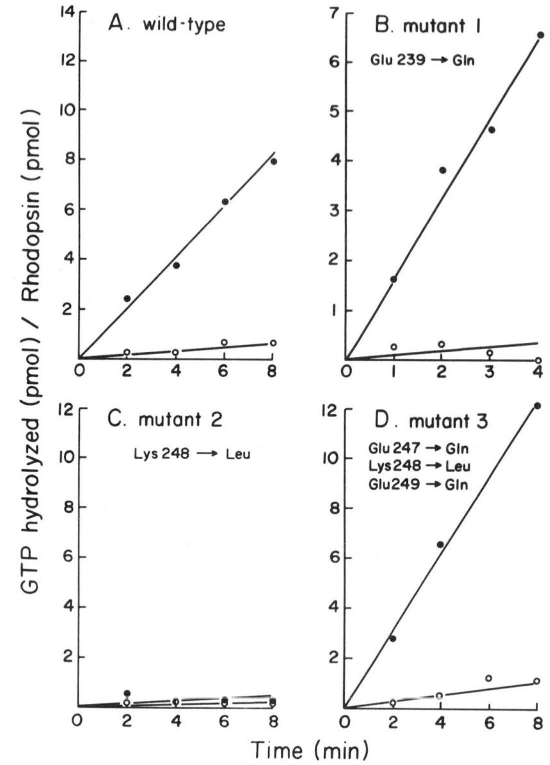

Figure 7. Rhodopsin mutant 2 (Lys-248→Leu) did not enhance the GTPase activity of transducin. A time course of GTP hydrolysis by transducin is shown for COS-1 rhodopsin and for each mutant. (*A*) Rhodopsin purified from COS-1 cells (wild-type); (*B*) mutant 1; (*C*) mutant 2; (*D*) mutant 3. Reactions were performed in the light and in the dark. Mutant 2 did not activate transducin. The specific activities of the other 2 mutants (1.6 min^{-1}) were similar to that of wild-type COS-1 cell rhodopsin (1.0 min^{-1}).

The mutant genes were expressed in COS cells, and the proteins were purified after the addition of retinal. Preliminary results show that none of the mutants bound 11-*cis*-retinal.

Removal of Phosphorylation Sites by Amino Acid Substitution

Rhodopsin undergoes light-dependent phosphorylation at threonine and serine residues near the carboxyl terminus (Wilden and Kuhn 1982). Some of the questions that need to be answered are (1) What is the role of phosphorylation at multiple sites? (2) Is there a temporal sequence in phosphorylation at the different sites and, if so, (3) what is the effect of blocking some or all of the phosphorylation sites with amino acids, such as alanine? Therefore, we are studying a series of mutants in which phosphorylation sites are systematically substituted by alanine residues singly or in groups.

TRANSDUCIN

Regions of Contact between β and γ Subunits

Transducin is a heterotrimer composed of α, β, and γ subunits with molecular weights of about 40K, 36K, and 8K, respectively. Although the primary structures of the three subunits via molecular cloning are known (Medynski et al. 1985; Sugimoto et al. 1985; Yatsunami and Khorana 1985; Yatsunami et al. 1985), no information is available concerning the topological organization of the functional domains and their interactions during the transduction process. We are employing chemical cross-linking techniques to identify specific amino acid residues that play a role in subunit-subunit interactions.

The ability of cupric phenanthroline (CuPh) to catalyze the formation of disulfide bonds in transducin was investigated. Incubation of transducin with CuPh produced a cross-linked species that migrated on electrophoresis under nonreducing conditions with an apparent molecular weight of 43K. The latter species was not observed when the sample was treated with β-mercaptoethanol before electrophoresis. Again, the 43K band was absent if the incubation mixture was treated with *N*-ethylmaleimide and EDTA prior to the addition of CuPh. Further experiments show that the 43K product results from disulfide bonds induced between T_β and T_γ. Identification of the specific cysteines involved is under investigation.

Synthesis and Expression of a Gene for the α Subunit of Transducin

A gene for the α subunit of transducin (T_α) was synthesized. By design, the synthetic gene contains 38 unique restriction sites, so as to facilitate site-specific mutagenesis by restriction fragment replacement. The 1076-bp gene was assembled by enzymatic joining of 44 synthetic oligonucleotides. The oligonucleotides were first joined to form four fragments of roughly equal lengths. These were purified and then cloned to assemble the entire gene (Sakmar and Khorana 1988). The synthetic T_α gene was expressed in transiently transfected COS cells using the expression vector (pMT3) described previously (Franke et al. 1988). The level of the synthetic T_α gene expression was comparable to the level of the corresponding cDNA expression, indicating that the codon changes introduced in the synthetic gene

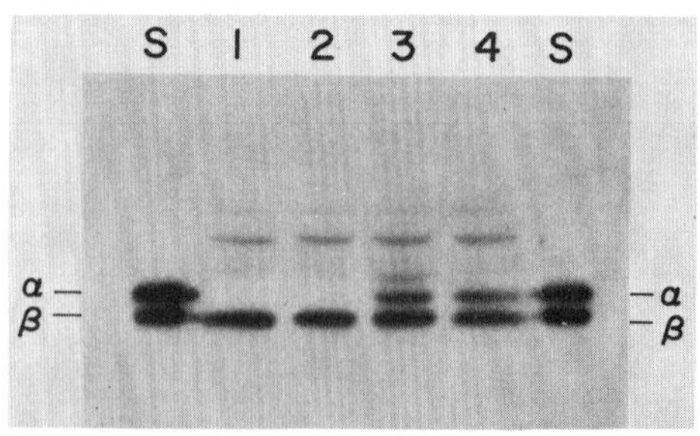

Figure 8. Immunoblot analysis of the α subunit of transducin was expressed in COS cells. (Lane *S*) Purified bovine holotransducin (100 ng); (lane *1*) mock transfected COS cells; (lane *2*) COS cells transfected with expression vector without insert; (lane *3*) COS cells transfected with vector containing synthetic gene; (lane *4*) COS cells transfected with vector containing cDNA. Endogenous COS cell G-protein 36 kD β subunit cross-reacts with anti-holotransducin antibody.

did not affect translation. Furthermore, levels of the endogenous COS cell G-protein 36-kD β subunit did not appear to be affected by overexpression of T_α. The 36-kD β subunit of various G proteins has been shown to be immunologically identical to T_β (Mumby et al. 1986). An immunoblot analysis using an anti-holotransducin antibody is shown in Figure 8. Purification and characterization of the T_α expressed in COS cells are in progress.

ACKNOWLEDGMENTS

This work was supported by grant GM-28289-07 from the National Institutes of Health and grant N-00014-82-K-0668 from the Office of Naval Research, Department of the Navy. M.B. was supported by a postdoctoral fellowship from the Helen Hay Whitney Foundation. S.L.F. and R.R.F. are the recipients of a fellowship and a graduate fellowship, respectively, from Deutscher Akademischer Austauschdienst. B.E.K. is the recipient of an NIH fellowship 1-F32-EY05916. T.M. is the recipient of a postdoctoral fellowship from the Swiss National Science Foundation. K.L.P. is the recipient of an American Cancer Society grant 03079. T.P.S. is the recipient of the NIH National Research Service award 1-F32-GM-11305. L.J.S. is an NIH predoctoral trainee. S.S. is the recipient of Damon Runyon-Walter Winchell Cancer Research Fund grant DRG-937. D.A.T. is the recipient of a postdoctoral fellowship from the NIH (F-32 EY-06024) and a Damon Runyon-Walter Winchell Cancer Fund fellowship (DRG-831). The assistance of Judy Carlin in preparing this manuscript is greatly appreciated.

REFERENCES

Ahl, P.L., L.J. Stern, D. During, T. Mogi, H.G. Khorana, and J.K. Rothschild. 1988. Effects of amino acid substitutions in the F-helix of bacteriorhodopsin: Low-temperature UV-visible difference spectroscopy. *J. Biol. Chem.* (in press).

Al-Saleh, S., M. Gore, and M. Akhtar. 1987. On the disulphide bonds of rhodopsins. *Biochem. J.* **246:** 131.

Applebury, M.L. and P.A. Hargrave. 1986. Molecular biology of the visual pigments. *Vision Res.* **26:** 1881.

Benovic, J.L., F. Mayor, Jr., R.L. Somers, M.G. Caron, and R.L. Lefkowitz. 1986. Light-dependent phosphorylation of rhodopsin by β-adrenergic receptor kinase. *Nature* **321:** 869.

Braiman, M.S., L.J. Stern, B.H. Chao, and H.G. Khorana. 1987. Structure-function studies on bacteriorhodopsin. IV. Purification and renaturation of bacterio-opsin polypeptide expressed in *Escherichia coli. J. Biol. Chem.* **262:** 9271.

Braiman, M.S., T. Mogi, L.J. Stern, N.R. Hackett, B.H. Chao, H.G. Khorana, and K.J. Rothschild. 1988. Vibrational spectroscopy of bacteriorhodopsin mutants. I. Tyrosine-185 deprotonates and protonates during the photocycle. *Proteins* **3:** 219.

Dascal, N. 1987. The use of *Xenopus* oocytes for the study of ion channels. *CRC Crit. Rev. Biochem.* **22:** 317.

Dratz, E.A. and P.A. Hargrave. 1983. The structure of rhodopsin and the rod outer segment disk membrane. *Trends Biochem. Sci.* **8:** 128.

Dunn, R., J. McCoy, M. Simsek, A. Majumdar, S.H. Chang, U.L. RajBhandary, and H.G. Khorana. 1981. The bacteriorhodopsin gene. *Proc. Natl. Acad. Sci.* **78:** 6744.

Dunn, R.J., N.R. Hackett, K.S. Huang, S.S. Jones, H.G. Khorana, D.-S. Lee, M.-J. Liao, K.-M. Lo, J. McCoy, S. Noguchi, R. Radhakrishnan, and U.L. RajBhandary. 1983. Studies on the light-transducing pigment bacteriorhodopsin. *Cold Spring Harbor Symp. Quant. Biol.* **48:** 853.

Ferretti, L., S.S. Karnik, H.G. Khorana, M. Nassal, and D.D. Oprian. 1986. Total synthesis of a gene for bovine rhodopsin. *Proc. Natl. Acad. Sci.* **83:** 599.

Franke, R.R., T.P. Sakmar, D.D. Oprian, and H.G. Khorana. 1988. A single amino acid substitution in rhodopsin (lysine 248→leucine) prevents activation of transducin. *J. Biol. Chem.* **263:** 2119.

Hackett, N.R., L.J. Stern, B.H. Chao, K.A. Kronis, and H.G. Khorana. 1987. Structure-function studies on bacteriorhodopsin. V. Effects of amino acid substitutions in the putative helix F. *J. Biol. Chem.* **262:** 9277.

Heyn, M.P., J. Westerhausen, I. Wallat, and F. Seiff. 1988. High-sensitivity neutron diffraction of membranes: Location of the Schiff base end of the chromophore of bacteriorhodopsin. *Proc. Natl. Acad. Sci.* **85:** 2146.

Hildebrandt, P. and M. Stockburger. 1984. Role of water in bacteriorhodopsin's chromophore: Resonance Raman study. *Biochemistry* **23:** 5539.

Huang, K.-S., H. Bayley, and H.G. Khorana. 1980. Delipidation of bacteriorhodopsin and reconstitution with exogeneous phospholipid. *Proc. Natl. Acad. Sci.* **77:** 323.

Huang, K.-S., R. Ramachandran, H. Bayley, and H.G. Khorana. 1982. Orientation of retinal in bacteriorhodopsin as studied by cross-linking using a photosensitive analog of retinal. *J. Biol. Chem.* **257:** 13616.

Huang, K.-S., H. Bayley, M.-J. Liao, E. London, and H.G. Khorana. 1981. Refolding of an integral membrane protein: Denaturation, renaturation and reconstitution of intact bacteriorhodopsin and two proteolytic fragments. *J. Biol. Chem.* **256:** 3802.

Karnik, S.S., M. Nassal, T. Doi, and H.G. Khorana. 1987. Structure-function studies on bacteriorhodopsin. II. Improved expression of the bacterio-opsin gene in *Escherichia coli. J. Biol. Chem.* **262:** 9255.

Kouyama, T., A.N. Kouyama, A. Ikegami, M.K. Mathew, and W. Stoeckenius. 1988. Bacteriorhodopsin photoreaction: Identification of a long-lived intermediate $N(P,R_{350})$ at high pH and its M-like photoproduct. *Biochemistry* **27:** 5855.

Kuhn, H., J.H. Cook, and W.J. Dreyer. 1973. Phosphorylation of rhodopsin in bovine photoreceptor membranes: A dark reaction after illumination. *Biochemistry* **12:** 2495.

Liao, M.-J., K.-S. Huang, and H.G. Khorana. 1984. Regeneration of native bacteriorhodopsin from fragments. *J. Biol. Chem.* **259:** 4200.

Liebman, P.A. and E.N. Pugh. 1980. ATP mediates rapid reversal of cyclic GMP phosphodiesterase activation in visual receptor membranes. *Nature* **287:** 734.

MacDonald, R.E. and J.K. Lanyi. 1975. Light-induced leucine transport in *Halobacterium halobium* envelope vesicles: A chemiosmotic system. *Biochemistry* **14:** 2882.

Medynski, D.C., K. Sullivan, D. Smith, C. Van Dop, F.-H. Chang, B.K.-K. Fung, P.H. Seeburg, and H.R. Bourne. 1985. Amino acid sequence of the α-subunit of transducin deduced from the cDNA sequence. *Proc. Natl. Acad. Sci.* **82:** 4311.

Mogi, T., L.J. Stern, N.R. Hackett, and H.G. Khorana. 1987. Bacteriorhodopsin mutants containing single tyrosine to phenylalanine substitutions are all active in proton translocation. *Proc. Natl. Acad. Sci.* **84:** 5595.

Mogi, T., L.J. Stern, T. Marti, B.H. Chao, and H.G. Khorana. 1988. Aspartic acid substitutions alter proton translocation by bacteriorhodopsin. *Proc. Natl. Acad. Sci.* **85:** 4148.

Mumby, S.M., R.A. Kahn, D.R. Manning, and A.G. Gilman. 1986. Antisera of designed specificity for subunits of guanine nucleotide-binding regulatory proteins. *Proc. Natl. Acad. Sci.* **83:** 265.

Nassal, M., T. Mogi, S.S. Karnik, and H.G. Khorana. 1987. Structure-function studies on bacteriorhodopsin. III. Total synthesis of a gene for bacterio-opsin and its expression in *Escherichia coli*. *J. Biol. Chem.* **262:** 9264.

Oprian, D.D., R.S. Molday, R.J. Kaufman, and H.G. Khorana. 1987. Expression of a synthetic bovine rhodopsin gene in monkey kidney cells. *Proc. Natl. Acad. Sci.* **84:** 8874.

Sakmar, T.P. and H.G. Khorana. 1988. Total synthesis and expression of a gene for the α-subunit of bovine rod outer segment guanine nucleotide-binding protein (transducin). *Nucleic Acids Res.* **16:** 6361.

Sibley, D.R., J.L. Benovic, M.G. Caron, and R.L. Lefkowitz. 1987. Regulation of transmembrane signaling by receptor phosphorylation. *Cell* **48:** 913.

Spudich, J.L. and R.A. Bogomolni. 1988. Sensory rhodopsins of halobacteria. *Annu. Rev. Biophys. Biophys. Chem.* **17:** 193.

Stoeckenius, W., R.H. Lozier, and R.A. Bogomolni. 1979. Bacteriorhodopsin and the purple membrane of *Halobacteria*. *Biochim. Biophys. Acta* **505:** 215.

Stryer, L. 1986. cGMP cascade of vision. *Annu. Rev. Neurosci.* **9:** 87.

Sugimoto, K., T. Nukada, T. Tanabe, H. Takahashi, M. Noda, N. Minamino, K. Kanagawa, H. Matsuo, T. Hirose, S. Inayama, and S. Numa. 1985. Primary structure of the β-subunit of bovine transducin deduced from the cDNA sequence. *FEBS Lett.* **191:** 235.

Wald, G. 1968. The molecular basis of visual excitation. *Nature* **219:** 800.

Wilden, U. and H. Kuhn. 1982. Light-dependent phosphorylation of rhodopsin: Number of phosphorylation sites. *Biochemistry* **21:** 3014.

Yatsunami, K. and H.G. Khorana. 1985. GTPase of bovine rod outer segments: The amino acid sequence of the α-subunit as derived from the cDNA sequence. *Proc. Natl. Acad. Sci.* **82:** 4316.

Yatsunami, K., B.Y. Pandya, D.D. Oprian, and H.G. Khorana. 1985. cDNA derived amino acid sequence of the α-subunit of GTPase from bovine rod outer segments. *Proc. Natl. Acad. Sci.* **82:** 1936.

Direct Coupling of G Proteins to Ionic Channels

A.M. Brown,* A. Yatani,* Y. Imoto,* G. Kirsch,* H. Hamm,‡
J. Codina,† R. Mattera,† and L. Birnbaumer,†
*Departments of Physiology and Molecular Biophysics and †Cell Biology, Baylor College of Medicine, Houston, Texas 77030; ‡Department of Physiology and Biophysics, University of Illinois College of Medicine at Chicago, Chicago, Illinois 60680

G proteins may act on membrane ionic channels indirectly by first stimulating membrane enzymes that alter the levels of second-messenger cytoplasmic mediators such as protein kinases, inositol trisphosphate, and Ca^{++}; the second messengers then act on the plasmalemmal ionic channels. G proteins can short-circuit the second-messenger pathway, however, and the short circuit is membrane-delimited. Although the possibility that a membrane protein closely linked to the channel is an intermediary cannot be excluded, it is assumed that the G protein acts directly on the ionic channel (Fig. 1). The functional significance of direct G-protein gating is that receptor activation can produce its effects more quickly. In the case of Ca^{++} channels, the same G protein, G_s, has two effectors, and both act on the Ca^{++} channel. Here, the direct effect will clearly be faster than the indirect effect via the cAMP cascade. Direct G-protein gating has been demonstrated for two different K^+ channels and two different Ca^{++} channels. Direct gating may be obligatory; it is required for the K^+ channels to open. Direct gating may also be modulatory; for Ca^{++} channels, depolarization is the necessary stimulus for opening to occur. We predict that direct G-protein gating of ionic channels will be widespread, and after summarizing the evidence for the four channels referred to, we propose some likely future candidates. The two major sets of ionic channels, voltage-gated and ligand-gated, are now joined by a third set, G-protein-gated ionic channels.

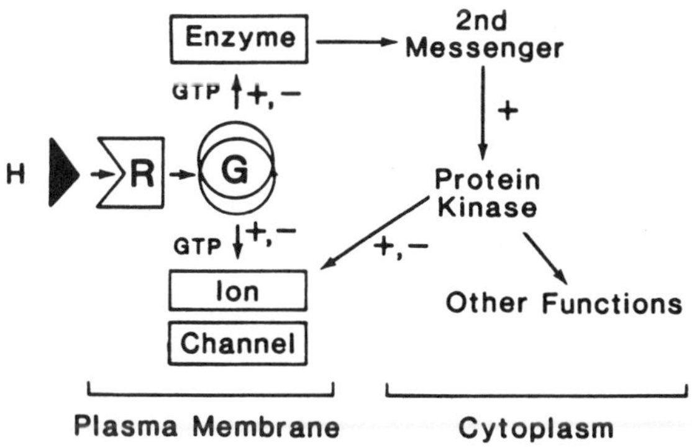

Figure 1. Diagram of the role of G proteins in receptor-mediated regulation of ionic channels. The indirect pathway is via a membrane enzyme and cytoplasmic second messenger. G, signal-transducing G protein with structure $\alpha\beta\gamma$, is shown as a double circle to denote the possibility that a single receptor may interact with more than one type of G protein (such as G_i and G_o) and/or that a single G protein may interact with more than one effector system (such as G_s with adenylyl cyclase and the Ca^{++} channel).

EXPERIMENTAL STRATEGY AND METHODS

Our approach was to isolate G-protein-gated channels from the cell and in this way exclude second messengers. This was accomplished either by using excised inside-out patches of membrane (Codina et al. 1987a,b; Logothetis et al. 1987; Yatani et al. 1987a,b) or by incorporating membrane vesicles into a planar lipid bilayer (Fig. 2) (Yatani et al. 1987c, 1988b). In either case, the single-channel currents were recorded using the high-gain current-voltage converter of a patch clamp amplifier. After analog to digital conversion, the single-channel currents were idealized as binary events, arranged into frequency distributions, and analyzed with standard statistical routines (Brown et al. 1984; Lux and Brown 1984). In general, we activated the target ionic channel via (1) its endogenous G-protein activator using either the appropriate ligand in combination with GTP or the nonhydrolyzable GTP analog, GTPγS, by itself, or (2) preactivated exogenous G proteins (Codina et al. 1984) purified from native membranes of human red blood cells or bovine brain. α subunits mediate direct G-protein gating of ionic channels (Codina et al. 1987a,b; Yatani et al. 1987a,b,c, 1988b; Imoto et al. 1988; Kirsch et al. 1988), and to sort out which α_i or which α_s was involved, we used the relevant recombinant α subunit after expressing it in the bacteria (Birnbaumer et al., this volume). The recombinant subunit was partially purified and preactivated prior to application.

By reconstituting the physiological response, the exogenous G proteins established the direct nature of the gating and specified the class of G proteins mediating receptor activation. We also studied G-protein coupling in situ by comparing the effects of GTP and GTPγS and by using a monoclonal antibody that was raised against the α subunit of transducin, α_t, and cross-reacted with α_k (Hamm and Bownds 1987; Hamm et al. 1987; Yatani et al. 1988a,b and in prep.).

RESULTS AND DISCUSSION

Acetylcholine (ACh) produces identical single-channel K^+ currents (K^+[ACh]) in cell-attached atrial

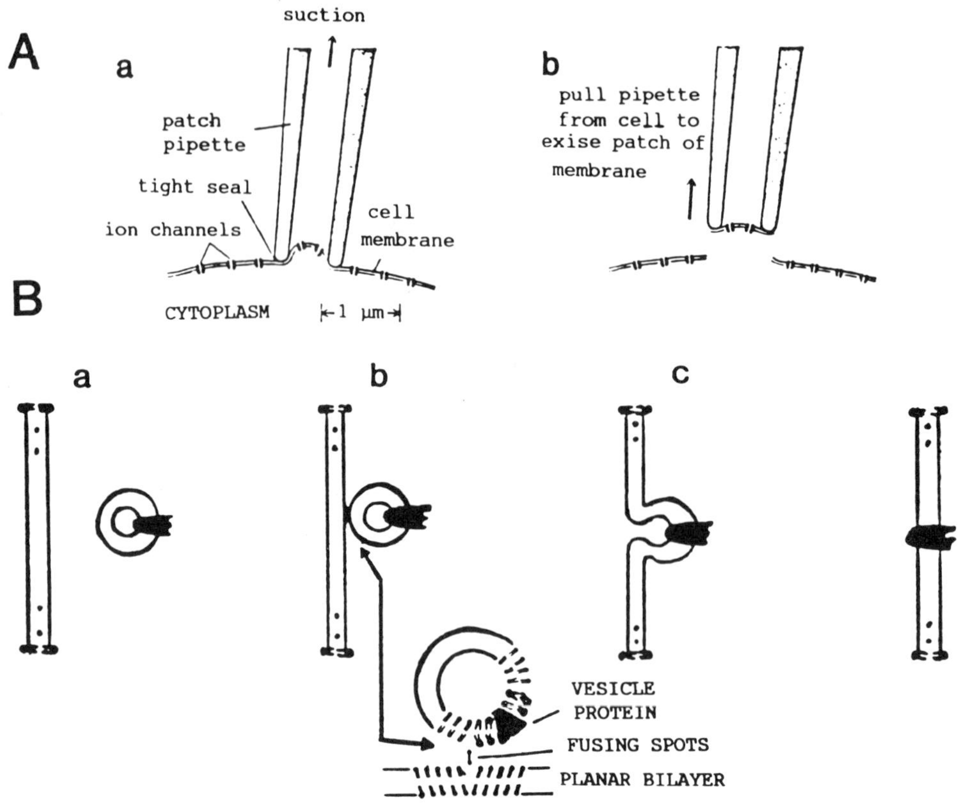

Figure 2. (*A*)Patch clamp method. Current enters or leaves the pipette by passing through the channels in the patch of membrane. Recordings of the current through these channels can be made with the patch still attached to the rest of the cell, as in (*a*), or excised, as in (*b*). *b* was the method we usually employed. (*B*) Method of incorporating membrane vesicles into planar lipid bilayers. The vesicles added to the *cis* chamber are carried to the bilayer by an osmotic gradient, and fusion begins at the fusing spots. The orientation of the Ca^{++} channels in the vesicle is usually right-side-out for cardiac sarcolemma and inside-out for skeletal muscle T tubules. The conventions for current recording are the same as those used with patch clamp: Positive current is outward.

membrane patches and in excised inside-out atrial membrane patches, provided GTP is present in the bath solution (Figs. 3 and 4; Table 1). At −80 mV, the single-channel currents in symmetrical isotonic K^+ solutions are inward, occur in bursts having a mean duration of 9 msec, have an open time frequency distribution that is a single exponential function with a mean value of 1.6 msec, and have a unitary conductance of 40 pS (Fig. 3; Table 1) (Brown and Birnbaumer 1988). Numerous simultaneous openings occur because there are usually between 4 and 6 K^+[ACh] channels per patch; the frequency of these simultaneous openings was predicted from binomial expectations. When GTP is removed from the bath, activity ceases. GTPγS at 100 μM produces an identical pattern of activity (Table 1; Fig. 4), but the presence of ligand is not essential, and the effect is not reversed by washing. As for GTP, Mg^{++} is required. A G protein purified from human red blood cells and preactivated with GTPγS can at 100 pM produce exactly the same effects as GTPγS (Table 1; Fig. 4). This exogenous G protein was called G_k (Yatani et al. 1987a) and when preactivated was des-

Table 1. Single-channel Properties of Agonist-activated K^+ Channel, GTPγS, G_k^*, α_k^*, or Recombinant α_{i-3}^*-activated K^+ Channels

Conditions		Amplitude (pA)	Mean open time (ms)	*n*
Agonist[a] + GTP	(100 μM)	2.2 ± 0.2	1.54 ± 0.6	14
GTPγS	(100 μM)	2.2 ± 0.3	1.70 ± 0.7	6
G_k^*	(20–50 M)	2.1 ± 0.4	1.60 ± 0.4	8
α_k^*	(10–50 pM)	2.1 ± 0.3	1.52 ± 0.6	12
α_{i-3}^*	(200–1000 pM)	2.2 ± 0.3	1.79 ± 0.3	4

Inside-out membrane patches from adult guinea pig were bathed in symmetrical isotonic K^+ solution; the holding potential was −80 mV. The slope conductance between −120 and −40 mV was 40 pS, and an offset current of ~1 pA was present.
[a]Carbachol or acetylcholine.

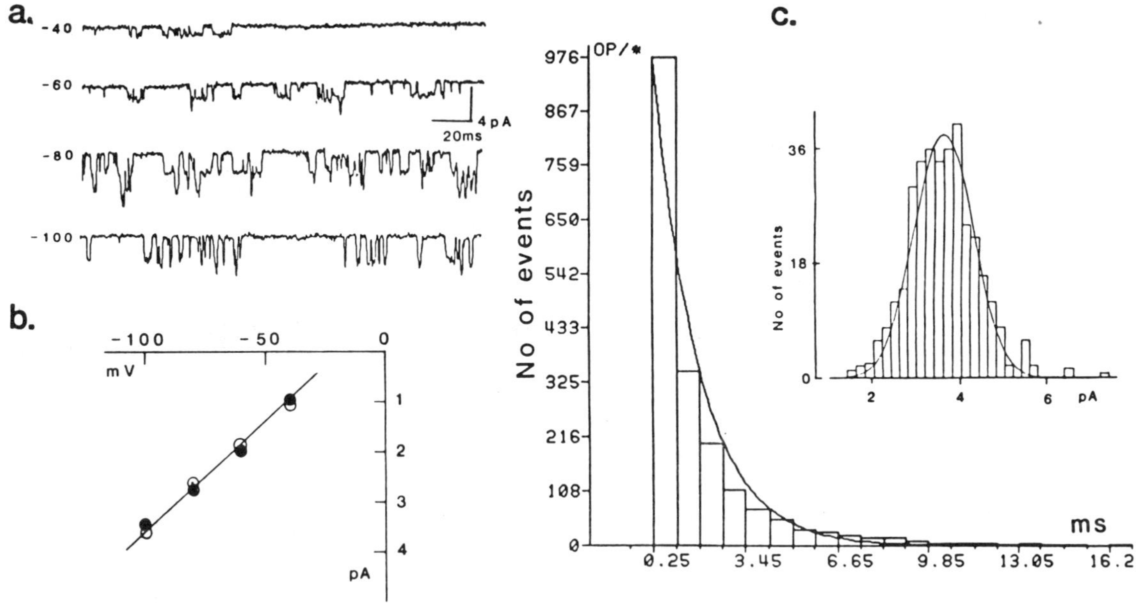

Figure 3. Conductance and open times of single muscarinic atrial K^+ channels. (*a*) Representative single-channel currents from a cell-attached patch with 1 μM carbachol in pipette at indicated holding potentials. Bathing solution contained isotonic K^+ and EGTA, which produced a resting potential of ~0 mV. There were at least two K^+[ACh] channels in this patch. (*b*) Current-voltage relationships for the channel in (*a*) (●) and for another channel from another patch under the same experimental conditions (○). Each point is the average value of at least 20 unitary openings. Slope conductance was 41 pS. (*c*) Frequency distribution of open times and amplitudes from another K^+[ACh] channel at −100 mV. Amplitude histogram (*inset*) was fit by a single Gaussian function with a mean value of 4.1 pA. Open-time histogram was fit with a single exponential function having a η of 1.57 ms.

ignated G_k^*. The putative endogenous G protein was referred to as G_k in an earlier publication (Breitwieser and Szabo, 1985). The preactivated α subunit, α_k^*, mimics the holo-G_k^* effect and is equipotent (Table 1; Fig. 4). The other G protein in human rbc, G_s, the stimulatory regulator of adenylyl cyclase, had no effect even at nanomolar concentrations. Since these G proteins share a common $\beta\gamma$ dimer, the specificity seems to lie in the α subunit, and as shown in Figure 6, this also applies to the physiological situation when the endogenous G protein is activated by ACh. The effects occurred in the absence of ATP, in the presence of AMP-P(NH)P at 100 μM, and were not influenced by phorbol ester. These results specify the target ionic channel and exclude activation of membrane-associated kinases such as protein kinase C. Activation of a phosphatase that might disinhibit a channel made quiescent when phosphorylated is excluded because withdrawal of GTP during ligand activation abolished rather than increased activity. To sum up, exogenous α_k^* acts in a membrane-delimited way to gate the K^+[ACh] channel directly, and we assume that an endogenous α_k is responsible for muscarinic activation.

α_k^* not only activates K^+[ACh] channels in adult guinea pig atria; it also does so in atria from neonatal rats and 12–14-day chick embryos (Table 2) (Krisch et al. 1988). Hence, the effects are independent of species, age, or tissue culture conditions. The general applicability of the result was important because in chick embryo, $\beta\gamma$ dimers were reported to activate K^+[ACh] channels, and α subunits were reported to have no effect (Logothetis et al. 1987). The difference between the two sets of results (Codina et al. 1987b; Logothetis et al. 1987; Yatani et al. 1987a) may otherwise have been attributed to differences of age, species, or tissue culture. The discrepancy may have resulted from improper activation of α subunits in the earlier experiments of Logothetis et al. (1987), since that laboratory has recently confirmed the effectiveness of α_k (Neer and Clapham 1988), refuting their own earlier conclusion that the $\beta\gamma$ and not the α subunits are responsible for activating K^+[ACh] channels. The effects of $\beta\gamma$ in our hands are unresolved. We have been unable to find a stimulatory effect because the hydrophobic dimer requires detergent buffers to prevent aggregation. At the required CHAPS concentration of 184 μM (Logothetis et al. 1987), the detergent itself has nonspecific effects on membrane channels (Wu et al. 1980; Kirsch et al. 1988), and at lower concentrations where the CHAPS has no effect, the $\beta\gamma$ is also ineffective. This result is complicated because at these concentrations, it is not certain that the dimer is still in solution. The best experiment we could do along these lines is shown in Figure 5. Here, we used a buffer containing Lubrol at 0.004% and bovine serum albumin at 0.004%, which by itself had no effect. The response could not be reconstituted with $\beta\gamma$ and GTP but was reconstituted with $\alpha\beta\gamma$ and GTP. In this case, the buffer was effective in keeping the holo-G protein and the $\beta\gamma$ dimers in solution, yet only the holoprotein was

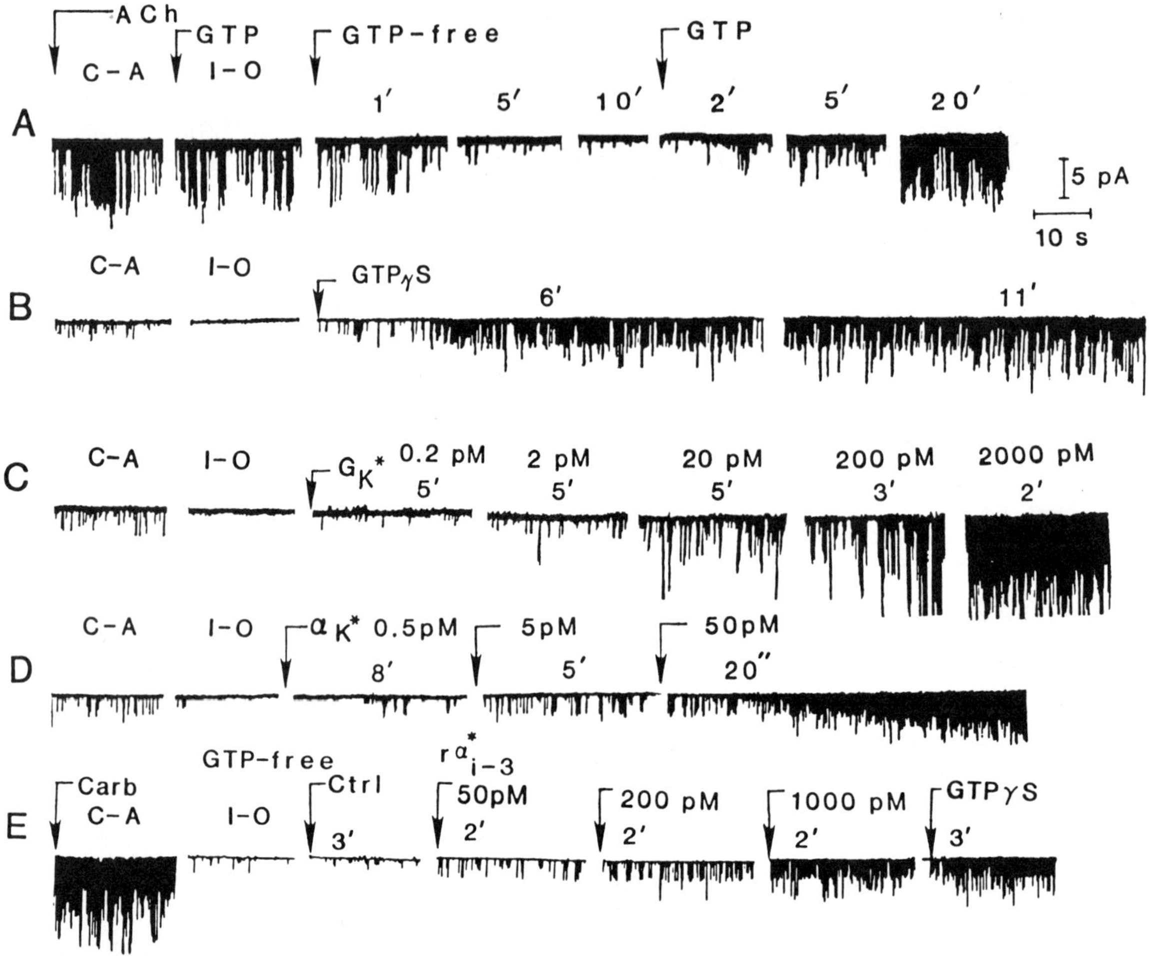

Figure 4. Activation of atrial K^+ channels in inside-out patches by ACh and GTP (*A*), GTPγS (*B*), GTPγS-treated, PTX-sensitive G protein G_k^* from human erythrocytes (*C*), GTPγS-treated α subunit of G_k, α_k^* (*D*), and recombinant α_i-3′rα_k (*E*). Single-channel K^+ currents from atrial in membrane patches were recorded in symmetrical isotonic K^+ solutions at −100 mV (*A*), −90 mV (*B*), and −80 mV (*C,D,E*), first in the cell-attached (CA) mode and then after patch excision under cell-free inside-out (I-O) conditions both before and after addition to the bathing media of (*A*) 100 μM GTP, (*B*) 100 μM GTPγS, (*C*) increasing concentrations of activated G_k^*, (*D*) increasing concentrations of activated α_k^*, and (*E*) increasing concentrations of rα_k. Times on top of each trace are minutes elapsed between the last addition and the recording of the trace shown. In *B*, GTPγS was added 7 min after patch excision, and the records were obtained 6 and 11 min after this addition. In *C*, *D*, and *E*, G_k^* and α_k^* additions were made at 25-min intervals, the first occurring 7 min after the excision of the patch.

effective. We conclude that the α subunit is the mediator of the G_k effect and that further experiments are required to establish the role of βγ.

The reconstitution of the response with G_k^* indicates that the coupling between endogenous G_k and the K^+[ACh] channel does not exclude access from other G proteins such as exogenous G_k. In other experiments, we examined whether the coupling between endogenous G_k and receptor was exclusive. Figure 5 shows that the exogenous G_k not preactivated with GTPγS can, in the presence of GTP at 100 μM, activate the K^+[ACh] channel when carbachol is applied to an atrial muscarinic receptor or when somatostatin is applied to a clonal anterior pituitary cell somatostatin receptor. Just as for the effector, the coupling between receptor and endogenous G_k does not exclude access from exogenous G_k.

Direct G-protein gating of other K^+ channels also occurs. In a clonal line of rat anterior pituitary cells, the GH_3 cells, somatostatin (SST), and ACh increase K^+ conductance in a cAMP-independent manner. The same strategy we used to establish direct gating of atrial K^+[ACh] and the putative activator G protein led to the same conclusions for K^+ channels in GH_3 cells (Table 3) (Codina et al. 1987a; Yatani et al. 1987b). Once again the α subunit mediates the G_k effect. However, the GH_3 K^+ channel differs from the atrial K^+[ACh] channel; it has a conductance of 55 pS versus the value of 40 pS found in atria of the same species, rat. This shows that direct G_k gating is not restricted to one type of K^+ channel.

The results on direct gating of K^+ channels have mainly dealt with reconstitution of the response using exogenous G proteins. To analyze functional coupling,

Table 2. Activation of Atrial K^+ Channels by α_k^*

Activator	Patch configuration	Slope conductance (pS)	Mean open time (ms)
		Guinea pig	
Carbachol	C-A	40.1 ± 1.5 (5)	1.6 ± 0.6 (7)
α_k^*	I-O	40.6 ± 1.7 (8)	1.4 ± 0.7 (9)
		Neonatal rat	
Carbachol	C-A	43.1 ± 1.2 (4)	1.7 ± 0.8 (4)
α_k^*	I-O	40.1 ± 1.1 (4)	1.8 ± 0.6 (6)
		Chick embryo	
Carbachol	C-A	44.5 ± 0.5 (4)	2.4 ± 0.3 (4)
α_k^*	I-O	42.4 ± 2.7 (4)	2.0 ± 0.4 (6)

Values are means ±S.D.; Number in parentheses is number of experiments; α_k^*, α_k activated with GTPγS; C-A, cell attached; I-O, inside-out. Carbachol was present in the pipette at 0.1 μM, and α_k^* was added at 5 pM to GTP-free bathing solution after patch excision. Slope conductances were measured between −120 and −40 mV. Mean open times were measured at −80 or −100 mV.

we used a monoclonal antibody to the α subunit of transducin, mAb 4A, which also reacts with the α_k subunit but not the βγ dimer (Fig. 6) (Yatani et al. 1988a and in prep.). mAb 4A blocked the atrial muscarinic response, and the effect was irreversible. However, after washing out the antibody, α_k^* could activate the K^+[ACh] channel. This indicated that the antibody could bind irreversibly to activated endogenous G protein but did not react with the channel. However, if the channel was activated irreversibly by GTPγS or by G_k^*, the antibody could not subsequently block activation. Another monoclonal antibody, 4H, which reacted with α_k, had no functional effects on the muscarinic response, and control IgG was also without effect.

The results suggested that mAb 4A reacted irreversibly with activated endogenous G_k. To confirm this, we did the experiments shown in Figure 6, B and C. When GTP is removed after ligand activation, the antibody no longer acts irreversibly, and subsequent application of GTPγS can activate endogenous G_k. Alternatively,

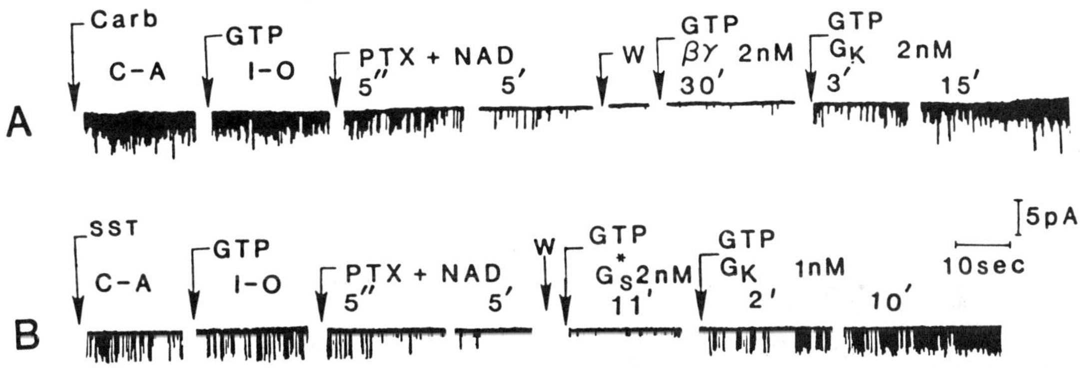

Figure 5. (*A*) Exogenous G_k activates single-channel atrial muscarinic K^+ currents in adult guinea pigs. Single-channel currents were recorded in symmetrical isotonic K^+ solutions. Holding potential was −80 mV, and single-channel K^+ currents are therefore inward. *A* begins with a cell-attached (C-A) recording and 1 μM carbachol (Carb) in patch pipette. Frequency of single-channel currents is high. Patch was then excised into inside-out (I-O) configuration with 100 μM GTP in the bath, and activity was sustained. Preactivated pertussis toxin (PTX) and NAD^+ were added to give 10 μg/ml and 1 mM, respectively. After 5 sec, there was little effect, but after about 5 min, activity was reduced. PTX and NAD^+ were then washed from the bath for 5 min, and at point W, there was no activity. Addition of βγ-dimer at 2 nM had no effect over a period of 30 min. Activity was restored when G_k was added at 2 nM. (*B*) Characteristics of the coupling system mediating K^+ channel activation by somatostatin (SST) receptors in GH_3 cell membranes. GH_3 cell K^+ currents in membrane patches were recorded in symmetrical isotonic K^+ solutions at a holding potential of −80 mV, first in the cell-attached mode (C-A) and then, after excision, in the inside-out configuration (I-O). Downward deflections denote channel openings and represent the current required to maintain the transmembrane potential at the holding value. Times on top of the traces refer to minutes (or seconds) elapsed between the last addition of test material and the recording of the trace shown. The first additions were made between 5 and 7 min after patch excision. Dependence of SST-stimulated K^+ currents on a PTX-sensitive step and reconstitution of receptor-K^+ channel coupling by readdition G_k, but not G_s. Throughout, the pipette contained 0.5 μM SST and the bath solutions contained 100 μM GTP. Preactivated PTX and NAD^+ were added to give 10 μg/ml and 1 mM, respectively; washing after PTX treatment (W) was for 5 min by perfusion of the chamber at 2 ml/min with the same GTP-containing medium but without PTX and NAD^+. Preactivated G_s^* and, 25 min afterwards, G_k were added to give final concentrations of 2000 pM each. Records were obtained 3 and 20 min after PTX plus NAD^+ addition, 15 min after G_s^* addition, and 2 and 10 min after G_k addition.

Table 3. Properties of GH_3 Somatostatin and Muscarinic K^+ Channels at −80 mV in Symmetrical Isotonic KMES Solution

GH_3	Mean amplitude (pA)	Mean open time (ms)	n
Agonist[a] + GTP (100 μM)	4.2 ± 0.3	1.77 ± 0.3	4
GTPγS (100 μM)	4.2 ± 0.2	1,54 ± 0.4	3
G_k^* (200 pM)	4.0 ± 0.3	1.90 ± 0.3	3
α_k^* (50 pM)	4.1 ± 0.1	1.70 ± 0.2	2

Slope conductance between −120 and −40 mV was 55 pS, and an offset current of ~0.6 pA was present.

[a]Somatostatin or carbachol.

if there was no prior activation by ligand, the effect of the antibody was also reversible. Moreover, in the presence of GDP, the unactivated G protein is dissociated from the receptor, and the antibody remains ineffective. This rules out a steric hindrance of the antibody by the receptor–G protein interaction. The results are compatible with one of two possibilities. In the first, the antibody binds to the activated, endogenous holo-G_k to produce block, in which case the holoprotein, or its mediator subunit, α (Codina et al. 1987b; Yatani et al. 1987a) or βγ (Logothetis et al. 1987), should be equipotent in reconstituting the response. However, only the holoprotein and the α subunit are equipotent; the βγ dimer, if it is effective, acts with, at best, 1/100th the potency of α_k (Logothetis et al. 1987). In this case, the K^+[ACh] channels are blocked at a time when βγ is released. If the antibody blocks by increasing the affinity of α for βγ, the two subunits would have to be equipotent, but as noted above, this is not the case. Hence, both possibilities are incompatible with βγ acting as an activator, and both support α_k as the mediator of activation of K^+[ACh] channels by ligand-activated endogenous G_k.

Direct G-protein gating of ionic channels can be explained by the interacting cycles shown in Figure 7. These are based on models proposed for the interaction of the activated α subunit with membrane enzymes (Birnbaumer et al. 1987; Gilman 1987), except that the ionic channel has replaced the enzyme as the target. In the case of the K^+[ACh] channel, the microscopic behavior of an αGTP with its effector can be observed directly, something that has not been achieved for α subunit interactions with membrane enzymes such as adenylyl cyclase and cGMP phosphodiesterase. The microscopic behavior can be interpreted by the sequential Markov reaction shown in Figure 7B. The rate constant for ACh activation is taken from whole-cell current measurements (Breitwieser and Szabo 1988), and activation ceases when αGTP is no longer bound to the channel. At low concentrations of ACh, bursting of single-channel currents is due to transitions among states 3, 2, and 1, and the mean open time (1.6 ms), mean burst duration (9 msec), and mean number of openings per burst (3) (Yatani et al. 1987b) specify the rate constants. Activation by αGTP or αGTPγS produced similar patterns of bursting (5000 bursts were

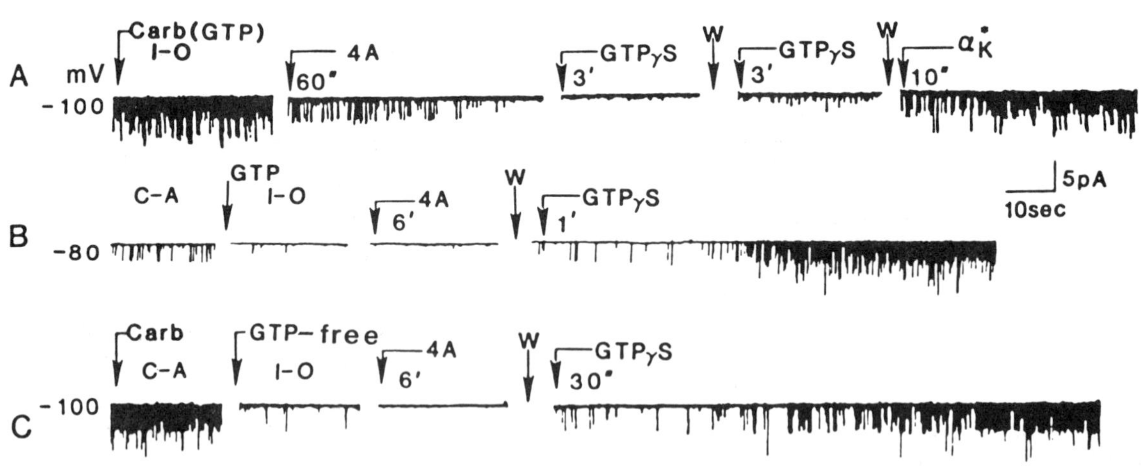

Figure 6. Monoclonal anti-transducin antibody 4A (mAb 4A) interacts with α_k and irreversibly blocks carbachol-stimulation of atrial single K^+ channel currents; but it does not interact with unactivated endogenous G_k. *A* shows that antibody block results after exposure of the cytoplasmic aspect of a membrane patch stimulated by carbachol in the pipette and GTP in the bath. *(B, C)* Exposure of the membrane patch to mAb 4A under conditions that do not lead to activation of endogenous G_k no longer leads to a persistent block of activation by GTPγS after washing out the antibody. Other experiments showed that the block is specific for mAb 4A; neither nonspecific IgG nor another anti-transducin α monoclonal antibody with a different epitope specificity has any effect.

A. Cycles for Direct α Activation

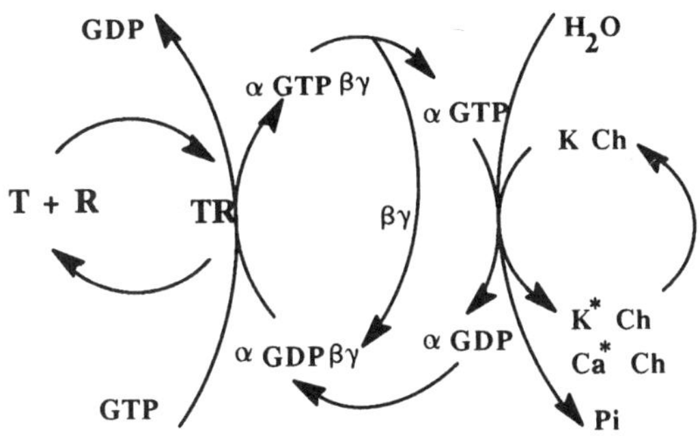

B. Markov Sequence for α Gating of K⁺[ACh] Channel.

$$\underset{(4)}{\alpha GDP\beta\gamma + GTP} \underset{3m^{-1}}{\overset{[ACh]10^9 m^{-1}M^{-1} + .3m^{-1}}{\rightleftharpoons}} \underset{(3)}{\beta\gamma + GDP + \alpha GTP + KCh} \underset{150s^{-1}}{\overset{150s^{-1}}{\rightleftharpoons}} \underset{(2)}{\alpha GTPKCh} \underset{660s^{-1}}{\overset{450s^{-1}}{\rightleftharpoons}} \underset{(1)}{\alpha GTPKCh^*}$$

Figure 7. (*A*) Cyclical scheme for direct G-protein gating of ionic channels. (T) Transmitter; (R) receptor; (KCh) potassium channel; (CaCh) calcium channel. Asterisks denote α-activated channels. (*B*) Minimum 4-state linear Markov model of a gated K^+[ACh] channel. State 1 is the open state, and state 2 is a preactivated open state.

compared) and open times (10,000 openings were compared). Since activation with αGTPγS is irreversible, this comparison shows that the α subunit neither hydrolyzes GTP nor leaves the channel while the channel is open or bursting. Otherwise, αGTP should produce shorter open times and burst durations than αGTPγS. This result is not unexpected, given that GTP hydrolysis occurs at about 3 min^{-1}. The difference also establishes a minimum of four states to explain the microscopic data (Fig. 7B). The longest closed state is on the order of 50 msec and sets this as an upper limit to the rate of GTP hydrolysis.

K^+ channels are not the only ionic channels directly gated by G proteins, and G_k is not the only G protein that gates ionic channels. G_s purified from human red blood cells and preactivated with GTPγS at picomolar concentrations directly activates Ca^{++} channels in skeletal muscle T-tubules (Yatani et al. 1988b) and cardiac sarcolemma (Fig. 8) (Yatani et al. 1987c). The effect is mediated by the α subunit, which is equipotent with the holoprotein, and the $\beta\gamma$ dimer is ineffective. Both types of Ca^{++} channels are dihydropyridine-sensitive, but they are distinguished by differences in conductance and kinetics (Yatani et al. 1988b). Unlike the atrial K^+[ACh] channels, the G-protein activation is not obligatory but is rather modulatory. Neither type of Ca^{++} channel will open unless the cell membrane is depolarized and G_s increases the frequency of opening. The cardiac myocyte experiments used excised patches, and survival required that the β-agonist isoproteronol be present in the patch pipette and GTPγS be present in the bath. This pointed to an involvement of G_s; the results with cholera toxin activation on skeletal muscle T-tubule Ca^{++} channels (Fig. 8C) support this conclusion. To have more stable recording conditions, we used the method of incorporating vesicles into planar lipid bilayers. In the case of cardiac sarcolemma, it was necessary to use the 1,4-dihydropyridine agonist, Bay K 8644, to obtain a reasonable opening frequency at the potentials where the single-channel currents were detectable. Nevertheless, this activity still decreased with time, and the G_s effect may have been to stabilize rather than to modulate the channel. In skeletal muscle, it was possible to demonstrate a stimulatory effect of G_s in the absence of Bay K 8644 or isoproteronol (Fig. 8). The G_s effects were independent of cAMP stimulation and occurred in the presence of ADPβS at 100 μM, 0 cAMP, and 0 ATP. In other experiments, we showed that the channel site activated by α_s was independent of the site that was phosphorylated by protein kinase A and the site at which Bay K 8644 acted (Yatani et al. 1988b).

As noted elsewhere in this book (Birnbaumer et al., this volume), the use of biochemically purified proteins always has an element of uncertainty due to the homology within subsets of α subunits, which prevents complete purification. This problem may be overcome by expressing individual recombinant α subunits in bacteria and testing the effects of each. We found that α_i-3 simulated the effects of human red blood cell α_k so that this α peptide could be α_k (Fig. 4E; Table 1). We have also observed that the short form of α_s simulated the

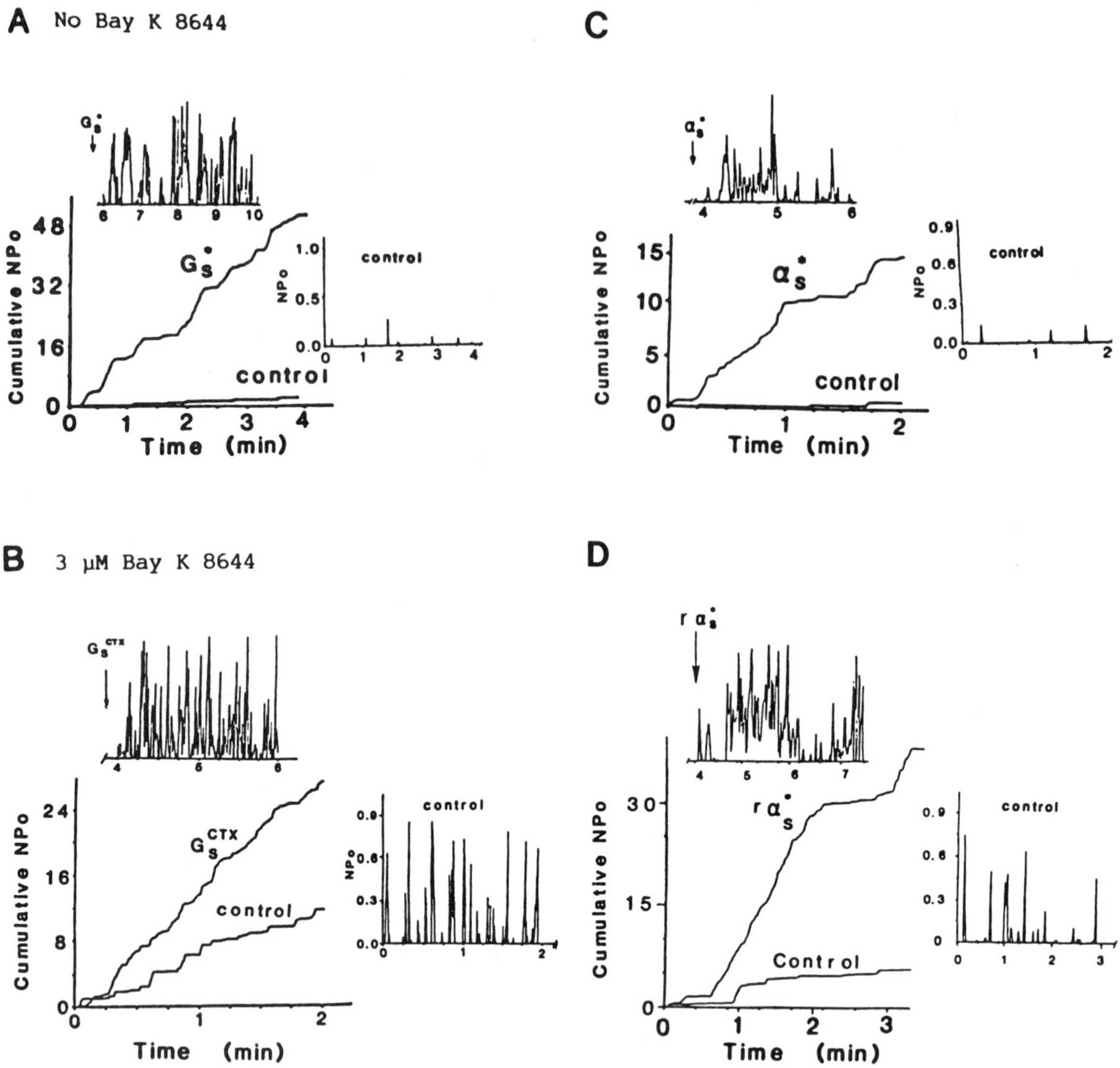

Figure 8. Representative experiments showing direct regulation of the skeletal muscle T-tubule dihydropyridine-sensitive Ca^{++} channel by purified human erythrocyte G_s. Experiments were carried out both in the absence (*A*) and in the presence (*B*) of the dihydropyridine agonist Bay K 8644 at 3 μM in the *cis* chamber. Shown are the NPo diaries and the cumulative NPo curves prior to and after addition of GTPγS-activated G_s, G_s^* GTPγ-activated α_s, α_s^* (*A*) and CTX-treated G_s (in the presence of GTP) and GTPγS-activated recombinant α_s of the short type (*B*). NPo is the probability of opening for the N channels in the patch. The holding potentials were either 0 or +20 mV, and all experiments were done with $BaCl_2$ in the *cis* chamber. Average stimulations of activity were between 10- and 20-fold in the absence and 2- to 3-fold in the presence of 3 μM Bay K 8644. The Ca^{++} channels incorporated into the bilayers can be stimulated either by exogenously activated G_s or by activation of an endogenous T-tubule G protein, presumably also G_s, co-incorporated with the Ca^{++} channel. In other experiments, it was established that, to be effective, GTPγS had to be added to the *cis* chamber, confirming the sidedness of the T-tubule membranes.

effects of human red blood cell α_s (Fig. 8D). In both cases, the concentrations required for clear effects were about 10× those required for the native peptide. This was also observed for the α_s effects on adenylyl cyclase (Graziano et al. 1987; Birnbaumer et al., this volume), which indicates that the expressed recombinants are not as potent as the native peptides in our assays.

The physiological importance of direct G_s gating of calcium channels in muscle remains to be elucidated, since the cAMP cascade, which indirectly modulates the channel, is activated by the same G protein. It is probably related to the fact that this pathway is faster than the second-messenger pathway. The significance of the result at present rests more with the fact that by identifying Ca^{++} channels as a direct target for G proteins, we have another indication that the phenomenon of direct G-protein gating of ionic channels is likely to be widespread. Other examples are now appearing: a G_i from bovine brain activates Ca^{++} currents in adrenal glomerulosa cells (Hescheler et al. 1988) in a cAMP-independent manner, and direct G-protein gating probably explains why intracellular injection of bovine brain G_o reconstituted the PTX blocked opioid inhibition of Ca^{++} channels in a neuroblastoma-glioma cell line (Hescheler et al. 1987). In preliminary experiments, we have demonstrated direct G_o effects on a variety of K^+ channels in hippocampal neurons (A.M.J. Van Dongen et al., in prep.). Overall, the possibilities are considerable. There are numerous distinct ionic channels, and many may be targets for G proteins. Because there are 10–15 G proteins, the permutations of direct G-protein gating of ionic channels becomes correspondingly greater. Could direct G-protein gating become the largest category of ionic channel gating?

ACKNOWLEDGMENTS

This research was supported by National Institutes of Health grants NS-23877, HL-36930, and HL-37044 to A.M.B.; and in part by U. S. Public Health Service research grants DK-19318, HD-09581, HL-31164, and HL-37044 to L.B.

REFERENCES

Birnbaumer, L., J. Codina, R. Mattera, A. Yatani, N. Scherer, M.-J. Toro, and A.M. Brown. 1987. Signal transduction by G protein. *Kidney Int.* **32:** 14.

Breitwieser, G.E. and G. Szabo. 1985. Uncoupling of cardiac muscarinic and β-adrenergic receptors from ion channels by a guanine nucleotide analogue. *Nature* **317:** 538.

———. 1988. Mechanism of muscarinic receptor-induced K^+ channel activation as revealed by hydrolysis-resistant GTP analogues. *J. Gen. Physiol.* **91:** 469.

Brown, A.M. and L. Birnbaumer. 1988. Direct G protein gating of ion channels. *Am. J. Physiol.* **23:** 401.

Brown, A.M., D.L. Wilson, and H.D. Lux. 1984. Activation and inactivation of single calcium channels in snail neurons. *J. Gen. Physiol.* **83:** 751.

Codina, J., G. Grenet, A. Yatani, L. Birnbaumer, and A.M. Brown. 1987a. Hormonal regulation of pituitary GH_3 cell K^+ channels by G_k is mediated by its *alpha* subunit. *FEBS Lett.* **216:** 104.

Codina, J., A. Yatani, D. Grenet, A.M. Brown, and L. Birnbaumer. 1987b. The *alpha* subunit of G_k opens atrial potassium channels. *Science* **236:** 442.

Codina, J., J.D. Hildebrandt, R.D. Sekura, M. Birnbaumer, J. Bryan, C.R. Manclark, R. Iyengar, and L. Birnbaumer. 1984. N_s and N_i: The stimulatory and inhibitory regulatory component of adenylyl cyclase. Purification of the human erythrocyte proteins without the use of activating regulatory ligands. *J. Biol. Chem.* **259:** 5871.

Gilman, A.G. 1987. G Proteins: Transducers of receptor-generated signals. *Annu. Rev. Biochem.* **56:** 615.

Graziano, M.P., P.J. Casey, and A.G. Gilman. 1987. Expression of cDNAs for G proteins in *Escherichia coli*. *J. Biol. Chem.* **262:** 11375.

Hamm, H.E. and M.D. Bownds. 1984. A monoclonal antibody to guanine nucleotide binding protein inhibits the light-activated cyclic GMP pathway in frog rod outer segments. *J. Biol. Physiol.* **84:** 265.

Hamm, H.E., D. Deretic, P. Hofmann, A. Schleicher, and B. Kohl. 1987. Mechanism of action of monoclonal antibodies that block the light-activation of the guanyl nucleotide binding protein, transducin. *J. Biol. Chem.* **262:** 10831.

Hescheler, J., W. Rosenthal, W. Trautwein, and G. Schultz. 1987. The GTP-binding protein G_o regulates calcium channels. *Nature* **325:** 445.

Hescheler, J., W. Rosenthal, K.-D. Hinsch, M. Wulfern, W. Trautwein, and G. Schultz. 1988. Angiotensin II-induced stimulation of voltage-dependent Ca^{2+} currents in an adrenal cortical cell line. *EMBO J.* **7:** 619.

Imoto, Y., A. Yatani, J.P. Reeves, J. Codina, L. Birnbaumer, and A.M. Brown. 1988. The α subunit of G_s directly activates cardiac calcium channels in lipid bilayers. *Am. J. Physiol.* (in press).

Kirsch, G.E., A. Yatani, J. Codina, L. Birnbaumer, and A.M. Brown. 1988. The α subunit of G_k activates atrial K^+ channels of embryonic chick, neonatal rat and adult guinea pig. *Am. J. Physiol.* **23:** 1200.

Logothetis, D.E., Y. Kurachi, J. Galper, E.J. Neer, and D.E. Clapham. 1987. The β subunits of GTP-binding proteins activate the muscarinic K^+ channel in heart. *Nature* **325:** 321.

Lux, H.D. and A.M. Brown. 1984. Patch and whole cell calcium currents recorded simultaneously in snail neurons. *J. Gen. Biochem.* **83:** 727.

Neer, E.J. and D.E. Clapham. 1988. Role of G protein subunits in transmembrane signalling. *Nature* **333:** 129.

Wu, C.H., P.J. Sides, and T. Narahashi. 1980. Interaction of deoxycholate with the sodium channel of squid axon membranes. *J. Gen. Physiol.* **76:** 355.

Yatani, A., J. Codina, A.M. Brown, and L. Birnbaumer. 1987a. Direct activation of mammalian atrial muscarinic potassium channels by GTP regulatory protein, G_k. *Science* **235:** 207.

Yatani, A., J. Codina, R.D. Sekura, L. Birnbaumer, and A.M. Brown. 1987b. Reconstitution of somatostatin and muscarinic receptor mediated stimulation of K^+ channels by isolated G_k protein in clonal rat anterior pituitary cell membranes. *Mol. Endocrinol.* **1:** 283.

Yatani, A., J. Codina, Y. Imoto, J.P. Reeves, L. Birnbaumer, and A.M. Brown. 1987c. A G protein directly regulates mammalian cardiac calcium channels. *Science* **238:** 1288.

Yatani, A., H.E. Hamm, J. Codina, L. Birnbaumer, and A.M. Brown. 1988a. Monoclonal antibody 4A to the α-subunit of transducin blocks activation of mammalian atrial K^+ channels. *FASEB J.* **2:** 1278.

Yatani, A., Y. Imoto, J. Codina, S. Hamilton, A.M. Brown, and L. Birnbaumer. 1988b. The stimulatory G protein of adenylyl cyclase, G_s, also stimulates dihydropyridine-sensitive Ca^{2+} channels: Evidence for direct regulation independent of phosphorylation by cAMP-dependent protein kinase or stimulation by a dihydropyridine agonist. *J. Biol. Chem.* **263:** 9887.

Role of G-protein-coupled Phosphatidylinositide System in Signal Transduction in Vertebrate Neurons: Experiments on Neuroblastoma Hybrid Cells and Ganglion Cells

D.A. Brown,* H. Higashida,† P.R. Adams,‡ N.V. Marrion,‡§ and T.G. Smart§

*MRC Neuropharmacology Research Group, Department of Pharmacology, University College London, London, WC1E6BT, United Kingdom; †Neuroinformation Research Institute, Department of Biophysics, Kanazawa University School of Medicine, Kanazawa, Ishikawa 920, Japan; ‡Howard Hughes Medical Institute Research Laboratories, Department of Neurobiology and Behavior, SUNY at Stony Brook, New York 11794; §MRC Neuropharmacology Research Group, Department of Pharmacology, University of London School of Pharmacy, London, WC1N1AX, United Kingdom

In the nervous system, a multiplicity of neurotransmitters are capable of activating the phosphatidylinositol (PtdIns) system, i.e., of accelerating the enzymatic breakdown of membrane phosphatidylinositide bisphosphate ($PtdInsP_2$) to inositol 1,4,5-trisphosphate ($Ins[1,4,5]P_3$) and diacylglycerol (DAG) (Fisher and Agranoff 1986; Berridge 1987). However, there is relatively little information on the consequences this has for neuronal function. We have studied the possible involvement of this system in mediating the regulation of membrane K currents by peptides and by acetylcholine in two types of neuronal cells: differentiated NG108-15 neuroblastoma × glioma hybrid cells, and amphibian and mammalian sympathetic neurons.

NEUROBLASTOMA HYBRID CELLS: RESPONSE TO BRADYKININ

NG108-15 cells show a vigorous PtdIns response to the peptide hormone bradykinin (BK). Thus, it accelerates the breakdown of ^{32}P-labeled phosphatidylinositide 4,5-bisphosphate at concentrations ≥ 1 nM (Yano et al. 1984) and induces a transient accumulation of $Ins(1,4,5)P_3$ and DAG (Yano et al. 1985). This is accompanied by a stimulation of GTPase activity (Grandt et al. 1986; Higashida et al. 1986). The nature of the G protein involved in the PtdIns response is not entirely clear, since these two reports yield conflicting information regarding the sensitivity of the GTPase to pertussis toxin (PTx), and the effect of PTx on the PtdIns response has not been directly measured. However, in the related cell line NG115-401L, the PtdIns response to BK is not inhibited by PTx (Jackson et al. 1987).

Electrophysiological Response

When applied by pressure or diffusion to NG108-15 cells, BK produces a biphasic change in membrane potential, consisting of an initial hyperpolarization, followed by a more sustained depolarization (Reiser and Hamprecht 1982; Yano et al. 1984). These potential changes are accompanied by a reduction and increase, respectively, in action potential discharges (Yano et al. 1984). Corresponding initial outward and subsequent inward membrane currents can be recorded under voltage clamp (see Fig. 1, top) (Higashida and Brown 1986; Brown and Higashida 1988a). These appear to result from changes in two different K currents: The initial outward current results from the activation of a voltage-insensitive K conductance, whereas the subsequent inward current results primarily from the inactivation of a voltage-dependent K current, the M current (I_M). These two currents can also be distinguished on pharmacological grounds: The initial outward current is blocked by the K-channel-blocking agent *d*-tubocurarine (dTC) (Fig. 2A), whereas the inward current is insensitive to dTC and instead is occluded by muscarine, which blocks the M current (see Brown and Adams 1980).

BK-induced Outward Current: The Role of $Ins(1,4,5)P_3$ and Ca

Responses to $Ins(1,4,5)P_3$. The duration of the initial outward current parallels the transient increase in $Ins(1,4,5)P_3$ (see Yano et al. 1985). To test the possible role of $Ins(1,4,5)P_3$ as a generator of the outward current, $Ins(1,4,5)P_3$ was injected into the cells from a second micropipette using brief iontophoretic current pulses. In a substantial proportion of cells, such injections yielded an appropriately brief outward current (Fig. 1, middle). This appeared to be mediated primarily by a K conductance, because the reversal potential (−87 mV) was near to E_K and shifted with changes in extracellular [K]. Like the BK-induced current, the conductance underlying the $Ins(1,4,5)P_3$-induced current was relatively insensitive to voltage, and the current was readily blocked by dTC (Fig. 2B) or by the bee venom apamin and was relatively insensitive to

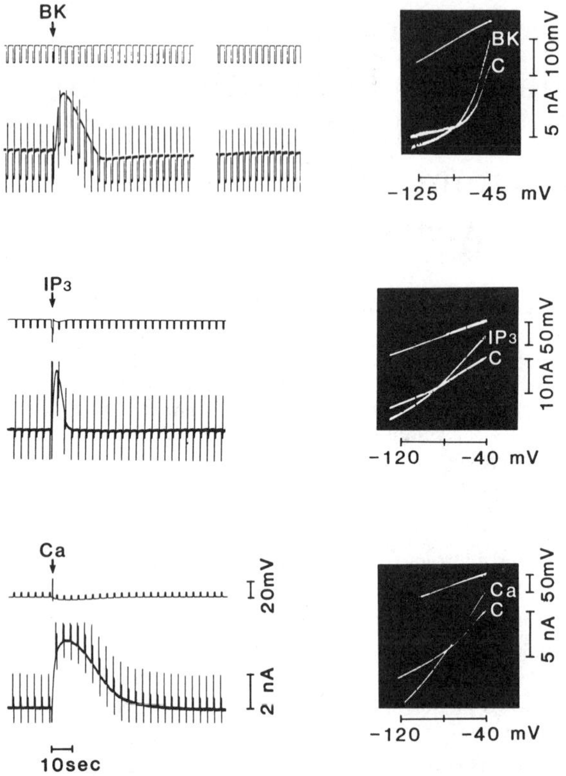

Figure 1. Membrane currents induced in three voltage-clamped NG108-15 cells by extracellular pressure application of BK (2 μl of a 10 μM solution) (*top*), intracellular iontophoretic injections of Ins(1,4,5)P_3 (IP_3; pipette solution 0.5 mM; current -12 nA for 1 sec) (*middle*), and Ca (pipette solution 0.5 M $CaCl_2$; current +20 nA for 0.5 sec) (*bottom*). Records at the left show time courses of membrane current changes (lower traces). Upper traces, voltage; holding potentials, -45 mV (*top*) and -40 mV (*middle* and *bottom*). Records at the right show superimposed ramp voltages (upper traces) and currents (lower traces) recorded before (C) and during the outward currents induced by the three agents. (Adapted from Higashida and Brown 1986.)

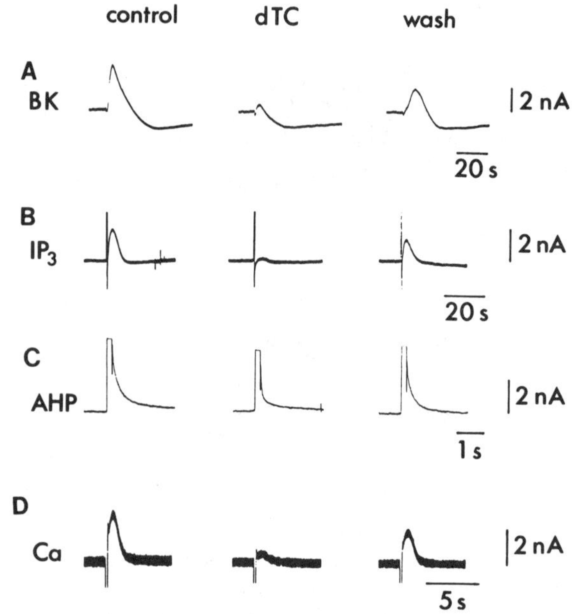

Figure 2. Effect of dTC (200 μM in *A*, *B*, and *C*; 500 μM in *D*) on membrane currents induced in NG108-15 cells by extracellular pressure application of BK (2 μl, 0.1 μM) (*A*), intracellular iontophoresis of Ins(1,4,5)P_3 (-30 nA, 0.5 sec) (*B*), a priming Ca current activated by a 50 msec, +50 mV depolarizing step (giving the Ca-activated K current designated AHP) (*C*), and intracellular iontophoresis of Ca ions (+58 nA, 0.2 sec) (*D*). Holding potentials, -40 mV in *A*, -50 mV in *B–D*. (Adapted from Brown and Higashida 1988a,b,c.)

tetraethylammonium. Thus, Ins(1,4,5)P_3 injections replicated the effect of external BK application.

There were, however, some differences between the responses to injected Ins(1,4,5)P_3 and those to BK. First, the response to Ins(1,4,5)P_3 was faster in onset (minimum latency about 0.4 sec, as opposed to ~1–1.5 sec for external iontophoretic application of BK) and shorter in duration. The shorter latency accords with the expectation for a second messenger; the faster offset suggests that the response to BK may be maintained by continued receptor activation. Second, the effects of Ins(1,4,5)P_3 could be repeated many times at injection intervals of 2 minutes or so, whereas the response to BK showed prominent fade at such repetition intervals. This suggests that the fade of the BK response results from receptor desensitization rather than depletion of Ca stores (see below). Third, in many cells, Ins(1,4,5)P_3 produced an inward current, either instead of or subsequent to the outward current, which differed from the secondary inward current produced by BK in that it resulted from an increased ionic conductance with a null potential of around -20 mV, instead of from a decreased K conductance. This appeared to result from activation of a nonspecific current, similar to that activated by other inositol phosphates (see below) and may have resulted from excessive local concentrations of Ins(1,4,5)P_3 or metabolism to other derivatives.

Responses to other inositol phosphates. In many cells, including neuroblastoma hybrid cells (Jackson et al. 1987), Ins(1,4,5)P_3 can be metabolized sequentially to inositol 1,3,4,5-tetrakisphosphate and then to inositol 1,3,4-trisphosphate (Batty et al. 1985; Irvine et al. 1986). When injected into NG108-15 cells, these compounds did not produce an outward current but instead, they produced an inward current. This reversed at about -20 mV and was accompanied by an increased membrane conductance, probably resulting from activation of a nonspecific ionic conductance. Since BK did not produce this effect, it probably does not contribute directly to the response to BK. Whether it contributed to the maintenance of the BK or Ins(1,4,5)P_3 effects, by promoting Ca influx and the refilling of Ins(1,4,5)P_3-depleted stores, as suggested previously by other investigators (Irvine and Moor 1986; Morris et al. 1987), remains to be determined.

Responses to Ca. The usual effect of Ins(1,4,5)P_3 is to release Ca from intracellular stores and thereby raise cytosolic Ca (Streb et al. 1983). In NG108-15 cells, BK increases Ca fluxes across the cell membrane (Yano et al. 1984), and both external application of BK (Reiser and Hamprecht 1985; Osugi et al. 1986) and intracellular injection of Ins(1,4,5)P_3 (Osugi et al. 1986) raise intracellular Ca concentrations. Hence, the activation of the K current by BK and Ins(1,4,5)P_3 might well be ultimately mediated by Ca. To discover whether this was plausible, we tested the effects of intracellular iontophoretic injections of Ca ions (Higashida and Brown 1986; Brown and Higashida 1988a). Such injections activated an outward current (Fig. 1, bottom) closely similar to that activated by either BK or Ins(1,4,5)P_3 in terms of both voltage insensitivity and pharmacological sensitivity to dTC (Fig. 2D) and apamin. A similar dTC and apamin-sensitive current could also be generated by a priming voltage-activated Ca current (Fig. 2C). The immediate source of Ca for the initial effects of BK and Ins(1,4,5)P_3 is probably intracellular, rather than extracellular. The responses to these agents declined in Ca-free solution (see Fig. 4), but this decline occurred over a few minutes, suggesting the depletion of some store of Ca, whereas the voltage-activated Ca current (and subsequent Ca-activated K current) showed an immediate reduction. However, a sustained smaller component of the elevated cytoplasmic Ca evoked by BK does appear to depend on an influx of extracellular Ca through voltage-activated Ca channels (Osugi et al. 1986), perhaps due to the delayed depolarization; the slow rundown of the responses to BK and Ins(1,4,5)P_3 in Ca-free solution does suggest a need for extracellular Ca in maintaining responses.

Single-channel currents. By applying a patch pipette to the outer surface of the cells, we have identified single ion channels responsible for the macroscopic K currents induced by BK, Ins(1,4,5)P_3, or Ca (Brown et al. 1987). Thus, as shown in Figure 3, intracellular iontophoretic injections of either Ins(1,4,5)P_3 or Ca evoked a burst of single-channel currents recorded through the cell-attached patch pipette. These coincided temporally with the membrane hyperpolarization induced by the macroscopic K current but were not secondary to the hyperpolarization, because they were not induced by an equivalent hyperpolarization effected by passing current through the voltage-recording microelectrode and were not inhibited when the cell hyperpolarization was reduced by bath-applied dTC. Two classes of channels were activated by Ins(1,4,5)P_3 injections, one with a conductance ranging from 32 to 44 pS (with isotonic KCl in the patch pipette) and a smaller, longer opening channel with a conductance of 10–15 pS; the former was the more frequent and more clearly correlated with the membrane hyperpolarization. Interestingly, Ins(1,4,5)P_3 or Ca injections only very occasionally activated large conductance channels in these cells. When Ins(1,4,5)P_3 was injected at different patch potentials, the single-channel currents reversed at a patch potential of about −50 mV with 150 mM KCl in

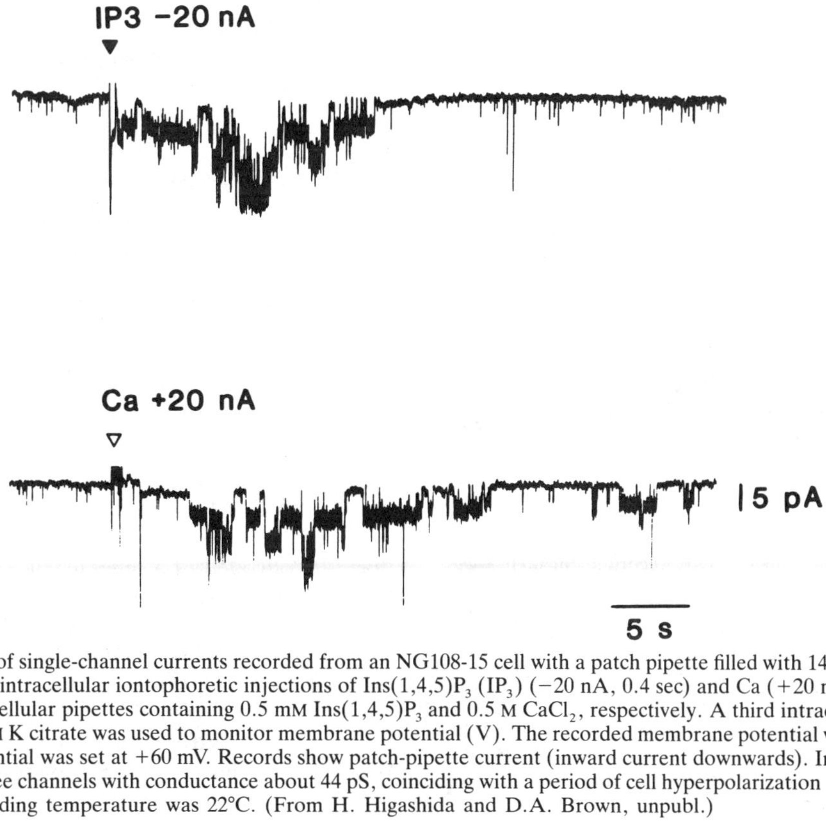

Figure 3. Clusters of single-channel currents recorded from an NG108-15 cell with a patch pipette filled with 140 mM KCl + 5 mM HEPES, following intracellular iontophoretic injections of Ins(1,4,5)P_3 (IP_3) (−20 nA, 0.4 sec) and Ca (+20 nA, 2 sec) through two separate intracellular pipettes containing 0.5 mM Ins(1,4,5)P_3 and 0.5 M $CaCl_2$, respectively. A third intracellular microelectrode filled with 1 M K citrate was used to monitor membrane potential (V). The recorded membrane potential was about −50 mV and the patch potential was set at +60 mV. Records show patch-pipette current (inward current downwards). Injections evoked a cluster of up to three channels with conductance about 44 pS, coinciding with a period of cell hyperpolarization (not shown on this record). The recording temperature was 22°C. (From H. Higashida and D.A. Brown, unpubl.)

the pipette (i.e., near 0 mV membrane potential, assuming a resting potential of about −50 mV), and the single-channel current voltage curve was linear. As expected from the macroscopic current, the probability of activating the channels did not obviously vary with holding potential. An important observation was that similar channels could be activated when BK was applied to the cell membrane outside the patch (Fig. 4), thus confirming that the effect of BK is mediated (in part at least) by a messenger system capable of working at a distance from the receptors.

BK-induced Inward Current: The Role of DAG

The secondary inward (depolarizing) current produced by BK results from a decreased ionic conductance (Fig. 5a). The conductance so affected differs in several respects from that activated during the initial outward current. First, it is not inhibited by dTC, so the inward current component of BK action persists in dTC solution at a time when the initial outward current is suppressed (Fig. 2A). Second, the conductance is voltage-sensitive, so that the inward current produced by BK diminishes markedly as the cell is hyperpolarized and does not invert to an outward current at potentials negative to E_K (Fig. 5c). This is because the current inhibited by BK is itself a voltage-dependent current I_M that is deactivated in a time-dependent manner when the membrane potential is hyperpolarized down to −70 mV or thereabouts (see Adams et al. 1982a). This current was also inhibited by muscarinic acetylcholine-receptor agonists (Fig. 5b) or by Ba ions, as in many other cells (see Brown 1988). Thus, these compounds produced an inward current similar to that produced by BK (albeit without the initial outward current) and thereby occluded the inward current response to BK.

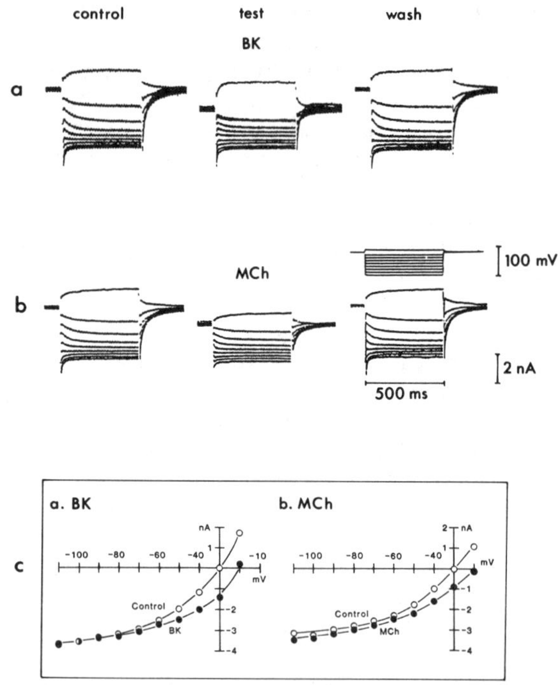

Figure 5. Inhibition of I_M in a voltage-clamped NG108-15 cell by BK (*a*) and methacholine (MCh) (*b*). BK and MCh were applied by pressure from pipettes containing 10 μM and 300 mM agonist, respectively. Records in *a* and *b* show families of currents evoked by +10 and −10 to −80 mV, 1-sec voltage steps from a holding potential of −30 mV. Note that hyperpolarizing steps produced slow inward current relaxations that accelerated with increasing hyperpolarization and inverted at between −80 and −90-mV command potential; these accord with the properties expected for deactivation of the I_M (see Adams et al. 1982b). BK and MCh (*middle* row) produced an inward current at the holding potential and reduced the amplitude of the deactivation relaxations. (The records in the middle row of *a* were obtained during the inward current phase of BK action, after the initial outward current [see Fig. 1] had subsided; the increased response to the +10-mV command may reflect a residuum of this outward current. MCh did not produce an initial outward current.) Graphs in *c* show final current levels plotted against command potential before (○) and after (●) applying agonist. Zero current is the initial holding current. Note that the curves show outward rectification and that both BK and MCh reduced this rectification; note also that the net current induced by the agonists does not reverse at command potentials beyond E_K (given by the inversion point for I_M relaxations), indicating that it results from the inhibition of a voltage-dependent conductance. (Adapted from Brown and Higashida 1988a,b.)

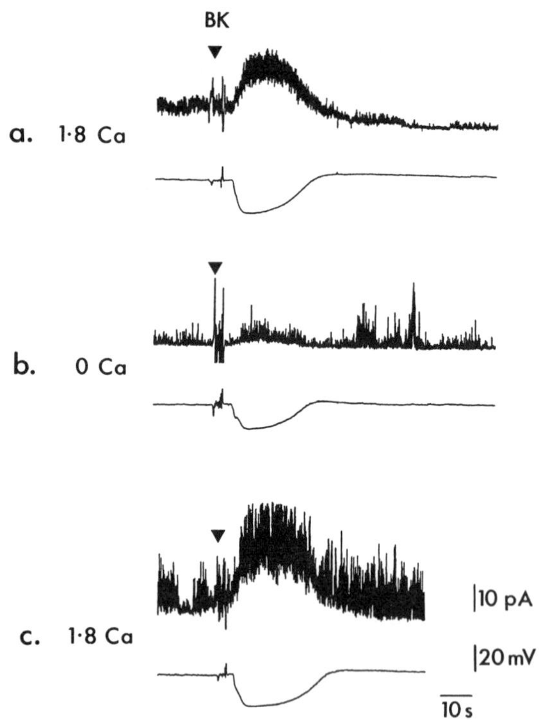

Figure 4. Multiple-channel activity recorded through a cell-attached KCl-filled patch pipette (pipette potential 0 mV) before and after pressure ejection of BK from a pipette filled with 20 μM BK onto the cell outside the patch in normal Dulbecco's modified Eagle's medium solution (*a* and *c*), and during perfusion with Ca-free solution (*b*). Upper traces are records of channel activity (inward current upwards); lower traces show membrane potential recorded through an additional intracellular micropipette (resting potential −35 mV). The recording temperature was 35°C. (From H. Higashida and D.A. Brown, unpubl.)

Intracellular injections of Ins(1,4,5)P_3, inositol 1,3,4,5-tetrakisphosphate, or Ca ions did not inhibit I_M: Even though these substances sometimes evoked an inward current, this was accompanied by an increased membrane conductance rather than a fall. Hence, we have to look for an alternative transduction mechanism for this component of BK action. One possibility is that it results from the action of the DAG formed from PtdInsP_2, along with the Ins(1,4,5)P_3 (Yano et al. 1985; Berridge 1987). The principal effect of DAG is to activate protein kinase C, an effect that can be imitated by exogenous DAGs or by tumor-promoting phorbol esters (Nishizuka 1984). Application of one such phorbol ester, phorbol 12,13-dibutyrate (PDBu), in concentrations from 10 nM to 1 μM, or of 1-oleoyl-2-acetylglycerol (OAG, 0.1–10 μM) consistently induced an inward current and inhibited I_M (Fig. 6), an effect consistent with a role for protein kinase C activation. However, we have not yet conclusively shown that formation of DAG is a necessary prelude to I_M inhibition, so we cannot eliminate alternative possibilities for the transduction of BK action.

One unexpected effect of phorbol esters was to increase the amplitude of the initial outward current produced either by BK itself or by the intracellular injection of either inositol phosphate or Ca ions (Brown and Higashida 1988c). This was also manifest at the single-channel level, where channel opening, induced by Ins(1,4,5)P_3 injections, was prolonged and additional channels were recruited. This was unexpected in that the usually reported effect of protein kinase C activation is to exert a negative feedback control of receptor activation or the subsequent step of G-protein activation. In this case, since the current response to Ca injections were also enhanced, C kinase activation must presumably modify the Ca-activated K channels.

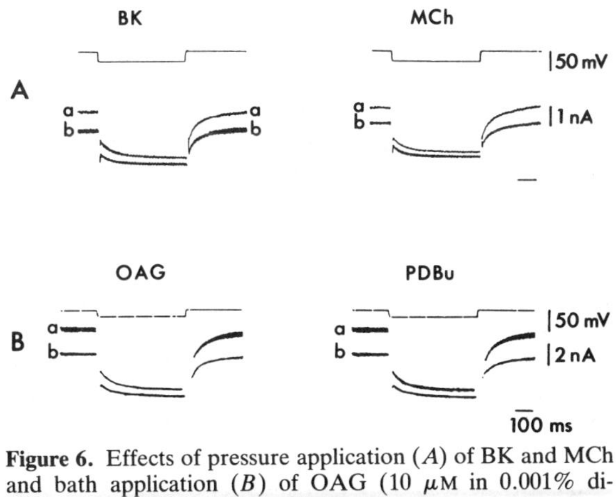

Figure 6. Effects of pressure application (*A*) of BK and MCh and bath application (*B*) of OAG (10 μM in 0.001% dimethylsulfoxide) and PDBu (0.33 μM) on current relaxations produced by 30 mV (*A*) and 20 mV (*B*) hyperpolarizing steps from a holding potential of −30 mV. Records show superimposed recordings of voltage (upper trace) and membrane current (lower traces) before (a) and after (b) applying drugs. Note that all drugs produced an inward shift in the holding current and reduced the amplitude of the current relaxations accompanying the voltage steps. (*A* and *B* are separate cells, adapted from Brown and Higashida 1988b and 1988c, respectively.)

SYMPATHETIC GANGLION CELLS: I_M CURRENT INHIBITION BY ACETYLCHOLINE AND PEPTIDES

As pointed out above, one effect of BK on neuroblastoma hybrid cells is to inhibit I_M, a voltage- and time-dependent K current. This current was first identified in frog sympathetic neurons (Brown and Adams 1980). In these cells, I_M is inhibited by muscarinic acetylcholine-receptor agonists and also by the peptide neurotransmitters luteinizing hormone-releasing hormone (LHRH and tLHRH) and substance P, and by uridine and adenosine nucleotides (for review, see Brown 1988). The analogous current in rat sympathetic neurons is also inhibited by muscarinic agonists and possibly by another peptide, angiotensin (Constanti and Brown 1981). The ability of several different agonists acting on different receptors to inhibit the same current in a mutually occlusive manner suggests an indirect form of coupling between receptor and ion channel, involving membrane-bound or soluble second messengers.

G-protein Coupling

Muscarinic and peptide receptors belong to a family of monomeric membrane receptors whose effects on cells are normally transduced through activation members of a family of membrane-located GTP-binding proteins (G proteins; see Gilman 1987). We have tested for the involvement of G proteins in I_M inhibition in dissociated cultured rat sympathetic neurons (see Marrion et al. 1987) by applying analogs of the natural ligands GTP and GDP to the cell interior through patch pipettes, using the approach pioneered by Pfaffinger et al. (1985) and Breitweiser and Szabo (1985) in their studies on muscarinic activation of K currents in cardiac cells.

Inclusion of the nonhydrolyzable GTP analogs, GTPγS (500 μM) or 5′ guanylylimidodiphosphate (Gpp[NH]p) (200 μM) had two effects on I_M (Fig. 7). First, they induced a slowly developing inhibition of the current in the absence of added agonist, with half-times of about 3 and 6 minutes, respectively, and second, they changed the normal rapidly reversible inhibition by the agonist muscarine (Fig. 7A) to an irreversible inhibition (Fig. 7B). These GTP analogs are very much less susceptible than native GTP to the GTPase action of the G protein. This is necessary for the reassociation and termination of the effect of the GTP-bound, activated α subunit of the G protein; hence, they trap the α subunit in the activated state, producing an irreversible effect. The slow onset of inhibition in the absence of agonist reflects the slow dissociation rate of GDP (0.1–0.2 min^{-1}) from the associated G protein in the ground state. Coupling of agonist-activated receptor accelerates the GDP dissociation rate to 3–4 min^{-1} (Gilman

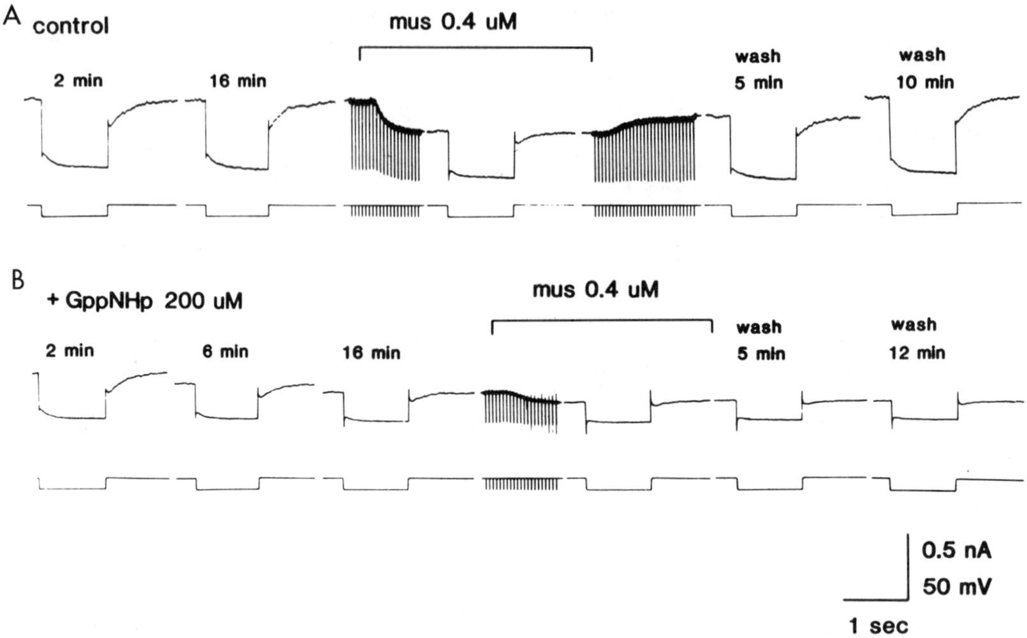

Figure 7. Effect of intracellular application of Gpp(NH)p on the action of muscarine on dissociated adult rat superior cervical sympathetic neurons. Neurons were dissociated, and the "whole cell" was clamped through a patch pipette filled with 150 mM K aspartate, 30 mM KCl, 1 mM $MgCl_2$, 1 mM Na_2ATP, 5 mM HEPES buffered to pH 6.7 with 3 mM NaOH, and 0.5 mM EGTA (135 nM free Ca^{++}) (see Marrion et al. 1987). Records show membrane current (upper traces) and membrane voltage (lower traces); times refer to time after pipette "breakthrough" or drug washout. The holding potential was −35 mV and −10 mV, 1-sec steps were superimposed at 10-sec intervals; these induced inward current relaxations resulting from I_M deactivation (see Constanti and Brown 1981). Addition of 0.4 μM muscarine (mus) to the bath fluid produced an inward current and reduced the amplitude of the I_M deactivation relaxations. In the control cell (*A*), this effect rapidly reversed on washing. (*B*) 200 μM Gpp(NH)p was added to the pipette solution. This led to a slowly developing inward current and a decline in the I_M deactivation relaxations, and the effect of muscarine could now not be reversed on washing. (From N.V. Marrion et al., unpubl.)

1987) and hence accelerates the trapping action of the nonhydrolyzable analogs.

In contrast, the stable GDP analog, GDPβS, competes with GTP for the G protein and hence inhibits activation and dissociation by GTP (Eckstein et al. 1979). In our experiments, inclusion of GDPβS (500 μM) in the pipette reduced the amount of I_M inhibition produced by a standard dose of muscarine by about 40% (Fig. 8).

These observations are compatible with the involvement of a G protein in the inhibition of I_M in rat ganglion cells by muscarine. The identity of the G protein is not yet clear. However, it appears to be neither G_i nor G_o (see Gilman 1987), since agonist-induced inhibition of I_M in rat ganglia is not inhibited by *Bordetella* PTx, which ADP-ribosylates G_i and G_o (Fig. 8) (Ui 1986). In fact, no substrate for ADP ribosylation by PTx could be detected in membrane preparations from these ganglia (Marrion 1987). Similar types of experiments have yielded comparable evidence for the role of a PTx-insensitive G protein in the inhibition of the I_M in frog ganglia by both muscarinic and peptide agonists (Jones 1987; Lopez et al. 1987; Pfaffinger 1987).

Second Messengers

The next question is, How does the activated G protein inhibit the I_M? One possibility is that it activates an enzyme to produce an amplifying second messenger. As in the neuroblastoma hybrid cells, a leading candidate for such an enzyme is the phosphodiesterase responsible for the hydrolysis of PtdInsP, since there have been several reports to the effect that muscarinic agonists can stimulate $PtdInsP_2$ hydrolysis in ganglia (see, e.g., Bone et al. 1984; Horwitz et al. 1984, 1985). To test this, we have recorded the responses of ganglion cells to the two principal second messengers generated by this reaction, Ins(1,4,5)P_3 and DAG (or alternative protein kinase C activators).

Ins(1,4,5)P₃. We have tried both iontophoretic injections of Ins(1,4,5)P_3 (in the manner used with the neuroblastoma hybrid cells) or addition of Ins(1,4,5)P_3 in concentrations of up to 100 μM to the cell interior via a patch pipette. These applications often produced a small outward current, much smaller than, although otherwise resembling, that evoked in the neuroblastoma hybrid cells (e.g., in their sensitivity to dTC). However, no clear inhibition of I_M was detectable, even at the highest concentrations used (many times greater than those likely to be generated by an agonist) (Fig. 9). Similarly, negative effects with both approaches have been reported in frog ganglion cells (Brown and Adams 1987; Pfaffinger et al. 1988). This, coupled with the fact that I_M is not consistently suppressed by elevation of intracellular Ca (Adams et al. 1982b; Pfaffinger et al. 1988), suggests that Ins(1,4,5)P_3

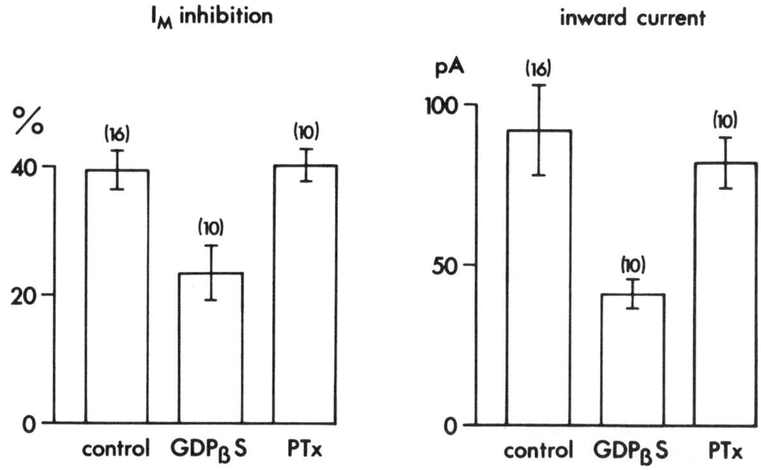

Figure 8. Effects of adding 500 μM GDPβS to the patch-pipette solution and 4–27-hr preincubation with 500 ng/ml^{-1} PTx on the inhibition of I_M and net inward current produced by 0.4 μM muscarine. Cells were voltage-clamped at −40 mV and subjected to −20-mV hyperpolarizing steps. Muscarine was applied by bath perfusion as in Fig. 7, and the amplitudes of the I_M deactivation relaxations before and during muscarine perfusion were measured to determine the percentage of inhibition by muscarine. The inward current produced by muscarine was measured at the holding potential. Bars show S.E.M.; numbers in brackets above blocks give number of cells. Note that inclusion of GDPβS in the pipette significantly ($p < 0.01$) reduced the effect of muscarine by 40–60%, but that preincubation with PTx had no effect. (Incubation for 4 hr with the same sample of PTx completely blocked the inhibition of the Ca current in NG108-15 cells produced by noradrenaline.) (From N.V. Marrion et al., unpubl.)

is unlikely to be the transducing agent in ganglion cells. This is in stark contrast to recent observations on hippocampal cells, where diffusion of Ins(1,4,5)P_3 from an inserted microelectrode sufficed to inhibit the hippocampal I_M (Dutar and Nicoll 1988).

Protein kinase C. In contrast to the negative effect of Ins(1,4,5)P_3, external application of a protein kinase C activator, such as PDBu, produced a consistent, although incomplete, inhibition of I_M in both rat (Marrion 1988) and frog (Brown and Adams 1987; Pfaffinger et al. 1988) ganglion cells (Fig. 10). However, it should be noted that there are some differences be-

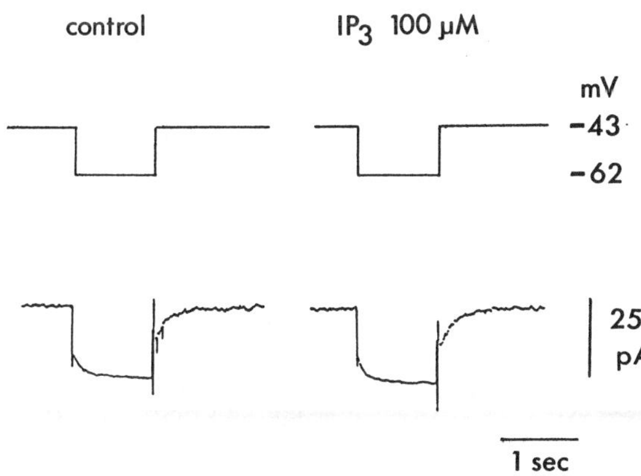

Figure 9. I_M relaxations in response to hyperpolarizing voltage steps recorded in two dissociated adult rat ganglion cells in the same culture dish whole-cell-clamped through patch pipettes, one containing the normal pipette solution (see Fig. 8) and the other with 100 μM Ins(1,4,5)P_3 (Li salt) added to the pipette solution. (*Upper*) Voltage (holding potential −43 mV); (*lower*) current. Note that the cell containing Ins(1,4,5)P_3 showed normal I_M deactivation relaxations during the hyperpolarizing step. (From N.V. Marrion et al., unpubl.)

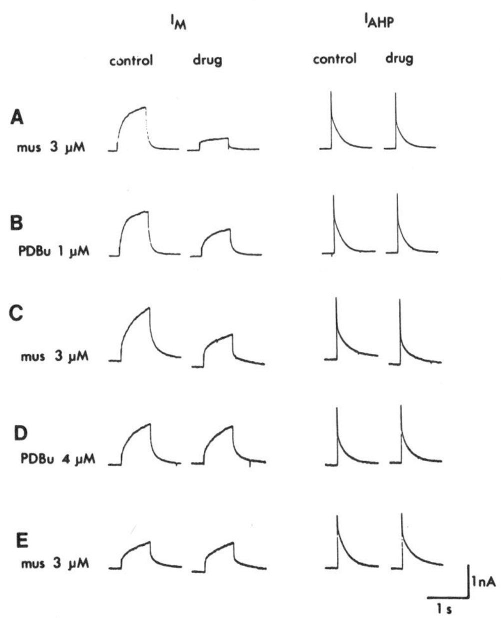

Figure 10. Effects of muscarine (mus) and PDBu on I_M and slow Ca-activated K current (I_{AHP}) recorded from a bullfrog lumbar sympathetic neuron impaled with a microelectrode containing 3 M KCl. Initial holding potential −59 mV. I_M was activated by a 700 msec, +15 mV depolarizing step; I_{AHP} was activated by a 50 msec, +35 mV depolarizing step, to induce a Ca current, and is revealed as a slowly declining outward tail current after repolarization. The initial application of 3 μM muscarine (*A*) suppressed I_M and slightly reduced I_{AHP}. These effects were partly replicated by 1 μM PDBu (*B*; 3 min application). Subsequent additions of 3 μM muscarine 15 min after washing out PDBu (*C*) produced less effect. A second application of 4 μM PDBu (*D*) produced no additional inhibition, but a further application of muscarine (*E*) was now ineffective. Thus, PDBu partly replicated but also reduced the effect of muscarine. (The holding potential was adjusted to −54 mV in *C–E*.) (From P.R. Adams and D.A. Brown, unpubl.; for technical details, see Brown and Adams 1987.)

tween the action of C kinase activators and the normal agonist. First, inhibition by C kinase activators was slow and incomplete; this does not seem to result from the requirement for an adjuvant increase in Ca (Nishizuka 1984), since the amount of inhibition was not increased by injecting Ins(1,4,5)P_3 (Brown and Adams 1987) or raising internal Ca directly (Pfaffinger et al. 1988). Second, PDBu reduced the sensitivity of the residual I_M to the normal agonist (Fig. 10). Third, in the rat ganglion (and indeed in neuroblastoma cells; Brown and Higashida 1988c), PDBu produced an additional inhibition of a "leak" component of membrane conductance that, although also susceptible to muscarine, was rather more profoundly affected by PDBu than by muscarine itself (Marrion 1988). Hence, although compatible in general terms with a role for DAG and consequent protein kinase C activation as a possible mechanism for agonist-induced I_M inhibition in ganglion cells, much needs to be done before this can be considered as the only or primary mechanism.

DISCUSSION

Figure 11 summarizes possible pathways for K-current control induced by agonist-stimulated PtdIns turnover in the two cell types studied in these experiments. As indicated above, there is good evidence that the principal agonists concerned (BK in NG108-15 cells and acetylcholine in rat ganglion cells) can stimulate PtdIns breakdown, so the main question is how far this latter effect is responsible for their electrophysiological effects.

The evidence in favor of a role for Ins(1,4,5)P_3 in triggering (via Ca release) the initial hyperpolarization and outward current produced by BK in NG108-15 cells is quite strong (although necessarily incomplete in the absence of appropriate blocking agents for the intracellular steps concerned). In view of the very widespread

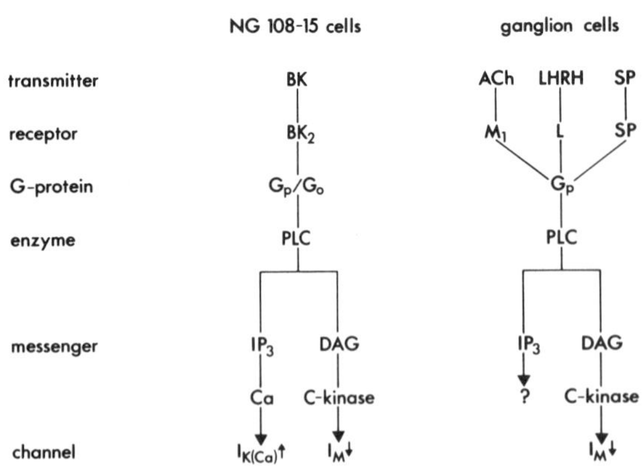

Figure 11. Possible pathways for K-current control by agonists operating through the PtdIns system in NG108-15 neuroblastoma × glioma hybrid cells and in sympathetic ganglion cells. (G) G protein; (PLC) phospholipase C; (IP_3) Ins(1,4,5)P_3; (DAG) diacylglycerol.

occurrence both of Ca-activated currents of the type seen in these cells and of transmitter-induced formation of Ins(1,4,5)P_3 within the nervous system, a hyperpolarization of this form might be expected to be a very common synaptic event. However, rather surprisingly, this appears not to be the case. Although many transmitters are capable of inducing an inhibitory synaptic potential mediated by an increased K conductance, in no other neuron so far have such potentials been shown to result from the activation of Ca-dependent channels or to involve the Ins(1,4,5)P_3/Ca release pathway. This contrasts sharply with the situation in secretory and neurosecretory cells, where secretagogue-induced changes in Ca-activated K currents mediated by the Ins(1,4,5)P_3 system are common (e.g., salivary cells [Evans and Marty 1986; Morris et al. 1987], liver cells [Capiod et al. 1987], and anterior pituitary cells [Dufy et al. 1987]). In this context, it should be noted that even NG108-15 cells behave as "neurosecretory" cells when cocultured with muscle cells, and both BK and intracellular Ins(1,4,5)P_3 increase this neurosecretory activity even at a time when the cell membrane is hyperpolarized and action potential discharges are inhibited (Higashida 1988). Thus, it may be that the membrane current responses mediated by the Ins(1,4,5)P_3 system in neural cells are in the nature of an epiphenomenon and that the primary purpose of the agonist-stimulated Ins(1,4,5)P_3 formation is not to modify electrical signaling but to control other aspects of cell function.

The role of DAG is less secure, relying as it does on the replicating effect of phorbol esters and exogenous DAG. Although there is good evidence that BK can induce DAG formation in NG108-15 cells (Yano et al. 1985) and that BK and PDBu can induce the phosphorylation of several proteins in common (thereby implying that BK probably activates C kinase; J. Sagara and H. Higashida, unpubl.), proof that this pathway is truly responsible for I_M inhibition requires additional tests with blocking agents. As with the response to Ins(1,4,5)P_3, it is possible that the principal raison d'etre of C kinase activation by BK has more to do with neurosecretion in these cells than with signaling per se, since like Ins(1,4,5)P_3, PDBu increased neurosecretory activity in NG108-15 muscle cocultures (Higashida 1988). Similar uncertainties apply to the role of DAG in mediating agonist-induced I_M inhibition in ganglion cells, an uncertainty exacerbated by the fact that phosphorylation studies do not provide unequivocal evidence for the ability of muscarinic agonists to activate C kinase in ganglion cells (Cahill and Perlman 1986, 1987). Notwithstanding, the ability of PDBu to inhibit I_M indicates the potential for receptor-triggered C kinase activation to modify electrical signaling in ganglion cells. In this regard, the I_M in ganglion and neuroblastoma cells joins both its counterpart in smooth muscle cells (Clapp et al. 1987) (but not in hippocampal cells; Malenka et al. 1986), and a number of other ionic currents (Kaczmarek 1987) in its sensitivity to C kinase activators.

ACKNOWLEDGMENTS

The work described in this paper was supported by grants from the United Kingdom Medical Research Council and Science and Engineering Research Council, the National Institutes of Health (grant NS-18579), and the Fogarty Centre of the National Institutes of Health.

REFERENCES

Adams, P.R., D.A. Brown, and A. Constanti. 1982a. M-currents and other potassium currents in bullfrog sympathetic neurones. *J. Physiol.* **330:** 537.

———. 1982b. Pharmacological inhibition of the M-current. *J. Physiol.* **332:** 223.

Batty, I.R., S.R. Nahorski, and R.F. Irvine. 1985. Rapid formation of inositol 1,3,4,5-tetrakisphosphate following muscarinic receptor stimulation of rat cerebral cortical slices. *Biochem. J.* **232:** 211.

Berridge, M.J. 1987. Inositol trisphosphate and diacylglycerol: Two interacting second messengers. *Annu. Rev. Biochem.* **56:** 159.

Bone, E.A., P. Fretten, S. Palmer, C.J. Kirk, and R.H. Michell. 1984. Rapid accumulation of inositol phosphates in isolated rat superior cervical ganglia exposed to V_1-vasopressin and muscarinic cholinergic stimulation. *Biochem. J.* **221:** 803.

Breitweiser, G.E. and G. Szabo. 1985. Uncoupling of cardiac muscarinic and β-adrenergic receptors from ion channels by a guanine nucleotide analogue. *Nature* **317:** 538.

Brown, D.A. 1988. M-currents. In *Ion channels* (ed. T. Narahashi), p. 55. Plenum Press, New York.

Brown, D.A. and P.R. Adams. 1980. Muscarinic suppression of a novel voltage-sensitive K^+-current in a vertebrate neurone. *Nature* **283:** 673.

———. 1987. Effects of phorbol dibutyrate on M currents and M current inhibition in bullfrog sympathetic neurons. *Cell. Mol. Neurobiol.* **7:** 255.

Brown, D.A. and H. Higashida. 1988a. Voltage- and calcium-activated potassium currents in mouse neuroblastoma × rat glioma hybrid cells. *J. Physiol.* **397:** 149.

———. 1988b. Membrane current responses of NG108-15 mouse neuroblastoma × rat glioma hybrid cells to bradykinin. *J. Physiol.* **397:** 167.

———. 1988c. Inositol 1,4,5-trisphosphate and diacylglycerol mimic bradykinin effects on mouse neuroblastoma × rat glioma hybrid cells. *J. Physiol.* **397:** 185.

Brown, D.A., K. Furuya, and H. Higashida. 1987. K channels in NG108-15 rodent neuroblastoma × glioma cells evoked by intracellular injections of inositol-1,4,5-trisphosphate (IP_3). *J. Physiol.* **390:** 217.

Cahill, A.L. and R.L. Perlman. 1986. Nicotinic and muscarinic agonists, phorbol esters, and agents which raise cyclic AMP levels phosphorylate distinct groups of proteins in the superior cervical ganglion. *Neurochem. Res.* **11:** 327.

———. 1987. Vasopressin stimulates the phosphorylation of an 83,000 M_r protein in the superior cervical ganglion. *Cell. Mol. Neurobiol.* **7:** 413.

Capiod, T., A.C. Field, D.C. Ogden, and C. Sandford. 1987. Internal perfusion of guinea-pig hepatocytes with buffered Ca^{2+} or with inositol 1,4,5-trisphosphate mimics noradrenaline activation of K^+ and Cl^- conductances. *FEBS Lett.* **217:** 247.

Clapp, L.H., M.H. Vivaudou, J.J. Singer, and J.V. Walsh. 1987. A diacylglycerol analogue mimics the action of acetylcholine and substance P on calcium currents in freshly dissociated smooth muscle cells. *Biophys. J.* **90:** 13a.

Constanti, A. and D.A. Brown. 1981. M-currents in voltage-clamped mammalian sympathetic neurones. *Neurosci. Lett.* **24:** 289.

Dufy, B., A.B. MacDermott, and J.L. Barker. 1986. Rundown of GH3 cell K^+ conductance response to TRH following patch recording can be obviated with a GH3 extract. *Biochem. Biophys. Res. Commun.* **137:** 388.

Dutar, P. and R.A. Nicoll. 1988. Stimulation of phosphatidyl inositol (PI) turnover may mediate the muscarinic suppression of the M-current in hippocampal pyramidal cells. *Neurosci. Lett.* **85:** 89.

Eckstein, R., D. Cassel, H. Levkovitz, M. Lowe, and Z. Selinger. 1979. Guanosine 5′0-(2-thiodiphosphate). An inhibitor of adenylate cyclase stimulation by guanine nucleotides and fluoride ions. *J. Biol. Chem.* **254:** 9829.

Evans, M.G. and A. Marty. 1986. Potentiation of muscarinic and α-adrenergic responses by an analogue of guanosine 5′ triphosphate. *Proc. Natl. Acad. Sci.* **83:** 4099.

Fisher, S.K. and B.W. Agranoff. 1986. Phosphoinositide turnover in the CNS and in neural-related tissues. In *Phosphoinositides and receptor mechanisms* (ed. J.W. Putney, Jr.), p. 219. A.R. Liss, New York.

Gilman, A.G. 1987. G-proteins: Transducers of receptor-generated signals. *Annu. Rev. Biochem.* **56:** 615.

Grandt, R., C. Greiner, P. Zubin, and K.H. Jakobs. 1986. Bradykinin stimulates GTP hydrolysis in NG108-15 membranes by a high-affinity, pertussis toxin-insensitive GTPase. *FEBS Lett.* **196:** 279.

Higashida, H. 1988. Acetylcholine release by bradykinin, inositol 1,4,5-trisphosphate and phorbol dibutyrate in rodent neuroblastoma cells. *J. Physiol.* **397:** 209.

Higashida, H. and D.A. Brown. 1986. Two polyphosphoinositide metabolites control two K^+ currents in a neuronal cell. *Nature* **323:** 333.

Higashida, H., R.A. Streaty, W.A. Klee, and M. Nirenberg. 1986. Bradykinin-activated transmembrane signals are coupled via N_o or N_i to production of inositol 1,4,5-trisphosphate, a second messenger in NG108-15 neuroblastoma-glioma hybrid cells. *Proc. Natl. Acad. Sci.* **83:** 942.

Horwitz, J., S. Tsymbalow, and R.L. Perlman. 1984. Muscarine increases tyrosine-3-monooxygenase activity and phospholipid metabolism in the superior cervical ganglion of the rat. *J. Pharmacol. Exp. Ther.* **229:** 577.

———. 1985. Muscarine stimulates the hydrolysis of inositol-containing phospholipids in the superior cervical ganglion. *J. Pharmacol. Exp. Ther.* **233:** 235.

Irvine, R.F. and R.M. Moor. 1986. Micro-injection of inositol 1,3,4,5-tetrakisphosphate activates sea-urchin eggs by a mechanism dependent on external Ca^{2+}. *Biochem. J.* **240:** 917.

Irvine, R.F., A.J. Letcher, J.P. Heslop, and M.J. Berridge. 1986. The inositol tris/tetrakisphosphate pathway—Demonstration on Ins(1,4,5)P_3 3-kinase activity in animal tissues. *Nature* **320:** 631.

Jackson, T.R., T.J. Hallam, C.P. Downes, and M.R. Hanley. 1987. Receptor coupled events in bradykinin action: Rapid production of inositol phosphates and regulation of cytosolic free Ca^{2+} in a neural cell line. *EMBO J.* **6:** 49.

Jones, S.W. 1987. GTP-γ-S inhibits the M-current of dissociated bullfrog sympathetic neurons. *Neurosci. Abstr.* **13:** 533.

Kaczmarek, L.K. 1987. The role of protein kinase C in the regulation of ion channels and neurotransmitter release. *Trends Neurosci.* **10:** 30.

Lopez, H.S., D.A. Brown, and P.R. Adams. 1987. Possible involvement of GTP-binding proteins in coupling of muscarinic receptors to M-current in bullfrog ganglion cells. *Neurosci. Abstr.* **13:** 532.

Malenka, R.C., D.V. Madison, R. Andrade, and R.A. Nicoll. 1986. Phorbol esters mimic some cholinergic actions in hippocampal pyramidal cells. *J. Neurosci.* **6:** 475.

Marrion, N.V. 1987. Probable role of a GTP-binding protein in mediating M-current inhibition by muscarine in rat sympathetic neurones. *J. Physiol.* **396:** 87P.

———. 1988. "Pharmacological control of M-currents in rat sympathetic neurones from the extracellular and intracellular milieus." Ph.D. Thesis, University of London.

Marrion, N.V., T.G. Smart, and D.A. Brown. 1987. Membrane currents in adult rat superior cervical ganglia in dissociated tissue culture. *Neurosci. Lett.* **77:** 55.

Morris, A.P., D.V. Gallagher, R.F. Irvine, and O.H. Petersen. 1987. Synergism of inositol trisphosphate and tetrakisphosphate in activating Ca^{2+}-dependent K^+ channels. *Nature* **330:** 653.

Nishizuka, Y. 1984. The role of protein kinase C in cell surface signal transduction and tumour promotion. *Nature* **308:** 693.

Osugi, T., S. Uchida, T. Imaizumi, and H. Yoshida. 1986. Bradykinin-induced intracellular Ca^{2+} elevation in neuroblastoma × glioma hybrid NG108-15 cells: Relationship to the action on inositol phospholipid metabolites. *Brain Res.* **379:** 84.

Pfaffinger, P.J. 1987. Control of M-current in dialyzed sympathetic ganglion neurons. *Neurosci. Abstr.* **13:** 152.

Pfaffinger, P.J., M.D. Liebowitz, M.M. Bosma, W. Almers, and B.Hille. 1988. M-current suppression by agonists: The role of the phospholipase C pathway. *Biophys. J.* **53:** 637a.

Pfaffinger, P.J., J.M. Martin, D.D. Hunter, N.M. Nathanson, and B. Hille. 1985. GTP-binding proteins couple cardiac muscarinic receptors to a K-channel. *Nature* **317:** 536.

Reiser, G. and B. Hamprecht. 1982. Bradykinin induces hyperpolarization in rat glioma cells and in neuroblastoma × glioma hybrid cells. *Brain Res.* **239:** 191.

———. 1985. Bradykinin causes a transient rise of intracellular Ca^{2+} activity in cultures neural cells. *Pfluegers Arch. Eur. J. Physiol.* **405:** 260.

Streb, H., R.F. Irvine, M.J. Berridge, and I. Schultz. 1983. Release of Ca^{2+} from a nonmitochondrial intracellular store in pancreatic acinar cells by inositol-1,4,5-trisphosphate. *Nature* **306:** 67.

Ui, M. 1986. Pertussis toxin as a probe of receptor coupling to inositol lipid metabolism. In *Phosphoinositides and receptor mechanisms* (ed. J.W. Putney, Jr.), p. 163. A.R. Liss, New York.

Yano, K., H. Higashida, H. Hattori, and Y. Nozawa. 1985. Bradykinin-induced transient accumulation of inositol trisphosphate in neuron-like cell line NG108-15 cells. *FEBS Lett.* **181:** 403.

Yano, K., H. Higashida, R. Inoue, and Y. Nozawa. 1984. Bradykinin-induced rapid breakdown of phosphatidylinositol 4,5-bisphosphate in neuroblastoma × glioma hybrid NG108-15 cells. A possible link to agonist-induced neuronal functions. *J. Biol. Chem.* **259:** 10201.

Functional Expression of the 5-HT$_{1c}$ Receptor in Neuronal and Nonneuronal Cells

D. JULIUS,* A.B. MACDERMOTT,*† T.M. JESSEL,*‡ K. HUANG,* S. MOLINEAUX,*
I. SCHIEREN,* AND R. AXEL*‡

‡*Department of Biochemistry and Molecular Biophysics,* †*Department of Physiology and Cellular Biophysics, and* **Howard Hughes Medical Institute, College of Physicians and Surgeons, Columbia University, New York, New York 10032*

Serotonin, 5-hydroxytryptamine (5-HT), is a neurotransmitter released by nerve terminals in most regions of the central and peripheral nervous system. Serotonin exerts its diverse physiological effects by binding to cell-surface receptors that activate several distinct second-messenger pathways that modulate the excitable properties of both central and peripheral neurons. Multiple serotonin receptor subtypes (including 5-HT$_{1a}$, 5-HT$_{1b}$, 5-HT$_{1c}$, 5-HT$_2$, and 5-HT$_3$) have been defined on the basis of their pharmacologic properties (Peroutka 1988). Individual receptor subtypes reveal characteristic ligand-binding profiles and couple to different intracellular second-messenger signaling systems. The 5-HT$_{1c}$ and 5-HT$_2$ receptors stimulate the phospholipase-C-mediated production of inositol phosphates (de Courcelles et al. 1985; Conn et al. 1986), whereas 5-HT$_{1a}$ and 5-HT$_{1b}$ receptors may regulate adenylate cyclase activity (DeVivo and Maayani 1986; Weiss et al. 1986; Hoyer and Schoekkter 1988) or couple to G proteins that directly activate ion channels (Aghajanian 1981; Andrade et al. 1986; Andrade and Nicoll 1987).

The diverse actions of serotonin in the nervous system are thought to be mediated by activation of these distinct receptor subtypes (for review, see Peroutka 1988). For example, the hallucinatory and perceptual disturbances associated with administration of lysergic acid diethylamide (LSD) and other psychedelic serotonin analogs (Leary et al. 1964) are probably elicited by activation of cortical 5-HT$_2$ receptors (Glennon et al. 1984). In contrast, the inflammatory and pain-producing effects of serotonin are mediated via 5-HT$_3$ receptors on peripheral sensory endings (Richardson et al. 1985).

The isolation of the genes encoding the multiple serotonin receptor subtypes and the ability to express these receptors in new cellular environments will help to elucidate the molecular mechanisms of action of serotonin in the mammalian brain. The cloning of most neurotransmitter receptors has required the purification of receptor, the determination of partial protein sequence, and the synthesis of oligonucleotide probes with which to obtain cDNA or genomic clones. However, the serotonin receptors have not been purified and antibodies have not been generated. We have therefore designed a cDNA expression system that permits the identification of functional cDNA clones encoding serotonin receptors in the absence of protein sequence information. We have combined cloning in RNA expression vectors with an electrophysiological assay in oocytes to isolate a functional cDNA clone encoding the entire 5-HT$_{1c}$ receptor. The sequence of this clone reveals that the 5-HT$_{1c}$ receptor belongs to a family of G-protein-coupled receptors that are thought to traverse the membrane seven times. Mouse fibroblasts transformed with this clone bind serotonergic ligands and respond to serotonin with an elevation in intracellular calcium. Moreover, in situ hybridization and Northern blot analysis indicate that the 5-HT$_{1c}$ receptor mRNA is expressed in a wide variety of neurons in the rat central nervous system, suggesting that this receptor plays a prominant role in neuronal function.

RESULTS

Isolation and Structure of the 5-HT$_{1c}$ Receptor

The expression of functional receptors in *Xenopus* oocytes has provided a sensitive assay for detection of mRNAs that encode receptors, including the 5-HT$_{1c}$ receptor, which couple via inositol phospholipid signaling systems (Gunderson et al. 1983; Lubbert et al. 1987a; Takahashi et al. 1987). High levels of serotonin 5-HT$_{1c}$ receptors (10^5 per cell) are expressed in the choroid plexus (Pazos and Palacios 1985; Yagaloff and Hartig 1985), a nonneuronal cell type in the central nervous system responsible for the production of cerebrospinal fluid. *Xenopus* oocytes injected with choroid plexus mRNA exhibit a serotonin-evoked inward current. Application of serotonin appears to liberate inositol phosphates that raise intracellular Ca^{++} levels, leading to the opening of Ca^{++}-dependent chloride channels (Gunderson et al. 1983; Lubbert et al. 1987a; Takahashi et al. 1987). Davidson and colleagues have used this oocyte assay to isolate a cDNA clone encoding the carboxy-terminal portion of the 5-HT$_{1c}$ receptor by hybrid depletion of choroid plexus mRNA (Lubbert et al. 1987b). We have combined cloning in RNA expression vectors with an electrophysiological assay in oocytes to isolate and characterize the expression of a functional cDNA clone encoding the entire 5-HT_{1c} receptor (Julius et al. 1988).

Briefly, a cDNA library was constructed from choroid plexus mRNA in λZAP, a vector permitting the in vitro transcription of cDNA inserts. RNAs transcribed from populations of cDNA clones were screened for their ability to induce serotonin-evoked currents in *Xenopus* oocytes. A single clone encoding a functional serotonin 5-HT_{1c} receptor was purified from a large population of clones by procedures of sibling selection.

Following this protocol, we isolated a single clone, pSR-1c, encoding an RNA that elicits serotonin sensitivity when injected into oocytes. Serial dilutions of RNA transcribed in vitro from pSR-1c demonstrated that injection of 0.5 pg of RNA was sufficient to generate a serotonin-evoked current of about 200 nA (Fig. 1). Pharmacological analysis of injected oocytes revealed that this cDNA encodes a serotonin receptor with properties similar to that of the 5-HT_{1c} subtype (Conn et al. 1986). Superfusion of oocytes with 1 μM

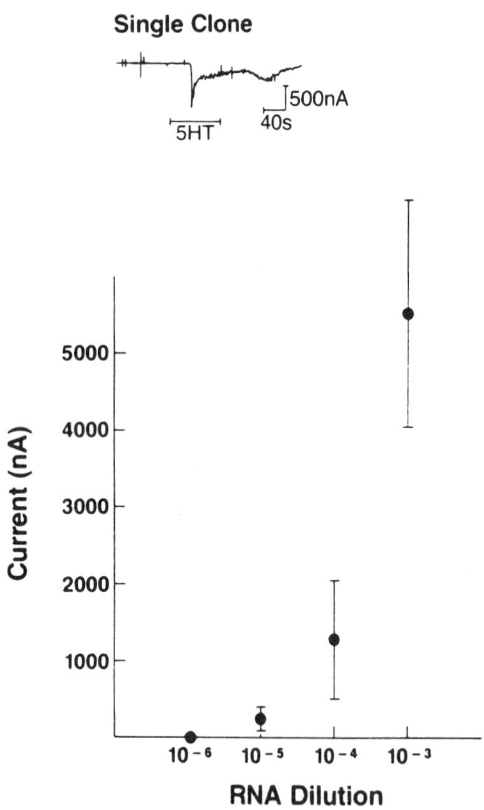

Figure 1. Serotonin sensitivity of *Xenopus* oocytes injected with RNA transcribed in vitro from λZAP cDNA clones. Inward current was evoked by serotonin (10^{-6} M) in *Xenopus* oocytes injected with RNA transcribed in vitro from a single λZAP cDNA clone (pSR-1c) at 10^{-4} dilution (5 pg) (trace). Serotonin-evoked currents after serial dilution of RNA transcribed in vitro from the λZAP cDNA clone pSR-1c. The undiluted RNA was at a concentration of 1 ng/nl. Each oocyte was injected with 50 nl of diluted RNA. Cells were incubated for 3 days and voltage clamped at −50 mV. Each point represents the mean and standard error of serotonin-evoked currents obtained from three to seven oocytes.

mianserin blocked the current evoked by 10 nM 5-HT in injected oocytes, whereas 1 μM spiperone produced only a small decrease in the serotonin-evoked current. The relative potency of these antagonists is therefore similar when assayed against 5-HT_{1c} receptors present in the choroid plexus or in oocytes injected with pSR-1c-derived RNA.

The deduced protein sequence of the 5-HT_{1c} receptor indicates that this protein shares numerous sequence and structural properties with the family of receptor molecules that has been predicted to span the lipid bilayer seven times (Fig. 2). This family includes rhodopsin and related opsins (Nathans and Hogness 1983), the α- and β-adrenergic receptors (Dohlman et al. 1987), the muscarinic cholinergic receptors (Kubo et al. 1986; Bonner et al. 1987; Peralta et al. 1987), the substance K neuropeptide receptor (Masu et al. 1987). the yeast mating-factor receptors (Burkholder et al. 1985; Nakayama et al. 1985; Hagan et al. 1986), and the oncogene c-*mas* (Young et al. 1986). Each of these receptors is thought to transduce extracellular signals by interaction with guanine-nucleotide-binding (G) proteins (for review, see Dohlman et al. 1987, 1988; O'Dowd et al. 1989). On the basis of structural homologies with rhodopsin, it is likely that the amino terminus of the 5-HT_{1c} receptor is located on the extracellular side of the plasma membrane, with the carboxyl terminus located intracellularly. In this scheme, three extracellular loops alternate with three intracellular loops to link the seven transmembrane domains.

Although the amino-terminal 65 amino acids of the 5-HT_{1c} receptor are thought to reside on the extracellular face of the membrane, the protein lacks a definable signal sequence. A potential N-linked glycosylation site is present at position 39. The receptor contains seven hydrophobic stretches 20–30 amino acids in length, which represent potential transmembrane domains. These domains constitute the regions of maximal sequence similarity with the other transmitter receptors of this class. The 5-HT_{1c} receptor shares 25% sequence identity with the β_2-adrenergic receptor and 20% identity with the muscarinic and substance K receptors (Fig. 3).

A conserved aspartate residue is present within the second putative transmembrane domain, and by analogy with the β_2-adrenergic receptor, this residue may participate in ligand binding. Mutations in this conserved aspartate in the β_2-adrenergic receptor reduce agonist binding (Strader et al. 1987b; Chung et al. 1988). In addition, amino acid residues in this domain can be cross-linked to affinity ligands of the β_2-adrenergic receptor (Dohlman et al. 1988; O'Dowd et al. 1989). These data, taken together with the observation that all ligands capable of associating with this family of receptors have amino groups, suggest that aspartate participates in stabilizing the ligand in the plane of the membrane.

The third cytoplasmic loop that is thought to interact with the different G proteins (Dixon et al. 1987a,b; Strader et al. 1987a) is of widely varying length in

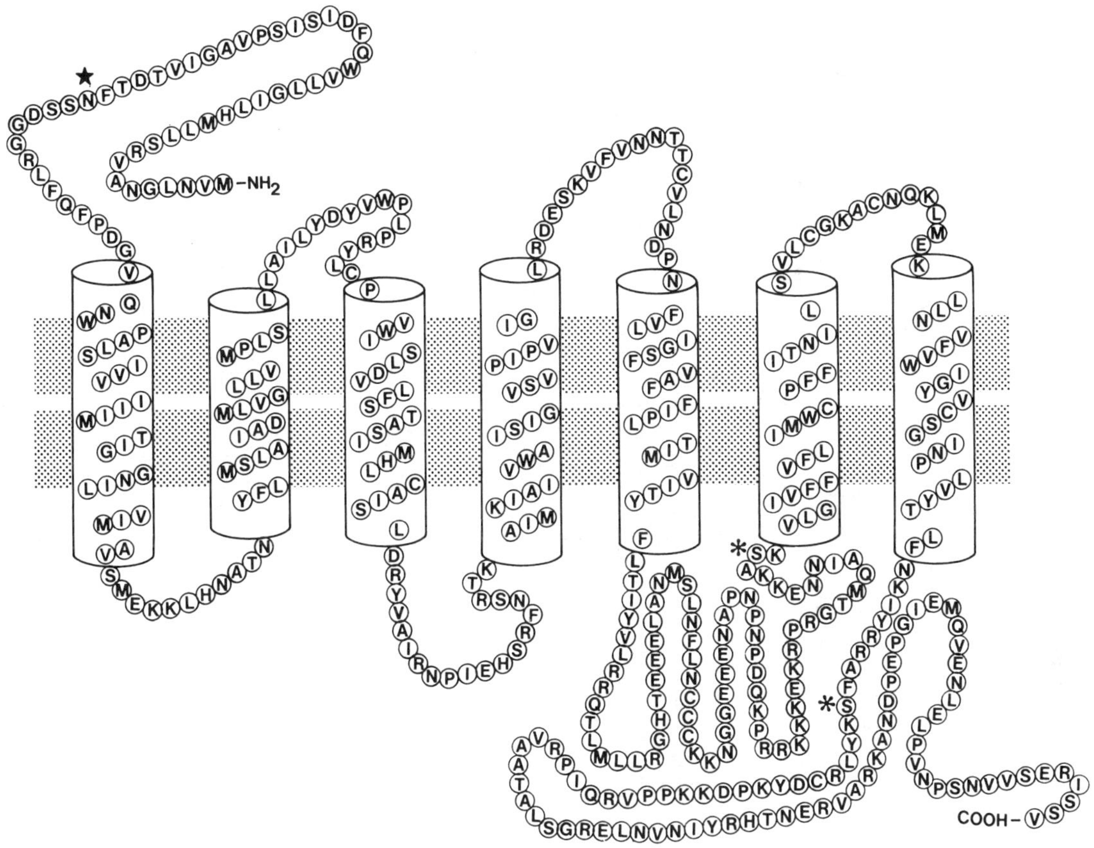

Figure 2. A model for the transmembrane topology of the rat 5-HT_{1c} receptor. The receptor is depicted as containing seven hydrophobic transmembrane regions. By analogy with the structure of rhodopsin, the amino terminus is located on the extracellular face of the lipid bilayer, and the carboxyl terminus is located on the cytoplasmic side. Asterisks indicate serine residues that represent potential phosphorylation sites. The star indicates an asparagine residue in the amino-terminal region that is a potential N-linked glycosylation site.

different receptors. No sequence similarities within this domain are apparent among receptors that couple to common signaling systems. In the 5-HT_{1c} receptor, this loop consists of 77 amino acids, a high proportion of which are basic, a feature characteristic of other receptors in this family.

Both the carboxy-terminal cytoplasmic domain and the third cytoplasmic loop contain sequences that may serve as substrates for phosphorylation by cAMP-dependent protein kinase or protein kinase C. In addition, there are four serine residues in the carboxy-terminal 12 amino acids which, by analogy with rhodopsin and the adrenergic receptors, may represent additional phosphorylation sites. These potential sites of phosphorylation may play a role in regulating the activity of the receptor molecule (Sibley et al. 1987).

5-HT_{1c} Receptor Expression in Mammalian Fibroblasts

The deduced protein sequence of pSR-1c indicates that we have cloned a new member of the gene family encoding G-protein-coupled neurotransmitter receptors. To establish further that this clone encodes a serotonin receptor, we have demonstrated that the introduction of this cDNA into mammalian fibroblasts renders these cells responsive to serotonin. The cDNA insert from pSR-1c was cloned into the expression vector pMV7 (Kirschmeir et al. 1988). This vector contains a murine leukemia virus long terminal repeat that serves as a promoter for expression of the serotonin cDNA, as well as an independent expression cassette encoding neomycin phosphotransferase. Transformed NIH-3T3 cells resistant to neomycin were isolated, and a single clone expressing significant levels of receptor mRNA (SR-3T3) was identified by RNA blot analysis.

Membranes prepared from SR-3T3 cells exhibited high-affinity binding sites for ligands that have been shown to interact with the 5-HT_{1c} receptor on choroid plexus cells (Kadan et al. 1984; Yagaloff and Hartig 1985). SR-3T3 cells expressed 10^3 to 10^4 high-affinity ^{125}I-labeled LSD-binding sites per cell, whereas no specific high-affinity sites were detected on the parental NIH-3T3 cell line. Moreover, the relative ability of specific antagonists to inhibit ^{125}I-labeled LSD binding to SR-3T3 cells paralleled their potency on choroid plexus cells (Fig. 4) (Conn et al. 1986). Mianserin

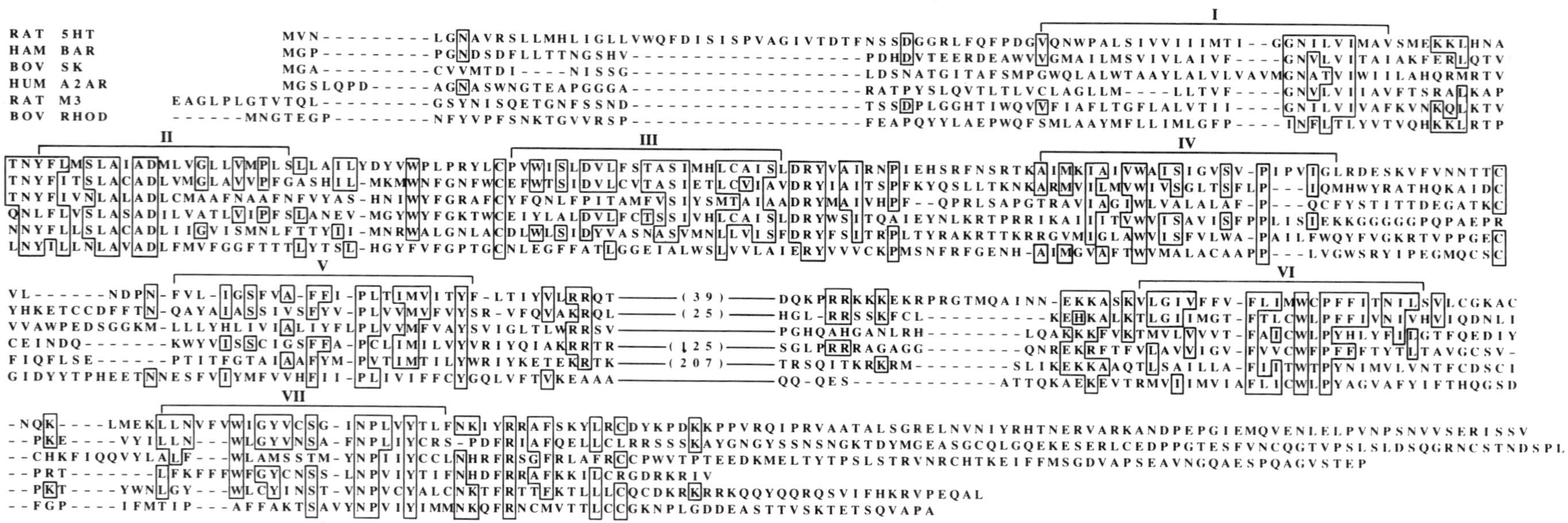

Figure 3. Alignment of the amino acid sequence of the rat 5-HT_{1c} receptor (RAT 5HT), hamster β_2-adrenergic receptor (HAM BAR), bovine substance K receptor (BOV SK); human α_2-adrenergic receptor (HUM A2AR); rat muscarinic M3 receptor (RAT M3); and bovine rhodopsin (BOV RHOD). The amino acid residues enclosed by solid lines represent residues that are identical in at least two of the five reference sequences when compared to the 5-HT_{1c} receptor. Sequences present in the highly variable third cytoplasmic loop are not included, but the number of residues present in this loop is indicated in parentheses. The first 25 amino acids of the rat M3 sequence have not been included here. Roman numerals and brackets denote the seven putative transmembrane domains.

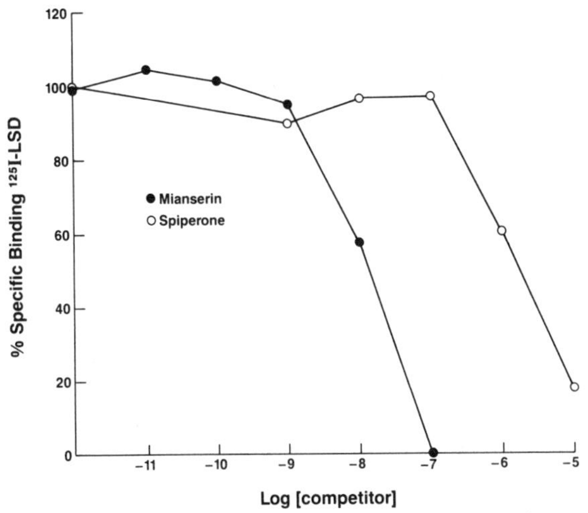

Figure 4. Pharmacological properties of the 5-HT_{1c} receptor expressed in transfected mouse fibroblasts, as determined by binding of ^{125}I-labeled LSD. Membranes were prepared from a fibroblast cell line (SR-3T3) transformed with an expression vector (pMV7) containing the pSR-1c cDNA insert. These membranes were incubated with ^{125}I-labeled LSD (1 nM final concentration) either alone or in the presence of the drugs mianserin or spiperone at the concentrations shown. Nonspecific binding of ^{125}I-labeled LSD was determined using an equivalent amount of membranes prepared from untransformed NIH-3T3 cells. This binding was not displaceable by 10^{-7} M mianserin and represented 45% of the total ligand bound.

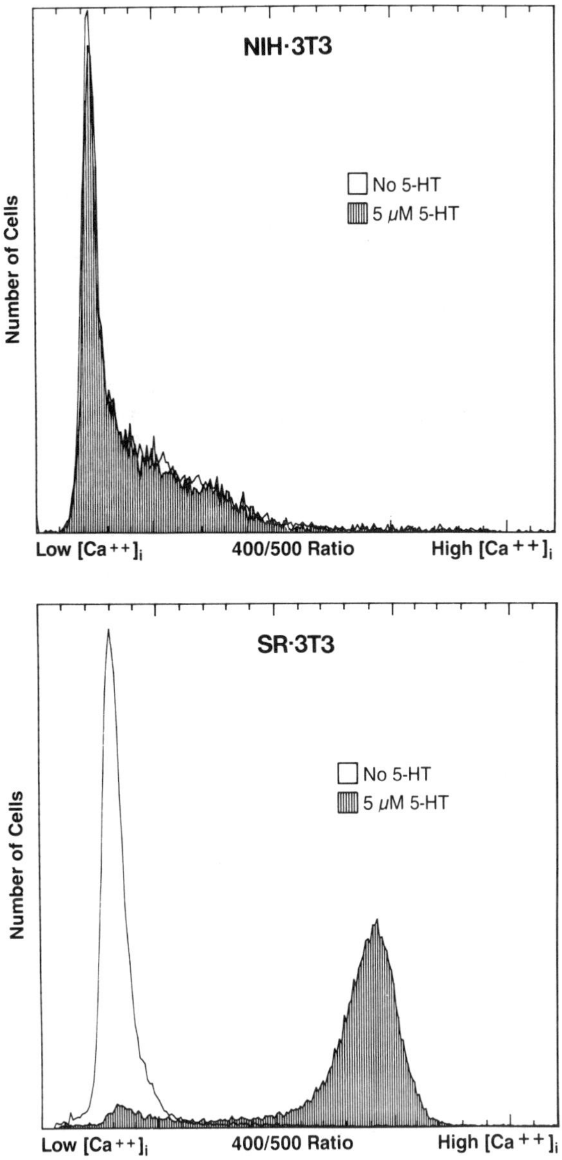

Figure 5. Activation of 5-HT_{1c} receptors in transformed fibroblasts elevates intracellular Ca^{++} concentration. Untransformed cells (NIH-3T3) or transformed cells expressing pSR-1c cDNA (SR-3T3) were loaded with the Ca^{++}-sensitive dye indo-1. Changes in the level of intracellular free Ca^{++} following exposure to 5 μM serotonin (hatched peak) were monitored using a flow cytometer to measure the 400/500-nm emission ratio. In each case, the resting intracellular Ca^{++} level in the absence of serotonin is also shown (open peak). For NIH-3T3 cells, the two peaks are coincident. For SR-3T3 cells, 95% of the population showed a marked elevation in intracellular Ca^{++} in the presence of serotonin. The mean resting Ca^{++} level in NIH-3T3 and SR-3T3 cells was similar.

(IC_{50} = 20 nM) was about two orders of magnitude more effective in displacing ^{125}I-labeled LSD binding than spiperone was (IC_{50} = 2 μM). Binding of ^{125}I-labeled LSD was also displaced by serotonin with an IC_{50} value of 20 nM. The expression of specific high-affinity binding sites for 5-HT_{1c}-selective ligands on transformed 3T3 cells provides independent confirmation that pSR-1c encodes a serotonin 1c receptor.

To ascertain whether the binding of serotonin activates intracellular signaling pathways in transformed fibroblasts, SR-3T3 cells were loaded with the Ca^{++}-sensitive dye indo-1 and analyzed in a fluorescence-activated cell sorter (Schieren and MacDermott 1988). Indo-1 undergoes a characteristic and quantitative shift in its fluorescence emission spectrum as a function of Ca^{++} concentration and serves as a quantitative measure of intracellular free Ca^{++} (Grynkiewicz et al. 1985). An increase in the 400/500-nm emission ratio is proportional to an increase in the intracellular Ca^{++} concentration. SR-3T3 and control NIH-3T3 cells loaded with indo-1 were exposed to serotonin immediately before fluorescence analysis to reduce the possibility of desensitization that may result from prolonged exposure of cells to agonist. Ninety-five percent of the indo-1-loaded SR-3T3 cells showed a marked increase in intracellular Ca^{++} when exposed to serotonin, whereas control, untransfected NIH-3T3 cells did not respond to serotonin (Fig. 5). A dose-response curve over a broad range of serotonin concentrations indicated that a threshold response is observed at 10^{-8} M 5-HT, and a maximal response occurs at 10^{-7} M 5-HT. These experiments indicate that the introduction of pSR-1c cDNA in mammalian fibroblasts leads to the expression of functional serotonin receptors. Thus, the 5-HT_{1c} receptor is capable of triggering the same transduction machinery, regardless of the cell type in which it is expressed.

We have also used this Ca^{++} assay to determine the relative potency of serotonin receptor antagonists in blocking the response of SR-3T3 cells to 10^{-6} M serotonin (Fig. 6). Mesulergine (10^{-6} M) is most effective in blocking the Ca^{++} response, followed by mianserin, whereas spiperone has little or no effect at the concentration used in this assay. This rank order of antagonist potency in this functional assay reflects the affinity of these drugs for the 5-HT_{1c} receptor in the choroid plexus (Conn et al. 1986), as well as in transformed 3T3 cells (Fig. 4), as determined by radioligand-binding assays.

Expression of 5-HT_{1c} Serotonin Receptor in the Nervous System

RNA blot analysis and in situ hybridization were performed to examine the expression of 5-HT_{1c} receptor mRNA in different brain regions and in peripheral tissues. In situ hybridization revealed a heavy grain density associated with epithelial cells of the choroid plexus in the third, fourth, and lateral ventricles. Interestingly, hybridization was not restricted to cells of the choroid plexus. Numerous neuronal cell groups throughout the central nervous system expressed 5-HT_{1c} receptor mRNA (S. Molineaux et al., in prep.). Many cortical areas were labeled with particularly intense hybridization observed in pyramidal neurons in the CA1 and CA2 region of the hippocampus (Fig. 7), although the CA3 and CA4 regions were not labeled. In cerebral cortex, hybridization was observed over cells in cortical layer five, although the intensity and number of labeled cells were much lower than in the hippocampus. Neurons in the cerebellar cortex did not express 5-HT_{1c} mRNA.

In subcortical regions, 5-HT_{1c} mRNA was detected in both monoaminergic neurons in the mesencephalon and brainstem. Dopaminergic neurons in the pars compacta of the substantia nigra and neurons in the raphe nuclei and adjacent nucleus paragigantocellularis expressed high levels of 5-HT_{1c} mRNA. These findings suggest that the 5-HT_{1c} receptor is involved in the pre- or postsynaptic regulation of monoaminergic neurons. Moreover, 5-HT_{1c} mRNA is located in many of the neuronal cell groups that comprise the descending pathway that functions to regulate incoming sensory information eventually perceived as pain. Thus, the 5-HT_{1c} receptor may be involved in a variety of neuronal functions, including pain transmission and analgesia, extrapyramidal motor coordination, and, potentially, learning and memory.

The distribution of 5-HT_{1c} receptor mRNA determined by in situ hybridization was supported by RNA blot analysis (Fig. 8). An intense 5.2-kb RNA was observed in choroid plexus and in a variety of other regions of the brain, including the basal ganglia, hypothalamus, hippocampus, pons medulla, and spinal cord. Receptor RNA was not detected in the cerebellum or in liver, kidney, intestine, heart, and lung. Titration studies with purified pSR-1c RNA indicate that the 5-HT_{1c} receptor RNA comprises about 0.02% of the choroid plexus message population and is therefore likely to be present at a level of about 40 copies per cell. The relative abundance of this RNA in other regions of the brain is at least tenfold lower. In situ hybridization, however, indicated that some neurons express receptor mRNA levels comparable to those in choroid plexus cells. These findings demonstrate that the 5-HT_{1c} receptor is not restricted to epithelial cells of the choroid plexus but is expressed in numerous discrete neuronal cell groups in many regions of the rat brain and spinal cord. This distribution of 5-HT_{1c} receptors suggests that this receptor subtype may mediate many of the central actions of serotonin.

CONCLUSIONS

We have cloned and characterized a functional cDNA encoding the 5-HT_{1c} subclass of serotonin receptor. RNA transcribed from this cDNA clone (pSR-

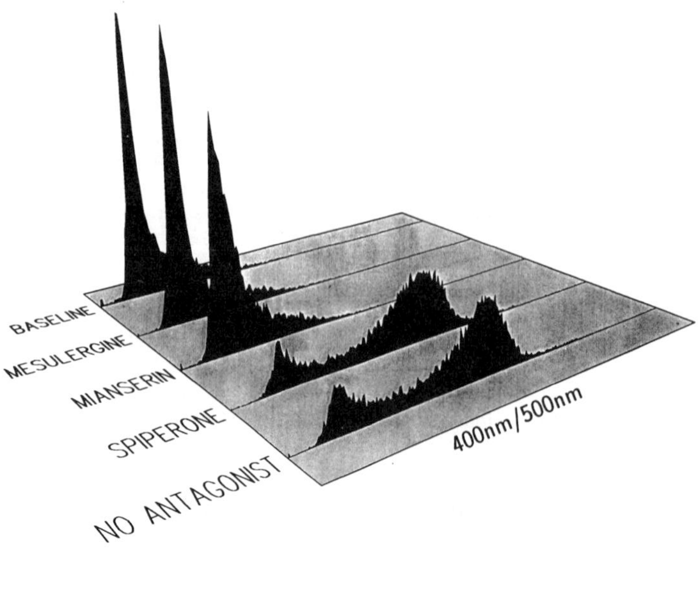

Figure 6. Pharmacological properties of the 5-HT_{1c} receptor expressed in transfected mouse fibroblasts, as determined by flow cytometry. SR-3T3 cells were loaded with the Ca^{++}-sensitive dye indo-1. Changes in the level of intracellular free Ca^{++} were monitored using a flow cytometer to measure the 400/500-nm emission ratio. The resting calcium level for these cells was determined in the absence of serotonin (baseline). Cells were exposed to 1 μM serotonin in the absence (no antagonist) or presence of 1 μM mesulergine, mianerin, or spiperone.

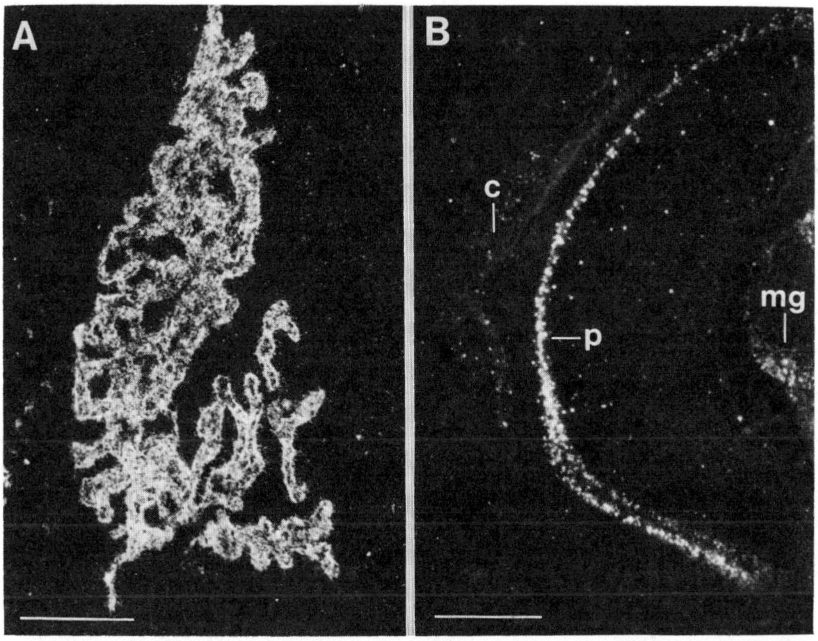

Figure 7. Localization of 5-HT_{1c} receptor mRNA in adult rat brain by in situ hybridization. ^{35}S-labeled antisense RNA probes were prepared by in vitro transcription of pSR-1c cDNA and hybridized with cryostat sections (15 μm) prepared from adult rat brain tissue fixed with 4% paraformaldehyde. (*A*) Hybridization of the antisense RNA probe to choroid plexus epithelial cells in the lateral ventricle. Most of the adjacent neural tissue exhibits low or background hybridization signals, with the exception of a few cells in the midbrain. (*B*) Hybridization of the antisense RNA probe to pyramidal neurons (P) in the CA1 and CA2 region of the hippocampus. Labeled cells are also detectable in deep cortical layers (c) and in the medial geniculate (mg). Bars: (*A*) 200 μm; (*B*) 250 μm.

1c) confers serotonin sensitivity to *Xenopus* oocytes. The expression of this cDNA in mouse fibroblasts results in the appearance of high-affinity serotonin-binding sites on the cell surface. Exposure of these transformants to serotonin increases intracellular Ca^{++} levels. Moreover, abundant expression of 5-HT_{1c} receptor mRNA is observed in subsets of neurons throughout the brain, suggesting that the 5-HT_{1c} receptor plays an important role in central neurotransmission.

The procedure used to isolate the functional 5-HT_{1c} receptor cDNA has combined cloning in RNA expression vectors with a sensitive electrophysiological assay for serotonin receptor function in *Xenopus* oocytes. Similar procedures have been used to isolate the cDNA encoding the receptor for the neuropeptide substance K (Masu et al. 1987) and the Na^{+}-glucose cotransporter (Hediger et al. 1987). In our experiments, the injection of as little as 5 pg of choroid plexus poly$(A)^{+}$ RNA (1 fg of receptor mRNA) into oocytes was adequate to generate serotonin-evoked membrane currents. The sensitivity of the oocyte expression assay presumably results from the signal amplification associated with the coupling of these receptors to second-messenger systems. In the oocyte, this assay is currently limited to receptors that activate second-messenger systems, particularly phospholipase C, which elevate intracellular Ca^{++} levels. This cloning strategy may therefore be generally applicable to the isolation of genes encoding neurotransmitter and growth factor receptors that initiate a similar signal amplification, even if their mRNAs are present in exceedingly small amounts in the total RNA population.

Pharmacological studies and, more recently, gene cloning have established that multiple receptor subtypes exist for most, if not all, neurotransmitters. The existence of multiple receptor subtypes provides one mechanism by which a single neurotransmitter can elicit distinct cellular responses. The variation in cellular response can be achieved by the association of individual receptor subtypes with different G proteins and different signaling systems. Further flexibility is provided by the ability of distinct receptors for the same ligand to activate or inhibit the same second-messenger system. For example, among the adrenergic receptors, β_1 and β_2 receptors activate adenylate cyclase, α_2 receptors inhibit adenylate cyclase, and α_1 receptors activate phospholipase C pathways (Strader et al. 1987b, Chung et al. 1988).

A similar array of cellular responses can be elicited by serotonin in cells bearing different receptor subtypes. 5-HT_{1c} and 5-HT_2 receptors stimulate the phospholipase-C-mediated production of inositol phosphates, whereas 5-HT_{1a} and 5-HT_{1b} receptors may regulate adenylate cyclase activity (DeVivo and Maayani 1986; Weiss et al. 1986; Hoyer and Schoekkter 1988) or couple to G proteins that directly activate ion channels (Aghajanian 1981; Siegelbaum et al. 1982; Andrade et al. 1986; Andrade and Nicoll 1987). The diverse neural actions of serotonin are thought to be mediated by

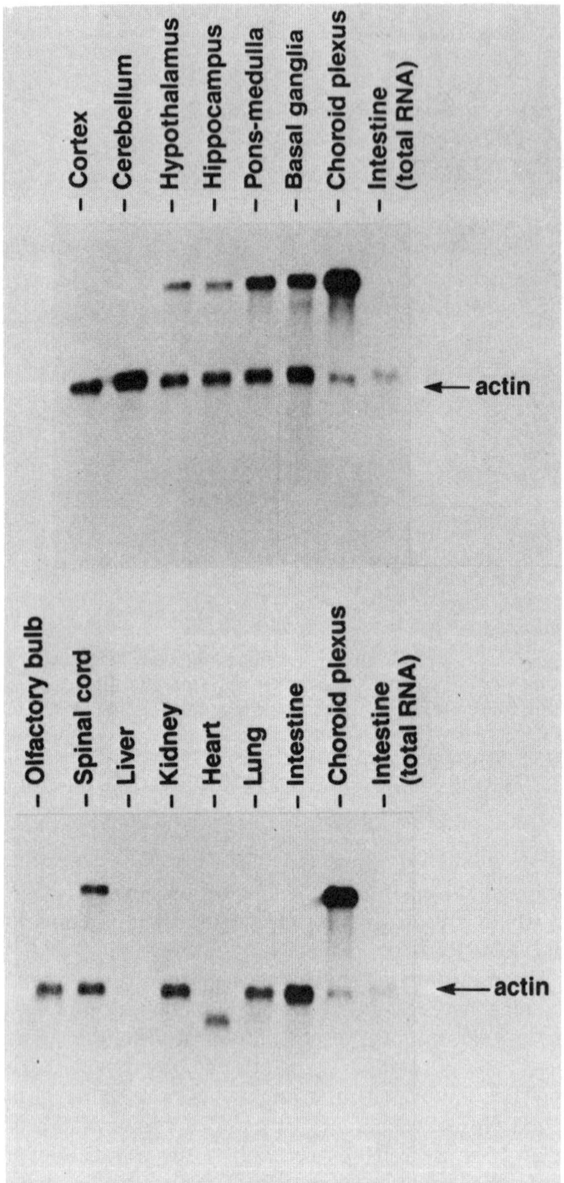

Figure 8. RNA blot analysis of the tissue distribution of 5-HT_{1c} receptor mRNA. Poly(A)$^+$ RNA (2–20 μg) prepared from different brain regions and peripheral tissues was subjected to Northern blot analysis. Blots were probed with a ^{32}P-labeled 3-kb *Eco*RI cDNA insert from pSR-1c. Tissue regions are indicated above each lane. Filters were rehybridized with ^{32}P-labeled human α-actin probe to assess relative amounts of RNA loaded in each lane. The 5-HT_{1c} mRNA was judged to be ~5.2 kb, using 18S and 28S RNAs as molecular-weight standards.

activation of these distinct receptor subtypes. In the cortex, for example, activation of 5-HT_2 receptors is thought to contribute to affective and perceptual states, and these receptors represent a major site of action of psychotropic drugs (Jacobs 1984). In contrast, in the spinal cord, the pain-producing effects of serotonin are mediated by 5-HT_3 receptors (Richardson et al. 1985), whereas control of sensory input is likely to be mediated by 5-HT_{1a} receptors (White and Neuman 1980; Fields 1984). The expression of 5-HT receptors in new cellular environments devoid of other receptor subtypes should permit the characterization of transduction systems associated with these receptor subtypes. We anticipate that the cloning of additional serotonin receptors will help to elucidate the molecular mechanisms of action of serotonin in the nervous system.

ACKNOWLEDGMENTS

This work was supported by the Howard Hughes Medical Institute (D.J., A.B.M., R.A., and T.M.J.) and by grants from the National Institutes of Health (R.A. and T.M.J.). D.J. was a fellow of the Jane Coffin Childs Memorial Fund for Medical Research.

REFERENCES

Aghajanian, G.K. 1981. The modulatory role of serotonin at multiple receptors in the brain. In *Serotonin neurotransmission and behavior* (ed. B.L. Jacobs and A. Gelperin), p. 156. MIT Press, Cambridge, Massachusetts.

Andrade, R. and R.A. Nicoll. 1987. Pharmacologically distinct actions of serotonin on single pyramidal neurons of the rat hippocampus recorded *in vitro*. *J. Physiol.* **394:** 99.

Andrade, R., R.C. Malenka, and R.A. Nicoll. 1986. A G protein couples serotonin and $GABA_B$ receptors to the same channels in hippocampus. *Science* **234:** 1261.

Bonner, T.I., N.J. Buckley, A.C. Loung, and M.R. Brann. 1987. Identification of a family of muscarinic acetylcholine receptor genes. *Science* **237:** 527.

Burkholder, A.C. and L.H. Hartwell. 1985. The yeast α-factor receptor: Structural properties deduced from the sequence of the *STE2* gene. *Nucleic Acids Res.* **13:** 8463.

Chung, F.-Z., C.-D. Wang, P.C. Potter, J.C. Venter, and C.M. Fraser. 1988. Site-directed mutagenesis and continuous expression of human β-adrenergic receptors. *J. Biol. Chem.* **263:** 4052.

Conn, P.J., E. Sanders-Bush, B.J. Hoffman, and P.R. Hartig. 1986. A unique serotonin receptor in choroid plexus is linked to phosphatidylinositol turnover. *Proc. Natl. Acad. Sci.* **83:** 4086.

de Courcelles, D., J. Leysen, F. De Clerck, H. Van Belle, and P. Janssen. 1985. Evidence that phospholipid turnover is the signal transducing system coupled to serotonin-S_2 receptor sites. *J. Biol. Chem.* **260:** 7603.

DeVivo, M. and S. Maayani. 1986. Characterization of the 5-hydroxytryptamine receptor-mediated inhibition of forskolin-stimulated adenylate cyclase activity in guinea pig and rat hippocampal membranes. *J. Pharmacol. Exp. Ther.* **238:** 248.

Dixon, T., I.S. Sigal, M.R. Candelore, B. Register, W. Scattergood, E. Rands, and C.D. Strader. 1987a. Structural features required for ligand binding to the β-adrenergic receptor. *EMBO J.* **6:** 3269.

Dixon, R., I.S. Sigal, E. Rands, B. Register, M.R. Candelore, A.D. Blake, and C.D. Strader. 1987b. Ligand binding to the β-adrenergic receptor involves its rhodopsin-like core. *Nature* **326:** 73.

Dohlman, H.G., M.G. Caron, and R.J. Lefkowitz. 1987. A family of receptors coupled to guanine nucleotide regulatory proteins. *Biochemistry* **26:** 2657.

Dohlman, H.G., M.G. Caron, C.D. Strader, N. Amlaiky, and R.J. Lefkowitz. 1988. Identification and sequence of a binding site peptide of the β-adrenergic receptor. *Biochemistry* **27:** 1813.

Fields, H.L. 1984. Brainstem mechanisms of pain modulation. In *Advances in pain research and therapy* (ed. L. Kruger and J.C. Liebeskind), p. 241. Raven Press, New York.

Glennon, R.A., M. Titeler, and J.D. McKenney. 1984. Evidence for 5-HT_2 involvement in the mechanism of action of hallucinogenic agents. *Life Sci.* **35:** 2505.

Grynkiewicz, G., M. Poenie, and R.Y. Tsien. 1985. A new generation of Ca^{2+} indicators with greatly improved fluorescence properties. *J. Biol. Chem.* **260:** 3440.

Gunderson, C.B., R. Miledi, and I. Parker. 1983. Serotonin receptors induced by exogenous messenger RNA in *Xenopus* oocytes. *Proc. R Soc. Lond. B. Biol. Sci.* **219:** 103.

Hagan, D.C., G. McCaffrey, and G.F. Sprague, Jr. 1986. Evidence that the yeast *STE3* gene encodes a receptor for the peptide pheromone α-factor: Gene sequence and implications for the structure of the presumed receptor. *Proc. Natl. Acad. Sci.* **83:** 1418.

Hediger, M.A., T. Ikeda, M. Coady, C.B. Gunderson, and E.M. Wright. 1987. Expression of size-selected mRNA encoding the intestinal Na/glucose cotransporter in *Xenopus laevis* oocytes. *Proc. Natl. Acad. Sci.* **84:** 2634.

Hoyer, D. and P. Schoekkter. 1988. 5-HT_{1D} receptor-mediated inhibition of forskolin-stimulated adenylate cyclase activity in calf substantia nigra. *Eur. J. Pharmacol.* **147:** 145.

Jacobs, B.L. 1984. Postsynaptic serotonergic action of hallucinogens. In *Hallucinogens: Neurochemical, behavioral, and clinical perspectives* (ed. B.L. Jacobs), p. 183. Raven Press, New York.

Julius, D., A.B. MacDermott, R. Axel, and T.M. Jessell. 1988. Molecular characterization of a functional cDNA encoding the serotonin 1c receptor. *Science* **241:** 558.

Kadan, M.J., A.M. Krohn, M.J. Evans, R.L. Waltz, and P.R. Hartig. 1984. Characterization of ^{125}I-lysergic acid diethylamide binding to serotonin receptors in rat frontal cortex. *J. Neurochem.* **43:** 601.

Kirschmeir, P., G.M. Housey, M. Johnson, A.S. Perkins, and I.B. Weinstein. 1988. Construction and characterization of a retroviral vector demonstrating efficient expression of cloned cDNA sequences. *DNA* (in press).

Kubo, T., K. Fukuda, A. Mikami, A. Maeda, H. Takahashi, M. Mishina, T. Haga, K. Haga, A. Ichiyama, K. Kangawa, M. Kojima, H. Matsuo, T. Hirose, and S. Numa. 1986. Cloning, sequencing and expression of complementary DNA encoding the muscarinic acetylcholine receptor. *Nature* **323:** 411.

Leary, T., R. Metzner, and R. Alpert. 1964. *The psychedelic experience.* University Books, New Hyde Park, New York.

Lubbert, H., T.P. Snutch, N. Dascal, H.A. Lester, and N. Davidson. 1987a. Rat brain 5-HT_{1c} receptors are encoded by a 5-6 K base mRNA size class and are functionally expressed in injected *Xenopus* oocytes. *J. Neurosci.* **4:** 1159.

Lubbert, H. B.J. Hoffman, T.P. Snutch, T. Van Dyke, A.J. Levine, P.R. Hartig, H.A. Lester, and N. Davidson. 1987b. cDNA cloning of a serotonin 5-HT_{1c} receptor by electrophysiological assays of mRNA-injected *Xenopus* oocytes. *Proc. Natl. Acad. Sci.* **84:** 4332.

Masu, Y., K. Nakayama, H. Tamaki, Y. Harada, K. Motoy, and S. Nakanishi. 1987. cDNA cloning of bovine substance-K receptor through oocyte expression system. *Nature* **329:** 836.

Nakayama, N., A. Miyajima, and K. Arai. 1985. Nucleotide sequences of *STE2* and *STE3*, cell type-specific sterile genes from *Saccharomyces cerevisiae*. *EMBO J.* **4:** 2643.

Nathans, J. and D.S. Hogness. 1983. Isolation sequence analysis, and intron-exon arrangement of the gene encoding bovine rhodopsin. *Cell* **34:** 807.

O'Dowd, B.F., R.J. Lefkowitz, and M.G. Caron. 1989. Structure of the adrenergic and related recepters. *Annu. Rev. Neurosci.* (in press).

Pazos, A. and J.M. Palacios. 1985. Quantitative autoradiographic mapping of serotonin receptors in the rat brain. Serotonin-1 receptors. *Brain Res.* **346:** 205.

Peralta, E., J. Winslow, G. Peterson, D. Smith, A. Ashkenazi, J. Ramachandran, M. Schimerlik, and D. Capon. 1987. Primary structure and biochemical properties of an M_2 muscarinic receptor. *Science* **236:** 600.

Peroutka, S.J. 1988. 5-Hydroxytryptamine receptor subtypes. *Annu. Rev. Neurosci.* **11:** 45.

Richardson, B.P., G. Engel, P. Donatsch, and P.A. Stadler. 1985. Identification of serotonin M-receptor subtypes and their specific blockade by a new class of drugs. *Nature* **316:** 126.

Schieren, I. and A.B. MacDermott. 1988. Flow cytometric identification and purification of cells by ligand-induced changes in intracellular calcium. *J. Neurosci. Methods* (in press).

Sibley, D.R., J.L. Benovic, M.G. Caron, and R.J. Lefkowitz. 1987. Regulation of transmembrane signalling by receptor phosphorylation. *Cell* **48:** 913.

Strader, C.D., R. Dixon, A.H. Cheung, M.R. Candelore, A.D. Blake, and I.S. Sigal. 1987a. Mutations that uncouple the β-adrenergic receptor from G_s and increase agonist affinity. *J. Biol. Chem.* **262:** 16439.

Strader, C.D., I.S. Sigal, R.B. Register, M.R. Candelore, E. Rands, and R. Dixon. 1987b. Identification of residues required for ligand binding to the β-adrenergic receptor. *Proc. Natl. Acad. Sci.* **84:** 4384.

Takahashi, T., E. Neher, and B. Sakmann. 1987. Rat brain serotonin receptors in *Xenopus* oocytes are coupled by intracellular calcium to endogenous channels. *Proc. Natl. Acad. Sci.* **84:** 5063.

Weiss, S., M. Sebben, D.E. Kemp, and J. Bockaert. 1986. Serotonin 5-HT_1 receptors mediate inhibition of cyclic AMP production in neurons. *Eur. J. Pharmacol.* **120:** 227.

White, S.R. and R.S. Neuman. 1980. Facilitation of spinal and motoneurone excitability by 5-hydroxytryptamine and noradrenaline. *Brain Res.* **188:** 119.

Yagaloff, K.A., and P.R. Hartig. 1986. ^{125}I lysergic acid diethylamide binds to a novel serotonergic site on rat choroid plexus epithelial cells. *J. Neuroscience* **5:** 3178.

Young, D., G. Waitches, C. Birchmeier, O. Fasano, and M. Wigler. 1986. Isolation and characterization of a new cellular oncogene encoding a protein with multiple potential transmembrane domains. *Cell* **45:** 711.

Molecular Convergence of Presynaptic Inhibition and Presynaptic Facilitation on Common Substrate Proteins of Individual Sensory Neurons of *Aplysia*

D. Sweatt, A. Volterra, S.A. Siegelbaum, and E.R. Kandel
Center for Neurobiology and Behavior and Howard Hughes Medical Institute, Columbia University College of Physicians & Surgeons, New York, New York 10032

The behavioral stimuli that serve as reinforcers for learning often have two components: a facilitatory component, important for sensitization and classical conditioning, and an inhibitory component, important for conditioned inhibition or as a break on the facilitation (Konorski 1967). We have encountered both of these components in *Aplysia* in the course of studying the effects of tail-reinforcing stimuli on the gill-withdrawal reflex of this animal (Carew et al. 1983; Frost et al. 1985; Mackey et al. 1987). These two opposing behavioral components are mediated, at the cellular level, by two opposing modulatory systems that act to regulate the strength of the connections between the sensory and the motor neurons of this reflex. Each of these two modulatory systems utilizes different transmitters, and these modulatory transmitters in turn activate distinct second-messenger systems. As a result, the anatomical pathways of the neuronal modulatory circuits in *Aplysia* are re-represented at the molecular level within individual nerve cells by the specific molecular circuitry of the second-messenger pathway. The facilitatory neurons utilize three transmitters, of which the best studied is serotonin (5-HT), and involve cAMP-dependent protein phosphorylation as a second-messenger system. The inhibition utilizes FMRFamide and dopamine as transmitters and the lipoxygenase metabolites of arachidonic acid as a second-messenger system.

The actions of both the facilitatory and inhibitory transmitters and their respective second-messenger systems converge on a number of common substrate proteins, including a common ion channel, the S-K^+ channel. 5-HT and cAMP lead to a decrease in current flow through these channels by reducing the number of open channels. FMRFamide and arachidonic acid lead to an increase in current flow by increasing the probability of channel opening. We have reason to believe that the S-type K^+ channels exist not only in the cell body where they regulate the excitability of the neurons, but also on the terminals where their modulation affects synaptic strength (Klein and Kandel 1978; Belardetti et al. 1986). Thus, this sensorimotor system allows us not only to examine the mechanisms whereby these two opposing transmitters acting through two different second messengers converge within the cell to regulate channel function, but it also allows us to explore how that interaction contributes to the up-and-down regulation of transmitter release (Fig. 1).

This interaction is of further interest because the opposing actions are not symmetrical. The modulatory effects of inhibition are more transient than those of facilitation, but when they occur together, this inhibition can fully override the facilitation (Fig. 2). The ability of the inhibitory system to override the facilitatory system also is reflected at the molecular level. In addition to increasing the opening probability of the S-type K^+ channel, FMRFamide can reverse the action of 5-HT and cAMP by reopening channels closed by them. This override also is evident in other substrate proteins. Quantitative two-dimensional gel studies reveal that 5-HT and cAMP lead to an increase in the level of phosphorylation of 17 proteins in the sensory neurons. In the absence of serotonin, FMRFamide acts on 10 of these proteins to reduce their basal level of phosphorylation. However, in the presence of serotonin, FMRFamide antagonizes the increase in phosphorylation of all 17 proteins normally phosphorylated by 5-HT. Thus, whereas 5-HT and cAMP activate a kinase (the cAMP-dependent protein kinase), as is characteristic of second-messenger pathways,

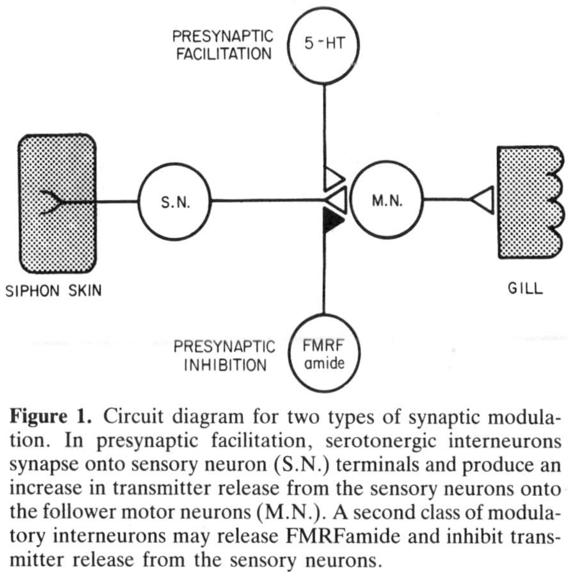

Figure 1. Circuit diagram for two types of synaptic modulation. In presynaptic facilitation, serotonergic interneurons synapse onto sensory neuron (S.N.) terminals and produce an increase in transmitter release from the sensory neurons onto the follower motor neurons (M.N.). A second class of modulatory interneurons may release FMRFamide and inhibit transmitter release from the sensory neurons.

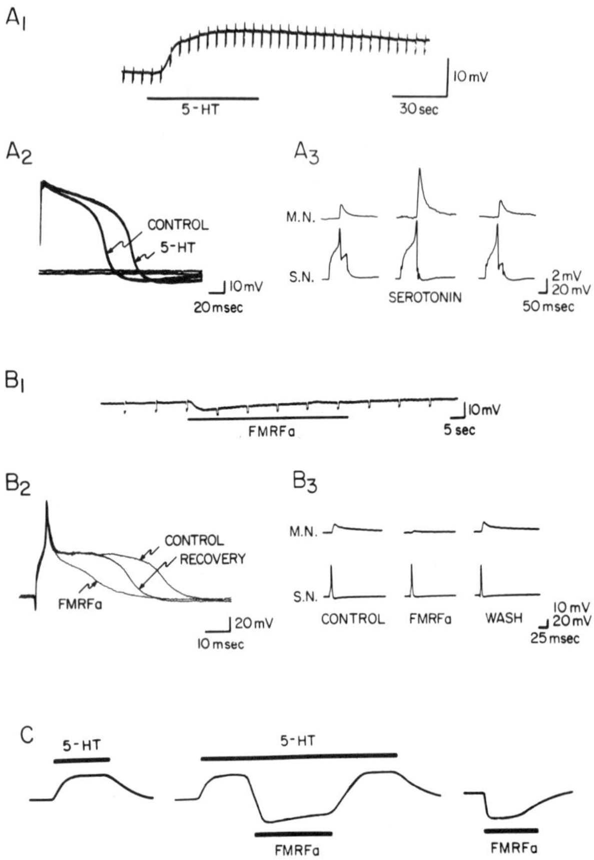

Figure 2. (*A*) 5-HT action on sensory neurons. (A_1) 5-HT produces a slow depolarization of the cell resting potential accompanied by a decrease in membrane conductance. Application of a 2 μM 5-HT is indicated by the solid line. Constant current hyperpolarizing pulses were delivered to the cell to monitor membrane conductance. (A_2) 5-HT causes an increase in action potential duration. Action potentials were recorded in the presence of 50 mM TEA. (A_3) 5-HT enhances synaptic transmission between sensory and motor neurons in the intact abdominal ganglion. Top trace shows fast excitatory postsynaptic potential (EPSP) in the motor neuron in response to stimulation of the sensory neuron (bottom trace). 5-HT produces a reversible increase in the size of the EPSP in the motor neuron. (*B*) FMRFamide action on sensory neurons. (B_1) FMRFamide produces a slow hyperpolarization of the cell resting potential accompanied by an increase in membrane conductance. The application of 2 μM FMRFamide is indicated by the solid line. Constant current hyperpolarizing pulses of 0.1 nA were delivered to the cell to monitor membrane conductance. (B_2) FMRFamide causes a reversible decrease in action potential duration. Action potentials recorded from sensory neuron in an intact abdominal ganglion in presence of 50 mM TEA (control). In absence of TEA, decrease in action potential duration is also observed with FMRFamide; however, the brief duration of the control action potential makes shortening difficult to measure accurately. Resting potential in B_2 was −43 mV before FMRFamide and −48 mV in presence of peptide. (B_3) FMRFamide inhibits synaptic transmission between sensory and motor neurons in culture. Top trace shows fast EPSP in the motor neuron in response to stimulation of the sensory neuron (bottom trace). FMRFamide produces a reversible decrease in the size of the EPSP in the motor neuron. (*C*) Representation of the interaction of 5-HT and FMRFamide at the level of resting membrane potential. When FMRFamide and 5-HT are applied simultaneously, the effect of FMRFamide predominates.

FMRFamide seems not to act through a kinase. Rather, it seems to initiate a novel pattern of action for a second-messenger system, and either activates a phosphatase (perhaps an arachidonic-acid-regulated phosphatase) or inhibits the cAMP-dependent protein kinase.

Presynaptic Facilitation

A strong electrical tail stimulus, presented alone, produces short-term behavioral sensitization of the siphon- and gill-withdrawal reflex. Sensitization leads to presynaptic facilitation, an increase in the synaptic strength at the connections between the sensory neurons and the interneurons and motor neurons mediating the withdrawal reflex. Changes also occur at other loci within the neural circuit, but we will not consider these here (see Frost et al. 1988). Rather, we will focus exclusively on the monosynaptic portion of the reflex circuit consisting of the sensory neuron, the motor neuron, and their connections. Sensitizing stimuli produce presynaptic facilitation of the sensory neuron to motor neuron connection by activating a facilitating system. Although we do not know all of the elements of this system, for example, all the transmitters these neurons use, we have identified three groups of facilitating cells, each of which uses a different transmitter: (1) 5-HT, released by the LC_B and RC_B cells; (2) an as-yet-unidentified transmitter released by the L29 cells, and (3) two small interrelated peptides, SCP_A and SCP_B, released by processes that come in close apposition to those of the sensory neuron in the neuropile. Their cell bodies have not yet been identified (see Mackey et al. 1987).

The presynaptic facilitation produced by these facilitating transmitters is mediated in the sensory neurons by increased levels of cAMP and by cAMP-dependent protein phosphorylation. The resulting presynaptic facilitation is associated with three characteristic changes in the sensory neurons: (1) a depolarization of the membrane potential associated with a decrease in membrane conductance; (2) a broadening of the action potential; and (3) an enhancement of transmitter release (Fig. 2A). On the basis of voltage-clamp and biochemical studies of the sensory neuron, both in the intact ganglion and in dissociated cell culture, we have outlined a molecular model for presynaptic facilitation (for review, see Kandel and Schwartz 1982). According to this view, 5-HT (as well as the other facilitating transmitters) released by the facilitating neurons activates a transmitter-sensitive adenylate cyclase in the membrane of the sensory neuron, including its presynaptic terminal, that increases the level of intracellular cAMP. cAMP then activates the cAMP-dependent protein kinase, which phosphorylates a number of substrate proteins including a K^+ channel protein or a protein associated with it (Castellucci et al. 1980; Klein and Kandel 1980; Shuster et al. 1985). This phosphorylation causes closure of the K^+ channels, reducing one of the K^+ conductances that normally repolarize the action potential. Reduction of this conductance produces three consequences: It depolarizes the cell, it increases the excitability of the neuron and prolongs the action potential so as to allow more Ca^{++} flow into the terminals, and it contributes to increased release of transmitter (Klein and Kandel 1978; Castellucci et al. 1980; Klein et al. 1982, 1986; Hochner et al. 1986a,b). Biochemical evidence for the role of cAMP in facilitation is consistent with the physiological effect of 5-HT and cAMP on transmitter release (see Bernier et al. 1982). 5-HT and intracellular injection of cAMP or the cAMP-dependent protein kinase simulate the effects of presynaptic facilitation on sensory neurons (Brunelli et al. 1975; Castellucci et al. 1980). Independent of K channel closure, 5-HT also alters the handling of Ca^{++} within the sensory neurons, which leads to a rise in free intracellular Ca^{++} (Boyle et al. 1984; Hochner et al. 1985). This rise in free Ca^{++} may be important for the mobilization of transmitter in the face of continued transmitter release (Hochner et al. 1986a,b). We here focus on only one of the effects of 5-HT and cAMP: their action on the K^+ channel.

5-HT Decreases the Number of Open S-type K^+ Channels

What are the properties of the K^+ channel modulated by 5-HT? Single-channel recordings from membranes of the sensory neurons revealed that the 5-HT-sensitive K^+ channel has novel properties (Siegelbaum et al. 1982). The channel is active at the resting potential. Its gating is not affected by the activity of intracellular calcium ions and shows only a moderate dependence on membrane potential. This initial analysis allowed us to address the next question: How does 5-HT modulate this channel? The total current flow (I) through populations of an ion channel is given by $I = N \times p \times i$. Does 5-HT affect the number of functional channels in the patch (N)? The probability of channel opening (p)? Or the elementary conductance of the channel (i)?

We have found that 5-HT reduces the *number* of functional channels in a patch of membrane and does so in an all-or-none manner (Fig. 3). Channels that still open in the presence of 5-HT appear to open and close normally, whereas channels once closed by 5-HT remain closed. The modulatory action of 5-HT does not involve either a change in the ionic specificity of the channel nor detectable changes in single-channel conductance over the lifetime of the channel in the open state. Rather, 5-HT greatly increases the lifetime of the channel in a closed state. This may simply represent the modification of the normal channel-gating process, whereby one of the normal rate constants of channel opening is dramatically decreased. Alternatively, 5-HT may induce new closed states of the channel (perhaps corresponding to the channel in a phosphorylated

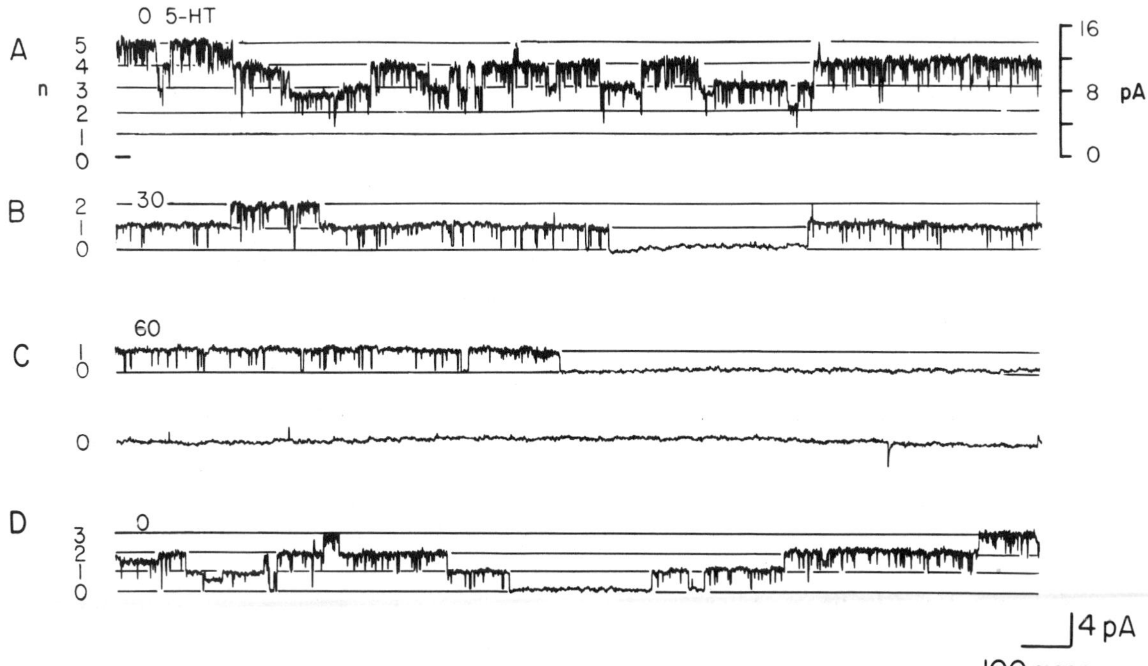

Figure 3. Action of 5-HT on single-channel current. Patch-clamp current recorded from mechanoreceptor sensory neurons in the abdominal ganglion of *Aplysia californica*. (Trace *A*) Current in the absence of serotonin. Left-hand ordinate shows the number of open channels; right-hand ordinate shows current magnitude. Individual current steps are 2.6 pA. The current record was well fitted by a binomial distribution assuming that five channels are active in the patch and that each channel opens with a probability of 0.84. (Trace *B*) Current obtained 2 min after addition of 30 μM 5-HT to the bath. (Trace *C*) 1 min after addition of a further dose of 5-HT, raising the total concentration to 60 μM. The two traces are a continuous recording.

form) in which the channel is not available to the normal gating process (Siegelbaum et al. 1982).

Intracellular injection of cAMP also causes channels to close in an all-or-none manner, similar to that caused by the action of 5-HT (Siegelbaum et al. 1982). Parallel effects are seen following application of the catalytic subunit of the cAMP-dependent protein kinase to the inside surface of an isolated membrane patch (Shuster et al. 1985). These results suggest that cAMP-dependent protein kinase leads to all-or-none closures of the S channels by acting on a membrane-associated substrate protein, either the S channel itself or a regulatory protein that acts on the channel.

Presynaptic Inhibition

At the level of the sensory neuron, the behavioral inhibition in part reflects presynaptic inhibition of transmitter release, a mirror image to the process of presynaptic facilitation. Although it is brief, behavioral inhibition has the interesting feature that it is capable of overriding the effects of facilitation.

Behavioral inhibition is mediated in part by the peptide FMRFamide. Small et al. (1988) have identified an interneuron, located within the left pleural ganglion, that shows FMRFamide immunoreactivity. This interneuron is activated by tail shock and, when stimulated with current pulses, this interneuron simulates the presynaptic inhibitory action produced by tail shock.

What are the molecular and ionic mechanisms of presynaptic inhibition? Stimulation of the inhibitory neuron or application of FMRFamide to the sensory neuron produces three actions: (1) a slow, transient hyperpolarization, associated with an increase in membrane conductance; (2) a narrowing of the action potential; and (3) a depression of transmitter release (Figs. 2, 4, and 6). To produce these three actions, FMRFamide alters several macroscopic currents. It increases a background K^+ current and decreases a voltage-dependent Ca^{++} current and a Ca^{++}-dependent K^+ current (Ruben et al. 1986; Belardetti et al. 1987; Brezina et al. 1987a,b; B. Edmonds et al., in prep.). We have here restricted our patch-clamp analysis to the actions on the background K current. Specifically, we have explored whether FMRFamide modulates the same class of K^+ channels closed by 5-HT. If so, which parameter of S-channel function is altered? Finally, how do 5-HT and FMRFamide interact in causing channel modulation, and what is at the single-channel level the mechanism of the inhibitory override?

FMRFamide Modulates the Same Channel Modulated by 5-HT

Single-channel recordings indicate that the slow hyperpolarization produced by FMRFamide is indeed associated with a modulation of the S-type K^+ channel (Fig. 4B). 5-HT decreases the macroscopic S current by causing prolonged all-or-none S-channel closures, resulting in an apparent decrease in the number of functional channels in the membrane. We therefore were interested in knowing: Does FMRFamide increase the mean S-channel current by increasing N, p, or i?

We found that FMRFamide produces no marked change in either the single-channel amplitude (i) or in the maximum number of channels that open simultaneously (N_f), suggesting that FMRFamide acts to increase p (Belardetti et al. 1987). The most convincing demonstration that FMRFamide acts to increase p was achieved when the patch contained only one active S channel (Belardetti et al. 1987). Within 1 minute after application of FMRFamide, there was even here a marked increase in channel opening with no recruitment of additional channels. Inasmuch as we never observed more than a single channel open in the presence of FMRFamide, despite the dramatic rise in channel-open probability, FMRFamide must act to increase p.

FMRFamide Overrides the Action of 5-HT at the Single-channel Level

The behavioral finding that inhibition can override facilitation and the cellular finding that both act on a common channel raises the question: How do the two modulatory inputs interact in modulating the S channel? We first examined the interaction of the two opposing transmitters at the level of membrane potential (Fig. 2C) and found that here also FMRFamide antagonizes and overrides the slow excitatory action of 5-HT (or cAMP). The single-channel basis for this override is shown in Figure 5. The patch of membrane illustrated in Figure 5 initially displayed four active S channels (control). On application of dibutyryl cAMP to the bath, two of the four active channels close (Fig. 5; cAMP), leading to a decrease in N_f as described earlier (Fig. 5). The two remaining active channels appear normal, with no change in single-channel amplitude or open probability. We then applied FMRFamide to the cell, still in the presence of cAMP (Fig. 5; cAMP-FMRFamide). Now, in addition to the normal increase in open probability, the peptide also produces an increase in the number of functional channels from two up to the original level of four. This is in contrast to the single action of FMRFamide to increase p in the absence of cAMP. The ability of FMRFamide to reopen channels closed by cAMP strengthens the argument that the peptide does indeed modulate the same class of S channels closed by cAMP and 5-HT. It also suggests that FMRFamide action does not require a decrease in adenylate cyclase activity.

These results show that the S channel can be modulated in opposite directions during presynaptic facilitation produced by 5-HT and during presynaptic inhibition produced by FMRFamide. However, the modes of action of these modulatory transmitters differ in fundamental ways. 5-HT causes prolonged channel closures, resulting in a decrease in the number of functional channels. In contrast, FMRFamide normally acts alone to increase channel opening without increasing

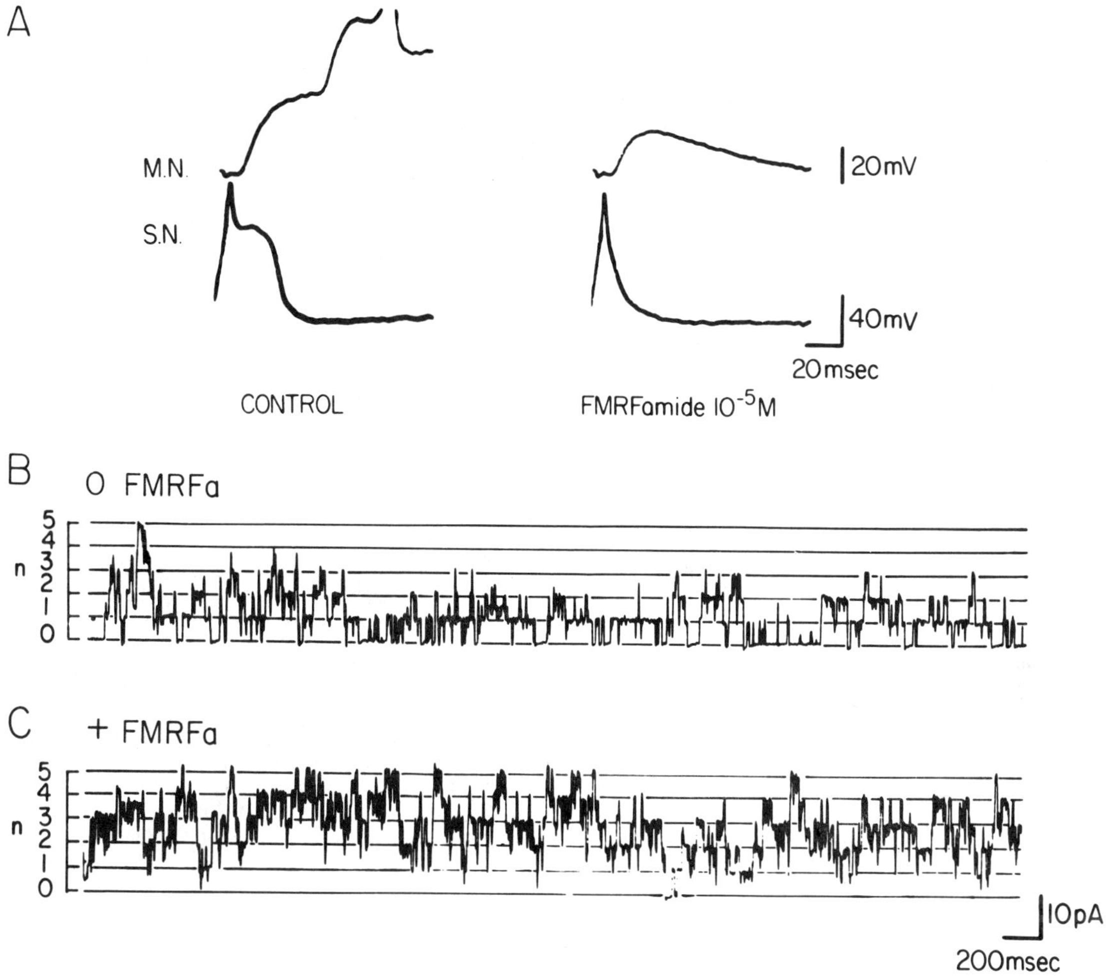

Figure 4. (*A*) Effect of FMRFamide on the duration of the presynaptic (sensory neuron) action potential and on the sensory-motor connection. Top traces show EPSP as recorded from motor neurons; the lower traces show sensory neuron action potential. The left side illustrates control condition; the right side, in the presence of FMRFamide (10^{-5} M). The bath contains 5 mM TEA. (*B*) Effect of FMRFamide on single S-channel current. Single-channel recordings from a cell-attached patch on a pleural sensory neuron in culture. (Control) Before application of any compounds. (FMRFa) +7 min after application of 5 μM FMRFamide to the bath. Binomial fit to data yields estimate of $n = 6$ for number of active S channels under all conditions. Records obtained at a patch potential depolarized by +80 mV above the resting potential.

the number of functional channels (Fig. 4B). Nevertheless, the two transmitter effects are not completely independent, since, in addition to increasing channel-open probability, FMRFamide also is able to override the action of 5-HT by reopening channels closed by cAMP (Fig. 5). This suggests that FMRFamide may have at least two sites of action on the S channel. In addition to an action at a site on the channel involved in gating, FMRFamide may also be able to stimulate phosphatase activity to reverse cAMP- and cAMP-dependent protein kinase-induced channel closures.

Receptors for Presynaptic Inhibition and Presynaptic Facilitation Activate Separate Intracellular Second-messenger Systems

Since application of FMRFamide to the bath is capable of opening channels under the patch pipette that have no direct access to the peptide, FMRFamide is most probably acting through a second-messenger system. What is this second messenger, and where does it act? Since 5-HT closes the S channels through the cAMP pathway, an obvious possibility is that FMRFamide produces an action comparable to the muscarinic action of vertebrate heart to activate a G_i-type protein that antagonizes the activation of the adenylate cyclase and decreases the resting level of cAMP. Biochemical measurements of Ocorr and Byrne (1985) and of T. Abrams and E.R. Kandel (unpubl.), however, show no change in levels of cAMP in sensory neurons in response to FMRFamide. In addition, our single-channel results also do not support a decrease in cAMP as the mode of FMRFamide action. This would be expected to lead to an increase in the number of functional channels (N_f) in the patch (i.e., the converse of the action of 5-HT) rather than to the increase in

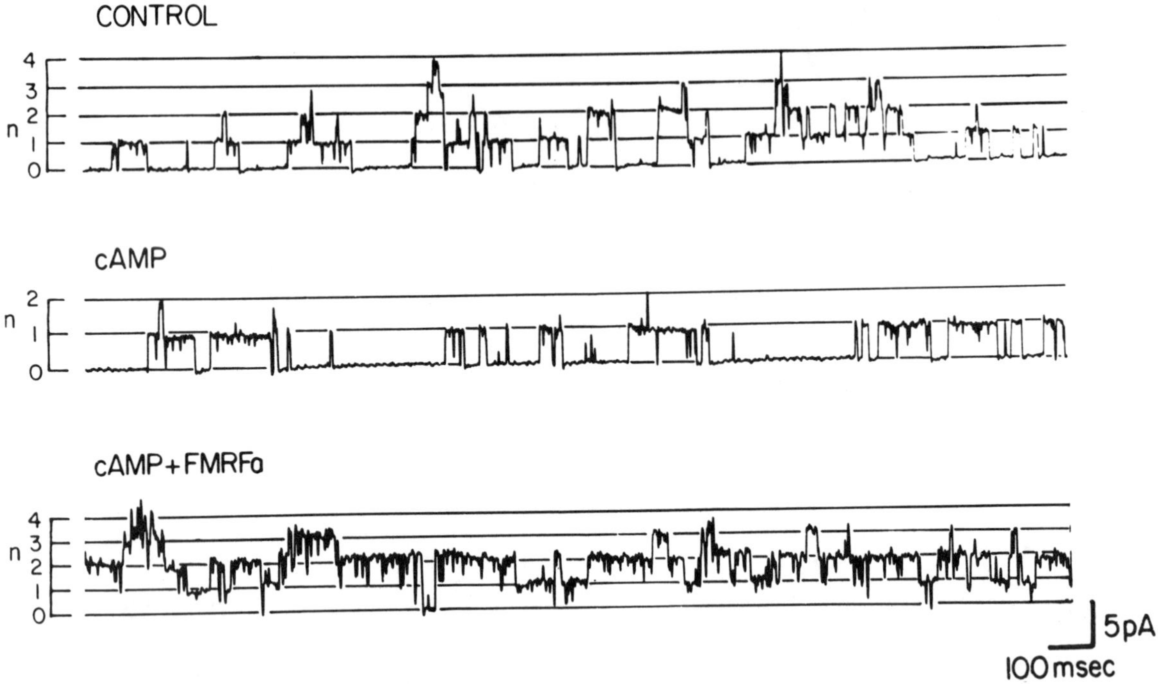

Figure 5. FMRFamide antagonizes the action of cAMP. On the level of single channels, FMRFamide antagonizes the closure of channels induced by cAMP. Top trace is a current record from patch containing four active S channels before application of cAMP. Middle trace is a current record obtained 10 min after addition of 0.1 mM dibutyryl cAMP (cAMP) to bath showing closure of two channels. Closures occurred within 5 min after addition of cAMP and persisted in maintained presence of cAMP until addition of FMRFamide. Lower trace is a current record 5 min after application of 2 μM FMRFamide (still in presence of cAMP), showing reopening of two channels and increase in open probability. Membrane potential was about 20 mV. Current records filtered at 1 kHz.

open probability observed. Thus, FMRFamide uses a second-messenger pathway that seems not to involve adenylate cyclase modulation. Studies from our laboratory and others also seem to rule out that several other known second messengers, such as $InsP_3$, diacylglycerol, or cGMP, mediate FMRFamide action on the S channel (Klein et al. 1986; Brezina et al. 1987b).

In collaborative experiments with Schwartz and Piomelli, we have recently found that presynaptic inhibition in the sensory neuron involves the receptor-mediated release of arachidonic acid from membrane phospholipids (Piomelli et al. 1987a,b). Release of arachidonic acid has been well characterized in non-neural cells, where free arachidonate is rapidly metabolized to eicosanoids, a family of active products that are thought to be both intracellular and intercellular messengers (for review, see Needleman et al. 1986). Although the metabolism of arachidonic acid in the brain has been studied for more than 20 years (Samuelsson 1964; Sautebin et al. 1978; Wolfe 1982; Lindgren et al. 1987; Shimizu et al. 1987), the complexity of the mammalian central nervous system has made it difficult to assign specific functions to the eicosanoids. In collaboration with J.H. Schwartz and D. Piomelli, we decided to study the arachidonic acid metabolism in *Aplysia*, taking advantage of the simpler nervous system of this marine mollusk, where it is possible to examine identified nerve cells of known behavioral function both in intact ganglia and in dissociated cells in culture. Our results indicate that the inhibitory synaptic actions of FMRFamide are mediated by lipoxygenase metabolites of arachidonic acid (Piomelli et al. 1987b; see also Shapiro et al., this volume).

Eicosanoid Pathways in *Aplysia*

Schwartz and his colleagues have found that arachidonate is a prominent component of *Aplysia* neural phospholipids (about 10% of total fatty acid content) and can be metabolized in *Aplysia* neurons through both the 12-lipoxygenase and the 5-lipoxygenase pathways, as well as through the cyclo-oxygenase pathway (Piomelli et al. 1987a). Two stable lipoxygenase products, the hydroxy acids 12-HETE and 5-HETE, have been identified in neural tissue from *Aplysia* (Piomelli et al. 1987a). In mammals, these hydroxy acids are formed by enzymatic reduction of the corresponding short-lived hydroperoxides (HPETE) (Needleman et al. 1986). The presence of hydroperoxy acids is only inferred in *Aplysia*, since we have not yet isolated these unstable precursors. Both 12-HPETE and 5-HPETE may be metabolized to products other than the hydroxy acids (Needleman et al. 1986). Finally, the cyclo-oxygenase pathway leads to the synthesis of various prostaglandins, which in *Aplysia* nervous tissue include PGE_2 and $PGF_{2\alpha}$ (Piomelli et al. 1987a).

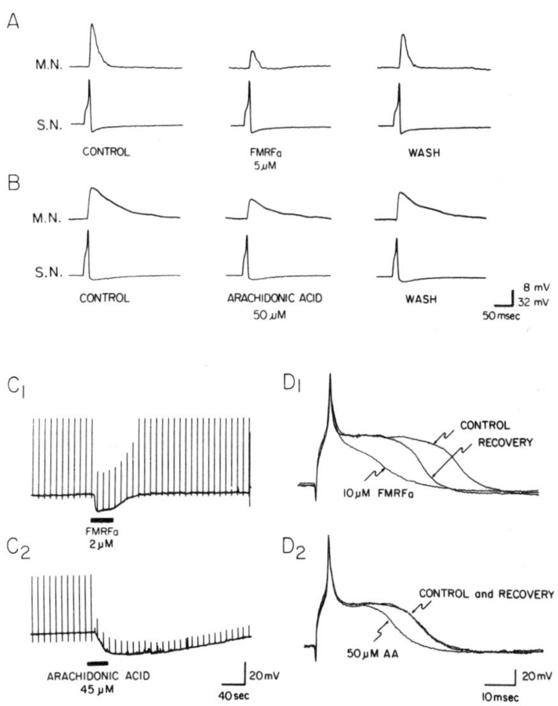

Figure 6. (*A*, *B*) Arachidonic acid simulates the effect of FMRFamide on postsynaptic potentials. Reduction in fast excitatory synaptic potential in follower motor neuron (upper traces) in response to firing an action potential in sensory neuron (lower traces). Records *A* (FMRFamide, 5 μm) and *B* (arachidonic acid, 50 μm) are from two different pairs of cells in intact abdominal ganglia. (*C*) Changes in membrane potential and spike firing of sensory neurons in culture in response to FMRFamide (2 μM) (C_1) or arachidonic acid (45 μM) (C_2). The solid bar indicates period of application. Intracellular records are from the same cell. Initial resting potential was −41 mV in A_1 and −35 mV in C_2. Upward spikes are action potentials elicited once every 10 sec by brief depolarizing (constant) current stimuli. The smaller spikes reflect action potential inhibition and are the passive depolarizing responses to the current step. The slower time course of the arachidonic-induced hyperpolarization most likely reflects the time required for diffusion (or transport) across the plasma membrane and metabolism. (*D*) FMRFamide and arachidonate cause a reversible decrease in action potential duration. Action potentials recorded from sensory neuron in an intact abdominal ganglion in presence of 50 mM TEA (control). In absence of TEA, decrease in action potential duration is also observed with FMRFamide and arachidonic acid; however, the brief duration of the control action potential makes shortening difficult to measure accurately. Resting potential in D_1 was −43 mV before FMRFamide and −48 mV in presence of peptide. In D_2, resting potential was −43 mV before arachidonate and −44.5 mV in presence of the fatty acid. In the intact ganglia, response to arachidonic acid is somewhat smaller and slower than in culture, probably due to the restricted access and diffusion of the fatty acid.

Arachidonate Mimics FMRFamide

To investigate the possible physiological role of arachidonic acid metabolites, Piomelli et al. first applied arachidonate extracellularly by pressure ejection onto the sensory neurons from a large pipette (Choi and Fischbach 1981). Arachidonic acid simulated the macroscopic actions of FMRFamide, producing a slow hyperpolarization of the membrane associated with an increase in membrane conductance, decreasing the duration of the action potential (Fig. 6) and inhibiting synaptic transmission between the sensory neuron and motor neuron (not shown). Eicosa-11-monoenoic acid, a 20-carbon chain fatty acid with one double bond, which constitutes 4.3% of *Aplysia* neural fatty acids (Piomelli et al. 1987a), had little effect on resting potentials, action potentials, or synaptic potentials.

To obtain more direct evidence that FMRFamide and arachidonic acid share a common mode of action, we studied single S-channel currents in cell-attached membrane patches (Piomelli et al. 1987b). FMRF amide caused a five- to sixfold increase in mean S-channel current over a range of membrane voltages with no significant effect on the single-channel current amplitude. Arachidonic acid caused a two- to fourfold increase in mean channel current with no significant effect on single-channel amplitude. Using a binomial analysis, we were able to confirm that the mode of action of arachidonic acid is similar to that of FMRFamide because it increases *p*, the channel open probability, with no marked effect on *N* or *i*. Thus, by all the criteria we have tested, the actions of arachidonic acid and FMRFamide are identical.

Although these experiments show that arachidonic acid simulates the action of FMRFamide, they do not prove that metabolites of endogenous arachidonate mediate the response to the peptide. Arachidonic acid could be acting on the K^+ channel through an independent parallel mechanism. We have obtained evidence for an obligatory role of arachidonic acid using two separate approaches, one pharmacological and the other biochemical.

Inhibitors of Arachidonic Acid Cascade Block the Action of FMRFamide

The first step in the receptor-mediated generation of eicosanoids is the release of free arachidonate from membrane phospholipids, a reaction that in mammals is most commonly catalyzed by a phospholipase A_2 or by the combined action of phospholipase C and diacylglycerol lipase (Van den Bosch 1980; Irvine 1982). Although selective inhibitors of either phospholipase are not yet available, both enzymes can be effectively blocked by 4-bromophenacyl bromide (Blackwell and Flower 1983). This inhibitor, at a concentration of 10 μM, completely suppresses the hyperpolarization produced by FMRFamide in cultured neurons (Piomelli et al. 1987b). The depolarizing response to 5-HT is not affected in the same cells, indicating that the drug does not interfere with S-channel function.

The experiments with the phospholipase inhibitor suggest that the release of arachidonic acid is a necessary step in the response to FMRFamide. To determine which metabolic pathway is responsible for producing the active eicosanoids, we tested the actions of indomethacin, which blocks cyclo-oxygenase (Shen and Winter 1977), and nordihydroguaiaretic acid (NDGA), which blocks lipoxygenases (Salari et al. 1984). Indomethacin, which we find inhibits formation of prostaglandins in *Aplysia* ganglia with a half-maximal inhibition concentration of 0.5 μM without affecting lipoxygenase activity, has no effect on the hyperpolarization induced by FMRFamide (Piomelli et al. 1987b). In contrast, NDGA, which inhibits lipoxygenase activity in *Aplysia* nervous tissue with an IC_{50} of 3 μM, suppresses the hyperpolarizing response of sensory cells in culture to FMRFamide (Fig. 8). The depolarization produced by 5-HT was not affected. These experiments suggest that a lipoxygenase enzyme rather than cyclo-oxygenase is involved in the intracellular signaling process.

To determine whether lipoxygenase metabolites of arachidonic acid were actually released in response to FMRFamide, we studied clusters of sensory neurons dissected from pleural ganglia and labeled by incubation with [^{3}H]arachidonic acid. FMRFamide stimulated the formation of radioactive material that migrated on reversed-phase HPLC with authentic 12-HETE and 5-HETE (Piomelli et al. 1987b). In control incubations or incubations with artificial seawater containing 5-HT, only background radioactivity was found in the 12-HETE and 5-HETE fractions.

12-HPETE Mimics the Action of FMRFamide

Independent support for the involvement of lipoxygenase metabolites comes from experiments where we directly tested the action of exogenously applied arachidonic acid metabolites on the sensory neurons. Lipoxygenases convert arachidonate into hydroperoxy acids; in mammals these are either reduced to the corresponding hydroxy acids or transformed into metabolites that are distinctive for each of the lipoxygenase pathways (Needleman et al. 1986). We tested the actions of both 12- and 5-hydroperoxy and hydroxy acids on the resting and action potentials of sensory neurons (Piomelli et al. 1987b). Application of 12-HPETE (1.5 μM) simulates the slow hyperpolarization of the resting potential and the decrease in excitability seen with FMRFamide. 5-HPETE, 5-HETE, and 12-HETE have little effect on resting potential. Both 12-HPETE and 5-HPETE simulate the decrease in action potential duration observed with FMRFamide. The average decrease in action potential duration with 12-HPETE, however, was two- to threefold greater than the decrease in response to 5-HPETE. Again, 12-HETE and 5-HETE had little effect on the duration of the action potential. These results confirm the biochemical and pharmacological experiments that implicate lipoxygenase metabolites in the actions of FMRFamide; they further suggest that the active second messenger derives from a hydroperoxy acid rather than from a hydroxy acid.

Molecular Sites of Interactions between 5-HT and FMRFamide

5-HT, acting through cAMP, causes S-channel closure and facilitation of transmitter release. FMRFamide, through lipoxygenase products, acts to open S channels and to cause presynaptic inhibition of neurotransmitter release. When 5-HT and FMRFamide are applied simultaneously, the effect of FMRFamide predominates, yielding a net presynaptic inhibition. How do these two competing cascades interact? What is the molecular mechanism whereby FMRFamide overrides presynaptic facilitation? Several points of interaction seem possible. First, there could be an interaction between G proteins involved in mediating the excitatory and inhibitory modulatory inputs on S-channel function (site 1 in Fig. 7). Second, FMRFamide acting through G_i could cause adenylate cyclase inhibition (site 2). Finally, FMRFamide, possibly acting through lipoxygenase products, could cause inhibition of the cAMP-dependent protein kinase (site 3) or activation of a phosphatase (site 4).

To investigate a possible site of interaction at the level of G proteins, we undertook studies on the effects of injection of nonhydrolyzable GTP analogs on S-channel function in sensory neurons (Volterra et al. 1987; Volterra and Siegelbaum 1988). As shown in Figure 8A, injection of GTPγS mimicked the effect of FMRFamide application; that is, it caused an increased-conductance membrane hyperpolarization. The ionic basis of this effect is consistent with GTPγS causing opening of S channels, and suggests that a G protein is probably involved in mediating the FMRFamide response. In fact, injection of pertussis toxin into sensory neurons completely blocks the ability of FMRFamide (but not arachidonic acid) to modulate S-channel function (data not shown). On the other hand, as shown in Figure 8B, the lipoxygenase inhibitor NDGA blocks GTPγS-induced hyperpolarization of

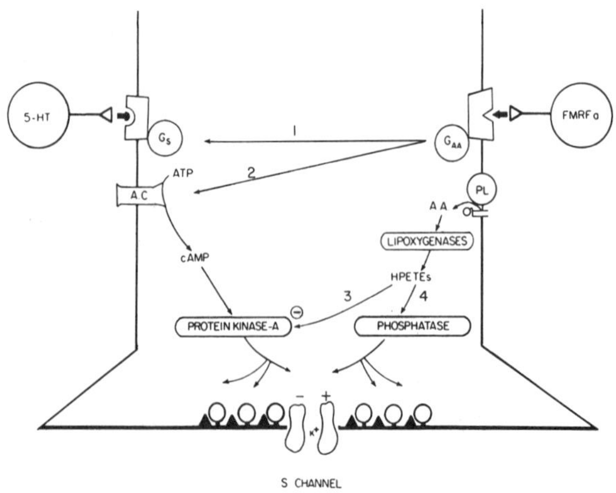

Figure 7. Model for the interaction between 5-HT and FMRFamide. Potential sites of interaction are shown. See text for discussion.

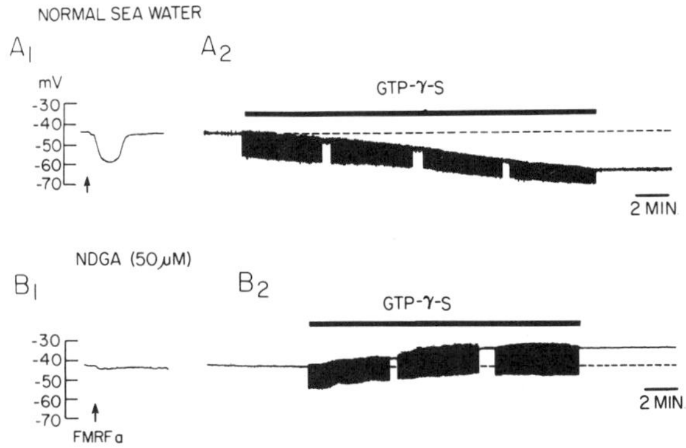

Figure 8. Effects of FMRFamide (*left*) and GTPγS (*right*) in absence (*A*) and presence (*B*) of the lipoxygenase inhibitor NDGA. (*A left*) Response to 10 μM FMRFamide in a cell impaled with a regular 3 M KCl electrode and bathed in artificial seawater. (*A right*) Slow, irreversible hyperpolarization of sensory cell membrane potential in response to iontophoretic injection from an electrode containing 25 mM GTPγS. The nucleotide was injected with 5-min periods of −0.2 nA current pulses (500 msec long at a frequency of 1 Hz). Input resistance given by amplitude of hyperpolarizing response to current pulses (regions with thickened trace). (*B*) Cell preincubated with the lipoxygenase inhibitor NDGA (50 μM). (*B left*) Response to 10 μM FMRFamide immediately after cell impalement with GTPγS (25 mM) containing electrode. (*B right*) GTPγS was then iontophoretically injected for 15 min. The blocker had no direct effect on membrane potential, input resistance, spike firing, or normal depolarizing response to 5-HT.

sensory neurons. This indicates that the effect of GTPγS on the S channel is indirect and mediated by lipoxygenase products. Taken together, these data strongly suggest that FMRFamide acts through a pertussis-toxin-sensitive G protein to cause phospholipase activation, and subsequent production of lipoxygenase metabolites mediates opening of the S channel.

Inspection of Figure 8B reveals that application of NDGA not only blocks the inhibitory effects of GTPγS, but also unmasks a component of the GTPγS effect that causes a decreased-conductance membrane depolarization. Thus, when lipoxygenase activity is blocked, GTPγS injection mimics 5-HT application and causes closure of the S channel. GTPγS injection therefore appears to be capable of activating both the facilitatory and the inhibitory mechanism, with the inhibitory mechanism normally predominating. Thus, there is a replication at the level of G proteins of the general finding that the inhibitory pathway overrides the stimulatory one. However, both the basic inhibitory mechanism and the inhibitory override appear to require lipoxygenase products as an obligatory intermediate. These data therefore rule out the existence in these cells of a G protein capable of *directly* interacting with the S channel to cause its opening, and they rule out the possibility of FMRFamide acting via such a mechanism.

The obligatory role for lipoxygenase products in mediating the inhibitory effects of GTPγS also rules out an interaction between G_i and G_s at the level of the adenylate cyclase. These data are consistent with those described previously by Ocorr and Byrne (1985) and by T. Abrams and E.R. Kandel (unpubl.), who found no effect of FMRFamide on cAMP levels in cells. Moreover, as described above (Fig. 5), FMRFamide can reopen S channels closed by cAMP, even in the continued presence of cAMP. Thus, FMRFamide must be able to act at a point distal to cAMP elevation.

Therefore, the findings suggest the possibility that 5-HT and FMRFamide interact at the level of protein phosphorylation (sites 3 and 4 in Fig. 7). To investigate this possibility, we developed a protein kinase assay using ^{32}P-labeled intact sensory neurons, two-dimensional gels, and computer-aided quantitational analysis. We first characterized the response of sensory neurons to application of 5-HT. We found that 5-HT or cell-permeable cAMP analog could cause an increase in the phosphorylation of at least 17 proteins in the sensory cells (data not shown). These proteins ranged in molecular weight from 16,000 to 60,000, and in pI from 4.5 to 7.5. The increased phosphorylation of a wide variety of proteins suggests that the effect of 5-HT on sensory cells is not limited to alteration of S-channel function but affects a variety of biochemical pathways.

Interestingly, application to the sensory cells of FMRFamide alone selectively caused a decrease in phosphorylation of 10 of these 17 proteins. Yet, when 5-HT and FMRFamide are presented together, FMRFamide completely overrides the effects of 5-HT. Not only do certain proteins have their phosphorylation decreased, but those proteins whose phosphorylation is normally increased by 5-HT also remain unchanged. Thus, the inhibitory override observed at the level of behavior, of neurotransmitter release, and of S-channel function, is also seen at the molecular level of phosphorylation of a variety of proteins in the cell.

These studies suggest that FMRFamide is able to antagonize the effects of 5-HT in a global fashion, at the level of substrate protein phosphorylation. At present, we do not know the precise molecular locus for this effect. However, these data are consistent with the hypothesis that sites of interaction for 5-HT and FMRFamide are at the level of kinase inhibition (site 3) or phosphatase activation (site 4), or both. We are presently undertaking studies to distinguish between these possibilities.

OVERALL VIEW

Whereas receptor-stimulated increase of cAMP mediates the facilitatory response of the sensory neurons to 5-HT, the receptor-stimulated release of arachidonic acid and its metabolism through a lipoxygenase pathway mediate the inhibitory synaptic response of FMRFamide. The results we summarize here and those presented by Shapiro et al. (this volume) thus strengthen the recent evidence of Piomelli et al. (1987b), which indicates that lipoxygenase products act as second messengers in nerve cells.

This function for the arachidonic acid cascade is not

unique to *Aplysia* sensory cells or to FMRFamide. Shapiro et al. have now found that the presynaptic inhibition produced by stimulating the putative histaminergic L32 cells on the L10-RB connection (Kretz et al. 1986) also operates through the same second-messenger mechanism (Piomelli et al. 1987b; Shapiro et al., this volume). It will, therefore, be of interest to see whether arachidonic acid participates in presynaptic inhibition in vertebrates, such as that produced by enkephalins (Mudge et al. 1979) and noradrenaline (Dunlap and Fischbach 1978) in the dorsal root ganglion cells that mediate response to painful stimuli.

Volterra and Siegelbaum (1988) and Shapiro et al. (this volume) have recently established that FMRFamide activation of the arachidonic acid cascade depends on a pertussis-toxin-sensitive G protein, which is likely to couple the peptide receptor to the phospholipase responsible for the release of arachidonate from sensory cell membrane phospholipids. From a functional point of view, this G protein seems not to belong to the family of G_i proteins. The inhibitory override does not reside at the level of the G proteins, and it does not have adenylate cyclase activity as a target (Ocorr and Byrne 1985; T. Abrams and E.R. Kandel, unpubl.). The failure to affect the cyclase adds further support to the earlier biochemical and biophysical evidence that, unlike the situation with transmitters utilizing G_i, the inhibitory override produced by FMRFamide lies at a later stage in the cAMP cascade. This override may involve either inhibition of the cAMP-dependent protein kinase or activation of a specific phosphatase capable of reversing the phosphorylation of the cAMP-dependent kinase. At present, we do not know whether the basic modulatory action of FMRFamide on the S channel (to increase the open probability) and its inhibition of 5-HT modulation (through reopening of the channels) share the same molecular mechanisms (e.g., phosphatase activation). Moreover, we have not yet identified the active lipoxygenase metabolite(s) directly responsible for S-channel modulation, although our results do suggest that the active molecule is closely related to a hydroperoxy acid, in particular to 12-HPETE. Whatever the answer to these questions, the inhibitory effect of FMRFamide acting through the eicosanoids is functionally antagonistic to the facilitatory action produced by 5-HT through cAMP. Since 5-HT and FMRFamide converge on the same K^+ channel, 5-HT acting to decrease and FMRFamide to increase channel activity, it is clear that an individual channel protein can serve as a final common effector for different second-messenger systems. This suggestion is supported by our finding with two-dimensional gels, which indicates that this convergence is not limited to one but extends to a whole family of substrate proteins.

Independent of the detailed biochemical mechanisms for modulation, the behavioral and cellular experiments on sensitization and inhibition in *Aplysia* combine to indicate that the animal's environmental input can produce substantial effects on behavior by activating different modulatory systems that act on nerve cells to regulate the excitability of the cells and to control synaptic strength. The modulatory systems, in turn, utilize two different families of transmitters and engage two different second-messenger systems within the sensory neurons. Therefore, our results suggest that these opposing behavioral events are represented within the nervous system of *Aplysia* by the activation of antagonistic modulatory systems and that these modulatory systems are re-represented within single sensory neurons by the balance of actions of competing second-messenger systems. Thus, these results begin to reveal an unexpected richness, at the cellular and molecular level, underlying the internal representation of external events.

REFERENCES

Belardetti, F., E.R. Kandel, and S. Siegelbaum. 1987. Neuronal inhibition by the peptide FMRFamide involves opening of S K^+ channels. *Nature* **325:** 153.

Belardetti, F., S. Schacher, E.R. Kandel, and S. Siegelbaum. 1986. The growth cones of *Aplysia* sensory neurons — Modulation by serotonin of action potential duration and single-potassium channel current. *Proc. Natl. Acad. Sci.* **83:** 7094.

Bernier, L., V.F. Castellucci, E.R. Kandel, and J.H. Schwartz. 1982. Facilitatory transmitter causes a selective and prolonged increase in adenosine 3′:5′-monophosphate in sensory neurons mediating the gill and siphon withdrawal reflex in *Aplysia. J. Neurosci.* **2:** 1682.

Blackwell, G.J. and R.J. Flower. 1983. Inhibition of phospholipase. *Br. Med. Bull.* **39:** 260.

Boyle, M.B., K. Klein, S.J. Smith, and E.R. Kandel. 1984. Serotonin increases intracellular Ca^{2+} transients in voltage-clamped sensory neurons of *Aplysia californica. Proc. Natl. Acad. Sci.* **81:** 7642.

Brezina, V., R. Eckert, and C. Erxleben. 1987a. Modulation of potassium conductances by an endogenous neuropeptide in neurones of *Aplysia californica. J. Physiol.* **382:** 267.

———. 1987b. Suppression of calcium current by an endogenous neuropeptide in neurones of *Aplysia californica. J. Physiol.* **388:** 565.

Brunelli, M., V. Castellucci, and E.R. Kandel. 1975. Serotonin mediated increase in cyclic AMP: A possible biochemical step in the increased transmitter output accompanying behavioral sensitization in *Aplysia. Soc. Neurosci. Abstr.* **1:** 568.

Carew, T.J., R.D. Hawkins, and E.R. Kandel. 1983. Differential classical conditioning at the defensive withdrawal reflex in *Aplysia californica. Science* **219:** 397.

Castellucci, V.F., E.R. Kandel, J.H. Schwartz, F.D. Wilson, A.C. Nairn, and P. Greengard. 1980. Intracellular injection of the catalytic subunit of cyclic AMP-dependent protein kinase simulates facilitation of transmitter release underlying behavioral sensitization in *Aplysia. Proc. Natl. Acad. Sci.* **77:** 7492.

Choi, D.W. and G.D. Fischbach. 1981. GABA conductance of chick spinal cord and dorsal root ganglion neurons in cell culture. *J. Neurophysiol.* **45:** 605.

Dunlap, K. and G.D. Fischbach. 1978. Neurotransmitters decrease the calcium component of sensory neuron action potentials. *Nature* **276:** 837.

Frost, W.N., G.A. Clark, and E.R. Kandel. 1988. Parallel processing of short-term memory for sensitization in *Aplysia. J. Neurobiol.* **19:** 297.

Frost, W.N., V.F. Castellucci, R.D. Hawkins, and E.R. Kandel. 1985. Monosynaptic connections made by the sensory neurons of the gill- and siphon-withdrawal reflex in *Aplysia* participate in the storage of long-term memory for sensitization. *Proc. Natl. Acad. Sci.* **82:** 8266.

Hochner, B., M. Klein, S. Schacher, and E.R. Kandel. 1986a. Action potential duration and modulation of transmitter release from the sensory neurons of *Aplysia* in presynaptic facilitation and behavioral sensitization. *Proc. Natl. Acad. Sci.* **83:** 8410.

———. 1986b. Additional component in the cellular mechanism of presynaptic facilitation contributes to behavioral dishabituation in *Aplysia. Proc. Natl. Acad. Sci.* **83:** 8794.

Hochner, B., S. Schacher, M. Klein, and E.R. Kandel. 1985. Presynaptic facilitation in *Aplysia* sensory neurons: A process independent of K^+ current modulation becomes important when transmitter release is depressed. *Soc. Neurosci. Abstr.* **11:** 29.

Irvine, R.F. 1982. How is the level of free arachidonic acid controlled in mammalian cells? *Biochem. J.* **204:** 3.

Kandel, E.R. and J.H. Schwartz. 1982. Molecular biology of an elementary form of learning: Modulation of transmitter release by cyclic AMP. *Science* **218:** 433.

Klein, M. and E.R. Kandel. 1978. Presynaptic modulation of voltage-dependent Ca^{2+} current: Mechanism for behavioral sensitization in *Aplysia californica. Proc. Natl. Acad. Sci.* **75:** 3512.

———. 1980. Mechanism of calcium current modulation and underlying synaptic facilitation and behavioral sensitization in *Aplysia. Proc. Natl. Acad. Sci.* **77:** 6912.

Klein, M., J.S. Camardo, and E.R. Kandel. 1982. Serotonin modulates a specific potassium current in the sensory neurons that show presynaptic facilitation in *Aplysia. Proc. Natl. Sci.* **79:** 5713.

Klein, M., B. Hochner, and E.R. Kandel. 1986. Facilitatory transmitters and cAMP can modulate accommodation as well as transmitter release in *Aplysia* sensory neurons: Evidence for parallel processing in a single cell. *Proc. Natl. Acad. Sci.* **83:** 7994.

Konorski, J. 1967. *Integrative activity of the brain: An interdisciplinary approach.* University of Chicago Press, Chicago.

Kretz, R., E. Shapiro, and E.R. Kandel. 1986. Presynaptic inhibition produced by an identified presynaptic inhibitory neuron. I. Physiological mechanisms. *J. Neurophysiol.* **55:** 131.

Lindgren, J.A., T. Hokfelt, S.E. Dahlen, C. Patrono, and B. Samuelsson. 1984. Leukotrienes in the rat central nervous system. *Proc. Natl. Acad. Sci.* **81:** 6212.

Mackey, S.L., D.L. Glanzman, S.A. Small, A.M. Dyke, E.R. Kandel, and R.D. Hawkins. 1987. Tail shock produces inhibition as well as sensitization of the siphon-withdrawal reflex of *Aplysia*: Possible behavioral role for presynaptic inhibition mediated by the peptide Phe-Met-Arg-Phe-NH_2. *Proc. Natl. Acad. Sci.* **84:** 8730.

Mudge, A., S. Leeman, and G.D. Fischbach. 1979. Enkephalin inhibits release of substance P from sensory neurons in culture and decreases action potential duration. *Proc. Natl, Acad. Sci.* **76:** 526.

Needleman, P., J. Turk, B.A. Jakschick, A.R. Morrison, and J.B. Lefkowith. 1986. Arachidonic acid metabolisms. *Annu. Rev. Biochem.* **55:** 69.

Ocorr, K.A. and J.H. Byrne. 1985. Membrane responses and changes in cAMP levels in *Aplysia* sensory neurons produced by serotonin, tryptamine, FMRFamide, and small cardioactive peptide B (SCP_B). *Neurosci. Lett.* **55:** 113.

Piomelli, D., E. Shapiro, S.J. Feinmark, and J.H. Schwartz. 1987a. Metabolites of arachidonic acid in the nervous system of *Aplysia*: Possible mediators of synaptic modulation. *J. Neurosci.* **7:** 3675.

Piomelli, D., A. Volterra, N. Dale, S.A. Siegelbaum, E.R. Kandel, J.H. Schwartz, and F. Belardetti. 1987b. Lipoxygenase metabolites of arachidonic acid as second messengers for presynaptic inhibition of *Aplysia* sensory cells. *Nature* **328:** 38.

Ruben, P., J.W. Johnson, and S. Thompson. 1986. Analysis of FMRF-amide effect on *Aplysia* bursting neurons. *J. Neurosci.* **6:** 252.

Salari, H., P. Braquet, and P. Borgeat. 1984. Comparative effects of indomethacin, acetylenic acids, 15-HETE, nordihydroguaiaretic acid and BW 755 C on the metabolism of arachidonic acid in human leukocytes and platelets. *Prostaglandins Leukotrienes Med.* **13:** 53.

Samuelsson, B. 1964. Identification of a smooth muscle-stimulating factor in bovine brain. *Biochem. Biophys. Acta* **84:** 218.

Sautebin, L., C. Spagnuolo, C. Galli, and G. Galli. 1978. A mass fragmentographic procedure for the simultaneous determination of HETE and $PGF_{2\alpha}$ in the central nervous system. *Prostaglandins* **16:** 985.

Shen, T.Y. and C.A. Winter. 1977. Chemical and biological studies on indomethacin, sulindac and their analogs. *Adv. Drug Res.* **12:** 89.

Shimizu, T., Y. Takusagawa, T. Izumi, N. Ohishi, and Y. Seyama. 1987. Enzymic synthesis of leukotriene B_4 in guinea pig brain. *J. Neurochem.* **48:**1541.

Shuster, M.J., J.S. Camardo, S.A. Siegelbaum, and E.R. Kandel. 1985. Cyclic AMP-dependent protein kinase closes the serotonin-sensitive K^+ channels of *Aplysia* sensory neurones in cell-free membrane patches. *Nature* **313:** 392.

Siegelbaum, S.A., J.S. Camardo, and E.R. Kandel. 1982. Serotonin and cyclic AMP close single K^+ channels in *Aplysia* sensory neurones. *Nature* **299:** 413.

Small, S.A., R.D. Hawkins, and E.R. Kandel. 1988. Activity-dependent enhancement of presynaptic inhibition in *Aplysia* sensory neurons. *Soc. Neurosci. Abstr.* **14:** (in press).

Van den Bosch, H. 1980. Intracellular phospholipase A. *Biochem. Biophys. Acta* **604:** 191.

Volterra, A. and S.A. Siegelbaum. 1988. Role of two different G proteins in the antagonistic modulation of the *S* K^+ channel by cAMP and arachidonic acid metabolites in *Aplysia* sensory neurons. *Proc. Natl. Acad. Sci.* (in press).

Volterra, A., J.D. Sweatt, and S.A. Siegelbaum. 1987. Involvement of G proteins in the inhibitory action of FMRFamide through lipoxygenase metabolities of arachidonic acid in *Aplysia* sensory neurons. *Soc. Neurosci. Abstr.* **13:** 1440.

Wolfe, L.S. 1982. Eicosanoids: Prostaglandins, thromboxanes, leukotrienes and other derivatives of carbon-20 unsaturated fatty acids. *J. Neurochem.* **38:** 1.

Electrical and Biochemical Properties of the cGMP-gated Cation Channel from Rod Photoreceptors

U.B. KAUPP, W. HANKE, R. SIMMOTEIT, AND H. LÜHRING

Abteilung Biophysik, Universität Osnabrück, D-4500 Osnabrück, Federal Republic of Germany

Vertebrate photoreceptors hyperpolarize in response to light due to the closure of light-sensitive channels in the plasma membrane. These channels have also been designated cGMP-gated channels because light controls their opening indirectly by changing the intracellular concentration of cGMP via a highly amplifying enzyme cascade. In this paper, we describe our present knowledge on the molecular properties of this sensory channel protein, the mechanism of its activation by cGMP, and its significance for the transduction process in the photoreceptor cell. In particular, we describe procedures to solubilize, purify, and reconstitute the cGMP-gated cation channel from vertebrate photoreceptors in a functional state. The activation by cGMP and the electrical properties of the channel in situ are compared with those of the purified and reconstituted protein.

Historical Review

Fensko et al. (1985) first described a conductance in excised patches of the plasma membrane of frog rods, which is directly and cooperatively activated by cGMP but not by cAMP. In addition, they provided initial evidence that the cGMP-activated conductance represents the light-sensitive conductance, implying that cGMP plays an important role as an intracellular transmitter in phototransduction. These results were confirmed by several other laboratories (Haynes and Yau 1985; Yau and Nakatani 1985; Zimmerman et al 1985; Stern et al. 1986).

Subsequent work has focused on several properties of the channel that are important for the normal physiology of the photoreceptor cell. One of the most urgent questions was to identify unambiguously the light-sensitive channel as the cGMP-gated channel in the plasma membrane. There is now ample evidence that these conductances are identical because both exhibit a similar ion selectivity (Yau and Nakatani 1984; Fesenko et al. 1985; Hodgkin et al. 1985; Hanke et al. 1988), a pronounced outward rectification in the presence of divalent cations (Bader et al. 1979; MacLeish et al. 1984; Baylor and Nunn 1986; Stern et al. 1986), and similar kinetics (Bodoia and Detwiler 1985; Gray and Attwell 1985; Matthews 1986, 1987) and comparable pharmacological (Stern et al. 1986; Nicol et al. 1987) properties.

Analysis of the current noise generated by the light-sensitive conductance in intact cells (Bodoia and Detwiler 1985; Gray and Attwell 1985) and by the cGMP-activated conductance (Fesenko et al. 1985) in the presence of millimolar concentrations of divalent cations suggested a single-channel conductance of a few femtoSiemens, a value that could also be due to a carrier-type mechanism of ion permeation. After the removal of Mg^{++} and Ca^{++} ions from the medium, the cGMP-activated macroscopic current across a patch of plasma membrane increases severalfold (Stern et al. 1986), and discrete, channel-like current events occur (Haynes et al. 1986; Zimmerman and Baylor 1986), suggesting (1) that the channel protein comprises an aqueous pore with a conductance of 25 pS in the presence of physiological Na^{+} concentration and (2) that the open channel may be blocked by divalent cations. Koch and Kaupp (1985) reported a cGMP-activated conductance to occur in a fraction of disk membranes, thereby confirming and extending previous results by Cavaggioni and collaborators (Caretta and Cavaggioni 1983; Caretta 1985; Caretta et al. 1985). Although the significance of a cGMP-gated channel in the disk membrane for the function of a photoreceptor cell remains uncertain, its properties were similar, if not identical, to those observed in excised patches of the plasma membrane (Kaupp and Koch 1986) and provided a useful system to examine the channel by chemical flux measurements.

Recently, the cGMP-gated channel protein has been identified, purified, and functionally incorporated into liposomes and planar lipid bilayers (Cook and Kaupp 1986; Cook et al. 1986, 1987; Hanke et al. 1988).

Isolation of the cGMP-gated Channel Protein

Purification and reconstitution scheme. The polypeptide that constitutes the channel protein has been purified from rod outer segments of the bovine retina by sequential column chromatography and has been functionally reincorporated into the membrane of liposomes and into planar lipid bilayers (Cook and Kaupp 1986; Cook et al. 1986, 1987; Hanke et al. 1988). Identification of the channel protein during purification was achieved by reconstituting column fractions into liposomes and subsequent testing for cGMP-stimulated Ca^{++}-flux activity, using the Ca^{++}-indicating dye arsenazo III (Koch and Kaupp 1985). This functional assay

  0-87969-055-0/88 $1.00

is particularly helpful in identifying hetero-oligomeric membrane receptors, which are composed of several types of subunits with different ligand-binding or pharmacological properties and is superior to photo-affinity-labeling techniques because it probes for the entire functional channel complex and not only for agonist- or toxin-binding subunit(s).

The reconstitution procedure is outlined schematically in Figure 1a. The most important factors that determine the activity of the reconstituted channel are (1) choice of detergent, (2) concentration of the detergent, and (3) presence of 0.1–0.2% (w/v) asolectin and 10 mM Ca^{++} during solubilization. Recovery of channel activity was maximal when photoreceptor membranes were solubilized with the detergent CHAPS, although Mega 9 and cholate are also suitable. Channel activity started to decrease at more than 10 mM CHAPS, suggesting that it is advisable to work at concentrations slightly above the critical micellar concentration value (8 mM for CHAPS) of the respective detergent.

Purification of the channel protein involved sequential fractionation of detergent-solubilized membrane proteins by an anion-exchange column, followed by an affinity column (Fig. 1b). The final purified extract contained only two polypeptide bands of 63 and 240 kD (Fig. 3, left). Although the 240-kD polypeptide often copurified with the 63-kD protein, its presence in the purified extract was highly variable, and it was even absent in some experiments. Channel activity observed after reconstitution of the purified protein was not correlated with the presence of the 240-kD band on the SDS-PAGE, indicating that it is not involved in forming a functional pore. Recently, it has been shown that a 240-kD polypeptide from bovine rod photoreceptors is antigenically related to spectrin of erythrocytes (Wong and Molday 1986), suggesting that the channel protein may be attached to this cytoskeletal element and that the protein-protein association is even maintained in detergent solutions. Such a stable interaction between the channel protein and the cytoskeleton could serve an important role for the architecture and physiology of the cell and requires careful attention in future research.

Subunit stoichiometry. Under the condition of denaturing gel electrophoresis, the channel protein appears as a single polypeptide of 63 kD. However, we must envisage the possibility that the functional channel in the membrane exists as an oligomeric complex, probably as a homotetramer (250 kD). First, it is cooperatively activated by at least three, and possibly four, molecules of cGMP (see below). A corollary of this hypothesis is that each subunit has one receptor site for cGMP and that cooperativity occurs between subunits rather than *within* a subunit. Alternatively, if the channel is composed of only one or two subunits, the number of cGMP-binding sites on each subunit would be two or four, respectively. Second, the two sets of channel proteins—voltage gated and chemically gated—that have been analyzed so far seem to either be composed of a single large subunit of 200,000 kD (Na^+-channel, Ca^{++}-channel) or constitute heterotetrameric (GABA and glycine receptors) or heteropentameric complexes (AchR), with a total molecular mass of 200–300 kD (see, e.g., Vandlen et al. 1979; Barnard et al. 1987; Betz 1987).

Channel density and subcellular distribution. The 63-kD polypeptide contributes approximately 0.5–0.8% (w/w) to the total protein content of photoreceptor membranes (Cook et al. 1987). Approximately 2.9×10^7 rhodopsin molecules exist in a rod outer segment from the bovine retina and 0.9×10^5 to 1.5×10^5 copies of the 63-kD subunit. The total number of channels per cell becomes fourfold smaller (2.3×10^4 to 3.8×10^4) if the channel is composed of four 63-kD subunits. A channel density between 365 μm^{-2} and 600 μm^{-2} is calculated if the channel is located exclusively in the plasma membrane (area 63 μm^{-2}). This number is in good agreement with estimates of channel density in the plasma membrane (100–400 μm^{-2}) from analysis of the light-sensitive current noise in whole rods (Bodoia and Detwiler 1985; Zimmerman and Baylor 1985) and the cGMP-activated current in excised membrane patches (Zimmerman and Baylor 1986; Haynes et al. 1986).

The existence of the cGMP-activated channel in the plasma membrane has been demonstrated by the patch-clamp technique and in the disk membrane by measuring the cGMP-stimulated ion flux from sealed disk vesicles. Both techniques are prone to misrepresent the true location and therefore are not particularly suitable for a quantitative determination of the subcellular distribution of the channel protein: The excision of a patch of plasma membrane by a glass pipette may be accompanied by fusion of pieces of the disk membrane with the plasma membrane because both membranes are only a few nanometers apart and are connected by filamentous structures. Similarly, outer segments shrink during purification by density gradient centrifugation in a hyperosmolar sucrose medium, which may cause the exchange of protein components between both membranes. Moreover, an exchange between the protein inventory of the plasma and disk membrane may occur during the lysis of intact rod outer segments, which is an obligatory step preceding the measurement of ion fluxes from disk vesicles. Only 10–20% of the intradiscal Ca^{++} can be released by cGMP, and we interpreted this result to indicate that it is only this fraction of disks that contained the channel protein (Koch and Kaupp 1985), possibly those vesicles that were freshly formed from invaginations of the plasma membrane at the ciliary base of the cell. Although the cellular origin of the disk membrane supports the notion that the cGMP-dependent channel may also occur in disks, it does not explain why only some disks would receive channel copies. Since the existence of the light-regulated channel in disks is unexpected and its function there at best is obscure, a final decision on this matter must await the availability of more appropriate techniques (i.e., immunohistochemistry or the preparative separation of disk and plasma membrane).

Irrespective of this unresolved issue, the cGMP-activated conductances examined by chemical flux measurements from sealed membrane vesicles and by electrophysiological techniques seem to behave similarly (Kaupp and Koch 1986). Both are activated directly and cooperatively by cGMP or analogs of cGMP that have been modified at the C-8 position. Both conductances are blocked by micromolar concentrations of *l-cis*-diltiazem (Koch and Kaupp 1985; Stern et al. 1986; see below). Both are weakly selective for alkali cations and even allow the passage of divalent cations (Caretta and Cavaggioni 1983; Fesenko et al. 1985; Koch and Kaupp 1985).

Pharmacology and Activation of the Channel by Analogs of cGMP

Pharmacology of the channel in situ. The cGMP-dependent channel is activated by cGMP and some analogs that have been modified at the C-8 position of the purine base but not by other analogs of cGMP or by cAMP, cIMP, and cXMP (Caretta and Cavaggioni 1983; Koch and Kaupp 1985; Zimmerman et al. 1985; Kaupp and Koch 1986). For example, the EC_{50} of channel activation by the R_p and S_p isomers of cGMPs is 100- and 1000-fold larger, respectively, than that for cGMP itself (Koch and Kaupp 1985; Zimmerman et al. 1985). Cyclic nucleotides exist in a conformational equilibrium about the glycosidic bond between a *syn* and an *anti* form. Bulky substituents at the C-8 position, e.g., bromide, shift the equilibrium to the *syn* form (Tavale and Sobell 1970; Fazakerley et al. 1977). Several analogs of cGMP that are modified at the C-8 position have a higher affinity for the channel-binding site (Caretta et al. 1985; Koch and Kaupp 1985) suggesting that the *syn* rather than the *anti* form binds to the cGMP-receptor of the channel protein.

Modifications of amino, carbonyl, or hydroxyl groups at the diester bond, the ribose residue, or the pyrimidine portion of the purine ring decrease the affinity of cGMP analogs by several orders of magnitude. Although we know little about the molecular structure of the cGMP-binding site, this result suggests that the formation of hydrogen bonds between those chemical groups and amino acids of the binding site may be important for the attachment of the cGMP molecule to the cGMP receptor. Alternatively, ineffective cGMP analogs may bind to the receptor without activating the channel. However, there is no evidence that these analogs may act as antagonists. A highly speculative model of the cGMP receptor that summarizes the above results is shown in Figure 2.

The *l* isomer of *cis*-diltiazem, a well-known Ca^{++} antagonist in heart muscle cells, blocks both the channel in isolated disk membranes and excised patches of the plasma membrane, whereas the *d* form is much less effective (Koch and Kaupp 1985; Stern et al. 1986). The drug is also moderately effective in suppressing the light-evoked current in whole cells when applied extracellularly but has no detectable effect on the cell's physiology when applied intracellularly via a patch pipette (Stern et al. 1986). We recently discovered two kinetically and pharmacologically distinct channel forms in washed disk membranes. One has a low EC_{50} of activation by cGMP (170 μM) and is effectively blocked by *l-cis*-diltiazem, whereas the other form has a high affinity for cGMP (20 μM) and is not blocked by *l-cis*-diltiazem (Koch et al. 1987).

Pharmacology of the purified and reconstituted channel. The properties of the reconstituted channel protein have been determined by measuring the cGMP-stimulated Ca^{++} efflux from liposomes with the indicator arsenazo III (Cook et al. 1986, 1987) and by recording single-channel activity in planar lipid bilayers (Hanke et al. 1988).

Significant cGMP-stimulated Ca^{++} fluxes (0.1–0.4 nmole Ca^{++}/sec/μg protein) were observed from liposomes that contained the 63-kD polypeptide (1–10 μg protein per milliliter of liposomes). The EC_{50} and the cooperativity of channel activation by cGMP were determined by analyzing the initial velocity, v_i, of the cGMP-stimulated Ca^{++} efflux. A series of Ca^{++}-efflux signals after injection of various amounts of cGMP into the cuvette is shown in Figure 3 (center), along with the corresponding Hill plot of the initial rates of the Ca^{++} efflux (Fig. 3, right). In this experiment, we found $EC_{50} = 27 \mu$M, and a Hill coefficient of $n = 2.4$. Surprisingly, the purified and reconstituted channel is blocked much less effectively by *l-cis*-diltiazem than the channel in situ. Concentrations of the drug approximately 10- to 20-fold higher are required to obtain an equivalent blocking action, and the stereo-selectivity seems to disappear at high concentrations. These observations suggest that the purified channel protein exists predominantly in the high-affinity form that is not blocked by *l-cis*-diltiazem. The above observations can be rationalized by proposing a cellular factor(s) or mechanism(s) that controls the sensitivity of the channel for cGMP. Apparently, this putative factor is removed during washing of photoreceptor membranes and isolating the channel along with the sensitivity to *l-cis*-diltiazem. A comparison of the channel properties in the reconstituted system and in situ is given in Table 1.

Single-channel Recording of the Purified and Reconstituted Protein

The electrical properties of the channel have been examined by analyzing single-channel activity after incorporation of the purified protein into planar lipid bilayers (Hanke et al. 1988).

Single-channel conductance. The predominant single-channel conductance is 26 pS (150 mM NaCl, in the absence of divalent cations), although channel events of smaller size (<6–10 pS) are also observed, similar to single-channel recordings in situ (Haynes et al. 1986; Zimmerman and Baylor 1986). Initial results suggest

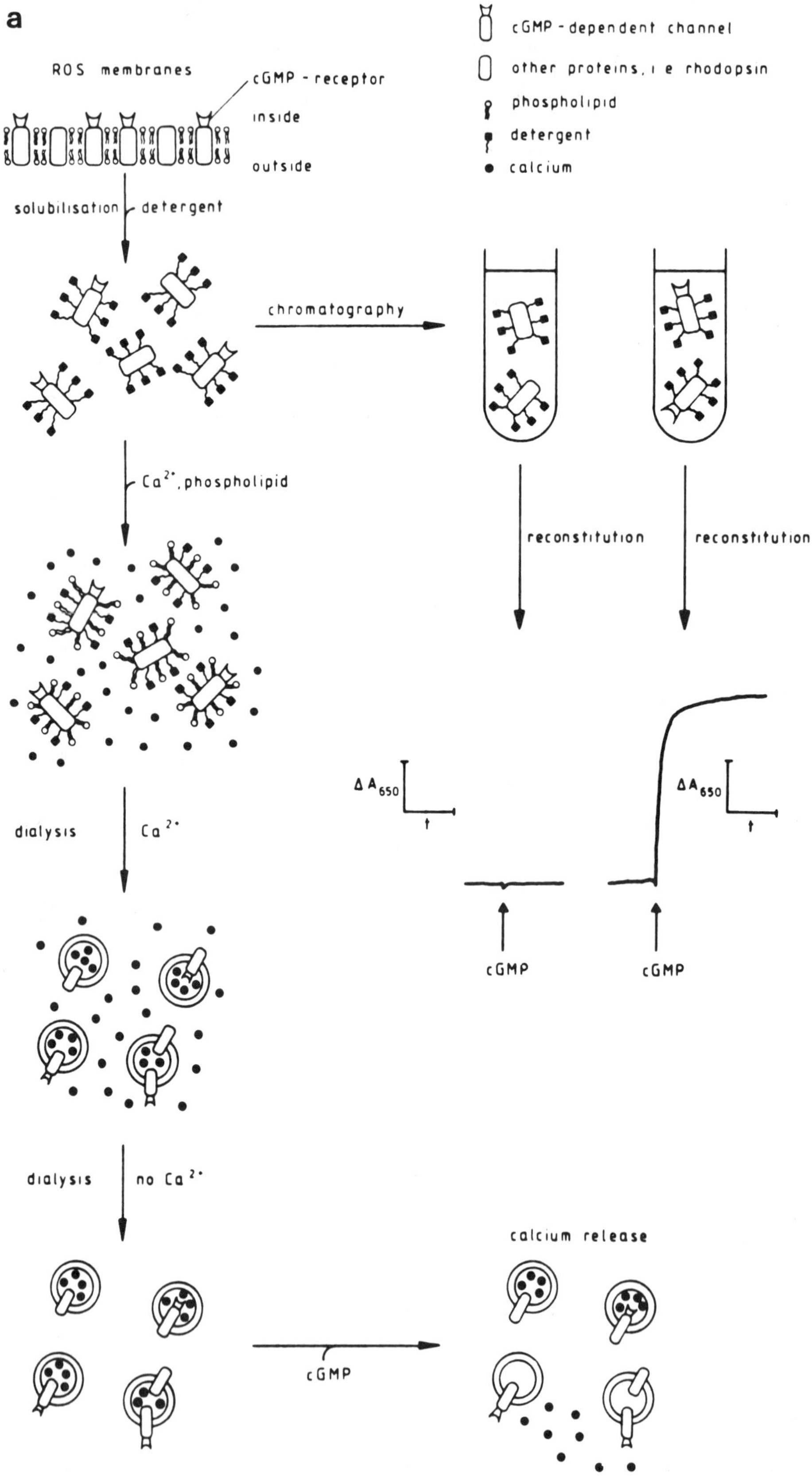

Figure 1. *(See facing page for part b and legend.)*

b

PURIFICATION SCHEME

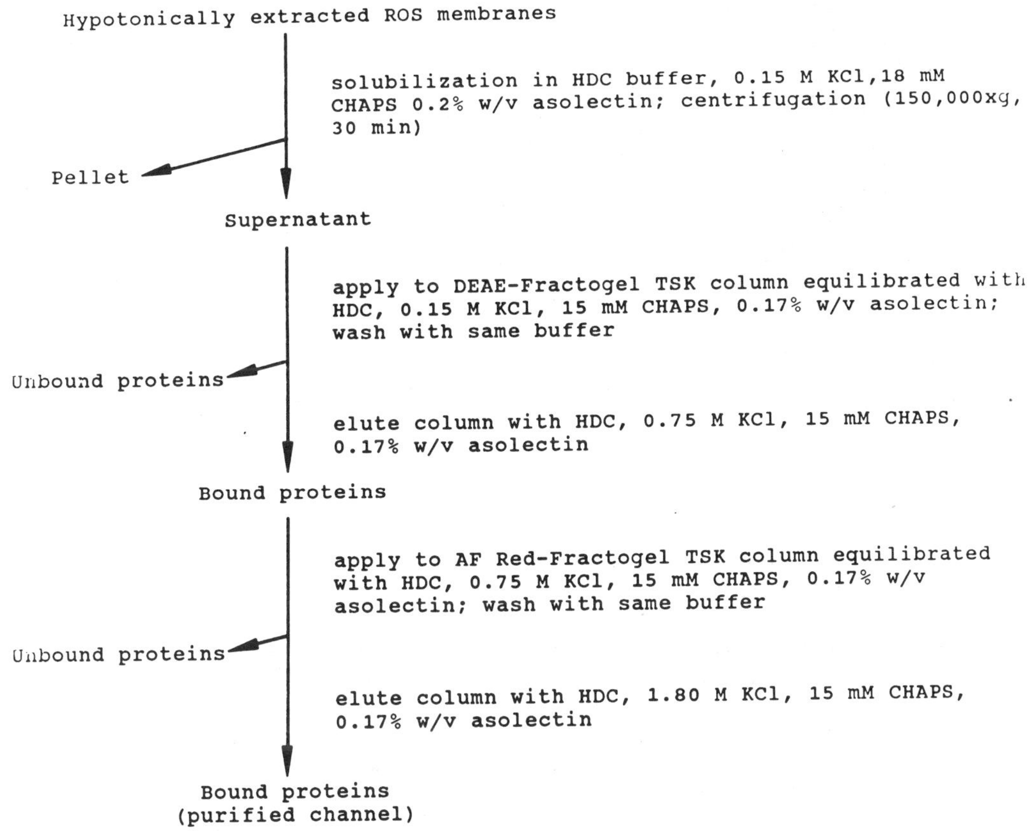

Figure 1. (*a*) Flow chart of the methods for the solubilization, reconstitution, and purification of the cGMP-dependent channel from ROS membranes. (*Left*) Steps involved when directly reconstituting the channel from solubilized membranes; (*right*) chromatographic methods were applied to the solubilized membrane extract, before reconstitution. Column fractions that did not contain the channel protein showed no cGMP-activated Ca^{++} release signal after reconstitution, whereas fractions that did contain the channel displayed a signal comparable to that shown in the figure (*middle, right side*). (*b*) Purification scheme of the cGMP-dependent channel from bovine rod outer segments. HDC (HEPES-DTT-Ca^{++}) buffer contained 10 mM HEPES-KOH (pH 7.4), 1 mM dithiothreitol, and 10 mM $CaCl_2$. Bound proteins refer to the protein fraction that bound to the columns at the salt concentrations present in the respective equilibration buffers (0.15 M KCl for the DEAE-Fractogel TSK column and 0.75 M for the AF Red-Fractogel TSK column) and that were eluted with high-salt buffer.

that channel-like current fluctuations of submaximal amplitude occur at intermediate concentrations of cGMP. An attractive hypothesis that could account for this observation is that partially liganded receptors (i.e., number of ligands less than four) also exist in an ion-conducting state and that the probability to observe this partially liganded state(s) is maximal at nonsaturating cGMP concentrations. Alternatively, the fully liganded channel may exist in more than one open state that differs in conductance.

Gating kinetics. The open time distribution of the channel at elevated cGMP concentrations (50 μM) can be described by a single exponential ($t_o = 6$ ms), whereas the closed time distribution can only be described by more than one exponential time constant. We interpret this result to indicate that channel activity occurs in bursts. The shorter time constant $t_{cib} = 7$ ms can be identified as the mean *c*losed time *i*n a *b*urst and the longer time constant $t_{cbb} = 37$ ms as the mean *c*losed time *b*etween *b*ursts. A corollary of this notion is that the channel has more than one closed state. The mean open time of the reconstituted channel is approximately one order of magnitude longer than that observed in situ (~1 ms; Haynes et al. 1986; Zimmerman and Baylor 1986) and probably reflects the different lipid environment in the artificial bilayer experiments.

Increasing the cGMP concentration affects the mean open time of channel events only weakly but decreases the mean closed times dramatically, i.e., the time between individual open events. This gating behavior is expected from a chemically gated channel, as the mean

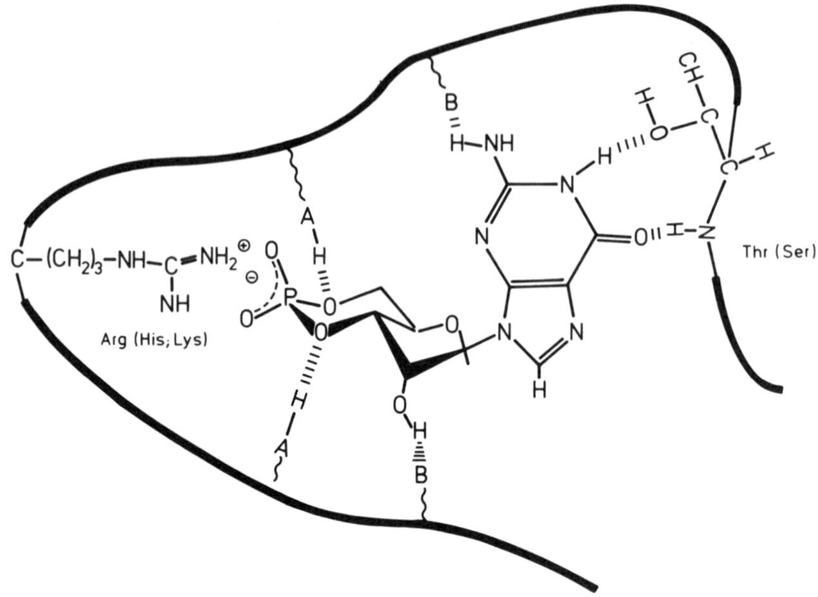

Figure 2. Hypothetical binding site of the cGMP receptor.

lifetime of the open channel in a simple two-state model only depends on the rate constant k_{oc} at which the agonist dissociates from the receptor (Hille 1984). The slight, but significant, increase in the mean open time with the cGMP concentration could result either from missed closing events in the analysis or from a second open state of the channel.

Voltage dependence and rectification. The current-voltage relation of the light-sensitive current in whole rods and of the cGMP-activated macroscopic current in excised membrane patches is nonlinear (Fig. 4a; MacLeish et al. 1984; Baylor and Nunn 1986; Stern et al. 1986; Yau et al. 1986). The current becomes almost independent of the voltage at negative potentials, and a strong outward rectification develops in the presence of divalent cations at the extracellular site of the membrane at positive potentials. If the divalent cation gradient is reversed, the current becomes inwardly rectifying (Matthews 1986; Yau et al. 1986), suggesting that diva-

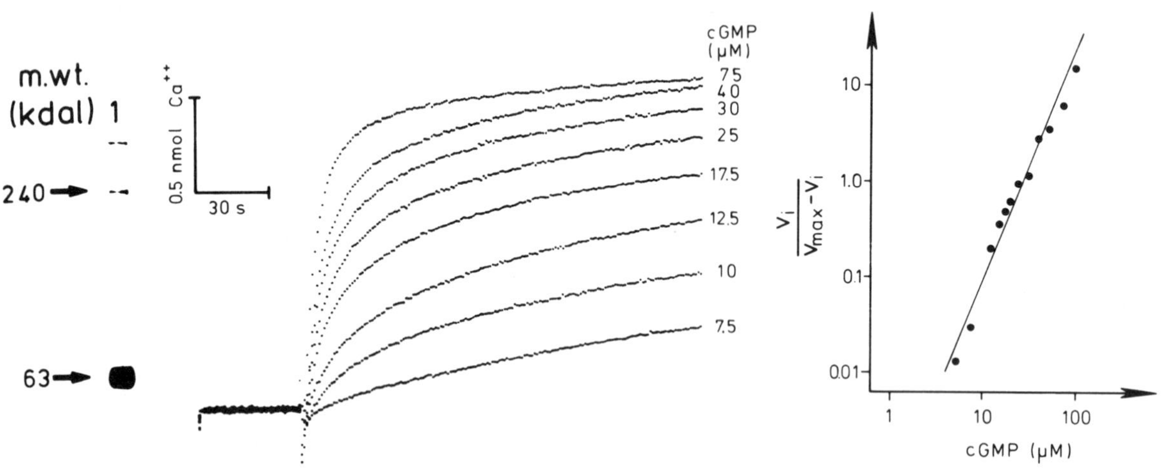

Figure 3. cGMP-dependent Ca^{++} efflux from liposomes containing purified channel protein. (*Left*) Gel electrophoresis of the purified channel protein was carried out under denaturing conditions. The lane contains 120 μl of a bound fraction that eluted from the AF Red-Fractogel TSK column containing 1–2 μg of protein. Details of the electrophoresis procedure are described by Cook et al. (1987) and Cook and Kaupp (1986). (*Middle*) Changes in absorbance after the injection of different concentrations of cGMP at time $t = 40$ sec into the cuvette. The difference in absorbance at $A = 650$ nm and $A = 730$ nm was recorded on an Aminco DW-2a spectrophotometer in the dual wavelength mode, stored on a transient recorder (Tracor TN-1500), and transferred to a PDP11/34 computer for further analysis (Koch and Kaupp 1985; Cook et al. 1986, 1987). The cuvette contained 0.3 ml of purified channel-containing liposomes (5 μg protein/ml, 10 mg phospholipid/ml) in a total volume of 2 ml (100 mM KCl, 10 mM HEPES-KOH [pH 7.4], 50 μM Arsenazo III and 5–10 μM free Ca^{++}). (*Right*) Hill plot of initial velocity v_i of the data shown in the middle part of the figure. The slope of the straight line is $n = 2.4$ and the $EC_{50} = 27$ μM.

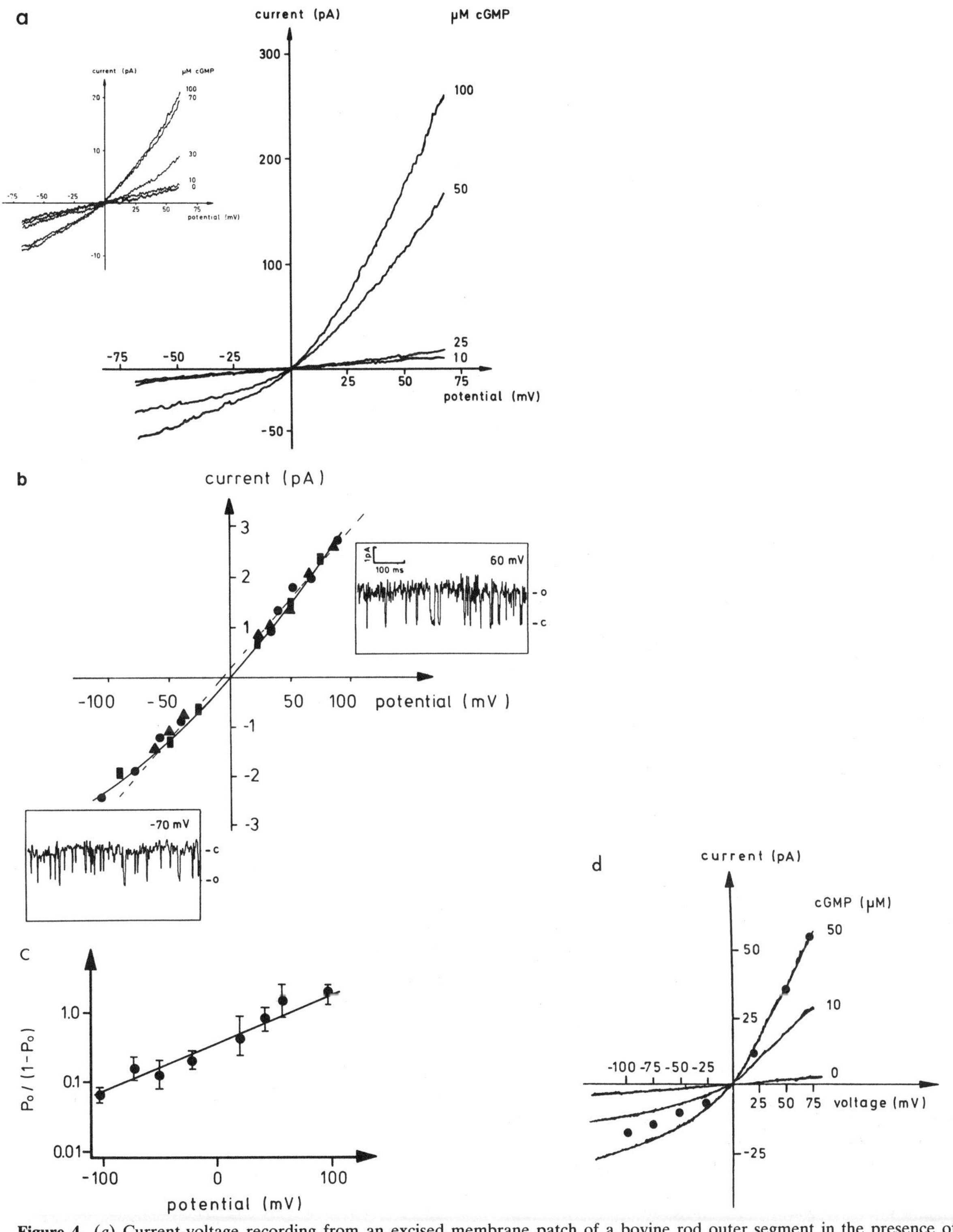

Figure 4. (*a*) Current-voltage recording from an excised membrane patch of a bovine rod outer segment in the presence of different concentrations of cGMP. Composition of the solution in the medium and in the pipette: 115 mM NaCl, 3 mM KCl, 3 mM $NaHCO_3$, and 10 mM HEPES-KOH (pH 7.4). (*Inset*) A series of current recordings in the presence of 1 mM Ca^{++} and Mg^{++}. (*b*) *i-V* relationship of the open state of a single reconstituted channel from three different bilayers as indicated by the various symbols in the absence of divalent cations. The solid line represents a best fit of the three sets of data. The reversal potential was imposed at 0 mV. The slope conductance at $V_m = 0$ mV was 27 pS. (*Insets*) Current fluctuations at two potentials V_m activated by 100 μM cGMP. (*c*) Voltage dependence of $P_o/(1 - P_o)$. The data represent mean values and the error bars indicate the standard deviation from five experiments at 50 μM cGMP. (*d*) *I-V* relationships of a bilayer membrane that contained several channel copies in the presence of the indicated cGMP concentrations. (●) Macroscopic current expected from the voltage dependence of the open probability and the *i-V* relationship of a single open channel.

Table 1. Properties of the cGMP-dependent Channel

	In situ	Reconstituted
Agonist	cGMP C-8-substituted analogs	cGMP C-8-substituted analogs
EC_{50} or K_m		
macroscopic current	17–67 μM	11–30 μM
single channel	5–45 μM	31 μM
Cooperativity		
macroscopic current	1.7–3.0	2.4–3.1
single channel	2.9–3.0	2.3
l-cis-diltiazem	blocks	little effect
Unit conductance	25 pS	26 pS
sublevels	8–10 pS	<10 pS
Mean open time	1–3 ms	>10 ms
Ion selectivity		
monovalent ions	Na > K > Rb > Cs	Na > K > Rb > Cs
divalent ions	Ca, Mg, Sr, Ba	Ca
I-V-relation	rectification	rectification
Divalent cation block	yes	yes

Properties of the channel in situ and in the reconstituted system. Data in the left column have been compiled from Fesenko et al. (1985), Zimmerman et al. (1985), Stern et al. (1986), Zimmerman and Baylor (1986), and Haynes et al. (1986). Data of the reconstituted channel have been taken from Cook et al. (1986), Cook and Kaupp (1986), Cook et al. (1987), and Hanke et al. (1988).

lent cations are responsible for the rectification, probably a mechanism that involves the entering and blocking of the channel. In the absence of divalent cations on both sides of the membrane, the *I-V* relationship in amphibian rods becomes less rectifying (Yau et al. 1986; Stern et al. 1987).

The nonlinearity that persists in the absence of divalent cations could reflect an intrinsic property of the conductive pathway through the channel protein, i.e., the macroscopic current *I* through many channels simply represents a multiple of the current *i* through a single open channel at all voltages. Consequently, we would expect a dependence of the single-channel conductance on the transmembrane voltage. One way to test this hypothesis involves recording of the microscopic current *i* through a single open channel as a function of the voltage (Hanke et al. 1988; Fig. 4b). This *i-V* relation of the open channel is almost linear and cannot account for the rectification of the macroscopic current. Alternatively, a change in gating kinetics and, consequently, of the open probability P_o could cause a nonlinear *I-V* relation. In fact, P_o of the channel decreases with hyperpolarizing potentials (Fig. 4c), suggesting that a significant portion of the nonlinearity may be accounted for by the voltage dependence of the gating process. After the incorporation of many channel copies into the planar lipid bilayer, macroscopic cGMP-activated currents can be recorded that exhibit a significant degree of outward directed rectification (Fig. 4d), thus demonstrating that this feature of the cGMP-activated current in excised membrane patches is also preserved in the reconstituted system.

The results presented in Figure 4 were performed in the absence of Ca^{++} and Mg^{++} and therefore suggest that a significant portion of the nonlinearity in the *I-V* relation is independent of divalent cations. Although these results are qualitatively similar to those obtained with excised patches from amphibian rods, it is conceivable that intrinsic properties of the ion pathway differ in mammalian and amphibian photoreceptor cells.

SUMMARY

The light-sensitive channel in the surface membrane of vertebrate photoreceptors is gated directly and cooperatively by cGMP, and the activation mechanism does not involve phosphorylation by a cGMP-dependent protein kinase. The channel protein most likely is composed of several copies of a single type of polypeptide, which can be removed from photoreceptor membranes by detergents and functionally reincorporated into the membrane of liposomes or planar bilayers. Most channel properties are preserved in the reconstituted system and provide a unique system to study the mechanisms of activation, regulation, and ion permeation in more detail.

ACKNOWLEDGMENT

This work was supported by the Deutsche Forschungsgemeinschaft SFB 171, Projekt C 6 and C 11.

REFERENCES

Bader, C.R., P.R. MacLeish, and E.A. Schwartz. 1979. A voltage-clamp study of the light response in solitary rods of the tiger salamander. *J. Physiol.* **296:** 1.

Barnard, E.A., M.G. Darlison, and P. Seeburg. 1987. Molecular biology of the $GABA_A$ receptor: The receptor/channel superfamily. *Trends Neurosci.* **10:** 502.

Baylor, D.A. and B.J. Nunn. 1986. Electrical properties of the light-sensitive conductance of rods of the salamander *Ambystoma tigrinum*. *J. Physiol.* **371:** 115.

Betz, H. 1987. Biology and structure of the mammalian glycine receptor. *Trends Neurosci.* **10:** 113.

Bodoia, R.D. and P.B. Detwiler. 1985. Patch-clamp recordings of light-sensitive dark noise in retinal rods from the lizard and frog. *J. Physiol.* **367:** 183.

Caretta, A. 1985. Effect of cGMP and cations on the permeability of cattle retinal disks. *Eur. J. Biochem.* **148:** 599.

Caretta, A. and A. Cavaggioni. 1983. Fast ionic flux activated by cyclic GMP in the membrane of cattle rod outer segments. *Eur. J. Biochem.* **132:** 1.

Caretta, A., A. Cavaggioni, and R.T. Sorbi. 1985. Binding stoichiometry of a fluorescent cGMP analogue to membranes of retinal rod outer segments. *Eur. J. Biochem.* **153:** 49.

Cook, N.J. and U.B. Kaupp. 1986. The cGMP-dependent cation channel from vertebrate photoreceptors: Purification and functional reconstitution. *Photobiochem. Photobiophys.* **13:** 331.

Cook, N.J., W. Hanke, and U.B. Kaupp. 1987. Identification, purification and functional reconstitution of the cGMP-dependent channel from rod photoreceptors. *Proc. Natl. Acad. Sci.* **84:** 585.

Cook, N.J., C. Zeilinger, K.-W. Koch, and U.B. Kaupp. 1986. Solubilization and functional reconstitution of the cGMP-dependent cation channel from bovine rod outer segments. *J. Biol. Chem.* **261:** 17033.

Fazakerley, G.V., J.C. Russell, and M.A. Wolfe. 1977. Determination of the *syn-anti* equilibrium of some purine 3′:5′-nucleotides by nuclear-magnetic relaxation perturbation in the presence of a lanthanide-ion probe. *Eur. J. Biochem.* **76:** 601.

Fesenko, E.E., S.S. Kolesnikov, and A.L. Lyubarsky. 1985. Induction by cyclic GMP of cationic conductance in plasma membrane of retinal rod outer segment. *Nature* **313:** 310.

Gray, M.P. and D. Attwell. 1985. Kinetics of light-sensitive changes in vertebrate receptors. *Proc. R. Soc. Lond. B Biol. Sci.* **223:** 379.

Hanke, W., N.J. Cook, and U.B. Kaupp. 1988. cGMP-dependent channel protein from photoreceptor membranes: Single-channel activity of the purified and reconstituted protein. *Proc. Natl. Acad. Sci.* **85:** 94.

Haynes, L.W. and K.-W. Yau. 1985. Cyclic GMP-sensitive conductance in outer segment membranes of catfish cones. *Nature* **317:** 61.

Haynes, L.W., A.R. Kay, and K.-W. Yau. 1986. Single cyclic GMP-activated channel activity in excised patches of rod outer segment membrane. *Nature* **321:** 66.

Hille, B. 1984. *Ionic channels of excitable membranes.* Sinauer Associates, Sunderland, Massachusetts.

Hodgkin, A.L., P.A. McNaughton, and B.J. Nunn. 1985. The ionic selectivity and calcium dependence of the light-sensitive pathway in toad rods. *J. Physiol.* **350:** 649.

Kaupp, U.B. and K.-W. Koch. 1986. Mechanism of photoreception in vertebrate vision. *Trends Biochem. Sci.* **11:** 43.

Koch, K.-W. and U.B. Kaupp. 1985. Cyclic GMP directly regulates a cation conductance in membranes of bovine rods by a cooperative mechanism. *J. Biol. Chem.* **260:** 6788.

Koch, K.-W., N.J. Cook, and U.B. Kaupp. 1987. The cGMP-dependent channel of vertebrate rod photoreceptors exists in two forms of different cGMP-sensitivity and pharmacological behavior. *J. Biol. Chem.* **262:** 14415.

MacLeish, P.R., E.A. Schwartz, and M. Tachibana. 1984. Control of the generator current of solitary rods of the *Ambystoma tigrinum* retina. *J. Physiol.* **348:** 645.

Matthews, G. 1986. Comparison of the light-sensitive and cGMP-sensitive conductances of the rod photoreceptor: Noise characteristics. *J. Neurosci.* **6:** 2521.

———. 1987. Single-channel recordings demonstrate that cGMP opens the light-sensitive ion channel of the rod photoreceptor. *Proc. Natl. Acad. Sci.* **84:** 299.

Nicol, G.D., P.P.M. Schnetkamp, Y. Saimi, E.J. Cragoe, Jr., and M.D. Bownds. 1987. A derivative of amiloride blocks both the light-regulated and cyclic GMP-regulated conductances in rod photoreceptors. *J. Gen. Physiol.* **90:** 651.

Stern, J.H., U.B. Kaupp, and P.R. MacLeish. 1986. Control of the light-regulated current in rod photoreceptors by cyclic GMP, calcium, and *l*-cis-diltiazem. *Proc. Natl. Acad. Sci.* **83:** 1163.

Stern, J.H., H. Knutsson, and P.R. MacLeish. 1987. Divalent cations directly affect the conductance of excised patches of rod photoreceptor membrane. *Science* **236:** 1674.

Tavale, S.S. and H.M. Sobell. 1970. Crystal and molecular structure of 8-bromoguanosine and 8-bromoadenosine, two purine nucleosides in the *syn* conformation. *J. Mol. Biol.* **48:** 109.

Vandlen, R.L., W.C.-S. Wu, J.C. Eisenach, and M.A. Raftery. 1979. Studies of the composition of purified *Torpedo californica* acetylcholine receptor and of its subunits. *Biochemistry* **18:** 1845.

Wong, S. and R.S. Molday. 1986. A spectrin-like protein in retinal rod outer segments. *Biochemistry* **25:** 6294.

Yau, K.-W and K. Nakatani. 1984. Cation selectivity of the light-sensitive conductance in retinal rods. *Nature* **309:** 352.

———. 1985. Light-suppressible, cGMP-sensitive conductance in the plasma membrane of a truncated rod outer segment. *Nature* **317:** 252.

Yau, K.-W., L.W. Haynes, and K. Nakatani. 1986. Roles of calcium and cyclic GMP in visual transduction. In *Membrane control of cellular activity* (ed. H.C. Lüttgau) p. 343. Gustav Fischer, Stuttgart.

Zimmerman, A.L. and D.A. Baylor. 1985. Electrical properties of the light-sensitive conductance of salamander retinal rods. *Biophys J.* **47:** 357a.

———. 1986. Cyclic GMP-sensitive conductance of retinal rods consists of aqueous pores. *Nature* **321:** 70.

Zimmerman, A.L., G. Yamanaka, F. Eckstein, D.A. Baylor, and L. Stryer. 1985. Interaction of hydrolysis-resistant analogs of cyclic GMP with the phosphodiesterase and light-sensitive channel of retinal rod outer segments. *Proc. Natl. Acad. Sci.* **82:** 8813.

Sevenless, a Gene Encoding a Putative Receptor for Positional Information

G.M. Rubin
Howard Hughes Medical Institute and Department of Biochemistry, University of California, Berkeley, California 94720

Development of specialized tissues in multicellular organisms requires the differentiation of distinct cell types and their assembly into integrated structures. The fate of a cell may be determined solely by lineage or may also require the reception and interpretation of positional information. A molecular description of the processes that allow cells to detect and respond to positional information is crucial to an understanding of cellular determination and differentiation.

The eye of *Drosophila melanogaster* is well suited to the study of cellular determination in response to positional information. The adult eye is an array of several hundred identical subunits, or ommatidia, each of which is a precise assembly of a few distinct cell types (Dietrich 1909). Extensive mosaic analyses have failed to detect any lineage relationships between cell types in the developing retina (Lawrence and Green 1979; Ready et al. 1976). Rather, progressive recruitment of undetermined cells in response to positional cues appears to be the primary mechanism of ommatidial assembly (Ready et al. 1976; Tomlinson 1985; Tomlinson and Ready 1987a). Aspects of this process can be readily observed in the eye imaginal disk of third instar larvae. Organization of the initally unpatterned epithelium of the disk is associated with the progression of the morphogenetic furrow anteriorly across the disk (Ready et al. 1976). Histological and immunochemical studies have demonstrated the progressive and stereotyped recruitment of cells posterior to the furrow into the cell clusters that will give rise to the adult ommatidia (Ready et al. 1976; Tomlinson 1985; Tomlinson and Ready 1987a). Each ommatidium is about 2 hours more mature than its anterior neighbor, so that by late third instar, over 24 hours of a smoothly graded sequence of development can be observed in a single disk preparation.

Analysis of mutants defective in ommatidial development provides a means of investigating the acquisition of cellular identity within the developing eye disk. The mutation *sevenless* causes the complete absence of the UV photoreceptor R7 from each ommatidium, with no other obvious aberration (Harris et al. 1976). Histological studies of eye disks from *sevenless* larvae have shown that a cell initially occupies the developmental position of the R7 cell but fails to differentiate as a photoreceptor, becoming a lens-secreting cone cell instead (Tomlinson and Ready 1986, 1987b). Analyses of ommatidia mosaic for *sevenless* have shown that *sevenless* is cell autonomous and that only the R7 cell appears to require *sevenless* function for its normal development (Harris et al. 1976; Campos-Ortega et al. 1979; Tomlinson and Ready 1987b). These studies suggest that *sevenless* is involved either in reading the positional information that specifies the R7 cell or in the subsequent differentiation of this cell type.

Isolation of the *sevenless* Gene

The *sevenless* gene has been cytogenetically mapped to the 10A1,2 polytene band on the X chromosome (Harris et al. 1976; Zhimulev et al. 1981). We (Hafen et al. 1987) used the microdissection cloning method (Scalenghe et al. 1981) to obtain chromosomal DNA from the vicinity of the *sevenless* gene. Starting from a DNA segment obtained by microdissection cloning that mapped to the interband region between 10A1,2 and 10B1,2, overlapping clones were isolated from genomic DNA libraries constructed in bacteriophage λ and cosmid vectors. Approximately 110 kb of chromosomal DNA derived from the interband between 10A1,2 and 10B1,2 and the proximal (with respect to the centromere) edge of 10A1,2 band were cloned. Two overlapping chromosomal deletions define the position of the gene (Zhimulev et al. 1981). The deletion *Df(1)v*^{64f29} removes part of the 10A1,2 band but does not affect *sevenless* gene function. However, the deletion *Df(1)rasv*^{17Cc8}, which removes slightly more of the 10A1,2 band, results in a *sev*$^{-}$ phenotype. The genomic region located between these two breakpoints is approximately 7 kb in length and must contain at least part of the *sevenless* gene. The *sevenless* gene has also been isolated by Banerjee et al. (1987a).

Rescue of the *sevenless* Phenotype by Germ-line Transformation

To determine the extent of the *sevenless* gene within the cloned region, we used P-element-mediated germ-line transformation to test the ability of a 16.3-kb genomic *Eco*RI DNA fragment to complement the *sevenless* mutant phenotype. This fragment includes the region between the deletion breakpoints, extends proximally approximately 8 kb beyond the *Df(1)rasv*^{17Cc8} breakpoint, and includes about 950 bp of the promoter region of this gene (see Fig. 1; Bowtell et al. 1988). We have generated a total of nine transformant lines using

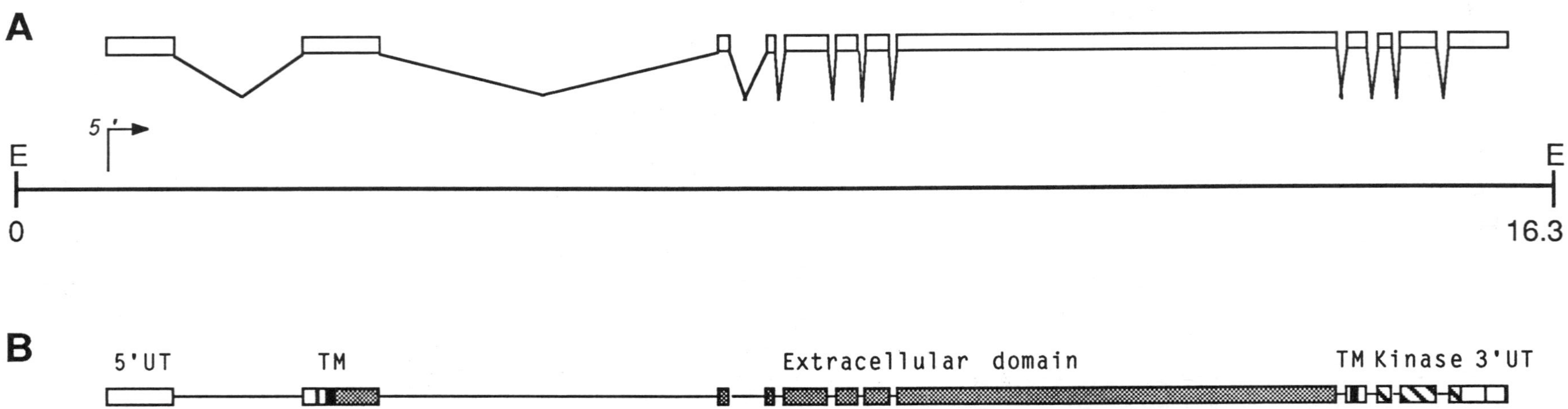

Figure 1. (*A*) Intron-exon structure of the *sevenless* gene deduced from the sequence of 11 cDNA fragments and most of the 16.3-kb genomic *Eco*RI (E) fragment. (*B*) Schematic representation of the various domains of the *sevenless* protein. (TM) Putative amino- and carboxy-terminal transmembrane domains; (UT) 5′- and 3′-untranslated regions; (kinase) putative tyrosine kinase domain.

this fragment (Hafen et al. 1987; Bowtell et al. 1988) and assayed *sevenless* function by three criteria (Bowtell et al. 1988). (1) The presence or absence of the R7 cell was judged by examining the pseudopupils of transformant flies (Fig. 2). (The pseudopupil image is formed by transmission of light through the rhabdomeres of the individual photoreceptors.) (2) R7 cell function and neuronal connection in the optic lobes was assayed in a behavioral test involving the response of the adult flies to UV light. All lines had apparently normal pseudopupils, and their behavioral preference for UV light was indistinguishable from that of wild-type flies. (3) The pattern of *sevenless* protein expression in these transformant lines was examined by staining eye imaginal disks with an antibody to the *sevenless* protein (Tomlinson et al. 1987; see below) and comparing the pattern to that of control wild-type flies. The pattern of *sevenless* expression in the transformant lines tested was the same as that in wild-type larvae. These experiments suggest that the 16.3-kb *Eco*RI genomic fragment encompasses the complete regulatory and transcriptional regions of the gene.

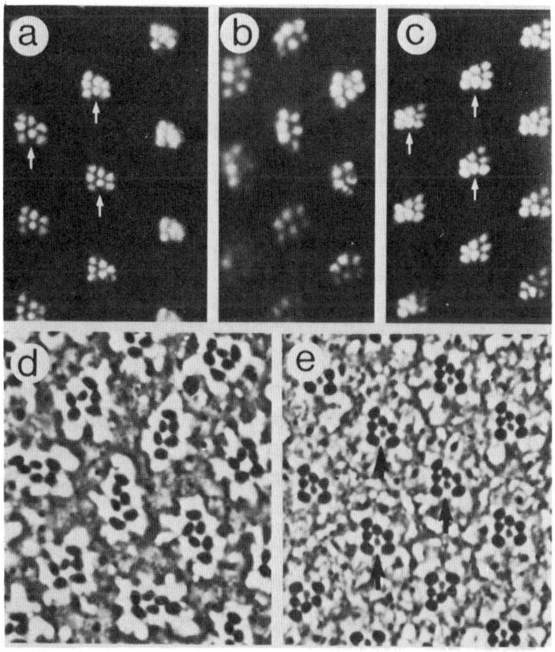

Figure 2. Rescue of the *sevenless* mutant phenotype by germline transformation. (*a–c*) The rhabdomere pattern of a wild-type fly, a *sev* fly, and a *sev* fly that also carries the 16.3-kb *Eco*RI fragment in a P transposon, as seen by the optical neutralization technique (Franceschini 1975). White arrows indicate the R7 cell rhabdomere that is present in wild-type (*a*) and transformant (*c*), but not *sev* (*b*) flies. The different arrangement of the trapezoids in *a* and *c* is due to the fact that the photographs have been taken from different hemispheres of the eye, which exhibit a mirror image symmetry and is not an effect of *sev*. (*b*) The mutant *sev* ommatidium at the top right shows seven instead of six rhabdomeres. Because of the vacant position of R7, the R8 cell can occasionally move up and occupy the R7 position (Campos-Ortega et al. 1979). (*d* and *e*) Radial plastic sections through the distal portion of the retina of a *sev* fly and a *sev*; P[sev] transformant fly, respectively. The arrows in *e* point to the rhabdomere of the R7 cell; no R7 cell rhabdomeres are seen in the *sev* eye (*d*). The line of mirror image symmetry can be seen passing through the bottom third of *e*. (Reprinted, with permission, from Hafen et al. 1987.)

Structure and Expression of the *sevenless* Gene

The structure of the transcribed region was determined by the isolation and sequencing of 11 overlapping cDNA fragments and the sequencing of the majority of the corresponding 16.3-kb genomic DNA fragment. The structure of the transcription unit deduced from these data is given in Figure 1A. The transcript of the *sevenless* locus is approximately 8.6 kb in length and can be detected in adult heads and bodies, as well as in the eye imaginal disk. One major and two minor transcription start sites were mapped by primer extension and RNase protection (Bowtell et al. 1988). These sites lie in a 100-bp region. Surrounding the transcriptional start sites are several regions with repeated sequence motifs. Repeated sequences have been found near the transcription start sites of a number of promoters, and in several cases, such repeated elements have been shown to bind factors that regulate transcription (for review, see McKnight and Tjian 1986). The role of the alternative *sevenless* start sites is unclear; there is no apparent tissue specificity to their use and no methionine codons appear between them and the major site that could represent alternate short translated sequences.

To determine the spatial distribution of *sevenless* transcripts, we (Hafen et al. 1987) examined the accumulation of the *sevenless* RNA by in situ hybridization to tissue sections of third instar larvae, using a tritium-labeled genomic restriction fragment that is homologous to the 8.6-kb *sevenless* transcript. Specific accumulation of *sevenless* transcripts was observed over the bilaterally paired eye imaginal disks. No other site of hybridization was detected in the larva. Because of the ordered arrangements of the developing photoreceptor cell clusters in the third instar eye imaginal disk, the different states of photoreceptor development are displayed spatially along the anterior-to-posterior axis of the disk. A horizontal section through an eye disk is shown in Figure 3. The silver grains were not distributed uniformly over the entire eye disk epithelium but were localized apically over the posterior two thirds of the disk epithelium. Cells in the anterior third of the disk did not contain significant levels of *sevenless* transcripts. The arrow in Figure 3a indicates the position of the morphogenetic furrow. Cells anterior to the morphogenetic furrow are undifferentiated and randomly organized, whereas cells posterior to the furrow become selected to different developmental pathways and become organized into ommatidial clusters. We detected *sevenless* transcripts around and posterior to the morphogenetic furrow. Therefore, it appears that the *sevenless* gene expression begins as soon as the developmental program for ommatidial assembly

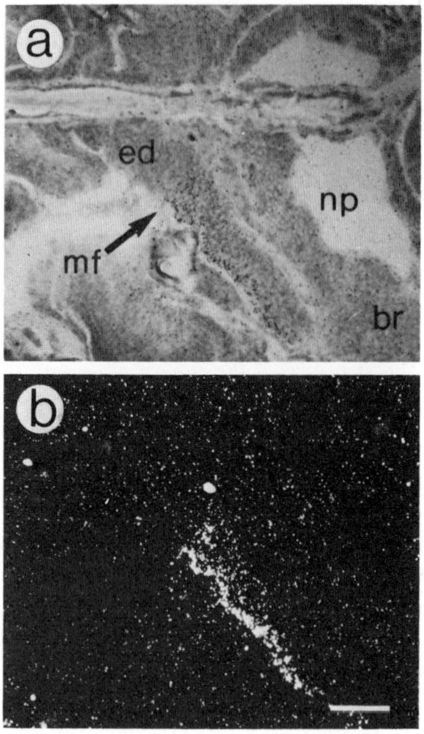

Figure 3. Spatial distribution of *sev* transcripts in third instar larvae. Frozen tissue sections of wild-type third instar larvae were hybridized with a 2.4-kb genomic *Hin*dIII restriction fragment labeled by nick translation. This probe is homologous to the 8.6-kb *sev* RNA. (*a* and *b*) Photomicrographs in bright- and dark-field illumination, respectively, of a horizontal section of an eye disk. The different developmental stages of ommatidial development are displayed spatially along the posterior-anterior axis. Arrow indicates the position of the morphogenetic furrow. Specific hybridization is observed over a subpopulation of cells in the apical region of the disk epithelium immediately posterior to the morphogenetic furrow. (br) Brain; (ed) eye imaginal disk; (mf) morphogenetic furrow; (np) neuropile. The horizontal bar in *b* represents 100 μm. Anterior is to the left. (Adapted from Hafen et al. 1987.)

is initiated. The resolution of the in situ hybridization technique is not sufficient to determine precisely which ommatidial precursor cells express the *sevenless* gene. However, this information could be obtained by electron microscopic immunocytochemistry using antibodies to the *sevenless* protein (see below).

Primary Structure of the Deduced *sevenless* Protein

Translation of the compiled cDNA sequence (Bowtell et al. 1988) revealed a long open reading frame of 7677 nucleotides (see Fig. 1B), extending to and encompassing a carboxy-terminal domain that has a high degree of homology with several protein tyrosine kinase receptors (Hafen et al. 1987). Examination of a hydrophobicity profile for the entire open reading frame revealed two distinct hydrophobic regions, one about 500 residues from the carboxyl terminus, representing the putative transmembrane region adjacent to the kinase domain, and a second region near the amino terminus. The average hydropathy of these sequences is +2.5 and +2.6, respectively; sequences with an average hydropathy of greater than +1.6 are considered to have a high probability of membrane insertion (Kyte and Doolittle 1982). The open reading frame extends for 106 amino acids upstream of this hydrophobic sequence and includes three, in-frame, methionine codons. However, the closest is 56 amino acids from the hydrophobic domain. We favor this residue as the most likely initiation codon, as the sequence surrounding this codon corresponds most closely to the *Drosophila* consensus sequence for translation initiation (Cavener 1987).

The amino-terminal hydrophobic sequence conforms to the comparison of amino-terminal anchor sequences (von Heijne 1986) by having a relatively long hydrophobic core compared with cleaved signal peptides and a positively charged residue at the amino-terminal end of the sequence. Assuming translation initiated at the third methionine codon, the amino-terminal region would be of similar length to other proteins anchored at their amino terminus. An amino-terminal anchor could result in the formation of a large loop in the extracellular domain, attached at the amino terminus and at the transmembrane domain adjacent to the kinase domain, with a short amino-terminal sequence and carboxy-terminal kinase domain on the cytoplasmic face of the cell.

The putative 56-amino-acid amino-terminal cytoplasmic sequence of the *sevenless* protein includes 12 serine and 4 threonine residues, which could act as sites of phosphorylation, perhaps modulating receptor function in a manner similar to that achieved by the phosphorylation of the EGF receptor by protein kinase C (Hunter et al. 1984). Although the amino-terminal hydrophobic sequence is suggestive of a transmembrane anchor, we cannot exclude the possibility that this sequence is cleaved, in which case we would expect the *sevenless* protein to adopt a conformation in the membrane similar to that of other tyrosine kinase receptors. We are currently investigating the topology of *sevenless* in the membranes of tissue-culture cells that express the protein under the control of heterologous promoters.

The unmodified protein is approximately 288 kD, considerably larger than other tyrosine kinase receptors. This predicted size is in good agreement with published estimates of the protein detected in eye disks and larval brains (Banerjee et al. 1987b) and in Schneider L2 cells expressing *sevenless* protein (M. Simon and D. Bowtell, unpubl.). The proposed extracellular domain contains 17 potential sites for amino-linked glycosylation. The subunit structure of the protein has not been determined. However, the presence of an additional 60-kD protein, specifically reactive with *sevenless* antisera on protein blots (Banerjee et al. 1987b; M. Simon and D. Bowtell, unpubl.), is consistent with *sevenless* being synthesized as a single polypeptide chain and then cleaved. If so, a stretch of arginine residues near the transmembrane domain (residues 2043-2051; Bowtell et al. 1988) may function as

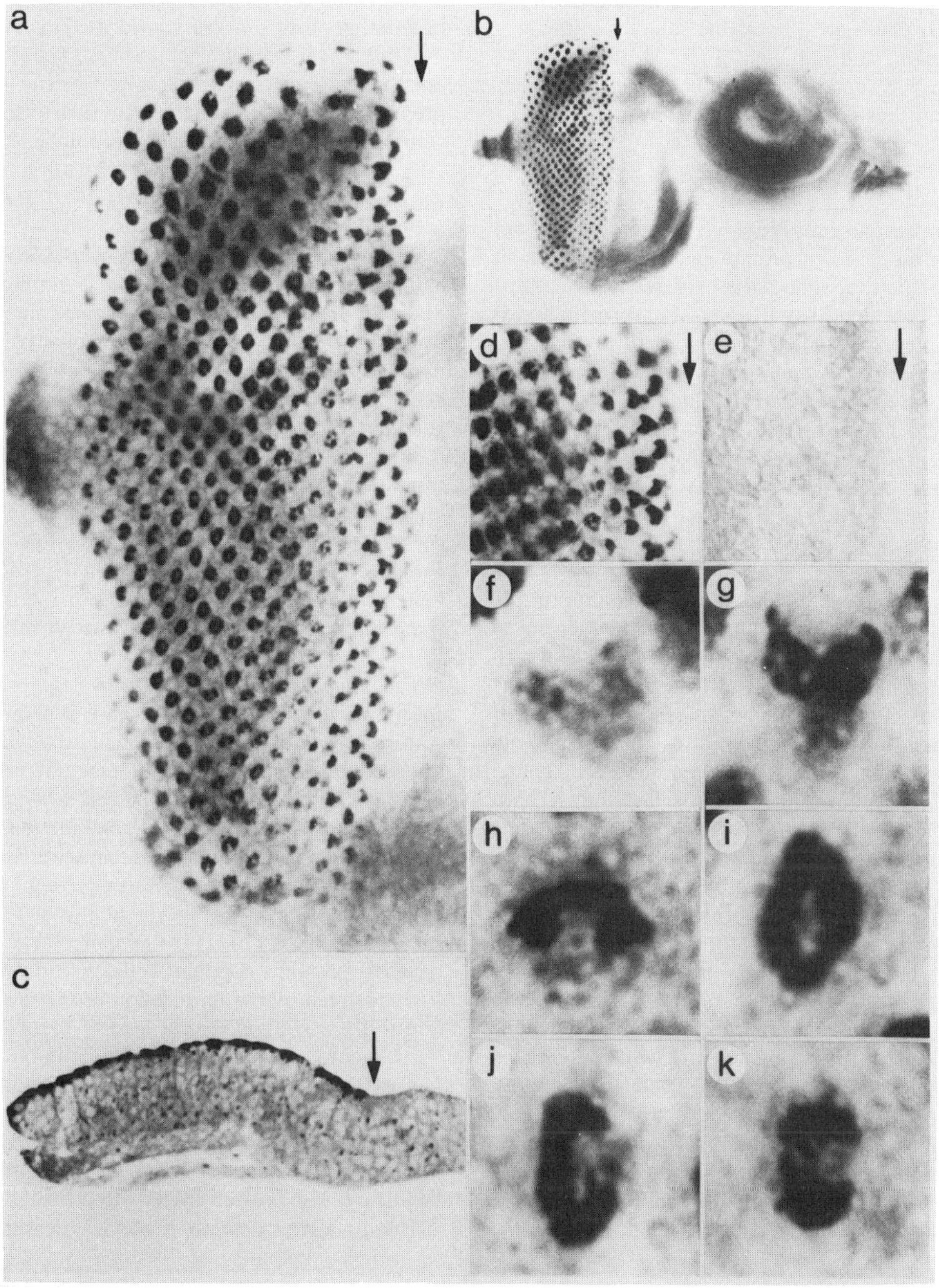

Figure 4. Light micrographs of eye imaginal disks from third instar larvae stained with *sevenless* antisera. Peripodial membranes of disks have been removed. Arrows mark the center of the morphogenetic furrow. All except *d* are stained with the antiserum raised against the carboxy-terminal peptide of *sevenless*. (*a–e*) Anterior is to the right. (*a* and *b*) Antibody staining of disk whole mounts; magnifications, 510× and 128×, respectively. Staining begins in the morphogenetic furrow and changes progressively toward the posterior of the disk. (*c*) Longitudinal section through a stained disk showing the onset of staining in the morphogenetic furrow and its apical position within the disk. The stain is in the region of the cells' microvilli, which project up into the lumen of the disk between the developing epithelium and the peripodial membrane, which has been removed; magnification, 446×. (*d*) Staining of a whole mount disk with an antiserum raised to a second *sevenless* peptide (see Tomlinson et al. 1987). An identical staining pattern is seen; magnification, 510×. (*e*) No detectable signal is seen in sev^{p3} control disks; magnification, 510×. (*f–k*) High power images (340×) of the progressive changes in staining pattern that occur. During disk development, the ommatidia undergo a 90° rotation. Initially, they are oriented directly posterior in the furrow, but they point toward the equator at the back of the disk (see Tomlinson and Ready 1987a). The shapes are shown in a standardized orientation, as though the rotation was complete, with anterior to the right. (*f*) The first staining seen is a diffuse butterfly shape. (*g*) Part of the butterfly darkens considerably. (*h*) The butterfly staining declines, and a horseshoe shape emerges. (*i*) The open end of the horseshoe is now filled, producing a complete ring of stain. (*j*) Reduction of stain in the anterior side of the circle produces another horseshoe shape. (*k*) Reduction in the posterior side now leaves two separated dots of stain. (Reprinted, with permission, from Tomlinson et al. 1987.)

the endopeptidase cleavage site, generating a subunit of 57.5 kD. There are 43 cysteine residues in the predicted protein, and although 10 are clustered in one region near the amino terminus, there does not appear to be an obvious cysteine-rich domain.

Expression of the *sevenless* Protein

We (Tomlinson et al. 1987) have generated antisera specific for the *sevenless* protein and have used them to examine the pattern of *sevenless* expression in wild-type eye disks. We find that the *sevenless* protein is expressed in each developing ommatidium prior to differentiation of the R7 cell and is found transiently at high levels in at least nine cells. Light micrographs of eye imaginal disks from third instar larvae stained with *sevenless* antisera are shown in Figure 4. To establish the identities of the cells expressing the *sevenless* protein, immunocytochemistry was performed using electron microscopy. The results of this analysis are summarized in Figure 5.

Within all cells expressing the *sevenless* antigen, the heaviest staining occurred in the microvilli at the apical surface of the epithelium, which lies below an extracellular matrix. Staining in the microvilli accounted for the shapes seen with light microscopy (Fig. 4f-k). Deeper into the cell, staining was found in large membrane-bound vesicle-like structures (an example is seen in Fig. 6b), which were found in all cells in which the *sevenless* protein was detected. A striking staining of plasma membranes occurred at the level of the adherens-type junctions in the first few microns of the tissue below the microvilli. Here, we saw an accumulation of stain within R3, R4, and R7 precisely at the position where they oppose R8 but not where they abut other cells (Fig. 6) or below the level of the adherens junctions.

Expression of *sevenless* protein is first detected in the morphogenetic furrow prior to the cell division that generates the presumptive R7 cell. The onset of expression in R3, R4, and other cells of the ommatidium closely follows the sequence of, but precedes, the maturation of these cells, as revealed by their expression of neural antigens (Tomlinson and Ready 1987a). In particular, *sevenless* staining is seen at high levels in the presumptive R7 cell some 8 hours before any overt differentiation of this cell is evident. However, unlike the staining revealed with antibodies to neuronal markers, the expression of *sevenless* is transient, and only a proportion of photoreceptors express detectable levels, with R2, R5, and R8 being the notable exceptions.

CONCLUSION

We have described the structure of the *sevenless* gene and its predicted protein product. The promoter region of the *sevenless* gene contains one major and two minor transcription start sites that are surrounded by a large number of repeated sequences. A number of lines were generated by P-element transformation with a 16.3-kb genomic fragment that included only about 950 bp of the sequences 5′ to the transcriptional start sites. Transformation with this fragment consistently rescued *sevenless* gene function, as determined by morphological and behavioral criteria, and was sufficient for the complex pattern of *sevenless* protein expression in eye imaginal disks. A transcription unit of 12 exons, which spans the majority of the genomic fragment, was identified.

The appearance of *sevenless* protein prior to the specification of R7 and its presence in many cells are consistent with a role for *sevenless* in receiving a signal required to induce the R7 developmental pathway. We believe that the identity of each cell in the developing ommatidium is specified by a combination of signals received from its neighbors and that *sevenless* mediates the reception of one of the signals required for R7 determination. It is unlikely that this signal alone is sufficient to specify the R7 developmental pathway; the *sevenless* protein's apparent activation in cells R3 and R4, suggested by the localization of the *sevenless* protein in these cells, does not lead to the R7 pathway. Furthermore, our data suggest that limitation of activation of *sevenless* protein to particular cells is accomplished by a combination of restricted temporal and spatial expression of the *sevenless* protein and a restricted distribution of its ligand. Our data suggest that this ligand is expressed by the R8 cell.

The putative *sevenless* protein differs from most

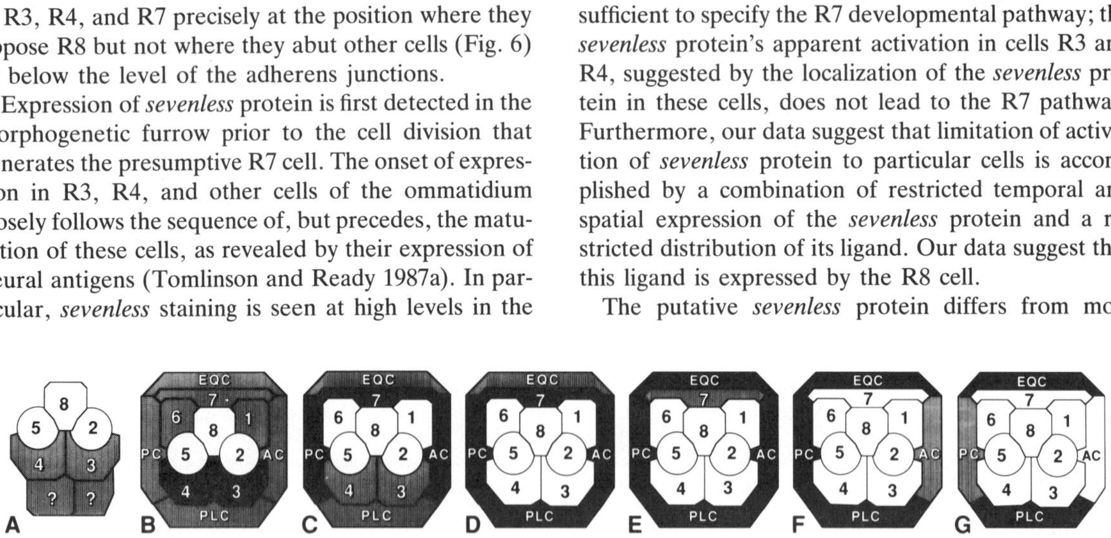

Figure 5. Schematic representation of the sequence of cells staining with antisera directed against the *sevenless* protein. Photoreceptors are numbered. (AC) Anterior cone cell; (PC) posterior cone cell; (EQC) equatorial cone cell; (PLC) polar cone cell; (?) cells present in the precluster but lost shortly thereafter (see Tomlinson et al. 1987). Three levels of staining are shown: white, which signifies a level indistinguishable from background; gray, which represents a low but definite level to a moderately high level; and black, which shows the cells heavily staining with the antibody. (*A*) Precluster stage at which the R3, R4, and ? cells stain. Following the precluster stage, cell divisions occur, generating all other cells of the ommatidium. (*B*) Immature eight-cell stage. (*C–E*) Progression through the symmetrical cluster stage to the two-cone-cell stage (*F* and *G*) Four-cone-cell stage. Cells are shown in the position they come to occupy, rather than their position at that developmental point. For example, EQC and PLC cannot be identified at stage *B*, but they are derived from cells surrounding the cluster, all of which show a low level of stain at this stage. (Adapted from Tomlinson et al. 1987.)

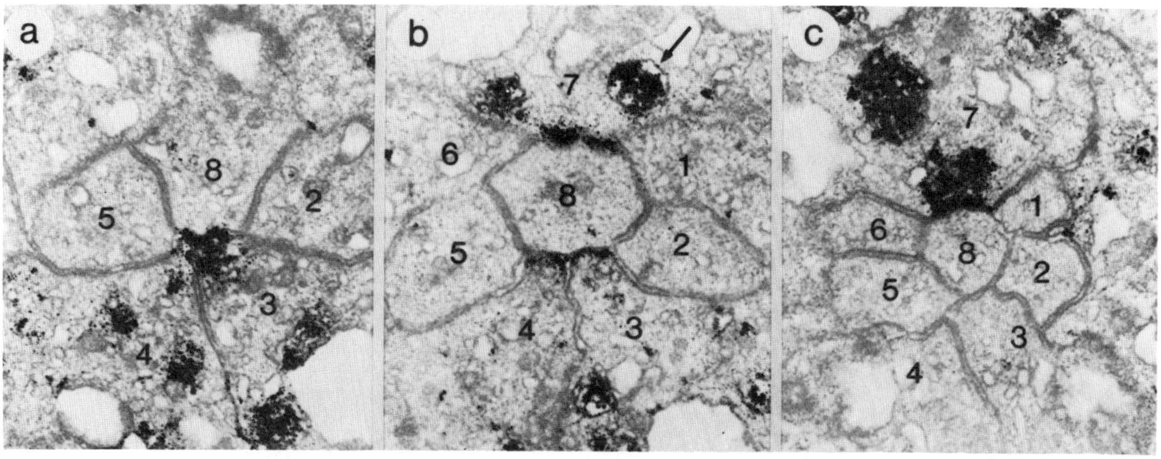

Figure 6. Intracellular localization of *sevenless* antibody. Magnification, 12,460×. (*a–c*) Cross sections through developing ommatidia at the level of the adherens-type junctions. Anterior is to the right. Photoreceptors are numbered. (*a*) At the immature eight-cell cluster stage, the staining is found on the membranes of R3 and R4 only where they contact R8. Note that intense staining also occurs in the cytoplasm of R3, R4, and R7 (see *b–c*) directly adjacent to where membrane staining is observed. (*b*) At the early symmetrical cluster stage, staining occurs simultaneously in R7, R3, and R4 and is localized in each of these cells at their points of contact with R8. Arrow indicates a membrane-bound vesicular structure that is strongly stained. Such structures are found in all cells in which expression is detected and account for the vast majority of intracellular staining. (*c*) Late symmetrical cluster stage in which staining in R3 and R4 has now reduced considerably and the stain in R7 where it contacts R8 has become more intense. (Adapted from Tomlinson et al. 1987.)

other tyrosine kinase cell-surface receptors in the unusually large size of its extracellular domain (~2000 amino acids) and the possible presence of two transmembrane domains in the mature protein. These structural features, as well as the limited pattern of *sevenless* expression, may relate to its function as a receptor for local positional information. The precise nature of this relationship awaits further study.

ACKNOWLEDGMENTS

I am grateful to my colleagues D. Bowtell, E. Hafen, M. Simon, and A. Tomlinson, whose results are summarized in this review.

REFERENCES

Banerjee, U., P.J. Renfranz, J.A. Pollock, and S. Benzer. 1987a. Molecular characterization and expression of *sevenless*, a gene involved in neuronal pattern formation in the *Drosophila* eye. *Cell* **49:** 281.

Banerjee, U., P.J. Renfranz, D.R. Hilton, B.A. Rabin, and S. Benzer. 1987b. The *sevenless*$^+$ protein is expressed apically in cell membranes of developing *Drosophila* retina: It is not restricted to cell R7. *Cell* **51:** 151.

Bowtell, D.D.L., M.A. Simon, and G.M. Rubin. 1988. Nucleotide sequence and structure of the *sevenless* gene of *Drosophila melanogaster*. *Genes Dev.* **2:** 620.

Campos-Ortega, J.A., G. Jürgens, and A. Hofbauer. 1979. Cell clones and pattern formation: Studies on *sevenless*, a mutant of *Drosophila melanogaster*. *Wilhelm Roux's Arch. Entwicklungsmech. Org.* **186:** 27.

Cavener, D.R. 1987. Comparison of the consensus sequence flanking translational start sites in *Drosophila* and vertebrates. *Nucleic Acids Res.* **15:** 1353.

Dietrich, W. 1909. Die Facettenaugen der Dipteren. *Z. Wiss. Zool.* **92:** 465.

Franceschini, N. 1975. Sampling of the visual environment by the compound eye of the fly: Fundamentals and applications. In *Photoreceptor optics* (ed. A.W. Snyder and R. Menzel), p. 98. Springer Verlag, Berlin.

Hafen, E., K. Basler, J.E. Edstroem, and G.M. Rubin. 1987. *sevenless*, A cell-specific homeotic gene of *Drosophila*, encodes a putative transmembrane receptor with a tyrosine kinase domain. *Science* **236:** 55.

Harris, W.A., W.S. Stark, and J.A. Walker. 1976. Genetic dissection of the photoreceptor system in the compound eye of *Drosophila melanogaster*. *J. Physiol.* **256:**415.

Hunter, T., N. Ling, and J.A. Cooper. 1984. Protein kinase C phosphorylation of the EGF receptor at a threonine residue close to the cytoplasmic face of the plasma membrane. *Nature* **311:**480.

Kyte, J. and R.F. Doolittle. 1982. A simple method for displaying the hydropathic character of protein. *J. Mol. Biol.* **157:**105.

Lawrence, R.A. and S.M. Green. 1979. Cell lineage in the developing retina of *Drosophila*. *Dev. Biol.* **71:**142.

McKnight, S. and R.Tjian. 1986. Transcriptional selectivity of viral genes in mammalian cells. *Cell* **46:** 795.

Ready, D.F., T.E. Hanson, and S. Benzer. 1976. Development of the *Drosophila* retina, a neurocrystalline lattice. *Dev. Biol.* **53:** 217.

Scalenghe, F., E. Turco, J.-E. Edstroem, V. Pirrotta, and M. Melli. 1981. Microdissection and cloning of DNA from a specific region of *Drosophila melanogaster* polytene chromosomes. *Chromosoma* **82:** 205.

Tomlinson, A. 1985. The cellular dynamics of pattern formation in the eye of *Drosophila*. *J. Embryol. Exp. Morphol.* **89:** 313.

Tomlinson, A. and D.F. Ready. 1986. *sevenless*: A cell specific homeotic mutation of the *Drosophila* eye. *Science* **231:** 400.

———. 1987a. Cell fate in the *Drosophila* ommatidium. *Dev. Biol* **123:** 264.

———. 1987b. Neuronal differentiation in the *Drosophila* ommatidium. *Dev. Biol.* **120:** 366.

Tomlinson, A., D.D.L. Bowtell, E. Hafen, and G.M. Rubin. 1987. Localization of the *sevenless* protein, a putative receptor for positional information, in the eye imaginal disc of *Drosophila*. *Cell* **51:** 143.

von Heijne, G. 1986. Towards a comparative anatomy of N-terminal topogenic protein sequences. *J. Mol. Biol.* **189:** 239.

Zhimulev, I.F., G.V. Pokholkova, A.V. Bgatov, V.F. Semeshin, and E.S. Belyaeva. 1981. Fine cytogenetical analysis of the band 10A1-2 and the adjoining regions in the *Drosophila melanogaster* X chromosome. *Chromosoma* **82:** 25.

The Role of Arachidonic Acid Metabolites in Signal Transduction in an Identified Neural Network Mediating Presynaptic Inhibition in *Aplysia*

E. Shapiro, D. Piomelli,* S. Feinmark,*† S. Vogel, G. Chin, and J.H. Schwartz
*Howard Hughes Medical Institute, Center for Neurobiology and Behavior, and Departments of *Pharmacology and †Medicine, Columbia University, New York, New York 10032*

Neuromodulation is a form of signal transduction that results in the biochemical control of neuronal excitability. Many neurotransmitters act through second messengers, and the examination of biochemical cascades initiated by neurotransmitter-receptor interaction has advanced the understanding of how information is acquired and stored in the nervous system (Kaczmarek and Levitan 1987). For example, as described by Sweatt et al. (this volume), 5-HT and other facilitory transmitters increase cAMP in sensory neurons of *Aplysia*, which enhances excitability and facilitates transmitter output. cAMP promotes protein phosphorylation, which closes K_s, a specific K^+ channel, thereby delaying membrane repolarization (Klein and Kandel 1980; Siegelbaum et al. 1982). Within sensory neurons, a novel second-messenger cascade using metabolites of arachidonic acid and activated by the peptide transmitter FMRF-amide has two actions on the same K^+ channel that is modulated by 5-HT to produce facilitation: (1) FMRF-amide through the arachidonic acid cascade increases the probability of channel opening (Piomelli et al. 1987b); (2) in the presence of 5-HT (or cAMP), FMRF-amide overrides the facilitation by reopening the channels closed by 5-HT (or cAMP), as well as by increasing channel opening. Sweatt et al. (this volume) suggest that this twofold inhibitory action might be produced by inhibiting the cAMP-dependent protein kinase or by stimulating a protein phosphatase.

It is important to test a new signal transduction mechanism in a variety of identified neurons, especially where it can be evoked by different transmitters and can influence different ion channels. Circuits involving histamine seem to be ideal to test whether the action of arachidonic acid metabolites is the same in every neuron, and specifically to test whether or not it operates by a mechanism involving protein dephosphorylation as proposed by Sweatt et al. (this volume). Neurons of the L32 circuit (Fig. 1) are well suited for this goal because in some histaminoceptive neurons in which the arachidonic acid cascade operates, e.g., L10, protein phosphorylation is not known to play any role, whereas the situation in others, e.g., L14 cells, is quite similar to that in sensory cells.

It is also important to examine how a novel second-messenger cascade can affect the function of neuronal circuits. An increasing number of circuits are being described in which a modulatory transmitter alters the output of a network of neurons by affecting the excitability of many target neurons in the circuit coordinately (Willard 1981; Harris-Warrick and Cohen 1985; Harris-Warrick and Flamm 1986; Dickinson et al. 1988; Frost et al. 1988).

We have examined the role of arachidonic acid metabolites in a neuronal circuit mediating presynaptic inhibition. First described by Byrne (1980a,b), L32 cells are a cluster of putative histaminergic neurons that each make dual-action synaptic potentials onto two follower neurons, L10 and L14 (Fig. 1A,B). The synaptic connections, biophysical properties, and roles in behavior of the L10 and L14 follower cells have been well studied (for L10, see Koester et al. 1974; Shapiro et al. 1980a,b; Connor et al. 1986; Koester and Koch 1987) (for L14, see Carew and Kandel 1977; Byrne et al. 1979; Shapiro et al. 1979; Byrne 1980c; Walsh and Byrne 1985). The types of ion channels causing each component of the L32-L10 and L32-L14 dual actions have been characterized (Byrne 1980a; Kretz et al. 1986a), and application of histamine mimics the effects of stimulating L32 in both L10 and L14 (Kretz et al. 1986b). L32 and histamine cause presynaptic inhibition of transmitter output from L10 by decreasing a Ca^{++} current and increasing a K^+ current, whose time courses overlap (Kretz et al. 1986a,b). The L32-L14 dual-action potential results from an early increase in a nonspecific cation current, which produces the fast excitatory component, and a slow increase in K^+ conductance, which gives rise to the late inhibitory phase (Byrne 1980a). In the L14 branch of this circuit, as in sensory neurons, metabolites of arachidonic acid oppose the actions of cAMP (for a description of the actions of serotonin and cAMP in L14, see Walsh and Byrne 1985). In the L10 branch of the circuit, however, cAMP-dependent protein phosphorylation has not been implicated in a synaptic response.

METHODS

Experiments were performed at room temperature on isolated desheathed abdominal ganglia of *Aplysia californica* pinned ventral side up in a chamber (5-ml vol) constantly perfused with seawater (460 mM NaCl, 10 mM KCl, 55 mM $MgCl_2$, 11 mM $CaCl_2$, 30 mM

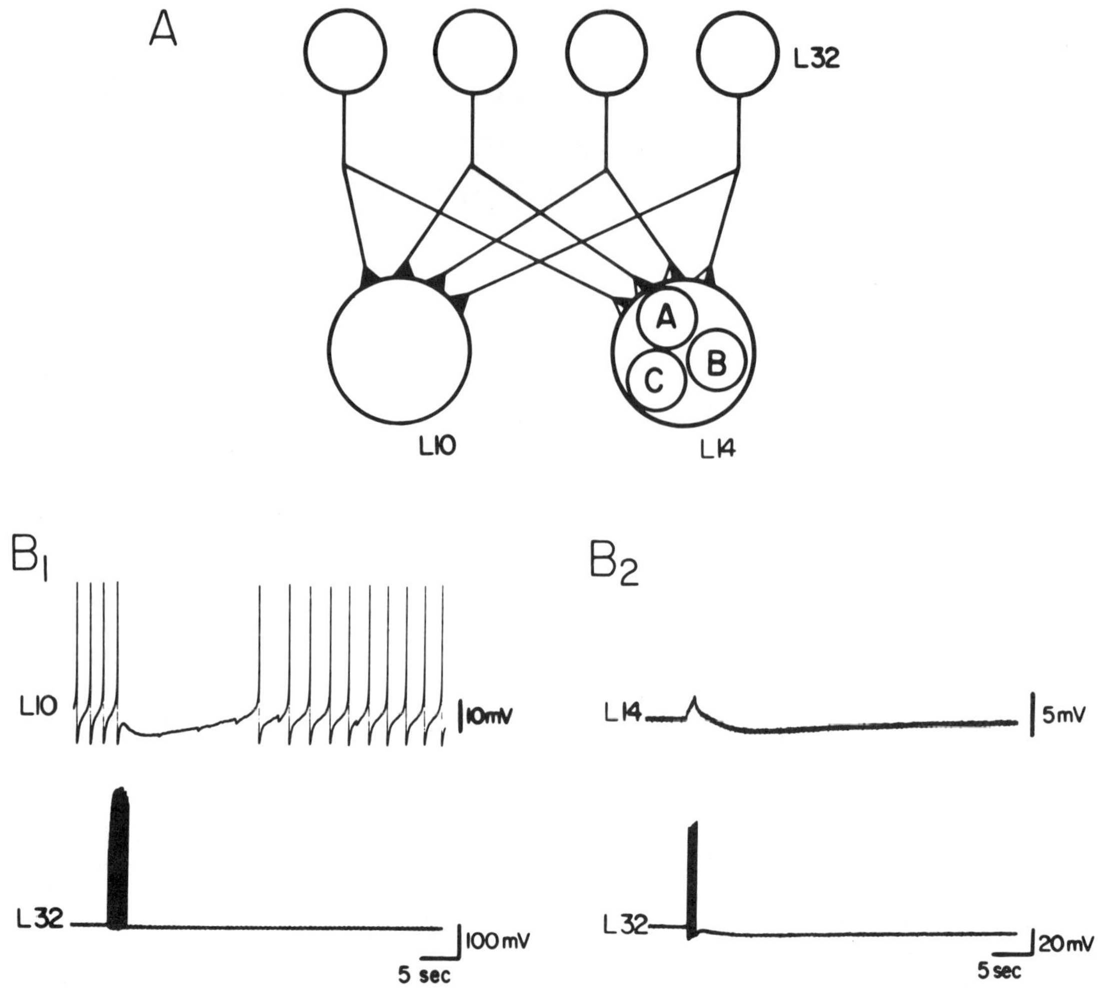

Figure 1. The L32 circuit. (*A*) Simplified schematic diagram of the neural network. L32 cells are a cluster of putative histaminergic neurons in the abdominal ganglion of *Aplysia*, which make inhibitory dual-action synaptic potentials onto L10 and mixed-action potentials on the three electrically coupled L14 neurons. For a more detailed description of this circuit, see Kretz et al. (1986a,b). Hyperpolarization (inhibition) is indicated by the filled-in synaptic endings. Conjoint action (depolarization/hyperpolarization) is indicated by synaptic terminal symbols (**▲**). (B_1) Intracellular stimulation of an L32 neuron for 2 sec at 10 Hz (lower trace) causes a slow hyperpolarization in L10 (upper trace). The bridge was unbalanced in the L32 neuron. (B_2) Intracellular stimulation of an L32 neuron for 1 sec at 10 Hz (upper trace) causes a dual-action, fast-excitatory, slow-inhibitory potential in L14 (upper trace). Records in B_1 and B_2 are from R. Kretz, E. Shapiro, and E.R. Kandel (unpubl.).

HEPES, at pH 7.4). Cells L32, L14, and L10 were identified as described previously (Kretz et al. 1986a). Transmitters and metabolites of arachidonic acid were administered from broken glass micropipettes (tip diameter 10–20 μm) placed about 1 mm above the cell body being tested. Positive pressure pulses of 1–10-second duration were applied to deliver 1–5-μl puffs. The concentrations given are those of test compounds in the pipette.

RESULTS

Arachidonic Acid Reduces Transmitter Release from Identified *Aplysia* Neurons

The first suggestion that the arachidonic acid cascade plays a role in synaptic modulation was the observation that bath application of two modulatory transmitters, histamine and FMRF-amide, stimulates the production of a variety of metabolites of arachidonic acid in *Aplysia* nervous tissue (Piomelli et al. 1986, 1987b,c). Because both of these transmitters had been implicated in presynaptic inhibition in *Aplysia* (Kretz et al. 1986b; Belardetti et al. 1987; Chiel et al. 1988), we examined the effects of arachidonic acid on the specific synapses in the abdominal ganglion presynaptically inhibited by histamine and FMRF-amide. As shown in Figure 2, the transmitter output of sensory neurons was reduced in the presence of arachidonic acid (Fig. 2A; Belardetti et al. 1987; Piomelli et al. 1987c). Arachidonic acid also reduced the L10-RB postsynaptic potential (PSP) in much the same manner as histamine (Fig. 2B). As with histamine, arachidonic acid produced mixed effects on RB follower cells. Even when no postsynaptic changes

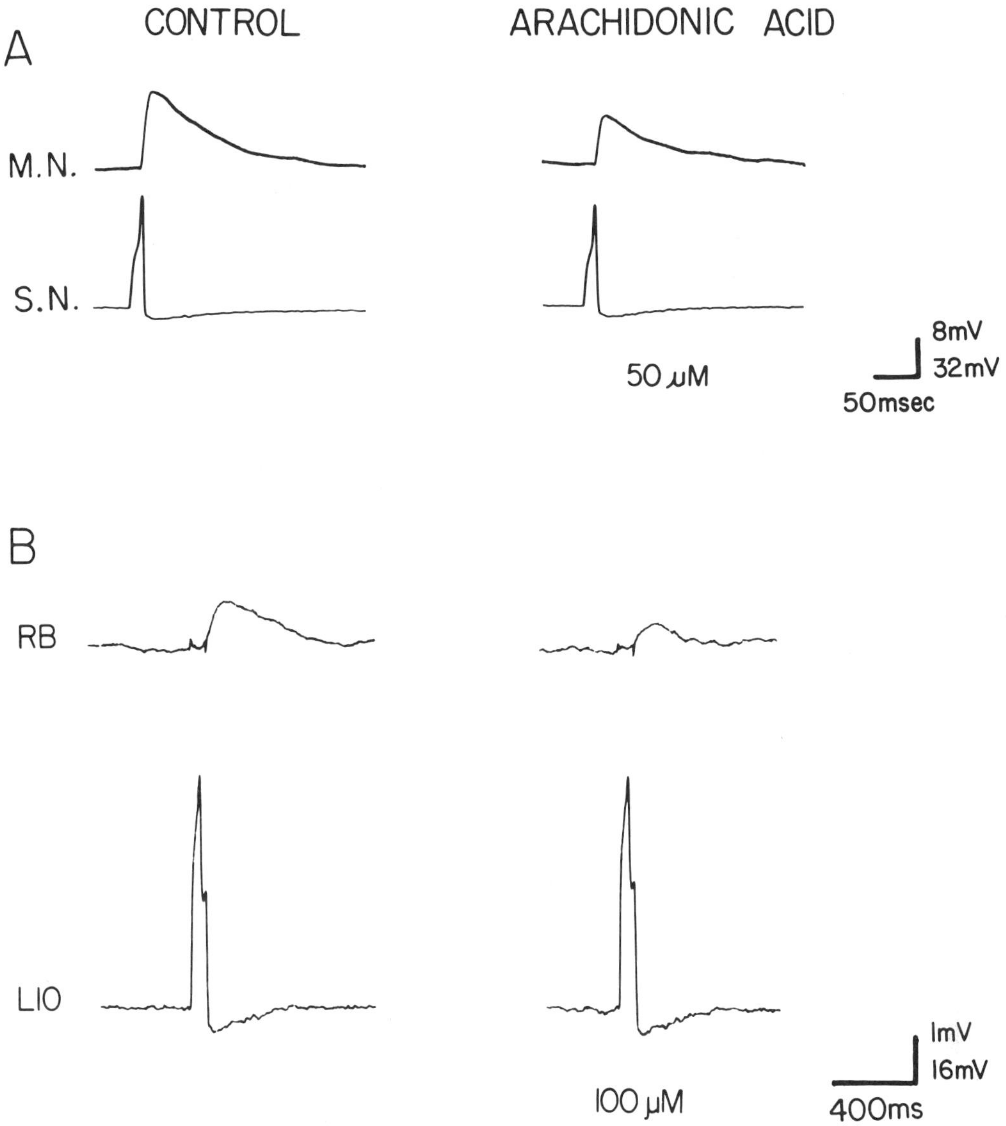

Figure 2. Arachidonic acid reduces transmitter output at two synapses in *Aplysia*. (*A*) Reduction of excitatory PSP in follower motor neuron (M.N. trace) in response to firing an action potential in sensory neurons (S.N. trace) (adapted from Piomelli et al. 1987c). (*B*) Reduction in PSP in follower RB neuron (RB trace) in response to firing an action potential in L10 (L10 trace). (*A*, *B*) Arachidonic acid was suspended in seawater by sonication just before application. In four experiments, the L10-RB PSP was maximally inhibited by $26 \pm 7\%$ (mean ± S.D.) during application of 100 μM arachidonic acid. Seawater or application of eicosamonoenoic acid were without effect on the L10-RB connection.

were observed, arachidonic acid reduced the size of the L10-RB connection (Fig. 2B). The reduction in size of the L10-RB connection in the presence of arachidonic acid was correlated with a slow increased conductance hyperpolarization in the presynaptic cell L10 (Shapiro et al. 1987), which is similar to the response of L10 to histamine and to stimulation of L32.

An Inhibitor of Phospholipase Blocks Responses to Transmitter

We then examined the effects of 4-bromophenacyl bromide (BPB) on the responses of sensory neurons to FMRF-amide and of L10 and L14 to histamine (Fig. 3). BPB, an irreversible inhibitor of phospholipase activity (Shen and Winter 1977) that would prevent the release of arachidonic acid from cell membranes, blocked the response of sensory neurons to FMRF-amide (Piomelli et al. 1987c) and the response of L10 and L14 to histamine. These results support the idea that the responses in all three cells to the applied transmitter require the activity of a phospholipase.

The response of L10 to histamine has been shown to be a dual-action conductance change (Kretz et al. 1986b); histamine also causes a biphasic response in L14 (Kretz et al. 1986b; Figs. 3 and 7), suggesting that it acts on more than one type of ion channel. BPB blocked the histamine response in these cells, suggest-

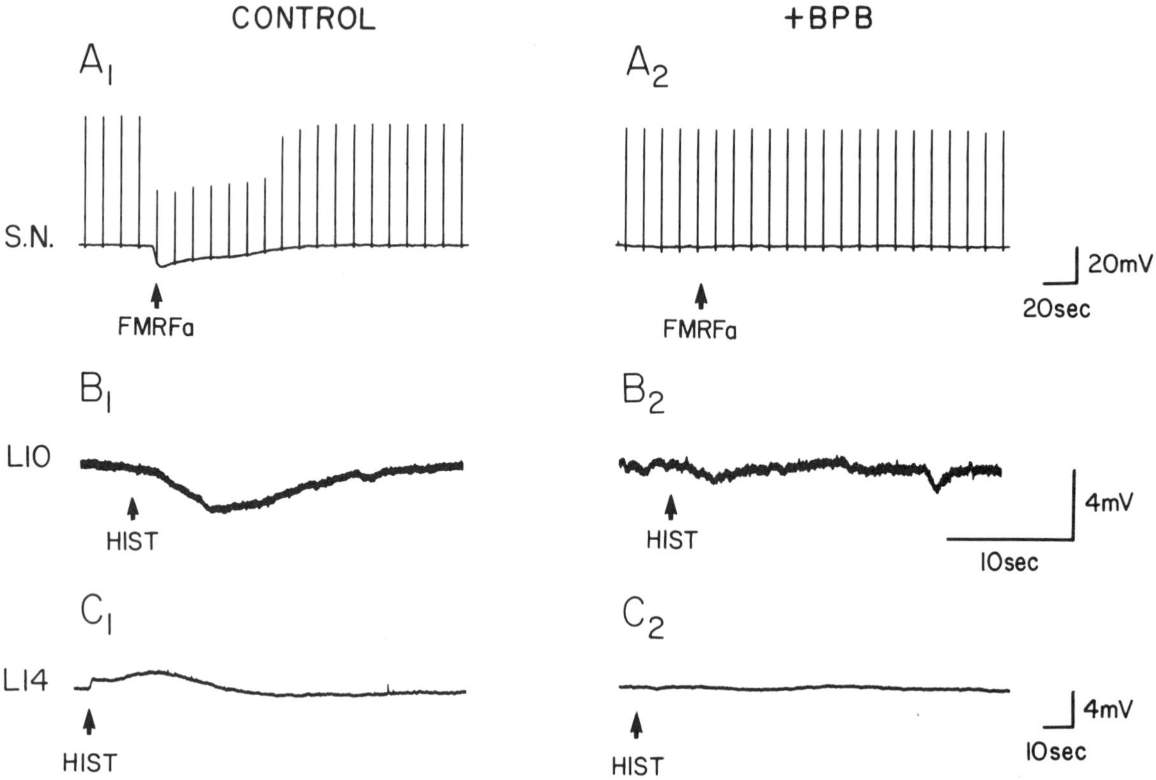

Figure 3. BPB, an inhibitor of phospholipase, blocks responses to transmitter in several identified *Aplysia* neurons. (A_1) Application of FMRF-amide (2 μM) (arrow) causes a slow hyperpolarization and decreases spike amplitude in sensory neurons (S.N.). (A_2) BPB (10 μM) abolishes the FMRF-amide response. (For experimental details, see Piomelli et al. 1987c.) (B_1) Application of histamine (10 μM) (arrow) causes a slow hyperpolarizing response in L10 in the isolated abdominal ganglion. (B_2) BPB (20 μM) blocks the histamine response ($n = 3$). Bath application of BPB (5 μM) reduced the histamine response by 50% in three other experiments with L10. (C_1) Histamine (100 μM) (arrow) causes a dual-action early-excitatory, late-inhibitory response in L14 neurons. (C_2) BPB (20 μM) abolishes both phases of the response ($n = 2$). Histamine was applied as described in Methods. BPB was gently added to the perfusing solution from a concentrated stock in dimethylsulfoxide (DMSO). The final bath concentration of DMSO was 0.1%. DMSO had no effect on histamine responses in L10 or L14 in control experiments.

ing that release of arachidonic acid participates in both components of the dual-action responses.

The Response of L10 to Histamine Involves a Pertussis-sensitive G Protein

Sasaki and Sato (1987) reported that histamine responses of unidentified neurons in another species of *Aplysia* are mediated through the action of a pertussis-sensitive G protein. L10 contains a protein with a relative molecular weight of 40,000 whose ADP-ribosylation is catalyzed by pertussis toxin. The protein from L10 is also detected by immunoblotting with an antibody that recognizes the α subunits of *Aplysia* G proteins (Vogel et al. 1987, 1988). We found that intracellular injection of pertussis toxin into L10 blocks the histamine response (Fig. 4). Injection of the inactivated toxin had no effect. Injection of toxin did not affect the input resistance or spike activity of L10. Sweatt et al. (this volume) and Volterra and Siegelbaum (1988) present evidence that the responses of sensory neurons to FMRF-amide require a pertussis-sensitive G protein.

Activity of L32 Neurons Stimulates the Formation of Metabolites of Arachidonic Acid

Thus far, we have presented pharmacological evidence implicating arachidonic acid in several responses to histamine. We next examined whether physiological stimulation of L32 neurons promotes the release of arachidonic acid in neurons. In these experiments, abdominal ganglia were prelabeled with [^{3}H]arachidonic acid by overnight incubation. L32 cells were then impaled and stimulated intracellularly with depolarizing pulses, and the bath samples were collected and analyzed for the appearance of labeled metabolites of arachidonic acid (Fig. 5). L32 stimulation caused both lipoxygenase and cyclo-oxygenase metabolites to appear (Piomelli et al. 1987b).

We have begun to study the 12-lipoxygenase pathway in more detail but have not yet examined the role of cyclo-oxygenase metabolites. The enzyme, 12-lipoxygenase, converts arachidonic acid to 12-hydroperoxy-5,8,10,14-eicosatetraenoic acid (12-HPETE), a reactive and unstable metabolite, which can be reduced to

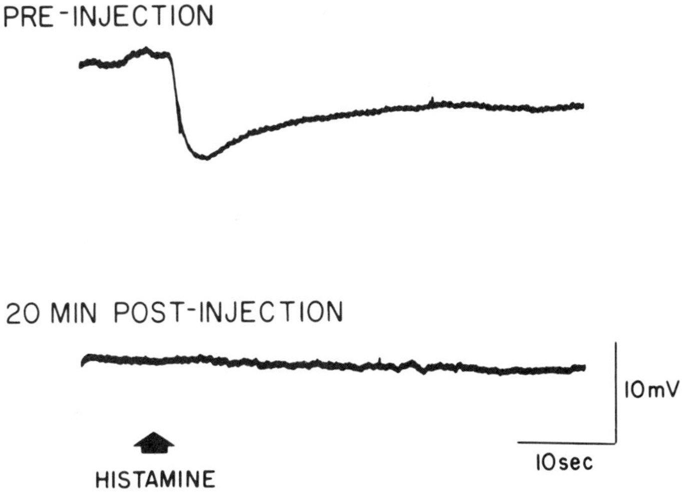

Figure 4. Histamine response of L10 blocked by injection of pertussis toxin. A 5-sec puff of histamine (100 μM) elicits a slow inhibitory response in L10 (*top* trace). After L10 was injected intracellularly with pertussis toxin (islet-activating protein; List Biochemical Lab.) at a concentration of 0.2 μg/μl in a buffer consisting of 10 mM dithiothreitol, 10 mM NAD^+, 10 mM $NaPO_4$, 50 mM NaCl, and 100 mM KCl (pH 7.0), the response to histamine was abolished. Injection of the toxin inactivated by heating at 80°C for 30 min did not affect the histamine response in L10. Details of the intracellular pressure-injection techniques are described in Eisenstadt et al. (1973).

the more stable, corresponding hydroxyl metabolite, 12-HETE, or metabolized further.

The products formed by the 12-lipoxygenase pathway in *Aplysia* nervous tissue are shown in Figure 6. In addition to 12-HETE, stimulating L32 evokes the production of the hydroxy epoxides 8-hydroxy-11,12,epoxy-5,9,14-eicosateraenoic acid (8-OH-EPETE) but not the 10-hydroxy-epoxy positional isomer, 10-OH-EPETE (Piomelli et al. 1987a). (These hydroxy-epoxide metabolites of 12-HPETE have previously been called hepoxilin A [8-OH-EPETE] and hepoxilin B [10-OH-EPETE]; Pace-Asciak and Martin

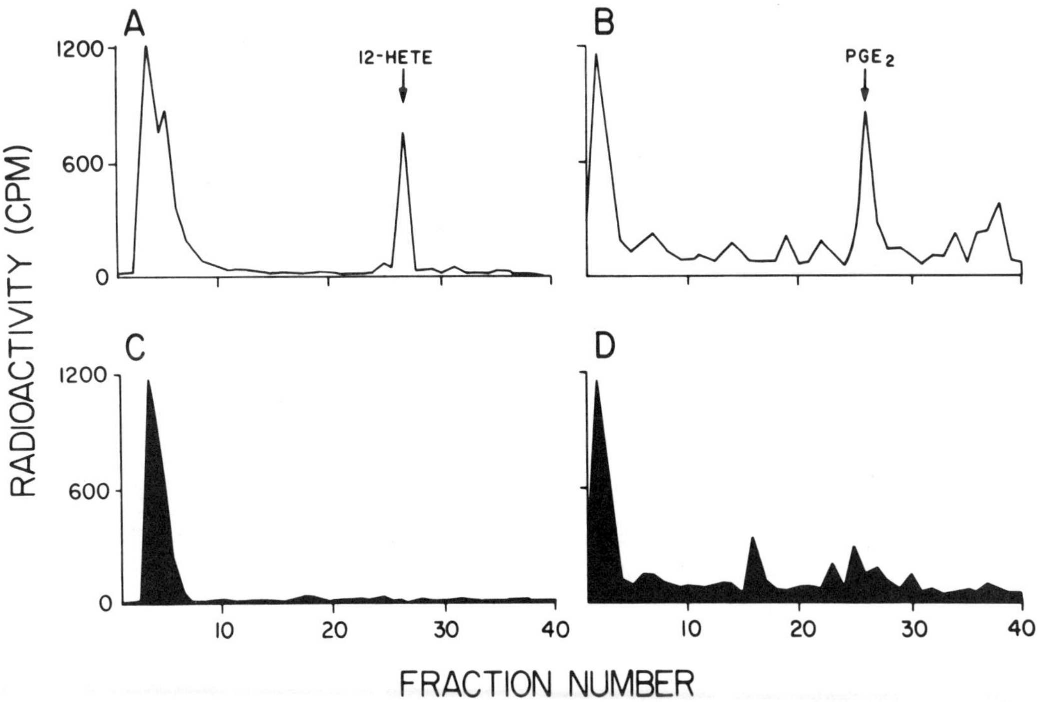

Figure 5. L32 stimulation causes the production of lipoxygenase and cyclo-oxygenase metabolites. Intact isolated abdominal ganglia were prelabeled with [^{3}H]arachidonic acid, L32 cells were stimulated, and bath samples were analyzed as described in Piomelli et al. (1987b). After the L32 cells were stimulated (*A*), reversed-phase high-performance liquid chromatography (HPLC) shows bath samples containing 12-HETE. No 12-HETE was detected before L32 stimulation (*C*). After stimulation, the ^{3}H-labeled 12-HETE fraction from the HPLC contained 450 ± 85 cpm per ganglion in five separate experiments. No counts above background were detected in 12-HETE from fractions taken before the stimulation. After L32 stimulation (*B*), normal-phase HPLC of bath sample reveals the presence of prostaglandin (PGE_2), a cyclo-oxygenase product. No PGE_2 was detected before L32 stimulation (*D*). In two experiments, intracellular stimulation of L32 resulted in the appearance of 377 and 619 cpm of ^{3}H-labeled PGE_2, with only background values in unstimulated controls. (Reprinted, with permission, from Piomelli et al. 1987b.)

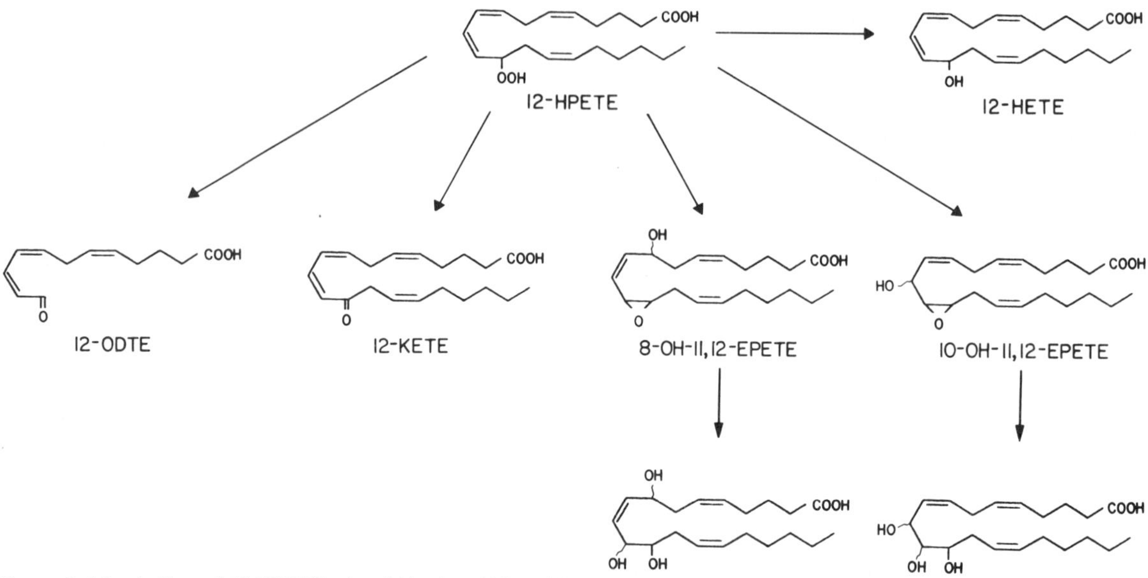

Figure 6. Metabolism of 12-HPETE. Arachidonic acid is oxidized to the hydroperoxy acid, 12-HPETE, by 12-lipoxygenase. In *Aplysia*, 12-HPETE can be further metabolized to the products shown (see text for details) (Piomelli et al. 1987a, 1988).

1984.) Application of histamine induces the formation of 12-keto-5,8,10,14-eicosatetraenoic acid (12-KETE); we also found that 12-oxo-5,8,10-dodecatrienoic acid (12-ODTE) is formed in nervous tissue incubated with arachidonic acid (Piomelli et al. 1988).

Products of the 12-lipoxygenase Pathway Mimic the Effects of L32 on Postsynaptic Targets

We have seen that the 12-lipoxygenase pathway in *Aplysia* is stimulated by histamine, FMRF-amide, and L32 activity. In addition, 12-HPETE was the most effective metabolite tested in mediating the actions of FMRF-amide in sensory neurons. Indomethacin, a cyclo-oxygenase inhibitor, did not block the FMRF-amide response of sensory neurons, whereas nordihydroguaiaretic acid, a lipoxygenase inhibitor, did (Piomelli et al. 1987c).

We tested the effects of 12-lipoxygenase products on cells of the L32 circuit. 12-HPETE can mimic the effects of histamine and L32 on L14 cells (Fig. 7). We administered 12-HPETE 34 times to L14 neurons in 15 experiments, and L14 responded to 12-HPETE with a dual-action response in 56% of these applications. An additional 24% of the applications of 12-HPETE re-

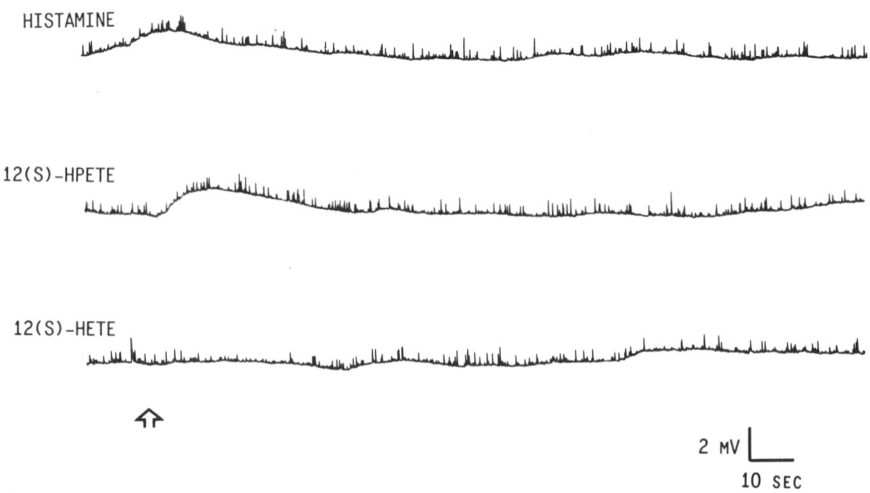

Figure 7. The lipoxygenase product 12-HPETE mimics the histamine response of L14. Application of 12-HPETE (Biomol, New Plymouth, PA) (50 μM in seawater in the ejection pipette; *middle* trace) elicits a response in L14 similar to that of histamine (100 μM; *top* trace). Application of 12(S)-HETE (Biomol) (50 μM; *lower* trace), a more stable metabolite, elicits no response in L14. To prepare 12-HPETE and 12(S)-HETE for application, samples from a stock solution kept in ethanol at −20°C were dried under a stream of nitrogen, reconstituted in seawater, and sonicated for 15 sec.

sulted in a response in L14 consisting only of the hyperpolarizing component of the dual-action response. In contrast, 20 applications in 7 experiments with 12(S)-HETE resulted in no response in 80% of the applications (Fig. 7). These results suggest that 12-HPETE, or a further product of its metabolism (Fig. 6), may be an active endogenous messenger at both histaminergic and FMRF-amidergic synapses.

Different branches of the L32 circuit may make use of other metabolites, however. When we tested the activity of 12-HPETE on cell L10 (in 9 experiments with 18 applications), we found that this cell did not respond in 78% of the trials.

DISCUSSION

From work on vertebrate nonneural tissues and cells, it is clear that many significant metabolites are produced from the arachidonic acid released from membranes by various extracellular stimuli, that these metabolites are produced through several biosynthetic pathways, and that many of them are highly active both as intercellular (first) messengers and as autacoids (see Needleman et al. 1986). The use of the invertebrate *Aplysia*, with its simple nervous system, should be suitable for analyzing the specific mechanisms by which these metabolites function in nerve cells.

Several important insights already have emerged from the work in *Aplysia* into the role of arachidonic acid metabolites in signal transduction in neurons. As with other second messengers, activation of receptors by neurotransmitters appears to be mediated by a G protein and is transduced through target conductance pathways, presumably specific ion channels. A critical step is the release of arachidonic acid from membrane phospholipids through the action of a phospholipase; metabolism of the arachidonate released produces several products with physiological activity.

We have described an identified cell circuit in *Aplysia* and two modulatory transmitters, histamine and FMRF-amide, that stimulate the release and metabolism of arachidonic acid. The major metabolic products produced by histamine, FMRF-amide, and L32 cell stimulation have been characterized, primarily those of the 12-lipoxygenase pathway. At two postsynaptic sites, L14 and sensory neurons, 12-HPETE mimics the actions of the appropriate transmitter. In addition to characterizing a novel possible second-messenger pathway, this information enables us to address three important questions about signal transduction mechanisms.

The first question concerns the complexity of interactions in regulatory processes produced by signal transduction. In many instances, two or more distinct second-messenger systems operate within the same cell (Burgoyne et al. 1987; Axelrod et al. 1988; Belardetti and Siegelbaum 1988). A case in point is cAMP and arachidonic acid in *Aplysia* sensory neurons (Sweatt et al., this volume). With the arachidonic acid cascade, however, it is possible that different metabolites produced from the same cascade have distinctive functional roles. Can more than one metabolite in an enzymatic cascade be physiologically active; e.g., is more than one metabolite involved in the sequential and overlapping dual-action PSPs? Are active metabolic pathways distributed functionally, for example, to mediate (presynaptic) inhibitory events, or are the pathways distributed as specific differentiated cellular properties?

Our recent experiments have shown that *Aplysia* neural tissue can metabolize 12-HPETE further (Fig. 6) and that some of these metabolites are active. In particular, we find that the epoxy alcohol, 8-OH-EPETE, produces a hyperpolarizing response in L14, and 12-KETE can produce its dual-action response (Piomelli et al. 1988 and in prep.). This is of particular interest because L14 cells respond to L32, histamine, and 12-HPETE with a sequential dual-action PSP. It will be important to examine the ion channels in L14 modulated by 12-HPETE and its products. It is also of interest that L14 neurons display a K^+ conductance that is inactivated by 5-HT acting through cAMP-dependent protein phosphorylation, just as sensory neurons do (Walsh and Byrne 1985). L14 neurons thus provide a second example of the functionally opposing interaction of the arachidonic acid and cAMP pathways.

Another question is whether arachidonic acid metabolites act in the same way in each neuron. As already mentioned, conductance mechanisms in L10 are not known to be modulated by protein phosphorylation, yet arachidonic acid produces inhibition in this cell. Do some metabolites act directly on ion channels while others operate by inhibiting third messengers (protein kinases or protein phosphatases)? This type of question can best by answered by examining the mechanisms of action of the metabolites in specific identified *Aplysia* neurons with the appropriate characteristics.

Are all the actions of the arachidonic acid cascade produced by metabolites acting as second messengers (intracellular signals) or do they also act as autacoids or as intercellular (first) messengers? The identification of presynaptic and postsynaptic neurons with synaptic connections at which the metabolites act should allow us to answer this question. As all of the cells in the L32 circuit are identified and can be monitored electrophysiologically, we have a chance to identify the specific components of the synapse where the metabolites act. We could determine whether a metabolite acts presynaptically or postsynaptically and assess whether there is presynaptic feedback mediated by metabolites generated postsynaptically. Pharmacological evidence has implicated lipoxygenase products in long-term potentiation in hippocampal slices (Bliss et al. 1988; Williams and Bliss 1988), where localized presynaptic and postsynaptic communication has been suggested as a mechanism (Piomelli et al. 1987c; Nicoll et al. 1988). As in the hippocampus, L32 cells converge on common postsynaptic follower neurons with varying synaptic strengths (Kretz et al. 1986a). We suspect that metabolites of arachidonic acid might modulate synaptic strength by acting both as second messengers within the follower cells and as local mediators (autacoids and intercellular messengers). Analysis of the strengths of specific synapses of L32 cells within the same ganglion would allow us to explore how the functioning of similar synapses are regulated by local signal transduction mechanisms.

ACKNOWLEDGMENTS

We are grateful to Dr. Robert Zipkin and Biomol (New Plymouth, PA) for supplying metabolites. We thank Dr. S. Mumby for providing antibody and Jillayn Lindahl for secretarial assistance.

REFERENCES

Axelrod, J., R.M. Burch, and C.L. Jelsema. 1988. Receptor-mediated activation of phospholipase A_2 via GTP-binding proteins: Arachidonic acid and its metabolites as second messengers. *Trends Neurosci.* **11:** 117.

Belardetti, F. and S.A. Siegelbaum. 1988. Up- and down-modulation of single K^+ channel function by distinct second messengers. *Trends Neurosci.* **11:** 232.

Belardetti, F., E.R. Kandel, and S.A. Siegelbaum. 1987. Neuronal inhibition by the peptide FMRF amide involves opening of S-K^+ channels in *Aplysia. Nature* **325:** 153.

Bliss, T.V.P., M.L. Errington, M.A. Lynch, and J.H. Williams. 1988. The lipoxygenase inhibitor nordihydroguaiaretic acid (NDGA) blocks the induction of both tetanus-induced and calcium-induced long-term potentiation in the hippocampus of the rat. *Pfluegers Arch. Eur. J. Physiol.* (in press).

Borgoyne, R.D., T.R. Cheek, and A.J. O'Sullivan. 1987. Receptor-activation of phospholipase A_2 in cellular signalling. *Trends Biochem. Sci.* **12:** 332.

Byrne, J.H. 1980a. Neural circuit for inking behavior in *Aplysia californica. J. Neurophysiol.* **43:** 896.

———. 1980b. Identification of neurons contributing to presynaptic inhibition in *Aplysia californica. Brain Res.* **199:** 235.

———. 1980c. Quantitative aspect of ionic conductance mechanisms contributing to firing pattern of motor cells mediated inking behavior in *Aplysia californica. J. Neurophysiol.* **43:** 651.

Byrne, J., E. Shapiro, N. Dieringer, and J. Koester. 1979. Biophysical mechanisms contributing to inking behavior in *Aplysia. J. Neurophysiol.* **42:** 1233.

Carew, T.J. and E.R. Kandel. 1977. Inking in *Aplysia californica.* I. Neural circuit of an all-or-none behavioral response. *J. Neurophysiol.* **40:** 692.

Chiel, H.J., I. Kupfermann, and K.R. Weiss. 1988. An identified histaminergic neuron can modulate the outputs of buccal-cerebral interneurons in *Aplysia* via presynaptic inhibition. *J. Neurosci.* **8:** 49.

Connor, J.A., R. Kretz, and E. Shapiro. 1986. Calcium levels measured in a presynaptic neurone of *Aplysia* under conditions that modulate transmitter release. *J. Physiol.* **375:** 625.

Dickinson, P.S., F. Nagy, and M. Moulins. 1988. Control of central pattern generators by an identified neurone in crustacea: Activation of the gastric mill motor pattern by a neurone known to modulate the pyloric network. *J. Exp. Biol.* **136:** 53.

Eisenstadt, M., J.E. Goldman, E.R. Kandel, H. Koike, J. Koester, and J.H. Schwartz. 1973. Intrasomatic injection of radioactive precursors for studying transmitter synthesis in identified neurons of *Aplysia californica. Proc. Natl. Acad. Sci.* **70:** 3371.

Frost, W.N., G.A. Clark, and E.R. Kandel. 1988. Parallel processing of short-term memory for sensitization in *Aplysia. J. Neurobiol.* **19:** 297.

Harris-Warrick, R.M. and A.H. Cohen. 1985. Serotonin modulates the central pattern generator for locomotion in the isolated lamprey spinal cord. *J. Exp. Biol.* **116:** 27.

Harris-Warrick, R.M. and R.E. Flamm. 1986. Chemical modulation of a small central pattern generator circuit. *Trends Neurosci.* **9:** 432.

Kaczmarek, L.K. and I.B. Levitan, eds. 1987. *Neuromodulation: The biochemical control of neuronal excitability.* Oxford University Press, New York.

Klein, M. and E.R. Kandel. 1980. Mechanism of calcium current modulation underlying presynaptic facilitation and behavioral sensitization in *Aplysia. Proc. Natl. Acad. Sci.* **77:** 6912.

Koester, J. and U. Koch. 1987. Neural control of the circulatory system in *Aplysia. Experientia* **43:** 972.

Koester, J., E. Mayeri, G. Liebeswar, and E.R. Kandel. 1974. Neural control of circulation in *Aplysia.* II. Interneurons. *J. Neurophysiol.* **37:** 476.

Kretz, R., E. Shapiro, and E.R. Kandel. 1986a. Presynaptic inhibition produced by an identified presynaptic inhibitory neuron. I. Physiological mechanisms. *J. Neurophysiol.* **55:** 113.

Kretz, R., E. Shapiro, C.H. Bailey, M. Chen, and E.R. Kandel. 1986b. Presynaptic inhibition produced by an identified presynaptic inhibitory neuron. II. Presynaptic conductance changes caused by histamine. *J. Neurophysiol.* **55:** 131.

Needleman, P., J. Turk, B.A. Jakschik, A.R. Morrison, and J.B. Lefkowith. 1986. Arachidonic acid metabolism. *Annu. Rev. Biochem.* **55:** 69.

Nicoll, R.A., J.A. Kauer, and R.C. Malenka. 1988. The current excitement in long-term potentiation. *Neuron* **1:** 97.

Pace-Asciak, C.R. and J. Martin. 1984. Hepoxilin, a new family of insulin secretagogues formed by intact rat pancreatic islets. *J. Prostagl. Leuk. Med.* **16:** 173.

Piomelli, D., S.J. Feinmark, and J.H. Schwartz. 1987a. 12-hydroperoxyeicosatetraenoic acid (12-HPETE), a lipoxygenase metabolite of arachidonic acid that produces presynaptic inhibition in *Aplysia*, is metabolized to hepoxylins A and B in neural tissue. *Soc. Neurosci. Abstr.* **13:** 598.

Piomelli, D., S.J. Feinmark, E. Shapiro, and J.H. Schwartz. 1986. Stimulation of C2, an identified histaminergic *Aplysia* neuron, results in increased release of lipoxygenase metabolites of arachidonic acid: Possible mediators of synaptic modulation. *Soc. Neurosci. Abstr.* **12:** 1340.

Piomelli, D., E. Shapiro, S.J. Feinmark, and J.H. Schwartz. 1987b. Metabolites of arachidonic acid in the nervous system of *Aplysia*: Possible mediators of synaptic modulation. *J. Neurosci.* **6:** 3675.

Piomelli, D., E. Shapiro, J.H. Schwartz, and S.J. Feinmark. 1988. Formation and biological activity of 12-keto-eicosatetraenoic acid in the nervous system of *Aplysia. J. Biol. Chem.* (in press).

Piomelli, D., A. Volterra, N. Dale, S.A. Siegelbaum, E.R. Kandel, J.H. Schwartz, and F. Belardetti. 1987c. Lipoxygenase metabolites of arachidonic acid as second messengers for presynaptic inhibition of *Aplysia* sensory cells. *Nature* **328:** 38.

Sasaki, K. and M. Sato. 1987. A single GTP-binding protein regulates K^+-channels coupled with dopamine, histamine and acetylcholine receptors. *Nature* **325:** 259.

Shapiro, E., V. Castellucci, and E.R. Kandel. 1980a. Presynaptic membrane potential affects transmitter release in an identified neuron in *Aplysia* by modulating the Ca^{2+} and K^+ currents. *Proc. Natl. Acad. Sci.* **77:** 629.

———. 1980b. Presynaptic inhibition in *Aplysia* involves a decrease in the Ca^{2+} current of the presynaptic neuron. *Proc. Natl. Acad. Sci.* **77:** 1185.

Shapiro, E., J. Koester, and J.H. Byrne. 1979. *Aplysia* ink release: Central locus for selective sensitivity to long-duration stimuli. *J. Neurophysiol.* **42:** 1223.

Shapiro, E., D. Piomelli, and J.H. Schwartz. 1987. Evidence that arachidonic acid cascade generates second messengers that mediate presynaptic inhibition in an identified cell circuit in *Aplysia. Soc. Neurosci. Abstr.* **13:** 597.

Shen, T.-Y. and C.A. Winter. 1977. Chemical and biological studies on indomethacin, sulendac, and their analogs. *Adv. Drug Res.* **12:** 89.

Siegelbaum, S., S.S. Camardo, and E.R. Kandel. 1982. Serotonin and cAMP close single K^+ channels in *Aplysia* sensory neurones. *Nature* **299:** 413.

Vogel, S.S., E. Shapiro, G.J. Chin, and J.H. Schwartz. 1987. Pertussis-sensitive G protein transduces the response to histamine that produces presynaptic inhibition in L10, an identified *Aplysia* interneuron. *Soc. Neurosci. Abstr.* **13:** 597.

Vogel, S.S., G.J. Chin, S.M. Mumby, M. Schonberg, and J.H. Schwartz. 1988. G proteins in *Aplysia*: Biochemical characterization and regional and subcellular distribution. *Brain Res.* (in press).

Volterra, A. and S.A. Siegelbaum. 1988. Role of two different guanine nucleotide-binding proteins in the antagonist modulation of the S-type K^+ channel by cAMP and arachidonic acid metabolites in *Aplysia* sensory neurons. *Proc. Natl. Acad. Sci.* **85:** 7810.

Walsh, J.P. and J.H. Byrne. 1985. Analysis of decreased conductance serotonergic response in *Aplysia* ink motor neurons. *J. Neurophysiol.* **53:** 590.

Willard, A.L. 1981. Effects of serotonin on the generation of the motor program for swimming by the medicinal leech. *J. Neurosci.* **1:** 936.

Williams, J.H. and T.V.P. Bliss. 1988. Induction but not maintenance of calcium-induced long-term potentiation in dentate gyrus and area CA1 of the hippocampal slice is blocked by nordihydroguaiaretic acid. *Neurosci. Lett.* **88:** 81.

Molecular Mechanisms of Phospholipid Signaling Pathways in Mammalian Nerve Cells

M.R. HANLEY, T.R. JACKSON, W.T. CHEUNG, M. DREHER, A. GATTI,
P. HAWKINS, S.I. PATTERSON, M. VALLEJO, A.P. DAWSON,* AND O. THASTRUP†

MRC Molecular Neurobiology Unit, University of Cambridge Medical School, Cambridge CB2 2QH, England;
**School of Biological Sciences, University of East Anglia, Norwich NR4 7TJ, England;*
†Department of Clinical Chemistry, University Hospital, Rigshospitalet, DK-2100 Copenhagen, Denmark

In addition to their structural roles, membrane phospholipids are now recognized to function as information-rich signal generators. This is particularly evident from recent demonstrations that plasma membrane phosphatidylinositol-4,5-bisphosphate ($PtdInsP_2$) is hydrolyzed by a surface-receptor-regulated phospholipase C to produce two molecules with intracellular activities, inositol-1,4,5-trisphosphate ($Ins[1,4,5]P_3$) and diacylglycerol (DAG) (Berridge 1987). The intracellular mediators $Ins(1,4,5)P_3$ and DAG can be regarded as authentic second messengers, comparable to the cyclic nucleotides, in that they function as chemical relays to downstream responses: the discharge of intracellular calcium (Berridge and Irvine 1984) and the activation of membrane-associated protein kinase C (PKC) (Nishizuka 1984), respectively. These events, in turn, are points of amplification and cell-specific diversification in the subsequent activation sequence. Little is known of how particular receptor inputs can generate distinctive responses or of how differentiated cell types modify the receptor-coupled cascade for distinct functional purposes. Accordingly, the detailed analysis of the gene products and their interactions in a characterized cell population is a necessary step in understanding the multi-effector pathways based on inositol lipid breakdown. The characterized cell population we have examined is a cloned mammalian cell line.

NG115-401L cells maintain many of the differentiated properties of neurons (Hatanaka and Amano 1981; Hanley 1987) and have distinct advantages over primary cultured neurons in genetic homogeneity and stability, and access to control of growth and maturation. Moreover, expression of functional and antigenic markers suggests that the 401L line can be regarded as a model for an identified population of neurons, the unmyelinated, small-diameter neurons of peripheral sensory ganglia. These neurons function in circumstances of tissue damage and respond to a variety of chemical stimulants, including endogenous mediators of acute inflammation. Thus, the 401L line can be regarded as a permanent clonal organism that can yield information applicable to the cellular physiology of a specialized mammalian cell population. We have examined the early events following activation of 401L cells by an inflammatory mediator, the peptide bradykinin, which is a recognized stimulant of the inositol lipid signaling pathways. The time course and cellular compartmentation of these events may give insights into physiological effects of kinins on peripheral sensory neurons and, more generally, illustrate how the individual gene products in phospholipid signal transduction are coordinated and regulated in the intact cell.

RESULTS

Inositol Phosphate Production and Cytosolic Calcium Regulation

The functional consequences of activating inositol lipid breakdown have now been described in great detail in a number of cell types. Figure 1A summarizes the identified species of inositol lipids that generate a variety of interconverting inositol phosphates (Fig. 1B). When monolayers of 401L cells are challenged with bradykinin, they exhibit a rapid breakdown of $PtdInsP_2$ with a concomitant production of inositol phosphates (Jackson et al. 1987a). Using the sensitive lithium-amplification assay of inositol monophosphate production (Berridge et al. 1982), the half-maximal effect of bradykinin can be seen at 2 nM (Jackson et al. 1987b), in good agreement with its physiological potency in activating sensory neurons (Burgess et al. 1988).

It is now appreciated that the complexity of inositol metabolites in mammalian cells vastly exceeds any simple functional context (Majerus et al. 1988). Accordingly, we have fractionated the water-soluble products of $[^3H]$inositol radiolabeling by high-performance liquid chromatography (HPLC) ion exchange, following bradykinin stimulation or in unstimulated controls, as shown in Figure 2. A number of the chromatographic peaks have been assigned to known inositol phosphate isomers by the use of external standards and more rigorous destructive methods (Stephens et al. 1987). One important conclusion that emerged from early experiments of this type (Irvine et al. 1984) was the existence of an unsuspected second isomer of inositol trisphosphate, having the structure $Ins(1,3,4)P_3$. $Ins(1,3,4)P_3$ is thought to be produced by a two-step

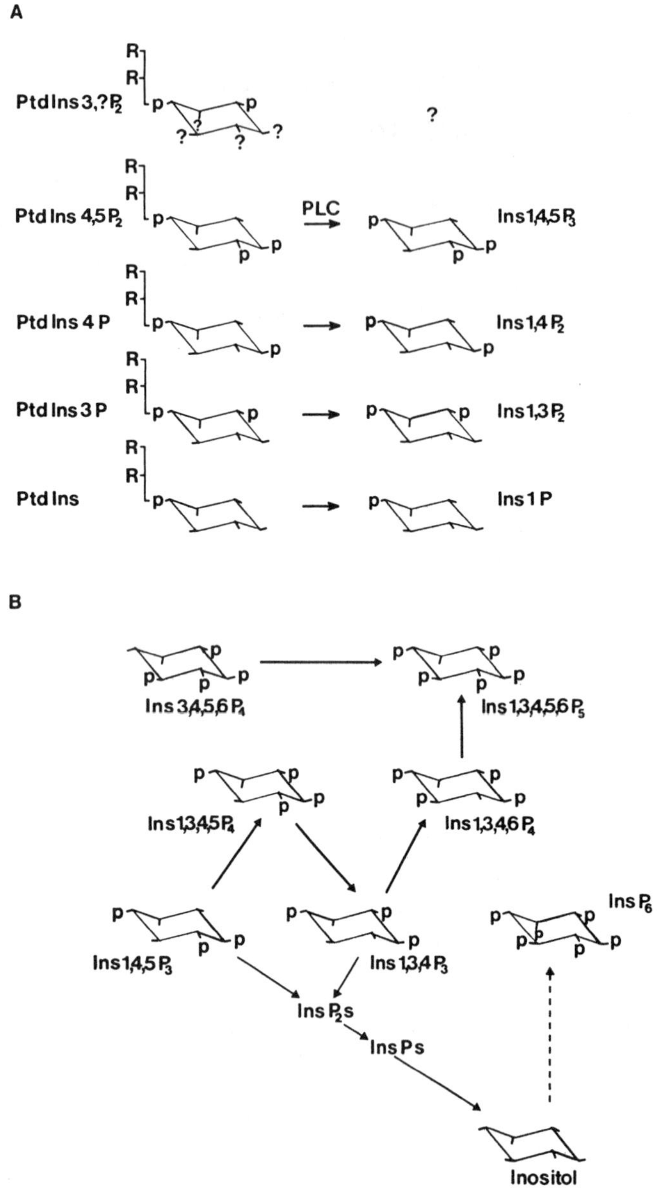

Figure 1. Major identified inositol lipids and inositol polyphosphates. (*A*) The known species of inositol-containing phospholipids from mammalian cells are shown with their predicted products following phospholipase C (PLC) hydrolysis. The possibility of one or more bisphospholipids containing the three-position ring phosphate is indicated by PtdIns3,? $P_2 \cdot$ PtdIns4, 5 P_2 (phosphatidylinositol-4, 5-bisphosphate); PtdIns4P; PtdIns3P (phosphatidylinositol-4- or 3-phosphate) (Whitman et al. 1988); PtdIns (phosphatidylinositol). (*B*) The known isomers of inositol tris-, tetrakis-, and pentakisphosphates, and documented metabolic conversions are illustrated (Stephens et al. 1987; Majerus et al. 1988).

process: kinase action on Ins(1,4,5)P_3 to produce Ins(1,3,4,5)P_4, followed by a phosphatase cleavage (see Fig. 1B). Using HPLC methods, the time course of appearance of these two trisphosphate isomers and their breakdown products, inositol bisphosphates, has been reported (Jackson et al. 1987a). Ins(1,4,5)P_3 and InsP_2 isomers exhibit rapid biphasic responses, whereas the Ins(1,3,4)P_3 isomer shows a steady, sustained increase after bradykinin stimulation. The slow kinetics of Ins(1,3,4)P_3 accumulation indicate that the isomer is relatively resistant to degradation and therefore does not exhibit the rapid, regulated changes in levels that suggest an intracellular messenger function. The temporal sequences of the appearance of Ins(1,4,5)P_3 and Ins(1,3,4,5)P_4 are compared to the changes in cytosolic free calcium concentration, as measured by the fluorescent indicator fura-2 (Cobbold and Rink 1987) in Figure 3. The first response is the appearance of authentic Ins(1,4,5)P_3 within 2 seconds, which parallels the initial calcium transient. This rapid calcium peak reaches calibrated levels approaching 700–800 nM, as has been seen in a variety of other cells (Hallam et al. 1984; Pozzan et al. 1986; McDonough et al. 1988), and remains unchanged upon removal of extracellular calcium, suggesting that it originates from an intracellular store. Much attention has been directed to initial calcium triggering, which is now considered to be due to Ins(1,4,5)P_3-mediated gating of an endoplasmic reticulum channel (Meyer et al. 1988). However, sustained elevation of cytoplasmic calcium levels follows bradykinin stimulation via activation, by unknown mechanisms, of a plasma membrane calcium channel (Jackson et al. 1987a), such as has been observed with other Ca^{++}-mobilizing receptors (Meldolesi and Poz-

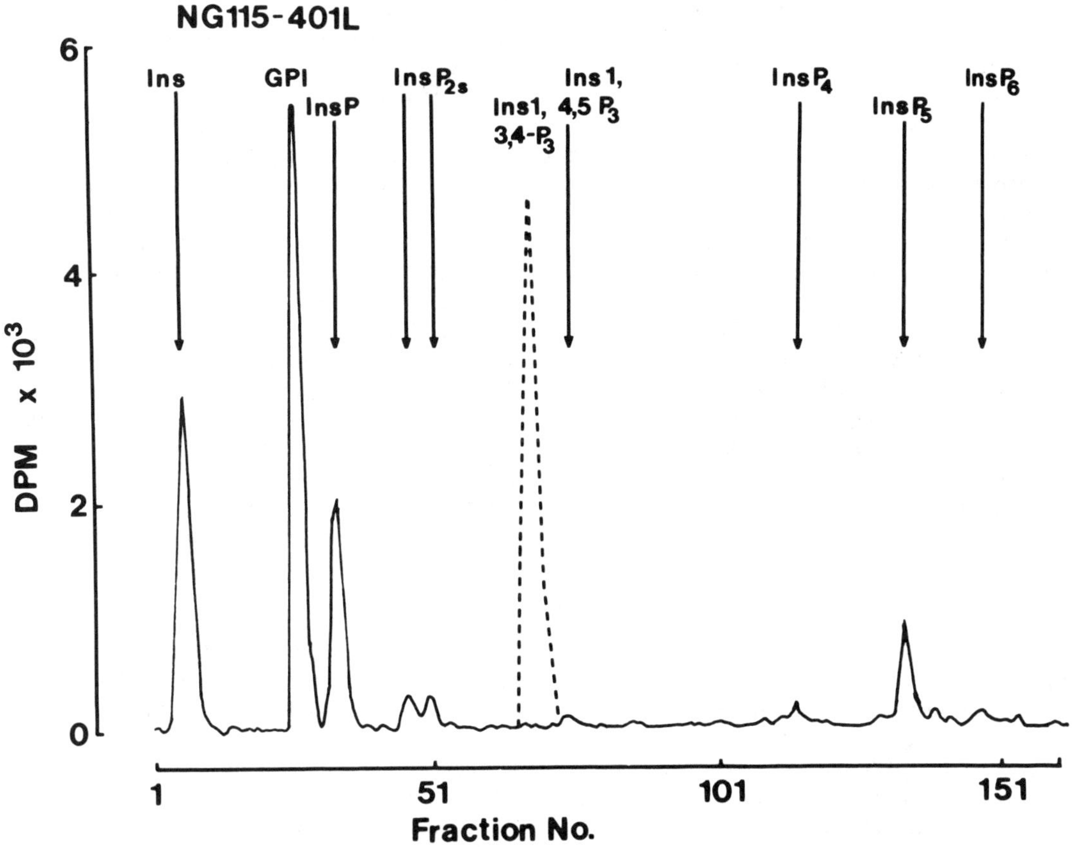

Figure 2. Separation and measurement of water-soluble [^{3}H]inositol-labeled compounds form NG115-401L cells by HPLC. For details of experimental procedures, see Jackson et al. (1987a). (Ins) Inositol; (GPI) glycerophosphoinositol; (InsP) inositol monophosphate; ($InsP_2$s) inositol bisphosphates; ($InsP_4$) inositol tetrakisphosphate; ($InsP_5$) inositol pentakisphosphate; ($InsP_6$) inositol hexakisphosphate (phytic acid). Note that the mobility of Ins(1,4,5)P_3, as well as the position of an external standard of ^{32}P-labeled Ins(1,3,4)P_3 (dotted line), are the only elution positions for explicitly identified structural isomers.

zan 1987). Several hypotheses have been considered for its origin. First, it has been suggested that it is directly linked to either the elevation in cytosolic calcium or the calcium loss from the endoplasmic reticulum store (Putney 1986). This "calcium-dependent" calcium influx model suggests that the calcium store communicates with the plasma membrane to indicate that discharge has taken place, initiating the transmembrane flux. Second, the calcium channel could be directly coupled to the receptor, as has been suggested for a smooth-muscle ATP receptor (Benham and Tsien 1987). Third, the calcium channel could be

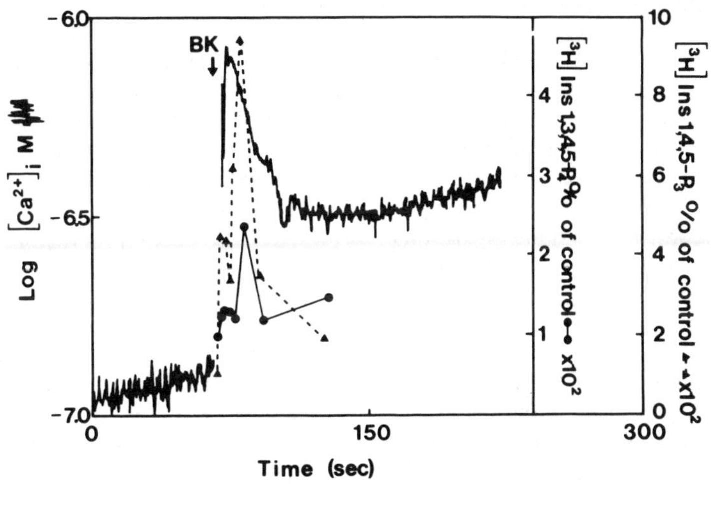

Figure 3. Time courses of [^{3}H]Ins(1,4,5)P_3, [^{3}H]Ins(1,3,4,5)P_4, and cytosolic free calcium in NG115-401L cells grown in monolayers. Calcium and inositol phosphate determinations were made under identical near-physiological conditions, as described previously (Jackson et al. 1987a). (BK) Bradykinin.

under indirect receptor control by production of an intracellular messenger (Meldolesi and Pozzan 1987). In this context, the signal stimulating the sustained calcium influx has been proposed to be Ins(1,3,4,5)P_4 (Irvine and Moor 1986), but experiments using direct microinjection of Ins(1,3,4,5)P_4 have given varying and often equivocal results (Higashida and Brown 1986). On the other hand, the coordinate administration of both Ins(1,4,5)P_3 and Ins(1,3,4,5)P_4 can elicit a protracted increase in calcium-dependent potassium current only when external Ca^{++} is present (Morris et al. 1987), suggesting that Ins(1,3,4,5)P_4 might require a specific priming stimulus in order to be active or that Ins(1,4,5)P_3 and Ins(1,3,4,5)P_4 may act in consort on certain cell types. For example, the release of intracellular calcium may be required to precede the appearance of Ins(1,3,4,5)P_4 because it may be an essential cofactor for the relevant channels or, alternatively, may initiate an obligate calcium-dependent phosphorylation.

To investigate the cause-and-effect interactions of inositol phosphates and calcium, it would be very useful to have pharmacological tools for the production of calcium transients in a hormone-independent manner. Thapsigargin, a sesquiterpene lactone irritant plant product, is a candidate for this type of pharmacological tool (Hanley et al. 1988 and references therein). In brief, thapsigargin has no ionophoric actions but can stimulate every vertebrate cell investigated so far to produce a rapid elevation of intracellular calcium. In 401L cells, this occurs without the production of inositol phosphates or the activation of PKC (Jackson et al. 1988), implying that the reagent causes the direct release of calcium by a novel mechanism. In 401L cells, the calcium discharge is followed by a return to prestimulus basal levels, even in the presence of extracellular calcium. Therefore, neither calcium discharge from the stores nor the rise in cytosolic calcium is sufficient to initiate the calcium influx. Similarly, the response to bradykinin is blocked by thapsigargin pretreatment, even though inositol phosphate production is unaffected (Jackson et al. 1988). This implies that the calcium channel cannot be receptor-operated directly. Accordingly, we are now left with the possibility that a second regulatory signal for calcium homeostasis is generated by the inositol lipid signaling sequence, for which the likeliest candidate is Ins(1,3,4,5)P_4 at present. An additional consideration is that the hypothetical messenger must be produced in a hormone-dependent manner; otherwise the response to bradykinin would be not blocked by thapsigargin pretreatment. This would be consistent with the regulation of Ins(1,3,4,5)P_4 conversion from Ins(1,4,5)P_3 by the elevation in cytosolic calcium concentration (Biden and Wollheim 1986; Rossier et al. 1987).

Comparison of thapsigargin responses in a variety of different cell types has established that only a limited subset shows a transient calcium elevation that returns to basal level. Other cell types, e.g., platelets or hepatocytes, respond to thapsigargin by shifting to a new elevated steady state, which may be as high as the peak of an initial hormone-stimulated transient. In platelets, maintained elevation appears to be due to calcium influx, as well as impaired sequestration (Thastrup et al. 1987). Thus, the differences between platelet and 401L cell responses to thapsigargin may imply cell-specific mechanisms for initiating calcium influx by a receptors. This perspective would predict that the regulation of calcium channels by a receptor-generated intracellular messenger may not be universal and may be limited to certain tissues and cell populations, such as neurons. Accordingly, more than one route for triggering sustained calcium entry may be used by different cells or in different domains of a single cell.

The use of thapsigargin and the interpretation of its effects would be extended greatly by the identification of its molecular target. Thapsigargin's potency and apparent universality of action argue for an interaction with a highly conserved component of the endoplasmic reticulum domain that stores and releases calcium. In isolated hepatic microsomes, thapsigargin causes a dose-dependent release of calcium, confirming that the endoplasmic reticulum is the likely subcellular site of action (O. Thastrup et al., in prep.). However, a complication exits because thapsigargin does not release only the Ins(1,4,5)P_3-sensitive calcium pool, but also a distinct microsomal calcium pool sensitive to GTP (Dawson et al. 1987). These results demonstrate that the molecular site at which thapsigargin acts cannot be the Ins(1,4,5)P_3-operated calcium channel. When the time course of accumulation and stimulated release of ^{45}Ca was examined in hepatic microsomes, thapsigargin both released ^{45}Ca stored previously and abolished the ATP-dependent loading process as well (O. Thastrup et al., in prep.). Indeed, the calcium-activated ATPase activity of microsomes was inhibited with a dose dependency very similar to that seen for stimulation of intracellular calcium discharge in intact cells. These results suggest that the endoplasmic reticulum calcium-activated ATPase, the presumptive calcium pump, is the site of thapsigargin action. In keeping with this, the effect of thapsigargin on inducing ^{45}Ca release from loaded microsomes can be mimicked by ATP depletion using hexokinase/glucose. Strikingly, neither hepatic plasma membrane nor skeletal muscle sarcoplasmic reticulum preparations exhibited any sensitivity to thapsigargin (O. Thastrup et al., in prep.). This specificity argues that the endoplasmic reticulum calcium-activated ATPase may have distinctive structural and functional properties and may predict more diverse roles in the control of calcium transients. For example, arrest of the calcium-activated ATPase may, in some circumstances, be coupled to Ins(1,4,5)P_3 gating of calcium release (however, see Muallem et al. 1985). Thapsigargin-induced calcium release requires a continuous or induced ionic leak of surprising magnitude to provide the Ins(1,4,5)P_3-independent calcium discharge following pump arrest. This pharmacological site may be worth considering as a potential point of physiological control. In this regard, a glucagon-derived peptide has been shown to cause a specific dose-dependent inhibition of the hepatic plasma membrane calcium pump (Mallat et al. 1987), suggesting that receptor-mediated regulation of calcium pump activity in several different membrane compartments might provide another level of tuning of the transient and sustained alterations in cytoplasmic calcium levels. The question of hormonal modulation of membrane Ca^{++}-activated ATPases remains unsettled, although both stimulation and inhibition of activity through second-messenger pathways have been invoked (Carofoli 1987).

These results are noteworthy in a different context. Thapsigargin has been shown to be a "non-TPA"-type second-stage tumor promoter, which does not activate PKC (Hakii et al. 1986; Hanley et al. 1988). With the identification of the microsomal calcium-activated ATPase as a thapsigargin target, this protein becomes correlated with non-TPA (12-*O*-tetradecanoylphorbol-13-acetate) mechanisms of tumor promotion. This correlation provides an appealing symmetry in that both wings of the inositol lipid pathways, previously proposed to be causal in mitogenesis (Berridge and Irvine 1984), could act independently in tumor promotion. More extravagantly, because thapsigargin apparently achieves its stimulant effects on cytoplasmic calcium elevation through an inactivating activity on the endoplasmic reticulum calcium-activated ATPase, an inactivating mutation of the same protein may contribute to multistage carcinogenesis.

In the investigation of intracellular signaling mechanisms, cloned cells such as the 401L line offer the advantage of comparison between individual cell and population measurement techniques, without the complications of genetic heterogeneity. Several methodologies for single-cell monitoring and "calcium imaging" have been introduced (Connor 1986; Tsien and Poenie 1986; Woods et al. 1986; Winiger et al. 1987) to extend the analysis of calcium responses to the level of individual cells. Using a video-based fura-2 imaging system, we have examined the single-cell patterns of calcium changes and have found, in keeping with other work, that there is considerable and unsuspected heterogeneity in individual cell responses (T.R. Jackson et al., in prep.). Some representative traces, including bradykinin stimulation, are shown in Figure 4. Note that each cell shows a different pattern of stimulation sensitivity. Furthermore, the calibrated levels of resting and stimulated calcium are consistently lower using these methods, giving basal values of 40–100 nM and peaks of no more than 400 nM. It is likely that these lower estimates are closer approximations to the physiological range of intracellular calcium concentrations. The origin of the discrepancy between the calibrated levels of intracellular Ca^{++} in single-cell and population measurements is unclear, but a contributing factor may be leakage or export of the indicator to the external medium, which would give anomalously high calcium readings in static population measurements (Malgaroli et al. 1987).

The strict temporal coordination of the production of inositol phosphates and intracellular calcium implies specialized domains for calcium sequestration and evoked discharge that, in view of the extremely short latency for induced calcium release, must be in close apposition to the plasma membrane. The calcium-storage portion of the bulk endoplasmic reticulum has been suggested to be physically distinct and potentially a discrete organelle, which can be isolated by subcellular fractionation (Volpe et al. 1988). It has been proposed that it is uniquely identified in nonmuscle cells by cross-reactivity with antigenic markers of muscle sarcoplasmic reticulum. These results motivate a more detailed physical mapping of the endoplasmic reticulum. Figure 5 shows the pattern observed in the 401L cell using an antibody to the luminal endoplasmic reticulum protein, endoplasmin (Koch et al. 1987). By using the enhanced resolving power of confocal microscopy (White et al. 1987), distinct, extensive regions of close contact between the endoplasmic reticulum and the surface can be discerned. The Ins(1,4,5)P_3-sensitive calcium store may be enriched in these zones. Additional domain-selective endoplasmic reticulum antigens are being investigated to enable the construction of more precise notions of calcium compartmentation within substructures of the rough and smooth endoplasmic reticulum.

PKC: Activation and Time Course of Membrane Effects

PKC, now recognized to be a complex multigene family of related structures (Carpenter et al. 1987; Kikkawa et al., this volume), is expressed by every differentiated cell type, and family members appear to share the property of activation through the stimulated breakdown of inositol lipids. Thus, it is to be expected that PKC stimulation must accompany the activation of 401L cells by bradykinin. However, the kinetics and duration of the activation are not known precisely in any cell nor are the functional consequences associated with specific molecular forms.

The starting point is the identification of the molecular forms of PKC expressed by the 401L cell line. In Figure 6, a Northern blot of poly(A)$^+$ mRNA from the 401L line shows two transcripts of 3.5 and 2.9 kb. The former has been assigned to the γ isoform of PKC, whereas the origin of the smaller transcript is unknown. Although the 2.9-kb mRNA has not been reported earlier, it could correspond to an alternatively processed form of the larger transcript or, more interestingly, it might correspond to a new neural form. Indeed, a novel neural isoform of PKC, with a single 5-kb transcript, has been described recently (Ohno et al. 1988). Thus, it should be considered that two molecular forms, one related to the identified γ form, may be present in this cloned cell population and may contribute differentially to PKC responses.

We have examined PKC activation in intact cells by several approaches, exploiting the potency and specificity of phorbol diesters to stimulate PKC selec-

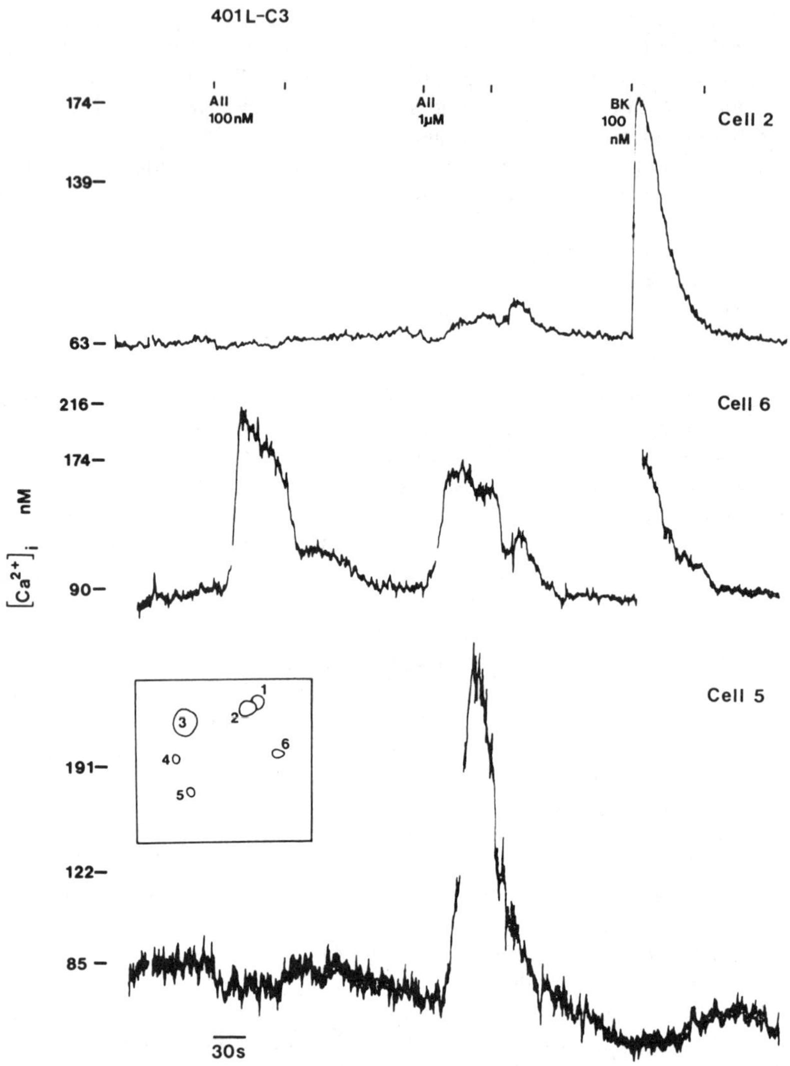

Figure 4. Single-cell determinations of cytosolic free calcium. Cells were plated on 22–mm-diameter glass coverslips and grown for 2 days prior to measurements. Fura-2-loaded cells were maintained at 37°C with superfusion of stimulants at the concentrations shown. Fluorescence images were recorded by alternate excitation at 340 and 380 nm, using an image-processing system interfaced with DEC MicroVAX II. The ratio image was obtained at video rate and filtered with a time constant of 200 msec. Cytosolic free calcium was calibrated by standard procedures (Cobbold and Rink 1987; Jackson et al. 1987a; Cheek et al. 1988).

tively and thereby identify responses mediated by targets of PKC. In 401L cells prelabeled with $[^{32}P]$orthophosphate, stimulation with phorbol-12,13-dibutyrate (PDBu) or TPA causes a rapid and reversible stimulation of the phosphorylation of several proteins, including the widely observed "80K" substrate (Blackshear et al. 1986). However, although these results establish that specific substrates are phosphorylated transiently by PKC in 401L cells, the onset, intensity, and duration of PKC-related effects can be examined more readily using noninvasive functional indicators of membrane effects.

In Figure 7, one such indicator is the fluorescent pH indicator BCECF. Under suitable conditions, the activation of the amiloride-sensitive sodium-proton antiporter can be monitored by the increase in BCECF fluorescence upon alkalinization of the cytoplasm (Hallam and Tashjian 1987). Upon stimulation with TPA, there is a long-lasting amiloride-sensitive increase in fluorescence in BCECF-loaded cells. The maximum TPA response is observed at 4 minutes after addition and appears slower than that elicited by bradykinin (Fig. 7). It is likely that this difference in latencies is set by the difference between inositol lipid hydrolysis and direct stimulation of the kinase activity of PKC. However, the pattern is precisely the opposite of what one might expect. A similar delay is observed with a separate noninvasive indicator, the extracellular release of choline metabolites (Fig. 7). Therefore, we examined the nature of inositol lipid coupling in more detail.

Such responses can be used as pharmacological tests to evaluate potential stimulants and blockers of PKC. Surprisingly, the widely held second-messenger candidate for activation of PKC, DAG, has little or no

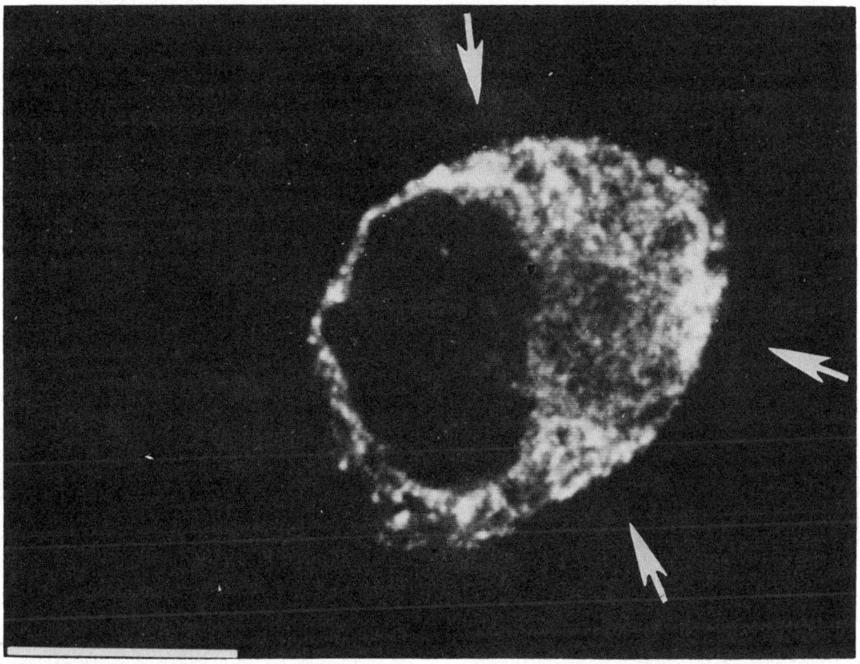

Figure 5. Visualization of the distribution of endoplasmic reticulum in NG115-401L cells. Fixed, permeabilized cells were stained on coverslips with a monospecific rabbit anti-endoplasmin antibody and photographed using scanning laser confocal fluorescence microscopy (Koch et al. 1987). Arrows call attention to dense staining in close proximity to the cell plasma membrane. Bar, 10 μm.

activity on 401L cells in these tests. We have used a variety of synthetic DAGs and analogs, including diolein, 1-oleyl,2-acetyl diglycerol (OAG), DiC8, and DiC10, and have found none to be active on these responses on intact 401L cells, although they are active on other cell lines, such as the AR42J rat pancreatic acinar cell line (Womack et al. 1985). The conclusion may be that not all molecular forms of PKC are controlled physiologically by the levels of DAG but rather respond to another signal (Dreher and Hanley 1988). However, this signal must be derived from early events preceding or following hydrolysis of inositol lipids because, as noted earlier, the activation of a receptor coupled to inositol lipid hydrolysis triggers these PKC-associated consequences rapidly. Candidates for coupling mediators are unsaturated fatty acids released by phospholipase action, such as arachidonic acid (Sekiguchi et al. 1987) or related metabolites. We do not as yet have any evidence to indicate whether these substances may act in coupling 401L bradykinin receptors to PKC, but it is pertinent to note that the γ-PKC isoform can be stimulated selectively by low concentrations of arachidonic acid (Sekiguchi et al. 1987).

The accelerated turnover of phosphatidylcholine has been recognized as a specific response to tumor promoters for some time (Guy and Murray 1982), but it has been reinvestigated only recently with the awareness that the tumor promoters are likely to be regulating this metabolic response through PKC (Besterman et al. 1986). It is found in a variety of cells, and, like inositol lipid hydrolysis, may be ubiquitous among vertebrate cell types. Consequently, the acceleration of phosphatidylcholine turnover may be a candidate for another signaling event in the temporal sequence following activation of PKC. The initial response has been proposed to be activation of an intracellular phospholipase C to produce DAG and phosphorylcholine (Besterman et al. 1986; Irving and Exton 1987; Miur and Murray 1987). This interpretation is heavily influenced by the awareness that a phosphatidylcholine-specific phospholipase C has been identified and purified (Sheikhnejad and Srivastava 1986). This accelerated turnover has been hypothesized to provide DAG, which, on activation of PKC, appears (Takuwa et al. 1987) to extend and amplify the response initiated by hydrolysis of inositol lipids, or to provide a mechanism to activate PKC independent of hydrolysis of inositol lipids.

The production of water-soluble choline metabolites from prelabeled 401L cells is not, however, in accord with these results and their conclusion. We have found that a variety of structurally related and structurally distinct tumor promoters activate the extracellular production of free choline without other choline metabolites (S. Patterson et al., in prep.). Extracellular choline is produced in the presence of a vast excess of unlabeled choline in the medium, indicating that it cannot be due to transport from the inside down a chemical gradient but is rather released directly from the surface. The pattern of metabolite production is unaffected by permeabilization, and measurements of intracellular choline products do not show substantial time-dependent changes. We conclude that although the activation

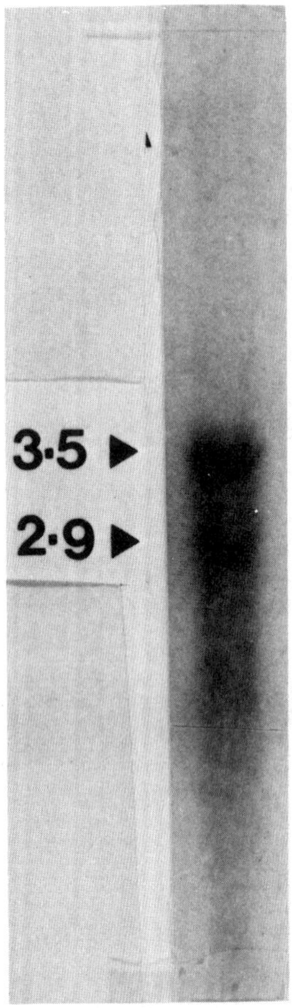

Figure 6. Northern blot of PKC-related transcripts in NG115-401L cells. Poly(A)$^+$ RNA (20 μg) was size separated on a glyoxal agarose gel, transferred to nylon filters, and hybridized with a random-primed ^{32}P-labeled probe. The probe was a 585-bp *Eco*RI-*Bgl*I fragment from the rat PKCI cDNA (Knopf et al. 1986), which recognizes a conserved region in identified PKC mRNAs. The blot was autoradiographed for 3 days at -70°C with an intensifying screen. λ *Hin*dIII/*Eco*RI fragments were used as markers to calibrate sizes. The major band was 3.5 kb, and a less intense band was detected at 2.9 kb.

by PKC is likely to be intracellular, the hydrolysis of phosphatidylcholine is a phospholipase D attack on substrate lipid on the surface leaflet of the plasma membrane, producing free choline and phosphatidic acid (PA). The possibility of a family of extracellularly acting phospholipases D is strengthened by the demonstration of a plasma-derived phospholipase D that is specific for the inositol glycan membrane protein anchor (Low and Prasad 1988). Whereas the latter enzyme may have a predicted role in coupling receptor activation events to release of anchored proteins, it is not clear what function cleavage of phosphatidylcholine could subserve. One point to consider is that the enzyme in effect provides a reverse form of transmembrane signaling to that provided by the inositol lipids, communicating to the exterior of the cell in response to intracellular stimulation. Although it is unlikely that the water-soluble product of this enzymatic response, choline, provides a messenger, the lipid-soluble product PA is a candidate for an informational molecule. Other work has shown that exogenous PA can activate cells (Moolenaar et al. 1986), and it has been suggested that there may be a surface "PA receptor" that may activate inositol lipid breakdown (Murayama and Ui 1987), cell proliferation, and gene regulation (Moolenaar et al. 1986; Tsai et al., this volume). If so, the physiological origin of the PA to activate this site is unknown, but cleavage of surface phosphatidylcholine by a kinase-regulated phospholipase D would provide a plausible mechanism. In a fashion comparable to another class of lipid-soluble mediators, the eicosanoids, perhaps the production of PA provides autocrine control to intensify the intracellular consequences of sustained stimulation or to integrate multiple cellular inputs that regulate PKC. Phosphatidylcholine hydrolysis may also function to remodel the properties of the cell surface, signaling the state of cell activity to ambient cells. Although phosphatidylcholine is not cleaved by a phospholipase C following PKC stimulation in 401L cells, one or more PKC-sensitive phospholipases, A, C, or D, may act on choline lipids to generate diverse signals in other cell types.

CONCLUSIONS

Figure 8 gives a schematic picture of the signaling sequence elicited by bradykinin in the NG115-401L clone. The early events triggered by receptor occupancy conform to a generic picture of signal transduction that has emerged in recent years: activation of a G protein, coupling to an enzymatic effector, and production of intracellular messengers. Because of the great similarity of this picture in different vertebrate cell types, it seems unlikely that specialization is achieved by deletion or addition of the major gene products; rather, several subtler routes may achieve specificity in output responses. First, many, if not all, of the identified gene products in Figure 8 (G proteins, phospholipases C, PKCs) are now recognized to belong to multigene families. This diversity provides specialized transduction components that can be expressed in a developmentally regulated and cell-specific manner. A corollary of this proposition is that different metabolic products can be recruited to act as intracellular or extracellular messengers, so that in addition to Ins(1,4,5)P_3/Ca^{++} in 401L cells, Ins(1,3,4,5)P_4, PA, and another mediator derived from inositol lipids, which is not DAG, may act as effectors. Moreover, other potential signals are derived from metabolites of these pathways, such as the inositol polyphosphates $InsP_5$ and $InsP_6$ (Vallejo et al. 1987), or derived from genetic induction, such as growth factors (Hanley 1988), which can add to the complexity of the acute and long-term signaling repertoire. Second, cell geometry itself can provide another form of specialization. For example, process-bearing cells, such as neurons, can segregate components so as to favor or eliminate certain interactions. This may be the ultimate reason for the apparently byzantine complexity of serially linked signaling events—to provide discrete steps that can be

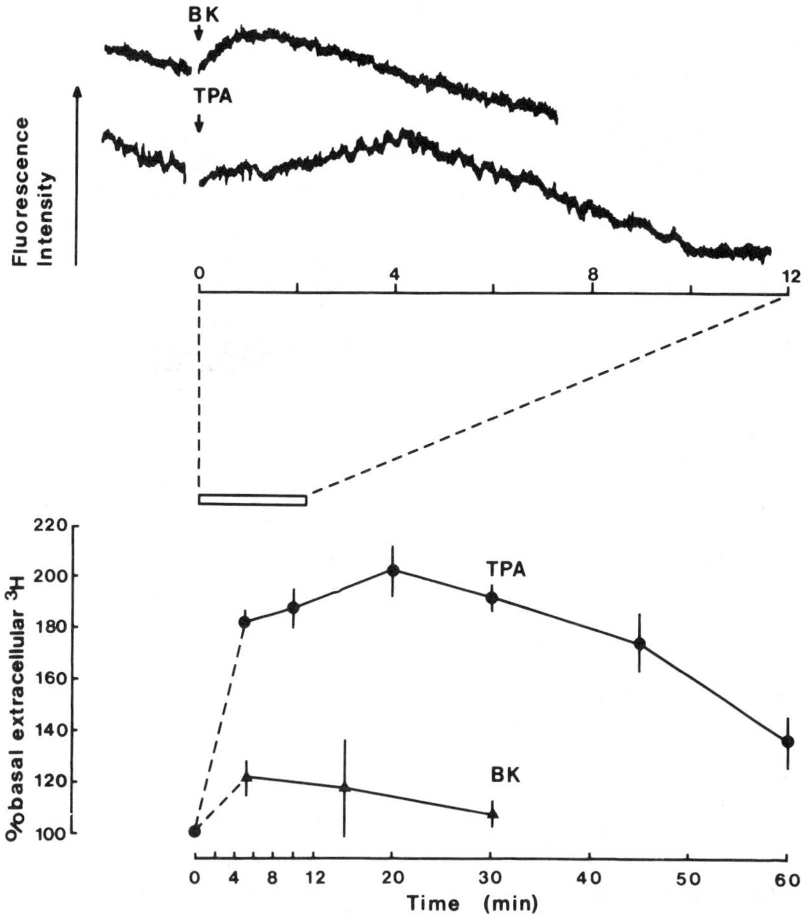

Figure 7. Time courses of cytoplasmic alkalinization (*upper* traces) and release of extracellular [^{3}H]choline (*lower* traces) in response to phorbol diester (TPA) or bradykinin (BK) stimulation, respectively. As elevation of cytosolic calcium has been shown previously to lower the basal choline release rate (Jackson et al. 1988), it was necessary to pretreat cells with thapsigargin before adding bradykinin. Without this manipulation, the bradykinin activation of choline release was masked by the inhibition of the basal release by increased cytoplasmic calcium levels. Measurements were made by procedures described in detail elsewhere (Hallam and Tashjian 1987; Jackson et al. 1988).

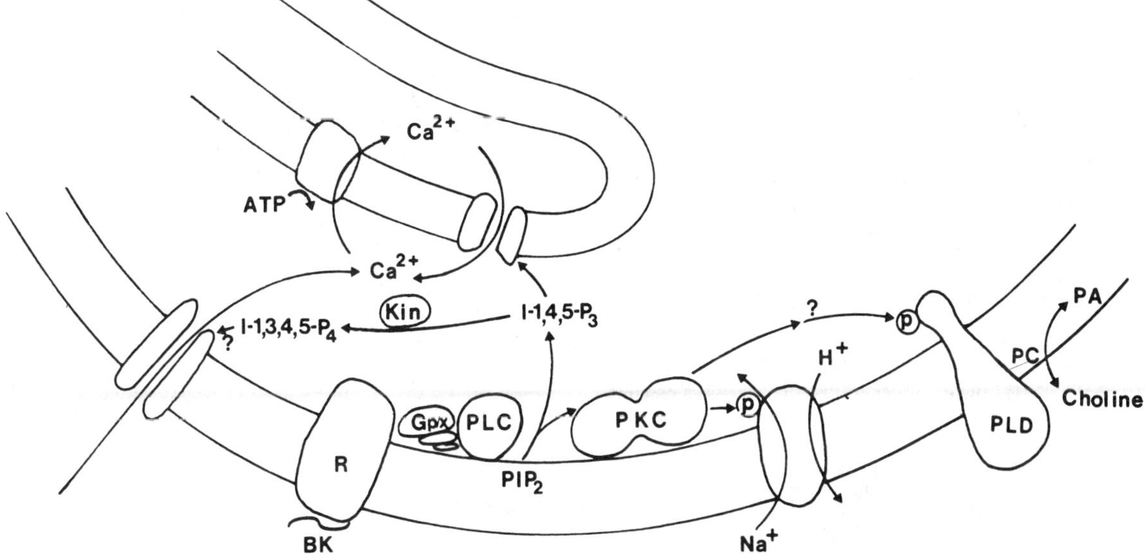

Figure 8. Early events coupled to activation of bradykinin (BK) receptors in the NG115-401L clone. Question mark (?) indicates that the evidence is speculative or incomplete. (Kin) Ins(1,4,5)P$_3$ 3-kinase; (PA) phosphatidic acid; (PC) phosphatidylcholine; (PIP$_2$) PtdInsP$_2$; (PLC) phospholipase C; (PLD) phospholipase D; (PKC) protein kinase C.

segregated physically. Third, the quantitative level of expression of transduction components can alter the efficiency of each stage in the signaling sequence. We have observed that receptors for the messengers generated by inositol lipid hydrolysis, the Ins(1,4,5)P_3 receptor and PKC, differ in levels by at least two orders of magnitude and that this stoichiometry varies from cell type to cell type. Fourth, response characteristics such as latency, duration, intensity, sensitivity to convergent inputs, desensitization, and recovery may provide a code identifying set points for activation of specific events. Thus, whether a receptor triggers secretion, motility, or entry into DNA synthesis may be governed by the quantitative spatial and temporal interplay of a limited number of informational steps in a signaling cascade.

ACKNOWLEDGMENTS

M.R.H. is a recipient of a Research Award from the International Life Sciences Research Institute. Support for the work was provided by the Medical Research Council (MRC) and MRC Partnership Award with SK & F Research (T.R.J.), a Croucher Foundation Studentship (W.T.C.), a Boehringer-Ingelheim Studentship (M.D.), a Senior Research Fellowship of the Lister Foundation (P.H.), an MRC Studentship (S.I.P.), and a long-term EMBO Fellowship (M.V.). We would like to express our appreciation to John Knopf (Genetics Institute) for provision of PKC cDNAs, Gordon Koch (MRC Laboratory of Molecular Biology, Cambridge) for anti-endoplasmin, and Mike Berridge, Roger Moreton, and Timothy Cheek (AFRC Unit, Department of Zoology, Cambridge) for assistance with single-cell calcium measurements.

REFERENCES

Benham, C.D. and R.W. Tsien. 1987. A novel receptor-operated Ca^{2+}-permeable channel activated by ATP in smooth muscle. *Nature* **328:** 275.

Berridge, M.J. 1987. Inositol trisphosphate and diacylglycerol: Two interacting second messengers. *Annu. Rev. Biochem.* **56:** 159.

Berridge,. M.J. and R.F. Irvine. 1984. Inositol trisphosphate, a novel second messenger in cellular signal transduction. *Nature* **312:** 315.

Berridge, M.J., C.P. Downes, and M.R. Hanley. 1982. Lithium amplifies agonist-dependent phosphatidylinositol responses in brain and salivary glands. *Biochem. J.* **206:** 587.

Besterman, J.M., V. Duronio, and P.M. Cuatrecasas. 1986. Rapid formation of diacylglycerol from phosphatidylcholine: A pathway for generation of a second messenger. *Proc. Natl. Acad. Sci.* **83:** 6785.

Biden, T.J. and C.B. Wollheim. 1986. Ca^{2+} regulates the inositol tris/tetrakisphosphate pathway in intact and broken preparations of insulin-secreting RINm5F cells. *J. Biol. Chem.* **261:** 11931.

Blackshear, P.J., L. Wen, B.P. Glynn, and L.A. Witters. 1986. Protein kinase C-stimulated phosphorylation in vitro of a M_r 80,000 protein phosphorylated in response to phorbol esters and growth factors in intact fibroblasts. *J. Biol. Chem.* **261:** 1459.

Burgess, G.M., M. McNeill, and I. Mullaney. 1988. Some actions of bradykinin in neonatal rat sensory neurones in culture are mediated by protein kinase C. *J. Physiol.* **398:** 17p.

Carofoli, E. 1987. Intracellular calcium homeostasis. *Annu. Rev. Biochem.* **56:** 395.

Carpenter, D., T.R. Jackson, and M.R. Hanley. 1987. Protein kinase Cs: Coping with a growing family. *Nature* **325:** 107.

Cheek, T.R., R.B. Moreton, M.J. Berridge, and O. Thastrup. 1988. The effects of the Ca^{2+}-mobilizing tumour promoter thapsigargin and an IP3-mobilizing agonist on cytosolic Ca^{2+} in bovine adrenal chromaffin cells. *Biochem. Soc. Trans.* (in press).

Cobbold, P.H. and T.J. Rink. 1987. Fluorescence and bioluminescence measurement of cytosolic free calcium. *Biochem. J.* **248:** 313.

Connor, J.A. 1986. Digital imaging of free calcium changes and of spatial gradients in growing processes in single mammalian central nervous system cells. *Proc. Natl. Acad. Sci.* **83:** 6179.

Dawson, A.P., G. Hills, and J.G. Comerford. 1987. The mechanism of action of GTP on Ca^{2+} efflux from rat liver microsomal vesicles. *Biochem. J.* **244:** 87.

Dreher, M. and M.R. Hanley. 1988. Multiple modes of protein kinase C regulation and their significance in signaling. *Trends Pharmacol. Sci.* **9:** 114.

Guy, G.R. and A.W. Murray. 1982. Tumor promoter stimulation of phosphatidylcholine turnover in HeLa cells. *Cancer Res.* **42:** 1980.

Hakii, H., H. Fujiki, M. Suganuma, M. Nakayasu, T. Tahira, T. Sugimura, P. Scheuer, and S.B. Christensen. 1986. Thapsigargin, a histamine secretogogue, is a non-12-o-tetradecanoylphorbol-13-acetate (TPA) type tumor promoter in two-stage mouse skin carcinogenesis. *Cancer Res. Clin. Oncol.* **111:** 177.

Hallam, T.J. and A.H. Tashjian. 1987. Thyrotropin-releasing hormone activates Na^+/H^+ exchange in rat pituitary cells. *Biochem. J.* **242:** 411.

Hallam, T.J., A. Sanchez, and T.J. Rink. 1984. Stimulus response coupling in human platelets. *Biochem. J.* **218:** 819.

Hanley, M.R. 1987. Analysis of receptor-coupled events in neuropeptide action using clonal cell lines. *Prog. Brain Res.* **72:** 189.

———. 1988. Proto-oncogenes in the nervous system. *Neuron* **1:** 175.

Hanley, M.R., T.R. Jackson, M. Vallejo, S.I. Patterson, O. Thastrup, S. Lightman, J. Rogers, G. Henderson, and A. Pini. 1988. Neural function: Metabolism and actions of inositol metabolites in mammalian brain. *Philos. Trans. Roy. Soc. B. Biol. Sci.* **320:** 381.

Hatanaka, H. and T. Amano. 1981. A mouse neuroblastoma × rat glioma produces immunoreactive substance P-like material. *Brain Res.* **215:** 305.

Higashida, H. and D.A. Brown. 1986. Membrane current responses to intracellular injection of inositol 1,3,4,5-tetrakisphosphate and inositol 1,3,4-trisphosphate in NG108-15 hybrid cells. *FEBS Lett.* **208:** 283.

Irvine, R.F. and R.M. Moor. 1986. Micro-injection of inositol 1,3,4,5-tetrakisphosphate activates sea urchin eggs by a mechanism dependent on external Ca^{2+}. *Biochem. J.* **240:** 917.

Irvine, R.F., A.J. Letcher, D.J. Lander, and C.P. Downes. 1984. Inositol trisphosphates in carbachol-stimulated rat parotid gland. *Biochem. J.* **223:** 237.

Irving, H.R. and J.H. Exton. 1987. Phosphatidylcholine breakdown in rat liver plasma membranes. *J. Biol. Chem.* **262:** 3440.

Jackson, T.R., T.J. Hallam, C.P. Downes, and M.R. Hanley. 1987a. Receptor-coupled events in bradykinin action: Rapid production of inositol phosphates and regulation of cytosolic free Ca^{2+} in a neural cell line. *EMBO J.* **6:** 49.

Jackson, T.R., S.I. Patterson, O. Thastrup, and M.R. Hanley. 1988. A novel tumour promoter, thapsigargin, transiently increases cytoplasmic free Ca^{2+} without generation of inositol phosphates in NG115-401L cells. *Biochem. J.* **253:** 81.

Jackson, T.R., S.I. Patterson, Y.H. Wong, and M.R. Hanley. 1987b. Bradykinin stimulation of inositol phosphate and calcium responses is insensitive to pertussis toxin in NG115-401L neuronal cells. *Biochem. Biophys. Res. Commun.* **148:** 412.

Knopf, J.L., L. Myung-Ho, L.A. Sultzman, R.W. Kriz, C.R. Loomis, R.M. Hewick, and R.M. Bell. 1986. Cloning and expression of multiple protein kinase C cDNAs. *Cell* **46:** 491.

Koch, G.L.E., D. Macer, and M.J. Smith. 1987. Visualization of the intact endoplasmic reticulum by immunofluorescence with antibodies to the major ER glycoprotein, endoplasmin. *J. Cell Sci.* **87:** 535.

Low, M.G. and A.R.S. Prasad. 1988. A phospholipase D specific for the phosphatidylinositol anchor of cell surface proteins is abundant in plasma. *Proc. Natl. Acad. Sci.* **85:** 980.

Majerus, P.W., T. Connolly, V. Bansal, R. Inhorn, T. Ross, and D. Lips. 1988. Inositol phosphates: Synthesis and degradation. *J. Biol. Chem.* **263:** 3051.

Malgaroli, A., D. Milani, J. Meldolesi, and T. Pozzan. 1987. Fura-2 measurements of cytosolic free Ca^{2+} in monolayers and suspensions of various types of animal cells. *J. Cell Biol.* **105:** 2145.

Mallat, A., C. Pavoine, M. Dufour, S. Lotersztajn, D. Bataille, and F. Pecker. 1987. A glucagon fragment is responsible for the inhibition of the liver Ca^{2+} pump by glucagon. *Nature* **325:** 620.

McDonough, P.M., J.H. Eubanks, and J.H. Brown. 1988. Desensitization and recovery of muscarinic and histaminergic Ca^{2+} mobilization in 1321N astrocytoma cells. *Biochem. J.* **249:** 135.

Meldolesi, J. and T. Pozzan. 1987. Pathways of Ca^{2+} influx at the plasma membrane: Voltage-, receptor-, and second messenger-operated channels. *Exp. Cell Res.* **171:** 271.

Meyer, T., D. Holowka, and L. Stryer. 1988. Highly cooperative opening of calcium channels by inositol 1,4,5-trisphosphate. *Science* **240:** 653.

Moolenaar, W.H., W. Kruijer, B.C. Tilly, I. Verlaan, A.J. Bierman, and S.W. de Laat. 1986. Growth factor-like action of phosphatidic acid. *Nature* **323:** 171.

Morris, A.P., D.V. Gallacher, R.F. Irvine, and O.H. Petersen. 1987. Synergism of inositol trisphosphate and tetrakisphosphate in activating Ca^{2+}-dependent K^+ channels. *Nature* **330:** 653.

Muallem, S., M. Schoeffield, S. Pandol, and G. Sachs. 1985. Inositol trisphosphate modification of ion transport in rough endoplasmic reticulum. *Proc. Natl. Acad. Sci.* **82:** 4433.

Muir, J.G. and A.W. Murray. 1987. Bombesin and phorbol ester stimulate phosphatidylcholine hydrolysis by phospholipase C: Evidence for a role of protein kinase C. *J. Cell. Physiol.* **130:** 382.

Murayama, T. and M. Ui. 1987. Phosphatidic acid may stimulate membrane receptors mediating adenylate cyclase inhibition and phospholipid breakdown in 3T3 fibroblasts. *J. Biol. Chem.* **262:** 5522.

Nishizuka, Y. 1984. The role of protein kinase C in cell surface signal transduction and tumour promotion. *Nature* **308:** 693.

Ohno, S., Y. Akita, Y. Konno, S. Imajoh, and K. Suzuki. 1988. A novel phorbol ester receptor/protein kinase, nPKC, distantly related to the protein kinase C family. *Cell* **53:** 731.

Pozzan, T., F. DiVirgilio, L. Vicentini, and J. Meldolesi. 1986. Activation of muscarinic receptors in PC12 cells. Stimulation of Ca^+ influx and redistribution. *Biochem. J.* **234:** 547.

Putney, J.W. 1986. A model for receptor-regulated calcium entry. *Cell Calcium* **7:** 1.

Rossier, M.F., A.M. Capponi, and M.B. Vallotton. 1987. Metabolism of inositol 1,4,5-trisphosphate in permeabilized rat aortic smooth-muscle cells. *Biochem. J.* **245:** 305.

Sekiguchi, K., M. Tsukuda, K. Ogita, U. Kikkawa, and Y. Nishizuka. 1987. Three distinct forms of rat brain protein kinase C: Differential response to unsaturated fatty acids. *Biochem. Biophys. Res. Commun.* **145:** 797.

Sheikhnejad, R.G. and P.N. Srivastava. 1986. Isolation and properties of a phosphatidylcholine-specific phospholipase C from bull seminal plasma. *J. Biol. Chem.* **261:** 7544.

Stephens, L., P.T. Hawkins, N. Carter, S.B. Chahwala, A.J. Morris, A.D. Whetton, and C.P. Downes. 1987. L-myo-inositol 1,4,5,6-tetrakisphosphate is present in both mammalian and avian cells. *Biochem. J.* **249:** 271.

Takuwa, N., Y. Takuwa, and H. Rasmussen. 1987. A tumour promoter 12-o-tetradecanoylphorbol 13-acetate increases cellular 1,2-diacylglycerol content through a mechanism other than phosphoinositide hydrolysis in Swiss-mouse 3T3 fibroblasts. *Biochem. J.* **243:** 647.

Thastrup, O., B. Foder, and O. Scharff. 1987. The calcium mobilizing and tumor promoting agent, thapsigargin, elevates the platelet cytoplasmic free calcium concentration to a higher steady state level. A possible mechanism of action for the tumor promoter. *Biochem. Biophys. Res. Commun.* **142:** 654.

Tsien, R.Y. and M. Poenie. 1986. Fluorescence ratio imaging: A new window into intracellular ionic signalling. *Trends Biochem. Sci.* **11:** 450.

Vallejo, M., T.R. Jackson, S.L. Lightman, and M.R. Hanley. 1987. Occurrence and extracellular actions of inositol pentakis- and hexakisphosphate in mammalian brain. *Nature* **330:** 656.

Volpe, P., K.-H. Krause, S. Hasimoto, F. Zorzato, T. Pozzan, J. Meldolesi, and D.P. Lew. 1988. "Calciosome," a cytoplasmic organelle: The inositol 1,4,5-trisphosphate-sensitive Ca^{2+} store of non-muscle cells? *Proc. Natl. Acad. Sci.* **85:** 1091.

White, J., W.B. Amos, and M. Fordham. 1987. An evaluation of confocal versus conventional imaging of biological structures by fluorescence light microscopy. *J. Cell Biol.* **105:** 41.

Whitman, M., C.P. Downes, M. Keeler, T. Keller, and L. Cantley. 1988. Type I phosphatidylinositol kinase makes a novel phospholipid, phosphatidylinositol-3-phosphate. *Nature* **332:** 644.

Winiger, B., F. Wuarin, G. Zahnd, C.B. Wollheim, and W. Schlegel. 1987. Single cell monitoring of cytosolic calcium reveals subtypes of rat lactotrophs with distinct responses to dopamine and thyrotropin-releasing hormone. *Endocrinology* **121:** 2222.

Womack, M., M.R. Hanley, and T.M. Jessell. 1985. Functional substance P receptors on a rat pancreatic acinar cell line. *J. Neurosci.* **5:** 3370.

Woods, N.M., K.S.R. Cuthbertson, and P. Cobbold. 1986. Repetitive transient rises in cytoplasmic free calcium in hormone-stimulated hepatocytes. *Nature* **319:** 600.

Purification and Characterization of Placental Membrane Phosphotyrosine Phosphatases

C.J. Pallen,* L. Sahlin,† G. Panayotou, and M.D. Waterfield
Ludwig Institute for Cancer Research, Courtauld Building, London, W1P 8BT, United Kingdom

The intrinsic tyrosine kinase activity of several growth factor receptors and retroviral oncogene products suggests that tyrosine phosphorylation plays a key role in the processes of cell growth and transformation (for review, see Hunter and Cooper 1985, 1986). Although viral tyrosine kinases may bring about transformation through their overexpression or expression in a form lacking regulatory features present in their cellular counterparts, the activities of the growth factor receptor and proto-oncogene product tyrosine kinases present in normal cells must be subject to strict control. By analogy with the well-characterized reversible phosphorylation systems involving serine- or threonine-specific kinases and phosphatases (Krebs 1986), the activation of tyrosine kinases implies the functional existence of compatible tyrosine phosphatases. Such reversible tyrosine phosphorylations have been observed, for example, in the case of temperature-sensitive mutants of the pp60^{src} tyrosine kinase, where the switch from a permissive to a nonpermissive temperature results in the loss of cellular phosphotyrosine and reversion of the transformed phenotype (Sefton et al. 1980) and in the dephosphorylation of the epidermal growth factor (EGF) receptor in membrane vesicles incubated at 30°C following phosphorylation at 0°C (Carpenter et al. 1979; Brautigan et al. 1981). The phosphotyrosine phosphatases responsible for these phenomena and other like enzymes potentially comprise a primary mechanism for modulation or termination of the activities and/or effects of their counterpart tyrosine kinases.

Several phosphotyrosine phosphatases have been purified or partially purified from the cytosolic fractions of a number of cell types and tissues (Horlein et al. 1982; Chernoff and Li 1983; Foulkes et al. 1983; Nelson and Branton 1984; Shriner and Brautigan 1984; Okada et al. 1986; Tung and Reid 1987). These enzymes exhibit a range of apparent molecular weights (23,000–95,000) and have been mainly identified using a variety of nonphysiological substrates. There appear to be three or more distinct phosphotyrosine phosphatases present in each cell type or tissue examined. All of these activities share a common set of responses to various effector compounds which define this family of enzymes and distinguish them from all other types of phosphatases. They are stimulated or unaffected by EDTA, inhibited by the divalent metal ions Mg^{++} or Mn^{++}, unaffected or only partially inhibited by sodium fluoride, and strongly inhibited by micromolar concentrations of Zn^{++} or vanadate.

In view of the membrane localization of the growth factor receptor tyrosine kinases, several of the nonreceptor tyrosine kinases, and a number of the substrates of these kinases, it is likely that there might coexist one or more phosphotyrosyl phosphatases that could play a role in the phosphorylation/dephosphorylation cycles occurring at or near the cell membrane. Until recently, no membrane tyrosine phosphatases had been purified, although activities with characteristics similar to those of the cytosolic tyrosine phosphatases have been reported to be associated with human erythrocyte membranes (Clari et al. 1987), rabbit kidney membranes (Rotenberg and Brautigan 1987), rat liver plasma membranes (Strout et al. 1988), and Rous-sarcoma-virus-transformed cell membranes (Gallis et al. 1981). Recently, Tonks et al. (1987) have purified two phosphatases from placental membranes that respond to various compounds in a manner similar to that of the cytosolic tyrosine phosphatases and therefore belong to this family of enzymes. We also chose to look for such enzymes in placental membranes where there are known to exist relatively high amounts of the EGF and insulin receptor-kinases. We report here the purification/partial purification of five tyrosine phosphatases from this source and the initial characterization of three of these enzymes. At least one of these phosphatases possesses activity toward the EGF receptor.

METHODS

Materials. The A-431 membranes were prepared according to the method of Thom et al. (1977). Casein kinase II and the catalytic subunit of the cAMP-dependent protein kinase were provided by P. Parker, and EGF was provided by J. Hsuan. Histone-type IIA-S, *p*-nitrophenyl phosphate (PNPP), phosphorylase kinase, and phosphorylase *b* were purchased from Sigma.

Phosphorylation of RR-*src *peptide. The RR-*src* peptide (RRLIEDAEYAARG) was phosphorylated by EGF receptor kinase in a 1.0-ml reaction containing 120 μg A-431 cell membranes, 0.2 μM EGF, 50 mM HEPES (pH 7.4), 150 mM NaCl, 0.02% Triton X-100,

Present addresses: *Institute of Molecular and Cell Biology, National University of Singapore, Singapore 0511, Republic of Singapore; †FFFE, Kliniska Forsknings Laboratoriet, Karolinska Sjukhuset, S-10401, Stockholm, Sweden.

5% glycerol, 2 mM $MnCl_2$, 100 μM Na_3VO_4, 1.4 mM RR-*src*, and 250 μM ATP (2000–3000 cpm/pmole). The mixture was incubated at room temperature for 1 hour in the absence of ATP, and then phosphorylation was initiated by the addition of ATP and continued for 16 hours at 30°C. The reaction was stopped with an equal volume of 1 N acetic acid and centrifuged to remove any denatured protein, and the supernatant was applied to an anion-exchange column (Bio-Rad AG 1X-2, 1.5 × 47 cm) that had been equilibrated in 0.5 N acetic acid. The phosphopeptide was eluted with 1 N acetic acid, while unreacted ATP remained bound to the column. Peak fractions of phosphopeptide were dried down by vacuum centrifugation. These fractions were resuspended for use in routine assays of phosphatase activity. To separate phospho- and dephosphopeptide, the dried fractions were resuspended in water and pooled, and the pool was applied to a 1.0-ml column of antiphosphotyrosine antibody conjugated to Sepharose 4B that had been equilibrated in 0.5 M ammonium acetate (pH 6.8). The column was washed in equilibration buffer, and then phosphopeptide was eluted with 0.5 M ammonium acetate (pH 6.8). Eluted fractions containing peak amounts of radioactivity and low (0.002 maximum) to undetectable absorbance at A_{280} were dried by vacuum centrifugation, resolved in 50 μl H_2O, and combined, and bovine serum albumin was added to a concentration of 1 mg/ml. Phosphopeptide concentration was calculated on the basis of the specific activity of the ATP in the phosphorylation reaction (only one tyrosine residue is present in the peptide substrate).

Phosphorylation of protein substrates. Phosphoseryl casein was prepared in a reaction mixture containing 0.5 mg/ml casein, 3 units/ml casein kinase II, 2.5 mM spermine, 25 mM Tris (pH 7.5), and [γ-^{32}P]ATP (100 μM, 1000 cpm/pmole). Phosphoseryl histone was prepared in a reaction mixture containing 2.0 mg/ml histone, 10 units/ml catalytic subunit of the cAMP-dependent protein kinase, 25 mM Tris (pH 7.5), 10 mM Mg^{++}, and 1 mM DTT. The reactions were carried out at 30°C for 30 minutes and stopped by the addition of cold trichloroacetic acid (TCA) at 20%. After sitting on ice for 20 minutes, the mixtures were centrifuged, and the supernatants were discarded. The pellets were washed twice more with 20% TCA and finally resuspended in 40 μl 1 N NaOH, and then diluted to 0.5 ml with 20 mM Tris (pH 7.5) and 10 mM 2-mercaptoethanol.

Phosphotyrosyl casein or histone was prepared in reaction mixtures containing 0.5 mg/ml casein or 2.0 mg/ml histone, 125 μg A-431 cell membranes, 2 μg/ml EGF, 12 mM Mg^{++}, 2 mM Mn^{++}, 20 mM HEPES (pH 7.4) and [γ-^{32}P]ATP (100 μM, 1000 cpm/pmole). The mixtures were preincubated at 25°C for 30 minutes in the absence of ATP, ATP was then added, and the reactions were incubated at 30°C for 60 minutes. The reactions were stopped with 15 mM EDTA and then microfuged to pellet A-431 membranes. The supernatants were adjusted to 20% TCA and incubated on ice for 20 minutes. This precipitation was repeated twice more, and the final pellets were resuspended as described above. All of the phosphoproteins were then dialyzed against several changes of 20 mM Tris (pH 7.5) and 10 mM 2-mercaptoethanol (48–72 hr) until only a background level of 1–2% free ^{32}P remained.

Phosphorylase *a* was prepared in a reaction mixture containing 0.5 mg/ml phosphorylase *b*, 0.5 mg/ml phosphorylase kinase, 75 mM Tris-HCl/75 mM β-glycerophosphate (pH 8.2), 2.8 mM magnesium acetate, and [γ-^{32}P]ATP (300 μm, 700 cpm/pmole). The mixture was incubated at 30°C for 60 minutes, and an equal volume of cold saturated $(NH_4)_2SO_4$ was then added. The mixture was incubated on ice for 60 minutes, and the resulting precipitate was then pelleted. The pellet was resuspended and dialyzed against 5 mM Tris-HCl (pH 7.0), 1 mM EDTA, and 50 mM 2-mercaptoethanol for 48 hours. The dialysate was microfuged, and the resulting pellet was resuspended and used in assays of phosphatase activity.

Assay of phosphatase activity. Phosphatase activity against the *src* peptide or protein substrates was assayed in reaction mixtures containing phosphosubstrates, 50 mM Mes (pH 6.0), 0.1 M NaCl, 2 mM EDTA, 1 mg/ml bovine serum albumin, 0.5 mM DTT, and 0.01% Chaps. The reaction was carried out at 30°C. In the case of the peptide substrates, the reaction was stopped by a tenfold excess volume of 5% TCA containing 100 mM potassium phosphate, and the phosphate released was quantitated as described by Chan et al. (1986). In the case of protein substrates, the reactions were stopped by the addition of cold 20% TCA, and bovine serum albumin was added (0.25 mg/ml) as a carrier. After incubation on ice for 10 minutes, the reactions were microfuged to pellet the precipitated proteins, and released ^{32}P in the supernatants was quantitated by scintillation counting.

RESULTS

Purification of Tyrosine Phosphatases

The following is a brief outline of the purification steps employed to obtain tyrosine phosphatases. A more detailed description of the purification procedure will be published elsewhere (C.J. Pallen et al., in prep.).

Preparation of placental membranes. Full-term human placentae obtained from normal delivery or Caesarian section were transported on ice to the laboratory. The cord, amnion, and chorion were removed, and the remaining tissue was cut into 1-inch pieces, rinsed twice in cold phosphate-buffered saline (Ca^{++}- and Mg^{++}-free), and homogenized in a Waring blender in 500 ml of Buffer A (25 mM HEPES [pH 7.6], 0.15 M NaCl, 2 mM EGTA, 0.1 mM PMSF, 10 μg/ml aprotinin). All subsequent steps were carried out at 4°C, unless otherwise indicated. The homogenate was centrifuged at 5700*g* for 20 minutes. The pellet was discarded, and the supernatant was centrifuged at 32,000*g* for 30 minutes. The pellet was resuspended in Buffer A using a Wheaton homogenizer and recentrifuged at 32,000*g* for 60 minutes. The pellet was resuspended in Buffer B (50 mM Tris-HCl [pH 7.6], 2% Triton X-100, 0.1 M NaCl, 2 mM EDTA, 0.05 mM PMSF, 10 μg/ml aprotinin) using a Wheaton homogenizer. The suspension was stirred for 60 minutes and then centrifuged at 100,000*g* for 60 minutes. The resulting supernatant either was processed immediately or could be stored at −70°C for at least 1 week until further use.

Heparin-Sepharose chromatography. An outline of the procedure is shown in Figure 1. The solubilized membrane fraction was diluted fourfold in Buffer C and applied to a heparin-Sepharose column equilibrated in Buffer C (50 mM Tris [pH 7.5], 20 mM 2-mercaptoethanol). After washing the column with Buffer C, bound protein was eluted with a 0–0.4 M NaCl gradient in Buffer C. A high level of phosphatase activity was detectable throughout the breakthrough fractions collected during column loading. These were pooled and processed further to obtain two phosphatases (PTP 2 and 3) as described below. A single peak of phosphatase activity eluted from the column at about 0.12 M NaCl (PTP 1).

PTP 1: FPLC gel-filtration chromatography. The active fractions eluting from the above column were pooled, and Chaps was added to a final concentration of 0.01%. The pool was concentrated to a volume of 1–2 ml by vacuum centrifugation. Phosphatase activity was unaffected by the high salt concentration resulting from this step. The concentrated sample was applied to two linked FPLC Superose 12 gel-filtration columns equilibrated in Buffer C containing 0.2 M NaCl and 0.1% Chaps. A single peak of phosphatase activity eluted with an estimated molecular weight of 50,000.

PTP 1: FPLC ion-exchange chromatography. Fractions comprising the above peak of activity were pooled, diluted tenfold with Buffer C containing 0.1% Chaps or dialyzed to remove salt, and applied to an FPLC Mono Q column. No phosphatase activity was detectable in the column breakthrough. Protein bound to the column was eluted with a 0–0.5 M NaCl gradient. A major peak of phosphatase activity eluted at about 1,7,100.1 M NaCl. These fractions were pooled, dialyzed against Buffer D (50 mM Tris [pH 7.5], 0.1 M NaCl, 50% glycerol, 0.5 mM DTT, 0.01% Chaps), and stored in small aliquots at −70°C. The activity was stable under these storage conditions for at least 6 months. This enzyme was termed PTP 1.

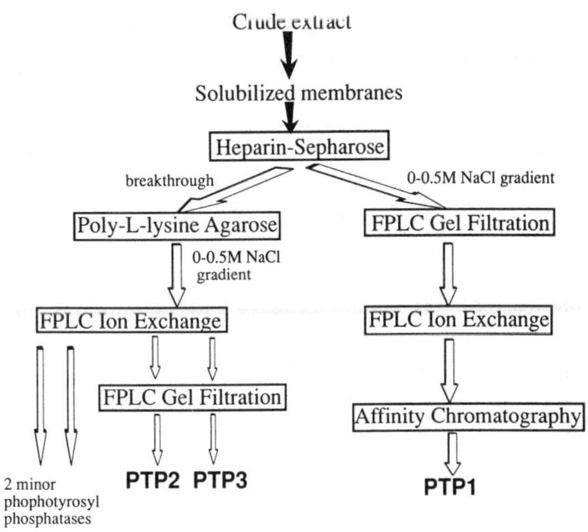

Figure 1. Phosphotyrosyl phosphatase purification scheme.

PTP 2 and 3: Poly-L-lysine agarose chromatography. The breakthrough fractions from the heparin-Sepharose column were pooled and applied to a poly-L-lysine agarose column. Approximately two thirds of the phosphatase activity did not bind to the column and this fraction was not processed further. After washing the column with Buffer C, bound protein was eluted with a 0–0.5 M NaCl gradient. A peak of phosphatase activity was eluted from the column at 0.45 M NaCl. Fractions containing this activity were pooled and dialyzed against Buffer C containing 0.1% Chaps.

PTP 2 and 3: FPLC ion-exchange chromatography. The dialyzed pool was applied to an FPLC Mono Q column equilibrated in Buffer C containing 0.1% Chaps, and protein was eluted with a gradient of 0–0.5 M NaCl. Two major and two minor peaks of phosphatase activity were detected in the eluted fractions. The two major peaks were termed PTP 2 and PTP 3, in order of their elution from the column.

PTP 2 and 3: FPLC gel-filtration chromatography. Active fractions of PTP 2 and PTP 3 were separately pooled and concentrated by vacuum centrifugation, and each pool was chromatographed on two linked FPLC Superose 12 columns. The estimated molecular weights of PTP 2 and PTP 3 were 51,000 and 32,000, respectively. The active fractions of each enzyme were pooled, dialyzed against Buffer D, and stored in small aliquots at −70°C.

Affinity chromatography of PTP 1. When analyzed by SDS-PAGE, the preparation of PTP 1 obtained from the last Mono Q chromatography step contained several protein bands as visualized by silver staining following SDS-PAGE. Therefore, a further step was necessary to improve the enzyme homogeneity. The best method to specifically isolate the protein at this stage of purification was judged to be immobilized substrate affinity chromatography. Certain serine-threonine phosphatases (Pato and Adelstein 1980; Tung et al. 1985) as well as another tyrosine phosphatase (Tonks et al. 1987) have been successfully purified to homogeneity using such a method involving chromatography on an immobilized thiophosphorylated substrate column. However, the stoichiometry of phosphorylation of the *src* peptide is low (see Methods), and the use of thiophosphorylated *src* as an affinity ligand would therefore be inefficent. Instead, we have used *p*-aminobenzylphosphonic acid (PABP) as a ligand. This compound has high structural homology with phosphotyrosine and has been used in a diazotized form as an immunogen for preparation of antiphosphotyrosine antibodies (Ross et al. 1981). When the PTP 1 preparation was applied to a column of immobilized PABP, protein but no phosphatase activity was detected in the breakthrough fractions. A very sharp peak of phosphatase activity could be eluted from the affinity column by a salt gradient. These active fractions were pooled and rerun on a regenerated column in an identical manner (Fig. 2). When the active fractions eluted from this column were analyzed by SDS-PAGE, a major band of protein with a molecular weight of 32,000–33,000 could be visualized by staining with Coomassie blue. In addition, one or two faint bands of protein in the molecular-weight range of 40,000–60,000 were observed (Fig. 2). We are currently in the process of attempting to identify conclusively which of these proteins is the tyrosine phosphatase by regenerating enzyme activity following SDS-PAGE.

Characterization of PTP 1, 2, and 3

To establish into which class(es) of phosphatases enzymes PTP 1, 2, and 3 might be related, we have characterized these enzymes in terms of activity requirements, susceptibility to various known phosphatase inhibitors/activators, and substrate specificity. This work was carried out prior to our recent employment of PABP as an affinity ligand, and therefore the preparations of PTP 1, 2, and 3 used to produce the results described below were obtained following chromatography on Mono Q (PTP 1) or Superose 12 (PTP 2 and 3).

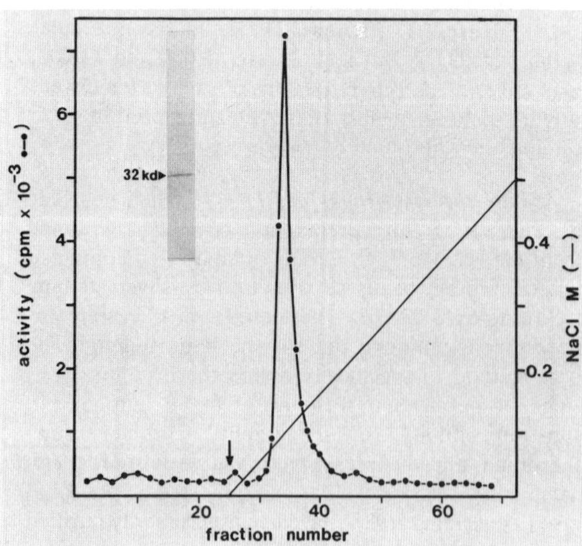

Figure 2. Purification of PTP 1 by affinity chromatography on immobilized PABP acid. Fractions were assayed for phosphatase activity as described in Methods. (*Inset*) Coomassie blue staining of a peak fraction after SDS-PAGE.

Activity requirements. The optimal pH for activity of PTP 1 toward the phospho-*src* peptide is 6.0, whereas both PTP 2 and 3 exhibit pH optima of 5.5. In all cases, the pH profiles of these enzymes are broad; e.g., PTP 1 is still 40% as active at pH 7.5 as it is at pH 6.0. As has been demonstrated for the majority of phosphotyrosine phosphatase activities identified (for review, see Foulkes 1983), PTP 1, 2, and 3 are stimulated by EDTA (Table 1). They are also stimulated to about an equal extent by EGTA, although in the case of other tyrosine phosphatases, this chelator is not as potent as EDTA (Chernoff and Li 1983). In contrast to the type 1 and 2 serine/threonine phosphatases (for review, see Ballou and Fischer 1986), PTP 1, 2, and 3 are inhibited by Mn^{++} and Mg^{++}. PTP 2 and 3 appear to be particularly susceptible to Mn^{++} inhibition as has been observed with some cytosolic tyrosine phosphatases from chicken brain (Foulkes et al. 1983). The divalent metal ion Ca^{++} has little or no effect on the activities of the three enzymes, whereas calmodulin alone or Ca^{++}/calmodulin in conjunction strongly inhibit the phosphatase activities. PTP 1, 2, and 3 are inhibited by small phosphocompounds such as ATP and ADP, and virtually no activities were observed in the presence of pyrophosphate. Due to the ability of sodium fluoride to inhibit serine/threonine phosphatases (Khatra and Soderling 1978), we examined the effect of this compound on PTP 1. In the presence of 2 mM EDTA and 50 mM NaF, only a partial inhibition (28%) was observed. In contrast, PTP 1 was subject to inhibition by the well-known tyrosine phosphatase inhibitors vanadate (Swarup et al. 1982) and Zn^{++} (Brautigan et al. 1981), with 50% inhibition observed at 0.3 μM and 95 μM, respectively.

Substrate specificity. Throughout the purification procedure, phosphatase activity was assayed against the phospho-*src* peptide, which was phosphorylated to a stoichiometry of only 0.05 mole phosphate/mole peptide. Therefore, the specific activities of the phosphatases toward this substrate must be considered as minimum values, since approximately 95% of the peptide added to the assay is already in the unphosphorylated or phosphatase reaction product form. To overcome this problem and obtain more accurate estimates of the abilities of the preparations of PTP 1, 2, and 3 to dephosphorylate the *src* peptide, we separated phospho- and nonphospho-*src* using an antiphosphotyrosine monoclonal antibody affinity column as described (Goris et al. 1988). A study of the kinetics of dephosphorylation of the resulting phospho-*src* (estimated 1 mole phosphate/mole peptide) resulted in observed K_m values of 4, 67, and 30 μM and V_{max} values of 70, 141, and 14 nmole/min/mg for PTP 1, 2, and 3 respectively. Higher affinities and maximal velocities may be expected with the affinity-purified phosphatases; this work is in progress. No activities of any of PTP 1, 2, or 3 toward free phosphotyrosine, phosphothreonine, or phosphoserine were observed.

Table 1. Effects of Various Compounds on Tyrosine Phosphatase Activity

Effector	Conc. (mM)	% Phosphatase activity PTP 1	PTP 2	PTP 3
None	—	100	100	100
EDTA	2	155	162	234
EGTA	2	168	165	227
$MnCl_2$	2	59	26	5
$MgCl_2$	20	47	53	66
$CaCl_2$	1	91	106	91
Calmodulin		5	11	1
Calmodulin + $CaCl_2$		19	18	8
ATP	1	33	30	37
ADP	1	33	36	10
PPi	2	0.4	2	1
50 mM NaF + 2 mM EDTA		72		

Reactions were carried out as described in Methods and stopped in the linear portion of the reaction time course. PTP 1 was present at 1.2 μg/ml, PTP 2 at 3.0 μg/ml, and PTP 3 at 9.0 μg/ml.

In view of the observed abilities of PTP 2 and 3 to dephosphorylate PNPP and also of the activity of alkaline phosphatase toward PNPP and phosphotyrosyl substrates (Swarup et al. 1981), we quantitated the specific activities of PTP 1, 2, and 3 toward PNPP to see how these would compare with that observed with placental alkaline phosphatase (324 μmole/min/mg; Harkness 1986). PTP 1, 2, and 3 had respective specific activities of 135, 95, and 22 nmole/min/mg when measured at pH 5.8, more than a 1000-fold lower order of magnitude than that of alkaline phosphatase. No activity was detectable when PTP 1, 2, and 3 were assayed against this substrate at pH 9.

To assess whether PTP 1, 2, and 3 are specific for phosphotyrosine residues, we examined the ability of these enzymes to dephosphorylate histone and casein phosphorylated on either tyrosine or serine/threonine. In all cases, only the phosphotyrosyl-containing proteins served as substrates, and little to no detectable activity was observed toward the phosphoseryl/phosphothreonyl-containing proteins (Fig. 3). Phosphorylase *a* (phosphorylated on serine residues), one of the best substrates for certain of the type-1 and type-2 serine/threonine-specific protein phosphatases (Ingebritsen et al. 1983), also was not dephosphorylated (<1% phosphate removed) by PTP 1, 2, or 3.

The presence of PTP 1, 2, and 3 in placental membranes suggests that these enzymes may play a role in the regulation of phosphotyrosine levels of membrane-associated proteins. Predominant among such phosphotyrosyl-containing proteins are the growth factor receptors, and because placental membranes are rich in EGF receptor, this phosphoprotein (obtained from A-431 cell membranes) was assayed as a potential substrate for PTP 1. The EGF receptor was found to be rapidly dephosphorylated by the tyrosine phosphatase, with 50% dephosphorylation reached in 10 minutes (Fig. 4). Interestingly, this was the maximum level of dephosphorylation observed even when the reaction was continued for a further 10 minutes.

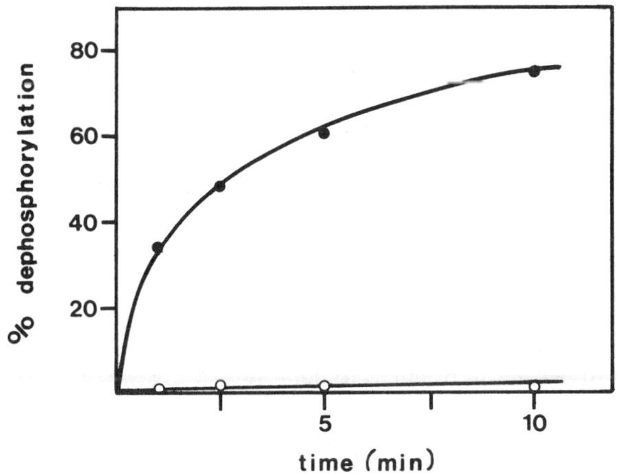

Figure 3. Dephosphorylation of P-Tyr (●) and P-Ser (○) casein by PTP 1. Reaction mixtures (250 μl) contained 4.5 μg/ml PTP 1 and 0.75 mg/ml Tyr-casein (total 4000 cpm) or 0.3 mg/ml Ser-casein (total 150,000 cpm). Aliquots of 40 μl were removed at the indicated times.

DISCUSSION

Three phosphatases have been purified and two minor phosphatases have been partially purified from human placental membranes using phosphotyrosyl *src*

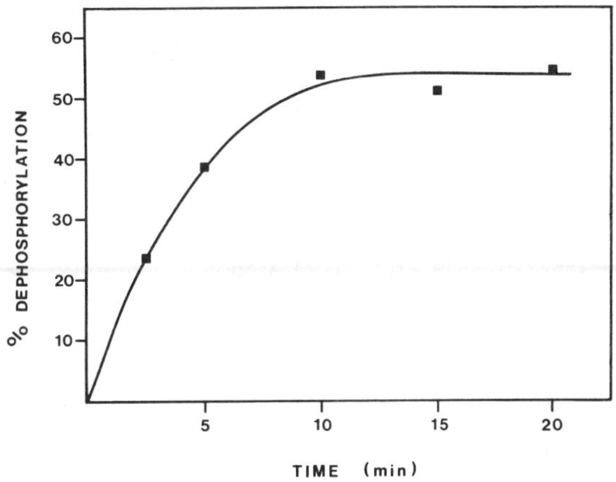

Figure 4. Dephosphorylation of EGF receptor by PTP 1.

peptide as a substrate. There appear to be a multiplicity of such enzymes in this tissue; several pools of activity were fractionated during the purification but not processed further. It is possible that still other tyrosine phosphatases with different substrate specificities are present that have remained undetectable by assay with the *src* peptide. Additionally, we did not attempt to purify any tyrosine phosphatases from the cytosolic fraction of this tissue. The number of distinct phosphatase activities identified in placental membrane is by no means unique; an equally wide variety of such activities have been described in a number of cell types and tissues, although the majority of these studies were concerned with cytosolic rather than particulate fractions. Recently, Tonks et al. (1987) have reported the purification of tyrosine phosphatases from both particulate and cytosolic fractions of placenta. Their studies indicate that the particulate phosphatases may have soluble counterparts. Rotenberg and Brautingan (1987) suggest that some cytosolic tyrosine phosphatases are derived by proteolysis of the membrane enzymes. The elucidation of the relationships between the various tyrosine phosphatases reported in the literature represents a very necessary future direction of research in this field but is presently complicated by the fact that these activities have been identified by separate groups, each using a different substrate.

The responses of the three tyrosine phosphatases PTP 1, 2, and 3 to various inhibitors and activators are characteristic of most of the other phosphotyrosyl protein phosphatase activities described to date, irrespective of enzyme source. No inhibition of activity by EDTA (or in certain cases, stimulation of activity such as observed with PTP 1, 2, and 3), a corresponding reduction of activity in the presence of the divalent metal ions Mn^{++} or Mg^{++}, and inhibition by micromolar concentrations of Zn^{++} or vanadate are responses that appear to define this family of enzymes. Most importantly, the responses of PTP 1, 2, and 3 to the above effectors, in conjunction with their phosphotyrosyl substrate specificity, distinguish them from alkaline and acid phosphatases and from the phosphoseryl/threonyl protein phosphatase types 2A and 2B, which have been shown also to possess phosphotyrosyl phosphatase activities (Swarup et al. 1981; Leis and Kaplan 1982; Chernoff et al. 1983, 1984; Foulkes et al. 1983; Pallen et al. 1985; Chan et al. 1986). The inhibitory effect of Mg^{++}, absence of activity at pH 9, and inability to dephosphorylate phosphoseryl histone or casein demonstrate that PTP 1, 2, and 3 cannot be alkaline phosphatases. Whereas several types of acid phosphatases show nonuniform responses to a variety of compounds, both bone and prostatic acid phosphatases differ from PTP 1, 2, and 3 in that they are strongly or completely inhibited by relatively low concentrations of fluoride (1 mM and 5 mM, respectively), can hydrolyze free phosphotyrosine, and, in the case of the skeletal enzyme, undergo activation in the presence of Mn^{++} and low concentrations of Zn^{++} (<1 mM) (Li et al. 1984; Lau et al. 1985, 1987). The effects of EDTA and Mn^{++} on the PTP activities are exactly opposite to their effects on protein phosphatase 2A (Foulkes et al. 1983). Furthermore, phosphorylase *a* is an excellent substrate for the latter enzyme (Ingebritsen et al. 1983), but it is not dephosphorylated by PTP 1, 2, or 3. It has been shown that ATP or inorganic pyrophosphate actually stimulates the phosphotyrosyl phosphatase activity of the type-2A enzymes (Goris et al. 1988), whereas these compounds are inhibitors of the PTP activities. Again, the observed effects of EDTA, Mg^{++}, and Mn^{++}, and in particular the inhibition of PTP 1, 2, and 3 by Ca^{++}/calmodulin, distinguish these enzymes from protein phosphatase 2B.

The results of our study to date indicate that PTP 1, 2, and 3 have an absolute specificity for phosphotyrosyl-containing substrates. However, within this category of substrates, we have not been able to detect a higher degree of specificity, excepting the inability of these three phosphatases to dephosphorylate free phosphotyrosine, suggesting a preference for protein or at least peptide substrate. Kidney cytosolic tyrosine phosphatases have been reported to be specific for acidic proteins, and it has consequently been suggested that the use of basic phosphotyrosyl-containing proteins such as histone or phosphotyrosyl peptides as substrates could only be expected to measure the activity of alkaline phosphatases and not tyrosine phosphatases (Sparks and Brautigan 1985). The demonstrated activity of PTP 1, 2, and 3 toward both acidic (casein) and basic (histone) protein substrates, as well as toward the *src* peptide, indicates that these membrane enzymes do not possess a substrate preference comparable in nature to that of the kidney cytosolic tyrosine phosphatases and also reveals potential pitfalls in assuming that all phosphotyrosyl histone phosphatase activity can be attributed to alkaline phosphatase. It is possible that some intrinsic features of the membrane environment or of the membrane-associated phosphatases themselves confer upon these enzymes the ability to dephosphorylate proteins irrespective of the net negative charge of the substrate. Other as yet undetermined properties of PTP 1, 2, and 3 may determine their substrate specificities, if any. The EGF receptor undergoes autophosphorylation on several tyrosine residues in the carboxy-terminal region in response to EGF (Downward et al. 1984). The observation that PTP 1 can only remove 50% of the tyrosine phosphate from the EGF receptor even after extended incubation suggests the possibility that the enzyme may possess a specificity for certain of the phosphorylation sites. Studies are in progress to determine if this is the case.

Perhaps the most important future direction in this area of research is the search for in vivo substrates of the tyrosine phosphatases. This is hampered by the lack of abundance of known phosphotyrosyl-containing proteins, by the frequently observed low stoichiometry of their phosphorylation, and by a lack of knowledge of the cellular pathways of tyrosine phosphorylation. The presence of PTP 1, 2, and 3 in solubilized membrane fractions and the identification/purification of other membrane tyrosine phosphatases (Gallis et al. 1981; Clari et al. 1987; Rotenberg and Brautigan 1987; Tonks et al. 1987; Strout et al. 1988) suggest that these enzymes may function to control the phosphotyrosine levels of membrane-associated proteins. A role in cell growth and transformation is indicated for the growth factor receptor tyrosine kinases of membranes and for oncogene-related tyrosine kinases, many of which are membrane-bound. PTP 1 can dephosphorylate the EGF receptor; Tonks et al. (1987) have purified a tyrosine phosphatase that dephosphorylates the insulin receptor, and on the basis of the abilities of PTP 1, 2, and 3 to dephosphorylate the *src* peptide in vitro, it is possible that they may also act on the protein tyrosine kinase pp60^{src}. The effects or defects (in the absence of dephosphorylation) of or in these activities at what are essentially control points in tyrosine phosphorylation pathways may prove decisive in the determination of subsequent cellular events.

ACKNOWLEDGMENT

C.J.P. was the recipient of a fellowship from the Medical Research Council of Canada.

REFERENCES

Ballou, L.M. and E.H. Fischer. 1986. Phosphoprotein phosphatases. In *The enzymes* (ed. P.D. Boyer and E.G. Krebs), vol. 17, p. 311. Academic Press, Orlando, Florida.

Brautigan, D.L., P. Bornstein, and B. Gallis. 1981. Phosphotryrosyl-protein phosphatase. *J. Biol. Chem.* **256:** 6519.

Carpenter, G., L. King, and S. Cohen. 1979. Rapid enhancement of protein phosphorylation in A-431 cell membrane preparations by epidermal growth factor. *J. Biol. Chem.* **254:** 4884.

Chan, C.P., B. Gallis, D.K. Blumenthal, C.J. Pallen, J.H. Wang, and E.G. Krebs. 1986. Characterization of the phosphotyrosyl protein phosphatase activity of calmodulin-dependent protein phosphatase. *J. Biol. Chem.* **261:** 9890.

Chernoff, J. and H.-C. Li. 1983. Multiple forms of phosphotyrosyl- and phosphoseryl-protein phosphatase from cardiac muscle: Partial purification and characterization of an EDTA-stimulated phosphotyrosyl-protein phosphatase. *Arch. Biochem. Biophys.* **226:** 517.

Chernoff, J., M.A. Sells, and H.-C. Li. 1984. Characterization of phosphotyrosyl-protein phosphatase activity associated with calcineurin. *Biochem. Biophys. Res. Commun.* **121:** 141.

Chernoff, J., H.-C. Li, Y.-S.E. Cheng, and L.B. Chen. 1983. Characterization of a phosphotyrosyl protein phosphatase activity associated with a phosphoseryl protein phosphatase of $M_r = 95{,}000$ from bovine heart. *J. Biol. Chem.* **258:** 7852.

Clari, G., A.M. Brunati, and V. Moret. 1987. Membrane-bound phosphotyrosyl-protein phosphatase activity in human erythrocytes. Dephosphorylation of membrane band 3 protein. *Biochem. Biophys. Res. Commun.* **142:** 587.

Downward, J., P. Parker, and M.D. Waterfield. 1984. Autophosphorylation sites on the epidermal growth factor receptor. *Nature* **311:** 483.

Foulkes, J.G. 1983. Phosphotyrosyl-protein phosphatases. *Curr. Top. Microbiol. Immunol.* **107:** 163.

Foulkes, J.G., E. Erikson, and R.L. Erikson. 1983. Separation of multiple phosphotyrosyl- and phosphoseryl-protein phosphatases from chicken brain. *J. Biol. Chem.* **258:** 431.

Gallis, B., P. Bornstein, and D.L. Brautigan. 1981. Tyrosylprotein kinase and phosphatase activities in membrane vesicles from normal and Rous sarcoma virus-transformed rat cells. *Proc. Natl. Acad. Sci.* **78:** 6689.

Goris, J., C.J. Pallen, P.J. Parker, J. Hermann, M.D. Waterfield, and J. Merlevede. 1988. Conversion of a phosphoseryl/threonyl phosphatase into a phosphotyrosyl phosphatase. *Biochem. J.* (in press).

Harkness, D.R. 1968. Studies on human placental alkaline phosphatase. II. Kinetic properties and studies on the apoenzyme. *Arch. Biochem. Biophys.* **126:** 513.

Horlein, D., B. Gallis, D.L. Brautigan, and P. Bornstein. 1982. Partial purification and characterization of phosphotyrosyl-protein phosphatase from Ehrlich ascites tumor cells. *Biochemistry* **21:** 5577.

Hunter, T. and J.A. Cooper. 1985. Protein-tyrosine kinases. *Annu. Rev. Biochem.* **54:** 897.

———. 1986. Viral oncogenes and tyrosine phosphorylation. In *The enzymes* (ed. P.D. Boyer and E.G. Krebs), vol. 17, p. 191. Academic Press, Orlando, Florida.

Ingebritsen, T.S., J.G. Foulkes, and P. Cohen. 1983. The protein phosphatases involved in cellular regulation 2. Glycogen metabolism. *Eur. J. Biochem.* **132:** 263.

Khatra, B.S. and T.R. Soderling. 1978. Reversible inhibition of skeletal muscle phosphoprotein phosphatase by ATP, phosphate and fluoride. *Biochem. Biophys. Res. Commun.* **85:** 647.

Krebs, E.G. 1986. The enzymology of control by phosphorylation. In *The enzymes* (ed. P.D. Boyer and E.G. Krebs), vol. 17, p. 3. Academic Press, Orlando, Florida.

Lau, K.-H.W., J.R. Farley, and D.L. Baylink. 1985. Phosphotyrosyl-specific protein phosphatase activity of a bovine skeletal acid phosphatase isoenzyme. *J. Biol. Chem.* **260:** 4653.

———. 1987. Bone acid phosphatase is a neutral pH phosphotyrosyl protein phosphatase. *Adv. Prot. Phos.* **4:** 165.

Leis, J.F. and N.O. Kaplan. 1982. An acid phosphatase in the plasma membranes of human astrocytoma showing marked specificity toward phosphotyrosine protein. *Proc. Natl. Acad. Sci.* **79:** 6507.

Li, H.C., J. Chernoff, L.B. Chen, and A. Kirschonbaum. 1984. A phosphotyrosyl-protein phosphatase activity associated with acid phosphatase from human prostate gland. *Eur. J. Biochem.* **138:** 45.

Nelson, R.L. and P.E. Branton. 1984. Identification, purification and characterization of phosphotyrosine-specific protein phosphatases from cultured chicken embryo fibroblasts. *Mol. Cell. Biol.* **4:** 1003.

Okada, M., K. Owada, and H. Nakagawa. 1986. [Phosphotyrosine] protein phosphatase in rat brain. *Biochem. J.* **239:** 155.

Pallen, C.J., K.A. Valentime, J.H. Wang, and M.D. Hollenberg. 1985. Calcineurin-mediated dephosphorylation of the human placental membrane receptor for epidermal growth factor urogastrone. *Biochemistry* **24:** 4727.

Pato, M.D. and R.S. Adelstein. 1980. Dephosphorylation of the 20,000-dalton light chain of myosin by two different phosphatases from smooth muscle. *J. Biol. Chem.* **255:** 6536.

Ross, A.H., D. Baltimore, and H.N. Eisen. 1981. Phosphotyrosine-containing proteins isolated by affinity chromatography with antibodies to a synthetic hapten. *Nature* **294:** 654.

Rotenberg, S.A. and D.L. Brautigan. 1987. Membrane protein phosphotyrosine phosphatase in rabbit kidney. *Biochem. J.* **243:** 747.

Sefton, B.M., T. Hunter, K. Beemon, and W. Eckhart. 1980. Evidence that the phosphorylation of tyrosine is essential for cellular transformation by Rous sarcoma virus. *Cell* **20:** 807.

Shriner, C.L. and D.L. Brautigan. 1984. Cytosolic protein phosphotyrosine phosphatases from rabbit kidney. *J. Biol. Chem.* **259:** 11383.

Sparks, J.W. and D.L. Brautigan. 1985. Specificity of protein phosphotyrosine phosphatases. *J. Biol. Chem.* **260:** 2042.

Strout, H.V., P.P. Vicario, R. Saperstein, and E. Slater. 1988. A protein phosphotyrosine phosphatase distinct from alkaline phosphatase with activity against the insulin receptor. *Biochem. Biphys. Res. Commun.* **151:** 633.

Swarup, G., S. Cohen, and D.L. Garbers. 1981. Selective dephosphorylation of proteins containing phosphotyrosine by alkaline phosphatases. *J. Biol. Chem.* **256:** 8197.

———. 1982. Inhibition of membrane phosphotyrosyl-protein phosphatase activity by vanadate. *Biochem. Biophys. Res. Commun.* **107:** 1104.

Thom, D., A.J. Powell, C.W. Lloyd, and D.A. Rees. 1977. Rapid isolation of plasma membranes in high yield from cultured fibroblasts. *Biochem. J.* **168:** 187.

Tonks, N.K., C.D. Diltz, and E.H. Fischer. 1987. A distinct class of phosphatases from human placenta which dephosphorylate tyrosyl residues in proteins. *Adv. Protein Phos.* **4:** 431.

Tung, H.Y.L. and L.J. Reed. 1987. Identification and purification of a cytosolic phosphotyrosyl protein phosphatase from bovine spleen. *Anal. Biochem.* **161:** 412.

Tung, H.Y.L., S. Alemany, and P. Cohen. 1985. The protein phosphatases involved in cellular regulation. 2. Purification, subunit structure and properties of protein phosphatases-2A, $2A_1$ and $2A_2$ from rabbit skeletal muscle. *Eur. J. Biochem.* **148:** 253.

Signal Transduction by the Platelet-derived Growth Factor Receptor

L.T. Williams, J.A. Escobedo, M.T. Keating, and S.R. Coughlin
Department of Medicine, Cardiovascular Research Institute, and Howard Hughes Medical Institute, University of California, San Francisco, California 94143-0724

When platelet-derived growth factor (PDGF) binds to its receptor on the surface of a mesenchymal cell, it stimulates almost every known type of hormone-activated signal transduction pathway. Within seconds, tyrosine kinase activity is stimulated (Nishimura et al. 1982; Ek et al. 1983; Frackelton et al. 1984), phosphatidylinositol (PtdIns) is hydrolyzed (Habenicht et al. 1981), the cellular pH changes (L'Allemain et al. 1984), and cytosolic calcium increases (Escobedo et al. 1988a). Other rapid responses include dramatic changes in the cytoskeleton, enhanced expression of a group of genes (Kelly et al. 1983; Greenberg and Ziff 1984), elevation in the level of cellular cAMP (Rozengurt et al. 1983), and internalization of the receptor. How these events culminate in DNA synthesis is still unknown. However, recent progress in understanding the structure and function of the PDGF receptor is providing mechanistic insight into the interrelationships among the pathways and their roles in mitogenesis.

The initial studies of PDGF were done in the context of its role in wound healing and atherogenesis. More recently, it has become apparent that many tissues express PDGF-like compounds and that PDGF is likely to be involved in embryonic development (Rizzino and Bowen-Pope 1985) and smooth muscle contraction (Berk et al. 1986). PDGF was first purified from outdated human platelets as a 32-kD disulfide-linked heterodimer consisting of two homologous polypeptide chains, A and B (Antoniades 1981). In addition to this A-B heterodimer, there are two homodimeric forms of PDGF, A-A and B-B, that are expressed by several nonplatelet tissues (Betsholtz et al. 1986; Heldin et al. 1986). In the interest of simplicity, all of these forms of PDGF are referred to as "platelet"-derived growth factor, indicating that they are structurally related to the form originally isolated from platelets.

At least three modes of action of PDGF have been found. First, it acts locally as a hormone released from activated platelets or macrophages at the site of tissue injury. Second, PDGF released by endothelial cells can act as a paracrine factor to stimulate the proliferation of nearby vascular smooth muscle cells (Barrett et al. 1984; Dicorleto and Bowen-Pope 1984). In this context, the expression of PDGF is transcriptionally regulated by both stimulatory (Daniel et al. 1986; Starksen et al. 1987; Kavanaugh et al. 1988) and inhibitory (Daniel et al. 1987; G.R. Harsh et al., in prep.) factors. Finally, PDGF can act as an autocrine factor, stimulating receptors on the same cells that produce the factor (Robbins et al. 1983; Johnsson et al. 1985; Habenicht et al. 1986; Keating and Williams 1988). The autocrine mode of action has been studied in most detail in cells transformed by the v-*sis* oncogene.

Whether PDGF acts through autocrine or paracrine processes, its mitogenic effect is mediated by the PDGF receptor. The 180-kD mouse PDGF receptor was recently purified on the basis of its ability to become tyrosine phosphorylated in response to the A-B human platelet form of PDGF (Daniel et al. 1985), and the receptor amino acid sequence was determined from a full-length cDNA clone (Yarden et al. 1986). The extracellular portion of this mouse receptor consists of a series of five immunoglobulin-like domains (J.F. Bazan et al., in prep.) that constitute the PDGF-binding region. The intracellular tyrosine kinase domain is split into two portions by a 104-amino-acid "kinase insert" (KI) region (Yarden et al. 1986). Recently obtained human cDNA clones (Fig. 1) confirmed that the human and mouse receptor sequences are highly conserved in the intracellular region (96% identity) and very similar in the extracellular domain (77% identity). The human receptor, like its murine homolog, contains five immunoglobulin-like domains in the extracellular region (J.F. Bazan et al., in prep.), as well as kinase and KI regions in the intracellular domain.

Both the human (Escobedo et al. 1988b) and mouse (Escobedo et al. 1988a) receptor cDNA sequences have been expressed in Chinese hamster ovary fibroblast (CHO) cells that normally lack PDGF receptors. When expressed in these cells, the receptors mediate most of the signal transduction pathways stimulated by PDGF in cells that normally respond to this factor. This paper summarizes recent results using this system to study signal transduction by the PDGF receptor. The findings show that (1) the KI region of the receptor plays an important role in mitogenesis but is not essential for many of the early responses to PDGF, including tyrosine kinase activation, PtdIns turnover, increase in cytoplasmic calcium, receptor internalization, and change in intracellular pH; (2) the tyrosine kinase activity of the receptor is essential for mitogenesis and for PtdIns hydrolysis but is not required for ligand-induced receptor down-regulation; (3) the transmembrane region of the receptor serves an important and specific role in signal transduction, other than simply providing a membrane anchor for the receptor; (4) when acti-

A

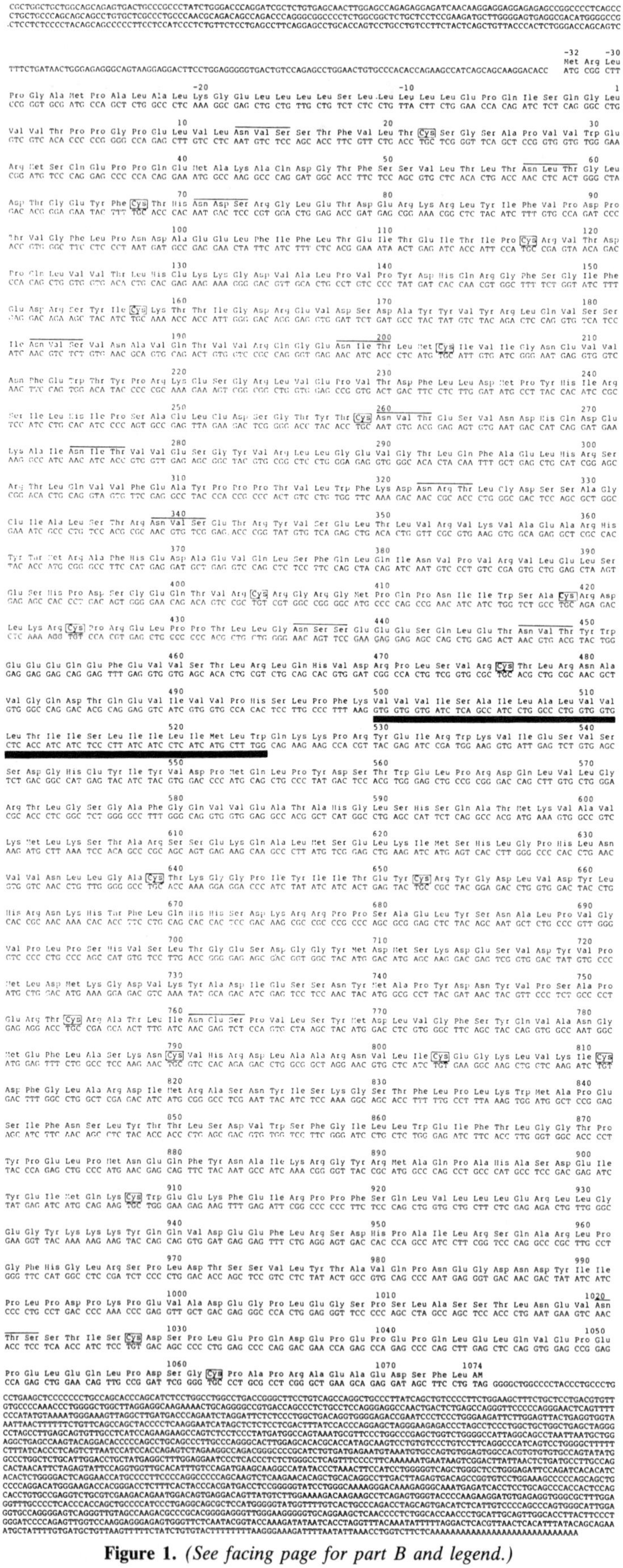

Figure 1. *(See facing page for part B and legend.)*

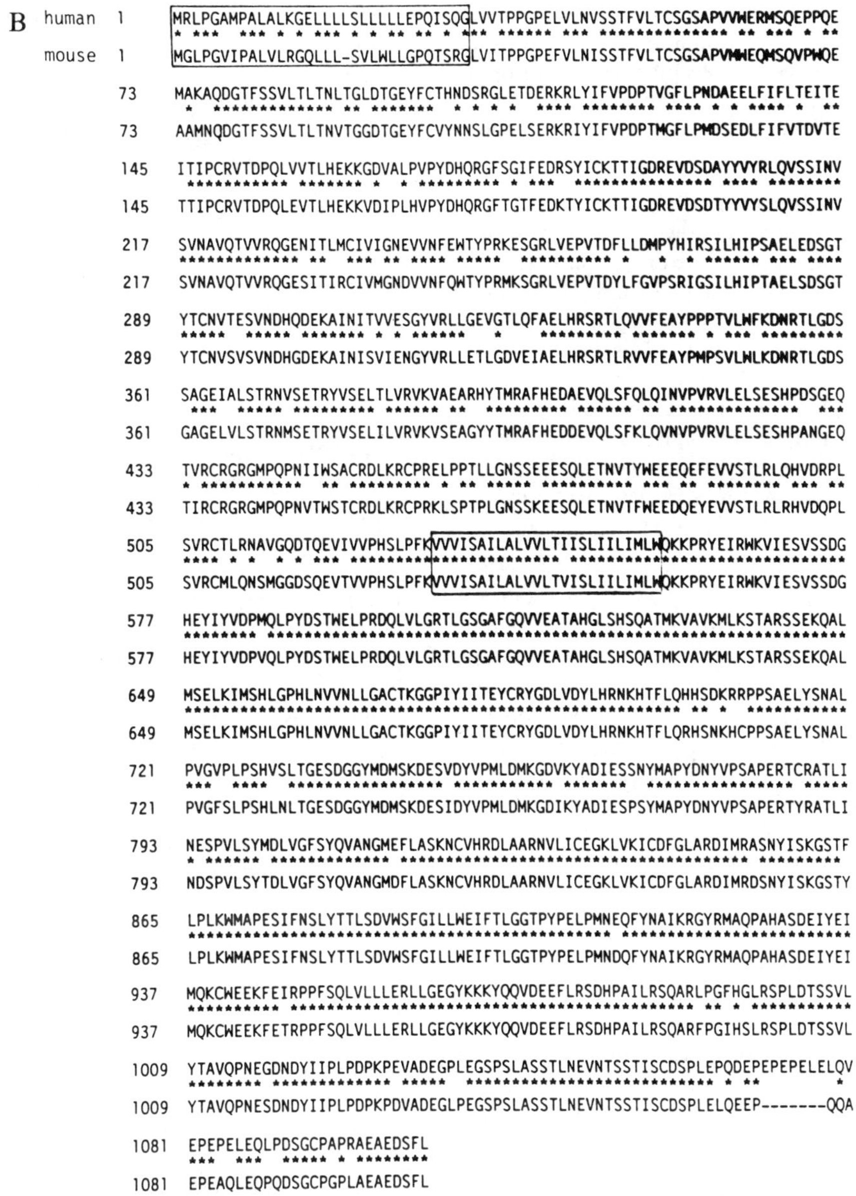

Figure 1. (*A*) Nucleotide and predicted amino acid sequence of a cDNA clone that encodes that human PDGF receptor. The sequence beginning at amino acid 1 corresponds to the amino terminus of the purified mouse PDGF receptor protein (Yarden et al. 1986). The first 32 amino acids (−32 to −1) encode the signal peptide sequence. The dark bar underlines the transmembrane sequence (amino acid residues 500–524). A line is placed over the potential *N*-glycosylation sites. The polyadenylation site in the 3′ end of the cDNA has been underlined. (*B*) Comparison of human and mouse PDGF receptor cDNA sequences. Asterisks indicate nucleotide identity. The signal sequence and transmembrane sequences are enclosed in boxes.

vated, the wild-type receptor associates with PDGF-sensitive PtdIns kinase, but nonmitogenic mutants of the receptor failed to associate with this kinase; (5) the receptor undergoes a conformational change when activated by PDGF.

METHODS

Construction and expression of PDGF receptor mutants. The details of the construction of the KI deletion mutant, a carboxy-terminal deletion (Δct) mutant, the ATP-binding site mutant (K602A), the transmembrane mutants, and the tyrosine-to-phenylalanine point mutations will be reported elsewhere (J.A. Escobedo and L.T. Williams; J.A. Escobedo et al.; both in prep.). In each case, the mutant cDNA clones were sequenced using dideoxy nucleotides and the Sequenase sequencing system. All of the mutated receptor cDNA sequences were cloned into an expression vector under the transcriptional control of the SV40 early promoter as described previously (Escobedo et al. 1988a). Stable transfectants were prepared by cotransfecting plasmids from each receptor mutant with a plasmid that confers resistance to neomycin (pSV2*neo*). Clones resistant to

G-418 were expanded and analyzed for the expression of the receptor using rabbit antibodies directed to the extracellular domain of the receptor. Goat anti-rabbit IgG conjugated to fluorescein isothiocyanate was used as an indicator in the fluorescence-activated cell sorter (FACS).

Early cellular responses to PDGF. Cells were plated in Ham's F-12 supplemented with 5% calf serum, penicillin, and streptomycin. After 12 hours, the cultures were changed to quiescing medium (Ham's F-12, 0.1% bovine serum albumin, and 5 μg/ml transferrin) for 48 hours. Cells were collected from confluent cultures by treatment with solution containing 2 mM EDTA in phosphate-buffered saline (PBS). [^{125}I]-PDGF-binding, antireceptor immunoblot, receptor tyrosine kinase activity, PtdIns turnover, and receptor down-regulation were measured as described previously (Escobedo et al. 1988a).

Rate of thymidine incorporation into DNA. Cells were seeded in Ham's F-12 medium containing 5% fetal calf serum. Twenty four hours later, the media were replaced with quiescing media. After 48 hours, confluent cultures were incubated with or without PDGF for 16 hours. The cells were then incubated with 0.75 μCi of [^{3}H]thymidine for 45 minutes at 37°C. Following incubation with thymidine, cells were rinsed with cold PBS and lysed in 1% SDS. DNA from the solubilized samples was precipitated with 10% trichloroacetic acid in the cold as described recently (Escobedo et al. 1988a).

PDGF-induced change in conformation. Cells were metabolically labeled with [^{35}S]methionine (0.25 mCi/ml) for 4 hours at 37°C and then incubated with PDGF at 4°C and solubilized as described previously (Keating and Williams 1987). Lysates were immunoprecipitated with antibodies against phosphotyrosine or against the PDGF receptor as described previously (Keating and Williams 1987). Antibody 83, which recognizes the conformational change in the PDGF receptor (M.T. Keating et al., in prep.), was generated against a synthetic peptide containing amino acid residues 934–951 of the PDGF receptor. The preparation of this antibody as well as antibody 88 (against 738–760) and antibody 77 (against external domain residues 425–446) has been described previously (Keating and Williams 1987).

PtdIns kinase measurements. Confluent quiescent cultures of BALB/c-3T3 cells in 150-cm^2 flasks were incubated in the presence of (+) or absence (−) of 10 ng/ml PDGF for 15 minutes at 37°C. The cells were washed once with cold PBS and lysed in situ as described previously (Frackelton et al. 1984). The lysates were clarified by centrifugation and incubated for 2 hours at 4°C with anti-receptor serum 77 or 88 at a 1:500 dilution in the presence or absence of the peptide (50 μg/ml) to which each antiserum was raised. After an additional 20-minute incubation at 4°C with protein A–Sepharose, the immunoprecipitates were recovered by centrifugation and washed three times with 1 ml of RIPA and then three times with 1 ml of 100 mM NaCl and 10 mM Tris (pH 7.4). The washed immunoprecipitates were resuspended in 100 μl of 20 mM HEPES (pH 7.2), 5 mM $MnCl_2$, and 0.2 mg/ml sonicated PtdIns, with 10 μCi [γ-^{32}P]ATP. The kinase reaction was allowed to continue for 20 minutes at room temperature; incorporation of ^{32}P into PtdIns-monophosphate was linear over this time. Labeled phospholipids were extracted, separated by thin-layer chromatography, and detected by autoradiography as described previously (Kaplan et al. 1987).

RESULTS AND DISCUSSION

A representation of the domains of the human and mouse PDGF receptors is shown in Figure 2. The extracellular portion of the receptor consists of a series of five immunoglobulin-like domains. There is a single transmembrane sequence of hydrophobic residues. The tyrosine-kinase-coding sequences, easily recognized by their homologies with other tyrosine kinases, are split into distinct regions separated by a 107-amino-acid sequence that we have termed the KI domain. The kinase domain that is closest to the membrane contains a lysine at position 602. This lysine is predicted to be the ATP-binding site of the receptor kinase, on the basis of consensus sequences for the ATP-binding region of other kinases (Escobedo et al. 1988b).

Deletion of the KI Domain

To determine the role of the KI region in receptor function, we constructed a mutant receptor in which 82 of the 107 amino acids of the KI region were deleted (residues 684–765). Plasmids containing the mutated receptor cDNA sequence under the transcriptional control of the SV40 early promotor were used to express the receptor in CHO cells that normally lack PDGF receptors. Stable transfectants that expressed either mutated or wild-type receptor cDNA were used to study the PDGF-induced responses. The levels of receptor expression in the ΔKI mutant transfectants were the same as or higher than the level of expression in wild-type transfectants (J.A. Escobedo and L.T. Williams, in prep.). The ΔKI mutant cells and CHO cells transfected with wild-type receptor cDNA had high-affinity binding sites for [^{125}I]PDGF, whereas the parent CHO cells had no specific PDGF binding (J.A. Escobedo and L.T. Williams, in prep.). Both mutant and wild-type receptors were down-regulated after 4 hours of PDGF treatment at 37°C (J.A. Escobedo and L.T. Williams, in prep.).

The ability of the ΔKI mutant to mediate PDGF-stimulated tyrosine phosphorylation of proteins in intact cells was assessed by immunoblot analysis using phosphotyrosine antibody to measure phosphotyrosine-containing protein in extracts of PDGF-stimulated cells (Fig. 3). The relative intensity of the signal in extracts of the PDGF-stimulated ΔKI transfectant was similar to the intensity of the signal from the wild-type transfectant. This was somewhat surprising, since

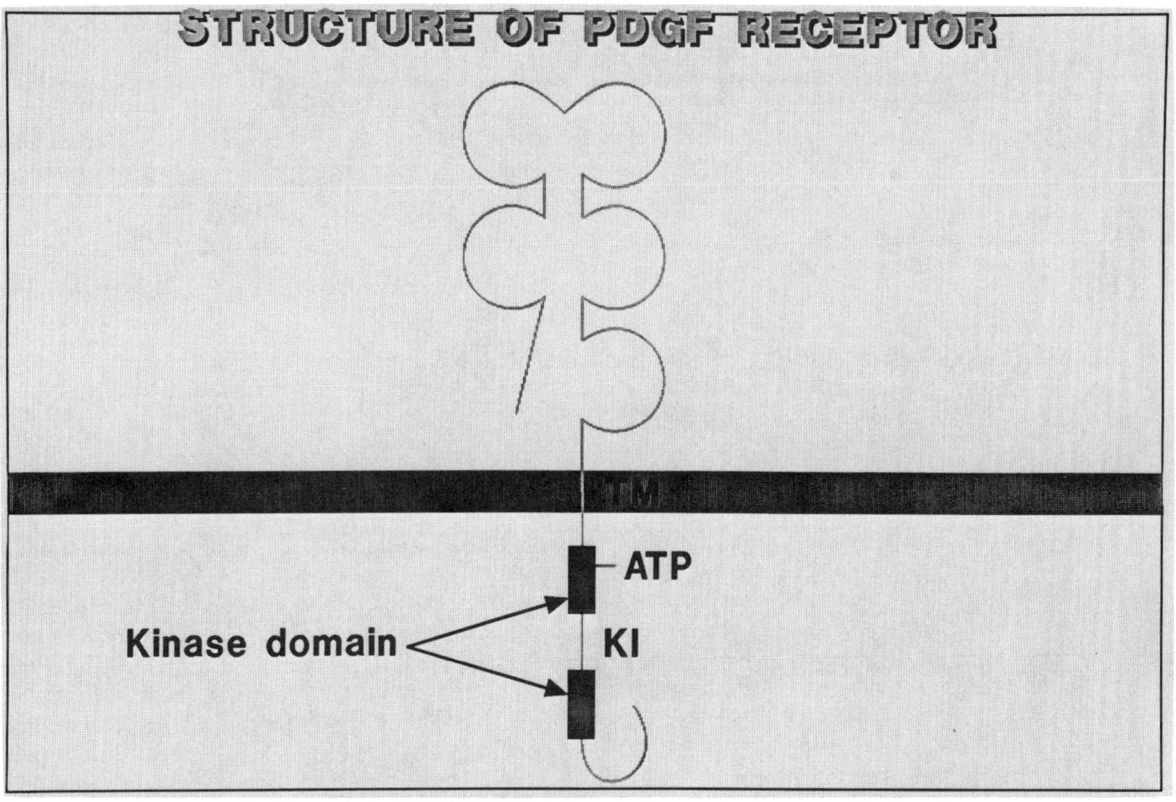

Figure 2. Domains of the PDGF receptor. The extracellular region consists of five immunoglobulin-like domains. There is a single transmembrane domain. The kinase domain is split into two parts separated by a KI region. The ATP-binding site is indicated.

we had anticipated that a large deletion of sequences in the intracellular domain might cause a loss of enzyme activity of the protein. Other early responses of PDGF, including hydrolysis of membrane PtdIns, a characteristic alteration of cellular pH, and an increase in intracellular calcium, were identical in the ΔKI mutant receptor and wild-type receptor transfectants (J.A. Escobedo and L.T. Williams, in prep.). Thus, the sequences in the KI region are not required for these PDGF-induced responses.

The analysis of the mitogenic response of the ΔKI mutant gave a surprising result. Despite the intact rapid PDGF-stimulated responses of the ΔKI transfectant (PDGF binding, down-regulation, tyrosine autophosphorylation, PtdIns hydrolysis, pH change, increase in intracellular calcium), the mitogenic responsiveness to PDGF was only 0–30% that of cells expressing the wild-type receptor. This loss of PDGF responsiveness in the ΔKI mutant was apparent in measurements of both DNA synthesis and cell number (J.A. Escobedo and L.T. Williams, in prep.).

These findings were surprising in two respects. First, a large change in the receptor primary sequence, a deletion of 82 amino acids, did not cause apparent loss of a number of rapid cellular responses to PDGF. Second, even though the rapid responses to PDGF were intact, the mitogenic response was significantly attenuated by deletion of the KI domain. These results suggest that this domain may be responsible for mediating the interaction of activated PDGF receptors with other molecules that are important in the mitogenic response to PDGF. Alternatively, the KI domain may impose a receptor confirmation that dictates the substrate specificity of tyrosine kinase. Another conclusion that can be made from these findings is that the stimulation of PtdIns turnover, pH change, and calcium flux change are not sufficient to explain PDGF-induced mitogenesis.

On the basis of the current data, it is not possible to define a specific role for the KI domain. The high degree of conservation between mouse and human KI sequences attests to the importance of this domain, and the reduction in mitogenic responsiveness in the KI deletion mutant supports this contention. The observations that the domain can be largely deleted and that some of the responses to PDGF appear to be unaffected suggest that this sequence functions as a distinct structural domain, perhaps as a "looped" structure that is not part of the kinase domain.

Receptor Mutants That Lack Tyrosine Kinase Activity

To investigate the role of tyrosine kinase in the cellular responses to PDGF, we have constructed mutants of the PDGF receptor that are defective in PDGF-stimulated tyrosine kinase activity. To ensure that the alterations in PDGF responsiveness observed in the

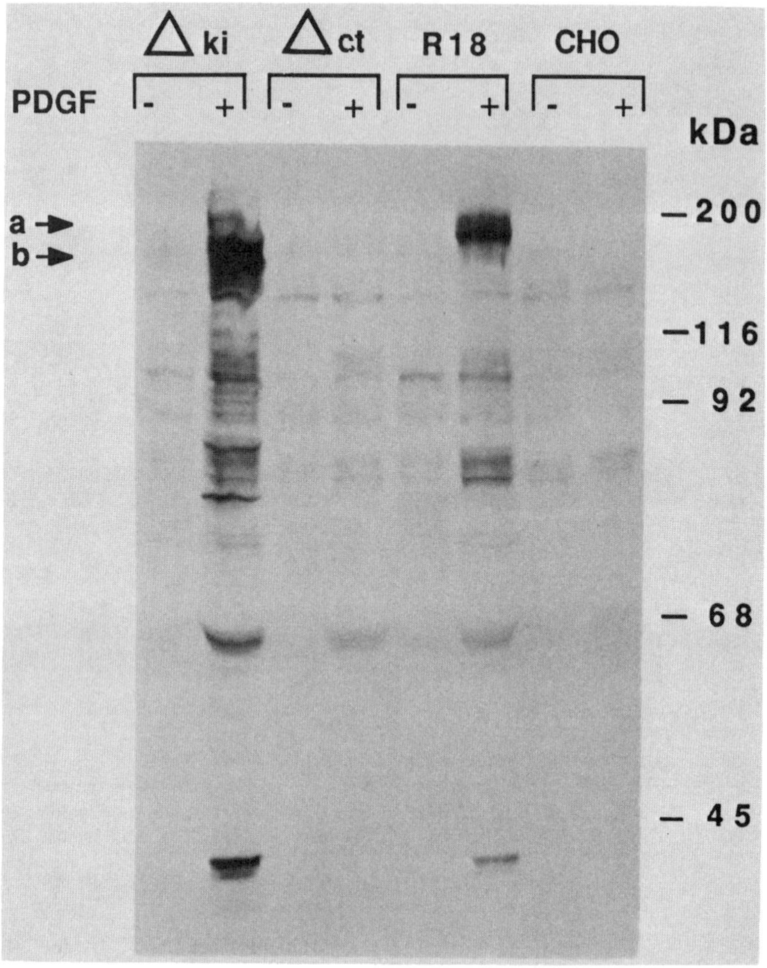

Figure 3. Receptor tyrosine kinase activity in intact cells. The pattern of PDGF-induced tyrosine phosphorylation in the KI deletion transfectant (ΔKI), wild-type receptor transfectant (R18), and a mutant receptor lacking the carboxy-terminal sequences (Δct) was determined by immunoblot using antibodies against phosphotyrosine. Arrows indicate the position of the receptor in the wild-type cells (*a*) and in the KI deletion mutant (*b*).

mutant receptors could be attributed to the loss of tyrosine kinase activity and not merely to conformational changes induced by the mutations, we inactivated the kinase by three independent mutations (Fig. 4). In the first mutation, the codon for the lysine at amino acid position 602, a site that corresponds to known ATP-binding sites of other tyrosine kinases, was converted to an alanine codon (K602A mutant). In the second mutation, the coding sequences for the carboxy-terminal 145 amino acids (amino acids 922–1067) were deleted, thus removing most of the sequences 3′ to the kinase domain. In the third group of mutations (transmembrane mutants), the sequences that encode the transmembrane region of the PDGF receptor were replaced by the corresponding transmembrane sequences of the low-density lipoprotein (LDL) receptor, the *neu* oncogene, or the cellular *neu* gene. Stable transfectants that expresssed each of the mutated receptor's sequence under the transcriptional control of the SV40 early promotor were prepared in CHO cells. The transfectants expressed surface receptor protein that was recognized by an antibody to the external domain of the PDGF receptor in a FACS. Analysis of these cell lines by immunoblot, using receptor antibody directed against the cytoplasmic domain sequence, showed that the mutant cell lines expressed receptor at levels comparable to those of the transfectant (R18) that expressed wild-type receptor. The binding of [^{125}I]PDGF to each of the cell lines was of high affinity (Escobedo et al. 1988a; J.A. Escobedo and L.T. Williams, in prep.). Half-maximal inhibition of binding by unlabeled PDGF occurred at 0.1–0.3 nM, which is the same concentration required to half-maximally inhibit [^{125}I]PDGF binding in 3T3 cells that have native PDGF receptors. Approximately 5000–8000 receptor sites per cell were expressed in the mutant and wild-type transfectants (J.A. Escobedo and L.T. Williams; J.A. Escobedo et al.; both in prep.).

The responsiveness of the mutant cells to PDGF was assessed by measurement of tyrosine kinase, PtdIns hydrolysis, change in cellular pH, increase in intracellular calcium, and PDGF-induced down-regulation (Es-

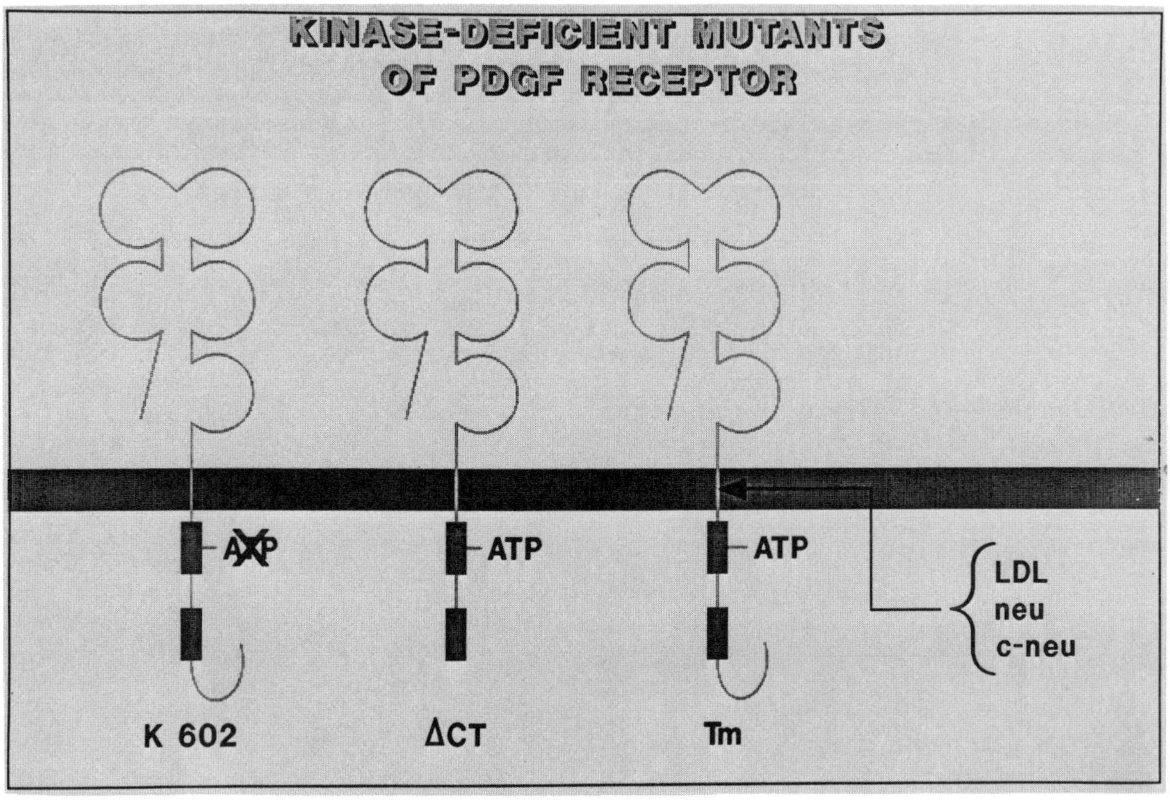

Figure 4. Kinase-deficient mutants of the PDGF receptor that have defective tyrosine kinase activity were prepared by altering the ATP-binding site (K602), deleting the carboxy-terminal domain (Δct), or substituting the transmembrane sequences of the LDL receptor, the *neu* oncogne, and the normal cellular *neu* gene as described in the text.

cobedo et al. 1988a). A summary of some of these findings is shown in Figure 5. PDGF-induced receptor autophosphorylation, measured by antiphosphotyrosine antibody immunoblot analysis, was defective in all of the mutant cell lines. Similarly, PDGF-induced changes in PtdIns turnover, pH, intracellular calcium levels, and DNA synthesis were also absent in the kinase-deficient mutants. In contrast, PDGF-induced receptor down-regulation was intact in all of the tyrosine-kinase-defective mutants. This result was surprising in the context of recently published experiments on ATP-binding-site mutants of the insulin and EGF receptors, which have shown that receptor down-regulation required an intact tyrosine kinase domain (Chou et al. 1987; Honegger et al. 1987).

Our findings that the kinase-deficient mutants cannot stimulate PtdIns hydrolysis suggest a direct link between receptor-mediated PtdIns hydrolysis and tyrosine kinase. Similarly, a link between the kinase activity and changes in calcium and pH can also be inferred from these findings. It seems likely either that the PDGF receptor phosphorylates an activator of these responses or that the autophosphorylation reaction alters the receptor so that it can interact with another molecule, which, in turn, mediates these early responses to PDGF. The loss of PDGF-induced mitogenesis in the kinase-deficient mutants was expected in the context of recently reported studies on ATP-binding-site mutants of the EGF and insulin receptors (Chou et al. 1987; Honegger et al. 1987). Our finding that three independent types of mutations cause a loss of receptor-mediated mitogenesis makes a compelling case for the role of kinase in growth-factor-stimulated cell proliferation. In addition, these data ruled out the possibility that internalization and down-regulation of this growth-factor receptor is a sufficient stimulus to induce DNA synthesis. In each of our kinase-deficient mutants, internalization and down-regulation appeared to be normal despite the fact that the other measured responses, including DNA synthesis, were lost.

Mutation of Putative Sites of Tyrosine Phosphorylation

There are 28 tyrosine residues in the cytoplasmic domain of the PDGF receptor. The specific tyrosines that become phosphorylated when the receptor binds PDGF have not been identified by direct phosphotyrosine measurements. As an alternative approach, we have used site-directed mutagenesis to alter six of these tyrosines, converting the coding sequences to phenylalanine codons. The only one of these mutants that showed a definitive loss of signal-transducing capability was the tyrosine at residue 825, which is within a sequence homologous to that of tyrosine 416 of the c-*src*-encoded protein. This mutant (Y825F) of the

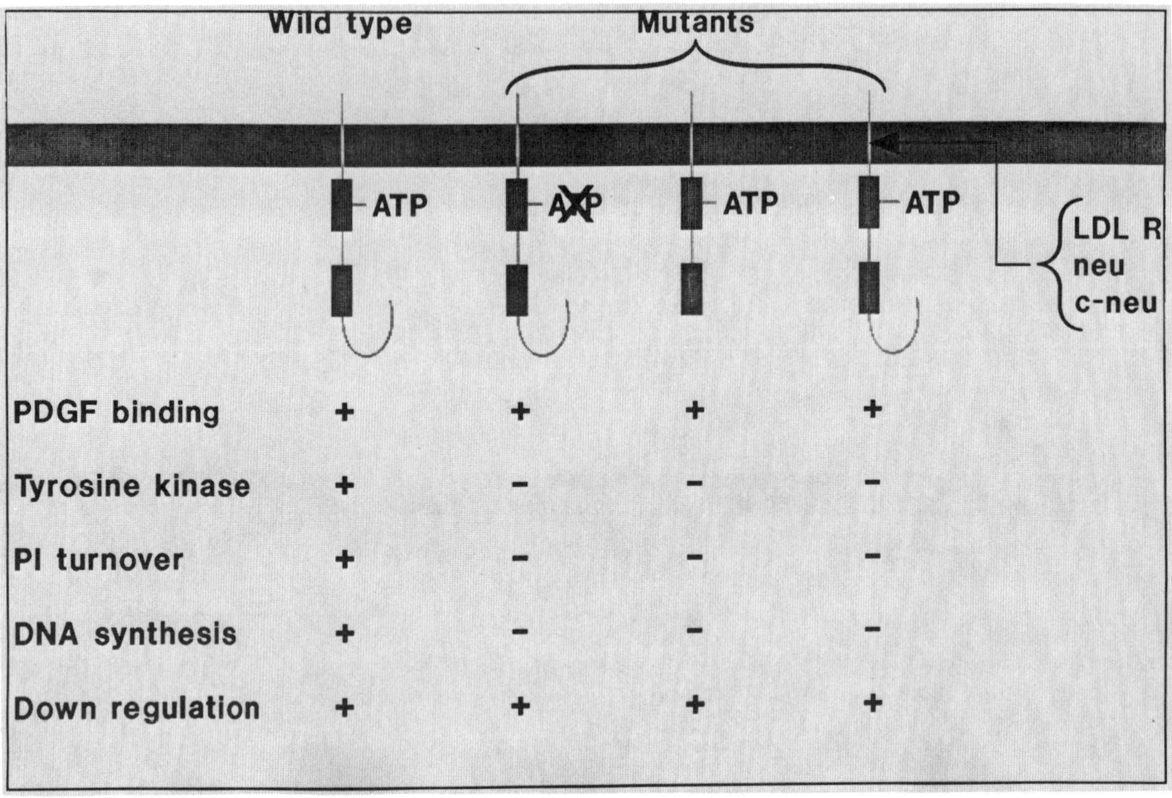

Figure 5. Responses of CHO cells transfected with wild-type or mutant PDGF receptor cDNA. + indicates a full response; − indicates that no responses were detected.

PDGF receptor bound [^{125}I]PDGF with high affinity and was down-regulated by chronic exposure to PDGF (W.J. Fantl et al., in prep.). However, its ability to mediate a PDGF increase in cell number was dramatically attenuated. It also failed to undergo PDGF-induced conformational change (see below) and had no measurable angiotensin phosphorylating activity (in vitro) in response to PDGF. The autophosphorylation response of the mutant receptor to PDGF in vivo was approximately 70% lower than the response of the wild-type receptor. These findings suggest that autophosphorylation of this tyrosine is involved in PDGF-induced change in receptor conformation and is important in determining the tyrosine kinase activity and mitogenic responsiveness to PDGF. However, further studies to determine sites of wild-type receptor protein phosphorylation should be performed before definitive conclusions can be made about the role of autophosphorylation in receptor function.

PDGF Induction of a Conformational Change in Its Receptor

To examine the effect of binding of PDGF on receptor conformation, we performed metabolic labeling and immunoprecipitation experiments using a panel of peptide antibodies directed against specific PDGF receptor domains (M.T. Keating et al., in prep.). Antibody 83, which was directed against the sequence in the carboxy-terminal domain of the receptor, immunoprecipitated ninefold more receptor protein from lysates of PDGF-activated 3T3 cells than from comparable lysates from unactivated 3T3 cells (Fig. 6). In contrast, antibodies directed to extracellular and other intracellular domains immunoprecipitated the stimulated and unstimulated forms of the receptor equally well (Fig. 6). These data suggest that ligand binding elicited a change in PDGF receptor conformation, exposing a cryptic epitope that was recognized by antibody 83.

The induction of receptor conformational change by PDGF could also be demonstrated in a solubilized cell-free system but only in the presence of ATP (M.T. Keating et al., in prep.). When AMP-PNP, a nonhydrolyzable analog of ATP, was added to the solubilized receptor, the conformational change was blocked, suggesting that phosphorylation was required for PDGF induction of the change in conformation (M.T. Keating et al., in prep.).

CHO transfectants that expressed mutant or wild-type receptor were also tested for PDGF-induced conformational change using antibody 83. The wild-type receptor transfectant readily underwent conformational change in response to PDGF. In contrast, none of the kinase-deficient mutants underwent a change in conformation, consistent with the notion that autophosphorylation was involved in conformational change (M.T. Keating et al., in prep.). Interestingly,

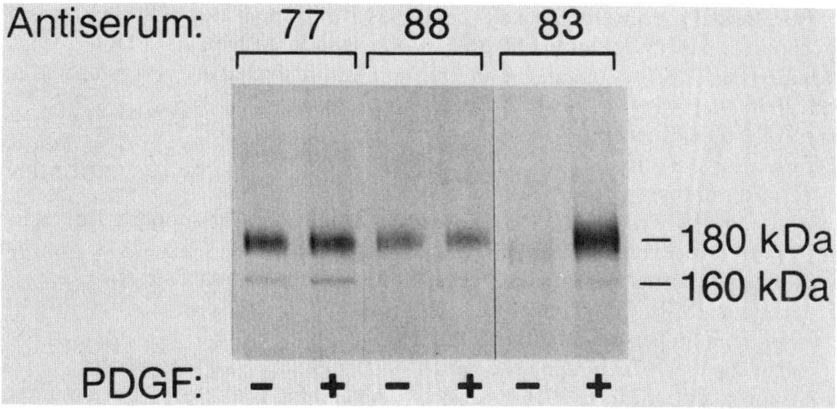

Figure 6. Selective immunoprecipitation of PDGF-stimulated receptors by antiserum 83. BALB/c-3T3 cells were labeled with [^{35}S]methionine and stimulated by PDGF (+) as described in Methods. Cell extracts were then immunoprecipitated using antibodies that recognize epitopes in the extracellular region (77), the insert region (88), or the carboxy tail region (83).

the conformational change could not be detected in the KI deletion mutant that did not mediate PDGF-stimulated mitogenesis (J.A. Escobedo and L.T. Williams, in prep.). Thus, the induction of conformational change may be an essential step in initiating the intracellular reactions required for DNA synthesis.

Association of a PDGF-sensitive PtdIns Kinase with the PDGF Receptor

Recently, Kaplan et al. (1987) reported that stimulation of intact cells with PDGF leads to the appearance of PtdIns kinase activity in antiphosphotyrosine immunoprecipitates, suggesting either that the PtdIns kinase is a substrate of PDGF-stimulated tyrosine kinase or that the PtdIns kinase is physically associated with the tyrosine-phosphorylated PDGF receptor. Previous studies by other investigators showed that a PtdIns kinase is known to associate with the v-*src*-encoded protein in cells transformed by Rous sarcoma virus (Sugimoto et al. 1984) and with the middle T–c–*src* complex in cells transformed by polyomavirus (Whitman et al. 1985). For these reasons, we tested the ability of wild-type and mutant PDGF receptors to associate physically with and regulate PtdIns kinase. In these experiments, antireceptor immunoprecipitates of PDGF-stimulated cells were assayed for PtdIns kinase activity. PtdIns kinase activity was specifically immunoprecipitated with antisera raised against peptides representing either extracellular (antibody 77) or intracellular (antibody 88) portions of the PDGF receptor (Fig. 7; S.R. Coughlin et al., in prep.). Immunoprecipitates from PDGF-stimulated cultures contained approximately tenfold more PtdIns kinase activity than those from unstimulated cultures. In parallel experiments, the recovery of the PDGF receptor in these immunoprecipitates was unaffected by PDGF treatment (S.R. Coughlin et al., in prep.). The peptides to which the

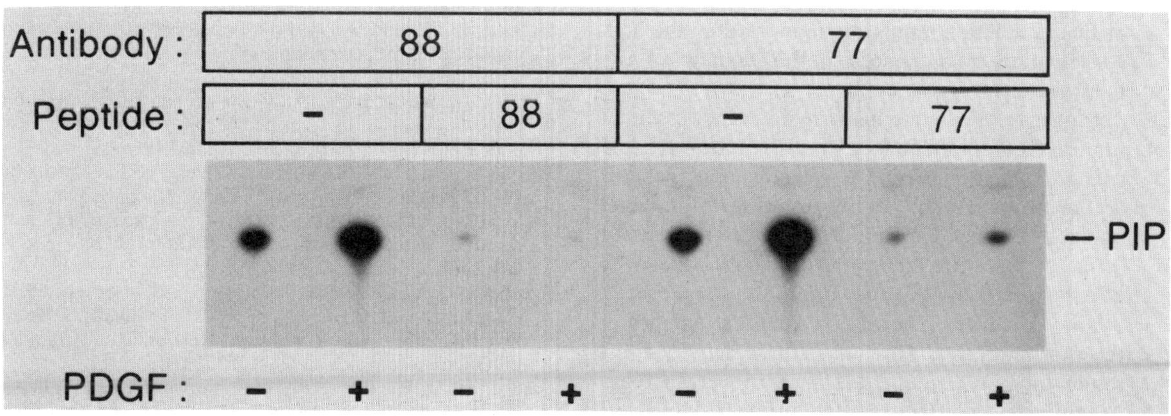

Figure 7. Phosphatidylinositol kinase activity and tyrosine kinase activity in antireceptor immunoprecipitates. Confluent quiescent cultures of BALB/c-3T3 cells in 150-cm^2 flasks were incubated in the presence (+) or absence (−) of 10 ng/ml PDGF for 15 min at 37°C. Cell lysates were immunoprecipitated with antipeptide sera directed against the PDGF receptor extracellular domain (77) or tyrosine kinase domain (88); competing peptides were included as indicated. The immunoprecipitates were incubated with sonicated PtdIns and [γ-^{32}P]ATP as described in Methods. Each PtdIns kinase reaction mixture was analyzed for ^{32}P incorporated into PtdIns-monophosphate, and labeled phospholipids were extracted, separated by thin-layer chromatography, and detected by autoradiography.

antireceptor sera were raised specifically blocked precipitation of the receptor and coprecipitation of PDGF-sensitive PtdIns kinase (Fig. 7; S.R. Coughlin et al., in prep.). These data show that a PtdIns kinase is physically associated with the PDGF receptor. In separate experiments, PDGF-sensitive PtdIns kinase was also coprecipitated from CHO fibroblasts transfected with cDNA encoding the mouse PDGF receptor (S.R. Couglin et al., in prep.). No PDGF-sensitive PtdIns kinase was coprecipitated from the parent CHO cells that lack PDGF receptors (S.R. Coughlin et al., in prep.). A number of nonmitogenic mutants of the PDGF receptor, including the ATP-binding-site mutant (K602A) (see above), did not have PDGF-sensitive PtdIns kinase in their immunoprecipitates (S.R. Coughlin et al., in prep.).

The PDGF-stimulated PtdIns kinase found in antireceptor immunoprecipitates utilized PtdIns in the micellar form of a substrate, not in detergent-solubilized form, and was insensitive to inhibition by adenosine (S.R. Coughlin et al., in prep.). These properties make PDGF-stimulated PtdIns kinase a type I PtdIns kinase, according to the classification by Cantley and colleagues (Kaplan et al. 1987). These investigators have reported that type-I PtdIns kinase converts PtdIns to PtdIns-3 monophosphate, a novel product that has potential, but as yet undocumented, roles in signal transduction. The role of this kinase in the stimulation of cell proliferation by PDGF remains to be determined.

CONCLUSIONS

These studies have provided the following insights into the mechanism of signal transduction by the PDGF receptor:

1. The KI domain of the PDGF receptor plays an important role in the stimulation of DNA synthesis by PDGF. This conclusion, derived from the studies of the KI deletion mutant, is supported by the high conservation of sequences in this domain of the human and mouse receptor.
2. Surprisingly, the kinase insert region is not essential for PDGF stimulation of PtdIns turnover, pH change, increase in cellular calcium, and receptor autophosphorylation. These responses stimulated by the kinase deletion mutant of the receptor are not sufficient for a full induction of DNA synthesis by PDGF.
3. Studies of three types of tyrosine-kinase-deficient receptor mutants suggest that kinase activity is essential for PDGF-stimulated pH turnover, increase in cellular calcium, and PtdIns turnover. However, kinase activity is clearly not required for PDGF-induced receptor down-regulation.
4. PDGF stimulates a conformational change in the receptor. This change appeared to require kinase activity and may be the result of autophosphorylation.
5. PDGF activates a PtdIns kinase that is physically associated with the PDGF receptor. This kinase is not activated by nonmitogenic mutants of the receptor.

ACKNOWLEDGMENTS

This work was supported by National Institutes of Health grant RO1-HL32898. We thank Betty Cheung for typing this manuscript.

REFERENCES

Antoniades, H.N. 1981. Human platelet-derived growth factor (PDGF): Purification of PDGF-I and PDGF-II and separation of their reduced subunits. *Proc. Natl. Acad. Sci.* **78:** 7314.

Barrett, T.B., C.M. Gajdusek, S.M. Schwartz, J.K. Mc Dougall, and E.P. Benditt. 1984. Expression of the *sis* gene by endothelial cells in culture and *in vivo*. *Proc. Natl. Acad. Sci.* **81:** 6772.

Berk, B.C., R.W. Alexander, T.A. Brock, M.A. Gimbrone, and C.R. Webb. 1986. Vasoconstriction: A new activity for platelet-derived growth factor. *Science* **232:** 87.

Betsholtz, C., A. Johnsson, C.-H. Heldin, B. Westermark, P. Lind, M.S. Urdea, R. Eddy, T.B. Shows, K. Philpott, A.L. Mellor, T.J. Knott, and J. Scott. 1986. cDNA sequence and chromosomal localization of human platelet-derived growth factor A-chain and its expression in tumor cell lines. *Nature* **320:** 695.

Chou, C.K., T.J. Dull, D.S. Russell, R. Gherzi, D. Lebwohl, A. Ullrich, and O.M. Rosen. 1987. Human insulin receptors mutated at the ATP-binding site lack protein tyrosine kinase activity and failed to mediate post receptor effects of insulin. *J. Biol. Chem.* **262:** 1842.

Daniel, T.O., V.C. Gibbs, D.F. Milfay, and L.T. Williams. 1987. Biosynthetic and glycosylation studies of cell surface PDGF receptors. *J. Biol. Chem.* **262:** 9778.

Daniel, T.O., P.M. Tremble, A.R. Frackelton, Jr., and L.T. Williams. 1985. Purification of the PDGF receptor by using an anti-phosphotyrosine antibody. *Proc. Natl. Acad. Sci.* **82:** 2684.

Daniel, T.O., V.C. Gibbs, D.F. Milfay, M.R. Garovoy, and L.T. Williams. 1986. Thrombin stimulates c-*cis* gene expression in microvascular endothelial cells. *J. Biol. Chem.* **261:** 9579.

DiCorleto, P.E. and D.F. Bowen-Pope. 1984. Cultured endothelial cells produce multiple growth factors for connective tissue cells. *Exp. Cell Res.* **153:** 167.

Ek, B., B. Westermark, A. Wasteson, and C.-H. Heldin. 1983. Stimulation of tyrosine specific phosphorylation by platelet-derived growth factor. *Nature* **195:** 419.

Escobedo, J.A., M.T. Keating, H.E. Ives, and L.T. Williams. 1988a. Platelet-derived growth factor receptors expressed by cDNA transfection couple to a diverse group of cellular responses associated with cell proliferation. *J. Biol. Chem.* **263:** 1482.

Escobedo, J.A., S. Navankasatussas, L.S. Cousens, S.R. Coughlin, G.I. Bell, and L.T. Williams. 1988b. Homodimeric A and B forms of PDGF activate a common PDGF receptor. *Science* **240:** 1532.

Frackelton, A.R., Jr., P.M. Tremble, and L.T. Williams. 1984. Evidence for the platelet-derived growth factor-stimulated tyrosine phosphorylation of the platelet-derived growth factor receptor *in vivo*. *J. Biol. Chem.* **259:** 7909.

Greenberg, M.E. and E.M. Ziff. 1984. Stimulation of 3T3 cells induces transcription of the c-*fos* proto-oncogene. *Nature* **311:** 43.

Habenicht, A., J. Glomset, W. King, C. Nist, C. Mitchell, and R. Ross. 1981. Early changes in phosphatidylinositol and arachidonic acid metabolism in quiescent Swiss 3T3 cells stimulated to divide by platelet-derived growth factor. *J. Biol. Chem.* **256:** 12329.

Habenicht, A.J.R., H.A. Dresel, M. Goerig, J.A. Weber, M. Stoehr, J.A. Glomset, R. Ross, and G. Schettler. 1986. Low-density lipo-protein receptor-dependent prostaglandin synthesis in Swiss 3T3 cells stimulated by platelet-derived growth factor. *Proc. Natl. Acad. Sci.* **83:** 1344.

Heldin, C.-H., A. Johnsson, S. Wennergren, C. Wernstedt, C. Betsholtz, and B. Westermark. 1986. A human osteosarcoma cell line secretes a growth factor structurally related to a homodimer of PDGF A-chains. *Nature* **319:** 511.

Honegger, A.M., T.J. Dull, S. Felder, E.V. Obberghen, F. Bellot, D. Szapary, A. Schmidt, A. Ullrich, and J. Schlessigner. 1987. Point mutation at the ATP binding site of EGF receptor abolishes protein-tyrosine kinase activity and alters cellular routing. *Cell* **51:** 199.

Johnsson, A., C. Bethsholtz, K. von der Heim, C.-H. Heldin, and B. Westermark. 1985. Platelet-derived growth factor agonist activity of a secreted form of the v-*sis* oncogene product. *Proc. Natl. Acad. Sci.* **82:** 1721.

Kaplan, D.R., M. Whitman, B. Schaffhausen, D.C. Pallas, M. White, L. Cantely, and T.M. Roberts. 1987. Common elements in growth factor stimulation and oncogenic transformation: 85 kd Phosphoprotein and phosphatidylinositol kinase activity. *Cell* **50:** 1021.

Kavanaugh, W.M., G.R. Harsh, N.F. Starksen, C.M. Rocco, and L.T. Williams. 1988. Transcriptional regulation of the A and B chain genes of platelet-derived growth factor in microvascular endothelial cells. *J. Biol. Chem.* **263:** 8470.

Keating, M.T. and L.T. Williams. 1987. Processing of the platelet-derived growth factor receptor. *J. Biol. Chem.* **262:** 7932.

———. 1988. Autocrine stimulation of intracellular PDGF receptors in v-*sis*-transformed cells. *Science* **239:** 914.

Kelly, K., B.H. Cochran, C.D. Stiles, and P. Leder. 1983. Cell-specific regulation of the c-*myc* gene by lymphocyte mitogens and platelet-derived growth factor. *Cell* **35:** 603.

L'Allemain, G., S. Paris, and J. Pouyssegur. 1984. Growth factor action and intracellular pH regulation in fibroblasts. *J. Biol. Chem.* **259:** 5809.

Nishimura, J., J.S. Huang, and T.F. Deuel. 1982. Platelet-derived growth factor stimulates tyrosine specific protein kinase activity in Swiss mouse 3T3 cell membranes. *Proc. Natl. Acad. Sci.* **79:** 4303.

Rizzino, A. and D.F. Bowen-Pope. 1985. Production of PDGF-like factors by embryonal carcinoma cells and response to PDGF by endoderm-like cells. *Dev. Biol.* **110:** 15.

Robbins, K.C., H.N. Antoniades, S.G. Devare, M.W. Hunkapiller, and S.A. Aaronson. 1983. Structural and immunological similarities between simian sarcoma virus gene product(s) and human platelet-derived growth factor. *Nature* **305:** 605.

Rozengurt, E., P. Stroobant, M.D. Waterfield, T.F. Deuel, and M. Keehan. 1983. Platelet-derived growth factor elicits cyclic AMP accumulation in Swiss 3T3 cells: Role of prostaglandin production. *Cell* **34:** 265.

Starksen, N.F., G.R. Harsh, V.C. Gibbs, and L.T. Williams. 1987. Regulated expression of the PDGF A chain gene in microvascular endothelial cells. *J. Biol. Chem.* **62:** 14381.

Sugimoto, Y., M. Whitman, L.C. Cantley, and R. Erikson. 1984. Evidence that the Rous sarcoma virus transforming gene product phosphorylates phosphatidylinositol and diaglycerol. *Proc. Natl. Acad. Sci.* **81:** 2117.

Whitman, M., D.R. Caplan, B.S. Schaffhausen, L. Cantley, and T.M. Roberts. 1985. Association of phosphatidylinositol kinase activity with polyoma middle-T competent for transformation. *Nature* **315:** 239.

Yarden, Y., J.A. Escobedo, W.-J. Kuang, T.L. Yang-Feng, T.O. Daniel, P.M. Tremble, E.Y. Chen, M.E. Ando, R.N. Harkins, U. Francke, V.A. Fried, A. Ullrich, and L.T. Williams. 1986. Structure of the receptor for the platelet-derived growth factor helps to identify a family of closely related growth factor receptors. *Nature* **323:** 226.

Role of Intrinsic Protein Tyrosine Kinase in Function and Metabolism of the Epidermal Growth Factor Receptor

G.N. GILL,* W.S. CHEN,*§, C.S. LAZAR,* J.R. GLENNEY, JR.,† H.S. WILEY,‡
H.A. INGRAHAM,† AND M.G. ROSENFELD*§

*Department of Medicine, Division of Endocrinology and Metabolism and Eukaryotic Regulatory Biology Program, University of California-San Diego, School of Medicine, La Jolla, California 92093; †Department of Biochemistry and Lucille P. Markey Cancer Center, University of Kentucky, Lexington, Kentucky 40536; ‡Department of Pathology, University of Utah School of Medicine, Salt Lake City, Utah 84132; §Howard Hughes Medical Institute, University of California-San Diego, School of Medicine, La Jolla, California 92093

Cellular responses to growth factors and peptide hormones depend on the concentration of both ligand and receptor by the mass action relation:

$$\underset{\text{inactive}}{[\text{ligand}]} + [\text{receptor}] \underset{k-1}{\overset{k1}{\rightleftharpoons}} \underset{\text{active}}{[\text{ligand} \cdot \text{receptor}]}$$

To maintain appropriate control of growth and metabolic processes, the concentrations of both ligand and receptor are extensively regulated at the levels of gene expression and metabolism. Expression at particular times during development may dictate crucial choices for the fate of precursor cells (Greenwald 1985; Wharton et al. 1985; Hafen et al. 1987; Kimelman and Kirschmer 1987; Weeks and Melton 1987). Constitutive or inappropriate expression either of growth factors or of their receptors can result in loss of regulated growth and neoplastic transformation (Bishop 1987; Friend et al. 1988).

Among the dominant oncogenes recognized as the transforming proteins of retroviruses is *erbB*, the transforming protein of the avian erythrocytosis virus (AEV) (Frykberg et al. 1983). *erbB* is derived from the avian epidermal growth factor (EGF)/tumor growth factor α receptor (Downward et al. 1984b; Lax et al. 1988). Insertional mutagenesis results in loss of the regulatory ligand-binding amino-terminal surface domain (Fung et al. 1983; Nilsen et al. 1985); progressive carboxy-terminal truncation of the cytoplasmic domain extends host range and tumor formation in recovered AEVs (Gamett et al. 1986). The proto-oncogene EGF receptor is amplified or overexpressed in several types of human neoplasms (Lin et al. 1984; Ullrich et al. 1984; Xu et al. 1984; Libermann et al. 1985; Hendler et al. 1985; Yamamoto et al. 1986). Experimentally, constitutive overexpression either of EGF (Stern et al. 1987) or of the EGF receptor (Di Fiore et al. 1987) results in transformation.

The EGF receptor is an 1186-amino-acid transmembrane glycoprotein with intrinsic protein tyrosine kinase activity (Ushiro and Cohen 1980; Cohen et al. 1982; Ullrich et al. 1984). Figure 1 shows a schematic view of the functional domains of the receptor. The cell-surface domain contains 2 cysteine-rich regions (Ullrich et al. 1984) and 12 asparagine-linked glycosylation sites (Mayes and Waterfield 1984; Weber et al. 1984b). This 621-amino-acid domain appears to bind a single EGF molecule (Weber et al. 1984a), perhaps via the fitting of the two antiparallel β sheets of EGF (Cooke et al. 1987) into a complementary cleft. Hydropathy calculations using the program described previously by Engelman et al. (1986) reveal a single transmembrane domain of hydrophobic amino acids between residues 621 and 644, with a characteristic stop-transfer sequence that begins at residue 644. The core protein kinase domain, which exhibits high homology

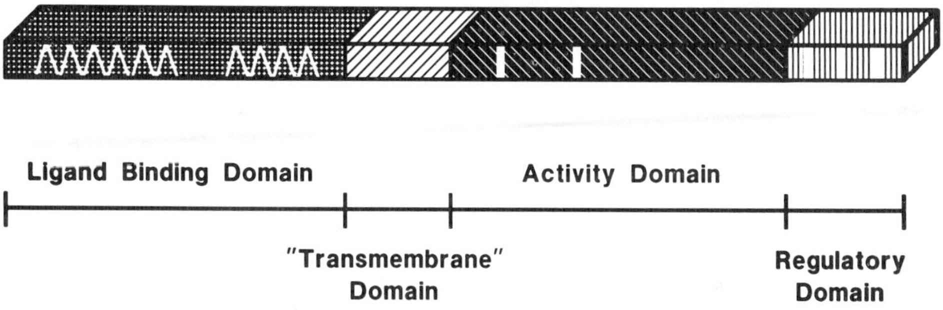

Figure 1. Schematic representation of the 1186-amino-acid EGF receptor. Cysteine-rich regions (zigzag lines) and the approximate boundaries of functional domains are indicated.

with the extended protein kinase gene family, extends to about residue 940 (Hunter and Cooper 1985). The carboxyl terminus distal to the kinase core contains the major identified sites of self-phosphorylation (Downward et al. 1984a) and functions as a regulatory domain (Bertics and Gill 1985).

Requirement for Intrinsic Protein Tyrosine Kinase Activity for Function of the EGF Receptor

Binding of EGF to its receptor activates several immediate biochemical processes, such as alterations of intracellular free calcium, pH, and increased transcription of several responsive genes, which usually culminates many hours later in DNA replication and cell division. Because several reports suggested that specific aspects of EGF receptor function were independent of its intrinsic protein tyrosine kinase activity, and because many transmembrane receptors interact with intrinsic membrane proteins to activate second-messenger systems, it was important to resolve experimentally whether mechanisms, in addition to the intrinsic protein tyrosine kinase activity, mediate some of the actions of EGF. The approach was to construct a mutant EGF receptor that lacked intrinsic protein tyrosine kinase activity and to compare the in vivo and in vitro activity of the expressed receptor with wild-type kinase-active receptor (Chen et al. 1987). To abolish protein tyrosine kinase activity with minimal structural alteration in the EGF receptor, the lysine at position 721 was mutated to a methionine. This amino acid, conserved among all the protein kinase family members, is predicted to provide a salt bridge to the β-phosphate of ATP in the nucleotide-binding site (Sternberg and Taylor 1984; Vogel et al. 1986). The methionine for lysine substitution was selected because, as these two residues have similar bond lengths and angles, it should optimally preserve the overall structure of the EGF receptor protein. A synthetic oligonucleotide (21-mer) was used to alter a single nucleotide (A to T) to generate a K to M substitution in amino acid 721 of a human EGF receptor transcription unit under control of the SV40 early promoter (Fig. 2a) (Chen et al. 1987). A mutant dihydrofolate reductase gene provided a dominant selectable marker and allowed for gene amplification (Simonsen and Levinson

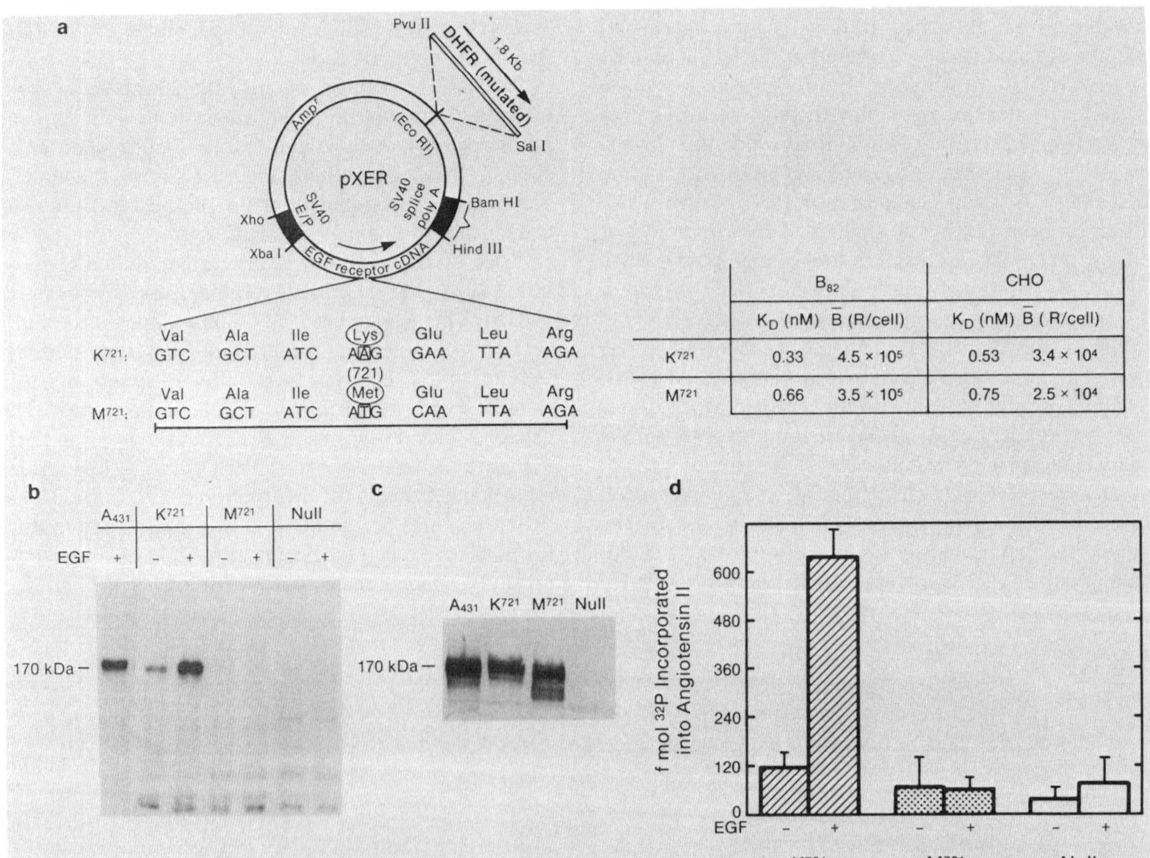

Figure 2. Construction, expression, ligand binding, and intrinsic protein tyrosine kinase activity of K^{721} and M^{721} receptors. (*a*) Diagram of the chimeric plasmid used to transfect K^{721} and M^{721} EGF receptor genes. The K_D and B_{max} of representative clonal mouse B82L and CHO cell transfectants using Scatchard analysis for a single binding site are indicated. (*b*) Self-phosphorylation of membranes prepared from the indicated cells; (*c*) Western blotting of EGF receptors in membranes prepared from the indicated cells; (*d*) phosphorylation of angiotensin II by membranes from the indicated cell lines. (Modified, with permission, from Chen et al. 1987.)

1983). The wild-type and mutant receptor cDNAs were transfected into recipient mouse L cells (B82) and Chinese hamster ovary (CHO) cells, which do not express endogenous EGF receptor mRNA or protein. Permanent transfectants were selected and analyzed. The K^{721} and mutant M^{721} EGF receptors were expressed to an equal extent, bound EGF with similar affinities, and were full-length 170-kD glycoproteins (Fig. 2a,c). The M^{721} EGF receptor was completely devoid of intrinsic protein tyrosine kinase activity, as assayed by self-phosphorylation and phosphorylation of substrates both in vivo and in vitro (Fig. 2b,d). Comparison of biological responses to EGF revealed that these were abolished by the point mutation that abolished intrinsic protein tyrosine kinase activity. Cells containing kinase-active EGF receptors exhibited rapid increases in intracellular $[Ca^{++}]$ as measured by computerized quantitation of fura-2 staining of individual cells (Poenie et al. 1985), by increases in gene transcription as measured by expression of a constructed response gene (*fos* enhancer, and thymidine kinase promoter placed in front of the luciferase reporter gene), and by cell replication. Cells expressing kinase-deficient receptors gave none of these biological responses to EGF (Table 1). Cells expressing kinase-deficient receptors also failed to undergo ligand-induced endocytosis and down-regulation. From these results, we conclude that the intrinsic protein tyrosine kinase activity of the EGF receptor is essential for its diverse biological effects.

Regulation of EGF Receptor Metabolism

A major property of cell-surface growth factor receptors is down-regulation in response to homologous ligand binding. On EGF binding, receptors, which are diffusely distributed on the cell surface, undergo rapid lateral mobility, clustering in coated pits, and endocytosis (Haigler et al. 1978, 1979; Schlessinger et al. 1978; Carpentier et al. 1987; Hertel and Perkins 1987). The endocytic vesicles carrying both EGF and its receptor ultimately fuse with lysosomes to result in degradation of both ligand and receptor (Carpenter and Cohen 1986; Krupp et al. 1982; Beguinot et al. 1984; Stoscheck and Carpenter 1984). The accelerated destruction of ligand · receptors results in loss of approximately 80% of cell-surface ligand binding, accounted for by an accelerated half-life and degradation of EGF receptor protein. This process of receptor desensitization necessitates higher concentrations of EGF to induce biological responses. It is an important mechanism in integrating appropriate biological responses and preventing constitutive overexpression that would perturb normal control. The process of endocytosis has also been suggested to be essential to signal transduction. In contrast to the endocytic pathway which results in receptor degradation, receptors for low-density lipoprotein, α_2- macroglobulin, asialoglycoprotein, and transferrin undergo ligand-induced endocytosis but follow an alternate intracellular pathway that uncouples ligand from receptor and recycles the receptor to the surface and delivers the ligand to the lysosomal degradative compartment (Willingham and Pastan 1980; Tycko and Maxfield 1982; Brown et al. 1983; Geuze et al. 1983). Components of receptor structure and of cellular machinery that decide between these alternate pathways of cellular trafficking are not completely understood.

Analysis of kinase-inactive EGF and insulin receptors revealed loss of ligand-induced receptor down-regulation (Chen et al. 1987; Hari and Roth 1987; Honegger et al. 1987; Russell et al. 1987). These studies suggested that the intrinsic protein tyrosine kinase activity of these transmembrane receptors regulated their metabolism either by controlling receptor-mediated endocytosis or by targeting them to a degradative pathway, rather than recycling pathway. Measurement of the endocytic rate constant (K_e), the cellular parameter for quantitating receptor-mediated endocytosis (Wiley and Cunningham 1982), revealed a sevenfold decrease in kinase-inactive M^{721} EGF receptors compared to kinase-active K^{721} EGF receptors (Fig. 3). The probability of an occupied EGF receptor being internalized in 1 minute was similar in B82L and CHO cells expressing transfected kinase-active human EGF receptors to that of endogenous EGF receptors in a variety of cell types from a variety of species. In contrast, the values for kinase-inactive EGF receptors were markedly reduced, indicating that they are unable to undergo normal ligand-induced internalization. Quantitation of EGF receptor mass revealed a loss that would quantitatively account for the loss of cell-surface

Table 1. Comparison of Biological Responses to EGF in Cells Expressing Kinase-active versus Kinase-inactive EGF Receptors

	EGF receptor	
	kinase active (K^{721})	kinase inactive (M^{721})
Ligand (EGF binding)	+	+
Increased cell $[Ca^{++}]$i	+	−
Induction of gene transcription	+	−
Ligand-induced receptor dimerization	+	+
Ligand-induced receptor endocytosis and down-regulation	+	−
Cell replication	+	−

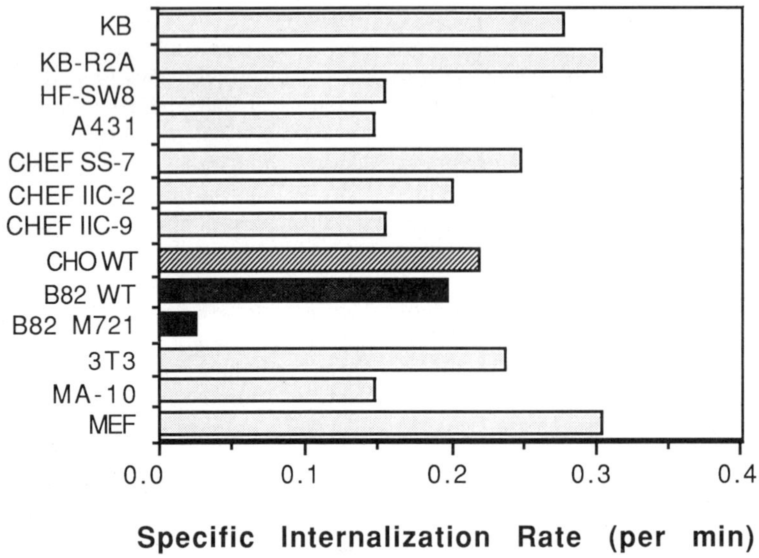

Figure 3. Endocytic rate constants (K_e) for EGF receptors measured in a variety of cell types. The specific internalization rate of the EGF receptor (probability of a given occupied receptor being internalized in 1 min of time) was determined by adding radiolabeled EGF to plates of cells maintained at 37°C and removing plates at 30-sec intervals. Unbound EGF was removed by rinsing, and the relative amount of labeled EGF associated with either the surface or interior of the cell was determined by the acid-stripping procedure (Haigler et al. 1980). Average specific internalization rates were calculated using a computer program during initial constant rate conditions (Wiley and Cunningham 1982). CHO WT and B82 WT indicate cells expressing transfected human K^{721} EGF receptors; B82 M^{721} indicates cells expressing human kinase-inactive EGF receptors. Other cell lines expressing their endogenous EGF receptors include human KB and a mutant defective in endosome acidification, KBR2A, HF-SW8 normal human foreskin fibroblasts, A431 human epidermoid carcinoma cells, CHEFSS-7, CHEFIIC-2, CHEFIIC-9, Chinese hamster embryo fibroblast clones, 3T3 mouse cells, MA10, mouse Leydig tumor cells, MEF, normal mouse embryo fibroblasts.

EGF binding in cells expressing kinase-active EGF receptors, whereas EGF receptor mass remained constant in cells expressing kinase-inactive receptors that did not undergo ligand-induced down-regulation (Glenney et al. 1988b).

The reduced rate of endocytosis of M^{721} EGF receptors measured by kinetics using ^{125}I-labeled EGF suggested that intrinsic protein tyrosine kinase activity was required for this process. This was examined directly by immunofluorescence microscopy using monoclonal and polyclonal anti-EGF receptor antibodies as well as monoclonal antiphosphotyrosine antibodies to identify proteins phosphorylated on tyrosine (Glenney et al. 1988b). EGF receptors are diffusely expressed on the surface of cells expressing both kinase-active and -inactive receptors (Fig. 4a,d). Addition of EGF resulted in endocytosis of kinase-active receptors into small endocytic vesicles that coalesced into larger vesicles by 10 minutes. Dual staining with antiphosphotyrosine antibodies confirmed that receptors were activated by EGF and that protein-containing phosphotyrosine colocalized with EGF receptors in both small and large endocytic vesicles (Fig. 4b,b′,c,c′). Kinase-inactive M^{721} EGF receptors remained diffusely distributed on the cell surface and did not internalize in response to EGF (Fig. 4d,e,f). Cells containing M^{721} EGF receptors did not stain with antiphosphotyrosine antibodies, confirming their lack of intrinsic protein tyrosine kinase activity (Fig. 4d′,e′,f′).

To confirm the requirement for intrinsic protein tyrosine kinase activity for ligand-induced endocytosis, cells expressing kinase-active EGF receptors were microinjected with two different antiphosphotyrosine monoclonal antibodies or with the same concentration of a control antibody and subsequently challenged with EGF under conditions where large vesicles were readily observed by fluorescence microscopy. Injected cells were identified by a fluorescein-labeled anti-mouse antibody (Fig. 5A,E). EGF receptors were identified by a rabbit anti-EGF receptor antibody visualized with rhodamine-labeled anti-rabbit IgG (Fig. 5B,F). All cells injected with antiphosphotyrosine antibodies were blocked in their ability to undergo endocytosis of the EGF receptor, exhibiting diffuse cell-surface staining with anti-EGF receptor antibodies, compared to surrounding uninjected cells that exhibited large EGF receptor-containing endocytic vesicles (Fig. 5B). Microinjection of the control antibody did not affect endocytosis (Fig. 5F). In vitro, the antiphosphotyrosine antibodies bound to self-phosphorylated EGF receptors became tyrosine phosphorylated on their heavy chain and blocked tyrosine phosphorylation of exogenous substrates (Glenney et al. 1988b).

Abolition of ligand-induced endocytosis by mutational inactivation of the intrinsic protein tyrosine kinase activity of the EGF receptor and by microinjection of monoclonal antiphosphotyrosine antibodies could be due to a requirement for phosphorylation of a cellular substrate(s) involved in endocytosis or of self-phosphorylation, which is proposed as essential to allosteric

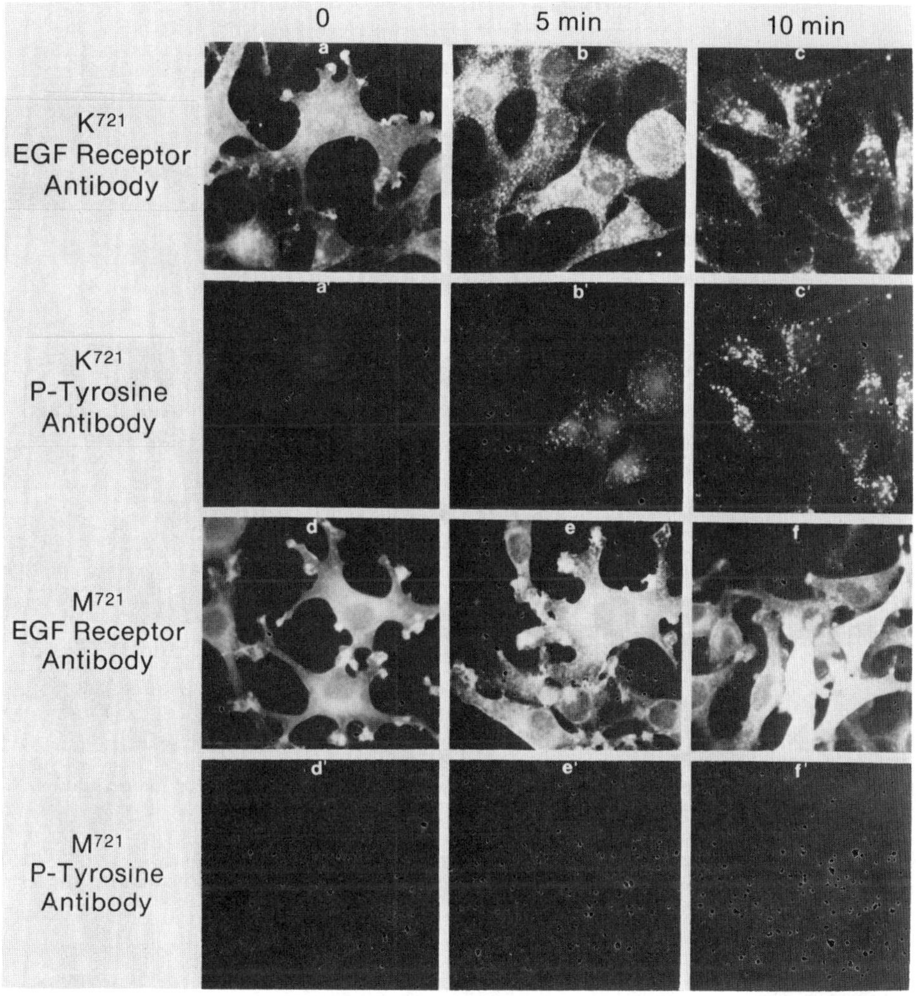

Figure 4. Effect of EGF on EGF receptors measured using immunofluorescence microscopy. B82 cells expressing A^{654} K^{721} kinase-active and M^{721} kinase-inactive EGF receptors were treated with 4 nM EGF at 37°C for the indicated times. Cells were fixed and directly stained using polyclonal EGF receptor antiserum (Weber et al. 1984b) and monoclonal antiphosphotyrosine antibody (Glenney et al. 1988a), with detection by rhodamine and fluorescein-conjugated second antibodies. (Reprinted, with permission, from Glenney et al. 1988b.)

activation of the EGF receptor (Bertics and Gill 1985; Bertics et al. 1985). Figure 6 presents a schematic model of the EGF receptor, diagraming these two possibilities. Domain motions in enzymes accompanying the binding of cofactors or substrates are documented to result in extensive conformational rearrangements from "open" with a deep cleft to a "closed" conformation in the presence of substrate, including rotation of helices with reduction of the radius of gyration (Bennett and Steitz 1980; Remington et al. 1982). By analogy, binding of EGF may cause a similar change in the conformation of the extracellular domain (Fig. 6). The induced allosteric change results in conformational changes in the intracellular kinase domain to cause rapid intramolecular self-phosphorylation (Weber et al. 1984a). This is proposed to remove an alternate substrate/competitive inhibitor, allowing tyrosine phosphorylation of cellular substrates (Bertics and Gill 1985). The requirement for intrinsic protein tyrosine kinase activity for ligand-induced endocytosis could reflect phosphorylation of cellular substrates involved in endocytosis (Fig. 6, center) or could reflect self-phosphorylation, which is an integral part of enzyme activation. This would place the EGF receptor in a favorable conformation to interact with cellular components dictating endocytosis (Fig. 6, right). These two models are not mutually exclusive because it is possible that the requirement for intrinsic protein tyrosine kinase activity could reflect both self-phosphorylation and phosphorylation of cellular components dictating endocytosis.

C′-terminal truncation into the conserved kinase core domain results in a kinase-inactive receptor that fails to undergo ligand-induced down-regulation (W.S. Chen et al., in prep.). Livneh et al. (1986) reported that an ~444-amino-acid C′-terminal truncated EGF receptor also failed to undergo ligand-induced down-regulation. Prywes et al. (1986) reported that EGF

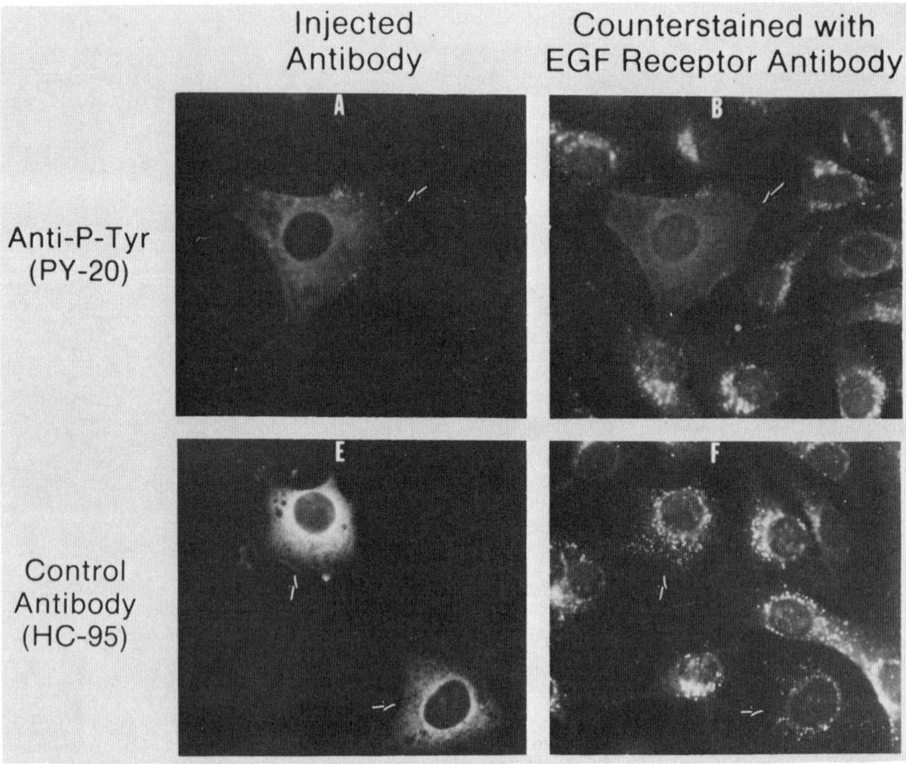

Figure 5. Effect of microinjected antiphosphotyrosine antibody on EGF-mediated receptor internalization. Cells expressing kinase-active EGF receptors were microinjected with monoclonal antiphosphotyrosine antibody P-Y 20 (*A*,*B*) or with a control monoclonal antibody against calpactin II (*E*,*F*). After a 1-hr recovery period, cells were treated with EGF for 20 min and processed for immunofluorescence microscopy. The EGF receptor was visualized with polyclonal anti-EGF receptor antibody and rhodamine anti-rabbit IgG. The injected antibodies were visualized with fluorescein anti-mouse IgG. (→) Injected cells. (Modified, with permission, from Glenney et al. 1988b.)

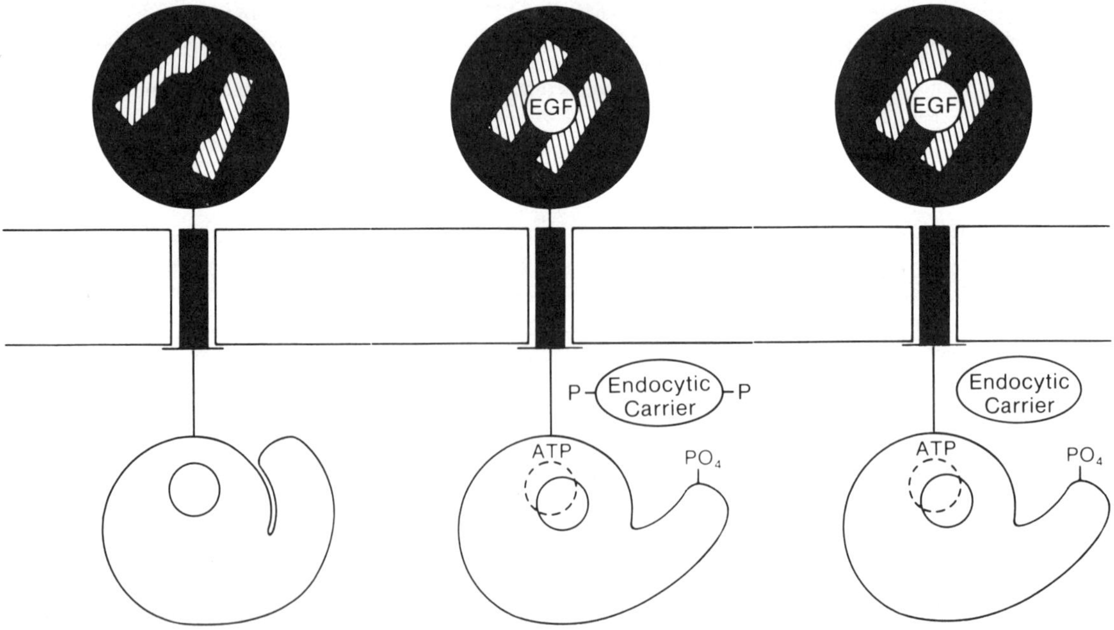

Figure 6. Schematic representation of the EGF receptor and two alternate mechanisms for control of ligand-induced endocytosis by intrinsic protein tyrosine kinase activity. The left panel shows the unstimulated receptor. The center and right panels visualize binding of EGF to the extracellular domain, with change from an open to a closed conformation, allosteric activation of the intracellular kinase domain, and self-phosphorylation involving the carboxyl terminus. The center panel presents mechanism 1, phosphorylation of substrates involved in the process of endocytosis. The right panel presents mechanism 2, a requirement for self-phosphorylation for full allosteric activation of the receptor permitting interaction with endocytic carriers.

failed to induce internalization of a chimeric receptor composed of the extracellular EGF-binding domain and a cytoplasmic v-*abl* kinase domain. These results suggest that sequences of the cytoplasmic domain of the EGF receptor are essential to interaction with the endocytic machinery of the cell and to the process of ligand-induced endocytosis.

At least three major sites of self-phosphorylation are located in the carboxyl terminus of the EGF receptor outside of the conserved protein kinase core (Downward et al. 1984a). This region does not appear to be essential for kinase activity because Cohen et al. (1982) reported that the ~150-kD C′-terminal truncated EGF receptor generated by calcium-dependent protease cleavage in vitro retained protein tyrosine kinase activity, and Clark et al. (1988) reported that an EGF receptor lacking the C′ terminus retained both protein tyrosine kinase and mitogenic activity in vivo. Analysis of a series of progressive C′-terminal truncations in the context of both kinase-active and kinase-inactive EGF receptors should allow selection between the models presented in Figure 6.

Structural Analysis of the Transmembrane Domain Using Site-directed Mutagenesis

An allosteric conformational change consequent to ligand EGF binding to the extracellular domain must be propagated via the transmembrane segment to the cytoplasmic protein tyrosine kinase domain. Helical structures have been confirmed for several transmembrane proteins whose structures have been solved (Leifer and Henderson 1983; Deisenhofer et al. 1985). Although 20 amino acids would be sufficient to span the 30-Å thickness of the hydrocarbon core of the membrane (Tanford 1978), studies by Adams and Rose (1985) indicate that 14 amino acids are sufficient to create a transmembrane domain, presumably as a 310 helix.

To investigate the structural requirements for signal transduction by the transmembrane domain of the EGF receptor, three types of mutants were constructed (H.A. Ingraham et al., in prep.). The first type of mutant was designed to shorten the transmembrane domain by deletion of amino acids from the amino terminus or by moving a stop-transfer sequence upstream of its original position (Table 2). One model of a propagation of conformational changes caused by binding of EGF predicts translational or "push" motion of the transmembrane helix to modify the conformation of the active site. Hydrophobicity calculations indicate that both of these mutational changes would make the amino-terminal portion less energetically favorable for partition in the membrane. Fewer residues may span the membrane, favoring a 310 helix or a mixed 310 and α helix conformation. The stop-transfer mutant would also extend the distance between the transmembrane helix and the ATP-binding site. The second type of mutant was designed to create single amino acid substitutions analogous to those present in transforming *neu* (c-*erbB*-2) (Bargmann et al. 1986). A single base change, T to A, that results in a valine to glutamic acid mutation was identified as the activating mutation converting c-*neu* to a transforming oncogene. Because of the close structural homology between c-*neu* and the EGF receptor, a series of point mutations was placed at residues 627 and 628 in the transmembrane domain of the EGF receptor (which correspond to valine and the spacing from the stop-transfer sequence, respectively) to determine whether constitutive activation would occur. The third type of mutation involved placing one, two, or three proline residues within the membrane-spanning helix. Proline, which tends to participate in β turns on the exterior and which causes a crystallographically demonstrated bend in the ion channel former alamethacin (Fox and Richards 1982), was considered because it might exert an effect on a helix, although this is less certain within the hydrophobic environment of the membrane (Kyte and Doolittle 1982).

Table 2 shows a partial list of EGF-dependent effects assayed in permanent transfected cells expressing EGF receptors mutated in their transmembrane domain. Group-1 mutants remained EGF-dependent for protein tyrosine kinase activity, EGF-induced receptor dimerization, and cell growth. Although not excluding an involvement of translational motion in signal transduc-

Table 2. EGF Effects on Cells Expressing EGF Receptors with Mutant Transmembrane Domains

Mutants	[^{125}I]EGF binding	EGF-dependent protein kinase: in vivo[a]	EGF-dependent protein kinase: in vitro[b]	EGF-induced dimer formation[c]	EGF-dependent cell growth[d]
Group 1 (deletions)	+[e]	+	+	+	+
Group 2 (point mutations) residue 627, 628	+	+	+	+	+
Group 3 (proline substitutions)	+	+	+	n.d.[f]	n.d.

[a]Assayed by Western blotting of total cell proteins using ^{125}I-labeled monoclonal antiphosphotyrosine antibody.

[b]Assayed in membranes as self-phosphorylation and phosphorylation of angiotensin II.

[c]Measured in membranes using cross-linking with 1-ethyl-3-[3-(dimethylamino)propyl] carbodiimide. Detection of monomers and dimers was via autoradiograms of self-phosphorylated receptors and by polyclonal anti-EGF receptor antibodies.

[d]Measured in both B82L and CHO recipients maintained in serum-free medium.

[e]+ = effective observed.

[f]n.d. indicates not done.

tion, these mutants, which are calculated to shorten the transmembrane domain, did not constitutively activate the EGF receptor. Analysis of mutations that shorten the transmembrane domain in an effort to mimic a "push" model thus provides no evidence for this occurring as an essential component of signal transduction.

Group-2 mutants, which were based on the observation that a V to E mutation in the transmembrane domain of the related protein, c-*erbB*-2, resulted in constitutive activation (Bargmann et al. 1986), also remained dependent on EGF for activation of protein tyrosine kinase activity, dimer formation, and cell proliferation. The EGF receptor thus differs from transforming *neu* in that it can tolerate a variety of amino acid substitutions at residues 627 and 628 without acquiring a constitutively active phenotype.

Group-3 mutations, which involved proline substitutions in the transmembrane domain, also failed to affect EGF dependence. Three prolines placed equidistant at approximately one helix turn apart neither constitutively activated the receptor nor inhibited EGF dependence.

The transmembrane segment that forms an essential link between the extracellular ligand-binding domain and the intracellular protein tyrosine kinase domain appears tolerant of a variety of amino acid substitutions without significantly affecting EGF-dependent signal transmission, neither mimicking EGF action by constitutive activation nor blocking ligand-induced signal transduction. The mechanism(s) of signal transduction, which must involve the transmembrane domain of the EGF receptor, remains uncertain.

ACKNOWLEDGMENTS

We thank our colleagues P.J. Bertics, C.D. Carpenter, J. Sowadski, G.M. Walton, and L.M. Zokas, who have made invaluable contributions to these studies. These studies were supported by research grants from the National Institutes of Health, the American Cancer Society, and the Council for Tobacco Research-U.S.A., Inc.

REFERENCES

Adams, G.A. and J.K. Rose. 1985. Structural requirements of a membrane-spanning domain for protein anchoring and cell surface transport. *Cell* **41:** 1007.

Bargmann, C.I., M.-C. Hung, and R.A. Weinberg. 1986. Multiple independent activations of the *neu* oncogene by a point mutation altering the transmembrane domain of p185. *Cell* **45:** 649.

Beguinot, L., R.M. Lyall, M.C. Willingham, and I. Pastan. 1984. Down-regulation of the epidermal growth factor receptor in KB cells is due to receptor internalization and subsequent degradation in lysosomes. *Proc. Natl. Acad. Sci.* **81:** 2384.

Bennett, W.S., Jr. and T.A. Steitz. 1980. Structure of a complex between yeast hexokinase A and glucose. I. Structure determination and refinement at 3.5 Å resolution. *J. Mol. Biol.* **140:** 183.

Bertics, P.J. and G.N. Gill. 1985. Self-phosphorylation enhances the protein-tyrosine kinase activity of the epidermal growth factor receptor. *J. Biol. Chem.* **260:** 14642.

Bertics, P.J., W. Weber, C. Cochet, and G.N. Gill. 1985. Regulation of the epidermal growth factor receptor by phosphorylation. *J. Cell. Biochem.* **29:** 195.

Bishop, J.M. 1987. The molecular genetics of cancer. *Science* **235:** 305.

Brown, M.S., R.G.W. Anderson, and J.L. Goldstein. 1983. Recycling receptors: The round-trip itinerary of migrant membrane proteins. *Cell* **32:** 663.

Carpenter, G. and S. Cohen. 1976. ^{125}I-labeled human epidermal growth factor: Binding internalization and degradation in human fibroblasts. *J. Cell Biol.* **71:** 159.

Carpentier, J.-L., M.R. White, L. Orci, and R.C. Kahn. 1987. Direct visualization of the phosphorylated epidermal growth factor receptor during its internalization in A-431 cells. *J. Cell Biol.* **105:** 2751.

Chen, W.S., C.S. Lazar, M. Poenie, R.Y. Tsien, G.N. Gill, and M.G. Rosenfeld. 1987. Requirement for intrinsic protein tyrosine kinase in the immediate and late actions of the EGF receptor. *Nature* **328:** 820.

Clark, S., D.J. Cheng, J.J. Hsuan, J.D. Haley, and M.D. Waterfield. 1988. Loss of three major auto phosphorylation sites in the EGF receptor does not block the mitogenic action of EGF. *J. Cell. Physiol.* **134:** 421.

Cohen, S., H. Ushiro, C. Stoscheck, and M. Chinkers. 1982. A native 170,000 epidermal growth factor receptor-kinase complex from shed plasma membrane vesicles. *J. Biol. Chem.* **257:** 1523.

Cooke, R.M., A.J. Wilkinson, M. Baron, A. Pastore, M.J. Tappin, I.D. Campbell, H. Gregory, and B. Sheard. 1987. The solution structure of human epidermal growth factor. *Nature* **324:** 339.

Deisenhofer, J., O. Epp, K. Miki, R. Huber, and H. Michel. 1985. Structures of the protein subunits in the photosynthetic reaction centre of *Rhodopseudomonas viridis* at 3 Å resolution. *Nature* **318:** 618.

Di Fiore, P.P., J.H. Pierce, T.P. Fleming, R. Hazan, A. Ullrich, C.R. King, J. Schlessinger, and S.A. Aaronson. 1987. Overexpression of the human EGF receptor confers an EGF-dependent transformed phenotype to NIH 3T3 cells. *Cell* **51:** 1063.

Downward, J., P. Parker, and M.D. Waterfield. 1984a. Autophosphorylation sites on the epidermal growth factor receptor. *Nature* **311:** 483.

Downward, J., Y. Yarden, E. Mayes, G. Scrace, N. Totty, P. Stockwell, A. Ullrich, J. Schlessinger, and M.D. Waterfield. 1984b. Close similarity of epidermal growth factor receptor and v-*erb-b* oncogene protein sequences. *Nature* **307:** 521.

Engelman, D.M., T.A. Steitz, and A. Goldman. 1986. Identifying nonpolar transbilayer helices in amino acid sequences of membrane proteins. *Annu. Rev. Biophys. Biophys. Chem.* **15:** 321.

Fox, R.O., Jr. and F.M. Richards. 1982. A voltage-gated ion channel model inferred from the crystal structure of alamethicin at 1.50Å resolution. *Nature* **300:** 325.

Friend, S.H., T.P. Dryja, and R.A. Weinberg. 1988. Oncogenes and tumor-suppressing genes. *N. Engl. J. Med.* **318:** 618.

Frykberg, L., S. Palmiere, H. Beug, T. Graf, M.J. Hayman, and B. Vennstorm. 1983. Transforming capacities of avian erythroblastosis virus mutants deleted in the *erbA* or *erbB* oncogenes. *Cell* **32:** 227.

Fung, Y.-K.T., W.G. Lewis, L.B. Crittenden, and H.-J. Kung. 1983. Activation of the cellular oncogene c-*erbB* by LTR insertion: Molecular basis for induction of erythroblastosis by avian leukosis virus. *Cell* **33:** 357.

Gamett, D.C., S.E. Tracy, and H.L. Robinson. 1986. Differences in sequences encoding the carboxyl-terminal domain of the epidermal growth factor receptor correlate with differences in the disease potential of viral *erbB* genes. *Proc. Natl. Acad. Sci.* **83:** 6053.

Geuze, H.J., J.W. Slot, G.J.A.M. Strous, H.F. Lodish, and A.L. Schwartz. 1983. Intracellular site of asialoglycoprotein receptor-ligand uncoupling: Double-label immunoelectron microscopy during receptor-mediated endocytosis. *Cell* **32:** 277.

Glenney, J.R., Jr., L.M. Zokas, and M. Kamps. 1988a. Monoclonal antibodies to phosphotyrosine. *J. Immunol. Methods* **109:** 277.

Glenney, J.R., Jr., W.S. Chen, C.S. Lazar, G.M. Walton, L.M. Zokas, M.G. Rosenfeld, and G.N. Gill. 1988b. Ligand-induced endocytosis of the EGF receptor is blocked by mutational inactivation and by microinjection of anti-phosphotyrosine antibodies. *Cell* **52:** 675.

Greenwald, I. 1985. *lin-12*, A nematode homeotic gene, is homologous to a set of mammalian proteins that includes epidermal growth factor. *Cell* **43:** 583.

Hafen, E., K. Basler, J.-E. Edstroem, and G.M. Rubin. 1987. *Sevenless*, a cell-specific homeotic gene of *Drosophila*, encodes a putative transmembrane receptor with a tyrosine kinase domain. *Science* **236:** 55.

Haigler, H.T., J.A. McKanna, and S. Cohen. 1979. Direct visualization of the binding and internalization of a ferritin conjugate of epidermal growth factor in human carcinoma cells A-431. *J. Cell Biol.* **81:** 382.

Haigler, H., J.F. Ash, S.J. Singer, and S. Cohen. 1978. Visualization by fluorescence of the binding and internalization of epidermal growth factor in human carcinoma cells A-431. *Proc. Natl. Acad. Sci.* **75:** 3317.

Haigler, H.T., F.R. Maxfield, M.C. Willingham, and I. Pastan. 1980. Dansylcadaverine inhibits internalization of ^{125}I-epidermal growth factor in BALB 3T3 cells. *J. Biol. Chem.* **255:** 1239.

Hari, J. and R.A. Roth. 1987. Defective internalization of insulin and its receptor in cells expressing mutated insulin receptors lacking kinase activity. *J. Biol. Chem.* **262:** 15341.

Hendler, F.J., C.S. Richards, A. Shum, D. Burns, S. Schaefer, and B. Ozanne. 1985. Nuclear mechanisms for the increase of epidermal growth factor receptor in squamous cell carcinoma. *Trans. Assoc. Am. Phys.* **98:** 189.

Hertel, C. and J.P. Perkins. 1987. Sequential appearance of epidermal growth factor in plasma membrane-associated and intracellular vesicles during endocytosis. *J. Biol. Chem.* **262:** 11407.

Honegger, A.M., T.J. Dull, S. Felder, E. Van Obberghen, F. Bellot, D. Szapary, A. Schmidt, A. Ullrich, and J. Schlessinger. 1987. Point mutation at the ATP binding site of EGF receptor abolishes protein-tyrosine kinase activity and alters cellular routing. *Cell* **51:** 199.

Hunter, T. and J.A. Cooper. 1985. Protein-tyrosine kinases. *Annu. Rev. Biochem.* **54:** 897.

Kimelman, D. and M. Kirschner. 1987. Synergistic induction of mesoderm by FGF and TGF-β and the identification of an mRNA coding for FGF in the early *Xenopus* embryo. *Cell* **51:** 869.

Krupp, M.N., D.T. Connolly, and M.D. Lane. 1982. Synthesis, turnover, and down-regulation of epidermal growth factor receptors in human A431 epidermoid carcinoma cells and skin fibroblasts. *J. Biol. Chem.* **257:** 11489.

Kyte, J. and R.F. Doolittle. 1982. A simple method for displaying the hydropathic character of a protein. *J. Mol. Biol.* **157:** 105.

Lax, I., A. Johnson, R. Howk, J. Sap, F. Bellot, M. Winkler, A. Ullrich, B. Vennstrom, J. Schlessinger, and D. Givol. 1988. Chicken epidermal growth factor (EGF) receptor: cDNA cloning, expression in mouse cells, and differential binding of EGF and transforming growth factor alpha. *Mol. Cell. Biol.* **8:** 1970.

Leifer, D. and R. Henderson. 1983. Three-dimensional structure of orthorhombic purple membrane at 6.5 Å resolution. *J. Mol. Biol.* **163:** 451.

Libermann, T.A., H.R. Nusbaum, N. Razon, R. Kris, I. Lax, H. Soreq, N. Whittle, M.D. Waterfield, A. Ullrich, and J. Schlessinger. 1985. Amplification enhanced expression and possible rearrangement of EGF receptor gene in primary human brain tumours of glial origin. *Nature* **313:** 144.

Lin, C.R., W.S. Chen, W. Kruijer, L.S. Stolarsky, W. Weber, R.M. Evans, I.M. Verma, G.N. Gill, and M.G. Rosenfeld. 1984. Expression cloning of human EGF receptor complementary DNA: Gene amplification and three related messenger RNA products in A431 cells. *Science* **224:** 843.

Livneh, E., R. Prywes, O. Kashles, N. Reiss, I. Sasson, Y. Mory, A. Ullrich, and J. Schlessinger. 1986. Reconstitution of human epidermal growth factor receptors and its deletion mutants in cultured hamster cells. *J. Biol. Chem.* **261:** 12490.

Mayes, E.L.V. and M.D. Waterfield. 1984. Biosynthesis of the epidermal growth factor receptor in A431 cells. *EMBO J.* **3:** 531.

Nilsen, T.W., P.A. Maroney, R.G. Goodwin, F.M. Rottman, L.B. Crittenden, M.A. Raines, and H.-J. Kung. 1985. c-*erbB* activation in ALV-induced erythroblastosis: Novel RNA processing and promoter insertion result in expression of an amino-truncated EGF receptor. *Cell* **41:** 719.

Poenie, M., J. Alderton, R.Y. Tsien, and R.A. Steinhardt. 1985. Changes of free calcium levels with stages of the cell division cycle. *Nature* **315:** 147.

Prywes, R., E. Livneh, A. Ullrich, and J. Schlessinger. 1986. Mutations in the cytoplasmic domain of EGF receptor EGF binding and receptor internalization. *EMBO J.* **5:** 2179.

Remington, S., G. Wiegand, and R. Huber. 1982. Crystallographic refinement and atomic models of two different forms of citrate synthase at 2.7 and 1.7 Å resolution. *J. Mol. Biol.* **158:** 111.

Russell, D.S., R. Gherzi, E.L. Johnson, C.-K. Chou, and O.M. Rosen. 1987. The protein-tyrosine kinase activity of the insulin receptor is necessary for insulin-mediated receptor down-regulation. *J. Biol. Chem.* **262:** 11833.

Schlessinger, J., Y. Shechter, M.C. Willingham, and I. Pastan. 1978. Direct visualization of binding, aggregation, and internalization of insulin and epidermal growth factor on living fibroblastic cells. *Proc. Natl. Acad. Sci.* **75:** 2659.

Simonsen, C.C. and A.D. Levinson. 1983. Isolation and expression of an altered mouse dihydrofolate reductase cDNA. *Proc. Natl. Acad. Sci.* **80:** 2495.

Stern, D.C., D.L. Hare, M.A. Cecchini, and R.A. Weinberg. 1987. Construction of a novel oncogene based on synthetic sequences encoding epidermal growth factor. *Science* **235:** 321.

Sternberg, M.J.E. and W.R. Taylor. 1984. Modelling the ATP-binding site of oncogene products, the epidermal growth factor receptor and related proteins. *FEBS Lett.* **175:** 387.

Stoscheck, C.M. and G. Carpenter. 1984. "Downregulation" of EGF receptors: Direct demonstration of receptor degradation in human fibroblasts. *J. Cell Biol.* **98:** 1048.

Tanford, C. 1978. The hydrophobic effect and the organization of living matter. *Science* **200:** 1012.

Tycko, B. and F.R. Maxfield. 1982. Rapid acidification of endocytic vesicles containing α_2-macroglobulin. *Cell* **28:** 643.

Ullrich, A., L. Coussens, J.S. Hayflick, T.J. Dull, A. Gray, A.S. Tam, J. Lee, Y. Yarden, T.A. Libermann, J. Schlessinger, J. Downward, E.L.V. Mayes, N. Whittle, M.D. Waterfield, and P.H. Seeburg. 1984. Human epidermal growth factor receptor cDNA sequence and aberrant expression of the amplified gene in A431 epidermoid carcinoma cells. *Nature* **309:** 418.

Ushiro, H. and S. Cohen. 1980. Identification of phosphotyrosine as a product of epidermal growth factor-activated protein kinase in A-431 cell membranes. *J. Biol. Chem.* **255:** 8363.

Vogel, S., W. Freist, and J. Hoppe. 1986. Assignment of conserved amino acid residues to the ATP site in the protein kinase domain of the receptor for epidermal growth factor. *Eur. J. Biochem.* **154:** 529.

Weber, W. P.J. Bertics, and G.N. Gill. 1984a. Immunoaffinity purification of the epidermal growth factor receptor: Stoichiometry of EGF binding and kinetics of self-phosphorylation. *J. Biol. Chem.* **259:** 14631.

Weber, W., G.N. Gill, and J. Spiess. 1984b. Production of an epidermal growth factor receptor-related protein. *Science* **224:** 294.

Weeks, D.L. and D.A. Melton. 1987. A maternal mRNA localized to the vegetal hemisphere in *Xenopus* eggs codes for a growth factor related to TGF-β. *Cell* **51:** 861.

Wharton, K.A., K.M. Johansen, T. Xu, and S. Artavanis-Tsakonas. 1985. Nucleotide sequence from the neurogenic locus notch implies a gene product that shares homology with proteins containing EGF-like repeats. *Cell* **43:** 567.

Wiley, H.S. and D.D. Cunningham. 1982. The endocytotic rate constant: A cellular parameter for quantitating receptor-mediated endocytosis. *J. Biol. Chem.* **257:** 4222.

Willingham, M.C. and I. Pastan. 1980. The receptosome: An intermediate organelle of receptor-mediated endocytosis in cultured fibroblasts. *Cell* **21:** 67.

Xu, Y.-H., S. Ishii, A.J.L. Clark, M. Sullivan, R.K. Wilson, D.P. Ma, B.A. Roe, G.T. Merlino, and I. Pastan. 1984. Human epidermal growth factor receptor cDNA is homologous to a variety of RNAs overproduced in A431 carcinoma cells. *Nature* **309:** 806.

Yamamoto, T., N. Kamata, H. Kawano, S. Shimizu, T. Kuroki, K. Toyoshima, K. Rikimaru, N. Nomura, R. Ishizaki, I. Pastan, S. Gamou, and N. Shimizu. 1986. High incidence of amplification of the epidermal growth factor receptor gene in human squamous carcinoma cell lines. *Cancer Res.* **46:** 414.

Molecular Analysis of the Nerve Growth Factor Receptor

B. Hempstead,* N. Patil,† K. Olson,† and M. Chao,*†

*Division of Hematology/Oncology and †Department of Cell Biology and Anatomy, Cornell University Medical College, New York, New York 10021

Survival and maintenance of sympathetic and sensory neurons and cholinergic neurons of the basal forebrain are crucially dependent on the action of nerve growth factor (NGF) (Thoenen and Barde 1980; Levi-Montalcini 1987). The biological effects of NGF are the result of its initial interaction with a specific cell-surface receptor, followed by internalization and retrograde transport to the perikaryon (Hendry et al. 1974; Johnson et al. 1978). Despite extensive characterization of the trophic and differentiative effects of NGF, its mechanism of action and mode of signal transduction remain largely unknown.

An essential molecule in the translocation of information by NGF to responsive cells is the cell-surface receptor for NGF. This receptor has been detected on a wide variety of both neuronal and nonneuronal derivatives of the neural crest, such as melanomas (Fabricant et al. 1977), PC12 pheochromocytoma cells (Grob et al. 1983), Schwann cells (Taniuchi et al. 1986a), and neurofibromas (Ross et al. 1984). Biochemical characterization of the receptor in rat sympathetic nerve cells and PC12 cells (Taniuchi et al. 1986b) and human melanoma cells (Grob et al. 1985) reveals a glycosylated transmembrane protein of 75–80 kD that is phosphorylated predominantly on serine residues. The phosphorylation of the receptor is unaffected by NGF treatment, and the receptor displays no intrinsic kinase activity.

Equilibrium binding of [^{125}I]NGF to cells reveals two distinct affinity states for the NGF receptor (Sutter et al. 1979; Landreth and Shooter 1980; Schechter and Bothwell 1981). In many responsive cells, approximately 10–15% of the receptors display high-affinity binding with a K_d of 10^{-11} M, with the remainder of the receptors possessing a K_d of 10^{-9} M. The difference in equilibrium binding is accounted for by a 100-fold difference in the rate of dissociation of NGF, thereby accounting for the designation of high- and low-affinity receptors as "slow" and "fast" receptors, respectively.

Several lines of evidence suggest that the high-affinity receptor is not only internalized upon NGF binding but is responsible for the actions of NGF. First, the SY5Y neuroblastoma cell line contains a small number of high-affinity sites and no low-affinity sites (Sonnenfeld and Ishii 1985) and yet remains responsive to NGF. Second, the concentration of NGF necessary to elicit responses in PC12 cells (Greene 1977) and in sympathetic and sensory ganglion neurons (Frazier et al. 1974; Hill and Hendry 1976) is consistent with mediation by the high-affinity receptor. However, the strongest evidence for a biological role for the high-affinity receptor comes from mutant PC12 cell lines that lack the high-affinity receptor but that retain low-affinity sites. These mutant cells are unable to internalize NGF or to differentiate morphologically in response to NGF (Green et al. 1986).

The biochemical basis for the two affinity forms is not understood. To date, only two established cell lines, the SY5Y and the PC12 cells, display high-affinity receptors and are responsive to NGF. The majority of cell lines that have been examined are unresponsive and have only low-affinity sites. It has been previously proposed that the two affinity forms are related in a precursor-product manner (Massagué et al. 1981). Alternatively, the two forms have been thought to be represented by two completely different molecules. The isolation of the gene encoding the NGF receptor by DNA-mediated gene transfer has provided a means to distinguish among these possibilities and afforded the opportunity to study the molecular requirements of the high-affinity, responsive NGF receptor.

Genomic Structure of the NGF Receptor Gene

The gene encoding the human NGF receptor was efficiently transfected into mouse fibroblasts by cotransformation with total genomic DNA and the herpesvirus thymidine kinase gene (Chao et al. 1986). Expression of NGF receptors in the transfected cells was detected by an in situ rosette assay employing a monoclonal antibody (ME20.4) that blocks NGF binding to NGF receptors on human melanoma cells (Ross et al. 1984). Biochemical characterization of the transfected receptor by affinity cross-linking with [^{125}I]NGF confirmed that the receptor was a 75-kD protein, identical to the protein on human melanoma cells and PC12 cells.

Human NGF receptor sequences were rescued from a transfected mouse L cell line using human Alu repetitive sequences. Genomic clones were isolated that were capable of expressing the gene in fibroblasts after gene transfer. Using genomic fragments as probes, NGF receptor cDNAs were isolated from a library made from the human melanoma A875 cell line (Johnson et al. 1986). A single mRNA species of 3.8 kb was detected in melanoma and neuroblastoma cells, as well as rat superior cervical and dorsal root neurons and pheochromocytoma cells (Chao et al. 1986; Buck et al. 1987).

The full-length cDNA predicts a 28-amino-acid signal sequence, an extracellular domain containing four 40-amino-acid repeats with six cysteine residues at conserved positions, followed by a serine/threonine-rich region, a single transmembrane domain, and a 155-amino-acid cytoplasmic domain. A large 3′-untranslated region exists in the receptor mRNA. The sequence of a full-length rodent NGF receptor cDNA (Radeke et al. 1987) indicated that the rat NGF receptor has considerable homology (92%) with the human NGF receptor from melanoma cells.

Isolation and DNA sequencing of coding segments of the receptor gene established that the human NGF receptor gene is divided into six exons, spanning 23 kb. Figure 1 outlines the exon-intron organization for the human NGF receptor. The 5′-untranslated and signal peptide sequences are contained in the first exon. The second and third exons encode the amino-terminal amino acid sequence of the mature receptor protein and a discrete domain of 161 amino acids, of which 24 are cysteine residues. It is of interest that the splice site for the third exon is located precisely after the 24th cysteine residue. Similarly, the transmembrane region found in the fourth exon contains an intron-exon boundary in the stop-transfer sequence directly following the hydrophobic-containing amino acids. Such precise divisions of the structural features of the receptor molecule are supportive of a functional role of these domains.

The mRNA for the NGF receptor contains a large 3′-untranslated region of 1989 bp. This sequence is uninterrupted and is contained in the largest receptor exon (2291 bp), together with sequences representing the intracellular domain of the receptor. In addition, the small fifth exon contains a portion of the putative cytoplasmic region of the receptor.

The coding sequence of the NGF receptor lacks significant homology with any other known receptor or protein. In particular, the primary sequences of the intracellular domains display no similarities to known growth factors, oncogenes, or the tyrosine kinase family of genes. This lack of homology with known tyrosine kinases agrees with the absence of tyrosine kinase activity by the NGF receptor from human melanoma (Grob et al. 1985) and from PC12 cells and rat superior cervical ganglia (Taniuchi et al. 1986b). The lack of similarity between the NGF receptor and the tyrosine kinase family of receptors is not unexpected since the primary target cells influenced by NGF do not undergo cell division in response to the hormone. Tyrosine kinase activity has been shown to be an integral step in the initiation of mitogenic activity by mesodermal growth factors (Moolenaar et al. 1988). The absence of such activity for NGF receptor suggests that the mechanism of action of NGF differs significantly from other peptide growth factors such as platelet-derived growth factor, epidermal growth factor (EGF), colony-stimulating factor-1, and insulin.

In experiments using intact cells, NGF induces changes in the patterns of protein serine/threonine phosphorylation (Halegoua and Patrick 1980), and the NGF receptor itself is known to be phosphorylated on serine residues (Grob et al. 1985; Taniuchi et al. 1986b). However, the protein kinase activities responsible for these phosphorylation events apparently are not intrinsic to the receptor, since the NGF receptor primary sequence lacks significant similarity with known serine/threonine protein kinases.

Structural Features of the NGF Receptor Gene

The initiation site for the NGF receptor gene in human A875 melanoma cells and in rat PC12 cells was localized by primer extension and S1 nuclease protection experiments (Sehgal et al. 1988). For both cell lines, the same start site for transcription was found. Isolation and sequencing of 5′ human (Sehgal et al. 1988) and rat genomic sequences were carried out by hybridization of 5′ cDNA fragments.

The sequence of the region upstream of the first coding exon for both the human and rat receptor genes is shown in Figure 2. The sequence composition of this genomic region contains a high G + C content but contains neither a characteristic "TATA" nor a "CAAT" consensus sequence. Several conserved GGGCGG sequences are present within 200 nucleotides of the initiator ATG codon. These GC-rich sequences represent potential binding sites for the transcription factor

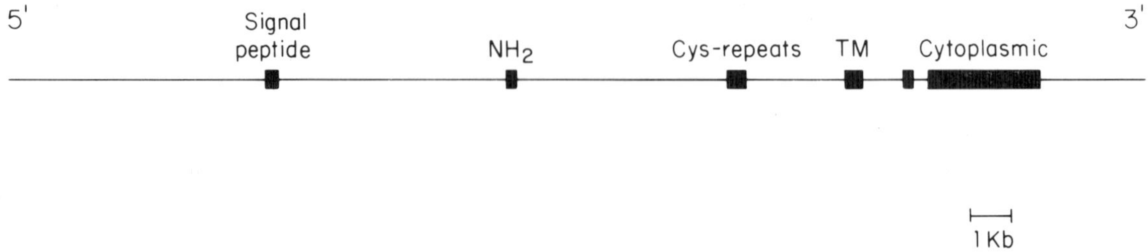

Figure 1. Genomic map of the human NGF receptor gene. The NGF receptor gene is shown in a 5′ to 3′ orientation, with exons represented as boxes. Nine bacteriophage λ clones were mapped by conventional techniques including oligomer mapping and Southern blot analysis with NGF receptor cDNA probes (Sehgal et al. 1988). Selected fragments were subcloned and the exon-intron junctions were determined.

```
Rat     GG GTATGTGGGA CAGCCCACAG AGAAGCCGCA GCGGGGCACA CGGGTGCGCT CCTCGCACGC
Human   GC GCAGTAGCGC CGGGAACTGG GTACCAGGGC GGGATGGGTG AGAGGCTCTA AGGGACAAGG
            -220                  -200                 -180

CTCACTTGGC TGCTTCGGAC TCCACACCCC AGGCAGCCGG GCGGGCGCGG TTCCGGACGG TTGGGGGGGG
CAGGGAGAAG GGAGAAGCGC ACGGGTGCGG GGAACCAGCC CTCCCTTTGC CTCTGCTTCC CACCCCGAGG
                     -140                  -120                  -100

GGGTAGTTGG GCGGGCTGGG CGGGGAGGAG GCGGGGCTGC TGCATTGCCT ACACCCAGCT GCTCCCGCCG
CGGCAGGGCG GGCGGGCAGG TTCCGGGGGT GGCGGGCTGG CGGGGCGGAG GCGGGGGCCG ACAGCTGGCT
         -80                  -60                  -40

CCAGCAGCCA GAGCGAGTGG AGCCTCGCGCAGCTCCGGCGGGCAGCAGGCGCTGGAGCGCATCGCAGTTCA
TCACCAGCCC TTCCGCCGTA GCAGAGCGAGCCGCGGCCAGCTCCGGCGGGCAGGGGGGGCGCTGGAGCGCA
-20

                                                                        MetGly
GCTCAGCGCAGCACCATCGGTCTGCGGAGCGGACTGAGCTAGAAGCGGAGCGCTGACGCCGGAGGCGTGCG ATGGGGGCA
GCGCAGCGCAGCCCCATCAGTCCGCAAAGCGGACCGAGCTGGAAGTCGAGCGCTGCCGCGCGGGAGGCGGGCG ATGGGGGCA
```

Figure 2. Promoter sequence of the human and rat NGF receptor genes. Rat genomic clones were originally isolated by virtue of cross-species hybridization with the human cDNA probe covering nucleotides 36–820 (called H1; Johnson et al. 1986). The 5′ end of the rat gene was further localized with a rat NGF receptor cDNA probe (Buck et al. 1988), and the site of initiation was determined by primer extension (Sehgal et al. 1988). The upstream sequence was determined and compared with the human sequence. The arrow specifies the site of initiation. Putative Sp1 binding sites are underlined and the beginning of the first exon is specified.

Sp1 (Dynan and Tjian 1983; Dynan et al. 1986). Two of the four potential Sp1 binding sites are conserved between the rat and human promoters. Transient expression analysis of the receptor promoter sequence from nucleotide +80 to −200 gave maximal promoter activity when this sequence was linked to the chloramphenicol acetyltransferase gene (Sehgal et al. 1988).

Several eukaryotic genes have promoters that are rich in GC content but lack the characteristic TATA and CAAT boxes. These include the hypoxanthine guanine phosphoribosyltransferase gene (Melton et al. 1984), the HMG-CoA reductase gene (Reynolds et al. 1984), the 3-phosphoglycerate kinase gene (Singer-Sam et al. 1984), the adenosine deaminase gene (Valerio et al. 1985), and the adenine phosphoribosyltransferase gene (Broderick et al. 1987), all of which encode enzymes with housekeeping functions. Expression of these genes is at a low constitutive level in diverse tissues. In addition, the promoters of several cellular growth control genes such as c-Harvey *ras* (Ishii et al. 1985a), c-Kirsten *ras* (Hoffman et al. 1987), the EGF receptor (Ishii et al. 1985b), and the insulin receptor (Araki et al. 1987) do not contain the typical TATA and CAAT transcriptional consensus sequence but have multiple GC box motifs. Hence, the promoter for the NGF receptor resembles those genes that are constitutively expressed or involved in growth regulation. An unusual feature of this promoter is that it appears to direct transcription primarily from a major site of initiation instead of using multiple sites of initiation as occurs with many constitutively expressed genes. No evidence was seen in PC12 RNA for multiple initiation sites (Sehgal et al. 1988). Thus, differential 5′ transcripts from the NGF receptor gene are not observed for a cell line displaying high- and low-affinity receptors.

NGF Receptor Models

Several other models could account for the two kinetic forms of the NGF receptor. First, the high-affinity and low-affinity receptors could be due to multiple receptor genes. As an example of a putative high-affinity receptor heterodimer, the interleukin-2 receptor is represented by two different proteins, each of which has the ability to bind interleukin-2 specifically (Sharon et al. 1986; Dukovich et al. 1987).

To investigate the possibility that there may be related NGF receptor genes, Southern blots were carried out at reduced stringency on human DNA. A set of parallel hybridizations were carried out under three different conditions. The receptor probe consisted of a cDNA that spans from nucleotide 36 to 820 (Johnson et al. 1986) and that included the entire extracellular portion of the receptor. This region of the receptor was chosen because the binding site for NGF is highly conserved, as demonstrated by the ability of mouse submaxillary NGF to bind to chicken, rat, and human NGF receptor. Hybridization at low stringency would be expected to yield any sequences that are similar to the human receptor. As demonstrated in Figure 3, no additional fragments are detected even under low stringency conditions. It should be emphasized that under similar conditions of hybridization, the rat (Buck et al. 1988) and chicken (E. Escandon, unpubl.) NGF re-

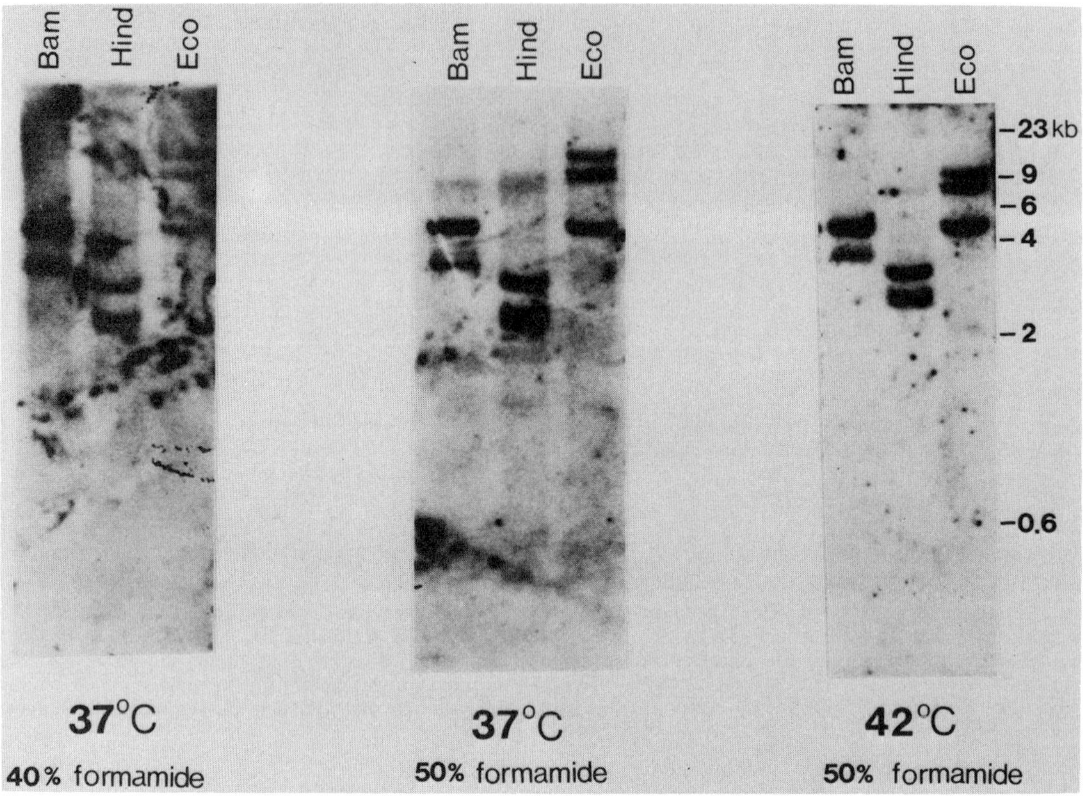

Figure 3. Southern blot of human NGF receptor sequences. Nitrocellulose filters containing transferred human DNA (Southern 1975) were hybridized at 5 × SSCPE, 1 × Denhardt's solution, 100 μg/ml denatured salmon sperm DNA, and 10% dextran sulfate at the indicated temperatures and formamide concentration with a human NGF receptor cDNA (H1). Filters were washed at 5°C to 1 × SSC.

ceptor genes have been isolated. These results indicate that within the limits of this hybridization, no other sequences are similar to the extracellular domain of the NGF receptor and are consistent with previous studies indicating that the NGF receptor gene is a single-copy gene (Chao et al. 1986; Huebner et al. 1986).

Alternative splicing is a common mechanism for generating multiple forms of the same protein. We have examined whether differential processing of the receptor gene occurs in PC12 cells and rat sympathetic neurons. Unique *Taq*I and *Mst*II sites within the rat receptor cDNA were end-labeled, and these restriction fragments were hybridized to RNA isolated from cells expressing both high- and low-affinity receptors, PC12 cells, and superior cervical ganglia (SCG). The hybridization reactions were digested by S1 nuclease and analyzed by a 6% polyacrylamide sequencing gel. Both probes gave a single protected fragment of the expected size (Fig. 4). No other protected fragments were detected in either PC12 RNA or SCG RNA.

Thus, alternative splicing from the NGF receptor gene in neurons and PC12 cells does not occur. Since both RNAs tested were isolated from cells that display both high- and low-affinity receptors, differential RNA processing does not appear to account for the generation of these kinetic forms.

Transfection of NGF Receptors

A full-length receptor cDNA, E1, was found to contain 130 bases of 5′-untranslated sequence, the entire coding region for the receptor, a polyadenylation signal, and a poly(A) tail (Johnson et al. 1986). It was introduced into a eukaryotic expression vector, pMV6, containing the Moloney leukemia virus long terminal repeats (LTRs) and the bacterial neomycin phosphotransferase gene under the control of the herpes simplex virus thymidine kinase promoter (Kirschmeier et al. 1988). A 3′ polyadenylation signal is provided by a 3′ LTR sequence. When inserted into this vector in a 5′-3′ orientation between the LTR sequences at a unique *Eco*RI site, the NGF receptor cDNA is under the control of the 5′-LTR promoter sequence.

We have used this vector to introduce the receptor into fibroblastic, epithelial, melanoma, pheochromocytoma, and neuroblastoma cells. Since the full-length receptor cDNA contained a significant amount of 3′-untranslated sequence, we examined whether the entire receptor cDNA sequence was required for expression in eukaryotic cells. A series of deletions in the 3′-untranslated segment of the human NGF receptor cDNA were generated (Fig. 5). Two different assays were used to detect the appearance of NGF receptors

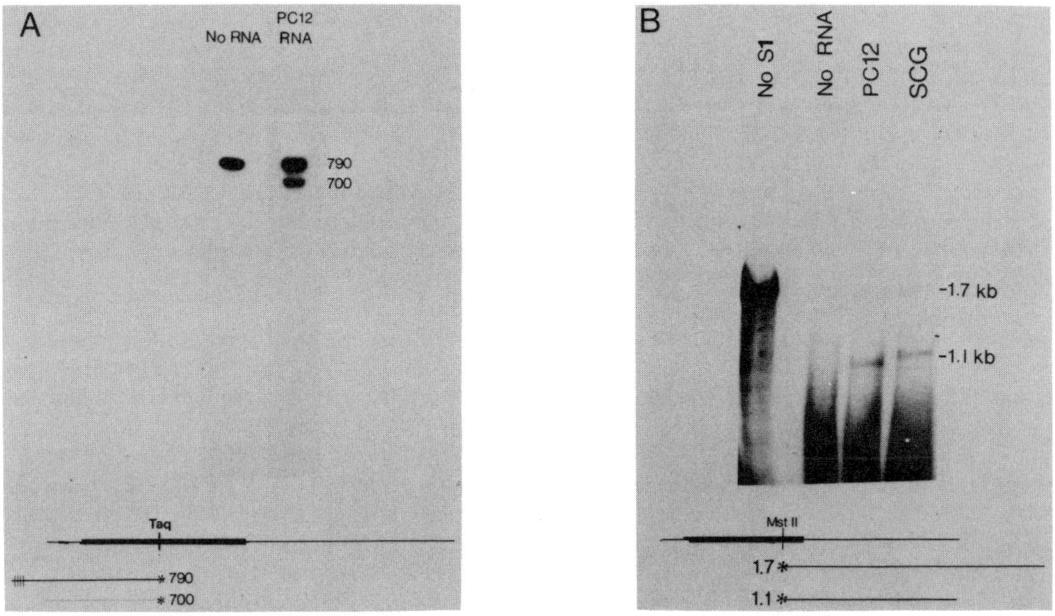

Figure 4. S1 nuclease analysis of the NGF receptor mRNA in rat PC12 cells and superior cervical ganglia. Mapping by S1 nuclease digestion was carried out according to the method of Weaver and Weissman (1979). Total RNA (50 μg) was mixed with 5′ end (*A*) or 3′ end (*B*) labeled restriction fragments from the rat NGF receptor cDNA (Buck et al. 1988) in 10 μl of hybridization buffer containing 80% formamide, 0.04 M PIPES (pH 6.5), 0.001 M EDTA, and 0.4 M NaCl. After annealing at 52°C for 12 hr, S1 digestion was carried out, followed by electrophoresis on 6% polyacrylamide, 8 M urea gels, and autoradiography. A schematic diagram of the expected protected fragments is shown at the bottom of the figure.

after gene transfer. Transient expression of NGF receptor was detected by indirect immunofluorescence using the monoclonal antibody against the human NGF receptor. In addition, stable transfectants were isolated after selection in neomycin and tested for receptor expression.

Removal of small segments of the 3′-untranslated sequence (deletions 0–5 and 18–4) or over 90% of the 3′-untranslated sequence (deletion B9) did not hinder receptor expression, as assayed by immunofluorescence and rosetting of stable clones. However, deletions that removed the authentic stop codon and altered the read-

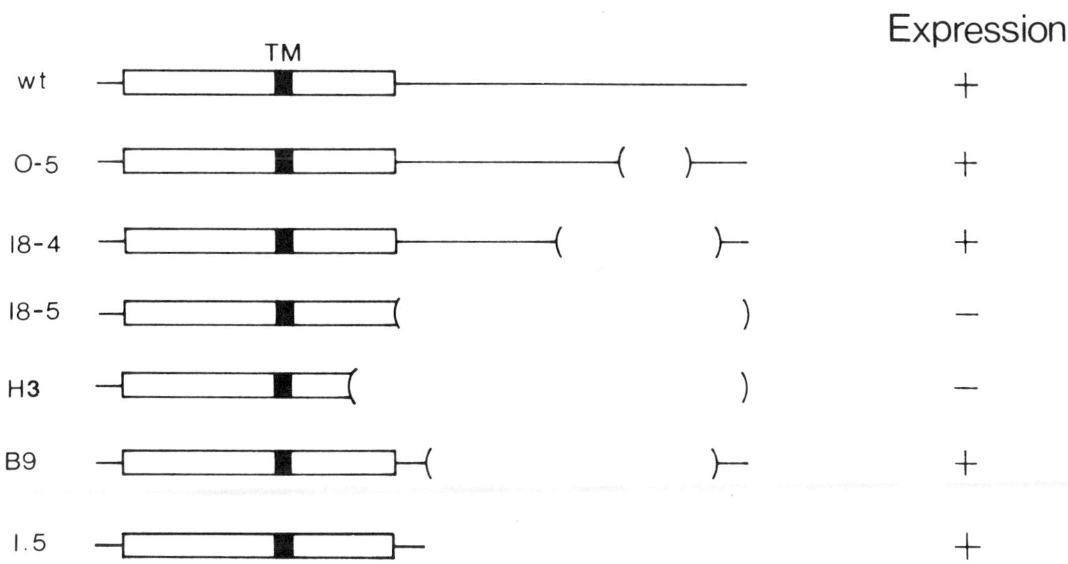

Figure 5. Deletions of the human NGF receptor cDNA. Schematic representation of the mutant human NGF receptor cDNAs derived from the cDNA E1. Mutant cDNAs were obtained by digestion of the plasmid, pMVE1 (Johnson et al. 1986) using key restriction sites. Mutants 0.4, 18.4, and 18.5 were constructed by linearization of pMVE1 at a unique *Xba*I site in the 3′-untranslated region and digestion with BAL 31. The plasmids were transfected into mouse fibroblasts (Wigler et al. 1979), and expression was assessed by rosetting or indirect immunofluorescence with the monoclonal anti-human receptor antibody ME20.4 (Ross et al. 1984).

ing frame of the receptor (deletions 18–5 and H3) resulted in no detectable expression of receptors at the cell surface. Elimination of the receptor polyadenylation signal resulted in a decrease in the level of expression of the receptor as assayed by immunofluoresence after transient expression and by RNA Northern blot analysis of stable transfectants.

These experiments indicate that a wide variety of cells can be transfected with the pMV6 expression plasmid bearing the human NGF receptor and that both transient and stable expression can be obtained. Although most of the 3′-untranslated sequence appears to be dispensable for expression, removal of the polyadenylation site appears to diminish the level of receptor expression (B. Hempstead, unpubl.). Maximal expression appears with the full-length cDNA, E1. The numbers of receptors on transfected cell lines can vary from 1000 to over 500,000 receptors per cell. However, the equilibrium dissociation constant measured for many transfected cells is consistent with the low-affinity form of the receptor.

A number of stable transformants abundantly expressing the human receptor have been examined for specific responses to NGF. These responses include (1) the acquisition of a differentiated phenotype with neurite outgrowth and cessation of cell division and (2) the induction of c-*fos* gene expression. Treatment of transfected fibroblasts, melanoma, and epithelial cells with NGF did not result in any induction of fiber formation or changes in growth pattern. Induction of c-*fos* transcription was also examined as an example of a distinct response to NGF. The c-*fos* gene is transiently stimulated by serum and a number of growth factors; additionally, c-*fos* transcription has been shown to be upregulated by NGF in PC12 cells (Greenberg et al. 1985).

Transfected mouse melanoma cells were grown in the presence and absence of 50 ng/ml NGF for 30 minutes, and total cellular RNA was isolated. A typical experiment is shown in Figure 6. Significant accumulation of c-*fos* RNA is observed in NGF-treated PC12 cells, but no expression is observed in similarly treated mouse melanoma cells expressing human NGF receptor. Similar experiments have been carried out using human melanoma cells and cell lines transfected with the human receptor, and no responses have yet been detected.

The most plausible explanation for these negative findings is that these cells lack the proper cellular environment to respond to NGF. Many of the cells that have been transfected are not of neural crest origin. In addition, the kinetic form of the NGF receptor found on the majority of transfected cells is predominately the low-affinity, fast form. These observations are consistent with the high-affinity receptor as the mediator of the biological responses of NGF, including c-*fos* induction (Greenberg et al. 1985; Green et al. 1986).

DISCUSSION

The human NGF receptor gene is a single-copy gene, consisting of six exons that span 23 kb. The receptor gene is capable of being transferred to fibroblast cells from human genomic DNA and expressed at high levels (Chao et al. 1986). The constitutive nature of the receptor promoter sequence is a partial explanation of why this tissue-specific gene is expressed efficiently in a variety of nonneuronal cells after genomic gene trans-

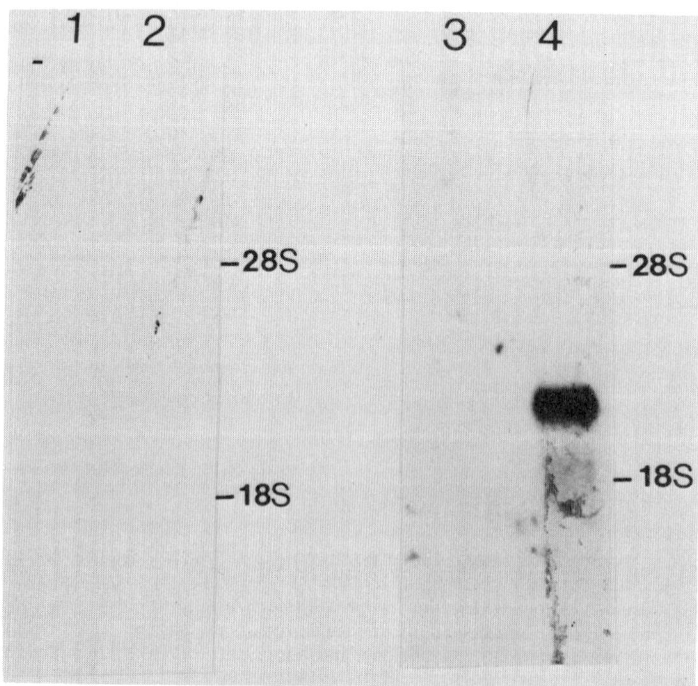

Figure 6. Lack of c-*fos* response in transfected cells. PC12 cells and mouse melanoma cells transfected with pMVE1 (designated H1c3) were treated with 50 ng/ml of NGF for 30 min. Total RNA was isolated and separated on a 1% agarose gel containing 2.2 M formaldehyde. Hybridization was carried out with a nick-translated v-*fos* probe (Curran et al. 1982). RNA samples were the following: (lane *1*) H1c3 cells; (lane *2*) H1c3 cells treated wtih NGF; (lane *3*) P12 cells; (lane *4*) PC12 cells treated with NGF.

fer. Significantly, receptors for EGF and insulin and the c-*ras* gene share promoter characteristics with the NGF receptor gene. Since these gene products are in some way involved with growth control and their promoter elements are similar, regulation of these genes may share common mechanisms.

Several lines of evidence strongly suggest that the cloned receptor is present in both high- and low-affinity species. First, a single receptor mRNA species is detected in nerve cells and in cultured cells expressing both high- and low-affinity receptors. Since the same mRNA species was observed in cells that express predominantly the low-affinity receptor species (Schwann and A875 melanoma cells) and in cells that express only the high-affinity receptor (SY5Y neuroblastoma cells), it appears that a single gene encodes both receptor forms (Chao et al. 1986). Moreover, both forms of the receptor must share common structural features, since the ME20.4 monoclonal antibody that blocks NGF binding to the human receptor recognizes the NGF receptor in both the low-affinity form (on A875 melanoma cells) and the high-affinity form (on SY5Y neuroblastoma cells). In addition, in the rodent system, the 192 monoclonal antibody specific for the rat NGF receptor can block certain biological responses such as c-*fos* induction (Milbrandt 1986) and neurite formation (Chandler et al. 1984). These data imply that the monoclonal antibodies against the NGF receptor can recognize common epitopes of both the high- and low-affinity forms of the receptor. These observations are further strengthened by the results of Green and Greene (1986), who have demonstrated by affinity cross-linking that both high- and low-affinity receptors are represented by the same receptor protein.

The lack of appropriate responses to NGF in many cells expressing abundant NGF receptors strongly suggests that most cells do not possess the proper cellular environment to respond to NGF. The lack of a response can also be correlated with the absence of the high-affinity form of the receptor, which mediates many of the biological responses of NGF (Green et al. 1986). Hence, clarification of the biochemical nature of the high-affinity form is crucial in understanding the mechanisms by which the receptor protein transmits its intracellular signal after binding to NGF. The data presented here indicate that differential splicing does not give rise to the two kinetic forms of the receptor, and, furthermore, no other evidence has been found for another gene related to the receptor that binds NGF. Although posttranslational modifications may account for a difference of affinity, there is presently no indication that a posttranslational event influences ligand affinity for NGF.

The most plausible explanation for the difference in receptor ligand affinity for the NGF receptor is that a distinct regulatory protein interacts with the receptor and dictates the affinity of the receptor for its ligand. A number of different observations makes this a tenable hypothesis. First, several groups, using affinity cross-linking agents to label NGF receptors, have described cross-linked species larger than the 100-kD receptor-NGF complex, which can be isolated from cells expressing the high-affinity form of the receptor (Massagué et al. 1981; Hosang and Shooter 1985). Second, there is circumstantial evidence for a regulatory site in the receptor sequence. Examination of the human and rat NGF receptor sequences indicates an extremely strong region of conservation of amino acid sequence is present in the transmembrane region and extending 50 amino acids into the cytoplasmic domain. It is attractive to speculate that this region of the receptor is so highly conserved because of important interactions with other proteins closely associated with the plasma membrane.

The two kinetic forms of the NGF receptor thus appear to be encoded by the same protein (Green and Greene 1986), which is the product of a single gene. Responses to NGF after gene transfer of receptors will probably be attained by the use of cells containing the correct intracellular machinery to transduce NGF's signal. The use of mutant cell lines derived from PC12 cells (Bothwell et al. 1980; Green et al. 1986) provides an important approach. Resolution of the molecular basis of the two kinetic forms of the NGF receptor will undoubtedly begin to unravel the complex series of events that result from NGF binding to responsive cells.

ACKNOWLEDGMENTS

We thank Tom Curran for the v-*fos* probe. We are grateful to Amita Sehgal, Randy Buck, Claudia Morgan, Todd Stukenburg, and Nancy Spickofsky for their contributions, and Lori Van Houton for photographic assistance. This work was supported by the National Institutes of Health (NS-21072 and HD-23315), the Cornell Scholars Program, and the Hirschl/Caulier Trust Fund, and a grant to B.L.H. by the Rose Kloss Foundation.

REFERENCES

Araki, E., F. Shimada, H. Uzawa, M. Mori, and Y. Ebina. 1987. Characterization of the promoter region of the human insulin receptor gene. *J. Biol. Chem.* **262:** 16186.

Bothwell, M.A., A.L. Schechter, and K.M. Vaughn. 1980. Clonal variants of PC12 pheochromocytoma cells with altered response to nerve growth factor. *Cell* **21:** 857.

Broderick, T.P., D.A. Schaff, A.M. Bertino, M.K. Dush, J.A. Tischfield, and P.J. Stambrook. 1987. Comparative anatomy of the human APRT gene and enzyme: Nucleotide sequence divergence and conservation of a nonrandom CpG dinucleotide arrangement. *Proc. Natl. Acad. Sci.* **84:** 3349.

Buck, C.R., H.J. Martinez, I.B. Black, and M.V. Chao. 1987. Developmentally regulated expression of the nerve growth factor receptor gene in the periphery and brain. *Proc. Natl. Acad. Sci.* **84:** 3060.

Buck, C.R., H.J. Martinez, M.V. Chao, and I.B. Black. 1988. Different expression of the nerve growth factor receptor gene in multiple brain areas. *Dev. Brain Res.* (in press).

Chandler, C.E., L.M. Parsons, M. Hosang, and E.M. Shooter. 1984. A monoclonal antibody modulates the interaction of nerve growth factor with PC12 cells. *J. Biol. Chem.* **259:** 6882.

Chao, M.V., M.A. Bothwell, A.H. Ross, H. Koprowski, A.A. Lanahan, C.R. Buck, and A. Sehgal. 1986. Gene transfer and molecular cloning of the human NGF receptor. *Science* **232:** 518.

Curran, T., G. Peters, C. Van Beveren, N.M. Teich, and I.M. Verma. 1982. BJ murine osteosarcoma virus: Identification and molecular cloning of biologically active proviral DNA. *J. Virol.*. **44:** 674.

Dukovich, M., Y. Wano, L.B. Thuy, P. Katz, B.R. Cullen, J.H. Kehrl, and W.C. Greene. 1987. A second human interleukin-2 binding protein that may be a component of high-affinity interleukin-2 receptors. *Nature* **327:** 518.

Dynan, W.S. and R. Tjian. 1983. The promoter-specific transcription factor Sp1 binds to upstream sequences in the SV40 early promoter. *Cell* **35:** 79.

Dynan, W.S., S. Sazer, R. Tjian, and R.T.Schimke. 1986. Transcription factor Sp1 recognizes a DNA sequence in the mouse dihydrofolate reductase promoter. *Nature* **319:** 246.

Fabricant, R.F., J.E. DeLarco, and G.J. Todaro. 1977. Nerve growth factor receptors on human melanoma cells in culture. *Proc. Natl. Acad. Sci.* **74:** 565.

Frazier, W.A., L.F. Boyd, and R.A. Bradshaw. 1974. Properties of the specific binding of ^{125}I-nerve growth factor to responsive peripheral neurons. *J. Biol. Chem.* **249:** 513.

Green, S.H. and L.A. Greene. 1986. A single $M_r = 103{,}000$ ^{125}I-β-nerve growth factor-affinity-labeled species respresents both the low and high affinity forms of the nerve growth factor receptor. *J. Biol. Chem.* **261:** 15316.

Green, S.H., R.E. Rydel, J.L. Connolly, and L.A. Greene. 1986. PC12 mutants that possess low- but not high-affinity nerve growth factor receptors neither respond to nor internalize nerve growth factor. *J. Cell Biol.* **102:** 830.

Greenberg, M., L.A. Greene, and E.B. Ziff. 1985. Nerve growth factor and epidermal growth factor induce rapid transient changes in proto-oncogene transcription in PC12 cells. *J. Biol. Chem.* **260:** 14101.

Greene, L.A. 1977. A quantitative bioassay for nerve growth factor (NGF) activity employing a clonal pheochromocytoma cell line. *Brain Res.* **133:** 350.

Grob, P.M., C.H. Berlot, and M.A. Bothwell. 1983. Affinity labeling and partial purification of nerve growth factor receptors from rat pheochromocytoma and human melanoma cells. *Proc. Natl. Acad. Sci.* **80:** 6819.

Grob, P.M., A.H. Ross, H. Koprowski, and M.A. Bothwell. 1985. Characterization of the human melanoma nerve growth factor receptor. *J. Biol. Chem.* **260:** 8044.

Halegoua, S. and J. Patrick. 1980. Nerve growth factor mediates phosphorylation of specific proteins. *Cell* **22:** 581.

Hendry, I.A., K. Stockel, H. Thoenen, and L.L. Iversen. 1974. The retrograde transport of nerve growth factor. *Brain Res.* **68:** 103.

Hill, C.E. and I.A. Hendry. 1976. Differences in sensitivity to nerve growth factor of axon formation and tyrosine hydroxylase induction in cultured sympathetic neurons. *Neuroscience* **1:** 489.

Hoffman, E.K., S.P. Trusko, N. Freeman, and D.L. George. 1987. Structural and functional characterization of the promoter region of the mouse c-Ki-*ras* gene. *Mol. Cell. Biol.* **7:** 2592.

Hosang, M. and E.M. Shooter. 1985. Molecular characteristics of nerve growth factor receptors on PC12 cells. *J. Biol. Chem.* **260:** 655.

Heubner, K., M. Isobe, M. Chao, M. Bothwell, A.H. Ross, J. Finan, J.A. Hoxie, A. Sehgal, C.R. Buck, A. Lanahan, P.C. Nowell, H. Koprowski, and C. Croce. 1986. The nerve growth factor receptor gene is at human chromosome region 17q12-12q22, distal to the chromosome 17 breakpoint in acute leukemias. *Proc. Natl. Acad. Sci.* **83:** 1403.

Ishii, S., G.T. Merlino, and I. Pastna. 1985a. Promoter region of the human Harvey *ras* proto-oncogene: Similarity to the EGF receptor proto-oncogene promoter. *Science* **230:** 1378.

Ishii, S., Y.H. Xu, R.H. Stratton, B.A. Roe, G.T. Merlino, and I. Pastan. 1985b. Characterization and sequence of the promoter region of the human epidermal growth factor receptor gene. *Proc. Natl. Acad. Sci.* **82:** 4902.

Johnson, D., A. Lanahan, C.R. Buck, A. Sehgal, C. Morgan, E. Mercer, M. Bothwell, and M. Chao. 1986. Expression and structure of the human NGF receptor. *Cell* **47:** 545.

Johnson, E.M., R.Y. Andres, and R.A. Bradshaw. 1978. Characterization of the retrograde transport of nerve growth factor (NGF) using high specific activity ^{125}I-NGF. *Brain Res.* **150:** 319.

Kirshmeier, P.T., G.M. Housey, M.D. Johnson, A.S. Perkins, and I.B. Weinstein. 1988. Construction and characterization of a retroviral vector demonstrating efficient expression of cloned cDNA sequences. *DNA* **7:** 219.

Landreth, G.E. and E.M. Shooter. 1980. Nerve growth factor receptors on PC12 cells: Ligand-induced conversion from low- to high-affinity states. *Proc. Natl. Acad. Sci.* **77:** 4751.

Levi-Montalcini, R. 1987. The nerve growth factor: Thirty-five years later. *Science* **237:** 1154.

Massagué, J., B.J. Guilette, M.P. Czech, C.J. Morgan, and R.A. Bradshaw. 1981. Identification of a nerve growth factor receptor protein in sympathetic ganglia membranes by affinity labeling. *J. Biol. Chem.* **256:** 9419.

Melton, D.W., D.S. Konecki, J. Brennand, and C.T. Caskey. 1984. Structure, expression, and mutation of the hypoxanthine phosphoribosyltransferase gene. *Proc. Natl. Acad. Sci.* **81:** 2147.

Milbrandt, J. 1986. Nerve growth factor rapidly induces c-*fos* mRNA in PC12 rat pheochromocytoma cells. *Proc. Natl. Acad. Sci.* **83:** 4789.

Moolenaar, W.H., A.J. Bierman, B.C. Tilly, I. Verlaan, L.H.K. Defize, A.M. Honegger, A. Ullrich, and J. Schlessinger. 1988. A point mutation at the ATP-binding site of the EGF-receptor abolishes signal transduction. *EMBO J.* **7:** 707.

Radeke, M.J., T.P. Misko, C. Hsu, L.A. Herzenberg, and E.M. Shooter. 1987. Gene transfer and molecular cloning of the rat nerve growth factor receptor. *Nature* **325:** 593.

Reynolds, G.A., S.K. Basu, T.F. Osborne, D.J. Chin, G. Gil, M.S. Brown, J.L. Goldstein, and K.L. Luskey. 1984. HMG CoA reductase: A negatively regulated gene with unusual promoter and 5′ untranslated regions. *Cell* **38:** 275.

Ross, A.H., P. Grob, M.A. Bothwell, D.E. Elder, C.S. Ernst, N. Marano, B.F.D. Ghrist, C.C. Slemp, M. Herlyn, B. Atkinson, and H. Koprowski. 1984. Characterization of nerve growth factor receptor in neural crest tumors using monoclonal antibodies. *Proc. Natl. Acad. Sci.* **81:** 6681.

Schechter, A. and M.A. Bothwell. 1981. NGF receptors on PC12 cells; evidence for two receptor classes with differing cytoskeletal association. *Cell* **24:** 867.

Sehgal, A., N. Patil, and M. Chao. 1988. A constitutive promoter directs expression of the NGF receptor gene. *Mol. Cell. Biol.* **8:** (in press).

Sharon, M, R.D. Klausner, B.R. Cullen, R. Chizzonite, and W.J. Leonard. 1986. Novel interleukin-2 receptor subunit detected by crosslinking under high-affinity conditions. *Science* **234:** 859.

Singer-Sam, J., D.H. Keith, K. Tani, R.L. Simmer, L. Shively, S. Lindsay, A. Yoshida, and A.D. Riggs. 1984. Sequence of the promoter region of the gene for human X-linked 3-phosphoglycerate kinase. *Gene* **32:** 409.

Sonnenfeld, K.H. and D.N. Ishii. 1985. Fast and slow nerve growth factor binding sites in human neuroblastoma and rat pheochromocytoma cell lines: Relationship of sites to each other and neurite formation. *J. Neurosci.* **5:** 1717.

Southern, E.M. 1975. Detection of specific sequences among DNA fragments separated by gel electrophoresis. *J. Mol. Biol.* **98:** 503.

Sutter, A., R.J. Riopelle, R.M. Harris-Warrick, and E.M. Shooter. 1979. NGF receptors: Characterization of two distinct classes of binding sites on chick embryo sensory ganglia cells. *J. Biol. Chem.* **254:** 5972.

Taniuchi, M., H.B. Clark, and E.M. Johnson. 1986a. Induction of nerve growth factor receptors in Schwann cells after axotomy. *Proc. Natl. Acad. Sci.* **83:** 4094.

Taniuchi, M., E.M. Johnson, P.J. Roach, and P.C. Lawrence. 1986b. Phosphorylation of nerve growth factor receptor proteins in sympathetic neurons and PC12 cells. *J. Biol. Chem.* **261:** 13342.

Thoenen, H. and Y.A. Barde. 1980. Physiology of nerve growth factor. *Physiol. Rev.* **60:** 1284.

Valerio, D., M.G.C. Doyvesteyn, B.M.M. Dekker, G. Weeda, T.M. Berkvens, L. van der Voorn, H. van Ormondt, and A.J. van der Eb. 1985. Adenosine deaminase: Characterization and expression of a gene with a remarkable promoter. *EMBO J.* **4:** 437.

Weaver, R.F. and C. Weissman. 1979. Mapping of RNA by a modification of the Berk-Sharp procedure: The 5′ termini of 15S β-globin mRNA have identical map coordinates. *Nucleic Acids Res.* **6:** 1175.

Wigler, M., R. Sweet, G.K. Sim, B. Wold, A. Pellicer, E. Lacy, T. Maniatis, S. Silverstein, and R. Axel. 1979. Transformation of mammalian cells with genes from procaryotes and eucaryotes. *Cell* **16:** 777.

Structure-Function Analysis of the β-Adrenergic Receptor

R.A.F. DIXON, I.S. SIGAL, AND C.D. STRADER*

*Departments of Molecular Biology and Biochemistry, Merck, Sharp, and Dohme Research Laboratories, West Point, Pennsylvania 19498 and *Rahway, New Jersey 07065*

A major goal of medicinal chemistry and pharmacology is to understand the molecular basis for the interactions of drugs with their targets. Many drugs act by binding to specific cell-surface receptors, either as agonists, which stimulate receptor function, or as antagonists, which block the action of endogenous agonists. In recent years, it has become clear that cell-surface receptors can be classified into groups on the basis of their ligand-binding characteristics, their mechanisms of signal transduction, and their structural properties (Dixon et al. 1988). One such class of receptors, which binds small molecule or peptide hormones and neurotransmitters, couples through guanine-nucleotide-binding regulatory proteins (G proteins) to modulate the intracellular levels of a variety of second-messenger molecules (Fig. 1). On binding agonists, receptors of this class catalyze the exchange of a bound GDP molecule for GTP on one of several G proteins in the cell membrane. Most receptors appear to associate with a single type of G protein, although recent reports suggest that under certain conditions, a receptor may interact with more than one type (Ashkenazi et al. 1987). The activated GTP–G-protein complex then interacts with an effector protein to modulate the level of a specific intracellular second messenger (Gilman 1987).

One of the best characterized of the G-protein-linked receptors is the β-adrenergic receptor (βAR), which binds the endogenous catacholamine epinephrine and couples to the G protein G_s to activate adenylyl cyclase, raising the intracellular level of cAMP (Lefkowitz et al. 1983). Recently, the cloning of the gene encoding the mammalian βAR has allowed predictions concerning the structure and function of the receptor protein to be tested by molecular genetic methods. Analysis of the amino acid sequence of the βAR by hydropathicity and hydrophobic moment plots and secondary structure prediction techniques has revealed that the protein contains seven regions of hydrophobic amino acids (20–30 residues in length) that could form membrane-spanning helices (Fig. 2) (Dixon et al. 1986; Yarden et al. 1986). These hydrophobic domains are connected by short hydrophilic segments, which would be predicted to form extramembranous loops. The tertiary structure predicted for the βAR is similar to that proposed for the visual pigment rhodopsin (Findlay and Pappin 1986; Nathans 1987), which acts through the G-protein transducin to stimulate cGMP phosphodiesterase (Stryer 1986). Further analysis has shown that the βAR shares significant amino acid similarity with rhodopsin (Dixon et al. 1986; Yarden et al. 1986). This primary sequence conservation is highest in the putative transmembrane regions that in opsin comprise the binding site for the ligand retinal (Findlay and Pappin 1986; Nathans 1987). In contrast, there is little or no similarity in sequence within the hydrophilic loop regions of the two proteins. Subsequently, several other members of the G-protein-coupled receptor family have been cloned and found to share similar primary structural characteristics (Kubo et al. 1986a,b; Frielle et al.

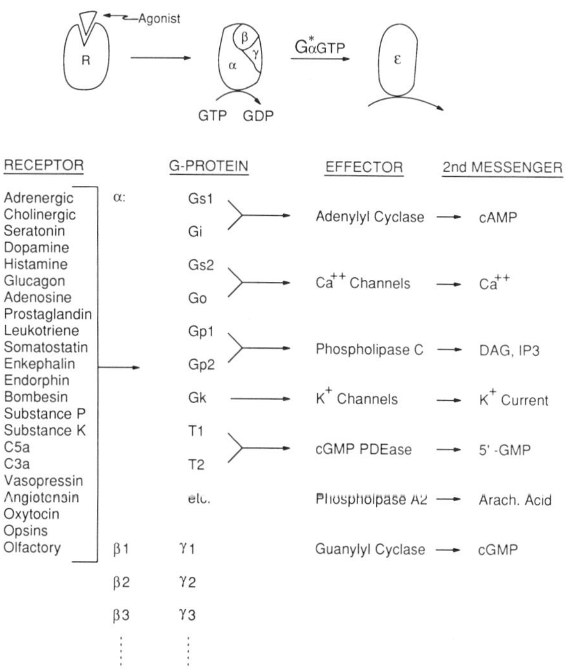

Figure 1. Components of the G-protein-coupled signal transduction pathways (Gilman 1987). The diagram illustrates the components and lists several members involved in each step of the G-protein-coupled pathway. The first component is the transmembrane receptor (R), which interacts with the ligand. Several ligands are listed that have been shown to utilize G-protein-coupled receptors. The second component is the G protein, which is composed of α, β, and γ subunits. Several subtypes of each subunit have been described and are listed. The third component is the effector (E). Several effector systems that are known to be modulated by G proteins and the second messengers generated by these effector proteins are listed.

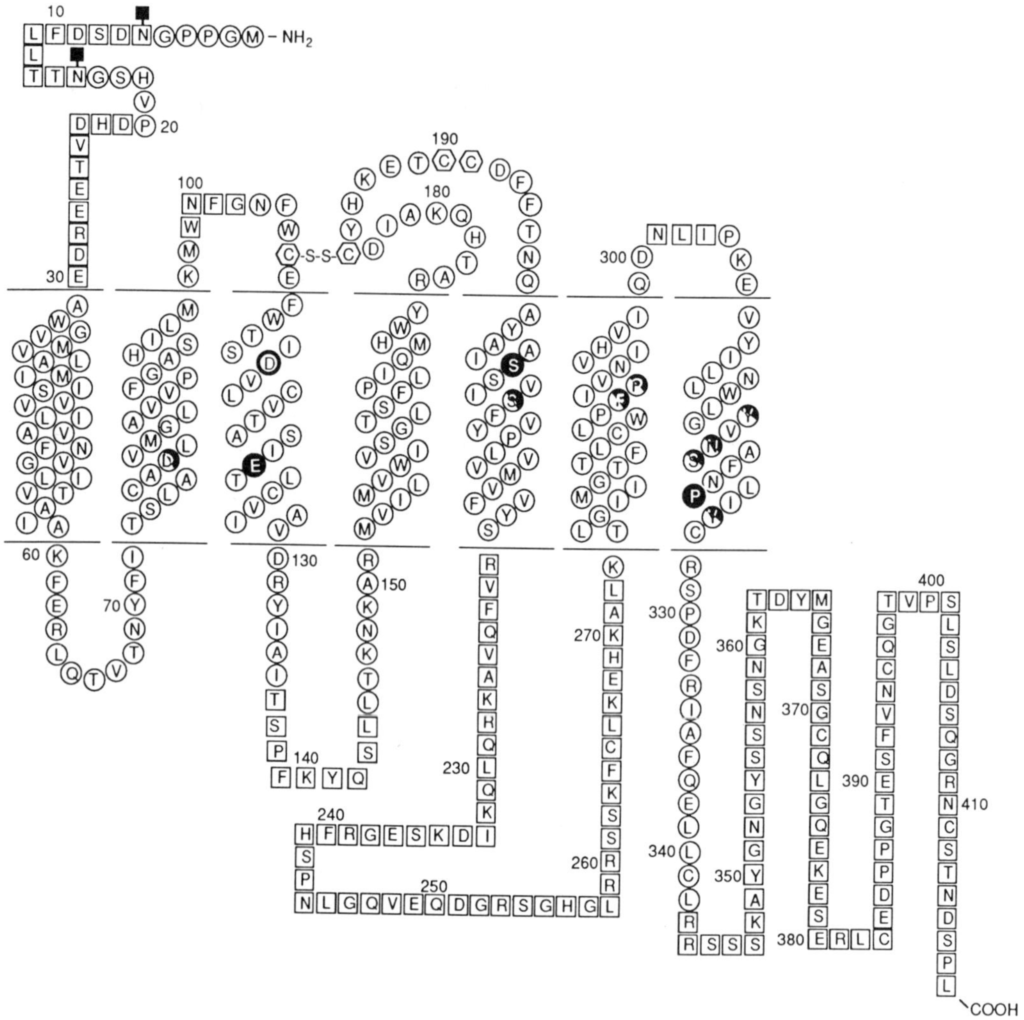

Figure 2. Proposed transmembrane topology and ligand-binding site of the βAR (Dixon et al. 1986; Yarden et al. 1986). The proposed topology of the βAR relative to the plasma membrane is indicated, with the top of the page representing the outside of the cell. The amino terminus is located on the extracellular surface, and the carboxyl terminus is located on the intracellular surface of the plasma membrane. The single-letter amino acid code is used to indicate individual residues. The deletion of the residues indicated in squares did not affect the accumulation of protein in the plasma membrane or ligand binding (Dixon et al. 1987a,b; Strader et al. 1987a,b, 1988). The circled residues either have not been tested or were determined to affect ligand binding. Asp-113, which is critical for the binding of all ligands of the receptor, is shown in the boldface circle. Single-residue substitutions that affect protein folding are shown in black. The partially shaded residues are those that are involved in agonist binding but not antagonist binding. The four cysteine residues that are critical for ligand binding are shown in hexagons. The solid squares above the two Asn residues near the amino terminus represent the attached N-linked carbohydrates.

1987; Kobilka et al. 1987; Masu et al. 1987; Peralta et al. 1987). On the basis of this similarity in protein structure and the common G-protein-based signal transduction mechanism shared by these receptors, we proposed that the interactions between all G-protein-linked receptors and their ligands should be similar to that between opsin and retinal (Dixon et al. 1986). Specifically, we hypothesized that the ligand-binding site lies in the transmembrane domain of the βAR.

To identify the regions of the receptor that are involved in ligand binding and G-protein interactions, specific mutations have been introduced into the βAR gene. Our structure-function study has relied on the expression of the mutant receptors in mammalian cells, the analysis of the levels and processing of the expressed protein by immunoblotting, the characterization of ligand interactions by radioligand-binding methodology, and, finally, the measurement of function by the determination of the ability of the mutant receptors to couple to adenylyl cyclase. As presented below, the results of this study have allowed us to map the ligand-binding site of the βAR and to determine regions of the receptor that are involved in G-protein coupling. Because of the high degree of structural similarity among the G-protein-linked receptors, an understanding of the mechanisms of ligand binding and G_s activation by the βAR should benefit drug development for this entire class of receptor proteins.

EXPERIMENTAL PROCEDURES

The cloning, mutagenesis, and expression of the hamster β_2AR have been described previously (Dixon et al. 1986, 1987b). Briefly, a full-length cDNA clone of the hamster β_2AR was inserted into the mammalian expression vector pSVL behind the SV40 early promoter. The vector contains a functional SV40 origin of DNA replication for use in a transient COS-7 cell expression system. Deletion and substitution mutations were constructed by oligonucleotide-directed mutagenesis. For transient expression experiments, the vectors were transfected into monolayers of COS-7 cells by the $CaPO_4$ precipitation method, followed by glycerol shock. The cultures were incubated at 37°C for 72 hours and harvested by scraping. Membranes were prepared from the transfected cells by freeze-thaw lysis of the cells, followed by centrifugation to remove soluble proteins, and assayed as described previously (Dixon et al. 1987b). Briefly, radioligand-binding assays were performed on the membranes by incubation with various concentrations of [^{125}I]cyanopindolol (ICYP) for 2 hours and collection of the membranes on glass-fiber filters. The nonspecific binding of ICYP to the membranes was determined in the presence of 10 μM alprenolol. Competition ligand binding was performed by incubation of the membranes with various concentrations of unlabeled ligand in the presence of 225 pM ICYP and assayed as described above. K_d values were determined by Scatchard and nonlinear regression analyses using the LIGAND program. For all transient expression experiments, only cultures that exhibited ICYP binding greater than fivefold over the background level of endogenous receptors were used in K_d determinations. The βAR protein produced during the transfection was visualized by immunoblot analysis, using an antipeptide sera specific for amino acids 404–418 of the β_2AR.

Stable clonal cell lines expressing the mutant βARs were constructed by cotransfection of the expression vectors and pRSV*neo* into mouse L cells, which are devoid of endogenous βAR. G-418-resistant colonies were isolated and examined for expression of the βAR by radioligand-binding analysis, immunoblotting, or indirect immunofluorescence (Dixon et al. 1987a; Strader et al. 1987c). Four to ten individual cell lines expressing each mutant receptor were examined. The ability of the receptor to couple to G proteins was determined by measuring the stimulation of adenylyl cyclase activity in response to agonists in membranes prepared from the transfected cell lines.

RESULTS AND DISCUSSION

The Ligand-binding Site Involves the Hydrophobic Core of the βAR

Preliminary determination of the regions of the βAR essential for protein folding and ligand binding involved the introduction of a series of in-frame deletion mutations into the βAR gene (Dixon et al. 1987a,b). The mutations resulted in the deletion of individual hydrophobic or hydrophilic domains of the receptor. In the discussion below, the hydrophobic regions of the βAR predicted to form transmembrane helices will be referred to as helices 1 through 7, numbered from the amino terminus to the carboxyl terminus. The connecting hydrophilic regions, which are predicted to extend intracellularly and extracellularly, will be termed inner and outer loops 1 through 3, respectively. The mutant βAR genes were transfected into COS-7 cells, and the βAR-related proteins that accumulated in the cell membranes were visualized by immunoblotting. The ability of the mutant βARs to bind agonists and antagonists was analyzed by radioligand binding. The results of these experiments are summarized in Figure 2. The amino- and carboxy-terminal hydrophilic regions, outer loops 1 and 3 and inner loops 2 and 3 could be truncated or removed completely without affecting the ability of the βAR to bind either the agonist isoproterenol or the antagonist ICYP. Most of these deletion mutations also did not affect the accumulation of the receptor protein in the cell membrane. Thus, these regions within the hydrophilic loops of the βAR are not essential either for protein folding and membrane insertion or for ligand binding. In contrast, other deletion mutations, most notably those involving regions immediately adjacent to each of the predicted transmembrane domains and those in inner loop 1, expressed significantly less protein which bound proportionally less ICYP than did the wild-type βAR. Removal of single transmembrane regions also resulted in the failure of the βAR to accumulate in the cell membrane or to bind ligands. Interestingly, when certain pairs of transmembrane regions were deleted, small amounts of receptor protein did accumulate in the membrane, but it did not appear to be properly glycosylated (see below) and was unable to bind antagonists. Thus, all seven of the transmembrane regions appear to be required for correct protein folding and ligand binding. As most of the hydrophilic loop regions of the βAR (approximately 40% of the total protein) could be removed without affecting the ligand-binding ability of the receptor, it appears that, like rhodopsin, the hydrophobic core of the βAR protein constitutes the ligand-binding site.

Conserved Cysteine Residues Are Required for Ligand Binding whereas Glycosylation Is Not

It was expected that certain residues of the βAR would play a major role in maintaining the overall tertiary structure of the receptor. There are 15 cysteine residues in the hamster βAR, of which 3 are completely conserved with all G-protein-coupled receptors and with opsin (Dixon et al. 1987a; Dohlman et al. 1987). Pharmacological studies have suggested that disulfide bonds and perhaps a rearrangement of disulfide bonds may be important for ligand binding and for agonist activation of the β_2AR (Moxham and Malbon 1985;

Pederson and Ross 1985). During the course of our structure-function analysis, mutations affecting 14 of the 15 cysteine residues have been introduced into the βAR (Dixon et al. 1987a). The effects of these mutations on ligand binding by the receptor are listed in Table 1. Only 4 of the 14 mutations, involving Cys-106 in outer loop 1 and Cys-184, Cys-190, and Cys-191 in outer loop 2, have any discernible effect on ligand-binding activity. Substitution of each of these cysteine residues results in the appearance of low-affinity binding states for both the agonist isoproterenol and the antagonist ICYP. Because Cys-106 and Cys-184 are completely conserved among the G-protein-linked receptors and removal of either of them or both together results in identical phenotypes for the mutant receptors, it was proposed that they were involved in an essential disulfide bridge (Dixon et al. 1987a). The identification of a second pair of cysteine residues that have similar effects suggests that a second essential disulfide bridge may be present in outer loop 2. Although the presence of additional disulfide bridges in the βAR cannot be excluded, no other cysteine residues appear to be essential for the receptor to accumulate in the membrane or to bind ligands.

Table 1. Effects of Substitutions of Cysteine Residues on Ligand Binding to the βAR

Mutant[a]	ICYP K_d (pM)[b]	Isoproterenol K_d (nM)[c]
βAR	60	200
[Val-77]βAR	135	140
[Val-106]βAR	100	50; 30000
[Val-116]βAR	130	60
[Val-184]βAR	70	3; 2000
[Val-190]βAR	n.d.	40; 9000
[Val-191]βAR	n.d.	20; 10000
D(265)	40	10
[Val-285]βAR	n.d.	200
[Val-327]βAR	n.d.	300
[Leu-341]βAR	120	200
D(371,383,395,411)	50	300

[a]The mutations indicated in brackets represent individual amino acid substitutions for cysteine at the position listed. Mutant D(265) refers to a deletion mutant that removes amino acid residues 258–270, including Cys-265. Mutant D(371,383,395,411) refers to a deletion mutation removing residues 354–418, including the four cysteines at the positions listed. Plasmids containing the mutated genes in the expression vector pSVL were transfected into COS-7 cells and incubated for 72 hr. Membranes were prepared from the transfected cells and assayed for ligand binding, as described in Experimental Procedures.

[b]The K_d values were determined by Scatchard analysis of radioligand-binding isotherms for ICYP. Membranes prepared for transfected cells were incubated with varying concentrations of ICYP for 90 min at 23°C, filtered through glass-fiber filters, washed, and counted. Nonspecific binding, determined with 10 μM alprenolol, was <10% of the total. The values presented are the averages of two determinations performed in triplicate with standard errors of <50%. n.d. indicates not determined.

[c]The K_d values were determined from nonlinear regression analysis of competition radioligand-binding assays. Membranes from the transfected cells were incubated in the presence of various concentrations of isoproterenol and 225 pM ICYP, and binding was measured as described in b. Results are listed for two classes of agonist affinity sites only when the fits of the data were significantly better ($p < .02$) than a fit using a single affinity site.

The βAR contains a significant (30% by apparent molecular weight) amount of N-linked carbohydrate (Stiles et al. 1984). Enzymatic removal of the carbohydrate region from purified receptors or biosynthesis of receptor in the presence of tunicamycin has little effect on the function of the βAR (Stiles et al. 1984; George et al. 1986). The βAR has four potential sites for N-linked glycosylation, two located near the amino terminus and two near the carboxyl terminus (Dixon et al. 1986). As summarized in Figure 2, truncations of the amino and carboxyl termini that would remove these sites had no effect on ligand binding. Immunoblot analysis showed that the truncation of the amino terminus, removing two potential glycosylation sites, resulted in the accumulation of a 43-kD protein in the cell membrane, rather than the 67-kD wild-type βAR (Dixon et al. 1987a). This 43-kD protein, which is the predicted molecular weight of the nonglycosylated receptor precursor, comigrates with the βAR produced in cells grown in the presence of tunicamycin (E. Rands et al., in prep.). Therefore, the removal of both of the amino-terminal consensus glycosylation sites was sufficient to prevent the glycosylation of the βAR, implicating the asparagine residues at positions 6 and 15 as the sites for βAR glycosylation. This assignment was confirmed by single-amino-acid substitutions at each of these positions (E. Rands et al., in prep.). Although the substitution of Asn-6 and Asn-15, independently or together, resulted in altered βAR glycosylation, these receptor mutants showed normal binding of both isoproterenol and ICYP. These results demonstrate that although glycosylation may be important for trafficking of the βAR through the cell or for receptor regulation, it plays a small role in ligand binding.

Asp-113 Is the Counterion for the Cationic Amine Group of Ligands

Having localized the ligand-binding domain of the βAR to its rhodopsin-like hydrophobic core, we further refined the map of the ligand-binding site by analyzing a series of mutant receptors that contained single-amino-acid substitutions within this region. The results of these studies are summarized in Figure 2. Structure-activity analysis of adrenergic ligands has shown that both the cationic amine moiety and the catechol ring are important for agonist activity (Lefkowitz et al. 1983; Main and Tucker 1985). Our initial series of experiments focused on identifying the residue of the receptor that functions as the counterion for the cationic amine group of the ligand (Strader et al. 1987b, 1988). Figure 3A presents a sequence comparison of transmembrane helices 2 and 3 for each subtype of G-protein-coupled receptor that has been cloned. Highlighted in the figure are four aspartate and glutamate residues, which, with the addition of Glu-268 in the dispensable third intracellular loop, are the only acidic residues in the βAR conserved among those receptors that bind cationic amine ligands (see Fig. 3B). Mutant receptors containing substitutions for each

A.

```
                 .                 II                 .     01             .            III                     .
                        79                                            106,107     113                      130
HAMBAR2   S L A C A D L V M G L A V V P F G A S H I L M K M W N F G N F W C E F W T S I D V L C V T A S I E T L C V I A V D R Y  1 3 2
HUMBAR1   S L A S A D L V M G L L V V P F G A T I V V W G R W E Y G S F F C E L W T S V D V L C V T A S I E T L C V I A L D R Y  1 5 7
HUMAAR2   S L A S A D I L V A T L V I P F S L A N E V M G Y W Y F G K T W C E I Y L A L D V L F C T S S I V H L C A I S L D R Y  1 3 2
HUMMAR1   S L A C A D L I I G T F S M N L Y T T Y L L M G H W A L G T L A C D L W L A L D Y V A S N A S V M N L L L I S F D R Y  1 2 4
HUMMAR2   S L A C A D L I I G V F S M N L Y T L Y T V I G Y W P L G P V V C D L W L A L D Y V V S N A S V M N L L I I S F D R Y  1 2 2
HUMMAR3   S L A C A D L I I G A F S M N L Y T V Y I I K G Y W P L G A V V C D L W L A L D Y V V S N A S V M N L L I I S F D R Y  1 3 1
HUMMAR4   S L A C A D L I I G V I S M N L F T T Y I I M N R W A L G N L A C D L W L A I D Y V A S N A S V M N L L V I S F D R Y  1 6 7
BOVSUBK   N L A L A D L C M A A F N A A F N F V Y A S H N I W Y F G R A F C Y F O N L F P I T A M F V S I Y S M T A I A A D R Y  1 3 2
HUMOPS    N L A V A D L F M V L G G F T S T L Y T S L H G Y F V F G P T G C N L E G F F A T L G G E I A L W S L V V L A I E R Y  1 3 6
HUMRED    N L A V A D L A E T V I A S T I S I V N Q V S G Y F V L G H P M C V L E G Y T V S L C G I T G L W S L A I I S W E R W  1 5 2
HUMGRE    N L A V A D L A E T V I A S T I S V V N Q V Y G Y F V L G H P M C V L E G Y T V S L C G I T G L W S L A I I S W E R W  1 5 2
HUMBLU    N V S F G G F L L C I F S V F P V F V A S C N G Y F V F G R H V C A L E G F L G T V A G L V T G W S L A F L A F E R Y  1 3 3
```

B.

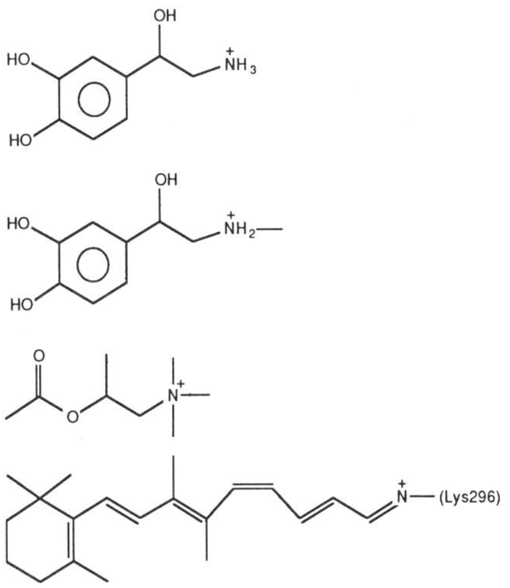

Ligand	Receptor
Norepinephrine	AAR2
Epinephrine	BAR1,2
Acetylcholine	MAR1-4
Retinal	OPSINS
NH_2 -HKTDSFVGLM - NH_2 Substance K	SUBK

Figure 3. (*A*) Comparison of amino acid sequences in hydrophobic regions 2 and 3 of each subtype of G-protein-coupled receptor whose sequence is known. The receptors are designated as follows: (HAMβAR$_2$) hamster β$_2$AR (Dixon et al. 1986); (HUMβAR$_1$) human β$_1$AR (Frielle et al. 1987); (HUMAAR$_2$) human α$_2$-adrenergic receptor (Kobilka et al. 1987); (HUMMAR$_{1-4}$) human M$_1$–M$_4$ muscarinic cholinergic receptors (Peralta et al. 1987); (BOVSUBK) bovine substance K receptor (Masu et al. 1987); (HUMOPS) human opsin (Nathans and Hogness 1984); (HUMRED, HAMGRE, and HUMBLU) human red, green, and blue opsins, respectively (Nathans et al. 1986). Numbers to the right of the row indicate the number for the last amino acid shown in the respective protein sequence. The dots above the comparison indicate the proposed border of the transmembrane region. O1 indicates the segment designated as outer loop 1. The numbers above the residues indicate the positions of the amino acids highlighted in the hamster β$_2$AR. (*B*) Structures of naturally occurring ligands for the receptors listed in *A*. The individual structures and their appropriate receptors are listed.

of these residues were expressed in COS-7 cells and assayed for protein accumulation and ligand binding. The protein accumulation and glycosylation patterns of all of the mutant receptors were similar to the wild-type βAR, suggesting that none of these mutations affected protein folding. The mutants [Ala-79]βAR, [Ala-107]βAR, and [Gly-130]βAR bound the antagonists ICYP, alprenolol, and propranolol with the same affinity as did the wild-type βAR. In contrast, [Asn-113]βAR failed to bind ICYP, even though normal levels of mature βAR protein were expressed. Although [Ala-107]βAR and [Gly-130]βAR also exhibited normal affinity for the agonists isoproterenol and epinephrine, [Ala-79]βAR displayed a tenfold reduction in affinity for these agonists.

Of the conserved acidic residues, Asp-113 alone appeared critical for both agonist and antagonist binding (Strader et al. 1987b). The role of Asp-113 in ligand binding was further explored by measuring agonist-promoted adenylyl cyclase activity in cell lines expressing either [Asn-113]βAR or [Glu-113]βAR (Strader et al. 1988). In membranes from cells expressing [Glu-113]βAR, 10^2-fold higher concentrations of isoproterenol were required for maximal adenylyl cyclase stimulation than those required for the wild-type receptor. The [Asn-113]βAR mutant required 10^4-fold higher concentration of the agonist for activation. As both mutant receptors responded to isoproterenol as a full agonist, the coupling between the agonist-occupied βAR and the G protein was apparently unaffected. Similar shifts in K_{act} for these mutant receptors were observed for the agonists epinephrine and norepinephrine, indicating that the β$_2$ subtype specificity of the receptor was not affected by the mutations. When the ability of the antagonist propranolol to inhibit the response to isoproterenol was measured, the K_i values for both receptor mutants were increased 10^4-fold. Thus, substitutions at Asp-113 dramatically affect the affinity of both agonists and antagonists for the βAR. On the basis of these results, we propose that Asp-113 acts as the counterion for the cationic amine group of the adrenergic ligands and that agonists and antagonists bind to overlapping sites on the receptor protein. All of the receptors whose sequences are known and which bind cationic amines as ligands contain an aspartate at a position analogous to that of Asp-113 of the βAR, whereas the substance K receptor and opsins do not, suggesting that the interaction between this aspartate and the amino group of the ligand is characteristic of this class of receptors.

Role of Conserved Hydrogen Bond Donors and Aromatic Residues in Ligand Binding

Those residues within the βAR-binding pocket that interact with the catechol group of the ligand might be expected to be conserved among the G-protein-linked receptors that bind catecholamines but not with the muscarinic acetylcholine receptor (MAR). Of particular interest are those residues that have side chains that could donate a hydrogen bond to the ligand. Several serine and threonine residues within the hydrophobic domain of the βAR meet these criteria (C.D. Strader et al., in prep.). Mutant receptors with amino acid substitutions for the serines at positions 42, 92, 161, and 220 and the threonines at positions 56, 164, and 283 were expressed and found to exhibit unaltered ICYP and isoproterenol binding. Mutant receptors with substitutions at Ser-204, Ser-207, and Ser-319 were expressed and found to bind ICYP with affinities identical to that of the wild-type βAR. However, these mutant receptors showed a 10–100-fold reduction in affinity for isoproterenol. Thus, the serine residues in helices 6 and 7 appear to affect the ability of the receptor to interact with agonists. At present, these results are insufficient to differentiate between an alteration in a direct ligand-binding interaction and a change in receptor conformation.

Interactions with the catechol ring of the ligand could also involve aromatic residues within the hydrophobic core of the βAR. Mutant receptors with substitutions for the tyrosines 132, 174, 219, and 308 displayed normal isoproterenol- and ICYP-binding characteristics. A mutant βAR with a leucine substituted for Tyr-326 bound ICYP with normal affinity but showed a tenfold reduction in the affinity for isoproterenol (C.D. Strader et al., in prep.). In addition, two phenylalanine residues in helix 6 seem to play a critical role in receptor folding and ligand binding (C.D. Strader et al., in prep.). Substitution of an alanine residue for Phe-289 resulted in normal ICYP binding, but a 10^3-fold reduction in agonist affinity. Substitution of Phe-290 with a methionine residue also decreased the affinity of the receptor for isoproterenol by tenfold, with no affect on ICYP binding. Therefore, the phenylalanines at positions 289 and 290 and the tyrosine at position 326 may be involved in the interaction of the receptor with agonists.

Proposed Structural Model for the Ligand-binding Site of the βAR

The structural model for rhodopsin serves as a paradigm for the entire class of G-protein-linked receptors. The structure proposed for rhodopsin, with its seven transmembrane helices and the ligand retinal intercalated within the helices, has been supported by biochemical and physicochemical studies (Findlay and Pappin 1986; Nathans 1987). The data presented above and biochemical evidence support a similar structure for the βAR (Dixon et al. 1988).

The genetic evidence reviewed in this paper demonstrates that the ligand-binding domain of the βAR involves residues within its rhodopsin-like hydrophobic core. A preliminary model for the ligand-binding site of the receptor that incorporates these results is shown in Figure 4. In this figure, the agonist isoproterenol is surrounded by several of the proposed transmembrane helices that contain residues demonstrated to be important for ligand binding. Although this representation is preliminary, it serves to generate predictions about the interactions of the βAR with ligands that can be tested experimentally. For example, most β-adrenergic antagonists have a longer chain length between the aromatic ring and the amine group than do agonists (Main and Tucker 1985). If, as proposed, agonists and antagonists both use Asp-113 as the counterion for the amine group, then the ring structures of the two classes of ligands should occupy different sites. Thus, future experiments should identify different contact points between the aromatic groups of agonists and antagonists and the receptor. The model for the ligand-binding domain should also stimulate the design of new classes of ligands that might utilize interactions with

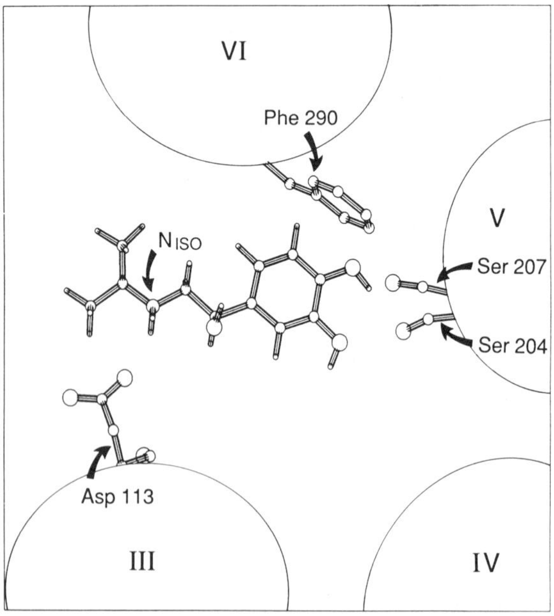

Figure 4. Proposed structural model for the ligand-binding site of the βAR, viewed from the outside of the cell (Strader et al. 1988). The ligand isoproterenol is shown surrounded by several of the proposed transmembrane helices, identified by Roman numerals. The proposed locations of the Asp-113, Phe-290, Ser-203, and Ser-207 side chains are indicated. The location of the cationic amine of iosproterenol in the binding pocket is indicated by N_{iso}.

other residues in the binding pocket that were not previously identified as being important.

Structural Requirements for Receptor Interaction with G_s

The topological model for the βAR predicts several regions of the receptor that might interact directly with G_s. Because G proteins are located on the cytoplasmic face of the plasma membrane, the internal hydrophilic loops and the lower portions of the transmembrane regions of the receptor would form the most likely sites of interaction with a G protein. To map the G-protein site on the receptor, we examined the ability of the mutant receptors that retained ligand binding to stimulate adenylyl cyclase (Fig. 5 and Table 2) (Dixon et al. 1987a,b; Strader et al. 1987a,c, 1988). No effect on G-protein coupling was observed for mutant receptors with intracellular truncations or deletions, including amino-terminal truncations and deletions within the three outer hydrophilic loops. Examination of the internal hydrophilic domains showed no effect on truncation of the carboxyl terminus up to position 345. The portion of the carboxy-terminal tail adjacent to the plasma membrane could not be investigated by deletion analysis, as this mutation resulted in the failure of the protein to be expressed. Therefore, this region (residues 327–336) was replaced with the analogous region from the M_1AR, which does not stimulate adenylyl cyclase but couples to G_p to stimulate phospholipase C (Nathanson 1987). The resulting mutant receptor coupled to adenylyl cyclase in response to isoproterenol but with a reduced efficiency (30–50%), reminiscent of the effects of a partial agonist. Deletions within inner loop 2 had no effect on receptor coupling to G_s. Likewise, mutant receptors in which this region was re-

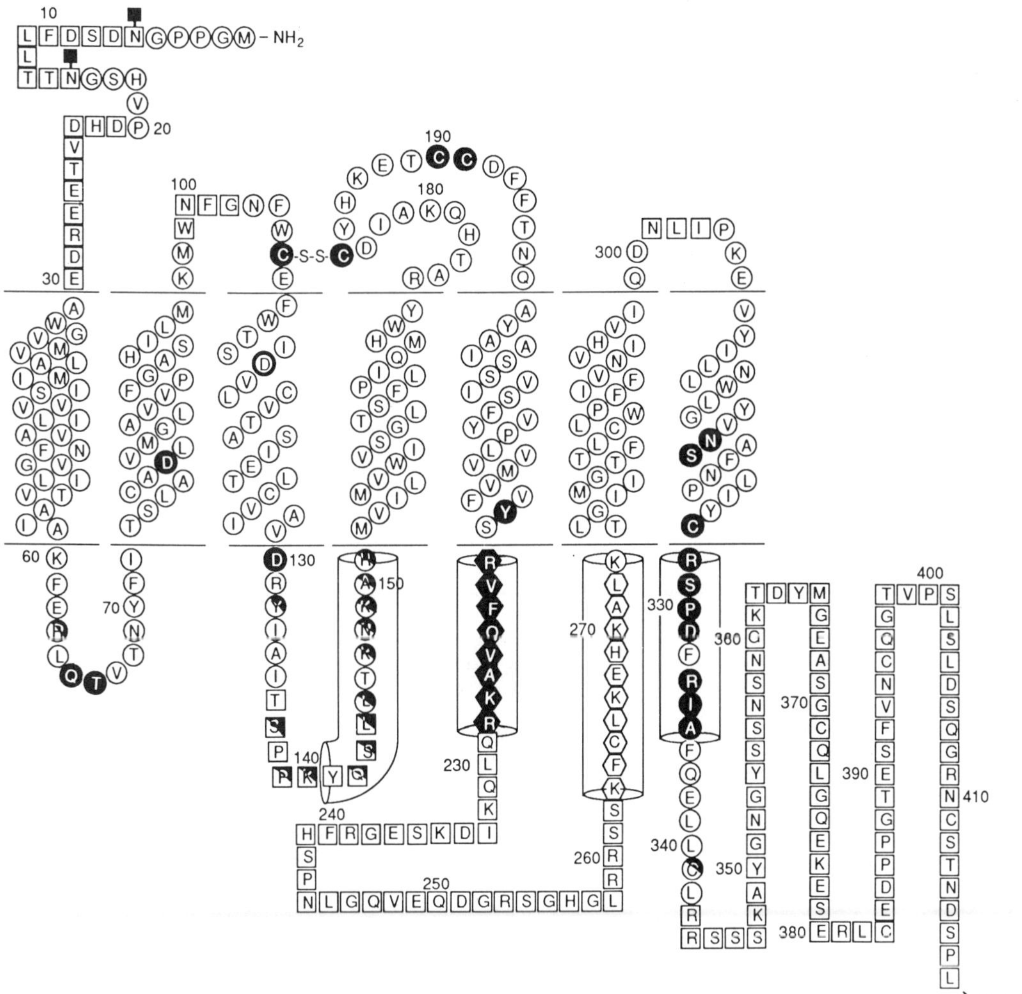

Figure 5. Residues required for interaction of the βAR with G_s (Dixon et al. 1987a,b; Strader et al. 1987a,c, 1988). The topology of the βAR relative to the membrane is the same as that shown in Fig. 2. Those residues shown as squares indicate the regions that can be deleted without affecting G_s coupling. The residues indicated in hexagons, when deleted, abolish the ability of the βAR to couple to G proteins. The fully shaded residues are those that, when substituted, render the receptor partially active in coupling, whereas the partially shaded residues can be substituted with no effect on coupling. The regions outlined by cylinders are those predicted to form amphiphilic α helices, as described in the text.

Table 2. Mutations That Affect the Coupling of the βAR with Adenylyl Cyclase

Mutant[a]	Isoproterenol[b] K_{act} (nM)	Cyclase stimulation[c] (% maximal)
L cell control	—	0
βAR	20	100
D(136–144)βAR	50	100
D(140–150)βAR	2000	50
D(222–229)βAR	—	0
D(229–236)βAR	5	100
D(230–262)βAR	20	50
D(238–251)βAR	10	100
D(239–272)βAR	—	0
D(250–259)βAR	10	100
D(258–270)βAR	5	30
D(345–418)βAR	20	100
[Gly-63]βAR	50	100
[Lys-65]βAR	200	50
[Gly-66]βAR	200	50
[Ala-79]βAR	200	50
[Val-106]βAR	8000	50
[Glu-113]βAR	10000	100
[Gly-130]βAR	20	50
[Leu-132]βAR	40	80
[Arg-137]βAR	100	100
[Leu-139]βAR	90	70
[Val-184]βAR	10000	50
[Val-190]βAR	4000	30
[Val-191]βAR	300	30
[Ala-227]βAR	20	80
[Arg-229]βAR	100	50
[Gly-268]βAR	n.d.	80
[Lys-318]βAR	300	50
[Ala-319]βAR	300	50
[Leu-341]βAR	n.d.	100
M(142–151)βAR	20	100
M(220–230)βAR	20	10
M(327–336)βAR	50	30

[a]The mutants designated as D(XX–XX)βAR refer to deletion mutations that remove the residues listed. The mutants listed in square brackets refer to amino acid substitutions that introduce the residue listed at the designated position. The mutants listed as M(XX–XX)βAR refer to substitution mutations in which the residues listed have been replaced by the analogous regions from the human M_1 muscarinic acetylcholine receptor. All mutant genes are contained on the plasmic pSVL, which was cotransfected with pRSV*neo* into mouse L cells that were devoid of endogenous βAR. Neomycin-resistant cell lines were isolated and analyzed for the expression of the βAR by radioligand-binding using ICYP. At least four different clonal cell lines were studied for each of the mutant βARs.

[b]For adenylyl cyclase stimulation, membranes were prepared from the cell lines and the amount of cAMP produced in response to varying concentrations of isoproterenol was determined as described in Experimental Procedures. The K_{act} reflects the concentration of isoproterenol required to achieve half-maximal stimulation of adenylyl cyclase. The values represent the average of at least two determinations. n.d. indicates not determined.

[c]The % maximal stimulation refers to the amount of the adenylyl cyclase activity mediated by a saturating concentration of isoproterenol, with the maximal level defined by 10 mM NaF. The value given is the median of the range with variance being ±20%. In those cases where 0% is listed, no stimulation in response to isoproterenol was observed.

placed with inner loop 2 from the M_1AR showed normal coupling to adenylyl cyclase. However, several single-residue substitutions in this region, including the replacement of Asp-130 with a glycine residue, resulted in a mutant receptor that displayed the partial agonist phenotype. Deletions within inner loop 1 resulted in the failure of the receptor protein to be expressed, although several single-residue substitutions within this region resulted in the expression of a protein with reduced efficiency of G_s coupling.

The most striking effects were observed with deletions of the third intracellular loop. Deletions within this loop did not affect expression of the receptor protein. However, a mutant receptor with a large deletion at the carboxy-terminal portion of this region, D(239–272)βAR, showed no agonist-mediated adenylyl cyclase stimulation, although there was significant stimulation with NaF or PGE_1, which utilize alternate mechanisms to activate the enzyme. A similarly large deletion from the middle portion of the loop resulted in a reduction in the efficiency of coupling with G_s but did not abolish it. Further investigation of the role of the third inner loop utilizing smaller deletions revealed that removal of this middle portion of the loop in smaller segments (8–10 residues each) did not affect the ability of the receptor to couple to G_s. However, the deletion of 8 residues from the amino terminus of this loop (222–229) resulted in a complete loss of agonist-mediated adenylyl cyclase activity, whereas the deletion of 12 residues from the carboxyl terminus of the loop (258–270) resulted in a receptor that stimulated adenylyl cyclase to only a fraction of the level of NaF stimulation. The coupling of D(222–229)βAR to adenylyl cyclase was only partially restored (0–15%) by replacement with the analogous region from the M_1AR. From this analysis, it appears that the regions at the amino and carboxyl termini of the third inner loop are critical for the interaction of the receptor with G proteins.

Sequence comparisons of other G-protein-coupled receptors within the regions that are essential for the activation of G_s by the βAR reveal strikingly little sequence identity among those receptors that couple to the same G protein (Dohlman et al. 1987; Strader et al. 1987b; Dixon et al. 1988). However, a key feature of all of these domains is their predicted ability to form amphiphilic α helices (Strader et al. 1987b). As illustrated in Figure 6 for the two regions of the βAR essential for G_s coupling, residues 221–228 would be predicted to form an almost ideal amphiphilic helix, whereas residues 258–273 would form a good but less ideal helix. The distribution of charge and the hydrophobic moment of the helices may determine G-protein specificity. Further support for this hypothesis comes from studies using mastoparan, which is a small peptide that assumes an amphiphilic helical conformation when in phospholipid micelles (Fig. 6) (Higashijima et al. 1988). Both in vivo and in vitro experiments have demonstrated that mastoparan and similar peptides can directly activate G proteins. In this activation process, these peptides may be mimicking the portions of the receptor that we have identified as being important for G-protein interaction. Further analysis of the structural requirements of both the receptor and the G protein for coupling could lead to the development of new drugs that will act as agonists or antagonists at sites distinct from the hormone-binding site.

CONCLUSIONS

The βAR serves as a prototype for the family of receptors that function by coupling to G proteins.

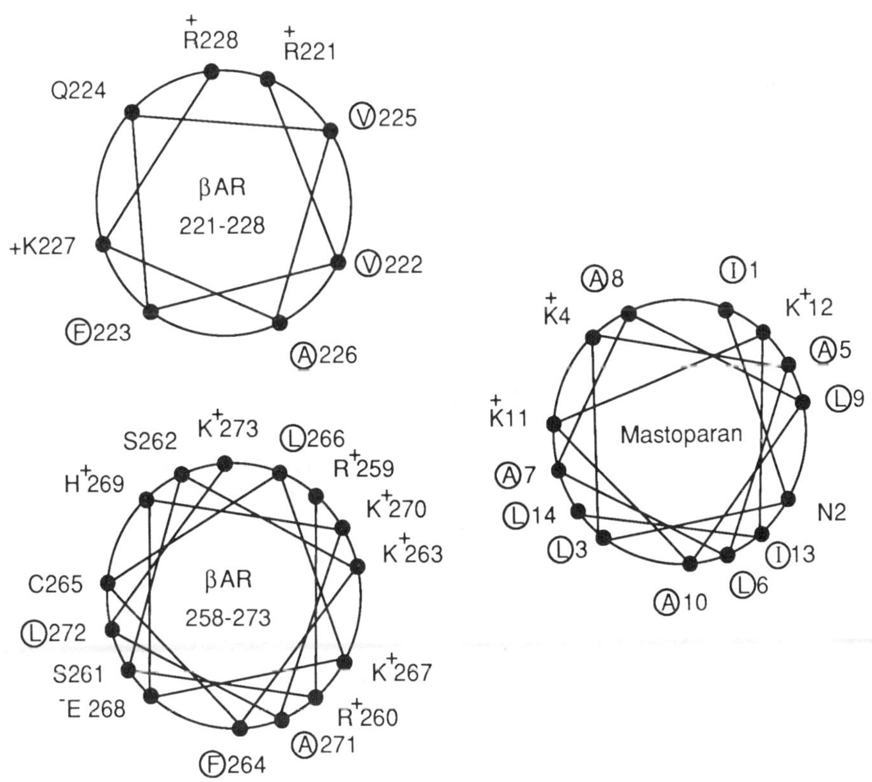

Figure 6. Proposed amphiphilic helical regions of the βAR (Strader et al. 1987b). The residues within regions 221–228 and 258–273 are represented on an Edmunson helical wheel. The charged residues are indicated with the appropriate charge. Hydrophobic residues are circled. The peptide mastoparan is also represented on a helical wheel for comparison (Higashijima et al. 1988).

Genetic and biochemical analysis of the βAR has resulted in a structural model for the receptor in which seven transmembrane helices comprise the ligand-binding site. Detailed mutagenesis studies of amino acid residues within the transmembrane regions have led to the identification of Asp-113 as the counterion for the catecholamine ligands. The conservation of this residue in the receptors that bind other amine ligands suggests that the presence of an aspartic acid at the analogous position in an unknown receptor will be predictive of the class of compounds it binds. The model allows the prediction of residues within the transmembrane domain that could form other contact points with the ligand. Our mutagenesis survey has also identified regions of the βAR that are critical for its coupling with G_s. These regions are proposed to form amphiphilic α helices, whose physicochemical properties may determine the specificity of G-protein interaction. Further genetic probing of the βAR, coupled with biophysical analysis of the protein, should lead to a full understanding of the mechanism of the agonist-receptor–G-protein interaction. This understanding should then stimulate the rational design of compounds that can modulate these interactions.

ACKNOWLEDGMENTS

The experiments from our laboratory, which have been reported elsewhere as referenced in the text, were performed by Drs. J. Allard, A. Blake, M. Candelore, A. Cheung, R. DeHaven, R. Diehl, S. Gould, W. Hill, E. Rands, B. Register, V. Strout, and B. Zemcik. We are grateful to Drs. E. Scolnick, P. Anderson, and E. Slater for their support throughout this study.

REFERENCES

Ashkenazi, A., J.W. Winslow, E.G. Peralta, G.L. Peterson, M.I. Schimerlik, D.J. Capon, and J. Ramachandran. 1987. A M_2 muscarinic receptor coupled to both adenylyl cyclase and phoshoinositide turnover. *Science* **238:** 672.

Dixon, R.A.F., C.D. Strader, and I.S. Sigal. 1988. Structure and function of G-protein coupled receptors. *Annu. Rep. Med. Chem.* **23:** (in press).

Dixon, R.A.F., I.S. Sigal, M.R. Candelore, R.B. Register, W. Scattergood, E. Rands, and C.D. Strader. 1987a. Structural features required for ligand binding to the β-adrenergic receptor. *EMBO J.* **6:** 3269.

Dixon, R.A.F., I.S. Sigal, E. Rands, R.B. Register, M.R. Candelore, A.D. Blake, and C.D. Strader. 1987b. Ligand binding to the β-adrenergic receptor involves its rhodopsin-like core. *Nature* **326:** 73.

Dixon, R.A.F., B.K. Kobilka, D.J. Strader, J.L. Benovic, H.G. Dohlman, T. Frielle, M.A. Bolanowski, C.D. Bennett, E. Rands, R.E. Diehl, R.A. Mumford, E.E. Slater, I.S. Sigal, M.G. Caron, R.J. Lefkowitz, and C.D. Strader. 1986. Cloning of the gene and cDNA for mammalian β-adrenergic receptor and homology with rhodopsin. *Nature* **321:** 75.

Dohlman, H.G., M.G. Caron, and R.J. Lefkowitz. 1987. A family of receptors coupled to guanine nucleotide regulatory proteins. *Biochemistry* **26:** 2657.

Findlay, J.B.C. and D.J.C. Pappin. 1986. The opsin family of proteins. *Biochem. J.* **238:** 625.

Freille, T., S. Collins, K.W. Daniel, M.G. Caron, R.J. Lefkowitz, and B.K. Kobilka. 1987. Cloning of the cDNA for the β_1-adrenergic receptor. *Proc. Natl. Acad. Sci.* **84:** 7920.

George, S.T., A.E. Ruoho, and C.C. Malbon. 1986. N-glycosylation in expression and function of β-adrenergic receptors. *J. Biol. Chem.* **261:** 16559.

Gilman, A.G. 1987. G proteins: Transducers of receptor-generated signals. *Annu. Rev. Biochem.* **56:** 615.

Higashijima, T., S. Uzu, T. Nakajima, and E.M. Ross. 1988. Mastoparan, a peptide from wasp venom, mimics receptors by activating GTP-binding regulatory proteins (G proteins). *J. Biol. Chem.* **262:** 6491.

Kobilka, B.K., H. Matsui, T.S. Kobilka, T.L. Yang-Feng, U. Francke, M.G. Caron, R.J. Lefkowitz, and J.W. Regan. 1987. Cloning, sequencing, and expression of the gene coding for the human platelet α_2-adrenergic receptor. *Science* **238:** 650.

Kubo, T., A. Maeda, K. Sugimoto, I. Akiba, A. Mikami, H. Takahashi, T. Haga, A. Ichiyama, K. Kangawa, H. Matsuo, T. Hirose, and S. Numa. 1986a. Primary structure of porcine cardiac muscarinic acetylcholine receptor deduced from the cDNA sequence. *FEBS Lett.* **209:** 367.

Kubo, T., K. Fukuda, A. Mikami, A. Maeda, H. Takahashi, M. Mishina, T. Haga, K. Haga, A. Ichiyama, K. Kangawa, H. Kojima, H. Matsuo, T. Hirose, and S. Numa. 1986b. Cloning, sequencing, and expression of complementary DNA encoding the muscarinic acetylcholine receptor. *Nature* **323:** 411.

Lefkowitz, R.J., J.M. Stadel, and M.G. Caron. 1983. Adenylate cyclase coupled β-adrenergic receptors: Structure and mechanisms of activation and desensitization. *Annu. Rev. Biochem.* **52:** 159.

Main, B.G. and H. Tucker. 1985. Recent advances in β-adrenergic blocking agents. *Prog. Med. Chem.* **22:** 122.

Masu, Y., K. Nakayama, H. Tamaki, Y. Harada, M. Kuno, and S. Nakanishi. 1987. cDNA cloning of bovine substance-K receptor through oocyte expression system. *Nature* **329:** 836.

Moxham, C.P. and C.C. Malbon. 1985. Fat cell β-adrenergic receptor: Structural evidence for existance of disulfide bridges essential for ligand binding. *Biochemistry* **24:** 6072.

Nathans, J. 1987. Molecular biology of visual pigments. *Annu. Rev. Neurosci.* **10:** 163.

Nathans, J. and D.S. Hogness. 1984. Isolation and nucleotide sequence of the gene encoding human rhodopsin. *Proc. Natl. Acad. Sci.* **81:** 4851.

Nathans, J., D. Thomas, and D.S Hogness. 1986. Molecular genetics of color vision: The genes encoding blue, green, and red pigments. *Science* **232:** 193.

Nathanson, N.M. 1987. Molecular properties of the muscarinic acetylcholine receptor. *Annu. Rev. Neurosci.* **10:** 195.

Pederson, S.E. and M.E. Ross. 1985. Functional activation of β-adrenergic receptors by thiols in the presence and absence of agonists. *J. Biol. Chem.* **260:** 14150.

Peralta, E.G., A. Ashkenazi, J.W. Winslow, D.H. Smith, J. Ramachandran, and D.J. Capon. 1987. Distinct primary structures, ligand-binding properties, and tissue-specific expression of four human muscarinic acetylcholine receptors. *EMBO J.* **6:** 3923.

Stiles, G.L., J.L. Benovic, M.G. Caron, and R.J. Lefkowitz. 1984. Mammalian β-adrenergic receptors: Distinct glycoprotein populations containing high mannose or complex type carbohydrate chains. *J. Biol. Chem.* **259:** 8655.

Strader, C.D., R.A.F. Dixon, A.H. Cheung, M.R. Candelore, A.D. Blake, and I.S. Sigal. 1987a. Mutations that uncouple the β-adrenergic receptor from G_s and increase agonist affinity. *J. Biol. Chem.* **262:** 16439.

Strader, C.D., I.S. Sigal, M.R. Candelore, E. Rands, W.S. Hill, and R.A.F. Dixon. 1988. Conserved aspartic acid residues 79 and 113 of the β-adrenergic receptor have different roles in receptor function. *J. Biol. Chem.* **263:** 10267.

Strader, C.D., I.S. Sigal, R.B. Register, M.R. Candelore, E. Rands, and R.A.F. Dixon. 1987b. Identification of residues required for ligand binding to the β-adrenergic receptor. *Proc. Natl. Acad. Sci.* **84:** 4384.

Strader, C.D., I.S. Sigal, A.D. Blake, A.H. Cheung, R.B. Register, E. Rands, B.A. Zemcik, M.R. Candelore, and R.A.F. Dixon. 1987c. The carboxyl terminus of the hamster β-adrenergic receptor expressed in mouse L cells is not required for receptor sequestration. *Cell* **49:** 855.

Stryer, L.C. 1986. Cyclic GMP cascade of vision. *Annu. Rev. Neurosci.* **9:** 87.

Yarden, Y., H. Rodriquez, S.K.-F. Wong, D.R. Brandt, D.C. May, J. Burnier, R.N. Harkins, E.Y. Chen, J. Ramachandran, A. Ullrich, and E.M. Ross. 1986. The avian β-adrenergic receptor: Primary structure and membrane topology. *Proc. Natl. Acad. Sci.* **83:** 6795.

Functional Domains in the β-Adrenergic Receptor

E.M. Ross, S.K.-F. Wong, R.C. Rubenstein, and T. Higashijima
Department of Pharmacology, University of Texas Southwestern Medical Center, Dallas, Texas 75235-9041

β-Adrenergic receptors (βAR) and numerous other cell-surface receptors communicate with their intracellular effector proteins indirectly, using a family of GTP-binding regulatory proteins, G proteins, as transducers. G-protein systems are unique in their ability to integrate multiple hormonal signals into a single intracellular second-messenger output or, alternatively, to propagate a single hormonal signal via two or more second messengers.

The βAR, an experimental prototype among G-protein-coupled receptors, selectively activates G_s. This G protein stimulates adenylate cyclase to increase the concentration of the second-messenger cAMP. Other receptors also stimulate G_s; such integration at the receptor–G protein level is observed in most cells. Integration at the G protein–effector level is also observed: At least two G proteins activate phospholipase C to release inositol phosphates. Signaling through multiple messengers provides further diversity: βAR, acting via G_s, can stimulate both adenylate cyclase and a Ca^{++} channel and inhibit transporters for Mg^{++} and glucose.

Receptors, G proteins, and effectors utilize only a few biochemical reactions to create this sophisticated signaling network. Only GTP-liganded G proteins can activate their effectors. Hydrolysis of bound GTP to GDP terminates G-protein activation and, consequently, the interaction with the effector. Thus, net activation of a signaling pathway reflects the steady-state concentration of the G protein–GTP complex. The receptor stimulates the system by catalyzing both the release of tightly bound GDP and the subsequent binding of GTP (for review, see Stryer and Bourne 1986; Gilman 1987).

The assembly of a G-protein-mediated signaling network takes advantage of the structural homology among different receptors, among G proteins, and (presumably) among effectors (Dohlman et al. 1987; Gilman 1987). Thus, the G_s-binding domain on a βAR is presumably similar in overall structure and function to the homologous domain on a receptor that regulates a different G protein but is different enough to display selectivity for G_s. Similarly, the receptor-binding domain on all G proteins presumably utilizes a single mechanism for responding to the signal that opens the nucleotide-binding site, but each G protein must retain selectivity for its appropriate receptor. The subunit interfaces on G proteins, and the G protein–effector interface must show the same sort of mechanistic unity combined with diverse patterns of selectivity. In an attempt to understand the protein-protein interactions that make up a G-protein-mediated signaling pathway, we have begun to dissect the structures of the βAR and of the receptor-G_s interface.

EXPERIMENTAL PROCEDURES

All studies reported here have been carried out using βAR from turkey erythrocytes. The avian erythrocyte βAR has been a well-studied model for G-protein-coupled receptors despite its unconventional origin. β-Adrenergic regulation of adenylate cyclase in erythrocyte membranes has been described in exquisite kinetic detail by Levitzki (1986). Our group has utilized purified βAR for detailed studies of its regulation of G_s after reconstituting both proteins, with or without purified adenylate cyclase, into large unilamellar vesicles (May et al. 1985; Brandt and Ross 1986). All of the procedures used in the experiments reported here have been described in detail elsewhere (Brandt and Ross 1986; Yarden et al. 1986; Rubenstein et al. 1987; Higashijima et al. 1988; Wong et al. 1988).

RESULTS AND DISCUSSION

The βAR is a relatively hydrophobic, integral membrane glycoprotein. The receptor from avian erythrocytes is a single polypeptide with an apparent molecular mass of about 52,000 daltons. This polypeptide is somewhat smaller than that of the mammalian receptor, but much of the difference appears to be due to glycosylation (Smigel et al. 1984; Dixon et al. 1986; Yarden et al. 1986). The avian βAR is commonly purified as an active 40,000-dalton proteolytic product that has apparently been cleaved at both termini and thereby partially deglycosylated (Jurss et al. 1985; Yarden et al. 1986; Rubenstein et al. 1987).

To explore the primary structure of the βAR in more detail, the cDNA that encodes the receptor was cloned from a fetal turkey erythrocyte library in collaboration with Yarden et al. (1986). The receptor cDNA was cloned using oligonucleotide probes based on its protein sequence and further identified according to the ability of antibodies raised against peptides encoded by the cDNA to cross-react with the purified receptor. Recombinant receptor has now been expressed in several heterologous cells, where it displays characteristic properties of the native protein (Fig. 1). Recombinant receptor binds β-adrenergic ligands with appropriate affinity, displaying a selectivity that is similar to, but

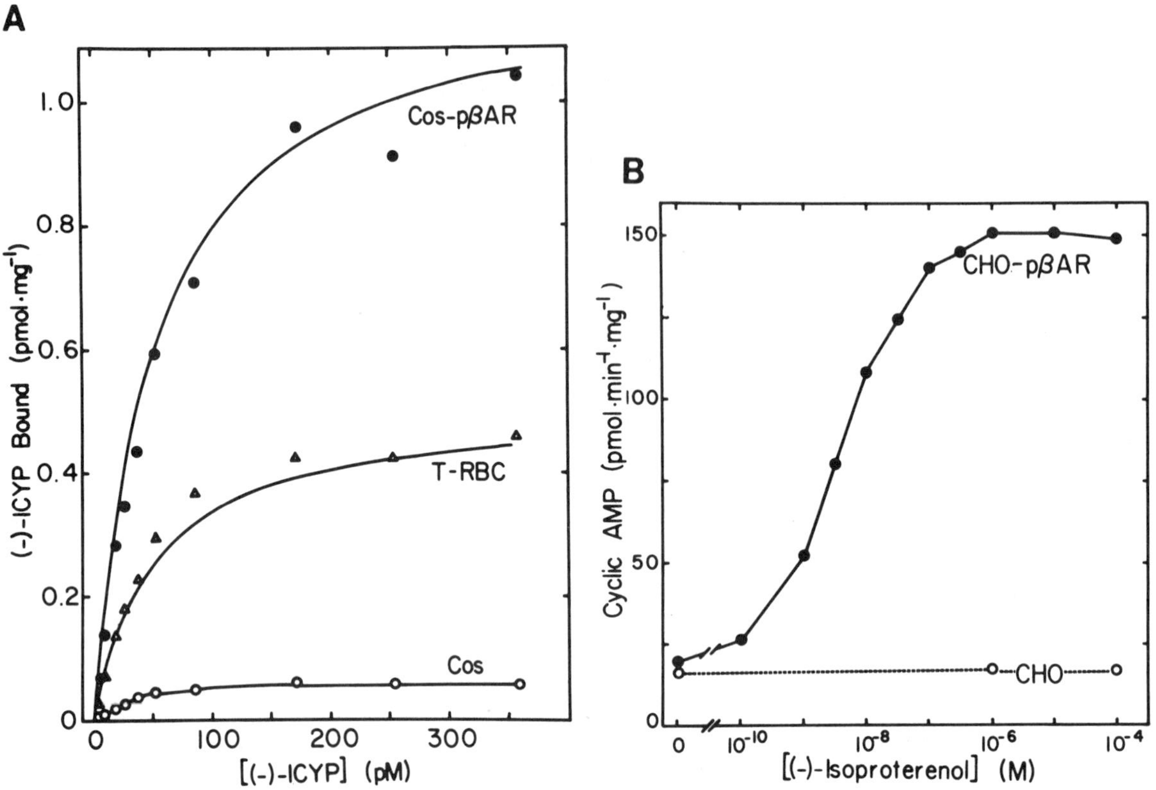

Figure 1. Activity of recombinant turkey erythrocyte βAR in mammalian cells. Full-length cDNA was prepared from the individual short clones described by Yarden et al. (1986) using standard procedures and introduced into the vectors pSVL (*A*) or pKSV-10 (*B*). (*A*) Binding of [^{125}I]iodocyanopindolol (ICYP). COS cells were transfected with βAR cDNA, and after 2 days, ligand-binding activity in a crude membrane fraction was measured using the β-adrenergic ligand, ICYP. Data shown are for cells transfected with βAR cDNA, cells transfected with salmon sperm DNA, and, as a referent, membranes from turkey erythrocytes. (*B*) Stimulation of adenylate cyclase. A crude membrane fraction was prepared from cells stably transfected with the βAR cDNA or from untransfected CHO cells. Adenylate cyclase activity was assayed in the presence of 100 μM GTP and increasing concentrations of (−)-isoproterenol.

not identical with the mammalian β_1-adrenergic receptor (see Minneman et al. 1980). The recombinant receptor also efficiently regulates mammalian G_s (Fig. 1), as predicted by earlier studies in which the purified avian receptor was successfully reconstituted with mammalian G_s (Brandt and Ross 1986).

The deduced primary structures of the avian βAR and of the receptor from hamster lung (Dixon et al. 1986) were immediately informative because these receptors share extensive homology with rhodopsin (for review, see Findlay and Pappin 1986). Rhodopsin is composed of seven largely helical membrane-spanning sequences, with its glycosylated amino terminus on the luminal side of the disk membrane and its carboxyl terminus on the cytoplasmic side. Homology among the receptors and rhodopsin was primarily conserved in six strongly hydrophobic regions and in a seventh, less hydrophobic span, which, in the case of rhodopsin, all cross the plasma membrane (see Fig. 2). Homology between the two βARs was also greatest in the putative membrane-spanning domains; homology was slight in the amino and carboxyl termini and in the large cytoplasmic loop that connects spans 5 and 6. This pattern has now been observed in comparisons among all of the G-protein-coupled receptors that have been sequenced thus far: three βARs, an α_2-adrenergic receptor, six muscarinic cholinergic receptors, a substance K receptor, the *mas* oncogene, and at least ten rhodopsins (for review, see Dohlman et al. 1987). The two nonconserved, intracellular segments of the receptor, located between spans 5 and 6 and at the carboxyl terminus, vary widely both in size and in amino acid composition. They are similar in only one odd respect: They tend to contain unusually high concentrations of carboxyl and hydroxyl amino acid residues and helix-breaking residues (proline and glycine) (see Fig. 2). The function of such regions in the receptors is unknown, but similar sequences have been implicated in targeting proteins for rapid turnover (Rogers et al. 1986).

The Hydrophobic Core of the Receptor Retains Regulatory Activity

The conservation of homology among G-protein-coupled receptors in their hydrophobic cores suggests the importance of this region in maintaining the three-dimensional structure of the receptors and in mediating

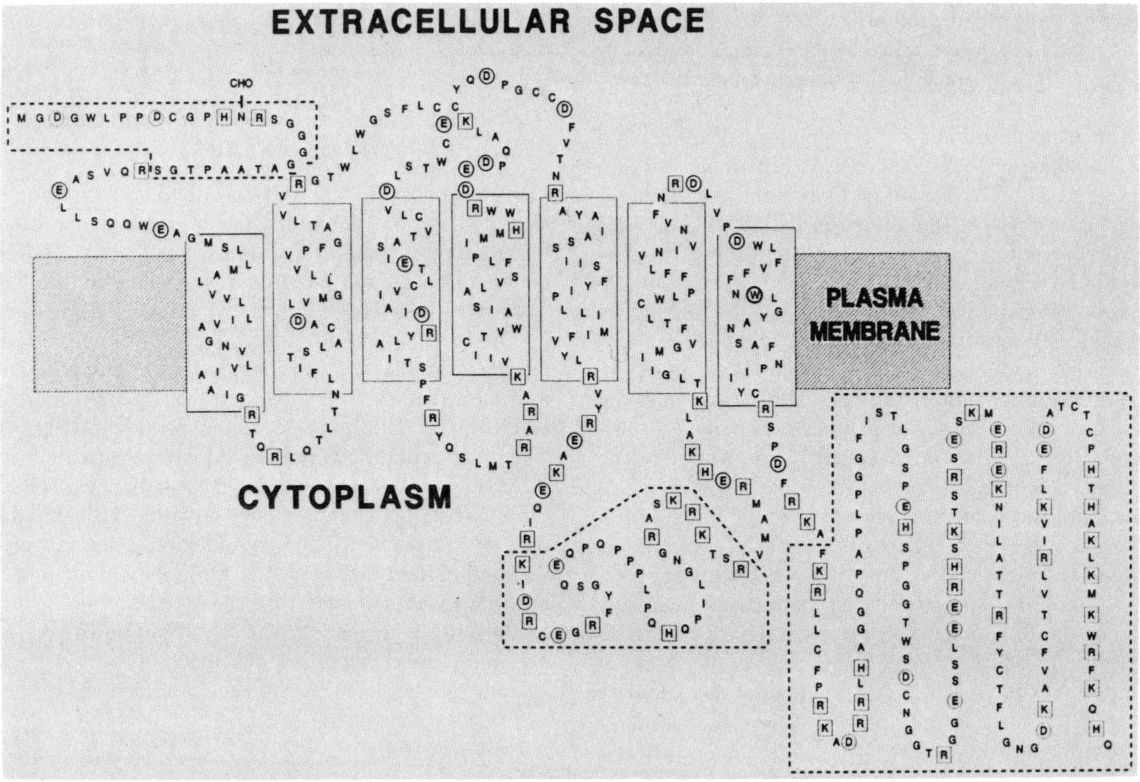

Figure 2. Proposed transmembrane structure of the turkey erythrocyte βAR. Assignment of transmembrane sequences was made to minimize the placement of hydrophilic or charged groups in the membrane and by analogy with the location of homologous membrane-spanning regions in rhodopsin (see Findlay and Pappin 1986). Those sequences in the dashed boxes can be proteolytically removed without loss of either ligand binding or G-protein-activating function (Rubenstein et al. 1987). The position of Trp-330 in the seventh membrane-spanning domain is indicated by a dark circle. Cationic amino acid residues are shown in boxes and anionic residues are shown in light circles. As an example of the odd amino acid composition of some of the nonconserved regions of G-protein-linked receptors, note the proline-rich sequence in the middle of the third cytoplasmic loop and the local concentration of glycine, serine, and glutamic acid residues in the middle of the carboxy-terminal domain. (Adapted from Yarden et al. 1986.)

signal transduction. Proteolytic studies in our laboratory (Rubenstein et al. 1987) and genetic deletion analysis (Dixon et al. 1987b; Strader et al. 1987) support this idea. When purified, detergent-solubilized βAR is treated with any of several proteases, a limit digest is formed that contains only two peptides with a relative molecular weight of >2000. The largest, which we refer to as peptide A, includes the first five membrane-spanning regions and four short connecting loops. The smaller, referred to as peptide B, contains essentially only spanning regions 6 and 7, their connecting loop, and short cytoplasmic sequences. (See Fig. 2 for termini of peptides A and B as prepared using trypsin.) Peptides A and B remain noncovalently bound to each other in detergent solution under nondenaturing conditions. The complex is nearly as stable as is the intact receptor.

Surprisingly, such a limit digest retains essentially unaltered β-adrenergic ligand-binding capacity. Even more surprising, the complex of peptide A and B retained its full ability to activate G_s in response to agonist. When the complex of peptides A and B was purified from a tryptic proteolytic digest by β-adrenergic affinity chromatography and reconstituted with purified G_s into unilamellar phospholipid vesicles, it catalyzed GTPγS binding to G_s as well as did the native receptor. The agonist-stimulated regulatory activity displayed by the tryptic core of the receptor indicates that large segments of the cytoplasmic domain of the receptor are not required for G-protein regulation. We are therefore forced to concentrate on the cationic regions near the cytoplasmic surface of the plasma membrane as possible G_s-interaction sites. The stability of the tryptic core of the receptor also suggests the importance of interactions among the membrane-spanning regions in stabilizing the overall conformation of the receptor.

The hydrophobic core of the receptor also retains the ability to be activated by thiols. Treatment of either avian or mammalian βAR with thiols to reduce disulfide bonds causes partial activation of unliganded receptor and markedly potentiates the stimulatory effect of agonist (Pedersen and Ross 1985; Moxham et al. 1988). When the hydrophobic core of the receptor was similarly reduced, agonist action was still potentiated by thiols (Rubenstein et al. 1987). Thiols did not activate the unliganded proteolyzed receptor, but activation might have been masked by the increased tendency of the unliganded receptor to denature (see Pedersen and Ross 1985).

The exact position of the regulatory disulfide that was modified in these experiments remains uncertain. Dixon et al. (1987a) found that mutation of either Cys-116 or Cys-191 (numbered in the avian sequence) yielded receptors with the same alteration in agonist binding, which argues that these cysteines are disulfide-linked and that this disulfide may be the regulatory disulfide. Because the proteolyzed receptor remains responsive to thiols and because peptides A and B are not covalently bound, our data also argue that the regulatory disulfide lies in peptide A. There is no evidence for a disulfide bond in peptide B. Note that the receptor contains at least two disulfides, not all of which are accessible to added thiols in the native state. As shown in Figure 3, the addition of dithiothreitol to native receptor caused only a partial shift in electrophoretic mobility, relative to that caused by reduction of the SDS-denatured receptor. Thus, it remains possible that there is more than one disulfide bond in peptide A of the receptor and that regulation of activity and binding may reflect the cleavage of one or more disulfides.

Location of the Catecholamine-binding Site on the βAR

In the context of the data described above, Wong et al. (1988) began to map the β-adrenergic ligand-binding site on the receptor by affinity labeling. Two different, highly selective β-adrenergic photoaffinity labels were used: [^{125}I]iodoazidobenzylpindolol (IABP) and [^{125}I]iodocyanopindolol-diazirine (ICYP-DZ). Peptide mapping of the receptor labeled with either ligand indicated that there are two sites of incorporation of label, one each in peptide A and peptide B (Fig. 4). The two peptides were labeled about equally with ICYP-DZ, whereas peptide B was labeled about threefold better than peptide A with IABP. Similar data were obtained if the tryptic core of the receptor was photoaffinity-labeled after proteolysis. Labeling of both peptides indicates that the β-adrenergic ligand-binding site is formed by the juxtaposition of two sequentially distal regions of the receptor.

Both labeled sites in the receptor have now been

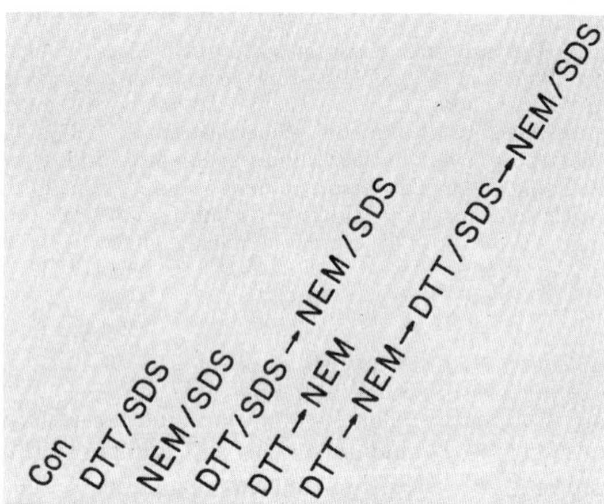

Figure 3. The presence of exposed and unexposed disulfides in the βAR. βAR was purified from turkey erythrocytes according to the method of Brandt and Ross (1986). Three samples were solubilized directly either in SDS solubilization buffer alone (CON), in sample buffer containing 10 mM dithiothreitol (DTT)→(DTT/SDS), or in sample buffer containing 5 mM *N*-ethylmaleimide (NEM)→(NEM/SDS). A fourth sample was dissolved in DTT-containing sample buffer heated at 45°C for 45 min and then exposed to 5 mM excess NEM (DDT/SDS→NEM/SDS). The fifth sample was incubated 1 hr with 5 mM DTT under nondenaturing conditions and then exposed to 5 mM excess NEM before exposure to sample buffer (DTT→NEM). The sixth sample was treated sequentially with DTT and NEM under nondenaturing conditions, then denatured in SDS in the presence of DTT, and finally alkylated with NEM in the presence of SDS. All samples finally were incubated at 45°C for 45 min before SDS-PAGE and autoradiography. The figure shows a silver-stained gel.

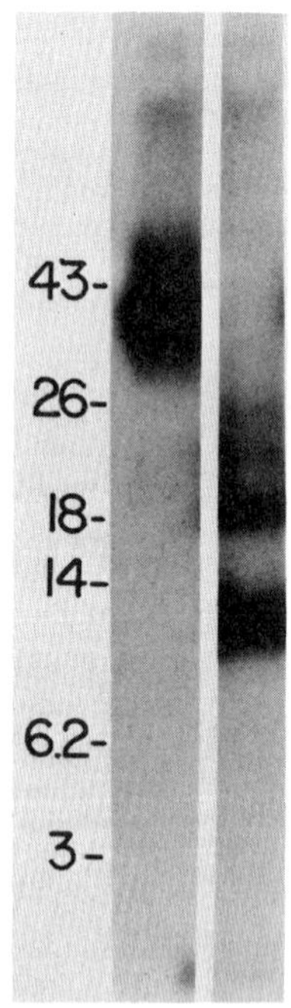

Figure 4. Limit proteolytic digest of avian erythrocyte βAR. Purified βAR (Brandt and Ross 1986) was labeled with IABP (Wong et al. 1988), and a portion was incubated at room temperature overnight with trypsin. Both samples were then analyzed by SDS-PAGE and autoradiography. The left lane shows untreated receptor, whose principal purified species electrophoreses at an equivalent molecular weight of approximately 41,000. The right-hand lane contained the limit digest. Peptides A ($M_r \sim 20{,}000$) and B ($M_r \sim 10{,}000$) are visible, along with some incompletely proteolyzed precursors to peptide A. Similar patterns are observed with silver staining, because the remainder of the receptor is proteolytically cleaved to very small fragments, although peptide B is only weakly stained by silver.

mapped in more detail. The site of labeling in peptide B was localized to the seventh membrane-spanning helix by further proteolysis, followed by the identification of peptides by Western blotting using sequence-directed antisera. When IABP-labeled peptide B was analyzed by gas-phase Edman degradation, we found that label was incorporated only at Trp-330 in span 7 (Wong et al. 1988). It is provocative that Trp-330 lies at a position in the βAR that is near that of Lys-296 in opsin, the residue to which retinal is covalently attached.

Further mapping of photoaffinity-labeled peptide A has localized the labeling site to an 8000-dalton fragment that contains a known antigenic site on the third extracellular loop (which connects spans 4 and 5). On the basis of the size of this fragment, the site of labeling in peptide A might lie in spans 3, 4, or 5, or in either of the first two extracellular loops. Unfortunately, attempts to isolate and sequence the 8000-dalton peptide to localize precisely the site of labeling have not yet been successful.

The photoaffinity-labeling data indicate that one face of the ligand-binding site is contributed by the extracellular half of the receptor's seventh membrane span at a position close to the site at which retinal is attached to opsin. The other face of the binding site lies in the amino-terminal half of the receptor somewhere between spans 2 and 5. Thus, the membrane-spanning domains of the receptor appear to be folded in a bundle, such that span 7 and spans 3–4 are juxtaposed. A similar conformation is generally accepted for rhodopsin (Findlay and Pappin 1986). We propose that β-adrenergic ligands may lie within the bundle of helices in roughly the same orientation as does retinal in opsin. A catecholamine agonist might promote activation of the βAR by altering its structure in the same way as *trans*-retinal alters the structure of opsin. It is interesting that catecholamines and the lysine-retinal conjugate of rhodopsin are both hydrophobic cations of roughly similar size and shape. Both are proposed to be ion paired with aspartyl or glutamyl residues in spans 2 or 3 (Findlay and Pappin 1986; Dixon et al. 1987a). It is plausible that these two regulatory ligands may act through similar molecular mechanisms. It will be fascinating to see if structurally dissimilar ligands of other G-protein-coupled receptors, such as prostaglandins or large peptides, bind to their receptors in similar or distinctively different fashions.

Mastoparans: Peptide Toxins as Models for the G-protein Regulatory Domain

The cytoplasmic surface of a G-protein-coupled receptor must, in response to agonist, selectively bind an appropriate G protein and alter its structure to catalyze guanine nucleotide exchange. In addition, this surface serves as a substrate for kinases and phosphatases (Benovic et al. 1987) and binds to whatever cytoskeletal components regulate receptor endocytosis. The G-protein interface is probably made up of more than one segment of the receptor's primary structure, and it is possible that different sites may be involved in selectivity for specific G proteins and for catalyzing nucleotide exchange. Moreover, it is not yet clear whether receptors bind only to the GTP-binding α subunit of G proteins or to the βγ subunits as well. The βγ subunits are required for receptor–G protein interaction (Fung 1983; Fung and Nash 1983; Im et al. 1987, 1988). They may have a regulatory role in this process in addition to anchoring α subunits to the plasma membrane (Sternweis 1986). Regulation of the receptor–G protein interface by agonists and nucleotides will be a complex area of study.

Recently, our laboratory has begun to exploit an experimental model for the receptors, G-protein regulatory domain, the toxic wasp venom peptides called mastoparans. Mastoparans are 14-amino-acid, tetracationic, amphiphilic peptides isolated from the venom of the genus *Vespa*. Mastoparan, the parent compound, was described by Nakajima and co-workers, first as the potent stimulator of histamine secretion from mast cells and subsequently as a promoter of secretion from a wide variety of secretory cells (cited in Wakamatsu et al. 1983; Higashijima et al. 1988). Secretion is frequently triggered by an increase in the cytoplasmic concentration of Ca^{++} that is itself caused by an increase in the intracellular second messenger inositol-1,4,5-triphosphate (Ins[1,4,5]P_3). Recently, such Ins(1,4,5)P_3-mediated increases of intracellular Ca^{++} have been shown to be controlled at the level of phospholipase C by G proteins (see Berridge 1987). Mastoparan-stimulated histamine release from mast cells is also mediated by an increase in cytoplasmic Ca^{++} (Okano et al. 1985). Higashijima et al. (1987) suggested that mastoparan might similarly act via a G-protein-mediated pathway, when they found that treating mast cells with pertussis toxin blocked mastoparan's ability to cause histamine release.

During the last year, we have demonstrated that mastoparan acts directly on G proteins to promote their activation by GTP by a mechanism strikingly similar to that utilized by agonist-liganded cell-surface receptors (Higashijima et al. 1988). Mastoparan increases the rate with which G_i, G_o, and G_s release GDP and bind GTP, thereby stimulating the overall rate of steady-state GTP hydrolysis. This receptor-mimetic activity also displays several properties that have become known as characteristic of receptor-regulated G-protein function: (1) The action of mastoparan on either G_i or G_o is blocked by a pertussis-toxin-catalyzed ADP-ribosylation of the α subunit. Such ADP-ribosylation generally does not alter the functions of an isolated G protein significantly but has been described as blocking the ability of the G protein to be regulated by receptor (Van Dop et al. 1984; Okajima et al. 1985). (2) Mastoparan displays its regulatory activity at low concentrations of Mg^{++} (Fig. 5). Significant stimulation is observed below 100 nM Mg^{++} and mastoparan stimulation is optimal in the presence of 10 μM Mg^{++}. Similar Mg^{++} dependence is noted for receptor-mediated stimulation of G proteins. (3) Mastoparan

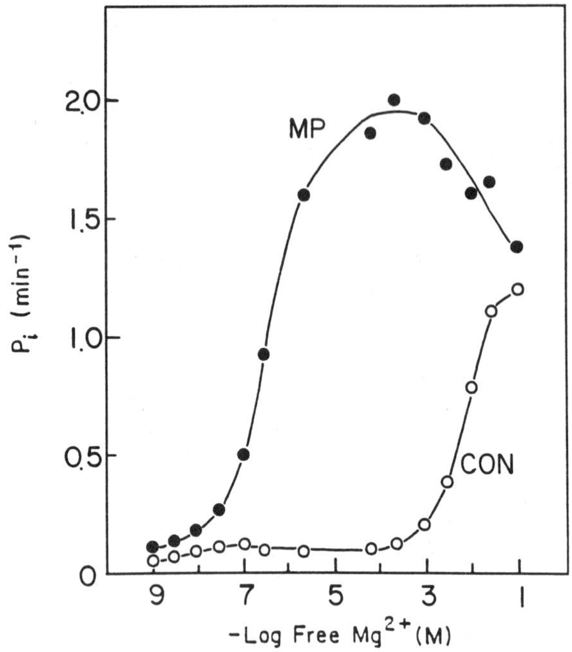

Figure 5. Stimulation of G_o-catalyzed GTP hydrolysis by mastoparan. G_o, purified from bovine brain, was incorporated into unilamellar phospholipid vesicles composed of phosphatidylcholine, phosphatidylserine, and phosphatidylethanolamine. The GTPase activity of this preparation was then assayed at increasing concentrations of free Mg^{++}, either in the presence (MP) or absence (CON) of 100 μM mastoparan. Stimulation of steady-state hydrolysis by mastoparan at 0.1 mM Mg^{++} is approximately 20-fold. (Reprinted, with permission, from Higashijima et al. 1988.)

does not alter the K_{cat} for hydrolysis of bound GTP. (4) Mastoparan is far more active when the G protein has been reconstituted into phospholipid vesicles, rather than when it is assayed in detergent solution. In general, receptors and G proteins do not interact productively in detergent solution.

The selectivity of mastoparan among G proteins is not as clear-cut as that displayed by most cell-surface receptors, but it is nonetheless impressive. For example, the potency and efficacy of mastoparan is greatest for G_o, somewhat less for G_i, and still less for G_s and transducin (Higashijima et al. 1988). It displays virtually no regulatory activity toward $p21^{ras}$ or the yeast RAS2 protein. Mastoparan can also distinguish somewhat among the three species of G_i. Other members of the mastoparan family and several synthetic cationic polypeptides display different patterns of selectivity among the G proteins. Selectivity appears to be at most tenfold, perhaps an order of magnitude less than that displayed by cell-surface receptors but certainly significant. Cationic antagonists of mastoparan are also under development. These compounds should allow us to discern the structural features necessary for G-protein binding and those necessary for activation.

Unfortunately, but not surprisingly, the potency of mastoparans is significantly lower than that of cell-surface receptors. Activation of G_o by mastoparan is half-maximally effective at around 10 μM, whereas receptors are frequently active in reconstituted systems at 1000-fold lower concentrations. The higher potency of receptors may reflect the fact that they are permanently anchored in membranes, thereby increasing their local concentration, whereas mastoparans equilibrate between the bilayer surface and the aqueous medium. It is also possible that more than one molecule of mastoparan is necessary to promote activation of a G protein. In the context of the relatively low potency of mastoparan, it should be noted that cellular effects of mastoparan on G proteins are probably central to their pharmacological activities. S. Uzu and T. Higashijima (in prep.) have demonstrated that the structure-activity relationships for mastoparan effects on G proteins and on histamine release are parallel. There is no correlation between the structure-activity relationship displayed for histamine release and that displayed for binding to phospholipid bilayers or for blocking the activity of calmodulin (both common nonspecific activities of basic peptides).

Despite the receptor-like activity of mastoparan, there is no significant similarity between its sequence and those of the G-protein-coupled receptors. What similarity exists must therefore lie at the level of tertiary structure. The three-dimensional structure of mastoparan when it is bound to a phospholipid bilayer has been determined by Higashijima et al. (1983; Wakamatsu et al. 1983) using circular dichroism measurements and transferred nuclear Overhauser effect (NOE) measurements of mastoparan bound to vesicles of perdeuterated phosphatidylcholine (Fig. 6). Mastoparans are short α helices when bound to the surface of bilayers, oriented such that the four positive charges are directed toward the aqueous solvent. It is thus likely that the surface of mastoparans that interacts with G proteins is such an ordered array of four positive charges. It is also likely that the G-protein-binding domain of receptors is an ordered array of positive charges. All of the potential candidate regions, the first and second intracellular loops, both origins of the large third intracellular loop, and the beginning of the cytoplasmic carboxy-terminal region, are strongly positively charged. We should now be in a position to use synthetic mastoparan analogs and specific mutants of the βAR to focus on what key features of these cationic arrays are important for G-protein binding and activation. It is reasonable to suppose that when a receptor is activated, the relevant conformational change on the intracellular face includes either the exposure or occlusion of positive charges or a change in the orientation of one charge cluster with respect to another. Probing the selectivity of chimeric receptors, mutated receptors, and novel synthetic mastoparans should help us approach this problem. With luck, these studies can lead to a clearer understanding of both the mechanism of G-protein activation by receptors and the selectivity of receptors for G proteins.

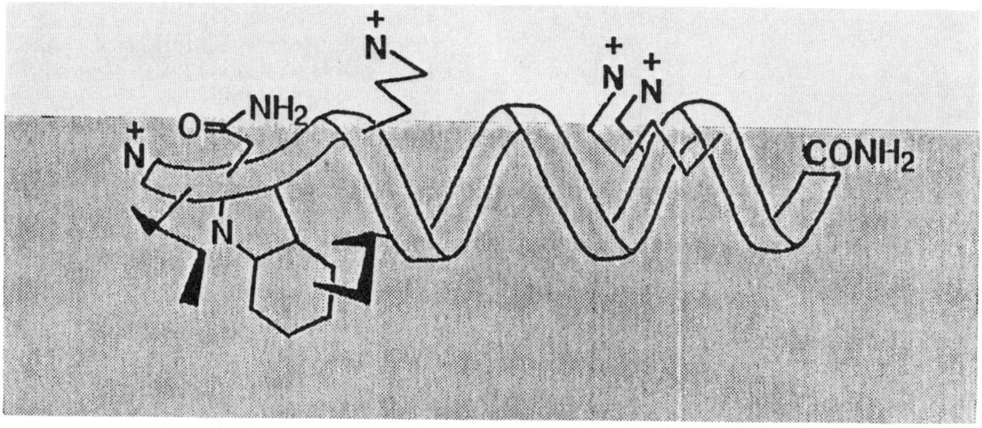

Figure 6. Structure of mastoparan-X. The structure of mastoparan-X when adsorbed to phospholipid vesicles was determined by circular dichroism and NMR spectroscopy using the transferred NOE technique (Higashijima et al. 1983; Wakamatsu et al. 1983). The shaded area represents the outer surface of the phospholipid bilayer. (Reprinted, with permission, from Wakamatsu 1986.)

ACKNOWLEDGMENTS

This work was supported by National Institutes of Health grant GM-30355, R.A. Welch Foundation grant I-982, and National Institutes of Health training grant GM-07062.

REFERENCES

Benovic, J.L., F. Mayor, Jr., C. Staniszewski, R.J. Lefkowitz, and M.G. Caron. 1987. Purification and characterization of the β-adrenergic receptor kinase. *J. Biol. Chem.* **262:** 9026.

Berridge, M.J. 1987. Inositol trisphosphate and diacylglycerol: Two interacting second messengers. *Annu. Rev. Biochem.* **56:**159.

Brandt, D.R. and E.M. Ross. 1986. Catecholamine-stimulated GTPase cycle: Multiple sites of regulation by β-adrenergic receptor and Mg^{2+} studied in reconstituted receptor-G_s vesicles. *J. Biol. Chem.* **261:** 1656.

Dixon, R.A.F., I.S. Sigal, M.R. Candelore, R.B. Register, W. Scattergood, E. Rands, and C.D. Strader. 1987a. Structural features required for ligand binding to the β-adrenergic receptor. *EMBO J.* **6:** 3269.

Dixon, R.A.F., I.S. Sigal, E. Rands, R.B. Register, M.R. Candelore, A.D. Blake, and C.D. Strader. 1987b. Ligand binding to the β-adrenergic receptor involves its rhodopsin-like core. *Nature* **326:** 73.

Dixon, R.A.F., B.K. Kobilka, D.J. Strader, J.L. Benovic, H.G. Dohlman, T. Frielle, M.A. Bolanowski, C.D. Bennett, E. Rands, R.E. Diehl, R.A. Mumford, E.E. Slater, I.S. Sigal, M.G. Caron, R.J. Lefkowitz, and C.D. Strader. 1986. Cloning of the gene and cDNA for mammalian β-adrenergic receptor and homology with rhodopsin. *Nature* **321:** 75.

Dohlman, H.G., M.G. Caron, and R.J. Lefkowitz. 1987. A family of receptors coupled to guanine nucleotide regulatory proteins. *Biochemistry* **26:** 2657.

Findlay, J.B.C. and D.J.C. Pappin. 1986. The opsin family of proteins. *Biochem. J.* **238:** 625.

Fung, B.K.-K. 1983. Characterization of transducin from bovine retinal rod outer segments. I. Separation and reconstitution of subunits. *J. Biol. Chem.* **258:** 10495.

Fung, B.K.-K. and C.R. Nash. 1983. Characterization of transducin from bovine retinal rod outer segments. II. Evidence for distinct binding sites and conformational changes revealed by limited proteolysis with trypsin. *J. Biol. Chem.* **258:** 10503.

Gilman, A.G. 1987. G proteins: Transducers of receptor-generated signals. *Annu. Rev. Biochem.* **56:** 615.

Higashijima, T., S. Uzu, T. Nakajima, and T. Miyazawa. 1987. Mechanism of histamine-release by mastoparan. In *Peptide chemistry 1986* (ed. T. Miyazawa), p. 75. Protein Research Foundation, Osaka, Japan.

Higashijima, T., S. Uzu, T. Nakajima, and E.M. Ross. 1988. Mastoparan, a peptide toxin from wasp venom, mimics receptors by activating GTP-binding regulatory proteins (G proteins). *J. Biol. Chem.* **263:** 6491.

Higashijima, T., K. Wakamatsu, M. Takemitsu, M. Fujino, T. Nakajima, and T. Miyazawa. 1983. Conformational change of mastoparan from wasp venom on binding with phospholipid membrane. *FEBS Lett.* **152:** 227.

Im, M.-J., A. Holzhöfer, H. Böttinger, T. Pfeuffer, and E.J. M. Helmreich. 1988. Interactions of pure βγ-subunits of G-proteins with purified β_1-adrenoceptor. *FEBS Lett.* **227:** 225.

Im, M.-J., A. Holzhöfer, A.K. Keenan, P. Gierschik, M. Hekman, E.J.M. Helmreich, and T. Pfeuffer. 1987. The role of β,γ-subunits of guanine nucleotide binding proteins in control of a reconstituted signal transmission chain containing purified components of the adenylate cyclase system. *J. Recept. Res.* **7:** 17.

Jurss, R., M. Hekman, and E.J.M. Helmreich. 1985. Proteolysis-associated deglycosylation of β_1-adrenergic receptor in turkey erythrocytes and membranes. *Biochemistry* **24:** 3349.

Levitzki, A. 1986. β-Adrenergic receptors and their mode of coupling to adenylate cyclase. *Physiol Rev.* **66:** 819.

May, D.C., E.M. Ross, A.G. Gilman, and M.D. Smigel. 1985. Reconstitution of catecholamine-stimulated adenylate cyclase activity using three purified proteins. *J. Biol. Chem.* **260:** 15829.

Minneman, K.P., G.A. Weiland, and P.B. Molinoff. 1980. A comparison of the beta-adrenergic receptor of the turkey erythrocyte with mammalian $beta_1$ and $beta_2$ receptors. *Mol. Pharmacol.* **17:** 1.

Moxham, C.P., E.M. Ross, S.T. George, and C.C. Malbon. 1988. β-Adrenergic receptors display intramolecular disulfide bridges *in situ*: Analysis by immunoblotting and functional reconstitution. *Mol. Pharmacol.* **33:** 486.

Okajima, F., T. Katada, and M. Ui. 1985. Coupling of the guanine nucleotide regulatory protein to chemotactic peptide receptors in neutrophil membranes and its uncoupling by islet-activating protein, pertussis toxin. *J. Biol. Chem.* **260:** 6761.

Okano, Y., H. Takagi, T. Tohmatsu, S. Nakashima, Y. Kuroda, K. Saito, and Y. Nozawa. 1985. A wasp venom mastoparan-induced polyphosphoinositide breakdown in rat peritoneal mast cells. *FEBS Lett.* **188:** 363.

Pedersen, S.E. and E.M. Ross. 1985. Functional activation of β-adrenergic receptors by thiols in the presence or absence of agonists. *J. Biol. Chem.* **260:** 14150.

Rogers, S., R. Wells, and M. Rechsteiner. 1986. Amino acid sequences common to rapidly degraded proteins: The PEST hypothesis. *Science* **234:** 364.

Rubenstein, R.C., S.K-F. Wong, and E.M. Ross. 1987. The hydrophobic tryptic core of the β-adrenergic receptor retains G_s-regulatory activity in response to agonists and thiols. *J. Biol. Chem.* **262:** 16655.

Smigel, M.D., E.M. Ross, and A.G. Gilman. 1984. Role of the β-adrenergic receptor in the regulation of adenylate cyclase. In *Cell membranes: Methods and reviews* (ed. E.L. Elson et al.), vol 2, p. 247. Plenum Press, New York.

Sternweis, P.C. 1986. The purified α subunits of G_o and G_i from bovine brain require βγ for association with phospholipid vesicles. *J. Biol. Chem.* **261:** 631.

Strader, C.D., I.S. Sigal, A.D. Blake, A.H. Cheung, R.B. Register, E. Rands, B.A. Zemcik, M.R. Candelore, and R.A.F. Dixon. 1987. The carboxyl terminus of the hamster β-adrenergic receptor expressed in mouse L cells is not required for receptor sequestration. *Cell* **49:** 855.

Stryer, L. and H.R. Bourne. 1986. G proteins: A family of signal transducers. *Annu. Rev. Cell Biol.* **2:** 391.

Van Dop, C., G. Yamanaka, F. Steinberg, R. Sekura, C.R. Manclark, L. Stryer, and H.R. Bourne. 1984. ADP-ribosylation of transducin by pertussis toxin blocks the light-stimulated hydrolysis of GTP and cGMP in retinal photoreceptors. *J. Biol. Chem.* **259:** 23.

Wakamatsu, K. 1986. "^{1}H-NMR analysis of the interaction of peptides with phospholipid bilayer: Relationships between interaction and physiological activity." Ph.D. thesis. Tokyo University, Japan.

Wakamatsu, K., T. Higashijima, M. Fujino, T. Nakajima, and T. Miyazawa. 1983. Transferred NOE analyses of conformations of peptides as bound to membrane bilayer of phospholipid; mastoparan X. *FEBS Lett.* **162:** 123.

Wong, S.K-F., C. Slaughter, A.E. Ruoho, and E.M. Ross. 1988. The catecholamine binding site of the β-adrenergic receptor is formed by juxtaposed membrane-spanning domains. *J. Biol. Chem.* **263:** 7925.

Yarden, Y., H. Rodriguez, S.K-F. Wong, D.R. Brandt, D.C. May, J. Burnier, R.N. Harkins, E.Y. Chen, J. Ramachandran, A. Ullrich, and E.M. Ross. 1986. The avian β-adrenergic receptor: Primary structure and membrane topology. *Proc. Natl. Acad. Sci.* **83:** 6795.

Molecular Biology of Adrenergic Receptors

R.J. Lefkowitz, B.K. Kobilka, J.L. Benovic, M. Bouvier, S. Cotecchia,
W.P. Hausdorff, H.G. Dohlman, J.W. Regan, and M.G. Caron
Departments of Medicine, Biochemistry, and Cell Biology, Howard Hughes Medical Institute, Duke University Medical Center, Durham, North Carolina 27710

The various receptors for catecholamines, termed adrenergic receptors, represent excellent model systems for the study of receptor-mediated transmembrane signaling because of their ubiquity, coupling to well-defined effector mechanisms, and the clinical importance of drugs that interact with them. The β_1- and β_2-adrenergic receptors stimulate adenylyl cyclase via the guanine nucleotide regulatory protein G_s (Lefkowitz and Caron 1988). The α_2-adrenergic receptor inhibits adenylyl cyclase via G_i. The α_2-adrenergic receptors also appear to activate a sodium/hydrogen antiporter (Isom et al. 1987). The α_1-adrenergic receptors stimulate hydrolysis of polyphosphoinositides by activating phospholipase C, thus generating inositol trisphosphate and diacylglycerol. Each of these systems is in turn analogous to the retinal light transduction system that consists of the prototypic receptor rhodopsin, a G-protein transducin, and an effector enzyme, which is a cGMP phosphodiesterase (Stryer 1986).

Over the past few years, our group has succeeded in cloning the genes for all of the known adrenergic receptors (Lefkowitz and Caron 1988). We have, in addition, cloned two other related genes, one for a previously unsuspected α_2-adrenergic receptor (Regan et al. 1988) and the other for an as yet unidentified receptor that we have termed G21 (Kobilka et al. 1987b). Our ability to clone these genes was based on our having developed means for purifying to homogeneity the various subtypes of adrenergic receptors (Caron et al. 1988). With pure protein in hand, we were able to obtain limited protein sequence information, design oligonucleotide probes, and clone the genes. In this paper, we summarize the structural features of the various adrenergic receptors and recent approaches for determining the structural basis for such receptor functions as ligand binding, G-protein activation, and receptor-mediated desensitization.

RESULTS AND DISCUSSION

Biochemical Properties of the Adrenergic Receptors

Some of the information that has emerged from the deduced sequences of the various adrenergic receptors is summarized in Table 1. As assessed by SDS-PAGE of the purified proteins, they all consist of single polypeptide chains with molecular weights of 64,000–80,000. They range in size from 413 amino acids for the human β_2-adrenergic receptor (Kobilka et al. 1987c) to 515 amino acids for the hamster α_1-adrenergic receptor (Cotecchia et al. 1988). All of these receptors share the characteristic and now well-known feature of having seven clusters of hydrophobic amino acids, 20–25 residues in length, long enough to represent membrane-spanning domains (Lefkowitz and Caron 1988). The same topology is shared by rhodopsin. All of the receptors have sites of N-linked glycosylation toward their amino termini, presumed to be extracellular, and with their carboxyl termini, presumed to be cytoplasmic. Although this organization remains largely speculative, recent results of limited proteolysis studies are quite consistent with such an organization (Dohlman et al. 1987b).

Sequence similarities among the adrenergic receptors, as well as all members of the G-protein-coupled receptor family, are greatest in these presumed membrane-spanning regions. Within these domains, the adrenergic receptors generally share 40–50% sequence identity (Lefkowitz and Caron 1988). The closest relationship is that between human β_1- and β_2-adrenergic

Table 1. Adrenergic Receptors

Subtype (species)	M_r = SDS-PAGE	Peptide length (amino acid)	Introns	Identity[a] (%)
β_2 (human)	64,000	413	0	100
β_1 (human)	64,000	477	0	71
α_{2A} (human)	64,000	450	0	42
α_{2B} (human)	80,000	461	?	40
α_1 (hamster)	80,000	515	1	42

[a]Sequence identity within the presumed membrane-spanning domains between a given receptor and the human β_2-adrenergic receptor.

receptors, which are identical in 70% of their residues within the membrane-spanning domains (Frielle et al. 1987).

The only features of the receptors that vary in size are the third cytoplasmic loop and the carboxyl terminus. The third loop is shortest in the β_2-adrenergic receptor, slightly longer in the α_1-adrenergic and β_1-adrenergic receptors, and longest in the α_2-adrenergic receptor. The carboxyl terminus is shortest in the α_2-adrenergic receptor, longer in the β-adrenergic receptors, and longest in the α_1-adrenergic receptor (Lefkowitz and Caron 1988). We do not know if these features are related to the specificity of G-protein coupling of the receptors.

Three of the five receptor genes listed in Table 1 have been found to lack introns within their coding regions. Only the α_1-adrenergic receptor contains a single intron within its coding block. Rhodopsin, of course, contains several introns.

Figure 1 is a schematic representation of the deduced sequence of a second α_2-adrenergic receptor that is found on a different chromosome (no. 4) than the human platelet α_2-adrenergic receptor gene originally cloned and localized to chromosome 10. As indicated in black dots, approximately 75% of the residues in the membrane-spanning domains are identical to those of the human platelet α_2-adrenergic receptor. The first and second cytoplasmic loops and the amino- and carboxy-terminal portions of the large third loop are also highly conserved with the other α_2-adrenergic receptor. Expression studies indicate that the pharmacology of the two α_2-adrenergic receptors is distinguishable (Regan et al. 1988) and is consistent with the recently suggested terminology of α_{2A} and α_{2B} receptor subsubtypes. The physiological roles and specific biochemical effector systems to which each of these receptor subtypes are coupled remain to be determined. Discovery of novel receptor subtypes such as this one, by receptor cloning, opens a whole new approach to development of new and potentially valuable therapeutic agents.

Structure-Activity Relationships

Plasma membrane receptors mediate two main functions. First, they bind ligands from the outside of the cell, and second, they activate an effector system leading to the generation of second messengers capable of activating various biochemical processes. In addition, processes by which drug and hormone responsiveness are regulated specifically involve modulation of receptor function. Thus, given the topographical organization of the G-protein-coupled receptors, it is possible that various regions of these receptors are involved in these different functions.

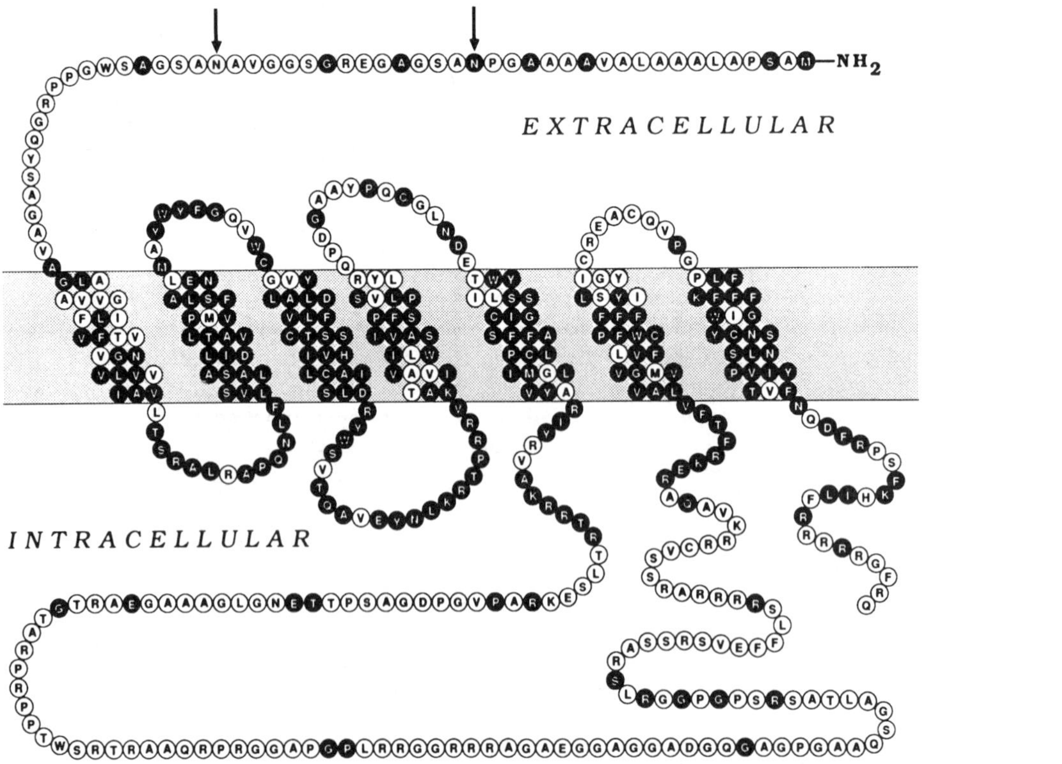

Figure 1. Primary structure of the human α_{2B}-adrenergic receptor. The protein is composed of 461 amino acid residues, which are represented by the one-letter code. The polypeptide chain is arranged according to the model that has been proposed for rhodopsin and other G-protein-coupled receptors (Lefkowitz and Caron 1988). The darkened residues represent amino acid identities between this receptor and the human α_{2A}-adrenergic receptor (Regan et al. 1988).

Chimeric receptors. One approach that we have used has been the construction and expression of chimeric α_2-β_2-adrenergic receptors. Over the past year or two, the structural basis for the various functional properties of this type of receptor has been probed by several groups (Dixon et al. 1987a,b; Kobilka et al. 1987a; Rubenstein et al. 1987). The major approaches that have been used are mutagenesis, especially deletion of specific sequences, and biochemical methods, for example, using proteases to cleave defined peptide segments from solubilized receptors. Although these methods have been useful in delineating regions of the receptors that do not influence its function, they suffer from difficulties in drawing compelling inferences about the role of specific domains on the basis of loss of function. This is so because deletion of specific sequences may adversely affect receptor function by influencing the conformation or folding of some distant region of the molecule. We wanted to circumvent such problems and establish a general approach to the study of G-protein-coupled receptors, one in which positive inferences could be drawn on the basis of acquisition of functions that were acquired by receptors in association with their acquisition of specific receptor domains. We were further encouraged to try this approach, because we had cloned a family of receptors, all of which bind epinephrine. We reasoned that chimeras made from various adrenergic receptors should all bind epinephrine and might be able to function.

The properties of the α_2- and β_2-adrenergic receptors are quite distinct. The α_2-receptor inhibits adenylyl cyclase through G_i, whereas the β_2-adrenergic receptor stimulates the enzyme via G_s. There are highly specific antagonists and a characteristic agonist order of potency of paraminoclonidine (PAC) > epinephrine > isoproterenol for the α_2 and isoproterenol > epinephrine > PAC for the β_2. In the next few figures, the α_2-adrenergic receptor sequence is represented as a solid line and the β_2-adrenergic receptor sequence is represented as an open line. In these studies, a series of 12 chimeric receptor genes containing coding sequences derived from both the α_2- and the β_2-adrenergic receptors was constructed. The chimeric receptor genes were expressed either by transfection into COS-7 cells or by injection of RNA into *Xenopus* oocytes (Kobilka et al. 1988).

Ligand-binding function. Figure 2 depicts 4 of the 12 chimeras that were constructed (Kobilka et al. 1988). This series can be used to delineate the importance of the transmembrane domains, but particularly the sixth and seventh transmembrane-spanning domains, in determining the binding specificity of these receptors. Shown in the upper left- hand corner of Figure 2 is the β_2-adrenergic receptor. In the chimera shown below it, the first two transmembrane helices of the β_2-receptor have been substituted with α_2-adrenergic receptor sequence. In chimeric receptor 2, the first four membrane helices are α_2, and by chimeric receptor 3, the first five helices and a good part of the third cytoplasmic loop are all donated by the α_2-adrenergic receptor. In chimeric receptor 3, only the last two

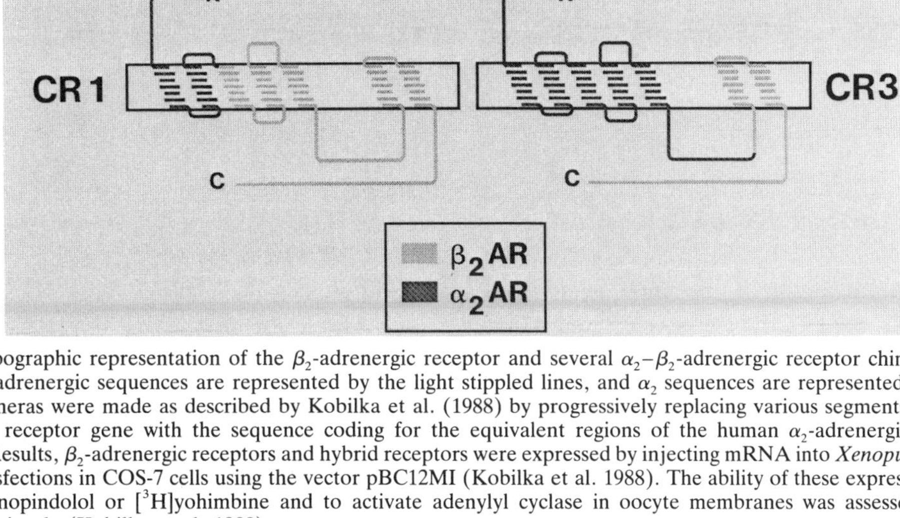

Figure 2. Topographic representation of the β_2-adrenergic receptor and several α_2–β_2-adrenergic receptor chimeras. In these chimeras, β_2-adrenergic sequences are represented by the light stippled lines, and α_2 sequences are represented by dark lines. Receptor chimeras were made as described by Kobilka et al. (1988) by progressively replacing various segments of the human β_2-adrenergic receptor gene with the sequence coding for the equivalent regions of the human α_2-adrenergic receptor. As described in Results, β_2-adrenergic receptors and hybrid receptors were expressed by injecting mRNA into *Xenopus* oocytes or by transient transfections in COS-7 cells using the vector pBC12MI (Kobilka et al. 1988). The ability of these expressed proteins to bind [^{125}I]cyanopindolol or [^{3}H]yohimbine and to activate adenylyl cyclase in oocyte membranes was assessed by methods described previously (Kobilka et al. 1988).

transmembrane regions are from the β_2-adrenergic receptor. As the α_2-sequence progressively replaces the β_2-sequence in this series, there is a progressive shift in agonist-binding affinity such that the β-agonist isoproterenol and the α_2-agonist PAC, which have vastly differing affinities for the β_2-receptor, become closer and closer in their affinities for the chimeras (data not shown). This suggests that many or all of the membrane-spanning domains influence agonist-binding specificity. Interestingly, somewhat of a different pattern was observed for the ability to bind antagonists. All three of these chimeras bind β-adrenergic antagonists with essentially wild-type affinity. None of them bind α_2-adrenergic antagonists. This suggests that the major determinants of antagonist-ligand-binding specificity of these receptors must reside in the sixth or seventh transmembrane-spanning regions.

Figure 3 shows the structure of chimeric receptor 2 next to that of another construction, chimeric receptor 8. The data obtained with these two chimeras emphasize further the importance of the seventh membrane-spanning domain for determining the specificity of ligand binding. As noted above, chimeric receptor 2 binds β-adrenergic antagonists and not α-adrenergic antagonists. Chimeric receptor 8 is made from chimeric receptor 2 simply by substituting the seventh membrane-spanning domain of the α_2-adrenergic receptor for that of the β_2-adrenergic receptor. This is the only difference between the two receptors. However, chimeric receptor 8 has lost the ability to bind β-antagonists and now binds α_2-adrenergic antagonists and other α-adrenergic ligands. Chimeric receptor 2 binds the β-antagonist alprenolol with relatively high affinity, and the β-agonist isoproterenol is more potent than the α_2-agonist PAC. In striking contrast, for chimeric receptor 8, PAC is most potent, alprenolol is about two orders of magnitude less potent than at chimeric receptor 2, and isoproterenol is very weak. Thus, the simple substitution of this one membrane-spanning domain (domain 7) completely switches the binding affinity of both agonists and antagonists.

G-protein-coupling function. Chimeric receptor 8 is also instructive in terms of determining that region of the receptors involved in determining the specificity of coupling to G_s. We had wondered about the minimum sequence of β_2-receptor that could be inserted into an α_2-adrenergic receptor to convey to it the ability to activate adenylyl cyclase. Chimeric receptor 8 appears to answer that question. Chimeric receptor 8, which contains largely the α_2-adrenergic receptor sequence, binds ligands with α_2-adrenergic specificity (Fig. 3). However, it is able to activate adenylyl cyclase. Moreover, as shown in Figure 4, this activation is mediated with a classic α_2-adrenergic specificity. Thus, PAC is more potent than epinephrine, which is much more potent and active than the β-adrenergic agonist isoproterenol in activating adenylyl cyclase through chimeric receptor 8. This construction is a true receptor chimera, binding ligands with an α-adrenergic specificity while performing the effector function of the β-adrenergic receptor, namely, activation of G_s and adenylyl cyclase. Another construction, chimeric receptor 9, which has a slightly smaller β-receptor insert

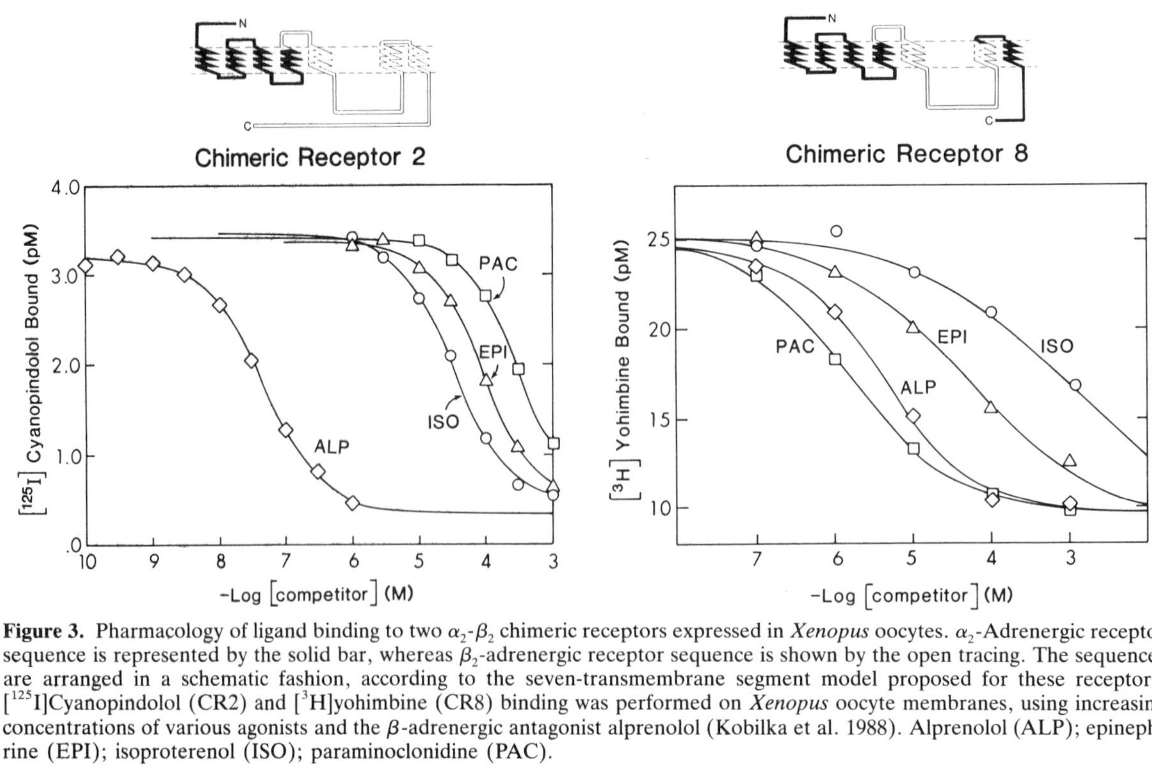

Figure 3. Pharmacology of ligand binding to two α_2-β_2 chimeric receptors expressed in *Xenopus* oocytes. α_2-Adrenergic receptor sequence is represented by the solid bar, whereas β_2-adrenergic receptor sequence is shown by the open tracing. The sequences are arranged in a schematic fashion, according to the seven-transmembrane segment model proposed for these receptors. [^{125}I]Cyanopindolol (CR2) and [^{3}H]yohimbine (CR8) binding was performed on *Xenopus* oocyte membranes, using increasing concentrations of various agonists and the β-adrenergic antagonist alprenolol (Kobilka et al. 1988). Alprenolol (ALP); epinephrine (EPI); isoproterenol (ISO); paraminoclonidine (PAC).

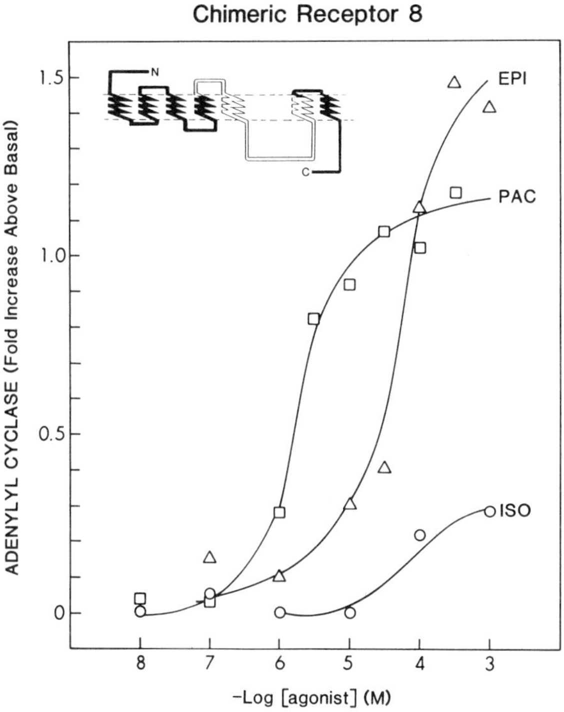

Figure 4. Stimulation of adenylyl cyclase activity by various adrenergic agonists in membranes of *Xenopus* oocytes injected with chimeric receptor 8 mRNA (Kobilka et al. 1988).

(amino acids 214–295; midpoint of the fifth transmembrane domain through the amino-terminal portion of the sixth transmembrane domain), also activates adenylyl cyclase but is significantly less effective (Kobilka et al. 1988).

In summary, the seventh membrane-spanning domain of the α_2- and β_2-adrenergic receptors is a major determinant of agonist and antagonist binding specificity. However, many of the membrane-spanning α-helices appear to play a role in forming the ligand-binding pocket. In accordance with this contention, we have found that a covalent β-adrenergic probe inserts into the second transmembrane helix of the β_2-adrenergic receptor (Dohlman et al. 1987a) and that a covalent α_2-adrenergic receptor probe inserts into the fourth transmembrane region of that receptor (H. Matsui et al., in prep.). The fifth and sixth membrane-spanning domains and the third cytoplasmic loop confer to the β-receptor the ability to couple to G_s. This last finding is consistent with data from Dixon et al. (1987b), who have shown that a large deletion in the third cytoplasmic loop leads to a loss of the ability to activate G_s. Recently, we have also shown (B.F. O'Dowd et al., in prep.) that deletions and/or substitutions, in the carboxy-terminal portion of the third cytoplasmic loop, as well as in the regions of the cytoplasmic tail of the receptor closest to the membrane, also lead to significant reduction in activation of G_s. Thus, these regions of the receptor may constitute part of a binding site that interacts with the G_s protein.

Receptor-mediated desensitization. Desensitization is a very general biological phenomenon characterized by the fact that the intensity of a response often wanes over time despite the presence of a stimulus of constant intensity (Sibley and Lefkowitz 1985). Adenylyl cyclase systems coupled to β-adrenergic receptors have been intensively studied, because they show several characteristic types of densensitization (Sibley and Lefkowitz 1985). One of these is termed homologous and refers to a phenomenon whereby only stimulation by a particular desensitizing hormone, such as a β-agonist, is attenuated. Heterologous desensitization means that diminished responsiveness to a variety of hormones and drugs may also follow stimulation by any one of them.

Over the past 5 years, we have been able to demonstrate that β-adrenergic receptor phosphorylation accompanies both homologous and heterologous forms of desensitization (Sibley et al. 1987). Our data suggested that phosphorylation of the receptor in heterologous densensitization is mediated by the cAMP-dependent protein kinase and accounts for a form of feedback regulation of receptor function (Benovic et al. 1985). In contrast, homologous desensitization, which is usually associated with receptor sequestration or internalization, is almost invariably a cAMP-independent process (Sibley and Lefkowitz 1985).

Several years ago, we characterized a novel protein kinase named the β-adrenergic receptor kinase, (βARK) that appears to be responsible for phosphorylating the β-adrenergic receptor during homologous desensitization (Benovic et al. 1986). This kinase has been purified to homogeneity, and its properties have been studied in detail. As summarized in Figure 5, βARK is an 80,000-dalton cytoplasmic protein (Benovic et al. 1987a). It is insensitive to all of the usual activators of protein kinases and does not phosphorylate such typical substrates as casein and histones. However, it does have the unique property of phosphorylating agonist-occupied β_2-adrenergic receptors as well as α_2-adrenergic receptors (Benovic et al. 1987c) to high stoichiometry. In this respect, it mimics the properties of another enzyme, rhodopsin kinase, which phosphorylates rhodopsin in a light-dependent way (Shichi et al. 1984).

In the case of rhodopsin, a very serine- and threonine-rich region at the carboxyl terminus of the protein has been demonstrated to serve as the major phosphate acceptor for rhodopsin kinase phosphorylation (Shichi et al. 1984). There may be one or two sites on the third cytoplasmic loop as well. Phosphorylation of these sites by the kinase leads to the binding of an additional protein, called 48K protein or arrestin, to the rhodopsin (Shichi et al. 1984; Stryer 1986). These interactions serve to attenuate markedly the coupling of rhodopsin to transducin. Several lines of evidence currently being developed in our laboratory suggest that a comparable serine- and threonine-rich region at the carboxyl terminus of the β-adrenergic receptor likely serves as the major locus for agonist-induced phosphorylation by βARK (Dohlman et al.

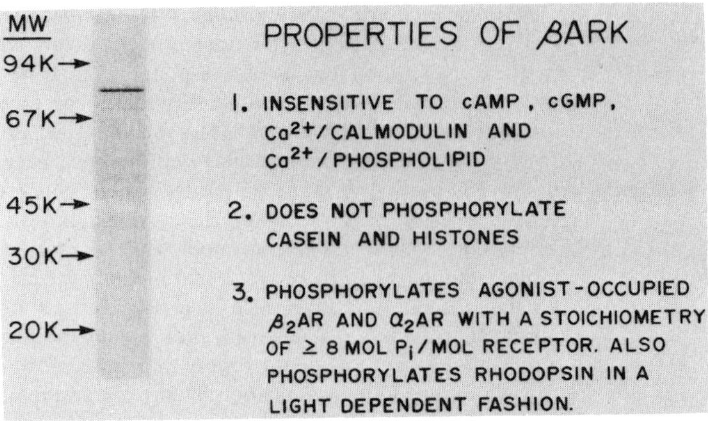

Figure 5. SDS-PAGE profile and properties of purified βARK. βARK was purified by various chromatographic procedures from bovine brain, according to the method of Benovic et al. (1987a). The SDS-PAGE shown on the left represents material purified through the hydroxyapatite step of purification, run on a 10% polyacrylamide gel, and visualized by Coomassie blue staining.

1987b; Bouvier et al. 1988; Lefkowitz and Caron 1988). There may also be additional sites on other regions. In addition, we have also shown that authentic retinal arrestin combines with the βARK-phosphorylated β-receptor and diminishes its interaction with G_s (Benovic et al. 1987b), suggesting the involvement of an analogous protein in the hormonal systems.

To test our hypotheses further about the location of these phosphorylation sites and the potential relevance of βARK-catalyzed phosphorylation of the β-receptor to the mechanism of homologous desensitization, we prepared the two β-adrenergic receptor mutants shown in Figure 6 (Bouvier et al. 1988). T365 is a truncation mutant in which the last 48 amino acids at the carboxyl terminus have been deleted. This serves to delete most of the serine and threonine residues in this region. The other mutant is a substitution mutant in which the 11 serine and threonine residues indicated with asterisks have been changed to either alanine or glycine, so that they can no longer serve as phosphate acceptors. Both

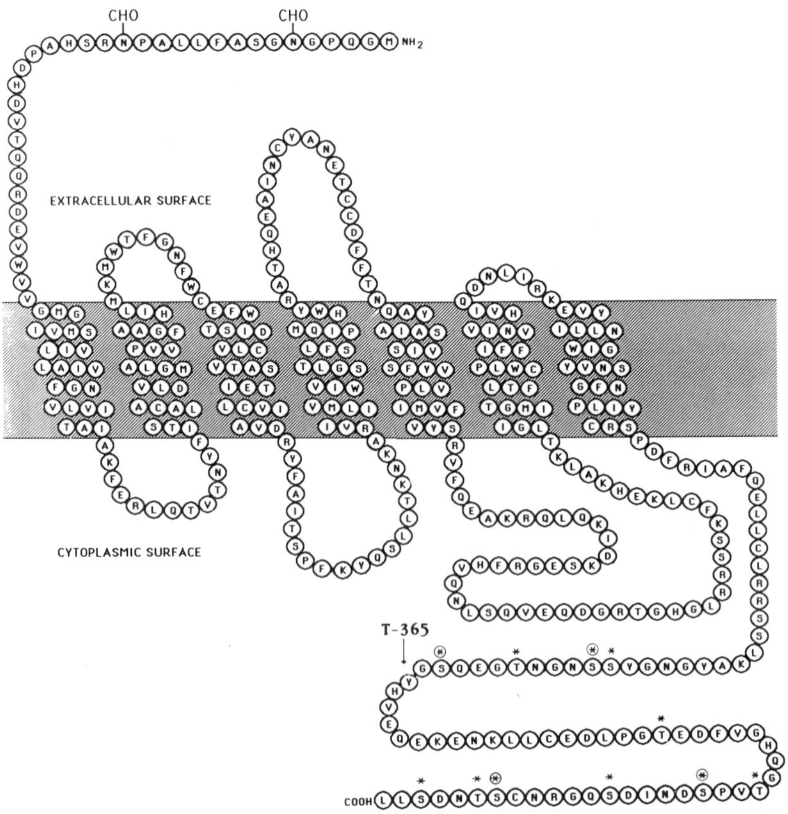

Figure 6. Schematic representation of the human β_2-adrenergic receptor sequence showing sites of two mutations that remove phosphorylation sites. Truncated mutant T365 (→) was obtained by introducing a stop codon in the β_2-adrenergic receptor sequence after residue G365 by site-directed mutagenesis. In S351, the various serine and threonine residues toward the carboxyl terminus of the protein were mutated to alanine (*) or glycine (⊛). Wild-type receptor and both mutant receptors were used to obtain stable transfected Chinese hamster fibroblast cell lines by methods described previously (Bouvier et al. 1988).

mutants were stably transfected into a mammalian fibroblast cell line and were shown to be capable of normal ligand binding and of normal activation of adenylyl cyclase (Bouvier et al. 1988). We demonstrated that when such cells were subjected to agonist-induced stimulation, the mutant receptors were phosphorylated to a much lesser extent than wild-type receptors.

We then examined the consequences of agonist stimulation of these different receptor constructs (Bouvier et al. 1988). Both showed normal or, in the case of the truncation mutant, enhanced agonist-induced receptor sequestration. However, the properties of the receptors with respect to desensitization were altered. Figure 7 shows the pattern of desensitization in the wild-type, truncation, and substitution mutants. The wild-type receptor shows a fairly typical and classic pattern of desensitization when the cells are exposed to the agonist isoproterenol (2 μM). Desensitization, which was apparent as a decrease in maximum isoproterenol-stimulated adenylyl cyclase, as well as an increase in half-maximal effective concentration, was apparent at the earliest time point studied, 2 minutes, and continued to evolve at 10, 60, and 180 minutes. In comparison, the evolution of desensitization was significantly slowed in the truncation mutant. There was effectively no desensitization at 2 minutes and 10 minutes. Desensitization at 60 minutes appeared to catch up with that in the wild type. The desensitization pattern in the substitution mutant was even more perturbed. Note that maximum catecholamine-stimulated adenylyl cyclase did not fall at all during the first 60 minutes of the incubation period. Interestingly, the twofold increase in EC_{50} did occur in the substitution mutant, just as it did in the wild-type receptor. Although the substitution mutant did show desensitization at 3 hours, this appears to be heterologous and equally involves basal and forskolin-stimulated adenylyl cyclase. These data represent direct evidence for a causal relationship between βARK-catalyzed phosphorylation on carboxy-terminal sites of the β-adrenergic receptor and the mechanism of homologous desensitization.

In summary, our data suggest an outline of the major domains of the adrenergic receptors that are involved in mediating the functions of ligand binding, G-protein coupling, and regulation of receptor function. As the structures of other members of the extended family of G-protein-coupled receptors become known, it should be possible to learn whether this preliminary map of functional domains can be generalized. Such information will ultimately lead to a clearer understanding of how hormonal signals are discriminated and transmitted across biological membranes and how the unique structural features of these receptor molecules have become specialized through evolution to accomplish these tasks.

ACKNOWLEDGMENTS

We thank Donna Addison for her expert secretarial assistance. This work was supported in part by National Institutes of Health grant HL-16037.

REFERENCES

Benovic, J.L., R.H. Strasser, M.G. Caron, and R.J. Lefkowitz. 1986. β-Adrenergic receptor kinase: Identification of a novel protein kinase which phosphorylates the agonist-occupied form of the receptor. *Proc. Natl. Acad. Sci.* **83:** 2797.

Benovic, J.L., F. Mayor, Jr., C. Staniszewski, R.J. Lefkowitz, and M.G. Caron. 1987a. Purification and characterization of the β-adrenergic receptor kinase. *J. Biol. Chem.* **262:** 9026.

Benovic, J.L., C. Staniszewski, F. Mayor, Jr., M.G. Caron, and R.J. Lefkowitz. 1988. β-Adrenergic receptor kinase: Activity of partial agonists for stimulation of adenylyl cyclase correlates with ability to promote receptor phosphorylation. *J. Biol. Chem.* **263:** 3893.

Benovic, J.L., H. Kuhn, I. Weyand, J. Codina, M.G. Caron, and R.J. Lefkowitz. 1987b. Functional desensitization of the isolated β-adrenergic receptor by the β-adrenergic receptor kinase: Potential role of an analog of the retinal protein arrestin (48K protein). *Proc. Natl. Acad. Sci.* **84:** 8879.

Benovic, J.L., J.W. Regan, H. Matsui, F. Mayor, Jr., S. Cotecchia, L.M.F. Leeb-Lundberg, M.G. Caron, and R.J. Lefkowitz. 1987c. Agonist-dependent phosphorylation of

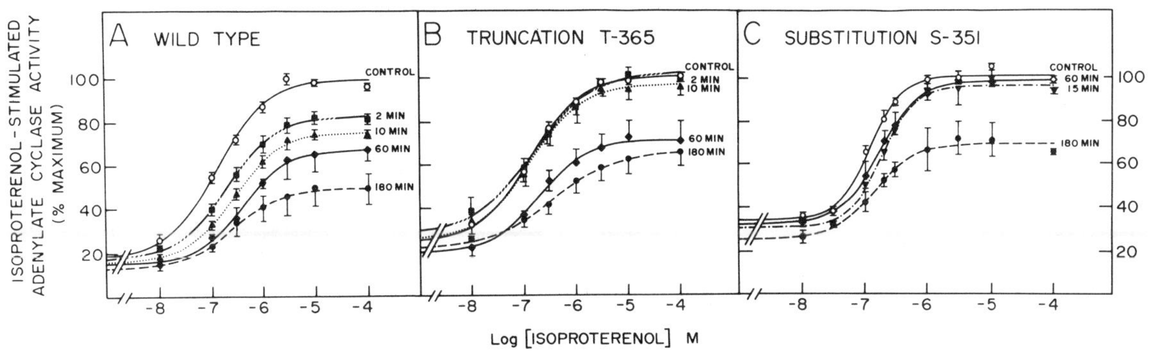

Figure 7. Comparison of the desensitization patterns of CHW cells expressing the wild-type β_2-human adrenergic receptor and two mutated forms of the receptor. β_2-Adrenergic receptor, T365, and S351-mutated receptor DNAs were transfected into CHW cells, and stable cell lines were established (Bouvier et al. 1988). Cells were exposed to 2 μM isoproterenol for the various times shown, after which membranes were prepared and the ability of various concentrations of isoproterenol to stimulate adenylyl cyclase was assessed in membranes (Bouvier et al. 1988).

the α_2-adrenergic receptor by the β-adrenergic receptor kinase. *J. Biol. Chem.* **262:** 17251.

Benovic, J.L., L.J. Pike, R.A. Cerione, C. Staniszewski, T. Yoshimasa, J. Codina, L. Birnbaumer, M.G. Caron, and R.J. Lefkowitz. 1985. Phosphorylation of the mammalian β-adrenergic receptor by cyclic AMP-dependent protein kinase: Regulation of the rate of receptor phosphorylation and dephosphorylation by agonist occupancy and effects on coupling of the receptor to the stimulatory guanine nucleotide regulatory protein. *J. Biol. Chem.* **260:** 7094.

Bouvier, M., W. Hausdorff, A. De Blasi, B.F. O'Dowd, B.K. Kobilka, M.G. Caron, and R.J. Lefkowitz. 1988. Removal of phosphorylation sites from the β_2-adrenergic receptor delays onset of agonist promoted densensitization. *Nature* **333:** 370.

Caron, M.G., B.K. Kobilka, T. Frielle, J.L. Benovic, J.W. Regan, S. Collins, C. McGregor, and R.J. Lefkowitz. 1988. Adrenergic receptors: Molecular and regulatory properties revealed by cloning of their genes. In *Recent advances in receptor chemistry* (ed. C. Melchiore and M. Giannella), p. 63. Elsevier, Amsterdam.

Cotecchia, S., D.A. Schwinn, R.R. Randall, R.J. Lefkowitz, M.G. Caron, and B.K. Kobilka. 1988. Molecular cloning and expression of the cDNA for the hamster α_1-adrenergic receptor. *Proc. Natl. Acad. Sci.* **85:** 7159.

Dixon, R.A.F., I.S. Sigal, M.R. Candelore, R.B. Register, W. Scattergood, E. Rands, and C.D. Strader. 1987a. Structural features required for ligand binding to the β-adrenergic receptor. *EMBO J.* **6:** 3269.

Dixon, R.A.F., I.S. Sigal, E. Rands, R.B. Register, M.R. Candelore, A.D. Blake, and C.D. Strader. 1987b. Ligand binding to the β-adrenergic receptor involves its rhodopsin-like core. *Nature* **326:** 73.

Dohlman, H.G., M.G. Caron, and R.J. Lefkowitz. 1987a. A family of receptors coupled to guanine nucleotide regulatory proteins. *Biochemistry* **26:** 2657.

Dohlman, H.G., M. Bouvier, J.L. Benovic, M.G. Caron, and R.J. Lefkowitz. 1987b. The multiple membrane spanning topography of the β_2-adrenergic receptor: Localization of the sites of binding, glycosylation and regulatory phosphorylation by limited proteolysis. *J. Biol. Chem.* **262:** 14282.

Frielle, T., S. Collins, K.W. Daniel, M.G. Caron, R.J. Lefkowitz, and B.K. Kobilka. 1987. Cloning of the cDNA for the human β_1-adrenergic receptor. *Proc. Natl. Acad. Sci.* **84:** 7920.

Ison, L.L., E.J. Cragoe, Jr., and L.E. Limbird. 1987. Multiple receptors linked to inhibition of adenylyl cyclase accelerate Na^+/H^+ exchange in neuroblastoma X glioma cells via a mechanism other than decreased cAMP accumulation. *J. Biol. Chem.* **262:** 17504.

Kobilka, B.K., T.S. Kobilka, J.W. Regan, M.G. Caron, and R.J. Lefkowitz, 1988. Chimeric α_2-β_2-adrenergic receptors: Delineation of domains involved in effector coupling and ligand binding specificity. *Science* **240:** 1310.

Kobilka, B.K., C. MacGregor, K. Daniel, T.S. Kobilka, M.G. Caron, and R.J. Lefkowitz. 1987a. Functional activity and regulation of human β_2-adrenergic receptors expressed in *Xenopus* oocytes. *J. Biol. Chem.* **262:** 15796.

Kobilka, B.K., T. Frielle, S. Collins, T. Yang-Feng, T.S. Kobilka, U. Francke, R.J. Lefkowitz, and M.G. Caron. 1987b. An intronless gene encoding a potential member of the family of receptors coupled to guanine nucleotide regulatory proteins. *Nature* **219:** 75.

Kobilka, B.K., R.A.F. Dixon, T. Frielle, H.G. Dohlman, M.A. Bolanowski, I.S. Sigal, T.L. Yang-Feng, U. Francke, M.G. Caron, and R.J. Lefkowitz. 1987c. cDNA for the human β_2-adrenergic receptor: A protein with multiple membrane spanning domains and a chromosomal location shared with the PDGF receptor gene. *Proc. Natl. Acad. Sci.* **84:** 46.

Lefkowitz, R.J. and M.G. Caron. 1988. Adrenergic receptors: Models for the study of receptors coupled to guanine nucleotide regulatory proteins. *J. Biol. Chem.* **263:** 4993.

Regan, J.W., T.S. Kobilka, T.L. Yang-Feng, M.G. Caron, R.J. Lefkowitz, and B.K. Kobilka. 1988. Cloning and expression of a human kidney cDNA for a novel α_2-adrenergic receptor subtype. *Proc. Natl. Acad. Sci.* **85:** 6301.

Rubenstein, R.C., S.K.-F. Wong, and E.M. Ross. 1987. The hydrophobic tryptic core of the β-adrenergic receptor retains G_s regulatory activity in response to agonists and thiols. *J. Biol. Chem.* **262:** 16655.

Shichi, H., K. Yamamoto, and R.L. Somers. 1984. GTP binding protein: Properties and the lack of activation by phosphorylated rhodopsin. *Vision Res.* **24:** 1523.

Sibley, D.R. and R.J. Lefkowitz. 1985. Molecular mechanisms of receptor desensitization using the β-adrenergic receptor-coupled adenylyl cyclase system as a model. *Nature* **317:** 124.

Sibley, D.R., J.L. Benovic, M.G. Caron, and R.J. Lefkowitz. 1987. Regulation of transmembrane signaling by receptor phosphorylation. *Cell* **48:** 913.

Stryer, L. 1986. Cyclic GMP cascade of vision. *Annu. Rev. Neurosci.* **9:** 87.

Signal Transduction by Epidermal Growth Factor Receptor

J. Schlessinger,*† A. Ullrich,‡ A.M. Honegger,† and W.H. Moolenaar§
†*Rorer Biotechnology, Inc., King of Prussia, Pennsylvania 19406;* ‡*Genentech, Inc., South San Francisco, California 94080;* §*The Netherlands Cancer Institute, 1066 CX Amsterdam, The Netherlands*

The proliferation of cells in vivo and in culture is tightly regulated by polypeptide growth factors. Like all polypeptide hormones, growth factors initiate their action by binding to specific, high-affinity receptor molecules on the cell surface. Interest in growth factors and their receptors has been dramatically intensified by the discovery that at least some of these molecules were identified as viral oncogene products. For example, the v-*sis* oncogene encodes the platelet-derived growth factor (PDGF)-B molecule, whereas the v-*erbB* oncogene product is a truncated form of the avian receptor for epidermal growth factor (EGF) (for review, see Schlessinger 1988; Yarden and Ullrich 1988). Inappropriate expression of the cellular counterparts of viral oncogenes is thought to play a role in the initiation and maintenance of malignant growth. Therefore, the study of the mode of action of growth factors will undoubtedly yield important insights into mechanisms underlying oncogenesis.

Mechanism of Receptor Activation

All polypeptide growth factors act by binding to specific cell-surface receptors that on activation transduce a broad range of signals, leading to cell growth and proliferation. One group of growth factor receptors contain intrinsic protein tyrosine kinase (TK) activity (for review, see Yarden and Ullrich 1988). The known growth factor receptors with protein TK activity belong to at least three distinct classes that have probably evolved from common ancestral receptor genes. All growth factor receptors with protein TK activity are composed of three major structural elements. An extracellular, glycosylated ligand-binding domain, a single transmembrane region, and a cytoplasmic protein TK domain (Schlessinger 1986, 1988; Yarden and Ullrich 1988). Moreover, for all growth factor receptors, ligand binding to the extracellular domain activates the cytoplasmic kinase domain, leading to autophosphorylation and to phosphorylation of various cellular substrates. Hence, growth-factor-induced activation of the intrinsic kinase activity of the various receptor molecules is probably mediated by a common mechanism. In principle, ligand control of receptor kinase activity could involve either an intramolecular or an intermolecular mechanism (Schlessinger 1986, 1988). An intramolecular mechanism for receptor activation would require the transduction of an allosteric signal from the extracellular ligand-binding domain, through the transmembrane region, to enhance the enzymatic activity of the cytoplasmic kinase domain (Bertics and Gill 1985; Staros et al. 1985). In an intermolecular model, receptor-receptor interactions stabilized by ligand binding would bring about the activation of the TK by allosteric subunit interaction (Schlessinger 1986, 1988; Yarden and Schlessinger 1987a,b).

An allosteric oligomerization model was proposed to explain how ligand binding to the extracellular domain can activate the TK domain which is on the cytoplasmic side of the plasma membrane (Schlessinger et al. 1983; Schlessinger 1986; Yarden and Schlessinger 1987a,b). According to this model, monomeric EGF receptors are in equilibrium with oligomeric receptors. It is postulated that the oligomeric receptors have higher ligand-binding affinity than receptor monomers, and therefore the binding of EGF will stabilize the oligomeric state (Fig. 1). This leads to the activation of the catalytic properties of the kinase domain by subunit interaction between neighboring cytoplasmic domains (Fig. 1). Such a mechanism bypasses the requirement for an

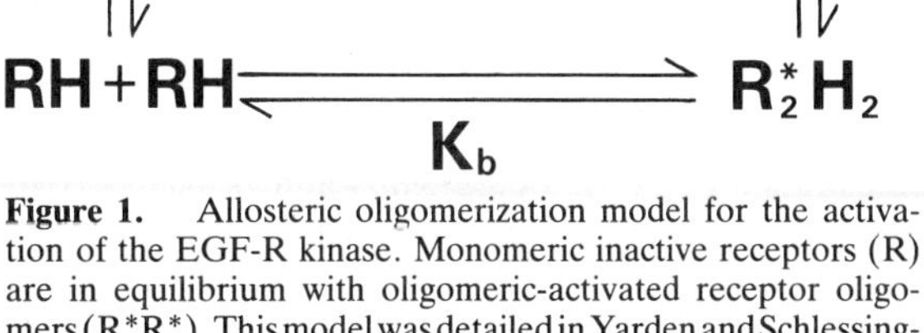

Figure 1. Allosteric oligomerization model for the activation of the EGF-R kinase. Monomeric inactive receptors (R) are in equilibrium with oligomeric-activated receptor oligomers (R*R*). This model was detailed in Yarden and Schlessinger (1987a). According to this scheme: $K_b/K_f = K_d^2/K_m^2$. Thus, higher binding affinities of the ligands to the oligomeric state $K_d > K_m$ implies that oligomerization is enhanced by the binding of the ligands ($K_b \gg K_f$). Assuming receptors are active only in the oligomeric state, ligand will activate the receptor because of its higher affinity toward the oligomeric state.

*On leave from the Weizmann Institute of Science, Rehovot, Israel.

energetically unfavorable conformational change to be transmitted through the transmembrane region connecting the two functional portions: an essential feature of every model, which assumes an intramolecular mechanism for the activation of the TK by the binding of EGF to the extracellular domain (Weber et al 1984; Bertics and Gill 1985). It is well established that EGF stimulates receptor oligomerization on the cell surface of living cells and in plasma membrane preparations. Receptor oligomerization induced by EGF has been demonstrated by various independent methods, including morphological (Haigler et al. 1979), biophysical (Zidovetzki et al. 1981, 1986; Hillman and Schlessinger 1982), and biochemical (Fanger et al. 1986; Boni-Schnetzler and Pilch 1987; Yarden and Schlessinger 1987a,b; Cochet et al. 1988) approaches. The capacity of EGF receptors to undergo oligomerization is an intrinsic property of the EGF receptor, since pure EGF receptor (EGF-R) undergoes rapid ligand-induced, temperature-sensitive, reversible receptor dimerization (Yarden and Schlessinger 1987b).

Mechanism and role of receptor autophosphorylation. Growth factor receptors with protein TK activity catalyze the phosphorylation both of exogenous substrates and of multiple intrinsic sites (Hunter and Cooper 1985). The autophosphorylation of the insulin receptor increases the V_{max} of the kinase activity, maintaining the receptor kinase in an activated state even in the absence of ligand binding. The known tyrosine autophosphorylation sites of the EGF-R are clustered in the carboxy-terminal end of the receptor in positions 1068, 1148, and 1173, respectively. It was reported that autophosphorylation of EGF-R increased (Bertics and Gill 1985) or had no effect (Downward et al. 1985) on the kinase activity of the EGF-R toward exogenous substrates. Hence, we have analyzed the properties of EGF-R mutants in which individual autophosphorylation sites were replaced by phenylalanine residues. These mutant receptors were expressed in NIH-3T3 cells lacking endogenous EGF-R, and both their enzymatic activities and biological properties were compared (Honegger et al. 1988a,b). It appears that alterations of individual autophosphorylation sites did not have a major effect on kinase activity. The V_{max} of the reaction was not affected, and the small increase in substrate affinity following autophosphorylation does not exceed the effect expected by a removal of a competitive substrate (Honegger et al. 1988a). The mutant receptors were as active as wild-type receptors in mediating various responses of EGF-R. Interestingly, the mutant receptors were slightly more sensitive (two- to fivefold) to EGF-induced mitogenesis than wild-type receptors expressed in the same cellular background (Honegger et al. 1988b).

Like the process of receptor activation, receptor autophosphorylation could proceed by two distinct mechanisms: by an intramolecular mechanism or by an intermolecular mechanism. In an intramolecular process, the catalytic domain of the EGF-R molecule phosphorylates the phosphorylation sites of the same molecule. Alternatively, the catalytic domain of one receptor may phosphorylate the phosphorylation sites of neighboring receptor molecules. We have studied the mechanism of EGF-R autophosphorylation by utilizing mixed populations of structurally distinguishable EGF-R mutants (A.M. Honegger et al., in prep.). The kinase-negative mutant was shown to be phosphorylated, by an enzymatically active EGF-R deletion mutant in an EGF-R-dependent manner, on sites virtually identical to the sites autophosphorylated by either wild-type or native human EGF-R (A.M. Honegger et al., in prep.). It therefore appears that autophosphorylation of solubilized EGF-R is mediated by intermolecular cross-phosphorylation (Fig. 2), which is probably facilitated by receptor oligomerization. According to this view, receptor oligomerization provides the mechanism for receptor activation and also sets the stage for receptor autophosphorylation. Figure 2 summarizes the current emerging concepts concerning the molecular steps involved in ligand-induced activation of the EGF-R kinase. There are still several open questions to be answered in the schematic diagram presented in Figure 2. Yet, this molecular scenario is based on numerous pieces of experimental evidence, and future studies will clarify the details and the generality of the concepts presented in this scheme.

Signal Pathways in Growth Factor Action

Understanding the mode of action of growth factors requires the identification of intracellular signals that are essential for the stimulation of DNA synthesis and cell division. Following growth factor binding, the activated receptor mediates a cascade of rapid biochemical and physiological changes in the cell, which ultimately (after 10–20 hr) lead to enhanced DNA synthesis. As a rule, the growth factor has to be present throughout the entire "prereplicative" G_0/G_1 phase (usually 8–10 hr) for commitment of DNA synthesis to occur. It appears that there are at least three potential signal pathways for mediating growth factor action: (1) activation of a protein TK activity that is intrinsic to the receptors for EGF, PDGF, or other growth factor receptors; (2) the phospholipase-C-mediated breakdown of inositol phospholipids generating various signal molecules; and (3) increases in cytoplasmic free Ca^{++} and pH (pH_i).

There is increasing evidence that in addition to the "classical" mitogens such as EGF and PDGF, certain Ca^{++}-mobilizing neurotransmitters can also function as growth factors. For example, the peptides bombesin, substance P, and vasopressin are mitogenic for quiescent fibroblasts, whereas bradykinin and histamine, acting through H-1 receptors, stimulate the growth of certain human carcinoma cells (Tilly et al. 1987 and unpubl.). Mitogen-induced alterations in the levels of cytoplasmic free Ca^{++}($[Ca^{++}]_i$) and H^+ are of special interest because these ions are thought to serve as second messengers in activated cells. During the past few years, our understanding of the ionic changes in stimulated cells has increased dramatically. The molecular mechanisms underlying ionic signaling have

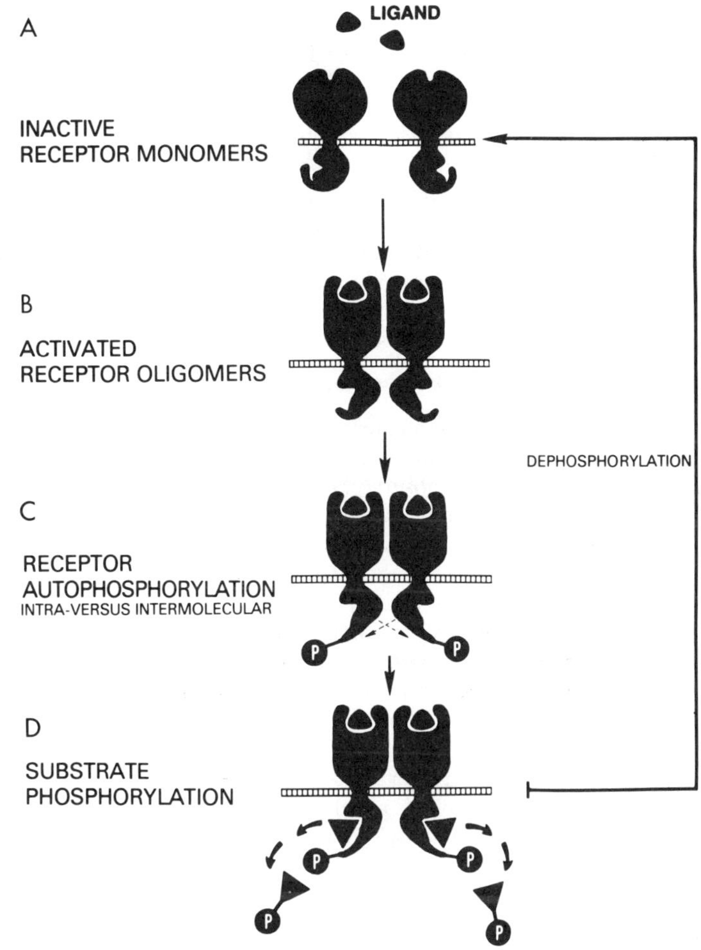

Figure 2. Molecular steps in the activation of the EGF-R kinase. (*A*) The binding of EGF stabilizes an oligomeric state with enhanced protein TK activity. (*B*) The activated receptor oligomers undergo autophosphorylation. Autophosphorylation is mediated by an intermolecular cross-phosphorylation mechanism (→). (*C*) The intrinsic autophosphorylation sites and the exogenous substrates compete for the substrate-binding region of the protein TK. Autophosphorylation renders the EGF-R kinase more accessible toward exogenous substrates. (*D*) A tyrosine-specific phosphatase removes the phosphate groups from the EGF-R.

been largely unraveled, and much effort is being devoted to determining how rapid increases in $[Ca^{++}]_i$ and pH_i may contribute to a proliferative response.

Activation of Na^+/H^+ Exchange

Most, if not all, mitogens activate an otherwise quiescent Na^+/H^+ exchanger in the plasma membrane of their target cells (for review, see Moolenaar 1986). This Na^+/H^+ exchanger, which is sensitive to the diuretic amiloride and its analogs, is normally involved in the close regulation of steady-state pH_i by virtue of its high sensitivity to cytoplasmic H^+. Most cells maintain their pH_i at 7.0–7.2, which is well above the electrochemical equilibrium value of 6.4 predicted by the Nernst equation from a transmembrane potential of −60 mV (at a pH of 7.4).

The functioning of the plasma membrane Na^+/H^+ exchanger and its normal housekeeping role in pH_i regulation are usually assessed by acutely acidifying the cytoplasm and monitoring the consequent recovery of pH_i back to its steady-state level. Although basal Na^+/H^+ exchange activity is normally very low and exactly balances the acidifying effects of H^+ influx and metabolic acid production, Na^+/H^+ exchange is dramatically accelerated as soon as the cytoplasm is acidified. Cytoplasmic H^+ is then rapidly extruded from the cell, causing pH_i to return to its basal value within a few minutes. Although Na^+/H^+ exchange is driven, in thermodynamic terms, by the steep transmembrane Na^+ gradient, the major determinant of the Na^+/H^+ exchange rate is cytoplasmic (H^+). It is generally assumed that cytoplasmic H^+ acts as an allosteric activator of the Na^+/H^+ exchanger by binding to a regulatory site. The protonation of this site sets the exchanger in motion, while a functionally separate H^+ transport site mediates H^+ efflux once the exchanger is activated. It appears that growth factors and activators of protein kinase C (phorbol esters) exploit this control mechanism to shift steady-state pH_i in alkaline direction. Growth factors act by increasing the sensitivity of the Na^+/H^+ exchanger for cytoplasmic H^+, thereby raising steady-state pH_i by 0.1–0.3 pH unit. This mitogen-induced shift in pH_i sensitivity presumably reflects an increase in the apparent affinity of the regulatory site for H^+. It is easy to imagine that some ionizable group at the regulatory site acquires a greater pK_a, because its immediate environment becomes more negatively charged, for example, by receptor-mediated phosphorylation, either via protein kinase C or through the intrinsic receptor TK.

A growth-factor-induced shift in pH_i of 0.2–0.3 units

would be expected to affect the role of numerous pH-sensitive processes in the cell. Perhaps the most convincing demonstration of a signaling role of pH_i in growth control has been made with mutant fibroblasts that lack a functional Na^+/H^+ exchanger (Pouysségur et al. 1984). It was shown that below a certain threshold value (around 7.0–7.2), pH_i becomes limiting for cell proliferation. Furthermore, one of the critical pH_i dependent steps in activated fibroblasts appears to be the initiation of protein synthesis (Chambard and Pouysségur 1986).

Ca^{++} Signaling and Phosphoinositide Hydrolysis

There is now general agreement that many, but not all, growth factors evoke the phospholipase-C-mediated breakdown of inositol phospholipids, resulting in the formation of diacylglycerol that activates protein kinase C and inositol 1,4,5-trisphosphate ($Ins[1,4,5]P_3$), which triggers the release of Ca^{++} from internal stores, as well as various other inositol polyphosphates. We have analyzed and compared the actions of EGF, on the one hand, and of Ca^{++}-mobilizing hormones (bradykinin, histamine), on the other hand, on both inositol phosphate formation and calcium signaling in responsive human carcinoma cells (Tilly et al. 1987, 1988). EGF is a rather weak inducer of phospholipase C activity, evoking a rather small and variable release of intracellularly stored Ca^{++}, whereas accumulation of $Ins(1,4,5)P_3$ is barely detectable. In contrast, bradykinin and histamine are among the most potent activators of phospholipase C, resulting in large Ca^{++} signals and substantial increases in inositol phosphate levels. Almost immediately after addition of these hormones to A431 or HeLa cells, the levels of $Ins(1,4,5)P_3$ start to increase, reaching peak values of ~10 times the basal level within 10–15 seconds, which coincides with the Ca^{++} peak. The formation of $Ins(1,4,5)P_3$ is a very transient phenomenon, and our results support the view that the (1,4,5) isomer is rapidly metabolized either to $InsP_2$ through phosphatase action or to $Ins(1,3,4)P_3$ via phosphorylation/dephosphorylation with $InsP_4$ as intermediate. The latter inositol polyphosphate may be involved in regulating Ca^{++} influx, but its precise role has not been fully clarified to date (Irvine and Moor 1986). Relatively large amounts of $InsP_5$ and $InsP_6$ are found in A431, HeLa, and neuroblastoma cells (Tilly et al. 1987), but $InsP_5/InsP_6$ levels remain virtually unaltered after receptor stimulation, suggesting that these inositol phosphates are not directly involved in cellular signaling.

The accumulation of inositol phosphate in response to bradykinin levels off in less than 2 minutes. This apparent desensitization is not observed with EGF as a stimulus; generally, the bradykinin-dependent pattern of stimulation is significantly different from that observed in EGF-stimulated A431 cells, suggesting that separate mechanisms of inositol lipid hydrolysis are involved.

Regardless of the distinct mechanisms by which growth factors raise $[Ca^{++}]_i$, it is widely accepted that Ca^{++} has an important role as a second messenger that regulates or modulates numerous cellular activities. The transient rise in $[Ca^{++}]_i$ is likely to trigger a sequence of early cellular changes occurring within minutes of receptor stimulation. Of special interest is the finding that artificial elevation of $[Ca^{++}]_i$ by means of ionophores mimics growth factors in rapidly inducing the expression of the c-*fos* and c-*myc* proto-oncogenes (Bravo et al. 1985), whereas suppression of the growth-factor-induced Ca^{++} signal can inhibit c-*fos* expression, at least in the action of bradykinin and histamine (W. Kruijer et al., in prep.).

Mutational Analysis of EGF-R-mediated Ionic Signaling

To explore the importance of the receptor TK in the action of EGF, we have used several different EGF-R mutants expressed in NIH-3T3 cells that contain undetectable amounts of endogenous EGF-R. The wild-type human EGF-R expressed in these cells behaves like the native EGF-R and is able to stimulate all the known responses to EGF, including inositol phosphate formation and ionic signaling (Moolenaar et al. 1988; Schlessinger 1988).

Of special interest is a point mutant receptor in which Lys-721, a key residue in the presumed ATP-binding region, is replaced by an alanine residue. This K721A mutant receptor, expressed in 3T3 cells, lacks protein TK activity but is normally processed and expressed on the cell surface. However, the kinase-deficient receptor is unable to stimulate both early responses such as activation of Na^+/H^+ exchange, Ca^{++} signaling, inositol phosphate formation, and late responses such as DNA synthesis and focus formation (Livneh et al. 1986, 1987; Chen et al. 1987; Honegger et al. 1987a,b). Hence, TK activity is essential, although not necessarily sufficient, for EGF-R signal transduction. In other words, EGF-induced phosphorylation of specific cellular substrates on tyrosine residues is a critical step in the pleiotropic response to EGF. The inability of the kinase-defective receptor to stimulate phosphoinositide hydrolysis is particularly intriguing, since it suggests that regulatory proteins in this pathway, such as G proteins or phospholipase C, may serve as substrates and thus be regulated by protein TK. Such a mechanism contrasts the accepted mode of action of "classical" Ca^{++}-mobilizing hormones, such as bradykinin and histamine, whose receptors are linked to G-protein intermediates without any evidence for tyrosine phosphorylation being required for the activation process.

In addition to the K721A EGF receptor mutant, we have also tested a truncated receptor that lacks 63 amino acid residues from the carboxy-terminal end, including two major autophosphorylation sites (Schlessinger 1986). This mutant receptor appeared to be indistinguishable from the wild-type receptor in terms of EGF-dependent ionic signaling, inositol phosphate formation, and DNA synthesis (W.H. Moolenaar et al., unpubl.; Schlessinger 1986). These results imply that receptor autophosphorylation is not essential for a normal signal-transducing response to EGF. It is obvious that the use of site-directed mutagenesis combined with transfections into cultured cells provides a promising, novel approach to the analysis of the various domains of the EGF-R and their role in signal transduction.

REFERENCES

Bertics, P.J. and G.N. Gill. 1985. Self-phosphorylation enhances the protein-tyrosine kinase activity of the epidermal growth factor receptor. *J. Biol. Chem.* **260:** 14642.

Boni-Schnetzler, M. and P.F. Pilch. 1987. Mechanism of epidermal growth factor receptor autophosphorylation and high affinity binding. *Proc. Natl. Acad. Sci.* **84:** 7832.

Bravo, R., J. Burckhardt, T. Curran, and R. Muller. 1985. Stimulation and inhibition of growth by EGF in different A 431 cell clones is accompanied by the rapid induction of c-*fos* and c-*myc* proto-oncogenes. *EMBO J.* **4:** 1193.

Chambard, J.C. and J. Pouysségur. 1986. Intracellular pH controls growth factor-induced ribosomal protein S6 phosphorylation and protein synthesis in the GO→G1 transition of fibroblasts. *Exp. Cell Res.* **164:** 282.

Chen, W.S., C.S. Lazar, M. Poenie, R.Y. Tsien, G.N. Gill, and M.G. Rosenfeld. 1987. Requirement for intrinsic protein-tyrosine kinase in the immediate and late actions of the EGF receptor. *Nature* **328:** 820.

Cochet, C., O. Kashles, E. Chambaz, I. Borrello, C.R. King, and J. Schlessinger. 1988. Demonstration of epidermal growth factor-induced receptor dimerization in living cells using a chemical covalent cross-linking agent. *J. Biol. Chem.* **263:** 3290.

Downward, J., M.D. Waterfield, and P.J. Parker. 1985. Autophosphorylation and protein kinase C phosphorylation of the epidermal growth factor receptor. *J. Biol. Chem.* **260:** 14538.

Fanger, B.O., K.S. Austin, S.H. Earp, and J.A. Cidlowski. 1986. Cross linking of epidermal growth factor receptors: Detection of initial stages of receptor clustering and determination of molecular weight of high-affinity receptors. *Biochemistry* **25:** 6414.

Haigler, H.T., J.A. McKanna, and S. Cohen. 1979. Direct visualization of the binding and internalization of a ferritin conjugate of epidermal growth factor in human carcinoma cells A-431. *J. Cell Biol.* **81:** 382.

Hillman, G.M. and J. Schlessinger. 1982. Lateral diffusion of epidermal growth factor complexed to its surface receptors does not account for the thermal sensitivity of patch formation and endocytosis. *Biochemistry* **21:** 1667.

Honegger, A.M., T.J. Dull, D. Szapary, A. Komoriya, R. Kris, A. Ullrich, and J. Schlessinger. 1988a. Kinetic parameter of the protein tyrosine kinase of EGF-receptor mutants with individually altered autophosphorylation sites. *EMBO J.* (in press).

Honegger, A.M., T.J. Dull, F. Bellot, E. Van Obberghen, D. Szapary, A. Schmidt, A. Ullrich, and J. Schlessinger. 1988b. Biological activities of EGF-receptor mutants with individually altered autophosphorylation sites. *EMBO J.* (in press).

Honegger, A.M., D. Szapary, A. Schmidt, R. Lyall, E. Van Obberghen, T.J. Dull, A. Ullrich, and J. Schlessinger. 1987a. A mutant epidermal growth factor receptor with defective protein tyrosine kinase is unable to stimulate proto-oncogene expression and DNA synthesis. *Mol. Cell. Biol.* **7:** 4567.

Honegger, A.M., T.J. Dull, S. Felder, E. Van Obberghen, F. Bellot, D. Szapary, A. Schmidt, A. Ullrich, and J. Schlessinger. 1987b. Point mutation at the ATP binding site of EGF receptor abolishes protein-tyrosine kinase activity and alters cellular routing. *Cell* **51:** 199.

Hunter, T. and J.A. Cooper. 1985. Protein-tyrosine kinases. *Annu. Rev. Biochem.* **54:** 894.

Irvine, R.F. and R.M. Moor. 1986. Microinjection of inositol 1,3,4,5 tetrakis inositol phosphate activates sea urchin eggs by a mechanism dependent on external Ca^{2+}. *Biochem. J.* **240:** 917.

Livneh, E., N. Reiss, E. Berent, A. Ullrich, and J. Schlessinger. 1987. An insertional mutant of epidermal growth factor receptor allows dissection of diverse receptor functions. *EMBO J.* **6:** 2669.

Livneh, E., R. Prywes, O. Kashles, N. Reiss, I. Sasson, A. Ullrich, and J. Schlessinger. 1986. Reconstitution of human epidermal growth factor receptors and its deletion mutants in cultured hamster cells. *J. Biol. Chem.* **261:** 12490.

Moolenaar, W.H. 1986. Effects of growth factors on intracellular pH regulation. *Annu. Rev. Physiol.* **48:** 363.

Moolenaar, W.H., A.J. Bierman, B.C. Tilly, I. Verlaan, A.M. Honegger, A. Ullrich, and J. Schlessinger. 1988. A point mutation at the ATP-binding site of the EGF-receptor abolishes signal transduction. *EMBO J.* **7:** 707.

Pouysségur, F., C. Sardet, A. Franchi, G. l'Allemain, and C. Paris. 1984. A specific mutation abolishing Na^+/H^+ antiport activity in hamster fibroblasts precludes growth at neutral and acidic pH. *Proc. Natl. Acad. Sci.* **81:** 4833.

Schlessinger, J. 1986. Allosteric regulation of the epidermal growth factor receptor kinase. *J. Cell Biol.* **103:** 2067.

———. 1988. The epidermal growth factor receptor as a multifunctional allosteric protein. *Biochemistry* **27:** 3119.

Schlessinger, J., A.B. Schreiber, A. Levi, I. Lax, T. Libermann, and Y. Yarden. 1983. Regulation of cell proliferation by epidermal growth factor. *CRC Crit. Rev. Biochem.* **14:** 93.

Tilly, B.C., P.A. van Paridon, I. Verlaan, S.W. de Laat, and W.H. Moolenar. 1988. Epidermal growth factor-induced formation of inositol phosphate in human A431 cells. Differences from the effect of bradykinin. *Biochem. J.* **252:** 857.

Tilly, B.C., P.A. van Paridon, I. Verlaan, K. Wirtz, S.W. de Laat, and W.H. Moolenaar. 1987. Inositol phosphate metabolism in bradykinin-stimulated human A431 carcinoma cells: Relationship to calcium signalling. *Biochem. J.* **244:** 129.

Weber, W., P.J. Bertics, and G.N. Gill. 1984. Immunoaffinity purification of the epidermal growth factor receptor, stoichiometry of binding and kinetics of self-phosphorylation. *J. Biol. Chem.* **259:** 14631.

Yarden, Y. and J. Schlessinger. 1987a. Self-phosphorylation of epidermal growth factor receptor: Evidence for a model of intermolecular allosteric activation. *Biochemistry* **26:** 1434.

———. 1987b. Epidermal growth factor induces rapid, reversible aggregation of the purified epidermal growth factor receptor. *Biochemistry* **26:** 1443.

Yarden, Y. and A. Ullrich. 1988. Molecular analysis of signal transduction by growth factors. *Biochemistry* **27:** 3113.

Zidovetzki, R., Y. Yarden, J. Schlessinger, and T.M. Jovin. 1981. Rotational diffusion of epidermal growth factor complexed to cell surface receptors reflects rapid microaggregation and endocytosis of occupied receptors. *Proc. Natl. Acad. Sci.* **78:** 6981.

———. 1986. Microaggregation of hormone-occupied epidermal growth factor receptors on plasma-membrane preparations. *EMBO J.* **5:** 247.

Macrophage Colony-stimulating Factor, CSF-1, and Its Proto-oncogene-encoded Receptor

C. J. SHERR, C.W. RETTENMIER, AND M.F. ROUSSEL
Department of Tumor Cell Biology, St. Jude Children's Research Hospital, and the Howard Hughes Medical Institute, Memphis, Tennessee 38101

The macrophage colony-stimulating factor, CSF-1, or M-CSF, is one of a family of hematopoietic growth factors that stimulates the proliferation of monocytes, macrophages, and their committed bone marrow progenitors (Stanley and Heard 1977; Stanley et al. 1983; Metcalf 1985). Unlike pluripotent hemopoietins such as granulocyte-macrophage colony-stimulating factor (GM-CSF) and interleukin-3 (IL-3 or multi-CSF), which affect the growth of myeloid cells of several different hematopoietic lineages (for review, see Metcalf 1985; Clark and Kamen 1987), CSF-1 acts only on cells of the mononuclear phagocyte series to stimulate their growth and enhance their survival (Stanley et al. 1983). CSF-1 is produced by mesenchymal cells, including fibroblasts and endothelium, and exerts its functions through its binding to a single class of high-affinity cell-surface receptors expressed primarily on mature mononuclear phagocytes and their immature bone marrow precursors (Guilbert and Stanley 1980, 1986; Byrne et al. 1981; Morgan and Stanley 1984). The CSF-1 receptor is the product of the c-*fms* proto-oncogene (Sherr et al. 1985) and is one of a family of growth factor receptors, including those for epidermal growth factor (Ushiro and Cohen 1980), platelet-derived growth factor (PDGF) (Ek et al. 1982; Nishimura et al. 1982), insulin (Kasuga et al. 1982), and insulin-like growth factor 1 (Jacobs et al. 1983), each of which exhibits an intrinsic kinase activity with specificity for protein tyrosine residues. Binding of CSF-1 to its receptor activates the receptor kinase and leads to autophosphorylation of the receptor on tyrosine (Sherr et al. 1985; Sacca et al. 1986; Yeung et al. 1987), internalization and degradation of receptor-ligand complexes (Guilbert and Stanley 1986; Wheeler et al. 1986; Downing et al. 1988), and phosphorylation of a series of cellular proteins (Downing et al. 1988), at least some of which are likely to be important in transducing growth-promoting signals from the plasma membrane to the cell nucleus (Müller et al. 1985; Bravo et al. 1987; Orlofsky and Stanley 1987).

Retroviral transduction of the feline c-*fms* gene in the Susan McDonough (SM) (McDonough et al. 1971) and Hardy Zuckerman-5 (HZ-5) (Besmer et al. 1986a) strains of feline sarcoma virus (FeSV) led to genetic alterations that endowed the recombined viral oncogene (v-*fms*) with the ability to transform cells in culture morphologically and to induce fibrosarcomas and hematopoietic neoplasms in susceptible animals. The v-*fms* oncogene product differs from the normal CSF-1 receptor in certain of its cardinal biochemical properties, most notably in exhibiting constitutively high basal levels of tyrosine kinase activity in the absence of its ligand (Sacca et al. 1986; Morrison et al. 1988). Comparative studies of the c-*fms* and v-*fms* genes coupled with analyses of engineered mutants and receptor chimeras have begun to pinpoint pertinent genetic alterations in the normal receptor gene that unmask its latent oncogenic potential. In addition, the availability of biologically active c-*fms*, v-*fms*, and CSF-1 cDNAs has allowed these genes to be mobilized and expressed in naive cells, thereby facilitating assays for receptor coupling with "downstream" components of the mitogenic pathway in diverse cell types.

RESULTS AND DISCUSSION

Biological Properties of CSF-1

CSF-1 is a glycosylated polypeptide growth factor produced by many mesenchymal cells (Stanley and Heard 1977; Das and Stanley 1982), but it is also transiently synthesized by monocytes and macrophages themselves after their induction with phorbol esters, γ-interferon, GM-CSF, or tumor necrosis factor-α (Horiguchi et al. 1986, 1987; Rambaldi et al. 1987a; Oster et al. 1987). Removal of carbohydrate from the disulfide-assembled homodimeric glycoprotein does not affect its biological activity, whereas reduction of disulfide bonds abrogates its potential to act as a mitogen (Das et al. 1981; Das and Stanley 1982). Unlike other hematopoietic colony-stimulating factors, CSF-1 normally circulates in the bloodstream, and its concentration is regulated through receptor-mediated uptake by circulating monocytes and tissue macrophages (Bartocci et al. 1987). In response to the growth factor, committed bone marrow progenitors plated in semisolid medium proliferate and differentiate, eventually giving rise to colonies containing only mature macrophages. In the presence of other hemopoietins, however, such as hemopoietin-1 (interleukin-1α), IL-3, and IL-6, CSF-1 can act on more immature and putatively multipotential cells, inducing their proliferation and differentiation to mononuclear phagocytes (Bartelmez and Stanley 1985; Stanley et al. 1986; Ikebuchi et al. 1987; Mochizuki et al. 1987; Warren and Moore 1987). Human CSF-1 binds to its specific receptors on human and murine mononuclear phagocytes with nearly equal affinity, whereas the murine growth factor is devoid of activity on human monocytes and macrophages. Paradoxically, the human hormone stimulates human bone marrow progenitors less than it stimulates mouse bone marrow progenitors (Ralph et al. 1986b), possibly reflecting unknown technical requirements to achieve optimal assay conditions for human macrophage progenitors, or, alternatively, the necessity for synergism with GM-CSF (Chen et al. 1988) or other interleukins (Ikebuchi et al. 1987; Warren and Moore 1987) for the stimulation of human macrophage colony formation.

In addition to its activities as a growth and survival factor, CSF-1 affects the functional integrity of mature monocytes and macrophages and is an important modulator of the inflammatory response. It enhances the production of other macrophage cytokines, such as interleukin-1, tumor necrosis factor, granulocyte colony-stimulating factor, and interferon, and potentiates the release of prostaglandins and biocidal oxygen metabolites (Metcalf and Nicola 1985; Ralph et al. 1986a; Warren and Ralph 1986). CSF-1 also protects macrophages from lytic infection by vesicular stomatitis virus (Lee and Warren 1987), enhances phagocytic activity (Tushinski et al. 1982), stimulates killing of microorganisms (Karbassi et al. 1987), and promotes antibody-dependent cytolysis of tumor cells (Ralph and Nakoinz 1987).

Although initially described only as a hemopoietin, it is likely that CSF-1 also functions outside the context of hematopoiesis. The growth factor is produced by uterine epithelium under control of ovarian steroids, and during pregnancy in mice, the local concentrations of CSF-1 increase over 1000-fold, being highest at parturition (Bartocci et al. 1986; Pollard et al. 1987). c-*fms* mRNA (Müller et al. 1983a,b) and authentic receptors for CSF-1 (Rettenmier et al. 1986) have been identified on placental trophoblasts and in malignant choriocarcinoma cell lines, suggesting that CSF-1 may have an important function in placental development. Transcripts of c-*fms* mRNA have also been detected in other organs, including spleen, brain, liver, lung, and bone marrow (Nienhuis et al. 1985; Rettenmier et al. 1985; Sariban et al. 1985), where their expression generally corresponds to the distribution of tissue macrophages (Byrne et al. 1981). However, sufficiently detailed studies have as yet not been performed to exclude the possibility that CSF-1 acts on other cell types, apart from its defined role in hematopoiesis and, possibly, placental development.

The CSF-1 Gene and Its Products

The CSF-1 gene has been mapped to the long arm of human chromosome 5 at band q33.1 and is linked to the c-*fms* proto-oncogene at 5q33.3 (Pettenati et al. 1987). To date, this is the only system in which genes encoding both a growth factor and its receptor were found to be closely linked on a single human chromosome. The adjacent region of chromosome 5q also includes genes encoding other hemopoietins (GM-CSF, IL-3, and IL-5) (Le Beau et al. 1986; Sutherland et al. 1988), the endothelial cell growth factor (acidic fibroblast growth factor) (Jaye et al. 1986), the receptor for PDGF (Yarden et al. 1986), the CD-14 myeloid differentiation antigen (Goyert et al. 1988), and a gene encoding a putative transcriptional regulatory factor (Sukhatme et al. 1988). Interstitial deletions in this region of human chromosome 5q have been frequently associated with therapy-related acute myeloid leukemia, myelodysplastic syndromes, and refractory anemias, suggesting that the functions of one or more clustered genes regulating hematopoiesis might be disrupted by such chromosomal rearrangements (for review, see Le Beau et al. 1986).

The human CSF-1 gene consists of ten exons spanning 21 kb encoding several alternatively spliced mRNAs (Kawasaki et al. 1985; Ralph et al. 1986a; Ladner et al. 1987), and two different human CSF-1 cDNAs of 4 kb and 1.6 kb have been molecularly cloned (Kawasaki et al. 1985; Wong et al. 1987). Expression of the 4-kb mRNA has been detected in many cell types (Wong et al. 1987), whereas the shorter mRNA, although inducible in one system (Ralph et al. 1986b), encodes a smaller form of the growth factor which has not been assigned a clearly defined physiological function. Two CSF-1 transcripts of the corresponding sizes are differentially expressed in normal mouse tissues (Rajavashisth et al. 1987), and a murine cDNA similar to the longer human clone has been isolated (DeLamarter et al. 1987).

Both human CSF-1 cDNAs have been expressed at high levels in mouse NIH-3T3 fibroblasts in which their synthesis and posttranslational processing have been studied in detail (Rettenmier et al. 1987; Rettenmier and Roussel 1988). Each encodes an integral transmembrane glycoprotein precursor (Fig. 1) that becomes oriented with its amino-terminal domain in the cisternae of the endoplasmic reticulum and its short carboxy-terminal tail in the cytoplasm. After cleavage of their 32-amino-acid signal peptides, both precursors undergo rapid assembly through disulfide bonds to form intracellular, membrane-bound homodimers.

The product of the 4-kb mRNA is a precursor of 554 amino acids (CSF-1^{554}) (Fig. 1) that acquires both N- and O-linked oligosaccharides during intracellular transport. The membrane-bound homodimer is proteolytically cleaved just prior to secretion to generate a soluble growth factor of 86 kD that is efficiently released from the cell. Since the polypeptide chain of each proteolyzed CSF-1^{554} subunit is about 220 amino acids in length (Wong et al. 1987), addition of carbohydrate contributes significantly to the apparent molecular mass of the soluble homodimer. On the basis of biochemical analyses of human urinary CSF-1 (Wong et al. 1987), it seems likely that the product of the 4-kb mRNA accounts for most, if not all, of the CSF-1 that appears in the circulation.

In contrast, the product of the 1.6-kb human cDNA

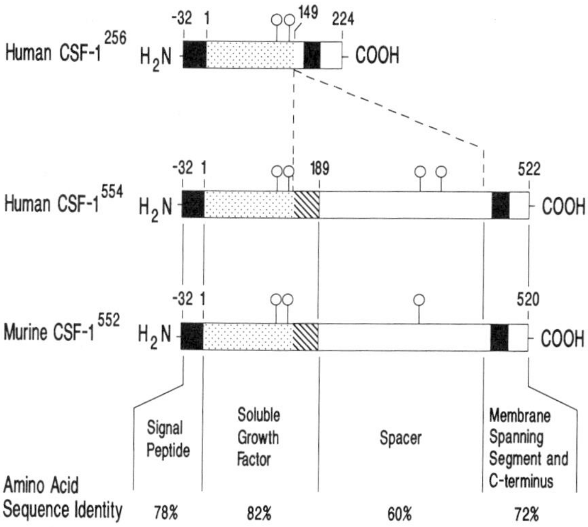

Figure 1. Predicted translation products encoded by human and mouse CSF-1 cDNAs. Each precursor contains a 32-amino-acid signal peptide (black bar), sequences representing the secreted growth factor (stippled and cross-hatched regions), a membrane-spanning segment (black bar), and a carboxy-terminal tail. Sites for N-linked oligosaccharide addition are indicated by circles. Human CSF-1^{554} and CSF-1^{256} differ by a 298-amino-acid residue spacer sequence (bracketed by dashed lines) that includes carboxy-terminal residues (cross-hatched) found in the soluble secreted growth factor derived from the larger precursor. Murine CSF-1^{552} is similar in structure to human CSF-1^{554}. (Data adapted from Kawasaki et al. 1985; DeLamarter et al. 1987; Wong et al. 1987.) (Reprinted, with permission, from Rettenmier et al. 1988.)

is a 256-amino-acid precursor (CSF-1^{256}) that contains only a portion of the amino-terminal domain of CSF-1^{554} as well as its hydrophobic membrane-spanning segment and short carboxy-terminal tail (Fig. 1). The shorter precursor lacks a "spacer" segment of 298 amino acids that contains the preferred site of proteolysis within CSF-1^{554}. Thus, it is transported as a membrane-bound homodimer to the cell surface and is slowly released by subsequent proteolytic cleavage of the externalized molecule that occurs within sequences close to the membrane-spanning segment. The proteases that cleave CSF-1^{554} or CSF-1^{256} have not been identified. In contrast to the intracellular proteolysis of the longer precursor, cleavage of the shorter polypeptide chains is inefficient, so that most of the homodimers remain cell-associated. The processed product of CSF-1^{256} that is released into the medium is only 44 kD in mass and consists of two glycosylated polypeptide chains, each containing approximately 160 amino acids. This carboxy-terminally truncated form of CSF-1 appears to be as active as the secreted product of CSF-1^{554} in binding to its receptor on macrophages and in stimulating their proliferation. Insertion of a termination codon amino-terminal to the transmembrane-spanning segment of CSF-1^{256} leads to the efficient and quantitative secretion of a soluble 44-kD homodimer that is not processed from membrane-bound intermediates (Heard et al. 1987a). Together, these data indicate that sequences encoding the biologically active growth factor moiety are included within residues 1–158 shared by both precursors. The fact that alternatively spliced CSF-1 mRNAs encode both membrane-bound and secreted forms of the growth factor raises the possibility that CSF-1 may stimulate receptor-bearing cells not only through endocrine or paracrine mechanisms, but also through direct cell-to-cell interactions.

The CSF-1 Receptor

The CSF-1 receptor (CSF-1R) is encoded by the c-*fms* proto-oncogene (Sherr et al. 1985) and is an integral transmembrane glycoprotein consisting of an extracellular ligand-binding domain connected by a single membrane-spanning segment to a carboxy-terminal tyrosine kinase domain (Coussens et al. 1986). CSF-1R is most closely related in its amino acid sequence to the PDGF receptor (Yarden et al. 1986) and to the c-*kit* oncogene product, a putative receptor for an as yet uncharacterized ligand (Besmer et al. 1986b; Yarden et al. 1987). An additional gene related to c-*fms* has also been detected in the genomes of vertebrates (Matsushime et al. 1987) but has a pattern of expression and chromosomal location in man distinct from c-*fms*.

In intact macrophages stimulated by CSF-1, CSF-1R is activated as a kinase, resulting in its immediate autophosphorylation and the phosphorylation of heterologous cellular substrates on tyrosine (Downing et al. 1988). Similarly, in membrane preparations or after receptor solubilization, incubation of CSF-1R preparations with [γ-^{32}P]ATP in the presence of purified CSF-1 and manganese activates the kinase and leads to receptor autophosphorylation on tyrosine (Sherr et al. 1985; Sacca et al. 1986; Yeung et al. 1987). In contrast, in in vitro assays performed with immune complexes formed with antibodies to the receptor, basal levels of receptor kinase activity appear to be increased, possibly due to receptor aggregation; autophosphorylation in the presence of [γ-^{32}P]ATP and manganese proceeds in the absence of ligand (Rettenmier et al. 1985) and is not enhanced by CSF-1 addition.

Ligand activation of the receptor kinase serves as the proximal signal that eventually elicits a mitogenic response. CSF-1 has both immediate and temporally prolonged effects on macrophages. Within seconds of binding to its receptor, it induces cytoplasmic alkalinization and secondary stimulation of Na^+, K^+ ATPase via activation of the Na^+, H^+ antiporter (Vairo and Hamilton 1988). CSF-1 also induces membrane vesiculation and enhancement of macrophage phagocytic activity (Tushinski et al. 1982). These events are followed by induction of a family of immediate early genes, including the proto-oncogenes c-*fos* (detectable within 15 min of CSF-1 stimulation and sustained for several hours) and c-*myc* (Müller et al. 1985; Bravo et al. 1987; Orlofsky and Stanley 1987). CSF-1 stimulation leads to an increase in overall protein synthesis within 2 hours (Tushinski and Stanley 1983) and to DNA synthesis within 8–12 hours and subsequent mitogenesis (Tushinski and Stanley 1985). The induction of immediate early genes is not sufficient for mitogenesis, since CSF-1 is required throughout the entire G_1 phase of the cell cycle (Tushinski and Stanley 1985). This suggests that different physiological substrates for the receptor kinase may be elaborated as the cells progress through G_1. At least some of the late effects of CSF-1 may be indirect, since the growth factor can induce the release of other cytokines and prostaglandins from macrophages (see above).

In the absence of ligand, CSF-1R turns over with a half-life of 2–3 hours (Rettenmier et al. 1987). Addition of saturating concentrations of CSF-1 to receptor-bearing cells leads to receptor internalization and the accelerated degradation of receptor-ligand complexes (receptor down-modulation), so that within 15 minutes of ligand addition, more than 90% of the cell-surface receptors are degraded to forms that are no longer recognized by specific antisera (Wheeler et al. 1986; Rettenmier et al. 1987). CSF-1 is itself degraded in lysosomes and released from cells as small acid-soluble peptides (Guilbert and Stanley 1986). After ligand-induced down-modulation, cells lacking CSF-1 binding sites are transiently refractory to further stimulation. In the subsequent absence of CSF-1, however, newly synthesized receptors reaccumulate on the plasma membrane so that a complete complement of binding sites is restored within several hours. Under physiological conditions where complete receptor occupancy is not required to generate a biological response, only a proportion of receptors undergo down-modulation in the presence of CSF-1 so that, at steady state, net receptor turnover is not so drastically accelerated.

CSF-1 receptors metabolically labeled with [^{32}P]-orthophosphate in the absence of CSF-1 are phosphorylated on serine residues (Woolford et al. 1985; Downing et al. 1988), and phorbol esters, which activate serine/threonine-specific protein kinase C, induce rapid receptor degradation (Chen et al. 1983; Wheeler et al. 1986). However, the mechanisms of ligand- and phorbol-ester-induced down-modulation are different, because under conditions in which protein kinase C is itself down-modulated by chronic treatment with 12-tetradecanoyl phorbol acetate (TPA), CSF-1 receptors reaccumulating on the plasma membrane after de novo synthesis can still be down-modulated by ligand (C.W. Rettenmier and C.J. Sherr, unpubl.). When a functional human c-*fms* gene was expressed in mouse NIH-3T3 cells, treatment of the cells with PDGF in medium containing low concentrations of serum resulted in down-modulation of CSF-1R (J.R. Downing and C.J. Sherr, unpubl.). These findings suggest that physiological inducers of protein kinase C can transmodulate CSF-1R and thereby inhibit their ability to respond to CSF-1.

Transduction of the human c-*fms* gene into mouse NIH-3T3 cells enables the cells to form colonies in semisolid medium in response to CSF-1 (Roussel et al. 1987). Thus, the human c-*fms* gene contains all the information necessary to render mouse fibroblasts responsive to the human hemopoietin, implying that downstream components of the mitogenic pathway include physiological targets for the receptor kinase, even in naive cells that do not otherwise express CSF-1R. Moreover, NIH-3T3 cells were morphologically transformed by an autocrine mechanism and rendered tumorigenic when the human c-*fms* gene was cotransfected with a second vector containing human CSF-1 cDNA (Roussel et al. 1987). In this cotransfection assay, CSF-1 cDNAs encoding either CSF-1^{554} or CSF-1^{256} elicited a transformed phenotype (Heard et al. 1987a), and CSF-1 receptors were persistently down-modulated in response to autocrine stimulation (Rettenmier et al. 1987). Under these conditions, neutralizing antibodies to CSF-1 did not revert the transformed phenotype, so that intracellular interactions between the ligand and its receptor could not be precluded. In conceptually analogous experiments, introduction of the human CSF-1 gene into murine SV40 immortalized, CSF-1-dependent macrophages induced factor independence and constitutive receptor down-modulation (Roussel et al. 1988a). However, the latter cells remained nontumorigenic in nude mice, suggesting that macrophages may be relatively more refractory to persistent CSF-1-induced mitogenic signals.

We have begun to introduce the c-*fms* gene into hematopoietic cell lines, which are dependent on other CSFs and interleukins for growth, in order to determine whether CSF-1R expression can alter their growth factor dependence and possibly affect their capacity to differentiate. The immature mouse myeloid cell line, FDC-P1, can be grown in either IL-3 or GM-CSF but cannot survive in CSF-1. These cells were infected with a helper-free retroviral vector containing both the human c-*fms* and neomycin resistance genes, grown in the presence of IL-3 and G-418, and then selected by fluorescence-activated flow cytometry for cells expressing c-*fms*-coded epitopes. When the receptor-positive cells were plated in semisolid medium containing either IL-3 or CSF-1 or neither of the growth factors, approximately 20% of the IL-3 responsive population formed colonies in the presence of CSF-1. However, about 3% also formed colonies in the absence of either growth factor. The majority of cell lines established from colonies proliferating in the presence of CSF-1 were able to grow continuously in either IL-3 or CSF-1 but did not survive in liquid culture in the absence of these growth factors. Although CSF-1 replaced their IL-3 requirement, the cells did not develop differentiated features of mononuclear phagocytes. An unresolved issue, then, is whether transduction of CSF-1R into pluripotent hematopoietic cell lines can alter their differentiation programs.

Properties of the v-*fms* Oncogene Product

The v-*fms* oncogene of SM-FeSV can transform fibroblasts (McDonough et al. 1971; Donner et al. 1982) and renders both CSF-1-dependent and IL-3-dependent mouse myeloid cell lines factor independent and tumorigenic by a nonautocrine mechanism (Wheeler et al. 1986, 1987). When mouse bone marrow cells infected with a helper-free retrovirus containing the v-*fms* gene were used to reconstitute lethally irradiated mice, recipient animals developed proliferative disorders involving multiple hematopoietic lineages, including frank leukemias (Heard et al. 1987b).

Unlike CSF-1R, which is tightly regulated as a kinase by its ligand, the v-*fms* gene product is constitutively phosphorylated on tyrosine (Woolford et al. 1985; Tamura et al. 1986; Downing et al. 1988; Morrison et al. 1988), shows high levels of basal kinase activity in membrane preparations in the absence of ligand (Sacca et al. 1986), and is relatively resistant to accelerated turnover induced by CSF-1 or TPA (Wheeler et al. 1986). The activity of the v-*fms* gene product as a constitutive kinase and its persistence at the cell surface in the presence of inducers of down-modulation may each contribute to ligand-independent signals for cell growth. Despite the fact that v-*fms* can transform cells by a nonautocrine mechanism, the gene product retains a functional CSF-1-binding domain (Sacca et al. 1986), and stimulation of cells with human CSF-1 leads to an increase in autophosphorylation of the v-*fms* gene product on tyrosine (Downing et al. 1988; Roussel et al. 1988b). In addition, the efficiency of colony formation of transformed cells in semisolid medium is increased when human (but not mouse) recombinant CSF-1 is added to the cultures (Roussel et al. 1988b). These observations imply that the v-*fms*-coded kinase can also be further up-regulated by feline CSF-1, perhaps explaining the propensity of both the SM and HZ-5 strains of FeSV to induce tumors of CSF-1-producing cells (i.e., fibrosarcomas) in domestic cats.

In contrast to normal CSF-1R, the posttranslational processing of immature forms of the v-*fms*-coded glycoprotein is relatively inefficient, so that at steady state, most of the molecules remain sequestered in the endoplasmic reticulum (ER), and only a minority are transported through the Golgi complex to the plasma membrane (Anderson et al. 1984). Studies of v-*fms* mutants that were further blocked in their transport from the ER to the Golgi complex (Roussel et al. 1984) or the use of inhibitors of glycosylation (Nichols et al. 1985; Hadwiger et al. 1986) revealed that transport of the v-*fms* product to the cell surface is required for cell transformation. Presumably, relevant substrates for the v-*fms*-coded kinase are located at the plasma membrane or in the endosome-lysosome compartment.

Transforming Potential of the c-*fms* Proto-oncogen

The feline v-*fms* and human c-*fms* gene products differ from one another by scattered point mutations, most of which are due to interspecies evolutionary divergence. In addition, the carboxy-terminal 40 amino acids of CSF-1R were replaced by 11 unrelated residues in the v-*fms* gene product, eliminating a single tyrosine residue (Tyr-969) four amino acids from the receptor carboxyl terminus (Hampe et al. 1984; Coussens et al. 1986) (Fig. 2). By analogy to studies with the trans-

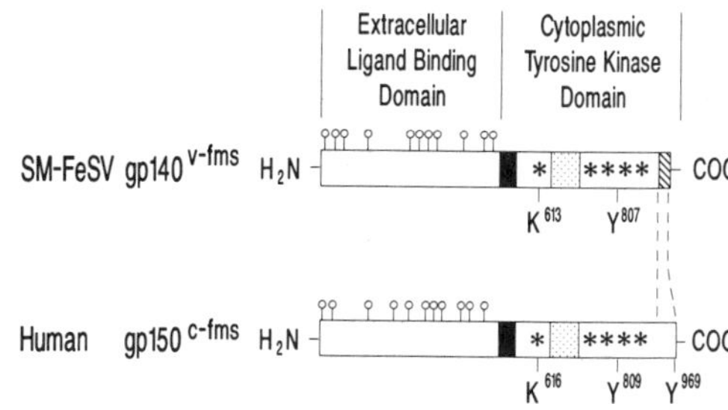

Figure 2. Comparative structures of the feline v-*fms* and human c-*fms* gene products. The membrane-spanning regions are indicated by a black bar and sites of N-linked glycosylation in the ligand-binding domains are shown by open circles. The carboxy-terminal 40 amino acids of human CSF-1R (gp $150^{c\text{-}fms}$) containing Tyr-969 (Y-969) were replaced by 11 unrelated amino acids in the v-*fms* gene product. The cytoplasmic kinase domains of both glycoproteins contain a "spacer" sequence of about 70 amino acids that shows no sequence similarity to analogous spacer regions in the kinase domains of the PDGF receptor and the c-*kit* oncogene product (Besmer et al. 1986b; Yarden et al. 1986, 1987). The predicted ATP-binding sites in each molecule (lysine residues: K-613 and K-616) are designated. Tyrosine residues (Y-807 and Y-809) located in positions analogous to the preferred sites of autophosphorylation in $pp60^{src}$ proteins are also shown. (Data adapted from Hampe et al. 1984; Coussens et al. 1986.) (Reprinted, with permission, from Rettenmier et al. 1988.)

forming protein encoded by $pp60^{v\text{-}src}$ and its normal c-*src*-coded progenitor (for review, see Hunter 1987), it was suggested that deletion of a negative regulatory site of tyrosine phosphorylation within the c-*fms*-coded carboxyl terminus might up-regulate its kinase activity and convert c-*fms* to an active oncogene (Besmer et al. 1986a; Coussens et al. 1986).

To test this hypothesis, human c-*fms* mutants containing either distal carboxy-terminal truncations or a Tyr-969 to Phe-969 mutation were cloned into retroviral vectors and transfected into mouse NIH-3T3 cells. Neither the wild-type nor mutant genes transformed cells when expressed at levels equal to those of v-*fms*. However, when c-*fms*(Tyr-969) and c-*fms*(Phe-969) were cotransfected into NIH-3T3 cells with a second vector encoding human CSF-1, cell transformation was induced by an autocrine mechanism, and the efficiency of transformation obtained with the mutant c-*fms* gene was reproducibly higher than that obtained with the wild-type cDNA. Thus, although Tyr-969 appears to regulate the activity of the c-*fms* gene product negatively, possibly by serving as a site of receptor phosphorylation, carboxy-terminal truncation of CSF-1R is insufficient to activate the transforming potential of the gene (Roussel et al. 1987). Since CSF-1R lacks phosphotyrosine unless it is stimulated by its ligand (Downing et al. 1988), phosphorylation of Tyr-969, if it indeed occurs, must follow ligand activation. Possibly, phosphorylation of this residue is catalyzed by another tyrosine kinase that recognizes the active CSF-1R conformation induced by CSF-1.

In a parallel series of experiments, chimeric constructs between the feline v-*fms* and human c-*fms* genes were engineered in which the truncated carboxyl terminus of the v-*fms* gene product was restored by carboxy-terminal residues derived from the normal receptor. The transforming efficiency of the v-*fms*/c-*fms* chimera was attenuated but not entirely eliminated (Browning et al. 1986; Roussel et al. 1987). When Tyr-969 in this chimeric construct was replaced by a phenylalanine residue, the efficiency of transformation was restored and approximated that of the v-*fms* gene (Roussel et al. 1987). Although these results confirmed the negative regulatory function of Tyr-969, the fact that the chimeric constructs were transforming, whereas both c-*fms* alleles were not, implied that a second "activating mutation" elsewhere in the v-*fms* gene was critical in enabling it to transform cells in a ligand-independent manner.

To map the location of the putative activating mutation in the v-*fms* gene, we used reciprocal recombination in bacteriophage λ to generate a series of c-*fms*/v-*fms* chimeras with crossovers at different homologous sites (Roussel et al. 1988b). One such chimera contained the distal amino-terminal domain of human CSF-1R (residues 1–308) fused to the remaining ligand-binding region, the transmembrane segment, and the complete carboxy-terminal kinase domain of the v-*fms* gene product (including its truncated carboxyl terminus). When expressed in NIH-3T3 cells, the chimeric gene product functioned as a normal human CSF-1 receptor. Unlike the v-*fms*-coded glycoprotein, the chimeric gene product was efficiently transported to the cell surface, was phosphorylated on tyrosine only in response to human CSF-1, and underwent normal ligand-induced down-modulation (Fig. 3). However, the chimeric gene product remained refractory to down-modulation by phorbol esters, confirming that the mechanisms of ligand- and TPA-induced receptor degradation differ from one another. Therefore, genetic alterations that distinguish the kinase domains of v-*fms* and c-*fms* are insufficient to account for the persistent phosphorylation of the v-*fms* gene product on tyrosine or its transforming activity in the absence of ligand. Rather, our data suggest that an activating mutation resides in the distal extracellular domain of the v-*fms* gene product, so that replacement of this region with sequences from normal CSF-1R eliminates its transforming activity and restores normal receptor function. Alternatively, activating mutations elsewhere in the gene might be phenotypically suppressed by substitution of the human CSF-1 ligand-binding domain. Recent studies using site-directed mutagenesis support the former interpretation (M.F. Roussel et al., unpubl.).

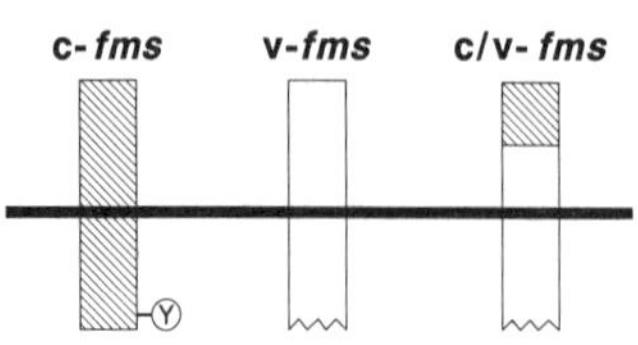

	c-fms	v-fms	c/v-fms
Transformation	No	Yes	No
Intracellular Transport	Fast	Slow	Fast
Autophosphorylation	Regulated	Constitutive	Regulated
Downmodulation (CSF-1)	Fast	Slow	Fast
Downmodulation (TPA)	Fast	Slow	Slow

Figure 3. Biological and biochemical characteristics of the human c-*fms*, feline v-*fms*, and chimeric c/v-fms^{308} molecules. (Data adapted from Roussel et al. 1988b.)

A Role for c-*fms* in Leukemogenesis

In mice, integration of the Friend strain of murine leukemia virus upstream of the c-*fms* locus is an early event that is etiologically associated with the development of myeloid leukemia in approximately 30% of infected animals (Gisselbrecht et al. 1987). Retroviral integration occurs in a head-to-head orientation with respect to sites of initiation of viral and c-*fms* transcripts and leads to aberrantly increased production of CSF-1R in immature myeloid progenitor cells. Cells having clonal retroviral insertions at the c-*fms* locus had a proliferative advantage over normal precursors and, although they initially required bone marrow stromal cell feeder layers for their propagation in culture, factor-independent variants emerged that were tumorigenic in vivo. In a percentage of leukemic lines, the overtly tumorigenic cells synthesized CSF-1 (Heard et al. 1984, 1986). Together, these findings suggest that abnormally increased expression of c-*fms* in early myeloid progenitors primes the cells to proliferate in response to exogenous CSF-1. Their subsequent expansion and leukemic evolution may potentially involve any of a number of secondary genetic events, but in some cases, constitutive coexpression of the CSF-1 gene is likely to be a factor contributing to malignant transformation.

Because interstitial deletions of human chromosome 5q are frequently associated with therapy-related acute myeloid leukemias in humans (see above), we assayed for rearrangements within c-*fms*-coding sequences (32 kb) and within the 50 kb of sequences flanking the 5′ end of the gene in 44 pediatric patients with acute myeloblastic leukemia (AML) (R.A. Ashmun et al., in prep.). No rearrangements were detected. In parallel, monoclonal antibodies to human CSF-1R were used to assay for receptor expression by flow cytometric techniques in 76 AML patients, both pediatric and adult. CSF-1R was detected in 30% of pediatric cases and in 16% of adult AMLs, but not in any of 19 cases of acute lymphocytic leukemia. In all cases, receptors expressed on human leukemic blasts were able to bind CSF-1 and underwent normal ligand- and TPA-induced down-modulation, suggesting that the receptors were functional.

Although in normal human blood and bone marrow, CSF-1R was detected only on monocytic cells, a significant subset of AML blasts without other morphologic, histochemical, or immunophenotypic evidence of monocytic differentiation expressed relatively high levels of CSF-1R. Such leukemic cells may have been derived from rare multipotential progenitor cells in the bone marrow that normally express low levels of CSF-1R but which cannot be readily detected in normal marrow by flow cytometric assays. This would account for the apparent lineage infidelity of CSF-1R expression observed in their leukemic counterparts.

Alternatively, expression of CSF-1R in these leukemias could reflect inappropriate gene expression and represent a factor contributing to the pathogenesis of AML. Characterization of the as yet unidentified c-*fms* promoter and the factors that regulate its transcription in both normal cells and myeloid leukemias will be a necessary prerequisite in distinguishing between these possibilities. Like Friend-virus-induced myeloid leukemia in mice, a subset of human leukemic blasts expressing CSF-1R also express CSF-1 mRNA, as measured by both Northern blotting analysis and in situ hybridization (Rambaldi et al. 1987b; Wakamiya et al. 1987). Thus, the processes of multistep disease progression in these human AMLs may be directly analogous to events observed in the animal model.

CONCLUSIONS

Although much has been learned about functions of CSF-1 that are mediated through its binding to its specific receptor, the identification of physiological substrates for the receptor kinase remains an elusive problem. It is presently unclear how receptor signals originating at the cell surface alter the regulation of CSF-1 responsive genes and elicit the pleiotropic actions ascribed to the growth factor. The problem is compounded by observations that CSF-1 receptors can be transmodulated by other growth factors and that CSF-1 induces the synthesis of other cytokines and prostaglandins that in turn act to modify cellular responses. Regardless of the exact molecular mechanisms of action, however, CSFs have recently been used clinically in the management of bone marrow aplasia following chemotherapy or transplantation and have likely applications in modifying the inflammatory response. The developing interest in hemopoietins on the part of the scientific and medical community and the availability of the purified recombinant growth factors for investigational and clinical use suggest that advances in receptor biology and in the therapy of blood disorders will be rapidly forthcoming.

ACKNOWLEDGMENTS

The authors thank the many investigators who worked closely with us on aspects of this problem and who co-authored papers with our group over the last 5 years. This work was supported by National Institutes of Health grants CA-38187 and CA-20180 and by ALSAC of St. Jude Children's Research Hospital. C.J.S. is Herrick Foundation Chairman of the Department of Tumor Cell Biology and an Investigator of the Howard Hughes Medical Institute.

REFERENCES

Anderson, S.J., M.A. Gonda, C.W. Rettenmier, and C.J. Sherr. 1984. Subcellular localization of glycoproteins encoded by the viral oncogene v-*fms*. *J. Virol.* **51:** 730.

Bartelmez, S.H. and E.R. Stanley. 1985. Synergism between hemopoietic growth factors (HGFs) detected by their effects on cells bearing receptors for a lineage specific HGF: Assay of hemopoietin-1. *J. Cell Physiol.* **122:** 370.

Bartocci, A., J.W. Pollard, and E.R. Stanley. 1986. Regulation of colony-stimulating factor 1 during pregnancy. *J. Exp. Med.* **164:** 956.

Bartocci, A., D.S. Mastrogiannis, G. Migliorati, R.J. Stockert, A.W. Wolkoff, and E.R. Stanley. 1987. Macrophages specifically regulate the concentration of their own growth factor in the circulation. *Proc. Natl. Acad. Sci.* **84:** 6179.

Besmer, P., E. Lader, P.C. George, P.J. Bergold, F.-H. Qiu, E.E. Zuckerman, and W.D. Hardy. 1986a. A new acute transforming feline retrovirus with *fms* homology specifies a C-terminally truncated version of the c-*fms* protein that is different from SM-feline sarcoma virus v-*fms* protein. *J. Virol.* **60:** 194.

Besmer, P., J.E. Murphy, P.C. George, F. Qiu, P.J. Bergold, L. Lederman, H.W. Snyder, Jr., D. Brodeur, E.E. Zuckerman, and W.D. Hardy. 1986b. A new acute transforming feline retrovirus and relationship of its oncogene v-*kit* with the protein kinase gene family. *Nature* **320:** 415.

Bravo, R., M. Neuberg, J. Burckhardt, J. Almendral, R. Wallich, and R. Müller. 1987. Involvement of common and cell type-specific pathways in c-*fos* gene control: Stable induction by cAMP in macrophages. *Cell* **48:** 251.

Browning, P.J., H.F. Bunn, A. Cline, M. Shuman, and A.W. Nienhuis. 1986. "Replacement" of COOH-terminal truncation of v-*fms* with c-*fms* sequences markedly reduces transformation potential. *Proc. Natl. Acad. Sci.* **83:** 7800.

Byrne, P.V., L.J. Guilbert, and E.R. Stanley. 1981. Distribution of cells bearing receptors for a colony-stimulating factor (CSF-1) in murine tissues. *J. Cell Biol.* **91:** 848.

Chen, B.D.-M., C.R. Clark, and T. Chou. 1988. Granulocyte-macrophage colony stimulating factor stimulates monocyte and tissue macrophage proliferation and enhances their responsiveness to macrophage colony stimulating factor. *Blood* **71:** 997.

Chen, B.D.-M., H.S. Lin, and S. Hsu. 1983. Tumor-promoting phorbol esters inhibit the binding of colony stimulating factor (CSF-1) to murine peritoneal exudate macrophages. *J. Cell Physiol.* **116:** 207.

Clark, S.C. and R. Kamen. 1987. The human hematopoietic colony-stimulating factors. *Science* **236:** 1229.

Coussens, L., C. Van Beveren, D. Smith, E. Chen, R.L. Mitchell, C.M. Isacke, I.M. Verma, and A. Ullrich. 1986. Structural alteration of viral homologue of receptor proto-oncogene *fms* at carboxyl terminus. *Nature* **320:** 277.

Das, S.K. and E.R. Stanley. 1982. Structure-function studies of a colony stimulating factor (CSF-1). *J. Biol. Chem.* **257:** 13679.

Das, S.K., E.R. Stanley, L.J. Guilbert, and L.W. Forman. 1981. Human colony-stimulating factor (CSF-1) radioimmunoassay: resolution of three subclasses of human colony-stimulating factors. *Blood* **58:** 630.

DeLamarter, J.F., C. Hession, D. Semon, N.M. Gough, R. Rothenbuhler, and J.-J. Mermod. 1987. Nucleotide sequence of a cDNA encoding murine CSF-1 (macrophage-CSF). *Nucleic Acids Res.* **15:** 2389.

Donner, L., L.A. Fedele, C.F. Garon, S.J. Anderson, and C.J. Sherr. 1982. McDonough feline sarcoma virus: Characterization of the molecularly cloned provirus and its feline oncogene (v-*fms*). *J. Virol.* **41:** 489.

Downing, J.R., C.W. Rettenmier, and C.J. Sherr. 1988. Ligand-induced tyrosine kinase activity of the colony stimulating factor-1 receptor in a murine macrophage cell line. *Mol. Cell. Biol.* **8:** 1795.

Ek, B., B. Westermark, A. Wasteson, and C.-H. Heldin.

1982. Stimulation of tyrosine-specific phosphorylation by platelet-derived growth factor. *Nature* **295:** 419.

Guilbert, L.J. and E.R. Stanley. 1980. Specific interaction of murine colony-stimulating factor with mononuclear phagocytic cells. *J. Cell Biol.* **85:** 153.

———. 1986. The interaction of ^{125}I-colony-stimulating factor-1 with bone marrow-derived macrophages. *J. Biol. Chem.* **261:** 4024.

Gisselbrecht, S., S. Fichelson, B. Sola, D. Bordereaux, A. Hampe, C. Andre, F. Galibert, and P. Tambourin. 1987. Frequent c-*fms* activation by proviral insertion in mouse myeloblastic leukaemias. *Nature* **329:** 259.

Goyert, S.M., E. Ferrero, W.J. Rettig, A.K. Yenamandra, F. Obata, and M.M. Le Beau. 1988. The CD14 monocyte differentiation antigen maps to a region encoding growth factors and receptors. *Science* **239**: 497.

Hadwiger, A., H. Niemann, A. Käbisch, H. Bauer, and T. Tamura. 1986. Appropriate glycosylation of the *fms* gene product is a prerequisite for its transforming potency. *EMBO J.* **5:** 689.

Hampe, A., M. Gobet, C.J. Sherr, and F. Galibert. 1984. The nucleotide sequence of the feline retroviral oncogene v-*fms* shows unexpected homology with oncogenes encoding tyrosine-specific protein kinase. *Proc. Natl. Acad. Sci.* **81:** 85.

Heard, J.M., M.F. Roussel, C.W. Rettenmier, and C.J. Sherr. 1987a. Synthesis, post-translational processing, and autocrine transforming activity of a carboxylterminal truncated form of colony stimulating factor-1. *Oncogene Res.* **1:** 423.

———. 1987b. Multilineage hematopoietic disorders induced by transplantation of bone marrow cells expressing the v-*fms* oncogene. *Cell* **51:** 663.

Heard, J.M., B. Sola, M.A. Martial, S. Fichelson, and S. Gisselbrecht. 1986. Long-term culture of bone marrow-derived preleukemic cells from F-MuLV-infected mice. *Blood* **68:** 193.

Heard, J.M., S. Fichelson, B. Sola, M.A. Martial, B. Varet, and J.P. Levy. 1984. Multistep virus-induced leukemogenesis in vitro: Description of a model specifying three steps within the myeloblastic malignant process. *Mol. Cell. Biol.* **4:** 216.

Horiguchi, J., M.K. Warren, and D. Kufe. 1987. Expression of the macrophage-specific colony-stimulating factor in human monocytes treated with granulocyte-macrophage colony-stimulating factor. *Blood* **69:** 1259.

Horiguchi, J., M.K. Warren, P. Ralph, and D. Kufe. 1986. Expression of the macrophage specific colony stimulating factor (CSF-1) during human monocytic differentiation. *Biochem. Biophys. Res. Commun.* **141:** 924.

Hunter, T. 1987. A tail of two srcs: Mutatis mutandis. *Cell* **49:** 1.

Ikebuchi, K., G.G. Wong, S.C. Clark, J.N. Ihle, Y. Hirai, and M. Ogawa. 1987. Interleukin 6 enhancement of interleukin-3 dependent proliferation of multipotential hematopoietic progenitors. *Proc. Natl. Acad. Sci.* **84:** 9035.

Jacobs, S., F.C. Kull, Jr., H.S. Earp, M.E. Svoboda, J.J. Van Wyk, and P. Cuatracasas. 1983. Somatomedin C stimulates the phosphorylation of the beta subunit of its own receptor. *J. Biol. Chem.* **258:** 9581.

Jaye, M., R. Howk, W. Burgess, G.A. Ricca, I.M. Chiu, M.W. Ravera, S.J. O'Brien, W.S. Modi, T. Maciag, and W.N. Drohan. 1986. Human endothelial cell growth factor: Cloning, nucleotide sequence, and chromosome localization. *Science* **233:** 541.

Karbassi, A., J.M. Becker, J.S. Foster, and R.N. Moore. 1987. Enhanced killing of *Candida albicans* by murine macrophages treated with macrophage colony stimulating factor: Evidence for augmented expression of mannose receptors. *J. Immunol.* **139:** 417.

Kasuga, M., Y. Zick, D.L. Blithe, M. Crettaz, and C.R. Kahn. 1982. Insulin stimulates tyrosine phosphorylation of the insulin receptor in a cell free system. *Nature* **298:** 667.

Kawasaki, E.S., M.B. Ladner, A.M. Wang, J. Van Arsdell, M.K. Warren, M.Y. Coyne, V.L. Schweickart, M.T. Lee, K.J. Wilson, A. Boosman, E.R. Stanley, P. Ralph, and D.F. Mark. 1985. Molecular cloning of a complementary DNA encoding human macrophage-specific colony stimulating factor (CSF-1). *Science* **230:** 2916.

Ladner, M.B., G.A. Martin, J.A. Noble, D.M. Nikoloff, R. Tal, E.S. Kawasaki, and T.J. White. 1987. Human CSF-1: Gene structure and alternative splicing of mRNA precursors. *EMBO J.* **6:** 2693.

Le Beau, M.M., M.J. Pettenati, R.S. Lemons, M.O. Diaz, C.A. Westbrook, R.A. Larson, C.J. Sherr, and J.D. Rowley. 1986. Assignment of the GM-CSF, CSF-1, and *fms* genes to human chromosome 5 provides evidence for linkage of a family of genes regulating hematopoiesis and for their involvement in the deletion (5q) in myeloid disorders. *Cold Spring Harbor Symp. Quant. Biol.* **51:** 899.

Lee, M.T. and M.K. Warren. 1987. CSF-1-induced resistance to viral infection in murine macrophages. *J. Immunol.* **138:** 3019.

Matsushime, H., M.C. Yoshida, M. Sasaki, and M. Shibuya. 1987. A possible new member of tyrosine kinase family, human *frt* sequence, is highly conserved in vertebrates and located on human chromosome 13. *Jpn. J. Cancer Res.* **78:** 655.

McDonough, S.K., S. Larsen, R.S. Brodey, N.D. Stock, and W.D. Hardy, Jr. 1971. A transmissible feline fibrosarcoma of viral origin. *Cancer Res.* **31:** 953.

Metcalf, D. 1985. The granulocyte-macrophage colony stimulating factors. *Science* **229:** 16.

Metcalf, D. and A. Nicola. 1985. Synthesis by mouse peritoneal cells of G-CSF, the differentiation inducer for myeloid leukemia cells: Stimulation by endotoxin, M-CSF, and multi-CSF. *Leuk. Res.* **9:** 35.

Mochizuki, D.Y., J.R. Eisenman, P.J. Conlon, A.D. Larsen, and R.J. Tushinski. 1987. Interleukin 1 regulates hematopoietic activity, a role previously ascribed to hemopoietin 1. *Proc. Natl. Acad. Sci.* **84:** 5267.

Morgan, C.J. and E.R. Stanley. 1984. Chemical crosslinking of the mononuclear phagocyte specific growth factor CSF-1 to its receptor at the cell surface. *Biochem. Biophys. Res. Commun.* **119:** 35.

Morrison, D.K., P.J. Browning, M.F. White, and T.M. Roberts. 1988. Tyrosine phosphorylations in vivo associated with v-*fms* transformation. *Mol. Cell. Biol.* **8:** 176.

Müller, R., T. Curran, D. Müller, and L. Guilbert. 1985. Induction of c-*fos* during myelomonocytic differentiation and macrophage proliferation. *Nature* **314:** 546.

Müller, R., J.M. Tremblay, E.D. Adamson, and I.M. Verma. 1983a. Tissue and cell type specific expression of two human c-*onc* genes. *Nature* **304:** 454.

Müller, R., D.J. Slamon, E.D. Adamson, J.M. Tremblay, D. Müller, M.J. Cline, and I.M. Verma. 1983b. Transcription of c-*onc* genes c-*ras*^{ki} and c-*fms* during mouse development. *Mol. Cell Biol.* **3:** 1062.

Nichols, E.J., R. Manger, S. Hakomori, A. Herscovics, and L.R. Rohrschneider. 1985. Transformation by the v-*fms* oncogene product: Role of glycosylational processing and cell surface expression. *Mol. Cell Biol.* **5:** 3467.

Nienhuis, A.W., H.F. Bunn, P.H. Turner, T.V. Gopal, W.G. Nash, S.J. O'Brien, and C.J. Sherr. 1985. Expression of the human c-*fms* proto-oncogene in hematopoietic cells and its deletion in the 5q$^-$ syndrome. *Cell* **42:** 421.

Nishimura, J., J.S. Huang, and T.F. Deuel. 1982. Platelet-derived growth factor stimulates tyrosine-specific protein kinase activity in Swiss mouse 3T3 cell membranes. *Proc. Natl. Acad. Sci.* **79:** 4303.

Orlofsky, A. and E.R. Stanley. 1987. CSF-1-induced gene expression in macrophages: Dissociation from the mitogenic response. *EMBO J.* **6:** 2947.

Oster, W., A. Lindemann, S. Horn, R. Mertelsmann, and F. Herrmann. 1987. Tumor necrosis factor (TNF)-alpha but not TNF-beta induces secretion of colony stimulating factor for macrophages (CSF-1) by human monocytes. *Blood* **70:** 1700.

Pettenati, M.J., M.M. Le Beau, R.S. Lemons, E.A. Shima, E.S. Kawasaki, R.A. Larson, C.J. Sherr, M.O. Diaz, and J.D. Rowley. 1987. Assignment of *CSF-1* to 5q33.1: Evidence for clustering of genes regulating hematopoiesis and for their involvement in the deletion of the long arm of chromosome 5 in myeloid disorders. *Proc. Natl. Acad. Sci.* **84:** 2970.

Pollard, J.W., A. Bartocci, R. Arceci, A. Orlofsky, M.B. Ladner, and E.R. Stanley. 1987. Apparent role of the macrophage growth factor, CSF-1, in placental development. *Nature* **330:** 484.

Rajavashisth, T.B., R. Eng, R.K. Shadduck, A. Waheed, C.M. Ben-Avram. J.E. Shively, and A.J. Lusis. 1987. Cloning and tissue-specific expression of mouse macrophage colony-stimulating factor mRNA. *Proc. Natl. Acad. Sci.* **84:** 1157.

Ralph, P. and I. Nakoinz. 1987. Stimulation of macrophage tumoricidal activity by the growth and differentiation factor CSF-1. *Cell Immunol.* **105:** 270.

Ralph, P., M.K. Warren, M.D. Ladner, E.S. Kawasaki, A. Boosman, and T.J. White. 1986a. Molecular and biological properties of human macrophage growth factor, CSF-1. *Cold Spring Harbor Symp. Quant. Biol.* **51:** 679.

Ralph, P., M.K. Warren, M.T. Lee, J. Csejtey, J.F. Weaver, H.E. Broxmeyer, D.E. Williams, E.R. Stanley, and E.S. Kawasaki. 1986b. Inducible production of human macrophage growth factor, CSF-1. *Blood* **68:** 633.

Rambaldi, A., D.C. Young, and J.D. Griffin. 1987a. Expression of the M-CSF (CSF-1) gene by human monocytes. *Blood* **69:** 1409.

Rambaldi, A., N. Wakamiya, E. Vallenga, J. Horiguchi, M.K. Warren, D. Kufe, and J.D. Griffin. 1987b. Expression of the macrophage colony stimulating factor and c-*fms* genes in human acute myeloblastic leukemia cells. *J. Clin. Invest.* **81:** 1030.

Rettenmier, C.W. and M.F. Roussel. 1988. Differential processing of colony-stimulating factor-1 precursors encoded by two human cDNAs. *Mol. Cell. Biol.* (in press).

Rettenmier, C.W., M.F. Roussel, and C.J. Sherr. 1988. The colony-stimulating factor 1 (CSF-1) receptor (c-*fms* proto-oncogene product) and its ligand. *J. Cell Sci.* (suppl. 9) (in press).

Rettenmier, C.W., J.H. Chen, M.F. Roussel, and C.J. Sherr. 1985. The product of the c-*fms* proto-oncogene: A glycoprotein with associated tyrosine kinase activity. *Science* **228:** 320.

Rettenmier, C.W., M.F. Roussel, R.A. Ashmun, P. Ralph, K. Price, and C.J. Sherr. 1987. Synthesis of membrane-bound colony-stimulating factor-1 (CSF-1) and downmodulation of CSF-1 receptors in NIH 3T3 cells transformed by cotransfection of the human CSF-1 and c-*fms* (CSF-1 receptor) genes. *Mol. Cell. Biol.* **7:** 2378.

Rettenmier, C.W., R. Sacca, W.L. Furman, M.F. Roussel, J.T. Holt, A.W. Nienhuis, E.R. Stanley, and C.J. Sherr. 1986. Expression of the human c-*fms* proto-oncogene product (colony-stimulating factor-1 receptor) on peripheral blood mononuclear cells and choriocarcinoma cell lines. *J. Clin. Invest.* **77:** 1740.

Roussel, M.F., C.W. Rettenmier, and C.J. Sherr. 1988a. Introduction of a human colony stimulating factor-1 gene into a mouse macrophage cell line induces CSF-1 independence but not tumorigenicity. *Blood* **71:** 1218.

Roussel, M.F., C.W. Rettenmier, A.T. Look, and C.J. Sherr. 1984. Cell surface expression of v-*fms*-coded glycoproteins is required for transformation. *Mol. Cell. Biol.* **4:** 1999.

Roussel, M.F., J.R. Downing, R.A. Ashmun, C.W. Rettenmier, and C.J. Sherr. 1988b. Colony-stimulating factor-1 mediated regulation of a chimeric c-*fms*/v-*fms* receptor containing the v-*fms*-coded tyrosine kinase domain. *Proc. Natl. Acad. Sci.* **85:** 5903.

Roussel, M.F., T.J. Dull, C.W. Rettenmier, P. Ralph, A. Ullrich, and C.J. Sherr. 1987. Transforming potential of the c-*fms* proto-oncogene (CSF-1 receptor). *Nature* **325:** 549.

Sacca, R., E.R. Stanley, C.J. Sherr, and C.W. Rettenmier. 1986. Specific binding of the mononuclear phagocyte colony-stimulating factor CSF-1 to the product of the v-*fms* oncogene. *Proc. Natl. Acad. Sci.* **83:** 3331.

Sariban, E., T. Mitchell, and D. Kufe. 1985. Expression of the c-*fms* proto-oncogene during human monocytic differentiation. *Nature* **316:** 64.

Sherr, C.J., C.W. Rettenmier, R. Sacca, M.F. Roussel, A.T. Look, and E.R. Stanley. 1985. The c-*fms* proto-oncogene product is related to the receptor for the mononuclear phagocyte growth factor, CSF-1. *Cell* **41:** 665.

Stanley, E.R. and P.M. Heard. 1977. Factors regulating macrophage production and growth. Purification and some properties of the colony stimulating factor from medium conditioned by mouse L cells. *J. Biol. Chem.* **252:** 4305.

Stanley, E.R., L.J. Guilbert, R.J. Tushinski, and S.H. Bartelmez. 1983. CSF-1—A mononuclear phagocyte lineage-specific hemopoietic growth factor. *J. Cell. Biochem.* **21:** 151.

Stanley, E.R., A. Bartocci, D. Patinkin, M. Rosendaal, and T.R. Bradley. 1986. Regulation of very primative, multipotent, hematopoietic cells by hemopoietin-1. *Cell* **45:** 667.

Sukhatme, V.P., X. Cao, L.C. Chang, C.-H. Tsai-Morris, D. Stamenkovich, P.C.P. Ferreira, D.R. Cohen, S.A. Edwards, T.B. Shows, T. Curran, M.M. Le Beau, and E.D. Adamson. 1988. A zinc finger-encoding gene coregulated with c-*fos* during growth and differentiation, and after cellular depolarization. *Cell* **53:** 37.

Sutherland, G.R., E. Baker, D.F. Callen, H.D. Campbell, I.G. Young, C.J. Sanderson, O.M. Garson, A.F. Lopez, and M.A. Vadas. 1988. Interleukin-5 is at 5q31 and is deleted in the $5q^-$ syndrome. *Blood* **71:** 1150.

Tamura, T., E. Simon, H. Niemann, G.T. Snoek, and H. Bauer. 1986. Gp140$^{v\text{-}fms}$ molecules expressed at the surface of cells transformed by the McDonough strain of feline sarcoma virus are phosphorylated in tyrosine and serine. *Mol. Cell. Biol.* **6:** 4745.

Tushinski, R.J. and E.R. Stanley. 1983. The regulation of macrophage protein turnover by a colony stimulating factor (CSF-1). *J. Cell Physiol.* **116:** 67.

———. 1985. The regulation of mononuclear phagocyte entry into S phase by the colony stimulating factor CSF-1. *J. Cell Physiol.* **122:** 221.

Tushinski, R.J., I.T. Oliver, L.J. Guilbert, P.W. Tynan, J.R. Warner, and E.R. Stanley. 1982. Survival of mononuclear phagocytes depends on a lineage-specific growth factor that the differentiated cells selectively destroy. *Cell* **28:** 71.

Ushiro, H. and S. Cohen. 1980. Identification of phosphotyrosine as a product of epidermal growth factor-activated protein kinase in A431 cell membranes. *J. Biol. Chem.* **255:** 8363.

Vairo, G. and J.A. Hamilton. 1988. Activation and proliferation signals in murine macrophages: Stimulation of Na^+, K^+-ATPase activity by hematopoietic growth factors and other agents. *J. Cell. Physiol.* **134:** 13.

Wakamiya, N., J. Horiguchi, and D. Kufe. 1987. Detection of c-*fms* and CSF-1 RNA by *in situ* hybridization. *Leukemia* **1:** 518.

Warren, D. and M.A.S. Moore. 1987. Synergy of interleukin 1 and granulocyte colony-stimulating factor: *In vivo* stimulation of stem cell recovery and hematopoietic regeneration following 5-fluorouracil treatment of mice. *Proc. Natl. Acad. Sci.* **84:** 7134.

Warren, M.K., and P. Ralph. 1986. Macrophage growth factor CSF-1 stimulates human monocyte production of interferon, tumor necrosis factor, and colony stimulating activity. *J. Immunol.* **137:** 2281.

Wheeler, E.F., C.W. Rettenmier, A.T. Look, and C.J. Sherr. 1986. The v-*fms* oncogene induces factor independence and tumorigenicity in CSF-1 dependent macrophage cell line. *Nature* **324:** 377.

Wheeler, E.F., D. Askew, S. May, J.N. Ihle, and C.J. Sherr. 1987. The v-*fms* oncogene induces factor-independent growth and transformation of the interleukin-3-dependent myeloid cell line FDC-P1. *Mol. Cell. Biol.* **7:** 1673.

Wong, G.G., P.A. Temple, A.C. Leary, J.S. Witek-Giannotti, Y.-C. Yang, A.B. Ciarletta, M. Chung, P. Murtha, R. Kriz, R.J. Kaufman, C.R. Ferenz, B.S. Sibley, K.J. Turner, R.M. Hewick, S.C. Clark, N. Yanai, H. Yokota, M. Yamada, M. Saito, K. Motoyoshi, and F. Takaku. 1987. Human CSF-1: Molecular cloning and expression of 4 kb cDNA encoding the human urinary protein. *Science* **235:** 1504.

Woolford, J., V. Rothwell, and L. Rohrschneider. 1985. Characterization of the human c-*fms* gene product and its expression in cells of the monocyte macrophage lineage. *Mol. Cell. Biol.* **5:** 3458.

Yarden, Y., W.J. Kuang, T. Yang-Feng, L. Coussens, T.J. Munemitsu, T.J. Dull, E. Chen, J. Schlessinger, U. Francke, and A. Ullrich. 1987. Human proto-oncogene c-*kit*: A new cell surface receptor tyrosine kinase for an unidentified ligand. *EMBO J.* **6:** 3341.

Yarden, Y., J.A. Escobedo, W.J. Kuang, T.L. Yang-Feng, T.O. Daniel, P.M. Tremble, E.Y. Chen, M.E. Ando, R.N. Harkins, U. Francke, V.A. Fried, A. Ullrich, and L.T. Williams. 1986. Structure of the receptor for platelet-derived growth factor helps define a family of closely related growth factor receptors. *Nature* **323:** 226.

Yeung, Y.G., P.T. Jubinsky, A. Sengupta, D.C.Y. Yeung, and E.R. Stanley. 1987. Purification of the colony-stimulating factor 1 receptor and demonstration of its tyrosine kinase activity. *Proc. Natl. Acad. Sci.* **84:** 1268.

Platelet-derived Growth Factor Generates at Least Two Distinct Intracellular Signals That Modulate Gene Expression

S.D. Jones, D.J. Hall, B.J. Rollins, and C.D. Stiles
Dana-Farber Cancer Institute, Boston, Massachusetts 02115

Platelet-derived growth factor (PDGF) binds to receptors located on the outer cell surface. From there, it stimulates transcription of between 10 and 100 genes within the cell nucleus. The proteins required for these genes to respond to PDGF already exist within the cell because the induction is not blocked (indeed, it is augmented) by inhibitors of protein synthesis. For this reason, Nathans and his colleagues named these PDGF-responsive genes "immediate early genes" (Lau and Nathans 1987). Baserga and his associates refer to them as "cell-division-cycle genes" (Hirschhorn et al. 1984). Our own laboratory refers to them as "competence genes," a term that will be used hereafter. The nuclear proto-oncogenes c-*myc*, c-*fos*, and c-*jun* are among those induced by PDGF (Kelly et al. 1983; Cochran et al. 1984; Greenberg and Ziff 1984; Kuijer et al. 1984; Ryder et al. 1988). A comprehensive review of the history, biology, and molecular biology of the competence genes is beyond the scope of this paper but can be found elsewhere (Rollins and Stiles 1988).

How does PDGF transduce a signal from the plasma membrane to the chromatin? Recent insights into the molecular biology of c-*fos* transcriptional control (reviewed elsewhere in this volume) have shed some light on this process. The laboratories of Treisman, Verma, Roeder, Ziff, and others identified *cis*-acting sequences, termed serum response elements, and *trans*-acting factors involved in activation of c-*fos* expression. Other elements in the PDGF signal transduction cascade may be illuminated through thoughtful exploitation of the *trans*-acting factors now in hand. The prospects for an orderly march backwards from the chromatin to the PDGF receptor would seem splendid, if the various other competence genes would follow the c-*fos* paradigm. Unfortunately, they may not.

The very first PDGF-responsive genes ever to be cloned and characterized as such are not coordinately regulated with c-*fos* in BALB/c-3T3 cells. The JE gene does not contain an obvious serum response element in its 5′ promoter domain, nor does c-*myc*. In this paper, we summarize evidence that PDGF generates at least two intracellular second signals and perhaps even more. We show, moreover, that the various competence genes respond selectively to one signal or to the other signals.

Not All Competence Genes Respond to Protein Kinase C Agonists

PDGF forms a complex with its cell-surface receptor, stimulating a variety of biochemical events, including an increase in the hydrolysis of phosphatidylinositol 4,5-bisphosphate to diacylglycerol and inositol trisphosphate (Glenn et al. 1982; Rozengurt 1986). Both products are involved in the subsequent intracellular events that lead to an increase in expression of specific genes (Berridge and Irvine 1984; Nishizuka 1984). Diacylglycerol is a potent activator of protein kinase C (PKC), as are the tumor-promoting phorbol esters, such as 12-*O*-tetradecanoylphorbol-13-acetate (TPA) (Nishizuka 1984). Prolonged treatment of cells with TPA causes a temporary loss of PKC activity. Hence, if PKC were the sole intracellular signal generated by PDGF, then pretreatment of cells with TPA would render PDGF ineffective at induction of gene expression. It has been reported that down-regulation of PKC diminished but did not eliminate the induction of c-*myc* by PDGF (Coughlin et al. 1985). A detailed examination of this phenomenon showed that when BALB/c-3T3 fibroblasts were pretreated with TPA, c-*fos* induction was completely blocked, c-*myc* induction was partially blocked, and both JE and KC inductions were unaffected (Fig. 1). When cells were treated with TPA for short times, expression of c-*fos* was induced to the same level and with the same kinetics as that of PDGF (Hall and Stiles 1987). c-*myc* expression was also induced but only to levels far less than those by PDGF (25%). Nuclear run-on experiments confirmed that gene expression measured by Northern blotting reflected changes in transcriptional activity. TPA did not significantly induce any expression of JE and KC genes. Therefore, the expression of at least two PDGF-inducible genes does not depend on a signal from PKC.

Two PDGF-inducible Genes Are Expressed Differentially in the Cell Cycle

An alternate approach to determine if the PDGF-inducible genes are coordinately regulated is to examine their expression during the cell cycle. Consistent with reports published previously, c-*myc* was expressed

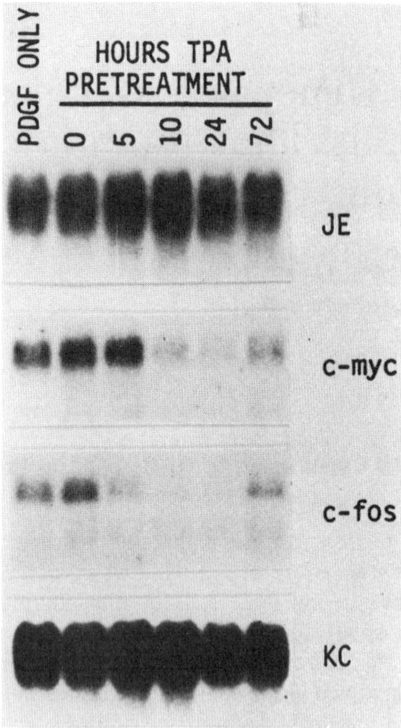

Figure 1. Down-regulation of PKC inhibits the response of c-*myc* and c-*fos* to PDGF. Cells were stimulated with TPA at 81 nM for 0–72 hr. PDGF at 1.7 nM was added for an additional 2 hr. PDGF and TPA were added simultaneously at zero time. As a control, PDGF was added alone to cells for 2 hr. Total RNA was isolated and analyzed by Northern blot. The blot was probed with c-*myc*, c-*fos*, KC, and JE cDNAs. Autoradiographs are shown. (Reprinted, with permission, from Hall and Stiles 1987.)

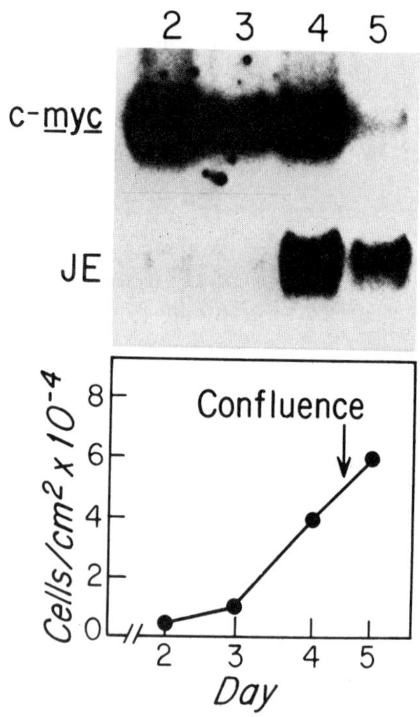

Figure 2. Effect of cell density on steady-state levels of c-*myc* and JE mRNAs. (*Top*) Cells were plated at 250,000 cells per 150-cm^2 dish in Dulbecco's modified Eagle's medium with 10% bovine calf serum, and the medium was replenished with fresh serum on days 2 and 4 after plating. RNA was collected daily (8 hr after refeeding on days 2 and 4) and analyzed by Northern blot. The same nitrocellulose filter was used to probe for both c-*myc* and JE. (*Bottom*) Cell number per dish was determined for each day. (→) The point at which confluence was reached. (Reprinted, with permission, from Rollins et al. 1987.)

at high levels during cell growth but was barely detectable once the cells were confluent (Fig. 2) (Hann et al. 1985; Thompson et al. 1985). If growing cells were deprived of serum growth factors, *myc* expression was diminished. Readdition of serum or PDGF induced its expression. Conversely, the JE gene was not expressed in growing cells until the cells approached confluence (Rollins et al. 1987). Furthermore, the JE gene was not inducible in sparsely plated cells, unless they were growth-arrested (G_0 phase) by extensive serum deprivation (72 hr). Other PDGF-inducible genes including KC and c-*fos* were only inducible by PDGF when the cells were in the G_0 phase (Rollins et al. 1987). Bravo et al. (1986) have shown that c-*fos* induction is not restricted to G_0 in a related cell line, NIH-3T3 fibroblasts. Consistent with our results, however, Kelly and Siebenlist (1988) found that there is a G_0 restriction on c-*fos* expression in macrophages. These apparently contradictory results may indicate differences in the signaling pathways in the different cell lines. The evidence points to at least two separate signals generated by PDGF in BALB/c-3T3 cells. One signal is only generated in G_0 cells and is responsible at least in part for the induction of JE, KC, and c-*fos*. The other signal is independent of the cell-cycle position and induces expression of c-*myc*.

The results presented previously suggested that activation of PKC is sufficient for full induction of c-*fos* expression, yet has no effect on expression of JE and KC. Perhaps *fos* activation requires both PKC activation and a G_0-specific signal, whereas JE and KC respond only to the G_0 signal. Studies of all these genes at the molecular level are in progress in our laboratory and others to determine how many separate transcription factors are needed for full transcriptional activation of each of these genes (Prywes and Roeder 1986; Treisman 1986; Sassone-Corsi and Verma 1987; Hayes et al. 1987). By working out from the nucleus, we may eventually understand which signals are involved in the expression of each gene.

Differential Response of Competence Genes to Interleukin

The genes that were initially cloned as members of the PDGF-inducible competence gene family respond to a variety of other agents. We reasoned that if stimuli could be found that induced only a subset of the competence genes, then additional features of the signal transduction pathways utilized by PDGF might be revealed. Interleukin-1 (IL-1) is synergistic with PDGF in inducing mitogenesis in BALB/c-3T3 fibroblasts (Singh et al. 1988). IL-1 does not induce the expression of c-*myc* or c-*fos* but induces JE and KC expression, suggesting that it generates a signal that discriminates between the JE transcriptional regulatory region and those of c-*myc* and c-*fos* (Fig. 3). Consistent with this result, IL-1 does not appear to activate PKC.

Poly(rI:rC) Differentially Regulates Competence Genes

A subset of the competence genes is induced by the synthetic double-stranded RNA poly(rI:rC) (poly-[IC]). Poly(IC) is a potent inducer of β-interferon expression in fibroblasts (Lengyel 1982). The induction of c-*myc* by poly(IC) has been described previously (Zullo et al. 1985). We found that poly(IC) induced c-*myc* and c-*fos* poorly compared to serum or PDGF (D. Hall et al., in prep.). However, poly(IC) induced both JE and KC as well as PDGF does. The c-*myc* and JE transcripts initiate at the same site, regardless of which stimulus is used. Hence, the addition of double-stranded RNA to BALB/c-3T3 fibroblasts does not have a global effect on the nature of transcription in the cells. The ability to induce a subset of the competence genes by either IL-1 or poly(IC) lends credence to the notion that more than one signal is necessary to elicit the induction of all competence genes.

Poly(IC) does not appear to accelerate the turnover of phosphatidylinositol and does not activate PKC (D. Hall et al., in prep.). This is consistent with the previous result that JE and KC induction is independent of PKC activation. Poly(IC) may enter the cell and be a signal for expression of these genes. Alternatively, it may bind to a cell-surface receptor and generate a signal separate from those generated by PDGF. Unlike PDGF, poly(IC) induces the expression of JE and KC in growing cells. Either its signal is distinct from that generated by PDGF or the cell-cycle constraint on this signal is overridden by poly(IC).

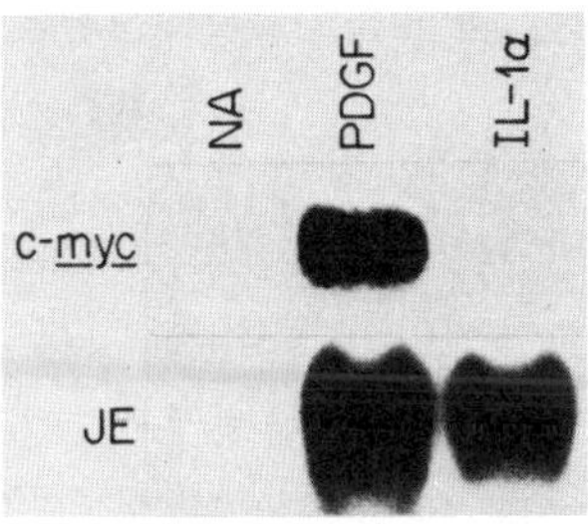

Figure 3. Induction of JE expression by IL-1α. Confluent, density-arrested monolayers of BALB/c-3T3 cells were prepared. Cells were made quiescent by placing them in 5% platelet-poor plasma and then treated with 300 μm/ml PDGF or 2.5 μm/ml recombinant IL-1α for 2 hr. RNA was collected for Northern analysis. (Reprinted, with permission, from Rollins et al. 1988.)

Preliminary Analysis of the JE Regulatory Region

Recent work in our laboratory has resulted in the cloning of more than 3000 bp of sequence upstream of the transcription initiation site of the JE gene. A preliminary examination of the sequence showed that the JE gene contains a TATA homology, a putative CCAAT box, and a perfect copy of a sequence found in some interferon-responsive genes (Rollins et al. 1988). JE is not induced by interferon, although pretreatment of BALB/c-3T3 fibroblasts with interferon causes a subsequent superinduction of JE by PDGF. No homologies with the known promoter and enhancer elements of the c-*fos* gene were found. This is consistent with the different kinetics of induction of these two genes and with the ability to separate their induction using a variety of techniques. It is likely that these two genes require different transcription factors to respond to PDGF. The JE upstream sequence contains two sites that are homologous to the binding site for factor NF-κB that binds to the immunoglobulin κ-light-chain gene enhancer. Experiments are in progress to determine if NF-κB plays a role in the regulation or transcription of the JE gene.

The Nature of the Signals

We have shown that PKC is involved in PDGF signal transduction (Hall and Stiles 1987). This kinase may have many substrates that can relay messages from the cell surface to the nucleus. The transcription factor AP-1, encoded by the c-*jun* proto-oncogene, is activated by treatment of cells with TPA (Angel et al. 1987; Bohman et al. 1987; Lee et al. 1987; Rauscher et al. 1988). Perhaps other transcription factors are also substrates for this kinase, and expression of genes whose promoter regions contain binding sites for these factors will increase in response to PKC activation. The PDGF receptor has an intrinsic tyrosine kinase activity (Ek et al. 1982; Bishayee et al. 1986), and substrates of this kinase may also mediate the cellular response to PDGF. Soluble kinases may also be activated by PDGF, and these might affect proteins whose ultimate site of action is the nucleus.

Two cellular proteins with molecular weights of 31,000 and 32,000 are phosphorylated by PDGF but not by TPA (Hall and Stiles 1987). In addition, at least three nuclear proteins were phosphorylated when BALB/c-3T3 cells were treated with PDGF or serum (Fig. 4). Further characterization of these proteins revealed that one, with an apparent molecular weight of 19,000, was soluble in 5% perchloric acid. This behavior is characteristic of the DNA-associated high-mobility group (HMG) proteins (Matthew et al. 1979). It is not known if HMG proteins are involved in the process of transcriptional activation of any genes. Experiments are in progress to determine which nuclear factors are necessary for the transcription of the competence genes. Perhaps some of the factors that we identify will be among those that are phosphorylated in response to PDGF.

Phosphorylation need not be the only method of transmitting signals from the cell surface to the nucleus. Many transcriptional activating proteins, including the steroid receptors, are activated by alterations in their structure. In the case of the steroid receptors, this alteration occurs as a result of the binding of a ligand (Holbrook et al. 1983). Another possibility, however, is that the structure of a factor can be altered simply by the change in pH or intracellular calcium. Alternately, the breakdown of some intracellular molecule could generate a ligand that could bind to a factor and alter its conformation. Resolution of these possibilities awaits the positive identification of the factors that are involved in the transcription of the competence genes.

SUMMARY AND CONCLUSIONS

Regulation of the genes by PDGF has some common features. All are primary response genes, and they can still be expressed in the presence of cycloheximide (Cochran et al. 1983; Kelly et al. 1983; Kruijer et al. 1984; Lau and Nathans 1985). In fact, many of the genes are superinduced when cells are treated with growth factors plus cycloheximide (Greenberg et al. 1986). The genes that have been characterized all contain a sequence motif in their 3′-noncoding sequences that appears to make the message labile (Meijlink et al. 1985; Treisman 1985; Shaw and Kamen 1986). All competence genes so far examined are controlled at least in part at the level of transcription (Cochran et al. 1983; Edwards et al. 1985; Almendral et al. 1988). Differences in regulation of the genes include variations in the time course of induction, ranging from 10 minutes to over 4 hours, and differences in the persistence of the mRNAs after their synthesis (Cochran et al. 1983; Muller et al. 1984; Lau and Nathans 1987). The data presented in this paper strongly suggest that multiple, distinct intracellular signals that lead to the expression of multiple genes are generated when cells are treated with growth factors such as PDGF. The variation in the time course of induction of PDGF-inducible genes suggests several models of signal transduction. Four such models are presented in Figure 5. One possibility (Fig. 5A) is that one signal is generated by the interaction of PDGF with its receptor and that this signal activates the very early genes, such as c-*fos*. Transcription of c-*fos* would then activate genes that are induced more slowly, such as c-*myc*. Experiments using antisense RNA to c-*fos* indicate that the fos protein is not necessary for induction of the later genes, such as c-*myc* (Nishikura and Murray 1987). The expression of the full set of early genes in the presence of protein synthesis inhibitors is consistent with this view.

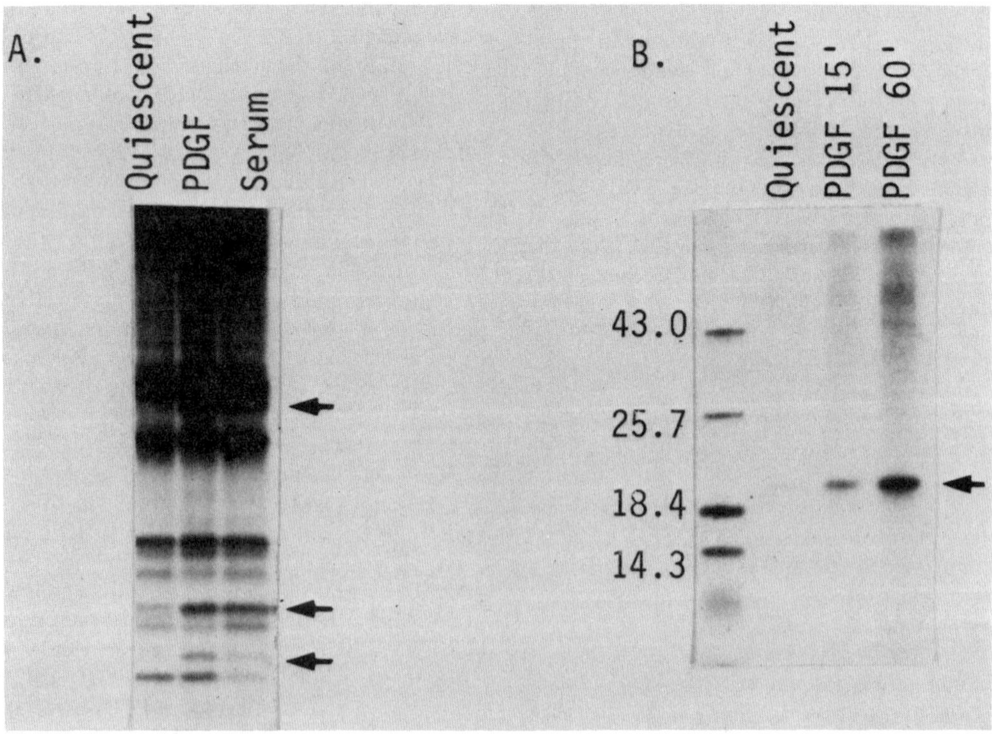

Figure 4. Induction of phosphorylation of nuclear proteins by PDGF. Quiescent, confluent BALB/c-3T3 cells were preincubated with 125 μm/ml [^{32}P]orthophosphate for 2 hr. Cells were then treated with 20% serum or 50 ng/ml PDGF. The cells were collected and lysed in NP-40 lysis buffer. Nuclei were spun at 800*g* for 1 min. (*A*) Total nuclear proteins isolated from cells treated with serum or PDGF for 60 min are shown. (→) Positions of proteins that are phosphorylated in response to these agents. (*B*) Nuclear proteins that are soluble in 5% perchloric acid.

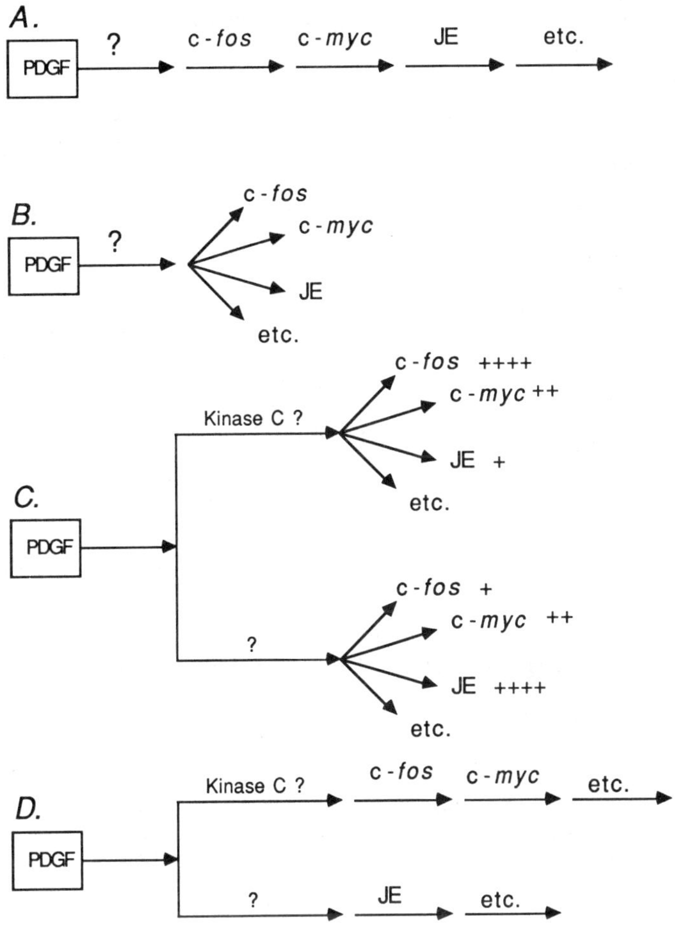

Figure 5. (*A–D*) Four models depicting possible pathways of signal transduction in PDGF-stimulated fibroblasts.

An alternate possibility is that the single signal generated by PDGF is sufficient to activate the entire cascade of early genes in a coordinate manner (Fig. 5B). If this model were accurate, all of the activated genes would require the same transcription factors. Then, solving the mechanisms of activation of one gene will solve all of them. A good understanding of c-*fos* regulation is emerging (Fisch et al. 1987; Hayes et al 1987; Sheng et al. 1988). We have shown that other genes are regulated differently from c-*fos* and therefore that model B is not accurate. The models presented in Figure 5, C and D, suggest the possibility that initial events occurring at the cell surface generate at least two signals that act independently, each turning on one or more genes. We believe that some variation of models C and D is likely to be correct.

The data presented in this paper indicate that the mechanisms controlling each gene will have to be ascertained individually. Perhaps subfamilies of genes with similar regulatory characteristics will emerge. Alternatively, each gene might respond to a combination of signals generated by the growth factor. Understanding more of these genes at a molecular level will allow us to grasp further the nature of the mechanism by which gene expression and cellular growth occurs in response to growth factors.

ACKNOWLEDGMENT

This work was supported by grants from the National Institutes of Health and the American Cancer Society.

REFERENCES

Almendral, J.M., D. Sommer, H. MacDonald-Bravo, J. Burckhardt, J. Perera, and R. Bravo. 1988. Complexity of the early genetic response to growth factors in mouse fibroblasts. *Mol. Cell. Biol.* **8:** 2140.

Angel, P., M. Imagawa, R. Chiu, B. Stein, R.J. Imbra, H.J. Rahmsdorf, C. Johat, P. Herrlich, and M. Karin. 1987. Phorbol ester-inducible genes contain a common *cis* element recognized by a TPA-modulated *trans*-acting factor. *Cell* **49:** 729.

Berridge, M.J. and R.F. Irvine. 1984. Inositol triphosphate, a novel second messenger in cellular signal transduction. *Nature* **312:** 315.

Bishayee, S., A.H. Ross, R. Womer, and C.D. Scher. 1986. Purified human platelet-derived growth factor receptor has ligand-stimulated tyrosine kinase activity. *Proc. Natl. Acad Sci.* **83:** 6756.

Bohmann, D., T.J. Bos, A. Admon, T. Nishimura, P.K. Vogt, and R. Tjian. 1987. Human proto-oncogene c-*jun* encodes a DNA binding protein with structural and functional properties of transcription factor AP1. *Science* **238:** 1386.

Bravo, R., J. Burckhardt, T. Curran, and R. Muller. 1986. Expression of c-*fos* in NIH3T3 cells is very low but inducible throughout the cell cycle. *EMBO J.* **5:** 695.

Cochran, B.H., A.C. Reffel, and C.D. Stiles. 1983. Molecular cloning of gene sequences regulated by platelet-derived growth factor. *Cell* **33:** 939.

Cochran, B.H., J. Zullo, I.M. Verma, and C.D. Stiles. 1984. Expression of the c-*fos* gene and of a *fos*-regulated gene is stimulated by platelet-derived growth factor. *Science* **226:** 1080.

Coughlin, S.R., W.M.F. Lee, P.W. Williams, G.M. Geiss, and L.T. Williams. 1985. c-*myc* gene expression is stimulated by agents that activate protein kinase C and does not account for the mitogenic effect of PDGF. *Cell* **43:** 243.

Edwards, D.R., C.L.J. Parfett, and D.T. Denhardt. 1985. Transcriptional regulation of two serum-induced RNAs in mouse fibroblasts: Equivalence of one species to B2 repetitive elements. *Mol. Cell. Biol.* **5:** 3280.

Ek, B., B. Westermark, A. Wasteson, and C.H. Heldin. 1982. Stimulation of tyrosine-specific phosphorylation by platelet-derived growth factor. *Nature* **295:** 419.

Fisch, T.M., R. Prywes, and R.G. Roeder. 1987. c-*fos* sequences necessary for basal expression and induction by epidermal growth factor, 12-*O*-tetradecanoyl phorbol-13-acetate, and the calcium ionophore. *Mol. Cel. Biol.* **7:** 3490.

Glenn, K., D.F. Bowen-Pope, and R. Ross. 1982. Platelet-derived growth factor. III. Identification of a PDGF-receptor by affinity labelling. *J. Biol. Chem.* **257:** 5172.

Greenberg, M.E. and E.B. Ziff. 1984. Stimulation of 3T3 cells induces transcription of the c-*fos* proto-oncogene. *Nature* **311:** 433.

Greenberg, M., A.L. Hermanowski, and E.B. Ziff. 1986. Effect on protein synthesis inhibitors on growth factor activation of c-*fos*, c-*myc*, and actin gene transcription. *Mol. Cell. Biol.* **6:** 1050.

Hall, D.J. and C.D. Stiles. 1987. Platelet-derived growth factor-inducible genes respond differentially to at least two distinct intracellular second messengers. *J. Biol. Chem.* **262:** 15302.

Hann, S.R., C.B. Thompson, and R.N. Eisenman. 1985. c-*myc* oncogene protein synthesis is independent of the cell cycle in human and avian cells. *Nature* **314:** 366.

Hayes, T.E., A.M. Kitchen, and B.H. Cochran. 1987. Inducible binding of a factor to the c-*fos* regulatory region. *Proc. Natl. Acad. Sci.* **84:** 1272.

Hirschhorn, R.R., P. Aller, Z.-A. Yuan, C.W. Gibson, and R. Baserga. 1984. Cell-cycle specific cDNAs from mammalian cells temperature sensitive for growth. *Proc. Natl. Acad. Sci.* **81:** 6004.

Holbrook, N.J., J.E. Bodwell, M. Jeffries, and A. Munck. 1983. Characterization of a nonactivated and activated glucocorticoid-receptor complexes from intact rat thymus cells. *J. Biol. Chem.* **258:** 6477.

Kelly, K. and U. Siebenlist. 1988. Mitogenic activation of normal T cells leads to increased initiation of transcription in the c-*myc* locus. *J. Biol. Chem.* **263:** 4828.

Kelly, K., B.H. Cochran, C.D. Stiles, and P. Leder. 1983. Cell specific regulation of c-*myc* gene by lymphocyte mitogens and platelet-derived growth factor. *Cell* **35:** 603.

Kruijer, W., J.A. Cooper, T. Hunter, and I.M. Verma. 1984. Platelet-derived growth factor induces rapid but transient expression of the c-*fos* gene and protein. *Nature* **312:** 711.

Lau, L.F. and D. Nathans. 1985. Identification of a set of genes expressed during the G0/G1 transition of cultured mouse cells. *EMBO J.* **4:** 3145.

———. 1987. Expression of a set of growth related immediate early genes in Balb/c-3T3 cells: Coordinated regulation with c-*fos* or c-*myc*. *Proc. Natl. Acad. Sci.* **84:** 1182.

Lee, W., P. Mitchell, and R. Tjian. 1987. Purified transcription factor AP-1 interacts with TPA-inducible enhancer element. *Cell* **49:** 741.

Lengyel, P. 1982. Biochemistry of interferons and their actions. *Annu. Rev. Biochem.* **51:** 251.

Matthew, C.G.P., G.H. Goodwin, and E.W. Johns. 1979. Studies on the association of the high mobility group nonhistone chromatin proteins with isolated nucleosomes. *Nucleic Acids Res.* **6:** 167.

Meijlink, F., T. Curran, A.D. Miller, and I.M. Verma. 1985. Removal of a 67-base-pair sequence in the non-coding region of a proto-oncogene *fos* converts it to a transforming gene. *Proc. Natl. Acad. Sci.* **82:** 4987.

Muller, R., R. Bravo, J. Burckhardt, and T. Curran. 1984. Induction of c-*fos* gene and protein by growth factors precedes activation of c-*myc*. *Nature* **312:** 716.

Nishikura, K. and J.M. Murray. 1987. Antisense RNA of proto-oncogene c-*fos* blocks renewed growth of quiescent 3T3 cells. *Mol. Cell. Biol.* **7:** 639.

Nishizuka, Y. 1984. The role of protein kinase C in cell surface signal transduction and tumor promotion. *Nature* **308:** 693.

Prywes, R. and R.G. Roeder. 1986. Inducible binding of a factor to the c-*fos* enhancer. *Cell* **47:** 777.

Rauscher, F.J., III, D.R. Cohen, T. Curran, T.J. Bos, P.K. Vogt, D. Bohmann, R. Tjian, and B.R. Franza, Jr. 1988. Fos-associated protein p39 is the product of the *jun* proto-oncogene. *Science* **240:** 1010.

Rollins, B.J. and C.D. Stiles. 1988. Serum-inducible genes. *Adv. Cancer Res.* (in press).

Rollins, B.J., E.D. Morrison, and C.D. Stiles. 1987. A cell cycle constraint on the regulation of gene expression by platelet-derived growth factor. *Science* **238:** 1269.

———. 1988. Cloning and expression of JE, a PDGF-inducible gene with cytokine-like properties. *Proc. Natl. Acad. Sci.* **85:** 3738.

Rozengurt, R. 1986. Early signals in the mitogenic response. *Science* **234:** 161.

Ryder, K., L.F. Lau, and D. Nathans. 1988. A gene activated by growth factors is related to the oncogene v-*jun*. *Proc. Natl. Acad. Sci.* **85:** 1487.

Sassone-Corsi, P. and I.M. Verma. 1987. Modulation of c-*fos* gene transcription by negative and positive cellular factors. *Nature* **326:** 507.

Shaw, G. and R. Kamen. 1986. A conserved AU sequence from the 3′ untranslated region of GM-CSF mRNA mediates selective mRNA degradation. *Cell* **46:** 659.

Sheng, M., S.T. Dougan, G. McFadden, and M.E. Greenberg. 1988. Calcium and growth factor pathways of c-*fos* transcriptional activation require distinct upstream regulatory sequences. *Mol. Cell. Biol.* **8:** 2787.

Singh, J.P., L.D. Adams, and P.D. Bonin. 1988. Mode of fibroblast growth enhancement by human IL-1. *J. Cell Biol.* **106:** 813.

Thompson, C.B., P.B. Challoner, P.E. Nieman, and M. Groudine. 1985. Levels of c-*myc* oncogene mRNA are invariant throughout the cell cycle. *Nature* **314:** 363.

Treisman, R. 1985. Transient accumulation of c-*fos* mRNA following serum stimulation requires a conserved 5′ element and c-*fos* 3′ sequences. *Cell* **42:** 889.

———. 1986. Identification of a protein-binding site that mediates transcriptional response of the c-*fos* gene to serum factors. *Cell* **46:** 567.

Zullo, J.N., B.H. Cochran, A.S. Huang, and C.D. Stiles. 1985. Platelet-derived growth factor and double-stranded ribonucleic acids stimulate expression of the same genes in 3T3 cells. *Cell* **43:** 793.

Insulin and Insulin-like Growth Factor Receptors and Responses

R.A. Roth,* G. Steele-Perkins,* J. Hari,* C. Stover,* S. Pierce,* J. Turner,† J.C. Edman,† and W.J. Rutter†

*Department of Pharmacology, Stanford University School of Medicine, Stanford, California 94305; †Hormone Research Laboratory, University of California, San Francisco, California 94143

Insulin is a member of a family of structurally related hormones with diverse physiological functions. The basic structure of these molecules is present even in hormones isolated from invertebrates (Thorpe and Duve 1988). In humans, the best-characterized members of this family include insulin, insulin-like growth factor (IGF)-I, and IGF-II. The physiological role of insulin is primarily to regulate rapid anabolic responses, such as glucose uptake into muscle and fat cells (Rosen 1987). IGF-I, on the other hand, appears to be primarily involved in regulating a long-term response, the growth of an organism (Rechler et al. 1987). The physiological role(s) of the third member of this family, IGF-II, is not known.

Each of these three polypeptide hormones has its own distinct receptor. The structures of each of these receptors have now been deduced from analyses of isolated cDNA clones (Ebina et al. 1985; Ullrich et al. 1985, 1986; Morgan et al. 1987). In agreement with prior biochemical data, the receptors for insulin and IGF-I have been found to be homologous. Both receptors are synthesized as a single precursor polypeptide and cleaved to generate an α subunit, which is completely extracellular, and a β subunit, which has both an extracellular and an intracellular domain. The intracellular domains of both receptors encode an intrinsic tyrosine-specific kinase activity. This region of the two receptors is most homologous, sharing 84% sequence identity. In contrast, the IGF-II receptor shows no sequence homology with either the IGF-I or insulin receptors. Most surprisingly, the deduced sequence of the human IGF-II receptor was 80% identical to the deduced sequence of the bovine receptor for mannose 6-phosphate (man-6-P) (Lobel et al. 1987), suggesting that these two proteins were the same. Additional studies have further documented that the IGF-II and man-6-P receptors are in fact the same protein (Roth 1988). Since the man-6-P receptor has been implicated in the targeting of lysosomal enzymes (Oshima et al. 1988), these studies have raised the possibility that the binding of IGF-II to its receptor is not a signal-transducing event.

To study further the responses mediated through these three different receptors, we have been studying cells expressing the proteins encoded by these three cDNAs. The isolated cDNAs have been transfected into Chinese hamster ovary (CHO) cells, and the resulting transfected cell lines have been characterized as to the ligand-binding activities and signal-transducing activities of the expressed proteins.

METHODS

Preparation of cell lines. Transfected CHO cells expressing the human insulin receptor and IGF-I receptor were prepared as described previously (Ellis et al. 1986; Ebina et al. 1987; Steele-Perkins et al. 1988).

Binding studies. Confluent 24-well culture plates of cells were washed and incubated with the indicated labeled ligand (50–150 pM) and unlabeled competitor for 5 hours at 4°C. The cells were washed twice at 4°C and then solubilized in 0.05% SDS and counted in a γ counter.

For measurements of binding to isolated receptors, a plate-binding assay was utilized (Morgan and Roth 1985). In brief, microtiter wells were sequentially coated with the appropriate antibody and receptor and then incubated with the iodinated ligand. After incubations at 4°C, wells were washed three times, cut out, and counted.

Biological studies. Confluent monolayers of the indicated cells in 24-well plates were incubated in serum-free Ham's F-12 medium for 36 hours at 37°C. Hormones and fresh medium were added, and the cells were incubated for an additional 8 hours. Cells were then pulsed with 0.75 μCi of [*methyl*-^{3}H]thymidine for 45 minutes, washed two times, and lysed with 0.075% SDS. Trichloroacetic acid (final concentration 10%) was added to the lysate, and the resulting precipitate was collected by filtration on Whatman glass-fiber filters, washed with 5% trichloroacetic acid, and counted.

Phosphorylation studies. Confluent 100-mm plates of cells were washed and then treated with hormone in serum-free Ham's F-12 media for 10 minutes at 37°C. In some experiments, the cells also received either 35 μM phenylarsene oxide or 0.5 mM vanadate. The cells were lysed, and the lysates were electrophoresed on 10% polyacrylamide SDS gels, transferred to nitrocellulose filters, and incubated with polyclonal rabbit antiphosphotyrosine antibodies. After 16 hours at 4°C, the filters were washed, and the bound rabbit immunoglobulin was detected by use of alkaline phosphatase-conjugated anti-rabbit immunoglobulin.

RESULTS

Binding Studies

The CHO cells expressing the transfected cDNAs for the human insulin and IGF-I receptors were tested for their ability to bind insulin, IGF-I, and IGF-II. These studies were complicated by the presence of endogenous insulin and IGF-I receptors in the parental CHO cells utilized in the transfected cells. However, the levels of these endogenous receptors are approximately tenfold lower than that of the expressed receptors (Ellis et al. 1986; Steele-Perkins et al. 1988). Thus, they were not found to affect these binding studies significantly. The expressed insulin receptor was found to bind insulin with the highest affinity (apparent $K_d \sim 0.3$ nM), IGF-II with approximately 100-fold weaker affinity, and IGF-I with approximately 500-fold weaker affinity (Fig. 1). In contrast, the expressed IGF-I receptor was found to bind IGF-I with highest affinity (apparent $K_d \sim 1$ nM) and insulin with approximately 100-fold weaker affinity (Fig. 2).

Surprisingly, the expressed IGF-I receptor was found to also bind IGF-II with high affinity (apparent $K_d \sim 3$ nM) (Fig. 2). To determine whether this high affinity for IGF-II was representative of other IGF-I receptors, binding studies were also performed with IGF-I receptors isolated from human fetal brain and rat hepatomas. As with the expressed IGF-I receptor, IGF-II was found to also bind to these other IGF-I receptors with high affinity (Fig. 3).

The IGF-II receptor purified from fetal brain was also tested for its ability to bind the three hormones. This receptor was found to bind IGF-II with high affinity (apparent $K_d \sim 1$ nM). However, this binding was not significantly inhibited by either 100 nM IGF-I or 100 nM insulin (Fig. 4). Additional studies with the IGF-II receptor from rat hepatoma cells also indicated that this receptor did not bind either insulin or IGF-I (data not shown).

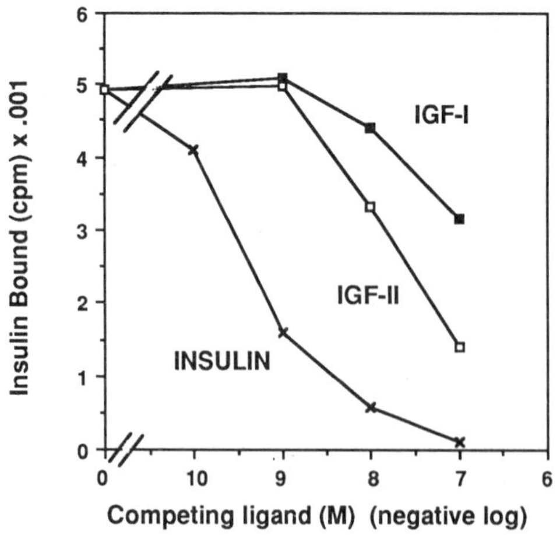

Figure 1. Specificity of the expressed human insulin receptor. [^{125}I]Insulin was incubated with CHO cells expressing the transfected human insulin receptor cDNA in the presence of the indicated concentrations of unlabeled ligands. Bound [^{125}I]insulin was determined as described in Methods.

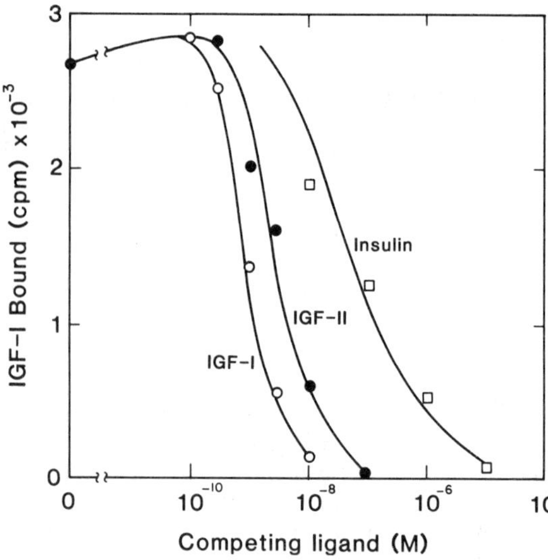

Figure 2. Specificity of the expressed IGF-I receptor. [^{125}I]IGF-I was incubated with CHO cells expressing the transfected human IGF-I receptor cDNA in the presence of the indicated concentrations of unlabeled ligands.

Biological Studies

The presence of endogenous insulin and IGF-I receptors in the parental CHO cells utilized in the transfection studies also complicated the biological studies. However, the expression of increased levels of a functional receptor would be expected to shift the dose response of the transfected cells to the left; i.e., the transfected cells should be more sensitive to low levels of hormone than the parental CHO cells.

Indeed, cells expressing the transfected IGF-I and insulin receptors were found to be more responsive to stimulation of a biological response by their respective ligands than the parental CHO cells (Ellis et al. 1986; Ebina et al. 1987; Steele-Perkins et al. 1988). For thymidine synthesis, a long-term response, the extent of shift in the dose response, and the maximal response to hormone were approximately the same in transfected cells expressing approximately the same number of either human IGF-I or insulin receptors (Fig. 5). These results suggested that there is no intrinsic difference in these two receptors to mediate a long-term response. This conclusion is also supported by the finding that the response of the parental CHO cells to these two hormones was also not significantly different (Fig. 5). Similarly, these two receptors were found to have comparable abilities to stimulate intermediate and rapid biological responses such as glucose uptake and glycogen synthesis (Steele-Perkins et al. 1988).

To verify that the human IGF-I receptor was responsible for the increased responsiveness of the transfected cells, we utilized a human-specific monoclonal

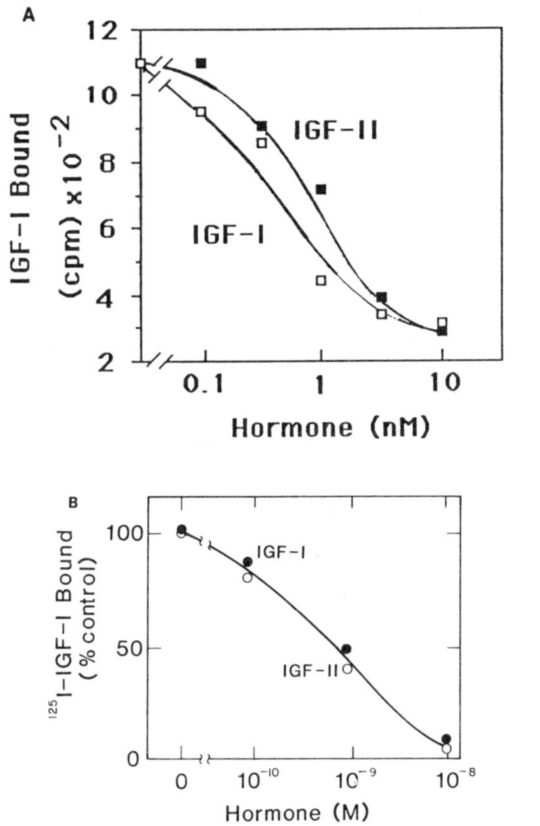

Figure 3. Specificity of the purified human and rat IGF-I receptors. [^{125}I]IGF-I was incubated with purified IGF-I receptor from either fetal human brain (*A*) or BRL-3A2 rat hepatomas (*B*). Binding was performed on receptor adsorbed to microtiter wells as described in Methods.

antibody to this receptor (Kull et al. 1983). This antibody does not recognize the endogenous hamster receptor (Steele-Perkins et al. 1988) and was found to stimulate a variety of responses in the transfected cells but not in the parental CHO cells, including the stimulation of thymidine incorporation (Table 1), glucose

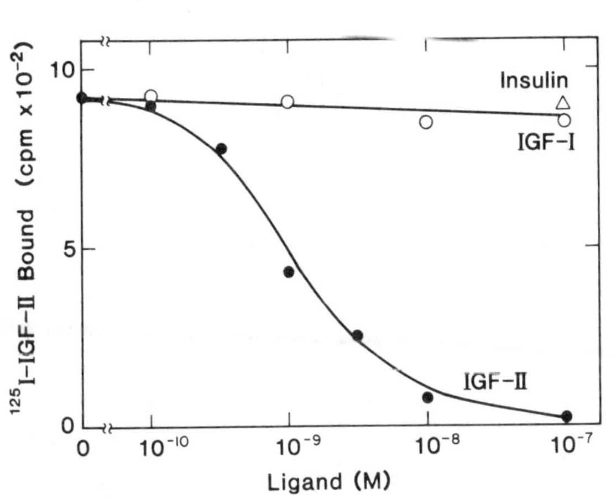

Figure 4. Specificity of the purified human IGF-II receptor. [^{125}I]IGF-II was incubated with purified IGF-II receptor from human brain in the presence of the indicated concentrations of unlabeled ligands. Binding was performed on receptor adsorbed to microtiter wells as described in Methods.

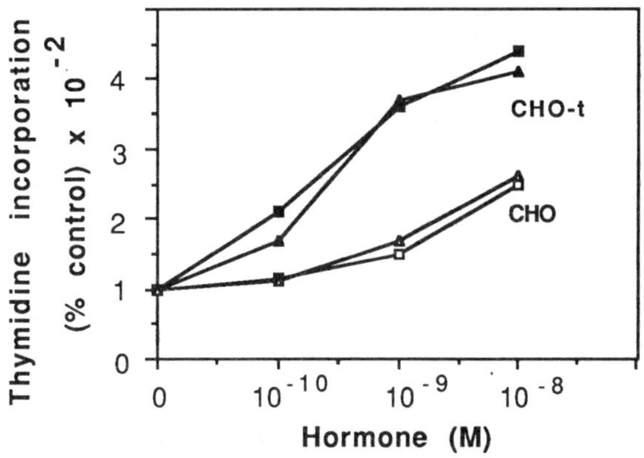

Figure 5. Stimulation of thymidine incorporation. Either the parental CHO cells (△, □), CHO cells expressing the human insulin receptor (■), or CHO cells expressing the human IGF-I receptor (▲) were treated with the indicated concentration of either insulin (□) or IGF-I (△) and tested for their incorporation of [^{3}H]thymidine as described in Methods. (CHO-t) Transfected CHO cells; (CHO) parental cells.

uptake, and glycogen synthesis (Steele-Perkins 1988). The effect of this monoclonal antibody on thymidine synthesis was observed with antibody purified from ascites fluid as well as antibody purified from hybridoma supernatants (Table 1). However, the maximal responses to the antibody were only one-third to one-half the maximal responses observed with IGF-I (Steele-Perkins et al. 1988).

Phosphorylation Studies

To examine the endogenous substrates of the intrinsic tyrosine kinase activities of the insulin and IGF-I receptors, the cells were treated with hormone and lysed, and the lysates were analyzed by Western blotting with a polyclonal antibody to phosphotyrosine. Insulin and IGF-I were both found to stimulate the tyrosine phosphorylation of proteins of $M_r \sim 95{,}000$ and 170,000 (Fig. 6). The proteins of $M_r \sim 95{,}000$ are presumably the β subunits of the receptors for these two ligands. The proteins of $M_r \sim 170{,}000$ may be the same as a protein of approximately the same molecular weight (first described by White et al. 1985) as an endogenous substrate for the insulin receptor in hepatoma cells. To examine other potential endogenous substrates for the two receptors, the cells were treated with hormone in the presence of phenylarsine oxide. Since this compound, like vanadate, can increase the detection of endogenous substrates of the insulin receptor kinase (Fig. 7), it may act like vanadate to inhibit phosphatases. Again, proteins of similar molecular weights were phosphorylated in response to either insulin or IGF-I in the presence of phenylarsine oxide (Fig. 6).

Since the monoclonal antibody to the human IGF-I receptor was found to stimulate biological responses, it was of interest to determine whether this antibody could stimulate the tyrosine phosphorylation of either the IGF-I receptor itself or any of the endogenous

Table 1. Stimulation of [^{3}H]Thymidine Incorporation into DNA by Antibodies to the Human IGF-I Receptor

Cells	Treatment	[^{3}H]Thymidine incorporated[a] (cpm) × 10^3
Parental CHO	none	2.27 ± .15
	10 nM IGF-I	6.96 ± .65
	1 nM αIR-3 purified from ascites	2.49 ± .17
	1 nM αIR-3 purified from hybridoma supernatants	2.24 ± .54
CHO cells expressing the human IGF-I receptor	none	2.59 ± .73
	10 nM IGF-1	6.78 ± .42
	1 nM αIR-3 purified from ascites	3.60 ± .14
	1 nM αIR-3 purified from hybridoma supernatants	3.99 ± .11

[a]The indicated cells were treated with either buffer, IGF-I, or the monoclonal antibody to the human IGF-I receptor (αIR-3). The amount of incorporated [^{3}H]thymidine was measured as described in Methods.

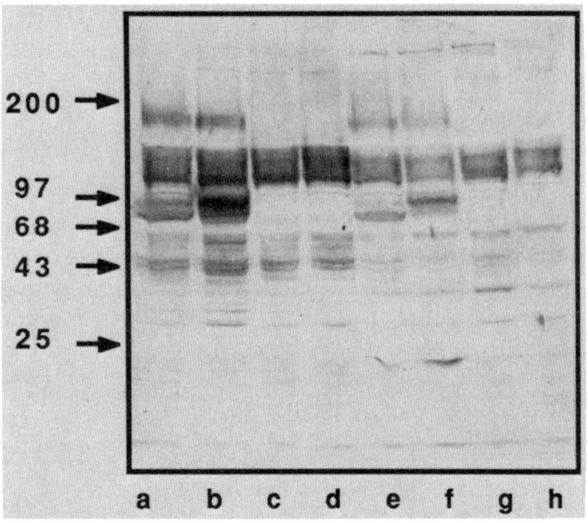

Figure 6. Endogenous substrates of the insulin and IGF-I receptors. CHO cells expressing the human insulin (lanes *a, c, e, g*) or IGF-I (lanes *b, d, f, h*) receptors were incubated for 10 min with either buffer (lanes *c, d, g, h*), 100 nM insulin (lanes *a, e*), or 100 nM IGF-I (lanes *b, f*). Cells were lysed, and the lysates were analyzed by Western blotting with antiphosphotyrosine antibodies. In some experiments, cells were pretreated with 35 μM phenylarsine oxide (lanes *a–d*).

substrates of the IGF-I receptor. Transfected cells expressing the human IGF-I receptor were therefore treated with the antibody, lysed, and analyzed by Western blotting with the antiphosphotyrosine antibody. This monoclonal antibody was not found to stimulate the tyrosine phosphorylation of either the receptor or any of the endogenous substrates, even in the presence of phenylarsine oxide (Fig. 8).

DISCUSSION

The presence in most cells of receptors for insulin, IGF-I, and IGF-II has complicated the assignment of particular responses to a specific receptor. In addition, the difficulties in isolating large amounts of pure IGF-I and IGF-II have led to problems in obtaining clear-cut data with these different receptors. Recently, recombinant DNA techniques have been utilized to solve these problems. Large amounts of homogeneous IGF-I and IGF-II can now be produced (Guler et al. 1987; Furman et al. 1987). Moreover, two of the three receptors have now been expressed from their cDNAs, and the third has been purified to homogeneity.

Binding studies with the recombinant hormones have clarified our understanding of the specificities of the three receptors. For example, initial studies suggested that the IGF-II receptor could bind IGF-I with approximately 1/100 the affinity of IGF-II (Rechler and Nissley 1985). However, studies with the recombinantly produced IGF-I with both whole cells (Ewton et al. 1987; Rosenfeld et al. 1987) and purified receptor (Fig. 4) have now shown that the IGF-II receptor binds IGF-I with less than 1/1000 the affinity of IGF-II. The earlier studies were probably incorrect because of the presence of contaminating IGF-II in the IGF-I preparations. Similarly, previous reports of the affinity of the IGF-I receptor for IGF-II are probably also incorrect. Several studies suggested that the IGF-I receptor binds IGF-II with approximately 1/50 the affinity of IGF-I. However, the present studies indicate that the IGF-I receptor binds IGF-II with almost equal affinity as IGF-I (Fig. 2). This was true for both the IGF-I receptor isolated from rat hepatomas and human brain and the IGF-I receptor expressed in CHO cells from its cDNA (Fig. 3). These studies cannot, however, rule out the existence of another more selective IGF-I receptor. Evidence for other IGF receptors has been presented recently (Jonas et al. 1986; Morgan and Roth 1986; Tollefsen et al. 1987).

The high affinity of the IGF-I receptor for IGF-II is consistent with a role for this receptor in mediating the responses of cells to IGF-II. Prior studies have found it difficult to demonstrate a biological response through the IGF-II receptor (Mottola and Czech 1984; Kiess et al. 1987). With the recent finding that the IGF-II receptor is in fact the receptor for man-6-P (Morgan et al. 1987), it is possible that the binding of IGF-II to its own receptor is not related to generating a biological response. Instead, the binding of IGF-II to the IGF-I receptor may be responsible for the initiation of the signal that causes the cells to respond to IGF-II. However, the presence of the IGF-II receptor could modulate the amount of IGF-II available for interaction with the IGF-I receptor. It is also still possible that the

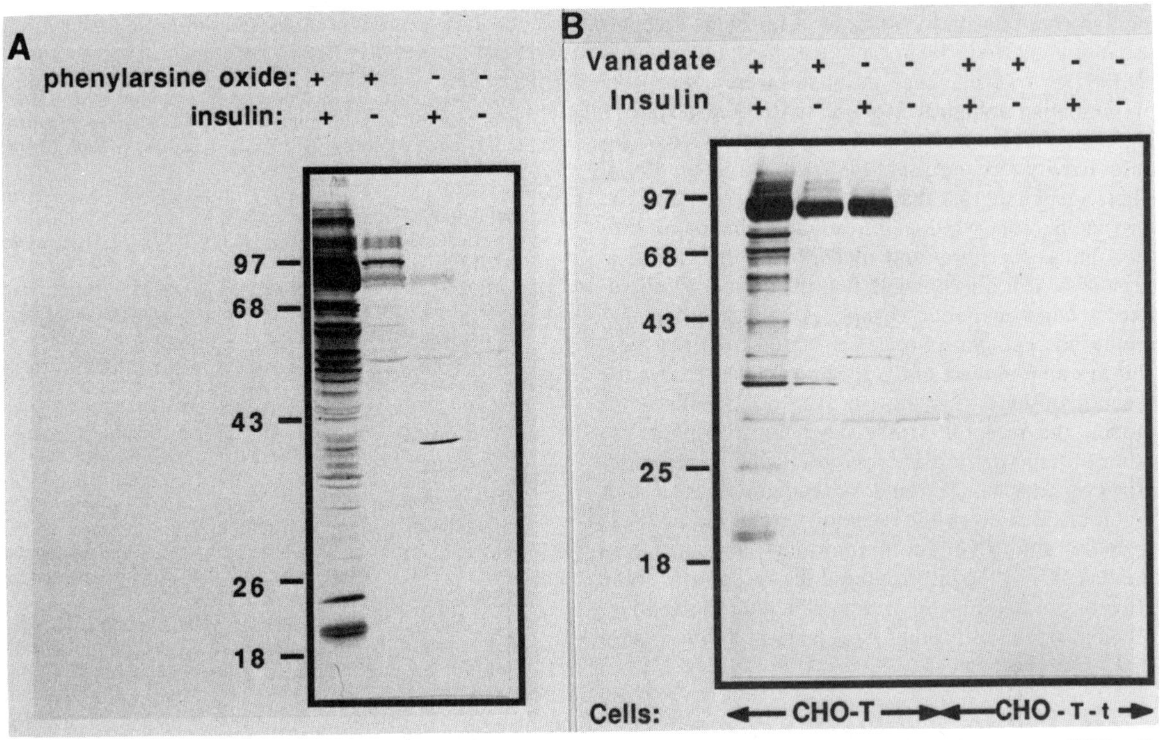

Figure 7. Effect of phenylarsine oxide and vanadate on the detection of endogenous substrates of the insulin receptor. CHO cells expressing the human insulin receptor (CHO-T) (*A*, *B*) or a mutated insulin receptor lacking kinase activity (CHO-T-t) (*B*) (Ellis et al. 1986) were treated with 100 nM insulin in the presence or absence of phenylarsine oxide (*A*) or vanadate (*B*), as indicated. Cells were lysed, and the lysates were analyzed by Western blotting with antiphosphotyrosine antibodies as described in Methods.

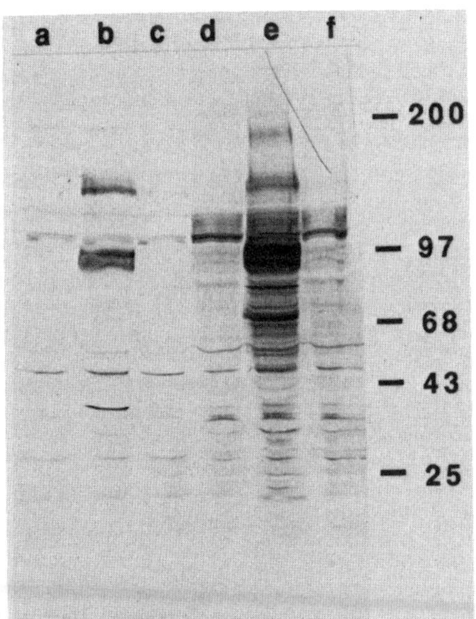

Figure 8. Effect of IGF-I and a monoclonal antibody on tyrosine phosphorylation of endogenous substrates and the IGF-I receptor. CHO cells expressing the human IGF-I receptor were treated with either buffer (lane *a*), 10 nM IGF-I (lane *b*), or 10 nM αIR-3 (lane *c*), lysed, and the lysates were analyzed by immunoblotting with antiphosphotyrosine antibodies. In another experiment, cells were treated with 35 μM phenylarsine oxide first, and then buffer (lane *d*), 10 nM IGF-I (lane *e*), or 10 nM αIR-3 (lane *f*) was added.

IGF-II receptor can initiate a response after binding IGF-II (Nishimoto et al. 1987; Hari et al. 1987).

The ability of the IGF-I receptor to stimulate thymidine incorporation was found to be comparable to the insulin receptor (Fig. 5). This was true for both the endogenous receptors present in the parental CHO cells and the expressed receptors in the transfected cells (Fig. 5). In addition, the insulin and IGF-I receptors were found to have similar potencies in stimulating short and intermediate responses (Steele-Perkins et al. 1988). These results would suggest that the different physiological roles of insulin and IGF-I in stimulating short- and long-term responses in vivo (Rechler and Nissley 1985; Rosen 1987) are dictated by the pharmacodynamics of the two hormones and/or the distribution of the two receptors on different cell types and not by any intrinsic difference in the signal-transducing activities of the two receptors. In support of this hypothesis was the finding that the endogenous substrates for the two receptors are very similar (Fig. 6).

Also surprising was the finding that a monoclonal antibody to the IGF-I receptor could stimulate several biological responses without stimulating tyrosine phosphorylation of the IGF-I receptor or any of the endogenous substrates (Steele-Perkins et al. 1988). Prior studies with several polyclonal and monoclonal antibodies to the insulin receptor have also indicated that these antibodies could act as agonists without stimulating tyrosine phosphorylation of the insulin receptor (Simpson and Hedo 1984; Zick et al. 1984; Forsayeth et al. 1987). However, these antibodies were not capable of stimulating biological responses in CHO cells expressing mutated receptors that were lacking tyrosine kinase activity (Ebina et al. 1987; Gherzi et al. 1987). These results suggest that the antibodies stimulate the tyrosine phosphorylation of a very low-abundance protein that is the important mediator of the biological responses. It is also possible that the antibodies stimulate a conformational change in the receptor that mimics the autophosphorylation state of the receptor, and this altered form of the receptor then activates the function of some other protein. According to this latter model, the receptor kinase need not phosphorylate other proteins on tyrosine residues to elicit a response. However, this model would also require that the point mutations that block the receptor kinase also interfere with the antibody-induced conformational change in the receptor or the interaction of the receptor with the effector protein. Finally, it is possible that the antibodies elicit responses through a mechanism different from that of the hormones.

ACKNOWLEDGMENTS

We are grateful to Dr. J. Merryweather (Chiron Corp.) for IGF-I, Dr. M. Smith (Eli Lilly) for IGF-II, Dr. J. Wang (U.C.-San Diego) for the antiphosphotyrosine antibodies, Dr. S. Jacobs (Wellcome Labs) for αIR-3, and Karen Bird for preparation of the manuscript. This work was supported by National Institutes of Health grants DK-34926 and DK-01393.

REFERENCES

Ebina, Y., E. Araki, M. Taira, F. Shimada, M. Mori, C.S. Craik, K. Siddle, S.B. Pierce, R.A. Roth, and W.J. Rutter. 1987. Replacement of lysine residue 1030 in the putative ATP-binding region of the insulin receptor abolishes insulin- and antibody-stimulated glucose uptake and receptor kinase activity. *Proc. Natl. Acad. Sci.* **84:** 704.

Ebina, Y., L. Ellis, K. Jarnagin, M. Edery, L. Graf, E. Clauser, J.-H. Ou, F. Masiarz, Y.W. Kan, I.D. Goldfine, R.A. Roth, and W.J. Rutter. 1985. The human insulin receptor cDNA: The structural basis for hormone-activated transmembrane signalling. *Cell* **40:** 747.

Ellis, L., E. Clauser, D.O. Morgan, M. Edery, R.A. Roth, and W.J. Rutter. 1986. Replacement of insulin receptor tyrosine residues 1162 and 1163 compromises insulin-stimulated kinase activity and uptake of 2-deoxyglucose. *Cell* **45:** 721.

Ewton, D.Z., S.L. Falen, and J.R. Florini. 1987. The type II insulin-like growth factor (IGF) receptor has low affinity for IGF-I analogs: Pleiotypic actions of IGFs on myoblasts are apparently mediated by the type I receptor. *Endocrinology* **120:** 115.

Forsayeth, J.R., A. Montemurro, B.A. Maddux, R. DePirro, and I.D. Goldfine. 1987. Effect of monoclonal antibodies on human insulin receptor autophosphorylation, negative cooperativity, and down-regulation. *J. Biol. Chem.* **262:** 4134.

Furman, T.C., J. Epp, H.M. Hsiung, J. Hoskins, G.L. Long, L.G. Mendelsohn, B. Schoner, D.P. Smith, and M.C. Smith. 1987. Recombinant human insulin-like growth factor II expressed in *Escherichia coli*. *Bio/Technology* **5:** 1047.

Gherzi, R., D.S. Russell, S.I. Taylor, and O.M. Rosen. 1987. Reevaluation of the evidence that an antibody to the insulin receptor is insulinmimetic without activating the protein tyrosine kinase activity of the receptor. *J. Biol. Chem.* **262:** 16900.

Guler, H.-P., J. Zapf, and E.R. Froesch. 1987. Short-term metabolic effects of recombinant human insulin-like growth factor I in healthy adults. *N. Engl. J. Med.* **317:** 137.

Hari, J., S.B. Pierce, D.O. Morgan, V. Sara, M.C. Smith, and R.A. Roth. 1987. The receptor for insulin-like growth factor II mediates an insulin-like response. *EMBO J.* **6:** 3367.

Jonas, H.A., J.D. Newman, and L.C. Harrison. 1986. An atypical insulin receptor with high affinity for insulin-like growth factors copurified with placental insulin receptors. *Proc. Natl. Acad. Sci.* **83:** 4124.

Kiess, W., J.F. Haskell, L. Nee, L.A. Greenstein, B.E. Miller, A.L. Aarons, M.M. Rechler, and S.P. Nissley. 1987. An antibody that blocks insulin-like growth factor (IGF) binding to the type II IGF receptor is neither an agonist nor an inhibitor of IGF-stimulated biologic responses in L6 myoblasts. *J. Biol. Chem.* **262:** 12745.

Kull, Jr., F.C., S. Jacobs, Y.-F. Su, M.E. Svoboda, J.J. Van Wyk, and P. Cuatrecasas. 1983. Monoclonal antibodies to receptors for insulin and somatomedin-C. *J. Biol. Chem.* **258:** 6561.

Lobel, P., N.M. Dahms, J. Breitmeyer, J.M. Chirgwin, and S. Kornfeld. 1987. Cloning of the bovine 215-kDa cation-independent mannose 6-phosphate receptor. *Proc. Natl. Acad. Sci.* **84:** 2233.

Morgan, D.O. and R.A. Roth. 1985. Plate binding assay for monoclonal anti-receptor antibodies. *Endocrinology* **116:** 1224.

———. 1986. Identification of a monoclonal antibody which can distinguish between two distinct species of the type I receptor for insulin-like growth factor. *Biochem. Biophys. Res. Commun.* **138:** 1341.

Morgan, D.O., J.C. Edman, D.N. Standring, V.A. Fried, M.C. Smith, R.A. Roth, and W.J. Rutter. 1987. Insulin-like growth factor II receptor as a multifunctional binding protein. *Nature* **329:** 301.

Mottola, C. and M.P. Czech. 1984. The type II insulin-like growth factor receptor does not mediate increased DNA synthesis in H-35 hepatoma cells. *J. Biol. Chem.* **259:** 12705.

Nishimoto, I., Y. Ohkuni, E. Ogata, and I. Kojima. 1987. Insulin-like growth factor II increases cytoplasmic free calcium in competent BALB/c 3T3 cells treated with epidermal growth factor. *Biochem. Biophys. Res. Commun.* **142:** 275.

Oshima, A., C.M. Nolan, J.W. Kyle, J.H. Grubb, and W.S. Sly. 1988. The human cation-independent mannose 6-phosphate receptor. *J. Biol. Chem.* **263:** 2553.

Rechler, M.M. and S.P. Nissley. 1985. The nature and regulation of the receptors for insulin-like growth factor. *Annu. Rev. Physiol.* **47:** 425.

Rechler, M.M., S.P. Nissley, and J. Roth. 1987. Hormonal regulation of human growth. *N. Engl. J. Med.* **316:** 941.

Rosen, O.M. 1987. After insulin binds. *Science* **237:** 1452.

Rosenfeld, R.G., C.A. Conover, D. Hodges, P.D.K. Lee, P. Misra, R.L. Hintz, and C.H. Li. 1987. Heterogeneity of insulin-like growth factor I affinity for the insulin-like growth factor-II receptor: Comparison of natural, synthetic and recombinant DNA-derived insulin-like growth factor-I. *Biochem. Biophys. Res. Commun.* **143:** 199.

Roth, R.A. 1988. Structure of the receptor for insulin-like growth factor II: The puzzle amplified. *Science* **239:** 1269.

Simpson, I.A. and J.A. Hedo. 1984. Insulin receptor phosphorylation may not be a prerequisite for acute insulin action. *Science* **223:** 1301.

Steele-Perkins, G., J. Turner, J.C. Edman, J. Hari, S.B. Pierce, C. Stover, W.J. Rutter, and R.A. Roth. 1988. Expression and characterization of a functional human insulin-like growth factor I receptor. *J. Biol. Chem.* **263:** 11486.

Thorpe, A. and H. Duve. 1988. Invertebrate neuroendocrinology: Insulin found at last? *Nature* **331:** 483.

Tollefsen, S.E., K. Thompson, and D.J. Petersen. 1987. Separation of the high affinity insulin-like growth factor I receptor from low affinity binding sites by affinity chromatography. *J. Biol. Chem.* **262:** 16461.

Ullrich, A., A. Gray, A.W. Tam, T. Yang-Feng, M. Tsubokawa, C. Collins, W. Henzel, T. Le Bon, S. Kathuria, E. Chen, S. Jacobs, U. Francke, J. Ramachandran, and Y. Fujita-Yamaguchi. 1986. Insulin-like growth factor I receptor primary structure: Comparison with insulin receptor suggests structural determinants that define functional specificity. *EMBO J.* **5:** 2503.

Ullrich, A., J.R. Bell, E.Y. Chen, R. Herrera, L.M. Petruzzelli, T.J. Dull, A. Gray, L. Coussens, Y.-C. Liao, M. Tsubokawa, A. Mason, P.M. Seeburg, C. Grunfeld, O.M. Rosen, and J. Ramachandran. 1985. Human insulin receptor and its relationship to the tyrosine kinase family of oncogenes. *Nature* **313:** 756.

White, M.F., R. Maron, and C.R. Kahn. 1985. Insulin rapidly stimulates tyrosine phosphorylation of a M_r-185,000 protein in intact cells. *Nature* **318:** 183.

Zick, Y., R.W. Rees-Jones, S.I. Taylor, P. Gorden, and J. Roth. 1984. The role of antireceptor antibodies in stimulating phosphorylation of the insulin receptor. *J. Biol. Chem.* **259:** 4396.

Response and Binding Elements for Ligand-dependent Positive Transcription Factors Integrate Positive and Negative Regulation of Gene Expression

M.G. ROSENFELD,*† C.K. GLASS,* S. ADLER,* E.B. CRENSHAW III,*‡ X. HE,*‡
S.A. LIRA,*¶ H.P. ELSHOLTZ,*† H.J. MANGALAM,*¶ J.M. HOLLOWAY,*§
C. NELSON,*‡ V.R. ALBERT,* AND H.A. INGRAHAM*

**Eukaryotic Regulatory Biology Program, †Howard Hughes Medical Institute, Departments of ‡Biology, §Chemistry, and ¶Physiology/Pharmacology, School of Medicine, University of California-San Diego, La Jolla, California 92093*

Accurate, regulated initiation of mRNA transcription by RNA polymerase II is dependent on the actions of a variety of positive and negative *trans*-acting factors that bind *cis*-acting promoter and enhancer elements (Serfling et al. 1985; Echols 1986; McKnight and Tjian 1986; Ptashne 1986; Struhl 1987). These transcription factors may exert their actions in a tissue-specific manner (see, e.g., Walker et al 1983; Ott et al. 1984; Wright et al. 1984; Ornitz et al. 1985; Hammer et al. 1987; Nelson et al. 1988) or function under control of plasma membrane or intracellular ligand-dependent receptors. A major goal in our laboratory has been to identify the molecular mechanisms responsible for the serial activation of hormone-encoding genes in the pituitary during development and the positive and negative regulation of their transcription.

The anterior pituitary gland, which is derived from a common primordium originating in Rathke's pouch, contains phenotypically distinct cell types, each of which expresses unique trophic hormones: adrenocorticotropic hormone, thyroid-stimulating hormone (TSH), prolactin, growth hormone, and follicle-stimulating hormone/luteinizing hormone. The structurally related prolactin and growth hormone genes are expressed in lactotrophs and somatotrophs, respectively, with their expression virtually limited to the pituitary gland. Prolactin and growth hormone are characteristically the last pituitary hormones expressed during development (Chetelain et al. 1979; Watanabe and Daikoku 1979; Oliver et al. 1980; Gash et al. 1982; Khorram et al. 1984; Hoeffler et al. 1985). The reported transient coexpression of these two structurally related neuroendocrine genes, within >95% prolactin-producing cells prior to the appearance of mature lactotrophs, raises the possibility that the prolactin and growth hormone genes are developmentally controlled by a common factor(s).

The prolactin and growth hormone genes are under multifactorial regulation. Prolactin gene transcription is rapidly, positively regulated by the thyrotropin-releasing hormone (TRH), estrogen, epidermal growth factor (EGF), and phorbol esters. Glucocorticoids result in inhibition of prolactin gene transcription. Conversely, the growth hormone gene is under positive regulation by thyroid hormone but not by phorbol esters or TRH. Regulation of prolactin and growth hormone genes by steroid and thyroid hormones permits analyses of the molecular mechanisms by which ligand-dependent transcription factors modulate the transcription rates of tissue-specific genes.

The steroid and thyroid hormone receptors are members of a super gene family of ligand-dependent transcription factors that contain a highly conserved domain mediating DNA binding (for review, see Evans 1988). Transcriptional activation requires that the ligand-activated receptors bind to *cis*-active regulatory elements in target genes. Because severely truncated steroid receptor mutants, containing little more than the DNA-binding domain, are usually much less effective at stimulating transcription than the wild-type receptor (see, e.g., Giguere et al. 1986; Miesfeld et al. 1986; Kumar et al. 1987), determinants required for the quantitative levels of transcriptional enhancement directed by the wild-type receptors are likely to reside outside of the DNA-binding domain.

In addition to their positive transcriptional effects, steroid and thyroid hormones negatively regulate gene expression. For example, thyroid hormone negatively regulates the transcription of the TSH gene (Shupnick et al. 1985) and several members of the myosin heavy-chain gene family (Izumo et al. 1986), whereas glucocorticoids negatively regulate prolactin gene expression (Camper et al. 1985). The general mechanisms involved in negative transcriptional control by thyroid and steroid hormones have not been defined.

The *cis*-active elements that mediate the positive transcriptional effects of estrogen, glucocorticoids, and progesterone are variations of a common DNA sequence motif (Payvar et al. 1981; Jost et al. 1984; Klein-Hipass et al. 1986; Glass et al. 1988). Response elements for the glucocorticoid receptor and progesterone receptor appear to be functionally indistinguishable; elements that mediate transcriptional activation in response to glucocorticoid also mediate progesterone responsiveness (Strähle et al. 1987). In contrast, estrogen and glucocorticoid response elements appear to be closely related but functionally distinct. A symmetric two-base change in a palindromic 15-bp sequence interconverts it between a glucocorticoid response element (GRE) and an estrogen response element (ERE) (Klock et al. 1987; Martinez et al. 1987).

In this paper, we review data that suggest that a tissue-specific positive transcription factor(s), referred to as pituitary transcriptional activator 1 (Pit-1), is responsible for pituitary-specific activation of the rat prolactin and growth hormone genes. The requirements for correct cell-type restriction have been defined using transgenic mice. Pit-1-binding *cis*-active elements also appear to mediate the transcriptional regulation of the prolactin gene by hypothalamic peptides and phorbol esters. The rat prolactin and growth hormone genes can be negatively regulated by steroid hormones. One mechanism for negative regulation reflects the unexpected property of estrogen and T_3 receptors to bind competitively to a series of identical sites at which only one of the receptors is in a conformation permitting transcriptional activation. A second type of inhibitory mechanism appears to involve an interaction between steroid hormone receptors and other transcription factor(s) responsible for pituitary expression of the prolactin gene. Binding of a transcription factor to sites in an inactive conformation can thus act to regulate transcription negatively as a consequence of the displacement of an active transcription factor. These events could account for cell-type-specific expression of related genes within an organ during development, as well as negative regulation of gene transcription by steroid hormones.

RESULTS

Mechanisms of Cell-specific Expression of Rat Prolactin and Growth Hormone Genes

The ontogeny of the anterior pituitary presents a developmental model system in which to understand the pathway of commitment to tissue and cellular phenotype. Tissue-specific enhancers in the 5′-flanking regions of both the prolactin and growth hormone genes appear to dictate pituitary-specific expression. On the basis of deletional analyses in vitro (Nelson et al. 1986), we suggest that cell-specific transcription of the rat prolactin gene depends on a distal enhancer (−1830 to −1530) accounting for 99% of activity and a proximal region (−422 to −36) accounting for 1–2% of maximal activity; both exert their effects in a cell-specific fashion. This proximal region contains several regulatory sequences, consistent with other evidence that has suggested a transcriptional role for the prolactin proximal region (Camper et al. 1985; Nelson et al. 1986; Cao et al. 1987; Gutierrez-Hartmann et al. 1987; Lufkin and Bancroft 1987).

The DNase I footprint analysis of the prolactin enhancer regions reveals four protected regions both in the distal enhancer (sites 1D–4D) and in the proximal region (1P–4P), none of which are observed with extracts prepared from HeLa or 208F fibroblast cell lines (Fig. 1B). Deletion of sites 1D, 2D, or 4D results in loss of most of the distal enhancer transcriptional activity, an indication of the importance of interactions between elements. Double-stranded oligonucleotides comprising each of the cell-specific elements in the prolactin distal enhancer and proximal enhancer specifically compete for binding to prolactin enhancer elements but with different affinities, suggesting that all sites bind a common (or similar) factor (Fig. 1B).

Cell-specific expression of the rat growth hormone gene depends on two tissue-specific transcriptional elements, designated −99 to −69 and −140 to −110 (Nelson et al. 1986, 1988; Bodner and Karin 1987; West et al. 1987; Ye and Samuels 1987), that are highly effective competitors of factors binding to prolactin tissue-specific elements (Fig. 1B). A comparison of the sequences of the individual rat prolactin and growth hormone elements (Fig. 1B) and their relative binding affinities suggested a consensus sequence with a core of A$\binom{AA}{TT}$TATNCAT. Mutational analysis of this sequence confirms the functional importance of the invariant bases within this core sequence in both Pit-1 binding and function (Nelson et al. 1988; H.P. Elsholtz et al., unpubl.).

The ability of the various sequences to compete in an in vitro transcription assay correlates well with the assignments of relative binding affinity determined by footprint competition analyses and their importance in transfectional analysis (Fig. 1B) (Nelson et al. 1988). Cell-specific elements mutated in the consensus sequence fail to compete enhancer-dependent transcription. From these in vitro transcription analyses, we conclude that a protein, referred to as Pit-1, binds to the tissue-specific elements of the prolactin genes and subserves a transcriptional function. To test this hypothesis further, we have used sequence-specific affinity chromatography to purify Pit-1 by a factor of more than 5000-fold and successfully complemented HeLa extract with affinity-purified material (<1 ng) specifically to stimulate transcription from both prolactin and growth hormone fusion genes (Nelson et al. 1988). Because purified Pit-1 is a doublet on denaturing gel electrophoresis, there may be differential activity for the growth hormone and prolactin elements for each form.

Inclusion of the rat growth hormone or prolactin proximal or distal enhancers in chimeric transcription units targets expression of these fusion genes to somatotrophs and lactotrophs, respectively, in transgenic mice (Lira et al. 1988; E.B. Crenshaw III et al., in prep.). As shown in Figure 2, both the prolactin proximal and distal regions target pituitary-specific expression. However, information outside the distal enhancer is required to restrict expression to lactotrophs; without this information, these chimeras also display ectopic expression in thyrotrophs, which express TSH. These data suggest the possibility that Pit-1 could be expressed in thyrotrophs. Similarly, 180 bp of 5′-flank-

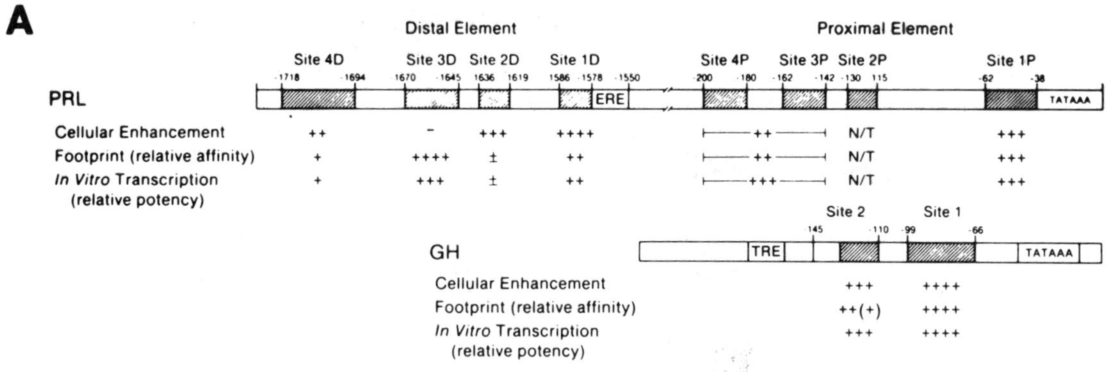

B

Factor Binding Sequences				Boundary
Higher Affinity	rGH1	antisense	CCTATC**ATTTATTCAT**GGC	-86
	p3D	sense	CTTCAT**TATTATTCAC**CAT	-1649
	rGH2	sense	CTTCTA**AATTATCCAT**CAG	-111
	p1P	sense	TTATAT**ATATATTCAT**GAA	-42
	p1D	sense	GCATTA**AAAAATGCAT**TTT	-1582
Lower Affinity	p3P	antisense	TTATTT**TCTTATTCAT**ATT	-157
	p4D	antisense	TTCGGG**ATTCATTCAT**CTG	-1707
	p2D	antisense	TTTTAG**AGATTTCCTA**AAT	-1639
	p4P	sense	TGCTGT**AATTAATCAA**AAT	-187
	consensus		$P_yP_y\frac{A}{T}N\frac{A}{T}NA\frac{A}{T}\frac{A}{T}$TATNCAT	
	CORE		A$\frac{A}{T}\frac{A}{T}$TATNCAT	
Factor Non-Binding Elements	p1P mutation	sense	TTATAT**ATATA**[**CGTGGT**]AA	-42
	Prl-TATAA	antisense	TTGACT**TTATAAACCT**TCG	-35

Figure 1. (*A*) Schematic representation of the rat prolactin and growth hormone *cis*-active elements showing the ability of each isolated element to compete for cell-specific footprint and for in vitro transcription, and its enhancer activity in vivo. (*B*) Sequences of the analyzed cell-specific elements from the prolactin distal and proximal enhancer regions and the consensus sequence are shown. Mutation of the consensus sequence abolished its activity.

ing information of the rat growth hormone gene, containing two Pit-1 elements, is sufficient to target pituitary (somatotroph) expression; at −45 bp, fusion genes fail to express (Fig. 2) (Lira et al. 1988). Thus, −180 and −45 bp contain all the information to target correctly growth hormone gene expression to pituitary.

Transcriptional enhancement of the rat prolactin gene by Pit-1 appears to be positively modulated by plasma-membrane-mediated peptide regulators, including TRH and EGF (Elsholtz et al. 1986). Because peptide hormone and 12-*O*-tetradecanoylphorbol-13-acetate (TPA)-mediated transcriptional regulation of fusion genes containing Pit-1 response elements occurs only in pituitary cells (Fig. 3), which exclusively express Pit-1, it appears that Pit-1 is among a series of different transcription factors that are under regulation by growth factors and phorbol esters (see, e.g., Angel et al. 1987; Lee et al. 1987; Treisman 1987).

Binding Elements for the T_3 Receptor That Mediate Positive or Negative Transcriptional Effects in Response to T_3

A deletional analysis of the rat growth hormone gene revealed that a minimal sequence sufficient for transcriptional activation by T_3 was located between 180 and 158 bp upstream of the transcriptional start site (Glass et al. 1987). An avidin-biotin complex DNA (ABCD)-binding assay was then established to demonstrate high-affinity, sequence-specific binding of the T_3 receptor to this element (Fig. 4A). Scatchard plots indicated a K_s of 1.23 to 10^{-9} M for the binding of pituitary T_3 receptors to the sequence.

Figure 4B compares the rat growth hormone T_3 response element with EREs present in the *Xenopus* vitellogenin A2 and rat prolactin genes and the metallothionein IIa GRE. When the sequences are aligned

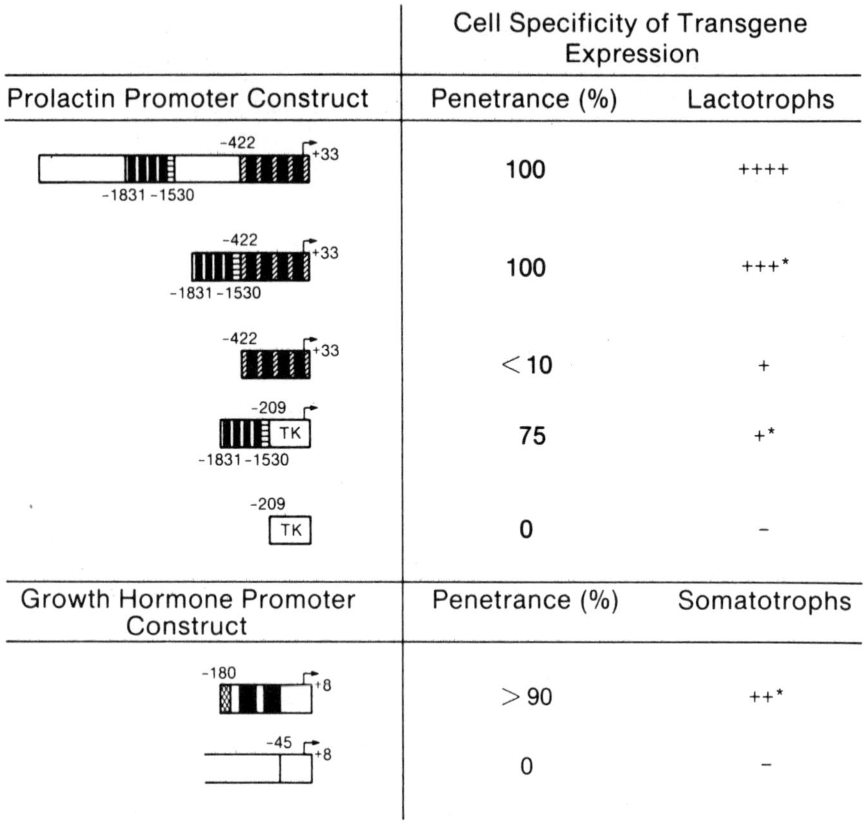

Figure 2. Expression of prolactin and growth hormone fusion genes in transgenic mice. Constructions containing the proximal or distal regions of the prolactin gene, or the growth hormone gene expressed exclusively in pituitary gland, in lactotrophs and somatotrophs, respectively. With removal of regions between the proximal and distal enhancers or in the context of a TK promoter, the prolactin distal enhancer also was expressed in thyrotrophs (*). (■) Pit-1-binding elements.

directly, all of the identical bases are clustered on one half of each response element. Introduction of a gap of three bases into the midst of the T_3 response element leads to a significant increase in the number of bases that are identical to those in the EREs and GREs. This suggests that the sequence determinants of binding might be quite similar or identical among the three classes of response elements, with the distinguishing feature between T_3 and steroid hormone response elements being a 3-bp gap.

T_3 receptors in pituitary cell nuclear extracts bind specifically and with high affinity to a biotinylated vitellogenin ERE (Fig. 3A). In contrast, no DNA-dependent binding was observed using a biotinylated oligonucleotide containing a *cis*-active element from the adenovirus-5 long terminal repeat (Fig. 4C). Surprisingly, the vitellogenin element binds more T_3 receptor than the growth hormone T_3 response element, whereas the MT IIa GRE bound only about one-fifth as much. On the basis of several types of kinetic analyses, we estimate the affinity of the T_3 receptor for the vitellogenin ERE to be 7×10^{-10} M, roughly equivalent to the affinity of the estrogen receptor for this element (Glass et al. 1988). Analysis of the effect of spacing mutations on the binding of the T_3 receptor to the rat growth hormone T_3 response element revealed that addition of between one and six bases into the midst of the T_3 receptor-binding site reduced the binding affinity by less than a factor of five (Fig. 4D).

To examine the sequence determinants of high-affinity binding, mutations were made within a 16-bp sequence of the rat growth hormone thyroid hormone response element (Fig. 5A). Most of the base substitutions within the growth hormone T_3 response element resulted in a decreased binding to T_3 receptors. In contrast, a limited number of mutations produced small but reproducible increases in the affinity of DNA binding. If the changes that resulted in increased binding affinity were to be simultaneously incorporated into the wild-type GH sequence, a near perfect palindrome would be obtained. The oligonucleotide corresponding to the predicted highest-affinity binding site (TCAGGTCA TGACCTGA) competed for T_3 receptor binding between three and four times more effectively than the wild-type growth hormone T_3 response element (Fig. 5B) and at least twice as effectively as any other palindromic sequence tested (Fig. 5B), and transferred 4–20-fold higher responsivity to T_3 than to the wild-type growth hormone T_3 response element. These experiments indicated that the highest-affinity binding

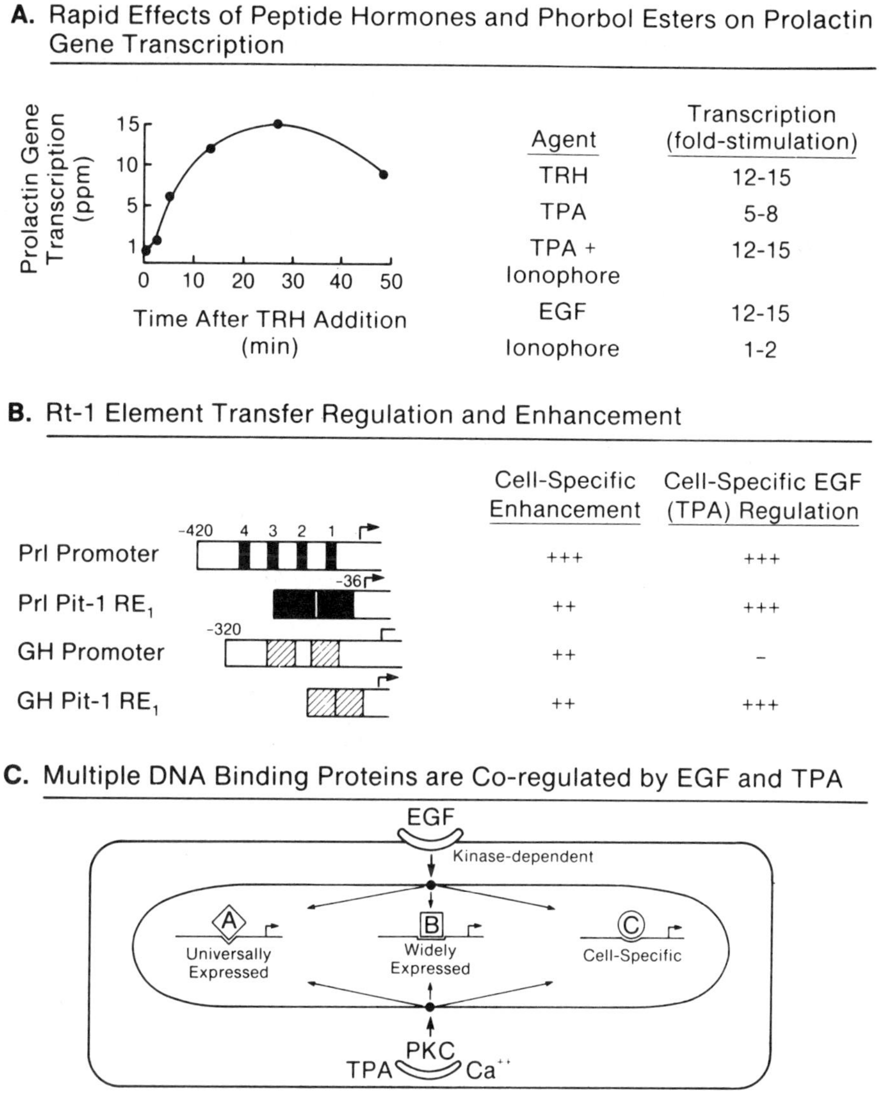

Figure 3. Transcriptional regulation by growth factors or phorbol esters is conferred by the Pit-1 *cis*-active sites. (*A*) Rapid transcriptional responses to TRH, EGF, and phorbol esters (TPA) are not mimicked by calcium ionophores alone. TPA and calcium ionophores together mimic the effects of TRH. (*B*) Both enhancement of gene transcription and regulation by EGF or TPA are transferred by Pit-1 elements from the prolactin or growth hormone gene but only in pituitary cells. In contrast, in the context of the normal rat growth hormone promoter, enhancement, but not EGF (TPA) dependent regulation, is conferred by Pit-1 elements. (*C*) Model of transcriptional activation by EGF and TPA. Regulated transcription is dependent on a number of widely distributed or tissue-specific transcription factors, binding to different *cis*-active sites, each regulated as a consequence of a modification to the factor itself and/or to an associated protein (e.g., c-*fos*) or other nuclear factors that modulate transcription.

sequence identified for the T_3 receptor is equivalent to the sequence of the vitellogenin ERE minus its three middle bases.

Methylation interference experiments demonstrate that the nucleotide contacts made between the response elements for T_3 or estrogen are very similar or identical (Fig. 5C). The symmetric pattern of contact on these two elements of different spacing also strongly suggests that the T_3 receptor is bound as a dimer, as do protein cross-linking experiments. The observation that the T_3 receptor could bind with high affinity to an ERE in vitro raises the question of whether or not the binding of the receptor to such an element in vivo would have any functional consequences. To examine this question, we mutated the thyroid hormone response element in the rat growth hormone promoter, as shown in Figure 5C, to a vitellogenin ERE or introduced a gap of three bases in the middle of the T_3 response element. Under T_3-stripped and low-estrogen conditions, T_3 stimulates gene expression from the wild-type T_3 response element, whereas mutation of the T_3 response element to either a "gapped" or vitellogenin ERE sequence confers estrogen-dependent regulation to this transcription unit. In contrast, addition of T_3 now results in a decrease in reporter gene expression to 20% of the base-line level. When estrogen concentration is

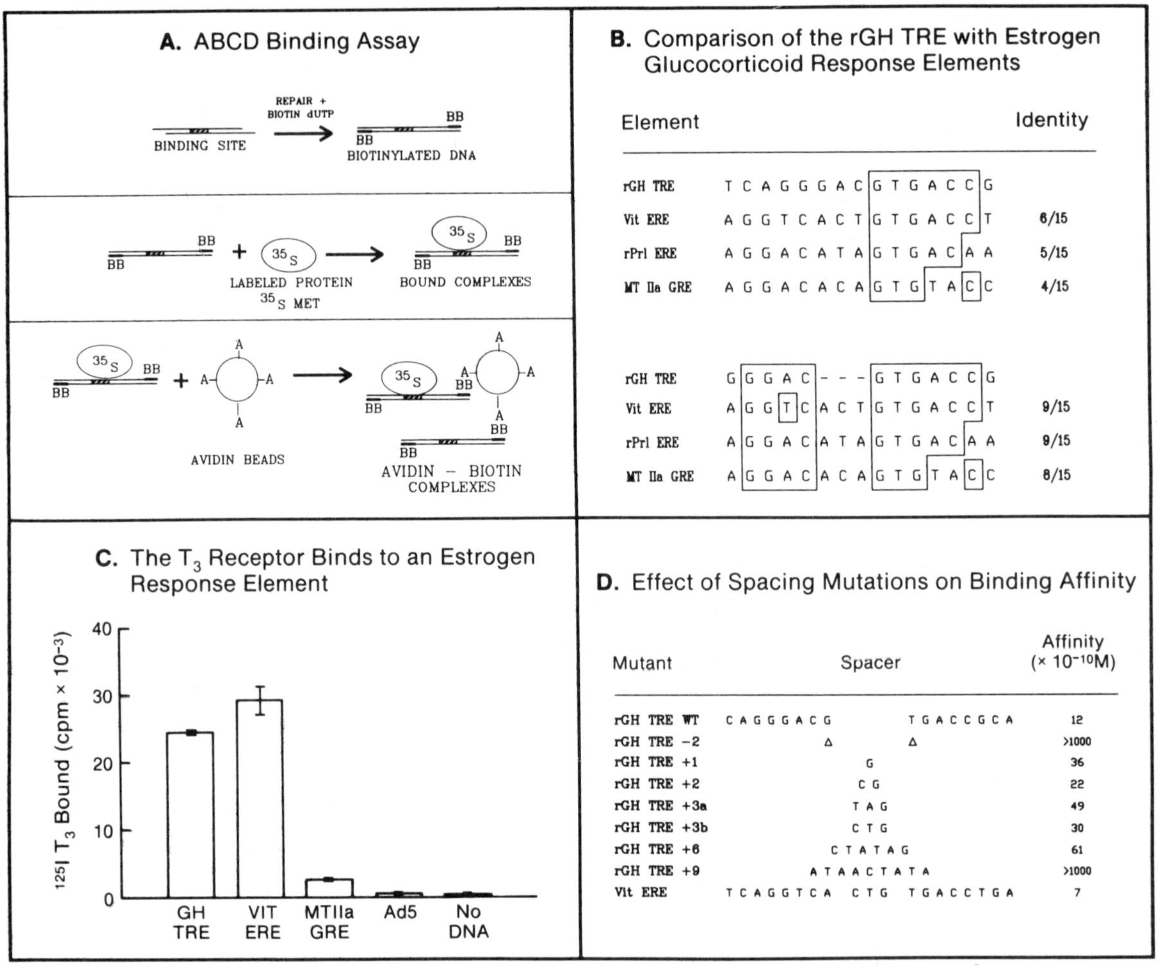

Figure 4. T_3 receptor regulation of gene transcription. (*A*) A schematic representation of the ABCD assay, used with ^{35}S-labeled protein or ^{125}I-labeled ligand. (*B*) Alternative alignment of the T_3 regulatory element against glucocorticoid regulatory element and estrogen regulatory element showing that optimal homology occurs with introduction of a 3-bp gap. (*C*) T_3 receptor effectively binds to the vitellogenin ERE, to a GRE, but not to a related sequence of the adenovirus E1A gene enhancer. (*D*) Effect of introduction of spacing between the consensus sequences on T_3 receptor binding.

further reduced ($<10^{-10}$ M), T_3 had little or no effect on gene expression. Similar results were obtained with EREs in several other chimeric transcription units, including the ERE of the prolactin distal enhancer. These results suggest that the negative transcriptional effects of T_3 on genes containing EREs are a consequence of T_3 receptor binding to these elements in vivo.

Steroid-hormone-dependent Inhibition of Prolactin Gene Expression

The major estrogen regulatory element responsible for estrogen-dependent stimulation of rat prolactin gene transcription has been identified as a 13-bp sequence located in the distal enhancer element (Fig. 6A) (Maurer and Notides 1987; Waterman et al. 1988). As shown in Figure 6B, under regulation of the intact 5′-flanking sequences, estrogen-dependent stimulation of fusion gene expression is observed, in both the presence and absence of a cotransfected estrogen receptor expression plasmid. When the estrogen regulatory element is deleted, estrogen-dependent stimulation of expression is no longer observed, and cotransfection of the estrogen receptor expression plasmid now produces a marked, estrogen-dependent inhibition of reporter gene expression. Conversely, mutation of the estrogen regulatory element to a glucocorticoid regulatory element converts the marked inhibitory actions of glucocorticoids into a four- to fivefold stimulation of prolactin gene expression (Fig. 6B), and a dramatic estrogen-dependent inhibition of the fusion gene is now observed. Therefore, both for estrogen and for glucocorticoids, a single appropriately located positive regulatory element can overcome the marked inhibitory effects conferred by either hormone.

As diagramed in Figure 6A, deletion to 422 bp of 5′-flanking information maintains both glucocorticoid and estrogen-dependent inhibition. Deletion of a sequence, at −173 to −161 bp, that has homology both

A

Element	Sequence	Relative Affinity
WT rGH TRE	CAGGGACG.TGACCGCA	1.0
G Pal TRE	TCAGGTCA.TGACCTGA	3.5
Vit GRE	TCAGGTCA TGACCTGA (insert CTG)	1.6

B Base Contacts of T_3 on T_3RE and ERE

G Pal TRE GATCTCAGGTCATGACCTGA
AGTCCAGTACTGGACTCTAG

Vit GRE GATCTCAGGTCACTGTGACCTGA
AGTCCAGTGACACTGGACTCTAG

C Responsiveness of Mutant rGH TREs to T_3 and Estrogen

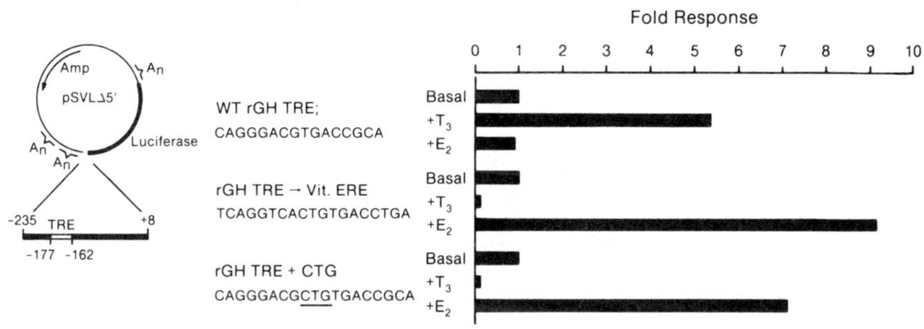

Figure 5. Effect of T_3 receptor on the ERE. (*A*) The optimal T_3 receptor-binding site is precisely identical to the palindromic vitellogenin ERE without the central gap of 3 bp. (*B*) Methylation interference analyses reveal identical symmetrical contacts of the T_3 receptor on both the estrogen and T_3 palindromic response elements. (*C*) An example of the effects of T_3 on expression of fusion genes under control of the rat growth promoter containing the wild-type T_3 response element, the element with a 3-bp gap, or the vitellogenin ERE. T_3 cells were maintained in T_3-stripped serum. T_3 causes stimulation of gene expression only with the wild-type thyroid hormone response element; in this context, estrogen has no effect. T_3 inhibits expression of the other two constructions, whereas estrogen is now stimulatory.

with glucocorticoid regulatory elements (8 out of 14 bp) and with estrogen regulatory elements (11 out of 14 bp), and which is also present in the bovine prolactin gene (Camper et al. 1985), does not eliminate the observed inhibitory effects of estrogens or glucocorticoids. Although further deletion of 5′-flanking sequences progressively decreases basal expression levels, steroid-dependent inhibition of fusion gene expression is reproducibly observed even with truncation to −78 bp. This sequence contains one cell-specific *cis*-active sequence spanning nucleotides −38 to −65 (Elsholtz et al. 1986; Nelson et al. 1986). Prolactin promoter fragments extending from −173 to −78 bp or from −135 to −205 bp and the upstream enhancer element extending from −1532 to −1831 bp also transfer negative regulation when placed distal to the herpes thymidine kinase (TK) promoter.

Each of the sites that transfer inhibition contain one or more of the Pit-1-binding elements. To test whether this effect is specific for Pit-1-binding sites, fusion genes containing isolated Pit-1, AP-1, and cAMP response elements adjacent to a TATAA element were analyzed. Both glucocorticoids and estrogen in the presence of overexpressed receptor selectively inhibited Pit-1-dependent gene expression but not expression conferred by AP-1 and cAMP elements (Fig. 6C). DNA-binding experiments fail to demonstrate high-affinity binding of the estrogen receptor to Pit-1-binding elements, suggesting that the estrogen receptor exerts its negative effects via a direct interaction with Pit-1.

Estrogen receptors (Green et al. 1986; Greene et al. 1986), truncated at amino acids 264 and 318, appear to be constitutively active (Adler et al. 1988; Waterman et al. 1988) in most cell types, stimulating all transcription units that contain an ERE (see Fig. 7A). However, a vitellogenin ERE fused to the herpes TK promoter, although stimulated by an estrogen holoreceptor, is not regulated by truncated receptors in HeLa cells, in concert with a report suggesting that the truncated receptor was not constitutively active as a transcription factor (Kumar et al. 1987; Maurer and Notides 1987). This appears to be a promoter-specific event, because the truncated receptor can constitutively stimulate tran-

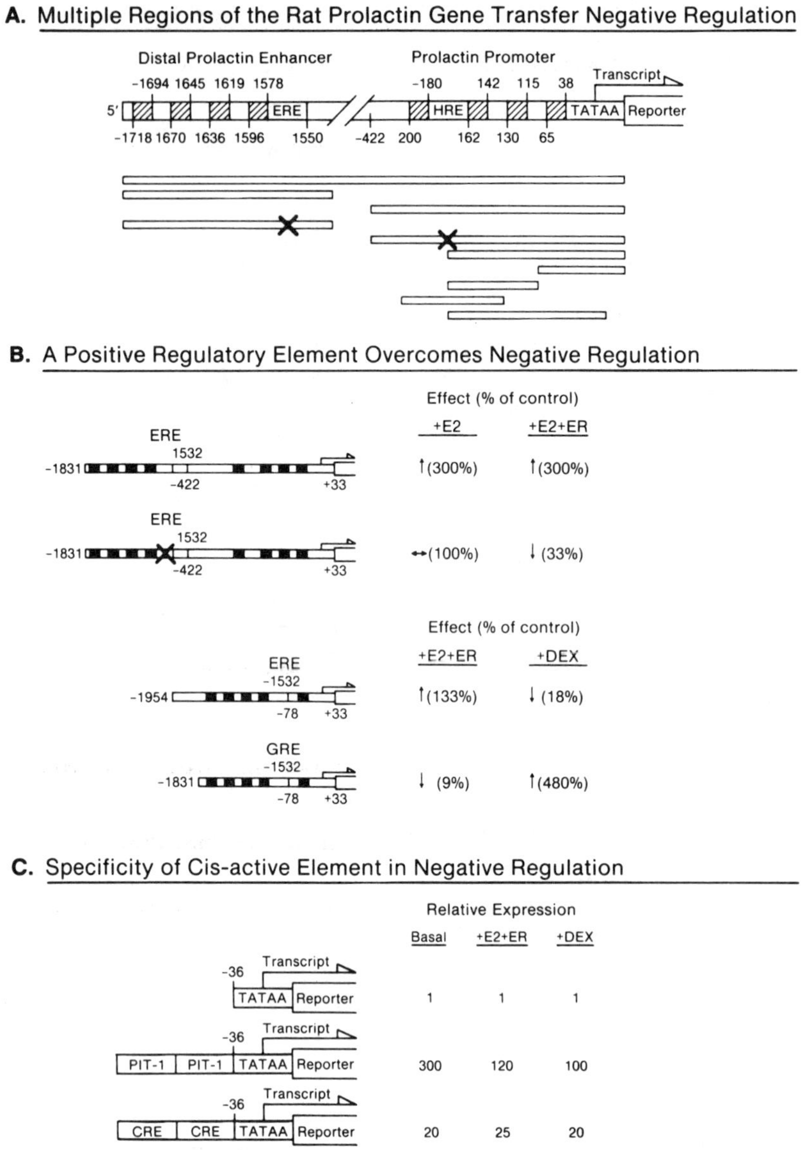

Figure 6. Inhibition of prolactin gene expression by overexpressed estrogen receptor or by glucocorticoids. (*A*) Multiple regions (as schematically shown) transfer negative regulation to glucocorticoids or overexpressed estrogen receptor. (*B*) One positive ERE or GRE converts estrogen or glucocorticoids from negative to positive regulators. (*C*) Pit-1 elements, but not AP-1 or cAMP regulatory sites, transfer negative regulation to glucocorticoids or overexpressed estrogen receptor.

scription of a second fusion gene under control of a vitellogenin estrogen regulatory element in HeLa cells (Fig. 7A). The availability of a constitutive estrogen receptor permitted mapping of the domains required for negative regulation by overexpression in GC cells, under conditions where the endogenous receptor present was not active (Adler et al. 1988). Deletion of the DNA-binding domain does not affect negative transcriptional regulation; however, receptors containing a deletion between amino acids 251 and 314 are not active as inhibitors of prolactin gene expression (Fig. 7A; Adler et al. 1988). Replacement of the deleted amino acids 251–314 of the human estrogen receptor with the corresponding region of the human glucocorticoid receptor (amino acids 486–531) restores inhibitory function to the chimeric receptor.

DISCUSSION

Positive and Restrictive Mechanisms in Pituitary Phenotypic Development

The ontogeny of the anterior pituitary involves sequential commitment to tissue and cellular phenotype. On the basis of our analyses, we propose that the prolactin and growth hormone *cis*-active elements can bind a common pituitary transcriptional activator that we have termed Pit-1 (Nelson et al. 1988). Cloning the

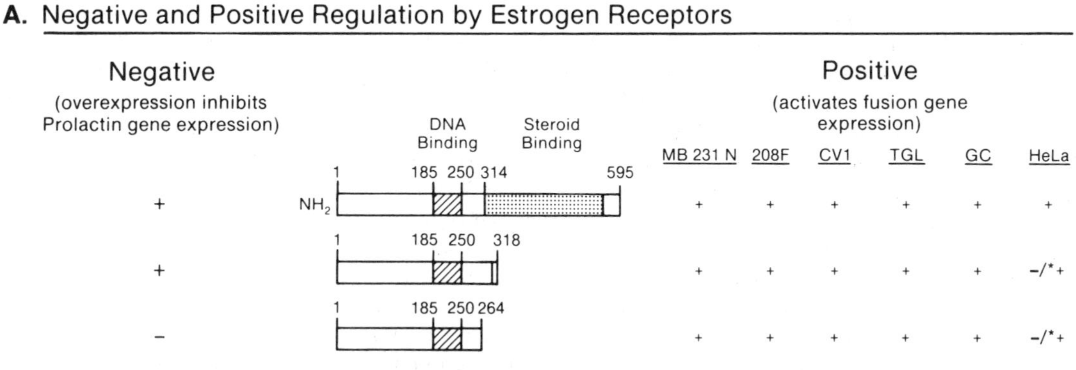

B. Estrogen Receptor Region Required for Negative Effect in Co-transfection Studies

Estrogen Receptor AA 251-314

MKGGIRKDRRGGRMLKHKRQRDDGEGRGEVGSAGDMRAANLWPSPLMIKRSKKNSLALSLTADQ

NLEARKTKKKIKGIQQATTGVSQETSENPGNKTIVPATLPQLTPT

Glucocorticoid Receptor AA 487-531

Figure 7. (*A*) Truncated estrogen receptors are constitutively active in a variety of cell types, except HeLa cells, using a vitellogenin ERE TK fusion gene. However, using a vitellogenin ERE prolactin TATA box (−36 to +34) fusion gene (∗), the truncated estrogen receptor is constitutively active in HeLa cells. (*B*) The region of the estrogen receptor that appears to be responsible for negative regulation, based on cotransfection assays with a prolactin fusion gene G/C cells. The comparable region of the glucocorticoid receptor is shown.

Pit-1-encoding gene has revealed that it encodes a 33-kD protein that contains a divergent homeodomain at the C′ terminus. This gene product transcriptionally activates prolactin and growth hormone fusion genes in heterologous cells, linking the homeodomain gene family to cell-specific activation of genes that define tissue and cellular phenotype.

The synergistic actions of multiple Pit-1-binding sites dispersed over a 2-kb region of the rat prolactin gene appear to be a required aspect of the developmental code dictating prolactin gene expression. If the prolactin, growth hormone, and possibly TSH genes require the same factor, Pit-1, for their activation, then restrictive mechanisms, whether repression or a requirement for additional interactive factors augmenting transcription, must account for their differential expression in mature lactotrophs, somatotrophs, and thyrotrophs. The data from analysis of fusion gene expression in transgenic mice suggest that, in the case of the prolactin gene, restriction from expression in thyrotrophs requires information located outside of the enhancer and promoter regions that are responsible for tissue-specific transcriptional activation in the pituitary gland. The growth hormone gene information between −180 and the cap site is sufficient to target expression selectively to somatotrophs and not to lactotrophs. Restriction could be explained, either by a model in which the growth hormone and prolactin *cis*-active elements serve as binding and response elements, respectively, for the two alternative forms of Pit-1 or by the actions of a second, selectively expressed factor binding to adjacent or overlapping sequences, exerting either synergistic or inhibitory actions. The steroid receptors are one candidate for such a factor.

The Pit-1 gene product also appears to serve as a homeostatic regulator in the mature lactotroph, under control of hypothalamic hormones (TRH). It will be of interest to determine whether regulation of Pit-1 is associated with other nuclear factors, such as c-*fos*, as has been reported for AP-1 (Rauscher et al. 1988). Activation of receptor mechanisms may profoundly influence pituitary cellular phenotype during development. The availability of the cloned gene encoding Pit-1 (H.A. Ingraham et al. in prep.) permits a direct assessment of the role of a tissue-specific positive transcription factor in developmental regulation of rat prolactin and growth hormone gene expression.

Binding Elements and Response Elements in Transcriptional Control

Although the precise mechanisms by which steroid and thyroid hormones exert negative transcriptional effects have not been defined, studies of other transcription factors indicate several possible mechanisms. A negative transcription factor may bind in close proximity to the TATA box, preventing access by RNA polymerase, as typified by prokaryotic repressor-operator interactions (see, e.g., Ptashne 1986). A second mechanism is the competition between positive and negative transcription factors for overlapping sequences, which may account for the inhibition of a series of transcription units by the α_2 gene product in yeast (see, e.g., Johnson and Herskowitz 1985; Nasmyth and Shore 1987). Additionally, negative regulation may result from protein-protein interactions rather than a direct DNA-binding event. For example, the negative transcriptional effects of *GAL80* (Ma and Ptashne 1987) and SV40 T antigen (Mitchell et al. 1987) appear to be a consequence of protein-protein interactions with *GAL4* and AP-2, respectively.

Our studies demonstrate that thyroid hormone can exert a negative effect on gene expression by decreasing estrogen-dependent transcription. We favor a mechanism involving T_3 receptor binding to EREs because the T_3 receptor binds to the vitellogenin A-2 ERE with an affinity higher than that of the wild-type rat growth hormone T_3 response element, even in the context of intact promoter elements. We propose that the model diagramed in Figure 8 accounts for the distinct actions of the T_3 receptor on T_3 and EREs. Because the binding affinities of the T_3 receptor and estrogen receptor for EREs are similar, the T_3 receptor can effectively compete for estrogen receptor binding. By binding in a transcriptionally inactive form, the net effect of the T_3 receptor is to decrease estrogen-dependent transcription (Fig. 8). Conversely, estrogen can decrease T_3-dependent transcription, consistent with the possibility that estrogen receptors bind to T_3 response elements in a transcriptionally inactive configuration. The intriguing implication of these findings is that a response element for one member of the steroid-receptor gene family can simultaneously serve as a binding (negative) element for a second member of this family of positive transcription factors (Fig. 8). The results suggest that in cells containing both T_3 and estrogen receptors, the net effect on the transcription of estrogen-regulated genes would depend on the ratio of the two activated receptors. These events are likely to operate for other members of the ligand-activated transcription factor family and for other classes of transcription factors, thereby providing the cell with an integrative mechanism to regulate, simultaneously positively and negatively, sets of genes utilizing identical *cis*-active elements, dependent on the ratios of activated positive transcription factors.

In contrast, negative regulation of the rat prolactin gene by overexpressed estrogen receptors does not appear to require the DNA-binding domain. The 63-amino-acid region of the estrogen receptor that is required for inhibition of prolactin gene expression is adjacent to the DNA-binding domain, and a region rich

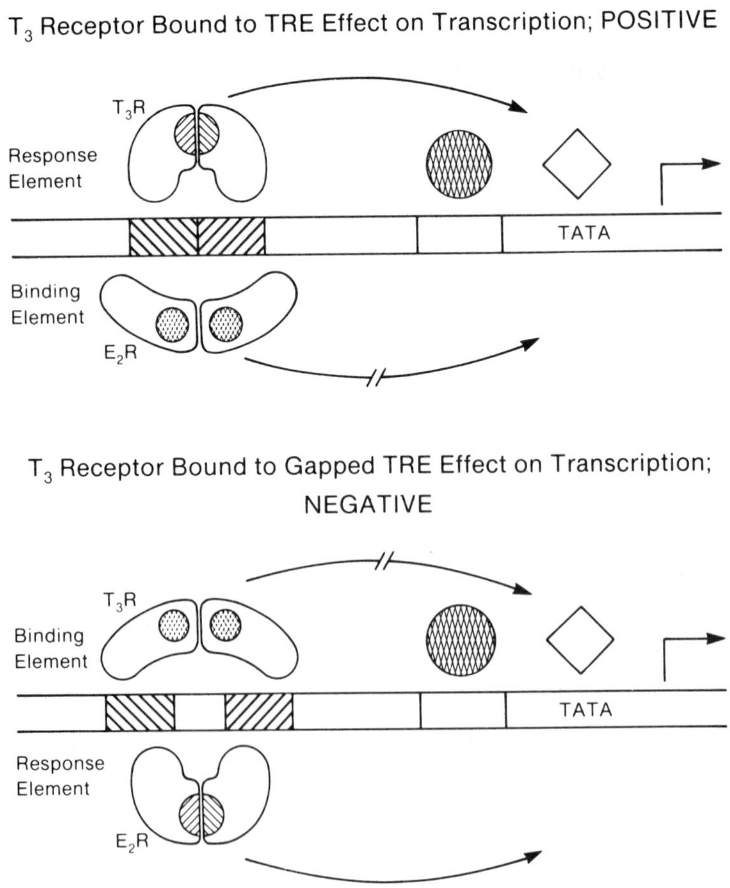

Figure 8. Model of binding (negative) and response (positive) elements in steroid receptor modulation of gene expression. As discussed in the text, we propose that response elements are a subset of the binding sites for the T_3 receptor, but that the T_3 receptor is active as a transcription factor only when bound to T_3 response elements. At binding sites, such as the ERE, T_3 receptor competes with estrogen receptor binding and can decrease gene transcription.

in basic amino acids appears to be critical for the observed effects. Inhibition by glucocorticoids or by overexpressed estrogen receptor can be transferred by a Pit-1-binding site. Furthermore, all of the prolactin genomic regions that have been shown to impart *cis*-active inhibition by steroid hormones contain the cell-specific Pit-1-binding sites. It is tempting to speculate that the overexpressed ligand-dependent transcription factors can inhibit the Pit-1-dependent stimulation of prolactin gene transcription based on protein-protein interactions and could inhibit either Pit-1 binding to DNA or its transcriptional function. Specificity is suggested because other *cis*-active elements, such as AP-1 or cAMP-regulatory elements, fail to exhibit negative regulation by glucocorticoids or by overexpressed estrogen receptor.

Additional mechanisms of negative regulation could operate in the prolactin gene, analogous to effects of T_3 receptor bound to "gapped" sites, by which glucocorticoid receptor could occupy binding sites in a transcriptionally inactive conformation, exerting negative transcriptional effects by displacing a transcription factor binding to an overlapped or to an identical regulatory site. This mechanism would not account, however, for the negative effects of overexpressed estrogen receptor.

Although additional experimentation will be required to prove the precise molecular interactions that are responsible for steroid-receptor-mediated inhibition of gene expression, it appears that steroid receptors can mediate positive and negative transcriptional regulation as a consequence of their association with structurally distinct binding and response elements and possibly also as a consequence of protein-protein interactions with other classes of positive transcription factors.

ACKNOWLEDGMENTS

We thank all members of the laboratory, past and present, for their contributions to the research presented, and we thank Dr. Gordon N. Gill for the helpful discussions. Research was conducted under support by grants from National Institutes of Health, American Cancer Society, and the Howard Hughes Medical Institute.

REFERENCES

Adler, S., M.W. Waterman, X. He, and M.G. Rosenfeld. 1988. Steroid receptor-mediated inhibition of rat prolactin gene expression does not require the receptor DNA-binding domain. *Cell* **52:** 685.

Angel, P. M. Imagawa, R. Chiu, B. Stein, R.J. Imbra, H.J. Rahmsdorf, C. Jonat, P. Herrlich, and M. Karin. 1987. Phorbol ester-inducible genes contain a common *cis* element recognized by a TPA-mediated *trans*-acting factor. *Cell* **49:** 729.

Bodner, M. and M. Karin. 1987. A pituitary-specific *trans*-acting factor can stimulate transcription from the growth hormone promoter in extracts of non-expressing cells. *Cell* **50:** 267.

Camper, S.A., Y.A.S. Yao, and F.M. Rothman. 1985. Hormonal regulation of the bovine prolactin promoter in rat pituitary tumor cells. *J. Biol. Chem.* **260:** 12246.

Cao, Z., E.A. Barron, A.J. Carrillo, and Z.D. Sharp. 1987. Reconstitution of cell-type-specific transcription of the rat prolactin gene in vitro. *Mol. Cell Biol.* **7:** 3402.

Chetelain, A., J.P. Dupuoy, and M.P. Dubois. 1979. Ontogenesis of cells producing polypeptide hormones (ACTH, MSH, LPH, GH, prolactin) in the fetal hypophysis of the rate: Influence of the hypothalamus. *Cell Tissue Res.* **196:** 409.

Echols, H. 1986. Multiple DNA protein interactions governing high precision DNA transactions. *Science* **233:** 1050.

Elsholtz, H.P., H.J. Mangalam, E. Potter, V.R. Albert, S. Supowit, R.M. Evans, and M.G. Rosenfeld. 1986. Two different *cis*-active elements transfer the transcriptional effects of both EGF and phorbol esters. *Science* **234:** 1552.

Evans, R.M. 1988. The steroid and thyroid hormone receptor superfamily. *Science* **240:** 889.

Gash, D., N. Ahmad, and J. Schechter. 1982. Comparison of gonadotroph, thyrotroph and mammotroph development in situ, in transplants and in organ culture. *J. Neuroendocrinol.* **34:** 222.

Giguere, V., S.M. Hollenberg, M.G. Rosenfeld, and R.M. Evans. 1986. Functional domains of the human glucocorticoid receptor. *Cell* **46:** 645.

Glass, C.K., J.M. Holloway, O.V. Devary, and M.G. Rosenfeld. 1988. The thyroid hormone receptor binds with opposite transcription effects to a common sequence motif in thyroid hormone and estrogen response elements. *Cell* **54:** 313.

Glass, C.K., R. Franco, C. Weinberger, V. Albert, R.M. Evans, and M.G. Rosenfeld. 1987. A c-*erb*-A binding site in the rat growth hormone gene mediates transactivation by thyroid hormone. *Nature* **329:** 738.

Green, S., P. Walter, V. Kumar, A. Krust, J.-M. Bornert, P. Argos, and P. Chambon. 1986. Human oestrogen receptor cDNA: Sequence, expression and homology to v-*erb*-A. *Nature* **320:** 134.

Greene, G.L., P. Gilna, M. Waterfield, A. Baker, Y. Hort, and J. Shine. 1986. Sequence and expression of human estrogen receptor complementary DNA. *Science* **231:** 1150.

Gutierrez-Hartmann, A., S. Siddiqui, and S. Loukin. 1987. Selective transcription and DNase I protection of the rat prolactin gene by GH_3 pituitary cell-free extracts. *Proc. Natl. Acad. Sci.* **84:** 5211.

Hammer, R.E., R. Krumlauf, S.A. Camper, R.L. Brinster, and S.M. Tilghman. 1987. Diversity of alpha-fetoprotein gene expression in mice is generated by a combination of separate enhancer elements. *Science* **235:** 53.

Hoeffler, J.P., F.R. Bookfor, and L.S. Frawley. 1985. Ontogeny of prolactin cells in neonatal rats: Initial prolactin secretors also release growth hormone. *Endocrinology* **117:** 187.

Izumo, S., B. Nadal-Ginard, and V. Mahdavi. 1986. All members of the MHC gene family respond to thyroid hormone in a highly tissue specific manner. *Science* **231:** 597.

Johnson, A.S., and I. Herskowitz. 1985. A repressor (MATα2 product) and its operator control expression of a set of cell-type-specific genes in yeast. *Cell* **42:** 237.

Jost, J.-P., M. Seldran, and M. Geiser. 1984. Preferential binding of estrogen-receptor complex to a region containing the estrogen-dependent hypomethylation site preceding the chicken vitellogenin II gene. *Proc. Natl. Acad. Sci.* **81:** 429.

Khorram, O., L.R. Depalatis, and S.M. McCann. 1984. Hypothalamic control of prolactin secretion during the perinatal period in the rat. *Endocrinology* **115:** 1698.

Klein-Hitpass, L., M. Schorpp, U. Wagner, and G.U. Ryffel. 1986. An estrogen-responsive element derived from the 5′ flanking region of the *Xenopus* vitellogenin A-2 gene functions in transfected human cell. *Cell* **46:** 1053.

Klock, G., U. Strahle, and G. Schutz. 1987. Oestrogen and glucocorticoid elements are closely related but distinct. *Nature* **329:** 734.

Kumar, V., S. Green, G. Stack, M. Berry, J. Jin, and P. Chambon. 1987. Functional domains of the human estrogen receptor. *Cell* **51:** 941.

Lee, W., P. Mitchell, and R. Tjian. 1987. Purified transcription factor AP_1 interacts with TPA-inducible enhancer element. *Cell* **49:** 741.

Lira, S.A., E.B. Crenshaw III, C.K. Glass, L.W. Swanson, and M.G. Rosenfeld. 1988. Identification of rat growth hormone genomic sequence targeting pituitary expression in transgenic mice. *Proc. Natl. Acad. Sci.* **85:** 4755.

Lufkin, T. and F.C. Bancroft. 1987. Identification by cell fusion of gene sequences that interact with positive *trans*-acting factors. *Science* **237:** 283.

Ma, J. and M. Ptashne. 1987. The carboxy terminal 30 amino acids of GAL4 are recognized by GAL80. *Cell* **51:** 137.

Martinez, E., F. Givel, and W. Wahli. 1987. The estrogen responsive element as an inducible enhancer: DNA sequence requirements and conversion to a glucocorticoid response element. *EMBO J.* **6:** 3719.

Maurer, R.A. and A.C. Notides. 1987. Identification of an estrogen-responsive element from the 5′-flanking region of the rat prolactin gene. *Mol. Cell. Biol.* **7:** 4247.

McKnight, R. and R. Tjian. 1986. Transcriptional selectivity of viral genes in mammalian cells. *Cell* **46:** 795.

Miesfeld, R., S. Rusconi, P.J. Godowski, B.A. Malcr, S. Okret, A.-C. Wikstrom, J.-A. Gustafsson, and K.R. Yamomoto. 1986. Genetic complementation of a glucocorticoid receptor deficiency by expression of cloned receptor cDNA. *Cell* **46:** 389.

Mitchell, P.J., C. Wang, and R. Tijian. 1987. Positive and negative regulation of transcription in vitro; enhancer binding protein AP2 is inhibited by SV40 T antigen. *Cell* **50:** 847.

Nasmyth, K., and D. Shore. 1987. Transcriptional regulation in the yeast life cycle. *Science* **237:** 1162.

Nelson, C., V. Albert, H.P. Elsholtz, L.-I. Lu, and M.G. Rosenfeld. 1988. Activation of cell-specific expression of rat growth hormone and prolactin genes by a common transcription factor. *Science* **239:** 1400.

Nelson, C., E.B. Crenshaw III, R. Franco, S.A. Lira, V.R. Albert, R.M. Evans, and M.G. Rosenfeld. 1986. Discrete *cis*-active genomic sequences dictate the pituitary cell type-specific expression of rat prolactin and growth hormone genes. *Nature* **322:** 557.

Oliver, C., R.L. Eskay, and J.C. Porter. 1980. Developmental changes in brain TRH and in plasma and pituitary TSH and prolactin levels in the rat. *Biol. Neonate* **37:** 145.

Ornitz, D.M., R.D. Palmiter, R.E. Hammer, R.L. Brinster, G.H. Swift, and R.J. MacDonald. 1985. Specific expression of an elastase-human growth hormone fusion gene in pancreatic acinar cells of transgenic mice. *Nature* **313:** 600.

Ott, M.-O., L. Sperling, P. Herbomel, M. Yaniv, and M.C. Weiss. 1984. Tissue-specific expression is conferred by a sequence from the 5′ end of the rat albumin gene. *EMBO J,* **3:** 2505.

Payvar, F., O. Wrange, J. Carlstedt-Duke, S. Okret, J.-A. Gustafsson, and K.R. Yamamoto. 1981. Purified glucocorticoid receptors bind selectively in vitro to a cloned DNA fragment whose transcription is regulated by glucocorticoids in vivo. *Proc. Natl. Acad. Sci.* **78:** 6628.

Ptashne, M. 1986. Gene regulation by proteins acting nearby and at a distance. *Nature* **322:** 697.

Rauscher, F.J., III, D.R. Cohen, T. Curran, T.J. Bos, P.K. Vogt, D. Bohmann, R. Tjian, and B.R. Franza, Jr. 1988. Fos-associated protein P39 is the product of the *inn* protooncogene. *Science* **240:** 1010.

Serfling, E., M. Jasin, and W. Schaffner. 1985. Enhancer and eukaryotic gene transcription. *Trends Genet.* **1:**224.

Shupnik, M.A., W.W. Chin, D.S. Ross, J.F. Habener, and E.C. Ridgway. 1985. Transcriptional regulation of the thyrotropin subunit genes by thyroid hormone. *J. Biol. Chem.* **260:** 2900.

Strähle, U., G. Klock, and G. Schütz. 1987. A DNA sequence of 15 base pairs is sufficient to mediate both glucocorticoid and progesterone induction of gene expression. *Proc. Natl. Acad. Sci.* **84:** 7871.

Struhl, K. 1987. Promoters, activator proteins and the mechanisms of transcriptional activation in yeast. *Cell* **49:** 295.

Treisman, R. 1987. Identification and purification of a polypeptide that binds to the c-*fos* serum response element. *EMBO J.* **6:** 2711.

Walker, M.D., T. Edlund, A.M. Boulet, and W.J. Rutter. 1983. Cell-specific expression controlled by the 5′-flanking region of insulin and chymotrypsin genes. *Nature* **306:** 557.

Watanabe, Y.G. and S. Daikoku. 1979. An immunohistochemical study on the cytogenesis of adenohypophysial cells in fetal rats. *Dev. Biol.* **68:** 557.

Waterman, M., S. Adler, C. Nelson, G.L. Greene, R.M. Evans, and M.G. Rosenfeld. 1988. A single domain of the estrogen receptor confers DNA binding and transcriptional activation of the rat prolactin gene. *Mol. Endocrinol* **2:** 14.

West, B.L., D.F. Catanzaro, S.H. Mellon, P.A. Cattini, J.D. Baxter, and T.L. Reudelhuber. 1987. Interaction of a tissue-specific factor with an essential rat growth hormone gene promoter element. *Mol. Cell. Biol.* **7:** 1193.

Wright, S., A. Rosenthal, R. Flavell, and J. Grosveld. 1984. DNA sequences required for regulated expression of beta-globin genes in murine erythroleukemia cells. *Cell* **38:** 265

Ye, Z.-S. and H.H. Samuels. 1987. Cell- and sequence-specific binding of nuclear proteins to 5′-flanking DNA of rat growth hormone gene. *J. Biol. Chem.* **262:** 6313.